College Board AP® Topic Outline	Friedland and Relyea: E
I. Earth Systems and Resources (10–15%)	
A. Earth Science Concepts	Chapter 1 Studying the Sta Chapter 2 Environmental Systems
B. The Atmosphere	Chapter 3 Ecosystem Ecology Chapter 4 Global Climates and Biomes
C. Global Water Resources and Use	Chapter 9 Water Resources
D. Soil and Soil Dynamics	Chapter 8 Earth Systems
II. The Living World (10–15%)	
A. Ecosystem Structure	Chapter 3 Ecosystem Ecology Chapter 5 Evolution of Biodiversity
B. Energy Flow	Chapter 3 Ecosystem Ecology
C. Ecosystem Diversity	Chapter 6 Population and Community Ecology
D. Natural Ecosystem Change	Chapter 3 Ecosystem Ecology Chapter 5 Evolution of Biodiversity
E. Natural Biogeochemical Cycles	Chapter 3 Ecosystem Ecology Chapter 4 Global Climates and Biomes
III. Population (10–15%)	
A. Population Biology Concepts	Chapter 6 Population and Community Ecology
B. Human Population	Chapter 7 The Human Population
IV. Land and Water Use (10–15%)	
A. Agriculture	Chapter 10 Land, Public and Private Chapter 11 Feeding the World
B. Forestry	Chapter 10 Land, Public and Private
C. Rangelands	Chapter 10 Land, Public and Private
D. Other Land Use	Chapter 10 Land, Public and Private
E. Mining	Chapter 8 Earth Systems
F. Fishing	Chapter 11 Feeding the World
G. Global Economics	Chapter 20 Sustainability, Economics, and Equity
V. Energy Resources and Consumption (10–15%)	
A. Energy Concepts	Chapter 12 Nonrenewable Energy Sources
B. Energy Consumption	Chapter 12 Nonrenewable Energy Sources
C. Fossil Fuel Resources and Use	Chapter 12 Nonrenewable Energy Sources
D. Nuclear Energy	Chapter 12 Nonrenewable Energy Sources
E. Hydroelectric Power	Chapter 12 Nonrenewable Energy Sources
F. Energy Conservation	Chapter 13 Achieving Energy Sustainability
G. Renewable Energy	Chapter 13 Achieving Energy Sustainability
VI. Pollution (25–30%)	
A. Pollution Types	Chapter 14 Water Pollution Chapter 15 Air Pollution and Stratospheric Ozone Depletion Chapter 16 Waste Generation and Waste Disposal
B. Impacts on the Environment and Human Health	Chapter 14 Water Pollution Chapter 15 Air Pollution and Stratospheric Ozone Depletion Chapter 16 Waste Generation and Waste Disposal Chapter 17 Human Health and Environmental Risks
C. Economic Impacts	Chapter 14 Water Pollution Chapter 15 Air Pollution and Stratospheric Ozone Depletion Chapter 16 Waste Generation and Waste Disposal
VII. Global Change (10–15%)	
A. Stratospheric Ozone	Chapter 15 Air Pollution and Stratospheric Ozone Depletion
B. Global Warming	Chapter 19 Global Change
C. Loss of Biodiversity	Chapter 18 Conservation of Biodiversity

TEACHER'S EDITION

FRIEDLAND and RELYEA

Environmental Science for AP®

SECOND EDITION

TEACHER'S EDITION

Nat Draper *Deep Run High School*

Elizabeth Jones *Sacred Heart Preparatory School*

Elisa McCracken *Brandeis High School*

FRIEDLAND and RELYEA

Environmental Science for AP® SECOND EDITION

Andrew Friedland

Dartmouth College

Rick Relyea

Rensselaer Polytechnic Institute

FREEMAN

W. H. Freeman and Company • New York

Publisher: Ann Heath
Sponsoring Editor: Jeffrey Dowling
Editorial Assistant: Matt Belford, Rachel Chlebowski
Marketing Manager: Julie Comforti
Marketing Assistant: Nont Pansringarm
Development Editor: Rebecca Kohn
Director, Content Management Enhancement: Tracey Kuehn
Managing Editor: Lisa Kinne
Project Editor: Kerry O'Shaughnessy
Design Manager & Cover Designer: Vicki Tomaselli
Text Designer: Patrice Sheridan
Photo Editor: Christine Buese
Illustration Coordinator: Matt McAdams
Art Development and Illustrations: Joseph BelBruno
Production Manager: Julia DeRosa
Composition: codeMantra
Printing and Binding: Willard

Cover Credit: Alice Cahill/Getty Images

Library of Congress Control Number: 2015931619
ISBN-13: 978-1-4641-5618-2
ISBN-10: 1-4641-5618-2

Printed in the United States of America

First printing

W. H. Freeman and Company
41 Madison Avenue
New York, NY 10010
Houndmills, Basingstoke RG21 6XS, England
www.whfreeman.com

What Is Included in the Teacher's Edition

Teacher's Edition Authors

Nat Draper

Nat Draper teaches AP® environmental science and Earth science at Deep Run High School in Glen Allen, Virginia. He has served as an AP® environmental science table leader for the past 4 years and was an AP® Annual Conference presenter in 2009 in San Antonio, Texas, and in 2013 in Las Vegas, Nevada. Nat was selected as a 2010 Toyota International Teacher and spent 3 weeks doing fieldwork in Costa Rica where he worked with other teachers to develop ways to make environmental science classes more engaging. Nat holds a BA in environmental studies from Randolph Macon College and a Master of Teaching from Virginia Commonwealth University.

Elizabeth Jones teaches AP® environmental science and biology at Sacred Heart Preparatory School in Atherton, California. She began teaching there in 2008 and introduced the AP® environmental science course in 2009. Elizabeth is also the faculty sponsor for the Green Team at Sacred Heart Preparatory School as well as cofounder of the Bay Area Green Council. Prior to teaching and attending graduate school, Elizabeth was a wildlife ecologist and worked for the Environmental and Energy Study Institute and the U.S. Environmental Protection Agency in Washington, D.C. Elizabeth holds a BA in environmental studies and biology from Dartmouth College, an MS in environmental science from Yale University, and a PhD in tropical ecology from Yale University. Elizabeth hopes to galvanize all students toward a greater environmental awareness that impels them to action and stewardship.

Jeff McCracken

Elisa McCracken is the science department coordinator at Brandeis High School in San Antonio, Texas. She earned her BS in genetics at Texas A&M and has taught high school science for 10 years. During her teaching career she has taught many science courses including AP® environmental science, chemistry, pre-AP® biology, and environmental systems. She currently serves as an AP® environmental science exam reader. Elisa was selected as the 2014 Brandeis High School Educator of the Year in recognition of the growth of the AP® Student Success Program. As a sponsor of the Brandeis High School environmental club, the EcoBroncos, Elisa directs over 100 student volunteers who participate in a campus-wide recycling program, a community garden, and various community service events designed to help make a difference in the local community and environment.

We would like to offer special thanks to the members of the first edition content advisory board whose advice and support were invaluable: Art Samel, Teri C. Balser, Dean Goodwin, Judith Treharne, Michael L. Denniston, and Jeffery A. Schneider. We would also like to thank the first edition supplements team who helped to create support materials for the first edition: Nat Draper, Courtney Mayer, Dora Barlaz, Joshua Castleberry, Maureen Goble, Connie L. Perreira Sergent, and Bishop Bosher.

About the Teacher's Edition

Our warm welcome into the AP® environmental science community with the first edition of *Friedland and Relyea Environmental Science for AP®* was a gratifying experience. Over the past few years we have spoken with and listened to many dedicated educators—some seasoned and others new to the AP® environmental science course. As authors of a textbook written specifically to meet the needs of AP® environmental science instructors, we wanted to make sure you have the best tools possible to make your class a success. So, with the second edition of *Friedland and Relyea Environmental Science for AP®* we have worked with a team of experienced and active AP® environmental science teachers to bring you this Teacher's Edition.

The Teacher's Edition includes:

- An AP® Primer that offers guidance on how to write, initiate, prepare, and teach an AP® environmental science course successfully
- The complete student text with wraparound content including Teaching Tips, Common Misconceptions, AP® Exam Tips, Exploring the Literature, More Math Practice, Teaching with Figures, Labs, Practicing Science—many of which are tied to the Teacher's Resource Materials (TRM)
- Introduction pages for each chapter that precede the wraparound student pages, providing an in-depth guide to the chapter content and support materials, including:
  - Chapter and module overviews featuring a discussion of the most important chapter ideas
  - A module alignment to the AP® environmental science course description
  - A complete list of chapter learning objectives
  - A pacing guide for both standard and block schedules
  - A comprehensive list of additional chapter resources available for download from the edAPtext Teacher's Edition (TE-book), the Book Companion Site (BCS), and the Teacher's Resource Flash Drive (TRFD)
  - A list of relevant free-response questions from previous AP® Environmental Science Exams
- Answers to all Do the Math problems, end-of-module AP® Review Questions, Critical Thinking Questions in Working Toward Sustainability, end-of-chapter and end-of-unit AP® Environmental Science Practice Exams, Science Applied questions, and the Cumulative AP® Environmental Science Practice Exam at the end of the book

Teacher's Resource Materials

The blue TRM button throughout the Teacher's Edition indicates that relevant Teacher's Resource Materials are available. Access these carefully developed resources in three ways: by clicking on the link buttons in the edAPtext Teacher's Edition (TE-book), opening the Teacher's Resource Flash Drive (TRFD), or logging onto the Book Companion Site (BCS), bcs.whfreeman.com/friedlandapes2e (teacher login required). Teacher's Resource Materials include:

- Chapter Overview Videos
- Chapter PowerPoint Presentations

- Optimized Art PowerPoint and JPEG Files
- Do the Math Videos
- Labs (instructor and student versions)
- Activities (instructor and student versions)
- More Math Practice (instructor and student versions)
- Measuring Your Impact (instructor and student versions) for each chapter
- Web Resources for each chapter
- Answers to end-of-chapter AP® Practice Exams, unit AP® Practice Exams, full-year Cumulative AP® Practice Exam, and Science Applied Free-Response Questions
- An additional Multiple-Choice AP® Practice Exam, with answers, for each chapter
- An additional Free-Response Question, with answer, for each unit exam
- Exploring the Literature references

Professional Development Videos

The green PD button appears at the beginning of each chapter. This is to let you know that a "just-in-time" professional development video presented by one of the authors or by an experienced AP® environmental science teacher is available for download. These videos offer tips and advice for teaching the chapter and will help you make the most of your Teacher's Resource Materials.

Features of the Teacher's Edition

Turn the page to take a tour through the Teacher's Edition.
See how this book will help you create the best course possible.

Chapter introduction pages show resources at a glance.

chapter

12

Nonrenewable Energy Resources

PD Overview

Chapter 12 explores the use of fossil fuels and nuclear energy resources, and how to project future supplies of these resources. This chapter offers an opportunity for students to focus on the advantages and disadvantages of the nonrenewable energy resources. For example, coal is abundant, energy-dense, easy to obtain, and relatively inexpensive. However, coal has impurities that are released into the atmosphere when it is burned, it leaves behind large deposits of ash, and it contributes to the increasing atmospheric concentrations of carbon dioxide, sulfur dioxide, and particulates. An understanding of energy is vital to the student's ability to understand environmental science because energy use impacts so many areas: air pollution, water pollution, food and agriculture, waste disposal, health, and climate change among others. This chapter on nonrenewable energy resources and the following chapter on renewable energy resources follow up on energy basics presented in Chapter 2.

Module 34: Patterns of Energy Use

This module examines patterns of energy use worldwide and in the United States. It demonstrates that evaluating energy efficiency helps us determine the best application for different energy sources. Finally, because electricity accounts for such a large percentage of our overall energy use, the module takes a close look at the ways in which electricity is generated.

Module 35: Fossil Fuel Resources

Fossil fuels provide most of the commercial energy in the world. They are also responsible for a good deal of the pollution that occurs, a topic that will be discussed in Chapter 15. This module describes the major fossil fuels—coal, petroleum, and natural gas—and discusses the advantages and disadvantages of each. It also considers the future use of fossil fuels.

Module 36: Nuclear Energy Resources

The combustion of fossil fuels releases large quantities of CO_2 into the atmosphere. Because nuclear energy has relatively low emissions of CO_2, it has received increasing interest as an energy source. This module explores how nuclear energy works and examines its advantages and disadvantages.

12-1

Alignment to AP® Environmental Science Course Description

AP® Outline	Module
V. Energy Resources and Consumption (10–15%)	
A. Energy Concepts	Module 34 Patterns of Energy Use
	Module 35 Fossil Fuel Resources
	Module 36 Nuclear Energy Resources
B. Energy Consumption	Module 34 Patterns of Energy Use
C. Fossil Fuel Resources and Use	Module 35 Fossil Fuel Resources
D. Nuclear Energy	Module 36 Nuclear Energy Resources

Chapter Learning Objectives

After completing this chapter students will be able t

- **describe the use of nonrenewable energy in t in the United States.**
- **explain why different forms of energy are bes certain purposes.**
- **understand the primary ways that electricity in the United States.**
- **discuss the uses of coal and its consequences**
- **discuss the uses of oil and its consequences.**
- **discuss the uses of natural gas and its conseq**
- **discuss the uses of oil sands and liquefied co consequences.**
- **describe future prospects for fossil fuel use.**
- **describe how nuclear energy is used to gene electricity.**
- **discuss the advantages and disadvantages of fuels to generate electricity.**

The **Chapter Overview** and **Module Summaries** highlight the key topics covered. Watch the PD **Overview Video** for just-in-time preparation.

The **Alignment Guide** to the AP® environmental science course description correlates module content to the College Board AP® environmental science outline.

Chapter 12 Pacing Guide

This pacing guide is based on a schedule with 120 sessions of 50 minutes each before the AP® Exam. If you have a different number of sessions before the exam, you can modify the pacing to suit your needs. If you have additional time, consider incorporating quizzes, released AP® Environmental Science free-response and multiple-choice questions, or additional activities.

Module	Standard Schedule Days	Block Schedule Days
Module 34	1	½
Module 35	3	1½
Module 36	2	1
Assessment	1	½

Chapter 12 Resources

The link to these resources can be found by clicking on the link buttons in the Teacher's e-Book (TE-book), opening the Teacher's Resource Flash Drive (TRFD), or logging onto the book's companion website bcs.whfreeman.com/friedlandapes2e (teacher login required).

- **PowerPoint Presentation 12**
- **Optimized Art PowerPoint and JPEG Files 12**
- **Do the Math Videos**
- **Measuring Your Impact 12:** Choosing a Car: Conventional or Hybrid?
- **Lab 12.1:** Coal Investigations
- **Exploring the Literature 12**
- **Answers to Chapter 12 AP® Exam**
- **Multiple-Choice AP® Practice Exam 12**

Free-Response Questions from Previous AP® Environmental Science Exams

Free-response questions from prior AP® Environmental Science Exams are available on the AP® course website: https://apstudent.collegeboard.org/apcourse/ap-environmental-science/exam-practice. Students should be able to answer all of the questions listed below with material learned in this and previous chapters. When a question requires students to understand material from multiple chapters, the question will be listed in the last chapter required to complete the entire question. Questions marked with an asterisk (*) are from exams with released multiple-choice questions. You may want to save these questions until the end of the year so you can give your students a complete released exam for practice. Questions marked with double asterisks (**) require math to calculate a problem. Look for references to these questions throughout the chapter.

Year	Question	Content
1998*	2	• The function of parts of a nuclear power plant • Environmental problems associated with nuclear power • The future of nuclear power use in the United States and the environmental implications
2005	3	• Reclamation of land formerly used for mining coal • Restoration of arid land • Environmental impacts of coal extraction and use • The future of coal consumption in the United States
2005	4 **	• Environmental costs of oil extraction from ANWR • Characteristics of a tundra biome • Use of oil in the United States and ways to use it more efficiently
2007	2 **	• Calculation of household water and electricity usage • Calculation of cost savings in using an energy-efficient hot water heater • Measures to reduce household energy and water use
2012	1	• Water-related environmental problems of hydraulic fracturing • The benefits of natural gas compared to coal • The economic benefits of fracking • Environmental problems associated with nuclear power
2014	1	• Alternative viewpoints on safety and impact of nuclear power • Environmental problems associated with construction of a nuclear power plant • Environmental impacts and ecological consequences of the pollution generated by a nuclear power plant • Reduction of electricity use

12-2

Learning Objectives for the chapter are clearly stated and the **Pacing Guide** helps you stay on track.

A complete list of **Chapter Resources** previews available Teacher Resource Materials for the chapter. Look for the blue TRM button throughout the chapter to find and download resources at point of use.

A list of **Released Free-Response Questions** from previous AP® Environmental Science Exams tells you which questions are related to topics in the chapter. More detailed summaries of each question are located throughout the chapter.

Margin content helps you prepare engaging lessons.

PD Chapter 12 Overview

Watch the video overview of Chapter 12 (for teachers) by clicking on the link buttons in the TE-book, opening the TRFD, or logging onto the book's companion website bcs.whfreeman.com /friedlandapes2e (teacher login required).

This massive structure is an oil refinery in Antwerp, Belgium. (Eberhard Streichan/Corbis)

TRM PowerPoint Presentation 12

Download the PowerPoint presentation for Chapter 12 by clicking on the link buttons in the TE-book, opening the TRFD, or logging onto the book's companion website bcs.whfreeman.com /friedlandapes2e (teacher login required).

396 Chapter 12 Nonrenewable Energy Resources

PD Chapter Overview Videos, presented by one of the authors or an experienced AP® environmental science instructor, offer tips for teaching the chapter and advice on how to make the most of the Teacher's Resource Materials.

PowerPoint Presentations available for download include relevant images and diagrams from the student textbook.

chapter 12 Nonrenewable Energy Resources

Module **34** Patterns of Energy Use

Module **35** Fossil Fuel Resources

Module **36** Nuclear Energy Resources

All Energy Use Has Consequences

The modern world is dependent on fossil and nuclear fuels for energy. Many of the benefits of our modern society—health care, comfortable living conditions, easy travel, abundant food—rely on readily accessible and relatively affordable fossil and nuclear fuels. But the costs to society are high.

For example, on April 20, 2010, an explosion and fire occurred at the British Petroleum Deepwater Horizon oil rig in the Gulf of Mexico. Oil gushed from the rig until it was capped 87 days later, on July 15. The BP accident killed 11 workers on the drilling platform, injured 17 others, and released more than 780 million liters (206 million gallons) of oil into the gulf. Scientists from the Center for Biological Diversity later estimated that at least 6,000 sea turtles, 26,000 marine mammals, and more than 82,000 birds were killed as a result of the spill. The oil spread through the Gulf of Mexico and washed up on the shores of Louisiana, Mississippi, Alabama, and Florida.

The modern world is dependent on fossil and nuclear fuels for energy but the costs to society are high.

Before the Deepwater Horizon incident, the largest spill in U.S. waters had been the March 1989 Exxon Valdez accident, when the Valdez, a supertanker carrying 200 million liters (53 million gallons) of oil, crashed into a reef in Prince William Sound, Alaska. Roughly 42 million liters (11 million gallons) of oil spilled into the sound, much of which washed up on shore, contaminating the coastline and killing perhaps half a million birds and thousands of marine mammals. The number of dead animals was so much greater in the Exxon Valdez accident because the spill occurred in a relatively enclosed sound rather than in open waters.

397

Teaching Tip: Chapter Opening Case

The chapter opening case, "All Energy Use Has Consequences," discusses some of the risks associated with use of nonrenewable fossil fuels. It mentions both the *Exxon Valdez* spill in 1989 and the BP *Deepwater Horizon* oil rig blowout in 2010, as well as the more recent events at the Fukushima nuclear power plant in Japan. Countries of the developed world consume more energy (EJ) per capita than undeveloped countries (see Figure 34.2 on page 400). The technological progress associated with increased energy consumption provides many benefits, but our dependence on fossil fuels for energy also has many long-term costs.

Teaching Tips to guide your planning are placed strategically throughout each chapter. Categories of teaching tips include: Warm-up, Discussion Starter, Activity, Engage, Remember, Video, Concept Map, Debate the Issue, Connections, Review, Beyond the Classroom, and Measuring Your Impact.

Teaching Tip: Beyond the Classroom

Ask students to research and write a short report on a major oil spill or coal mining disaster. Possible topics include:

- Santa Barbara blowout (1969)
- *Exxon Valdez* oil spill (1989)
- BP oil refinery explosion (2005)
- BP *Deepwater Horizon* oil well blowout (2010)
- West Virginia coal mining accident (2010)

Students should address the following in their report:

- Where and when did the event occur?
- How did it happen?
- What were the environmental implications?
- What were the economic implications?
- Did any laws or policies change as a result of the event?

Chapter 12 Nonrenewable Energy Resources 397

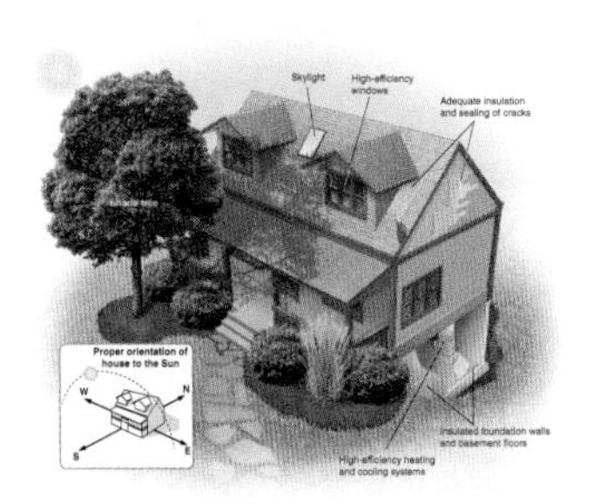

Teaching Tip: Discussion Starter

Ask students: How can building design contribute to energy conservation and efficiency? *A sustainable building design considers a home as a system and attempts to minimize energy inputs and losses from that system. It incorporates proper solar orientation and landscaping as well as insulated windows, walls, and floors.*

Practicing Science: Demo

In this demonstration, students compare incandescent, compact fluorescent, and LED light bulbs. Find light bulbs that can work in the same light fixture. You will also need an electricity usage monitor that plugs into a wall outlet and provides a socket for your lamp.

For each bulb, measure and record the wattage. Have students note the cost and life expectancy of the bulb, which should be listed on the package. Finally, use a thermometer to record the temperature of the bulb, holding the thermometer 3 cm away from the bulb for 5 minutes. Record the data in a table like the one below. (Sample answers are provided in italics.)

Ask students to answer the following questions about the three different types of light bulbs:

- Which light bulb uses the most wattage? Does the temperature of the bulbs tell us anything about the efficiency of the light bulb used? *Incandescent light bulbs use the most watts. They will also have the highest temperature reading on the thermometer because the filament that produces the light also produces wasted heat energy. You should notice that LED light bulbs do not produce much heat and use only about 9 watts of electricity to generate the same amount of light as a 60-watt incandescent bulb.*
- Considering the purchase price of the light bulb, its life expectancy, and the cost of electricity used to run the bulb during its lifetime, do you think LED bulbs save consumers money? Explain your answer. *Although an LED bulb costs more to purchase than other types of bulbs, it uses less electricity over its lifetime and provides the same amount of light. Even when you include the purchase price of the LED bulb, the consumer ends up spending less money over the lifetime of the LED bulb.*

Type and brand	Incandescent bulb Philips	Compact fluorescent EcoSmart	LED Cree
Wattage	*60 watts*	*17 watts*	*9.5 watts*
Estimated life expectancy	*0.9 years*	*9.1 years*	*22.8 years*
Estimated yearly energy cost based on 3 hours per day	*$7.23*	*$1.69*	*$1.14*
Cost of individual bulb	*$0.80*	*$1.50*	*$7.00–$10.00*
Brightness (lumens)	*860 lumens*	*800 lumens*	*800 lumens*

Module 37 Conservation, Efficiency, and Renewable Energy **435**

Practicing Science includes suggestions for in-class demos and hands-on science experiences to help students understand the concepts.

Many full-scale **Labs** are available for download in the Teacher's Resource Materials. Labs and field work develop good science practice and help students understand complex relationships in the natural world.

lab
Solar Energy

This lab allows students to experiment with solar cells and to compare energy output with different variables in order to learn how passive solar systems work.

Download this lab by clicking the link buttons in the TE-book, opening the TRFD, or logging onto the book's companion website bcs.whfreeman.com /friedlandapes2e (teacher login required).

TRM Lab 13.1 Solar Energy

Teaching Tip: Review

Ask students the following questions about solar energy:

- In what ways do humans capture solar energy for their use? *Humans capture solar energy in four main ways: passive energy systems, active energy systems, photovoltaic energy systems, and solar thermal power plants. Some humans also capture energy from the Sun to cook food in a solar cooker.*
- Why are active solar energy systems not feasible everywhere? *The average amount of solar energy available varies geographically. In the continental United States, average daily solar radiation ranges from 3 kWh of energy per square meter in the Pacific Northwest to almost 7 kWh per square meter in parts of the Southwest.*

Teaching Tip: Video

How Solar Panels Work
Use this short video clip from Science 360 to help students understand how solar panels work.

The link to this video may be found by clicking the link buttons in the TE-book, opening the TRFD, or logging onto the book's companion website bcs.whfreeman.com/friedlandapes2e (teacher login required).

TRM Chapter 13 Web Resources

Module 39 Solar, Wind, Geothermal, and Hydrogen **453**

◀ TEACHING with FIGURES

Show students Figure 34.6 and ask: Will there be a trend to purchase gas-efficient compact cars or will Americans continue to purchase larger, less fuel-efficient vehicles? *If prices rise and gas is more expensive for an extended period, overall efficiency of U.S. automobiles that are purchased will increase. People will begin to buy smaller, more compact cars that achieve a higher mpg. If the price per gallon falls, people will not have an incentive to economize on fuel use and may continue to purchase vehicles that have a lower mpg.*

AP® Exam Tip

Review kinetic and potential energy before discussing how coal or other fossil fuels produce electricity. This is a good time to review the first and second laws of thermodynamics. Use these concepts to describe what happens in a coal-fired electricity generation plant. In such plants, the potential energy in coal is converted to the kinetic energy of water that is converted into steam, which, in turn, spins a turbine. The second law of thermodynamics tells us that when energy is transformed, the quantity of energy remains the same, but its ability to do work diminishes.

AP® Exam Tip

Household electricity and water usage is covered on the 2007 AP® Exam, Question 2. To answer this question, students must

- use provided data to calculate the amount of water a family uses and the yearly cost of electricity for the family to shower
- calculate the time needed to recoup the investment in an energy-efficient hot water heater
- describe how a family can reduce overall water use
- list measures to reduce total energy use

Teaching Tip: Discussion Starter

Ask students: What is the basic process by which the energy in a fuel is converted into electricity? *Electricity is a secondary source of energy, meaning that we obtain it from the conversion of a primary source such as coal, oil, or natural gas. By the nature of being a secondary source, electricity is an energy carrier—something that can move and deliver energy in a convenient, usable form to end users.*

Module 34 Patterns of Energy Use **405**

Teaching with Figures provides suggestions and questions to help teach students how to read and analyze graphs and figures.

AP® Exam Tip offers additional hints for your students to succeed on the exam and summarizes relevant free-response questions from past exams that you can use with your students.

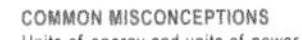

COMMON MISCONCEPTIONS
Units of energy and units of power

Units of energy and power can often be difficult for students to understand. Review the difference between energy and power and make sure students know the units used for each.

Energy is the ability to do work or transfer heat. Below are some common units of energy that might appear on the exam. See Table 5.1 on page 44 for common usages.

- **joule (J):** The amount of energy used when a 1-watt electrical device is turned on for 1 second.
- **British thermal unit (Btu):** The amount of energy required to heat 1 pound of water by 1 degree Fahrenheit.
- **calorie:** The amount of energy it takes to heat 1 gram of water by 1 degree Celsius.
- **kilowatt hour (kWh):** The amount of energy expended by using 1 kilowatt of electricity for 1 hour.

Power is the rate at which work is done. The units for power describe the amount of work done over time. Below are some common units of power that may appear on the exam.

- **watt (W):** 1 watt is equal to 1 joule per second (1 J/s).
- **horsepower (hp):** One horsepower is equal to 746 watts.

COMMON MISCONCEPTIONS
Calories and calories

Students often confuse calories and Calories. Remind them that calorie is a basic scientific unit for energy. A food calorie, like that listed on the packaging of a product, is written as Calorie (uppercase C) and is equal to 1,000 calories, or 1 kilocalorie (kcal).

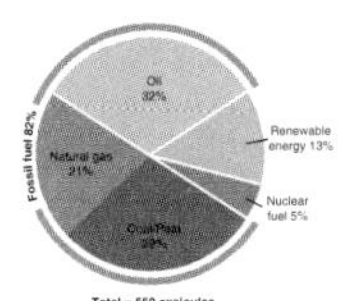

more math practice
Conversions

You can convert from one unit of energy to another:
1 calorie = 4.184 J
1 Btu = 1,055 J
1 kWh = 3,600,000 J = 3.6 Megajoules (MJ)
1 hp = 746W

Here are two problems to try with your class:

1. A runner burns 600 Calories in track practice. How many joules (J) did the runner burn?

600 Calories × (1,000 calories/Calorie) × (4.184 J/calorie) = 2,510,400 J

2. A lawnmower has 5 hp. How many watts would that equal?
5 hp × (746 W/hp) = 3,730 W

400 Chapter 12 Nonrenewable Energy Resources

Common Misconceptions alerts you to topics that may require additional attention.

Step-by-step **Answers** to **Do the Math** problems are easy to find.

For an additional math challenge, **More Math Practice** includes math teaching tips and worked problems to help students practice the math skills they need to succeed on the exam.

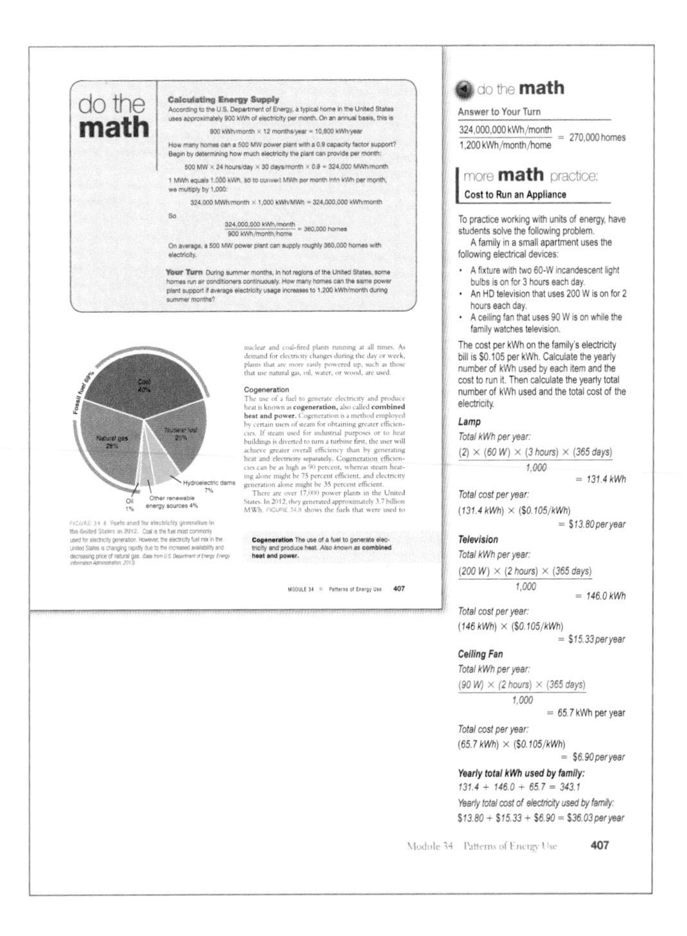

do the math
Answer to Your Turn

$\frac{324{,}000{,}000 \text{ kWh/month}}{1{,}200 \text{ kWh/month/home}} = 270{,}000$ homes

more math practice:
Cost to Run an Appliance

To practice working with units of energy, have students solve the following problem.

A family in a small apartment uses the following electrical devices:

- A fixture with two 60-W incandescent light bulbs is on for 3 hours each day.
- An HD television that uses 200 W is on for 2 hours each day.
- A ceiling fan that uses 90 W is on while the family watches television.

The cost per kWh on the family's electricity bill is \$0.105 per kWh. Calculate the yearly number of kWh used by each item and the cost to run it. Then calculate the yearly total number of kWh used and the total cost of the electricity.

Lamp
Total kWh per year:
$\frac{(2) \times (60\ W) \times (3\ hours) \times (365\ days)}{1{,}000} = 131.4$ kWh
Total cost per year:
$(131.4\ kWh) \times (\$0.105/kWh) = \$13.80$ per year

Television
Total kWh per year:
$\frac{(200\ W) \times (2\ hours) \times (365\ days)}{1{,}000} = 146.0$ kWh
Total cost per year:
$(146\ kWh) \times (\$0.105/kWh) = \$15.33$ per year

Ceiling Fan
Total kWh per year:
$\frac{(90\ W) \times (2\ hours) \times (365\ days)}{1{,}000} = 65.7$ kWh per year
Total cost per year:
$(65.7\ kWh) \times (\$0.105/kWh) = \$6.90$ per year

Yearly total kWh used by family:
131.4 + 146.0 + 65.7 = 343.1
Yearly total cost of electricity used by family:
\$13.80 + \$15.33 + \$6.90 = \$36.03 per year

Module 34 Patterns of Energy Use 407

Teaching Tip: Measuring Your Impact

Choosing a Light Bulb
Measuring Your Impact 13 asks students to calculate the costs of using incandescent and fluorescent light bulbs. The link to this resource may be found by clicking on the link buttons in the TE-book, opening the TRFD, or logging onto the book's companion website bcs.whfreeman.com/friedlandapes2e (teacher login required).

Measuring Your Impact 13: Choosing a Light Bulb

Exploring the Literature

Friedland, A. J., and K. T. Gillingham. 2010. Carbon accounting a tricky business (letter). *Science* 327:410–411.

International Energy Agency: http://www.iea.org.

Kruger, P. 2006. *Alternative Energy Resources*. John Wiley & Sons.

Randolph, J., and G. M. Masters. 2008. *Energy for Sustainability*. Island Press.

Smith, T. P. 2014. Fish, tides and turbines. *Natural History* 121:32–39.

U.S. Department of Energy, National Renewable Energy Laboratory: http://www.nrel.gov.

U.S. Energy Information Administration, Independent Statistics and Analysis: http://www.eia.doe.gov.

Wolfson, R. 2011. *Energy, Environment, and Climate*. 2nd ed. W. W. Norton.

Download a printable version of this list by clicking on the link buttons in the TE-book, opening the TRFD, or logging onto the book's companion website bcs.whfreeman.com/friedlandapes2e (teacher login required).

Exploring the Literature 13

Chapter 13 Review 467

Measuring Your Impact offers students an opportunity to explore the environmental impact of decisions made in the real world. These activities also provide additional math practice.

Exploring the Literature offers a list of recent and classic sources for students who want to delve deeper into a chapter topic.

Answers are located where you need them.

Answers to the Module AP® Review Questions, Working Toward Sustainability Critical Thinking Questions, and Chapter AP® Practice Exams are placed beside the questions. Answers for Free-Response Questions continue in the Answer Appendix.

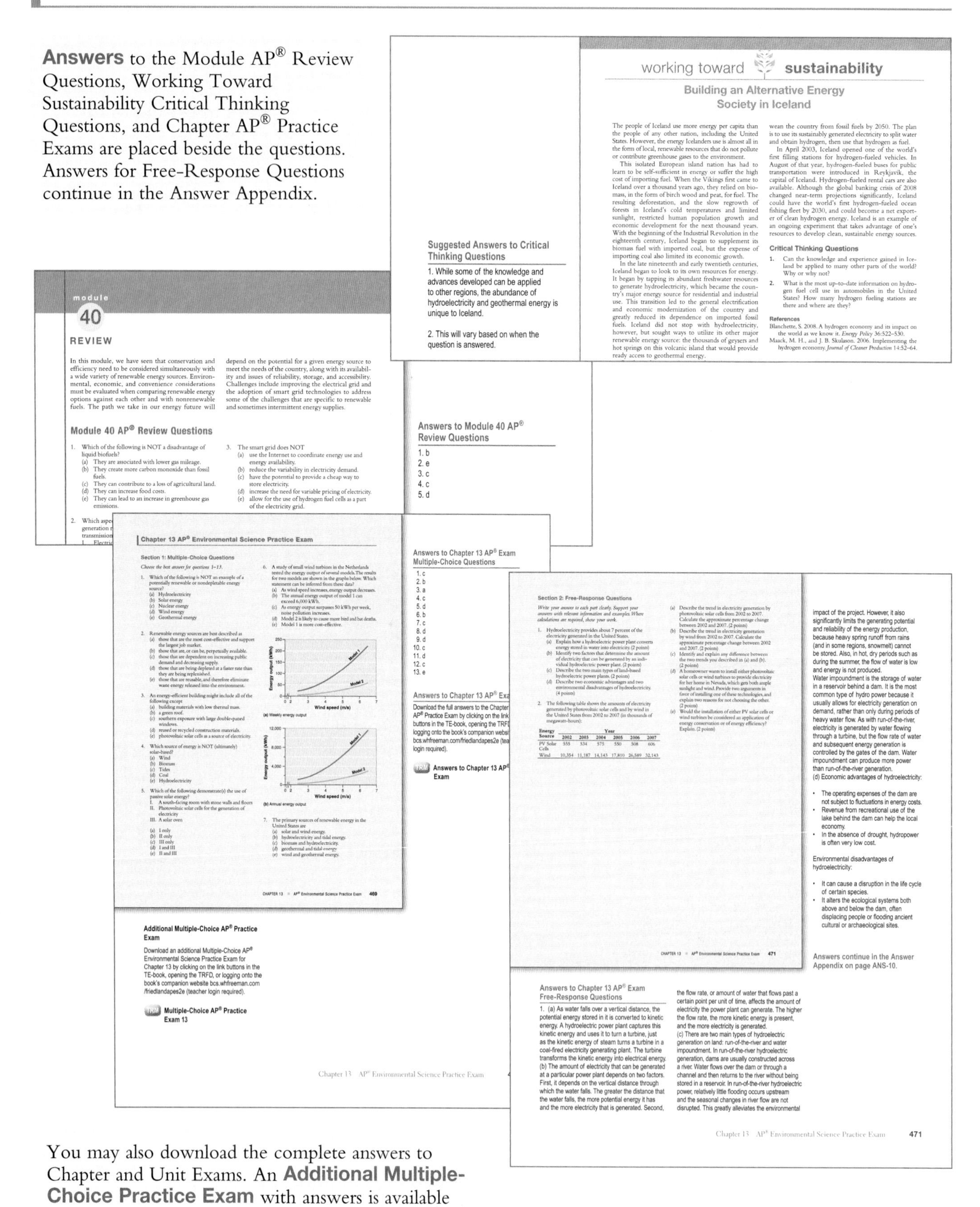

working toward sustainability
Building an Alternative Energy Society in Iceland
Critical Thinking Questions
Suggested Answers to Critical Thinking Questions
1. While some of the knowledge and advances developed can be applied to other regions, the abundance of hydroelectricity and geothermal energy is unique to Iceland.
2. This will vary based on when the question is answered.
module 40 REVIEW
Module 40 AP® Review Questions
Answers to Module 40 AP® Review Questions
1. b
2. e
3. c
4. c
5. d
Chapter 13 AP® Environmental Science Practice Exam
Section 1: Multiple-Choice Questions
Answers to Chapter 13 AP® Exam Multiple-Choice Questions
1. c
2. b
3. a
4. c
5. d
6. b
7. c
8. d
9. d
10. c
11. d
12. c
13. e
Additional Multiple-Choice AP® Practice Exam
Download an additional Multiple-Choice AP® Environmental Science Practice Exam for Chapter 13 by clicking on the link buttons in the TE-book, opening the TRFD, or logging onto the book's companion website bcs.whfreeman.com/friedlandapes2e (teacher login required).
Multiple-Choice AP® Practice Exam 13
Section 2: Free-Response Questions
Answers to Chapter 13 AP® Exam Free-Response Questions
Answers continue in the Answer Appendix on page ANS-10.

You may also download the complete answers to Chapter and Unit Exams. An **Additional Multiple-Choice Practice Exam** with answers is available for each chapter.

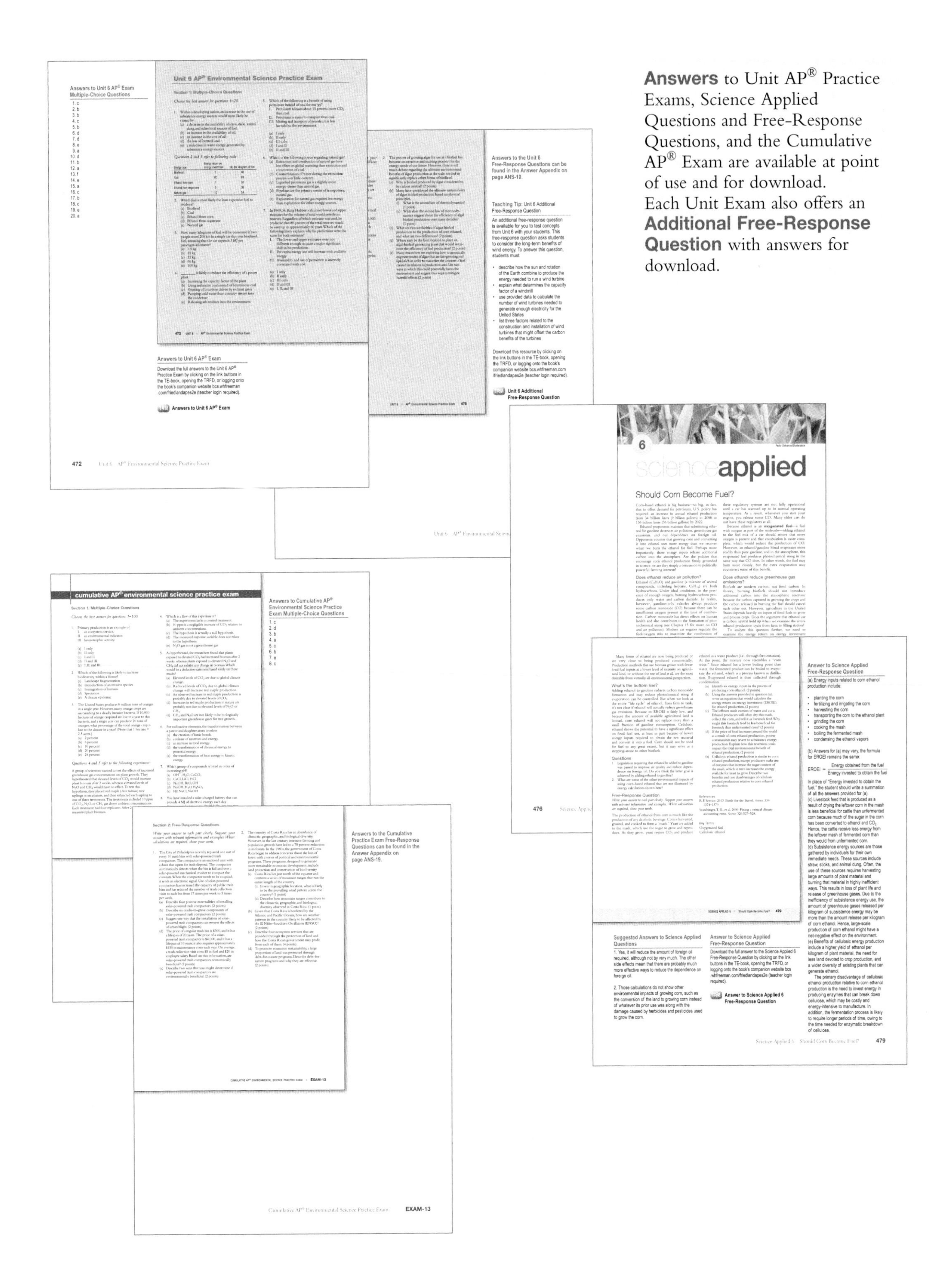

Answers to Unit AP® Practice Exams, Science Applied Questions and Free-Response Questions, and the Cumulative AP® Exam are available at point of use and for download.
Each Unit Exam also offers an **Additional Free-Response Question** with answers for download.

Available Resources

For Teachers

edAPtext Teacher's Edition

This innovative new digital product gives you full access to the Teacher's Edition for *Friedland and Relyea Environmental Science for AP®*, Second Edition, when and where you want it. The TE-book offers page fidelity, so that it matches the print page exactly. You may read the student textbook and teacher content and take notes on the computer at school, and then later you can pick up where you left off on your home computer or in a downloaded version on your iPhone/Android smartphone or iPad. All of your notes sync once you are connected to the Internet and log in. Your edAPtext TE-book is the ultimate integrator for all of your *Friedland and Relyea Environmental Science for AP®* resources. While you are in the connected mode, you may access all of the PD Videos, TRM resources, and other links at point of use. If you want to provide assignments, notes, or give quizzes, you may do so using the social-media function of edAPtext and the teacher dashboard. The edAPtext TE-book is designed to make classroom management and communication easier so you will be able to focus on teaching.

Teacher's Resource Flash Drive (TRFD) (ISBN: 1-4641-5619-0)

Supporting teacher resources can be found on the TRFD, the Book Companion Site, as well as through the direct links embedded in the edAPtext TE-book. The teacher's resources on the TRFD include:

- Chapter Overview Videos
- Chapter PowerPoint Presentations
- Optimized Art PowerPoint and JPEG Files
- Do the Math Videos
- Labs (instructor and student versions)
- Activities (instructor and student versions)
- More Math Practice (instructor and student versions)
- Measuring Your Impact (instructor and student versions)
- Web Resources for each chapter
- Answers to end-of-chapter AP® Practice Exams, unit AP® Practice Exams, full-year Cumulative AP® Practice Exam, and Science Applied Free-Response Questions
- Additional Multiple-Choice AP® Practice Exams, with answers, for each chapter
- An additional Free-Response Question, with answer, for each unit exam
- Exploring the Literature references

Book Companion Site (BCS)

The Book Companion Site is a free website that offers valuable tools for teachers and students. The teacher's view of the website features the extensive Teacher's Resource Materials for use with the course. This is a password-protected site found at bcs.whfreeman.com/friedlandapes2e (login required).

ExamView® Assessment Suite (ISBN: 1-4641-5612-3)

The ExamView® Assessment Suite allows teachers to create online and paper tests and quizzes quickly and easily. Users may select from the extensive bank of more than a thousand questions or may write their own. Questions can be sorted according to various metadata fields and scrambled to create different versions of tests. Tests may be printed or administered online using the ExamView® Player. The ExamView® Assessment Suite can automatically grade and send results of multiple-choice questions as well as create reports for teachers.

Chapter PD Videos

Chapter overview videos give teachers a quick guide to what is covered in each chapter. Presented by one of the authors or by an experienced AP® environmental science teacher, these videos offer tips and advice for teaching the chapter and help you make the most of your Teacher's Resource Materials.

Do the Math: Foundations Videos

Each of these short (2–5 minutes) videos presents a teacher working step-by-step through an AP® environmental science problem that requires the student to recall and use basic math skills. Some of the math concepts and skills covered are scientific notation, dimensional analysis, and exponential growth.

PowerPoint Presentations

PowerPoint presentations for each chapter in *Friedland and Relyea Environmental Science for AP®*, Second Edition, include relevant images and diagrams from the student textbook for use during class lectures.

Student Resources

edAPtext Student e-Book

This innovative new digital product gives students full access to *Friedland and Relyea Environmental Science for AP®*, Second Edition, when and where they want it. They may read the textbook and take notes on the computer at school and then later pick up where they left off using a downloaded version on an iPhone/Android smartphone or iPad. Notes taken offline will sync once the user is connected to the Internet and logs in. In addition to accessing student notes, the social media function allows students to receive group or individual messages from the teacher, to work in study groups, to take quizzes, and to complete other assignments, all at the point of use, within the edAPtext program.

Strive for a 5: Preparing for the AP® Environmental Science Examination (ISBN: 1-4641-5616-6)

Strive for a 5 is designed to help students evaluate their understanding of the material covered in the student textbook, to reinforce key concepts, and to prepare students for success on the AP® Environmental Science Exam. The book has a study guide section that corresponds to each textbook chapter, and a test preparation section.

Your AP® Environmental Science Primer

Preparing to Teach AP® Environmental Science

Teaching AP® environmental science is an exciting and rewarding experience. Students find the curriculum interesting and appreciate the numerous practical applications they can make to their daily lives. The course encourages them to think about situations in their own communities and about important global issues. The course also lends itself to many engaging hands-on activities.

If you are like many AP® environmental science teachers, you are a converted biology, Earth science, or chemistry teacher. You may have taken an environmental science class in college that was different from the AP® environmental science course you are preparing to teach. Whether you are new to the course or an experienced instructor, be sure to utilize the many opportunities and resources available to the AP® environmental science community.

In addition to reviewing the materials in this Teacher's Edition and the supplementary items in the Teacher's Resource Materials, we recommend you take advantage of as many of the following as possible:

- Make sure your school has completed the AP® Course Audit. See the AP® coordinator at your school or go to www.collegeboard.com/html/apcourseaudit/ for more information.
- Review the Environmental Science Course Home Page at the AP® Central website: http://apcentral.collegeboard.com/apc/public/courses/teachers_corner/2128.html. Here you can find many of the materials you will need to begin planning your curriculum, including:
  - Course descriptions
  - Course overview
  - Teacher's guide
  - Free released exam
  - Past exam free-response questions
  - Information on scoring questions
  - Classroom resources
- Set up your free account on the AP® Environmental Science Teacher Community website: https://apcommunity.collegeboard.org/web/apenvscience. This website will enable you to:
  - Find and share class materials for your course
  - Connect with other AP® environmental science instructors
  - Participate in topic discussions
  - Sign up for email notifications

- If possible, attend a workshop approved by the College Board. These week-long workshops are typically offered in the summer in a number of locations and are conducted by consultants the College Board has endorsed. Shorter workshops are offered throughout the year.
- Consider attending the College Board AP® national conference in July. This conference is an excellent opportunity to learn what your peers are doing, get inspiring new ideas, and share strategies for teaching AP® environmental science.

An Introduction to the AP® Environmental Science Exam

The AP® Environmental Science Exam consists of 100 multiple-choice questions and 4 free-response questions. The multiple-choice section of the exam accounts for 60 percent of the score and the free-response section accounts for 40 percent. The exam is 3 hours long, with 90 minutes allotted to each of the two sections. Note that calculators are *not* permitted.

The free-response questions are divided as follows:

- One question based on a provided document such as a newspaper excerpt or research report
- One question requiring analysis of a provided data set
- Two questions requiring synthesis and evaluation

Students must answer all 100 of the multiple-choice questions. This section of the exam is graded by computer. Free-response questions are graded by a group of high school and college instructors hired to score the exams.

To make the grading as fair and objective as possible, each free-response question is assigned a grading rubric that allocates points for scoring the exam. To help familiarize students with this system, all free-response questions in *Friedland and Relyea Environmental Science for AP®* indicate a point allocation.

The Educational Testing Service typically sends students their scores in early July. Depending upon each student's choice, the scores also may be sent directly to colleges and universities. Teachers can access student scores through an Educational Testing Service website. Your school's testing coordinator will help you to set up a password and obtain your school's access code.

Exam results fall into one of the following categories on a five-point scale:

5	Extremely well qualified
4	Well qualified
3	Qualified
2	Possibly qualified
1	No recommendation

Most colleges accept a score of 4 or 5 for credit and placement, while some colleges offer credit for a score of 3. Students may go online to the college they will be attending and find the score that institution accepts, or they may contact that school's admissions office for more information.

Topics Covered on the AP® Environmental Science Exam

The AP® Environmental Science Exam is divided into seven major topics areas, shown below with a brief description of each. The emphasis given to the topic areas on the exam is noted in parentheses. The teaching guide at the beginning of each chapter of the Teacher's Edition indicates how the chapter modules correspond to these topics.

Topic area	Subtopics
Earth systems and resources (10–15%)	• Earth science concepts • Earth's atmosphere • global water resources and use • soil and soil dynamics
The living world (10–15%)	• ecosystem structure • energy flow • ecosystem diversity • natural ecosystem change • biogeochemical cycles
Population (10–15%)	• population ecology • human population
Land and water use (10–15%)	• agriculture • pesticides • forestry • rangelands • mining • other land use • fishing and aquaculture • global economics
Energy resources and consumption (10–15%)	• energy concepts • energy consumption • fossil fuel resources and use • nuclear energy • energy conservation • renewable energy
Pollution (25–30%)	• air pollution • noise pollution • water pollution • solid waste reduction and disposal • pollution hazards to human health • hazardous chemicals in the environment • economic impacts of pollution
Global change (10–15%)	• stratospheric ozone • global warming • loss of biodiversity

Pacing

Pacing is often one of the biggest challenges for teachers of an AP® environmental science course. This suggested pacing guide for use with the textbook is based on a schedule with 120 sessions of 50 minutes each before the AP® Exam. If you have a different number of sessions before the exam, modify the pacing to suit your needs. If you have additional time, consider incorporating quizzes, released AP® environmental science free-response and multiple-choice questions, or additional activities.

Chapter	Standard schedule days	Block schedule days
1	3	1½
2	2	1
3	8	4
4	8	4
5	5	2½
6	8	4
7	6	3
8	6	3
9	7	3½
10	5	2½
11	5	2½
12	7	3½
13	8	4
14	8	4
15	8	4
16	6	3
17	6	3
18	5	2½
19	6	3
20	3	1½

Preparing Students for the AP® Environmental Science Exam

Although there is no foolproof way to guarantee your students success on the AP® Environmental Science Exam, by choosing *Friedland and Relyea Environmental Science for AP®* you are off to an excellent start. This textbook program has been developed and written specifically for a high school AP® environmental science course in consultation with experienced high school and college instructors who understand what teachers need to help students do their best and realize their potential. Here are some of the features of *Friedland and Relyea Environmental Science for AP®* that are designed to help you help your students excel on the AP® Exam:

- Every topic on the official AP® Environmental Science Topic Outline is covered in detail in this book.

- Throughout the Teacher's Edition you will find AP® Exam Tips.
- Each module ends with AP® Review Questions for practice.
- Critical Thinking Questions in Working Toward Sustainability help students to think and write critically about current events.
- Each chapter ends with an AP® Environmental Science Practice Exam. An Additional Multiple-Choice Practice Exam is available in the Teacher's Resource Materials for each chapter.
- Each unit concludes with an AP® Environmental Science Practice Exam that includes at least 20 multiple-choice questions and 2 free-response questions covering topics from the entire unit. An additional practice free-response question is available in the Teacher's Resource Materials for each unit.
- The Science Applied feature includes review questions and a practice free-response question related to the topic covered in the article.
- The book concludes with a Cumulative AP® Environmental Science Practice Exam containing 100 multiple-choice questions and 4 free-response questions, just like the actual exam, described in the sections below.

In addition to using the Teacher's Edition and all the available resources, experienced AP® environmental science teachers recommend a number of strategies for helping students achieve success on the exam, described in the sections below.

AP® Topic Outline

Give students a copy of the AP® Topic Outline and get in the habit of alerting students when you have finished one topic and are starting a new one. We have also placed the AP® Topic Outline in the front endpapers of both the student text and this Teacher's Edition.

Preparing for Free-Response Questions

Make reading and writing an integral part of your course work. Many free-response questions require students to use these skills. In addition to the practice free-response questions in the text and those from past exams, you might require students to read articles about current environmental issues or events every week and to write a brief summary of them in their journals.

Assessments

Although you will find a multitude of assessments and test items with this program, you probably will not have time to use all of them. How each teacher assesses students is a matter of personal preference. Some teachers find time for quizzes, while many teachers only formally assess their students at the end of each chapter or unit.

No matter how you decide to assess student progress, you should use as many actual AP® environmental science questions as possible. There are a number of ways to incorporate previous free-response questions into your course. Some teachers include one or two AP® free-response questions on each test. Other teachers assign them for homework or in cooperative groups during class. If you assign them for homework, be aware that students can access the solutions online. Because there are more than a hundred released free-response questions, it is probably safe to include one or two on an exam. In the Teacher's Edition we list released questions that are appropriate for each chapter. You can find the list in the introductory pages that precede each chapter, with more detailed descriptions throughout the chapter. Here are some additional tips for assessment:

- Don't be afraid to give your students difficult questions. Try to be a fair but tough grader, because the rubrics on the actual AP® Exam are often quite challenging.
- Offer cumulative assessments to help students retain material from earlier parts of the course. Some teachers include cumulative questions on each exam, while others give cumulative exams every few chapters.
- Make sure students know how the AP® Environmental Science Exam is graded. Show them the scoring guidelines and examples of student responses after you answer old AP® questions in class. The rubrics and examples of student responses (with commentary) are available on AP® Central.
- Shortly before they take the actual exam, have students come in after school or on a weekend to take a full-length practice exam. You can use the exam at the end of the text or one of the released past exams from the College Board. It is highly beneficial for students to experience a real exam, given under actual conditions, so they can get a sense of how fast they need to work and what topics they should review. If possible, administer the practice exam in the room in which students will take the actual AP® Exam.

Labs and Field Activities

The AP® environmental science course is designed to have a strong laboratory component. You can find links to resources for labs and field activities on AP® Central. The Practicing Science feature in the Teacher's Edition offers ways to help bring the practice of science into your classroom. Beyond the Classroom, also found in the Teacher's Edition, suggests numerous field trip and guest speaker ideas for both large and small classrooms. In addition, the Teacher's Resource Materials include a number of full-length labs.

Math Practice

The AP® Environmental Science Exam requires students to answer questions that involve math. This can sometimes be a stumbling block for students who may be intimidated by mathematical problem solving. To help students prepare to solve the math problems on the exam, the student textbook includes Do the Math boxes in most chapters. These boxes review core math skills and calculations with which students must be familiar to work problems or answer questions on numerous environmental science topics. The Do the Math boxes are a great way for teachers to help students work through some tough math problems without the use of calculators. Where possible, math practice is provided in the AP® Practice Exams that fall at the ends of modules, chapters, and units, as well as at the end of the year. The more students practice, the more comfortable they will be with math problems on the actual exam.

Look for more math support tools in the Teacher's Edition and Teacher's Resource Materials. More Math Practice in the Teacher's Edition includes math teaching tips and additional worked problems to help students practice the math skills they need to succeed on the exam. The Teacher's Resource Materials also include additional downloadable More Math Practice problems with complete solutions. Look for references to these additional problems in the margins of your Teacher's Edition. Teaching with Figures provides suggestions and questions to help you teach students how to read and analyze graphs and figures in the text.

Homework

How each teacher assigns and grades homework is a matter of personal preference. The number of questions and activities available in the student textbook and Teacher's Edition should offer ample opportunity for homework assignments in any given school year.

Activities and Projects

Many teachers assign activities or more involved projects to help promote student engagement and synthesize learning across topics. The Teacher's Edition includes multiple suggestions for every chapter on ways to enhance your class with videos, visuals, debates, discussions, projects, activities, and field trips. Although you may feel overwhelmed by the content you must cover in the course, experienced teachers agree that getting students engaged in activities and projects is time well spent. Activities and projects promote critical thinking, facilitate connections between concepts, build teamwork, and help students put their knowledge to work. Although it is probably not possible to assign a project for each chapter, many teachers try to assign one project per semester.

Technology

Technology is a vital part of AP® environmental science. Today students are expected to use computers to help create tables and graphs and to perform calculations. Many teachers encourage the use of spreadsheet programs to organize and then graph data. There are also numerous websites available that allow students the opportunity to learn with digital support tools. But don't forget that you can have an effective and exciting class with mini white boards and poster paper as well.

Technology can enhance learning experiences in many ways. However, you should be aware that at present, on the actual AP® Environmental Science Exam, students are not allowed to use calculators when solving problems. Make sure that you give students plenty of calculator-free practice time throughout the year so they are adept at solving math problems on paper.

What to Do After the Exam

If you still have time with your students after the AP® Exam, there are several options:

- **Projects** This is an ideal time for students to bring together all of the topics from the course. Have students research a question about environmental policies that have multiple perspectives and topics to consider.
- **Debates** There are opportunities throughout the course for students to research and debate topics and ideas related to environmental science. At the end of the year, you might offer students an opportunity to work on public speaking and debating skills.
- **Labs and field activities** If you did not have time to complete certain labs or field activities, the period after the exam could be a chance to offer these experiences as enrichment.

- **Movies and documentaries** The Teacher's Resource Materials recommend many films and documentaries your students might enjoy. You can add a critical component to watching a movie in class by having students reflect and write about environmental science topics addressed in the film.
- **Guest speakers** If you did not have time during the year, the period after the exam is an excellent opportunity to invite environmental science professionals to speak to the class about their careers and how they are making a difference.

Recruiting Students for Next Year

Students may have heard about AP® environmental science at your school without fully understanding what is covered in the course. You have the opportunity to promote the hands-on learning and real-world applications that make an AP® environmental science class worth taking. Consider recruiting new students by visiting appropriate science classes and giving a short presentation about the class. When talking to prospective students, you might try to hit three main themes:

- AP® environmental science is not a typical science class. However, this doesn't mean the class will be easy. Students will need good critical thinking and communication skills to succeed.
- AP® environmental science offers great preparation for college. Many liberal arts colleges and universities require students to take at least two science courses. Taking an AP® environmental science course might count toward this prerequisite. It also demonstrates that a student can handle college-level work.
- AP® environmental science covers issues of pressing interest and relevance to the world around us. It relies on real-world studies in a variety of fields. A student who has taken this course will be able to think critically and discuss with confidence claims made in the media about issues related to the environment.

Brief Contents

Contents

UNIT 4
Earth Systems and Resources

UNIT 5
Land Use

UNIT 6
Energy Resources and Consumption

UNIT 8
Global Change and a Sustainable Future

About the Authors

Nancy Nutile-McMenemy

Brian Mattes

Andrew Friedland is Richard and Jane Pearl Professor in Environmental Studies and former chair of the Environmental Studies Program at Dartmouth College. He was the founding chair of the Advanced Placement Test Development Committee (College Board) for Environmental Science. He has a strong interest in high school science education, and in the early years of AP® environmental science he participated in many trainer and teacher workshops. For more than 10 years, Andy has been a guest lecturer at the St. Johnsbury Academy Advanced Placement Institute for Secondary Teachers. He has also served on the College Board AP® Environmental Science Curriculum Development and Assessment Committee.

Andy regularly teaches introductory environmental science and energy courses at Dartmouth and has taught courses in forest biogeochemistry, global change, and soil science, as well as foreign study courses in Kenya. Beginning in 2015, Andy brings his introductory environmental science course to the massive, open, online course format through the DartmouthX platform.

Andy received a BA degree in both biology and environmental studies, and a PhD in earth and environmental science, from the University of Pennsylvania. For more than three decades, Andy has been investigating the effects of air pollution on the cycling of carbon, nitrogen, and lead in high-elevation forests of New England and the Northeast. Recently, he has been examining the impact of increased demand for wood as a fuel, and the subsequent effect on carbon stored deep in forest soils.

Andy has served on panels for the National Science Foundation, USDA Forest Service, and Science Advisory Board of the Environmental Protection Agency. He has authored or coauthored more than 65 peer-reviewed publications and one book, *Writing Successful Science Proposals* (Yale University Press).

Andy is passionate about saving energy and has pursued many energy efficiency endeavors in his home. Recently, he installed a 4 kW solar photovoltaic tracker that follows the Sun during the day.

Rick Relyea is the David Darrin Senior '40 Endowed Chair in Biology and the executive director of the Darrin Freshwater Institute at Rensselaer Polytechnic Institute. Rick teaches courses in ecology, evolution, and animal behavior at the undergraduate and graduate levels. He received a BS in environmental forest biology from the State University of New York College of Environmental Science and Forestry, an MS in wildlife management from Texas Tech University, and a PhD in ecology and evolution from the University of Michigan.

Rick is recognized throughout the world for his work in the fields of ecology, evolution, animal behavior, and ecotoxicology. He has served on multiple scientific panels for the National Science Foundation and has been an associate editor for the journals of the Ecological Society of America. For two decades, he has conducted research on a wide range of topics, including predator-prey interactions, phenotypic plasticity, eutrophication of aquatic habitats, sexual selection, disease ecology, long-term dynamics of populations and communities across the landscape, and pesticide impacts on aquatic ecosystems. He has authored more than 110 scientific articles and book chapters, and has presented research seminars throughout the world. Rick recently moved to Rensselaer from the University of Pittsburgh, where he was named the Chancellor's Distinguished Researcher in 2005 and received the Tina and David Bellet Teaching Excellence Award in 2014.

Rick has a strong interest in high school education. High school science teachers conduct research in his laboratory and he offers summer workshops for high school teachers in the fields of ecology, evolution, and ecotoxicology. Rick also works to bring cutting-edge research experiments into high school classrooms.

Rick's commitment to the environment extends to his personal life. He lives in a home constructed with a passive solar building design and equipped with active solar panels on the roof. The solar panels generate so much electricity that he sells the extra electricity back to the local electric utility every month.

Acknowledgments

We would like to thank the many people at Bedford, Freeman, and Worth who helped guide us through the publication process in both the first and second editions of this book. They have taught us a great deal and have been crucial to our book becoming greatly appreciated by so many people. We especially want to acknowledge:

Ann Heath, Jeffrey Dowling, Becky Kohn, Fred Burns, Janie Pierce-Bratcher, Kerry O'Shaughnessy, Julia DeRosa, Matt McAdams, Joseph BelBruno, Anna Skiba-Crafts, Aaron Stoler, Lucas Sanford-Long, Christine Buese, Vicki Tomaselli, Lee Wilcox, Jerry Correa, Beth Howe, Cindi Weiss, Karen Misler, Deborah Goodsite, Ted Szczepanski, and Cathy Murphy. We thank David Courard-Hauri, Ross Jones, and Susan Weisberg for contributions to the first edition of this book.

We also wish to convey our appreciation to the dozens of reviewers who constantly challenged us to write a clear, correct, and philosophically balanced textbook.

From Andy Friedland . . .

A large number of people have contributed to this book in a variety of ways. I would like to thank all of my teachers, students, and colleagues. Professors Robert Giegengack and Arthur Johnson introduced me to environmental science as an undergraduate and graduate student. My current and previous colleagues in the Environmental Studies Program at Dartmouth and elsewhere have contributed in a variety of ways. I thank Doug Bolger, Michael Cox, Rich Howarth, Anne Kapuscinski, Karol Kawiaka, Rosi Kerr, Nick Reo, Bill Roebuck, Jack Shepherd, Chris Sneddon, Scott Stokoe, Ross Virginia, and D.G. Webster for all sorts of contributions to my teaching and scholarship and to this book. Graduate students Chelsea Petrenko and Justin Richardson have also contributed. Emily Lacroix and Jacob Ebersole, Dartmouth undergraduates who have taken courses from me, provided excellent editorial, proofreading, and writing assistance. Many other colleagues have had discussions with me or evaluated sections of text including William Schlesinger, Ben Carton, Jon Kull, Nat Draper, Bob Hawley, Jim Labelle, Tim Smith, Charlie Sullivan, Jenna Pollock, Jim Kaste, Carol Folt, Celia Chen, Matt Ayres, Kathy Cottingham, and Mark McPeek. Since the time when AP® Environmental Science was just an idea at a College Board workshop, Beth Nichols, Tom Corley, and many others, especially teachers I have since met at meetings and workshops, have introduced me to the world of Advanced Placement® teaching.

I wish to acknowledge Dana Meadows and Ned Perrin, both of whom have since passed away, for contributions during the early stages of this work. Terry Tempest Williams has been a tremendous source of advice and wisdom about topics environmental, scientific, and practical.

I am grateful to Dick and Janie Pearl for friendship and support through the Richard and Jane Pearl Professorship in Environmental Studies. Finally, I thank Katie, Jared, and Ethan Friedland, and my mother Selma.

From Rick Relyea . . .

I would like to thank my family—my wife Christine and my children Isabelle and Wyatt. Too many nights and weekends were taken from them and given to this textbook and they never complained. Their presence and patience continually inspired me to push forward and complete the project.

I am also grateful to the many people at Bedford, Freeman, and Worth who helped guide me and taught me a great deal about the publication process. I would like to especially thank Jerry Correa for convincing me to join the first edition of this book.

Reviewers

High School Focus Group Participants and Reviewers

Our deep appreciation and heartfelt thanks are due to the experienced AP® teachers who participated in focus groups and/or reviewed the manuscript during the development of this book. Their contributions have been invaluable.

Cynthia Ahmed, *Signature School, IN*
Timothy Allen, *Thomas A. Edison Preparatory High School, OK*
Julie Back, *Kecoughtan High School, VA*
Maureen Bagwell, *Collierville High School, TN*
Fredrick Baldwin, *Kendall High School, NY*
Lisa Balzas, *Indian Springs School, AL*
Debra Bell, *Montgomery High School, TX*
Melinda Bell, *Flagstaff Arts and Leadership Academy, AZ*
Karen Benton, *South Brunswick High School, NJ*
Richard Benz, *Wickliffe High School, OH*
Cindy Birkner, *Webber Township High School, IL*
Christine Bouchard, *Milford Public Schools, CT*
Gail Boyarsky, *East Chapel Hill High School, NC*
Rebecca Bricen, *Johnsonburg High School, PA*
Deanna Brunlinger, *Elkhorn Area High School, WI*
Kevin Bryan, *Woodrow Wilson Senior High School, CA*
Tanya Bunch, *Carter High School, TN*
Diane Burrell, *Starr's Mill High School, GA*
Teri Butler, *New Hanover High School, NC*
Charles Campbell, *Russellville High School, AR*
Sande Caton, *Concord High School, DE*
Andrea Charles, *West Side Leadership Academy, IN*
Linda Charpentier, *Xavier High School, CT*
Blanca Ching, *Fort Hamilton High School, NY*
Ashleigh Coe, *Bethesda-Chevy Chase High School, MD*
Bethany Colburn, *Randolph High School, MA*
Jonathan D. Cole, *Holmdel High School, NJ*
Robert Compton, *Walled Lake Northern High School, MI*
Ann Cooper, *Oseola High School, AR*
Thomas Cooper, *The Walker School, GA*
Joyce Corriere, *Hampton High School, VA*
Stephanie Crow, *Milford High School, MI*
Stephen Crowley, *Winooski High School, VT*
Linda D'Apolito, *Trinity School, NY*
Brygida DeRiemaker, *Eisenhower High School, MI*
Chand Desai, *Martin Luther King Magnet High School, TN*
Michael Douglas, *Bronx Prep Charter School, NY*
Nancy Dow, *A. Crawford Mosley High School, FL*
Nat Draper, *Deep Run High School, VA*
Denis DuBay, *Leesville High School, NC*
John Dutton, *Shaw High School, OH*
Heather Earp, *West Johnston High School, NC*
Kim Eife, *Academy of Notre Dame, PA*
Brian Elliot, *San Dimas High School, CA*
Christina Engen, *Crescenta Valley High School, CA*
Mary Anne Evans, *Allendale Columbia School, NY*
Kay Farkas, *Rush-Henrietta High School, NJ*
Tim Fennell, *LASA at LBJ High School, TX*
Michael Finch, *Greene County Tech High School, AR*
Robert Ford, *Fairfield College Preparatory School, CT*
Paul Frisch, *Fox Lane High School, NY*
Bob Furhman, *The Covenant School, VA*
Nivedita (Nita) Ganguly, *Oak Ridge High School, TN*
Mike Gaule, *Ladywood High School, MI*
Billy Goodman, *Passaic Valley High School, NJ*
Amanda Graves, *Mt. Tahoma High School, WA*
Barbara Gray, *Richmond Community High School, VA*
Jack Greene, *Logan High School, UT*
Jeannie Kornfeld, *Hanover High School, NH*
Jen Kotkin, *St. Philip's Academy, NJ*
Pat Kretzer, *Timber Creek High School, FL*
Michelle Krug, *Coral Springs High School, FL*
Jim Kuipers, *Chicago Christian High School, IL*

Claire Kull, *Career Center, NC*
Jay Kurima, *O. D. Wyatt High School, TX*
Tom LaHue, *Aptos High School, CA*
Cathy Larson, *Patuxent High School, MD*
Michael Lauer, *Danville High School, KY*
Sonia Laureni, *West Orange High School, NJ*
Amy Lawson, *Naples High School, FL*
Jim Lehner, *The Taft School, CT*
Dr. Avon Lewis, *Lexington High School, MA*
Marie Lieberman, *Ravenscroft School, NC*
John Ligget, *Conestoga High School, PA*
Ann Linsley, *Bellaire High School, TX*
Mark Little, *Broomfield High School, CO*
Leyana Lloyd, *Washington Senior Academy, GA*
Larry Lollar, *Alice High School, TX*
Stephanie Longfellow, *Deltona High School, FL*
Sue Ellen Lyons, *Holy Cross School, LA*
Theresa Lyster, *Camden County High School, GA*
John F. Madden, *Ashley Hall School, SC*
Jeremy Magee, *Sandy High School, OR*
Mike Mallon, *James I. O'Neill High School, NY*
Scott Martin, *Deer Creek High School, OK*
Kristi Martinez, *Eastlake High School, WA*
Christeena Mathews, *The Philadelphia High School for Girls, PA*
Courtney Mayer, *Winston Churchill High School, TX*
Monica Maynard, *Schurr High School, CA*
James McAdams, *Center Grove High School, IN*
Kristen McClellen, *Grand Junction High School, CO*
Sandy McDonough, *North Salem Middle/High School, NY*
Diane Medford, *Los Alamos High School, NM*
Leslie Miller, *Flintridge Sacred Heart Academy, CA*
Lonnie Miller, *El Diamante High School, CA*
Melody Mingus, *Breckinridge County High School, KY*
Myra Morgan, *National Math & Science Initiative, AP® Environmental Consultant*
Tammy Morgan, *Lake Placid High School, NY*
David Moscarelli, *Ponaganset High School, RI*
Terri Mountjoy, *Greene County Career Center, OH*
Bill Mulhearn, *Archmere Academy, DE*
Sharna Murphy, *Millikan High School, CA*
Jeanine Musgrove, *Oakton High School, VA*
Anna Navarro, *Veterans Memorial High School, TX*
Barbara Nealon, *Southern York County School District, PA*
Dara Nix-Stevenson, *American Hebrew Academy, NC*
Bennett O'Connor, *Dallas ISD, TX*
Robert Oddo, *Horace Greeley High School, NY*
Kate Oitzinger, *El Molino High School, CA*
Paul Olson, *Redwood High School, CA*
Janet Ort, *Hoover High School, AL*
Roger Palmer, *Bishop Dunne High School, TX*
Annetta Pasquarello, *Triton Regional High School, NJ*
Lynn Paulsen, *Mayde Creek High School, TX*
Judy Perrella, *Academy of the Holy Names, FL*
Carolyn Phillips, *Southeastern High School, IL*
Pam Phillips, *Hayden High School, AL*
Alanna Piccillo, *Palisade High School, CO*
Julie Quinn Kiernan, *Cretin-Durham Hall, NC*
Jenny Ramsey, *Charlotte Christian School, NC*
Susan Ramsey, *VASS, VA*
Cristen Rasmussen, *Costa Mesa High School, CA*
Alesa Rehmann, *Coral Shores High School, FL*
Mark Reilly, *Jefferson High School, IΩ*
Kimbell Reitz, *Penn High School, IN*
Cheryl Rice, *Howard High School, MD*
Sharon Riley, *Springfield High School, OH*
Chris Robson, *Ironwood Ridge High School, AZ*
James Rodewald, *Shaker High School, NY*
Kurt Rogers, *Northern Highlands Regional High School, NJ*
Kris Rohrbeck, *Almont High School, MI*
David Rouby, *Hall High School, AR*
Rebecca Rouch, *East Bay High School, FL*
Jennifer Roy, *TrekNorth Junior & Senior High School, MN*
Reva Beth Russell, *Lehi High School, UT*
Sheila Scanlan, *Highland High School, AZ*
Kristi Schertz, *Saugus High School, CA*
Greg Schiller, *James Monroe High School, CA*
Amy Schwartz, *Aragon High School, CA*
Shashi Sharma, *Henry Snyder High School, NJ*
Tonya Shires, *Edgewood High School, MD*
Pamela Shlachtman, *South Dade Senior High School, FL*
Julie Smiley, *Winchester Community High School, IN*
Amy Snodgrass, *Central High School, AR*
Bill Somerlot, *New Albany High School, OH*
Anne Soos, *Stuart Country Day School of the Sacred Heart, NJ*
Joan Stevens, *Arcadia High School, CA*
Marianne Strickhart, *Henry Snyder High School, NJ*
Timothy Strout, *Jericho High School, NY*

Robert Summers, *A+ College Ready, AL*
Jeff Sutton, *The Harker School, CA*
Dave Szaroleta, *Salesianum School, DE*
Kristen Thomson, *Saratoga High School, CA*
James Timmons, *Carrboro High School, NC*
Thomas Tokarski, *Woodlands High School, NY*
Susan Tully, *Salem Academy Charter School, MA*
Debra Tyson, *Brooke Pointe High School, VA*
Melissa Valentine, *Elizabeth Seton High School, MD*
Dirk Valk, *McKeel Academy, FL*
Gene Vann, *Head-Royce School, CA*
Rebecca Van Tassell, *Herron High School, IN*
Marc Vermeire, *Friday Harbor High School, WA*
Naomi Volain, *Springfield Central High School, MA*
Betty Walden, *Merritt Island High School, FL*
Craig Wallace, *North Oldham High School, KY*
Abbie Walston, *North Haven High School, CT*
Annette Weeks, *Battle Ground High School, WA*
Pamela Weghorst, *Ardrey Kell High School, NC*
Michelle Whitehurst, *Powhatan High School, VA*
Jane Whitelock, *Easton High School, MD*
Laurie Whitesell, *Eli Whitney Middle School, OK*
Robert Whitney, *Westview High School, CA*
Carol Widegren, *Lincoln Park High School, IL*
Sarrah Williams, *Hamden Hall Country Day School, CT*
Robert Willis, *Lakeside High School,GA*

College Reviewers

We are also indebted to numerous college instructors, many of whom are also involved in AP® Environmental Science, for their insights and suggestions through various stages of development. The content experts who carefully reviewed Chapters in their area of expertise are designated with an asterisk (*).

M. Stephen Ailstock, PhD, *Anne Arundel Community College*
Deniz Z. Altin-Ballero, *Georgia Perimeter College*
Daphne Babcock, *Collin County Community College District*
Jay L. Banner, *University of Texas at San Antonio*
James W. Bartolome, *University of California, Berkeley*
Ray Beiersdorfer, *Youngstown State University*
Grady Price Blount, *Texas A&M University, Corpus Christi*
Dr. Edward M. Brecker, *Palm Beach Community College, Boca Raton*
Anne E. Bunnell, *East Carolina University*
Ingrid C. Burke, *Colorado State University*
Anya Butt, *Central Alabama Community College*
John Callewaert, *University of Michigan**
Kelly Cartwright, *College of Lake County*
Mary Kay Cassani, *Florida Gulf Coast University*
Young D. Choi, *Purdue University Calumet*
John C. Clausen, *University of Connecticut**
Richard K. Clements, *Chattanooga State Technical Community College*
Thomas Cobb, *Bowling Green State University, OH*
Stephen D. Conrad, *Indiana Wesleyan University*
Terence H. Cooper, *University of Minnesota, Saint Mary's Winona Campus*
Douglas Crawford-Brown, *University of North Carolina at Chapel Hill*
Wynn W. Cudmore, *Chemeketa Community College*
Katherine Kao Cushing, *San Jose State University*
Maxine Dakins, *University of Idaho*
Robert Dennison, *Heartland Community College*
Michael Denniston, *Georgia Perimeter College*
Roman Dial, *Alaska Pacific University*
Robert Dill, *Bergen Community College*
Michael L. Draney, *University of Wisconsin, Green Bay*
Anita I. Drever, *University of Wyoming**
James Eames, *Loyola University New Orleans*
Kathy Evans, *Reading Area Community College*
Mark Finley, *Heartland Community College*
Dr. Eric J. Fitch, *Marietta College*
Karen F. Gaines, *Northeastern Illinois University*
James E. Gawel, *University of Washington, Tacoma*
Carri Gerber, *Ohio State University Agricultural Technical Institute*
Julie Grossman, *Saint Mary's University of Minnesota, Saint Mary's Winona Campus*

Lonnie J. Guralnick, *Roger Williams University*
Sue Habeck, *Tacoma Community College*
Hilary Hamann, *Colorado College*
Dr. Sally R. Harms, *Wayne State College*
Floyd Hayes, *Pacific Union College*
Keith R. Hench, *Kirkwood Community College*
William Hopkins, *Virginia Tech*★
Richard Jensen, *Hofstra University*
Sheryll Jerez, *Stephen F. Austin State University*
Shane Jones, *College of Lake County*
Caroline A. Karp, *Brown University*
Erica Kipp, *Pace University, Pleasantville/Briarcliff*
Christopher McGrory Klyza, *Middlebury College*★
Frank T. Kuserk, *Moravian College*
Matthew Landis, *Middlebury College*★
Kimberly Largen, *George Mason University*
Larry L. Lehr, PhD, *Baylor University*
Zhaohui Li, *University of Wisconsin, Parkside*
Thomas R. MacDonald, *University of San Francisco*
Robert Stephen Mahoney, *Johnson & Wales University*
Bryan Mark, *Ohio State University, Columbus Campus*
Paula J.S. Martin, *Juniata College*
Robert J. Mason, *Tennessee Temple University*
Michael R. Mayfield, *Ball State University*
Alan W. McIntosh, *University of Vermont*
Dr. Kendra K. McLauchlan, *Kansas State University*★
Patricia R. Menchaca, *Mount San Jacinto Community College*
Dr. Dorothy Merritts, *Franklin and Marshall College*★
Bram Middeldorp, *Minneapolis Community and Technical College*
Tamera Minnick, *Mesa State College*
Mark Mitch, *New England College*
Ronald Mossman, *Miami Dade College, North*
William Nieter, *St. John's University*
Mark Oemke, *Alma College*
Victor Okereke, PhD, PE, *Morrisville State College*
Duke U. Ophori, *Montclair State University*
Chris Paradise, *Davidson College*
Dr. Clayton A. Penniman, *Central Connecticut State University*
Christopher G. Peterson, *Loyola University Chicago*
Craig D. Phelps, Rutgers, *The State University of New Jersey, New Brunswick*
F.X. Phillips, PhD, *McNeese State University*
Rich Poirot, *Vermont Department of Environmental Conservation*★
Bradley R. Reynolds, *University of Tennessee, Chattanooga*
Amy Rhodes, *Smith College*★
Marsha Richmond, *Wayne State University*
Sam Riffell, *Mississippi State University*
Jennifer S. Rivers, *Northeastern Illinois University*
Ellison Robinson, *Midlands Technical College*
Bill D. Roebuck, *Dartmouth Medical School*★
William J. Rogers, *West Texas A&M University*
Thomas Rohrer, *Central Michigan University*
Aldemaro Romero, *Arkansas State University*
William R. Roy, *University of Illinois at Urbana-Champaign*
Steven Rudnick, *University of Massachusetts, Boston*
Heather Rueth, *Grand Valley State University*
Eleanor M. Saboski, *University of New England*
Seema Sah, *Florida International University*
Shamili Ajgaonkar Sandiford, *College of DuPage*
Robert M. Sanford, *University of Southern Maine*
Nan Schmidt, *Pima Community College*
Jeffery A. Schneider, *State University of New York at Oswego*
Bruce A. Schulte, *Georgia Southern University*
Eric Shulenberger, *University of Washington*
Michael Simpson, *Antioch University New England*★
Annelle Soponis, *Reading Area Community College*
Douglas J. Spieles, *Denison University*
David Steffy, *Jacksonville State University*
Christiane Stidham, *State University of New York at Stony Brook*
Peter F. Strom, Rutgers, *The State University of New Jersey, New Brunswick*
Kathryn P. Sutherland, *University of Georgia*
Christopher M. Swan, *University of Maryland, Baltimore County*★
Melanie Szulczewski, *University of Mary Washington*
Jamey Thompson, *Hudson Valley Community College*
John A. Tiedemann, *Monmouth University*
Conrad Toepfer, *Brescia University*
Todd Tracy, *Northwestern College*
Steve Trombulak, *Middlebury College*
Zhi Wang, *California State University, Fresno*
Jim White, *University of Colorado, Boulder*
Rich Wolfson, *Middlebury College*★
C. Wesley Wood, *Auburn University*
David T. Wyatt, *Sacramento City College*

Getting the Most from This Book

Daily life is filled with decisions large and small that affect our environment. From the food we eat, to the cars we drive or choose not to drive, to the chemicals we put into the water, soil, and air. The impact of human activity is wide-ranging and deep. And yet making decisions about the environment is often not easy or straightforward. Is it better for the environment if we purchase a new, energy-efficient hybrid car or should we continue using the older car we already own? Should we remove a dam that provides electricity for 70,000 homes because it interferes with the migration of salmon? Are there alternatives to fossil fuel for heating our homes?

The purpose of this book is to give you a working knowledge of the big ideas of environmental science and help you to prepare for the AP® Environmental Science Exam. The book is designed to provide you with a strong foundation in the scientific fundamentals, to introduce you to the policy issues and conflicts that emerge in the real world, and to offer you an in-depth exploration of all the topics covered on the advanced placement exam in environmental science.

Like the first edition, *Friedland and Relyea Environmental Science for AP®*, Second Edition, is organized to closely follow the AP® environmental science course description. Every item on the College Board's "Topic Outline" is covered thoroughly in the text. Look inside the front cover for a detailed alignment guide. The textbook offers comprehensive coverage of all required AP® course topics and will help you prepare for success on the exam by:

- providing chapter opening case studies that will help you to see how environmental science is grounded in your daily life and in the world around you
- dividing each chapter into manageable modules that will help you to be organized and keep up with the challenging pace of the AP® environmental science course
- using the same terminology, language, and formulas that you will see on the AP® environmental science exam
- using expertly selected and artistically rendered figures, photographs, graphs, and visuals that will help you to understand and remember the big ideas and important concepts that will be on the exam
- providing you with many opportunities to practice for the exam throughout the year, including end-of-module AP® review questions, chapter AP® practice exams, unit AP® practice exams, and a cumulative AP® practice exam at the end

The next few pages offer you a brief tour of the features of this book that have been designed to help you succeed in the course and on the exam.

Explore the world around you through science.

chapter

8

Earth Systems

Module 24 Mineral Resources and Geology

Module 25 Weathering and Soil Science

Are Hybrid Electric Vehicles as Environmentally Friendly as We Think?

Many people in the environmental science community believe that hybrid electric vehicles (HEV) and all-electric vehicles are some of the most exciting innovations of the last decade. Cars that run on electric power or on a combination of electricity and gasoline are much more efficient in their use of fuel than similarly sized internal combustion (IC) automobiles. Some of these cars use no gasoline at all, while others are able to run as much as twice the distance as a conventional IC car on the same amount of gasoline.

Although HEV and all-electric vehicles reduce our consumption of liquid fossil fuels, they do come with environmental trade offs. The construction of HEV vehicles uses scarce metals, including neodymium, lithium, and lanthanum. Neodymium is needed to form the magnets used in the electric motors, and lithium and lanthanum are used in the compact high-performance batteries the vehicles require. At present,

Although HEV and all-electric vehicles reduce our consumption of liquid fossil fuels, they do come with environmental trade offs.

there appears to be enough lanthanum available in the world to meet the demand of the Toyota Motor Corporation, which has manufactured more than 3 million Prius HEV vehicles. Toyota obtains its lanthanum from China. There are also supplies of lanthanum in various geologic deposits in California, Australia, Bolivia, Canada, and elsewhere, but most of these deposits have not yet been developed for mining. Until this hap some scientists believe that the pr tion of HEVs and all-electric ve will eventually be limited b availability of lanthanum.

In addition to the scarci metals needed to make HEV all-electric vehicles, we ha consider how we acquire metals. Wherever mining occu has a number of environm consequences. Material extra leaves a landscape fragmente holes, and road construction nece for access to and from the minin further alters the habitat. Erosio water contamination are also cor results of mining.

A typical Toyota Prius HEV approximately 1 kg (2.2 pound

Chapter Opening Case Study

Read the intriguing case study that begins each chapter and think about the environmental challenges and trade-offs that are introduced. The subjects of these studies often will spark spirited class discussion.

As you can see from case studies like this one from Chapter 8, it's not always easy to make sustainable choices.

module

34

Patterns of Energy Use

In this module we begin our study of nonrenewable energy sources by looking at patterns of energy use throughout the world and in the United States. We will see how evaluating energy efficiency can help us determine the best application for different energy sources. Finally, because electricity accounts for such a large percentage of our overall energy use, we will examine the ways in which electricity is generated.

Learning Objectives

After reading this module, you should be able to

- describe the use of nonrenewable energy in the world and in the United States.
- explain why different forms of energy are best suited for certain purposes.
- understand the primary ways that electricity is generated in the United States.

Nonrenewable energy is used worldwide and in the United States

Fossil fuels are fuels derived from biological material that became fossilized millions of years ago. Fuels from this source provide most of the energy used in both developed and developing countries. The vast majority of the fossil fuels we use—coal, oil, and natural gas—come from deposits of organic matter that were formed 50 million to 350 million years ago. As we saw in Chapter 3 (see Figure 7.2 on page 83), when organisms die, decomposers break down most of the dead biomass aerobically, and it quickly reenters the food web. However, in an anaerobic environment—for example in places such as swamps, river deltas, and the ocean floor—a large amount of detritus may build up quickly. Under these conditions, decomposers cannot break down all of the detritus. As this material is buried under succeeding layers of sediment and exposed to heat and pressure, the organic compounds within it are chemically transformed into high-energy solid, liquid, and gaseous components that are easily combusted. Because fossil fuel cannot be replenished once it is used up, it is known as a **nonrenewable energy resource. Nuclear fuel,** derived from radioactive materials that give off energy, is another major source of nonrenewable energy on which we depend. The supplies of these energy types are finite.

Every country in the world uses energy at different rates and relies on different energy resources. Factors that determine the rate at which energy is used include the resources that are available and affordable. In the past few decades, people have also begun to consider environmental impacts in some energy-use decisions.

Fossil fuel A fuel derived from biological material that became fossilized millions of years ago.

Nonrenewable energy resource An energy source with a finite supply, primarily the fossil fuels and nuclear fuels.

Nuclear fuel Fuel derived from radioactive materials that give off energy.

MODULE 34 ■ Patterns of Energy Use 399

Module Structure

Chapters are divided into short Modules to help keep you on pace. Each module opens with a brief description of what topics will be covered.

Learning Objectives

A list key ideas at the beginning of the module help to keep you focused as you read.

Running glossary

Important key terms are set in bold type in the text and defined at the bottom of the page on which they are introduced. Key terms are also defined in the glossary at the end of the book.

Math practice makes perfect.

Do the Math

Among the biggest challenges on the AP® Environmental Science Exam are questions that ask you to solve environmental science math problems. "Do the Math" problems help you practice the math skills that you'll need to tackle these problems on the exam.

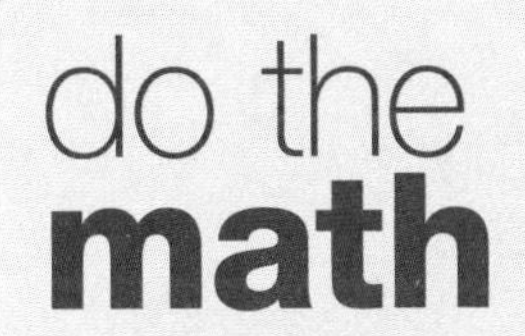

Calculating Energy Supply

According to the U.S. Department of Energy, a typical home in the United States uses approximately 900 kWh of electricity per month. On an annual basis, this is

$$900 \text{ kWh/month} \times 12 \text{ months/year} = 10{,}800 \text{ kWh/year}$$

How many homes can a 500 MW power plant with a 0.9 capacity factor support? Begin by determining how much electricity the plant can provide per month:

$$500 \text{ MW} \times 24 \text{ hours/day} \times 30 \text{ days/month} \times 0.9 = 324{,}000 \text{ MWh/month}$$

1 MWh equals 1,000 kWh, so to convert MWh per month into kWh per month, we multiply by 1,000:

$$324{,}000 \text{ MWh/month} \times 1{,}000 \text{ kWh/MWh} = 324{,}000{,}000 \text{ kWh/month}$$

So

$$\frac{324{,}000{,}000 \text{ kWh/month}}{900 \text{ kWh/month/home}} = 360{,}000 \text{ homes}$$

On average, a 500 MW power plant can supply roughly 360,000 homes with electricity.

Your Turn During summer months, in hot regions of the United States, some homes run air conditioners continuously. How many homes can the same power plant support if average electricity usage increases to 1,200 kWh/month during summer months?

Your Turn

Each "Do the Math" box has a "Your Turn" practice problem to help you review and practice the math skills introduced.

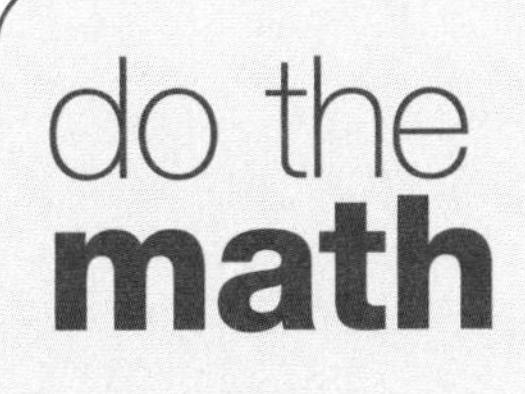

Converting Between Hectares and Acres

In the metric system, land area is expressed in hectares. A hectare (ha) is 100 meters by 100 meters. In the United States, land area is most commonly expressed in acres. There are 2.47 acres in 1 ha. The conversion from hectares is relatively easy to do without a calculator; rounding to two significant figures gives us 2.5 acres in 1 ha. If a nature preserve is 100 ha, what is it size in acres?

$$100 \text{ ha} \times 2.5 \text{ acres} = 250 \text{ acres}$$

Your Turn A particular forest is 10,000 acres. Determine its size in hectares.

Prepare for the Exam

Once you are comfortable with the math skills introduced, you'll be prepared for quantitative problems on the exam.

Analyze and interpret visual data.

(a) Random distribution

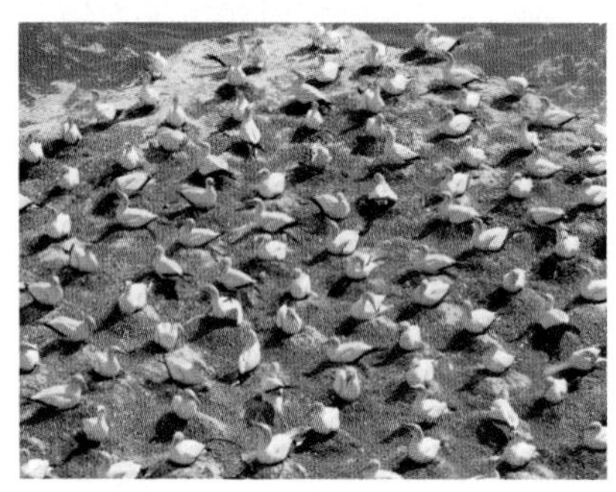

(b) Uniform distribution

Photos and Illustrations

The photos and illustrations in this book are more than just pretty pictures. They have been carefully chosen and developed to help you comprehend and remember the key ideas.

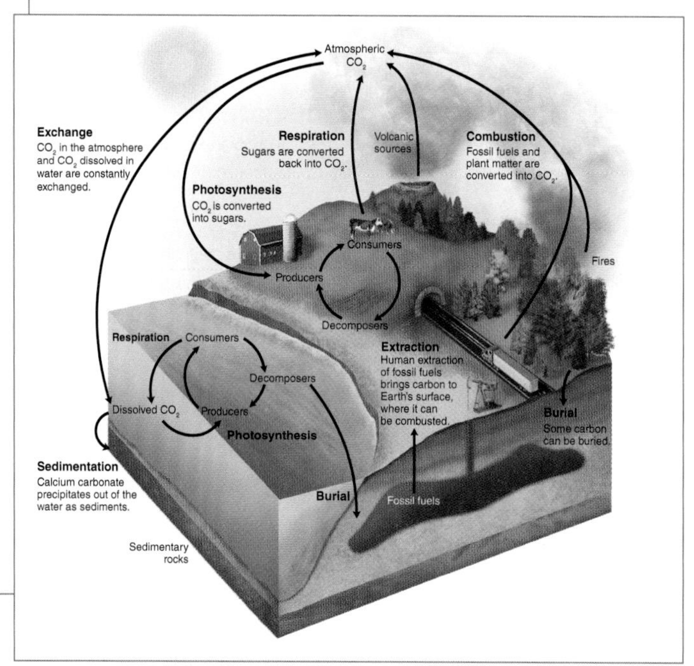

TABLE 36.1 Comparison of nonrenewable energy fuels

Energy Type	Advantages	Disadvantages	Pollutant and greenhouse gas emissions	Electricity (cents/kWh)	Energy return on energy investment*
Oil/gasoline	• Ideal for mobile combustion (high energy/mass ratio) • Quick ignition/turn-off capability • Cleaner burning than coal	• Significant refining required • Oil spill potential effect on habitats near drilling sites • Significant dust and emissions from fossil fuels used to power earth-moving equipment • Human rights/ environmental justice issues in developing countries that export oil • Will probably be much less available in the next 40 years or so	• Second highest emitter of CO_2 among fossil fuels • Hydrocarbons • Hydrogen sulfide	• Relatively little electricity is generated from oil	4.0 (gasoline) 5.7 (diesel)
Coal	• Energy-dense and abundant—U.S. resources will last at least 200 years • No refining necessary • Easy, safe to transport • Economic backbone of some small towns	• Mining practices frequently risk human lives and dramatically alter natural landscapes • Coal power plants are slow to reach full operating capacity • A large contributing factor to acid rain in the United States	• Highest emitter of CO_2 among energy sources • Sulfur • Trace amounts of toxic metals such as mercury	5 cents/kWh	14
Natural Gas	• Cogeneration power plants can have efficiencies up to 60% • Efficient for cooking, home heating, etc. • Fewer impurities than coal or oil	• Risk of leaks/ explosions • Twenty-five times more effective as a greenhouse gas than CO_2 • Not available everywhere because it is transported by pipelines	• Methane • Hydrocarbons • Hydrogen sulfide	6–8 cents/kWh	
Nuclear Energy	• Emits no CO_2 once plant is operational • Offers independence from imported oil • High energy density, ample supply	• Very unpopular; generates protests • Plants are very expensive to build because of legal challenges • Meltdown could be catastrophic • Possible target for terrorist attacks	• Radioactive waste is dangerous for hundreds of thousands of years • No long-term plan currently in place to manage radioactive waste • No air pollution during production	12–15 cents/kWh	

*Estimates vary widely.

Tables and Graphs

To understand environmental science and succeed on the exam, you need to engage in the scientific practice of analyzing and interpreting a variety of tables, graphs, and charts.

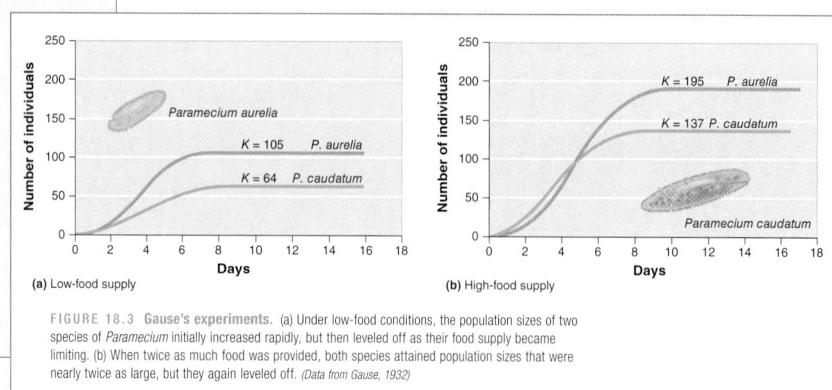

FIGURE 18.3 **Gause's experiments.** (a) Under low-food conditions, the population sizes of two species of *Paramecium* initially increased rapidly, but then leveled off as their food supply became limiting. (b) When twice as much food was provided, both species attained population sizes that were nearly twice as large, but they again leveled off. *(Data from Gause, 1932)*

Review and practice for quizzes and tests.

module

3

REVIEW

In this module, we have seen how specific aspects of the scientific method are used to conduct field and laboratory evaluations of how human activity affects the natural environment. The scientific method follows a process of observations and questions, testable hypotheses and predictions, and data collection. Results are interpreted and shared with other researchers. Experiments can be either controlled (manipulated) experiments or natural experiments that make use of natural events. There are often challenges in environmental science including the lack of baseline data and the interactions with social factors such as human preferences.

Module 3 AP® Review Questions

1. The first step in the scientific process is
   (a) collecting data.
   (b) observations and questions.
   (c) forming a hypothesis.
   (d) disseminating findings.
   (e) forming a theory.

Use the following information for questions 2 and 3:
Two new devices for measuring lead contamination in water are tested for accuracy. Scientists test each device with seven samples of water known to contain 400 ppm of lead. Their data is shown below. Concentration is in parts per billion.

Water Sample	1	2	3	4	5	6	7
Device 1	415	417	416	417	415	416	416
Device 2	398	401	400	402	398	400	399

2. The data from device 1 is
   (a) accurate, but not precise.
   (b) precise, but not accurate.
   (c) both accurate and precise.
   (d) neither accurate nor precise.
   (e) not clear enough to support any conclusion about accuracy or precision.

3. Assuming the devices were used correctly, and assuming we want to choose a device that accurately reflects the true concentration of lead in the water samples, which conclusion does the data support?
   (a) Device 1 is superior to device 2 because it is more precise.
   (b) Device 2 is superior to device 1 because it is more precise.
   (c) Device 1 is superior to device 2 because it is more accurate.
   (d) Device 2 is superior to device 1 because it is more accurate.
   (e) Both devices are equally effective at measuring contaminates.

4. Challenges in the study of environmental science include all of the following except
   (a) dangers of studying natural systems.
   (b) lack of baseline data.
   (c) subjectivity of environmental impacts.
   (d) complexity of natural systems.
   (e) complex interactions between humans and the environment.

5. A control group is
   (a) a group with the same conditions as the experimental group.
   (b) a group with conditions found in nature.
   (c) a group with a randomly assigned population.
   (d) a group with the same conditions as the experimental group except for the study variable.
   (e) a group that is kept at the same conditions throughout the experiment

Module Review

Solidify your understanding by reviewing the main ideas in each module review.

Exam Prep All Year

Each module ends with multiple-choice questions similar to those on the AP® exam. Practicing your test-taking strategies for multiple-choice questions throughout the year will pay off when you take the exam.

chapter

REVIEW

Throughout this chapter, we have outlined principles, techniques, and methods that will allow us to approach environmental science from an interdisciplinary perspective as we evaluate the current condition of Earth and the ways that human beings have influenced it. We identified that we can use environmental indicators to show the status of specific environmental conditions in the past, at present, and, potentially, into the future. These indicators and other environmental metrics must be measured using the same scientific process used in other fields of science. Environmental science does contain some unique challenges because there is no undisturbed baseline—humans began manipulating Earth long before we have been able to study it.

Key Terms

Fracking
Environment
Environmental science
Ecosystem
Biotic
Abiotic
Environmentalist
Environmental studies
Ecosystem services
Environmental indicators
Biodiversity
Genetic diversity
Species
Species diversity
Speciation
Background extinction rate
Greenhouse gases
Anthropogenic
Development
Sustainability
Sustainable development
Biophilia
Ecological footprint
Scientific method
Hypothesis
Null hypothesis
Replication
Sample size
Accuracy
Precision
Uncertainty
Theory
Control group
Natural experiment

Learning Objectives Revisited

Module 1 Environmental Science

- **Define the field of environmental science and discuss its importance.**
  Environmental science is the study of the interactions among human-dominated systems and natural systems and how those interactions affect environments. Studying environmental science helps us identify, understand, and respond to anthropogenic changes.
- **Identify ways in which humans have altered and continue to alter our environment.**
  The impact of humans on natural systems has been significant since early humans hunted some large animal species to extinction. However, technology and population growth have dramatically increased both the rate and the scale of human-induced change.

Module 2 Environmental Indicators and Sustainability

- **Identify key environmental indicators and their trends over time.**
  Five important global-scale environmental indicators are biological diversity, food production, average global surface temperature and atmospheric CO_2 concentrations, human population, and resource depletion. Biological diversity is decreasing as a result of human actions, most notably habitat destruction and habitat degradation. Food production appears to be leveling off and may be decreasing. Carbon dioxide concentrations are steadily increasing as a result of fossil fuel combustion and land conversion. Human population continues to increase and probably will continue to do so throughout this century. Resource depletion for most natural resources continues to increase.

Chapter Review

At the end of each chapter, take time to review the main ideas and key terms.

Learning Objectives Revisited

Check your notes against summaries of the learning objectives for each module in the chapter.

Prepare and practice for the AP® Environmental Science Exam.

Chapter AP® Environmental Science Practice Exam

When you finish a chapter take the practice exam to check your understanding of the main ideas. The practice exam will help you become familiar with the style of questions on he AP® Environmental Science Exam.

Chapter 1 AP® Environmental Science Practice Exam

Section 1: Multiple-Choice Questions

Choose the best answer for questions 1–11.

1. Which of the following events has increased the impact of humans on the environment?
   I. advances in technology
   II. reduced human population growth
   III. use of tools for hunting

   (a) I only
   (b) I and II only
   (c) II and III only
   (d) I and III only
   (e) I, II, and III

2. As described in this chapter, environmental indicators
   (a) always tell us what is causing an environmental change.
   (b) can be used to analyze the health of natural systems.
   (c) are useful only when studying large-scale changes.
   (d) do not provide information regarding sustainability.
   (e) take into account only the living components of ecosystems.

3. Which statement regarding a global environmental indicator is NOT correct?
   (a) Concentrations of atmospheric carbon dioxide have been rising quite steadily since the Industrial Revolution.
   (b) World grain production has increased fairly
   (c) For the past 130 years, average global surface temperatures have shown an overall increase that seems likely to continue.
   (d) World population is expected to be between 8.1 billion and 9.6 billion by 2050.
   (e) Some natural resources are available in finite amounts and are consumed during a one-time use, whereas other finite resources can be used multiple times through recycling.

4. Figure 2.5 (on page 12) shows atmospheric carbon dioxide concentrations over time. The measured concentration of CO_2 in the atmosphere is an example of
   (a) a sample of air from over the Antarctic.
   (b) an environmental indicator.
   (c) replicate sampling.
   (d) calculating an ecological footprint.
   (e) how to study seasonal variation in Earth's temperatures.

5. Environmental metrics such as the ecological footprint are most informative when they are considered along with other environmental indicators. Which indicator, when considered in conjunction with the ecological footprint, would provide the most information about environmental impact?
   (a) biological diversity
   (b) food production
   (c) human population
   (d) CO_2
   (e) wate

6. In science

Multiple-Choice Questions

Each chapter exam begins with multiple-choice questions modeled after those you'll see on the exam. Many of the questions ask you to analyze or interpret tables, graphs, or figures.

Section 2: Free-Response Questions

Write your answer to each part clearly. Support your answers with relevant information and examples. Where calculations are required, show your work.

1. Your neighbor has fertilized her lawn. Several weeks later, she is alarmed to see that the surface of her ornamental pond, which sits at the bottom of the sloping lawn, is covered with a green layer of algae.
   (a) Suggest a feasible explanation for the algal bloom in the pond. (2 points)
   (b) Design an experiment that would enable you to validate your explanation. Include and label in your answer:
      (i) a testable hypothesis (2 points)
      (ii) the variable that you will be testing (1 point)
      (iii) the data to be collected (1 point)
      (iv) a description of the experimental procedure (2 points)
      (v) a description of the results that would validate your hypothesis (1 point)
   (c) Based on the data from your experiment and your explanation of the problem, think of and suggest one action that your neighbor could take to help the pond recover. (1 point)

Free-Response Questions

Chapter exams include two free-response questions. Points are assigned to indicate how a complete, correct answer would be scored on the AP® exam. The more practice you have in writing answers to free-response questions, the better you will do on the exam.

Unit AP® Environmental Science Practice Exam

The textbook is divided into 8 major units. At the end of each unit, you are provided with a longer practice exam containing 20 multiple-choice questions and 2 free-response questions. These exams give you a chance to review material across multiple chapters and to practice your test-taking skills.

Unit 1 AP® Environmental Science Practice Exam

Section 1: Multiple-Choice Questions

Choose the best answer for questions 1–20.

1. Which best describes how humans have altered natural systems?
   I. Overhunted many large mammals to extinction.
   II. Created habitat for species to thrive.
   III. Emitted greenhouse gases.
   (a) I only
   (b) I and II only
   (c) II and III only
   (d) I and III only
   (e) I, II, and III

2. Which does NOT describe a benefit of biodiversity?
   (a) Genetic biodiversity improves the ability of a population to cope with environmental change.
   (b) Ecosystems with higher species diversity are more productive.
   (c) Species serve as environmental indicators of global-scale problems.
   (d) Speciation reduces natural rates of species extinction.
   (e) Humans rely on ecological interactions among species to produce ecosystem services.

3. Which of the following is NOT a consequence of human population growth?
   (a) Depletion of natural resources

6. The greatest value of the scientific method is best stated as:
   (a) The scientific method permits researchers a rapid method of disseminating findings.
   (b) The scientific method removes bias from observation of natural phenomenon.
   (c) The scientific method allows findings to be reproduced and tested.
   (d) The scientific method promotes sustainable development.
   (e) The scientific method reduces the complexity of experimental results.

7. Researchers conducted an experiment to test the hypothesis that the use of fertilizer near wetlands is associated with increased growth of algae. An appropriate null hypothesis would be:
   (a) The use of fertilizer near wetlands is associated with an increase in fish biomass.
   (b) Growth of algae in wetlands is never associated with increased fertilizer use.
   (c) Application of fertilizers near wetlands is always associated with increased growth of algae.
   (d) Fertilizer use near wetlands has no association with growth of algae.
   (e) Fertilizer use near wetlands leads to increased growth of algae as a result of elevated nutrient concentrations.

Cumulative AP® Environmental Science Practice Exam

At the end of the text you will find a cumulative exam with 100 multiple-choice questions and 4 free-response questions. This exam matches the actual AP® Environmental Science exam in length and scope.

cumulative AP® environmental science practice exam

Section 1: Multiple-Choice Questions

Choose the best answer for questions 1–100.

1. Primary production is an example of
   I. an ecosystem service.
   II. an environmental indicator.
   III. heterotrophic activity.
   (a) I only
   (b) II only
   (c) I and II
   (d) II and III
   (e) I, II, and III

2. Which of the following is likely to increase biodiversity within a biome?
   (a) Landscape fragmentation
   (b) Introduction of an invasive species
   (c) Immigration of humans
   (d) Speciation
   (e) A disease epidemic

3. The United States produces 8 million tons of oranges in a single year. However, many orange crops are succumbing to a deadly invasive bacteria. If 10,000 hectares of orange cropland are lost in a year to this bacteria, and a single acre can produce 20 tons of oranges, what percentage of the total orange crop is lost to the disease in a year? (Note that 1 hectare = 2.5 acres.)
   (a) 2 percent
   (b) 6 percent
   (c) 10 percent
   (d) 20 percent
   (e) 24 percent

4. Which is a flaw of this experiment?
   (a) The experiment lacks a control treatment.
   (b) 10 ppm is a negligible increase of CO_2 relative to ambient concentrations.
   (c) The hypothesis is actually a null hypothesis.
   (d) The measured response variable does not relate to the hypothesis.
   (e) N_2O gas is not a greenhouse gas.

5. As hypothesized, the researchers found that plants exposed to elevated CO_2 had increased biomass after 2

Section 2: Free-Response Questions

Write your answer to each part clearly. Support your answers with relevant information and examples. Where calculations are required, show your work.

1. The City of Philadelphia recently replaced one out of every 10 trash bins with solar-powered trash compactors. The compactor is an enclosed unit with a door that opens for trash disposal. The compactor automatically detects when the bin is full and uses a solar-powered mechanical crusher to compact the contents. When the compactor needs to be emptied, it sends an electronic signal. Use of solar-powered compactors has increased the capacity of public trash bins and has reduced the number of trash collection visits to each bin from 17 times per week to 5 times per week.
   (a) Describe four positive externalities of installing solar-powered trash compactors. (2 points)
   (b) Describe six cradle-to-grave components of solar-powered trash compactors. (2 points)
   (c) Suggest one way that the installation of solar-powered trash compactors can reverse the effects of urban blight. (2 points)

2. The country of Costa Rica has an abundance of climactic, geographic, and biological diversity. However, in the last century intensive farming and population growth have led to a 75 percent reduction in its forests. In the 1980s, the government of Costa Rica began to address concerns about the loss of forest with a series of political and environmental programs. These programs, designed to generate more sustainable economic development, include land protection and conservation of biodiversity.
   (a) Costa Rica lies just north of the equator and contains a series of mountain ranges that run the entire length of the country.
      (i) Given its geographic location, what is likely to be the prevailing wind pattern across the country? (1 point)
      (ii) Describe how mountain ranges contribute to the climactic, geographic, and biological diversity observed in Costa Rica. (1 point)
   (b) Given that Costa Rica is bordered by the Atlantic and Pacific Oceans, how are weather patterns in the country likely to be affected by the El Niño–Southern Oscillation (ENSO)? (2 points)

Be inspired by individuals making a difference.

working toward  **sustainability**

A New Cook Stove Design

In China, India, and sub-Saharan Africa, people in 80 to 90 percent of households cook food using wood, animal manure, and crop residues as their fuel. Since women do most of the cooking, and young children are with the women of the household for much of the time, it is the women and young children who receive the greatest exposure to carbon monoxide and particulate matter. When biomass is used for cooking, concentrations of particulate matter in the home can be 200 times higher than the exposure limits recommended by the EPA. A wide range of diseases has been associated with exposure to smoke from cooking. Earlier in this chapter, we described that indoor air pollution is responsible for 4 million deaths annually around the world, and indoor cooking is a major source of indoor air pollution.

There are hundreds of projects underway around the world to enable women to use more efficient cooking stoves, ventilate cooking areas, cook outside whenever possible, and change customs and practices that will reduce their exposure to indoor air pollution. The use of an efficient cook stove will have the added benefit of consuming less fuel. This improves air quality and reduces the amount of fuel needed, which has

Two innovators from the United States developed a cook stove for backpackers and other outdoor enthusiasts who needed to cook a hot meal with little impact on the environment. They described their stove as needing no gasoline and no batteries, both desirable features for people carrying all their belongings on their backs. They soon realized that their stove, which could burn wood, animal manure, or crop residue, could make an important contribution in the developin physically sepa form when th burn the gases. inside the stov

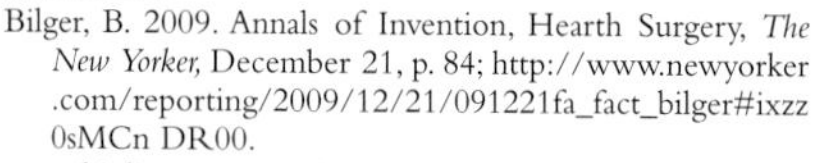

Critical Thinking Questions

1. Why are women and children often the ones most exposed to indoor air pollution in developing countries?
2. How can technology offer solutions to cooking over open fires?

References

Bilger, B. 2009. Annals of Invention, Hearth Surgery, *The New Yorker*, December 21, p. 84; http://www.newyorker.com/reporting/2009/12/21/091221fa_fact_bilger#ixzz0sMCn DR00.

www.biolitestove.com, homepage of BioLite stove.

Working Toward Sustainability

At the end of each chapter read about people and organizations that are making a difference.

Critical Thinking Questions

Working Toward Sustainability provides questions that give you a chance to hone your critical thinking and writing skills.

Science in the real world.

Tracy Packer Photography/Getty Images

1

science**applied**

What Happened to the Missing Salt?

At the beginning of the twentieth century, the City of Los Angeles needed more water for its inhabitants. As we saw at the beginning of Chapter 2, in 1913 the city designed a plan to redirect water away from Mono Lake in California. Before the Los Angeles Aqueduct was built, approximately 120 billion liters of stream water (31 billion gallons) flowed into Mono Lake in an average year. The City of Los Angeles altered the water balance of Mono Lake and at the same time caused a series of changes to the Mono Lake system that led to an increase in the salt concentration in Mono Lake.

water from streams must be equal to the outpu through evaporation.

How did the salt balance change at Mono Lake?

Although we can make the assumption that in Mono Lake is in steady state in a typical salt balance in the lake is not. By applying so principles we have learned in the first two ch can make observations and draw conclusi what has probably happened at Mono L

Free-Response Question

Water that flows into Mono Lake contains a much smaller concentration of salt than the water already in the lake. This inflow tends to stratify, or float on top of existing water, because fresh water is less dense that salt water. As salt from the lower layer dissolves into the upper layer, nutrients from the bottom of the lake also rise to the surface. This exchange of nutrients is critical for the growth of algae in the surface waters. Recent research suggests that the reduction of water diversion from Mono Lake had unexpected results:

> In 1995, the reduction of stream diversions from Mono Lake, combined with greater than average quantities of fresh water from snowmelt runoff, led to a rapid rise in water level. The large volume

Science Applied

At the end of each unit, the "Science Applied" feature offers you an opportunity to read about how the science you are learning is used to make decisions about environmental issues.

Practice Free-Response Questions

Science Applied includes a free-response question related to the topic in the article.

TEACHER'S EDITION

FRIEDLAND and RELYEA

Environmental Science for AP® SECOND EDITION

chapter 1

Environmental Science: Studying the State of Our Earth

PD Overview

This chapter describes the study of environmental science, how we define sustainability, and what constitutes the scientific method. The chapter contains vocabulary terms and details about the scientific method that establish important foundations for the chapters to come. You should also use this chapter to get the students excited about the material so they feel a sense of personal engagement.

Module 1: Environmental Science

Humans are dependent on Earth's air, water, and soil for existence. However, we have altered the planet in many ways, both large and small. This module describes how humans have changed the planet and identifies ways humans can respond to those changes.

Module 2: Environmental Indicators and Sustainability

To study the natural world and the ways in which humans have altered it, students will need techniques to measure and quantify change. This module examines the environmental indicators used to assess the impact of humans on Earth. These indicators allow us to determine whether or not the quality of the natural environment is improving and they inform discussions on the sustainability of humans on the planet.

Module 3: Scientific Method

All scientific endeavors use reproducible scientific methods. In this module, students are encouraged to become scientists themselves as well as good consumers of science by learning the steps of the scientific method and its essential application to environmental science.

Alignment to AP® Environmental Science Course Description

AP® Outline	Module	
I. Earth Systems and Resources (10–15%)		
A. Earth Science Concepts	Module 1	Environmental Science
	Module 2	Environmental Indicators and Sustainability
	Module 3	Scientific Method

Chapter Learning Objectives

After completing this chapter students will be able to

- define the field of environmental science and discuss its importance.
- identify ways in which humans have altered and continue to alter our environment.
- identify key environmental indicators and their trends over time.
- define sustainability and explain how it can be measured using the ecological footprint.
- explain the scientific method and its application to the study of environmental problems.
- describe some of the unique challenges and limitations of environmental science.

Chapter 1 Pacing Guide

This pacing guide is based on a schedule with 120 sessions of 50 minutes each before the AP® Exam. If you have a different number of sessions before the exam, you can modify the pacing to suit your needs. If you have additional time, consider incorporating quizzes, released AP® Environmental Science free-response and multiple-choice questions, or additional activities.

Module	Standard Schedule Days	Block Schedule Days
Module 1	½	¼
Module 2	½	¼
Module 3	1	½
Assessment	1	½

Chapter 1 Resources

These resources can be found by clicking on the link buttons in the Teacher's e-Book (TE-book), opening the Teacher's Resource Flash Drive (TRFD), or logging onto the book's companion website, bcs.whfreeman.com/friedlandapes2e (teacher login required).

- **PowerPoint Presentation 1**
- **Optimized Art PowerPoint and JPEG files 1**
- **Do the Math Videos**
- **More Math Practice 1.1:** Converting Square Miles to Hectares
- **Measuring Your Impact 1:** Exploring Your Footprint
- **Chapter 1 Web Resources**
- **Exploring the Literature 1**
- **Answers to Chapter 1 AP® Exam**
- **Multiple-Choice AP® Practice Exam 1**

PD Chapter 1 Overview

Watch the video overview of Chapter 1 (for teachers) by clicking on the link buttons in the TE-book, opening the TRFD, or logging onto the book's companion website bcs.whfreeman.com /friedlandapes2e (teacher login required).

TRM PowerPoint Presentation 1

Download the PowerPoint presentation for Chapter 1 by clicking on the link buttons in the TE-book, opening the TRFD, or logging onto the book's companion website bcs.whfreeman.com /friedlandapes2e (teacher login required).

A hydraulic fracturing site like this one near Canton, Pennsylvania, can contain many features that are seen prominently here including a concrete pad, a drilling rig, and many storage containers. *(Les Stone/Corbis)*

chapter 1

Environmental Science: Studying the State of Our Earth

To Frack, Or Not to Frack

The United States—like other developed countries—is highly dependent on fuels such as coal and oil that come from the remains of ancient plants and animals. However, the use of these fossil fuels is responsible for many environmental problems that include land degradation and the release of pollutants into the air and water. Natural gas, also known as methane, is the least harmful producer of air pollution among the fossil fuels; it burns more completely and cleanly than coal or oil, and it contains fewer impurities.

Due to advances in technology, oil and mining companies have recently increased their reliance on *fracking.* **Fracking**, short for hydraulic fracturing, is a method of oil and gas extraction that uses high-pressure fluids to force open existing cracks in rocks deep underground. This technique allows extraction of natural gas from locations that were previously so difficult to reach that extraction was economically unfeasible. As a result, large quantities of natural gas are now available in the United States at a lower cost than before. A decade ago, 40 percent of energy in the United States was used to generate electricity with half of that energy coming from coal. As a result of fracking, electricity generation now uses less coal and more natural gas. Since coal emits more air pollutants—including carbon dioxide—than does natural gas, increased fracking initially appeared to be beneficial to the environment.

> Footage of flames shooting from kitchen faucets became popular on YouTube.

Fracking Hydraulic fracturing, a method of oil and gas extraction that uses high-pressure fluids to force open cracks in rocks deep underground.

Teaching Tip: Chapter Opening Case

The chapter opening case, *To Frack, Or Not to Frack*, introduces students to the costs and benefits of natural gas fracking. This case demonstrates how human activities that are initially perceived as causing little harm to the environment may in fact have substantial adverse effects. It also illustrates the controversial side of issues that environmental scientists explore and the difficulty in obtaining absolute answers to environmental problems and questions.

Teaching Tip: Warm-up

Have students read the chapter opening case, *To Frack, Or Not to Frack,* which discusses the various environmental benefits as well as the costs of natural gas fracking. Fracking is a topic widely publicized, so ask these questions to start a discussion about fracking and broader environmental issues.

- Have you heard of hydraulic fracturing, or fracking?
- Where have you heard about it?
- What do you know, or think you know, about fracking?

The dilemma of fracking illustrates the inherent tradeoffs that are part of every environmental issue. This is also a good opportunity to point out that information in the popular press is not always reliable or objective. Emphasize that this course takes a scientific, objective approach that will empower them to evaluate the validity of publicly disseminated information.

Teaching Tip: Video

Communities Divided Over Natural Gas Drilling

After discussing the opening case, show students this Associated Press video.

Ask students to list the arguments made for and against fracking in the movie, and then have them share and discuss their answers with the class. You may also want to hold a class debate about whether fracking should be allowed in your community. Issues to address during the debate include:

- the advantages of burning natural gas over other fossil fuels
- global energy demand
- the use of unknown chemicals
- the contamination of drinking-water wells
- the escape of natural gas
- the economic feasibility of extraction

In class, select groups to debate the issue. Allow each side 2 minutes to prepare, 3 minutes to present, and 1 minute for rebuttal. Have the remainder of the class take notes, ask questions, and decide which side has the most persuasive argument.

The link to this video can be found by clicking on the link buttons in the TE-book, opening the TRFD, or logging onto the book's companion website bcs.whfreeman.com/friedlandapes2e (teacher login required).

Chapter 1 Web Resources

However, reports soon began appearing both in the popular press and in scientific journals about the negative consequences of fracking. Large amounts of water are used in the fracking process with millions of gallons of water taken out of local streams and rivers and pumped down into each gas well. A portion of this water is later removed from the well and must be properly treated after use to avoid contaminating local water bodies.

A variety of chemicals are added to the fracking fluid to facilitate the release of natural gas. Mining companies are not required to publicly identify all of these chemicals. Environmental scientists and concerned citizens began to wonder if fracking was responsible for chemical contamination of underground water and, in one case, the poisoning of livestock. Some drinking-water wells near fracking sites became contaminated with natural gas, and homeowners and public health officials asked if fracking was the culprit. Water with high concentrations of natural gas can be flammable, and footage of flames shooting from kitchen faucets after someone ignited the water became popular on YouTube, in documentaries, and in feature films. However, it wasn't clear if fracking caused natural gas to contaminate well water or if some of these wells contained natural gas long before fracking began. Several reputable studies showed that drinking-water wells near some fracking sites were contaminated, with natural gas concentrations in the nearby wells being much higher than in more distant wells. These issues need further study, which may take years.

Scientists have begun to assess how much natural gas escapes during the fracking and gas extraction process. As we will learn in Chapter 19, methane is a greenhouse gas and is much more efficient at trapping heat from Earth than carbon dioxide, which is the greenhouse gas most commonly produced by human activity. As the number of potential environmental issues associated with fracking began to increase, environmental scientists and activists began to ask whether fracking was making the greenhouse problem and other environmental problems worse. By 2014, it appeared that opponents of fracking were as numerous as supporters.

Certainly, using natural gas is better for the environment than coal, though using less fossil fuel—or using no fossil fuel at all—would be even better. However, at present it is difficult to know whether the benefits of using natural gas outweigh the problems that extraction causes. Many years may pass before the extent and nature of harm from fracking is known.

The story of natural gas fracking provides a good introduction to the study of environmental science. It shows us that human activities that are initially perceived as causing little harm to the environment can in fact have adverse effects, and that we may not recognize these effects until we better understand the science surrounding the issue. It also illustrates the difficulty in obtaining absolute answers to questions about the environment and demonstrates that environmental science can be controversial. Finally, it shows us that making assessments and choosing appropriate actions in environmental science are not always as clear-cut as they first appear.

Sources:
S. G. Osborn et al., Methane contamination of drinking water accompanying gas-well drilling and hydraulic fracturing, *Proceedings of the National Academy of Sciences* 108 (2011): 8172–8176; Drilling down. Multiple authors in 2011 and 2012. *New York Times*, viewed at: http://www.nytimes.com/interactive/us/DRILLING_DOWN_SERIES.html.

The process of scientific inquiry builds on previous work and careful, sometimes lengthy, investigations. For example, we will eventually accumulate a body of knowledge on the effects of hydraulic fracturing of natural gas, but until we have this knowledge, we will not be able to make a fully informed decision about the policies of energy extraction. In the meantime, we may need to make interim decisions based on incomplete information. This uncertainty is one feature—and an exciting aspect—of environmental science.

To investigate important topics such as the extraction and use of fossil fuels, environmental science relies on a number of indicators, methodologies, and tools. This chapter introduces you to the study of the environment and outlines some of the important foundations and assumptions you will use throughout your study.

module

1

Environmental Science

Humans are dependent on Earth's air, water, and soil for our existence. However, we have altered the planet in many ways, both large and small. The study of environmental science can help us understand how humans have changed the planet and identify ways of responding to those changes.

Learning Objectives

After reading this module you should be able to

- define the field of environmental science and discuss its importance.
- identify ways in which humans have altered and continue to alter our environment.

Environmental science offers important insights into our world and how we influence it

Stop reading for a moment and look up to observe your surroundings. Consider the air you breathe, the heating or cooling system that keeps you at a comfortable temperature, and the natural or artificial light that helps you see. Our **environment** is the sum of all the conditions surrounding us that influence life. These conditions include living organisms as well as nonliving components such as soil, temperature, and water. The influence of humans is an important part of the environment as well. The environment we live in determines how healthy we are, how fast we grow, how easy it is to move around, and even how much food we can obtain. One environment may be strikingly different from another—a hot, dry desert versus a cool, humid tropical rainforest, or a coral reef teeming with marine life versus a crowded city street.

We are about to begin an examination of **environmental science,** the field of study that looks at interactions among human systems and those found in nature. By *system* we mean any set of interacting components that influence one another by exchanging energy or materials. We have already seen that a change in one part of a system—for example, fracking in a particular geologic formation—can cause changes throughout the entire system, such as in a nearby well that supplies drinking water.

An environmental system may be completely human-made, like a subway system, or it may be natural, like weather. The scope of an environmental scientist's work can vary from looking at a small population of individuals, to multiple populations that make up a species, to a community of interacting species, or to even larger systems, such as the global climate system. Some environmental scientists are interested in regional problems. The specific case of fracking at a particular location in the United States, for example, is a regional problem. Other environmental scientists

Environment The sum of all the conditions surrounding us that influence life.

Environmental science The field of study that looks at interactions among human systems and those found in nature.

Teaching Tip: Warm-up

Show students a picture of an animal in its natural environment, such as a frog in the rainforest or a lion in the savanna. Ask students to name as many conditions surrounding the animal as they can, including both living organisms and nonliving components. This exercise should acquaint your students with the components of natural habitats and should develop their observational skills. *Answers will vary. As an example, consider a frog's surroundings in the rainforest: leaves, ants, wasps, snakes, vines, soil, water, and rocks.*

Teaching Tip: Connections

Ask students to work with a partner and list five ways they interact with the environment daily. Have each pair share their results with the class. *Answers will vary but should help the class begin to understand the importance of the subject and the benefit of further study. Answers that discuss activist activity may help you lead into a discussion of the difference between environmental science and environmentalism. Answers involving the popular press or possible misinformation might help you raise awareness about the scientific method.*

Teaching Tip: Venn Diagram

Have students make a Venn diagram, as shown below, to compare environmental scientists with environmentalists. Note that the left side shows characteristics of environmental scientists, the right side shows characteristics of environmentalists, and the center shows areas in which the two overlap. After they have written down the similarities and differences between the two, have them discuss their thoughts with the class. *Answers will vary but may include items listed in the sample diagram below.*

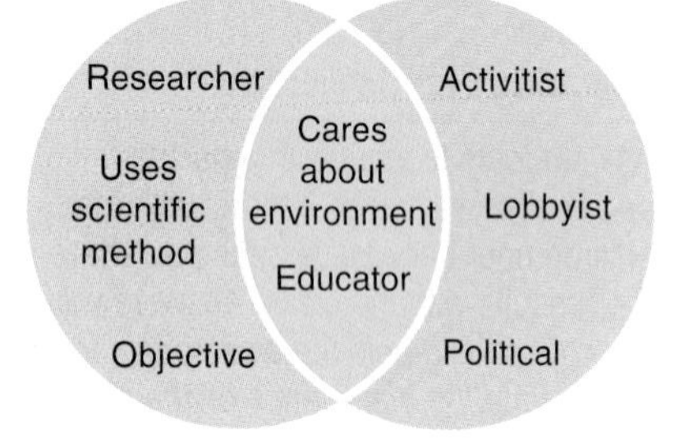

Teaching Tip: Discussion Starter

Ask students: In what ways is the field of environmental studies interdisciplinary? *The field of environmental studies includes many disciplines, such as environmental policy, economics, literature, and ethics. Environmental science is only a subset of the environmental studies field.*

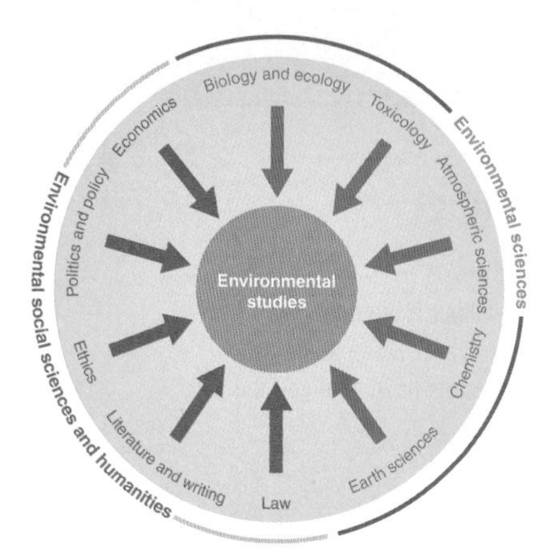

FIGURE 1.1 **Environmental studies.** The study of environmental science uses knowledge from many disciplines.

work on global issues, such as species extinction and climate change.

Many environmental scientists study a specific type of natural system known as an *ecosystem*. An **ecosystem** is a particular location on Earth with interacting components that include living, or **biotic,** components and nonliving, or **abiotic,** components.

As a student of environmental science, you should recognize that environmental science is different from *environmentalism,* which is a social movement that seeks to protect the environment through lobbying, activism, and education. An **environmentalist** is a person who participates in environmentalism. In contrast, an environmental scientist, like any scientist, follows the process of observation, hypothesis testing, and field and laboratory research. We'll learn more about the process of science later in this chapter.

Ecosystem A particular location on Earth with interacting biotic and abiotic components.

Biotic Living.

Abiotic Nonliving.

Environmentalist A person who participates in environmentalism, a social movement that seeks to protect the environment through lobbying, activism, and education.

Environmental studies The field of study that includes environmental science and additional subjects such as environmental policy, economics, literature, and ethics.

So what does the study of environmental science actually include? As FIGURE 1.1 shows, environmental science encompasses topics from many scientific disciplines, such as chemistry, biology, and Earth science. Environmental science is itself a subset of the broader field known as **environmental studies,** which includes additional subjects such as environmental policy, economics, literature, and ethics. Throughout the course of this book you will become familiar with these and many other disciplines.

We have seen that environmental science is a deeply interdisciplinary field. It is also a rapidly growing area of study. As human activities continue to affect the environment, environmental science can help us understand the consequences of our interactions with our planet and help us make better decisions about our actions.

Humans alter natural systems

Think of the last time you walked in a wooded area. Did you notice any dead or fallen trees? Chances are that even if you did, you were not aware that living and nonliving components were interacting all around you. Perhaps an insect pest killed the tree you saw and many others of the same species. Over time, dead trees in a forest lose moisture. The increase in dry wood makes the forest more vulnerable to intense wildfires. But the process doesn't stop there. Wildfires trigger the germination of certain tree seeds, some of which lie dormant until after a fire. And so what began with the activity of insects leads to a transformation of the forest. In this way, biotic factors interact with abiotic factors to influence the future of the forest. All of these factors are part of a system.

Systems can vary in size. A large system may contain many smaller systems within it. FIGURE 1.2 shows an example of complex, interconnecting systems that operate at multiple space and time scales: the fisheries of the North Atlantic. A physiologist who wants to study how codfish survive in the North Atlantic's freezing waters must consider all the biological adaptations of the cod that enable it to be part of one system. In this case, the fish and its internal organs are the system being studied. In the same environment, a marine biologist might study the predator-prey relationship between cod and herring. That relationship constitutes another system, which includes two fish species and the environment they live in. At an even larger scale, a scientist might examine a system that includes all of these systems as well as people, fishing technology, policy, and law. The global environment is composed of both small-scale and large-scale systems.

Teaching Tip: Engage

Ask students: In what ways do humans change the environment? *Answers will vary.*

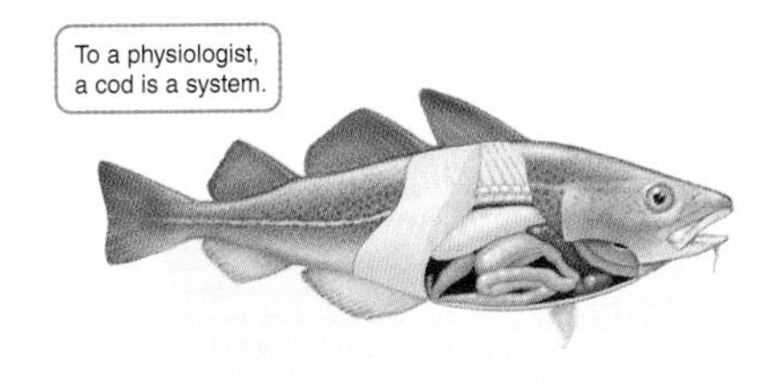

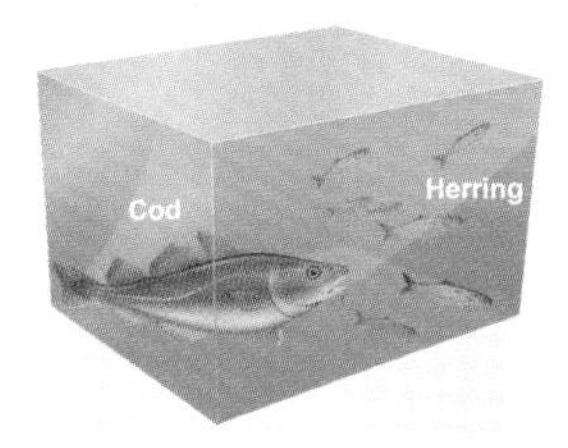

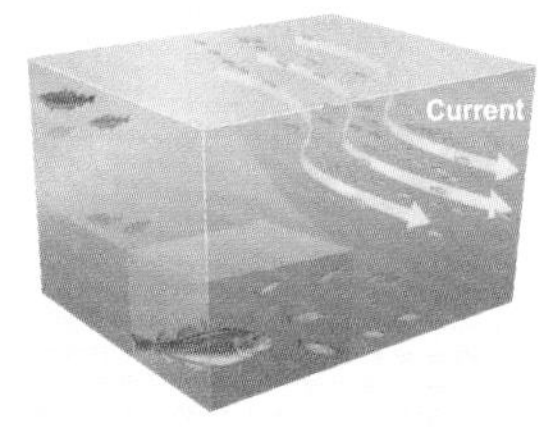

FIGURE 1.2 **Systems within systems.** The boundaries of an environmental system may be defined by the researcher's point of view. Physiologists, marine biologists, oceanographers, and fisheries managers would all describe the North Atlantic Ocean fisheries system differently.

Humans manipulate the systems in their environment more than any other species. We convert land from its natural state into urban, suburban, and agricultural areas. We change the chemistry of our air, water, and soil, both intentionally—for example, by adding fertilizers—and unintentionally—for example, by our activities that generate pollution. Even where we don't manipulate the environment directly, the simple fact that there are so many of us affects our surroundings.

Humans and our direct ancestors (other members of the genus *Homo*) have lived on Earth for about 2.5 million years. During this time, and especially during the last 10,000 to 20,000 years, we have shaped and influenced our environment. As tool-using, social animals, we have continued to develop a capacity to directly alter our environment in substantial ways. *Homo sapiens*—genetically modern humans—evolved to be successful hunters; when they entered a new environment, they often hunted large animal species to extinction. In fact, early humans are thought to be responsible for the extinction of mammoths, mastodons, giant ground sloths, and many types of birds. More recently, hunting in North America led to the extinction of the passenger pigeon (*Ectopistes migratorius*) and nearly caused the loss of the American bison (*Bison bison*).

But the picture isn't all bleak. Human activities have also created opportunities for certain species to thrive. For example, for thousands of years Native Americans on the Great Plains used fire to capture animals for food. The fires they set kept trees from encroaching on the plains, which in turn created a window for an entire ecosystem to develop. Because of human activity, this ecosystem—the tallgrass prairie—is now home to numerous unique species.

During the last two centuries, the rapid and widespread development of technology, coupled with dramatic human population growth, has substantially increased both the rate and the scale of our global environmental impact. Modern cities with electricity, running water, sewer systems, Internet connections, and public transportation systems have improved human well-being, but they have come at a cost. Because cities cover land that was once natural habitat, species that relied on that habitat must adapt, relocate, or go extinct. Human-induced changes in climate—for example, in patterns of temperature and precipitation—affect the health of natural systems on a global scale. Current changes in land use and climate are rapidly outpacing the rate at which natural systems can evolve. Some species have not "kept up" and can no longer compete in the human-modified environment.

Moreover, as the number of people on the planet has grown, their effect has multiplied. Six thousand people can live in a relatively small area with only minimal effects on the environment. But when roughly 4 million people live in a modern city like Los Angeles, their combined activity will cause environmental damage that will inevitably pollute the water, air, and soil as well as introduce other adverse consequences (FIGURE 1.3).

Teaching Tip: Activity

Have students examine Figure 1.2, and then in pairs or groups, ask them to create their own "systems within systems" diagram. To get them started, you can assign each group a starting organism, such as a deer or grizzly bear. *Answers will vary. One example could be: deer → deer and wolves → deer and wolves in the Yellowstone ecosystem → Yellowstone ecosystem and Yellowstone Park rangers.*

Teaching Tip: Discussion Starter

Ask students: How does human development impact natural systems, and how has this changed over the last 2 centuries? *Answers will vary, but should include the increasing magnitude of our impact on land use and climate.*

▲ TEACHING with FIGURES

Have students examine Figure 1.2, and then ask them to think of two more systems that could be included in the diagram. *Answers may include references to the ocean floor, plate tectonics, heat exchange, and pollutants.*

Teaching Tip: Beyond the Classroom

Have students replicate Figure 1.3 for another city. Ask students to search the Web for a current photograph of a city, and then one of the same city when it was significantly smaller. Students should share and discuss their findings with each other.

Answers to Module 1 AP® Review Questions

1. d
2. b
3. c
4. c

(a)

(b)

FIGURE 1.3 **Human impact on Earth.** It is impossible for millions of people to inhabit an area without altering it. (a) In 1880, fewer than 6,000 people lived in Los Angeles. (b) In 2013, Los Angeles had a population of 3.9 million people, and the greater Los Angeles metropolitan area was home to nearly 13 million people.
(a: The Granger Collection, New York; b: LA/AeroPhotos/Alamy)

module 1

REVIEW

In this module we have seen that the study of environmental science helps us understand the role humans have played in the natural environment, and how that role has changed over time. There are specific approaches to the study of environmental science, some of which utilize terms and concepts from other disciplines. To study environmental science, we utilize specific techniques and environmental indicators, the focus of the next module.

Module 1 AP® Review Questions

1. Impacts of fracking include
   I contamination of ground water.
   II increased use of coal.
   III lower natural gas prices.
   (a) I only
   (b) I and II only
   (c) II and III only
   (d) I and III only
   (e) I, II, and III

2. Which of the following is an abiotic component?
   (a) an eagle
   (b) a rock
   (c) a tree
   (d) a human
   (e) a virus

3. Which of the following is NOT true about ecosystems?
   (a) They include biotic components.
   (b) They can be a wide range of sizes.
   (c) They include no human components.
   (d) Many interactions among species occur in them.
   (e) They include abiotic components.

4. Each of the following is an example of how humans have negatively affected the environment except
   (a) hunting large mammals.
   (b) conversion of arid land to agricultural use.
   (c) the use of fire to create the Great Plains.
   (d) slash-and-burn forest clearing.
   (e) fertilizer additions to lakes and rivers.

module

2

Environmental Indicators and Sustainability

As we study the way humans have altered the natural world, it is important to have techniques for measuring and quantifying human impact. Environmental indicators allow us to assess the impact of humans on Earth. The use of these indicators help us determine whether or not the quality of the natural environment is improving and inform discussions on the sustainability of humans on the planet.

Learning Objectives

After reading this module you should be able to

- identify key environmental indicators and their trends over time.
- define sustainability and explain how it can be measured using the ecological footprint.

Teaching Tip: Discussion Starter

Ask students: What is an environmental indicator and what does it tell us? *An environmental indicator describes the current state of an environmental system. It tells us when we might need to look more deeply into a particular issue.*

Environmental scientists monitor natural systems for signs of stress

One critical question that environmental scientists investigate is whether the planet's natural life-support systems are being degraded by human-induced changes. Natural environments provide what we refer to as **ecosystem services**—the processes by which life-supporting resources such as clean water, timber, fisheries, and agricultural crops are produced. Although we often take a healthy ecosystem for granted, we notice when an ecosystem is degraded or stressed because it is unable to provide the same services or produce the same goods. To understand the extent of our effect on the environment, we need to be able to measure the health of Earth's ecosystems.

To describe the health and quality of natural systems, environmental scientists use *environmental indicators*. Just as body temperature and heart rate can indicate whether a person is healthy or sick, **environmental indicators** describe the current state of an environmental system. These indicators do not always tell us what is causing a change, but they do tell us when we might need to look more deeply into a particular issue. Environmental indicators provide valuable information about natural systems on both small and large scales. Some of these indicators and the chapters in which they are covered are listed in **TABLE 2.1**.

In this book we will focus on the five global-scale environmental indicators listed in **TABLE 2.2**: biological diversity, food production, average global surface temperature and carbon dioxide concentrations in the atmosphere, human population, and resource depletion. Throughout the text we will cover each of these five indicators in greater detail. Here we take a first look.

Ecosystem services The processes by which life-supporting resources such as clean water, timber, fisheries, and agricultural crops are produced.

Environmental indicator An indicator that describes the current state of an environmental system.

Teaching Tip: Warm-up

Ask students: What services and products does nature provide for us? Which of these can we live without? *Answers will vary, but may include oxygen production, water purification, food production, pollination, fuel production, soil formation, flood regulation, and climate regulation.*

TABLE 2.1 Some common environmental indicators

Environmental indicator	Unit of measure	Chapter
Human population	Individuals	7
Ecological footprint	Hectares of land	1
Total food production	Metric tons of grain	11
Food production per unit area	Kilograms of grain per hectare of land	11
Per capita food production	Kilograms of grain per person	11
Carbon dioxide	Concentration in air (parts per million)	19
Average global surface temperature	Degrees centigrade	19
Sea level change	Millimeters	19
Annual precipitation	Millimeters	4
Species diversity	Number of species	5, 18
Fish consumption advisories	Present or absent; number of fish allowed per week	17
Water quality (toxic chemicals)	Concentration	14
Water quality (conventional pollutants)	Concentration; presence or absence of bacteria	14
Deposition rates of atmospheric compounds	Milligrams per square meter per year	15
Fish catch or harvest	Kilograms of fish per year or weight of fish per effort extended	11
Extinction rate	Number of species per year	5
Habitat loss rate	Hectares of land cleared or "lost" per year	18
Infant mortality rate	Number of deaths of infants under age 1 per 1,000 live births	7
Life expectancy	Average number of years an infant born today can be expected to live under current conditions	7

Teaching Tip: Beyond the Classroom

Divide students into five groups and assign to each group one of the five key global indicators listed in Table 2.2. Ask them to use the Web or other resources to find data that support the recent trend in each indicator. For example, for biological diversity, "large number of extinctions" can be supported by the estimate of more than 1,000 species going extinct each year.

TABLE 2.2 Five key global indicators

Indicator	Recent trend	Outlook for the future	Overall impact on environmental quality
Biological diversity	Large number of extinctions, extinction rate increasing	Extinctions will continue	Negative
Food production	Per capita production possibly leveling off	Unclear	May affect the number of people Earth can support
Average global surface temperature and CO_2 concentration	CO_2 concentrations and temperatures increasing	Probably will continue to increase, at least in the short term	Effects are uncertain and varied but probably detrimental
Human population	Still increasing, but growth rate slowing	Population leveling off; resource consumption rates also a factor	Negative
Resource depletion	Many resources being depleted at rapid rate, but human ingenuity develops "new" resources, and efficiency of resource use is increasing in many cases	Unknown	Increased use of most resources has negative effects

Biological Diversity

Biological diversity, or **biodiversity,** is the diversity of life forms in an environment. It exists on three scales: *ecosystem, species,* and *genetic,* illustrated in FIGURE 2.1. Each level of biodiversity is an important indicator of environmental health and quality.

Genetic Diversity

Genetic diversity is a measure of the genetic variation among individuals in a population. Populations with high genetic diversity are better able to respond to environmental change than populations with lower genetic diversity. For example, if a population of fish possesses high genetic diversity for disease resistance, at least some individuals are likely to survive whatever diseases move through the population. If the population declines in number, however, the amount of genetic diversity it can possess is also reduced, and this reduction increases the likelihood that the population will decline further when exposed to a disease.

Species Diversity

A **species** is defined as a group of organisms that is distinct from other groups in its morphology (body form and structure), behavior, or biochemical properties. Individuals within a species can breed and produce fertile offspring. Scientists have identified and cataloged approximately 2 million species on Earth. Estimates of the total number of species on Earth range between 5 million and 100 million, with the most common estimate at 10 million. This number includes a large array of organisms with a multitude of sizes, shapes, colors, and roles.

Species diversity indicates the number of species in a region or in a particular type of habitat. Scientists have observed that ecosystems with more species—that is, higher species diversity—are more productive and resilient—that is, better able to recover from disturbance. For example, a tropical forest with a large number of plant species growing in the understory is likely to be more productive, and better able to withstand change, than a nearby tropical forest plantation with one crop species growing in the understory.

Environmental scientists often focus on species diversity as a critical environmental indicator. The number of frog species, for example, is used as an indicator of regional environmental health because frogs are exposed to both the water and the air in their ecosystem. A decrease in the number of frog species in a particular ecosystem may be an indicator of environmental problems there. Species losses in several ecosystems can indicate environmental problems on a larger scale. Not all species losses are indicators of environmental problems, however. Species arise and others go extinct as part of the natural evolutionary process. The

(a) Ecosystem diversity

(b) Species diversity

(c) Genetic diversity

FIGURE 2.1 **Levels of biodiversity.** Biodiversity exists at three scales. (a) Ecosystem diversity is the variety of ecosystems within a region. (b) Species diversity is the variety of species within an ecosystem. (c) Genetic diversity is the variety of genes among individuals of a species.

Biodiversity The diversity of life forms in an environment.

Genetic diversity A measure of the genetic variation among individuals in a population.

Species A group of organisms that is distinct from other groups in its morphology (body form and structure), behavior, or biochemical properties.

Species diversity The number of species in a region or in a particular type of habitat.

Teaching Tip: Discussion Starter

Ask students: Which species do you think has higher genetic diversity, domestic dogs or domestic cats? Where do humans fall on this scale? *Of the three species, domestic dogs have the most genetic diversity. Domestic cats have the least genetic diversity. Humans fall in between dogs and cats in genetic diversity.*

Teaching Tip: Beyond the Classroom

Ask each student to bring in a picture of an endangered species, with information on its habitat, population status, major threats, and any current related conservation activity. You may even want to hang pictures of these species in the classroom for future reference.

(a)

(b)

(c)

(d)

FIGURE 2.2 **Species on the brink.** Humans have saved some species from the brink of extinction, such as (a) the American bison and (b) the peregrine falcon. Other species, such as the (c) snow leopard and (d) the West Indian manatee, continue to decline. *(a: Richard A. McMillin/Shutterstock; b: Jim Zipp/Science Source; c: AlanCarey/Science Source; d: Douglas Faulkner/Science Source)*

evolution of new species, known as **speciation,** typically happens very slowly—perhaps on the order of one to three new species per year worldwide. The average rate at which species go extinct over the long term is referred to as the **background extinction rate.** The background extinction rate is also very slow: about one species in a million every year. So with 2 million identified species on Earth, the background extinction rate should be about two species per year.

Under conditions of environmental change or biological stress, species may go extinct faster than new ones evolve. Some scientists estimate that more than 1,000 species are currently going extinct each year—which is about 500 times the background rate of extinction. Habitat destruction and habitat degradation are the major causes of species extinction today, although climate change, overharvesting, and pressure from introduced species also contribute to species loss. Human intervention has saved certain species, including the American bison, peregrine falcon (*Falco peregrinus*), bald eagle (*Haliaeetus leucocephalus*), and American alligator (*Alligator mississippiensis*). But other large animal species, such as the Bengal tiger (*Panthera tigris*), snow leopard (*Panthera uncia*), and West Indian manatee (*Trichechus manatus*), remain endangered and may go extinct if present trends are not reversed. Overall, the number of species has been declining (FIGURE 2.2).

Speciation The evolution of new species.

Background extinction rate The average rate at which species become extinct over the long term.

Ecosystem Diversity

Ecosystem diversity is a measure of the diversity of ecosystems or habitats that exist in a given region. A greater number of healthy and productive ecosystems means a healthier environment overall. As an environmental

Practicing Science: Analyze

After students read the discussion of speciation and background extinction rates, instruct them to calculate the average net gain or loss for species on Earth, without human-driven biological stress. Students should use data given in the text.

Speciation = an addition of 1–3 species/year = average 2 additional species/year

Background extinction = loss of 1 species per million species/year × 2 million total species = loss of 2 species/year. Net = 2 additional species + 2 species lost = 0

Teaching Tip: Connections

Ask students: What factors other than human activities cause species extinction? *Climate fluctuations, natural disasters, meteorites, competition among species.*

Teaching Tip: Discussion Starter

Ask students: Do you think Earth would really be at zero net species gain without human impacts? *Answers will vary.*

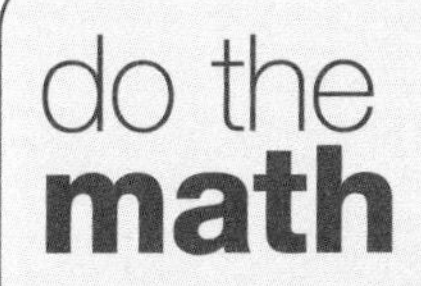

Converting Between Hectares and Acres

In the metric system, land area is expressed in hectares. A hectare (ha) is 100 meters by 100 meters. In the United States, land area is most commonly expressed in acres. There are 2.47 acres in 1 ha. The conversion from hectares is relatively easy to do without a calculator; rounding to two significant figures gives us 2.5 acres in 1 ha. If a nature preserve is 100 ha, what is it size in acres?

$$100 \text{ ha} \times 2.5 \text{ acres} = 250 \text{ acres}$$

Your Turn A particular forest is 10,000 acres. Determine its size in hectares.

indicator, the current loss of biodiversity tells us that natural systems are facing strains unlike any in the recent past. We will look at this important topic in greater detail in Chapters 5 and 18.

Some measures of biodiversity are given in terms of land area, so becoming familiar with measurements of land area is important to understanding them. A hectare (ha) is a unit of area used primarily in the measurement of land. It represents 100 meters by 100 meters. In the United States we measure land area in terms of square miles and acres. However, the rest of the world measures land in hectares. "Do the Math: Converting Between Hectares and Acres" shows you how to do the conversion.

Food Production

The second of our five global indicators is food production: our ability to grow food to nourish the human population. Just as a healthy ecosystem supports a wide range of species, a healthy soil supports abundant and continuous food production. Food grains such as wheat, corn, and rice provide more than half the calories and protein humans consume. Still, the growth of the human population is straining our ability to grow and distribute adequate amounts of food.

In the past we have used science and technology to increase the amount of food we can produce on a given area of land. World grain production has increased fairly steadily since 1950 as a result of expanded irrigation, fertilization, new crop varieties, and other innovations. At the same time, worldwide production of grain *per person*, also called *per capita* world grain production, has leveled off. FIGURE 2.3 shows what might be a slight downward trend in wheat production since about 1985.

In 2008, food shortages around the world led to higher food prices and even riots in some places. Why did this happen? The amount of grain produced worldwide is influenced by many factors. These factors include climatic conditions, the amount and quality of land under cultivation, irrigation, and the human labor and energy required to plant, harvest, and bring the grain to market. Grain production is not keeping up with population growth because in some areas the productivity of agricultural ecosystems has declined as a result of soil degradation, crop diseases, and unfavorable weather conditions such as drought or flooding. In addition, demand is outpacing supply. While the rate of human population growth has outpaced increases in food production, humans currently use more grain to feed livestock than they consume themselves. Finally, some government policies discourage food production by making it more profitable for land to remain uncultivated or by encouraging farmers to grow crops for fuels such as ethanol and biodiesel instead of food.

Will there be sufficient grain to feed the world's population in the future? In the past, whenever a shortage of food has loomed, humans have discovered and employed technological or biological innovations to increase production. However,

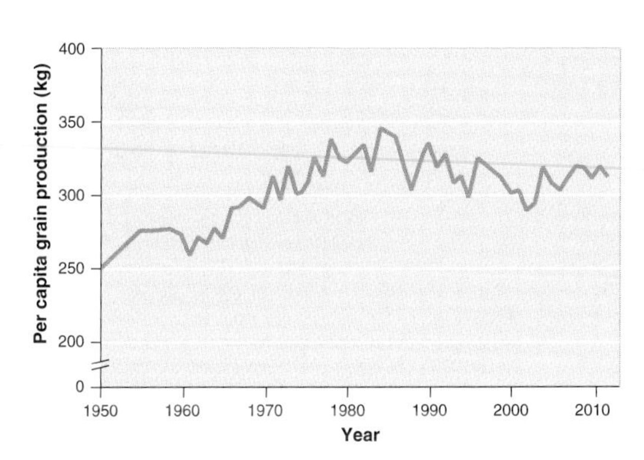

FIGURE 2.3 **World grain production per person.** Grain production has increased since the 1950s, but it has recently begun to level off. *(After http://www.earth-policy.org/index.php?/indicators/C54)*

Answer to Your Turn

There are 2.5 acres in a hectare (ha); the size of the forest is

$$\frac{10{,}000 \text{ acres}}{2.5 \text{ acres/ha}} = 4{,}000 \text{ ha}$$

Teaching Tip: Engage

Ask students: What are the three most-consumed food grains in the world? *Wheat, corn, and rice are the three most-consumed grains in the world.*

Teaching Tip: Beyond the Classroom

Ask students to explore agricultural production patterns in the United States on the Web. Students can also look for information about agricultural production in their particular state. You may want to have students make a map of where their favorite crops are produced, find the top five agricultural producing states, or find nearby farms in their own state.

For a link to a USDA website that contains current data, click on the link buttons in the TE-book, open the TRFD, or log onto the book's companion website bcs.whfreeman.com/friedlandapes2e (teacher login required).

TRM **Chapter 1 Web Resources**

COMMON MISCONCEPTIONS
The Greenhouse Effect

Students often misunderstand the greenhouse effect. This chapter introduces the concept of the greenhouse effect, but it will be covered in greater detail in Chapter 4. We suggest you defer a lengthy discussion of this topic until then.

these innovations often put a strain on the productivity of the soil. If we continue to overexploit the soil, its ability to sustain food production may decline dramatically. We will take a closer look at soil quality in Chapter 8 and food production in Chapter 11.

Average Global Surface Temperature and Carbon Dioxide Concentrations

We have seen that biodiversity and abundant food production are necessary for life. One of the things that makes them possible is a stable climate. Earth's temperature has been relatively constant since the earliest forms of life began, about 3.5 billion years ago. The temperature of Earth allows the presence of liquid water, which is necessary for life.

What keeps Earth's temperature so constant? As FIGURE 2.4 shows, our thick planetary atmosphere contains many gases. Some of these atmospheric gases, known as **greenhouse gases,** trap heat near Earth's surface. The most important greenhouse gas is carbon dioxide (CO_2). During most of the history of life on Earth, greenhouse gases have been present in the atmosphere at fairly constant concentrations for relatively long periods. They help keep Earth's surface within the range of temperatures at which life can flourish.

In the past 2 centuries, however, the concentrations of CO_2 and other greenhouse gases in the atmosphere have risen. Today, atmospheric CO_2 concentrations are greater than 400 parts per million (ppm). During roughly the same period, as the graph in FIGURE 2.5 shows, while global temperatures have fluctuated considerably, they have displayed an overall increase. (Note that this graph has two *y* axes. See the appendix "Reading Graphs" if you'd like to learn more about reading a graph like this one.) Many scientists believe that the increase in atmospheric CO_2 during the last two centuries is **anthropogenic**—that is, the increase is derived from human activities. The two major sources of anthropogenic CO_2 are the combustion of fossil fuels and the net loss of forests and other habitats that would otherwise take up and store CO_2 from the atmosphere. We will discuss climate in Chapter 4 and global climate change in Chapter 19.

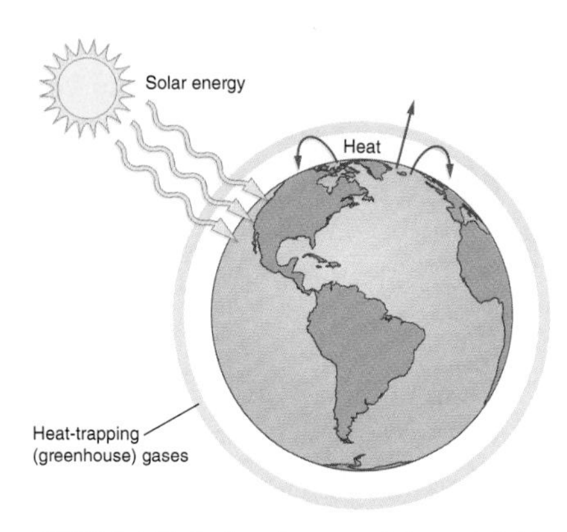

FIGURE 2.4 **The Earth-surface energy balance.** As Earth's surface is warmed by the Sun, it radiates heat outward. Heat-trapping gases absorb the outgoing heat and reradiate some of it back to Earth. Without these greenhouse gases, Earth would be much cooler.

Human Population

In addition to biodiversity, food production, and global surface temperature, the size of the human population can tell us a great deal about the health of our global environment. The human population is currently 7.2 billion and growing. The increasing world population places additional demands on natural systems, since each new person requires food, water, and other resources. In any given 24-hour period, 387,000 infants are born

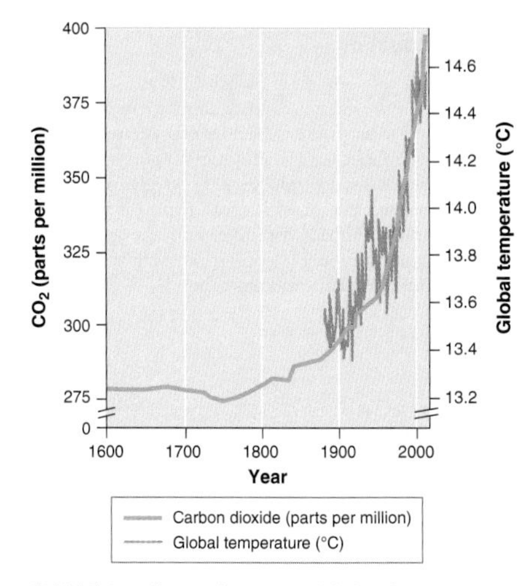

FIGURE 2.5 **Changes in average global surface temperature and in atmospheric CO_2 concentrations.** Earth's average global surface temperature has increased steadily for at least the past 100 years. Carbon dioxide concentrations in the atmosphere have varied over geologic time, but have risen steadily since 1960. *(Data from http://data.giss.nasa.gov/gistemp/graphs_v3/ and http://www.esrl.noaa.gov/gmd/ccgg/trends/#mlo_full)*

Greenhouse gases Gases in Earth's atmosphere that trap heat near the surface.

Anthropogenic Derived from human activities.

Teaching Tip: Discussion Starter

Ask students: Do the data shown in Figure 2.5 surprise you? Why or why not? *Answers will vary.*

FIGURE 2.6 **The effect of crowded cities on natural and human systems.** The human population will continue to grow for at least 50 years. Unless humans can devise ways to live more sustainably, these population increases will put additional strains on natural systems. Here, a street scene in Kolkata. *(Deshakalyan Chowdhury/AFP/Getty Images)*

and 155,000 people die. The net result is 232,000 new inhabitants on Earth each day, or over a million additional people every 5 days. Although the rate of population growth has been slowing since the 1960s, world population size will still continue to increase for at least another 50 to 100 years. Most population scientists project that the human population will be somewhere between 8.1 billion and 9.6 billion in 2050 and will stabilize between 7.1 billion and 10.5 billion by 2100.

Can the planet sustain so many people (FIGURE 2.6)? Even if the human population eventually stops growing, the billions of additional people will create a greater demand on Earth's finite resources, including food, energy, and land. Unless humans work to reduce these pressures, the human population will put a rapidly growing strain on natural systems for at least the first half of this century. We discuss human population issues in Chapter 7.

Resource Depletion

Natural resources provide the energy and materials that support human civilization but, as the human population grows, the resources necessary for our survival become increasingly depleted. In addition, extracting these natural resources can affect the health of our environment in many ways. Pollution and land degradation caused by mining, waste from discarded manufactured products, and air pollution from fossil fuel combustion are just a few of the negative environmental consequences of resource extraction and use.

Some natural resources, such as coal, oil, and uranium, are finite and cannot be renewed or reused. Others, such as aluminum or copper, also exist in finite quantities but can be used multiple times through reuse or recycling. Renewable resources, such as timber, can be grown and harvested indefinitely, but in some locations these resources are being used faster than they can be naturally replenished. "Do the Math: Rates of Forest Clearing" on page 14 provides an opportunity to calculate rates of one type of resource depletion.

Sustaining the global human population requires vast quantities of resources. However, in addition to the total amounts of resources used by humans, we must consider per capita resource use.

Patterns of resource consumption vary enormously among nations depending on their level of development. What exactly do we mean by *development?* **Development** is defined as improvement in human well-being through economic advancement. Development influences personal and collective human lifestyles—things such as automobile use, the amount of meat in the diet, and the availability and use of technologies such as cell phones and personal computers. As economies develop, resource consumption also increases: People drive more automobiles, live in larger homes, and purchase more goods. These increases can often have implications for the natural environment.

According to the United Nations Development Programme, people in developed nations—including the United States, Canada, Australia, most European countries, and Japan—use most of the world's resources. FIGURE 2.7 shows that the 20 percent of the global population that lives in developed nations owns 87 percent of the world's automobiles and consumes 58

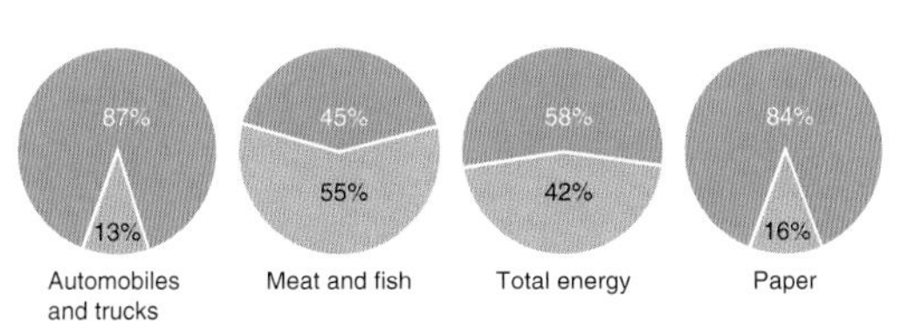

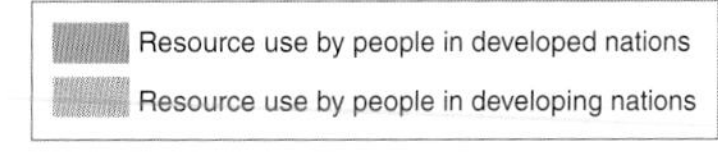

FIGURE 2.7 **Resource use in developed and developing countries.** Only 20 percent of the world's population lives in developed countries, but that 20 percent uses most of the world's resources. The remaining 80 percent of the population lives in developing countries and uses far fewer resources per capita.

Development Improvement in human well-being through economic advancement.

Teaching Tip: Engage

Draw the graphs below on the board, and ask students to identify which one most closely resembles human population growth over the past 2,000 years. Then ask students how they think the human population will change in the future. *The middle graph most closely resembles human population growth.*

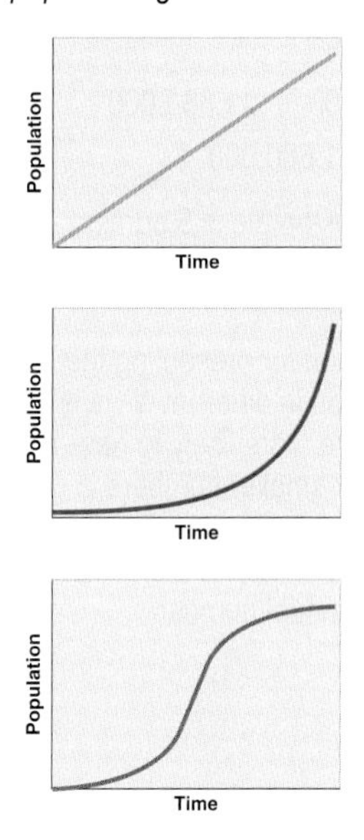

Teaching Tip: Discussion Starter

Ask students: Are resources always either renewable or nonrenewable? How might a renewable resource become nonrenewable? *Resources are not always either renewable or nonrenewable. As an example, in the case of timber production, if harvesting techniques are not sustainable, production rates would decline over time and eventually reach zero.*

Teaching Tip: Engage

Have students complete the following table comparing renewable versus nonrenewable resources using these or other terms: biofuel, coal, copper, oil, tidal energy, timber, uranium, wind energy. (Answers are provided in italics.)

Renewable resources	Nonrenewable resources
Biofuel	*Coal*
Tidal energy	*Copper*
Timber	*Oil*
Wind energy	*Uranium*

Teaching Tip: Beyond the Classroom

Have students research a resource not included in Figure 2.7 and create a similar pie chart to compare use of the resource in developing versus developed countries. You may want to assign each student or group of students a resource to work with. These might include copper, aluminum, coal, oil, natural gas, water, or land. Have students share and discuss their findings with each other.

do the **math**

Answer to Your Turn

Estimate 2: 32,000 ha/day $\times$ 365 days/year = 11,680,000 ha cleared per year

Estimate 1: 12,614,400 ha cleared per year

Difference:

$$\frac{12{,}614{,}400 - 11{,}680{,}000}{11{,}680{,}000} = 8 \text{ percent}$$

The first estimate is 8 percent larger than the second and third estimates.

One reason for presenting the information in different ways is that many people are more likely to have an idea of 1 acre, or smaller numbers in general. Therefore, the 1 acre per second number could be easier to understand than a daily or annual number.

more **math** practice

Conversions

More Math Practice 1.1 offers an additional problem for students to practice converting land areas. To download this problem, click on the link buttons in the TE-book, open the TRFD, or log onto the book's companion website bcs.whfreeman.com/friedlandapes2e (teacher login required).

More Math Practice 1.1: Converting Square Miles to Hectares

do the **math**

Rates of Forest Clearing

A Web search of environmental organizations yielded a range of estimates of the amount of forest clearing that is occurring worldwide:

Estimate 1: 1 acre per second
Estimate 2: 80,000 acres per day
Estimate 3: 32,000 ha per day

Convert the first two estimates into hectares per year and compare them.

There are 2.47 acres per hectare (see "Do the Math: Converting Between Hectares and Acres"). Therefore, 1 acre = 0.40 ha.

Estimate 1: 1.0 acre/second $\times$ 0.40 ha/acre
= 0.40 ha/second 0.40 ha/second $\times$ 60 seconds/minute
$\times$ 60 minutes/hour $\times$ 24 hours/day $\times$ 365 days/year
= 12,614,400 ha cleared per year

Estimate 2: 80,000 acres/day $\times$ 0.40 ha/acre = 32,000 ha cleared per day

Your Turn Notice that Estimate 2, when converted to hectares, is identical to Estimate 3. Now convert the estimate of 32,000 ha/day into the amount cleared per year. How much larger is Estimate 1 than Estimate 2? Why might environmental organizations, or anyone else, choose to present similar information in different ways?

percent of all energy, 84 percent of all paper, and 45 percent of all fish and meat. The poorest 20 percent of the world's people consume 5 percent or less of these resources. Thus, even though the number of people in the developing countries is much larger than the number in the developed countries, their total consumption of natural resources is relatively small.

So while it is true that a larger human population has greater environmental impacts, a full evaluation requires that we look at economic development and consumption patterns as well. We will take a closer look at resource depletion and consumption patterns in Chapters 7, 12, and 13.

Human well-being depends on sustainable practice

The five key environmental indicators that we have just discussed help us analyze the health of the planet. We can use this information to guide us toward **sustainability,** by which we mean living on Earth in a way that allows us to use its resources without depriving future generations of those resources. Many scientists maintain that achieving sustainability is the single most important goal for the human species. It is also one of the most challenging tasks we face.

The Impact of Consumption on the Environment

We have seen that people living in developed nations consume a far greater share of the world's resources than do people in developing countries. What effect does this consumption have on our environment? It is easy to imagine a very small human population living on Earth without degrading its environment because there simply would not be enough people to do significant damage. Today, however, Earth's population is 7.2 billion people and growing. Many environmental scientists ask how we will be able to continue to produce sufficient food, build needed infrastructure, and process pollution and waste. Our current attempts to sustain the human population have already modified many environmental systems. Can we continue our current level of resource consumption without jeopardizing the well-being of future generations?

Teaching Tip: Warm-up

Ask students: What are the five global-scale environmental indicators we focus on in this book, and how do they help us monitor the health of the environment? *Biological diversity, food production, average global surface temperature and CO_2 production, human population, and resource depletion. These environmental indicators provide valuable information about the health of natural systems, such as our forests, oceans, and agricultural fields. They allow us to assess the impact of humans on Earth and determine whether or not the quality of the natural environment is improving.*

Teaching Tip: Discussion Starter

Ask students: How do human activities contribute to changes in the five global-scale environmental indicators? *Human population growth and development deplete habitats for other species and require greater food production. Our increasing use of fossil fuels produces greater amounts of CO_2, which increases global surface temperatures. Lastly, human population growth leads to greater consumption and therefore increases resource depletion.*

FIGURE 2.8 **The cautionary story from Easter Island.** The overuse of resources by the people of Easter Island is probably the primary cause for the demise of that civilization. *(Hubertus Kanus/Science Source)*

Easter Island, in the South Pacific, provides a cautionary tale (FIGURE 2.8). This island, also called Rapa Nui, was once covered with trees and grasses. When humans settled the island many hundreds of years ago, they quickly multiplied in its hospitable environment. They cut down trees to build homes and to make canoes for fishing, but they overused the island's soil and water resources. By the 1870s, almost all of the trees were gone. Without the trees to hold the soil in place, massive erosion occurred, and the loss of soil caused food production to decrease. While other forces, including diseases introduced by European visitors, were also involved in the destruction of the population, the unsustainable use of natural resources on Easter Island appears to be the primary cause for the collapse of its civilization.

Most environmental scientists believe that there are limits to the supply of clean air and water, nutritious foods, and other life-sustaining resources our environment can provide. They also believe there is a point at which Earth will no longer be able to maintain a stable climate. We must meet several requirements in order to live sustainably:

- Environmental systems must not be damaged beyond their ability to recover.
- Renewable resources must not be depleted faster than they can regenerate.
- Nonrenewable resources must be used sparingly.

Sustainable development is development that balances current human well-being and economic advancement with resource management for the benefit of future generations. This is not as easy as it sounds. The issues involved in evaluating sustainability are complex, in part because sustainability depends not only on the number of people using a resource but also on how that resource is used. For example, eating chicken is sustainable when people raise their own chickens and allow them to forage for food on the land. However, if all people, including city dwellers, wanted to eat chicken six times a week, the amount of resources needed to raise that many chickens would probably make the practice of eating chicken unsustainable.

Living sustainably means *acting in a way such that activities that are crucial to human society can continue.* It includes practices such as conserving and finding alternatives to nonrenewable resources as well as protecting the capacity of the environment to continue to supply renewable resources (FIGURE 2.9).

Consider iron, a nonrenewable resource derived from ore removed from the ground. Iron is the major constituent of steel, which we use to make many things, including automobiles, bicycles, and strong frames for tall buildings. Historically, our ability to

FIGURE 2.9 **Living sustainably.** Sustainable choices such as bicycling to work or school can help protect the environment and conserve resources for future generations. *(Jim West/The Image Works)*

Sustainability Living on Earth in a way that allows humans to use its resources without depriving future generations of those resources.

Sustainable development Development that balances current human well-being and economic advancement with resource management for the benefit of future generations.

Teaching Tip: Engage

Ask students: How could the people on Rapa Nui have lived sustainably? What actions could they have taken to prevent the collapse of their civilization? *Answers will vary, but may include planting trees to replace those they cut down, having fewer children, or making homes of nontimber resources.*

Teaching Tip: Journal Prompt

Have students make a list of five personal actions and choices that are sustainable and a list of five that are not sustainable. Then ask them to write a brief reflection on how they might be able to minimize or change their nonsustainable actions. *Answers will vary.*

smelt iron for steel limited our use of that resource, but as we have improved steel manufacturing technology, steel has become more readily available and the demand for it has grown. Because of this increased demand, our current use of iron is unsustainable. What would happen if we ran out of iron? While not too long ago the depletion of iron ore might have been a catastrophe, today we have developed materials that can substitute for certain uses of steel—for example, carbon fiber—and we also know how to recycle steel. Developing substitutes and recycling materials are two ways to address the problem of resource depletion and to increase sustainability.

The example of iron leads us to a question that environmental scientists often ask: How do we determine the importance of a given resource? If we use up a resource such as iron for which substitutes exist, it is possible that the consequences will not be severe. However, if we are unable to find an alternative to the resource—for example, something to replace fossil fuels—people in the developed nations may have to make significant changes in their consumption habits.

FIGURE 2.10 **Central Park, New York City.** New Yorkers have set aside 341 ha (843 acres) in the center of the largest city in the United States—a testament to the compelling human need for interactions with nature. *(ExaMediaPhotography/Shutterstock)*

Teaching Tip: Discussion Starter

Ask students: What are basic human needs? *Basic human needs are the essentials that sustain human life, including air, water, food, and shelter.*

Teaching Tip: Journal Prompt

Ask students to list the items they actually need in order to survive. Be sure to emphasize the difference between *wanting* something and *needing* something. Then ask students to write a brief reflection on the advantages and disadvantages of reducing the amount of unnecessary items in their lives. You may ask students to share and discuss their lists with each other.

Defining Human Needs

We have seen that sustainable development requires us to determine how we can meet our current needs without compromising the ability of future generations to meet their own needs. Let's look at how environmental science can help us achieve that goal. We will begin by defining *needs*.

If you have ever experienced an interruption of electricity to your home or school, you know how frustrating it can be. Without the use of lights, computers, televisions, air-conditioning, heating, and refrigeration, many people feel disconnected and uncomfortable. Almost everyone in the developed world would insist that they need—indeed, cannot live without—electricity. In other parts of the world, however, people have never had these modern conveniences. So, when we speak of *basic needs,* we are referring only to the essentials that sustain human life, including air, water, food, and shelter.

But humans also have more complex needs. Many psychologists have argued that we require meaningful human interactions in order to live a satisfying life, and so a community of some sort might be considered a human need. Biologist Edward O. Wilson wrote that humans exhibit **biophilia**—that is, love of life—which is a need to make "the connections that humans subconsciously seek with the rest of life." Thus our needs for access to natural areas, for beauty, and for social connections can be considered as vital to our well-being as our basic physical needs and must be considered as part of our long-term goal of global sustainability (FIGURE 2.10).

Biophilia Love of life.

The Ecological Footprint

We have begun to see the multitude of ways in which human activities affect the environment. As countries prosper, their populations use more resources. Economic development can sometimes improve environmental conditions. For instance, wealthier countries may have the resources to implement pollution controls and invest money to protect native species. So although people in developing countries do not consume the same quantity of resources as those in developed nations, they may be less likely to use environmentally friendly technologies or to have the financial resources to implement environmental protections.

How do we determine what lifestyles have the greatest environmental impact? This is an important question for environmental scientists if we are to understand the effects of human activities on the planet and develop sustainable practices. Calculating sustainability, however, is more difficult than one might think because we must consider the impacts of our activities and lifestyles on different aspects of our environment. We use land to grow food and to build structures on, and for parks and recreation. We require water for drinking, for cleaning, and for manufacturing products such as paper, and we need clean air to breathe. Yet these goods and services are all interdependent: Using or protecting one has an effect on the others. For example, using land for conventional agriculture may require water for irrigation, fertilizer to promote plant growth, and pesticides to reduce crop damage. This use of land reduces the amount of water available for human use: The plants consume it and the pesticides pollute it.

Teaching Tip: Warm-up

Ask each student or groups of students to bring an everyday object they use to class. Have them describe the resources required to produce it and project how this can be expressed in an area of land. They don't need to estimate actual footprint area, but rather discuss the item's connection to the land. Once everyone is finished, point out that every object in our life originated from Earth, so everything we use affects our ecological footprint.

One method used to assess whether we are living sustainably is to measure the impact of a person or country on world resources. The tool many environmental scientists use for this purpose, the *ecological footprint,* was developed in 1995 by Professor William E. Rees and his graduate student Mathis Wackernagel. An individual's **ecological footprint** is a measure of how much that person consumes, expressed in area of land—that is, the output from the total amount of land required to support a person's lifestyle represents that person's ecological footprint (FIGURE 2.11).

Rees and Wackernagel maintained that if our lifestyle demands more land than is available, then we must be living unsustainably—using up resources more quickly than they can be produced, or producing wastes more quickly than they can be processed. For example, each person requires a certain number of food calories every day. We know the number of calories in a given amount of grain or meat. We also know how much farmland or rangeland is needed to grow the grain to feed people or livestock such as sheep, chickens, or cows. If a person eats only grains or plants, the amount of land needed to provide that person with food is simply the amount of land needed to grow the plants they eat. If that person eats meat, however, the amount of land required to feed that person is greater, because we must also consider the land required to raise and feed the livestock that ultimately become meat. Thus one factor in the size of a person's ecological footprint is the amount of meat in the diet. Meat consumption is a lifestyle choice, and per capita meat consumption is much greater in developed countries. We can calculate the ecological footprint of the food we eat, the water and energy we use, and even the activities we perform that contribute to climate change. Other metrics for calculating our impact on Earth exist as well.

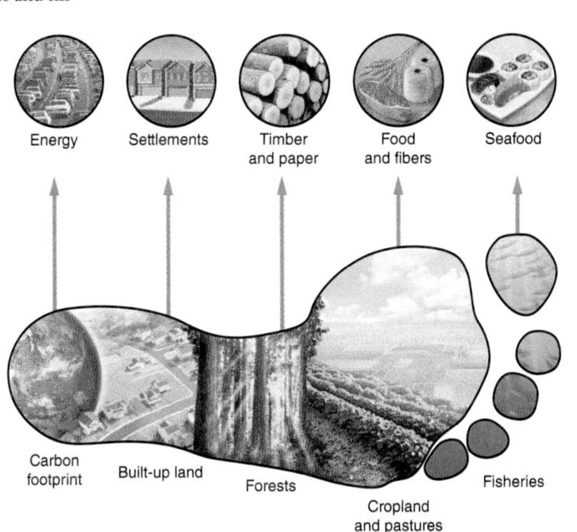

FIGURE 2.11 **The ecological footprint.** An individual's ecological footprint is a measure of how much land is needed to supply the goods and services that individual uses. Only some of the many factors that go into the calculation of the footprint are shown here. (The actual amount of land used for each resource is not drawn to scale.)

Ecological footprint A measure of how much an individual consumes, expressed in area of land.

module
2
REVIEW

In this module we have identified global-scale indicators of environmental health that allow us to monitor specific parameters over time. Ultimately, these indicators contribute to a picture of the sustainability of human activities on Earth. We have identified that biodiversity is decreasing and that food production has leveled off. Atmospheric carbon dioxide concentrations are steadily increasing and global temperatures fluctuate, although the overall change is toward an increase. The human population continues to increase in size but the rate of increase has been declining. One measurement that allows us to assess the sustainability of these different parameters and how they change over time is the ecological footprint. The final module in this chapter introduces us to some of the scientific methods and techniques we will use in order to measure these indicators and make assessments about sustainability.

Teaching Tip: Activity

Have students go to the website "My Footprint" and calculate their ecological footprint. The questions, which will take10 to 15 minutes to answer, ask students for household information. If students are not sure about how to respond, consider providing some data for an average American family. After they have completed the questions, ask students:

- How many Earths would be needed if everyone on the planet had your lifestyle?
- Which of your consumption categories have the largest and smallest footprints?
- Is your footprint bigger or smaller than the country average? Does this surprise you? Why or why not?
- Which biome contributes the most to your footprint? Why do you think that is?
- What are three ways that you can reduce your ecological footprint?

For a link to "My Footprint," click on the link buttons in the TE-book, open the TRFD, or log on to the book's companion website bcs.whfreeman.com /friedlandapes2e (teacher login required).

 **Chapter 1 Web Resources**

Teaching Tip: Discussion Starter

Ask students: What does an ecological footprint tell us? Why is it important to calculate? *An ecological footprint tells us how much land is needed to supply the goods and services that an individual uses. This ecological footprint is important because it helps us assess whether we are living sustainably.*

Answers to Module 2 AP® Review Questions

1. e
2. a
3. e
4. c
5. d

Module 2 AP® Review Questions

1. Common global-scale environmental indicators include all of the following except
   (a) atmospheric carbon dioxide concentrations.
   (b) human population.
   (c) natural resource depletion.
   (d) ocean fish harvest.
   (e) pollution in a local stream.

2. How many hectares of land is a 500-acre park? (1 acre = 0.405 ha)
   (a) 200 ha
   (b) 250 ha
   (c) 500 ha
   (d) 750 ha
   (e) 1,250 ha

3. Refer to Figure 2.7 (on page 13). How does fish and meat consumption in developed and developing countries compare?
   (a) Developing countries consume slightly more meat and fish per capita.
   (b) Developed countries consume slightly more meat and fish per capita.
   (c) Developing and developed countries consume about the same amount of meat and fish per capita.
   (d) Developing countries consume about four times more meat and fish per capita.
   (e) Developed countries consume about four times more meat and fish per capita.

4. In 2011, 640,000 ha of the Amazon rainforest were cleared. Approximately how many hectares is that each hour?
   (a) 1.2 ha
   (b) 29 ha
   (c) 73 ha
   (d) 178 ha
   (e) 1,752 ha

5. A person's ecological footprint is
   (a) the land that a person lives on.
   (b) the amount of carbon dioxide a person contributes to climate change.
   (c) the land required to produce a person's food.
   (d) the land needed to support all of a person's activities.
   (e) the amount of fossil fuel a person uses.

module 3

Scientific Method

Environmental indicators are important for understanding human impacts on Earth systems and the sustainability of those systems. In order to evaluate environmental indicators, we need to use reproducible scientific methods. An understanding of the *scientific method* is essential for environmental science.

Learning Objectives

After reading this module you should be able to

- explain the scientific method and its application to the study of environmental problems.
- describe some of the unique challenges and limitations of environmental science.

Science is a process

Humans during the past century have learned a lot about the impact of their activities on the natural world. Scientific inquiry has provided great insights into the challenges we are facing and has suggested ways to address those challenges. For example, a hundred years ago, we did not know how significantly or rapidly we could alter the chemistry of the atmosphere by burning fossil fuels. Nor did we understand the effects of many common materials, such as lead and mercury, on human health. Much of our knowledge comes from the work of researchers who study a particular problem or situation to understand why it occurs and to determine how we can fix or prevent it from occurring. In this section we will look at the process scientists use to ask and answer questions about the environment.

The Scientific Method

To investigate the natural world, scientists, such as those who examined the effects of fracking as described at the beginning of this chapter, have to be as objective and methodical as possible. They must conduct their research in such a way that other researchers can understand how their data were collected and agree on the validity of their findings. To do this, scientists follow a process known as the **scientific method,** which is an objective way to explore the natural world, draw inferences from it, and predict the outcome of certain events, processes, or changes. The scientific method is used in some form by scientists in all parts of the world and is a generally accepted way to conduct science.

As we can see in FIGURE 3.1, the scientific method has a number of steps, including observing and questioning, forming hypotheses, collecting data, interpreting results, and disseminating findings.

Observing and Questioning

Homeowners and scientists noticed, in areas where fracking occurred, that certain household wells contained high methane concentrations and they wanted to know why this was occurring. Such observing and questioning is where the process of scientific research begins.

Forming Hypotheses

Observing and generating questions lead a scientist to formulate a *hypothesis*. A **hypothesis** is a testable conjecture about how something works. It may be an idea, a proposition, a possible mechanism of interaction, or a statement about an effect. For example, we might hypothesize that when the air temperature rises over time, certain plant species will be more likely, and others less likely, to persist.

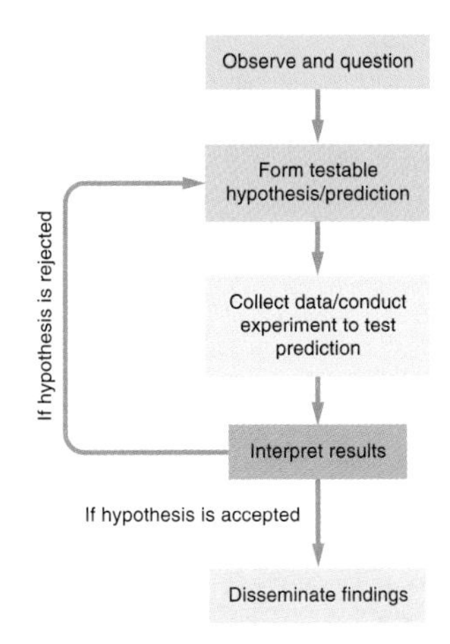

FIGURE 3.1 **The scientific method.** In an actual investigation, a researcher might reject a hypothesis and investigate further with a new hypothesis, several times if necessary, depending on the results of the experiment.

What makes a hypothesis testable? We can test the idea about the relationship between air temperature and plant species by growing plants in a greenhouse at different temperatures. "Fish kills are caused by something in the water" is a testable hypothesis because it speculates that there is an interaction between something in the water and the observed dead fish.

Sometimes it is easier to prove something wrong than to prove it is true beyond doubt. In this case, scientists use a *null hypothesis*. A **null hypothesis** is a prediction that there is no difference between groups or conditions, or a statement or idea that can be falsified, or proved wrong. The statement "Fish deaths have no relationship to something in the water" is an example of a null hypothesis.

Scientific method An objective method to explore the natural world, draw inferences from it, and predict the outcome of certain events, processes, or changes.

Hypothesis A testable conjecture about how something works.

Null hypothesis A prediction that there is no difference between groups or conditions, or a statement or an idea that can be falsified, or proved wrong.

Teaching Tip: Warm-up

Ask students to make an observation about their current surroundings. Have them discuss their observation with a partner or share it with the rest of the class.

Practicing Science: Hypothesize and Collect Data

Have students practice formulating a hypothesis and explain how they might test it. For example, they might hypothesize that the temperature inside the room would decrease if all the windows were opened. Ask them to think about an experimental procedure that would allow them to test this hypothesis. For example, they might want to measure the temperature of the room before and after the windows were opened. What instruments might be needed for this? How might the data be recorded and analyzed? Would testing one classroom on one day be adequate to collect enough data to support their hypothesis? Would the data be different at different times of the day or year?

Practicing Science: Analyze

Write the following statements on the board. Ask students to determine which demonstrate inductive reasoning and which demonstrate deductive reasoning:

- Every cat you have observed has meowed. Therefore, all cats must meow. *Inductive*
- All whales are mammals and all mammals have hair. Therefore all whales have hair. *Deductive*
- John is a teacher. Every teacher you know is funny. Therefore, John must be funny. *Inductive*
- Lizards are reptiles and reptiles are cold blooded. Therefore, lizards are cold blooded. *Deductive*
- Sarah and Kate are friends. Sara likes swimming, running, and rock climbing. Kate likes swimming and rock climbing. Kate must also like running. *Inductive*
- The oak is a tree and all trees have bark. Oaks must have bark. *Deductive*

Collecting Data

Scientists typically take several sets of measurements—a procedure called **replication.** The number of times a measurement is replicated is the **sample size** (sometimes referred to as n). A sample size that is too small can cause misleading results. For example, if a scientist chose three men out of a crowd at random and found that they all had size 10 shoes, she might conclude that all men have a shoe size of 10. If, however, she chose a larger sample size—100 men—it is very unlikely that all 100 individuals would happen to have the same shoe size.

Proper procedures yield results that are accurate and precise. They also help us determine the possible relationship between our measurements or calculations and the true value. **Accuracy** refers to how close a measured value is to the actual or true value. For example, an environmental scientist might estimate how many songbirds of a particular species there are in an area of 1,000 ha by randomly sampling 10 ha and then projecting or extrapolating the result up to 1,000 ha. If the extrapolation is close to the true value, it is an accurate extrapolation. **Precision** is how close to one another the repeated measurements of the same sample are. In the same example, if the scientist counted birds five times on five different days and obtained five results that were similar to one another, the estimates would be precise. **Uncertainty** is an estimate of how much a measured or calculated value differs from a true value. In some cases, it represents the likelihood that additional repeated measurements will fall within a certain range. Looking at FIGURE 3.2, we see that high accuracy and high precision is the most desirable result.

FIGURE 3.2 **Accuracy and precision.** Accuracy refers to how close a measured value is to the actual or true value. Precision is how close repeated measurements of the same sample are to one another.

Interpreting Results

We have followed the steps in the scientific method from making observations and asking questions, to forming a hypothesis, to collecting data. What happens next? Once results have been obtained, analysis of data begins. A scientist may use a variety of techniques to assist with data analysis, including summaries, graphs, charts, and diagrams.

As data analysis proceeds, scientists begin to interpret their results. This process normally involves two types of reasoning: inductive and deductive. Inductive reasoning is the process of making general statements from specific facts or examples. If the scientist who sampled a songbird species in the preceding example made a statement about all birds of that species, she would be using inductive reasoning. It might be reasonable to make such a statement if the songbirds that she sampled were representative of the whole population. Deductive reasoning is the process of applying a general statement to specific facts or situations. For example, if we know that, in general, air pollution kills trees, and we see a single, dead tree, we may attribute that death to air pollution. But a conclusion based on a single tree might be incorrect, since the tree could have been killed by something else, such as a parasite or fungus. Without additional observations or measurements, and possibly experimentation, the observer would have no way of knowing the cause of death with any degree of certainty.

The most careful scientists always maintain multiple working hypotheses—that is, they entertain many possible explanations for their results. They accept or reject certain hypotheses based on what the data show or do not show. Eventually, they determine that certain explanations are the most likely, and they begin to generate conclusions based on their results.

Replication The data collection procedure of taking repeated measurements.

Sample size (n) The number of times a measurement is replicated in data collection.

Accuracy How close a measured value is to the actual or true value.

Precision How close the repeated measurements of a sample are to one another.

Uncertainty An estimate of how much a measured or calculated value differs from a true value.

Disseminating Findings

A hypothesis is never confirmed by a single experiment. That is why scientists not only repeat their experiments themselves, but also present papers at conferences and publish the results of their investigations. This dissemination of scientific findings allows other scientists to repeat the original experiment and verify or challenge the results. The process of science involves ongoing discussion among scientists, who frequently disagree about hypotheses, experimental conditions, results, and the interpretation of results. Two investigators may even obtain different results from similar measurements and experiments, as hap-

AP® Exam Tip

Students should understand how to do a lab write-up using the scientific method. On past AP® Exams they have been asked in the free-response section to design an experiment that covers each step of the scientific method.

pened with investigations of fracking. Only when the same results are obtained over and over by different investigators can we begin to trust that those results are valid. In the meantime, the disagreements and discussion about contradictory findings are a valuable part of the scientific process. They help scientists refine their research to arrive at more consistent, reliable conclusions.

Like any scientist, you should always read reports of "exciting new findings" with a critical eye. Question the source of the information, consider the methods or processes that were used to obtain the information, and draw your own conclusions. This process, essential to all scientific endeavors, is known as critical thinking.

A hypothesis that has been repeatedly tested and confirmed by multiple groups of researchers and has reached wide acceptance becomes a **theory.** Current theories about how plant species distributions change with air temperature, for example, are derived from decades of research and evidence. Notice that this sense of theory is different from the way we might use the term in everyday conversation (such as, "But that's just a theory!"). To be considered a theory, a hypothesis must be consistent with a large body of experimental results. A theory cannot be contradicted by any replicable tests.

Scientists work under the assumption that the world operates according to fixed, knowable laws. We accept this assumption because it has been successful in explaining a vast array of natural phenomena and continues to lead to new discoveries. When the scientific process has generated a theory that has been tested multiple times, we can call that theory a natural law. A natural law is a theory to which there are no known exceptions and which has withstood rigorous testing. Familiar examples include the law of gravity and the laws of thermodynamics, which we will look at in the next chapter. These theories are accepted as facts by the scientific community, but they remain subject to revision if contradictory data are found.

Scientific Method in Action: The Chlorpyrifos Investigation

Let's look at what we have learned about the scientific method in the context of an actual scientific investigation. In the 1990s, scientists suspected that organophosphates—a group of chemicals commonly used in insecticides—might have serious effects on the human central nervous system. By the early part of the decade, scientists suspected that organophosphates might be linked to problems such as neurological disorders, birth defects, ADHD, and palsy. One of these chemicals, chlorpyrifos (klor-PEER-i-fos), was of particular concern because it is among the most widely used pesticides in the world, with large amounts applied in homes in the United States and elsewhere.

The Hypothesis

The researchers investigating the effects of chlorpyrifos on human health formulated a hypothesis: Chlorpyrifos causes neurological disorders and negatively affects human health. Because this hypothesis would be hard to prove conclusively, the researchers also proposed a null hypothesis: Chlorpyrifos has no observable negative effects on the central nervous system. We can follow the process of their investigation in FIGURE 3.3.

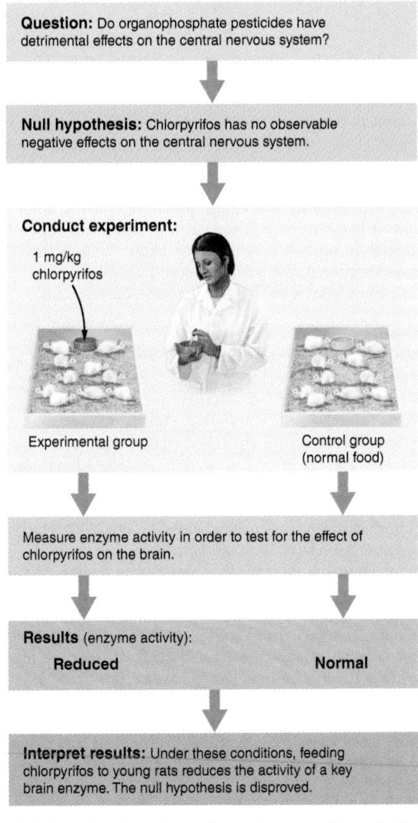

FIGURE 3.3 **A typical experimental process.** An investigation of the effects of chlorpyrifos on the central nervous system illustrates how the scientific method is used.

Theory A hypothesis that has been repeatedly tested and confirmed by multiple groups of researchers and has reached wide acceptance.

Teaching Tip: Connections

Have students research a scientific theory. Ask them to write down the name of the theory in large letters, as well as one sentence explaining it. Then have the students tape their theories to the board, and encourage them to read everyone else's work. You may use the discussion to talk about how scientists arrive at a theory using the scientific method. *Answers may include cell theory, theory of evolution, germ theory, atomic theory, Big Bang theory, theory of relativity, plate tectonics theory, or others.*

Practicing Science: Communicate

Have students read and evaluate the methodology of the chlorpyrifos investigation. Then ask them how this experiment might be improved for more robust results. Have them either write down their answers or discuss them with a partner. *Answers may include using more rats, running the experiment for a longer period of time, or using different concentrations of the drug in multiple experimental groups.*

Teaching Tip: Debate the Issue

The Ethics of Animal Testing
The use of animals in medical research has been a subject of ethical debate for many decades. Have your class debate the question: "Was it ethical to use the rats in the chlorpyrifos investigation?"

Students should be prepared to discuss these topics in their debate:

- Does human well-being justify harming animals?
- Would it have been possible to understand the effects of chlorpyrifos without animal testing?
- How do scientists determine which animal species to use in testing? Ethically are some animal species more acceptable to use for testing than others?

In class, select groups to debate the issue. Allow each side 2 minutes to prepare, 3 minutes to present, and 1 minute for rebuttal. Have the remainder of the class take notes, ask questions, and decide which side has the most persuasive argument.

Teaching Tip: Discussion Starter

Ask students: What is the difference between controlled and natural experiments? Why do we need each type? *A controlled experiment is conducted under controlled conditions, such as in a laboratory, whereas a natural experiment occurs when a natural event takes place or when natural processes unfold over time. A natural experiment is not controlled and many variables can change at once. Controlled experiments enable scientists to collect the most reliable data, whereas natural experiments allow scientists to study processes that cannot be examined in a lab.*

Testing the Hypothesis

To test the null hypothesis, the scientists designed experiments using rats. One experiment used two groups of rats, with 10 individuals per group. The first group—the experimental group—was fed small doses of chlorpyrifos for each of the first 4 days of life. No chlorpyrifos was fed to the second group. That second group was a **control group**: a group that experiences exactly the same conditions as the experimental group, except for the single variable under study. In this experiment, the only difference between the control group and the experimental group was that the control group was not fed any chlorpyrifos. By designating a control group, scientists can determine whether an observed effect is the result of the experimental treatment or of something else in the environment to which all the subjects are exposed. For example, if the control rats—those that were not fed chlorpyrifos—and the experimental rats—those that were exposed to chlorpyrifos—showed no differences in their brain chemistry, researchers could conclude that the chlorpyrifos had no effect. If the control group and experimental group had very different brain chemistry after the experiment, the scientists could conclude that the difference must have been due to the chlorpyrifos.

The Results

At the end of the experiment, the researchers found that the rats exposed to chlorpyrifos had much lower levels of the enzyme choline acetyltransferase in their brains than the rats in the control group. But without a control group for comparison, the researchers would never have known whether the chlorpyrifos or something else caused the change observed in the experimental group.

The discovery of the relationship between ingesting chlorpyrifos and a single change in brain chemistry might seem relatively small. But that is how most scientific research works: Very small steps establish that an effect occurs and, eventually, how it occurs. In this way, we progress toward a more thorough understanding of how the world works. This particular research on chlorpyrifos, combined with numerous other experiments testing specific aspects of the chemical's effect on rat brains, demonstrated that chlorpyrifos was capable of damaging developing rat brains at fairly low doses. The results of this research have been important for our understanding of human health and toxic substances in the environment.

Control group In a scientific investigation, a group that experiences exactly the same conditions as the experimental group, except for the single variable under study.

Natural experiment A natural event that acts as an experimental treatment in an ecosystem.

Controlled Experiments and Natural Experiments

The chlorpyrifos experiment we have just described was conducted in the controlled conditions of a laboratory. However, not all experiments can be done under such controlled conditions. For example, it would be difficult to study the interactions of wolves and caribou in a controlled setting because both species need large amounts of land and because their behavior changes in captivity. Other reasons that a controlled laboratory experiment may not be possible include prohibitive costs and ethical concerns.

Under these circumstances, investigators look for a natural experiment. A **natural experiment** occurs when a natural event acts as an experimental treatment in an ecosystem. For example, a volcano that destroys thousands of hectares of forest provides a natural experiment for understanding large-scale forest regrowth (FIGURE 3.4). We would never destroy that much forest just to study regrowth, but we can study such natural disasters when they occur. Still other cases of natural experiments do not involve disasters. For example, we can study the process of ecological succession by looking at specific areas where forests have been growing for different amounts of time and comparing them. We can study the effects of species invasions by comparing uninvaded ecosystems with invaded ones.

Because a natural experiment is not controlled, many variables can change at once, and results can be difficult to interpret. Ideally, researchers compare multiple examples of similar systems in order to exclude the influences of different variables. For example, after a forest fire, researchers might not only observe how a burned forest responds to the disturbance but also compare it with a nearby forest that did not burn. In this case, the researchers are comparing similar forests that differ in only one variable, fire. If, however, they tried to compare the burned forest with a different type of forest, perhaps one at a different elevation, it would be difficult to separate the effects of the fire from the effects of elevation. Still, because they may be the only way to obtain vital information, natural experiments are indispensable.

Let us return to the study of chlorpyrifos. Researchers wanted to know if human brains that were exposed to the chemical would react in the same way as rat brains. Because researchers would never feed pesticides to humans to study their effects, for obvious ethical reasons, they conducted a natural experiment. They looked for groups of people who were similar in most ways—for example, income, age, level of education—but who varied in their exposure to chlorpyrifos. To gather data on variation of exposure they looked at how often people in each group used pesticides that contained the chemical, the brand they used, and the frequency and location of use. Researchers found that

Teaching Tip: Beyond the Classroom

Ask students to find an example of a natural or anthropogenic event that could have acted as an experimental treatment in its ecosystem. *Answers may include various volcanic eruptions, the Fukushima nuclear disaster, the BP oil spill, various hurricanes, or any forest fire.*

(a)

(b)

(c)

FIGURE 3.4 **A natural experiment.** The Mount St. Helens eruption in 1980 created a natural experiment for understanding large-scale forest regrowth. (a) A pre-eruption forest near Mount St. Helens in 1979; (b) the same location, post-eruption, in 1982; (c) the same location in 2009 begins to show forest regrowth. *(U.S. Forest Service)*

tissue concentrations of chlorpyrifos were highest in groups that were exposed to the chemical in their jobs and among poor urban families whose exposure to residential pesticides was high. Among these populations, a number of studies connected exposure to chlorpyrifos with low birth weight and other developmental abnormalities.

Science and Progress

The chlorpyrifos experiment is a good example of the process of science. Based on observations, the scientists proposed a hypothesis and null hypothesis. The null hypothesis was tested and rejected. Multiple rounds of additional testing gave researchers confidence in their understanding of the problem. Moreover, as the research progressed, the scientists informed the public, as well as the scientific community, about their results. Finally, in 2000, as a result of the step-by-step scientific investigation of chlorpyrifos, the U.S. Environmental Protection Agency (EPA) decided to prohibit its use for most residential applications. It also prohibited agricultural use on fruits that are eaten without peeling, such as apples and pears, and those that are especially popular with children, such as grapes.

Environmental science presents unique challenges

Environmental science has many things in common with other scientific disciplines. However, it presents a number of challenges and limitations that are not usually found in most other scientific fields. These challenges and limitations are a result of the nature of environmental science and the way research in the field is conducted.

Lack of Baseline Data

The greatest challenge to environmental science is the fact that there is no undisturbed baseline—no "control planet"—with which to compare contemporary Earth. Virtually every part of the globe has been altered by humans in some way (FIGURE 3.5). Even though some remote regions appear to be undisturbed, we can still find quantities of lead in the Greenland ice sheet, traces of the anthropogenic compound PCB in the fatty tissue of penguins in Antarctica, and invasive species from many locations carried by ship to remote tropical islands. This situation makes it difficult to know the original levels of contaminants or numbers of species that existed before humans began to alter the planet. Consequently, we can only speculate about how the current conditions deviate from those of prehuman activity.

Teaching Tip: Engage

Ask students: What are some of the challenges environmental scientists face? *Environmental science research is often conducted in the field, which introduces many challenges such as a lack of baseline data and the presence of complex interactions.*

Teaching Tip: Discussion Starter

Ask students: Why (or when) is the lack of baseline data a problem in environmental science? *Since every region of the planet has been affected by human activity, we cannot compare a disturbed area of Earth with an undisturbed area. Because there is no "control planet" with which to compare contemporary Earth, humans can't fully understand our impact on the planet.*

Teaching Tip: Discussion Starter

Ask students: How do scientists use the scientific method to address environmental problems? *Scientists use the scientific method to explore the natural world objectively, draw inferences from it, and predict the outcome of certain events, processes, or changes. They must conduct their research in such a way that other researchers can understand how their data were collected and agree on the validity of their findings. In the case of fracking, as discussed in the chapter opener, scientists evaluate its environmental impact by testing groundwater quality, monitoring the escape of natural gas, documenting water and chemical use, and monitoring the health of workers and nearby community members. Comparing these data with those of sites without fracking indicates whether fracking has an adverse effect on the environment.*

Teaching Tip: Beyond the Classroom

Have students research the advantages and disadvantages of paper bags versus plastic bags. They can also make a list of cities in the United States that have ordinances concerning the use of plastic or paper bags.

Teaching Tip: Discussion Starter

Ask students: What makes environmental systems so complex? *Environmental systems have many interacting parts, so the results of a study of one system cannot always be applied to similar systems. Additionally, human behavior, which also affects natural systems, cannot always be accounted for or predicted.*

FIGURE 3.5 **The global nature of human impacts.** The trash that washed up onto the beach of this remote Pacific island vividly demonstrates the difficulty of finding any part of Earth unaffected by human activities. *(Ashley Cooper/Alamy)*

Subjectivity

A second challenge unique to environmental science lies in the dilemmas raised by subjectivity. For example, when you go to the grocery store, the bagger may ask, "Paper or plastic?" How can we know for certain which type of bag has the least environmental impact? There are techniques for determining what harm may come from using the petrochemical benzene to make a plastic bag and from using chlorine to make a paper bag. However, different substances tend to affect the environment differently: Benzene may pose more of a risk to people, whereas chlorine may pose a greater risk to organisms in a stream. It is difficult, if not impossible, to decide which is better or worse for the environment overall. There is no single measure of environmental quality. Ultimately, our assessments and our choices involve value judgments and personal opinions.

Interactions

A third challenge is the complexity of natural and human-dominated systems. All scientific fields examine interacting systems, but those systems are rarely as complex and as intertwined as they are in environmental science. Because environmental systems have so many interacting parts, the results of a study of one system cannot always be easily applied to similar systems elsewhere.

There are also many examples in which human preferences and behaviors affect environmental systems as much as the natural laws that describe them. For example, many people assume that if we built more efficient automobiles, the overall consumption of gasoline in the United States would decrease. To decrease gas consumption, however, it is necessary not only to build more efficient automobiles, but also to get people to purchase those vehicles and use them in place of less efficient ones. During the 1990s and early 2000s, even though there were many fuel-efficient cars available, the majority of buyers in the United States continued to purchase larger, heavier, and less fuel-efficient cars, minivans, light trucks, and sport-utility vehicles. Environmental scientists thought they knew how to reduce gasoline consumption, but they neglected to account for consumer behavior.

Human Well-Being

As we continue our study of environmental science, we will see that many of its topics touch on human well-being. In environmental science, we study how humans impact the biological systems and natural resources of the planet. We also study how changes in natural systems and the supply of natural resources affect humans.

We know that people who are unable to meet their basic needs are less likely to be interested in or able to be concerned about the state of the natural environment. The principle of environmental equity—the fair distribution of Earth's resources—adds a moral issue to questions raised by environmental science. Pollution and environmental degradation are inequitably distributed, with the poor receiving much more than an equal share. Is this a situation that we, as fellow humans, can tolerate? Environmental justice is a social movement and field of study that works toward equal enforcement of environmental laws and the elimination of disparities, whether intended or unintended, in how pollutants and other environmental harms are distributed among the various ethnic and socioeconomic groups within a society (FIGURE 3.6).

FIGURE 3.6 **Electronic waste recycling.** The poor are exposed to a disproportionate amount of pollutants and other hazards. The people shown here, located in a small village on the outskirts of New Delhi, India, are recycling circuit boards from discarded electronics products. *(Peter Essick/Aurora Photos/Alamy)*

24 CHAPTER 1 ■ Environmental Science: Studying the State of Our Earth

Teaching Tip: Journal Prompt

Pollution and environmental degradation are inequitably distributed in many societies, with the poor more exposed than others. Ask students to discuss why this might happen in a society and to suggest what might be done to address such disparities.

module

3

REVIEW

In this module, we have seen how specific aspects of the scientific method are used to conduct field and laboratory evaluations of how human activity affects the natural environment. The scientific method follows a process of observations and questions, testable hypotheses and predictions, and data collection. Results are interpreted and shared with other researchers. Experiments can be either controlled (manipulated) experiments or natural experiments that make use of natural events. There are often challenges in environmental science including the lack of baseline data and the interactions with social factors such as human preferences.

Module 3 AP® Review Questions

1. The first step in the scientific process is
   (a) collecting data.
   (b) observations and questions.
   (c) forming a hypothesis.
   (d) disseminating findings.
   (e) forming a theory.

Use the following information for questions 2 and 3:
Two new devices for measuring lead contamination in water are tested for accuracy. Scientists test each device with seven samples of water known to contain 400 ppm of lead. Their data is shown below. Concentration is in parts per billion.

Water Sample	1	2	3	4	5	6	7
Device 1	415	417	416	417	415	416	416
Device 2	398	401	400	402	398	400	399

2. The data from device 1 is
   (a) accurate, but not precise.
   (b) precise, but not accurate.
   (c) both accurate and precise.
   (d) neither accurate nor precise.
   (e) not clear enough to support any conclusion about accuracy or precision.

3. Assuming the devices were used correctly, and assuming we want to choose a device that accurately reflects the true concentration of lead in the water samples, which conclusion does the data support?
   (a) Device 1 is superior to device 2 because it is more precise.
   (b) Device 2 is superior to device 1 because it is more precise.
   (c) Device 1 is superior to device 2 because it is more accurate.
   (d) Device 2 is superior to device 1 because it is more accurate.
   (e) Both devices are equally effective at measuring contaminates.

4. Challenges in the study of environmental science include all of the following except
   (a) dangers of studying natural systems.
   (b) lack of baseline data.
   (c) subjectivity of environmental impacts.
   (d) complexity of natural systems.
   (e) complex interactions between humans and the environment.

5. A control group is
   (a) a group with the same conditions as the experimental group.
   (b) a group with conditions found in nature.
   (c) a group with a randomly assigned population.
   (d) a group with the same conditions as the experimental group except for the study variable.
   (e) a group that is kept at the same conditions throughout the experiment.

Answers to Module 3 AP® Review Questions

1. b
2. b
3. d
4. a
5. d

Teaching Tip: Activity

Have your students develop a sustainability plan for your school. Their plan should focus on 7 to 10 environmental concerns, such as biodiversity, energy use, water use, food consumption, hazardous material use, student health, open spaces, air quality, waste production, recycling rates, and transportation use. For each of the environmental concerns, have students set out long-term objectives, as well as the actions required to achieve those objectives. Finally, ask students to choose two to three specific indicators they can use to monitor the effectiveness of the proposed plan to address each of the 10 environmental concerns. For example, student health could be evaluated by the total number of student sick days and the number of visits to the school nurse. Students can work collaboratively on the report through a shared document system and then present their work to school administrators.

Suggested Answers to Critical Thinking Questions

1. The indicators used by a metropolitan region often focus on those factors that more directly impact humans in that immediate area. For example, hazardous materials and solid waste may have much larger impacts at a local level than worldwide because they become much more dispersed when measured globally. Indicators will also be more directly related to human activity and human life because of the concentration of people living in a metropolitan area.

2. Some potential features for a sustainability plan could include:

- reducing the use of cars and increasing the use of public transportation
- converting abandoned lots into parks or natural areas with native species
- starting recycling and composting programs to reduce the amount of waste
- using water conservation programs in dry areas, limiting the use of water for watering lawns, and using low-flow shower heads and toilets
- increasing the availability of food and organizing farmers' markets or community gardens

working toward  **sustainability**

Using Environmental Indicators to Make a Better City

We have seen that environmental indicators can be used to monitor conditions across a range of scales, from local to global. They are also being used by people looking for ways to apply environmental science to the urban planning process in countries as diverse as China, Brazil, and the United States.

San Francisco, California, is one example. In 1997 the city adopted a sustainability plan to go along with its newly formed Department of the Environment. The San Francisco Sustainability Plan focuses on 10 environmental concerns including air quality; biodiversity; energy, climate change, and ozone depletion; food and agriculture; hazardous materials; human health; parks, open spaces, and streetscapes; solid waste; transportation; and water and wastewater.

Although some of these topics may not seem like components of urban planning, the drafters of the plan recognized that the everyday choices of city dwellers can have wide-ranging environmental impacts, both in and beyond the city. For example, purchasing local produce or organic food affects the environments and economies of both San Francisco and the agricultural areas that serve it.

For each of the 10 environmental concerns, the sustainability plan sets out a series of 5-year and long-term objectives as well as specific actions required to achieve them.

To monitor the effectiveness of the various actions, San Francisco chose specific environmental indicators for each of the 10 environmental concerns. These indicators had to indicate a clear trend toward or away from environmental sustainability, demonstrate cost effectiveness, be understandable to the nonscientist, and be easily presented to the media. For example, to evaluate biodiversity, San Francisco uses four indicators:

1. Number of volunteer hours dedicated to managing, monitoring, and conserving San Francisco's biodiversity
2. Number of square feet of the worst non-native species removed from natural areas
3. Number of surviving native plant species planted in developed parks, private landscapes, and natural areas
4. Abundance and species diversity of birds, as indicated by the Golden Gate Audubon Society's Christmas bird counts

Together, these indicators provide a relatively inexpensive and simple way to summarize the level of biodiversity, the threat to native biodiversity from non-native species, and the amount of effort going into biodiversity protection.

More than 15 years later, what do the indicators show? In general, there has been a reasonable amount of improvement. For example, in the category of solid waste, San Francisco has increased the amount of waste recycled from 30 to 70 percent, with a goal of 75 percent by 2020, and it now has the largest urban composting program in the country. San Francisco has also improved its air quality, reducing the number of days in which fine particulate matter exceeded the EPA air quality safe level from 27 days in 2000 to 8 days in 2011. These and other successes have won the city numerous accolades: It has been selected as one of "America's Top Five Cleanest Cities" by *Reader's Digest* and as one of the "Top 10 Green Cities" by *The Green Guide*. In 2013, San Francisco was named the most sustainable city in the United States by a Toronto-based research company.

Critical Thinking Questions

1. How do the indicators used by a city or metropolitan region differ from the global indicators we described earlier in the chapter?
2. Think about a city you have lived in or visited. What features might you recommend for a sustainability plan for that location?

References

http://www.thegreenguide.com
http://www.sf-planning.org

A "green" city. San Francisco's adoption of environmental indicators has helped it achieve many of its sustainability goals. *(Richard H/FeaturePics)*

chapter 1

REVIEW

Throughout this chapter, we have outlined principles, techniques, and methods that will allow us to approach environmental science from an interdisciplinary perspective as we evaluate the current condition of Earth and the ways that human beings have influenced it. We identified that we can use environmental indicators to show the status of specific environmental conditions in the past, at present, and, potentially, into the future. These indicators and other environmental metrics must be measured using the same scientific process used in other fields of science. Environmental science does contain some unique challenges because there is no undisturbed baseline—humans began manipulating Earth long before we have been able to study it.

Key Terms

Fracking
Environment
Environmental science
Ecosystem
Biotic
Abiotic
Environmentalist
Environmental studies
Ecosystem services
Environmental indicators
Biodiversity
Genetic diversity
Species
Species diversity
Speciation
Background extinction rate
Greenhouse gases
Anthropogenic
Development
Sustainability
Sustainable development
Biophilia
Ecological footprint
Scientific method
Hypothesis
Null hypothesis
Replication
Sample size
Accuracy
Precision
Uncertainty
Theory
Control group
Natural experiment

Learning Objectives Revisited

Module 1 Environmental Science

- **Define the field of environmental science and discuss its importance.**
  Environmental science is the study of the interactions among human-dominated systems and natural systems and how those interactions affect environments. Studying environmental science helps us identify, understand, and respond to anthropogenic changes.
- **Identify ways in which humans have altered and continue to alter our environment.**
  The impact of humans on natural systems has been significant since early humans hunted some large animal species to extinction. However, technology and population growth have dramatically increased both the rate and the scale of human-induced change.

Module 2 Environmental Indicators and Sustainability

- **Identify key environmental indicators and their trends over time.**
  Five important global-scale environmental indicators are biological diversity, food production, average global surface temperature and atmospheric CO_2 concentrations, human population, and resource depletion. Biological diversity is decreasing as a result of human actions, most notably habitat destruction and habitat degradation. Food production appears to be leveling off and may be decreasing. Carbon dioxide concentrations are steadily increasing as a result of fossil fuel combustion and land conversion. Human population continues to increase and probably will continue to do so throughout this century. Resource depletion for most natural resources continues to increase.

Teaching Tip: Measuring Your Impact

Exploring Your Footprint
Measuring Your Impact 1 asks students to make a list of the activities they did in one day and attempt to describe their impact on the five global environmental indicators described in this chapter. To download this resource, click on the link buttons in the TE-book, open the TRFD, or log onto the book's companion website bcs.whfreeman.com/friedlandapes2e (teacher login required).

Measuring Your Impact 1: Exploring Your Footprint

Exploring the Literature

Ehrenfeld, J.R., and A.J. Hoffman. 2013. *Flourishing: A Frank Conversation About Sustainbility.* Stanford Business Books, 2013.

Meadows, D. H., J. Randers, and D. L. Meadows. 2004. *Limits to Growth: The 30-Year Update.* Chelsea Green. See also http://www.sustainer.org/.

United Nations Development Programme. 2013. *Human Development Report 2013.* Accessed at: http://hdr.undp.org/en/

Wackernagel, M., et al. 2004. *Ecological Footprint and Biocapacity Accounts 2004: The Underlying Calculation Method.* Global Footprint Network. www.footprintnetwork.org

Wheelan, C. *Naked Statistics: Stripping the Dread from the Data.* W.W. Norton, 2013.

Download a printable version of this list by clicking on the link buttons in the TE-book, opening the TRFD, or logging onto the book's companion website bcs.whfreeman.com/friedlandapes2e (teacher login required).

Exploring the Literature 1

- **Define sustainability and explain how it can be measured using the ecological footprint.**
  Sustainability is the use of Earth's resources to meet our current needs without jeopardizing the ability of future generations to meet their own needs. The ecological footprint is the land area required to support a person's (or a country's) lifestyle. We can use that information to say something about how sustainable that lifestyle would be if it were adopted globally.

Module 3 Scientific Method

- **Explain the scientific method and its application to the study of environmental problems.**
  The scientific method is a process of observation, hypothesis generation, data collection, analysis of results, and dissemination of findings. Repetition of measurements or experiments is critical if one is to determine the validity of findings. Hypotheses are tested and often modified before being accepted.
- **Describe some of the unique challenges and limitations of environmental science.**
  We lack an undisturbed "control planet" with which to compare conditions on Earth today. Assessments and choices are often subjective because there is no single measure of environmental quality Environmental systems are so complex that they are poorly understood, and human preferences and policies may affect them as much as do natural laws.

Answers to Chapter 1 AP® Exam Multiple-Choice Questions

1. d
2. b
3. b
4. b
5. c
6. c
7. e
8. b
9. e
10. d
11. a

Answers to Chapter 1 AP® Exam

Download the full answers to the Chapter 1 AP® Practice Exam by clicking on the link buttons in the TE-book, opening the TRFD, or logging onto the book's companion website bcs.whfreeman.com/friedlandapes2e (teacher login required).

Answers to Chapter 1 AP® Exam

Chapter 1 AP® Environmental Science Practice Exam

Section 1: Multiple-Choice Questions

Choose the best answer for questions 1–11.

1. Which of the following events has increased the impact of humans on the environment?
   I. advances in technology
   II. reduced human population growth
   III. use of tools for hunting
   (a) I only
   (b) I and II only
   (c) II and III only
   (d) I and III only
   (e) I, II, and III
2. As described in this chapter, environmental indicators
   (a) always tell us what is causing an environmental change.
   (b) can be used to analyze the health of natural systems.
   (c) are useful only when studying large-scale changes.
   (d) do not provide information regarding sustainability.
   (e) take into account only the living components of ecosystems.
3. Which statement regarding a global environmental indicator is NOT correct?
   (a) Concentrations of atmospheric carbon dioxide have been rising quite steadily since the Industrial Revolution.
   (b) World grain production has increased fairly steadily since 1950, but worldwide production of grain per capita has decreased dramatically over the same period.
   (c) For the past 130 years, average global surface temperatures have shown an overall increase that seems likely to continue.
   (d) World population is expected to be between 8.1 billion and 9.6 billion by 2050.
   (e) Some natural resources are available in finite amounts and are consumed during a one-time use, whereas other finite resources can be used multiple times through recycling.
4. Figure 2.5 (on page 12) shows atmospheric carbon dioxide concentrations over time. The measured concentration of CO_2 in the atmosphere is an example of
   (a) a sample of air from over the Antarctic.
   (b) an environmental indicator.
   (c) replicate sampling.
   (d) calculating an ecological footprint.
   (e) how to study seasonal variation in Earth's temperatures.
5. Environmental metrics such as the ecological footprint are most informative when they are considered along with other environmental indicators. Which indicator, when considered in conjunction with the ecological footprint, would provide the most information about environmental impact?
   (a) biological diversity
   (b) food production
   (c) human population
   (d) CO_2 concentration
   (e) water quality
6. In science, which of the following is the most certain?
   (a) hypothesis
   (b) idea
   (c) natural law
   (d) observation
   (e) theory

Additional Multiple-Choice AP® Practice Exam

To download an additional Multiple-Choice AP® Environmental Science Practice Exam for Chapter 1, click on the link buttons in the TE-book, open the TRFD, or log onto the book's companion website bcs.whfreeman.com/friedlandapes2e (teacher login required).

Multiple-Choice AP® Practice Exam 1

7. All of the following would be exclusively caused by anthropogenic activities except
(a) combustion of fossil fuels.
(b) overuse of resources such as uranium.
(c) forest clearing for crops.
(d) air pollution from burning oil.
(e) forest fires.

8. Use Figure 2.3 (on page 11) to calculate the approximate percentage change in world grain production per person between 1950 and 2000.
(a) 10 percent (d) 40 percent
(b) 20 percent (e) 50 percent
(c) 30 percent

9. The populations of some endangered animal species have stabilized or increased in numbers after human intervention. An example of a species that is still endangered and needs further assistance to recover is the
(a) American bison. (d) American alligator.
(b) peregrine falcon. (e) snow leopard.
(c) bald eagle.

Questions 10 and 11 refer to the following experimental scenario:

An experiment was performed to determine the effect of caffeine on the pulse rate of five healthy 18-year-old males. Each was given 250 mL of a beverage with or without caffeine. The men had their pulse rates measured before they had the drink (time 0 minutes) and again after they had been sitting at rest for 30 minutes after consuming the drink. The results are shown in the following table.

Subject	Beverage	Caffeine content (mg/mL)	Pulse rate at time 0 minutes	Pulse rate at time 30 minutes
1	Water	0	60	59
2	Caffeine-free soda	0	55	56
3	Caffeinated soda	10	58	68
4	Coffee, decaffeinated	3	62	67
5	Coffee, regular	45	58	81

10. Before the researchers began the experiment, they formulated a null hypothesis. The best null hypothesis for the experiment would be that caffeine
(a) has no observable effect on the pulse rate of an individual.
(b) will increase the pulse rates of all test subjects.
(c) will decrease the pulse rates of all test subjects.
(d) has no observable effects on the pulse rates of 18-year-old males.
(e) from a soda will have a greater effect on pulse rates than caffeine from coffee.

11. After analyzing the results of the experiment, the most appropriate conclusion would be that caffeine
(a) increased the pulse rates of the 18-year-old males tested.
(b) decreased the pulse rates of the 18-year-old males tested.
(c) will increase the pulse rate of any individual that is tested.
(d) increases the pulse rate and is safe to consume.
(e) makes drinks better than decaffeinated beverages.

Section 2: Free-Response Questions

Write your answer to each part clearly. Support your answers with relevant information and examples. Where calculations are required, show your work.

1. Your neighbor has fertilized her lawn. Several weeks later, she is alarmed to see that the surface of her ornamental pond, which sits at the bottom of the sloping lawn, is covered with a green layer of algae.
(a) Suggest a feasible explanation for the algal bloom in the pond. (2 points)
(b) Design an experiment that would enable you to validate your explanation. Include and label in your answer:
(i) a testable hypothesis (2 points)
(ii) the variable that you will be testing (1 point)
(iii) the data to be collected (1 point)
(iv) a description of the experimental procedure (2 points)
(v) a description of the results that would validate your hypothesis (1 point)
(c) Based on the data from your experiment and your explanation of the problem, think of and suggest one action that your neighbor could take to help the pond recover. (1 point)

2. The study of environmental science sometimes involves examining the overuse of environmental resources.
(a) Identify one general effect of overuse of an environmental resource. (3 points)
(b) For the effect you listed above, describe a more sustainable strategy for resource utilization. (3 points)
(c) Describe how the events from Easter Island can be indicative of environmental issues on Earth today. (4 points)

Answers to Chapter 1 AP® Exam Free-Response Questions

1. (a) It is most likely that too much fertilizer was applied to the lawn. When it rained, some of the runoff flowed down the sloping lawn and entered the ornamental pond. This increased the nutrient supply in the pond and resulted in an algal bloom a few weeks later.
(b) (i) The experiment will test the hypothesis that fertilizer will promote the growth of pond algae in water.
(ii) The variable that will be tested is the amount of fertilizer that will be added to each experimental group.
(iii) The data that will be collected is the amount of algae growth in each of the experimental groups.
(iv) The experiment will be performed using the following procedure: A sample of the pond water is obtained and filtered through a paper towel. The algae left on the paper towel is transferred to a container of distilled water (1 L) and mixed thoroughly. Fifty mL of the diluted algae solution is measured into five identical small clear glass containers. The following amounts of fertilizer are put into the containers and mixed thoroughly: 0.00 g into container 1, 0.25 g into container 2, 0.50 g into container 3, 0.75 g into container 4, 1.00 g into container 5. The above steps are repeated another three times to use up all of the 1 L of the algae mixture, providing a total of four experimental groups; that is, the experiment is repeated four times. The container with no fertilizer represents the control group. Each of the labeled glass containers is placed on a sunny windowsill and monitored on a daily basis. The glass containers are compared for algal growth by assessing the green coloration of each container. (This could be done qualitatively as described above and/or compared to the color of 50 mL of water taken directly from the surface of the pond with the algal bloom *or* quantitatively by using a colorimeter, if one is available.)
(v) It is expected that within 3 to 7 days, depending on the amount of sunlight, more algal growth will be evident in the containers that have fertilizer added to them. The amount of algae will depend on the amount of fertilizer that was added—the more fertilizer, the more algae. This result validates the hypothesis that fertilizer will promote the growth of pond algae in water. This also helps to validate the explanation in (a) above.
(c) It may be difficult to get the pond to recover immediately due to the high level of fertilizer that was added to the lawn. The algae growth may continue until all the fertilizer has run off from the lawn. The neighbor should not apply any more fertilizer to the lawn. She needs to use a soil test kit and assess the nutrient levels of the lawn. Using a mulching lawn mower will help restore nutrient levels naturally. The algae in the pond could be harvested or scooped off when it is in a quantity that enables this to occur. This could also be done in conjunction with drawing down the level of the pond water and replacing it with fresh water. This will help dilute the nutrient level in the pond. She could also plant vegetation on the side of the pond that is facing up the slope. This will help reduce nutrient input from runoff water when it rains, as the plants will take up the nutrients before runoff enters the pond.

Answers continue in the Answer Appendix on page ANS-1.

chapter

Environmental Systems

PD Overview

This chapter reviews the concept of systems in nature, the different properties of matter, and the various forms of energy. It offers an overview of key ideas from chemistry and physics that scientists use to understand the environment and to measure human impact on the environment. If your students have already taken chemistry and physics, you may choose to review this material quickly. However, if you teach environmental science to first-year students, you will probably need to spend more time on this chapter.

Module 4: Systems and Matter

All questions about the environment involve matter, the atoms and molecules that compose materials, and the systems in which the matter circulates. Module 4 examines the basic building blocks of matter and how matter moves among systems. Students will take a detailed look at water, an important component of most environmental systems. Finally, students explore how matter is conserved in chemical and biological systems.

Module 5: Energy, Flows, and Feedbacks

This module discusses the various ways that energy flows within and among systems. Comparing the transfer of energy from sunlight to plants and then from plants to animals shows students that some transfers occur more efficiently than others. Some transfers occur within a system and others occur between different systems.

Alignment to AP® Environmental Science Course Description

AP® Outline	Module
I. Earth Systems and Resources (10–15%)	
A. Earth Science Concepts	Module 4 Systems and Matter Module 5 Energy, Flows, and Feedbacks

Chapter Learning Objectives

After completing this chapter students will be able to

- **describe how matter comprises atoms and molecules that move among different systems.**
- **explain why water is an important component of most environmental systems.**
- **discuss how matter is conserved in chemical and biological systems.**
- **distinguish among various forms of energy and understand how they are measured.**
- **discuss the first and second laws of thermodynamics and explain how they influence environmental systems.**
- **explain how scientists keep track of energy and matter inputs, outputs, and changes to environmental systems.**

Chapter 2 Pacing Guide

This pacing guide is based on a schedule with 120 sessions of 50 minutes each before the AP® Exam. If you have a different number of sessions before the exam, you can modify the pacing to suit your needs. If you have additional time, consider incorporating quizzes, released AP® Environmental Science free-response and multiple-choice questions, or additional activities.

Module	Standard Schedule Days	Block Schedule Days
Module 4	½	¼
Module 5	½	¼
Assessment	1	½

Chapter 2 Resources

These resources can be found by clicking on the link buttons in the Teacher's e-Book (TE-book), opening the Teacher's Resource Flash Drive (TRFD), or logging onto the book's companion website bcs.whfreeman.com/friedlandapes2e (teacher login required).

- **PowerPoint Presentation 2**
- **Optimized Art PowerPoint and JPEG Files 2**
- **Do the Math Videos**
- **Measuring Your Impact 2:** Bottled Water versus Tap Water
- **Activity 2.1:** Word Splash
- **Lab 2.1:** Who's Got the Power?
- **Exploring the Literature 2**
- **Answers to Chapter 2 AP® Exam**
- **Multiple-Choice AP® Practice Exam 2**
- **Answers to Unit 1 AP® Exam**
- **Unit 1 Additional Free-Response Question**
- **Answer to Science Applied 1 Free-Response Question**

Free-Response Questions from Previous AP® Environmental Science Exams

Free-response questions from prior AP® Environmental Science Exams are available on the AP® course website, https://apstudent.collegeboard.org/apcourse/ap-environmental-science/exam-practice. Students should be able to answer all of the questions listed below with material learned in this and previous chapters. When a question requires students to understand material from multiple chapters, the question will be listed in the last chapter required to complete the entire question. Questions marked with an asterisk (*) are from exams with released multiple-choice questions. You may want to save these questions until the end of the year so you can give your students a complete released exam for practice. Questions marked with double asterisks (**) require math to calculate a problem. Look for references to these questions throughout the chapter.

Year	Question	Content
1998*	3	• The pH scale
1999	1	• System dynamics • Scientific method

PD Chapter 2 Overview

Watch the video overview of Chapter 2 (for teachers) by clicking on the link buttons in the TE-book, opening the TRFD, or logging onto the book's companion website bcs.whfreeman.com/friedlandapes2e (teacher login required).

TRM PowerPoint Presentation 2

Download the PowerPoint presentation for Chapter 2 by clicking on the link buttons in the Teacher's e-Book (TE-book), opening the TRFD, or logging onto the book's companion website bcs.whfreeman.com/friedlandapes2e (teacher login required).

Tufa towers rise out of the salty water of Mono Lake. *(Dan Kosmayer/Shutterstock)*

chapter 2

Environmental Systems

Module 4 Systems and Matter

Module 5 Energy, Flows, and Feedbacks

A Lake of Salt Water, Dust Storms, and Endangered Species

Located between the deserts of the Great Basin and the mountains of the Sierra Nevada, California's Mono Lake is an unusual site. It is characterized by eerie towers of limestone rock known as tufa, unique animal species, glassy waters, and frequent dust storms. Mono Lake is a terminal lake, which means that water flows into it but does not flow out. As water moves through the mountains and desert soil, it picks up salt and other minerals, which it deposits in the lake. As the water evaporates, these minerals are left behind. Over time, evaporation has caused a buildup of salt concentrations so high that the lake is actually saltier than the ocean, and no fish can survive in the lake's water.

The Mono brine shrimp (*Artemia monica*) and the larvae of the Mono Lake alkali fly (*Ephydra hians*) are two of only a few animal species that can tolerate the conditions of the lake. The brine shrimp and the fly larvae consume microscopic algae, millions of tons of which grow in the lake each year. In turn, large flocks of migrating birds, such as sandpipers, gulls, and flycatchers, use the lake as a stopover, feeding on the brine shrimp and fly larvae to replenish their energy stores. The lake is an oasis on the migration route for these birds and they have come to depend on its food and water resources. The health of Mono Lake is therefore critical for many species.

> Just when it appeared that Mono Lake would never recover, circumstances changed.

In 1913, the City of Los Angeles drew up a controversial plan to redirect water away from Mono Lake and its neighbor, the larger and shallower Owens Lake. Owens Lake was diverted first, via a 359-km (223-mile) aqueduct that drew water away from the springs and streams that kept Owens Lake full. Soon, the lake began to dry up, and by the 1930s only an empty salt flat remained. Today the dry lake bed covers roughly 44,000 ha (109,000 acres). It is one of the nation's largest sources of windblown dust, which lowers visibility in the area's national parks. Even worse, because of the local geology, the dust contains high concentrations of arsenic, which is a major threat to human health.

31

Teaching Tip: Chapter Opening Case

The chapter opening case, "A Lake of Salt Water, Dust Storms, and Endangered Species," shows us that the activities of humans, the lives of other organisms, and abiotic processes in the environment are interconnected. Humans, water, animals, plants, and the desert environment all interact at Mono Lake to create a complex environmental system. The case also demonstrates how a single change made to an ecosystem often has wide-ranging and unexpected consequences. A more detailed discussion of Mono Lake is given in the Science Applied section at the end of this chapter.

Teaching Tip: Warm-up

The opening case is a great way to demonstrate how a single change made to an ecosystem can have ripple effects that cause more changes. Have students read the case and then discuss other examples of ripple effects that come from altering ecosystems. Ask them to think of local examples. One example might be how deforestation causes droughts, landslides, and species loss.

Teaching Tip: Beyond the Classroom

After discussing the case of Mono Lake, have students find other examples of terminal lakes. Ask students: Do these lakes face the same challenges as Mono Lake? Why or why not? *Examples of other lakes include the Caspian Sea, the Aral Sea, the Great Salt Lake, and Crater Lake.*

In 1941, despite the environmental degradation at Owens Lake, Los Angeles extended the aqueduct to draw water from the streams feeding Mono Lake. By 1982, with less fresh water feeding the lake, its depth had decreased by half, to an average of 14 m (45 feet), and the salinity of the water had doubled to more than twice that of the ocean. The salt killed the lake's algae and, without the algae to eat, the Mono brine shrimp also died. Most birds stayed away, and newly exposed land bridges allowed coyotes from the desert to prey on the colonies of nesting birds that remained.

However, just when it appeared that Mono Lake would never recover, circumstances changed. In 1994, after years of litigation led by the National Audubon Society and tireless work by environmentalists, the Los Angeles Department of Water and Power finally agreed to reduce the amount of water it diverted and to allow the lake to refill to about two-thirds of its historical depth. By summer 2009, lake levels had risen to just short of that goal, and the ecosystem was slowly recovering. In 2013, water levels remained close to the targeted goals. The brine shrimp are thriving, and many birds are returning to Mono Lake.

Water is a scarce resource in the Los Angeles area, and demand is particularly high. To decrease the amount of water diverted from Mono Lake, the City of Los Angeles had to reduce its water consumption. The city converted grass lawns requiring a great deal of water to native shrubs that were drought-tolerant, and it imposed new rules requiring low-flow showerheads and water-saving toilets. Through these seemingly small, but effective, measures, the inhabitants of Los Angeles were able to cut their water consumption and, in turn, protect nesting birds, Mono brine shrimp, and algae populations, and restore the Mono Lake ecosystem.

Sources:
J. Kay, It's rising and healthy, *San Francisco Chronicle*, July 29, 2006; Mono Lake Committee, *Mono Lake* (2013), http://www.monolake.org/.

The story of Mono Lake shows us that the activities of humans, the lives of other organisms, and abiotic processes in the environment are interconnected. Humans, water, animals, plants, and the desert environment all interact at Mono Lake to create a complex environmental system. The story also demonstrates a key principle of environmental science: a change in any one factor often has unexpected effects.

In Chapter 1, we learned that a system is a set of interacting components connected in such a way that a change in one part of the system affects one or more other parts of the system. The Mono Lake system is relatively small. Other complex systems exist on a much larger scale. The largest system that environmental science considers is Earth. Many of the most important current environmental issues—including human population growth and climate change—exist at the global scale. Throughout this book we will define a given system in terms of the environmental issue we are studying and the scale in which we are interested.

Organisms, nonliving matter, and energy all interact in natural systems. Taking a systems approach to an environmental issue decreases the chance of overlooking important components of that issue. Whether investigating ways to reduce pollution, increase food supplies, or find alternatives to fossil fuels, environmental scientists must have a thorough understanding of matter and energy and how these components interact within and across systems. In this chapter, we lay the foundation for the systems approach in environmental science. We will begin by exploring the properties of matter, and we will then discuss the various types of energy and how they influence and limit systems.

module 4

Systems and Matter

All questions about the environment involve *matter,* the atoms and molecules that compose materials, and the systems in which the matter circulates. In this module we will look at the basic building blocks of matter and how matter moves among systems. We will look in detail at water, an important component of most environmental systems. Finally, we will explore how matter is conserved in chemical and biological systems.

Learning Objectives

After reading this module you should be able to

- describe how matter comprises atoms and molecules that move among different systems.
- explain why water is an important component of most environmental systems.
- discuss how matter is conserved in chemical and biological systems.

Matter comprises atoms and molecules that move among different systems

What do rocks, water, air, the book in your hands, and the cells in your body have in common? They are all forms of *matter.* **Matter** is anything that occupies space and has *mass.* The **mass** of an object is a measurement of the amount of matter it contains. Note that the words "mass" and "weight" are often used interchangeably, but they are not the same thing. Weight is the force that results from the action of gravity on mass. Your own weight, for example, is determined by the amount of gravity pulling you toward the planet's center. Whatever your weight on Earth, you would weigh less on the Moon where the action of gravity is weaker. In contrast, mass stays the same under any gravitational influence. So although your weight would change on the Moon, your mass would remain the same because the amount of matter you are made of would be the same. In this section we will look at important properties of matter, starting with the building blocks.

Atoms and Molecules

All matter is composed of tiny particles that cannot be broken down into smaller pieces. The basic building blocks of matter are known as *atoms.* An **atom** is the

Matter Anything that occupies space and has mass.

Mass A measurement of the amount of matter an object contains.

Atom The smallest particle that can contain the chemical properties of an element.

Teaching Tip: Activity

If your students have already taken biology, chemistry, or physics, you may want to play a review game using the key terms from this chapter (see page 57). Divide the class into two teams. Ask one student to come to the front of the class and face his or her team, back to the board. Using either a projected slide or a flashcard, display a review term to the student's team. Instruct the team to offer clues about the term until the student guesses it. When the student correctly identifies the term, display another term and continue as before, until 1 minute has passed. When the minute is over, tally and record the number of terms the student correctly identified. It is now the other team's turn. Switch back and forth between teams, making sure that as many students as possible get a turn. When all the terms have been used, tally the final score and announce the results.

Teaching Tip: Venn Diagram

Have students compare and contrast protons, neutrons, and electrons using a Venn diagram as shown below. They should place features that are unique to each term in the outermost part of the circle, with features common to two terms in the area where two circles overlap, and features that are common to all three terms in the central area where all three circles overlap. One example for each part of the diagram is provided.

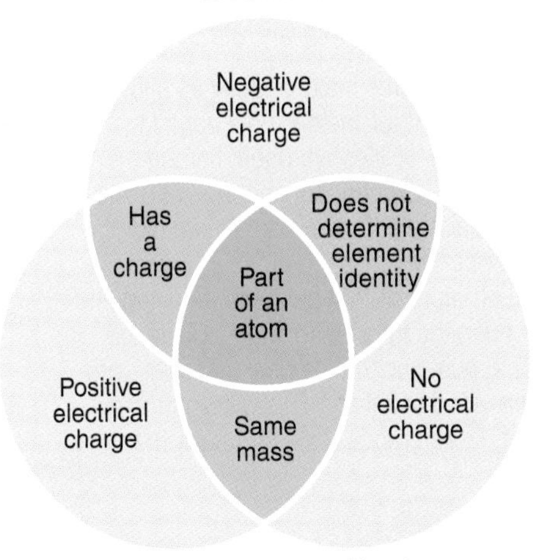

TEACHING with FIGURES ▶

Show students Figure 4.1, or a similar figure, with the labels removed. Have them redraw the model in their notebook, with special attention to the number of protons, neutrons, and electrons. Ask them to label each part of the figure, and then ask whether they can identify the element. For an alternative exercise, ask students to draw the structure of nitrogen or another atom without the aid of a figure.

smallest particle that can contain the chemical properties of an *element*. An **element** is a substance composed of atoms that cannot be broken down into smaller, simpler components. At Earth's surface temperatures, elements can occur as solids (such as gold), liquids (such as mercury), or gases (such as helium). Atoms are so small that a single human hair measures about a few hundred thousand carbon atoms across.

Ninety-four elements occur naturally on Earth, and another 24 have been produced in laboratories. The **periodic table** lists all of the elements currently known, organized by their properties. The full periodic table is reproduced at the end of this book. As you can see, each element is identified by a one- or two-letter symbol; for example, the symbol for carbon is C, and the symbol for oxygen is O. These symbols are used to describe the atomic makeup of **molecules,** which are particles that contain more than one atom. Molecules that contain more than one element are called **compounds.** For example, a carbon dioxide molecule (CO_2) is a compound composed of one carbon atom (C) and two oxygen atoms (O_2). Let's take a closer look at atoms and how they behave.

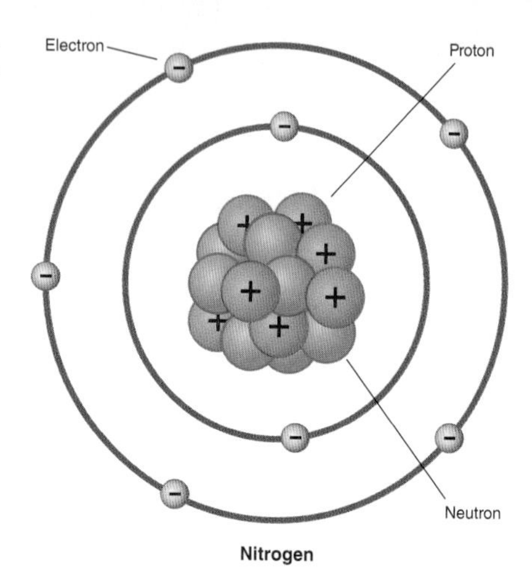

FIGURE 4.1 **Structure of the atom.** An atom is composed of protons, neutrons, and electrons. Neutrons and positively charged protons make up the nucleus. Negatively charged electrons surround the nucleus.

The Structure of an Atom

FIGURE 4.1 shows a simplified model of single atom of nitrogen. Like the nitrogen atom, every atom consists of a nucleus, or core, surrounded by electrons. The nucleus consists of protons and neutrons. Protons and neutrons have roughly the same mass—both minutely small. Protons have a positive electrical charge, like the "plus" side of a battery. The number of protons in the nucleus of a particular element—called the **atomic number**—is unique to that element. The periodic table lists the atomic number; in the periodic table at the back of this book, it is the whole number next to each element symbol. Neutrons have no electrical charge, but they are critical to the stability of nuclei because they keep the positively charged protons together. Without them, the protons would repel one another and separate.

Turning back to Figure 4.1 you can see that the space around the nucleus is occupied by electrons. Electrons are negatively charged, like the "minus" side of a battery, and have a much smaller mass than protons or neutrons. In the molecular world, opposites always attract, so negatively charged electrons are attracted to positively charged protons. This attraction binds the electrons to the nucleus. In a neutral atom, the numbers of protons and electrons are equal.

The total number of protons and neutrons in an element is known as its **mass number.** Because the mass of an electron is insignificant compared with the mass of a proton or neutron, we do not include electrons in mass number calculations. The periodic table lists the mass number of each element under the element name.

Although the number of protons in a chemical element is constant, atoms of the same element may have different numbers of neutrons and, therefore, different mass numbers. Atoms of the same element with different numbers of neutrons are called **isotopes.** Isotopes of the element carbon, for example, have six protons, but can occur with six, seven, or eight neutrons, which yield mass numbers of 12, 13, or 14, respectively. In nature, carbon occurs as a mixture of carbon isotopes. All carbon isotopes behave the same chemically. However, biological processes sometimes favor one isotope over

Element A substance composed of atoms that cannot be broken down into smaller, simpler components.

Periodic table A chart of all chemical elements currently known, organized by their properties.

Molecule A particle that contains more than one atom.

Compound A molecule containing more than one element.

Atomic number The number of protons in the nucleus of a particular element.

Mass number A measurement of the total number of protons and neutrons in an element.

Isotopes Atoms of the same element with different numbers of neutrons.

Teaching Tip: Activity

Have students make a table, as shown to the right, of the six most abundant elements of life: hydrogen, carbon, nitrogen, oxygen, phosphorus, and sulfur. For each element, record its atomic number, which is the number of protons in an atom of that element. Ask students to use the Internet to find a substance that contains the element. (Answers are provided in italics.) This will become a useful reference table for future chapters.

Element	Atomic number	Number of protons	Example
Hydrogen	*1*	*1*	H_2O
Carbon	*6*	*6*	CO_2
Nitrogen	*7*	*7*	N_2
Oxygen	*8*	*8*	O_2
Phosphorus	*15*	*15*	H_3PO_4
Sulfur	*16*	*16*	H_2SO_4

another. Certain isotopic signatures, or ratios of isotopes, can be left behind by different biological processes. Environmental scientists use these signatures to learn about certain processes or to evaluate environmental conditions such as air pollution. Because modern carbon in wood and fossil carbon in oil or coal have different carbon isotope signatures, the proportions of different isotopes in polluted air can tell us whether the pollution comes from a forest fire or combustion of fossil fuels.

Radioactivity

The nuclei of isotopes can be stable or unstable, depending on the mass number of the isotope and the number of neutrons it contains. Unstable isotopes are radioactive. Radioactive isotopes undergo **radioactive decay,** the spontaneous release of material from the nucleus. Radioactive decay changes the radioactive element into a different element. For example, uranium-235 (^{235}U) decays to form thorium-231 (^{231}Th). The original atom (uranium) is called the parent and the resulting decay product (thorium) is called the daughter. The radioactive decay of ^{235}U and certain other elements emits a great deal of energy that can be captured as heat. Nuclear power plants use this heat to produce steam that turns turbines to generate electricity.

We measure radioactive decay by recording the average rate of decay of a quantity of a radioactive element. This measurement is commonly stated in terms of the **half-life** of the element, which is the time it takes for one-half of the original radioactive parent atoms to decay. An element's half-life is a useful parameter to know because some elements that undergo radioactive decay emit harmful radiation. Knowledge of the half-life allows scientists to determine the length of time that a particular radioactive element may be dangerous. For example, using the half-life allows scientists to calculate the period of time that people and the environment must be protected from depleted nuclear fuel, like that generated by a nuclear power plant. As it turns out, many of the elements produced during the decay of ^{235}U have half-lives of tens of thousands of years and more. From this we can see why long-term storage of radioactive nuclear waste is so important. We will discuss this further in Chapter 12.

The measurement of isotopes has many applications in environmental science as well as in other scientific fields. For example, carbon in the atmosphere exists in a known ratio of the isotopes carbon-12 (99 percent), carbon-13 (1 percent), and carbon-14 (which occurs in trace amounts, on the order of one part per trillion). Carbon-14 is radioactive and has a half-life of 5,730 years. Carbon-13 and carbon-12 are stable isotopes. Living organisms incorporate carbon into their tissues at roughly the known atmospheric ratio. But after an organism dies, it stops incorporating new carbon into its tissues. Over time, the radioactive carbon-14 in the organism decays to nitrogen-14. By calculating the proportion of carbon-14 in dead biological material—a technique called carbon dating—researchers can determine how many years ago an organism died.

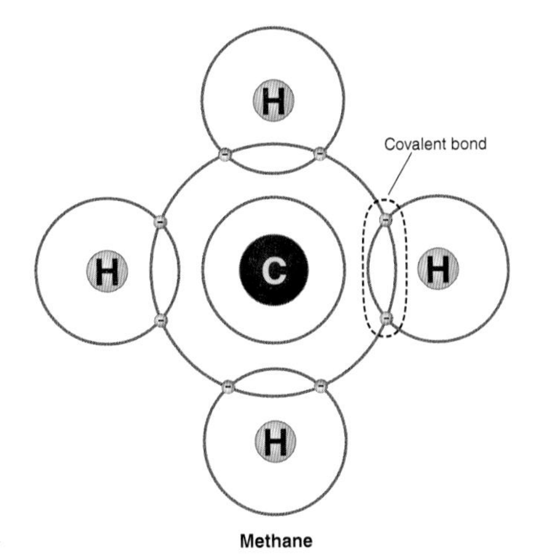

FIGURE 4.2 **Covalent bonds.** Molecules such as methane (CH_4) are associations of atoms held together by covalent bonds, in which electrons are shared between the atoms. As a result of the four hydrogen atoms sharing electrons with a carbon atom, each atom has a complete set of electrons in its outer shell—two for the hydrogen atoms and eight for the carbon atom.

Chemical Bonds

We have seen that matter is composed of atoms, which form molecules or compounds. In order to form molecules or compounds, atoms must be able to interact or join together. This happens by means of chemical bonds of various types. Chemical bonds fall into three categories: *covalent bonds, ionic bonds,* and *hydrogen bonds.*

Covalent Bonds

Elements that do not readily gain or lose electrons form compounds by sharing electrons. Compounds formed by sharing electrons are said to be held together by **covalent bonds.** FIGURE 4.2 illustrates the covalent

Radioactive decay The spontaneous release of material from the nucleus of radioactive isotopes.

Half-life The time it takes for one-half of an original radioactive parent atom to decay.

Covalent bond The bond formed when elements share electrons.

Teaching Tip: Discussion Starter

Ask students: What is the difference between an element that has a half-life measured in years and an element that has a half-life measured in days? *An element with a half-life measured in years will stay radioactive much longer than one measured in days, since it takes years, rather than days, to lose half its radioactivity.*

more **math** practice

Graphing

Have students fill in a table like the one below, to track the radioactive decay of carbon-14. Complete the rows 0 to 3, but leave the table blank for rows 4, 5, and 6,and ask students to fill in the missing information. (Answers are provided in italics.) After students have completed the table, ask them to construct a graph with "percent C-14 remaining" on the *y* axis and "years" on the *x* axis. A sample answer graph is shown below.

Number of half-lives	Years	Percent of C-14 remaining
0	0	100
1	5,730	50
2	11,460	25
3	17,190	12.5
4	*22,920*	*6.25*
5	*28,650*	*3.13*
6	*34,380*	*1.56*

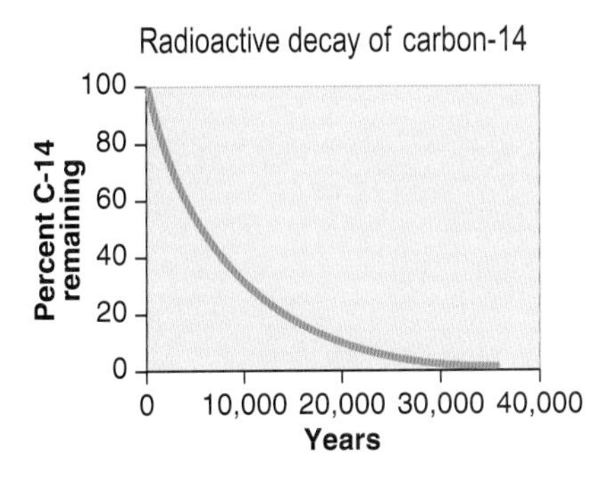

Teaching Tip: Discussion Starter

Ask students to describe how hydrogen bonding differs from covalent and ionic bonding. *Hydrogen bonding occurs between molecules. Hydrogen bonding is a weak bond that forms when a hydrogen atom on one molecule is attracted to an atom on another molecule. Covalent and ionic bonding occur when atoms join together to form compounds.*

Teaching Tip: Review

Ask students: What is the difference between covalent bonding and ionic bonding? *In a covalent bond, electrons are shared between atoms. In an ionic bond, an electron is transferred to another atom.*

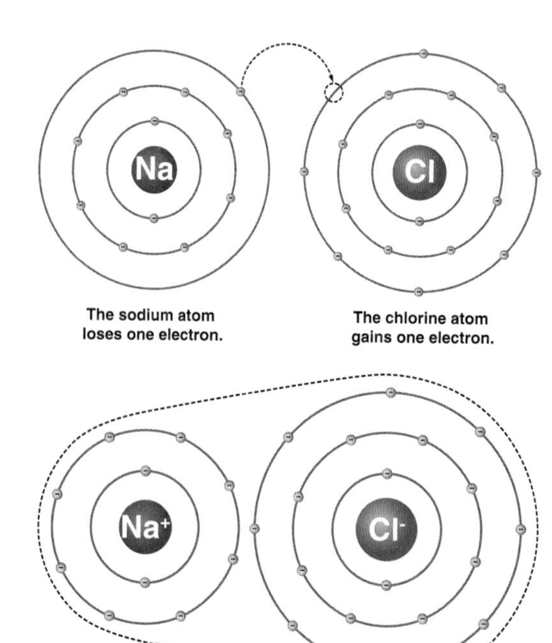

Na
Cl
The sodium atom loses one electron.
The chlorine atom gains one electron.
Na+
Cl-
The sodium ion and the chloride ion form an ionic bond: NaCl.

FIGURE 4.3 **Ionic bonds.** To form an ionic bond, the sodium atom loses an electron and the chlorine atom gains one. As a result, the sodium atom becomes a positively charged ion (Na^+) and the chlorine atom becomes a negatively charged ion (Cl^-, known as chloride). The attraction between ions of opposite charges—an ionic bond—forms sodium chloride (NaCl), or table salt.

bonds in a molecule of methane (CH_4, also called natural gas). A methane molecule is made up of one carbon (C) atom surrounded by four hydrogen (H) atoms. Covalent bonds form between the single carbon atom and each hydrogen atom. Covalent bonds also hold the two hydrogen atoms and the oxygen atom in a water molecule together.

Ionic bond A chemical bond between two ions of opposite charges.

Hydrogen bond A weak chemical bond that forms when hydrogen atoms that are covalently bonded to one atom are attracted to another atom on another molecule.

Polar molecule A molecule in which one side is more positive and the other side is more negative.

Ionic Bonds

In a covalent bond, atoms share electrons. Another kind of bond between two atoms involves the transfer of electrons. When such a transfer happens, one atom becomes electron deficient (positively charged), and the other becomes electron rich (negatively charged). The charged atoms are called ions. The charge imbalance holds the two atoms together and the attraction between ions of opposite charges forms a chemical bond known as an **ionic bond.** FIGURE 4.3 shows an example of this process. Sodium (Na) donates one electron to chlorine (Cl), which gains one electron, to form sodium chloride (NaCl), or table salt.

An ionic bond is not usually as strong as a covalent bond. This means that the compound can readily dissolve. For example, as long as sodium chloride remains in a salt shaker, it remains in solid form. But if you shake some into water, the salt dissolves into sodium and chloride ions (Na^+ and Cl^-).

Hydrogen Bonds

The third type of chemical bond is weaker than both covalent bonds and ionic bonds. A **hydrogen bond** is a weak chemical bond that forms when hydrogen atoms that are covalently bonded to one atom are attracted to an atom on another molecule. When atoms of different elements form bonds, their electrons may be shared unequally; that is, shared electrons may be pulled closer to one atom than to the other. In some cases, the strong attraction of the hydrogen electron to other atoms creates a charge imbalance within the covalently bonded molecule.

Looking at FIGURE 4.4a, we see that water is an excellent example of this type of unequal electron distribution. Each water molecule as a whole is neutral; that is, it carries neither a positive nor a negative charge. But water has unequal covalent bonds between its two hydrogen atoms and one oxygen atom. Because of these unequal bonds and the angle formed by the H—O—H bonds, water is known as a *polar molecule.* In a **polar molecule,** one side is more positive and the other side is more negative. We can see the result in Figure 4.4b where a hydrogen atom in one water molecule is attracted to the oxygen atom in another nearby water molecule. That attraction forms a hydrogen bond between the two molecules.

Hydrogen bonds also occur in nucleic acids such as DNA, the biological molecule that carries the genetic code for all organisms.

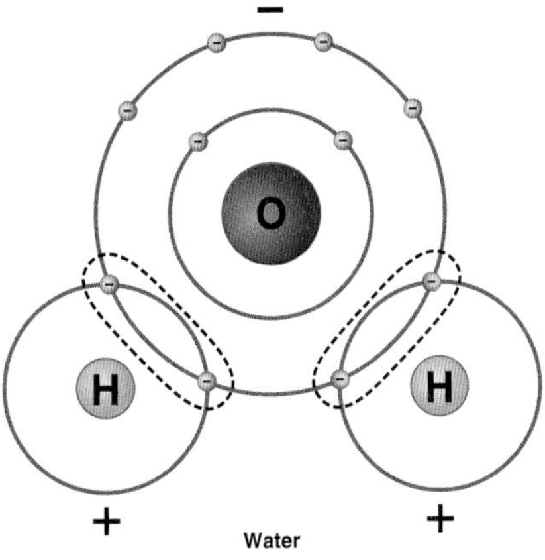

(a) Water molecule

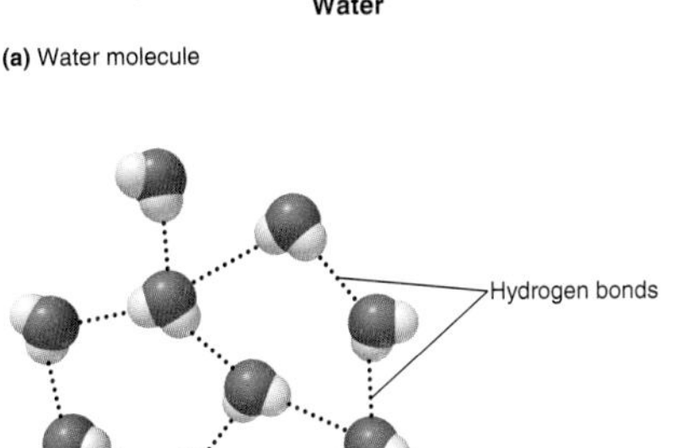

(b) Hydrogen bonds between water molecules

FIGURE 4.4 **The polarity of the water molecule.** (a) Water (H_2O) consists of two hydrogen atoms covalently bonded to one oxygen atom. Water is a polar molecule because its shared electrons spend more time near the oxygen atom than near the hydrogen atoms. The hydrogen atoms thus have a slightly positive charge, and the oxygen atom has a slightly negative charge. (b) The slightly positive hydrogen atoms are attracted to the slightly negative oxygen atom of another water molecule. The result is a hydrogen bond between the two molecules.

Water is a vital component of most environmental systems

Because water is often the vehicle for transferring chemical elements and compounds from one system to another, it is vital for environmental scientists to understand how water behaves. The molecular structure of water gives it unique properties that support the conditions necessary for life on Earth. Among these properties are surface tension, capillary action, a high boiling point, and the ability to dissolve many different substances. Each of these properties is essential to physiological functioning and the movement of elements through systems.

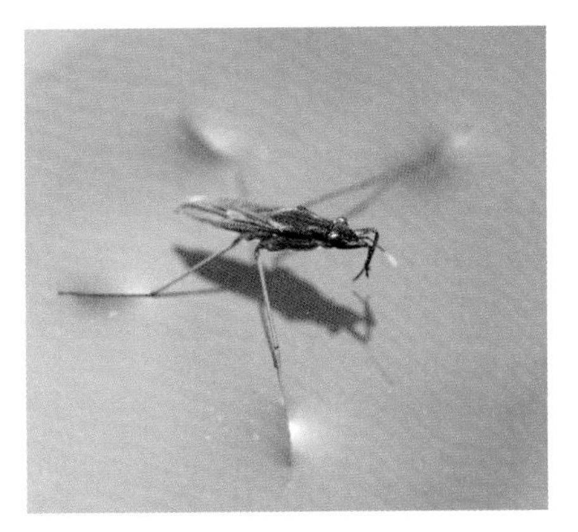

FIGURE 4.5 **Surface tension.** Hydrogen bonding between water molecules creates the surface tension necessary to support this water strider. *(optimarc/Shutterstock)*

Surface Tension and Capillary Action

Although we don't generally think of water as being sticky, hydrogen bonding makes water molecules stick strongly to one another in an action known as cohesion. Hydrogen bonding also makes water molecules stick strongly to certain other substances, an action known as adhesion. The ability to cohere or adhere underlies two unusual properties of water: *surface tension* and *capillary action*.

Surface Tension

Surface tension, which results from the cohesion of water molecules at the surface of a body of water, creates a sort of skin on the water's surface. Have you ever seen an aquatic insect, such as a water strider, walk across the surface of the water? This is possible because of surface tension (FIGURE 4.5). Surface tension also makes water droplets smooth and more or less spherical as they cling to a water faucet before dropping.

Surface tension A property of water that results from the cohesion of water molecules at the surface of a body of water and that creates a sort of skin on the water's surface.

Practicing Science: Demo

Show students some or all of the following demos to exhibit the properties of water. After each demo, ask students to name the property of water exhibited. (Answers are provided in italics.)

- Pour water into a glass until it forms a slight dome on top. Then carefully rest a paper clip on the water's surface. *Surface tension.*
- Drop ice into a clear bottle of water and seal the top. Then turn the bottle upside down until the ice rises to the water's surface. *Expands when freezes.*
- Drop a couple drops of dark food coloring into a beaker of water. Watch as the color dissolves in the water. You may want to place the beaker on a sheet of white paper so it's easier to see. *Universal solvent.*
- Dip the edge of a paper towel into a shallow bowl containing water in which food coloring has been added. Lay the rest of the paper on the table and watch as the water creeps up the paper. *Capillary action.*

Teaching Tip: Discussion Starter

Ask students: How do the unique properties of water make life on Earth possible? *Water facilitates the transfer of chemical elements and compounds from one system to another. The molecular structure of water gives it unique properties that support the conditions necessary for life on Earth. These properties are essential to physiological functioning of plants and animals and the movement of elements through systems.*

Teaching Tip: Discussion Starter

Ask students to explain why a water molecule is a polar molecule. *A water molecule is a polar molecule because its shared electrons spend more time near the oxygen atom than the hydrogen atoms. Therefore the hydrogen end of the molecule is slightly positive and the oxygen end is slightly negative.*

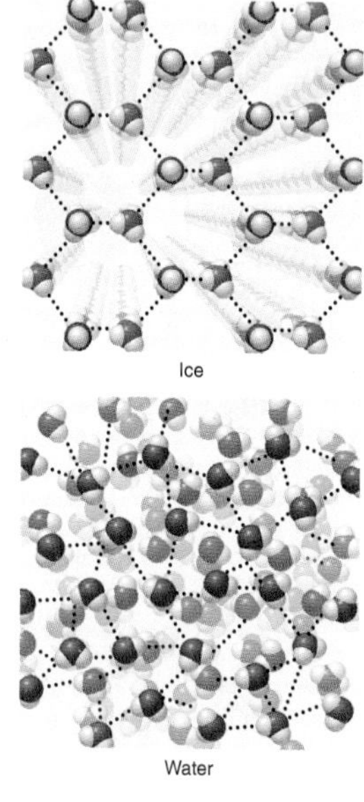

FIGURE 4.6 **The structure of water.** Below 4°C, water molecules realign into a crystal lattice structure. With its molecules farther apart, solid water (ice) is less dense than liquid water. This property allows ice to float on liquid water. *(Patrick Poendl/Shutterstock)*

Capillary Action

Capillary action happens when adhesion of water molecules to a surface is stronger than cohesion between the molecules. The absorption of water by a paper towel or a sponge is the result of capillary action. This property is important in thin tubes, such as the water-conducting vessels in tree trunks, and in small pores in soil. It is also important in the transport of underground water, as well as dissolved pollutants, from one location to another.

Boiling and Freezing

At the atmospheric pressures found at Earth's surface, water boils (becomes a gas) at 100°C (212°F) and freezes (becomes a solid) at 0°C (32°F). If water behaved like structurally similar compounds such as hydrogen sulfide (H_2S), which boils at −60°C (−76°F), it would be a gas at typical Earth temperatures and life as we know it could not exist. Because of cohesion, however, water can be a solid, a gas, or—most importantly for living organisms—a liquid at Earth's surface temperatures. In addition, the hydrogen bonding between water molecules means that it takes a great deal of energy to change the temperature of water. Thus the water in organisms protects them from wide temperature swings. Hydrogen bonding also explains why geographic areas near large lakes or oceans have moderate climates. The water body holds summer heat, which releases slowly as the atmosphere cools in the fall. Similarly, the water body warms slowly in spring, which prevents the adjacent land area from heating up too quickly.

Water has another unique property: It takes up a larger volume in solid form than it does in liquid form. FIGURE 4.6 illustrates the difference in molecular structure between liquid water and ice. As liquid water cools, it becomes denser, until it reaches 4°C (39°F), the temperature at which it reaches maximum density. As it cools from 4°C down to freezing at 0°C, its molecules realign into a crystal lattice structure, and its volume expands. You can see the result any time you add an ice cube to a drink: Ice floats on liquid water.

What does this unique property of water mean for life on Earth? Imagine what would happen if water acted like most other liquids. As it cooled, it would continue to become denser. Its solid form (ice) would

Capillary action A property of water that occurs when adhesion of water molecules to a surface is stronger than cohesion between the molecules.

Teaching Tip: Review

Ask students the following questions:

- Does frozen water float or sink? *Frozen water floats.*
- What happens when you pour water into a glass just past the rim, just before it overflows? *The water forms a slight dome.*
- Can you think of any interesting characteristics of water? *Answers will vary, but may include the ability of water to form drops, to dissolve many substances, or to move against gravity in plants.*

Teaching Tip: Activity

Have students create a table with two columns, one headed "Property of water" and the other "Function in environment," as shown below. First ask students to list the properties of water. Then ask them to brainstorm with a partner about the function of each water property in the environment. Reconvene as a class and fill in the "Function in environment" column together, using the input from students. (Suggested answers are provided in italics.)

Property of water	Function in environment
Surface tension	*Water forms droplets in clouds and falls as rain.*
Capillary action	*Water is transported from plant roots to leaves.*
High boiling and melting points	*Liquid water exists on Earth's surface.*
Expands when freezes	*Lakes remain liquid under ice, which allows aquatic organisms to survive during winter.*
Good solvent	*Water transfers dissolved substances through environmental and biological systems.*

AP® Exam Tip

The pH scale is covered in the 1998 AP® Exam, Question 3. To answer this question, students must

- understand pH ion concentration
- graph the type of organisms that can live in certain pH ranges
- know that pesticides, herbicides, and fertilizers do not necessarily acidify bodies of water
- understand the concept of neutralization
- know how lime can be used as a means of remediation for acidified bodies of water

sink, and lakes and ponds would freeze from the bottom up. As a result, very few aquatic organisms would be able to survive in temperate and cold climates.

Water as a Solvent

In our table salt example, we saw that water makes a good solvent. Many substances, such as table salt, dissolve well in water because their polar molecules bond easily with other polar molecules. This explains the high concentrations of dissolved ions in seawater as well as the capacity of living organisms to store many types of molecules in solution in their cells. Unfortunately, many toxic substances also dissolve well in water, which makes them easy to transport through the environment. Fertilizers, human waste, and road deicers such as road salt are all pollutants that dissolve easily in water and so are transported far from their sources.

Acids, Bases, and pH

Another important property of water is its ability to dissolve hydrogen or hydroxide-containing compounds known as *acids* and *bases*. An **acid** is a substance that contributes hydrogen ions to a solution. A **base** is a substance that contributes hydroxide ions to a solution. Both acids and bases typically dissolve in water.

When an acid is dissolved in water, it dissociates into positively charged hydrogen ions (H^+) and negatively charged ions. Two important acids we will discuss in this book are nitric acid (HNO_3) and sulfuric acid (H_2SO_4), the primary constituents of acid deposition, one form of which is acid rain.

Bases, on the other hand, dissociate into negatively charged hydroxide ions (OH^-) and positively charged ions. Some examples of bases are sodium hydroxide ($NaOH$) and calcium hydroxide ($Ca(OH)_2$), which can be used to neutralize acidic emissions from power plants.

The **pH** indicates the relative strength of acids and bases in a substance. A pH value of 7—the pH of pure water—is neutral, meaning that the number of hydrogen ions is equal to the number of hydroxide ions. Anything above 7 is basic, or alkaline, and anything below 7 is acidic. The lower the number, the stronger the acid, and the higher the number, the more basic the substance. The pH scale is logarithmic, meaning that each number on the scale changes by a factor of 10. For example, a substance with a pH of 5 has 10 times the hydrogen ion concentration of a substance with a pH of 6—it is 10 times more acidic. Water in equilibrium with Earth's atmosphere typically has a pH of 5.65 because carbon dioxide from the atmosphere dissolves in it, which makes it weakly acidic. FIGURE 4.7 lists the pH of many familiar substances on the pH scale, which ranges from 0 to 14.

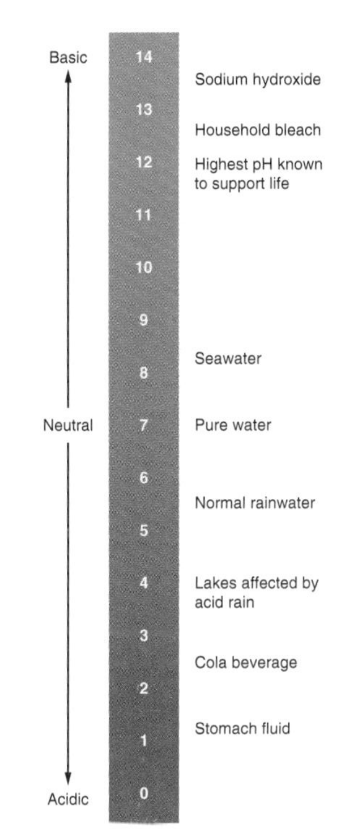

FIGURE 4.7 **The pH scale.** The pH scale shows how acidic or how basic a solution is.

Environmental systems contain both chemical and biological reactions

The chemical principles that we have described play an important role in many environmental systems through chemical and biological reactions. Although we will look at biological and chemical reactions

Acid A substance that contributes hydrogen ions to a solution.

Base A substance that contributes hydroxide ions to a solution.

pH The number that indicates the relative strength of acids and bases in a substance.

◀ TEACHING with FIGURES

Examine Figure 4.7 and then show students the figure below. These data are used in the 1998 AP® Exam, Question 3. Ask students the following questions:

- At what pH range can all selected species survive? *6.0 – 7.0.* Is this acidic, basic, or neutral? *Neutral to slightly acidic.*
- At what pH range do the fewest species survive? *4.0 – 4.5.* Is this acidic, basic, or neutral? *Acidic.*
- Which species are the most sensitive to acidic environments? How can you tell? *Clams and snails are more sensitive, because they are the first species to show an intolerance to increasing acidity.*
- Which species are the most tolerant of acidic environments? How can you tell? *Perch and frogs are the most tolerant because they are the two species able to survive at the most acidic levels.*
- Do you think pH is an important factor in determining the health of an ecosystem? Why or why not? *Yes, pH is an important factor in determining the health of an ecosystem. As shown in this figure, pH directly affects the species composition. Acidic environments are intolerable for some species, and therefore limit species diversity in a given ecosystem.*

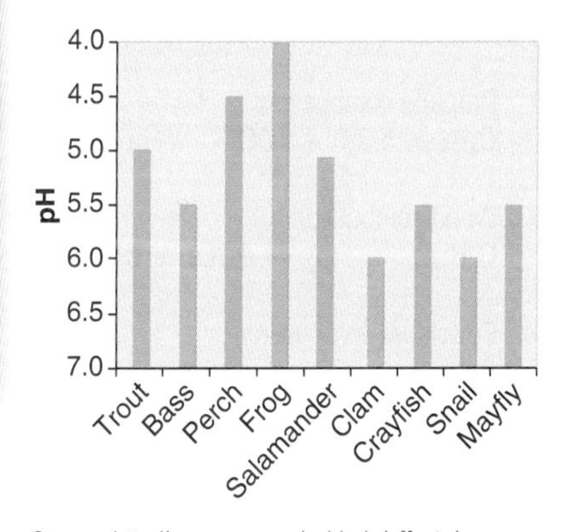

Source: http://www.epa.gov/acidrain/effects/surface_water.html

Practicing Science: Minilab

Set up beakers containing different household liquids around the classroom, and place pH indicators next to each beaker. Give students the list of liquids and have them first predict the pH of each one. Then instruct them to rotate through the classroom and record the pH of each liquid. When the class has finished, ask students which liquids they were most and least accurate in predicting the pH. Liquids that you may want to use include milk, cola, lemon juice, bleach, vinegar, tomato juice, soapy water, and pure water. Remind students that even though these are household substances, they need to wear safety goggles in the lab and follow all the appropriate science safety rules for your facility.

Practicing Science: Demo

The following demo illustrates the interaction between acid and limestone. Prepare two small beakers, one with crushed chalk or limestone (both contain calcium carbonate, $CaCO_3$) and one with 1 mL of vinegar. Take the pH of the vinegar and announce the result to the class. Then pour the vinegar into the beaker with the limestone. Observe how the limestone reacts. After 10 minutes, take the pH of the limestone and vinegar solution. The pH of the combined solution should be higher than that of the original vinegar. Discuss the implications of this interaction with your class, including the effect of acid rain on marble buildings, the effect of acidic oceans on shelled organisms, and the ability of limestone to neutralize acidic emissions from power plants.

Practicing Science: Chemical Reactions

Give students the following chemical reactions for the combustion of various fuels, substituting a blank line space for the numbers in red italics. Ask them to balance the equations and check their answers with a partner.

- Methane combustion:
  $CH_4 + 2O_2 \rightarrow CO_2 + 2H_2O$
- Propane combustion:
  $C_3H_8 + 5O_2 \rightarrow 3CO_2 + 4H_2O$
- Wood combustion:
  $C_6H_{10}O_5 + 6O_2 \rightarrow 6CO_2 + 5H_2O$
- Petroleum combustion:
  $2C_8H_{18} + 25O_2 \rightarrow 16CO_2 + 18H_2O$

FIGURE 4.8 **The law of conservation of matter.** Even though this forest seems to be disappearing as it burns, all the matter it contains is conserved in the form of water vapor, carbon dioxide, and solid particles. *(Exactostock/SuperStock)*

separately in Chapter 3, here we shall see that chemical and biological components interact in most environmental systems.

Chemical Reactions and the Conservation of Matter

A **chemical reaction** occurs when atoms separate from molecules or recombine with other molecules. In a chemical reaction, no atoms are ever destroyed or created, although the bonds between particular atoms may change. For example, when methane (CH_4) is burned in air, it reacts with two molecules of oxygen (2 O_2) to create one molecule of carbon dioxide (CO_2) and two molecules of water (2 H_2O):

$$CH_4 + 2\,O_2 \rightarrow CO_2 + 2\,H_2O$$

Notice that the number of atoms of each chemical element is the same on each side of the reaction.

Chemical reactions can occur in either direction. For example, during the combustion of fuels, nitrogen gas (N_2) combines with oxygen gas (O_2) from the atmosphere to form two molecules of nitrogen oxide (NO), which is an air pollutant:

$$N_2 + O_2 \rightarrow 2\,NO$$

This reaction can also proceed in the opposite direction:

$$2\,NO \rightarrow N_2 + O_2$$

The observation that no atoms are created or destroyed in a chemical reaction leads us to the **law of conservation of matter,** which states that matter cannot be created or destroyed; it can only change form. For example, when paper burns, it may seem to vanish, but no atoms are lost. In this case, the carbon and hydrogen that make up the paper combine with oxygen in the air to produce carbon dioxide, water vapor, and other materials, which either enter the atmosphere or form ash. Combustion converts most of the solid paper into gases, but all of the original atoms remain. The same process occurs in a forest fire, but on a much larger scale (FIGURE 4.8). The only known exception to the law of conservation of matter occurs in nuclear reactions, in which small amounts of matter change into energy.

The law of conservation of matter explains why we cannot easily dispose of hazardous materials. If something is hazardous, it typically will remain hazardous even if we try to convert it through combustion. For example, when we burn material that contains heavy metals, such as an automotive battery, the atoms

Chemical reaction A reaction that occurs when atoms separate from molecules or recombine with other molecules.

Law of conservation of matter A law of nature stating that matter cannot be created or destroyed; it can only change form.

Practicing Science: Ask Questions

Make sure students understand the law of conservation of matter. Have each student make a list of questions they have about the topic. Questions can be general or specific. Have students volunteer to share their questions with a partner and then with the class. Select specific questions to start a discussion. *Answers will vary but may include questions such as: What happens to wood matter when it burns? What happens to organic matter when an animal decomposes? What's the source of plant matter as it grows?*

of the metals in the battery do not disappear. They turn up elsewhere in the environment, where they may harm humans and other organisms. For this and other reasons, understanding the law of conservation of matter is crucial to the study of environmental science.

Biological Molecules and Cells

As we mentioned earlier, chemical and biological reactions happen simultaneously in most environmental systems. We have seen how chemical compounds form and how they respond to various processes such as burning and freezing. To understand biological processes, we must first look at the distinction between compounds that are *inorganic* and those that are *organic*. **Inorganic compounds** are compounds that either do not contain the element carbon or contain carbon that is bound to elements other than hydrogen. Examples include ammonia (NH_3), sodium chloride (NaCl), water (H_2O), and carbon dioxide (CO_2). **Organic compounds** are compounds that have carbon-carbon and carbon-hydrogen bonds. Examples of organic compounds include glucose ($C_6H_{12}O_6$) and fossil fuels, such as natural gas (CH_4).

Organic compounds are the basis of the biological molecules that are important to life: *carbohydrates, proteins, nucleic acids,* and *lipids*. Because these four types of molecules are relatively large, they are also known as macromolecules.

Carbohydrates

Carbohydrates are compounds composed of carbon, hydrogen, and oxygen atoms. Glucose ($C_6H_{12}O_6$) is a simple sugar (a monosaccharide, or single sugar) easily used by plants and animals for quick energy. Sugars can link together in long chains called complex carbohydrates, or polysaccharides (many sugars). For example, plants store energy as starch, which is made up of long chains of covalently bonded glucose molecules. The starch can also be used by animals that eat the plants. Cellulose, a component of plant leaves and stems, is another polysaccharide consisting of long chains of glucose molecules. Cellulose is the raw material for cellulosic ethanol, a type of fuel that has the potential to replace or supplement gasoline.

Proteins

Proteins, a critical component of living organisms, are made up of long chains of nitrogen-containing organic molecules called amino acids. Proteins play a role in structural support, energy storage, internal transport, and defense against foreign substances. Enzymes are proteins that help control the rates of chemical reactions. The antibodies that protect us from infections are also proteins.

Nucleic Acids

Nucleic acids are organic compounds found in all living cells. Long chains of nucleic acids form *DNA* and *RNA*. **DNA (deoxyribonucleic acid)** is the genetic material organisms pass on to their offspring; it contains the code for reproducing the components of the next generation. **RNA (ribonucleic acid)** translates the code stored in DNA, which makes possible the synthesis of proteins.

Lipids

Lipids are smaller biological molecules that do not mix with water. Fats, waxes, and steroids are all lipids. Lipids form a major part of the membranes that surround cells.

Cells

We have looked at the four types of macromolecules required for life. But how do they work as part of a living organism? The smallest structural and functional component of organisms is known as a *cell*.

Inorganic compound A compound that does not contain the element carbon or contains carbon bound to elements other than hydrogen.

Organic compound A compound that contains carbon-carbon and carbon-hydrogen bonds.

Carbohydrate A compound composed of carbon, hydrogen, and oxygen atoms.

Protein A critical component of living organisms made up of a long chain of nitrogen-containing organic molecules known as amino acids.

Nucleic acid Organic compounds found in all living cells.

DNA (deoxyribonucleic acid) A nucleic acid, the genetic material that contains the code for reproducing the components of the next generation, and which organisms pass on to their offspring.

RNA (ribonucleic acid) A nucleic acid that translates the code stored in DNA, which makes possible the synthesis of proteins.

Lipid A smaller organic biological molecule that does not mix with water.

Teaching Tip: Discussion Starter

Ask students: How are the four types of biological molecules different from each other? *Carbohydrates, proteins, nucleic acids, and lipids differ in their biological functions and structures. Carbohydrates are composed of chains made of carbon, hydrogen, and oxygen. Proteins are composed of amino acid chains. Nucleic acids have a double helix structure composed of nucleic acids. Lipids contain fatty acid chains attached to a glycerol group.*

Teaching Tip: Activity

Divide students into pairs or groups and have them start a table with the column heads, as shown at the right. Then provide them with a shuffled list of the terms in the table and ask them to place each term in the correct column. (Answers are provided in italics.) Ask students to research and identify images that show the structure of a carbohydrate, protein, nucleic acid, and lipid. Use the terms from the table to discuss each image in more detail.

Carbohydrates	Proteins	Nucleic acids	Lipids
monosaccharide	*amino acid*	*DNA*	*fats*
sugar	*peptide*	*RNA*	*waxes*
glucose	*enzyme*	*double helix*	*steroids*
cellulose	*antibodies*	*nucleotide*	*cholesterol*
starch	*muscle*	*code of life*	*triglyceride*

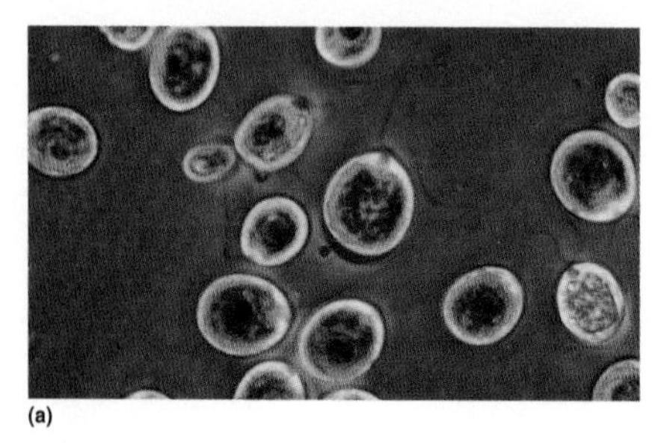

(a)

(b)

FIGURE 4.9 **Cellular composition of organisms.** (a) Some organisms, such as these green algae, consist of a single cell. (b) More complex organisms, such as the Mono Lake brine shrimp, are made up of millions of cells. *(a: Biophoto Associates/Science Source; b: Stuart Wilson/Science Source)*

A **cell** is a highly organized living entity that consists of the four types of macromolecules and other substances in a watery solution, surrounded by a membrane. Some organisms, such as most bacteria and some algae, consist of a single cell. This cell contains all of the functional structures, or organelles, needed to keep the cell alive and allow it to reproduce (FIGURE 4.9a). Larger and more complex organisms, such as Mono Lake's brine shrimp, are multicellular (Figure 4.9b). Throughout this book, we will be examining biological reactions and the resulting effects on the environment. Some effects are at the cellular level, while others are at the organismal level. Each is important to understanding environmental science.

Cell A highly organized living entity that consists of the four types of macromolecules and other substances in a watery solution, surrounded by a membrane.

module 4

REVIEW

In this module, we saw that all environmental systems consist of matter. Matter has specific properties that can be described and measured. Properties of water include surface tension, capillary action, a high boiling point, and the ability to dissolve many different substances. These properties make water a critical part of most environmental systems. Matter is conserved in chemical reactions and the components of biological reactions. In the next module we will expand our view to look at how the movement of matter in environmental systems is strongly influenced by energy inputs, flows, and outputs.

Answers to Module 4 AP® Review Questions

1. c
2. d
3. e
4. b
5. a
6. c

Module 4 AP® Review Questions

1. If two atoms of an element are isotopes, then they have a different
   (a) atomic symbol.
   (b) number of protons.
   (c) number of neutrons.
   (d) number of electrons.
   (e) atomic number.

2. The chemical bond that forms from the attraction of sodium ions and chlorine atoms in table salt (NaCl) is called
   (a) a covalent bond.
   (b) a polar bond.
   (c) a hydrogen bond.
   (d) an ionic bond.
   (e) a nucleic bond.

3. Which of the following is NOT a property of water that allows it to support life?
   (a) Surface tension
   (b) Capillary action
   (c) Solvent ability
   (d) High boiling point
   (e) High viscosity

4. Which of the following has the highest pH?
   (a) Pure water
   (b) Bleach
   (c) Cola beverage
   (d) Seawater
   (e) Acid rain

5. Which of the following is an organic compound?
   (a) CH_4
   (b) NH_3
   (c) NaCl
   (d) H_2O
   (e) CO_2

6. Which of the following is NOT a macromolecule?
   (a) Carbohydrates
   (b) Nucleic acids
   (c) Organelles
   (d) Proteins
   (e) Lipids

module 5

Energy, Flows, and Feedbacks

Energy flows within and among systems. Plants and other photosynthetic organisms such as the algae in Mono Lake absorb solar energy and use it in photosynthesis to convert carbon dioxide and water into sugars they need to survive, grow, and reproduce. Animals such as the brine shrimp in Mono Lake eat those plants and the energy is transferred. When migrating gulls use Mono Lake as a stopover, they consume the brine shrimp and transfer the energy and nutrients elsewhere. Some transfers occur more effectively than others. Some transfers occur within a given system while others, like those of migrating gulls, result in transfers of material and energy to another system.

Learning Objectives

After reading this module you should be able to

- distinguish among various forms of energy and understand how they are measured.
- discuss the first and second laws of thermodynamics and explain how they influence environmental systems.
- explain how scientists keep track of energy and matter inputs, outputs, and changes to environmental systems.

Teaching Tip: Activity

Word Splash
Divide students into pairs and pass out a copy of the Word Splash worksheet to each pair. Ask students to first circle all the different types of energy on the sheet (*thermal, chemical, kinetic, potential, nuclear, electrical, mechanical, solar*). Then have students use the terms on the sheet to create two true statements about energy, and one false statement about energy. Have students share their statements with the class, and write the best statements on the board for further discussion.

Download this resource by clicking on the link buttons in the TE-book, opening the TRFD, or logging onto the book's companion website bcs.whfreeman.com /friedlandapes2e (teacher login required).

Activity 2.1 Word Splash

TABLE 5.1	Common units of energy and their conversion into joules		
Unit	Definition	Relationship to joules	Common uses
calorie	Amount of energy it takes to heat 1 gram of water 1°C	1 calorie = 4.184 J	Energy expenditure and transfer in ecosystems; human food consumption
Calorie	Food Calorie (always shown with a capital "C")	1 Calorie = 1,000 calories = 1 kilocalorie (kcal) = 4,184 J	Food labels; human food consumption
British thermal unit (Btu)	Amount of energy it takes to heat 1 pound of water 1°F	1Btu = 1,055 J	Energy transfer in air conditioners and home water heaters
Kilowatt-hour (kWh)	Amount of energy expended by using 1 kilowatt of electricity for 1 hour	1 kWh = 3,600,000 J = 3.6 megajoules (MJ)	Energy use by electrical appliances; often given in kWh per year

Energy is a fundamental component of environmental systems

Earth's systems cannot function, and organisms cannot survive, without *energy*. **Energy** is the ability to do work, or transfer heat. Water flowing into a lake has energy because it moves and can move other objects in its path.

All living systems absorb energy from their surroundings and use it to organize and reorganize molecules within their cells and to power movement. The sugars in plants are also an important energy source for many animals. Humans, like other animals, absorb the energy they need for cellular respiration from food. This provides the energy for our daily activities, from waking to sleeping to walking, and everything in between. Constructed human systems also utilize energy. If you took mass transportation or an automobile to get to school, you most likely utilized fossil fuel energy that was converted to the energy of motion of your vehicle.

The basic unit of energy in the metric system is the *joule* (J). A **joule** is the amount of energy used when a 1-watt light bulb is turned on for 1 second—a very small amount. Although the joule is the preferred energy unit in scientific study, many other energy units are commonly used. Conversions between these units and joules are given in **TABLE 5.1**.

Although we often use the words "energy" and "power" interchangeably, they are not the same thing. We have seen that energy is the ability to do work. **Power** is the rate at which work is done:

$$\text{energy} = \text{power} \times \text{time}$$
$$\text{power} = \text{energy} \div \text{time}$$

When we talk about generating electricity, we often hear about kilowatts and kilowatt-hours. The kilowatt (kW) is a unit of power while the kilowatt-hour (kWh) is a unit of energy. Therefore, the capacity of a turbine is given in kW because that measurement refers to the turbine's power. Your monthly home electricity bill reports energy use—the amount of energy from electricity that you have used in your home—in kWh. "Do the Math: Calculating Energy Use and Converting Units" (see page 46) gives you an opportunity to practice working with these units.

Forms of Energy

Energy exists in different forms and can be converted from one form to another. Potential energy, kinetic energy, light energy, chemical energy, and sound energy are all important energy forms in the environmental sciences.

Electromagnetic Radiation

Ultimately, most energy on Earth derives from the Sun. The Sun emits **electromagnetic radiation,** a form of energy that includes, but is not limited to, visible light, ultraviolet light, and infrared energy, which we perceive as heat. The scale at the top of FIGURE 5.1 shows these and other types of electromagnetic radiation.

Electromagnetic radiation is carried by **photons,** massless packets of energy that travel at the speed of light and can move even through the vacuum of space. The amount of energy contained in a photon depends on its wavelength—the distance between two peaks or troughs in a wave, as shown in the inset in Figure 5.1.

Energy The ability to do work or transfer heat.

Joule The amount of energy used when a 1-watt electrical device is turned on for 1 second.

Power The rate at which work is done.

Electromagnetic radiation A form of energy emitted by the Sun that includes, but is not limited to, visible light, ultraviolet light, and infrared energy.

Photon A massless packet of energy that carries electromagnetic radiation at the speed of light.

Teaching Tip: Engage

Ask students: What is the difference between power and energy? *Energy is the ability to do work, whereas power is the rate at which work is done.*

lab
Who's Got the Power?

This lab offers students an opportunity to explore the relationship between work and power. Students will first learn how to differentiate between the two, and then will calculate how much work is required to use up the energy in a snack.

Download the lab by clicking on the link buttons in the TE-book, opening the TRFD, or logging onto the book's companion website bcs.whfreeman.com/friedlandapes2e (teacher login required).

 **Lab 2.1** Who's Got the Power?

Photons with long wavelengths, such as radio waves, have very low energy, while those with short wavelengths, such as X-rays, have high energy. Photons of different wavelengths are used by humans for different purposes. For example, high-energy, short-wavelength X-rays are used for diagnostic medical purposes while lower-energy, long-wavelength infrared rays are used to identify heat loss from buildings during an environmental energy audit.

Potential Energy

Many stationary objects possess a large amount of **potential energy**—energy that is stored but has not yet been released. For example, water impounded behind a dam contains a great deal of potential energy. Potential energy stored in chemical bonds is known as **chemical energy.** The energy in food is a familiar example. By breaking down the high-energy bonds in the salad you had for lunch, your body obtains energy to power its activities and functions. Likewise, an automobile engine combusts gasoline and releases its chemical energy to propel the car.

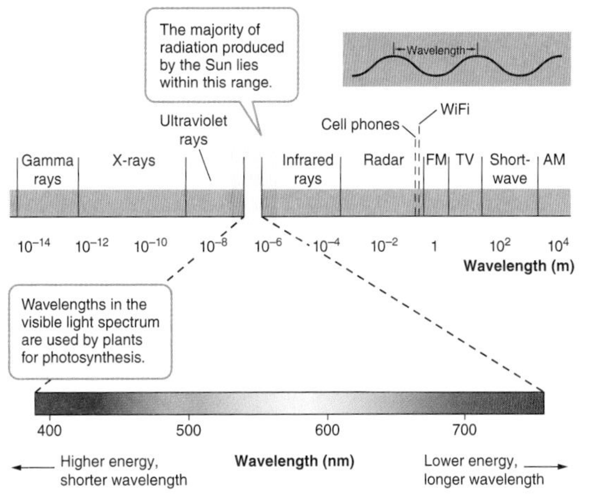

FIGURE 5.1 **The electromagnetic spectrum.** Electromagnetic radiation can take numerous forms, depending on its wavelength. The Sun releases photons of various wavelengths, but primarily between 250 and 2,500 nanometers (nm).

Kinetic Energy

We noted that water impounded behind a dam contains a great deal of potential energy. When the water is released and flows downstream, that potential energy becomes **kinetic energy,** the energy of motion (FIGURE 5.2). The kinetic energy of moving water can be captured at a dam and transferred to a turbine and generator, and ultimately to the energy in electricity. Can you think of other common examples of kinetic energy? A car moving down the street, a flying honeybee, and a football traveling through the air all have kinetic energy. Sound also has kinetic energy because it travels in waves through the coordinated motion of atoms. Systems can contain potential energy, kinetic energy, or some of each.

All matter, even the frozen water in the world's ice caps, contains some energy. When we say that energy moves matter, we mean that it is moving the molecules within a substance. The measure of the average kinetic energy of a substance is its **temperature.**

Changes in temperature—and, therefore, in energy—can convert matter from one state to another such as liquid water freezing and becoming ice. At a certain temperature, the molecules in a solid substance start moving so fast that they begin to flow, and the substance melts into a liquid. At an even higher temperature, the molecules in the liquid move still faster, with increasing amounts of energy. Finally the molecules move with such speed and energy that they overcome the forces holding them together and become gases.

FIGURE 5.2 **Potential and kinetic energy.** The water stored behind this dam in Arizona has potential energy. The potential energy is converted into kinetic energy as the water flows through the gates. *(Richard Kolar/Earth Scenes/Animals Animals)*

Potential energy Stored energy that has not been released.

Chemical energy Potential energy stored in chemical bonds.

Kinetic energy The energy of motion.

Temperature The measure of the average kinetic energy of a substance.

Teaching Tip: Activity

Divide students into groups and give each group a list (or flashcards) of the types of electromagnetic radiation shown in Figure 5.1. Also include purple light, yellow light, and red light. The list should be in a random order. Ask students to put the terms in order from longest wavelength to shortest wavelength. Evaluate the accuracy of each group's work, clarifying where needed.

Teaching Tip: Connections

Ask students: How do potential energy and kinetic energy differ? What is chemical energy? *Potential energy is stored energy that has not yet been released, whereas kinetic energy is the energy of motion, often generated when potential energy is released. Chemical energy is a form of potential energy that is stored in chemical bonds, such as those in food.*

do the **math**

Answer to Your Turn

We are told to assume that we can save 76 L per month (L/month) by repairing our car and are asked to convert this quantity to joules per year (J/year). We are then asked to compare this to the quantity of electricity that we would save, also converted to joules per year.

- Car

$(76\ \text{L/month}) \times (12\ \text{months/year}) \times (10\ \text{kWh/L}) \times (3.6 \times 10^6\ \text{J/kWh}) = 3.28 \times 10^{10}\ \text{J/year}$

To the correct number of significant figures this is 3.3×10^{10} J/year.

- Refrigerator
  From the Do the Math box we see that we save 1,790 kWh per year, therefore

$(1790\ \text{kWh/year}) \times (3.6 \times 10^6\ \text{J/kWh}) = 6.44 \times 10^9\ \text{J/year}$

We would save more by making repairs to our car.

more **math** practice

Metric Conversion

Students should be able to understand metric conversion with area, length, mass, temperature, power, energy, and volume. Some of the most common conversions are listed in the table. Several free-response questions from past exams have asked students to calculate the number of acres or hectares of land needed to feed a population or to grow a certain crop. Students should also be able to convert all of the metric units by moving the decimal to the left or right.

Area Conversions

Acres × 0.4047 = Hectares
Hectares × 2.47 = Acres

Length Conversions

Feet × 0.3048 = Meters
Meters × 3.2808 = Feet
Kilometers × 0.62 = Miles
Miles × 1.609347 = Kilometers

do the math

Calculating Energy Use and Converting Units

Your electricity bill shows that you use 600 kWh of electricity each month. Your refrigerator, which is 15 years old, could be responsible for up to 25 percent of this electricity consumption. Newer refrigerators are more efficient, meaning that they use less energy to do the same amount of work. If you wish to conserve electrical energy and save money, should you replace your refrigerator? How can you compare the energy efficiency of your old refrigerator with that of more-efficient newer models?

Your refrigerator uses 500 watts when the motor is running. The motor runs for about 30 minutes per hour (or a total of 12 hours per day). How much energy in kilowatt-hours per year will you save by using the best new refrigerator instead of your current one? How long will it take you to recover the cost of the new appliance?

1. Start by calculating the amount of energy your current refrigerator uses.

$$0.5\ \text{kW} \times 12\ \text{hours/day} = 6\ \text{kWh/day}$$
$$6\ \text{kWh/day} \times 365\ \text{days/year} = 2{,}190\ \text{kWh/year}$$

2. How much more efficient is the best new refrigerator compared with your older model?
   The best new model uses 400 kWh per year. Your refrigerator uses 2,190 kWh per year.

$$2{,}190\ \text{kWh/year} - 400\ \text{kWh/year} = 1{,}790\ \text{kWh/year}$$

3. Assume that you are paying, on average, \$0.10 per kilowatt-hour for electricity. A new refrigerator would cost you \$550. You will receive a rebate of \$50 from your electric company for purchasing an energy-efficient refrigerator. If you replace your refrigerator, how long will it be before your energy savings compensate you for the cost of the new appliance? You will save

$$1{,}790\ \text{kWh/year} \times \$0.10/\text{kWh} = \$179/\text{year}$$

Dividing \$500 by \$179 indicates that in less than 3 years, you will recover the cost of the new appliance.

Your Turn Environmental scientists must often convert energy units in order to compare various types of energy. For instance, you might want to compare the energy you would save by purchasing an energy-efficient refrigerator with the energy you would save by driving a more fuel-efficient car. Assume that for the amount you would spend on the new refrigerator (\$500), you can make repairs to your car engine that would save you 20 gallons (76 liters) of gasoline per month. (Note that 1 L of gasoline contains the energy equivalent of about 10 kWh.) Using this information and Table 5.1 on page 44, convert the quantities of both gasoline and electricity into joules and compare the energy savings. Which decision would save the most energy?

Energy Conversions

Individual organisms rely on a continuous input of energy in order to survive, grow, and reproduce. But interactions beyond the organism can also be seen as a process of converting energy into organized structures such as leaves and branches. Consider a forest ecosystem. Trees absorb water through their roots and carbon dioxide through their leaves. By combining these compounds in the presence of sunlight, they convert water and carbon dioxide into sugars that will provide them with the energy they need. But then a deer grazes on tree leaves, and later a mountain lion eats the deer. At each step, energy is converted by organisms into work.

46 CHAPTER 2 ■ Environmental Systems

Examples:

56.8 m = how many millimeters?

Move the decimal three spaces to the right (millimeter).
56.8 m = 56,800 mm

4,885 cm = how many kilometers?

Move the decimal five spaces to the left (kilometer).
4,885 cm = 0.04885 km

Metric Prefixes

	Prefixes	Means	Abbreviation
Largest	kilo	1,000	k
	hecto	100	h
	deca	10	da
Base Unit	meter, liter, gram	1	m, L, g
	deci	0.1	d
	centi	0.01	c
Smallest	milli	0.001	m

The form and amount of energy available in an environment determines what kinds of organisms can live there. Plants thrive in tropical rainforests where there is plenty of sunlight and water. Many food crops, not surprisingly, can be planted and grown in temperate climates that have a moderate amount of sunlight. Life is much more sparse at high latitudes, toward the North and South Poles, where less solar energy is available to organisms. These landscapes are populated mainly by small plants and shrubs, insects, and migrating animals. Plants cannot live at all on the deep ocean floor, where no solar energy penetrates. The animals that live there, such as eels, anglerfish, and squid, get their energy by feeding on dead organisms that sink from above. Chemical energy, in the form of sulfides emitted from deep-ocean vents, supports a plantless ecosystem that includes sea spiders, 2.4-m (8-foot) tube worms, and bacteria (FIGURE 5.3).

The laws of thermodynamics describe how energy behaves

In Chapter 1 we saw that some theories have no known exceptions. Two theories concerning energy fall in this category. The laws of thermodynamics are among the most significant principles in all of science.

First Law of Thermodynamics

Impounded water behind a dam is quiet, still, and unmoving. It doesn't seem like it contains a lot of energy. But in fact that water contains a great deal of potential energy. If you release that water by opening a gate in the dam, the water rushes out. The potential energy of the impounded water becomes the kinetic energy of the water rushing through the gates of the dam. This is an illustration of the **first law of thermodynamics,** which states that energy is neither created nor destroyed but it can change from one form to another.

The first law of thermodynamics dictates that you can't get something from nothing. When an organism needs biologically usable energy, it must convert it from an energy source such as the Sun or food. The potential energy contained in firewood never goes away but is transformed into heat energy permeating a room when the wood is burned in a fireplace. Sometimes it may be difficult to identify where the energy is going, but it is always conserved.

Look at FIGURE 5.4, which uses a car to show the first law in action through a series of energy conversions. Think of the car, including its fuel tank, as a system. The potential energy of the fuel (gasoline) is converted into kinetic energy when the battery supplies a spark in the presence of gasoline and air. The gasoline combusts, and the resulting gases expand, pushing the pistons in the engine—converting the chemical energy

(a)

(b)

(c)

FIGURE 5.3 **The role of energy in a natural system.** The amount of energy in a natural system determines which organisms can live in it. (a) A tropical rainforest such as this one in Costa Rica has abundant energy available from the Sun and enough moisture for plants to make use of that energy. (b) Arctic tundra, for example this area in Denali National Park, Alaska, has much less energy available, so plants grow more slowly there and do not reach large sizes. (c) The energy supporting this deep-ocean vent community in the Pacific Ocean comes from chemicals emitted from the vent. Bacteria convert the chemicals into forms of energy that other organisms, such as tube worms, can use. *(a: Steffen Foerster/Shutterstock; b: NancyS/Shutterstock; c: Emory Kristof/National Geographic Stock)*

First law of thermodynamics A physical law which states that energy can neither be created nor destroyed but can change from one form to another.

Teaching Tip: Discussion Starter

Ask students to describe how animals convert energy from one form into another. *When an animal eats, it converts the chemical energy of its food to kinetic energy generated by the movement of its muscles. When the organism moves, some of this kinetic energy is converted to thermal energy, which the animal releases as heat.*

Teaching Tip: Discussion Starter

Ask students: In what ways does energy determine the suitability of an environment for growing food? *When food is grown, solar energy is converted to chemical energy stored in plants. Therefore the amount of solar energy available in an environment determines the amount of chemical energy available through food production. If solar energy is initially limited, the amount of food production will be limited and there will be less chemical energy available.*

Teaching Tip: Activity

Ask students to create a drawing that illustrates the fundamental concept that energy can neither be created nor destroyed but can change from one form to another. Students may discuss their drawings with a partner first and then volunteer to share their drawings with the class.

AP® Exam Tip

Make sure students can convert units. For example, can they go from millimeters to meters or from grams to kilograms? These conversions show up frequently on the free-response questions and often cause students to lose points.

Teaching Tip: Engage

Ask students: If Figure 5.4 were a human instead of a car, what would be the energy inputs and outputs? *Energy inputs: potential (chemical) energy in food, drinks, and vitamins. Energy outputs: kinetic energy in muscle movement, thermal energy from heat released during muscle movement, chemical energy in feces.*

Teaching Tip: Activity

Divide students into groups and ask each group to create a jingle or rhyme that conveys the concept that when energy is transformed, the quantity of energy remains the same, but its ability to do work diminishes. When students are finished, have each group share their jingle or rhyme with the class.

more **math** practice

Calculating Energy Efficiency

Ask students to calculate the overall energy efficiency of converting coal into compact fluorescent lighting, which is 20 percent efficient. How does this compare to the overall efficiency of an incandescent bulb?

Coal to electricity × electricity transport × fluorescent light efficiency = overall efficiency
$0.35 \times 0.90 \times 0.20 = 0.063$ (6.3% efficiency)

Using a compact fluorescent bulb is four times more efficient than using an incandescent bulb.

FIGURE 5.4 Conservation of energy within a system. In a car, the potential energy of gasoline is converted into other forms of energy. Some of that energy leaves the system, but all of it is conserved.

in the gasoline into the kinetic energy of the moving pistons. Energy is transferred from the pistons to the drivetrain, and from there to the wheels, which propel the car. The combustion of gasoline also produces heat, which dissipates into the environment outside the system. The kinetic energy of the moving car is converted into heat and sound energy as the tires create friction with the road and the body of the automobile moves through the air. When the brakes are applied to stop the car, friction between brake parts releases heat energy. No energy is ever destroyed in this example, but chemical energy is converted into motion, heat, and sound. Notice that some of the energy stays within the system and some energy, for example the heat from burning gasoline, leaves the system.

Second Law of Thermodynamics

We have seen how the potential energy of gasoline is transformed into the kinetic energy of moving pistons in a car engine. But as Figure 5.4 shows, some of that energy is converted into a less usable form—in this case, heat. The heat that is created is called waste heat, meaning that it is not used to do any useful work. And it is inevitable—it is a natural law—that any time there is a conversion of energy from one form to another, some of that energy will be lost as heat. This is one of the implications of another very important law: The **second law of thermodynamics** tells us that when energy is transformed, the quantity of energy remains the same, but its ability to do work diminishes.

Energy Efficiency

To quantify the second law of thermodynamics, we use the concept of *energy efficiency*. **Energy efficiency** is the ratio of the amount of energy expended in the desired form to the total amount of energy that is introduced into the system. Two machines or engines that perform the same amount of work, but use different amounts of energy to do that work, have different energy efficiencies. Consider the difference between modern woodstoves and traditional open fireplaces. A woodstove that is 70 percent efficient might use 2 kg of wood to heat a room to a comfortable 20°C (68°F), whereas a fireplace that is 10 percent efficient would require 14 kg of wood to achieve the same temperature—a sevenfold greater energy input (FIGURE 5.5).

We can also calculate the energy efficiency of transforming one form of energy into other forms of energy. Let's consider what happens when we convert the chemical energy of coal into the electricity that provides light from a reading lamp and the heat that the lamp releases. FIGURE 5.6 shows the process.

A modern coal-burning power plant can convert 1 metric ton of coal, containing 24,000 megajoules (MJ; 1 MJ = 1 million joules) of chemical energy into about 8,400 MJ of electricity. Since 8,400 is 35 percent of 24,000, this means that the process of turning coal into electricity is about 35 percent efficient. The rest of the energy from the coal—65 percent—is lost as waste heat.

In the electrical transmission lines between the power plant and the house, 10 percent of the electrical energy from the plant is lost as heat and sound, so the transport of energy away from the plant is about 90 percent efficient. We know that the conversion of electrical energy into light in an incandescent bulb is 5 percent efficient; again, the rest of the energy is lost as heat. From beginning to end, we can calculate the energy efficiency of converting coal into incandescent lighting by multiplying all the individual efficiencies:

Calculating energy efficiency:

$$\text{Coal to electricity} \times \text{transport of electricity} \times \text{light bulb efficiency} = \text{overall efficiency}$$

$$0.35 \times 0.90 \times 0.05 = 0.016$$

(1.6% efficiency)

Second law of thermodynamics The physical law stating that when energy is transformed, the quantity of energy remains the same, but its ability to do work diminishes.

Energy efficiency The ratio of the amount of energy expended in the form you want to the total amount of energy that is introduced into the system.

AP® Exam Tip

Make sure students understand the difference between the first law of thermodynamics and the second law of thermodynamics.

FIGURE 5.5 **Energy efficiency.** (a) The energy efficiency of a traditional fireplace is low because so much heated air can escape through the chimney. (b) A modern woodstove, which can heat a room using much less wood, is considerably more energy efficient. *(a: Sergey Karpov/Shutterstock; b: Andrew Brookes/Corbis)*

Energy Quality

Most of us have an intuitive sense about the relative effectiveness of various energy sources. For example, we realize that gasoline is a more useful source of energy than paper. This difference is a function of each material's **energy quality,** the ease with which an energy source can be used for work. A high-quality energy source has a convenient, concentrated form so that it does not take too much energy to move it from one place to another.

Gasoline, for example, is a high-quality energy source because its chemical energy is concentrated (about 44 MJ/kg), and because we have technology that can conveniently transport it from one location to another. In addition, it is relatively easy to convert gasoline energy into work and heat. Wood, on the other hand, is a lower-quality energy source. It has less than half the energy concentration of gasoline (about 20 MJ/kg) and is more difficult to use to do work. Imagine using wood to power an automobile. Clearly, gasoline is a higher-quality energy source than wood. Energy quality is one important factor humans must consider when they make energy choices. Considering that wood has less than half the energy content of gasoline, could you imagine driving a car powered by wood? It would not be practical for many reasons, including its lower energy quality.

Entropy

The second law of thermodynamics also tells us that all systems move toward randomness rather than toward order. This randomness in a system, called **entropy,** is

Energy quality The ease with which an energy source can be used for work.

Entropy Randomness in a system.

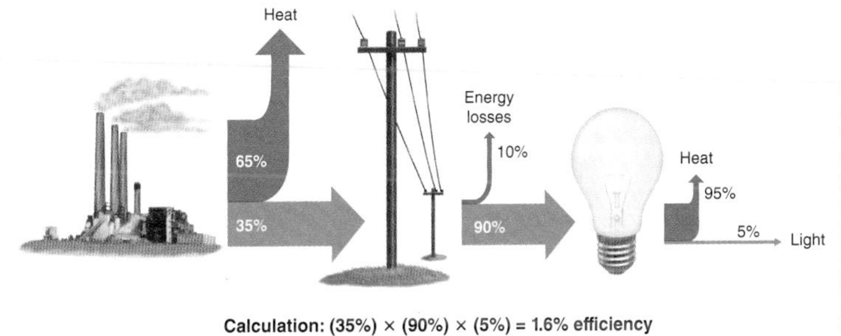

FIGURE 5.6 **The second law of thermodynamics.** Whenever one form of energy is transformed into another, some of that energy is converted into a less usable form of energy, such as heat. In this example, we see that the conversion of coal into the light of an incandescent bulb is only 1.6 percent efficient.

Teaching Tip: Discussion Starter

Ask students: What are some sources of energy other than gasoline and wood? *Answers include but are not limited to coal, natural gas, the Sun, nuclear energy, food, water, and wind.*

Teaching Tip: Connections

Have students name an example of entropy in a system. *Answers will vary, but may include a locker getting messy, a binder getting disorganized, or an individual getting cold in the winter without a warm coat.*

Teaching Tip: Engage

As class is ending, ask students to take out a sheet of paper and write down the following:

- three things that interested them about today's class
- two things about today's class that they would like to know more about
- one question that they still have about today's class

Collect their papers and review them after class to better understand students' grasp of the material.

AP® Exam Tip

System dynamics are covered in the 1999 AP® Exam, Question 1. To answer this question, students must

- use the scientific method to compare two local ponds
- name and describe three water quality tests that could be conducted on a pond, and explain how the water quality tests could be used to assess the distribution of aquatic organisms
- design a controlled experiment assessing the larvae of certain insects found at both ponds, and include a hypothesis, *x* and *y* variables, outline of field work procedures, and a conclusion
- define the term "indicator species" and explain how an indicator species is used to assess environmental quality

(a)

(b)

FIGURE 5.7 **Energy and entropy.** Entropy increases in a system unless an input of energy from outside the system creates order. (a) In order to reduce the entropy of this messy room, a human must expend energy, which comes from food. (b) A tornado has increased the entropy of this forest system in Wisconsin. *(a: Norm Betts/ Landov; b: AP Photo/The Post Crescent, Dan Powers)*

always increasing unless new energy from outside the system is added to create order.

Think of your bedroom as a system. At the start of the week, your books may be in the bookcase, your clothes may be in the dresser, and your shoes may be lined up in a row in the closet. But what happens if, as the week goes on, you don't expend energy to put your things away (FIGURE 5.7)? Unfortunately, your books will not spontaneously line up in the bookcase, your clothes will not fall folded into the dresser, and your shoes will not pair up and arrange themselves in the closet. Unless you bring energy into the system to put things in order, your room will slowly become more and more disorganized.

The energy you use to pick up your room comes from the energy stored in food. Food is a relatively high-quality energy source because the human body easily converts it into usable energy. The molecules of food are ordered rather than random. In other words, food is a low-entropy energy source. Only a small portion of the energy in your digested food is converted into work, however; the rest becomes body heat, which may or may not be needed. This waste heat has a high degree of entropy because heat is the random movement of molecules. Thus, in using food energy to power your body to organize your room, you are decreasing the entropy of the room, but increasing the entropy in the universe by producing waste body heat.

Another example of the second law can be found in the observation that energy always flows from hot to cold. A pot of water will never boil without an input of energy, but hot water left alone will gradually cool as its energy dissipates into the surrounding air. This application of the second law is important in many of the global circulation patterns that are powered by the energy of the Sun.

Matter and energy flow in the environment

Why do environmental scientists study whole systems rather than focusing on the individual plants, animals, or substances within a system? Imagine taking apart your cell phone and trying to understand how it works simply by focusing on the microphone. You wouldn't get very far. Similarly, it is important for environmental scientists to look at the whole picture and not just the individual parts of a system in order to understand how that system works. With a working knowledge of how a system functions, we can predict how changes to any part of the system—for example, changes in the water level at Mono Lake—will change the entire system.

Studying systems allows scientists to think about how matter and energy flow in the environment. In this way, researchers can learn about the complex relationships between organisms and the environment. In this section we will explore system dynamics and changes in systems across space and over time. In each case we will focus on how energy and matter flow in the environment.

System Dynamics

As we suggested at the beginning of this chapter, the Mono Lake ecosystem changes over time. Some years there is more algae growing in the lake and that feeds

Teaching Tip: Discussion Starter

Ask students: Why is it important to look at a whole system rather than only the parts of a system? *In order to understand how a given system works and functions, we need to know how all the parts interact with each other. A system is not only a sum of its parts, but is defined by the nature of the interactions among its various components.*

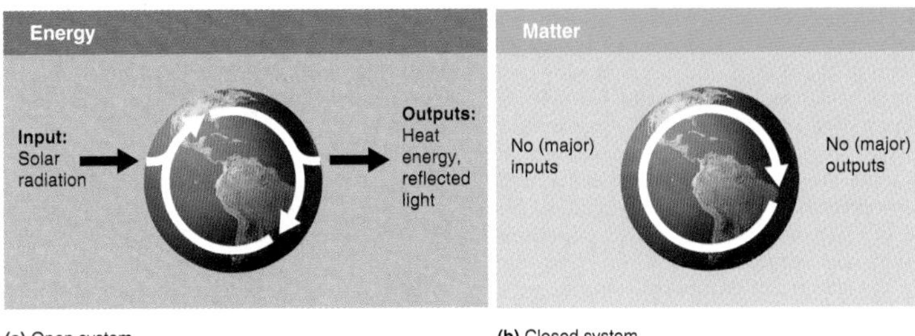

FIGURE 5.8 **Open and closed systems.** (a) Earth is an open system with respect to energy. Solar radiation enters the Earth system, and energy leaves it in the form of heat and reflected light. (b) Earth is essentially a closed system with respect to matter because very little matter enters or leaves Earth's system. The white arrows indicate the cycling of energy and matter.

more brine shrimp. Some years fewer migrating gulls stop over and feed, removing less matter and energy from the system, while in other years, more gulls stop over. These changing parameters describe the system dynamics of the Mono Lake ecosystem. There are a number of terms used to describe systems and we will present some of them in this section.

Open and Closed Systems

Systems can be either *open* or *closed*. In an **open system,** exchanges of matter or energy occur across system boundaries. Most systems are open. Even at remote Mono Lake, water flows in and birds fly to and from the lake. The ocean is also an open system. Energy from the Sun enters the ocean, warming the waters and providing energy to plants and algae. Energy and matter are transferred from the ocean to the atmosphere as energy from the Sun evaporates the water, giving rise to meteorological events such as tropical storms in which clouds form and send rain back to the ocean surface. Matter, such as sediment and nutrients, enters the ocean from rivers and streams and leaves it through geologic cycles and other processes.

In a **closed system,** matter and energy exchanges do not occur across system boundaries. Closed systems are less common than open systems. Some underground cave systems are almost completely closed systems.

As FIGURE 5.8 shows, Earth is an open system with respect to energy. Solar radiation enters Earth's atmosphere, and heat and reflected light leave it. But because of its gravitational field, Earth is essentially a closed system with respect to matter. Only an insignificant amount of material enters or leaves the Earth system. All important material exchanges occur within the system.

Inputs and Outputs

By now you have seen numerous examples of both **inputs,** which are additions to a given system, and **outputs,** which are losses from the system. People who study systems often conduct a **systems analysis,** in which they determine inputs, outputs, and changes in the system under various conditions. For instance, researchers studying Mono Lake might quantify the inputs to that system—such as water and salts—and the outputs—such as water that evaporates from the lake and brine shrimp removed by migratory birds. Because no water flows out of the lake, salts are not removed, and even without the aqueduct, Mono Lake, like other terminal lakes, would slowly become saltier.

Steady States

In any given period at Mono Lake, the same amount of water that enters the lake eventually evaporates. In many cases, the most important aspect of conducting a systems analysis is determining whether your system is in **steady state**—that is, whether inputs equal outputs, so that the system is not changing over time. This information is particularly useful in the study of environmental science. For example, it allows us to know whether the amount of a valuable resource or a harmful pollutant is increasing, decreasing, or staying the same.

Open system A system in which exchanges of matter or energy occur across system boundaries.

Closed system A system in which matter and energy exchanges do not occur across boundaries.

Input An addition to a system.

Output A loss from a system.

Systems analysis An analysis to determine inputs, outputs, and changes in a system under various conditions.

Steady state A state in which inputs equal outputs, so that the system is not changing over time.

Teaching Tip: Activity

Divide students into pairs or groups, then distribute the name of a system, such as one of the following, to each group: pond, beehive, beaver lodge, cave, river, farm, power plant, human or other organism, household, car, or bank. Have students sketch the system, and then list the inputs coming in from the left, and the outputs going out to the right. Their diagrams should look something like the one shown below. Students can share their diagrams.

You can extend this activity by asking students what they should measure and compare in order to find out whether the system they drew is in a steady state. *In the example of a cow, one could measure and compare the water the cow drinks against total water released in urine, sweat, and milk, or nutrients eaten in grass against nutrients lost in waste.*

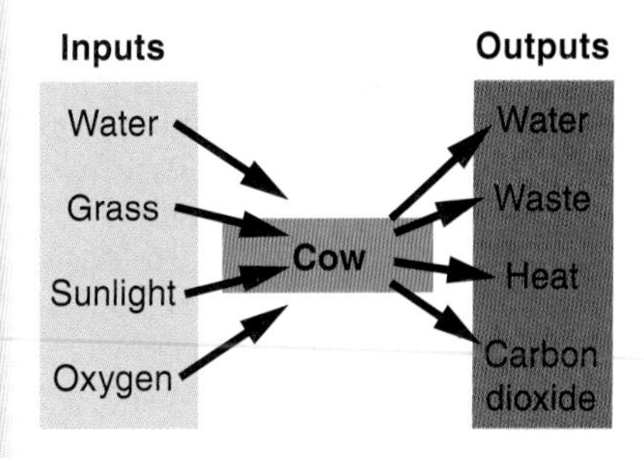

Teaching Tip: Discussion Starter

Ask students: What is a steady state? *A steady state occurs when inputs equal outputs in a system, so that the system is not changing over time.*

Teaching Tip: Discussion Starter

Ask students: What is the difference between an open system and a closed system? What is an example of each? *In an open system, exchanges of matter or energy occur across system boundaries; for example, a lake interacts with the surrounding land and atmosphere. In a closed system—for example, an underground cave—matter and energy exchanges do not occur across boundaries.*

Teaching Tip: Discussion Starter

Ask students: What are feedback loops? What is the difference between a positive feedback loop and a negative feedback loop? *Feedback loops occur when the results of a process return back to the system and change the rate of that process. Negative feedback loops provide a mechanism by which organisms or ecological systems can return to a steady state when faced with stresses or disturbances. Positive feedback loops amplify a change in the system.*

COMMON MISCONCEPTIONS
Positive and negative feedback loops

Students often assume that positive feedback loops must be good and negative feedback loops must be bad. This is a topic that will come back many times during the year, so be sure students understand these concepts now. You may want to explain feedback loops in relation to the human body. When they exercise hard, their bodies sweat. This sweating cools the body off and returns it to homeostasis, or to the original state. This response is an example of a negative feedback loop that is beneficial. A positive feedback loop occurs when a change to a system is amplified. For example, when a woman goes into labor, the first contraction causes the next contraction to get stronger, which then causes the next contraction to get stronger, and so on. This is a positive feedback loop and usually ends in a stronger or added result.

The first step in determining whether a system is in steady state is to measure the amount of matter and energy within it. If the scale of the system allows, we can perform these measurements directly. Consider the leaky bucket shown in FIGURE 5.9. We can measure the amount of water going into the bucket and the amount of water flowing out through the holes in the bottom. However, some properties of systems, such as the volume of a lake or the size of an insect population, are difficult to measure directly, so we must calculate or estimate the amount of energy or matter stored in the system. We can then use this information to determine the inputs to and outputs from the system to determine whether it is in steady state.

Many aspects of natural systems, such as the water vapor in the global atmosphere, have been in steady state for at least as long as we have been studying them. The amount of water that enters the atmosphere by evaporation from oceans, rivers, and lakes is roughly equal to the amount that falls from the atmosphere as precipitation. Until recently, the oceans have also been in steady state: The amount of water that enters from rivers and streams has been roughly equal to the amount that evaporates into the air. One concern about the effects of global climate change is that some global systems, such as the system that includes water balance in the oceans and atmosphere, may no longer be in steady state.

It's interesting to note that one part of a system can be in steady state while another part is not. Before the Los Angeles Aqueduct was built, the Mono Lake system was in steady state with respect to water but not with respect to salt. The inflow of water equaled the rate of water evaporation but salt was slowly accumulating, as it does in all terminal lakes.

Feedbacks

Most natural systems are in steady state. Why? A natural system can respond to changes in its inputs and outputs. For example, during a period of drought, evaporation from a lake will be greater than combined precipitation and stream water flowing into the lake. Therefore, the lake will begin to dry up. Soon there will be less surface water available for evaporation, so the evaporation rate will continue to fall until it matches the new, lower precipitation rate. When this happens, the system returns to steady state, and the lake stops shrinking.

Negative feedback loop A feedback loop in which a system responds to a change by returning to its original state, or by decreasing the rate at which the change is occurring.

Positive feedback loop A feedback loop in which change in a system is amplified.

Of course, the opposite is also true. In very wet periods, the size of the lake will grow, and evaporation from the expanded surface area will continue to increase until the system returns to a steady state at which inputs and outputs are equal.

Adjustments in input or output rates caused by changes to a system are called feedbacks; the results of a process feed back into the system to change the rate of that process. Feedbacks, which can be diagrammed as loops or cycles, are found throughout the environment.

Feedback can be either negative or positive. In natural systems, scientists most often observe **negative feedback loops,** in which a system responds to a change by returning to its original state, or by decreasing the rate at which the change is occurring. FIGURE 5.10a shows how the negative feedback loop for Mono Lake works: When water levels drop, there is less lake surface area, so evaporation decreases. With less evaporation, the water in the lake slowly returns to its original volume.

Positive feedbacks also occur in the natural world. Figure 5.10b shows an example of how births in a population can give rise to a **positive feedback loop** in which change in a system is amplified. The more members of a species that can reproduce, the more births there will be, creating even more of the species to give birth, and so on.

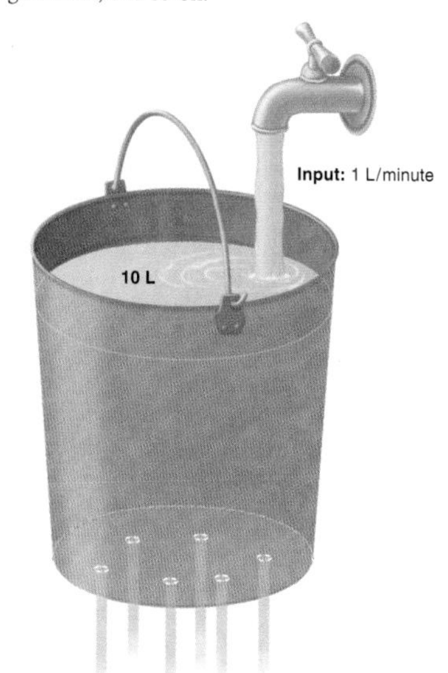

FIGURE 5.9 **A system in steady state.** In this leaky bucket, inputs equal outputs. As a result, there is no change in the total amount of water in the bucket; the system is in steady state.

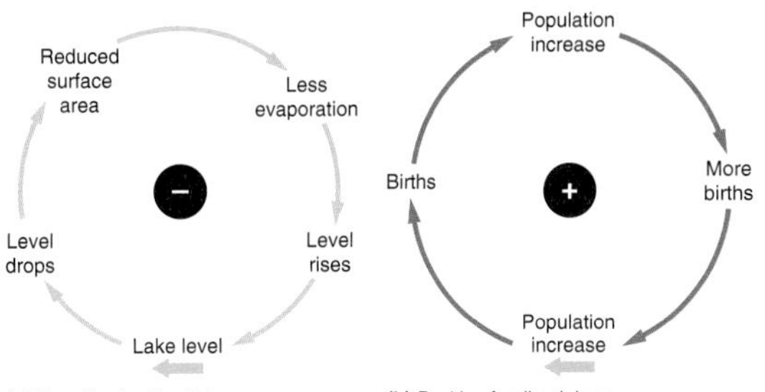

(a) Negative feedback loop (b) Positive feedback loop

FIGURE 5.10 **Negative and positive feedback loops.** (a) A negative feedback loop occurs at Mono Lake: When the water level drops, the lake surface area is reduced and evaporation decreases. As a result of the decrease in evaporation, the lake level rises again. (b) Population growth is an example of positive feedback. As members of a species reproduce, they create more offspring that will be able to reproduce in turn, creating a cycle that increases the population size. The green arrow indicates the starting point of each cycle.

It's important to note that positive and negative here do not mean good and bad; instead, positive feedback amplifies changes, whereas negative feedback resists changes. People often talk about the balance of nature. That balance is the logical result of systems reaching a state at which negative feedbacks predominate—although positive feedback loops play important roles in environmental systems as well.

Environmental scientists are especially concerned with the extent to which Earth's temperature is regulated by feedback loops. Understanding the role of feedback loops in temperature regulation, as well as the types of feedbacks and their scale, can help us make better predictions about climatic changes in the coming decades.

In general, warmer temperatures at Earth's surface increase the evaporation of water. The additional water vapor that enters the atmosphere by evaporation causes two kinds of clouds to form. Low-altitude clouds reflect sunlight back into space. The result is less heating of Earth's surface, less evaporation, and less warming—a negative feedback loop. High-altitude clouds, on the other hand, absorb terrestrial energy that might otherwise have escaped the atmosphere, leading to higher temperatures near Earth's surface, more evaporation of water, and more warming—a positive feedback loop. In the absence of other factors that compensate for or balance the warming, this positive feedback loop will continue making temperatures warmer, driving the system further away from its starting point. This and other potential positive feedback loops may play critical roles in climate change.

The health of many environmental systems depends on the proper operation of feedback loops. Sometimes, natural or anthropogenic factors lead to a breakdown in a negative feedback loop and drive an environmental system away from its steady state. As you study the exploitation of natural resources, try to determine what factors may be disrupting the negative feedback loops of the systems that provide those resources.

Change Across Space and over Time

Differences in environmental conditions affect what grows or lives in an area, which creates geographic variation among natural systems. Variations in temperature, precipitation, or soil composition across a landscape can lead to vastly different numbers and types of organisms. In Texas, for example, sycamore trees grow in river valleys where there is plenty of water available, whereas pine trees dominate mountain slopes because they can tolerate the cold, dry conditions there. Paying close attention to these natural variations may help us predict the effect of any change in an environment. We know that if the rivers that support the sycamores in Texas dry up, then the trees will probably die.

Natural systems are also affected by the passage of time. Thousands of years ago, when the climate of the Sahara was much wetter than it is today, it supported large populations of Nubian farmers and herders. Small changes in Earth's orbit relative to the Sun, along with a series of other factors, led to the disappearance of monsoon rains in northern Africa. As a result of these changes, the Sahara—now a desert nearly the size of the continental United States—became one of Earth's driest regions (FIGURE 5.11). Other, more dramatic changes have occurred on the planet. In the last few million

FIGURE 5.11 **The Sahara.** The Sahara desert was once a lush grassland that dried up over time. *(Rachel Carbonell/Getty Images)*

Teaching Tip: Connections

Have students examine the positive and negative feedback loops in Figure 5.10, and ask them to create their own diagram using another example. Alternatively, provide students with the following examples and ask them to determine whether the feedback is positive or negative. Examples may include the following:

- **Negative**: body temperature, blood sugar, blood pressure, predation pressure on prey populations
- **Positive**: population growth, ice melting when the atmospheric temperature rises, labor contractions, melting of permafrost that releases methane

Teaching Tip: Discussion Starter

Ask students: Why is it important to study variations in natural systems over space and time? *Differences in environmental conditions affect what grows or lives in an area. Studying these natural variations helps scientists understand correlations between species distribution and landscape variation. With this understanding, scientists can better predict the effect of changes in a given ecosystem. Natural systems are also affected by the passage of time. Studying how natural systems respond to changes in the global environment over time can help scientists predict impacts from the alterations humans are making to the world today.*

Teaching Tip: Discussion Starter

Ask students: What are some examples of environmental conditions that might vary among natural systems? *Examples include temperature, precipitation, soil composition, topography, and altitude.*

years, Earth has moved in and out of several ice ages; 70 million years ago, central North America was covered by a sea; 240 million years ago, Antarctica was warm enough for 2-meter-long (6.6-foot) salamander-like amphibians to roam its swamps. Natural systems respond to such changes in the global environment with migrations and extinctions of species as well as the evolution of new species.

Throughout Earth's history, small natural changes have had large effects on complex systems, but human activities have increased both the pace and the intensity of these natural environmental changes, as they did at Mono Lake. Studying variations in natural systems over space and time can help scientists learn more about what to expect from the alterations humans are making to the world today.

module 5

REVIEW

In this module, we have seen that energy is a fundamental component of environmental systems and that there are different forms of energy. The first and second laws of thermodynamics describe energy behavior. Matter and energy flow within and between systems and are subject to feedbacks that regulate and influence the behavior of systems. Natural systems change, sometimes quickly, sometimes slowly, and change can be natural or caused by human beings.

Module 5 AP® Review Questions

1. If a solar photovoltaic panel produces 1,000 watts of electrical energy and is active for 12 hours each day, how many kWh of electricity will be produced in a week?
   (a) 63 kWh
   (b) 12 kWh
   (c) 84 kWh
   (d) 70 kWh
   (e) 7 kWh

2. A car traveling down the highway represents
   (a) kinetic energy.
   (b) electromagnetic radiation.
   (c) potential energy.
   (d) chemical energy.
   (e) entropy.

3. The concept of energy efficiency is used to quantify
   (a) the first law of thermodynamics.
   (b) the second law of thermodynamics.
   (c) conservation of matter.
   (d) energy quality.
   (e) the third law of thermodynamics.

4. A terminal lake like Mono Lake is an example of
   (a) a closed system.
   (b) an open system with only inputs.
   (c) an open system with only outputs.
   (d) an open system with both inputs and outputs.
   (e) a closed system in steady state.

5. Which of the following will be most likely to return to steady state after a disturbance?
   (a) A system with mostly positive feedback loops
   (b) A system with mostly negative feedback loops
   (c) A system with the same number of positive and negative feedback loops
   (d) An open system with many inputs and outputs
   (e) A closed system

6. Entropy is
   (a) energy of a system that is stored in molecular movement.
   (b) the amount of heat in a system.
   (c) the lowest level of energy quality.
   (d) the ease with which an energy source can be used for work.
   (e) the randomness of a system.

Answers to Module 5 AP® Review Questions

1. c
2. a
3. b
4. d
5. b
6. e

Managing Environmental Systems in the Florida Everglades

South Florida's vast Everglades ecosystem extends over 5,000,000 ha (12,400,000 acres). The region, which includes the Everglades and Biscayne Bay national parks and Big Cypress National Preserve, is home to many threatened and endangered bird, mammal, reptile, and plant species, including the Florida panther (*Puma concolor coryi*) and the Florida manatee (*Trichechus manatus latirostris*). The 400,000 ha (988,000 acre) subtropical wetland area for which the region is best known has been called a "river of grass" because a thin sheet of water flows constantly through it, allowing tall water-tolerant grasses to grow.

A hundred years of rapid human population growth, and the resulting need for water and farmland, have had a dramatic impact on the region. Flood control, dams, irrigation, and the need to provide fresh water to Floridians have led to a 30 percent decline in water flow through the Everglades. Much of the water that does flow through the region is polluted by phosphorus-rich fertilizer and waste from farms and other sources upstream. Cattails thrive on the input of phosphorus, choking out other native plants. The reduction in water flow and water quality is, by most accounts, destroying the Everglades. Can we save this natural system while still providing water to the people who need it?

The response of scientists and policy makers has been to treat the Everglades as a set of interacting systems and to manage the inputs and outputs of water and pollutants to those systems. The Comprehensive Everglades Restoration Plan of 2000 is a systems-based approach to the region's problems. It covers 16 counties and 46,600 km^2 (11,500,000 acres) of South Florida. The plan is based on three key steps: increasing water flow into the Everglades, reducing pollutants coming in, and developing strategies for dealing with future problems.

The first step—increasing water flow—will counteract some of the effects of decades of drainage by local communities. Its goal is to provide enough water to support the Everglades' aquatic and marsh organisms. The plan calls for restoring natural water flow as well as natural hydroperiods (seasonal increases and decreases in water flow). Its strategies include removal of over 390 km (240 miles) of inland levees, canals, and water control structures that have blocked this natural water movement.

Water conservation will also be a crucial part of reaching this goal. New water storage facilities and restored wetlands will capture and store water during rainy seasons for use during dry seasons, redirecting much of the 6.4 billion liters (1.7 billion gallons) of fresh water that currently flow to the ocean every day. About 80 percent of this fresh water will be redistributed back into the ecosystem via wetlands and aquifers. The remaining water will be used by cities and farms. The federal and state governments also hope to purchase nearby irrigated cropland and return it to a more natural state. In 2009, for example, the state of Florida purchased 29,000 ha (71,700 acres) of land from the United States Sugar Corporation, the first of a number of actions that will allow engineers to restore the natural flow of water from Lake Okeechobee into the Everglades. Florida is currently negotiating to purchase even more land from United States Sugar. In 2013, pilot projects for water storage in Lake Okeechobee were underway.

To achieve the second goal—reducing water pollution—local authorities will improve waste treatment facilities and place restrictions on the use of agricultural chemicals. Marshlands are particularly effective at absorbing nutrients and breaking down toxins. Landscape engineers have designed and built more than 21,000 ha (52,000 acres) of artificial marshes upstream of the Everglades to help clean water before it reaches Everglades National Park. Although not all of the region has seen water quality improvements, phosphorus concentrations in runoff from farms south of Lake Okeechobee are lower, meaning that fewer pollutants are reaching the Everglades.

The third goal—to plan for addressing future problems—requires an **adaptive management plan:**

River of grass. The subtropical wetland portion of the Florida Everglades has been described as a river of grass because of the tall water-tolerant grasses that cover its surface. *(Philip Lange/Shutterstock)*

Adaptive management plan A plan that provides flexibility so that managers can modify it as changes occur.

Teaching Tip: Beyond the Classroom

Have students pick two cars and look up the miles per gallon (mpg) ratings of each. One could be a large SUV and the other a hybrid. Then, calculate the amount of gasoline that would be needed in each vehicle for a trip from Los Angeles to New York City (approximately a distance of 3,000 miles). Discuss the amount of energy used and see what ideas they have about gasoline and the impacts it has on the environment.

a strategy that provides flexibility so that managers can modify it as future changes occur. Adaptive management is an answer to scientific uncertainty. In a highly complex system such as the Everglades, any changes, however well intentioned, may have unexpected consequences. Management strategies must adapt to the actual results of the restoration plan as they occur. In addition, an adaptive management plan can be changed to meet new challenges as they come. One such challenge is global warming. As the climate warms, glaciers melt and sea levels rise, so much of the Everglades could be inundated by seawater, which would destroy freshwater habitat. Adaptive management essentially means paying attention to what works and adjusting methods accordingly. The Everglades restoration plan will be adjusted along the way to take the results of ongoing observations into account, and it has put formal mechanisms in place to ensure that this will occur.

The Everglades plan has its critics. Some people are concerned that control of water flow and pollution will restrict the use of private property and affect economic development, possibly even harming the local economy. Yet other critics fear that the restoration project is underfunded or moving too slowly, and that current farming practices in the region are inconsistent with the goal of restoration.

In spite of its critics, the Everglades restoration plan is, historically speaking, a milestone project, not least because it is based on the concept that the environment is made up of interacting systems.

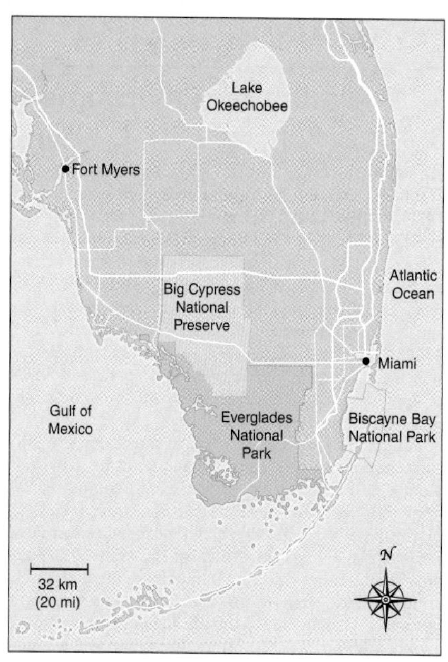

The Florida Everglades Ecosystem. This map shows the locations of Lake Okeechobee and the broader Everglades ecosystem, which includes Everglades and Biscayne Bay national parks and Big Cypress National Preserve.

Critical Thinking Questions

1. Why are the Florida Everglades environmentally significant?
2. How does your understanding of the Florida Everglades change when you think of the Everglades as a set of interacting systems?
3. What are some adaptive management strategies utilized in the Florida Everglades?

References

Kiker, C., W. Milon, and A. Hodges. 2001. South Florida: The reality of change and the prospects for sustainability. Adaptive learning for science-based policy: The Everglades restoration. *Ecological Economics* 37:403–416.

The Comprehensive Everglades Restoration Plan (CERP) Website. http://www.evergladesplan.org/. Accessed 25 September 2013.

Suggested Answers to Critical Thinking Questions

1. The Everglades region is a unique habitat and is home to many endangered and threatened species. It also serves many important functions as a large subtropical wetland area.

2. In thinking of the Everglades as a system, one considers how water use and water pollution in the surrounding areas of Florida have such a large impact on the Everglades and why managing the whole area is necessary for the restoration of the Everglades.

3. Adjusting the plan to account for rising sea levels due to global warming is one example of an adaptive management strategy.

chapter 2 REVIEW

Throughout this chapter, we have examined environmental systems. Earth is one large interconnected system. Components of the system follow basic principles of chemistry and biology. Energy is an important component of these systems. Energy conversions are frequently used in systems analysis. Natural systems change over space and time and humans are sometimes major actors in causing system change.

Teaching Tip: Measuring Your Impact

Bottled Water versus Tap Water
Measuring Your Impact 2 asks students to compare the energy use and per capita cost of bottled water versus tap water in the United States. Download this resource by clicking on the link buttons in the TE-book, opening the TRFD, or logging onto the book's companion website bcs.whfreeman.com/friedlandapes2e (teacher login required).

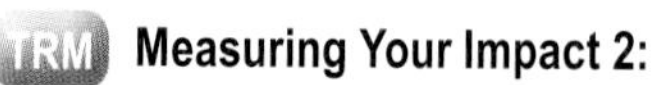

Measuring Your Impact 2:
Bottled Water versus Tap Water

Key Terms

Matter
Mass
Atom
Element
Periodic table
Molecule
Compound
Atomic number
Mass number
Isotopes
Radioactive decay
Half-life
Covalent bond
Ionic bond
Hydrogen bond
Polar molecule
Surface tension
Capillary action
Acid
Base
pH
Chemical reaction
Law of conservation of matter
Inorganic compound
Organic compound
Carbohydrate
Protein
Nucleic acid
DNA (deoxyribonucleic acid)
RNA (ribonucleic acid)
Lipid
Cell
Energy
Electromagnetic radiation
Photon
Potential energy
Kinetic energy
Chemical energy
Joule (J)
Power
Temperature
First law of thermodynamics
Second law of thermodynamics
Energy efficiency
Energy quality
Entropy
Open system
Closed system
Input
Output
Systems analysis
Steady state
Negative feedback loop
Positive feedback loop

Learning Objectives Revisited

Module 4 Systems and Matter

- **Describe how matter comprises atoms and molecules that move among different systems.**
  Matter is composed of atoms, which are made up of protons, neutrons, and electrons. Atoms and molecules can interact in chemical reactions in which the bonds between particular atoms may change.
- **Explain why water is an important component of most environmental systems.**
  Water facilitates the transfer of chemical elements and compounds from one system to another. The molecular structure of water gives it unique properties that support the conditions necessary for life on Earth. These properties are essential to physiological functioning of plants and animals and the movement of elements through systems.
- **Discuss how matter is conserved in chemical and biological systems.**
  Matter cannot be created or destroyed, but its form can be changed within chemical and biological systems. This is part of the reason we cannot easily dispose of certain chemical compounds, such as hazardous materials.

Module 5 Energy, Flows, and Feedbacks

- **Distinguish among various forms of energy and understand how they are measured.**
  Energy can take various forms, including energy that is stored (potential energy) and the energy of motion (kinetic energy). Joules and calories are two important energy units.
- **Discuss the first and second laws of thermodynamics and explain how they influence environmental systems.**
  The first law of thermodynamics states that energy cannot be created or destroyed, but it can be converted from one form into another. The second law of thermodynamics states that in any conversion of energy, some energy is converted into unusable waste energy, and the entropy of the universe is increased. The quantities and forms of energy present in various systems influence the types of organisms in those systems.
- **Explain how scientists keep track of energy and matter inputs, outputs, and changes to environmental systems.**
  Systems can be open or closed to exchanges of matter, energy, or both. A systems analysis determines what goes into, what comes out of, and what has changed within a given system. Environmental scientists use systems analysis to calculate inputs to and outputs from a system and its rate of change. If there is no overall change, the system is in steady state. Changes in one input or output can affect the entire system.

Exploring the Literature

Hart, J. 1996. *Storm Over Mono: The Mono Lake Battle and the California Water Future.* University of California Press.

Kendall, K. E., and J. E. Kendall. 2008. *Systems Analysis and Design.* 7th ed. Prentice Hall.

Meadows, D. 2008. *Thinking in Systems: A Primer.* Chelsea Green.

Moore, J. T. E. 2005. *Chemistry Made Simple.* Broadway Press.

Zolli, A., and A.M. Healy. 2013. *Resilience: Why Things Bounce Back.* Simon & Schuster.

Download a printable version of this list by clicking on the link buttons in the TE-book, opening the TRFD, or logging onto the book's companion website bcs.whfreeman.com/friedlandapes2e (teacher login required).

Exploring the Literature 2

Answers to Chapter 2 AP® Exam Multiple-Choice Questions

1. b
2. d
3. e
4. c
5. a
6. b
7. e
8. c
9. b
10. e
11. d
12. c
13. a
14. e

Answers to Chapter 2 AP® Exam

Download the full answers to the Chapter 2 AP® Practice Exam by clicking on the link buttons in the TE-book, opening the TRFD, or logging onto the book's companion website bcs.whfreeman.com /friedlandapes2e (teacher login required).

TRM **Answers to Chapter 2 AP® Exam**

Chapter 2 AP® Environmental Science Practice Exam

Section 1: Multiple-Choice Questions

Choose the best answer for questions 1–14

1. Which statement about atoms and molecules is correct?
   (a) The mass number of an element is always less than its atomic number.
   (b) Isotopes are the result of varying numbers of neutrons in atoms of the same element.
   (c) Ionic bonds involve electrons while covalent bonds involve protons.
   (d) Inorganic compounds never contain the element carbon.
   (e) Protons and electrons have roughly the same mass.

2. Which of the following does NOT demonstrate the law of conservation of matter?
   (a) $CH_4 + 2\,O_2 \rightarrow CO_2 + 2\,H_2O$
   (b) $NaOH + HCl \rightarrow NaCl + H_2O$
   (c) $2\,NO_2 + H_2O \rightarrow HNO_3 + HNO_2$
   (d) $PbO + C \rightarrow 2\,Pb + CO_2$
   (e) $C_6H_{12}O_6 + 6\,O_2 \rightarrow 6\,CO_2 + 6\,H_2O$

3. Pure water has a pH of 7 because
   (a) its surface tension equally attracts acids and bases.
   (b) its polarity results in a molecule with a positive and a negative end.
   (c) its ability to dissolve carbon dioxide adjusts its natural pH.
   (d) its capillary action attracts it to the surfaces of solid substances.
   (e) its H^+ concentration is equal to its OH^- concentration.

4. Which of the following is NOT a type of organic biological molecule?
   (a) Lipids
   (b) Carbohydrates
   (c) Salts
   (d) Nucleic acids
   (e) Proteins

5. A wooden log that weighs 1.00 kg is placed in a fireplace. Once lit, it is allowed to burn until there are only traces of ash, weighing 0.04 kg, left. Which of the following best describes the flow of energy?
   (a) The potential energy of the wooden log was converted into the kinetic energy of heat and light.
   (b) The kinetic energy of the wooden log was converted into 0.04 kg of ash.
   (c) The potential energy of the wooden log was converted into 1.00 J of heat.
   (d) Since the ash weighs less than the wooden log, matter was converted directly into energy.
   (e) The burning of the 1.00 kg wooden log produced 0.96 kg of gases and 0.04 kg of ash.

6. Consider a power plant that uses natural gas as a fuel to generate electricity. If there are 10,000 J of chemical energy contained in a specified amount of natural gas, then the amount of electricity that could be produced would be
   (a) greater than 10,000 J because electricity has a higher energy quality than natural gas.
   (b) something less than 10,000 J, depending on the efficiency of the generator.
   (c) greater than 10,000 J when energy demands are highest; less than 10,000 J when energy demands are lowest.
   (d) greater than 10,000 J because of the positive feedback loop of waste heat.
   (e) equal to 10,000 J because energy cannot be created or destroyed.

7. A lake that has been affected by acid rain has a pH of 4. How many more times acidic is the lake water than seawater? (See Figure 4.7 on page 39.)
   (a) 4
   (b) 10
   (c) 100
   (d) 1,000
   (e) 10,000

8. An automobile with an internal combustion engine converts the potential energy of gasoline (44 MJ/kg) into the kinetic energy of the moving pistons. If the average internal combustion engine is 10 percent efficient and 1 kg of gasoline is combusted, how much potential energy is converted into energy to run the pistons?
   (a) 39.6 MJ
   (b) 20.0 MJ
   (c) 4.4 MJ
   (d) Depends on the capacity of the gas tank
   (e) Depends on the size of the engine

9. If the average adult woman consumes approximately 2,000 kcal per day, how long would she need to run in order to utilize 25 percent of her caloric intake, given that the energy requirement for running is 42,000 J per minute?
   (a) 200 minutes
   (b) 50 minutes
   (c) 5 minutes
   (d) 0.05 minutes
   (e) 0.012 minutes

10. The National Hurricane Center studies the origins and intensities of hurricanes over the Atlantic and Pacific oceans and attempts to forecast their tracks, predict where they will make landfall, and assess what damage will result. Its systems analysis involves
    (a) changes within a closed system.
    (b) inputs and outputs within a closed system.
    (c) outputs only within an open system.
    (d) inputs from a closed system and outputs in an open system.
    (e) inputs, outputs, and changes within an open system.

Additional Multiple-Choice AP® Practice Exam

Download an additional Multiple-Choice AP® Environmental Science Practice Exam for Chapter 2 by clicking on the link buttons in the TE-book, opening the TRFD, or logging onto the book's companion website bcs.whfreeman.com /friedlandapes2e (teacher login required).

Multiple-Choice AP® Practice Exam 2

11. Based on the graph below, which of the following is the best interpretation of the data?

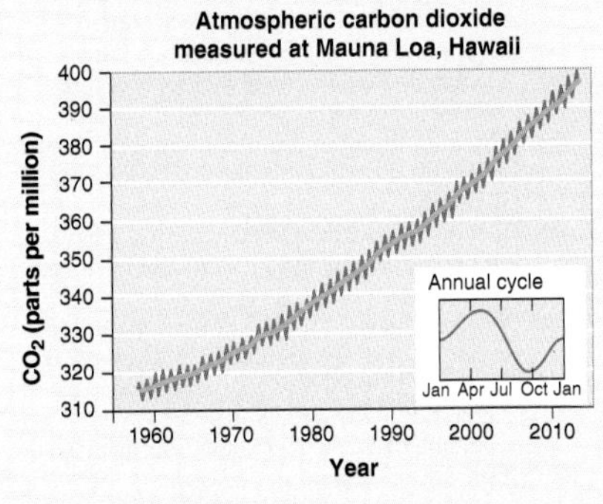

(a) The atmospheric carbon dioxide concentration is in steady state.
(b) The output of carbon dioxide from the atmosphere is greater than the input into the atmosphere.
(c) The atmospheric carbon dioxide concentration appears to be decreasing.
(d) The input of carbon dioxide into the atmosphere is greater than the output from the atmosphere.
(e) The atmospheric carbon dioxide concentration will level off due to the annual cycle.

12. Study the diagram below and select the concept it represents.

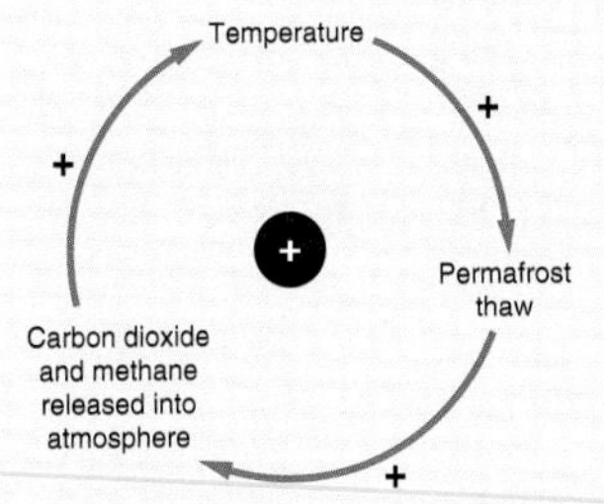

(a) A negative feedback loop, because melting of permafrost has a negative effect on the environment by increasing the amounts of carbon dioxide and methane in the atmosphere.
(b) A closed system, because only the concentrations of carbon dioxide and methane in the atmosphere contribute to the permafrost thaw.
(c) A positive feedback loop, because more carbon dioxide and methane in the atmosphere result in greater permafrost thaw, which releases more carbon dioxide and methane into the atmosphere.
(d) An open system that resists change and regulates global temperatures.
(e) Steady state, because inputs and outputs are equal.

13. Which of the following would represent a system in steady state?
I. The birth of chameleons on the island of Madagascar equals their death rate.
II. Evaporation from a lake is greater than precipitation and runoff flowing into the lake.
III. The steady flow of the Colorado River results in more erosion than deposition of rock particles.

(a) I only
(b) II only
(c) III only
(d) I and II
(e) I and III

14. Which of the following statements about the Comprehensive Everglades Restoration Plan is NOT correct?
(a) Human and natural systems interact because feedback loops lead to adaptations and changes in both systems.
(b) Water conservation will alter land uses and restore populations of aquatic and marsh organisms.
(c) Improvements in waste treatment facilities and restrictions on agricultural chemicals will reduce the nutrients and toxins in the water that reaches the Everglades.
(d) Adaptive management will allow for the modification of strategies as changes occur in this complex system.
(e) The Florida Everglades is a closed system that includes positive and negative feedback loops and is regulated as such.

Chapter 2 AP® Environmental Science Practice Exam 59

Answers to Chapter 2 AP® Exam Free-Response Questions

1. (a) (i) (nuclear decay to electricity) × (transport of electricity) × (fluorescent light efficiency) = (overall efficiency)

$(0.35 \times 0.90) \times 0.22 = 0.069$
$= 6.9\%$ efficiency

(ii) 6.7×10^{13} J/kg × 0.069 (1 kg U-235) $= 4.6 \times 10^{12}$ J

(iii) The efficiency of this system would be improved by using the waste heat or decreasing the amount of heat and sound lost as the energy is transported from the power plant to the home or office using the light. It could also be improved by using a more efficient light bulb.

(b) The first law of thermodynamics states that energy cannot be created or destroyed, but it can be converted from one form to another. This explains how it is possible to take the energy from the fission of uranium-235 and convert it to the energy used in fluorescent lighting. The second law of thermodynamics states that when energy is transformed, some of that energy is converted to a less usable form, usually heat, and the capacity of that energy to do work diminishes. This explains why no conversion from one energy form to another is 100 percent efficient and why less usable energy is available for applications such as the fluorescent light.

(c) 10 half-lives × 704 million years/half-life = 7,040 million years or 7.04 billion years

Answers continue in the Answer Appendix on page ANS-1.

Section 2: Free-Response Questions

Write your answer to each part clearly. Support your answers with relevant information and examples. Where calculations are required, show your work.

1. The atomic number of uranium-235 is 92, its half-life is 704 million years, and the radioactive decay of 1 kg of ^{235}U releases 6.7×10^{13} J. Radioactive material must be stored in a safe container or buried deep underground until its radiation output drops to a safe level. Generally, it is considered "safe" after 10 half-lives.
   (a) Assume that a nuclear power plant can convert energy from ^{235}U into electricity with an efficiency of 35 percent, the electrical transmission lines operate at 90 percent efficiency, and fluorescent lights operate at 22 percent efficiency.
      (i) What is the overall efficiency of converting the energy of ^{235}U into fluorescent light? (2 points)
      (ii) How much energy from 1 kg of ^{235}U is converted into fluorescent light? (2 points)
      (iii) Name one way in which you could improve the overall efficiency of this system. Explain how your suggestion would improve efficiency. (2 points)
   (b) What are the first and second laws of thermodynamics? (2 points)
   (c) How long would it take for the radiation from a sample of ^{235}U to reach a safe level? (2 points)

2. U.S. wheat farmers produce, on average, 3,000 kg of wheat per hectare. Farmers who plant wheat year after year on the same fields must add fertilizers to replace the nutrients removed by the harvested wheat. Consider a wheat farm as an open system.
   (a) Identify two inputs and two outputs of this system. (4 points)
   (b) Using one input to one output from (a), diagram and explain one positive feedback loop. (2 points)
   (c) Identify two adaptive management strategies that could be employed if a drought occurred. (2 points)
   (d) Wheat contains about 2.5 kcal per gram, and the average U.S. male consumes 2,500 kcal per day. How many hectares of wheat are needed to support one average U.S. male for a year, assuming that 30 percent of his caloric intake is from wheat? (2 points)

Unit 1 AP® Environmental Science Practice Exam

Section 1: Multiple-Choice Questions

Choose the best answer for questions 1–20.

1. Which best describes how humans have altered natural systems?
   I. Overhunted many large mammals to extinction.
   II. Created habitat for species to thrive.
   III. Emitted greenhouse gases.
   (a) I only
   (b) I and II only
   (c) II and III only
   (d) I and III only
   (e) I, II, and III

2. Which does NOT describe a benefit of biodiversity?
   (a) Genetic biodiversity improves the ability of a population to cope with environmental change.
   (b) Ecosystems with higher species diversity are more productive.
   (c) Species serve as environmental indicators of global-scale problems.
   (d) Speciation reduces natural rates of species extinction.
   (e) Humans rely on ecological interactions among species to produce ecosystem services.

3. Which of the following is NOT a consequence of human population growth?
   (a) Depletion of natural resources
   (b) Background extinction
   (c) Emission of greenhouse gases
   (d) Rise in sea level
   (e) Reduction in per capita food supply

4. An example of sustainable development is
   (a) harvesting enough crops to provide the basic needs of all humans.
   (b) increasing the price of vegetables.
   (c) reducing the use of all major modes of transportation.
   (d) creating renewable sources of construction material.
   (e) enforcing laws that stop future development of cities.

5. The ecological footprint of a human is
   (a) a measure of how much a human consumes, expressed in joules.
   (b) a measure of human consumption, expressed in area of land.
   (c) a measure of biodiversity loss stemming from industrial processes.
   (d) a measure of plant biomass removed by a farmer.
   (e) a measurement calculated through statistical methods.

6. The greatest value of the scientific method is best stated as:
   (a) The scientific method permits researchers a rapid method of disseminating findings.
   (b) The scientific method removes bias from observation of natural phenomenon.
   (c) The scientific method allows findings to be reproduced and tested.
   (d) The scientific method promotes sustainable development.
   (e) The scientific method reduces the complexity of experimental results.

7. Researchers conducted an experiment to test the hypothesis that the use of fertilizer near wetlands is associated with increased growth of algae. An appropriate null hypothesis would be:
   (a) The use of fertilizer near wetlands is associated with an increase in fish biomass.
   (b) Growth of algae in wetlands is never associated with increased fertilizer use.
   (c) Application of fertilizers near wetlands is always associated with increased growth of algae.
   (d) Fertilizer use near wetlands has no association with growth of algae.
   (e) Fertilizer use near wetlands leads to increased growth of algae as a result of elevated nutrient concentrations.

Questions 8 and 9 refer to the following experiment:

Researchers designed an experiment to test the hypothesis that air pollution positively correlates with the number of asthma-related problems among humans. To test this hypothesis, they compared medical records obtained from large hospitals in 10 major U. S. cities.

8. This experiment is an example of a
   (a) controlled study.
   (b) manipulative experiment.
   (c) laboratory experiment.
   (d) replication.
   (e) natural experiment.

9. Results of the study indicated that cities with more air pollution had a higher number of patients with asthma. The most appropriate conclusion from this study is that
   (a) air pollution causes asthma in humans.
   (b) air pollution is a cause of asthma in humans.
   (c) air pollution is associated with asthma in humans.
   (d) there is no association between air pollution and asthma in humans.
   (e) confounding variables make the results difficult to interpret.

Answers to Unit 1 AP® Exam Multiple-Choice Questions

1. e
2. d
3. b
4. d
5. b
6. c
7. d
8. e
9. c
10. a
11. d
12. b
13. a
14. b
15. a
16. e
17. d
18. d
19. a
20. c

Answers to Unit 1 AP® Exam

Download the full answers to the Unit 1 AP® Practice Exam by clicking on the link buttons in the TE-book, opening the TRFD, or logging onto the book's companion website bcs.whfreeman.com/friedlandapes2e (teacher login required).

Answers to Unit 1 AP® Exam

10. Which of the following constitutes baseline data on the effects of humans on natural ecosystems?
    (a) Concentrations of atmospheric CO_2 before humans existed
    (b) Current rates of species extinction
    (c) Global rate of freshwater consumption from 1900 to 2010
    (d) Current rate of human population growth
    (e) Average plant productivity on a remote island uninhabited by humans

11. During radioactive decay
    I. there is a release of material from the nucleus of unstable isotopes.
    II. there is a change in the half-life of an element.
    III. an element is changed into a different element.

    (a) I only
    (b) II only
    (c) I and II only
    (d) I and III only
    (e) I, II, and III

12. Which of the following statements about hydrogen bonding of water is NOT correct?
    (a) The positive charge of the hydrogen atom results from a covalent bond of hydrogen with oxygen.
    (b) Hydrogen atoms are strongly bonded to other hydrogen atoms.
    (c) Hydrogen atoms in one molecule are weakly bonded to oxygen atoms in other molecules.
    (d) Polarity causes weak attraction between molecules.
    (e) Hydrogen bonding causes atoms to align in a crystal structure at cold temperatures.

13. The upward movement of water through soil is an example of
    (a) capillary action.
    (b) ionic bonding.
    (c) covalent bonding.
    (d) surface tension.
    (e) evapotranspiration.

14. Which list contains only organic material?
    (a) Proteins, lipids, salts
    (b) Dead trees, decomposing leaves, earthworms
    (c) Water, ash, CO_2 gas
    (d) Cellulose, ethanol, calcium chloride
    (e) NH_3, NaOH, NO_2^-

15. A grasshopper can extract energy from ingested food at an efficiency of 10 percent. If the insect consumes 10 Calories of food and uses the food's energy during 1 minute, how much energy did it exert per second?
    (a) 70 J
    (b) 700 J
    (c) 7,000 J
    (d) 10,000 calories
    (e) 10 kWh

16. Temperature may be best described as
    (a) a measure of the average potential energy of a substance.
    (b) a measure of the average chemical energy of a substance.
    (c) a measure of the average energy efficiency of a chemical reaction.
    (d) a measure of the capacity for potential energy to be converted into kinetic energy.
    (e) a measure of the average kinetic energy of a substance.

17. When the seed pods of a pea plant dry in the sun, the skin of the pods exert inward pressure on the encased seeds. This provides the seeds with potential energy that is converted to kinetic energy when the pod is ruptured and the seeds shoot far distances. A researcher claims that the seed's potential energy is converted to kinetic energy with 100 percent efficiency. This result would violate
    (a) the law of conservation of matter.
    (b) the law of conservation of energy.
    (c) the first law of thermodynamics.
    (d) the second law of thermodynamics.
    (e) the law of energy quality.

For questions 18 and 19, refer to the following table that documents the material inputs and outputs to a 100-m sectio n of forest stream

Type of matter	Headwater inputs	Downstream outputs
Leaf litter	250 g/m^2	100 g/m^2
Woody debris	100 g/m^2	90 g/m^2
Dead insects	1 g/m^2	0.5 g /m^2
Stream sediment	10 g/m^2	3.5 g/m^2
Fish	30 g/m^2	150 g/m^2
Insects	5 g/m^2	52 g/m^2

18. Which of the following terms apply to the overall flow of organic material in this stream?
    I. A system with open boundaries
    II. Steady-state system
    III. Closed ecosystem

    (a) I only
    (b) II only
    (c) III only
    (d) I and II
    (e) II and III

19. When leaf litter inputs to the stream decrease, the amount of fish and insect biomass leaving the downstream section decreases by a similar amount. This represents
    (a) a negative feedback loop.
    (b) conservation of potential energy.
    (c) a positive feedback loop.
    (d) a decrease in entropy.
    (e) a positive feedback loop and an increase in entropy.

20. The reaction of sodium hydroxide, NaOH, and hydrochloric acid, HCl, results in the following reaction: $NaOH + HCl \rightarrow NaCl + H_2O$.
This product represents
(a) an acidic product.
(b) a basic product.
(c) a pH-neutral product.
(d) the formation of both an inorganic and organic compound.
(e) the loss of matter.

Section 2: Free-Response Questions

Write your answer to each part clearly. Support your answers with relevant information and examples. Where calculations are required, show your work.

1. The planned development site of a residential neighborhood includes 4 km^2 of forest habitat and 5,000 m^2 of stream habitat. Engineers plan to clear-cut the forest and construct a culvert that will bury the stream underground. To mitigate the loss of natural habitat, the developers will construct a 0.75 km^2 wetland at the outlet of the culvert.
   (a) Calculate the hectares of natural habitat that will be lost as result of this development. (3 points)
   (b) Describe two ecosystem services that will be lost as a result of burying the stream and two ecosystem services that will be gained by construction of the wetland. (3 points)
   (c) Using water quality as an environmental indicator, design a natural experiment that could test the impacts of this residential development. In your answer, include and label:

      (i) A testable hypothesis. (1 point)
      (ii) The data to be collected. (1 point)
      (iii) A description of the experimental procedure. (1 point)
      (iv) A description of the results that would validate your hypothesis. (1 point)

2. Approximately 72 billion liters of milk are produced each year in the United States, from 8 million cows. On average, a single cow consumes 13,500 kg of corn feed each year. It requires 40 MJ to produce a kilogram of corn feed, which contains 20 MJ of energy. There are 15 MJ of energy in a single liter of milk.
   (a) Calculate the energy efficiency of growing corn and converting it into milk. (4 points)
   (b) Describe two processes that reduce efficiency of milk production. Consider the entire process of milk production from the growth of cattle feed to the collection of milk. (2 points)
   (c) To increase the energy efficiency of milk production farmers can harvest the fecal waste (manure) from cows and use the gas it produces as a source of energy.

      (i) What is the main chemical in gas produced by cow manure that can be used as a source of energy? (1 point)
      (ii) At the molecular level, how is energy derived from this compound? (1 point)
      (iii) If 10 percent of the food energy not used by cows could be captured as chemical energy from gas released by manure, what would be the energy efficiency of converting corn into milk? (2 points)

Answers to the Unit 1 Free-Response Questions can be found in the Answer Appendix on page ANS-1.

Teaching Tip: Unit 1 Additional Free-Response Question

An additional practice free-response question covering concepts from Unit 1 is available. The question presents a short discussion of the process of carbon fractionation. To answer this question students must

- explain why C-13 and C-14 are more massive than C-12
- explain the presence of different carbon isotope ratios in fossils of the same species
- name two environmental systems in which a researcher could study isotopic ratios
- describe the process of carbon dating

Download this resource by clicking on the link buttons in the TE-book, opening the TRFD, or logging on to the book's companion website bcs.whfreeman.com/friedlandapes2e (teacher login required).

Unit 1 Additional Free-Response Question

Teaching Tip: Beyond the Classroom

This Science Applied section guides students through a thorough investigation and calculation of the missing salt in Mono Lake. In conjunction with the opening case about Mono Lake, have students investigate the flora and fauna that live in and around Mono Lake. Ask them to consider the adaptations that enable these organisms to survive in their salty environment. Alternatively, have students research the history of Mono Lake and its relationship with humans. You may want to assign each student a decade to report on, starting with the present and going back as far as your class size allows.

AP® Exam Tip

The Mono Lake investigation and the free-response question are a great way to practice test-taking skills with students before the AP® Exam or a unit test.

Tracy Packer Photography/Getty Images

1 science applied

What Happened to the Missing Salt?

At the beginning of the twentieth century, the City of Los Angeles needed more water for its inhabitants. As we saw at the beginning of Chapter 2, in 1913 the city designed a plan to redirect water away from Mono Lake in California. Before the Los Angeles Aqueduct was built, approximately 120 billion liters of stream water (31 billion gallons) flowed into Mono Lake in an average year. The City of Los Angeles altered the water balance of Mono Lake and at the same time caused a series of changes to the Mono Lake system that led to an increase in the salt concentration in Mono Lake.

To understand the problems at Mono Lake, ecosystem scientists had to examine water and chemical flows in natural waterways. Looking at the water and salt budgets of Mono Lake gave rise to observations, conclusions, and new studies on how human activities influence lakes. In a way, the City of Los Angeles conducted an experiment of what happens if you stop the flow of water into a terminal lake.

What is a terminal lake?

Mono Lake is a terminal lake because it is at the lowest point of the landscape: Water flows into the lake from rivers and streams and from precipitation, but does not flow out. However, in a typical year before Los Angeles began diverting water, the water level did not rise or fall at Mono Lake. The water exiting a terminal lake must balance with the water entering. If it does not, the lake will eventually either dry out or overflow its banks. But if the water level stays constant, and since Mono Lake is a terminal lake with no surface exits for liquid water, how is the water in balance? Mono Lake provides an excellent lesson in the mass balance of water: If the size of the pool does not change, then outputs must equal inputs. In this case, roughly the same amount of water that enters the lake must leave the lake. The only way this is possible is through evaporation. The input of water from streams must be equal to the output of water through evaporation.

How did the salt balance change at Mono Lake?

Although we can make the assumption that the water in Mono Lake is in steady state in a typical year, the salt balance in the lake is not. By applying some of the principles we have learned in the first two chapters, we can make observations and draw conclusions about what has probably happened at Mono Lake. The stream water that entered Mono Lake contained salt, as all natural waters do. The salt content of this water flowing into Mono Lake varied, but a typical liter of lake water averaged 50 mg of salt. Note that 50 mg/L is equivalent to 50 parts per million.

To calculate the total amount of salt that entered Mono Lake each year, we can multiply the concentration of salt, 50 mg per liter of water, by the number of liters of water flowing into the lake, before it was diverted by the City of Los Angeles: 120 billion liters per year:

$$50 \text{ mg/L salt} \times 120 \text{ billion L/year} = 6 \text{ trillion mg salt/year}$$

$$6 \text{ trillion mg salt/year} \times \frac{1 \text{ million kg}}{1 \text{ trillion mg}} = 6 \text{ million kg salt/year}$$

This is the annual input of salt by weight to Mono Lake. The lake today contains about 285 billion kilograms of dissolved salt, based on measurements and estimates conducted recently.

At the yearly rate of salt input we have just calculated, how long would it take to accumulate that much salt, starting with no salt in the lake? We have just determined that the salt concentration of Mono Lake increases by 6 million kilograms per year. Mono Lake contains

Answer to Science Applied 1 Free-Response Question

Download the full answer to the Science Applied 1 Free-Response question by clicking on the link buttons in the TE-book, opening the TRFD, or logging onto the book's companion website bcs.whfreeman.com/friedlandapes2e (teacher login required).

Answer to Science Applied 1 Free-Response Question

approximately 285 billion kilograms of dissolved salts today, so at the rate of stream flow before the diversion, it would have taken about 47,500 years to accumulate that much salt:

285 billion kg ÷ 6 million kg/year = 47,500 years

Does our calculation agree with the salt in Mono Lake?

Earth scientists believe that no water has flowed out of the Mono Lake basin since it was formed about 120,000 years ago. Assuming that Earth's climate hasn't changed significantly over that time and that water inputs to Mono Lake have not changed drastically over that time period, what can we calculate about how much salt should be in the water of Mono Lake?

At today's input rate, how much salt should be in the water of Mono Lake today?

6 million kg/year × 120,000 years = 720 billion kg of dissolved salt

versus 285 billion kg estimated recently.

The calculated salt contents do not match. How can we explain the discrepancy?

The lake's towering tufa formations, prominently featured in the photograph at the beginning of Chapter 2, hold the answer: Many of the salts that entered Mono Lake over time (including calcium, sodium, and magnesium) have precipitated—that is, solidified—out of the water to form the tufa rock. In this way, the salts have been removed from the water, but not from the Mono Lake system as a whole. Our analysis of salts in Mono Lake is complete when we account for the salts removed from the lake as tufa. FIGURE SA1.1 summarizes these inputs to and outputs from the Mono Lake system. And they show us how we can apply environmental science to learn about natural processes in systems, and understand how humans impact natural systems, in this case by diverting water (FIGURE SA1.2).

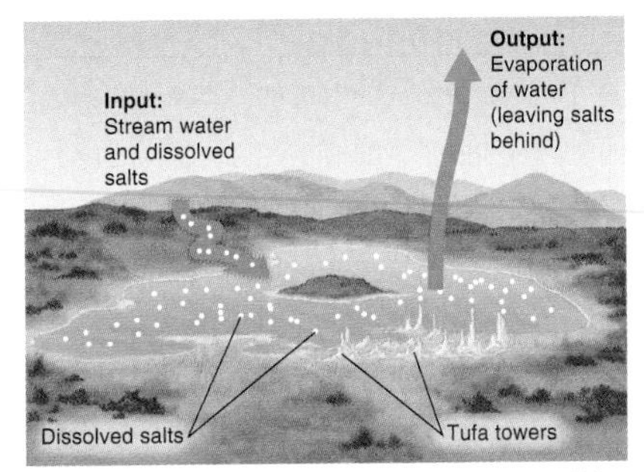

FIGURE SA1.1 **The Mono Lake System.** In this terminal lake system, inputs are from stream water while outputs are evaporated water only. All salts remain in the lake.

FIGURE SA1.2 **Research at Mono Lake.** This photo shows a scientist collecting a water sample at Mono Lake. *(Henry Bortman/NASA)*

Questions

1. How did Los Angeles inadvertently conduct an experiment at Mono Lake?
2. What chemical principle causes terminal lakes to become more salty?
3. What is the reason for the discrepancy between the two calculations of salt content in Mono Lake?

Free-Response Question

Water that flows into Mono Lake contains a much smaller concentration of salt than the water already in the lake. This inflow tends to stratify, or float on top of existing water, because fresh water is less dense that salt water. As salt from the lower layer dissolves into the upper layer, nutrients from the bottom of the lake also rise to the surface. This exchange of nutrients is critical for the growth of algae in the surface waters. Recent research suggests that the reduction of water diversion from Mono Lake had unexpected results:

In 1995, the reduction of stream diversions from Mono Lake, combined with greater than average quantities of fresh water from snowmelt runoff, led to a rapid rise in water level. The large volume of fresh water from streams led to a long-term stratification of the lake, with fresh water on the surface and salt water on the bottom. Relative to baseline data taken before the initial stream diversions, stratification has severely reduced the rate at which nutrients rise from the bottom of the lake. Long-term projections based on mathematical models suggest that the current degree of stratification will persist for decades.

(a) List three potential consequences of reduced lake mixing. (3 points)
(b) Describe two adaptive management strategies that could reduce lake stratification in Mono Lake. (3 points)
(c) What is the chemical property of water that allows salt to dissolve? (2 points)
(d) Why would the mixing of salt water with fresh water be considered an example of increased entropy? (2 points)

Suggested Answers to Science Applied Questions

1. Los Angeles redirected the flow of water into Mono Lake, thereby reducing the amount of water that was entering. This is a natural experiment that shows what happened to the lake as the inputs changed.

2. The principle that outputs must equal inputs applies.

3. Salts (calcium, magnesium, and sodium) precipitate out of the water and into the tufa rock formations.

Answer to Science Applied Free-Response Question

(a) As a result of reduced lake mixing, fewer nutrients are transported from the bottom of the lake to the surface waters, and, as a consequence, the algae that use these nutrients to grow and reproduce are likely to decrease in abundance. As primary producers, algae are at the bottom of the food web and their reduction may have numerous consequences throughout the food web. Hence, potential consequences of reduced lake mixing may be:

- a reduction in primary consumers, such as brine shrimp and alkali flies, that are typically abundant in Mono Lake
- a reduction in the growth and abundance of algae
- a reduction in the presence of secondary consumers, such as migratory shorebirds that consume the brine shrimp and alkali flies
- a reduction in total photosynthesis as well as a reduction in oxygen production in the lake because of the reduction in algae production; this may have further consequences for primary consumers that rely on oxygen production for respiration
- stagnancy of nutrients on the bottom of the lake because of their long-term burial, resulting in reduced lake production in future years

(b) Appropriate adaptive management strategies would include any strategy that actively monitors the condition of the lake and continually adjusts to improve the functioning of the ecosystem. One potential adaptive management strategy could be to incrementally reduce stream diversions, so the inflow of freshwater to the lake does not substantially alter the rate of water mixing. Rates of lake mixing could be monitored yearly to adjust the rate of stream diversion until stream diversions are completely eliminated. A second adaptive management strategy might be to add nutrients to stream inflows, so nutrient concentrations in surface water are similar to baseline concentrations. Nutrient concentrations could be monitored yearly to adjust the amount of nutrients added.
(c) Water molecules are polar, meaning that one side of the molecule has a negative charge whereas the other side has a positive charge. In the case of water, the oxygen atom has a net negative charge whereas hydrogen atoms have a net positive charge. Since most salts are composed of atoms ionically bonded to each other, the atoms readily break apart and adhere to the polar water molecules.
(d) Entropy is defined as randomness in a system. When saltwater is mixed with freshwater, the ions of salt increase their random placement in the water as they dissolve into the freshwater. Additionally, entropy is always increasing unless new energy from outside the system is added to create order. Since no energy is added to a solution of saltwater and freshwater, the system must be increasing in entropy.

chapter

3 Ecosystem Ecology

PD Overview

This chapter examines the movement of water, energy, and nutrients that organisms need to grow and reproduce. Key concepts include food webs, the water (hydrologic) cycle, and the cycling of the macronutrients carbon, nitrogen, phosphorous, and sulfur. The AP® exam frequently contains numerous multiple-choice questions about biogeochemical cycles and they also appear in free-response questions, so you will want to review this material thoroughly. Students must understand how nutrients cycle, the parts and names of each cycle, and the phases of matter in which each occurs.

Module 6: The Movement of Energy

To examine the movement of energy in ecosystems, students must begin with an understanding of ecosystem boundaries. With this foundation established, the text goes on to review how photosynthesis and respiration capture and release energy. The discussion then looks at the movement of energy through the different components of an ecosystem, touching on the concepts of food chains, trophic levels, and ecosystem productivity.

Module 7: The Movement of Matter

This module explores the cycling of matter with a focus on the relevant major biotic and abiotic processes. It begins with the hydrologic cycle—the movement of water—then explore the cycles of carbon, nitrogen, and phosphorus. Finally, it takes a brief look at the cycles of calcium, magnesium, potassium, and sulfur.

Module 8: Responses to Disturbances

In this module, students learn how scientists study disturbances, how ecosystems are affected by disturbances, and how quickly ecosystems can bounce back to their pre-disturbance condition. The module concludes with a discussion of the intermediate disturbance hypothesis, which describes the relationship between ecosystem disturbance and biodiversity.

Alignment to AP® Environmental Science Course Description

AP® Outline	Module
I. Earth Systems and Resources (10–15%)	
B. The Atmosphere	Module 6 The Movement of Energy Module 7 The Movement of Matter
II. The Living World (10–15%)	
A. Ecosystem Structure	Module 6 The Movement of Energy Module 7 The Movement of Matter Module 8 Responses to Disturbances
B. Energy Flow	Module 6 The Movement of Energy
D. Natural Ecosystem Change	Module 8 Responses to Disturbances
E. Natural Biogeochemical Cycles	Module 7 The Movement of Matter

Chapter Learning Objectives

After completing this chapter students will be able to

- explain the concept of ecosystem boundaries.
- describe the processes of photosynthesis and respiration.
- distinguish among the trophic levels that exist in food chains and food webs.
- quantify ecosystem productivity.
- explain energy transfer efficiency and trophic pyramids.
- describe how water cycles within ecosystems.
- explain how carbon cycles within ecosystems.

- describe how nitrogen cycles within ecosystems.
- explain how phosphorus cycles within ecosystems.
- discuss the movement of calcium, magnesium, potassium, and sulfur within ecosystems.
- explain the insights gained from watershed studies.
- distinguish between ecosystem resistance and ecosystem resilience.
- explain the intermediate disturbance hypothesis.

Chapter 3 Pacing Guide

This pacing guide is based on a schedule with 120 sessions of 50 minutes each before the AP® Exam. If you have a different number of sessions before the exam, you can modify the pacing to suit your needs. If you have additional time, consider incorporating quizzes, released AP® Environmental Science free-response and multiple-choice questions, or additional activities.

Module	Standard Schedule Days	Block Schedule Days
Module 6	3	1½
Module 7	3	1½
Module 8	1	½
Assessment	1	½

Chapter 3 Resources

The link to these resources can be found by clicking on the link buttons in the Teacher's e-Book (TE-book), opening the Teacher's Resource Flash Drive (TRFD), or logging onto the book's companion website bcs.whfreeman.com/friedlandapes2e (teacher login required).

- PowerPoint Presentation 3
- Optimized Art PowerPoint and JPEG Files 3
- Do the Math Videos
- **Measuring Your Impact 3:** Human Influence on the Carbon Cycle
- **Lab 3.1:** Ecosystem Fieldwalk
- Chapter 3 Web Resources
- Exploring the Literature 3
- Answers to Chapter 3 AP® Exam
- Multiple-Choice AP® Practice Exam 3

Free-Response Questions from Previous AP® Environmental Science Exams

Free-response questions from prior AP® Environmental Science exams are available on the AP® course website, https://apstudent.collegeboard.org/apcourse/ap-environmental-science/exam-practice. Students should be able to answer all of the questions listed below with material learned in this and previous chapters. When a question requires students to understand material from multiple chapters, the question will be listed in the last chapter required to complete the entire question. Questions marked with an asterisk (*) are from exams with released multiple-choice questions. You may want to save these questions until the end of the year so you can give your students a complete released exam for practice. Questions marked with double asterisks (**) require math to calculate a problem. Look for references to these questions throughout the chapter.

Year	Question	Content
2001	2	• Food webs
2003*	1	• Cycling of organic matter
2011	1	• Ecosystem disturbance
2014	4	• Carbon cycle

PD Chapter 3 Overview

Watch the video overview of Chapter 3 (for teachers) by clicking on the link buttons in the TE-book, opening the TRFD, or logging onto the book's companion website bcs.whfreeman.com /friedlandapes2e (teacher login required).

PowerPoint Presentation 3

Download the PowerPoint presentation for Chapter 3 by clicking on the link buttons in the TE-book, opening the TRFD, or logging onto the book's companion website bcs .whfreeman.com/friedlandapes2e (teacher login required).

The boundary between the Dominican Republic, on the left, and Haiti, on the right, shows massive deforestation in Haiti which has disrupted natural cycles in the ecosystem. *(James P. Blair/National Geographic/Getty Images)*

chapter

3 Ecosystem Ecology

Reversing the Deforestation of Haiti

Even before the devastating earthquake of 2010, life in Haiti was hard. On the streets of the capital city, Port-au-Prince, people would line up to buy charcoal to cook their meals. According to the United Nations, 76 percent of Haitians lived on less than $2.00 a day. Because other forms of cooking fuel, including oil and propane, were too expensive, people turned to the forests, cutting trees to make charcoal from firewood.

Relying on charcoal for fuel has had a serious impact on the forests of Haiti. In 1923, 60 percent of this mountainous country was covered in forest. However, as the population grew and demand for charcoal increased, the amount of forest shrunk. By 2012, with more than 9 million people living in this small nation, less than 2 percent of its land remained forested. Today, most trees in Haiti are cut before they grow to more than a few centimeters in diameter. This rate of deforestation is not sustainable for the people or for the forest.

Deforestation disrupts the ecosystem services that living trees provide. In Chapter 1 we saw how events on Easter Island demonstrated some of the consequences of subjecting land to massive deforestation. When Haitian forests are clear-cut, the land becomes much more susceptible to erosion. When trees are cut, their roots die and can no longer stabilize the soil. Without roots to anchor it, the soil is eroded away by the heavy rains of tropical storms and hurricanes. Unimpeded by vegetation, the rainwater runs quickly down the mountainsides, dislodging the topsoil that is so important for forest growth. In addition, this oversaturation of the soil causes massive mudslides that can destroy entire villages.

> By 2012, with more than 9 million people living in this small nation, less than 2 percent of its land remained forested.

But the news from Haiti is not all bad. Since the 1980s, the U.S. Agency for International Development, in cooperation with other groups, has provided funds to plant 60 million trees in Haiti. Unfortunately, simply planting trees is not enough; the local people can't afford to let trees grow when they are in desperate need of firewood and charcoal. One solution has been to plant mango trees (*Mangifera indica*).

Teaching Tip: Chapter Opening Case

The chapter opening case, "Reversing the Deforestation of Haiti," shows us how economic thinking can be used to solve an environmental problem. Economic concepts are essential to the decisions we make about the environment and should be threaded into everything we teach. Students should understand that keeping a mango tree that can provide between $70 and $150 each year is better than cutting down a tree for use as firewood. Use this case to introduce the idea of how many problems can be solved by using economic principles.

Teaching Tip: Warm-up

After students read the opening case, divide them into small groups to discuss another location where keeping the habitat intact provides more economic benefit than using up all its resources. Then ask students to share group discussion points with the class. *Answers will vary but may include the Great Barrier Reef or any of the national parks.*

Teaching Tip: Beyond the Classroom

After discussing the story of reforestation in Haiti, have students do a research project to learn more about the country. Divide students into pairs or groups and assign each group one of the following tasks:

- Describe Haiti's natural landscape and ecology including vegetation, animals, climate, topography, and biome.
- Identify two endangered species native to Haiti. Find the population sizes of each species and the main threat to each.
- Find the size of Haiti's human population and how it compares to all other countries. Find the population growth rate of Haiti and determine whether this growth rate is faster or slower than the U.S. growth rate.
- Identify the main agricultural products and agricultural imports of Haiti.
- Identify the main source for the generation of electricity in Haiti and the percentage of electricity that comes from this source.
- Identify minerals extracted in Haiti and describe what is done with them.
- Describe two of the biggest environmental problems in Haiti.
- Describe two initiatives, projects, or programs that are working to fix Haiti's environmental problems.

Students can write a report or make a poster. Have students present their findings to the rest of the class. Discuss the challenges that developing nations face as they strive to advance in an environmentally sustainable manner. Emphasize the ecology topics covered in this chapter, while referring to the material in future chapters, such as energy, agriculture, and pollution. This will spark curiosity about future topics and provide students with a reference point when you reach those chapters.

Because a mature mango tree can provide $70 to $150 worth of mangoes annually, this value provides an economic incentive to let the trees grow to maturity. The deforestation problem is also being addressed through efforts to develop alternative fuel sources, such as discarded paper that is processed into dried cakes that can be burned. In 2013, the president of Haiti announced a new program to plant 50 million trees every year, with the goal of moving the country from the current 2 percent forest cover to 29 percent forest cover 50 years from now.

Extensive forest removal is a problem in many developing nations. Widespread removal of trees on mountains causes rapid soil erosion and substantial disruptions of the natural cycles of water and soil nutrients. This leads to long-term degradation of the environment. The results not only illustrate the connectedness of ecological systems, but also show how forest ecosystems, like all ecosystems, can be influenced by human decisions.

Sources:
"Haitians seek remedies for environmental ruin," National Public Radio, July 15, 2009, http://www.npr.org/templates/story/story.php?storyId=104684950; "Haiti to plant millions of trees to boost forests and help tackle poverty," *The Guardian*, March 28, 2013, http://www.theguardian.com/world/2013/mar/28/haiti-plant-millions-trees-deforestation.

The story of deforestation in Haiti reminds us that all the components of an ecosystem are interrelated. As we noted in Chapter 1, an ecosystem is a particular location on Earth distinguished by its particular mix of interacting biotic and abiotic components. A forest, for example, contains many interacting biotic components, such as trees, wildflowers, birds, mammals, insects, fungi, and bacteria, that are quite distinct from those found in a grassland. Ecosystems also have abiotic components such as sunlight, temperature, soil, water, pH, and nutrients. The abiotic components of the ecosystem help determine which organisms can live there. In this chapter, we will see that ecosystems control the movement of the energy, water, and nutrients organisms must have to grow and reproduce. Understanding the processes that determine these movements is the goal of ecosystem ecology.

module

6

The Movement of Energy

To understand how an ecosystem functions and the relationship of its biotic and abiotic components, we must first study how energy moves through the ecosystem. In this module we will examine how we delineate ecosystems so that we can examine the movement of energy. We will then consider how photosynthesis and respiration capture and release energy. Finally, we will look at how energy moves through the different components of the ecosystem.

Learning Objectives

After reading this module you should be able to

- explain the concept of ecosystem boundaries.
- describe the processes of photosynthesis and respiration.
- distinguish among the trophic levels that exist in food chains and food webs.
- quantify ecosystem productivity.
- explain energy transfer efficiency and trophic pyramids.

Ecosystem boundaries are not clearly defined

The characteristics of any given ecosystem are highly dependent on the climate that exists in that location on Earth. For example, ecosystems in the dry desert of Death Valley, California, where temperatures may reach 50°C (120°F), are very different from those on the continent of Antarctica, where temperatures may drop as low as −85°C (−120°F). Similarly, water can range from being immeasurable in deserts to being the defining feature of the ecosystem in lakes and oceans. On less extreme scales, small differences in precipitation and the ability of the soil to retain water can favor different terrestrial ecosystem types. Regions with greater quantities of water in the soil can support trees, whereas regions with less water in the soil can support only grasses.

The biotic and abiotic components of an ecosystem provide the boundaries that distinguish one ecosystem from another. Some ecosystems have well-defined boundaries, whereas others do not. A cave, for example, is a well-defined ecosystem (FIGURE 6.1). It contains identifiable biotic components, such as animals

FIGURE 6.1 A cave ecosystem. Cave ecosystems, such as this one in Kenya with emerging bats, typically have distinct boundaries and are home to highly adapted species. *(Ivan Kuzmin/age fotostock)*

Teaching Tip: Activity

Create a handout titled "Ecosystems" with a list of the following related terms, as well as others that you choose:

- Abiotic
- Biotic
- Open
- Closed
- Boundary
- Energy
- Water
- Soil
- Atmosphere
- Plants
- Animals
- Microorganisms
- Temperature
- Salinity
- Precipitation
- Aquatic
- Terrestrial
- Physical
- Chemical
- Biological
- Components
- Large
- Small
- Interact
- Define
- Contain
- Provide
- Distinguish
- Adapt
- Live

In pairs or groups, have students use the words to create three true statements and two false statements about ecosystems. Encourage them to use as many words on the page as they can, as well as others as needed. Then have groups share their statements with the class. Award "bonus points" to the group that uses the most terms from the handout. *Answers will vary but may include:*

- *Ecosystems contain abiotic components like water and biotic components like plants (True).*
- *Ecosystems can be large or small and vary in temperature and precipitation (True).*
- *Ecosystems always have well-defined boundaries (False).*
- *An animal that lives in one ecosystem cannot live in another ecosystem (False).*

Teaching Tip: Discussion Starter

Ask students: What is an ecosystem and what are its components? *An ecosystem is a natural unit distinguished by the nonliving, physical environment and the community of all organisms living in a given area. It includes nonliving, or abiotic, components such as water, soil, and the atmosphere, as well as living, or biotic, components such as animals, plants, and microorganisms.*

and microorganisms that are specifically adapted to live in a cave environment, as well as distinctive abiotic components, including temperature, salinity, and water that flows through the cave as an underground stream. Roosting bats fly out of the cave each night and consume insects. When the bats return to the cave and defecate, their feces provide energy that passes through the relatively few animal species that live in the cave. In many caves, for example, small invertebrate animals consume bat feces and are in turn consumed by cave salamanders.

The cave ecosystem is relatively easy to study because its boundaries are clear. With the exception of the bats feeding outside the cave, the cave ecosystem is easily defined as everything from the point where the stream enters the cave to the point where it exits. Likewise, many aquatic ecosystems, such as lakes, ponds, and streams, are relatively easy to define because the ecosystem's boundaries correspond to the boundaries between land and water. Knowing the boundaries of an ecosystem makes it easier to identify the system's biotic and abiotic components and to trace the cycling of energy and matter through the system.

In most cases, however, determining where one ecosystem ends and another begins is difficult. For this reason, ecosystem boundaries are often subjective. Environmental scientists might define a terrestrial ecosystem as the range of a particular species of interest, such as the area where wolves roam, or they might define it by using topographic features, such as two mountain ranges enclosing a valley. The boundaries of some managed ecosystems, such as national parks, are set according to administrative rather than scientific criteria. Yellowstone National Park, for example, was once managed as its own ecosystem until scientists began to realize that many species of conservation interest, such as grizzly bears (*Ursus arctos horribilis*), spent time both inside and outside the park, despite the park's massive area of 898,000 ha (2.2 million acres). To manage these species effectively, scientists had to think much more broadly; they had to include nearly 20 million ha (50 million acres) of public and private land outside the park. This larger region, shown in FIGURE 6.2a, was named the Greater Yellowstone Ecosystem. As the name suggests, the actual ecosystem extends well beyond the administrative boundaries of the park.

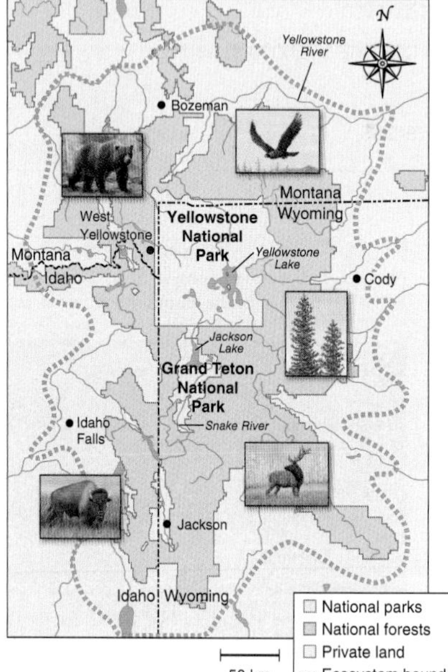

(a) The Greater Yellowstone Ecosystem

(b) A small ecosystem

FIGURE 6.2 Large and small ecosystems. (a) The Greater Yellowstone Ecosystem includes the land within Yellowstone National Park and many adjacent properties. (b) Some ecosystems are very small, such as a rain-filled tree hole that houses a diversity of microbes and aquatic insects.

Teaching Tip: Engage

Ask students: How do you know when you leave one ecosystem and enter another? *It depends. Sometimes it is easy to know when you are moving from one ecosystem to another, such as when you enter a body of water from land or walk into a cave. In other cases, ecosystems do not have well-defined boundaries and you may enter a transition zone when going from one to the other. This transition zone might include an overlap of biotic and abiotic components from both ecosystems, such as vegetation type, soil substrate, or topography.*

Teaching Tip: Activity

Have students name one large ecosystem and identify three abiotic and three biotic components. Then ask students to name a smaller ecosystem within the first and describe what defines the second ecosystem. Continue asking students to name an ecosystem within their previous choice until the scale becomes microscopic. For example:

- I. *Greater Yellowstone Ecosystem*
  - A. *Lamar Valley*
    - 1. *Large tree*
      - a. *Squirrel*
        - i. *Tick*

When students have identified at least three components of the ecosystem, ask them to explain how the abiotic and biotic components differ at various scales. *Abiotic components in Yellowstone and Lamar Valley include rocks, water, and soil substrate. The abiotic components of a tree might be pebbles, sand, and water. A squirrel or tick may have dust and water as part of their abiotic components. The biotic components of Yellowstone and Lamar Valley include all large organisms such as bears, wolves, and elk, as well as small organisms that are found on the tree, such as squirrels, birds, and insects. The biotic community of a squirrel includes parasites such as fleas and ticks, as well as the microscopic community of bacteria that also inhabit ticks.*

FIGURE 6.3 **The flow of energy in the Serengeti ecosystem of Africa.** The Serengeti ecosystem has more plants than herbivores, and more herbivores than carnivores. *(Michel & Christine Denis-Huot/Science Source)*

Teaching Tip: Discussion Starter

Ask students: How do boundaries created by humans differ from natural boundaries? *Boundaries created by humans are often set according to administrative rather than scientific criteria.*

As we saw in Chapter 2, not all ecosystems are as vast as the Greater Yellowstone Ecosystem. Some can be quite small, such as a water-filled hole in a fallen tree trunk or an abandoned car tire that fills with rainwater (Figure 6.2b). Such tiny ecosystems include all the physical and chemical components necessary to support a diverse set of species, such as microbes, mosquito larvae, and other insects.

Although it is helpful to divide locations on Earth into distinct ecosystems, it is important to remember that each ecosystem interacts with surrounding ecosystems through the exchange of energy and matter. Organisms such as bats—which fly in and out of caves—move across ecosystem boundaries, as do chemical elements, such as carbon or nitrogen dissolved in water. As a result, changes in any one ecosystem can ultimately have far-reaching effects on the global environment.

The combination of all ecosystems on Earth forms the **biosphere,** which is the region of our planet where life resides. It forms a 20-km (12-mile) thick layer around Earth between the deepest ocean bottom and the highest mountain peak.

Photosynthesis captures energy and respiration releases energy

To understand how ecosystems function and how best to protect and manage them, ecosystem ecologists also study the processes that move energy and matter within an ecosystem. To understand energy relationships, we need to look at the way energy flows across an ecosystem.

Consider the Serengeti Plain in East Africa (FIGURE 6.3). Plants, such as grasses and acacia trees, absorb energy directly from the Sun. That energy spreads throughout an ecosystem as plants are eaten by animals, such as gazelles, and the animals are subsequently eaten by predators, such as cheetahs. There are millions of herbivores, such as zebras and wildebeests, in the Serengeti ecosystem, but far fewer carnivores, such as lions (*Panthera leo*) and cheetahs (*Acinonyx jubatus*), that feed on herbivores. In accordance with the second law of thermodynamics, when one organism consumes another, not all of the energy in the consumed organism is transferred to the consumer. Some of that energy is lost as heat. Therefore, all the carnivores in an area contain less energy than all the herbivores in the same area because all the energy going to the carnivores must come from the animals they eat. Let's trace this energy flow in more detail by looking at the processes of *photosynthesis* and *cellular respiration.*

Photosynthesis

Nearly all of the energy that powers ecosystems comes from the Sun as solar energy, which is a form of kinetic energy. Plants, algae, and some bacteria that use the Sun's energy to produce usable forms of energy are called **producers,** or **autotrophs.** As you can see

Biosphere The region of our planet where life resides, the combination of all ecosystems on Earth.

Producer An organism that uses the energy of the Sun to produce usable forms of energy. *Also known as* **Autotroph.**

Teaching Tip: Discussion Starter

Ask students: What is photosynthesis and what is its significance? *Photosynthesis converts energy from the Sun to chemical energy for organisms to consume. Without photosynthesis, there would be no usable energy for heterotrophs, and food webs would not exist. Photosynthesis also produces the oxygen that organisms need to perform cellular respiration.*

> **AP® Exam Tip**
>
> Students should know how photosynthesis and cellular respiration work, including the various inputs and outputs of each process. It is also important for students to understand their role in biogeochemical cycles, particularly the carbon cycle.

in the top half of FIGURE 6.4, these producers use **photosynthesis,** which means they use solar energy to convert carbon dioxide (CO_2) and water (H_2O) into glucose ($C_6H_{12}O_6$). Glucose is a form of potential energy that can be used by a wide range of organisms. The photosynthesis process also produces oxygen (O_2) as a waste product. That is why plants and other producers are beneficial to our atmosphere; they produce the oxygen we need to breathe.

Cellular Respiration

Producers use the glucose they produce by photosynthesis to store energy and to build structures such as leaves, stems, and roots. Other organisms, such as the herbivores on the Serengeti Plain, eat the tissues of producers and gain energy from the chemical energy contained in those tissues. They do this through **cellular respiration,** a process by which cells unlock the energy of chemical compounds. **Aerobic respiration,** which is shown in the bottom half of Figure 6.4, is the opposite of photosynthesis; cells convert glucose and oxygen into energy, carbon dioxide, and water. In essence, organisms conducting aerobic respiration run photosynthesis backward to recover the solar energy stored in glucose. Some organisms, such as bacteria that live in the mud underlying a swamp where oxygen is not available, conduct **anaerobic respiration,** a process by which cells convert glucose into energy in the absence of oxygen. Anaerobic respiration does not provide as much energy as aerobic respiration.

Many organisms—including producers—carry out aerobic respiration to fuel their own metabolism and growth. Thus producers both produce and consume oxygen. When the Sun is shining and photosynthesis occurs, producers generate more oxygen through photosynthesis than they consume through respiration. At night, when producers only respire, they consume oxygen without generating it. Overall, producers photosynthesize more than they respire. The net effect is an excess of oxygen released into the air and an excess of carbon stored in the tissues of producers.

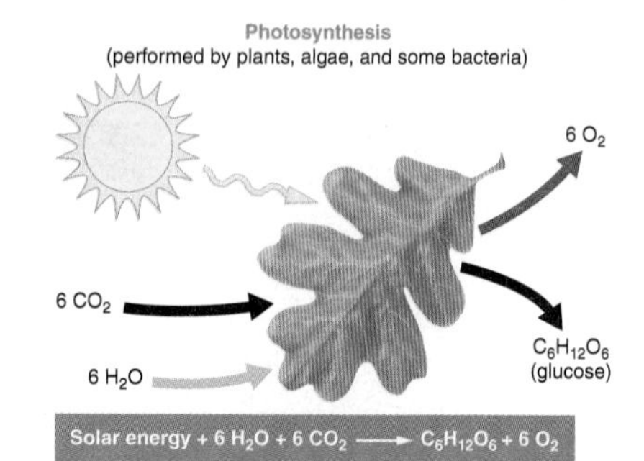

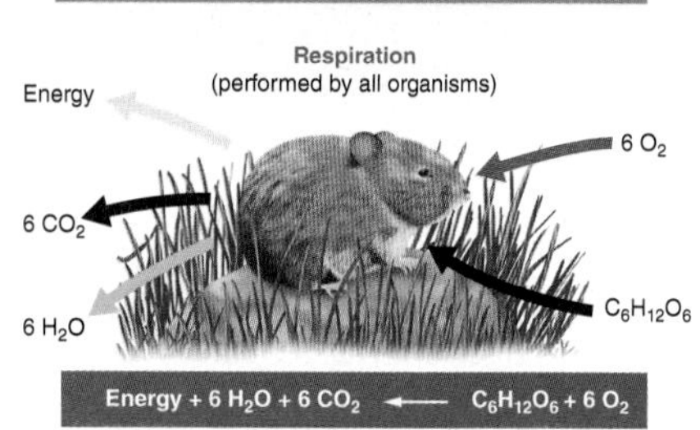

FIGURE 6.4 **Photosynthesis and respiration.** Photosynthesis is the process by which producers use solar energy to convert carbon dioxide and water into glucose and oxygen. Respiration is the process by which organisms convert glucose and oxygen into water and carbon dioxide, releasing the energy needed to live, grow, and reproduce. All organisms, including producers, perform respiration.

> **Photosynthesis** The process by which producers use solar energy to convert carbon dioxide and water into glucose.
>
> **Cellular respiration** The process by which cells unlock the energy of chemical compounds.
>
> **Aerobic respiration** The process by which cells convert glucose and oxygen into energy, carbon dioxide, and water.
>
> **Anaerobic respiration** The process by which cells convert glucose into energy in the absence of oxygen.
>
> **Consumer** An organism that is incapable of photosynthesis and must obtain its energy by consuming other organisms. *Also known as* **Heterotroph.**
>
> **Herbivore** A consumer that eats producers. *Also known as* **Primary consumer.**

Energy captured by producers moves through many trophic levels

We have seen that producers make their own food. However, **consumers,** or **heterotrophs,** are incapable of photosynthesis and must obtain their energy by consuming other organisms. In FIGURE 6.5, we can see that consumers in both terrestrial and aquatic ecosystems fall into different categories. Consumers that eat producers are called **herbivores** or **primary consumers.** Primary consumers include a variety of familiar plant- and

Teaching Tip: Discussion Starter

Ask students: Why is cellular respiration an important process? *Organisms need cellular respiration to convert the chemical energy found in food to ATP, a form of energy that fuels their metabolism and growth. Without cellular respiration, organisms would not be able to convert chemical energy into a usable form.*

Teaching Tip: Concept Map

Have students compare and contrast photosynthesis and cellular respiration using a concept map like the one shown to the right. Features that are unique to photosynthesis are placed in boxes on the left, features common to both are placed in boxes in the center, and features that are unique to cellular respiration are placed in boxes on the right. (One example for each is provided.)

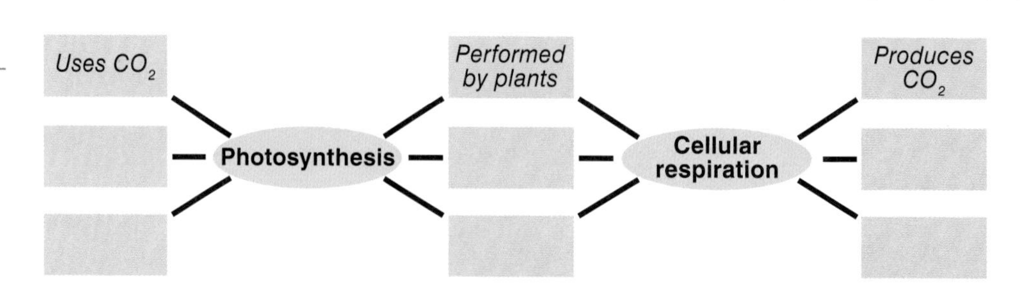

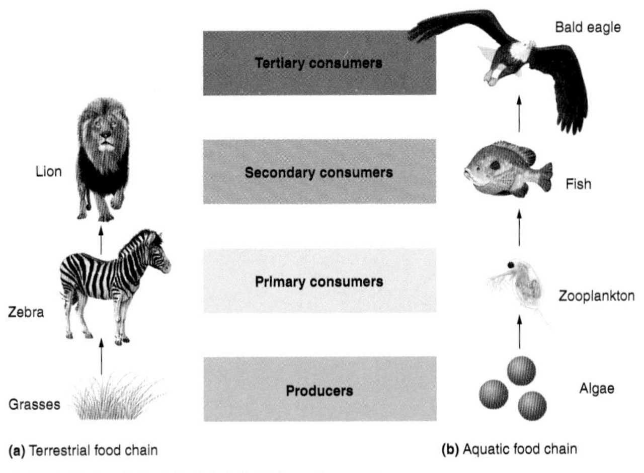

FIGURE 6.5 **Simple food chains.** A simple food chain that links producers and consumers in a linear fashion illustrates how energy and matter move through the trophic levels of an ecosystem. (a) An example of a terrestrial food chain. (b) An example of an aquatic food chain.

alga-eating animals, such as zebras, grasshoppers, tadpoles, and zooplankton. Consumers that eat other consumers are called **carnivores.** Carnivores that eat primary consumers are called **secondary consumers.** Secondary consumers include creatures such as lions, hawks, and rattlesnakes. Carnivores that eat secondary consumers are called **tertiary consumers.** As you can see on the right side of Figure 6.5, animals such as bald eagles can be tertiary consumers. In this food chain, the algae (producers) living in lakes convert sunlight into glucose, zooplankton (primary consumers) eat the algae, fish (secondary consumers) eat the zooplankton, and eagles (tertiary consumers) eat the fish.

The successive levels of organisms consuming one another are known as **trophic levels** (from the Greek word *trophe,* which means "nourishment"). The sequence of consumption from producers through tertiary consumers is known as a **food chain,** where energy moves from one trophic level to the next. A food chain helps us visualize how energy and matter move between trophic levels.

Species in natural ecosystems are rarely connected in such a simple, linear fashion. A more realistic type of model, shown in FIGURE 6.6, is known as a *food web.* A **food web** is a complex model of how energy and matter move through trophic levels. Food webs illustrate one of the most important concepts of ecology: All species in an ecosystem are connected to one another.

Not all organisms fit neatly into a single trophic level. Some organisms, called *omnivores,* operate at several trophic levels. Omnivores include grizzly bears, which eat berries and fish, and the Venus flytrap (*Dionaea muscipula*), which can photosynthesize as well as digest insects that become trapped in its leaves.

Each trophic level eventually produces dead individuals and waste products. Three groups of organisms feed on this dead organic matter: *scavengers, detritivores,* and *decomposers.* **Scavengers** are organisms,

Carnivore A consumer that eats other consumers.

Secondary consumer A carnivore that eats primary consumers.

Tertiary consumer A carnivore that eats secondary consumers.

Trophic levels The successive levels of organisms consuming one another.

Food chain The sequence of consumption from producers through tertiary consumers.

Food web A complex model of how energy and matter move between trophic levels.

Scavenger An organism that consumes dead animals.

Teaching Tip: Activity

Ask students to draw the pathway of energy as it travels from the Sun to a hawk, similar to the sample pathway shown below.

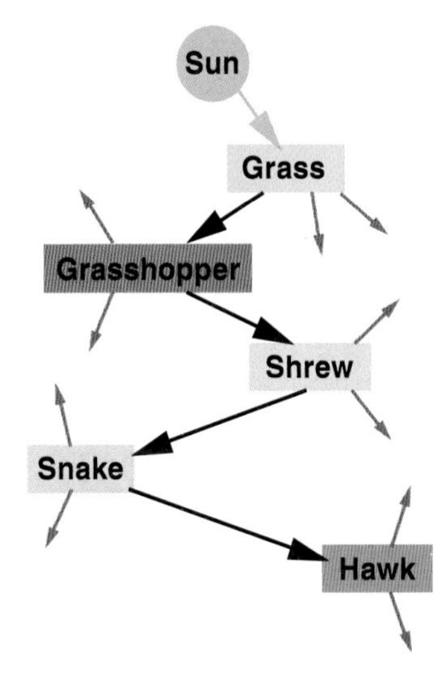

Make sure students indicate the direction of energy flow and label the types of energy in the diagram. The orange arrow from the Sun represents solar energy and the black arrows represent the transfer of chemical energy. Dark orange arrows indicate the loss of heat from each organism as it respires. When they have completed their diagram ask the following two questions:

- Why does only about 10 percent of the energy available at a given trophic level get passed on to the next trophic level? *Much of the energy consumers obtain from a lower trophic level is either indigestible or is used for the organism's day-to-day activities, growth, and the production of heat.*
- What happens to the energy flow beyond the hawk? *Energy will be transferred to the organisms that consume the hawk. This would likely be a scavenger, detritivore, or decomposer. Energy will also be lost as heat when organisms consume the hawk.*

If students need to review types of energy, direct them to Module 5 in Chapter 2 beginning on p. 43.

COMMON MISCONCEPTIONS

Heterotroph and herbivore

Students often confuse heterotroph and herbivore. To help clarify the hierarchy of feeding classifications, ask them to create a concept map (like the one shown to the right) using these terms: carnivore, consumer (heterotroph), herbivore, omnivore, producer (autotroph), and scavenger.

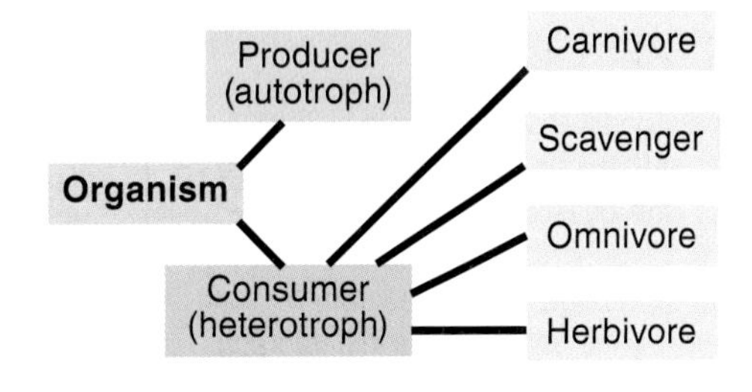

TEACHING with FIGURES ▶

Have students examine Figure 6.6, and ask them to write three different food chains within the food web depicted. Ask volunteers to share their findings with the class. *Answers may include but are not limited to: grass → zebra → hyena; grass → hare → cheetah; acacia tree → giraffe → lion.*

Teaching Tip: Activity

Create a diverse collection of 25 to 30 organism cards, each containing the name of an organism and a picture of it. Have students work in groups to sort the cards based on these feeding classification categories: producer, herbivore, carnivore, omnivore, scavenger, detritivore, and decomposer. Have students share their results with the class.

FIGURE 6.6 **A simplified food web.** Food webs are more realistic representations of trophic relationships than simple food chains. They include scavengers, detritivores, and decomposers, and they recognize that some species feed at multiple trophic levels. Arrows indicate the direction of energy movement. This is a real but somewhat simplified food web; in an actual ecosystem, many more organisms are present. In addition, there are many more energy movements.

such as vultures, that consume dead animals. **Detritivores** are organisms, such as dung beetles, that specialize in breaking down dead tissues and waste products (referred to as detritus) into smaller particles. These particles can then be further processed by **decomposers,** which are the fungi and bacteria that complete the breakdown process by converting organic matter into small elements and molecules that can be recycled back into the ecosystem. Without scavengers, detritivores, and decomposers, there would be no way of recycling organic matter and energy, and the world would rapidly fill up with dead plants and animals.

Detritivore An organism that specializes in breaking down dead tissues and waste products into smaller particles.

Decomposers Fungi and bacteria that convert organic matter into small elements and molecules that can be recycled back into the ecosystem.

Some ecosystems are more productive than others

The amount of energy available in an ecosystem determines how much life the ecosystem can support. For example, the amount of sunlight that reaches a lake surface determines how much algae can live in the lake. In turn, the amount of algae determines the number of zooplankton the lake can support, and the size of the zooplankton population determines the number of fish the lake can support.

To understand where the energy in an ecosystem comes from and how it is transferred through food

AP® Exam Tip

Food webs are covered in 2001 AP® Exam, Question 2. To answer this question, students must

- create a food web from information given in paragraph form
- design a controlled experiment regarding acorn production and gypsy moth cycles

webs, environmental scientists measure the ecosystem's productivity. The **gross primary productivity (GPP)** of the ecosystem is a measure of the total amount of solar energy that the producers in the system capture via photosynthesis over a given amount of time. Note that the term *gross,* as used here, indicates the total amount of energy captured by producers. In other words, GPP does not subtract the energy lost when the producers respire. The energy captured minus the energy respired by producers is the ecosystem's **net primary productivity (NPP)**:

net primary productivity =
gross primary productivity − respiration by producers

You can think of GPP and NPP in terms of a paycheck: GPP is the total amount your employer pays you while NPP is the actual amount you take home after taxes are deducted.

GPP is essentially a measure of how much photosynthesis is occurring over some amount of time. Determining GPP is a challenge for scientists because a plant rarely photosynthesizes without simultaneously respiring. However, if we can determine the rate of photosynthesis and the rate of respiration, we can use this information to calculate GPP.

We can determine the rate of photosynthesis by measuring the compounds that participate in the reaction. So, for example, we can measure the rate at which CO_2 is taken up during photosynthesis and the rate at which CO_2 is produced during respiration. A common approach to measuring GPP is to first measure the production of CO_2 in the dark. Because no photosynthesis occurs in the dark, this measure eliminates CO_2 uptake by photosynthesis. Next, we measure the uptake of CO_2 in sunlight. This measure gives us the net movement of CO_2 when respiration and photosynthesis are both occurring. By adding the amount of CO_2 produced in the dark to the amount of CO_2 taken up in the sunlight, we can determine the gross amount of CO_2 that is taken up during photosynthesis:

CO_2 taken up during photosynthesis =
CO_2 taken up in sunlight + CO_2 produced in the dark

In this way, we can derive the GPP of an ecosystem per day within a given area. We can give our answer in units of kilograms of carbon taken up per square meter per day (kg $C/m^2/day$).

Converting sunlight into chemical energy is not an efficient process. As FIGURE 6.7 shows, only about 1 percent of the total amount of solar energy that reaches the producers in an ecosystem—the sunlight on a pond surface, for example—is converted into chemical energy via photosynthesis. Most of that solar energy is lost from the ecosystem as heat that returns to the atmosphere. Some of the lost energy consists of wavelengths of light that producers cannot absorb.

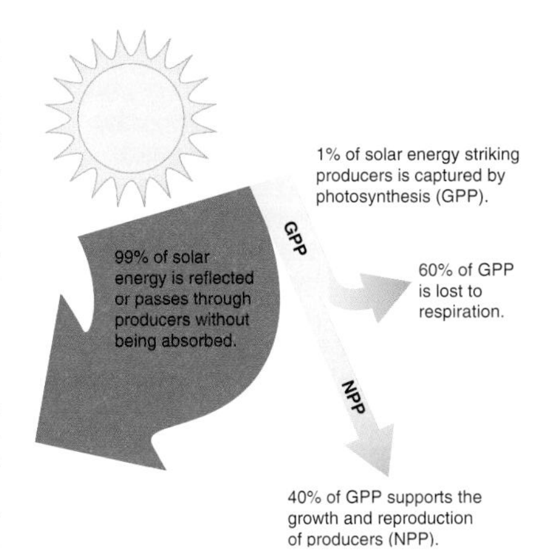

FIGURE 6.7 **Gross and net primary productivity.** Producers typically capture only about 1 percent of available solar energy via photosynthesis. This is known as gross primary productivity, or GPP. About 60 percent of GPP is typically used for respiration. The remaining 40 percent of GPP is used for the growth and reproduction of the producers. This is known as net primary productivity, or NPP.

Those wavelengths are either reflected from the surfaces of producers or pass through their tissues.

The NPP of ecosystems ranges from 25 to 50 percent of GPP, or as little as 0.25 percent of the solar energy striking the plant. Clearly, it takes a lot of energy to conduct photosynthesis. Let's look at the math. Recall from Figure 6.7 that on average, of the 1 percent of the Sun's energy that is captured by a producer, about 60 percent is used to fuel the producer's respiration. The remaining 40 percent can be used to support the producer's growth and reproduction. A forest in North America, for example, might have a GPP of 2.5 kg $C/m^2/year$ and lose 1.5 kg $C/m^2/year$ to respiration by plants. Because NPP = GPP − respiration, the NPP of the forest is 1 kg $C/m^2/year$ (1.8 pounds $C/yard^2/year$). This means that the plants living in 1 m^2 of forest will add 1 kg of carbon to their tissues every year by means of growth and reproduction. So, in this example, NPP is 40 percent of GPP.

Gross primary productivity (GPP) The total amount of solar energy that producers in an ecosystem capture via photosynthesis over a given amount of time.

Net primary productivity (NPP) The energy captured by producers in an ecosystem minus the energy producers respire.

Teaching Tip: Connections

Help students understand GPP and NPP by drawing an analogy to our own personal energy balance. Explain that GPP is similar to the total amount of calories we take in from food that we eat. Like producers, we "burn" many of these calories during cellular respiration (moving, digesting, etc.). The calories that aren't lost to respiration are stored in our tissues as fat. This stored energy is similar to NPP, the energy that producers store in their plant tissues.

more **math** practice

Calculating GPP

Have students solve the following problem:

A forest has an NPP of 1.4 kg $C/m^2/year$ and the rate of cellular respiration is 2.4 kg $C/m^2/year$. What is the GPP?

$$NPP = GPP - respiration$$

$$1.4 = GPP - 2.4$$

$$GPP = 1.4 + 2.4 = 3.8 \text{ kg } C/m^2/year$$

Teaching Tip: Discussion Starter

Ask students: What determines the productivity of an ecosystem? *The productivity of an ecosystem is largely determined by the amount of sunlight and water available. The more sunlight and water, the more productive the ecosystem can be. Other limitations to ecosystem productivity include low nutrient availability, poor soil quality, rugged topography, and frequent disturbances from natural disasters such as fires or volcanoes.*

TEACHING with FIGURES ▶

Show students Figure 6.8 and ask them the following questions:

- What accounts for the large difference in NPP among different ecosystems? *Ecosystems with abundant sunlight, water, and nutrients have the highest productivity because producers grow best under these environmental conditions. Ecosystems that are cold and/or dry, such as the Arctic and deserts, have low productivity because plants grow poorly there.*
- Which ecosystems would vary the most over the course of a year? Why? *The most variable ecosystems are those that have seasonal climates, such as temperate forests. These regions have periods of abundant sunlight and rainfall, and therefore high productivity, coupled with periods of lower sunlight and rainfall, and therefore lower productivity. Tropical regions with continual precipitation have high productivity year-round.*
- Are the rates of cellular respiration the same in each ecosystem? Why or why not? *No, the rates are not the same. Because rates of respiration are affected by both temperature and the biomass of living organisms in the ecosystem, different ecosystems will have different rates of respiration.*

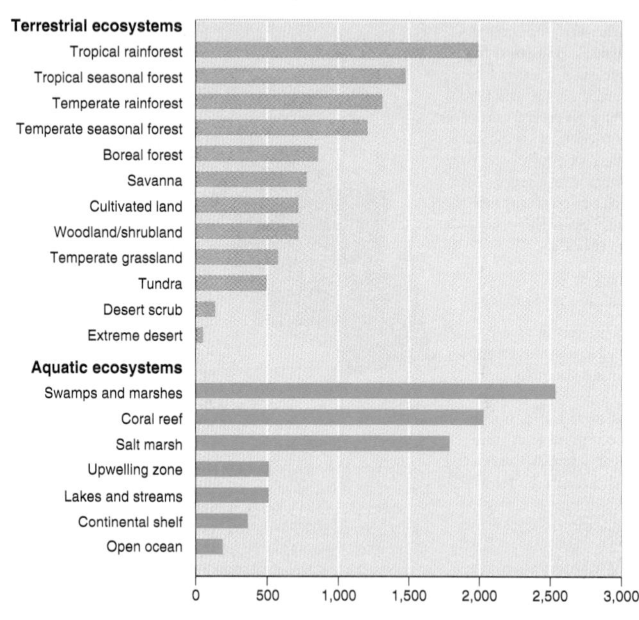

FIGURE 6.8 Net primary productivity. Net primary productivity varies among ecosystems. Productivity is highest where temperatures are warm and water and solar energy are abundant. As a result, NPP varies tremendously among different areas of the world. (Data from R. H. Whittaker and G. E. Likens, Primary production: The biosphere and man, *Human Ecology* 1 (1973): 357–369.)

Measurement of NPP allows us to compare the productivity of different ecosystems, as shown in FIGURE 6.8. As you can see, producers grow best in ecosystems where they have plenty of sunlight, lots of available water and nutrients, and warm temperatures, such as tropical rainforests and salt marshes, which are the most productive ecosystems on Earth. Conversely, producers grow poorly in the cold regions of the Arctic, dry deserts, and the dark regions of the deep sea. In general, the greater the productivity of an ecosystem, the more primary consumers can be supported.

Measuring NPP is also a useful way to measure change in an ecosystem. For example, after a drastic change alters an ecosystem, the amount of stored energy (NPP) tells us whether the new system is more or less productive than the previous system.

Biomass The total mass of all living matter in a specific area.

Standing crop The amount of biomass present in an ecosystem at a particular time.

The efficiency of energy transfer affects the energy present in each trophic level

The net primary productivity of an ecosystem establishes the rate at which **biomass**—the total mass of all living matter in a specific area—is produced over a given amount of time. The amount of biomass present in an ecosystem at a particular time is its **standing crop.** It is important to differentiate standing crop, which measures the amount of energy in a system at a given time, from productivity, which measures the rate of energy production over a span of time. For example, slow-growing forests have low productivity; the trees add only a small amount of biomass through growth and reproduction each year. However, the standing crop of long-lived trees—the biomass of trees that has accumulated over hundreds of years—is quite high. In contrast, the high growth rates of algae living in the ocean make them extremely productive. But because primary consumers eat these algae so rapidly, the standing crop of algae at any particular time is relatively low.

Not all of the energy contained in a particular trophic level is in a usable form. Some parts of plants are not

Lab
Ecosystem Field Walk

This lab offers students an opportunity to observe an ecosystem in nature.

Download the lab by clicking on the link buttons in the TE-book, opening the TRFD, or logging onto the book's companion website bcs.whfreeman.com/friedlandapes2e (teacher login required).

TRM **Lab 3.1** Ecosystem Field Walk

digestible and are excreted by primary consumers. Secondary consumers such as owls consume the muscles and organs of their prey, but they cannot digest bones and hair. Of the food that is digestible, some fraction of the energy it contains is used to power the consumer's day-to-day activities, including moving, eating, and (for birds and mammals) maintaining a constant body temperature. That energy is ultimately lost as heat. Any energy left over may be converted into consumer biomass for growth and reproduction and thus becomes available for consumption by organisms at the next higher trophic level. The proportion of consumed energy that can be passed from one trophic level to another is referred to as **ecological efficiency.**

Ecological efficiencies are fairly low; they range from 5 to 20 percent and average about 10 percent across all ecosystems. In other words, of the total biomass available at a given trophic level, only about 10 percent can be converted into energy at the next higher trophic level. We can represent the distribution of biomass among trophic levels using a **trophic pyramid,** like the one for the Serengeti ecosystem shown in FIGURE 6.9. Trophic pyramids tend to look similar across ecosystems. Most of the energy and biomass are found at the producer level, and they commonly decrease as we move up the pyramid.

The Serengeti ecosystem offers a good example of a trophic pyramid. The biomass of producers (such as grasses and shrubs) is much greater than the biomass of primary consumers (such as gazelles, wildebeests, and zebras) for which the producers serve as food. Likewise, the biomass of primary consumers is much greater than the biomass of secondary consumers (such as lions and cheetahs). The flow of energy between trophic levels helps to determine the population sizes of the various species within each trophic level. As we saw earlier in this chapter, the number of primary consumers in an area is generally higher than that of the carnivores they sustain.

The principle of ecological efficiency also has implications for the human diet. For example, if all humans were to act only as primary consumers—that is, become vegetarians—we would harvest much more energy from any given area of land or water. How would this work?

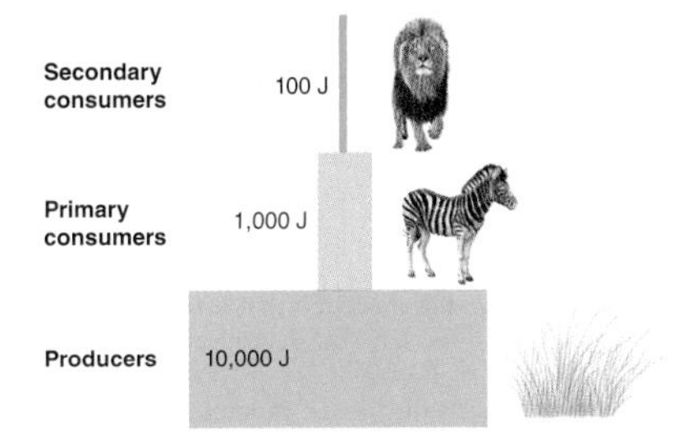

FIGURE 6.9 Trophic pyramid for the Serengeti ecosystem. This trophic pyramid represents the amount of energy that is present at each trophic level, measured in joules (J). While this pyramid assumes 10 percent ecological efficiency, actual ecological efficiencies range from 5 to 20 percent across different ecosystems. For most ecosystems, graphing the numbers of individuals or biomass within each trophic level would produce a similar pyramid.

Suppose a hectare of cropland could produce 1,000 kg of soybeans. This food could feed humans directly. Or, if we assume 10 percent ecological efficiency, it could be fed to cattle to produce approximately 100 kg of meat. In terms of biomass, there would be 10 times more food available for humans acting as primary consumers by eating soybeans than for humans acting as secondary consumers by eating beef. However, 1 kg of soybeans actually contains about 2.5 times as many calories as 1 kg of beef. Therefore, 1 ha of land would produce 25 times more calories when used for soybeans than when used for beef. In general, when we act as secondary consumers, the animals we eat require land to support the producers they consume. When we act as primary consumers, we require only the land necessary to support the producers we eat.

Ecological efficiency The proportion of consumed energy that can be passed from one trophic level to another.

Trophic pyramid A representation of the distribution of biomass, numbers, or energy among trophic levels.

◀ TEACHING with FIGURES

Show students Figure 6.9 and ask them the following questions:

- Where does the energy go? *Energy is ultimately lost to the surrounding environment as organisms undergo metabolism and consumers fail to digest all of the organic matter that they consume.*
- How do these data correspond to the number of plants, zebras, and lions that you would expect to see on a grassland? *On a grassland you would expect to see an abundance of grasses and herbaceous plants, followed by herds of grazing herbivores, such as zebras. The populations of carnivores, such as lions, are smaller than those of the herbivores, as depicted in the diagram.*
- What organisms might be on the fourth level? *Organisms that eat lions are likely to be scavengers, detritivores, or decomposers.*

Practicing Science:
Synthesize, Analyze, and Hypothesize

Using the pyramid and table to the right, ask students to calculate the amount of energy that moves through each trophic level. Have students duplicate the table to fill in their answers. (Answers are provided in italics.) Students should use the given energy efficiency values to make their calculations. Start with the first trophic level—grass at 100,000 kcals—and determine how much energy moves to the second trophic level, rabbits. Once they calculate the amount of energy transferred to the rabbits, students should calculate how much energy moves to the foxes and finally to the hawks.

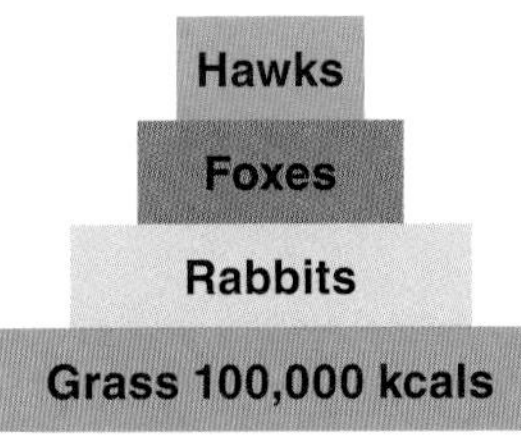

Transfer of energy through trophic levels

Between trophic levels	Percentage of energy efficiency	Energy in kcals that moves to the next trophic level?	
		Calculation	Rounded answer
Grass to rabbits	12% energy efficiency	*12,000*	*12,000 kcals*
Rabbits to foxes	14% energy efficiency	*1,680*	*1,700 kcals*
Foxes to hawks	8% energy efficiency	*134.4*	*130 kcals*

module 6

REVIEW

In this module, we have learned that ecosystems can range widely in size and have boundaries that are either natural or defined by humans. Energy from the Sun is captured by producers during the process of photosynthesis and released during the process of respiration. The energy captured by producers helps to determine the productivity of different ecosystems around the world. This energy moves from producers through the many trophic levels. The efficiency of such energy transfers determines the biomass found in different trophic levels. In the next module we will apply this understanding of how energy moves through an ecosystem to consider how matter cycles around an ecosystem.

Answers to Module 6 AP® Review Questions

1. a
2. d
3. a
4. c
5. b
6. e

Module 6 AP® Review Questions

1. Autotrophs
   (a) use photosynthesis.
   (b) are able to survive without oxygen.
   (c) are primary consumers.
   (d) are at the top of the food chain.
   (e) cannot assimilate carbon.

2. A zebra is an example of
   (a) a secondary consumer.
   (b) a producer.
   (c) a detritivore.
   (d) a primary consumer.
   (e) a scavenger.

3. If gross primary productivity in a wetland is 3 kg C/m^2/year and respiration is 1.5 kg C/m^2/year, what is the net primary productivity of the wetland?
   (a) 1.5 kg C/m^2/year
   (b) 2 kg C/m^2/year
   (c) 3 kg C/m^2/year
   (d) 4.5 kg C/m^2/year
   (e) Impossible to determine from the given information

4. The average efficiency of energy transfer between trophic levels is approximately
   (a) 1 percent.
   (b) 4 percent.
   (c) 10 percent.
   (d) 40 percent.
   (e) 50 percent.

5. The gross primary productivity of an ecosystem is
   (a) the total amount of biomass.
   (b) the total energy captured by photosynthesis.
   (c) the energy captured after accounting for respiration.
   (d) the energy available to primary consumers.
   (e) the biomass of the producers.

6. Ecosystem boundaries are
   (a) based primarily on topographic features.
   (b) boundaries to nutrient flows.
   (c) never based on human created features.
   (d) are only used for ecosystems smaller than a few square hectares.
   (e) depend on many subjective factors.

module

7

The Movement of Matter

As we saw in Chapter 2, Earth is an open system with respect to energy, but a closed system with respect to matter. In the previous module, we examined how energy flows through the biosphere; it enters as energy from the Sun, moves among the living and nonliving components of ecosystems, and is ultimately emitted back into space. In contrast, matter—such as water, carbon, nitrogen, and phosphorus—does not enter or leave the biosphere, but cycles within the biosphere in a variety of forms. In this module, we will explore the cycling of matter with a focus on the major biotic and abiotic processes that cause this cycling.

The specific chemical forms that elements take determine how they cycle within the biosphere. We will look at the elements that are the most important to the productivity of photosynthetic organisms. We will begin with the hydrologic cycle—the movement of water. Then we will explore the cycles of carbon, nitrogen, and phosphorus. Finally, we will take a brief look at the cycles of calcium, magnesium, potassium, and sulfur.

Learning Objectives

After reading this module you should be able to

- describe how water cycles within ecosystems.
- explain how carbon cycles within ecosystems.
- describe how nitrogen cycles within ecosystems.
- explain how phosphorus cycles within ecosystems.
- discuss the movement of calcium, magnesium, potassium, and sulfur within ecosystems.

Teaching Tip: Engage

Before teaching any of the biogeochemical cycles, discuss the following questions:

- Why does matter on Earth go through cycles? What would happen if it didn't? *Without nutrient cycles, Earth's available nutrients would quickly be exhausted and unavailable to support life.*
- What are the most important cycles? *While all nutrients that are essential to life cycle through nature, the cycles that are most important to plant productivity are the cycles of water, carbon, nitrogen, phosphorus, and sulfur.*
- How and why are these cycles different from energy flow? *Energy is in a constant, one-direction flow, starting from the Sun, through food webs, and dissipating back into the atmosphere and space. Unlike energy that comes from the Sun, all the nutrients on Earth come from Earth and therefore must be recycled in order to replenish supplies.*
- Why are they called biogeochemical cycles? *The term biogeochemical refers to the biological, geological, and chemical processes that occur during nutrient cycling.*

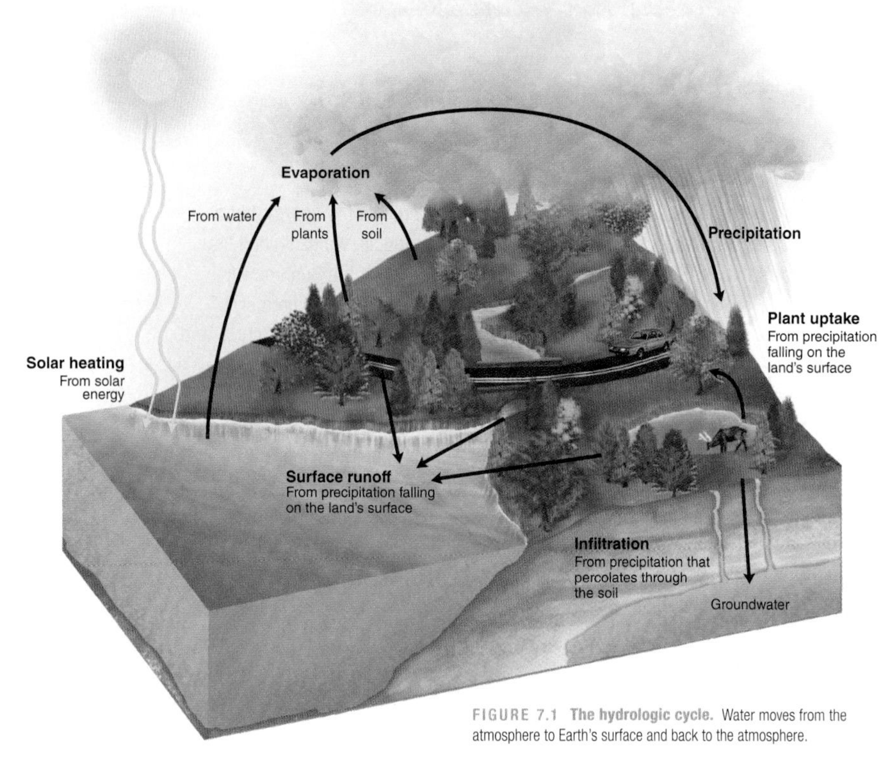

FIGURE 7.1 **The hydrologic cycle.** Water moves from the atmosphere to Earth's surface and back to the atmosphere.

COMMON MISCONCEPTIONS
The water cycle and the hydrologic cycle

Make sure students know that the water cycle and the hydrologic cycle are the same thing.

The hydrologic cycle moves water through the biosphere

Because the movement of matter within and between ecosystems involves cycles of biological, geological, and chemical processes, these cycles are known as **biogeochemical cycles.** To keep track of the movement of matter in biogeochemical cycles, we refer to the components that contain the matter—including air, water, and organisms—as "pools." Processes that move matter between pools are known as "flows." We will begin our study of biogeochemical cycles with water.

Biogeochemical cycle The movements of matter within and between ecosystems.

Hydrologic cycle The movement of water through the biosphere.

Water is essential to life. It makes up over one-half of a typical mammal's body weight, and no organism can survive without it. Water allows essential molecules to move within and between cells, draws nutrients into the leaves of trees, dissolves and removes toxic materials, and performs many other critical biological functions. On a larger scale, water is the primary agent responsible for dissolving and transporting the chemical elements necessary for living organisms. The movement of water through the biosphere is known as the **hydrologic cycle.**

The Hydrologic Cycle

FIGURE 7.1 shows how the hydrologic cycle works. Heat from the Sun causes water to evaporate from oceans, lakes, and soils. Solar energy also provides the energy for photosynthesis, during which plants release water from their leaves into the atmosphere—a process

do the **math**

Raising Mangoes

As we saw at the beginning of this chapter, farmers in Haiti are being encouraged to plant mango trees to help reduce runoff and increase the uptake of water by the soil and the trees. Consider a group of Haitian farmers that decides to plant mango trees. Mango saplings cost $10 each. Once the trees become mature, each tree will produce $75 worth of fruit per year. A village of 225 people decides to pool its resources and set up a community mango plantation. Their goal is to generate a per capita income of $300 per year for everyone in the village.

1. How many mature trees will the village need to meet the goal?
   Total annual income desired:

   $$\$300/\text{person} \times 225 \text{ persons} = \$67{,}500$$

   Number of trees needed to produce $67,500 in annual income:

   $$\$67{,}500 \div \$75/\text{tree} = 900 \text{ trees}$$

2. Each tree requires 25 m^2 of space. How many hectares must the village set aside for the plantation?

   $$900 \text{ trees} \times 25 \text{ m}^2 = 22{,}500 \text{ m}^2 = 2.25 \text{ ha}$$

Your Turn

Each tree requires 20 L of water per day during the 6 hot months of the year (180 days). The water must be pumped to the plantation from a nearby stream. How many liters of water are needed each year to water the plantation of 900 trees?

known as **transpiration.** The water vapor that enters the atmosphere eventually cools and forms clouds, which, in turn, produce precipitation in the form of rain, snow, and hail. Some precipitation falls back into the ocean and some falls on land.

When water falls on land, it may take one of three distinct routes. First, it may return to the atmosphere by evaporation or, after being taken up by plant roots, by transpiration. The combined amount of evaporation and transpiration, called **evapotranspiration,** is often used by scientists as a measure of the water moving through an ecosystem. Alternatively, water can be absorbed by the soil and percolate down into the groundwater. Finally, water can move as **runoff** across the land surface and into streams and rivers, eventually reaching the ocean, which is the ultimate pool of water on Earth. As water in the ocean evaporates, the cycle begins again.

The hydrologic cycle is instrumental in the cycling of elements. Many elements are carried to the ocean or taken up by organisms in dissolved form. As you read about biogeochemical cycles, notice the role that water plays in these processes.

Human Activities and the Hydrologic Cycle

Because Earth is a closed system with respect to matter, water never leaves it. Nevertheless, human activities can alter the hydrologic cycle in a number of ways. For example, harvesting trees from a forest can reduce evapotranspiration by reducing plant biomass. If evapotranspiration decreases, then runoff or percolation will increase. On a moderate or steep slope, most water will leave the land surface as runoff. That is why, as we saw at the opening of this chapter, clear-cutting a mountain slope can lead to erosion and flooding. Similarly, paving over land surfaces to build roads, businesses, and homes reduces the amount of percolation that can take place in a given area, increasing runoff and evaporation. Humans can also alter the hydrologic cycle by diverting water from one area to another to provide water for drinking, irrigation, and industrial uses. In "Do the Math: Raising Mangoes" you can see how citizens in Haiti are working to reforest the country and reduce the impacts of humans on the hydrologic cycle.

Transpiration The release of water from leaves during photosynthesis.

Evapotranspiration The combined amount of evaporation and transpiration.

Runoff Water that moves across the land surface and into streams and rivers.

Answer to Your Turn

Total Water Needed:

20 L/tree/day × 180 days/year × 900 trees = 3,240,000 L

Teaching Tip: Activity

Students often have a prior knowledge of the hydrologic and carbon cycles. If so, you might want to create blank diagrams of each with the associated terms (condensation, evaporation, photosynthesis, respiration, etc.) listed on the sides. Ask students to label the diagram and write a brief explanation for each term. The hydrologic cycle is illustrated in Figure 7.1 on p. 80 and the carbon cycle is illustrated in Figure 7.2 on p. 83. They can then switch with a partner to check their answers. Have a class discussion about what was easiest and most difficult to label.

AP® Exam Tip

The cycling of organic matter is covered in the 2003 AP® Exam, Question 1. In order to answer this question, students must

- discuss the role that leaf litter plays in the forest floor ecosystem
- name three abiotic factors that would change if leaf litter disappeared
- describe how introducing exotic earthworms into the ecosystem changes abiotic conditions in the forest floor
- design a controlled experiment with exotic worm species

The carbon cycle moves carbon between air, water, and land

Carbon is the most important element in living organisms; it makes up about 20 percent of their total body weight. Carbon is the basis of the long chains of organic molecules that form the membranes and walls of cells, constitute the backbones of proteins, and store energy for later use. Other than water, few molecules in the bodies of organisms do not contain carbon. To understand the **carbon cycle,** which is the movement of carbon around the biosphere, we need to examine the flows that move carbon and the major pools that contain carbon.

The Carbon Cycle

FIGURE 7.2 illustrates the seven processes that drive the carbon cycle: *photosynthesis, respiration, exchange, sedimentation, burial, extraction,* and *combustion.* These processes can be categorized as either fast or slow. The fast part of the cycle involves processes that are associated with living organisms. The slow part of the cycle involves carbon that is held in rocks, in soils, or as petroleum hydrocarbons (the materials we use as fossil fuels). Carbon may be stored in these forms for millions of years.

Photosynthesis and Respiration

Let's take a closer look at Figure 7.2, which depicts the carbon cycle. When producers photosynthesize, whether on land or in the water, they take in CO_2 and incorporate the carbon into their tissues. Some of this carbon is returned as CO_2 when organisms respire. It is also returned after organisms die. When organisms die, carbon that was part of the live biomass pool becomes part of the dead biomass pool. Decomposers break down the dead material, which returns CO_2 to the water or air via respiration and continues the cycle.

Exchange, Sedimentation, and Burial

As you can see on the left side of Figure 7.2, carbon is exchanged between the atmosphere and the ocean. The amount of carbon released from the ocean into the atmosphere roughly equals the amount of atmospheric CO_2 that diffuses into ocean water. Some of the CO_2 dissolved in the ocean enters the food web via photosynthesis by algae.

Another portion of the CO_2 dissolved in the ocean combines with calcium ions in the water to form calcium carbonate ($CaCO_3$), a compound that can precipitate out of the water and form limestone and dolomite rock via sedimentation and burial. Although sedimentation is a very slow process, the small amounts of calcium carbonate sediment formed each year have accumulated over millions of years to produce the largest carbon pool in the slow part of the carbon cycle.

A small fraction of the organic carbon in the dead biomass pool is buried and incorporated into ocean sediments before it can decompose into its constituent elements. This organic matter becomes fossilized and, over millions of years, some of it may be transformed into fossil fuels. The amount of carbon removed from the food web by this slow process is roughly equivalent to the amount of carbon returned to the atmosphere by weathering of rocks containing carbon (such as limestone) and by volcanic eruptions, so the slow part of the carbon cycle is in steady state.

Carbon cycle The movement of carbon around the biosphere.

Extraction and Combustion

The final processes in the carbon cycle are extraction and combustion, as are shown in the middle of Figure 7.2. The extraction of fossil fuels by humans is a relatively recent phenomenon that began when human society started to rely on coal, oil, and natural gas as energy sources. Extraction by itself does not alter the carbon cycle; it is the subsequent step of combustion that makes the difference. Combustion of fossil fuels by humans and the natural combustion of carbon by fires or volcanoes release carbon into the atmosphere as CO_2 or into the soil as ash.

As you can see, combustion, respiration, and decomposition operate in very similar ways: All three processes cause organic molecules to be broken down to produce CO_2, water, and energy. However, respiration and decomposition are biotic processes, whereas combustion is an abiotic process.

Human Impacts on the Carbon Cycle

In the absence of human disturbance, the exchange of carbon between Earth's surface and atmosphere is in steady state. Carbon taken up by photosynthesis eventually ends up in the soil. Decomposers in the soil gradually release that carbon at roughly the same rate it is added. Similarly, the gradual movement of carbon into the buried or fossil fuel pools is offset by the slow processes that release it. Before the Industrial Revolution, the atmospheric concentration of carbon dioxide had changed very little for 10,000 years (see Figure 2.5 on page 12). So, until recently, carbon entering any of these pools was balanced by carbon leaving these pools.

Since the Industrial Revolution, however, human activities have had a major influence on carbon cycling.

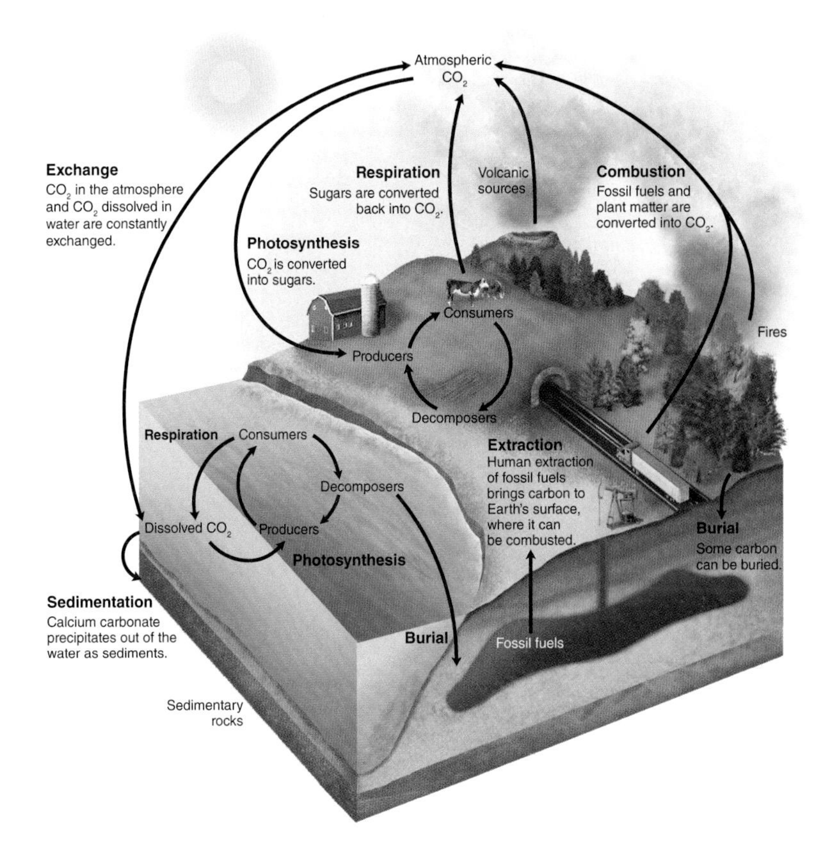

FIGURE 7.2 **The carbon cycle.** Producers take up carbon from the atmosphere via photosynthesis and pass it on to consumers and decomposers. Some inorganic carbon sediments out of the water to form sedimentary rock while some organic carbon may be buried and become fossil fuels. Respiration by organisms returns carbon to the atmosphere and water. Combustion of fossil fuels and other organic matter returns carbon to the atmosphere.

The best-known and most significant human alteration of the carbon cycle is the combustion of fossil fuels. This process releases fossilized carbon into the atmosphere, which increases atmospheric carbon concentrations and upsets the balance between Earth's carbon pools and the atmosphere. The excess CO_2 in the atmosphere acts to increase the retention of heat energy in the biosphere. The result, global warming, is a major concern among environmental scientists and policy makers.

Tree harvesting is another human activity that can affect the carbon cycle. Trees store a large amount of carbon in their wood, both above and below ground. The destruction of forests by cutting and burning increases the amount of CO_2 in the atmosphere. Unless enough new trees are planted to recapture the carbon, the destruction of forests will upset the balance of CO_2. To date, large areas of forest, including tropical forests as well as North American and European temperate forests, have been converted into pastures, grasslands, and croplands. In addition to destroying a great deal of biodiversity, this destruction of forests has added large amounts of carbon to the atmosphere. The increases in

Teaching Tip: Activity

Have students draw an illustrated diagram of the carbon cycle. Tell them to include the following processes, but not to label them: burial, combustion, exchange, extraction, photosynthesis, respiration, and sedimentation. Ask students to switch diagrams with a partner who will label the diagram with the seven components of the carbon cycle. Diagrams should resemble Figure 7.2.

AP® Exam Tip

The carbon cycle is covered in the 2014 AP® Exam, Question 4. In order to answer this question, students must

- describe the biological processes by which carbon is removed from the atmosphere, converted to organic molecules, and then converted back to a gas and returned to the atmosphere
- understand that oceans and terrestrial systems are important carbon reservoirs
- discuss human activities, aside from burning fossil fuels, that increase atmospheric carbon concentrations, and identify environmental problems associated with these activities
- contrast the carbon cycle with the phosphorus cycle and explain why phosphorus is necessary for organisms

Teaching Tip: Activity

The nitrogen cycle is the most complex biogeochemical cycle covered. After teaching the cycle, break students into groups and ask them to write a jingle or a rap to describe the process. Explain that they must use the terms for all five steps: ammonification, assimilation, denitrification, nitrification, and nitrogen fixation (which all happen to rhyme). Their composition should explain what occurs at each step, starting and ending with nitrogen (N_2). When complete, ask groups to perform their composition for the class.

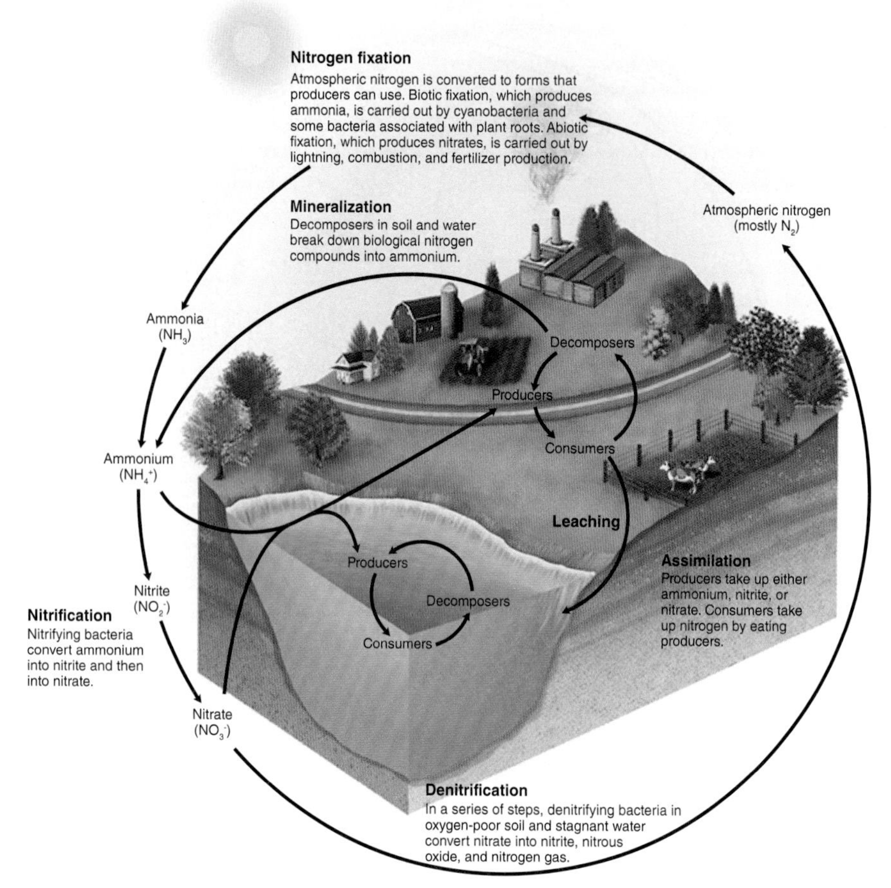

Nitrogen fixation
Atmospheric nitrogen is converted to forms that producers can use. Biotic fixation, which produces ammonia, is carried out by cyanobacteria and some bacteria associated with plant roots. Abiotic fixation, which produces nitrates, is carried out by lightning, combustion, and fertilizer production.
Atmospheric nitrogen (mostly N_2)
Mineralization
Decomposers in soil and water break down biological nitrogen compounds into ammonium.
Ammonia (NH_3)
Decomposers
Producers
Consumers
Ammonium (NH_4^+)
Leaching
Producers
Decomposers
Consumers
Assimilation
Producers take up either ammonium, nitrite, or nitrate. Consumers take up nitrogen by eating producers.
Nitrite (NO_2^-)
Nitrification
Nitrifying bacteria convert ammonium into nitrite and then into nitrate.
Nitrate (NO_3^-)
Denitrification
In a series of steps, denitrifying bacteria in oxygen-poor soil and stagnant water convert nitrate into nitrite, nitrous oxide, and nitrogen gas.

FIGURE 7.3 **The nitrogen cycle.** The nitrogen cycle moves nitrogen from the atmosphere and into soils through several fixation pathways, including the production of fertilizers by humans. In the soil, nitrogen can exist in several forms. Denitrifying bacteria release nitrogen gas back into the atmosphere.

atmospheric carbon due to human activities have been partly offset by an increase in carbon absorption by the ocean. Still, the loss of forest remains a concern.

Macronutrient One of six key elements that organisms need in relatively large amounts: nitrogen, phosphorus, potassium, calcium, magnesium, and sulfur.

The nitrogen cycle includes many chemical transformations

There are six key elements, known as **macronutrients,** that are needed by organisms in relatively large amounts: nitrogen, phosphorus, potassium, calcium, magnesium, and sulfur.

Nitrogen is an element that organisms need in relatively high amounts. Nitrogen is used to form amino acids, the building blocks of proteins, and nucleic acids,

Teaching Tip: Remember

Help students remember the processes and products of the nitrogen cycle by using the mnemonic FixNAAD ANPAN, which stands for nitrogen fixation, nitrification, assimilation, ammonification, denitrification, ammonium, nitrates, proteins, ammonium, and nitrogen. Have students set up a table like the one shown to the right.

Process	Product
Fix-nitrogen fixation	Ammonium
Nitrification	Nitrates
Assimilation	Proteins
Ammonification	Ammonium
Denitrification	Nitrogen

the building blocks of DNA and RNA. Because so much of it is required, nitrogen is often a *limiting nutrient* for producers. A **limiting nutrient** is a nutrient required for the growth of an organism but available in a lower quantity than other nutrients. Therefore, a lack of nitrogen constrains the growth of the organism. Adding other nutrients, such as water or phosphorus, will not improve plant growth in nitrogen-poor soil.

The Nitrogen Cycle

The **nitrogen cycle** is the movement of nitrogen around the biosphere. As nitrogen moves through an ecosystem, it experiences many chemical transformations. As shown in FIGURE 7.3, there are five major transformations in the nitrogen cycle: *nitrogen fixation, nitrification, assimilation, mineralization,* and *denitrification.*

Nitrogen Fixation

Although Earth's atmosphere is 78 percent nitrogen by volume, the vast majority of that nitrogen is in a gas form that most producers cannot use. **Nitrogen fixation** is the process that converts nitrogen gas in the atmosphere (N_2) into forms of nitrogen that producers can use. As you can see in Figure 7.3, nitrogen fixation can occur through biotic or abiotic processes. In the biotic process, a few species of bacteria can convert N_2 gas directly into ammonia (NH_3), which is rapidly converted to ammonium (NH_4^+), a form that is readily used by producers. Nitrogen-fixing organisms include cyanobacteria (also known as blue-green algae) and certain bacteria that live within the roots of legumes, which include plants such as peas, beans, and a few species of trees. Nitrogen-fixing organisms use the fixed nitrogen to synthesize their own tissues, then excrete any excess. Cyanobacteria, which are primarily aquatic organisms, excrete excess ammonium ions into the water, where they can be taken up by aquatic producers. Nitrogen-fixing bacteria that live within plant roots excrete excess ammonium ions into the plant's root system; the plant, in turn, supplies the bacteria with sugars it produces via photosynthesis.

Nitrogen fixation can also occur through two abiotic pathways. N_2 can be fixed in the atmosphere by lightning or during combustion processes such as fires and the burning of fossil fuels. These processes convert N_2 into nitrate (NO_3^-), which is usable by plants. The nitrate is carried to Earth's surface in precipitation. Humans have developed techniques for nitrogen fixation into ammonia or nitrate to be used in plant fertilizers. Although these processes require a great deal of energy, humans now fix more nitrogen than is fixed in nature. The development of synthetic nitrogen fertilizers has led to large increases in crop yields, particularly for crops such as corn that require large amounts of nitrogen.

Nitrification

Another step in the nitrogen cycle is **nitrification,** which is the conversion of ammonium (NH_4^+) into nitrite (NO_2^-) and then into nitrate (NO_3^-). These conversions are conducted by specialized species of bacteria. Although nitrite is not used by most producers, nitrate is readily used.

Assimilation

Once producers take up nitrogen in the form of ammonia, ammonium, nitrite, or nitrate, they incorporate the element into their tissues in a process called **assimilation.** When primary consumers feed on the producers, some of the producer's nitrogen is assimilated into the tissues of the consumers while the rest is eliminated as waste products.

Mineralization

Eventually, organisms die and their tissues decompose. In a process called **mineralization,** fungal and bacterial decomposers break down the organic matter found in dead bodies and waste products and convert organic compounds back into inorganic compounds. In the nitrogen cycle, the process of mineralization is sometime called **ammonification** because organic nitrogen compounds are converted into the inorganic ammonium (NH_4^+). The ammonium produced by this process can either be taken up by producers in the ecosystem or be converted into nitrite (NO_2^-) and nitrate (NO_3^-) through the process of nitrification.

Limiting nutrient A nutrient required for the growth of an organism but available in a lower quantity than other nutrients.

Nitrogen cycle The movement of nitrogen around the biosphere.

Nitrogen fixation A process by which some organisms can convert nitrogen gas molecules directly into ammonia.

Nitrification The conversion of ammonia (NH_4^+) into nitrite (NO_2^-) and then into nitrate (NO_3^-).

Assimilation The process by which producers incorporate elements into their tissues.

Mineralization The process by which fungal and bacterial decomposers break down the organic matter found in dead bodies and waste products and convert it into inorganic compounds.

Ammonification The process by which fungal and bacterial decomposers break down the organic nitrogen found in dead bodies and waste products and convert it into inorganic ammonium (NH_4^+).

Teaching Tip: Activity

Ask each student to draw a diagram of the nitrogen cycle. Encourage students to highlight the main forms and processes of the nitrogen cycle and to include the main nitrogen reservoirs. You may wish to provide the following more explicit instructions:

- Box the main nitrogen reservoirs: the atmosphere, plants, and animals.
- Circle the main forms of nitrogen: N_2, ammonium, and nitrate.
- Title arrows with the main processes of the nitrogen cycle: ammonification, assimilation, denitrification, nitrification, and nitrogen fixation.

Student diagrams should resemble the one shown below.

Denitrification
Atmosphere N_2
Nitrate (NO_3^-)
Nitrogen fixation
Nitrification
Nitrite (NO_2^-)
Ammonium (NH_4^+)
Assimilation
Plants
Animals
Mineralization

Practicing Science: Demo

Bring to class a live legume such as a pea, alfalfa, or bean plant. Carefully remove the plant from its soil and show the class the nodules on its roots. (You may want to check the plant in advance to make sure that it has visible nodules.) Enhance the demonstration by projecting a close-up picture of root nodules, and then pass the plant around the class as you discuss nitrogen fixation.

Teaching Tip: Connections

When covering excesses of nitrogen and phosphorus in the nitrogen cycle and phosphorus cycle, refer to the topic of water pollution covered in Chapter 14. To spark student curiosity about future topics and provide them with a reference point when you reach Chapter 14, you might note the following:

- An excess of nutrients in water bodies, particularly nitrogen and phosphorus, causes an explosion of algae growth.
- When algae growth increases, the populations of consumers that eat algae, such as zooplankton and bacteria, also increase.
- The overpopulation of decomposers that occurs when algae die depletes oxygen levels and therefore threatens survival of aquatic life.

Denitrification

The final step that completes the nitrogen cycle is **denitrification,** which is the conversion of nitrate (NO_3^-) in a series of steps into the gases nitrous oxide (N_2O) and, eventually, nitrogen gas (N_2), which is emitted into the atmosphere. Denitrification is conducted by specialized bacteria that live under anaerobic conditions, such as waterlogged soils or the bottom sediments of oceans, lakes, and swamps.

Human Impacts on the Nitrogen Cycle

Nitrogen is a limiting nutrient in most terrestrial ecosystems, so excess inputs of nitrogen can have consequences in these ecosystems. For example, nitrate is readily transported through the soil with water through **leaching,** a process in which dissolved molecules are transported through the soil via groundwater. In addition, adding nitrogen to soils in fertilizers ultimately increases atmospheric concentrations of nitrogen in regions where the fertilizer is applied. This nitrogen can be transported through the atmosphere and deposited by rainfall in natural ecosystems that have adapted over time to a particular level of nitrogen availability. The added nitrogen can alter the distribution or abundance of species in those ecosystems.

In one study of nine different terrestrial ecosystems across the United States, scientists added nitrogen fertilizer to some plots and left other plots unfertilized as controls. They found that adding nitrogen reduced the number of species in a plot by up to 48 percent because some species that could survive under low-nitrogen conditions could no longer compete against larger plants that thrived under high-nitrogen conditions. Other studies have documented cases in which plant communities that have grown on low-nitrogen soils for millennia are now experiencing changes in their species composition. An influx of nitrogen due to human activities has favored colonization by new species that are better adapted to soils with higher fertility.

The observation that nutrients can have unintended effects on ecosystems highlights an important principle of environmental science: In ecosystems containing species that have adapted to their environments over thousands of years or longer, changes in conditions are likely to cause changes in biodiversity as well as in the movement of energy through, and the cycling of matter within, those ecosystems.

Denitrification The conversion of nitrate (NO_3^-) in a series of steps into the gases nitrous oxide (N_2O) and, eventually, nitrogen gas (N_2), which is emitted into the atmosphere.

Leaching The transportation of dissolved molecules through the soil via groundwater.

Phosphorus cycle The movement of phosphorus around the biosphere.

The phosphorus cycle moves between land and water

Organisms need phosphorus for many biological processes. Phosphorus is a major component of DNA and RNA as well as ATP (adenosine triphosphate), the molecule cells use for energy transfer. Required by both producers and consumers, phosphorus is a limiting nutrient second only to nitrogen in its importance for successful agricultural yields. Thus phosphorus, like nitrogen, is commonly added to soils in the form of fertilizer.

The Phosphorus Cycle

The **phosphorus cycle** is the movement of phosphorus around the biosphere. As shown in FIGURE 7.4, it primarily operates between land and water. There is no gas phase, although phosphorus does enter the atmosphere in very small amounts when dust is dissolved in rainwater or sea spray. Unlike nitrogen, phosphorus rarely changes form; it is typically found in the form of phosphate (PO_4^{3-}).

Assimilation and Mineralization

The biotic processes that affect the phosphorus cycle are not complex. Producers on land and in the water take up inorganic phosphate and assimilate the phosphorus into their tissues as organic phosphorus. The waste products and eventual dead bodies of these organisms are decomposed by fungi and bacteria, which causes the mineralization of organic phosphorus back to inorganic phosphate.

Sedimentation, Geologic Uplift, and Weathering

The abiotic processes of the phosphorus cycle involve movements between the water and the land. In water, phosphorus is not very soluble, so much of it precipitates out of solution in the form of phosphate-laden sediments in the ocean. You can see this process in the left corner of Figure 7.4. Over time, geologic forces can lift these ocean layers up and they become mountains. The phosphate rocks in the mountains are slowly weathered by natural forces including rainfall and this weathering brings phosphorus to terrestrial and aquatic habitats. Phosphorus is tightly held by soils, so it is not easily leached from soils and into water bodies. Because so little phosphorus leaches into water bodies and because much of what enters water precipitates out of solution, very little dissolved phosphorus is naturally available in streams, rivers, and lakes. As a result, phosphorus is a limiting nutrient in many aquatic systems.

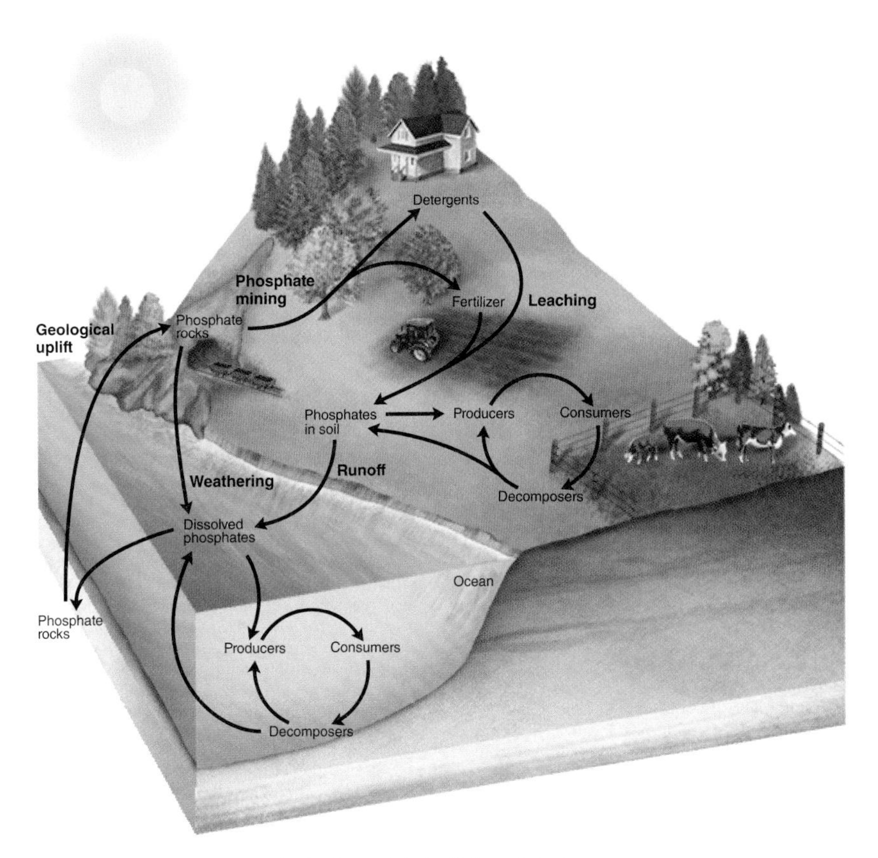

FIGURE 7.4 **The phosphorus cycle.** The phosphorus cycle begins with the weathering or mining of phosphate rocks and use of phosphate fertilizer, which releases phosphorus into the soil and water. This phosphorus can be used by producers and subsequently moves through the food web. In water, phosphorus can precipitate out of solution and form sediments, which over time are transformed into new phosphate rocks.

Human Impacts on the Phosphorus Cycle

Humans have had a dramatic effect on the phosphorus cycle. For example, humans mine the phosphate sediments from mountains to produce fertilizer. When these fertilizers are applied to lawns, gardens, and agricultural fields, excess phosphorus can leach into water bodies. Because aquatic systems are commonly limited by a low availability of phosphorus, even small inputs of leached phosphorus into these systems can greatly increase the growth of producers. Phosphorus inputs can cause a rapid increase in the algal population of a waterway, known as an **algal bloom,** that can quickly increase the biomass of

Algal bloom A rapid increase in the algal population of a waterway.

Teaching Tip: Discussion Starter

Ask students: What role does water play in nutrient cycling? *Water is critical to nutrient cycles because it enables nutrients to be leached from soil, transported by streams and rivers, and precipitate out of solution in the ocean. Nutrients would have a difficult time moving through ecosystems without water.*

COMMON MISCONCEPTIONS
The phosphorus cycle

Students often think there is an atmospheric form of phosphorus. Point out that this is not the case. To clarify this concept, it can be helpful for students to connect the phosphorus cycle to the sulfur cycle, since the major natural source for both is rocks. Be sure they understand that the two cycles differ because of one significant fact: Sulfur has a gaseous component and phosphorus does not.

FIGURE 7.5 **Algal bloom.** When excess phosphorus enters waterways, it can stimulate a sudden and rapid growth of algae that turns the water bright green, like this area along the Susquehanna River in Pennsylvania. The algae eventually die, and the resulting increase in decomposition can reduce dissolved oxygen to levels that are lethal to fish and shellfish. *(Michael P. Gadomski/Science Source)*

algae in the ecosystem (FIGURE 7.5). As the algae die, decomposition consumes large amounts of oxygen. As a result, the water becomes low in oxygen, or **hypoxic.** Hypoxic conditions kill fish and other aquatic animals. Hypoxic dead zones occur around the world, including where the Mississippi River empties into the Gulf of Mexico.

A second major source of phosphorus in waterways is from the use of household detergents. From the 1940s through the 1990s, laundry detergents in the United States contained phosphates to make clothes cleaner. The water discharged from washing machines inadvertently fertilized streams, rivers, and lakes. Because ecological dead zones caused by excess phosphorus represent substantial environmental and economic damage, manufacturers stopped adding phosphates to laundry detergents in 1994 and 16 states banned phosphates in dishwashing detergents in 2010. In 2013, 2 more states imposed a similar ban.

Hypoxic Low in oxygen.

Sulfur cycle The movement of sulfur around the biosphere.

In Europe, the European Union recently passed a ban on phosphates in laundry detergents starting in 2013 and a ban on phosphates in dishwashing detergents starting in 2017.

In addition to causing algal blooms, increases in phosphorus concentrations can alter plant communities. We have already seen one example in Chapter 2, where we discussed the deterioration of the environment in the Florida Everglades. Because of agricultural expansion in southern Florida, the water that flows through the Everglades has experienced elevated phosphorus concentrations. This change in nutrient cycling has altered the ecosystem of the Everglades. For example, over time, cattails have become more common and sawgrass has declined. Animals that depend on sawgrass are now experiencing reduced food and habitat from this plant.

Calcium, magnesium, potassium, and sulfur also cycle in ecosystems

Calcium, magnesium, and potassium play important roles in regulating cellular processes and in transmitting signals between cells. Like phosphorus, these macronutrients are derived primarily from rocks and decomposed vegetation. All three can be dissolved in water as positively charged ions: Ca^{2+}, Mg^{2+}, and K^{+}. None is present in a gaseous phase, but all can be deposited from the air in small amounts as dust.

Because of their positive charges, calcium, magnesium, and potassium ions are attracted to the negative charges present on the surfaces of most soil particles. Calcium and magnesium occur in high concentrations in limestone and marble. Because Ca^{2+} and Mg^{2+} are strongly attracted to soil particles, they are abundant in many soils overlying these types of rock. In contrast, K^{+} is only weakly attracted to soil particles and is therefore more susceptible to being leached away by water moving through the soil. Leaching of potassium can lead to potassium-deficient soils that constrain the growth of plants and animals.

The final macronutrient, sulfur, is a component of proteins and plays an important role in allowing organisms to use oxygen. The **sulfur cycle,** which is the movement of sulfur around the biosphere, is shown in FIGURE 7.6. Most sulfur exists in rocks and is released into soils and water as the rocks weather over time. Producers absorb sulfur through their roots in the form of sulfate ions (SO_4^{2-}), and the sulfur then cycles through the food web. The sulfur cycle also has a gaseous component. Volcanic eruptions are a natural source of atmospheric sulfur in the form of sulfur dioxide (SO_2). Human activities also add sulfur

Teaching Tip: Activity

To review all of the biogeochemical cycles, divide students into groups and assign each group one of the biogeochemical cycles. Instruct groups to diagram their cycle on the board or on a large piece of poster paper. When groups are about halfway done with the cycle they are working on, interrupt them and ask them to switch places with another group. Students will then finish and perhaps edit the previous group's work. Once all cycles are complete, rotate groups once more to a different cycle. Each group should then present the cycle they end up with.

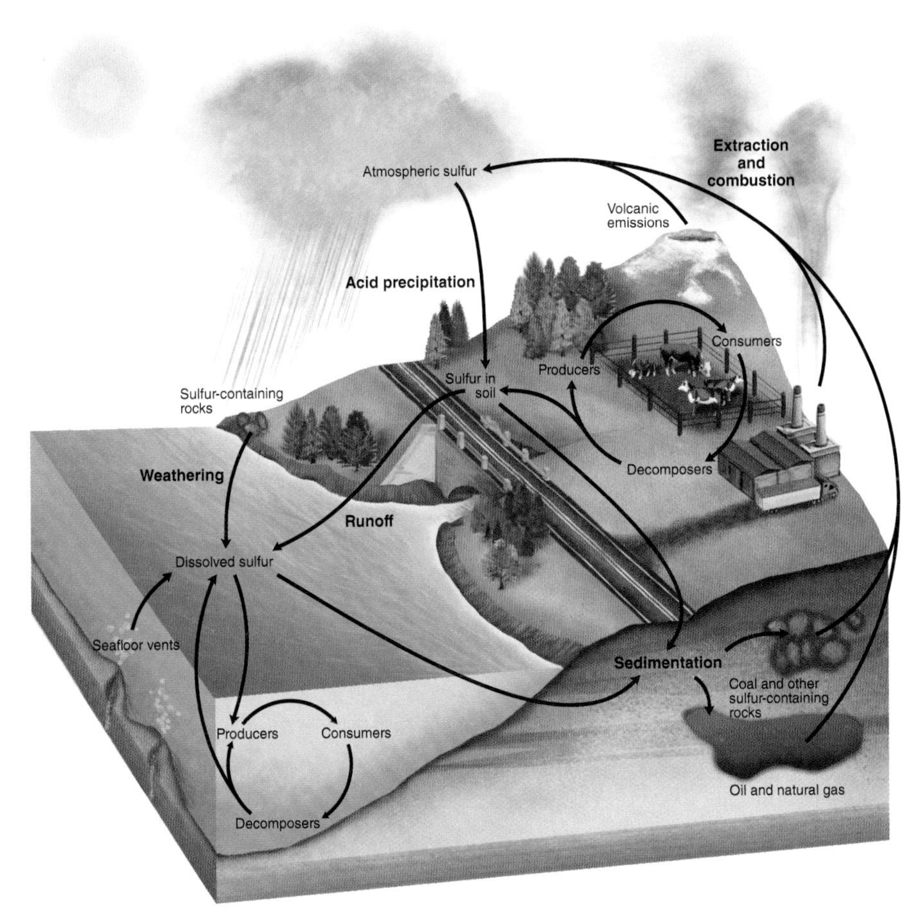

FIGURE 7.6 **The sulfur cycle.** Most sulfur exists as rocks. As these rocks are weathered over time, they release sulfate ions (SO_4^{2-}) that producers can take up and assimilate. This assimilated sulfur then passes through the food web. Volcanoes, the burning of fossil fuels, and the mining of copper put sulfur dioxide (SO_2) into the atmosphere. In the atmosphere, sulfur dioxide combines with water to form sulfuric acid (H_2SO_4). This sulfuric acid is carried back to Earth when it rains or snows.

dioxide to the atmosphere, especially the burning of fossil fuels and the mining of metals such as copper. In the atmosphere, SO_2 is converted into sulfuric acid (H_2SO_4) when it mixes with water. The sulfuric acid can then be carried back to Earth when it rains or snows. As humans add more sulfur dioxide to the atmosphere, we cause more acid precipitation, which can negatively affect terrestrial and aquatic ecosystems. Although anthropogenic deposition of sulfur remains an environmental concern, clean air regulations in the United States have significantly lowered these deposits since 1995.

Teaching Tip: Review

Ask students: What are the main similarities and differences among the carbon, nitrogen, and phosphorus cycles? *Carbon, nitrogen, and phosphorus are all needed by organisms for survival, and animals acquire all three nutrients by consuming plants. Carbon and phosphorus are taken up directly by plants, whereas nitrogen must be converted to ammonium or nitrate before plants can assimilate it. Carbon and nitrogen have an atmospheric component, while phosphorus does not. Nitrogen and phosphorus are used in fertilizers, but carbon is not.*

module

7

REVIEW

In this module, we have learned that water, carbon, and the six macronutrients cycle around ecosystems. These elements become available to organisms in different ways, but once they are obtained by producers from the atmosphere or as ions dissolved in water, the elements cycle through trophic levels in similar ways. Consumers then obtain these elements by eating producers. Finally, decomposers absorb these elements from dead producers and consumers and their waste products. Through the process of decomposition, they convert the elements into forms that are once again available to producers. In the next module, we will explore how ecosystems respond to disturbances in these cycles.

Answers to Module 7 AP® Review Questions

1. d
2. b
3. c
4. a
5. e
6. d
7. c

Module 7 AP® Review Questions

1. Which one of the following is fixed from the atmosphere by bacteria?
   (a) Magnesium
   (b) Phosphorus
   (c) Sulfur
   (d) Nitrogen
   (e) Potassium

2. Human construction of buildings and pavement affect the hydrological cycle by
   I. increasing runoff.
   I. increasing evaporation.
   III. increasing percolation.

   (a) I only
   (b) I and II only
   (c) I and III only
   (d) II and III only
   (e) III only

3. The largest carbon pool is found in
   (a) oceans.
   (b) the atmosphere.
   (c) sedimentary rock.
   (d) living organisms.
   (e) fossil fuels.

4. Phosphorus
   (a) is a limiting nutrient in many aquatic systems.
   (b) has an important gaseous phase.
   (c) is easily lost from soils due to leaching.
   (d) is often produced by volcanic eruptions.
   (e) changes chemical form often during its biogeochemical cycle.

5. Acid rain is associated with which geochemical cycle?
   (a) Potassium
   (b) Calcium
   (c) Carbon
   (d) Phosphorus
   (e) Sulfur

6. Which of the following processes is also known as ammonification?
   (a) Nitrogen fixation
   (b) Nitrification
   (c) Assimilation
   (d) Mineralization
   (e) Denitrification

7. Haitian farmers are buying mango tree saplings for $10 each, and the full-grown trees produce $75 of fruit each year. If a farmer wishes to earn $1,500 per year when the trees are grown, how much will the farmer have to spend on saplings?
   (a) $100
   (b) $150
   (c) $200
   (d) $750
   (e) $1,000

module

8

Responses to Disturbances

As we have seen in the previous modules, flows of energy and matter in ecosystems are essential to the species that live in them. However, sometimes ecosystems experience major disturbances that alter how they operate. Disturbances can occur over both short and long time scales, and ecosystem ecologists are often interested in how disturbances affect the flow of energy and matter through an ecosystem. More specifically, they are interested in whether an ecosystem can resist the impact of a disturbance and whether a disturbed ecosystem can recover its original condition. In this module, we will look at how scientists study disturbances, how ecosystems are affected by disturbances, and how quickly these ecosystems can bounce back to their pre-disturbance condition. Finally, we will apply our knowledge to an important theory about how systems respond to disturbances.

Learning Objectives

After reading this module you should be able to

- distinguish between ecosystem resistance and ecosystem resilience.
- explain the insights gained from watershed studies.
- explain the intermediate disturbance hypothesis.

Teaching Tip: Discussion Starter

Ask students: What is the difference between resistance and resilience in an ecosystem? *An ecosystem that demonstrates resistance is able to withstand a disturbance with minimal effect on its flow of energy and matter. An ecosystem that demonstrates resilience may be severely affected by a disturbance, with significant changes to its flow of energy and matter, but is able to return to its original state relatively rapidly.*

Ecosystems are affected differently by disturbance and in how well they bounce back after the disturbance

In 2012, a major hurricane came to shore in New York and New Jersey and caused a tremendous amount of damage to local ecosystems, not to mention local homes and businesses (FIGURE 8.1). Hurricanes, ice storms, tsunamis,tornadoes, volcanic eruptions, and forest fires can all be classified as a

FIGURE 8.1 **Ecosystem disturbance.** Large disturbances can have major effects on ecosystems, such as this area of New Jersey where Hurricane Sandy devastated shorelines and the residences along these shorelines. *(The Star-Ledger/Andy Mills/The Image Works)*

AP® Exam Tip

Ecosystem disturbance is covered in the 2011 AP® Exam, Question 1. In order to answer this question, students must

- explain why fire-suppression may lead to increased pine beetle activity
- describe two ways that beetle activity might enhance climate change
- identify two physical changes that occur in forest ecosystems when mature trees die, and describe how those changes impact the forest ecosystem
- describe two economic consequences of the collapse of managed honeybee colonies

Teaching Tip: Discussion Starter

Ask students: What are the characteristics of resilient ecosystems? *Resilient ecosystems contain many generalist species, have a high net primary productivity, contain fast-growing producers, and have moderate to high species diversity.*

(a)

(b)

FIGURE 8.2 **Wetland restoration.** (a) Once a forested wetland, this property in Maryland was cleared and drained for agricultural use in the 1970s. (b) Beginning in 2003, efforts began to plug the drainage ditches, remove undesirable trees, and plant wetland plants to restore the property to a wetland habitat. *(Rich Mason, USFWS)*

disturbance because they are events caused by physical, chemical, or biological agents that results in changes in population size or community composition in ecosystems. Disturbances also can be due to anthropogenic causes, such as human settlements, agriculture, air pollution, forest clear-cutting, and the removal of entire mountaintops for coal mining.

Not every ecosystem disturbance is a disaster. For example, a low-intensity fire might kill some plant species, but at the same time it might benefit fire-adapted species that can use the additional nutrients released from the dead plants. So, although the population of a particular producer species might be diminished or even eliminated, the net primary productivity of all the producers in the ecosystem might remain the same. When this is the case, we say that the productivity of the system is *resistant*. The **resistance** of an ecosystem is a measure of how much a disturbance can affect the flows of energy and matter. When a disturbance influences populations and communities, but has no effect on the overall flows of energy and matter, we say that the ecosystem has high resistance.

Disturbance An event, caused by physical, chemical, or biological agents, resulting in changes in population size or community composition.

Watershed All land in a given landscape that drains into a particular stream, river, lake, or wetland.

Resistance A measure of how much a disturbance can affect flows of energy and matter in an ecosystem.

Resilience The rate at which an ecosystem returns to its original state after a disturbance.

Restoration ecology The study and implementation of restoring damaged ecosystems.

When the flows of energy and matter of an ecosystem are affected by a disturbance, environmental scientists often ask how quickly and how completely the ecosystem can recover its original condition. The rate at which an ecosystem returns to its original state after a disturbance is termed **resilience.** A highly resilient ecosystem returns to its original state relatively rapidly; a less resilient ecosystem does so more slowly. For example, imagine that a severe drought has eliminated half the species in an area. In a highly resilient ecosystem, the flows of energy and matter might return to normal in the following year. In a less resilient ecosystem, the flows of energy and matter might not return to their pre-drought conditions for many years.

An ecosystem's resilience often depends on specific interactions of the biogeochemical and hydrologic cycles. For example, as human activity has led to an increase in global atmospheric CO_2 concentrations, terrestrial and aquatic ecosystems have increased the amount of carbon they absorb. In this way the carbon cycle as a whole has offset some of the changes that we might expect from increases in atmospheric CO_2 concentrations, including global climate change. Conversely, when a drought occurs, the soil may dry out and harden so much that when it eventually does rain, the soil cannot absorb as much water as it did before the drought. The soil changes in response to the drought, which leads to further drying and intensifies the drought damage. In this case, the hydrologic cycle does not relieve the effects of the drought; instead, a positive feedback in the system makes the situation worse.

Many anthropogenic disturbances—for example, housing developments, clear-cutting, or draining of wetlands—are so large that they eliminate an entire ecosystem. In some cases, however, scientists can work to reverse these effects and restore much of the original function of the ecosystem (FIGURE 8.2). Growing interest in restoring damaged ecosystems has led to the creation of a new scientific discipline called **restoration ecology.** Restoration ecologists are currently working on two high-profile ecosystem restoration projects, in the Florida Everglades and in the Chesapeake Bay, to restore water flows and nutrient inputs that are closer to historic levels so that the functions of these ecosystems can be restored.

Watershed studies help us understand how disturbances affect ecosystem processes

Understanding the natural rates and patterns of biogeochemical cycling in an ecosystem provides a basis for determining how a disturbance has changed the system. Because it is difficult to study biogeochemical cycles on a global scale, most of this research takes place on a smaller scale where scientists can measure all of the ecosystem processes. Scientists commonly conduct such studies in a *watershed.* As shown in FIGURE 8.3, a **watershed** is all of the land in a given landscape that drains into a particular stream, river, lake, or wetland.

One of the most thorough studies of disturbance at the watershed scale has been ongoing for more than 50 years in the Hubbard Brook ecosystem of New Hampshire. Since 1962, investigators have monitored the hydrological and biogeochemical cycles of six watersheds at Hubbard Brook, ranging in area from 12 to 43 ha (30 to 106 acres). The soil in each watershed is underlain by impenetrable bedrock, so there is no deep percolation of water; all precipitation that falls on the watershed leaves it either by evapotranspiration or by runoff. Scientists measure precipitation throughout each watershed, and a stream gauge at the bottom of the main stream that drains a given watershed allows them to measure the amounts of water and nutrients leaving the system.

Researchers at Hubbard Brook investigated the effects of clear-cutting and subsequent suppression of plant regrowth. The researchers cut down the forest in one watershed and used herbicides to suppress the regrowth of vegetation for several years. A nearby watershed that was not clear-cut served as a control (FIGURE 8.4). The concentrations of nitrate in stream

FIGURE 8.3 **Watershed.** A watershed is the area of land that drains into a particular body of water.

FIGURE 8.4 **Studying disturbance at the watershed scale.** In the Hubbard Brook ecosystem, researchers clear-cut one watershed to determine the importance of trees in retaining soil nutrients. They compared nutrient runoff in the clear-cut watershed with that in a control watershed that was not clear-cut. (The two other watersheds shown in the photo received other experimental treatments). *(U.S. Forest Service, Northern Research Station)*

Practicing Science: Analyze

Review the Hubbard Brook experiment in the framework of the scientific method. Have students determine the following:

- **Hypothesis:** *Intact forests affect the retention of nutrients in their soils.*
- **Experimental design and data collection:** *Identify six watersheds in the Hubbard Brook system. Set up a stream gauge at the bottom of each stream that drains a given watershed and use it to measure levels of water and nutrients leaving the system. Monitor the system for several years. Clear-cut and suppress tree growth in one watershed. Compare the stream nitrate concentrations of the clear-cut watershed to that of a nearby control that was not clear-cut.*
- **Results:** *The clear-cut watershed showed significant increases in stream nitrate concentrations when compared to the control area.*
- **Conclusion:** *When trees are no longer present to take up nitrate from the soil, nitrate leaches out of the soil and ends up in the stream that drains the watershed.*

Ask students: Are there any ways this experiment could have been improved? *The experimental manipulation lacks replication, as only one watershed was clear-cut. A more robust design would have multiple treatment groups and multiple control groups.*

Teaching Tip: Review

Ask students the following questions:

- Why is Hubbard Brook valuable as a study area? *The Hubbard Brook study demonstrates the importance of plants in regulating the cycling of nutrients.*
- What does the Hubbard Brook study teach us? *It teaches us a great deal about biogeochemical cycles, particularly that large amounts of nutrients accumulate in vegetation and soil where forests and grasslands grow. Without intact plant communities, nutrients would cycle through systems and end up in the ocean.*

TEACHING with FIGURES ▶

Show students Figure 8.5a and ask why species richness is low on the left side and on the right side of the graph. Is it low on both sides for the same reason? *Species richness is low on opposite sides of the graph for different reasons. Species richness is low on the left because competitively dominant species take over when a lack of disturbance doesn't regulate their population. Species richness is low on the right because the high level of disturbance prevents many species from establishing stable populations.*

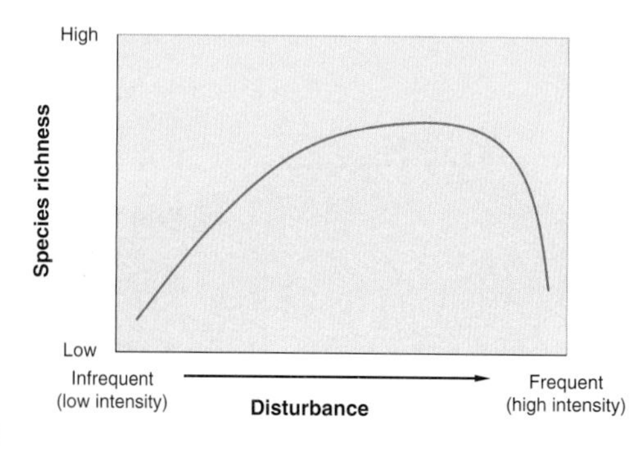

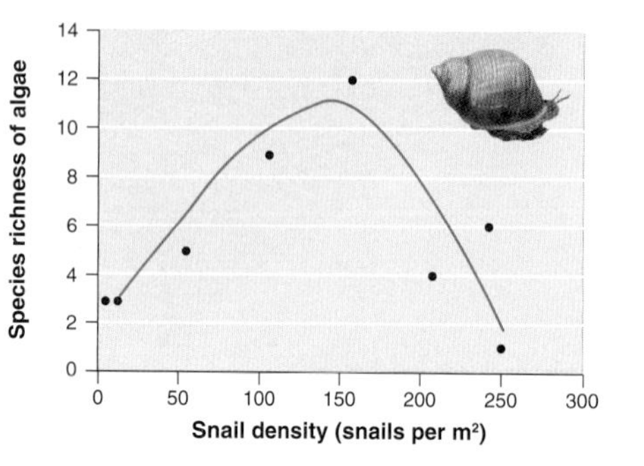

FIGURE 8.5 **Intermediate disturbance hypothesis.** (a) In general, we expect to see the highest species diversity at intermediate levels of disturbance. Rare disturbances favor the best competitors, which outcompete other species. Frequent disturbances eliminate most species except those that have evolved to live under such conditions. At intermediate levels of disturbance, species from both extremes can persist. (b) An example of the intermediate disturbance in the number of algal species observed in response to different amounts of herbivory by marine snails. When few or many snails are present, there is a low diversity of algal species, but when an intermediate density of snails are consuming algae, the snails cause an intermediate amount of disturbance and a higher diversity of algal species can persist in the ecosystem.

water were similar in the two watersheds before the clear-cutting. Within 6 months after the cutting, the clear-cut watershed showed significant increases in stream nitrate concentrations. With this information, the researchers were able to determine that when trees are no longer present to take up nitrate from the soil, nitrate leaches out of the soil and ends up in the stream that drains the watershed. This study and subsequent research have demonstrated the importance of plants in regulating the cycling of nutrients, as well as the consequences of not allowing new vegetation to grow when a forest is cut.

Studies such as the one done at Hubbard Brook allow investigators to learn a great deal about biogeochemical cycles. We now understand that as forests and grasslands grow, large amounts of nutrients accumulate in the vegetation and in the soil. The growth of forests allows the terrestrial landscape to accumulate nutrients that would otherwise cycle through the system and end up in the ocean. Forests, grasslands, and other terrestrial ecosystems increase the retention of nutrients on land. This is an important way in which ecosystems directly influence their own growing conditions.

Intermediate disturbance hypothesis The hypothesis that ecosystems experiencing intermediate levels of disturbance are more diverse than those with high or low disturbance levels.

Intermediate levels of disturbance favor high species diversity

We have seen that not all disturbance is bad. In fact, some level of ecosystem disturbance is natural, and may even be necessary to maintain species diversity. The **intermediate disturbance hypothesis** states that ecosystems experiencing intermediate levels of disturbance will favor a higher diversity of species than those with high or low disturbance levels. The graph in FIGURE 8.5a illustrates this relationship between ecosystem disturbance and species diversity. Ecosystems in which disturbances are rare experience intense competition

Teaching Tip: Beyond the Classroom

Have students find a local watershed at the EPA website. After the students find the name of a local watershed, divide them into groups to research more about it. You may want to assign each group different tasks, such as drawing the watershed boundaries, finding all cities and towns supplied by the watershed, learning the names of streams and tributaries in the watershed, determining the quality of water in the watershed, and identifying dams located in the watershed. Have each group present their findings to the class.

Find a link to the EPA website by clicking on the link buttons in the TE-book, opening the TRFD, or logging onto the book's companion website bcs.whfreeman.com/friedlandapes2e (teacher login required).

TRM **Chapter 3 Web Resources**

Teaching Tip: Engage

Have students work with a partner to write a clear and simple explanation of the intermediate disturbance hypothesis. Have volunteers share their work with the class.

among species. Because of this, populations of only a few highly competitive species eventually dominate the ecosystem. In places where disturbances are frequent, population growth rates must be high enough to counter the effects of frequent disturbance and prevent species extinction.

An example of the intermediate disturbance hypothesis can be found in marine algae that spend their lives attached to rocks along the rocky coast of New England. In areas containing low densities of common periwinkle snails (*Littorina littorea*), which cause a low disturbance through low amounts of herbivory, just a few algal species dominate the community, as shown in Figure 8.5b. In areas containing high densities of sails, the disturbance from their high amount of herbivory caused only the most herbivore-resistant algal species to persist. However, when snails were present at an intermediate density, which represents an intermediate disturbance, the populations of best competitors never reach a size at which they can dominate, and populations of other species are never driven too close to zero. As a result, we see the highest diversity of species in ecosystems that experience an intermediate frequency of disturbance.

Teaching Tip: Review

Ask students to name different ecosystem disturbances. *Disturbances include natural disasters such as floods, fires, hurricanes, landslides, and droughts. Disturbances can also be biological, as seen with herbivory, predation, and disease.*

module

8 REVIEW

In this module, we learned that disturbances are events that can alter population sizes and community compositions of ecosystems. Watershed studies help scientists investigate how disturbances affect ecosystems; by manipulating watersheds, they can observe how the cycling of water and elements is altered. When a disturbance occurs, some ecosystems exhibit high resistance to the disturbance. Other ecosystems show high resilience, which allows them to bounce back quickly to pre-disturbance conditions. Finally, an ecosystem disturbance of intermediate magnitude favors a high diversity of species.

Module 8 AP® Review Questions

1. Which is NOT true about disturbances?
   (a) They are caused by natural events such as hurricanes.
   (b) They occur only on short time scales.
   (c) They can cause complete destruction of an ecosystem.
   (d) Many result in no change in ecosystem productivity.
   (e) Some are due to anthropogenic causes.

2. An ecosystem that rapidly returns to its original state after a disturbance is
   (a) resistant.
   (b) vigorous.
   (c) resilient.
   (d) stable.
   (e) adaptable.

3. The intermediate disturbance hypothesis states that intermediate levels of disturbance will
   (a) increase runoff.
   (b) increase ecosystem nutrient cycling.
   (c) decrease primary productivity.
   (d) increase species diversity.
   (e) decrease biomass.

4. The Hubbard Brook experiment showed that
   (a) the intermediate disturbance hypothesis is plausible.
   (b) freshwater aquatic ecosystems are often very resilient.
   (c) river restoration can take many years to complete.
   (d) evapotranspiration increases with more vegetation cover.
   (e) deforestation increases nutrient runoff.

5. Which is a measure of how much a disturbance can affect the flows of energy and matter in an ecosystem?
   (a) Diversity
   (b) Intensity
   (c) Resistance
   (d) Resilience
   (e) Homeostasis

Answers to Module 8 AP® Review Questions

1. b
2. c
3. d
4. e
5. c

working toward sustainability

Can We Make Golf Greens Greener?

Though golf is played outdoors on open green courses designed around the contours of the natural landscape, golf courses have a poor environmental reputation. Golf courses are highly managed ecosystems that cover over 3 million ha (7.5 million acres) worldwide—an area about the size of Belgium. About two-thirds of this area consists of closely mowed turfgrass. Closely mowed grass has short leaves that cannot gain enough energy from photosynthesis to grow deep roots. This causes the grass to dry out easily and makes it difficult to obtain soil nutrients. As a result, the grass is susceptible to challenges from weeds, grubs that feed on grass roots, and fungal diseases that can weaken or kill the grass.

To combat these challenges, golf courses use disproportionate amounts of water, fertilizer, and pesticides. Because humans expect to see green, well-manicured golf courses no matter where in the world they are located, golf courses collectively use 9.5 billion liters (2.5 billion gallons) of water annually to keep their grasses green. Much of this water is used in regions where water is scarce. In addition, providing the grass with sufficient nutrients requires a large amount of fertilizer. For example, the putting greens that surround each of the 18 holes in a golf course require as much nitrogen per hectare as corn, the heaviest nitrogen user of all major food crops. If the course requires irrigation soon after the application of fertilizer, or if it rains, up to 60 percent of the fertilizer can be leached into nearby waterways. To maintain a uniform texture on the greens, golf courses use about six times the amount of agricultural pesticides per hectare as do conventional farms. These chemicals include herbicides to remove weeds, insecticides to kill soil-dwelling grubs, and fungicides to control disease.

Since 1991, the Audubon Cooperative Sanctuary Program (ACSP), a partnership between Audubon International and the U.S. Golf Association, has been working to improve the environmental management of golf courses. The ACSP encourages golf course managers to develop courses that perform more like natural ecosystems, with nutrient and water recycling to reduce waste and biodiversity to increase ecosystem resilience. It also educates golf course managers about low-impact pest management, water conservation, and water quality management.

The golf course of the Palisades Country Club in North Carolina was constructed with natural ecosystem processes in mind. To prevent the runoff of nutrients and lawn-care chemicals into nearby waters, the course directs all runoff water through a treatment system. The course was designed to reduce the amount of closely mowed turfgrass. Deep-rooted native grasses surrounding the greens and fairways soak up nutrients and help to direct water underground. As a result, maintenance costs, chemical applications, and the time spent using machinery have all declined. Smaller areas of turfgrass also leave space for more native vegetation of various heights, providing better habitats for birds and predatory insects. These consumers keep pest populations low, which reduces the need for pesticides. When pesticides are used, they are chosen to protect nontarget wildlife and applied on wind-free days to keep them from spreading beyond where they are needed.

By 2012, more than 2,400 golf courses worldwide were participating in the ACSP. Audubon International found that over 80 percent of the courses in the program reduced the amounts and toxicity of pesticides applied, improved nutrient retention within the course, and used less water for irrigation. The average course in the program saved about 7 million liters (1.9 million gallons) of water per year, and the amount of land area devoted to providing wildlife habitat increased by about 50 percent, from 18 to 27 ha (45 to 67 acres) per 60-ha (150-acre) golf course. Moreover, 99 percent of managers reported that playing quality and golfer satisfaction were maintained or improved.

Even with these changes, golf courses still require large amounts of water, nutrients, fossil fuel energy, and upkeep. Highly managed ecosystems cannot be made input free. However, within these limits, a growing number of courses are attempting to reduce their ecological footprint and make their greens greener.

Making golf more sustainable. The Palisades Country Club in North Carolina is making its golf course more environmentally friendly by considering the important roles of natural ecosystem processes. *(Courtesy The Palisades Country Club)*

Teaching Tip: Discussion Starter

The developed nations of the world burn many more fossil fuels in the form of coal, oil, and natural gas than do the developing nations. Ask students to discuss the carbon cycle and the activities in which developed nations participate that add carbon to our atmosphere. Then ask students to discuss individual ways that will allow all of us to lower our carbon footprint.

Teaching Tip: Measuring Your Impact

Human Influence on the Carbon Cycle Measuring Your Impact 3 asks students to calculate their carbon footprint and compare their results to the U.S. average. This item can be downloaded by clicking on the link buttons in the TE-book, opening the TRFD, or logging onto the book's companion website bcs.whfreeman.com/friedlandapes2e (teacher login required).

Measuring Your Impact 3: Human Influence on the Carbon Cycle

Critical Thinking Questions

1. In addition to helping the environment, what factors do you think help motivate the golf course operators to join the Audubon Cooperative Sanctuary Program?
2. Is the high use of water, fertilizers, and pesticides on golf courses justified by the fact that it brings people outside to enjoy and appreciate nature?

References

Audubon Cooperative Sanctuary Program. http://www.auduboninternational.org/acspgolf.

Beecham, Tara. 2007. How green is your tee? *Stormwater Features*. http://www.stormh2o.com/september-2007/audubon-program-golf.aspx.

chapter 3 REVIEW

In this chapter, we have learned how ecosystems function by looking at how energy moves through an ecosystem and how matter cycles around an ecosystem. This energy and matter form the basis of the trophic groups that exist in ecosystems and they are responsible for the abundance of each group in nature. The typical movement of energy and matter can be altered by ecosystem disturbances, although the magnitude of the impact depends on the resistance of a particular ecosystem and its resilience after the disturbance.

Key Terms

Biosphere
Producer
Autotroph
Photosynthesis
Cellular respiration
Aerobic respiration
Anaerobic respiration
Consumer
Heterotroph
Herbivore
Primary consumer
Carnivore
Secondary consumer
Tertiary consumer
Trophic levels
Food chain
Food web
Scavenger
Detritivore
Decomposers
Gross primary productivity (GPP)
Net primary productivity (NPP)
Biomass
Standing crop
Ecological efficiency
Trophic pyramid
Biogeochemical cycle
Hydrologic cycle
Transpiration
Evapotranspiration
Runoff
Carbon cycle
Limiting nutrient
Macronutrient
Nitrogen cycle
Nitrogen fixation
Nitrification
Assimilation
Mineralization
Ammonification
Denitrification
Leaching
Phosphorus cycle
Algal bloom
Hypoxic
Sulfur cycle
Disturbance
Resistance
Resilience
Watershed
Restoration ecology
Intermediate disturbance hypothesis

Learning Objectives Revisited

Module 6 The Movement of Energy

- **Explain the concept of ecosystem boundaries.**

  Ecosystem boundaries distinguish one ecosystem from another. Although boundaries can be well-defined, often they are not. Boundaries are commonly defined either by topographic features, such as mountain ranges, or are subjectively set by administrative criteria rather than biological criteria.

Suggested Answers to Critical Thinking Questions

1. The reduced cost of maintaining the golf courses is one motivator for the golf course operators. The decreased use of pesticides and the decreased maintenance and labor costs lead to large cost savings. In addition, golfers will be attracted to courses that are more environmentally friendly, which will increase the popularity of golf courses that are members of the program.

2. (Answers will vary.) Some students may feel it is not justified because there are many other activities that allow people to enjoy nature without harming the environment. Other students may feel that the cost to the environment is worthwhile because golfing is an activity that brings many people outside to appreciate the beauty of the landscape.

Exploring the Literature

Bernal, S., et al. 2012. Complex response of the forest nitrogen cycle to climate change. *PNAS* 109:3406–3411.

Crowder, L. B., et al. 2008. The impacts of fisheries on marine ecosystems and the transition to ecosystem-based management. *Annual Review of Ecology, Evolution, and Systematics* 39:259–278.

Elser, J. J., et al. 2007. Global analysis of nitrogen and phosphorus limitation of primary producers in freshwater, marine, and terrestrial ecosystems. *Ecology Letters* 10:1135–1142.

Ricklefs, R. E., and R. A. Relyea. 2014. *Ecology: The Economy of Nature*. 7th ed. W. H. Freeman.

Svensson, J. R., et al. 2012. Disturbance-diversity models: What do they really predict and how are they tested? *Proceedings of the Royal Society*. Series B. 279:2163–2170.

Download a printable version of this list by clicking on the link buttons in the TE-book, opening the TRFD, or logging onto the book's companion website bcs.whfreeman.com/friedlandapes2e (teacher login required).

TRM **Exploring the Literature 3**

- **Describe the processes of photosynthesis and respiration.**

  Photosynthesis captures the energy of the Sun to convert CO_2 and water into carbohydrates. Respiration, whether aerobic or anaerobic, unlocks the chemical energy stored in the cells of organisms.

- **Distinguish among the trophic levels that exist in food chains and food webs.**

  The trophic levels consist of producers that convert solar energy into producer biomass through photosynthesis, primary consumers that eat the producers, secondary consumers that eat the primary consumers, and tertiary consumers that eat the secondary consumers. Omnivores eat individuals from more than one trophic group. Trophic groups that eat waste products and dead organisms are scavengers, detritivores, and decomposers.

- **Quantify ecosystem productivity.**

  Ecosystem productivity can be quantified by measuring the total amount of solar energy that producers capture, which is gross primary productivity, or by measuring the total amount of solar energy captured minus the amount of energy used for respiration, which is net primary productivity.

- **Explain energy transfer efficiency and trophic pyramids.**

  The energy present in one trophic level can be transferred to a higher trophic level and the efficiency of this transfer is generally about 10 percent. Because of this low energy transfer efficiency, the amount of energy present in each trophic level declines as we move to higher trophic levels. We can represent the energy in each trophic level as a rectangular block in a pyramid, with a size that is proportional to the energy found in the trophic level. Low ecological efficiency results in a large biomass of producers, but a much lower biomass of primary consumers and an even lower biomass of secondary consumers.

Module 7 The Movement of Matter

- **Describe how water cycles within ecosystems.**

  In the water cycle, water evaporates from water bodies and transpires from plants. The resulting water vapor cools and forms clouds, which ultimately drop water back to Earth in the form of precipitation. When the water falls onto the land, it can evaporate, be taken up by plants and transpired, percolate into the groundwater, or run off along the soil surface and ultimately return to lakes and oceans.

- **Explain how carbon cycles within ecosystems.**

  In the carbon cycle, producers take up CO_2 for photosynthesis and transfer the carbon to consumers and decomposers. Some of this carbon is converted back into CO_2 by respiration, while the rest is lost to sedimentation and burial. The extraction and combustion of fossil fuels, as well as the destruction of forests, returns CO_2 to the atmosphere.

- **Describe how nitrogen cycles within ecosystems.**

  In the nitrogen cycle, nitrogen fixation by organisms, lightning, or human activities converts nitrogen gas into ammonium (NH_4^+) or nitrate (NO_3^-). Nitrification is a process that converts ammonium into nitrite (NO_2^-) and then into nitrate (NO_3^-). Once producers take up nitrogen as ammonia, ammonium, nitrite, or nitrate, they incorporate it into their tissues in a process called assimilation. Eventually, organisms die and their tissues decompose and are converted to ammonium in a process called mineralization. Finally, denitrification returns nitrogen to the atmosphere.

- **Explain how phosphorus cycles within ecosystems.**

  The phosphorus cycle involves a large pool of phosphorus in rock that is formed by the precipitation of phosphate onto the ocean floor. Geologic forces can lift these sediments and form mountains. The phosphorus in the mountains can be made available to producers either by weathering or by mining. Producers assimilate phosphorus from the soil or water and consumers assimilate it when they eat producers. The waste products and dead bodies of organisms experience mineralization, which returns phosphorus to the environment where it can be ultimately transferred back to the ocean.

- **Discuss the movement of calcium, magnesium, potassium, and sulfur within ecosystems.**

  Calcium, magnesium, and potassium are derived from rock and can be held by soils. Producers can assimilate these elements, and mineralization of waste products and dead organisms returns the elements back to the environment. Most sulfur exists in the form of rocks and is released through the process of weathering, which makes it available for plant assimilation. Some sulfur exists as a gas in the form of sulfur dioxide (SO_2), which can be produced by volcanic eruptions and the burning of fossil fuels. In the atmosphere, SO_2 is converted into sulfuric acid (H_2SO_4) when it mixes with water. The sulfuric acid can then be carried back to the ground when it rains or snows.

Module 8 Responses to Disturbances

- **Distinguish between ecosystem resistance and ecosystem resilience.**

  The resistance of an ecosystem is a measure of how much a disturbance can affect its flows of energy and matter. In contrast, the resilience of an ecosystem is the rate at which an ecosystem returns to its original state after a disturbance has occurred.

- **Explain the insights gained from watershed studies.**
  Because watersheds contain all of the land in a given landscape that drains into a particular water body, experimental manipulations such as logging allow scientists to determine how a disturbance to an ecosystem alters the flow of energy and matter.
- **Explain the intermediate disturbance hypothesis.**
  The intermediate disturbance hypothesis states that ecosystems experiencing intermediate levels of disturbance are more diverse than those with high or low disturbance levels.

Chapter 3 AP® Environmental Science Practice Exam

Section 1: Multiple-Choice Questions

Choose the best answer for questions 1–18.

1. Which of the following is NOT an example of an abiotic component of an ecosystem?
   (a) Water
   (b) Minerals
   (c) Sunlight
   (d) Fungi
   (e) Air

2. Which of the following is NOT characteristic of most ecosystems?
   (a) Biotic components
   (b) Abiotic components
   (c) Recycling of matter
   (d) Distinct boundaries
   (e) A wide range of sizes

3. The waste product in photosynthesis is
   (a) carbon dioxide.
   (b) oxygen.
   (c) glucose.
   (d) water.
   (e) energy.

4. Which of the following could be a cause of decreased evapotranspiration?
   (a) Increased precipitation
   (b) Decreased runoff
   (c) Increased percolation
   (d) Increased vegetation
   (e) Increased sunlight

For questions 5, 6, and 7, select from the following choices:
   (a) Producers
   (b) Decomposers
   (c) Primary consumers
   (d) Secondary consumers
   (e) Tertiary consumers

5. At which trophic level are eagles that consume fish that eat algae?

6. At which trophic level do organisms use a process that produces oxygen as a waste product?

7. At which trophic level are dragonflies that consume mosquitoes that feed on herbivorous mammals?

8. Beginning at the lowest trophic level, arrange the following food chain found on the Serengeti Plain of Africa in the correct sequence.
   (a) Shrubs–gazelles–cheetahs–decomposers
   (b) Shrubs–decomposers–gazelles–cheetahs
   (c) Shrubs–decomposers–cheetahs–gazelles
   (d) Gazelles–decomposers–cheetahs–shrubs
   (e) Decomposers–cheetahs–shrubs–gazelles

9. Which macronutrient is required by humans in the largest amounts?
   (a) Calcium
   (b) Nitrogen
   (c) Sulfur
   (d) Potassium
   (e) Magnesium

10. Roughly what percentage of incoming solar energy is converted into chemical energy by producers?
   (a) 99
   (b) 80
   (c) 50
   (d) Between 5 and 20
   (e) 1

11. The net primary productivity of an ecosystem is 1 kg C/m^2/year, and the energy needed by the producers for their own respiration is 1.5 kg C/m^2/year. The gross primary productivity of such an ecosystem would be
   (a) 0.5 kg C/m^2/year.
   (b) 1.0 kg C/m^2/year.
   (c) 1.5 kg C/m^2/year.
   (d) 2.0 kg C/m^2/year.
   (e) 2.5 kg C/m^2/year.

Answers to Chapter 3 AP® Exam Multiple-Choice Questions

1. d
2. d
3. b
4. c
5. d
6. a
7. e
8. a
9. b
10. e
11. e
12. b
13. c
14. a
15. a
16. b
17. e
18. b

Additional Multiple-Choice AP® Practice Exam

Download an additional Multiple-Choice AP® Environmental Science Practice Exam for Chapter 3 by clicking on the link buttons in the TE-book, opening the TRFD, or logging onto the book's companion website bcs.whfreeman.com/friedlandapes2e (teacher login required).

Multiple-Choice AP® Practice Exam 3

Answers to Chapter 3 AP® Exam

Download the full answers to the Chapter 3 AP® Practice Exam by clicking on the link buttons in the TE-book, opening the TRFD, or logging onto the book's companion website bcs.whfreeman.com/friedlandapes2e (teacher login required).

Answers to Chapter 3 AP® Exam

12. An ecosystem has an ecological efficiency of 10 percent. If the producer level contains 10,000 kilocalories of energy, how much energy does the tertiary consumer level contain?
(a) 1 kcal
(b) 10 kcal
(c) 100 kcal
(d) 1,000 kcal
(e) 10,000 kcal

13. Which biogeochemical cycle(s) does NOT have a gaseous component?
I. Potassium
II. Sulfur
III. Phosphorus
(a) II only
(b) I and II only
(c) III only
(d) II and III only
(e) I and III only

14. Which of the following statements about the carbon cycle is true?
(a) Carbon transfer from photosynthesis is in steady state with respiration and death.
(b) The majority of dead biomass is accumulated in sedimentation.
(c) Combustion of carbon is equivalent in mass to sedimentation.
(d) Most of the carbon entering the oceans is from terrestrial ecosystems.
(e) Carbon exchange between the atmosphere and terrestrial ecosystems is primarily an abiotic process.

15. Human interaction with the nitrogen cycle is primarily due to
(a) the leaching of nitrates into terrestrial ecosystems.
(b) the breakdown of ammonium into ammonia for industrial uses.
(c) the interruption of the mineralization process in urban areas.
(d) the acceleration of the nitrification process in aquatic ecosystems.
(e) the decreased assimilation of ammonium and nitrates.

16. Research at Hubbard Brook showed that stream nitrate concentrations in two watersheds were ________ before clear-cutting, and that after one watershed was clear-cut, its stream nitrate concentration was ________.
(a) similar/decreased
(b) similar/increased
(c) similar/the same
(d) different/increased
(e) different/decreased

17. Small inputs of this substance, commonly a limiting factor in aquatic ecosystems, can result in algal blooms and dead zones.
(a) Dissolved carbon dioxide
(b) Sulfur
(c) Dissolved oxygen
(d) Potassium
(e) Phosphorus

18. After a severe drought, the productivity in an ecosystem took many years to return to pre-drought conditions. This observation indicates that the ecosystem has
(a) high resilience.
(b) low resilience.
(c) high resistance.
(d) low resistance.
(e) equal resilience and resistance.

Section 2: Free-Response Questions

Write your answer to each part clearly. Support your answers with relevant information and examples. Where calculations are required, show your work.

1. Nitrogen is crucial for sustaining life in both terrestrial and aquatic ecosystems.
   (a) Draw a fully labeled diagram of the nitrogen cycle. (4 points)
   (b) Describe the following steps in the nitrogen cycle:
      (i) Nitrogen fixation (1 point)
      (ii) Ammonification (1 point)
      (iii) Nitrification (1 point)
      (iv) Denitrification (1 point)
   (c) Describe one reason why nitrogen is crucial for sustaining life on Earth. (1 point)
   (d) Describe one way that the nitrogen cycle can be disrupted by human activities. (1 point)

2. Read the following article written for a local newspaper and answer the questions below.

Neighbors Voice Opposition to Proposed Clear-Cut

A heated discussion took place last night at the monthly meeting of the Fremont Zoning Board. Local landowner Julia Taylor has filed a request that her 150-acre woodland area be rezoned from residential to multiuse in order to allow her to remove all of the timber from the site.

"This is my land, and I should be able to use it as I see fit," explained Ms. Taylor. "In due course, all of the trees will return and everything will go back to the same as it is now. The birds and the squirrels will still be there in the future. I have to sell the timber because I need the extra revenue to supplement my retirement as I am on a fixed income. I don't see what all the fuss is about," she commented.

A group of owners of adjacent properties see things very differently. Their spokesperson, Ethan Jared, argued against granting a change in the current zoning. "Ms. Taylor has allowed the community to use these woods for many years, and we thank her for that. But I hope that the local children will be able to hike and explore the woods with their children as I have done with mine. Removing the trees in a clear-cut will damage our community in many ways, and it could lead to contamination of the groundwater and streams and affect many animal and plant species. Like the rest of us property owners, Ms. Taylor gets her drinking water from a well, and I do not think she has really looked at all the ramifications should her plan go through. We strongly oppose the rezoning of this land—it has a right to be left untouched."

After more than 2 hours of debate between Ms. Taylor and many of the local residents, the chair of the Zoning Board decided to research the points raised by the neighbors and report on his findings at next month's meeting.

   (a) Name and describe the ecosystem value(s) that are being expressed by Ms. Taylor in her proposal to clear-cut the wooded area. (2 points)
   (b) Name and describe the ecosystem value(s) that Mr. Jared is placing on the wooded area. (2 points)
   (c) Provide three realistic suggestions for Ms. Taylor that could provide her with revenue from the property but leave the woods intact. (3 points)
   (d) Identify and then discuss the validity of the environmental concerns that were raised by Mr. Jared. (3 points)

Answers to Chapter 3 AP® Exam Free-Response Questions

1. (a) See Figure 7.3 on page 84.
(b) (i) Nitrogen fixation is the conversion of nitrogen gas (N_2) to ammonia (NH_3) that can be utilized by producers. Certain bacteria perform the process, and legumes such as peas and beans contain such bacteria and fungi on their roots. Nitrogen fixation can also occur from lightning and the process of combustion.
(ii) When an organism dies, decomposer microorganisms that use nitrogen-containing materials for food convert organic matter to ammonium ions through the process of ammonification.
(iii) Nitrification is the conversion of ammonium ions through a two-step process, first into nitrite (NO_2^-) and then into nitrate (NO_3^-), which can be absorbed by producers.
(iv) Denitrification is the natural conversion of nitrate to nitrous oxide (N_2O), a gas that is released into the atmosphere where it can be converted into nitrogen gas.
(c) Possible answers include:

- Nitrogen is found in amino acids, which are the building blocks of proteins, found in all plant and animal species.
- As a macronutrient, nitrogen is required in high concentrations for living organisms, including humans; it is often considered a limiting factor.
- Specific amounts of nitrogen are required by plants for healthy growth.

(d) The burning of fossil fuels can fix more nitrogen into nitrate, which comes to Earth in precipitation, and could lead to greater nitrate input than ecosystems have historically experienced.

2. (a) Ms. Taylor is referring to the instrumental value of an ecosystem, where an ecosystem provides a direct benefit to humans, such as providing cash revenue from the trees in the 150-acre woodland. Specifically, this is the provisions category.
(b) Mr. Jared is referring to a number of ecosystem values. He refers to the instrumental value of the ecosystem filtering the water, providing the residents with a supply of good quality water in their wells. He is also referring to the cultural benefits, because the local residents use the woods to enjoy the natural beauty of the area as well as the aesthetic pleasure that spending time in the area with their children provides them. He also notes the intrinsic value of the woods when he states, "They have a right to be left untouched," because that does not refer to any direct human benefit.
(c) Possible answers include:

- Estimate the total revenue that Ms. Taylor might get from the logging operation and then set up a group of local residents who will raise funds if she agrees not to clear-cut in the future.
- Have local residents pay a yearly fee to Ms. Taylor for using the woods and the natural services that they provide. The amount would be based on matching the income from logging.
- Money could be raised by the local residents or the town to buy the land from Ms. Taylor or to purchase the rights to preserve the land, thereby providing money for Ms. Taylor in her retirement.

(d) Mr. Jared is concerned that the clear-cut could cause a decline in ecosystem services provided by the woodland. For example, if the land were clear-cut, a number of consequences might result, including elevation of nitrogen (nitrate) levels in nearby streams, loss of habitat leading to loss of biodiversity, erosion of soil into nearby streams, and loss of aesthetic value of the natural environment.

chapter 4

Global Climates and Biomes

PD Overview

This chapter explores the factors that affect the distribution of heat and precipitation around the globe, including the unequal heating of Earth, air currents, and ocean currents. After looking at the pattern of climate differences around the world, the text describes how similar climates support similar types of plants and details characteristics of the nine terrestrial biomes. The chapter closes with a description of different aquatic biomes.

Many topics in this chapter have been covered on previous AP® exams. Make sure students know the layers of the atmosphere and temperature trends in each layer. Discuss the seasons and what causes them. Students should be able to describe the Coriolis effect and ocean currents. The rain shadow effect and the El Niño–Southern Oscillation are also topics that should be covered in detail.

Module 9: The Unequal Heating of Earth

The unequal heating of Earth by the Sun is a major driver of the world's climates. To explain how this unequal heating occurs, the module begins with the composition of Earth's atmosphere. Other important topics include the angle at which sunlight strikes Earth, the area over which the Sun's energy is spread, and the ability of different regions of the world to absorb this energy. Finally the tilt of Earth on its axis affects the energy received from one season to another throughout the year.

Module 10: Air Currents

This module explores how air currents affect world climates. It begins with a discussion of how the properties of air affect the amount of moisture it can carry. The unequal heating of Earth causes air to circulate along the ground and up into the atmosphere. The direction of this air movement is also affected by the rotation of Earth. Air currents that travel over mountain ranges can cause different climates to exist on opposite sides of the mountains.

Module 11: Ocean Currents

This module examines the circulation of ocean waters, both at the surface and deep in the ocean. Ocean circulation affects the transport of heat around the globe. This movement of heat affects the climates of continents. Disruptions to these circulation patterns can dramatically alter global climates.

Module 12: Terrestrial Biomes

Regional differences in climate help determine the species of plants and animals that live in each region. Such patterns help scientists categorize these regions by their dominant growth forms. This module introduces climate diagrams and the concept of biomes. It moves on to cover the nine terrestrial biomes in detail.

Module 13: Aquatic Biomes

This module looks at aquatic biomes. Aquatic biomes fall into two broad categories: freshwater and marine. Freshwater biomes include streams, rivers, lakes, and wetlands. Marine biomes include estuaries, coral reefs, and the open ocean.

Alignment to AP® Environmental Science Course Description

AP® Outline	Module	
I. Earth Systems and Resources (10–15%)		
A. Earth Science Concepts	Module 9	The Unequal Heating of Earth
B. The Atmosphere	Module 9	The Unequal Heating of Earth
	Module 10	Air Currents
	Module 11	Ocean Currents
II. The Living World (10–15%)		
A. Ecosystem Structure	Module 12	Terrestrial Biomes
	Module 13	Aquatic Biomes

Chapter Learning Objectives

After completing this chapter students will be able to

- identify the five layers of the atmosphere.
- discuss the factors that cause unequal heating of Earth.
- describe how Earth's tilt affects seasonal differences in temperatures.
- explain how the properties of air affect the way it moves in the atmosphere.
- identify the factors that drive atmospheric convection currents.
- describe how Earth's rotation affects the movement of air currents.
- explain how the movement of air currents over mountain ranges affects climates.
- describe the patterns of surface ocean circulation.
- explain the mixing of surface and deep ocean waters from thermohaline circulation.
- identify the causes and consequences of the El Niño–Southern Oscillation.
- explain how we define terrestrial biomes.
- interpret climate diagrams.
- identify the nine terrestrial biomes.
- identify the major fresh water biomes.
- identify the major marine biomes.

Chapter 4 Pacing Guide

This pacing guide is based on a schedule with 120 sessions of 50 minutes each before the AP® Exam. If you have a different number of sessions before the exam, you can modify the pacing to suit your needs. If you have additional time, consider incorporating quizzes, released AP® Environmental Science free-response and multiple-choice questions, or additional activities.

Module	Standard Schedule Days	Block Schedule Days
Module 9	1	½
Module 10	2	1
Module 11	1	½
Module 12	2	1
Module 13	1	½
Assessment	1	½

Chapter 4 Resources

The link to these resources can be found by clicking on the link buttons in the Teacher's e-Book (TE-book), opening the Teacher's Resource Flash Drive (TRFD), or logging onto the book's companion website bcs.whfreeman.com/friedlandapes2e (teacher login required).

- **PowerPoint Presentation 4**
- **Optimized Art PowerPoint and JPEG Files 4**
- **Do the Math Videos**
- **Measuring Your Impact 4:** How Much Paper Do You Use?
- **Activity 4.1:** Aquatic Biomes Card Sort
- **Chapter 4 Web Resources**
- **Exploring the Literature 4**
- **Answers to Chapter 4 AP® Exam**
- **Multiple-Choice AP® Practice Exam 4**

Free-Response Questions from Previous AP® Environmental Science Exams

Free-response questions from prior AP® Environmental Science Exams are available on the AP® course website at https://apstudent.collegeboard.org/apcourse/ap-environmental-science/exam practice. Students should be able to answer all of the questions listed below with material learned in this and previous chapters. When a question requires students to understand material from multiple chapters, the question will be listed in the last chapter required to complete the entire question. Questions marked with an asterisk (*) are from exams with released multiple-choice questions. You may want to save these questions until the end of the year so you can give your students a complete released exam for practice. Questions marked with double asterisks (**) require math to calculate a problem. Look for references to these questions throughout the chapter.

Year	Question	Content
2002	4	• The El Niño–Southern Oscillation
2005	4	• The tundra habitat of the Alaskan National Wildlife Refuge
2011	2	• Coral reefs

PD **Chapter 4 Overview**

Watch the video overview of Chapter 4 (for teachers) by clicking on the link buttons in the TE-book, opening the TRFD, or logging onto the book's companion website bcs.whfreeman.com/friedlandapes2e (teacher login required).

TRM **PowerPoint Presentation 4**

Download the PowerPoint presentation for Chapter 4 by clicking on the link buttons in the TE-book, opening the TRFD, or logging onto the book's companion website bcs.whfreeman.com/friedlandapes2e (teacher login required).

Grapes used for fine wines grow best in only a few regions of the world that have a similar climate. *(swetta/Getty Images)*

chapter

4

Global Climates and Biomes

Growing Grapes to Make a Fine Wine

Winemaking has its origin in the Mediterranean region, in places such as Egypt, Greece, Italy, France, and Spain. As the European nations began to colonize other parts of the world, they brought grape vines with them. Today wine is made throughout the world, but the regions known for the finest wines are the Mediterranean, California, Chile, South Africa, and southwestern Australia. What is it about these regions that favors the production of great wines?

Wine making critically depends on having proper growing conditions for grapes. The best conditions are mild, moist winters and hot, dry summers. Mild winters are important because temperatures that fall below freezing can damage the grape vines. Hot, dry summers are important because they are moderately stressful for the plants, which causes them to create the perfect balance of sugars and acids in the grapes. The dry climate also reduces outbreaks of many grape vine diseases that are more prevalent in humid environments.

> Because wine grapes grow best under a narrow range of climatic conditions, global climate change is causing great concern among winegrowers.

The five regions with the best growing conditions are all situated between 30° and 50° latitude, next to the ocean, and typically on the western side of continents. Their similarity in geographic position causes them to have air and water currents that produce comparable climates that provide similar growing conditions. Because of this, these regions also contain plants that look quite similar. Although the plant species in these five locations are not closely related, they consist of drought-tolerant grasses, wildflowers, and shrubs. In

103

Teaching Tip: Chapter Opening Case

The chapter opening case, "Growing Grapes to Make a Fine Wine," illustrates that different regions of the world contain distinct climates and that these climates affect the species that can live in each region. Students learn that certain regions have comparable climates and therefore support similar plant and animal communities. The example of wine grapes further demonstrates that as climate changes, we can expect changes in the species that live in these regions as well as changes in the way humans use these ecosystems.

Teaching Tip: Warm-up

Use the opening case to demonstrate how climatic conditions determine the plant composition of a given ecosystem. Bring bunches of grapes to class, pass them out if you wish, and ask students the following:

- In what regions of the world do wine grapes grow? *The Mediterranean, California, Chile, South Africa, and southwestern Australia.*
- What type of climate do these regions share? *Mild, moist winters and hot, dry summers.*
- Why do grapes thrive in this type of environment? *Mild winters are important because freezing temperatures can damage the grape vines. Hot, dry summers are important because they are moderately stressful for the plants, which causes them to create the perfect balance of sugars and acids in the grapes. The dry climate also reduces outbreaks of many grape vine diseases that are more prevalent in humid environments.*
- Why don't wine grapes grow in other regions of the world? *Wine grapes grow best under a narrow range of climatic conditions. Other regions of the world are either too cold, too dry, or too wet.*

Teaching Tip: Beyond the Classroom

After discussing the opening case on wine grapes, ask students to choose one of their favorite fruits. Then have them use the Internet to research the regions of the world where their fruit is produced. Have students address the following:

- In what parts of the world is your fruit grown?
- To what biome or biomes do these regions belong?
- What climatic conditions in the biome(s) enable your fruit to grow so well?
- How might you expect the production of your fruit to change in the face of climate change?

short, the plants that grow naturally in these areas, much like grapes, are well adapted to the local climate.

Because wine grapes grow best under a narrow range of climatic conditions, global climate change is causing great concern among winegrowers. For example, scientists in 2013 evaluated the current and future climates in the wine region of California, where 90 percent of U.S. wine is produced. They predicted that many vineyards will have to shift northward to Oregon and Washington by 2040 because the ideal climate for winegrowing will shift northward. Similarly, winemakers in France have experienced a decade of exceptionally hot and dry summers, which has made it difficult to grow the unique varieties of French wine grapes. In contrast, English wineries, which are located farther north and have not previously had ideal conditions for growing wine grapes, are now producing some of their best wines because the current English climate is starting to resemble the historic French climate. As winemakers face the reality of global warming and changing climates, they will have to decide whether to move their vineyards to more hospitable climates, modify the growing conditions by adding irrigation, or plant varieties of grapes that are more tolerant to the changing climate, and therefore produce different types of wines.

The story of wine grapes illustrates the point that different regions of the world contain distinct climates and that these climates affect the species that can live in each region. It further demonstrates that when these climates change, we can expect changes in the species that live in the regions and changes in the way humans use these ecosystems.

Sources:
J. T. Iverson, How global warming could change the winemaking map, *Time*, December 3, 2009, at http://www.time.com/time/specials/packages/article/0,28804,1929071_1929070_1945282,00.html;
L. Hannah et al., Climate change, wine and conservation, *Proceedings of the National Academy of Sciences* 110 (2013): 6907–6912.

As we have seen with wine grapes, annual patterns of temperature and precipitation help to determine the types of plants and animals that can live in aquatic and terrestrial habitats around the world. These annual patterns represent a region's **climate,** which is the average weather that occurs in a given region over a long period—typically over several decades. We can contrast climate with the concept of **weather,** which is the short-term conditions of the atmosphere in a local area that include temperature, humidity, clouds, precipitation, and wind speed. To understand how climate differences arise around the world, we need to look at the factors that affect the distribution of heat and precipitation across the globe, which include the unequal heating of Earth, air currents, and ocean currents. Once we understand the pattern of climate differences around the world, we can explore how similar climates support similar types of plants. We can also look at characteristics of different aquatic environments.

Climate The average weather that occurs in a given region over a long period of time.

Weather The short-term conditions of the atmosphere in a local area, which include temperature, humidity, clouds, precipitation, and wind speed.

module

9

The Unequal Heating of Earth

The unequal heating of Earth by the Sun is a major driver of the climates found around the world. To understand how this unequal heating occurs, we need to examine the makeup of Earth's atmosphere. We can then investigate how the angle of sunlight striking Earth affects the area over which the Sun's energy is spread, and the ability of different regions of the world to absorb this energy. Finally, we must consider how the tilt of Earth on its axis affects the energy received from one season to another throughout the year.

Learning Objectives

After reading this module you should be able to

- identify the five layers of the atmosphere.
- discuss the factors that cause unequal heating of Earth.
- describe how Earth's tilt affects seasonal differences in temperatures.

Earth's atmosphere is composed of layers

As FIGURE 9.1 shows, Earth's atmosphere consists of five layers of gases. The pull of gravity on the gas molecules keeps these layers of gases in place. Because each layer of gas has mass, the layers closest to Earth have a greater mass of air above them. This causes the layers closest to Earth to have more densely packed molecules, which causes higher air pressure.

The atmospheric layer closest to Earth's surface, the **troposphere,** extends roughly 16 km (10 miles) above Earth. It is the densest layer of the atmosphere and it is the layer where most of the atmosphere's nitrogen, oxygen, and water vapor occur. The troposphere experiences a great deal of circulation of liquids and gases, and it is the layer where Earth's weather occurs. Air temperature in the troposphere decreases with distance from Earth's surface and varies with latitude. Temperatures can fall as low as −52°C (−62°F) near the top of the troposphere.

Above the troposphere is the **stratosphere,** which extends roughly 16 to 50 km (10–31 miles) above Earth's surface. The stratosphere is less dense than the troposphere. Ozone, a pale blue gas composed of molecules made up of three oxygen atoms (O_3), forms a layer within the stratosphere. This ozone

Troposphere A layer of the atmosphere closest to the surface of Earth, extending up to approximately 16 km (10 miles).

Stratosphere The layer of the atmosphere above the troposphere, extending roughly 16 to 50 km (10–31 miles) above the surface of Earth.

Teaching Tip: Video

Layers of the Atmosphere
Show your class this tutorial. As they watch, have students record the information given for each of the atmospheric layers in the following categories: atmospheric layer name, altitude range, and distinguishing characteristics.

Find the link to this video by clicking on the link buttons in the TE-book, opening the TRFD, or logging onto the book's companion website bcs.whfreeman.com/friedlandapes2e (teacher login required).

Chapter 4 Web Resources

Teaching Tip: Engage

Ask students: Why do airplanes usually fly in the stratosphere and not in the troposphere? *It is difficult for planes to fly in the troposphere because it is dense with air and water molecules; the troposphere is where Earth's weather occurs. The stratosphere is much less dense and therefore a better place for planes to fly.*

Teaching Tip: Remember

An easy way for students to remember the thermosphere is to refer to this layer as the Earth's blanket. It is the thickest and hottest of the atmospheric layers, and the prefix "therm" can help students remember this.

more **math** practice:

Graphing

This activity works best if students use a computer with a plotting application. Present the data in the table below, which were collected from a high-altitude weather balloon. Ask students to create an altitude and temperature graph. This graph (see below) shows the four major layers of Earth's atmosphere, which are easily identified by the changes in temperature. In the troposphere the temperature decreases. In the stratosphere the temperature increases because the ozone layer absorbs the UV rays from the sun. In the mesosphere the temperature decreases, and in the thermosphere the temperature increases.

Have students print their graphs and label the four layers of the atmosphere.

Altitude (km)	Temperature (°C)
0	32
5	0
10	−70
15	−80
20	−72
25	−50
30	−32
35	−20
40	−15
45	−12
50	−10
55	−22
60	−35
65	−51
70	−60
75	−75
80	−85
85	−92
90	−100
95	−95
100	−90
105	−62
110	−5
115	40
120	80

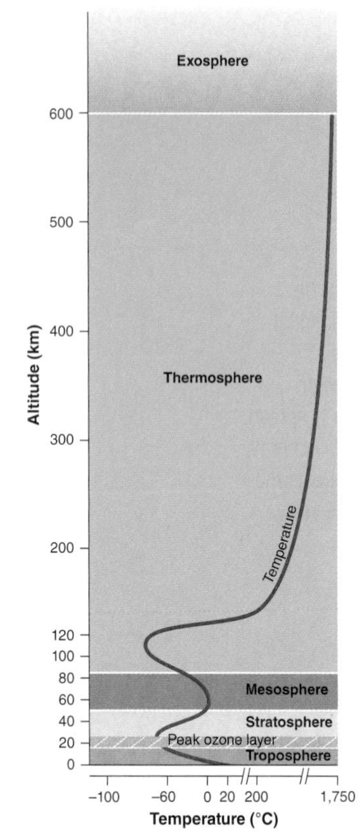

FIGURE 9.1 **The layers of Earth's atmosphere.** The troposphere is the atmospheric layer closest to Earth. Because the density of air decreases with altitude, the troposphere's temperature also decreases with altitude. Temperature increases with altitude in the stratosphere because the Sun's UV-B and UV-C rays warm the upper part of this layer. Temperatures in the thermosphere can reach 1,750°C (3,182°F). *(After http://www.nasa.gov/audience/forstudents/9-12/features/912_liftoff_atm.html)*

layer absorbs most of the Sun's ultraviolet-B (UV-B) radiation and all of its ultraviolet-C (UV-C) radiation. UV radiation can cause DNA damage and cancer in organisms, so the stratospheric ozone layer provides critical protection for our planet. The upper layers in the stratosphere absorb the UV radiation and convert it to infrared radiation, which is released as heat. Because UV radiation from the Sun reaches the higher altitudes of the stratosphere, where much of the UV radiation is absorbed first, there is less UV radiation remaining to be absorbed in the lower stratosphere. As a result, the upper stratosphere is warmer than the lower stratosphere.

Beyond the stratosphere are the mesosphere, the thermosphere, and the exosphere. The atmospheric pressure and density in each of these layers continues to decrease as we move toward the outer layers of the atmosphere. The thermosphere is particularly important to organisms on Earth's surface because of its ability to block harmful X-ray and UV radiation from reaching our planet. The thermosphere is also interesting because it contains charged gas molecules that, when hit by solar energy, begin to glow and produce light, in the same way that a light bulb glows when electricity is applied. Because this interaction between solar energy and gas molecules is driven most intensely by magnetic forces at the North Pole and South Pole, the best places to view the phenomenon are at high latitudes. In the northern United States, Canada, and northern Europe, these glowing gases are known as the northern lights, or aurora borealis. In Australia and southern South America, they are called the southern lights, or aurora australis (FIGURE 9.2).

The amount of solar energy reaching Earth varies with location

Now that we know something about Earth's atmosphere, we can take a closer look at the processes that affect heat and precipitation distribution. As the Sun's energy passes through the atmosphere and strikes land and water, it warms the planet's surface. But this warming does not occur evenly across the planet. This uneven warming pattern has three primary causes.

The first cause of unequal warming is variation in the angle at which the Sun's rays strike Earth. As we

FIGURE 9.2 **Northern lights.** The glowing, moving lights that are visible at high latitudes in both hemispheres are the product of solar radiation energizing the gases of the thermosphere. *(Stephen Mcsweeny/Shutterstock)*

106 CHAPTER 4 ■ Global Climates and Biomes

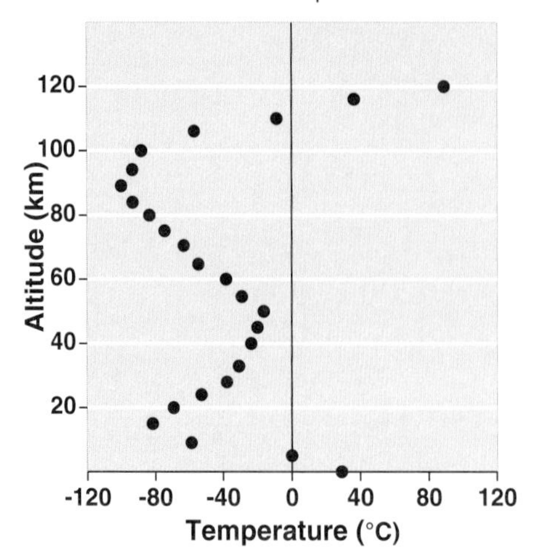

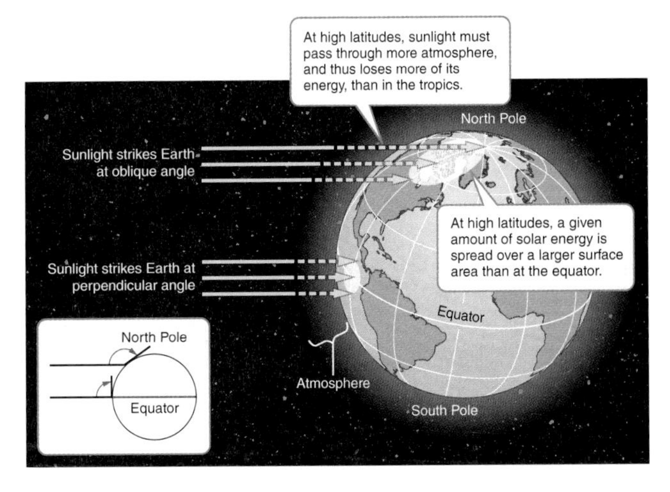

FIGURE 9.3 Differential heating of Earth. Tropical regions near the equator receive more solar energy than mid-latitude and polar regions, where the Sun's rays strike Earth's surface at an oblique angle.

can see in FIGURE 9.3, in the region nearest to the equator—the tropics—the Sun strikes at a perpendicular, or right, angle. In the mid-latitude and polar regions, the Sun's rays strike at a more oblique angle. As a result, the Sun's rays travel a relatively short distance through the atmosphere to reach Earth's surface in the tropics but they must travel a longer distance through the atmosphere to reach Earth's surface near the poles. Because solar energy is lost as it passes through the atmosphere, more solar energy reaches the equator than the mid-latitude and polar regions.

The second cause of the uneven warming of Earth is variation in the amount of surface area over which the Sun's rays are distributed. As you may know, the Sun's rays strike Earth at different angles in different places on the globe, and the angle can change with the time of year. When the Sun's rays strike near the equator, the solar energy is distributed over a smaller surface area than near the poles. Thus, regions near the equator receive more solar energy per square meter than mid-latitude and polar regions. You can replicate this phenomenon by shining a flashlight onto a round object, such as a basketball, in a dark room. If you shine the light perpendicular to the surface of the ball, you will create a small circle of bright light. If you shine the flashlight at an oblique angle, you will create a large oval pool of dim light because the light is distributed over a larger area.

Finally, some areas of Earth reflect more solar energy than others. The percentage of incoming sunlight that is reflected from a surface is called its **albedo.** The higher the albedo of a surface, the more solar energy it reflects and the less it absorbs. A white surface has a higher albedo than a black surface, so it tends to stay cooler. FIGURE 9.4 shows albedo values for various surfaces on Earth. Although Earth has an average albedo of 30 percent, tropical regions with dense green foliage have albedo values of 10 to 20 percent, whereas the snow-covered polar regions have values of 80 to 95 percent.

Earth's tilt causes seasonal changes in climate

As we saw in the previous section, differences in the amount of solar energy striking various latitudes on Earth depend on the angle of the Sun's rays. For the same reason, the amount of solar energy reaching various latitudes shifts over the course of the year. Because Earth's axis of rotation is tilted 23.5°, Earth's orbit around the Sun causes most regions of the world to experience seasonal changes in temperature and precipitation. Specifically, when the Northern Hemisphere is tilted toward the Sun, the Southern Hemisphere is tilted away from the Sun, and vice versa.

FIGURE 9.5 will help us visualize how this works. The Sun's rays strike the equator directly twice a year: once during the March equinox, on March 20 or 21, and again during the September equinox, on September 22 or 23. On those days, virtually all regions of Earth (except those nearest the poles) receive 12 hours of daylight and 12 hours of darkness. For the 6 months

Albedo The percentage of incoming sunlight reflected from a surface.

Teaching Tip: Video

Earth's Tilt 1: The Reason for the Seasons

This video demonstrates Earth's rotation and the Sun's effect at different angles. You may want to have students practice the flashlight demonstration after they see the video.

Find the link to this video by clicking on the link buttons in the TE-book, opening the TRFD, or logging onto the book's companion website bcs.whfreeman.com/friedlandapes2e (teacher login required).

 **Chapter 4 Web Resources**

Teaching Tip: Engage

Ask students the following questions:

- What surface types have a high albedo? *Deserts, white roofs, and new concrete have high albedo.*
- What surface types have a low albedo? *New asphalt, dark soil, and dark buildings have low albedo.*

Teaching Tip: Discussion Starter

Ask students: Why do you think it is hotter at the equator than at the poles? *At high latitudes, sunlight strikes Earth at an oblique angle, passing through more of the atmosphere and therefore losing more solar energy than at the equator. The oblique angle of the Sun at the poles also distributes solar energy over a greater area than at the equator, which causes lower energy per square meter than at the equator.*

TEACHING with FIGURES ▶

Show students Figure 9.4, concealing the percentage of incoming solar energy that each surface type reflects. Ask students to rank the following surfaces from highest to lowest albedo: clouds, cropland/grassland, forest, fresh snow, sea ice, and water. *From highest to lowest albedo: Fresh snow, sea ice, clouds, water, cropland/grassland, forest, and asphalt.*

Teaching Tip: Activity

Before showing students Figure 9.5, ask them to draw a picture of Earth as it orbits around the Sun on its axis. Encourage them to draw Earth's position during each of the four seasons. When they have finished, have each student share his or her drawing with a partner, then be sure to clarify any misconceptions that you observe in their work.

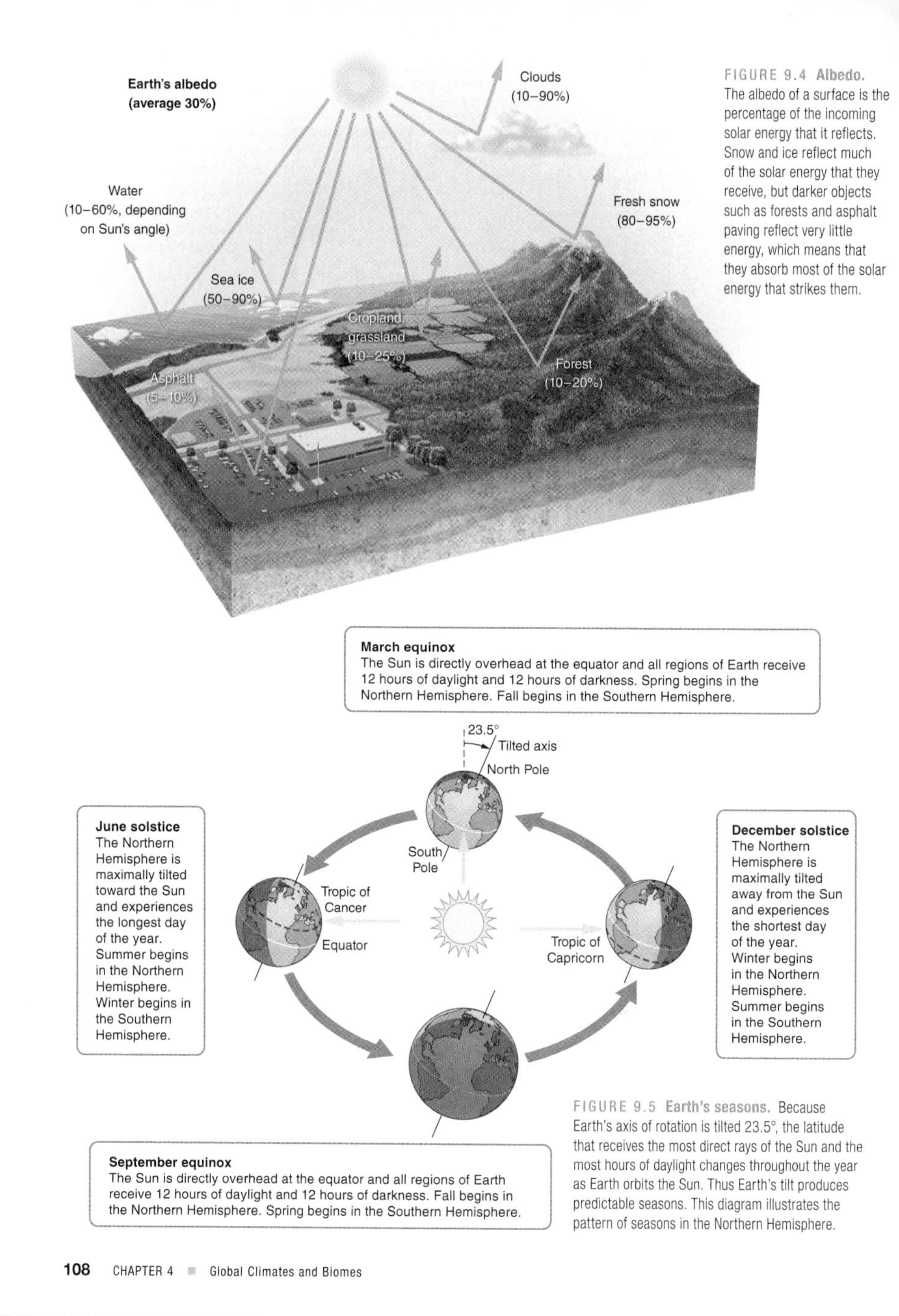

FIGURE 9.4 **Albedo.** The albedo of a surface is the percentage of the incoming solar energy that it reflects. Snow and ice reflect much of the solar energy that they receive, but darker objects such as forests and asphalt paving reflect very little energy, which means that they absorb most of the solar energy that strikes them.

FIGURE 9.5 **Earth's seasons.** Because Earth's axis of rotation is tilted 23.5°, the latitude that receives the most direct rays of the Sun and the most hours of daylight changes throughout the year as Earth orbits the Sun. Thus Earth's tilt produces predictable seasons. This diagram illustrates the pattern of seasons in the Northern Hemisphere.

▲ TEACHING with FIGURES

Show students Figure 9.5 with the labels removed. Ask them to match each Earth image with the season that it represents. When a student identifies an Earth image correctly, ask students to explain why it is correct.

between the March and September equinoxes, the Northern Hemisphere tilts toward the Sun, experiencing more hours of daylight than darkness. The opposite is true in the Southern Hemisphere. On June 20 or 21, the Sun is directly above the Tropic of Cancer at 23.5° N latitude. On this day—the June solstice—the Northern Hemisphere experiences more daylight hours than on any other day of the year. For the 6 months between the September and March equinoxes, the Northern Hemisphere tilts away from the Sun, experiencing fewer hours of daylight than darkness. On December 21 or 22—the December solstice—the Sun is directly over the Tropic of Capricorn at 23.5° S latitude. On this day, the Northern Hemisphere experiences its shortest daylight period of the year, and the Southern Hemisphere experiences its longest daylight period of the year.

module

9

REVIEW

In this module, we have learned that the atmosphere has five layers, each with different properties. These layers absorb most of the UV radiation and X-rays that are emitted toward Earth from the Sun, while much of the Sun's remaining energy passes through the atmosphere to strike Earth's surface. The amount of solar energy that strikes the surface of Earth at any location depends on the angle of the Sun's rays and the albedo. Earth's tilted axis causes the planet's unequal heating to shift throughout the year, which produces seasonal changes in the amount of solar energy that strikes different regions of the world. In the next module, we will examine how this unequal heating of Earth affects the circulation of air currents, which, in turn, help determine global climates.

Module 9 AP® Review Questions

1. Which list is in the correct order of atmospheric layers starting from Earth's surface?
   (a) Thermosphere, stratosphere, troposphere, mesosphere, exosphere
   (b) Exosphere, troposphere, mesosphere, stratosphere, thermosphere
   (c) Mesosphere, stratosphere, thermosphere, exosphere, troposphere
   (d) Thermosphere, troposphere, stratosphere, exosphere, mesosphere
   (e) Troposphere, stratosphere, mesosphere, thermosphere, exosphere

2. Which of the following contributes to the unequal warming of Earth?
   I. Variation in the angle of sunlight that reaches Earth
   II. Variation in the amount of surface area over which the Sun's rays are distributed
   III. Variation in the amount of sunlight reflected from clouds

   (a) I only
   (b) I and II only
   (c) II and III only
   (d) III only
   (e) I, II, and III

3. Which of the following has the highest albedo?
   (a) Soil
   (b) Snow
   (c) Water
   (d) Tropical rainforest
   (e) Pavement

4. Summer in the Northern Hemisphere is warmer primarily because of
   I. increased atmospheric absorption of solar radiation.
   II. the tilt of Earth's axis.
   III. reduced albedo.

   (a) I only
   (b) I and II only
   (c) II only
   (d) I and III only
   (e) II and III only

5. Which atmospheric layer contains the protective ozone layer?
   (a) Exosphere
   (b) Mesosphere
   (c) Stratosphere
   (d) Thermosphere
   (e) Troposphere

Teaching Tip: Activity

To do this activity you will need compasses, protractors, a globe, and a flashlight. On a sunny day, go outside with your class and follow the steps listed below.

- Before you begin, be sure to warn students that they must never look directly at the Sun.
- Locate the Sun and take note of the date and the time.
- Have students identify north, south, east, and west, and ask them to predict the path of the Sun as it moves across the sky during the day. *The Sun starts in the east, low on the horizon, and then moves higher in the sky as it travels west, where it approaches the horizon again.*
- Have students use a protractor to estimate the Sun's current angle off the horizon, and ask them to predict what that angle is at midday. *Answers will vary. On the June solstice, the Northern Hemisphere is maximally tilted toward the Sun, and in the United States the Sun appears to be directly overhead. This angle would be close to 90° off the horizon. On the December solstice, the Northern Hemisphere is maximally tilted away from the Sun, and in the United States the Sun appears to be lower in the sky. This angle would be closer to 60° off the horizon, depending on your specific geographic location.*
- Ask students to explain why the Sun is lower in the sky during winter than during the summer. *During the summer, the Northern Hemisphere is tilted toward the Sun, so the Sun appears directly overhead in the sky. During the winter, the Northern Hemisphere is tilted away from the Sun, so the Sun appears to be lower in the sky.*

(Note that you can use a globe and flashlight to simulate this occurrence.)

Answers to Module 9 AP® Review Questions

1. e
2. b
3. b
4. c
5. c

module

10 Air Currents

The unequal heating of Earth has wide-ranging effects on the climates of our planet not only because it determines regional temperatures, but also because it drives the circulation of air around the planet. These air currents bring warm and cold air to different regions. They also move moisture around the planet, which causes some regions to receive more precipitation than others. In this module, we will explore how air currents have a major impact on the climates of the world. We will begin by examining how the properties of air affect the amount of moisture it can carry. We will then explore how the unequal heating of Earth causes air to circulate along the ground and up into the atmosphere. As we will see, the direction of this air movement is also affected by the rotation of Earth. Finally, we will discuss how air currents that travel over mountain ranges can cause different climates to exist on opposite sides of the mountains.

Learning Objectives

After reading this module you should be able to

- explain how the properties of air affect the way it moves in the atmosphere.
- identify the factors that drive atmospheric convection currents.
- describe how Earth's rotation affects the movement of air currents.
- explain how the movement of air currents over mountain ranges affects climates.

Teaching Tip: Engage

Ask students: Why does warm air have a lower density than cold air? *As air becomes warmer, air molecules gain energy and therefore move more quickly. This causes warm air molecules to be less densely packed than cold air molecules, which results in a lower density of warm air.*

Air has several important properties that determine how it circulates in the atmosphere

Air has four properties that determine how it circulates in the atmosphere: density, water vapor capacity, adiabatic heating or cooling, and latent heat release.

The first property is air density, which is the mass of all molecules in the air in a given volume. The density of air determines its movement: Less dense air rises, whereas denser air sinks. At a constant atmospheric pressure, warm air has a lower density than cold air. Because of this density difference, warm air rises, whether in a room in your house or in the atmosphere.

The second property that determines how air circulates is its capacity to contain water vapor. For example, warm air is not only less dense than cold air, it also has a higher capacity for water vapor. In regions of the world that receive substantial amounts of precipitation, hot summer days are associated with high humidity because the warm air contains a lot of water vapor. The maximum amount of water vapor that can be in the air at a given temperature is called

Teaching Tip: Remember

Help students remember the relationship between temperature and the capacity of air to contain water vapor by using the phrases "hot and humid" and "cool and crisp."

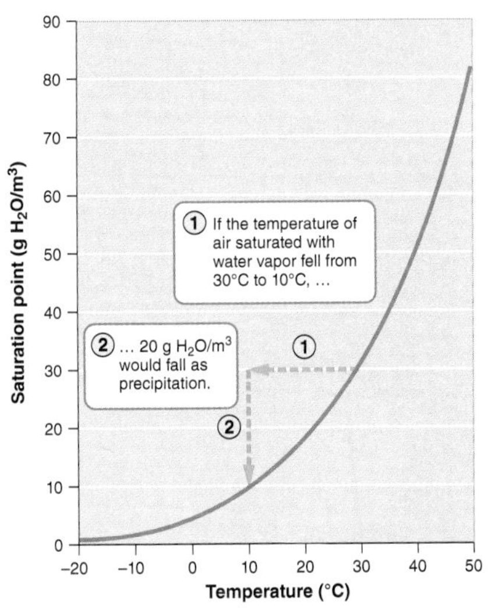

FIGURE 10.1 **The saturation point of air.** When air cools and its saturation point drops, water vapor condenses into liquid water that forms clouds. These clouds are ultimately the source of precipitation.

its **saturation point.** FIGURE 10.1 shows the relationship between the temperature of air and its saturation point. When the temperature of air falls, its saturation point decreases, water vapor condenses into liquid water, clouds form, and precipitation occurs.

A third property of air is its response to changes in pressure. As air rises higher in the atmosphere, the pressure on it decreases. The lower pressure allows the rising air to expand in volume, and this expansion lowers the temperature of the air. The cooling effect of reduced pressure on air as it rises in the atmosphere and expands is called **adiabatic cooling.** Conversely, when air sinks toward Earth's surface, the pressure on it increases. The higher pressure forces the air to decrease in volume, and this decrease raises the temperature of the air. The heating effect of increased pressure on air as it sinks toward the surface of Earth and decreases in volume is called **adiabatic heating.**

The final property of air that determines how it circulates is the production of heat when water vapor condenses from a gas to a liquid. As you may know, the Sun provides the energy necessary to evaporate water on Earth's surface and convert it into water vapor, which enters the atmosphere. In the reverse process, when water vapor in the atmosphere condenses into liquid water, energy is released as heat. The release of energy when water vapor in the atmosphere condenses into liquid is known as **latent heat release.** Because of latent heat release, whenever water vapor in the atmosphere condenses, the air will become warmer and rise in the atmosphere.

Atmospheric convection currents move air and moisture around the globe

A major contributor to the different climates of the world is *atmospheric convection currents.* **Atmospheric convection currents** are global patterns of air movement that are initiated by the unequal heating of Earth. These air currents are called convection currents because they involve the movement of air that absorbs and releases heat as the air moves up into the atmosphere and then back to Earth. Now that we understand the four properties of air, we can look at how these properties contribute to atmospheric convection currents. In the next section we will consider how these atmospheric currents are deflected by Earth's rotation.

We can examine how atmospheric currents develop in FIGURE 10.2. The warming of humid air at Earth's surface decreases the density of the air. As a result, the air begins to rise and, as it does, it begins to experience lower atmospheric pressures and adiabatic cooling. The cooling causes the air to reach its saturation point, which leads to condensation that causes cloud formation and precipitation. Condensation also causes latent heat release, which offsets some of the adiabatic cooling and becomes a strong driving force to make the air expand further and rise higher. Collectively, these processes cause air to rise continuously from Earth's surface near the equator, forming a river of air flowing upward into the troposphere.

As the air reaches the top of the troposphere, it is chilled by adiabatic cooling. This air now contains relatively little water vapor. As warmer air continually

Saturation point The maximum amount of water vapor in the air at a given temperature.

Adiabatic cooling The cooling effect of reduced pressure on air as it rises higher in the atmosphere and expands.

Adiabatic heating The heating effect of increased pressure on air as it sinks toward the surface of Earth and decreases in volume.

Latent heat release The release of energy when water vapor in the atmosphere condenses into liquid water.

Atmospheric convection current Global patterns of air movement that are initiated by the unequal heating of Earth.

Practicing Science
Understanding Processes

The processes that create atmospheric currents are complex. Help students clarify the various steps by asking them first to read carefully the explanation in the text, and then to create a flow chart outlining each step in the process. Their flow charts should include the sequence shown below. *Air warms → pressure decreases → air rises → lower atmospheric pressure → adiabatic cooling → condensation and precipitation → latent heat release → air warms → air rises more → lower atmospheric pressure → adiabatic cooling → horizontal displacement and sinking → higher atmospheric pressure → adiabatic warming*

Teaching Tip: Connections

Understanding the processes of adiabatic cooling and heating is often difficult for students. If students have taken physics or chemistry, they may find it useful to refer to the ideal gas law: $PV = nRT$. This helps explain that when pressure (P) decreases, temperature (T) also decreases to keep the equation balanced. (The change in volume is less significant than the change in pressure.) Conversely, when pressure (P) increases, temperature (T) also increases.

Teaching Tip: Video

Latent Heat Captured on Video!
Show your class this video demonstrating latent heat release in action. The video also explains how to perform the demonstration, in case you prefer to do it live in front of your class.

Find the link to this video by clicking on the link buttons in the TE-book, opening the TRFD, or logging onto the book's companion website bcs.whfreeman.com/friedlandapes2e (teacher login required).

Chapter 4 Web Resources

Teaching Tip: Engage

Show students *Global Maps* from the NASA Space Observatory, an animated vegetation map of Earth. Point out the dense rainforests that occur along the equator and the desert regions to their north and south, at approximately 30° N and 30° S. Then ask: Using what we learned about Hadley cells, how do you explain why this pattern occurs? *The region where solar energy strikes the Earth most intensely causes warm air to rise, which subsequently cools, condenses, and falls as precipitation. This region of intense sunlight occurs around the equator between 23.5° N and 23.5° S, which is where most of the tropical rainforests are located. After the tropical air loses its moisture, the air sinks back to Earth at approximately 30° N and 30° S, where it becomes warm and dry. As a result, regions at 30° N and 30° S are typically hot, dry deserts.*

Find a link to the vegetation map by clicking on the link buttons in the TE-book, opening the TRFD, or logging onto the book's companion website bcs.whfreeman.com/friedlandapes2e (teacher login required).

Chapter 4 Web Resources

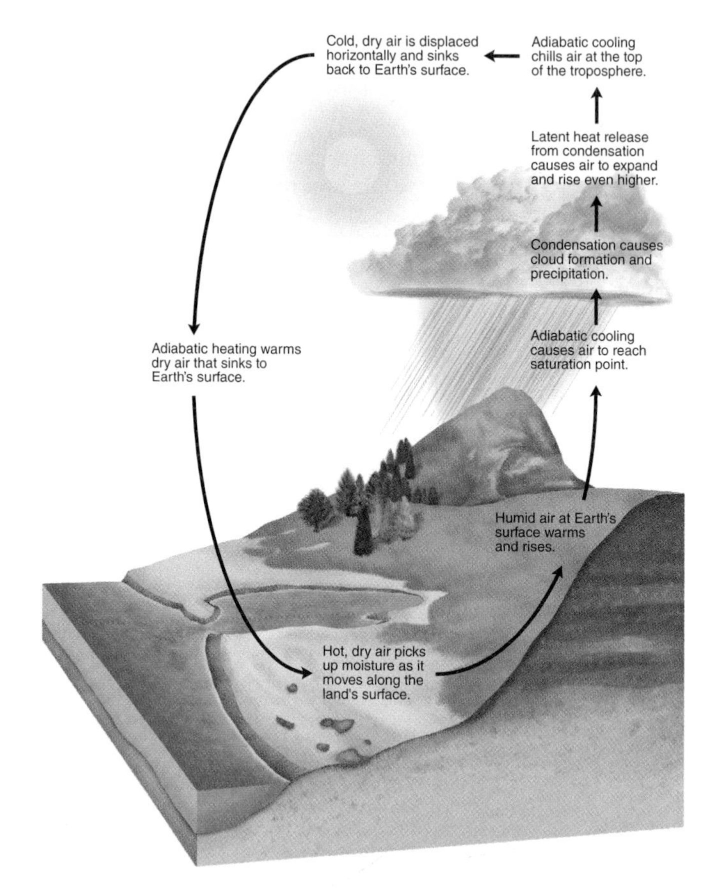

FIGURE 10.2 **Atmospheric currents.** Warming at Earth's surface causes air to rise up into the atmosphere where it experiences lower pressures, adiabatic cooling, and latent heat release. The cool air near the top of the atmosphere is then displaced horizontally before it sinks back to Earth. As it sinks, the air experiences adiabatic heating and then moves horizontally along the surface of Earth to complete the cycle.

rises from below, the cold, dry air at the top of the troposphere is displaced horizontally and eventually begins to sink back to Earth's surface. As the air sinks, it experiences higher atmospheric pressures, and its reduction in volume causes adiabatic heating. By the time the air reaches Earth's surface, it is hot and dry. This hot and dry air circulates back to the starting point and picks up moisture along the way.

Atmospheric convection currents on Earth are found at specific locations. For example, **Hadley cells** are convection currents that cycle between the equator and approximately 30° N and 30° S. In a Hadley cell, illustrated in FIGURE 10.3, the unequal warming of the tropics causes the air near the equator to expand and rise. This rising air cools and the condensed water falls as precipitation over the tropics. After being chilled by adiabatic cooling near the top of the troposphere, the cold dry air is displaced horizontally north and south of the equator. This air descends back to Earth at approximately 30° N and 30° S. As it sinks, adiabatic heating causes the air to warm. As a result, regions at 30° N and 30° S are typically hot, dry deserts. Once the warm, dry air reaches Earth, it flows back toward the equator to complete the cycle.

Hadley cell A convection current in the atmosphere that cycles between the equator and 30° N and 30° S.

The circulation of the Hadley cells is driven by the intense solar energy that strikes Earth near the equator.

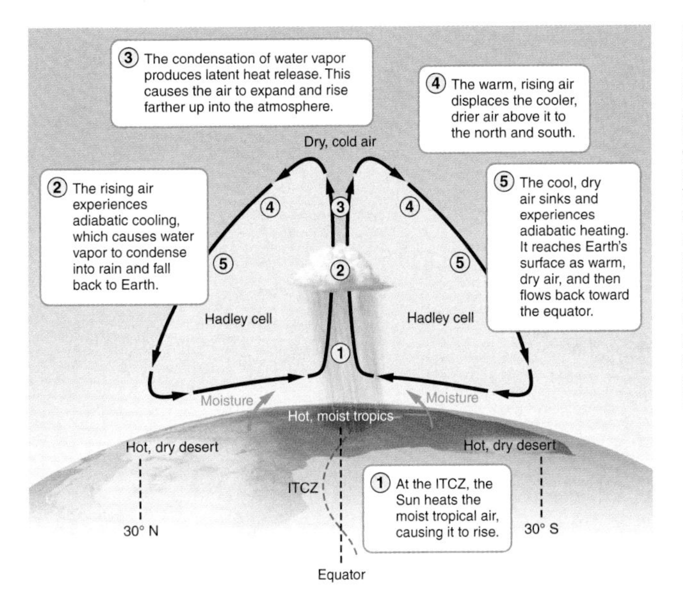

FIGURE 10.3 **Hadley cells.** Hadley cells are atmospheric convection currents that operate between the equator and 30° N and 30° S. Solar energy warms humid air in the tropics. The warm air rises and eventually cools below its saturation point. The water vapor it contains condenses into clouds and precipitation. The air, which now contains little moisture, sinks to Earth's surface at approximately 30° N and 30° S. As the air descends, it is warmed by adiabatic heating. This descent of hot, dry air causes desert environments to develop at those latitudes.

The latitude that receives the most intense sunlight, which causes the ascending branches of the two Hadley cells to converge, is called the **intertropical convergence zone (ITCZ).** Because humid air rises at this latitude and precipitation falls, the ITCZ is typified by dense clouds and intense thunderstorm activity.

The latitude of the ITCZ is not fixed. Because Earth's axis of rotation is tilted, the area receiving the most intense sunlight shifts between approximately 23.5° N and 23.5° S as Earth orbits the Sun. Because the ITCZ occurs at the latitudes that receive the most intense sunlight, the ITCZ moves north and south of the equator over the course of a year. As a result, the tropics experience seasons of high and low precipitation.

Another set of atmospheric convection currents occur near the poles. **Polar cells** are convection currents formed by air that rises at 60° N and 60° S and sinks at the poles (90° N and 90° S). At 60° N and 60° S, the rising air cools and the water vapor condenses into precipitation. The air dries as it moves toward the poles, where it sinks back to Earth's surface. At the poles, the air moves back toward 60° N and 60° S, completing the cycle.

A third convection current, known as **Ferrell cells,** lies between Hadley cells and polar cells. Air currents at these latitudes do not form distinct convection cells, but are driven by the circulation of the neighboring Hadley cells and polar cells. At Earth's surface, some of the warmer air from the Hadley cells moves toward the poles from 30° N and 30° S, and some of the cooler air from the polar cells moves toward the equator from 60° N and 60° S. This movement not only helps to distribute warm air away from the tropics and cold air away from the poles, but also allows a wide range of warm and cold air currents to circulate between 30° and 60°. In this latitudinal range, which includes most of the United States, wind direction can be quite variable, both at Earth's surface and at the top of the troposphere.

Collectively, these convection currents slowly move the warm air of the tropics toward the mid-latitude and polar regions. Because these currents determine the patterns of temperature and precipitation around the world, they are largely responsible for the locations of rainforests, deserts, and grasslands on Earth.

Intertropical convergence zone (ITCZ) The latitude that receives the most intense sunlight, which causes the ascending branches of the two Hadley cells to converge.

Polar cell A convection current in the atmosphere, formed by air that rises at 60° N and 60° S and sinks at the poles, 90° N and 90° S.

Ferrell cell A convection current in the atmosphere that lies between Hadley cells and polar cells.

◀ TEACHING with FIGURES

Copy Figure 10.3 with the bubble caption text covered. (Leave the numbers in place.) Distribute the figure and have students write their own captions for processes 1 to 5 with a partner. If students are struggling with the concept of convection currents, write the text captions on the side of the figure for them, and have students match the captions to the correct stage in the cycle.

Teaching Tip: Engage

Ask students: Based on the patterns of convection currents alone, where on Earth would you most want to live? *Answers will vary.*

Teaching Tip: Activity

To help demonstrate the Coriolis effect, simulate the example of a ball being thrown from the North Pole. Divide students into pairs and give each pair an inflated balloon and a felt marker. Have one student hold the balloon while the other student touches the marker to the "North Pole." Then ask students to simultaneously start rotating the balloon to the "east" and move the pen to the "equator." When the pen reaches the equator, tell students to stop and observe the drawn line. It should curve to the "west" as shown in Figure 10.4a.

Teaching Tip: Remember

A good way for students to remember how the Coriolis effect influences wind patterns is to explain that in the Northern Hemisphere, winds move right from the point of origin. Note: Northeast trade winds move right as they move south from their northern point of origin. In the Southern Hemisphere, winds move left from the point of origin.

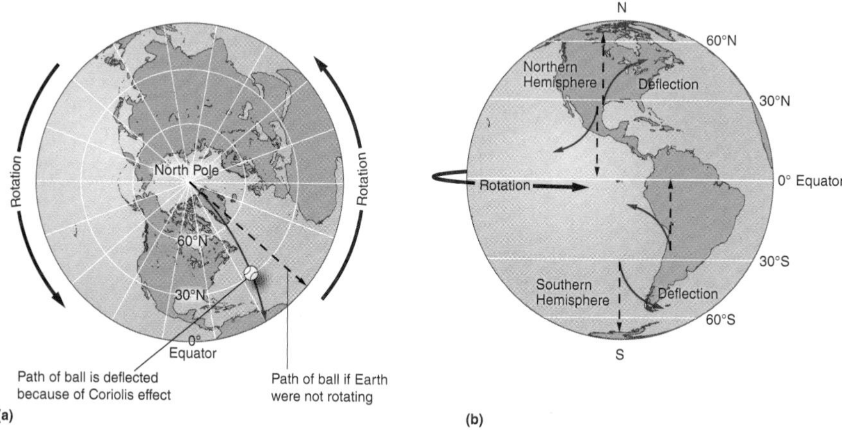

FIGURE 10.4 **The Coriolis effect.** (a) A ball thrown from the North Pole toward the equator would be deflected to the west by the Coriolis effect. (b) The different rotation speeds of Earth at different latitudes cause a deflection in the paths of traveling objects.

Earth's rotation causes the Coriolis effect

The atmospheric convection currents cause air to move directly north or south. However, the actual air currents at Earth's surface are deflected to the east or west because our planet is rotating. The rotation of Earth causes the deflection of objects that are moving directly north or south, a phenomenon that we call the **Coriolis** (core-ee-oh-lis) **effect.** Look at the left side of FIGURE 10.4a and imagine that you can stand at the North Pole and throw a ball directly south, all the way down to the equator (0° latitude). If Earth did not rotate, the ball would travel south to the equator in a straight line. But because Earth rotates to the east while the ball is traveling through the air, the ball will land in a location that is west of its intended target. This happens because Earth's surface beneath the ball is moving faster and faster as the ball moves toward the equator. In other words, the path of the ball is deflected with respect to a given location on the globe because Earth is rotating. Now imagine you are standing at 30° N or 30° S latitude. As shown in Figure 10.4b, throwing a ball toward the equator results in a deflection of the ball's path to the west. However, throwing a ball toward the nearest pole results in a deflection of the ball's path to the east.

Coriolis effect The deflection of an object's path due to the rotation of Earth.

The deflection of objects moving north or south occurs because the planet's surface moves much faster at the equator than at higher latitudes. The planet's circumference is 40,000 km (25,000 miles) at the equator, but decreases to 7,000 km (4,350 miles) at 80° N or 80° S latitude. Now imagine traveling all the way around Earth in 24 hours at the equator versus at 80° N latitude. FIGURE 10.5 will help you visualize this journey. At the equator, you must travel 40,000 km; at 80° N latitude you must travel only 7,000 km. To make the trips in 24 hours, you must travel much faster at the

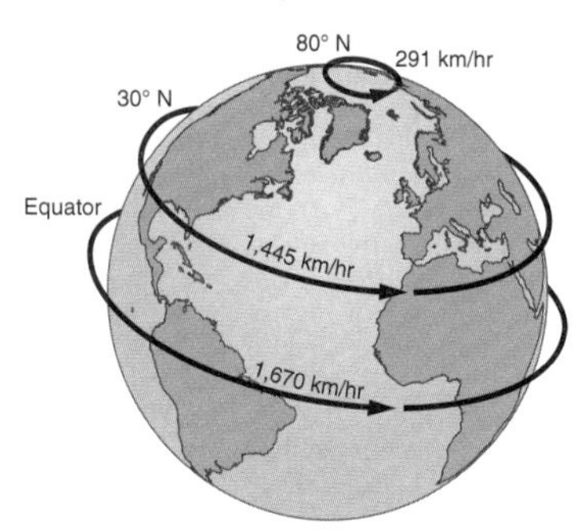

FIGURE 10.5 **The speed of Earth's rotation.** Because all locations on Earth complete one revolution every 24 hours, and because Earth has a greater circumference near the equator than near the poles, its speed of rotation is much faster at the equator than near the poles.

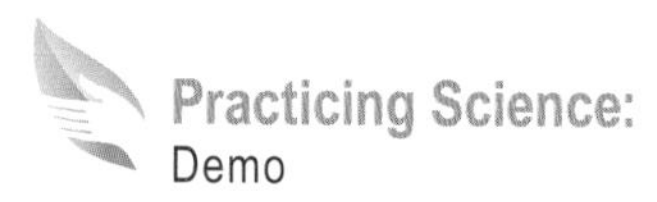

Practicing Science: Demo

Demonstrate how rotation speed increases with an increase in circumference.
Take a paper plate and place a drop of water-soluble paint at its center, a drop halfway to the edge, and another drop close to the edge. Then spin the plate and watch the paint drops move. The center drop should stay relatively stable, while the drop on the edge should form the longest streak.

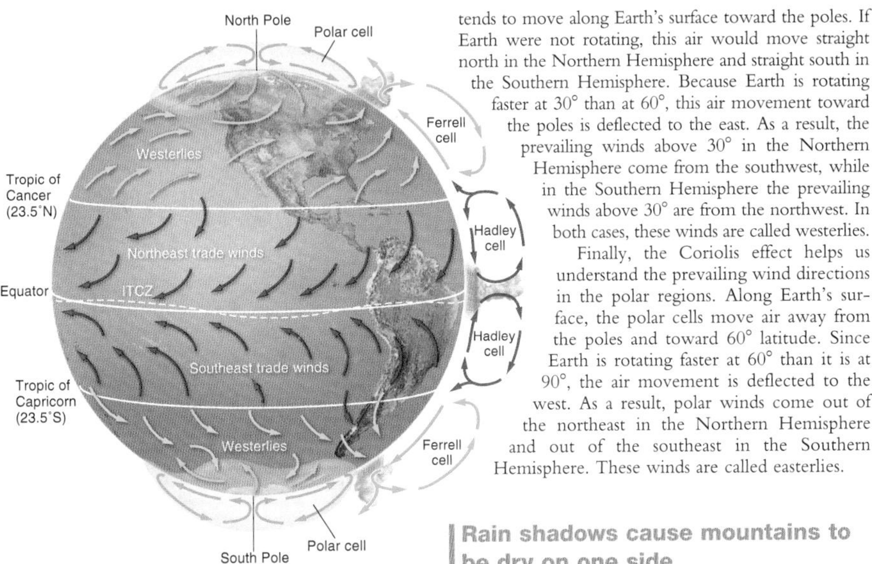

FIGURE 10.6 **Prevailing wind patterns.** Prevailing wind patterns around the world are produced by a combination of atmospheric convection currents and the Coriolis effect.

equator. Indeed, Earth's surface moves at 1,670 km per hour (1,038 miles per hour) at the equator but only 291 km per hour (181 miles per hour) at 80° N or 80° S.

If Earth did not rotate, the air in the convection cells that we discussed earlier would simply move directly north or south and cycle back again. For example, the air in a Hadley cell sinks to Earth's surface at approximately 30° latitude and travels along Earth's surface toward the equator. However, the Coriolis effect produced by Earth's rotation deflects the path of atmospheric convection currents that move toward or away from the equator, as we can see in FIGURE 10.6. Because Earth's speed of rotation is faster near the equator, the path of the air current moving toward the equator is deflected to the west, just as we saw in our example of throwing a ball toward the equator. This deflection causes the Hadley cell north of the equator to produce prevailing winds along Earth's surface that come from the northeast (called northeast trade winds), whereas the cell south of the equator produces prevailing winds that come from the southeast (called southeast trade winds).

The Coriolis effect also explains the prevailing wind directions in the mid-latitudes (between 30° and 60°) where we find the Ferrell cells. These winds can be quite variable as a result of the mixing of air currents from the adjacent Hadley cells and polar cells. Closer to 30°, air tends to move along Earth's surface toward the poles. If Earth were not rotating, this air would move straight north in the Northern Hemisphere and straight south in the Southern Hemisphere. Because Earth is rotating faster at 30° than at 60°, this air movement toward the poles is deflected to the east. As a result, the prevailing winds above 30° in the Northern Hemisphere come from the southwest, while in the Southern Hemisphere the prevailing winds above 30° are from the northwest. In both cases, these winds are called westerlies.

Finally, the Coriolis effect helps us understand the prevailing wind directions in the polar regions. Along Earth's surface, the polar cells move air away from the poles and toward 60° latitude. Since Earth is rotating faster at 60° than it is at 90°, the air movement is deflected to the west. As a result, polar winds come out of the northeast in the Northern Hemisphere and out of the southeast in the Southern Hemisphere. These winds are called easterlies.

Rain shadows cause mountains to be dry on one side

Although many processes that affect weather and climate operate on a global scale, local features, such as mountain ranges, can also play a role. Air moving inland from the ocean often contains a large amount of water vapor. As shown in FIGURE 10.7, when this air meets the windward side of a mountain range—the side facing the wind—it rises and begins to experience adiabatic cooling. Because water vapor condenses as air cools, clouds form and precipitation falls. As is the case in Hadley cells, this condensation causes latent heat release, which helps to accelerate the upward movement of the air. Thus the presence of the mountain range causes large amounts of precipitation to fall on its windward side. The cold, dry air then travels to the other side of the mountain range—called the leeward side—where it descends and experiences higher pressures, which cause adiabatic heating. This now warm, dry air produces arid conditions on the leeward side of the range. It is common to see lush vegetation growing on the windward side of a mountain range while very dry conditions prevail on the leeward side. The dry region formed on the leeward side of the mountain as a result of the humid winds from the ocean that cause precipitation on the windward side is known as a **rain shadow.**

Rain shadow A region with dry conditions found on the leeward side of a mountain range as a result of humid winds from the ocean causing precipitation on the windward side.

◀ Teaching with Figures

Show students Figure 10.6 and ask: Is the ITCZ line shown at the equator always at that location? Why or why not? *No, the ITCZ line shifts throughout the year between 23.5° N and 23.5° S. This occurs as Earth moves around the Sun, which causes the area where solar energy strikes the Earth most intensely to shift north and south.*

Teaching Tip: Discussion Starter

How does the altitude of a mountain range affect its rain shadow? *The altitude of a mountain range determines how high the prevailing winds rise before reaching their peak. As winds rise higher, the air becomes cooler, condenses more, and therefore loses more moisture. As prevailing winds cross a mountain range of lower altitude, the air rises less, cools less, and therefore loses relatively less moisture. As a result, the rain shadow of a lower altitude mountain range may be less distinct.*

Teaching Tip: Beyond the Classroom

Have students research pictures of various rain shadows around the world. Ask them to compare rain shadows in tropical areas and mid-latitude zones, and discuss the difference with the class. *In tropical areas, rain shadows tend to be on the western sides of mountain ranges because of the prevailing trade winds moving from east to west. In mid-latitude zones, such as North America, rain shadows are commonly on the eastern sides of mountain ranges because the prevailing westerlies move from west to east.*

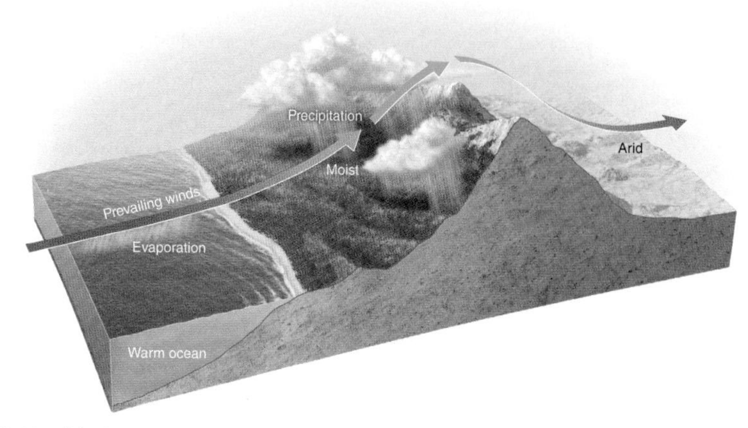

FIGURE 10.7 **Rain shadow.** Rain shadows occur where humid winds blowing inland from the ocean meet a mountain range. On the windward (wind-facing) side of the mountains, air rises and cools, and large amounts of water vapor condense to form clouds and precipitation. On the leeward side of the mountains, cold, dry air descends, warms via adiabatic heating, and causes much drier conditions.

module 10

REVIEW

In this module, we have seen that the properties of air affect how it moves in the atmosphere and how much water vapor it contains. These properties help us understand how the intense sunlight at latitudes near the equator drives the formation of Hadley cells, which are responsible for also producing the polar cells and Ferrell cells. These air convection currents, combined with the Coriolis effect, cause the predominant wind directions that exist around the planet. Air convection currents and the effects of rain shadows help to determine the distribution of heat and precipitation around the globe. In the next module, we will see that many of these processes also drive ocean currents, which also affect climates throughout the world.

Module 10 AP® Review Questions

1. Temperature change in adiabatic heating occurs due to
   (a) water vapor condensing.
   (b) a pressure increase.
   (c) absorption of sunlight.
   (d) a pressure decrease.
   (e) a volume increase.

2. Latent heat release
   (a) increases air temperature.
   (b) decreases air temperature.
   (c) increases the amount of water vapor.
   (d) increases air pressure.
   (e) decreases air pressure.

3. Which of the following does NOT contribute to atmospheric convection currents?
   (a) Hadley cells
   (b) Westerlies
   (c) Gyres
   (d) Polar cells
   (e) The intertropical convergence zone

4. The maximum amount of water vapor air can hold at a given temperature is called the
   (a) saturation point.
   (b) latent heat capacity.
   (c) absolute humidity.
   (d) dew point.
   (e) vapor pressure.

Answers to Module 10 AP® Review Questions

1. b
2. a
3. c
4. a
5. d
6. a
7. d

5. The Coriolis effect is responsible for
   (a) the convergence of two ascending Hadley cells.
   (b) the location of atmospheric currents caused by the unequal heating of Earth.
   (c) the cooling effect of air as it rises and expands.
   (d) the deflection of atmospheric currents.
   (e) the counterclockwise air circulation in the Northern Hemisphere due to Earth's tilt.

6. Which of the following does NOT play a role in causing a rain shadow?
   (a) Cold polar air
   (b) Prevailing trade winds
   (c) A mountain range
   (d) Adiabatic cooling
   (e) Humid ocean air

7. Which is the correct order of the convection cells, starting from the equator?
   (a) Polar, Ferrell, Hadley
   (b) Ferrell, Hadley, Polar
   (c) Polar, Hadley, Ferrell
   (d) Hadley, Ferrell, Polar
   (e) Hadley, Polar, Ferrell

module 11

Ocean Currents

We have seen that the unequal heating of Earth together with the Coriolis effect drive air currents around the world. These two phenomena also affect ocean currents. In this module, we will explore the circulation of ocean waters, both at the surface and in the deep ocean. We will see how ocean circulation affects the transport of heat around the globe, how this heat movement affects the climates of continents, and how disruptions to these circulation patterns can dramatically alter global climates.

Learning Objectives

After reading this module you should be able to

- describe the patterns of surface ocean circulation.
- explain the mixing of surface and deep ocean waters from thermohaline circulation.
- identify the causes and consequences of the El Niño–Southern Oscillation.

Surface ocean currents move warm and cold water around the globe

The flow of ocean water is an important factor for global climates because it moves warm and cold waters to different parts of the globe. In doing so, ocean currents affect the primary productivity found in different ocean regions as well as the climate of the adjacent continents. When warm ocean currents flow from the tropics to higher latitudes, the neighboring continents receive warm air from the ocean. In this section, we will examine the major ocean currents that occur near the oceans and then

Teaching Tip: Connections

Show Figure 10.6 side-by-side with Figure 11.1. Ask students: What would ocean currents look like if there were no continents? *Without continents, ocean currents would more closely resemble prevailing wind patterns. Instead of having clockwise and counterclockwise gyres, water at the equator would move continuously west, and water in mid-latitude regions would move continuously east. This could result in the formation of gyres at the poles.*

Teaching Tip: Engage

Ask students: How would you feel about swimming in an area where upwelling occurs? *Answers will vary, but students would most likely not want to swim in an area of upwelling. Not only are these regions cold, but the abundance of nutrients attracts many fish and other aquatic life, which make the waters not desirable for swimming, except perhaps for snorkeling or scuba diving.*

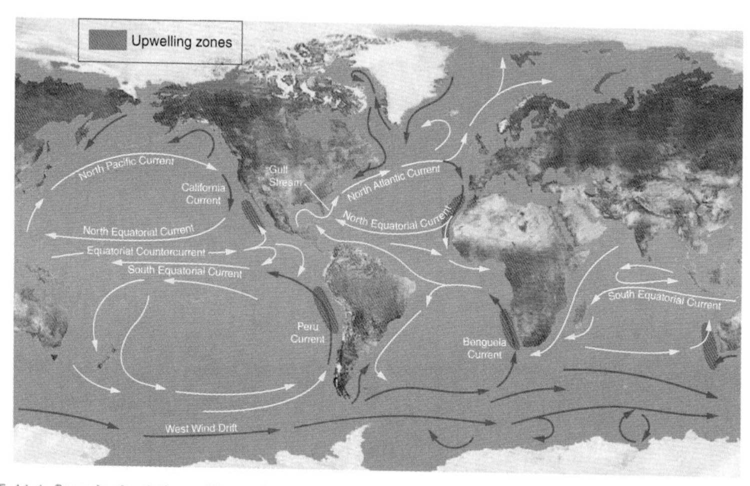

FIGURE 11.1 **Oceanic circulation patterns.** Oceanic circulation patterns are the result of differential heating, gravity, prevailing winds, the Coriolis effect, and the locations of continents. Each of the five major ocean basins contains a gyre driven by the trade winds in the tropics and the westerlies at mid-latitudes. The result is a clockwise circulation pattern in the Northern Hemisphere and a counterclockwise circulation pattern in the Southern Hemisphere. Along the west coasts of many continents, currents diverge and cause the upwelling of deeper and more fertile water.

consider currents that mix the surface water with much deeper water.

Ocean currents are driven by a combination of temperature, gravity, prevailing winds, the Coriolis effect, salinity, and the locations of continents. As we have already observed, the tropics receive the most direct sunlight throughout the year, which generally makes tropical waters warm. Warm water, like warm air, expands and rises. This process raises the tropical water surface about 8 cm (3 inches) higher than mid-latitude waters. While this difference might seem trivial, the slight slope is sufficient for the force of gravity to make water flow away from the equator. As we will see, this movement away from the equator due to gravity combines with other forces to make the surface waters circulate.

Gyres

As the warm tropical waters are pushed away from the equator by gravity, prevailing wind patterns around the globe also play a role in determining the direction in which ocean surface water moves. In the Northern Hemisphere, for example, the trade winds near the equator push water from the northeast to the southwest. However, the Coriolis effect deflects this wind-driven current so that water actually moves from east to west at the equator. Similarly, when winds in northern mid-latitude regions push water from the southwest to the northeast, the Coriolis effect deflects this current so that water actually moves from west to east. FIGURE 11.1 shows the overall effect: Ocean surface currents between the equator and the mid-latitudes rotate in a clockwise direction in the Northern Hemisphere and in a counterclockwise direction in the Southern Hemisphere. These large-scale patterns of water circulation that move clockwise in the Northern Hemisphere and counterclockwise in the Southern Hemisphere are called **gyres.**

Gyres redistribute heat in the ocean, just as atmospheric convection currents redistribute heat in the atmosphere. Cold water from the polar regions moves along the west coasts of continents, and the cool air immediately above these waters brings cooler temperatures to the adjacent continents. For example, the California Current, which flows south from the North Pacific along the coast of California, causes coastal areas of California to have cooler temperatures than areas at similar latitudes on the east coast of the United States. Similarly, warm water from the tropics moves along the east coasts of continents, and the warm air immediately above these waters causes warmer temperatures on land.

Gyre A large-scale pattern of water circulation that moves clockwise in the Northern Hemisphere and counterclockwise in the Southern Hemisphere.

Teaching Tip: Discussion Starter

Ask students: What effect does Earth's rotation have on ocean currents? Why do the gyres move clockwise in the Northern Hemisphere and counterclockwise in the Southern Hemisphere? *Earth's rotation affects ocean currents in much the same way that it affects atmospheric convection currents. Trade winds move water from northeast to southwest along the equator, but the Coriolis effect deflects this wind-driven current so that water actually moves from east to west. Similarly, when winds in northern mid-latitude regions push water from the southwest to the northeast, the Coriolis effect deflects this current so that water actually moves from west to east. This results in ocean surface currents that rotate in a clockwise direction in the Northern Hemisphere and in a counterclockwise direction in the Southern Hemisphere.*

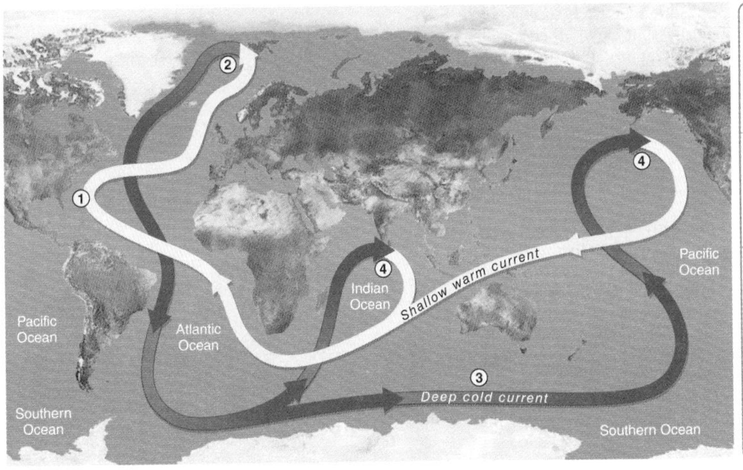
1 Warm water flows from the Gulf of Mexico to the North Atlantic, where some of it freezes and evaporates.
2 The remaining water, now saltier and denser, sinks to the ocean bottom.
3 The cold water travels along the ocean floor, connecting the world's oceans.
4 The cold, deep water eventually rises to the surface and circulates back to the North Atlantic.
Pacific Ocean
Atlantic Ocean
Indian Ocean
Shallow warm current
Deep cold current
Southern Ocean

FIGURE 11.2 **Thermohaline circulation.** The sinking of dense, salty water in the North Atlantic drives a deep, cold current that moves slowly around the world.

Upwelling

Ocean currents also help explain why some regions of the ocean support highly productive ecosystems. Along the west coasts of most continents, for example, the surface currents diverge, or separate from one another, causing deeper waters to rise and replace the water that has moved away. This upward movement of water toward the surface, shown as dark blue regions in Figure 11.1, is called **upwelling.** The deep waters bring with them nutrients from the ocean bottom that support large populations of producers. The producers, in turn, support large populations of fish that have long been important to commercial fisheries.

Deep ocean currents circulate ocean water over long time periods

Another oceanic circulation pattern, **thermohaline circulation,** drives the mixing of surface water and deep water. This process is crucial for moving heat and nutrients around the globe. Thermohaline circulation appears to be driven by surface waters that contain unusually large amounts of salt. As Figure 11.1 shows, warm currents flow from the Gulf of Mexico to the very cold North Atlantic. Some of this water freezes or evaporates, and the salt that remains behind increases the salt concentration of the water. This cold, salty water is relatively dense, so it sinks to the bottom of the ocean, mixing with deeper ocean waters. Two processes—the sinking of cold, salty water at high latitudes and the rising of warm water near the equator—create the movement necessary to drive a deep, cold current that slowly moves past Antarctica and northward to the northern Pacific Ocean, where it returns to the surface and then makes its way back to the Gulf of Mexico. This global round trip, shown in FIGURE 11.2, can take hundreds of years to complete. As you can see in the figure, thermohaline circulation helps to mix the water of all the oceans.

Ocean currents associated with thermohaline circulation affect the temperature of nearby landmasses much as the major gyres affect surface waters. For example, the ocean current known as the Gulf Stream originates in the tropics near the Gulf of Mexico and flows toward western Europe. This movement of warm tropical waters brings warmer temperatures to latitudes that would otherwise be much colder. For instance, England's average winter temperature is 20°C (36°F) warmer than that of Newfoundland, Canada, which is located at a similar latitude but receives cold ocean currents from the North Atlantic.

Scientists are concerned that global warming could affect thermohaline circulation. If increased air temperatures accelerate the melting of glaciers in the Northern Hemisphere, the waters of the North Atlantic

Upwelling The upward movement of ocean water toward the surface as a result of diverging currents.

Thermohaline circulation An oceanic circulation pattern that drives the mixing of surface water and deep water.

◀ TEACHING with FIGURES

Use Figure 11.2 to help students understand thermohaline circulation. Copy the figure with the descriptions in boxes 1 to 4 covered up. Distribute the figure and ask students to write their own captions for boxes 1 to 4 with a partner. If students are struggling with the concept of thermohaline circulation, write the text captions on the side of the figure without using numbers, and have them match the captions to the correct stage in the cycle.

Teaching Tip: Video

The Thermohaline Circulation (The Great Ocean Conveyor Belt)
Show students the short, silent video animating thermohaline circulation. After they have watched the video, divide students into groups and allow each group to access the video on a computer, tablet, or smartphone. Instruct them to watch the video and write a script to accompany the animation, being sure to explain thermohaline circulation thoroughly. Clarify that the dialogue should last the entire video. When finished, have each group present its work to the class.

Find a link to this video by clicking on the link buttons in the TE-book, opening the TRFD, or logging onto the book's companion website bcs.whfreeman.com/friedlandapes2e (teacher login required).

 **Chapter 4 Web Resources**

Practicing Science: Demo

Demonstrate how cold water at the poles and warm water at the equator drive ocean circulation patterns. Before the demo, make ice cubes with blue food coloring dissolved in the water. Fill a clear plastic box (about the size of a shoebox) with tap water, and place the box on top of white paper. To start the demo, carefully drop a blue ice cube on one side of the box and place a few drops of red food coloring on the other side of the box. Watch as the blue food coloring sinks to the bottom and the red food coloring flows on top.

Teaching Tip: Connections

Ask students: In what ways are atmospheric and oceanic circulation patterns similar? How are they different? *Differences in temperature and density play a major role in both atmospheric and oceanic circulation patterns. The Coriolis effect also deflects both wind and ocean currents in a similar way. However, salinity and the location of the continents have a greater effect on oceanic circulation patterns than they do on atmospheric circulation patterns.*

could become less salty and thus less likely to sink. Such a change could potentially shut down thermohaline circulation and stop the transport of warm water to western Europe, making it a much colder place.

The El Niño–Southern Oscillation is caused by a shift in ocean currents

Our discussion of ocean currents has focused on the commonly occurring ocean currents. In the southern Pacific Ocean, for example, trade winds from South America and the Coriolis effect causes the equatorial water to flow from east to west. As we have discussed, this current produces an upwelling of cold, nutrient-rich water along the west coast of South America, which produces large populations of fish.

Every 3 to 7 years, however, the tropical current moves in the opposite direction. Because this phenomenon often begins around the December 25 Christmas holiday, it is has been named El Niño ("the baby boy"). However, since the cause of the phenomenon is due to a reversal of wind and water currents in the South Pacific, the preferred name for the phenomenon is the **El Niño–Southern Oscillation (ENSO)**.

FIGURE 11.3 shows this process in action. First, the trade winds that normally blow from South America and push the surface waters to the west weaken or reverse direction. This change in the wind allows warm equatorial water from the western Pacific to move eastward toward the west coast of South America. The movement of warm water and air toward South America suppresses upwelling off the coast of Peru. The effect can last from a few weeks to a few years.

El Niño–Southern Oscillation (ENSO) A reversal of wind and water currents in the South Pacific.

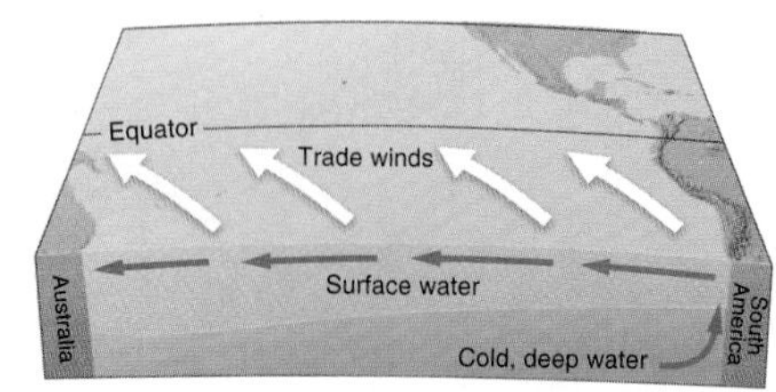

(a) Normal year

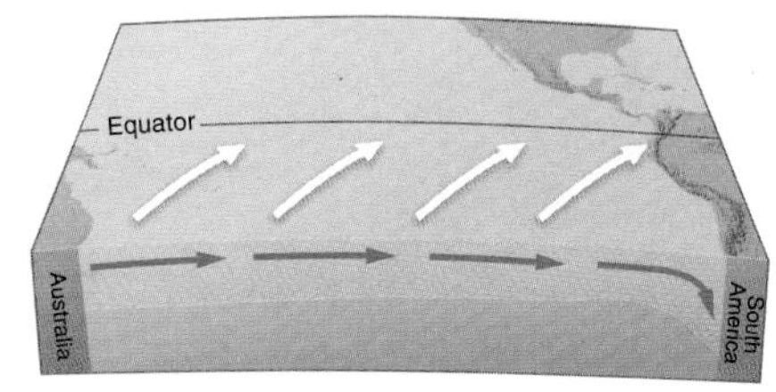

(b) El Niño year

FIGURE 11.3 **The El Niño–Southern Oscillation.** (a) In a normal year, trade winds push warm surface waters away from the coast of South America and promote the upwelling of water from the ocean bottom. (b) In an El Niño year, trade winds weaken or reverse direction, so warm waters build up along the west coast of Peru.

Because ocean currents are so important to global climates, the ENSO has several consequences. As we have noted, it reduces the upwelling off the South American coast, which decreases productivity and dramatically reduces fish populations. It also has widespread effects around the world, including cooler and wetter conditions in the southeastern United States and unusually dry weather in southern Africa and Southeast Asia.

module 11

REVIEW

In this module, we have learned that the circulation of surface ocean currents and deep ocean currents is driven by the unequal heating of Earth, air currents, Coriolis forces, and differences in salt concentrations. These currents help to transport heat to different regions of the world and therefore help determine the climates around the globe. The disruption of these currents—due to the ENSO, and potentially because of glacier melting—can have major impacts on these climates. In the next module, we will examine how differences in climate help to determine the plants and animals that live in different parts of the world.

Teaching Tip: Activity

Review with your class the processes that cause the El Niño–Southern Oscillation, referring to Figure 11.3 as needed. Then divide students into groups and have each group fill in the table below. (Answers are provided in italics.) Keep Figure 11.3 projected for students to use as a reference.

	Normal conditions	El Niño conditions
Wind direction	*East to west*	*West to east*
Wind strength	*Medium*	*Low*
Precipitation location	*Western Pacific*	*Eastern Pacific*
Water temperature in eastern Pacific	*Cold*	*Warm*
Water temperature in western Pacific	*Warm*	*Warm*
Change from normal conditions	*N/A*	*Surface winds change direction and warm water flows to eastern Pacific*

Module 11 AP® Review Questions

1. Which of the following does NOT drive ocean currents?
   (a) Salinity
   (b) Prevailing winds
   (c) Gravity
   (d) Temperature
   (e) Precipitation
2. High productivity and nutrient availability in the ocean occurs in areas with
   (a) gyres.
   (b) upwelling.
   (c) warm waters.
   (d) El Niño.
   (e) prevailing westerlies.
3. Climate change could potentially disrupt which one of the following phenomena?
   (a) El Niño–Southern Oscillation (ENSO)
   (b) Thermohaline circulation
   (c) Rotation of Southern Hemisphere gyres
   (d) Upwelling off the coasts of Australia and New Zealand
   (e) The polar current along Antarctica
4. Which of the following is NOT true of an El Niño–Southern Oscillation (ENSO) event?
   (a) It causes increased upwelling on the coast of South America.
   (b) It causes wetter conditions in the southern United States.
   (c) It can last from several weeks to several years.
   (d) It includes a reversal in tropical ocean currents.
   (e) It often begins near Christmas.
5. Gyres do all of the following EXCEPT
   (a) redistribute heat in the ocean.
   (b) result from the Coriolis effect.
   (c) rotate clockwise in the Northern Hemisphere.
   (d) redistribute nutrients from the deep ocean.
   (e) affect the temperatures of coastal areas.

Answers to Module 11 AP® Review Questions

1. e
2. b
3. b
4. a
5. d

module

12 Terrestrial Biomes

Climate affects the distribution of species around the globe. For example, only species that are well adapted to hot and dry conditions can survive in deserts. A very different set of organisms survives in cold, snowy places. In this module, we will examine how differences in climate help determine the species of plants and animals that live in each region and how scientists categorize these regions.

AP® Exam Tip

The El Niño–Southern Oscillation is covered in the 2002 AP® Exam, Question 4. To answer this question, students must

- make a connection between atmosphere and ocean currents (ENSO)
- understand how climate is linked to human disease
- describe environmental problems associated with ENSO

Learning Objectives

After reading this module you should be able to

- explain how we define terrestrial biomes.
- interpret climate diagrams.
- identify the nine terrestrial biomes.

(a)

(b)

FIGURE 12.1 **Similar growth forms.** Cacti and euphorbs are not closely related, but they have evolved many similar adaptations that allow them to live in hot, dry environments. (a) Organ pipe cactus (*Stenocereus thurberi*) in Arizona. (b) Euphorbia (*E. damarana*) in Namibia. *(a: Dhoxax/Shutterstock; b: Patti Murray /Earth Scenes/Animals Animals)*

Terrestrial biomes are defined by the dominant plant growth forms

Scientists have long recognized that organisms possess distinct growth forms, many of which represent adaptations to local temperature and precipitation patterns. For example, if we were to examine the plants from all the deserts of the world, we would find many cactus-like species. Plants that look like cacti in North American deserts are indeed members of the cactus family, but those in the Kalahari Desert are members of the euphorb family. These two distantly related groups of species look similar because they have evolved similar adaptations to hot, dry environments—including the ability to store large amounts of water in their tissues and a waxy coating to reduce water loss (FIGURE 12.1).

Despite these common adaptations to desert conditions, these two plant families have distinctive flowers and spines, and only the euphorb family produces a milky sap. These differences help to confirm that while the two groups may look similar, genetically they are not closely related. Instead, exposure to similar selective pressures over long periods of time favors the evolution of many similar adaptations. As a result, scientists can categorize terrestrial regions of the world into **terrestrial biomes,** which are geographical regions that each have a particular combination of average annual temperature and precipitation and contain distinctive plant growth forms that are adapted to that climate. In the next module we will examine **aquatic biomes**, which are categorized by particular combinations of salinity, depth, and water flow. FIGURE 12.2 shows the range of terrestrial biomes on Earth in the context of precipitation and temperature. For example, boreal and tundra biomes have average annual temperatures below 5°C (41°F), whereas temperate biomes have average annual temperatures between 5°C and 20°C (68°F), and tropical biomes have average annual temperatures above 20°C. Within each of these temperature ranges, we can observe a wide range of precipitation. FIGURE 12.3 shows the distribution of biomes around the world.

Note that although terrestrial biomes are categorized by plant growth forms, the animal species living in different biomes are often quite distinctive as well. For example, rodents inhabiting deserts around the world have a number of adaptations for hot, dry climates, including highly efficient kidneys that allow very little water loss through urination.

Terrestrial biome A geographic region categorized by a particular combination of average annual temperature, annual precipitation, and distinctive plant growth forms on land.

Aquatic biome An aquatic region characterized by a particular combination of salinity, depth, and water flow.

122 CHAPTER 4 ■ Global Climates and Biomes

Teaching Tip: Activity

Have students organize the information in this module by creating a table like the one shown below. Encourage them to fill in the relevant information as they read about each biome. They may want to use additional resources for information on characteristic animals. (Answers are provided in italics)

Biome	Locations	Average annual temperature (°C)	Precipitation range (cm)	Indicator plants	Indicator animals
Tundra	*Northernmost Northern Hemisphere, Antarctica*	*−10–5*	*0–100*	*Woody shrubs, mosses, heaths, lichens*	*Hare, fox, lemming, caribou, polar bear*
Boreal forest	*50°N–60°N in Europe, Russia, and North America*	*−5–10*	*25–200*	*Coniferous trees: spruce, pine, fir; some deciduous trees: birch, maple, aspen*	*Moose, beaver, bear, wolf*
Temperate rainforest	*West coast of North America, southern Chile, eastern Australia, western New Zealand*	*5–20*	*150–350*	*Coniferous trees: fir, spruce, cedar, hemlock, redwood*	*Blue grouse, Clark's nutcracker, banana slug*
Temperate seasonal forest	*Eastern U.S., Japan, Chile, China, Europe, eastern Australia*	*5–20*	*50–225*	*Deciduous trees: oak, maple, beech, hickory; some coniferous trees*	*Toad, squirrel, chipmunk*
Woodland/shrubland	*Southern California, southern South America, southern Africa, southwestern Australia, Mediterranean*	*−5–20*	*25–125*	*Drought-resistant shrubs: yucca, sagebrush, scrub oak*	*Small nocturnal animals*
Temperate grassland/cold desert	*North America (prairies), South America (pampas), Central Asia, and eastern Europe (steppes)*	*−5–20*	*0–50*	*Grasses, nonwoody flowering plants*	*Bison, zebra, rhinoceros, kangaroo*
Tropical rainforest	*20°N–20°S Central and South America, Africa, southeast Asia, northeastern Australia*	*20–30*	*250–450*	*Lush vegetation including two-thirds of Earth's terrestrial species; cocoa, coffee, cassava, bamboo, strangler figs*	*Great ape, monkey, sloth, toucan*
Tropical seasonal forest/Savanna	*Central America, southern Asia, coastal South America, sub-Saharan Africa, northwestern Australia*	*20–30*	*50–275*	*Dense stands of shrubs and trees: broadleaf grasses, acacia, baobab*	*Scorpion, giraffe, southern white rhinoceros*
Subtropical desert	*30°N and 30°S Mojave Desert (U.S.), Sahara (Africa), Arabian Desert (Middle East), Great Victoria Desert (Australia)*	*18–31*	*0–100*	*Drought-resistant plants: cacti, euphorbs, succulents*	*Bactrian camel, rock hyrax*

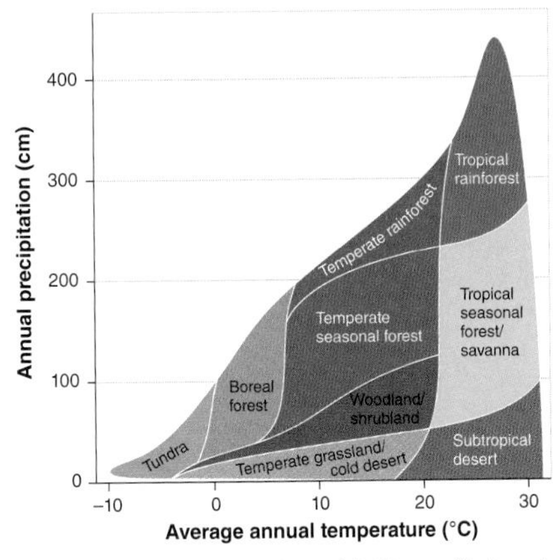

FIGURE 12.2 **Locations of the world's biomes.** Tundra and boreal biomes generally exist when the average annual temperature is less than 5°C (41°F). In contrast, temperate biomes exist in regions that experience average annual temperatures between 5°C and 20°C (41°F and 68°F). Tropical biomes exist in regions that experience average annual temperatures above 20°C.

Climate diagrams illustrate patterns of annual temperature and precipitation

Climate diagrams, such as those shown in FIGURE 12.4, are a helpful way to visualize regional patterns of temperature and precipitation. By graphing average monthly temperature and precipitation, these diagrams illustrate how the conditions in a biome vary during a typical year. They also indicate when the temperature is warm enough for plants to grow—that is, the months when it is above 0°C (32°F), known as the *growing season*. In Figure 12.4a, we can see that the growing season is mid-March through mid-October.

In addition to the growing season, climate diagrams can show the relationship between precipitation, temperature, and plant growth. For every 10°C (18°F) temperature increase, plants need 20 mm (0.8 inches) of additional precipitation each month to supply the extra water demand that warmer temperatures cause. As a result, plant growth can be limited either by temperature or by precipitation. In Figure 12.4a, the precipitation line is above the temperature line during all months. This means that water supply of plants exceeds the water demand, so plant growth is more constrained by temperature than by precipitation.

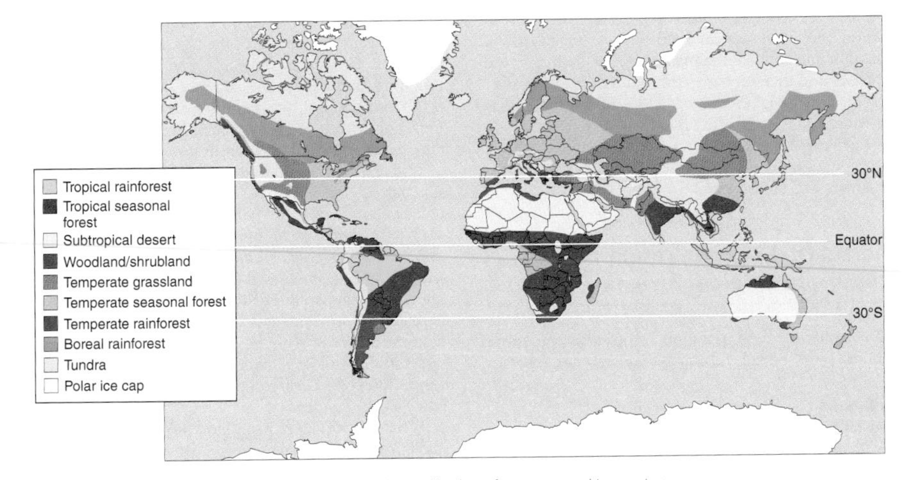

FIGURE 12.3 Biomes. Biomes are categorized by particular combinations of average annual temperature and annual precipitation. *(Data from http://www.biome-explorer.net/Biomes-Map.html)*

◀ TEACHING with FIGURES

On Figure 12.2, remove the biome labels and list the biomes in alphabetical order on the side of the figure. Copy and distribute. Have students work with a partner to match each biome to its location on the graph.

Teaching Tip: Discussion Starter

Ask students: What characteristics are used to identify terrestrial biomes? *Terrestrial biomes are characterized by their plant growth forms, which are in turn determined by various abiotic factors, including average annual precipitation, temperature, and soil type.*

◀ TEACHING with FIGURES

On Figure 12.3, remove the biome labels in the key and list the biomes in alphabetical order on the side of the figure. Copy and distribute. Have students work with a partner to match each biome to its correct location. Ask students to name the biome in which your school is located.

more **math** practice:

Graphing

Ask students to make a climate diagram using the data in the table below. The graph generated by your students should look similar to the graph below the table. After students complete their graphs, ask them which biome the climate diagram represents and how they arrived at their conclusion. *This biome is a tropical rainforest. First, this is clearly a biome along the equator because the temperature is consistently warm year-round. Second, the high levels of precipitation can only be that of a rainforest; the precipitation is too high for the other tropical biomes, deserts and savannas.*

Month	Average temperature (°C)	Average precipitation (mm)
January	26	214.3
February	27	204.14
March	26	204.14
April	24	237.16
May	23	186.36
June	22	181.28
July	21	155.88
August	22	150.8
September	22	186.36
October	23	188.9
November	24	196.52
December	25	170.18

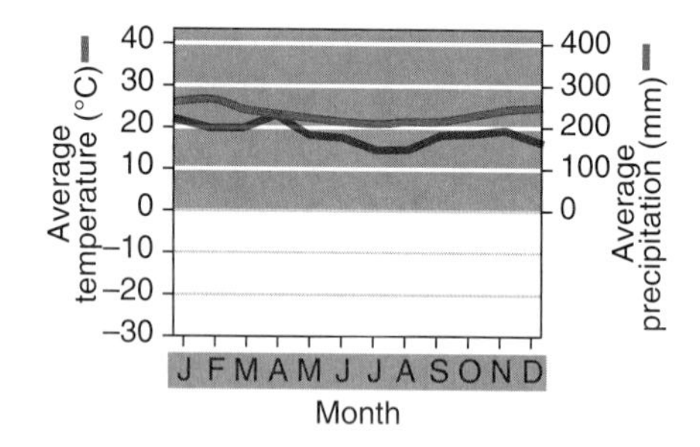

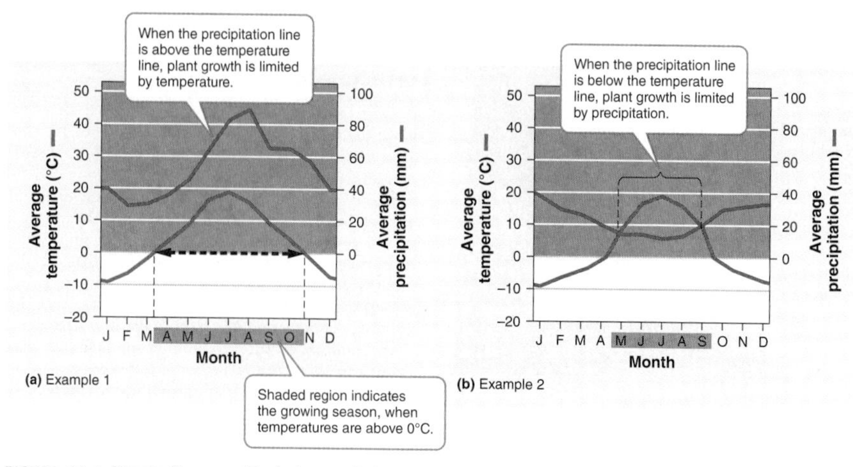

FIGURE 12.4 Climate diagrams. Climate diagrams display monthly temperature and precipitation values, which help determine the productivity of a biome.

In Figure 12.4b, we see a different scenario. When the precipitation line intersects the temperature line, the amount of precipitation available to plants equals the amount of water lost by plants through evapotranspiration. At any point where the precipitation line is below the temperature line, water demand exceeds supply. In this situation, plant growth will be constrained more by precipitation than by temperature.

Climate diagrams also help us understand how humans use different biomes. For example, areas of the world that have warm temperatures, long growing seasons, and abundant rainfall are generally highly productive and therefore are well suited to growing many crops. Warm regions that have less abundant precipitation are suitable for growing grains such as wheat and for grazing domesticated animals, including cattle and sheep. Colder regions are often best used to grow forests that will be harvested for lumber.

Terrestrial biomes range from tundra to tropical forests

We can divide terrestrial biomes into three categories: tundra and boreal forest, temperate, and tropical. These three categories contain a total of nine biomes. We will examine each of these biomes in turn, looking at temperature and precipitation patterns, geographic distribution, and typical plant growth forms.

Tundra A cold and treeless biome with low-growing vegetation.

Permafrost An impermeable, permanently frozen layer of soil.

Tundra

The **tundra,** shown in FIGURE 12.5, is a cold and treeless biome, with low-growing vegetation. In winter, the soil is completely frozen. Arctic tundra is found in the northernmost regions of the Northern Hemisphere in Russia, Canada, Scandinavia, and Alaska. Antarctic tundra is found along the edges of Antarctica and on nearby islands. At lower latitudes, alpine tundra can be found on high mountains, where high winds and low temperatures prevent trees from growing.

The growing season in the tundra is very short, usually only about 4 months during summer, when the polar region is tilted toward the Sun and the days are very long. During this time the upper layer of soil thaws, creating pools of standing water that are ideal habitat for mosquitoes and other insects. The underlying subsoil, known as **permafrost,** is an impermeable, permanently frozen layer that prevents water from draining and roots from penetrating. Permafrost, combined with the cold temperatures and a short growing season, prevents deep-rooted plants such as trees from living in the tundra.

While the tundra receives little precipitation, there is enough to support some plant growth. The characteristic plants of this biome, such as small woody shrubs, mosses, heaths, and lichens, can grow in shallow, waterlogged soil

Teaching Tip: Engage

Ask students: Which biomes have the longest periods during which water demand exceeds supply? *Biomes with the longest periods during which water supply exceeds water demand include: Woodland/shrubland, subtropical desert, temperate grassland/cold desert, and tropical seasonal forest/savanna.*

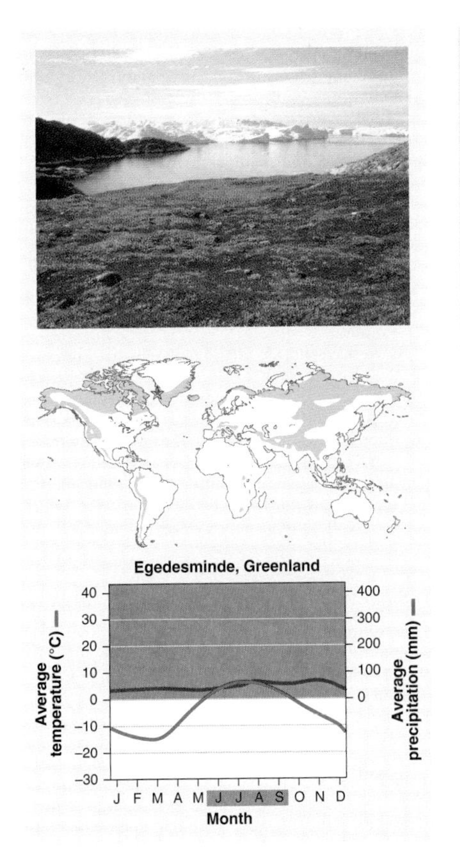

FIGURE 12.5 **Tundra biome.** The tundra is cold and treeless, with low-growing vegetation. *(Cusp/SuperStock)*

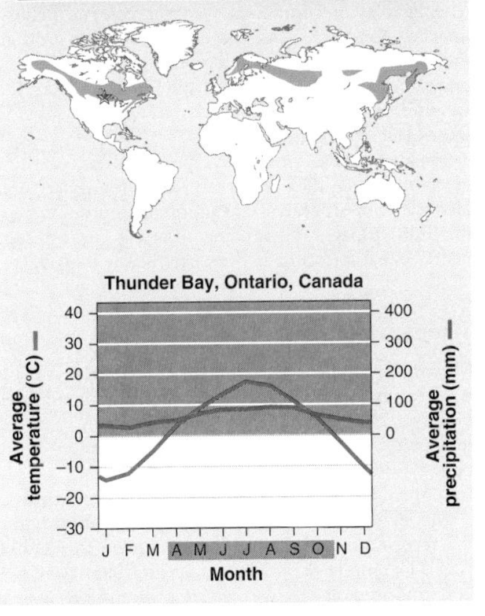

FIGURE 12.6 **Boreal forest biome.** Boreal forests are made up primarily of coniferous evergreen trees that can tolerate cold winters and short growing seasons. *(Bill Brooks/Alamy)*

and can survive short growing seasons and bitterly cold winters. At these cold temperatures, chemical reactions occur slowly, and as a result, dead plants and animals decompose slowly. This slow rate of decomposition results in the accumulation of organic matter in the soil over time with relatively low levels of soil nutrients.

Boreal Forest

The **boreal forest** biome, shown in FIGURE 12.6, consists primarily of coniferous (cone-bearing) evergreen trees that can tolerate cold winters and short growing seasons. Boreal forests are sometimes called taiga. Evergreen trees appear green year-round because they drop only a fraction of their needles each year. Boreal forests are found between about 50° N and 60° N in Europe, Russia, and North America. This subarctic biome has a very cold climate, and plant growth is more constrained by temperature than by precipitation.

Boreal forest A forest biome made up primarily of coniferous evergreen trees that can tolerate cold winters and short growing seasons.

Teaching Tip: Activity

Copy Figures 12.5 through 12.13, and cut off the caption of each so the biomes are not identified. If your copies are black and white, make the temperature line clearly dashed so that it can be distinguished from the precipitation line. Divide students into groups or pairs, and provide each with a list of the nine terrestrial biomes. Have them match each figure to the correct biome. For a more challenging activity, separate the picture, graph, and map in each figure, and have students match the separate images to the correct biome.

◀ TEACHING with FIGURES

Have your class compare the climate diagrams of the tundra (Figure 12.5) and boreal forest biomes (Figure 12.6). Ask students to work in pairs and to find two similarities and two differences between the graphs. *Answers may include: The monthly precipitation range in both is between about 30 mm and less than 100 mm; they both have temperatures reaching below −10°C; the tundra biome has a shorter growing season; and the boreal forest biome reaches higher temperatures in the summer.*

Teaching Tip: Connections

When covering the tundra's growing season and permafrost, refer to the topic of global warming, which is covered in Chapter 19. To spark your students' curiosity about future topics and provide them with a reference point when you reach Chapter 19, you might note the following:

- As atmospheric concentrations of CO_2 increase, Arctic regions become warmer and permafrost starts to thaw.
- Thawing permafrost causes anaerobic decomposition and the release of methane, a potent greenhouse gas.
- The release of methane leads to even more global warming and therefore initiates a positive feedback loop where permafrost melts at an increasing rate.

Teaching Tip: Measuring Your Impact

How Much Paper Do You Use? Measuring Your Impact 4 asks students to evaluate how much paper they use and to calculate the impact of this use on the environment. Download this item by clicking on the link buttons in the TE-book, opening the TRFD, or logging onto the book's companion website bcs.whfreeman.com/friedlandapes2e (teacher login required).

Measuring Your Impact 4: How Much Paper Do You Use?

AP® Exam Tip

The tundra habitat of the Alaskan National Wildlife Refuge is covered in the 2005 AP® Exam, Question 4. To answer this question, students must

- describe two characteristics of Arctic tundra that make it fragile
- explain how the two characteristics make the tundra susceptible to damage from human impacts

As in the tundra, cold temperatures and relatively low precipitation make decomposition in boreal forests a slow process. In addition, the waxy needles of evergreen trees contain compounds that are resistant to decomposition. As a result of the slow rate of decomposition and the low nutrient content of the needles, boreal forest soils are covered in a thick layer of organic material, but are poor in nutrients.

The climate characteristics of boreal forests—cold temperatures, low precipitation, and nutrient-poor soil—determine the species of plants that can survive in them. In addition to coniferous trees such as pine, spruce, and fir, some deciduous trees, such as birch, maple, and aspen, can also be found in this biome. The needles of coniferous trees can tolerate below-freezing conditions, but the deciduous trees drop all their leaves in autumn before the subfreezing temperatures of winter can damage them. When the weather warms, the deciduous trees produce new leaves and grow rapidly.

Because boreal forests have poor soils and a short growing season, they are poorly suited for agriculture. However, they serve as an important source of trees for pulp, paper, and building materials. As a result, many boreal forests have been extensively logged.

Temperate Rainforest

Moving to the mid-latitudes, we find that the climate is more temperate, with average annual temperatures between 5°C and 20°C (41°F and 68°F). A range of temperate biomes exists in this area including temperate rainforest, temperate seasonal forest, woodland/shrubland, and temperate grassland/cold desert.

Moderate temperatures and high precipitation typify the **temperate rainforest** biome, shown in FIGURE 12.7. The temperate rainforest is a coastal biome and can be found in relatively narrow areas around the world. Temperate rainforests exist along the west coast of North America from northern California to Alaska, in southern Chile, on the east coast of Australia and in neighboring Tasmania, and on the west coast of New Zealand. Ocean currents along these coasts help to moderate temperature fluctuations, and ocean water provides a source of water vapor. The result is relatively mild summers and winters, compared with other biomes at similar latitudes, and a nearly 12-month growing season. In the temperate rainforest, winters are rainy and summers are foggy.

Temperate rainforest A coastal biome typified by moderate temperatures and high precipitation.

Temperate seasonal forest A biome with warm summers and cold winters with over 1 m (39 inches) of precipitation annually.

The combination of mild temperatures and high precipitation supports the growth of very large trees. In North America, the most common temperate rainforest trees are coniferous species, including fir, spruce, cedar, and hemlock as well as some of the world's tallest trees: the coastal redwoods (*Sequoia sempervirens*). These immense trees can live hundreds to thousands of years and achieve heights of 90 m (295 feet) and diameters of 8 m (26 feet). Because many of these large tree species are attractive sources of lumber, much of this biome has been logged and subsequently converted into single-species tree plantations.

As we have already seen, coniferous trees produce needles that are slow to decompose. The relatively cool temperatures in the temperate rainforest also favor slow decomposition, although it is not nearly as slow as in boreal forest and tundra. The nutrients released are rapidly taken up by the trees or leached down through the soil by the abundant rainfall, which leaves the soil low in nutrients. Ferns and mosses, which can survive in nutrient-poor soil, are commonly found living under the enormous trees.

Temperate Seasonal Forest

The **temperate seasonal forest** biome, shown in FIGURE 12.8, experiences warm summers and cold winters with over 1 m (39 inches) of annual precipitation. It is found in the eastern United States, Japan, China, Europe, Chile, and eastern Australia. Away from the moderating influence of the ocean, these forests experience much warmer summers and colder winters than temperate rainforests. They are dominated by broadleaf deciduous trees such as beech, maple, oak, and hickory, although some coniferous tree species may also be present. Because of the predominance of deciduous trees, these forests are also called temperate deciduous forests.

The warm summer temperatures in temperate seasonal forests favor rapid decomposition. Because the leaves shed by broadleaf trees are more readily decomposed than the needles of coniferous trees, the soils of temperate seasonal forests generally contain more nutrients than those of boreal forests. Their higher soil fertility, combined with their longer growing season, means that temperate seasonal forests have greater productivity than boreal forests.

Because the temperate seasonal forest is so productive, it has historically been one of the first biomes to be converted to agriculture on a large scale. When European settlers arrived in North America, they cleared large areas of the eastern forests for agriculture.

Teaching Tip: Engage

Ask students to research the biome of a favorite city, anywhere in the world.

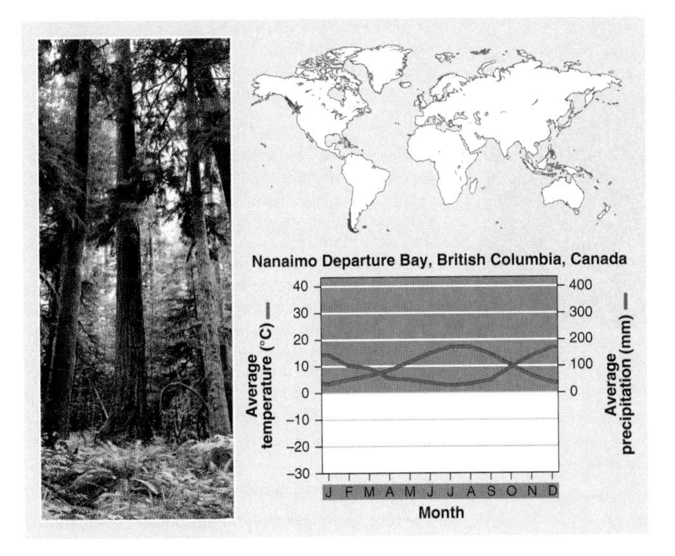

FIGURE 12.7 **Temperate rainforest biome.** Temperate rainforests have moderate mean annual temperatures and high precipitation that supports the growth of very large trees. *(BGSmith/Shutterstock)*

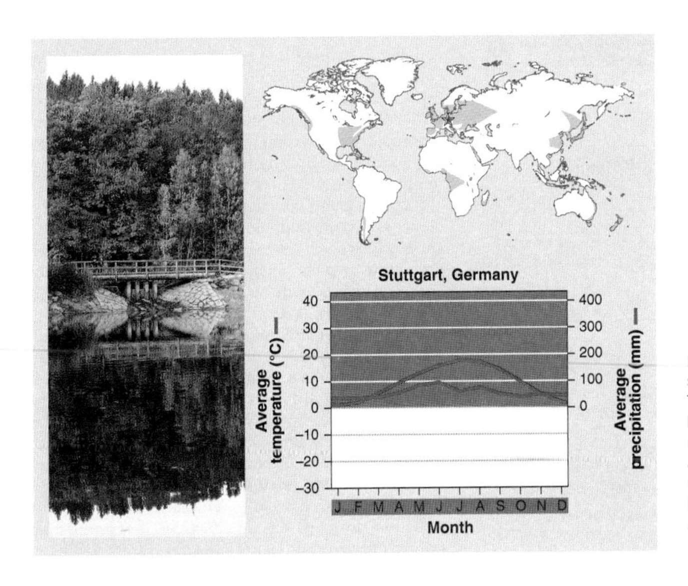

FIGURE 12.8 **Temperate seasonal forest biome.** Temperate seasonal forest biomes have moderate mean annual temperatures and moderate amounts of precipitation that support broadleaf deciduous trees such as beech, maple, oak, and hickory. *(Ursula Sander/Getty Images)*

Teaching Tip: Discussion Starter

Ask students: What abiotic factor distinguishes temperate rainforests from temperate seasonal forests? *Both biomes have similar annual temperatures, but temperate rainforests receive significantly more rain.*

Teaching Tip: Activity

What's My Biome?

To prepare for the game: Randomly assign each student one of the nine terrestrial biomes. There will likely be more than one student per biome. Then allow students from 5 to 10 minutes to write a short description of the assigned biome, which should include climatic conditions, global location, and characteristic flora and fauna. Give students from 10 to 15 blank paper tickets (depending on class size), and have them write the name of their biome on each ticket.

To play the game, have students face each other in two even lines. (If there is an odd number of students in your class, join the game and play along as a student.) Each student reads his or her biome description (without naming the biome) to the person directly across. After listening to the description, the student guesses the name of the biome. Students who guess correctly receive a ticket from their reader; otherwise the reader simply corrects the guesser's mistake. Partners then switch roles. After the pairs have finished, rotate one space in the line and begin again. The end of the line should loop around. Continue playing until all students have been matched together. When the game is over, have each student count the tickets he or she has won and declare winners for first, second, and third place.

Teaching Tip: Engage

Ask students: Why do you think early European settlers cut down so much of the primary forests of the eastern United States? *Early European settlers cut down much of the temperate forests in North America for agriculture and lodging.*

Teaching Tip: Beyond the Classroom

What other crops are grown in the woodland/shrubland biome? Have students investigate the answer for homework and report back to the class. *Answers may include but are not limited to olives, oranges, lemons, avocados, almonds, and figs.*

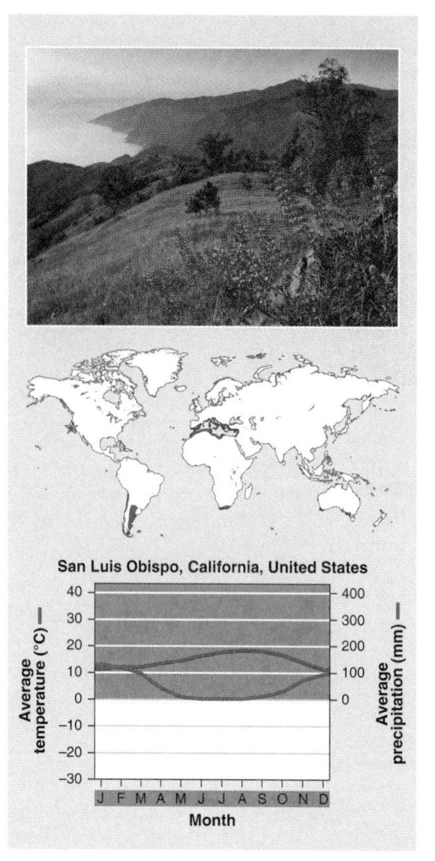

FIGURE 12.9 **Woodland/shrubland biome.** The woodland/shrubland biome is characterized by hot, dry summers and mild, rainy winters. *(Gary Crabbe/Enlightened Images)*

Woodland/Shrubland

The **woodland/shrubland** biome, illustrated in FIGURE 12.9, is characterized by hot, dry summers and mild, rainy winters. This biome is found on the coast of southern California (where it is called chaparral), in southern South America (matorral), in southwestern Australia (mallee), in southern Africa (fynbos), and in a large region surrounding the Mediterranean Sea (maquis). There is a 12-month growing season, but plant growth is constrained by low precipitation in summer and by relatively low temperatures in winter.

The hot, dry summers of the woodland/shrubland biome favor the natural occurrence of wildfires. Plants of this biome are well adapted to both fire and drought. Many plants quickly resprout after a fire, while others produce seeds that open only upon exposure to the intense heat of a fire. Typical plants of this biome include drought-resistant shrubs such as yucca, scrub oak, and sagebrush. Soils in this biome are low in nutrients because of leaching by the winter rains. As a result, the major agricultural uses of this biome are for grazing by animals and growing drought-tolerant, deep-rooted crops, such as grapes to make wine.

Woodland/shrubland A biome characterized by hot, dry summers and mild, rainy winters.

Temperate grassland/cold desert A biome characterized by cold, harsh winters, and hot, dry summers.

Temperate Grassland/Cold Desert

The **temperate grassland/cold desert** biome, shown in FIGURE 12.10, is characterized by cold, harsh winters and hot, dry summers. Temperate grasslands are found in the Great Plains of North America (where they are called prairies), in South America (pampas), and in central Asia and eastern Europe (steppes). Cold, harsh winters and hot, dry summers characterize this biome. Thus, as in the woodland/shrubland biome, plant growth is constrained by insufficient precipitation in summer and cold temperatures in winter. Fires are common, as the dry and frequently windy conditions fan flames ignited by lightning. Although estimates vary, it is thought that, historically, large wildfires occurred in this biome every few years, sometimes burning as much as 10,000 ha (nearly 25,000 acres) in a single fire.

Typical plants of temperate grasslands include grasses and nonwoody flowering plants. These plants are generally well adapted to wildfires and frequent grazing by animals. Their deep roots store energy to enable quick regrowth. Within this biome, the amount of rainfall determines which plants can survive in a region. In the North American prairies, for example, nearly 1 m (39 inches) of rain falls per year on the eastern edge of the biome, supporting grasses that can grow up to 2.5 m (8 feet) high. Although these tallgrass prairies receive sufficient rainfall for trees to grow, frequent wildfires keep trees from encroaching. In fact, the Native American people are thought to have intentionally kept the eastern prairies free of trees by using controlled burning. To the west, annual precipitation drops to 0.5 m (20 inches), favoring the growth of grasses less than 0.5 m (20 inches) tall. These shortgrass prairies are simply too dry to support trees

Teaching Tip: Activity

Make a biome worksheet using the lists below. Have students match each biome to the correct vegetation type and climatic condition. (Answers are provided in italics.) When students are finished, give them the answer key and have them correct their own work.

Biome	Characteristic vegetation	Climate condition
Boreal forest *(4b)*	1. Acacia trees	a. Cold, harsh winters and hot, dry summers
Subtropical desert *(2d)*	2. Cacti	b. Cold, snowy winters, short growing seasons
Temperate grassland/cold desert *(6a)*	3. Coastal redwoods	c. Coldest region on Earth
Temperate rainforest *(3f)*	4. Coniferous trees	d. Extremely hot and dry all year
Temperate seasonal forest *(5i)*	5. Deciduous trees	e. Hot, dry summers and mild, rainy winters
Tropical rainforest *(7g)*	6. Grasses and flowers	f. Moderate temperature and high rainfall
Tropical seasonal forest/savanna *(1h)*	7. Lianas	g. Very warm and very wet
Tundra *(8c)*	8. Lichen	h. Very warm with distinct wet and dry seasons
Woodland/shrubland *(9e)*	9. Scrub oak	i. Warm summers, cold winters, high rainfall

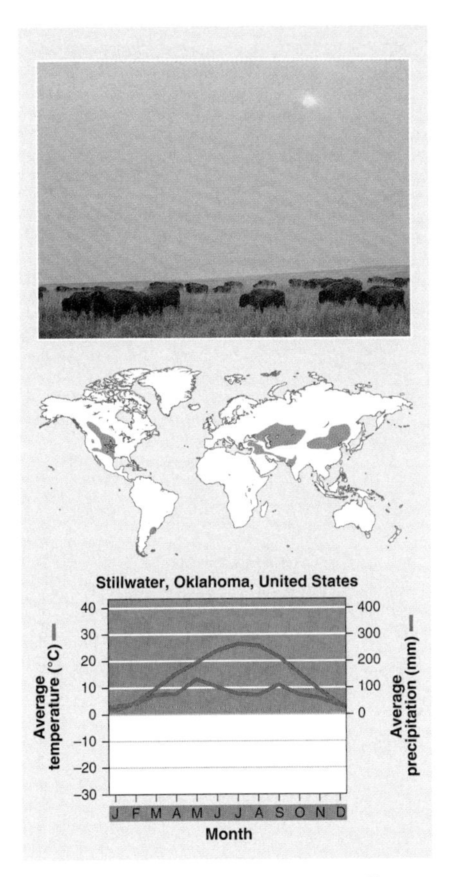

FIGURE 12.10 **Temperate grassland/cold desert biome.** The temperate grassland/cold desert biome has cold, harsh winters and hot, dry, summers that support grasses and nonwoody flowering plants. *(AP Photo/Brandi Simons)*

or tall grasses. Farther west, in the rain shadow of the Rocky Mountains, annual precipitation continues to decline to 0.25 m (10 inches). In this region, the shortgrass prairie gives way to cold desert.

Cold deserts, also known as temperate deserts, have even sparser vegetation than shortgrass prairies. Cold deserts are distinct from subtropical deserts in that they have much colder winters and do not support the characteristic plant growth forms of hot deserts, such as cacti and euphorbs.

The combination of a relatively long growing season and rapid decomposition that adds large amounts of nutrients to the soil makes temperate grasslands very productive. More than 98 percent of the tallgrass prairie in the United States has been converted to agriculture, while the less productive shortgrass prairie is predominantly used for growing wheat and grazing cattle.

Tropical Rainforest

In the tropics, average annual temperatures exceed 20°C. Here we find the tropical biomes: tropical rainforests, tropical seasonal forests/savannas, and subtropical deserts.

The **tropical rainforest** biome, shown in FIGURE 12.11, is a warm and wet biome that lies within approximately 20° N and 20° S of the equator. This biome is found in Central and South America, Africa, Southeast Asia, and northeastern Australia. It is also found on large tropical islands, where the oceans provide a constant source of atmospheric water vapor.

The tropical rainforest biome is warm and wet, with little seasonal temperature variation. Precipitation occurs frequently, although there are seasonal patterns in precipitation that depend on when the ITCZ passes overhead. Because of the warm temperatures and abundant rainfall, productivity is high, and decomposition is extremely rapid. The lush vegetation takes up nutrients quickly, leaving few nutrients to accumulate in the soil. Because of its high productivity, approximately 24,000 ha (59,500 acres) of tropical rainforest are cleared each year for agriculture. However, the high rate of decomposition causes the soils to lose their fertility quickly. As a result, farmers growing crops on tropical soils often have to keep moving to newly deforested areas.

Tropical rainforests contain more biodiversity per hectare than any other terrestrial biome. As much as two-thirds of Earth's terrestrial species are found in this biome. These forests have several distinctive layers of vegetation. Large trees form a forest canopy that shades the underlying vegetation. Several layers of successively shorter trees make up the subcanopy, also known as the understory. Attached to the trunks and branches of the trees are epiphytes, plants that hold small pools of water that support small aquatic ecosystems far above the forest floor. Numerous species of woody vines (also called lianas) are rooted in the soil, but climb up the trunks of trees and often into the canopy.

Tropical rainforest A warm and wet biome found between 20° N and 20° S of the equator, with little seasonal temperature variation and high precipitation.

Teaching Tip: Engage

Ask students: How do humans use rainforests? *Humans use rainforests in various ways. People who live in rainforests eat the wide variety of native plants and animals, and use lianas and trees to build shelters. Rainforests are also a valuable resource for medicinal compounds. Humans also clear many tropical rainforests, using the trees for lumber and the cleared land for agriculture. Once the nutrients in the soil have been depleted, the agricultural fields are converted to rangelands for raising livestock.*

Teaching Tip: Discussion Starter

Ask students the following questions:

- How do humans use grasslands? *Humans use grasslands predominately for agriculture and for grazing livestock.*
- Why do humans use grasslands in this way? *Grasslands are ideal for these uses because the combination of a relatively long growing season and rapid decomposition adds large amounts of nutrients to the soil and makes temperate grasslands very productive. Additionally, grasslands are usually flat landscapes, which makes it relatively easy to work the land.*

Teaching Tip: Beyond the Classroom

Assign a biome to each student and ask them to find a movie that takes place there. *Answers will vary but may include: Tundra – Snow Dogs; Temperate seasonal forest – Last of the Mohicans; Temperate grassland – Dances with Wolves; Tropical rainforest – The Jungle Book; Savanna – The Lion King; Boreal forest – Frozen.*

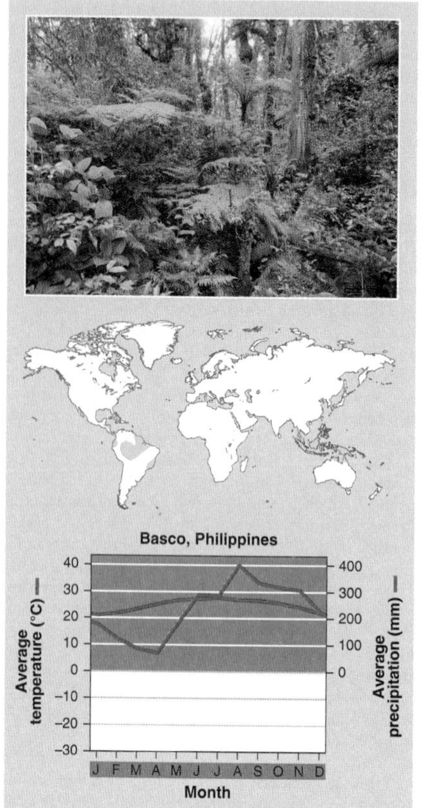

FIGURE 12.11 **Tropical rainforest biome.** Tropical rainforests are warm and wet, with little seasonal temperature variation. These forests are highly productive with several distinctive layers of vegetation. *(Doug Weschler/Earth Scenes/Animals Animals)*

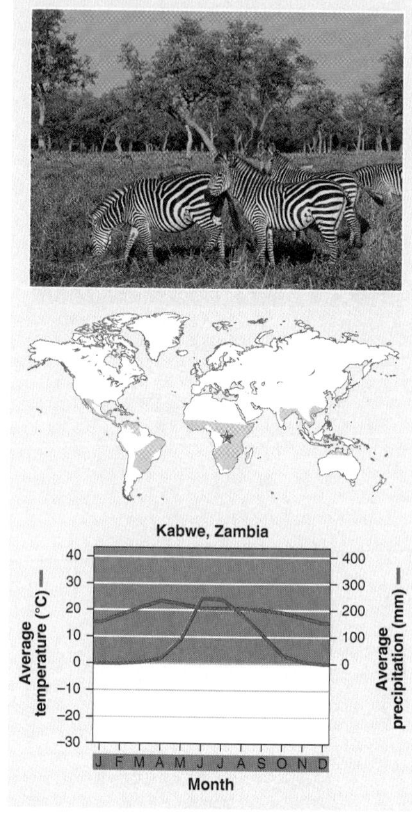

FIGURE 12.12 **Tropical seasonal forest/savanna biome.** Tropical seasonal forest and savannas have warm temperatures and distinct wet and dry seasons. Vegetation ranges from dense stands of shrubs and trees to relatively open landscapes dominated by grasses and scattered deciduous trees. *(John Warburton-Lee/DanitaDelimont.com)*

Tropical Seasonal Forest/Savanna

The **tropical seasonal forests/savanna** biome, shown in FIGURE 12.12, is marked by warm temperatures and distinct wet and dry seasons. This seasonal pattern is caused by the seasonal movement of the ITCZ, which, because it tracks the seasonal movement of the most intense sunlight, passes overhead and drops precipitation only during summer. The trees drop their leaves during the dry season as an adaptation to survive the drought conditions and produce new leaves during the wet season. Thus these forests are also called tropical deciduous forests.

Tropical seasonal forest/savanna A biome marked by warm temperatures and distinct wet and dry seasons.

Tropical seasonal forests are common in much of Central America, on the Atlantic coast of South America, in southern Asia, in northwestern Australia, and in sub-Saharan Africa. Areas with moderately long dry seasons support dense stands of shrubs and trees. In areas with the longest dry seasons, the tropical seasonal climate leads to the formation of savannas, relatively open landscapes dominated by grasses and scattered deciduous trees. Common plants in this biome include acacia and baobab trees. Grazing and fire discourage the growth of many smaller woody plants and keep the savanna landscape open. The presence of trees and a warmer average annual temperature distinguish savannas from grasslands.

The warm temperatures of the tropical seasonal forest/savanna biome promote decomposition, but the low amounts of precipitation constrain plants from using the soil nutrients that are released. As a result, the soils of this biome are fairly fertile and can be farmed. Their fertility has resulted in the conversion of large areas of tropical seasonal forest and savanna into agricultural fields and grazing lands. For example, over 99 percent of the tropical seasonal forest of Pacific Central America and the Atlantic coast of South America has been converted to human uses, including agriculture and grazing.

Subtropical Desert

The **subtropical desert** biome, shown in FIGURE 12.13, is located at roughly 30° N and 30° S, and is characterized by hot temperatures, extremely dry conditions, and sparse vegetation. Also known as hot deserts, this biome includes the Mojave Desert in the southwestern United States, the Sahara Desert in Africa, the Arabian Desert of the Middle East, and the Great Victoria Desert of Australia. Cacti, euphorbs, and succulent plants are well adapted to this biome. To prevent water loss, the leaves of desert plants may be small, nonexistent, or modified into spines, and the outer layer of the plant is thick, with few pores for water and air exchange. Most photosynthesis occurs along the plant stem, which stores water so that photosynthesis can continue even during very dry periods. To protect themselves from herbivores, desert plants have developed defense mechanisms such as spines to discourage grazing.

When rain does fall, the desert landscape is transformed. Annual plants—those that live for only a few months, reproduce, and die—grow rapidly during periods of rain. In contrast, perennial plants—those that live for many years—experience spurts of growth when it rains, but then exhibit little growth during the rest of the year. The slow overall growth of perennial plants in subtropical deserts makes them particularly vulnerable to disturbance, and they have long recovery times.

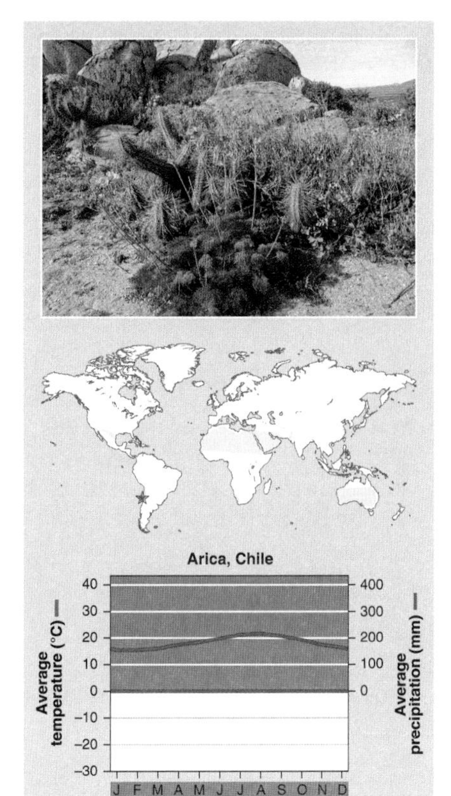

FIGURE 12.13 **Subtropical desert biome.** Subtropical deserts have hot temperatures, extremely dry conditions, and sparse vegetation. *(Topham/The Image Works)*

Subtropical desert A biome prevailing at approximately 30° N and 30° S, with hot temperatures, extremely dry conditions, and sparse vegetation.

Teaching Tip: Engage

Show pictures of a baobab tree and an acacia tree to your class. Ask students: What adaptations enable these trees to survive in their environments? *Baobab trees have very thick trunks, which store water and protect against desiccation in their dry environments. Acacia trees have thorns on their branches and leaves that grow high above the ground, which protect against herbivory.*

Teaching Tip: Discussion Starter

Ask students: What adaptations allow desert plants and animals to survive in their hot and dry environment? *Answers may include:*

- *Succulents store large amounts of water in their tissues and have a waxy coating to reduce water loss.*
- *Succulents have spines to protect against animals who try to access their valuable water stores.*
- *Animals avoid the heat by being active at night and burrowing underground.*
- *Animals reduce water loss by concentrating their urine and excreting dry feces.*

Teaching Tip: Activity

Have students make a brochure advertising their favorite biome. It should include pertinent information for people visiting the biome and highlight its special features. The cities that one can visit in order to see the biome should also be included. Encourage students to use colorful pictures or photos in their brochures. Have students present their brochures to the class.

module

12

REVIEW

In this module, we have learned that terrestrial biomes are categorized by the dominant plant growth forms that exist in a region. These dominant plants coincide with the climates of a region, which we can illustrate graphically using climate diagrams. Based on patterns of climate and plant growth forms, we can categorize nine terrestrial biomes. In the next module, we will see that aquatic biomes can also be categorized, though the criteria are quite different from terrestrial biomes.

Module 12 AP® Review Questions

1. In addition to temperature, a terrestrial biome is defined by
   I. annual precipitation.
   II. distinctive animal species.
   III. distinctive plant species.

   (a) I only
   (b) I and II only
   (c) I and III only
   (d) II and III only
   (e) I, II, and III

2. The precipitation line below the temperature line in a climate diagram shows
   (a) the primary growing season.
   (b) when plants will grow the least.
   (c) the biome is a desert or tundra.
   (d) when plant growth will be limited by precipitation.
   (e) the seasons in which droughts are most likely to occur.

3. Permafrost is an important factor in which of the following biomes?
   I. Tundra
   II. Boreal forest
   III. Cold desert

   (a) I only
   (b) I and II only
   (c) II only
   (d) I and III only
   (e) III and III only

4. Plant growth in which of the following biomes is primarily constrained by precipitation?
   (a) Boreal forest
   (b) Temperate seasonal forest
   (c) Temperate grassland
   (d) Tropical rainforest
   (e) Tundra

5. Which of the following biomes has the highest soil nutrient levels?
   (a) Tropical rainforest
   (b) Temperate rainforest
   (c) Boreal forest
   (d) Woodland/shrubland
   (e) Temperate seasonal forest

Answers to Module 12 AP® Review Questions

1. c
2. d
3. a
4. c
5. e

Teaching Tip: Activity

Divide students into groups and assign a biome to each group. If you don't use all nine terrestrial biomes, the best biomes for this activity are tropical rainforests, tundra, subtropical desert, tropical seasonal forest/savanna, and boreal forest. Tell students that they will be making a survival guide for their biome. The survival guides should outline items needed for a person to survive a camping trip in each biome. Students should consider possible weather conditions that occur in different times of year, animals and plants that they may encounter, and ability to find shelter. *Answers will vary, but should resemble the following for tropical rainforests: poncho or umbrella, light and quick-dry clothing, floppy hat, bug spray, guide for poisonous snakes and frogs, EpiPen® (for reaction to bee stings or other allergens), mosquito net, tent with extra water protection, light sleeping bag, water filter, and plastic bags to keep everything dry.*

Teaching Tip: Review

Ask students: What characteristics of a terrestrial biome determine its productivity? *The productivity of a biome is characterized by its average annual temperature and precipitation level. The most productive biomes have both high temperatures and high precipitation, such as tropical rainforests. Areas with high temperatures but low precipitation, such as subtropical deserts, have low productivity.*

module

13

Aquatic Biomes

Whereas terrestrial biomes are categorized by dominant plant growth forms, aquatic biomes are categorized by physical characteristics such as salinity, depth, and water flow. Temperature is an important factor in determining which species can survive in a particular aquatic habitat, but it is not a factor used to categorize aquatic biomes. Aquatic biomes fall into two broad categories: freshwater and marine. Freshwater biomes include streams, rivers, lakes, and wetlands. Saltwater biomes, also known as marine biomes, include shallow marine areas such as estuaries and coral reefs as well as the open ocean.

Learning Objectives

After reading this module you should be able to

- identify the major freshwater biomes.
- identify the major marine biomes.

Freshwater biomes have low salinity

Freshwater biomes can be categorized as streams and rivers, lakes and ponds, or freshwater wetlands.

Streams and Rivers

Streams and rivers are characterized by flowing fresh water that may originate from underground springs or as runoff from rain or melting snow (FIGURE 13.1). Streams (also called creeks) are typically narrow and carry relatively small amounts of water. Rivers are typically wider and carry larger amounts of water. It is not always clear, however, at what point a particular stream, as it combines with other streams, becomes large enough to be called a river.

As water flow changes, biological communities also change. Most streams and many rapidly flowing rivers have few plants or algae to act as producers. Instead, inputs of organic matter from terrestrial biomes, such as fallen leaves, provide the base of the food web. This organic matter is consumed by insect larvae and crustaceans such as crayfish, which then provide food for secondary consumers such as fish. As fast-moving streams combine to form rivers, the water flow

FIGURE 13.1 **Streams and rivers.** Streams and rivers are freshwater aquatic biomes that are characterized by flowing water. This photo shows Berea Falls on the Rocky River near Cleveland, Ohio. *(Jim West/The Image Works)*

Teaching Tip: Warm-up

Ask students: How are aquatic biomes categorized? Why are they categorized differently from terrestrial biomes? *Aquatic biomes are categorized by physical characteristics such as salinity, depth, and water flow. The two broad categories of aquatic biomes are freshwater and marine. In terrestrial biomes, differences in temperature and precipitation determine the dominant plant forms, which in turn determine the composition of the animal community. Because the composition of the aquatic animal community is influenced more by salinity, water flow, and depth than by plant forms, aquatic biomes are categorized by these abiotic factors.*

Teaching Tip: Activity

Aquatic Biomes Card Sort
In this activity, students match descriptive cards to the correct aquatic biome. This activity should be done in groups.

Download the activity by clicking on the link buttons in the TE-book, opening the TRFD, or logging onto the book's companion website bcs.whfreeman.com /friedlandapes2e (teacher login required).

Activity 4.1 Aquatic Biomes Card Sort

FIGURE 13.2 **Lakes and ponds.** Lakes, such as Lake George in New York State, are characterized by standing water and a central zone of water that is too deep for emergent vegetation. *(Frank Paul/Alamy)*

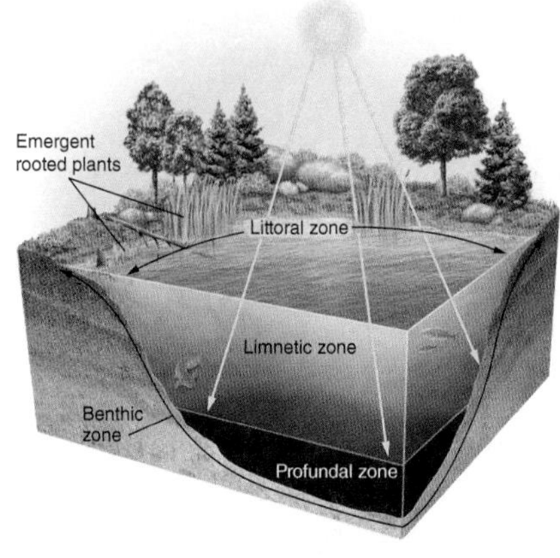

FIGURE 13.3 **Lake zones.** The littoral zone consists of shallow water with emerging, rooted plants whereas the limnetic zone is the deeper water where plants do not emerge. The deepest water, where oxygen can be limiting because little sunlight penetrates to allow photosynthesis by producers, is the profundal zone. The sediments that lie beneath the littoral, limnetic, and profundal zones constitute the benthic zone.

typically slows, sediments and organic material settle to the bottom, and rooted plants and algae are better able to grow.

Fast-moving streams and rivers typically have stretches of turbulent water called rapids, where water and air are mixed together. This mixing allows large amounts of atmospheric oxygen to dissolve into the water. Such high-oxygen environments support fish species such as trout and salmon that need large amounts of oxygen. Slower-moving rivers experience less mixing of air and water. These lower-oxygen environments favor species such as catfish that can better tolerate low-oxygen conditions.

Lakes and Ponds

Lakes and ponds contain standing water, at least some of which is too deep to support emergent vegetation (plants that are rooted to the bottom and emerge above the water's surface). Lakes are larger than ponds, but as with streams and rivers, there is no clear point at which a pond is considered large enough to be called a lake (FIGURE 13.2).

As FIGURE 13.3 shows, lakes and ponds can be divided into several distinct zones. The **littoral zone** is the shallow area of soil and water near the shore where algae and emergent plants such as cattails grow. Most photosynthesis occurs in this zone. In the open water, or **limnetic zone,** rooted plants can no longer survive; floating algae called **phytoplankton** are the only photosynthetic organisms. The limnetic zone extends as deep as sunlight can penetrate. Very deep lakes have a region of water below the limnetic zone, called the **profundal zone.** Because sunlight does not reach the profundal zone, producers cannot survive there, so nutrients are not easily recycled into the food web. Bacteria decompose the detritus that reaches the profundal zone, but they consume oxygen in the process. As a result, dissolved oxygen concentrations are not sufficient to support many large organisms. The muddy bottom of a lake or pond beneath the limnetic and profundal zones is called the **benthic zone.**

Lakes are classified by their level of primary productivity. Lakes that have low productivity due to low amounts of nutrients such as phosphorus and nitrogen in the water are called **oligotrophic** lakes. In contrast,

Littoral zone The shallow zone of soil and water in lakes and ponds where most algae and emergent plants grow.

Limnetic zone A zone of open water in lakes and ponds.

Phytoplankton Floating algae.

Profundal zone A region of water where sunlight does not reach, below the limnetic zone in very deep lakes.

Benthic zone The muddy bottom of a lake, pond, or ocean.

Oligotrophic Describes a lake with a low level of productivity.

134 CHAPTER 4 ■ Global Climates and Biomes

Teaching Tip: Concept Map

To help clarify the categorization of aquatic biomes, ask students to create a concept map like the one shown below using these terms:

- coral reefs
- freshwater biomes
- freshwater wetlands
- intertidal zones
- lakes and ponds
- mangrove swamps
- marine biomes
- salt marshes
- streams and rivers
- open ocean

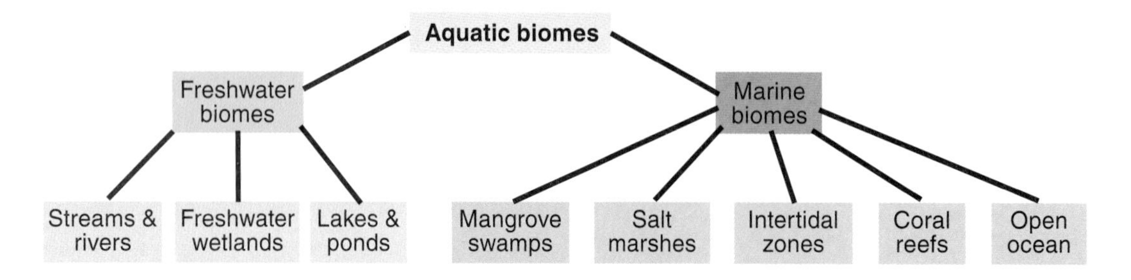

(b)

(a) (c)

FIGURE 13.4 **Freshwater wetlands.** Freshwater wetlands have soil that is saturated or covered by fresh water for at least part of the year and are characterized by particular plant communities. (a) In this swamp in southern Illinois, bald cypress trees emerge from the water. (b) This marsh in south central Wisconsin is characterized by cattails, sedges, and grasses growing in water that is not acidic. (c) This bog in northern Wisconsin is dominated by sphagnum moss as well as shrubs and trees that are adapted to acidic conditions. *(Lee Wilcox)*

lakes with a moderate level of productivity are called **mesotrophic** lakes, and lakes with a high level of productivity are called **eutrophic** lakes.

Freshwater Wetlands

Freshwater wetlands are aquatic biomes that are submerged or saturated by water for at least part of each year, but shallow enough to support emergent vegetation. They support species of plants that are specialized to live in submerged or saturated soils.

Freshwater wetlands include swamps, marshes, and bogs. Swamps are wetlands that contain emergent trees, such as the Great Dismal Swamp in Virginia and North Carolina and the Okefenokee Swamp in Georgia and Florida (FIGURE 13.4a). Marshes are wetlands that contain primarily non-woody vegetation, including cattails and sedges (Figure 13.4b). Bogs, in contrast, are very acidic wetlands that typically contain sphagnum moss and spruce trees (Figure 13.4c).

Freshwater wetlands are among the most productive biomes on the planet, and they provide several critical ecosystem services. For example, wetlands can take in large amounts of rainwater and release it slowly into the groundwater or into nearby streams, thus reducing the severity of floods and droughts. Wetlands also filter pollutants from water, recharging the groundwater with clean water. Many bird species depend on wetlands during migration or breeding. As many as one-third of all endangered bird species in the United States spend some part of their lives in wetlands, even though this biome makes up only 5 percent of the nation's land area. More than half of the freshwater wetland area in the United States has been drained for agriculture or development or to eliminate breeding grounds for mosquitoes and various disease-causing organisms.

Mesotrophic Describes a lake with a moderate level of productivity.

Eutrophic Describes a lake with a high level of productivity.

Freshwater wetlands An aquatic biome that is submerged or saturated by water for at least part of each year, but shallow enough to support emergent vegetation.

Teaching Tip: Engage

Ask students: Where in the United States are wetlands located? What are the ecological consequences if wetlands continue to disappear? *Wetlands occur largely in the southeastern United States, particularly Florida, Georgia, Alabama, Mississippi, Louisiana, and the Carolinas. The Chesapeake Bay region of the East Coast also has extensive wetland habitats. The disappearance of wetlands would increase the severity of floods in coastal regions and inhibit proper filtration of groundwater. Additionally, the loss of wetlands would disrupt the breeding and migration patterns of many bird species, potentially leading to their extinction.*

Teaching Tip: Activity

Divide students into groups and have each group create a way to remember the various zones of a lake: littoral zone, limnetic zone, profundal zone, and benthic zone. Encourage students to be creative and suggest that they use a rhyme, rap, song, or jingle as aids to remembering. When students finish, have each group present its memory aid to the class.

Teaching Tip: Beyond the Classroom

Take students on a field trip to an aquatic biome close to your school. Before the trip, ask students to make a list of the plants, animals, and abiotic components that they expect to find in the aquatic habitat. Then during the field trip, have students check off the organisms and features from their lists that they observe, while they add those that aren't on their lists. Be sure to bring along field guides for identifying plants and animals that you see.

Teaching Tip: Activity

Prepare a slideshow of photographs from the eight categories of aquatic biomes. Number each picture, but provide no other identification. Have students write down the type of aquatic biome represented by each picture. After completing the slideshow, run through it once more, giving students the answers. Then ask students which biomes were the easiest and hardest to identify.

COMMON MISCONCEPTIONS
Swamps and marshes

Students often confuse marshes and swamps. Explain that the main difference is that swamps have trees, while marshes do not.

Teaching Tip: Activity

Divide students into groups and give each group the list of terms for aquatic biomes shown below. Have students put the terms into categories of their own making. They may choose to use broad categories, such as freshwater and marine biomes, or to use the narrower categories of each aquatic biome. Go around to each group and monitor their progress, giving clarifications where needed.

- Littoral zone
- Limnetic zone
- Phytoplankton
- Profundal zone
- Benthic zone
- Oligotrophic
- Mesotrophic
- Eutrophic
- Freshwater wetland
- Salt marsh
- Mangrove swamp
- Bog
- Intertidal zone
- Coral reef
- Coral bleaching
- Open ocean
- Photic zone
- Aphotic zone
- Chemosynthesis
- Turbulent water
- Standing water
- Cattails and sedges
- Estuaries
- Trees with submerged roots
- Nonwoody emergent vegetation
- Barnacles and mussels
- Limestone skeleton
- Deep water
- Pelagic fish

FIGURE 13.5 **Salt marsh.** The salt marsh is a highly productive biome typically found in temperate regions where fresh water from rivers mixes with salt water from the ocean. This salt marsh is in Plum Island Sound in Massachusetts. *(Jerry and Marcy Monkman)*

Marine biomes have high salinity

Marine biomes contain salt water and can be categorized as salt marshes, mangrove swamps, intertidal zones, coral reefs, and the open ocean.

Salt Marshes

Like freshwater marshes, **salt marshes**—found along the coast in temperate climates—contain nonwoody emergent vegetation (FIGURE 13.5). The salt marsh is one of the most productive biomes in the world. Many salt marshes are found in estuaries, which are areas along the coast where the fresh water of rivers mixes with salt water from the ocean. Because rivers carry large amounts of nutrient-rich organic material, estuaries are extremely productive places for plants and algae, and the abundant plant life helps filter contaminants out of the water. Salt marshes provide important habitat for spawning fish and shellfish; two-thirds of marine fish and shellfish species spend their larval stages in estuaries.

Mangrove Swamps

Mangrove swamps occur along tropical and subtropical coasts and, like freshwater swamps, contain trees whose roots are submerged in water (FIGURE 13.6). Unlike most trees, however, mangrove trees are salt tolerant. They often grow in estuaries, but they can also be found along shallow coastlines that lack inputs of fresh water. The trees help to protect those coastlines from erosion and storm damage. Falling leaves and trapped organic material produce a nutrient-rich environment. Like salt marshes, mangrove swamps provide sheltered habitat for fish and shellfish.

Intertidal Zones

The **intertidal zone** is the narrow band of coastline that exists between the levels of high tide and low tide (FIGURE 13.7). Intertidal zones range from steep, rocky areas to broad, sloping mudflats. Environmental conditions in this biome are relatively stable when submerged during high tide. However, conditions can become quite harsh during low tide when organisms are exposed to direct sunlight, high temperatures, and desiccation. Moreover, waves crashing onto shore can make it a challenge for organisms to hold on and not get washed away. Intertidal zones are home to a wide variety of organisms that have adapted to these conditions, including barnacles, sponges, algae, mussels, crabs, and sea stars.

Salt marsh A marsh containing nonwoody emergent vegetation, found along the coast in temperate climates.

Mangrove swamp A swamp that occurs along tropical and subtropical coasts, and contains salt-tolerant trees with roots submerged in water.

Intertidal zone The narrow band of coastline between the levels of high tide and low tide.

FIGURE 13.6 **Mangrove swamp.** Salt-tolerant mangrove trees, such as these in Everglades National Park, are important in stabilizing tropical and subtropical coastlines and in providing habitat for marine organisms. *(Biosphoto/T. & S. Allofs)*

136 CHAPTER 4 ■ Global Climates and Biomes

Teaching Tip: Discussion Starter

Ask students: What adaptations allow intertidal organisms to survive in their harsh and variable environment? *Answers may include: shells that protect organisms from crashing waves, the ability to close shells during low tide to prevent water loss, adaptations that hold organisms to rocks in strong currents, and a higher tolerance for variability in salinity and temperatures than other aquatic organisms.*

FIGURE 13.7 **Intertidal zone.** Organisms that live in the area between high and low tide, such as these giant green sea anemones (*Anthopleura xanthogrammica*), goose barnacles (*Lepas anserifera*), and ochre sea stars (*Pisaster ochraceus*), must be highly tolerant of the harsh, desiccating conditions that occur during low tide. This photo was taken at Olympic National Park, Washington. *(Jim Zipp/Science Source)*

FIGURE 13.8 **Coral reef.** The skeletons of millions of corals build reefs that serve as home to a great variety of other marine species. Sea goldies (*Pseudanthias squamipinnis*) and other animals inhabit this reef of soft coral in the Indian Ocean. *(Helmut Corneli/imageb/imagebroker.net/SuperStock)*

Coral Reefs

Coral reefs, which are found in warm, shallow waters beyond the shoreline, represent Earth's most diverse marine biome (FIGURE 13.8). Corals are tiny animals that secrete a layer of limestone (calcium carbonate) to form an external skeleton. The animal living inside this tiny skeleton is essentially a hollow tube with tentacles that draw in plankton and detritus. Corals live in water that is relatively poor in nutrients and food, which is possible because of their relationship with single-celled algae that live within the tissues of the corals. When a coral digests the food it captures, it releases CO_2 and nutrients. The algae use the CO_2 during photosynthesis to produce sugars and the nutrients stimulate the algae to release their sugars to the coral. The coral gains energy in the form of sugars, and the algae obtain CO_2, nutrients, and a safe place to live within the coral's tiny limestone skeleton. But this association with photosynthetic algae means that corals can live only in shallow waters where light can penetrate.

Although each individual coral is tiny, most corals live in vast colonies. As individual corals die and decompose, their limestone skeletons remain. Over time, these skeletons accumulate and develop into coral reefs, which can become quite massive. The Great Barrier Reef of Australia, for example, covers an area of 2,600 km^2 (1,600 $miles^2$). A tremendous diversity of other organisms, including fish and invertebrates, use the structure of the reef as both a refuge in which to live and a place to find food. At the Great Barrier Reef there are more than 400 species of coral, 1,500 species of tropical fish, and 200 species of birds.

Coral reefs are currently facing a wide range of challenges, including pollutants and sediments that make it difficult for the corals to survive. Coral reefs also face the growing problem of **coral bleaching,** a phenomenon in which the algae inside the corals die. Without the algae, the corals soon die as well, and the reef turns white. Scientists believe that the algae are dying from a combination of disease and environmental changes, including lower ocean pH and abnormally high water temperatures. Coral bleaching is a serious problem: Without the corals, the entire coral reef biome is endangered.

The Open Ocean

The **open ocean** contains deep ocean water that is located away from the shoreline where sunlight can no longer reach the ocean bottom. The exact depth of penetration by sunlight depends on a number of factors,

Coral reef The most diverse marine biome on Earth, found in warm, shallow waters beyond the shoreline.

Coral bleaching A phenomenon in which algae inside corals die, causing the corals to turn white.

Open ocean Deep ocean water, located away from the shoreline where sunlight can no longer reach the ocean bottom.

AP® Exam Tip

Coral reefs are covered in the 2011 AP® Exam, Question 2. To answer this question, students must

- explain how an increase in the amount of dissolved CO_2 in ocean water results in a decrease in the pH of ocean water
- calculate the annual global increase of calcium carbonate in coral reefs
- identify and describe one likely negative environmental consequence of coral reef loss
- identify one environmental problem (other than one due to ocean acidification or loss of coral reefs) that affects marine ecosystems on a global scale

Teaching Tip: Engage

Ask students: Why are coral reefs sometimes referred to as the rainforests of the sea? How and why are coral reefs similar to tropical rainforests? *Coral reefs are comparable to tropical rainforests because both are among the most productive and diverse ecosystems in the world. They are similar in this way because both biomes occur along the equator, and therefore receive more solar energy year-round than any other region on Earth. Coral reefs also have warm temperatures because they reside in shallow areas of the ocean.*

Teaching Tip: Discussion Starter

Ask students: What are the different zones of the open ocean and what defines them? *The zones of the ocean are largely defined by the amount of sunlight they receive. The upper layer of water, which receives enough sunlight to allow photosynthesis, is called the photic zone. The deeper layer of water that lacks sufficient sunlight for photosynthesis is called the aphotic zone. There is also the benthic zone, which is the area of ocean floor.*

including the amounts of sediment and algae suspended in the water, but it generally does not exceed 200 m (approximately 650 feet).

Like a pond or lake, the ocean can be divided into zones. These zones are shown in FIGURE 13.9. The upper layer of ocean water that receives enough sunlight to allow photosynthesis is the **photic zone,** and the deeper layer of water that lacks sufficient sunlight for photosynthesis is the **aphotic zone.** The ocean floor is called the benthic zone.

In the photic zone, algae are the major producers. They form the base of a food web that includes tiny zooplankton, fish, and whales. In the aphotic zone, because of the lack of light, there are no photosynthetic producers. However, there are some species of bacteria that can use the energy contained in the bonds of methane and hydrogen sulfide, which are both found in the deep ocean, to generate energy via **chemosynthesis** rather than photosynthesis. These bacteria form the base of a deep-ocean food web that includes animals such as tube worms (see Figure 5.3c). The aphotic zone also contains a variety of organisms that can generate their own light to help them feed in the dark waters. These organisms include several species of crustaceans, jellyfish, squid, and fish.

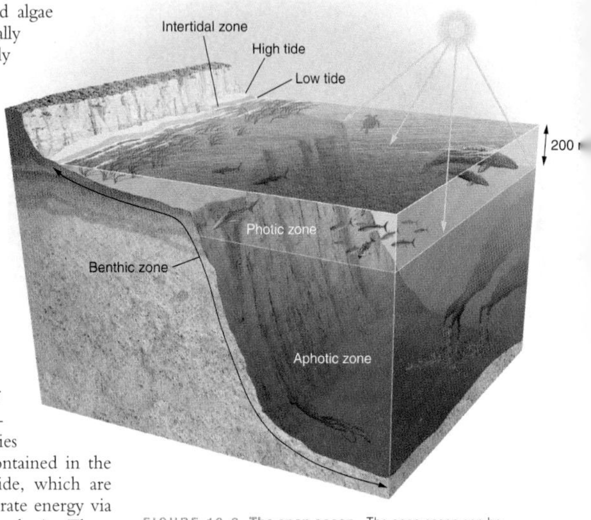

Intertidal zone
High tide
Low tide
200 m
Photic zone
Benthic zone
Aphotic zone

FIGURE 13.9 **The open ocean.** The open ocean can be separated into several distinct zones.

Photic zone The upper layer of ocean water in the ocean that receives enough sunlight for photosynthesis.

Aphotic zone The deeper layer of ocean water that lacks sufficient sunlight for photosynthesis.

Chemosynthesis A process used by some bacteria in the ocean to generate energy with methane and hydrogen sulfide.

module 13

REVIEW

In this module, we have learned that aquatic biomes are characterized by physical features such as salinity, depth, and water flow. Freshwater biomes include streams and rivers, which have flowing water, and lakes, ponds, and wetlands, which have standing water. Marine biomes contain salt water and include salt marshes, mangrove swamps, intertidal zones, coral reefs, and the open ocean. Differences in water flow, depth, and salinity help us understand why different species of producers and consumers, including commercially important species of fish and shellfish, live in different aquatic regions of the world.

Practicing Science: Analyze and Predict

The table below includes hypothetical data from a sample lake, marsh, open ocean, and river. Have students work in pairs to match each of the given biomes to its corresponding data. When completed, have volunteers explain how they determined their answers.

Match the following aquatic biomes to the data below: lake, marsh, open ocean, and river.

	Biome A	Biome B	Biome C	Biome D
pH	7.3	6.8	7.5	8.1
Flow rate (cubic feet per second)*	430	5	3	10
Salinity (parts per thousand)	0.5	18	0.5	35
Surface temperature (°C)	21	25	24	19

*Note: Large, deep bodies of water, such as oceans, are not typically measured in cubic feet per second, but hypothetical measurements are given here for the sake of comparison.

Answers: *Biome A = River; Biome B = Marsh; Biome C = Lake; Biome D = Open ocean.*

Teaching Tip: Engage

Ask students: What is your favorite aquatic animal and in which biome does it live? Give students a few minutes to think this over, and then ask each to share the answer with a partner. When finished, have students volunteer to share their answers with the class.

Module 13 AP® Review Questions

1. Which of the following ecosystems experiences harsh conditions due to conditions from tides?
   (a) Coral reef
   (b) Freshwater wetlands
   (c) Open ocean
   (d) Intertidal zone
   (e) Ponds and lakes

2. Most of the photosynthesis in lakes and ponds occurs in the
   (a) benthic zone.
   (b) littoral zone.
   (c) limnetic zone.
   (d) profundal zone.
   (e) aphotic zone.

3. Which of the following is NOT an important ecosystem service provided by wetlands?
   (a) Flood control
   (b) Breeding habitat for birds
   (c) Migratory habitat for birds
   (d) Water filtration
   (e) Seed dispersal

4. Aquatic biomes are categorized by which of the following?
   I. Dominant plant growth forms
   II. Depth
   III. Salinity

   (a) I and II only
   (b) I and III only
   (c) II and III only
   (d) III only
   (e) I, II, and III

5. Which biome contains the aphotic zone?
   (a) Coral reefs
   (b) Mangrove swamps
   (c) Streams and rivers
   (d) Freshwater wetlands
   (e) Open ocean

Answers to Module 13 AP® Review Questions

1. d
2. b
3. e
4. c
5. e

working toward  sustainability

Is Your Coffee Made in the Shade?

Around the world, people enjoy drinking coffee. In the United States alone, 54 percent of adults drink coffee every day, at an average of 3 cups per day. Worldwide, people buy 7.7 billion kilograms (16.9 billion pounds) of coffee beans each year. In short, coffee is an important part of many people's lives. But have you ever thought about where your coffee comes from?

Coffee beans come from several species of shrubs that historically grew in Ethiopia under the shade of the tropical rainforest canopy. In the fifteenth century, coffee was brought to the Middle East and eventually spread throughout the world. Because of its popularity, coffee is now farmed in many places around the world, including South America, Africa, and Southeast Asia.

As farmers began cultivating coffee, they grew it like many other crops by clearing large areas of rainforest and planting coffee bushes close together in large open fields. Because the coffee plant's native habitat is a shady forest, coffee farmers found that they had to construct shade over the plants to prevent them from becoming sunburned in the intense tropical sunlight. Over the past several decades, however, breeders have developed more sunlight-tolerant plants that cannot only handle intense sunlight but can also produce many more coffee beans per plant.

As coffee was transformed from a plant that was naturally scattered throughout a diverse rainforest to one that was grown as a single species in large numbers in open fields, the coffee fields became attractive targets for insect pests and diseases. Farmers have applied a variety of pesticides to combat these pests, which has increased the cost of farming coffee, poisoned workers, and polluted the environment. Given the world's demand for coffee, what other options do coffee farmers have?

Some coffee farmers thought back to the natural environment in which coffee grows and wondered if they could farm coffee under more natural conditions. Such coffee, called shade-grown coffee, is grown in one of three ways: by planting coffee bushes in an intact rainforest, by planting the bushes in a rainforest that has had some of the trees removed, or by planting the bushes in a field alongside trees that produce other

Teaching Tip: Beyond the Classroom

Using resources in one way requires giving up alternative uses. After learning about the different biomes around the planet, have students do the activity "Site Selection: A Land Use Simulation," which teaches about trade-offs. In this activity, students role-play common environmental dilemmas as community members try to decide where to locate a new school. You can have each group of students locate your school in a different biome of the world where they will face different decisions according to the characteristics of that biome.

Find a link to this resource by clicking on the link buttons in the TE-book, opening the TRFD, or logging onto the book's companion website bcs.whfreeman.com/friedlandapes2e (teacher login required).

Chapter 4 Web Resources

Suggested Answers to Critical Thinking Questions

1. Because the shade-grown coffee plants produce less coffee per hectare, farmers must sell it at a higher price. Although some people are willing to pay more for this coffee, many are not. Because the demand for expensive coffee is limited, not all coffee sold can be shade-grown.

2. If three times as much coffee can be grown in the Sun, then producing shade-grown coffee will result in much more land being used for growing coffee; however, this land would experience less deforestation than would land used for sun-grown coffee.

marketable products, including fruit. Coffee bushes grown in this way attract fewer pests, so less money is needed to buy and apply pesticide, and there is less risk to workers and the nearby soil and water. Using these methods, coffee can be grown while still preserving some of the plant diversity of the rainforest. And the coffee often tastes better. The density of coffee plants is lower in these more diverse landscapes, however, which means that only about one-third as much coffee is produced per hectare. So, while there are cost savings, the yield is lower. Economically, this means that owners of shade-grown coffee farms need to charge higher prices to match the profits of other farms.

How can farmers producing shade-grown coffee stay in business? A number of environmental groups that want to preserve biodiversity in tropical rainforests have stepped in to help. Researchers found that shade-grown coffee farms provided habitat for approximately 150 species of rainforest birds, whereas open-field coffee farms provided habitat for only 20 to 50 bird species. Not surprisingly, researchers also found that other groups of animals were more diverse on shade-grown coffee farms. In response to these findings, the Smithsonian Migratory Bird Center in Washington, D.C., developed a program to offer a "Bird Friendly" seal of approval to coffee farmers who were producing shade-grown coffee. Combined with an advertising campaign that explained the positive effects of shade-grown coffee on biodiversity, this seal of approval alerted consumers to make a conscious choice about the impact that their favorite beverage was having on rainforests. From 2005 to 2011, the sales of the shade-grown coffee with the Smithsonian seal of approval increased by an amazing 250 percent. The Arbor Day Foundation, an environmental organization that promotes the planting of trees, joined the effort by selling its own brand of shade-grown coffee. Over the past 20 years, it has become clear that when consumers are informed about how coffee is grown, many people are willing to choose the shade-grown varieties, even if it requires spending more money to reduce adverse impacts on the tropical rainforest biome.

Shade-grown coffee in Honduras. Coffee grown in the shade requires less pesticide, helps to preserve the plant diversity of the rainforest, and even tastes better. *(AP Photo/Ginnette Riquelme)*

A bird-friendly certification label. Bird-friendly labels were introduced by the Smithsonian Migratory Bird Center to inform consumers that the coffee was grown in the shade in a manner that improves bird habitat. *(Francesca Slater)*

Critical Thinking Questions

1. If shade-grown coffee produces less coffee per hectare, what economic factor might prevent all coffee from being grown this way?
2. If three times as much coffee can be grown in the Sun than in the shade, what are the trade offs in terms of the amount of land used for growing coffee under these two alternative agricultural practices?

References

Philpott, S. M., et al. 2008. Biodiversity loss in Latin American coffee landscapes: Review of the evidence on ants, birds, and trees. *Conservation Biology* 22:1093–1105.

Smithsonian Migratory Bird Center. *Coffee Drinkers and Bird Lovers.* http://nationalzoo.si.edu/SCBI/MigratoryBirds/Coffee/lover.cfm.

chapter 4

REVIEW

In this chapter we have examined how global processes such as air and water currents determine regional climates and how these regional climates have a major effect on the types of organisms that can live in different parts of the world. Among the terrestrial biomes, temperature and precipitation affect the rate of decomposition of dead organisms and the productivity of the soil. Understanding these patterns helps us understand how humans have come to use the land in different ways: growing crops in regions with enough water and a sufficient growing season, grazing domesticated animals in drier areas, and harvesting lumber from forests. Among the aquatic biomes, differences in flow, salinity, and depth help to determine the aquatic species that can live in different aquatic regions of the world.

Key Terms

Climate
Weather
Troposphere
Stratosphere
Albedo
Saturation point
Adiabatic cooling
Adiabatic heating
Latent heat release
Atmospheric convection current
Hadley cell
Intertropical convergence zone (ITCZ)
Polar cell
Ferrell cell
Coriolis effect
Rain shadow
Gyres
Upwelling
Thermohaline circulation
El Niño–Southern Oscillation (ENSO)
Terrestrial biome
Aquatic biome
Tundra
Permafrost
Boreal forest
Temperate rainforest
Temperate seasonal forest
Woodland/shrubland
Temperate grassland/cold desert
Tropical rainforest
Tropical seasonal forest/savanna
Subtropical desert
Littoral zone
Limnetic zone
Phytoplankton
Profundal zone
Benthic zone
Oligotrophic
Mesotrophic
Eutrophic
Freshwater wetland
Salt marsh
Mangrove swamp
Intertidal zone
Coral reef
Coral bleaching
Open ocean
Photic zone
Aphotic zone
Chemosynthesis

Learning Objectives Revisited

Module 9 The Unequal Heating of Earth

- **Identify the five layers of the atmosphere.**
  Above Earth's surface, the first layer of atmosphere is the troposphere, followed by the stratosphere, mesosphere, thermosphere, and the exosphere.
- **Discuss the factors that cause unequal heating of Earth.**
  The unequal heating of Earth is caused by differences in the angle of the Sun's rays that strike Earth, the amount of atmosphere that the Sun's rays must pass through before striking Earth's surface, and how much of the solar energy that reaches Earth is reflected rather than absorbed.
- **Describe how Earth's tilt affects seasonal differences in temperatures.**
  Earth's central axis is tilted at 23.5°, which causes seasonal changes in the latitudes that receive the most intense sunlight.

Exploring the Literature

Marimi, M. A., et al. 2009. Predicted climate-driven bird distribution changes and forecasted conservation conflicts in a neotropical savanna. *Conservation Biology* 23: 1558–1567.

National Science Teachers Association. *World biomes.com.* http://www.worldbiomes.com/.

U.S. Environmental Protection Agency. *Wetlands.* http://water.epa.gov/type/wetlands/.

Download a printable version of this list by clicking on the link buttons in the TE-book, opening the TRFD, or logging onto the book's companion website bcs.whfreeman.com/friedlandapes2e (teacher login required).

 **Exploring the Literature 4**

Module 10 Air Currents

- **Explain how the properties of air affect the way it moves in the atmosphere.**

  Air rises when it becomes less dense and sinks when it becomes more dense. Warm air has a higher saturation point for water vapor than cold air. Changes in air pressure result in adiabatic cooling or heating; when water condenses it emits heat, which is known as latent heat release.

- **Identify the factors that drive atmospheric convection currents.**

  Atmospheric convection currents are driven by the intense sunlight that strikes Earth near the tropics. This solar energy warms the surface of Earth, which causes moist air to rise, cool, and release water as precipitation. As the air continues to rise, it reaches the top of the troposphere. The air, which is now cold and dry, moves toward the poles until it descends at approximately 30° N or 30° S latitude. As it descends back to Earth's surface, the air warms and then moves back toward the equator.

- **Describe how Earth's rotation affects the movement of air currents.**

  Because the surface of Earth travels faster near the equator than near the poles, the Coriolis effect causes convection currents traveling north and south to be deflected, thereby creating trade winds, westerlies, and easterlies.

- **Explain how the movement of air currents over mountain ranges affects climates.**

  When moist air from the ocean moves up a mountain, the air cools and releases water as precipitation, which results in a moist environment on the windward side. On the other side of the mountain, the cool, dry air descends, which results in a dry environment on the leeward side of the mountain.

Module 11 Ocean Currents

- **Describe the patterns of surface ocean circulation.**

  Ocean currents are driven by a combination of temperature, gravity, prevailing winds, the Coriolis effect, and the locations of continents. Together, prevailing winds and ocean currents distribute heat and precipitation around the globe.

- **Explain the mixing of surface and deep ocean waters from thermohaline circulation.**

  As ocean water flows from the Gulf of Mexico to the North Atlantic, water evaporates or freezes, and this causes the remaining water to have a high salt concentration and therefore a high density. This dense water sinks to the bottom of the ocean and later comes back to the surface near the equator.

- **Identify the causes and consequences of the El Niño–Southern Oscillation.**

  The El Niño–Southern Oscillation occurs when the typical trade winds from South America weaken or reverse, which allows the equatorial current that usually flows from east to west to reverse direction. When this happens, the upwelling along the western coast of South America is impeded, which affects climates around the world.

Module 12 Terrestrial Biomes

- **Explain how we define terrestrial biomes.**

  Terrestrial biomes are categorized by the dominant plant forms that exist in a region.

- **Interpret climate diagrams.**

  Climate diagrams illustrate monthly patterns of temperature and precipitation during the year. They also illustrate the growing season of a biome and the months during which plants are more constrained by temperature or precipitation.

- **Identify the nine terrestrial biomes.**

  The nine terrestrial biomes are tundra, boreal forests, temperate rainforests, temperate seasonal forests, woodland/shrublands, temperate grasslands/cold deserts, tropical rainforests, tropical seasonal forests/savannas, and subtropical deserts.

Module 13 Aquatic Biomes

- **Identify the major freshwater biomes.**

  There are three types of freshwater biomes. Streams and rivers have flowing fresh water. Lakes and ponds have standing water, at least some of which is too deep to support emergent vegetation. Freshwater wetlands are submerged or saturated by water for at least part of the year, but shallow enough to support emergent vegetation.

- **Identify the major marine biomes.**

  There are five types of marine biomes. Salt marshes are found along the coast in temperate climates and contain nonwoody emergent vegetation. Mangrove swamps occur along tropical and subtropical coasts and contain trees that have roots submerged in the water. The intertidal zone is the narrow band of coastline that exists between the levels of high tide and low tide. Coral reefs are found in warm, shallow waters beyond the shoreline and represent Earth's most diverse marine biome. The open ocean is characterized by deep water where sunlight can no longer reach the ocean bottom.

Chapter 4 AP® Environmental Science Practice Exam

Section 1: Multiple-Choice Questions

Choose the best answer for questions 1–13.

1. In which layer of Earth's atmosphere does most weather occur?
   (a) Troposphere
   (b) Stratosphere
   (c) Mesosphere
   (d) Thermosphere
   (e) Lithosphere

2. Which statement best explains why polar regions are colder than tropical regions?
   (a) Polar regions have lower albedo values.
   (b) Polar regions receive less solar energy per unit of surface area.
   (c) Tropical regions receive less direct sunlight throughout the year.
   (d) Sunlight travels through more atmosphere and loses more energy in tropical regions.
   (e) Tropical regions rotate at a faster speed than polar regions.

3. Which statement about patterns of air convection is NOT correct?
   (a) The air in a Hadley cell rises where sunlight strikes Earth most directly.
   (b) The greatest amount of precipitation occurs at the intertropical convergence zone.
   (c) The air in a Hadley cell descends near 30° N and 30° S, causing the formation of deserts.
   (d) The air of a polar cell rises near 60° latitude.
   (e) Along Earth's surface, the air of a Hadley cell moves away from the equator.

4. An increase in evaporation near the equator would most likely cause
   (a) increased precipitation at the ITCZ.
   (b) decreased precipitation at the ITCZ.
   (c) increased precipitation in Ferrell cells.
   (d) increased precipitation at 30° N and 30° S.
   (e) decreased precipitation in Ferrell cells.

5. The high heat capacity of water causes what effect when combined with ocean circulation?
   (a) The high salinity of deep polar water in the thermohaline cycle
   (b) Warm temperatures in continental coastal areas
   (c) The suppression of upwelling during an ENSO event
   (d) Increased elevation of warm tropical waters, driving surface currents
   (e) The heat transfer between the ocean and atmosphere due to evaporation along the equator

6. Which of the following processes is NOT characteristic of oceanic circulation?
   (a) Counterclockwise gyres in the Northern Hemisphere
   (b) Slow thermohaline circulation of surface and deep ocean waters
   (c) Unequal heating of tropical versus polar ocean waters
   (d) El Niño–Southern Oscillation
   (e) Coriolis effect

7. Which statement about rain shadows is correct?
   (a) They occur on the western sides of mountain ranges in the Northern Hemisphere.
   (b) Air gains water vapor as it rises.
   (c) As air rises over a mountain range, water vapor condenses into precipitation.
   (d) They occur on the eastern sides of mountain ranges in the Southern Hemisphere.
   (e) The rain shadow side of a mountain range receives the most rain.

8. Why do scientists use dominant plant growth forms to categorize terrestrial biomes?
   (a) Plants with similar growth forms are always closely related genetically.
   (b) Different plant growth forms indicate climate differences, whereas different animal forms do not.
   (c) Plants from similar climates evolve different adaptations.
   (d) Similar plant growth forms are found in climates with similar temperatures and amounts of precipitation.
   (e) Similar plant growth forms exist in both terrestrial and aquatic biomes.

9. Which information is NOT found in climate diagrams?
   (a) Average annual temperature
   (b) Seasonal changes in temperature
   (c) Average annual humidity
   (d) The months when plant growth is limited by precipitation
   (e) The length of the growing season

10. Which statement about tundras and boreal forests is correct?
   (a) Both are characterized by slow plant growth, so there is little accumulation of organic matter.
   (b) Tundras are warmer than boreal forests.
   (c) Boreal forests have shorter growing seasons than tundras.
   (d) Plant growth in both biomes is limited by precipitation.
   (e) Boreal forests have larger dominant plant growth forms than tundras.

Answers to Chapter 4 AP® Exam Multiple-Choice Questions

1. a
2. b
3. e
4. a
5. b
6. a
7. c
8. d
9. c
10. e
11. a
12. b
13. d

Additional Multiple-Choice AP® Practice Exam

Download an additional Multiple-Choice AP® Environmental Science Practice Exam for Chapter 4 by clicking on the link buttons in the TE-book, opening the TRFD, or logging onto the book's companion website bcs.whfreeman.com/friedlandapes2e (teacher login required).

Multiple-Choice AP® Practice Exam 4

Answers to Chapter 4 AP® Exam

Download the full answers to the Chapter 4 AP® Practice Exam by clicking on the link buttons in the TE-book, opening the TRFD, or logging onto the book's companion website bcs.whfreeman.com/friedlandapes2e (teacher login required).

Answers to Chapter 4 AP® Exam

11. Which of the following statements about temperate biomes is NOT correct?
    (a) Temperate biomes have average annual temperatures above 20°C.
    (b) Temperate rainforests receive the most precipitation, whereas cold deserts receive the least precipitation.
    (c) Temperate rainforests can be found in the northwestern United States.
    (d) Temperate seasonal forests are characterized by trees that lose their leaves.
    (e) Temperate shrublands are adapted to frequent fires.

12. Which statement about tropical biomes is correct?
    (a) Tropical seasonal forests are characterized by evergreen trees.
    (b) Tropical rainforests have the highest precipitation due to the proximity of the ITCZ.
    (c) Savannas are characterized by the densest forests.
    (d) Tropical rainforests have the slowest rates of decomposition due to high rainfall.
    (e) Subtropical deserts have the highest species diversity.

13. Which statement about aquatic biomes is correct?
    (a) They are characterized by dominant plant growth forms.
    (b) They can be categorized by temperature and precipitation.
    (c) Lakes contain littoral zones and intertidal zones.
    (d) Freshwater wetlands have emergent plants in their deepest areas, whereas ponds and lakes do not.
    (e) Coral reefs have the lowest diversity of species.

144 Chapter 4 Global Climates and Biomes

Section 2: Free-Response Questions

Write your answer to each part clearly. Support your answers with relevant information and examples. Where calculations are required, show your work.

1. As the greenhouse effect continues to warm the planet slowly, the glaciers of Greenland are melting at a rapid rate. Scientists are concerned that this melting may dilute the salt water in that region of the ocean enough to shut down thermohaline circulation. Use what you know about climate to answer the following questions.
   (a) Explain how shutting down thermohaline circulation would affect the temperature of western Europe. (2 points)
   (b) Explain the possible consequences for agriculture in western Europe. (2 points)
   (c) Why might the populations of fish along the west coasts of most continents increase if thermohaline circulation shuts down? (3 points)
   (d) How would shutting down thermohaline circulation affect the transport of nutrients among the oceans of the world? (3 points)

2. A number of Earth's features determine the locations of biomes around the world.
   (a) Explain why the tropical rainforests are found in regions of the world that receive the most direct sunlight. (4 points)
   (b) Describe how movement of the ITCZ over the year influences the location of seasonal forests in tropical regions. (2 points)
   (c) Identify the mechanisms by which albedo and the angle of the Sun's rays cause colder temperatures to occur on Earth near the North and South Poles. (2 points for each mechanism)

Answers to Chapter 4 AP® Exam Free-Response Questions

1. (a) Shutting down thermohaline circulation would reduce the amount of warm surface water flowing from the Gulf of Mexico to Europe, and this would cause western Europe to experience colder temperatures.
(b) Colder temperatures in western Europe would cause shorter growing seasons and reduced crop production.
(c) Surface currents would separate from one another, causing deeper waters to rise and replace the water that has moved away. These deep waters bring with them nutrients from the ocean bottom that support large populations of producers. The producers then support large populations of fish.
(d) Shutting down the thermohaline circulation would reduce the amount of nutrient flow among the world's oceans.

2. (a) Sunlight that is most direct produces the warmest temperatures. The most direct sunlight is also the driving force behind the primary Hadley cells, which release large amounts of precipitation at the intertropical convergence. Thus, these locations have both very warm temperatures and very high precipitation, and these two conditions favor the growth of tropical rain forests.
(b) The intertropical convergence zone is typified by dense clouds and intense thunderstorm activity. In tropical regions that are north and south of the equator, the ITCZ passes overhead twice each year, producing two seasons of high rainfall and two seasons of low rainfall. The location and amount of this rainfall determines the location and productivity of the seasonal forests in tropical regions.
(c) The angle of the Sun is least direct near the poles. This provides the least amount of solar energy per unit area, which causes lower temperatures. The high albedo effect near the poles, largely due to the high reflectance of snow, allows little absorption of solar energy and further reinforces the cold temperatures that favor the growth of tundra and boreal plants.

chapter

5 Evolution of Biodiversity

PD Overview

Chapter 5 describes the biodiversity of Earth, how it came to be, and how environmental factors can cause it to decline. Biodiversity can be described at the genetic level, the species level, and the ecosystem level. In any given location, biodiversity can be quantified by evaluating species richness and species evenness. Evolution is the underlying mechanism of biodiversity. Populations change over time in response to natural and artificial selection as well as random processes. Changes in the genetic composition of a population allow species to adapt to changing environmental conditions. Speciation occurs through allotropic and sympatric processes. Species evolve to exist in a unique niche that determines their geographic distribution. Environmental change can alter species distribution and lead to extinctions.

Module 14: The Biodiversity of Earth

Biodiversity occurs at three different scales, as shown in Figure 2.1 on page 9. Within a given region, for example, the variety of ecosystems is a measure of ecosystem diversity. Within a given ecosystem, the variety of species constitutes species diversity. Within a given species, the variety of genes can be used as a measure of genetic diversity. Every individual organism is distinguished from every other organism, at the most basic level, by the differences in the information coded in their genes. Because genes form the blueprint for an organism's traits, the diversity of genes on Earth ultimately helps determine the species diversity and ecosystem diversity on Earth. In other words, all three scales of biodiversity contribute to the overall biodiversity of the planet. In this module, students will examine how scientists estimate the number of species on Earth and how they quantify biodiversity. The module concludes by introducing the concept of a phylogeny as a way to illustrate relatedness among species.

Module 15: How Evolution Creates Biodiversity

This module examines the processes of evolution, which is the source of biodiversity. Because the evolution of biodiversity depends on genetic diversity, the module begins by exploring the sources of genetic variation. It then considers how humans and the natural world select from this variation to favor particular individuals that go on to reproduce in the next generation. The module closes with a discussion of how random processes can also cause evolution.

Module 16: Speciation and the Pace of Evolution

Over time, speciation has given rise to the millions of species present on Earth today. Beyond determining how many species exist, environmental scientists are also interested in understanding how quickly existing species can change, how quickly new species can evolve, and how quickly species can go extinct. This module examines the processes that produce new species and the factors that determine how rapidly species can evolve in response to changes in the environment.

Module 17: Evolution of Niches and Species Distributions

Because evolution alters the traits that species express and because different traits perform well in some environments but not in others, the evolution of species affects where species are able to live on Earth. This module explores how species have responded to environmental change over millions of years by altering their distributions. This response also suggests that environmental changes in the future will continue to affect the distributions of species. Species that can move easily will likely persist whereas species that cannot adjust or move will likely go extinct.

Alignment to AP® Environmental Science Course Description

AP® Outline	Module	
II. The Living World (10–15%)		
C. Ecosystem Structure	Module 14	The Biodiversity of Earth
D. Natural Ecosystem Change	Module 15	How Evolution Creates Biodiversity
	Module 16	Speciation and the Pace of Evolution
	Module 17	Evolution of Niches and Species Distributions

Chapter Learning Objectives

After completing this chapter students will be able to

- understand how we estimate the number of species living on Earth.
- quantify biodiversity.
- describe patterns of relatedness among species using a phylogeny.
- identify the processes that cause genetic diversity.
- explain how evolution can occur through artificial selection.
- explain how evolution can occur through natural selection.
- explain how evolution can occur through random processes.
- explain the processes of allopatric and sympatric speciation.
- understand the factors that affect the pace of evolution.
- explain the difference between a fundamental and a realized niche.
- describe how environmental change can alter species distributions.
- discuss how environmental change can cause species extinctions.

Chapter 5 Pacing Guide

This pacing guide is based on a schedule with 120 sessions of 50 minutes each before the AP® Exam. If you have a different number of sessions before the exam, you can modify the pacing to suit your needs. If you have additional time, consider incorporating quizzes, released AP® Environmental Science free-response and multiple-choice questions or additional activities.

Module	Standard Schedule Days	Block Schedule Days
Module 14	2	1
Module 15	½	¼
Module 16	½	¼
Module 17	1	½
Assessment	1	½

Chapter 5 Resources

The link to these resources can be found by clicking on the link buttons in the Teacher's e-Book (TE-book), opening the Teacher's Resource Flash Drive (TRFD), or logging onto the book's companion website bcs.whfreeman.com/friedlandapes2e (teacher login required).

- **PowerPoint Presentation 5**
- **Optimized Art PowerPoint and JPEG Files 5**
- **Do the Math Videos**
- **Measuring Your Impact 5:** The True Cost of a Green Lawn
- **Lab 5.1:** Calculating the Diversity of Trail Mix Using Shannon's Index
- **Chapter 5 Web Resources**
- **Exploring the Literature 5**
- **Answers to Chapter 5 AP® Exam**
- **Multiple-Choice AP® Practice Exam 5**
- **Answers to Unit 2 AP® Exam**
- **Unit 2 Additional Free-Response Question**
- **Answer to Science Applied 2 Free-Response Question**

Free-Response Questions from Previous AP® Environmental Science Exams

Free-response questions from prior AP® Environmental Science Exams are available on the AP® course website, https://apstudent.collegeboard.org/apcourse/ap-environmental-science/exam-practice. Students should be able to answer all of the questions listed below with material learned in this chapter. When a question requires students to understand material from multiple chapters, the question will be listed in the last chapter required to complete the entire question. Questions marked with an asterisk (*) are from exams with released multiple-choice questions. You may want to save these questions until the end of the year so you can give your students a complete released exam for practice. Questions marked with double asterisks (**) require math to calculate a problem. Look for references to these questions throughout the chapter.

Year	Question	Content
2013	4	• Characteristics of biodiverse ecosystems • How humans affect biodiverse ecosystems • Benefits of biodiversity

PD Chapter 5 Overview

Watch the video overview of Chapter 5 (for teachers) by clicking on the link buttons in the TE-book, opening the TRFD, or logging onto the book's companion website bcs.whfreeman.com /friedlandapes2e (teacher login required).

TRM PowerPoint Presentation 5

Download the PowerPoint presentation for Chapter 5 by clicking on the link buttons in the TE-book, opening the TRFD, or logging onto the book's companion website bcs.whfreeman.com /friedlandapes2e (teacher login required).

The plant known as the Dung of the Devil, discovered as a treatment for the Spanish flu of 1918, may also be a remedy for the H1N1 virus. *(Rstudio/Alamy)*

chapter 5

Evolution of Biodiversity

The Dung of the Devil

From 1918 to 1920, the world experienced a flu outbreak of unprecedented scale. Known as the Spanish flu, the disease had a devastating effect. Mortality estimates from that time vary, but somewhere between 20 million and 100 million people died worldwide, including more than 600,000 people in the United States. During the height of the outbreak, reports stated that some people in China had found the roots of a particular plant beneficial in fighting the flu. The plant (*Ferula assafoetida*) had a pleasant smell when cooked, but the raw sap from the roots had a foul smell that inspired the plant's common name, the Dung of the Devil.

The Dung of the Devil story does not end in 1920. It turns out that Spanish flu was caused by an H1N1 virus that is closely related to the H1N1 virus that caused the worldwide "swine flu" outbreak of 2009–2010. Scientists in China recalled that people had used the plant to fight the Spanish flu 80 years ago, so they decided to explore its potential to combat the modern H1N1 flu virus. They found that extracts from the plant had strong antiviral properties, stronger even than those of contemporary antiviral drugs. Thus the Dung of the Devil has the potential to produce a new pharmaceutical drug to fight future H1N1 flu epidemics.

The Dung of the Devil has the potential to produce a new pharmaceutical drug to fight future H1N1 flu epidemics.

The Dung of the Devil is just one of the organisms from which humans have extracted life-saving drugs. Willow trees from temperate forests were the original source of salicylic acid,

Teaching Tip: Chapter Opening Case

The chapter opening case, "The Dung of the Devil," illustrates one important reason to protect biodiversity: the use of resources for the development of new pharmaceutical drugs. Humans have already extracted life-saving drugs from a variety of species. With increasing rates of deforestation and habitat loss, many species that have never been researched for medical use are at risk of extinction. This introductory case helps students understand the significance of biodiversity and the underlying mechanisms that allow organisms to adapt to their ever-changing environments.

Teaching Tip: Warm-up

One reason to protect biodiversity is the potential development of future medicines found in living organisms. Ask students to brainstorm other reasons to protect biodiversity. *Ecosystems provide natural capital such as lumber. Biodiversity supports ecosystem services that include water purification and pollination. Humans derive recreational and economic benefits from highly diverse ecosystems. Ecosystem stability is supported by ecosystems that are highly diverse. Increased genetic diversity exists in highly diverse ecosystems, which protects the overall health of populations.*

from which aspirin is derived. More recently, wild plants have provided several important medicines for treating a variety of cancers. For example, the rosy periwinkle (*Catharanthus roseus*), found only in the tropical forests of Madagascar, is the source of two drugs used to treat childhood leukemia and Hodgkin's disease. The mayapple (*Podophyllum peltatum*), a common herb of the eastern United States, is the source of two other anticancer drugs. Many new medicines, including anti-inflammatory, antiviral, and antitumor drugs, have come from a variety of invertebrate animals that inhabit coral reefs, including sponges, corals, and sea squirts. Of the most promising current candidates for new drugs, 70 percent were first discovered in plants, animals, and microbes. Unfortunately, many species that are either known or suspected sources of drugs are being lost to deforestation, agriculture, and other human activities. At the same time, indigenous people with knowledge about medicinal uses of the natural drugs in their environment are being forced to relocate, and their knowledge may soon be lost.

Only a small fraction of the millions of species on Earth has been screened for useful drugs. It is likely that many more medicines could be found in living organisms. The continual discovery of new drugs in organisms around the world, including the Dung of the Devil, makes yet another convincing argument for conserving Earth's biodiversity.

Sources:
C. L. Lee et al., Influenza A (H1N1) antiviral and cytotoxic agents from *Ferula assafoetida*, *Journal of Natural Products* 72 (2009): 1568–1572; D. Newman and G. M. Cragg, Natural products as sources of new drugs over the last 25 years, *Journal of Natural Products* 70 (2007): 461–477.

The use of plants for drugs that can help humans fight diseases is just one of many reasons that we want to protect the biodiversity of the planet. In general, biodiversity is an important indicator of environmental health, so a rapid decline of biodiversity in an ecosystem indicates that it is under stress. The biodiversity on Earth today is the result of evolution and extinction. Knowledge of these processes helps us to understand past and present environmental changes and their effects. In this chapter, we will examine how scientists quantify biodiversity and then look at how the process of evolution creates biodiversity. We will also examine the processes of speciation and extinction and the ways species have evolved unique ways of life that affect the abiotic and biotic conditions under which they can live.

module 14

The Biodiversity of Earth

As you will recall from Chapter 1, we can think about biodiversity at three different scales (see Figure 2.1 on page 9). Within a given region, for example, the variety of ecosystems is a measure of ecosystem diversity. Within a given ecosystem, the variety of species constitutes species diversity. Within a given species, we can think about the variety of genes as a measure of genetic diversity. Every individual organism is distinguished from every other organism, at the most basic level, by the differences in the information coded by their genes. Because genes form the blueprint for an organism's traits, the diversity of genes on Earth ultimately helps determine the species diversity and ecosystem diversity on Earth. In other words, all three scales of biodiversity contribute to the overall biodiversity of the planet. In this module, we will examine how we estimate the number of species on Earth and how scientists quantify biodiversity. We will then examine how scientists illustrate the relatedness among species.

Learning Objectives

After reading this module you should be able to

- understand how we estimate the number of species living on Earth.
- quantify biodiversity.
- describe patterns of relatedness among species using a phylogeny.

Teaching Tip: Discussion Starter

Ask students the following questions:

- What is the range of current estimates for the number of species on Earth? *The range is 5 to 100 million species.*
- What estimate do most scientists agree upon? *Most scientists agree that there are approximately 10 million species.*

It is difficult to estimate the number of species on Earth

A short walk through the woods, a corner lot, or a city park makes one thing clear: Life comes in many forms. A small plot of untended land or a tiny pond contains dozens, perhaps hundreds, of different kinds of plants and animals visible to the naked eye as well as thousands of different kinds of microscopic organisms. In contrast, a carefully tended lawn or a commercial timber plantation usually supports only a few types of grasses or trees (FIGURE 14.1). The total number of organisms in the plantation or lawn may be the same as the number in the pond or in the untended plot, but the number of species will be far smaller.

Recall from Chapter 1 that a species is defined as a group of organisms that is distinct from other such

Teaching Tip: Warm-up

Knowing the number of species on Earth is not a useful indicator of how many species live in a particular location. Ask students to explain why. *Species are not uniformly distributed across the planet. To accurately assess diversity, it must be measured at the local or regional level.*

(a)

(b)

FIGURE 14.1 **Species diversity and ecosystems.** (a) Natural forests contain a high diversity of tree species. (b) In forest plantations, in which a single tree species has been planted for lumber and paper products, species diversity is low. *(a: Ron and Patty Thomas/Getty Images; b: Brent Waltermire/Alamy)*

Teaching Tip: Journal Prompt

Ask students to answer the following questions:

- What is the difference between species richness and species evenness? *Species richness is the number of species in a given area. Species evenness describes species abundance, which is the relative proportion of individuals within the different species in a given location.*
- Why are they both important measures? *Higher numbers of species means greater diversity if there is equal distribution of each of those species in an area. If one species dominates the community, diversity is not considered to be as high even if there are other species present. Scientists use both of these measures to establish a baseline to measure changes in diversity.*

groups in terms of size, shape, behavior, or biochemical properties, and that can interbreed with other individuals in its group to produce viable offspring. This last requirement is important because sometimes individuals from different species can mate, but they do not produce offspring that survive.

The number of species in any given place is the most common measure of biodiversity, but estimating the total number of species on Earth is a challenge. Many species are easy to find, such as the birds or small mammals you might see in your neighborhood. Others are not so easy to find. Some species are active only at night, live in inaccessible locations such as the deep ocean, or cannot be seen without a microscope. To date, scientists have named approximately 2 million species, which means the total must be larger than that.

The insects represent a group that contains more species than most other groups, so scientists reason that if we could get a good estimate for the number of insect species in the world, we would have a much better sense of the total number of species. In one study, researchers fumigated the canopies of a single tree species in the tropical rainforest and then collected all the dead insects that fell from the trees onto a tarp on the ground. From this collection of dead insects, they counted the number of beetle species that fed on only the one tree species they fumigated. By multiplying this number of beetle species by the total number of tropical tree species, they estimated that in the tropics there were perhaps 8 million species of beetles that feed on a single species of tree. Because beetles make up about 40 percent of all insect species, and because insect species in the forest canopy tend to be about twice as numerous as insect species on the forest floor, the researchers suggested that a reasonable estimate for the total number of tropical insect species might be 30 million. More recent work has indicated that this number is probably too high. Current estimates for the total number of species on Earth range between 5 million and 100 million, but most scientists estimate that there are about 10 million species.

We can measure biodiversity in terms of species richness and evenness

Because species are not uniformly distributed, the number of species on Earth is not a useful indicator of how many species live in a particular location. Often we desire to know the number of species at a given location to determine whether a region is being affected by

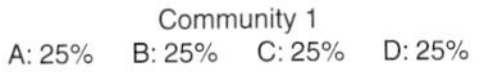

Community 2
A: 70% B: 10% C: 10% D: 10%

FIGURE 14.2 **Measures of species diversity.** Species richness and species evenness are two different measures of species diversity. Although both communities contain the same number of species, community 1 has a more even distribution of species and is therefore more diverse than community 2.

human activities. To measure species diversity at local or regional scales, scientists have developed two measures: *species richness* and *species evenness*.

The number of species in a given area, such as a pond, the canopy of a tree, or a plot of grassland, is known as **species richness.** Species richness is used to give an approximate sense of the biodiversity of a particular place. However, we may also want to know the **species evenness,** which is the relative proportion of individuals within the different species in a location. Species evenness tells us whether a particular ecosystem is numerically dominated by one species or whether all of its species have similar abundances. An ecosystem has high species evenness if its species are all represented by similar numbers of individuals. An ecosystem has low species evenness if one species is represented by many individuals whereas other species are represented by only a few individuals. In this case, there is effectively less diversity.

Scientists evaluating the biodiversity of an area must often look at both species richness and species evenness. Consider the two forest communities, community 1 and community 2, shown in FIGURE 14.2. Both forests contain 20 trees that are distributed among four species. In community 1, each species is represented by 5 individuals. In community 2, one species is represented by 14 individuals and each of the other three species is represented by 2 individuals. Although the species richness of the two forests is identical, the four species are more evenly represented in community 1. That forest therefore has greater species evenness and is considered to be more diverse.

Because species richness or evenness often declines after a human disturbance, knowing the species richness and species evenness of an ecosystem gives environmental scientists a baseline they can use to determine how much that ecosystem has changed. "Do the Math: Measuring Species Diversity" on page 108 demonstrates one common way of calculating species diversity.

The evolutionary relationship among species can be illustrated using a phylogeny

Scientists organize species into categories that indicate how closely related they are to one another. The branching pattern of evolutionary relationships is called

Species richness The number of species in a given area.

Species evenness The relative proportion of individuals within the different species in a given area.

◀ TEACHING with FIGURES

Have students evaluate both communities in Figure 14.2. Ask them to vote on which community appears to be more diverse. Then have students explain their reasoning using the terms "species richness" and "species evenness."

Practicing Science: Evaluate

Use the Howard Hughes Medical Institute (HHMI) interactive called "Using DNA to Explore Lizard Phylogeny" to reinforce concepts pertaining to phylogenic trees. Students will sort lizard species by appearance, and then create a phylogenetic tree using the lizards' DNA sequences to evaluate whether species that appear to be the same are closely related to each other.

The link to this activity can be found by clicking on the link buttons in the TE-book, opening the TRFD, or logging onto the book's companion website bcs.whfreeman .com/friedlandapes2e (teacher login required).

Chapter 5 Web Resources

Teaching Tip: Engage

To illustrate the difference between species richness and species evenness, ask two students to flip over their desks. Look at the gum that has been stuck to the bottom of the desk. Then, discuss which desk has greater species diversity and which has greater species richness. If your desks are remarkably gum-free, you can create two different bags with varying mixtures of candy or dried beans. Make sure that one bag represents "species richness" with a great variety of items and the other represents "species evenness" with just a few varieties.

Lab

Calculating the Diversity of Trail Mix

Shannon's index is used by environmental scientists to compare diversity of ecosystems and to quantify species evenness and species richness. This lab asks students to apply Shannon's index to communities of trail mix. It should take your class approximately 50 minutes to complete.

Download the lab by clicking on the link buttons in the TE-book, opening the TRFD, or logging on to the book's companion website bcs.whfreeman.com/friedlandapes2e (teacher login required).

Lab 5.1 Calculating the Diversity of Trail Mix

Answer to Your Turn

$H = -[0.5 \times \ln(0.5) + 0.5 \times \ln(0.5)]$
$= 0.69$

This value is between the two Shannon's index values for the examples with four species.

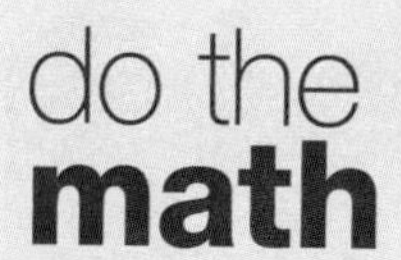

Measuring Species Diversity

Environmental scientists are often interested in evaluating both species richness and species evenness, so they have come up with indices of species diversity that take both measures into account. One commonly used index is Shannon's index of diversity. To calculate this index, we must know the total number of species in a community (n) and, for each species, the proportion of the individuals in the community that represent that species (p_i). Once we have this information, we can calculate Shannon's index (H) by taking the product of each proportion (p_i) and its natural logarithm [ln (p_i)] and then summing these products, as indicated by the summation symbol (Σ):

$$H = -\sum_{i=1}^{n} p_i \ln(p_i)$$

The minus sign makes the index a positive number. Higher values of H indicate higher diversity.

Imagine a community of 100 individuals that are evenly divided among four species, so that the proportions (p_i) of the species all equal 0.25. We can calculate Shannon's index as follows:

$H = -[(0.25 \times \ln 0.25) + (0.25 \times \ln 0.25) + (0.25 \times \ln 0.25) \times (0.25 \times \ln 0.25)]$
$H = -[(-0.35) + (-0.35) + (-0.35) + (-0.35)]$
$H = 1.40$

Now imagine another community of 100 individuals that also contains four species, but in which one species is represented by 94 individuals and the other three species are each represented by 2 individuals. We can calculate Shannon's index to see how this difference in species evenness affects the value of the index:

$H = -[(0.94 \times \ln 0.94) + (0.02 \times \ln 0.02) + (0.02 \times \ln 0.02) + (0.02 \times \ln 0.02)]$
$H = -[(-0.06) + (-0.08) + (-0.08) + (-0.08)]$
$H = 0.30$

Because this value of H is lower than the value we calculated for the first community, we can conclude that the second community has lower diversity. Note that the total number of individuals does not affect Shannon's index of diversity; only the number of species and the proportion of individuals within each species matter.

Your Turn Imagine a third community of 100 individuals in which those individuals are distributed evenly among all the species, but there are only two species, not four. Calculate Shannon's index to see how this difference in species richness affects the value of the index.

a **phylogeny.** Phylogenies can be described with a diagram like the one shown in FIGURE 14.3, called a phylogenetic tree.

Phylogeny The branching pattern of evolutionary relationships.

The relatedness of the species in a phylogeny is determined by similarity of traits: The more similar the traits of two species, the more closely related the two species are assumed to be. Historically, scientists used mostly morphological traits, including a large number of bone measurements, to measure similarity. Today, scientists base phylogenies on a variety of characteristics, including morphology, behavior, and genetics.

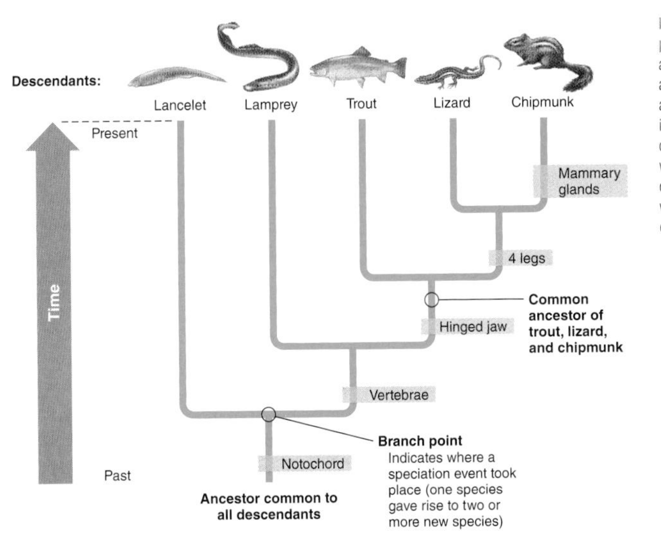

FIGURE 14.3 A phylogenetic tree. Phylogenies are based on the similarity of traits among species. Scientists can assemble phylogenetic trees that indicate how different groups of organisms are related and show where speciation events have occurred. The brown boxes indicate when major morphological changes evolved over evolutionary time.

◀ TEACHING with FIGURES

Have students describe how scientists determine phylogenic trees. Using Figure 14.3 as an example, ask students to identify the changes that distinguished each species from its common ancestor. *Scientists base phylogenies on morphology, behavior, and genetics. Based on Figure 14.3, differences in morphology such as the presence of a notochord, vertebrae, hinged jaw, four legs, and mammary glands are major morphological changes that evolved over time.*

module

14

REVIEW

In this module, we learned that scientists are not able to determine the exact number of species on Earth. Most estimates agree on approximately 10 million species, of which we have identified approximately 2 million. We saw that biodiversity can be quantified in terms of species richness and evenness. Species can be arranged on a phylogenetic tree that illustrates the evolutionary steps that gave rise to current species. In the next module, we will examine how evolution has produced such a large diversity of species.

Module 14 AP® Review Questions

1. How many species are estimated to exist on Earth?
   (a) 2 million
   (b) 8 million
   (c) 10 million
   (d) 30 million
   (e) 100 million

2. Two savanna communities both contain 15 plant species. In community A, each of the 15 species is represented by 20 individuals. In community B, 10 of the species are each represented by 12 individuals; the remaining 5 species are each represented by 3 individuals. Which statement best describes the two communities?
   (a) Community A has the same biodiversity as community B.
   (b) Both communities have the same species evenness.
   (c) Community B has a higher species richness.
   (d) Community A has a higher species evenness.
   (e) Community B has a lower species richness.

Teaching Tip: Activity

Assign "Science Applied 2: How Should We Prioritize the Protection of Species Diversity?" on page 184. This article introduces students to biodiversity hotspots, which are isolated areas that contain a high proportion of all the species found on Earth. Students will learn about the characteristics that define a hotspot. They will discover areas that have been identified as hotspots around the world. Students must also consider the costs as well as the benefits of conserving these areas. A quick activity students may enjoy after reading the article is to create a postcard for an identified hotspot. Use the Critical Ecosystem Partnership Fund (CEPF) website to identify hotspots. On one side, students must include pictures and descriptions of two animal and three plant or fungi species endemic to the hotspot. On the other side students should describe the biome and give the climate diagram. They should include threats to the species living in this hotspot and the conservation efforts being made to protect the area.

The link to the CEPF website can be found by clicking on the link buttons in the TE-book, opening the TRFD, or logging onto the book's companion website bcs.whfreeman.com /friedlandapes2e (teacher login required).

 **Chapter 5 Web Resources**

Answers to Module 14 AP® Review Questions

1. c
2. d
3. c
4. b

Teaching Tip: Video

Myths and Misconceptions About Evolution

This video clip from TED-Ed introduces evolution and helps students understand the mechanisms of evolution. The link to this resource can be found by clicking on the link buttons in the TE-book, opening the TRFD, or logging onto the book's companion website bcs.whfreeman.com/friedlandapes2e (teacher login required).

 **Chapter 5 Web Resources**

Teaching Tip: Journal Prompt

Ask students the following:

- Define "evolution." *Evolution is a change in the genetic composition of a population over time.*
- Differentiate between microevolution and macroevolution. *Microevolution occurs within a species, creating a variety of characteristics seen among individuals. Macroevolution involves genetic changes that give rise to new species, genera, families, classes, or phyla.*
- Describe how microevolution and macroevolution contribute to biodiversity on Earth. *Both of these processes contribute to biodiversity by creating genetic variation at the species level and greater species diversity in communities. Greater variation allows for increased survival and stability as populations and communities have greater adaptability to changing conditions.*

3. Phylogeny is
   (a) the number of evolutionarily related species in an ecosystem.
   (b) the study of morphological traits.
   (c) the branching pattern of evolutionary relationships.
   (d) the process of evolution that creates new species.
   (e) the genetic biodiversity of a species.

4. Which of the following is used to calculate Shannon's index of diversity?
   I. The proportion of individuals in each species
   II. The total number of species
   III. The number of individuals in each species

   (a) I only
   (b) I and II only
   (c) II only
   (d) II and III only
   (e) I, II, and III

module 15

How Evolution Creates Biodiversity

We have seen the importance of biodiversity. In this module we will look at the processes of *evolution*, which is the source of biodiversity. Because the evolution of biodiversity depends on genetic diversity, we will examine how genetic diversity is created. We begin by exploring the sources of genetic variation and then consider how humans and the natural world select from this variation to favor particular individuals that go on to reproduce in the next generation. Finally, we will examine how several random processes can also cause evolution.

Learning Objectives

After reading this module you should be able to

- identify the processes that cause genetic diversity.
- explain how evolution can occur through artificial selection.
- explain how evolution can occur through natural selection.
- explain how evolution can occur through random processes.

Genetic diversity is created through mutation and recombination

Earth's biodiversity is the product of **evolution,** which can be defined as a change in the genetic composition of a population over time. Evolution can occur at multiple levels. Evolution below the species level, such as the evolution of different varieties of apples or potatoes, is called **microevolution.** In contrast, when genetic changes give rise to new species, or to new genera, families, classes, or phyla—larger categories of organisms into which species are organized—we call the process **macroevolution.** Among these many levels of macroevolution, the term speciation is restricted to the evolution of new species.

To understand how genetic diversity is created, we first need to understand *genes*. **Genes** are physical locations on chromosomes within each cell of an organism. A given gene has DNA that codes for a particular trait, such as body size, but the DNA can take different forms known as alleles. An organism's genes determine the range of possible traits (physical or behavioral characteristics) that it can pass down to its offspring. The complete set of genes in an individual is called its **genotype.** In this section, we will discuss how genotypes help to determine the traits of individuals and the two processes that can create genetic diversity in a population: *mutation* and *recombination.*

Genotypes versus Phenotypes

An individual's genotype serves as the blueprint for the complete set of traits that organism may potentially possess. An individual's **phenotype** is the actual set of traits expressed in that individual. Among these traits are the individual's anatomy, physiology, and behavior. The color of your eyes, for example, is your phenotype, whereas the genes that code for eye color are a part of your genotype. Changes in genotypes can produce important changes in phenotypes.

In some cases, an individual's phenotype is determined almost entirely by genes. For instance, a person who inherits the genes for brown eyes will have brown eyes, regardless of where that person lives. Most phenotypes, however, are the product of an individual's environment as well as its genotype. For example, in many turtle and crocodile species, the temperature of eggs during incubation determines whether the offspring will hatch as males or females. The water flea, a tiny animal that lives in ponds and lakes, offers another interesting example. The body shape of the water flea depends on whether or not a young individual smells predators in its environment (FIGURE 15.1). If predators are absent, the water flea develops a relatively small head and tail spine. If predators are present, however, the water flea develops a much larger head and a long tail spine. Although the larger head and longer tail spine help prevent the water flea from being eaten, they come at the cost of slower reproduction. Therefore, it is beneficial that the water flea not produce these defenses unless they are needed. By being able to respond to changing environmental conditions, organisms such as the water flea can improve their ability to survive and reproduce in any environment.

FIGURE 15.1 **Environmental effects on phenotype.** Water fleas raised in the absence of predators produce relatively small heads and short tail spines (left), whereas individuals raised in the presence of predators produce relatively large heads and long tail spines (right). *(Christian Laforsch/Science Photo Library/Photo Researchers)*

Evolution A change in the genetic composition of a population over time.

Microevolution Evolution below the species level.

Macroevolution Evolution that gives rise to new species, genera, families, classes, or phyla.

Gene A physical location on the chromosomes within each cell of an organism.

Genotype The complete set of genes in an individual.

Phenotype A set of traits expressed by an individual.

Teaching Tip: Warm-up

Ask students to differentiate between genotype and phenotype. *The complete set of genes in an individual is its genotype. These genes, combined with influences from the environment, determine an individual's phenotype, which is the set of traits expressed by an individual.*

Teaching Tip: Discussion Starter

Phenotype can be influenced by both genotype and environmental factors. Ask students: How do variations in phenotype benefit the organism? *By being able to respond to changing environmental conditions, organisms can improve their ability to survive and reproduce.*

Teaching Tip: Warm-up

Ask students to identify and describe the three primary ways in which evolution occurs.

- ***Artificial selection:*** *Humans determine which individuals will breed to select for certain traits.*
- ***Natural selection:*** *The environment determines which individuals survive and reproduce.*
- ***Random processes:*** *Random processes include mutation, gene flow, genetic drift, bottleneck effect, and founder effect. These mechanisms change a population composition but the changes are not related to differences in fitness among individuals.*

Mutation

DNA is copied millions of times during an organism's lifetime as cells grow and divide. An occasional mistake in the copying process produces a random change in the genetic code, which is known as a **mutation.** Environmental factors, such as ultraviolet radiation from the Sun, can also cause mutations. When mutations occur in cells responsible for reproduction, such as the eggs and sperm of animals, those mutations can be passed on to the next generation.

Most mutations are detrimental, and many cause the offspring that carry them to die while they are embryos. The effects of some mutations are less severe, but can still be detrimental. For example, some dusky-headed conures (*Aratinga weddellii*) have a mutation that makes these normally green-feathered parrots produce feathers that appear to be blue (FIGURE 15.2). In the wild, individuals with this mutation have a poor chance of survival because blue feathers stand out against the green vegetation and make them conspicuous to predators.

Sometimes a mutation improves an organism's chances of survival or reproduction. If such a mutation is passed along to the next generation, it adds new genetic diversity to the population. Some mosquitoes, for example, possess a mutation that makes them less vulnerable to insecticides. In areas that are sprayed with insecticides, this mutation improves an individual mosquito's chance of surviving and reproducing.

Recombination

Genetic diversity can also be created through *recombination.* In plants and animals, genetic **recombination** occurs as chromosomes are duplicated during reproductive cell division and a piece of one chromosome breaks off and attaches to another chromosome. This process does not create new genes, but it does bring together new combinations of alleles on a chromosome and can therefore produce novel traits. For example, the human immune system must battle a large variety of viruses and bacteria that regularly attempt to invade the body. Recombination allows new allele combinations to come together, and this provides new immune defenses that may prove to be effective against the invading organisms.

Mutation A random change in the genetic code produced by a mistake in the copying process.

Recombination The genetic process by which one chromosome breaks off and attaches to another chromosome during reproductive cell division.

Evolution by artificial selection The process in which humans determine which individuals breed, typically with a preconceived set of traits in mind.

FIGURE 15.2 **Mutations.** A mutation in the genetic code of the dusky-headed conure causes these normally green-feathered parrots to develop feathers that appear blue. In nature the mutation makes individuals more conspicuous and prone to predation. *(Howard Voren at www.Voren.com)*

Evolution can occur through artificial selection

Evolution occurs in three primary ways: *artificial selection, natural selection,* and *random processes.* In this section we will look at artificial selection.

Humans have long influenced evolution by breeding plants and animals for desirable traits. For example, all breeds of domesticated dogs belong to the same species as the gray wolf (*Canis lupus*), yet dogs exist in an amazing variety of sizes and shapes, ranging from toy poodles to Siberian huskies. FIGURE 15.3 shows the phylogenetic relationships among the wolf and different breeds of domestic dogs that were bred from the wolf by humans. Beginning with the domestication of wolves, dog breeders have selectively bred individuals that had particular qualities they desired, including body size, body shape, and coat color. After many generations of breeding, the selected traits became more and more exaggerated until breeders felt satisfied that the desired characteristics of a new dog breed had been achieved. As a result of this carefully controlled breeding, we have a tremendous variety of dog sizes, shapes, and colors today. Yet dogs remain a single species: All dog breeds can still mate with one another and produce viable offspring.

When humans determine which individuals to breed, typically with a preconceived set of traits in mind, we call the process **evolution by artificial selection.** Artificial selection has produced numerous breeds of horses, cattle, sheep, pigs, and chickens with traits that humans find useful or aesthetically pleasing. Most of our modern agricultural crops are also the result of many years of careful breeding. For example, starting with a single species of wild mustard, *Brassica oleracea,* plant breeders have produced a variety of food

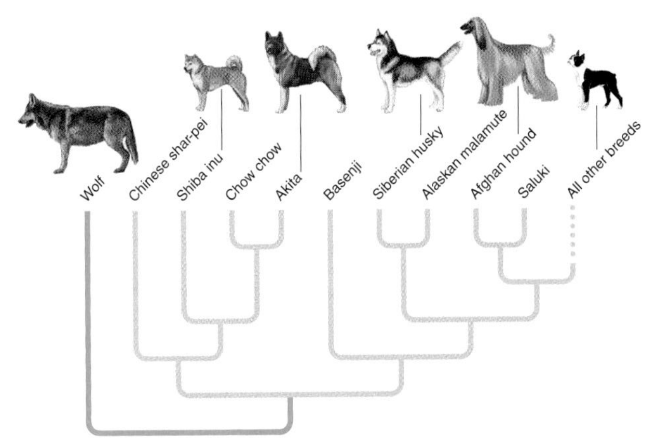

FIGURE 15.3 **Artificial selection on animals.** The diversity of domesticated dog breeds is the result of artificial selection on wolves. The wolf is the ancestor of the various breeds of dogs. It is illustrated at the same level as the dogs in this phylogeny because it is a species that is still alive today. *(Data from H. G. Parker et al., Science 304 (2004): 1160–1164.)*

crops, including cabbage, cauliflower, broccoli, Brussels sprouts, kale, and kohlrabi, shown in FIGURE 15.4.

As useful as artificial selection has been to humans, it can also produce a number of unintended results. For example, farmers often use herbicides to kill weeds. However, as we cover larger and larger areas with herbicides, there is an increasing chance that at least one weed will possess a mutation that allows it to survive the herbicide application. If that one mutant plant passes on its herbicide resistance to its offspring, we will have artificially selected for herbicide resistance in that weed. This process is occurring in many parts of the world where increased use of the popular herbicide Roundup (chemical name: glyphosate) has led to the evolution of several species of Roundup-resistant weeds. A similar process has occurred in hospitals, where the use of antibiotics and antibacterial cleaners has caused artificial selection of harmful drug-resistant bacteria. These examples underscore the importance of understanding the mechanisms of evolution and the ways in which humans can either purposefully or inadvertently direct the evolution of organisms.

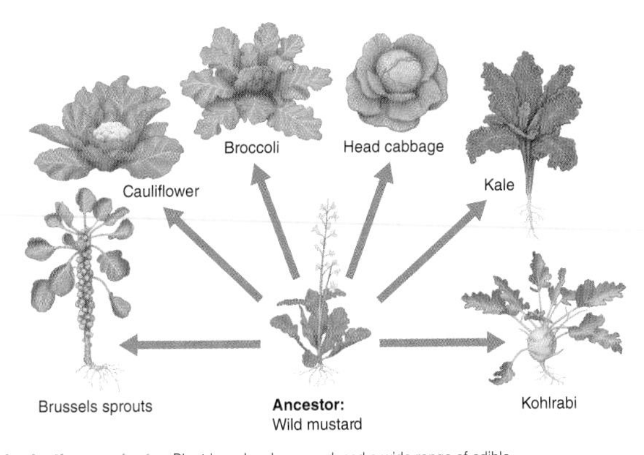

FIGURE 15.4 **Artificial selection on plants.** Plant breeders have produced a wide range of edible plants from a single species of wild mustard.

Teaching Tip: Discussion Starter

Humans use artificial selection to select for desirable traits. Ask students to give an example of human actions that result in unintentional selection. *Humans use herbicides to kill weeds. If one plant possesses a mutation that allows it to survive, it will pass on the beneficial mutation to future generations. The herbicide-resistant weeds continue to thrive and reproduce, which represents an unintentional selection for herbicide-resistant weeds. Another example is the use of antibiotics and antibacterial cleaners, which have led to artificial selection for drug-resistant bacteria.*

Evolution can occur through natural selection

Evolution also takes place through natural mechanisms. In **evolution by natural selection,** the environment determines which individuals survive and reproduce. Members of a population naturally vary in their traits, and certain combinations of those traits make individuals better able to survive and reproduce. As a result, the genes that produce those traits are more common in the next generation.

Prior to the mid-nineteenth century, the idea that species could evolve over time had been suggested by a number of scientists and philosophers. However, the concept of evolution by natural selection did not become synthesized into a unifying theory until two scientists, Alfred Wallace (1823–1913) and Charles Darwin (1809–1882), independently put the various pieces together.

Of the two scientists, Charles Darwin is perhaps the better known. At age 22, he became the naturalist on board HMS *Beagle,* a British survey ship that sailed around the world from 1831 to 1836. During his journey, Darwin made many observations of trait variation across a tremendous variety of species. In addition to observing living organisms, he found fossil evidence of a large number of extinct species. He also recognized that organisms produce many more offspring than are needed to replace the parents, and that most of these offspring do not survive. Darwin questioned why, out of all the species that had once existed on Earth, only a small fraction had survived. Similarly, he wondered why, among all the offspring produced in a population in a given year, only a small fraction survived to the next year. During the decades following his voyage, he developed his ideas into a robust theory. His *On the Origin of Species by Means of Natural Selection,* published in 1859, changed the way people thought about the natural world.

The key ideas of Darwin's theory of evolution by natural selection are the following:

- Individuals produce an excess of offspring.
- Not all offspring can survive.
- Individuals differ in their traits.
- Differences in traits can be passed on from parents to offspring.
- Differences in traits are associated with differences in the ability to survive and reproduce.

Evolution by natural selection The process in which the environment determines which individuals survive and reproduce.

FIGURE 15.5 shows how this process works using the example of body size in a group of crustaceans known as amphipods. We can begin with parents producing offspring that vary in their body size. The largest offspring are consumed by fish because fish prefer to eat large prey rather than small prey. As a result, the smaller offspring are left to reproduce. Because body

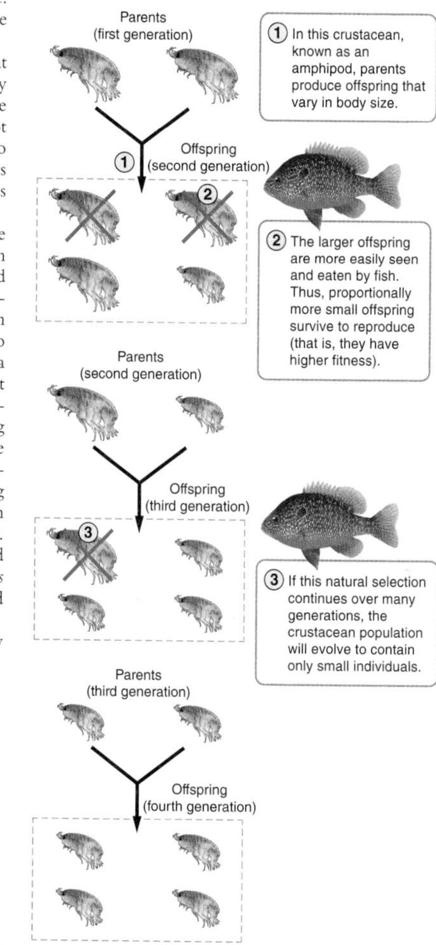

FIGURE 15.5 Natural selection. All species produce an excess number of offspring. Only those offspring with the fittest genotypes will pass on their genes to the next generation.

▲ TEACHING with FIGURES

Have students consider Figure 15.5. Larger amphipods are easily seen and eaten by fish. Help students understand predation as an environmental pressure that will limit amphipod population.
Ask students: Why, after many generations, will the amphipod population contain only small individuals? *If larger offspring are more easily seen by predators, they are more likely to be eaten than smaller offspring. Proportionally more offspring that are smaller survive to reproduce. Over many generations, those offspring with genotypes for smaller size will continue to be the most successful—have the highest fitness—and the population will evolve to contain only small individuals.*

size is, in part, determined by an individual's genes, the next generation of amphipods will be smaller. This process can continue over many generations and over time the fish will cause the evolution of smaller body sizes in amphipods.

Both artificial and natural selection begin with the requirement that individuals vary in their traits and that these variations are capable of being passed on to the next generation. In both cases, parents produce more offspring than necessary to replace themselves, and some of these offspring either do not survive or do not reproduce. But in the case of artificial selection, humans decide which individuals will breed, based on those individuals that possess the traits that tend toward some predetermined goal, such as a curly coat or large size. Natural selection does not select for specific traits that tend toward some predetermined goal. Rather, natural selection favors any combination of traits that improves an individual's **fitness**—its ability to survive and reproduce, as we saw in the case of the smallest amphipods surviving predation by fish. Traits that improve an individual's fitness are called **adaptations.**

Natural selection can favor multiple solutions to a particular environmental challenge, as long as each solution improves an individual's ability to survive and reproduce. For example, while all plants living in the desert face the challenge of low water availability in the soil, different species have evolved different solutions to this common challenge. Some species have evolved large taproots to draw water from deep in the soil. Other species have evolved the ability to store excess water during infrequent rains. Still other species have evolved waxy or hairy leaf surfaces that reduce water loss. Each of these very different adaptations allows the plants to survive and reproduce in a desert environment (FIGURE 15.6).

(a)

(b)

(c)

FIGURE 15.6 **Adaptations.** Desert plants have evolved several different adaptations to their desert environment. (a) The wedgeleaf draba (*Draba cuneifolia*) has leaf hairs that reduce water loss. (b) The Leuchtenbergia cactus (*Leuchtenbergia principis*) has a large taproot to draw water from deep in the soil. (c) The waxy outer layers of *Aloe vera* reduce water loss. (Inset) Close-up of waxy layer on an aloe that has been scraped with a knife. (a: FPI/Alamy; b: Ian Nartowicz; c: Mark Hannaford/JWL/Aurora Photos; inset: Lee Wilcox)

Evolution can also occur through random processes

Artificial and natural selection are important mechanisms of evolution, but evolution can also occur by random, or nonadaptive, processes. In these cases, the genetic composition of a population changes over time, but the changes are not related to differences in fitness among individuals. There are five random processes: *mutation, gene flow, genetic drift, bottleneck effects,* and *founder effects.*

Mutation

If a random mutation is not lethal, it can add to the genetic variation of a population. As shown in FIGURE 15.7, the larger the population, the more opportunities there will be for mutations to appear within it. As the number of mutations accumulates in the population over time, evolution occurs.

Gene flow

Gene flow is the process by which individuals move from one population to another and thereby alter the genetic composition of both populations. Populations

Fitness An individual's ability to survive and reproduce.

Adaptation A trait that improves an individual's fitness.

Gene flow The process by which individuals move from one population to another and thereby alter the genetic composition of both populations.

Teaching Tip: Concept Map

To review the five random processes that drive evolution, ask each student to create a concept map that includes the terms shown below. Divide the class into five groups. Assign each group an evolutionary process to describe. What is the significance of this mechanism to biodiversity? Have each group present their information and ask students to complete their concept maps.

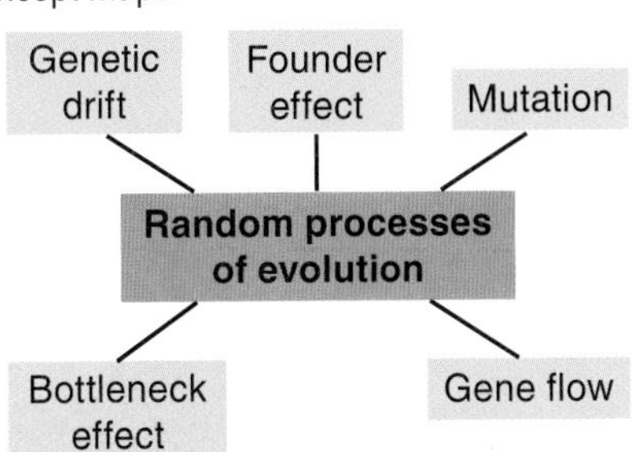

- **Mutation:** *A random change in the genetic code produced by a mistake in the copying process. If a mutation is not lethal it might increase the organism's chance for survival and reproduction. Organisms with a mutation that improves fitness can pass on the mutation to future generations and the mutation will increase in frequency. Mutations add genetic variation to a population, increasing genetic diversity.*
- **Gene Flow:** *The process by which individuals move from one population to another and thereby alter the genetic composition of both populations. New alleles may be introduced into a population by immigrating individuals or some alleles may be removed from a population by emigrating individuals. Either way migrating individuals alter the allele frequency in a population. Gene flow can be helpful in bringing genetic variation to a population that lacks it.*
- **Genetic Drift:** *A change in the genetic composition of a population over time as a result of random mating. Genetic drift has a particularly significant effect on the genetic composition of small populations. Random mating among individuals can eliminate some individuals with rare genotypes simply because they did not find a mate in a given year. Because their genes will not be passed on, the genetic composition of the population will change.*
- **Bottleneck Effect:** *A reduction in the genetic diversity of a population caused by a reduction in its size. When a population becomes smaller, genetic variation might also be reduced. Low genetic variation in a population can cause problems such as low fertility and increased susceptibility to disease. Small populations with low genetic diversity are less able to adapt to changing environmental conditions.*
- **Founder Effect**: *A change in the genetic composition of a population as a result of descending from a small number of colonizing individuals. For example, when individuals from the mainland colonize a new island, these founding individuals possess all the alleles for the new population. Because they may not bring all the diversity of the mainland with them, the founders may give rise to a population with a more limited genetic composition.*

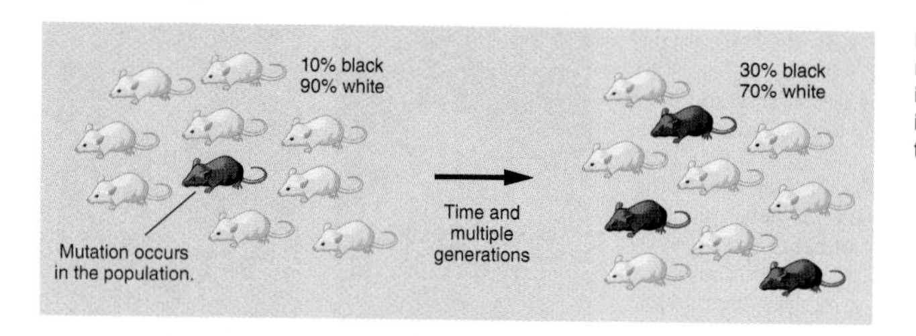

FIGURE 15.7 Evolution by mutation. A mutation can arise in a population and, if it is not lost, it may increase in frequency over time.

can experience an influx of migrating individuals with different alleles. The arrival of these individuals from adjacent populations alters the frequency of alleles in the population. High gene flow between two populations can cause the two populations to become very similar in genetic composition. In a population that is experiencing natural or artificial selection, high gene flow from outside can prevent the population from responding to selection.

Gene flow can be helpful in bringing in genetic variation to a population that lacks it. For example, the Florida panther is a subspecies of panther that once roamed throughout much of the southeastern United States and likely experienced gene flow with other subspecies of panthers. By 1995, the Florida panther only lived in southern Florida, occupying only 5 percent of its original habitat. Moreover, the number of panthers declined to only about 30 individuals, and because the Florida subspecies was isolated from other subspecies, it did not experience gene flow. As a result, shown in FIGURE 15.8, the small population had low genetic variation and the remaining individuals became very inbred. Being inbred causes individuals to express homozygous, harmful alleles. In the case of the panthers, these deleterious alleles caused a high prevalence of kinked tails, heart defects, and low sperm counts. In response, the U.S. Fish and Wildlife Service captured eight panthers from the Texas subspecies and introduced them to Florida with the hope that this gene flow

TEACHING with FIGURES ▶

Show the class Figure 15.8 and describe how the Florida panther population experienced low genetic variation and showed signs of inbreeding. Ask students: Why can inbreeding lead to an increase in detrimental phenotypes such as heart defects? *Inbreeding increases the likelihood that recessive harmful alleles will increase in a gene pool and will more likely be paired and therefore expressed.*

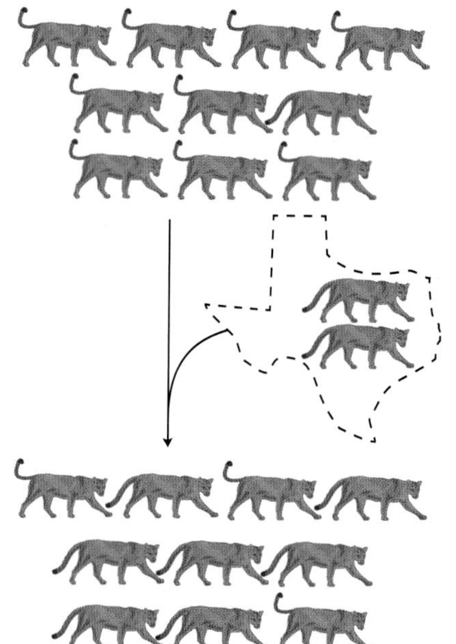

Florida
As the population of Florida panthers declined to very low numbers, the percentage of kinked tails increased to approximately 90 percent.
Texas
Several Texas panthers were brought to Florida.
Florida
After breeding with the Texas panthers, the Florida panther population was less inbred and the percentage of kinked tails declined.

FIGURE 15.8 Evolution by gene flow. As the Florida panther declined in population size, the animals experienced low genetic variation and showed signs of inbreeding, which lead to kinky tails, heart defects, and low sperm counts. With the introduction of eight panthers from Texas, the Florida population experienced a decline in the prevalence of defects and a growth in population from 30 to 160 individuals.

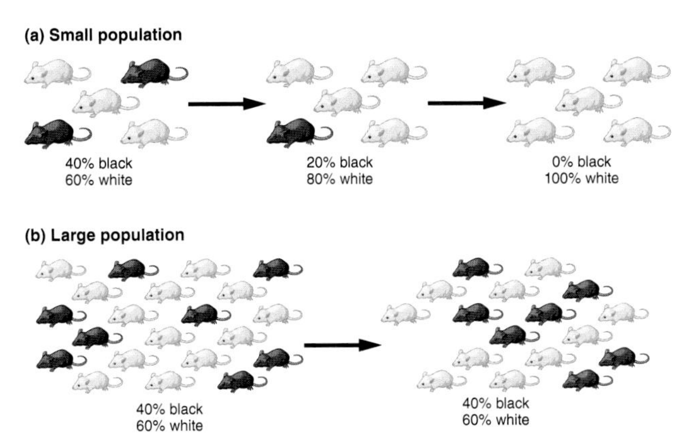

FIGURE 15.9 **Evolution by genetic drift.** (a) In a small population, some less-common genotypes can be lost by chance as random mating among a small number of individuals can result in the less-common genotype not mating. As a result, the genetic composition can change over time. (b) In a large population, it is more difficult for the less-common genotypes to be lost by chance because the absolute number of these individuals is large. As a result, the genetic composition tends to remain the same over time in larger populations.

would increase the genetic variation and allow the population to grow. By 2011, the prevalence of defects previously seen from inbreeding had declined and the Florida panther population had increased to 160 individuals.

Genetic Drift

Genetic drift is a change in the genetic composition of a population over time as a result of random mating. Like mutation and gene flow, genetic drift is a nonadaptive, random process. It can have a particularly important role in altering the genetic composition of small populations, as illustrated in FIGURE 15.9. In small populations, shown in Figure 15.9a, random mating among individuals can eliminate some of the rare individuals simply because they did not find a mate in a given year. For example, imagine a small population of five animals, in which two individuals carry genes that produce black hair and three individuals carry genes that produce white hair. If, by chance, the individuals that carry the genes for black hair fail to find a mate, those genes will not be passed on. The next generation will be entirely white-haired, and the black-haired phenotype will be lost. In this case, the genetic composition of the population has changed, and the population has therefore evolved. The cause underlying this evolution is random; the failure to find a mate has nothing to do with hair color. In contrast, a large population that has the same proportion of black-haired mice, shown in Figure 15.9b, has a greater absolute number of mice. As a result, it is less likely that random mating events will cause all of the black-haired mice to not find a mate, so their genes are passed on to the next generation and genetic drift is less likely to occur.

Bottleneck Effect

A drastic reduction in the size of a population that reduces genetic variation—known as a **bottleneck effect**—is another random process that can change a population's genetic composition. A population might experience a drastic reduction in its numbers for many reasons, including habitat loss, a natural disaster, harvesting by humans, or changes in the environment. FIGURE 15.10 illustrates the bottleneck effect using the example of cheetahs and spots. When the size of a population is reduced, the amount of genetic variation that can be present in the population is also reduced. With fewer individuals there are fewer unique genotypes remaining in the population.

Low genetic variation in a population can cause several problems, including increased risk of disease and low fertility. In addition, species that have been through a population bottleneck are often less able to adapt to future changes in their environment. In some cases, once a species has been forced through a bottleneck, the

Genetic drift A change in the genetic composition of a population over time as a result of random mating.

Bottleneck effect A reduction in the genetic diversity of a population caused by a reduction in its size.

Teaching Tip: Video

Five Fingers of Evolution
Show this video clip from TED-Ed to help students remember the 5 random processes that drive evolution.

The link to this video can be found by clicking on the link buttons in the TE-book, opening the TRFD, or logging onto the book's companion website bcs.whfreeman.com/friedlandapes2e (teacher login required).

 **Chapter 5 Web Resources**

TEACHING with FIGURES ▶

Ask students to evaluate Figure 15.10 and suggest a potential consequence of losing one of the coat variations among the cheetah population. *With fewer variations in coat phenotype, the population will have fewer options to adapt to changing environmental conditions. Fewer numbers may be able to successfully reproduce and this may lead to extinction.*

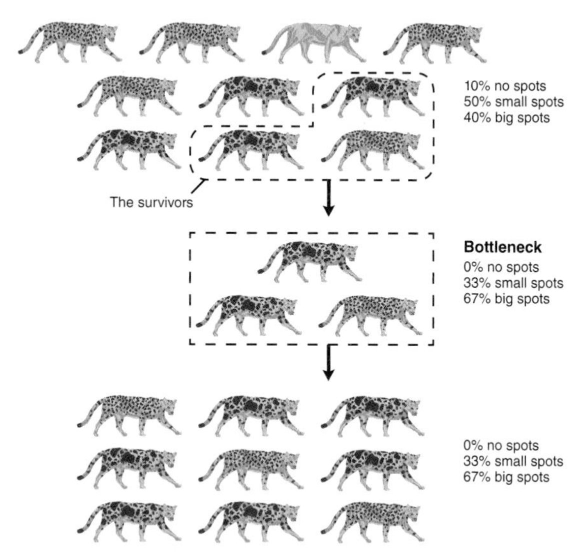

FIGURE 15.10 **Evolution by the bottleneck effect.** If a population experiences a drastic decrease in size (goes through a "bottleneck"), some genotypes will be lost, and the genetic composition of the survivors will differ from the composition of the original group.

resulting low genetic diversity causes it to decline to **extinction,** which occurs when the last member of a species dies. Such declines are thought to be occurring in a number of species today. The cheetah, for example, has relatively little genetic variation due to a bottleneck that appears to have occurred 10,000 years ago.

Founder Effect

Imagine that a few individuals of a particular bird species happen to be blown off their usual migration route and land on a hospitable oceanic island, as illustrated in FIGURE 15.11. These two individuals will have been drawn at random from the mainland population, and the genotypes they possess are only a subset of those in the original mainland population. These colonizing individuals, or founders, will give rise to an island population that has a genetic composition very different from that of the original mainland population. A change in the genetic composition of a population as a result of descending from a small number of colonizing individuals is known as the **founder effect.** Like mutation, genetic drift, and the bottleneck effect, the founder effect is a random process that is not based on differences in fitness.

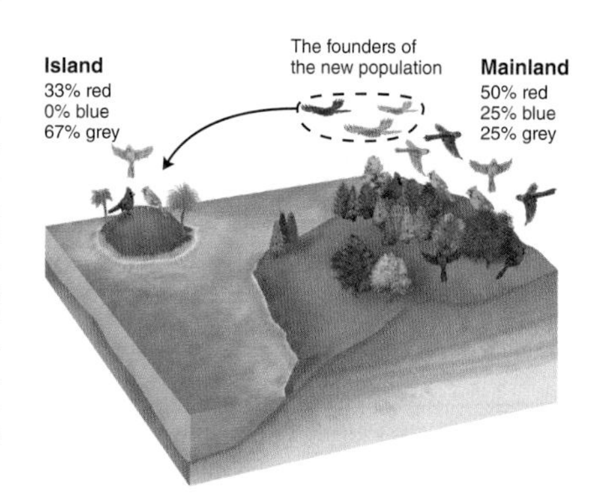

FIGURE 15.11 **Evolution by the founder effect.** If a few individuals from a mainland population colonize an island, the genotypes on the island will represent only a subset of the genotypes present in the mainland population. As with the bottleneck effect, some genotypes will not be present in the new population.

Extinction The death of the last member of a species.

Founder effect A change in the genetic composition of a population as a result of descending from a small number of colonizing individuals.

▲ TEACHING with FIGURES

After showing students Figure 15.11, ask the following:

- Predict the possible phenotypes of the new island. *Phenotypes should include red and gray birds.*
- Why might a blue bird suddenly appear on the island? *If a blue bird were to suddenly appear, it might be because the blue allele is recessive and was therefore included in the gene pool of the colonizing population. Or, a blue bird could have migrated to the new island and introduced blue alleles into the gene pool.*

We can see an example of the founder effect in the Amish communities of Pennsylvania. The Amish population was founded by a relatively small number of individuals—about 200 people from Germany. This group happened to carry a mutation for Ellis-van Creveld syndrome, a condition that causes a variety of malformations including extra fingers. The mutation is rare in humans around the world, but by chance the frequency of the mutation was higher in the early Amish colonists and has remained higher because the population is an isolated group with little gene flow from outside its community.

Teaching Tip: Engage

Ask students: How does evolution lead to biodiversity? *The mechanisms of evolution—artificial selection, natural selection, and random processes—change the genetic composition of populations over time. With variation in genetic code, greater variability is gained creating diversity at the genetic level, the species level, and the ecosystem level. Biodiversity is the product of evolution.*

module 15 REVIEW

In this module, we learned that genetic variation helps to determine the traits that individuals express. We also learned that artificial selection, natural selection, and random processes can all cause the evolution of populations. In the next module, we will examine how these evolutionary processes can cause the evolution of new species, which increases biodiversity, and how environmental change that is too rapid for selection can result in extinctions that cause declines in biodiversity.

Module 15 AP® Review Questions

1. Which evolutionary effect results in reduced genetic variation in a community?
   (a) Natural selection
   (b) The founder effect
   (c) Artificial selection
   (d) Gene flow
   (e) Mutation

2. A phenotype is
   (a) an adaptation that creates a new species.
   (b) the genes of a particular individual.
   (c) a result of genetic recombination.
   (d) the set of traits expressed in an individual.
   (e) a genetic mutation passed from parent to offspring.

3. Which of the following processes create genetic diversity in a population?
   I. Mutation
   II. Allele division
   III. Recombination

   (a) I only
   (b) I and II only
   (c) I and III only
   (d) II and III only
   (e) I, II, and III

4. Evolved resistance to a pesticide is an example of
   (a) a nonadaptive process.
   (b) the bottleneck effect.
   (c) natural selection.
   (d) artificial selection.
   (e) range of tolerance.

5. Which is the best definition of an adaptation?
   (a) A mutation that creates a new species
   (b) A trait that improves an individual's fitness
   (c) A trait that is passed on to the next generation
   (d) A trait that has no effect on an individual's fitness
   (e) A trait created by natural selection

6. The change in the genetic composition of a population over time due to random mating is called
   (a) the bottleneck effect.
   (b) gene flow.
   (c) genetic drift.
   (d) mutation.
   (e) phenotype adaption.

7. In a particular zoo the population of spider monkeys has a higher proportion of individuals with light golden brown fur than spider monkeys in the wild. If the monkeys were recently captured from the wild and if fur color is largely determined by genetics, what evolutionary process is at work?
   (a) The founder effect
   (b) The bottleneck effect
   (c) Microevolution
   (d) Artificial selection
   (e) Genetic drift

Answers to Module 15 AP® Review Questions

1. b
2. d
3. c
4. d
5. b
6. c
7. a

Teaching Tip: Journal Prompt

Ask students to describe the process of allopatric speciation. *Allopatric speciation creates new species though geographic isolation. When individuals are separated, the genetic composition of the populations that are isolated from each other might diverge over time either because of random processes or natural selection. If individuals cannot move between the populations, then the two populations will become even more genetically distinct. Eventually they will not only be separated by geographic barriers but will be reproductively isolated. At this point the populations will be distinct species.*

module 16

Speciation and the Pace of Evolution

Over time, speciation has given rise to the millions of species present on Earth today. Beyond determining how many species exist, environmental scientists are also interested in understanding how quickly existing species can change, how quickly new species can evolve, and how quickly species can go extinct. In this section we will examine the processes that produce new species and the factors that determine how rapidly species can evolve in response to changes in the environment.

Learning Objectives

After reading this module you should be able to

- explain the processes of allopatric and sympatric speciation.
- understand the factors that affect the pace of evolution.

Speciation can be allopatric or sympatric

Microevolution is happening all around us, from the breeding of agricultural crops, to the unintentional evolution of drug-resistant bacteria in hospitals, to the bottleneck that reduced genetic variation in the cheetah. But how do we move from the evolution of genetically distinct populations of a species to the evolution of genetically distinct species? That is, how do we move from microevolution to macroevolution?

Geographic isolation Physical separation of a group of individuals from others of the same species.

Allopatric speciation The process of speciation that occurs with geographic isolation.

Two common processes are *allopatric speciation* and *sympatric speciation*.

Allopatric speciation

One common way in which evolution creates new species is through **geographic isolation,** which means the physical separation of a group of individuals from others of the same species. The process of speciation that occurs with geographic isolation is known as **allopatric speciation** (from the Greek *allos,* meaning "other," and *patris,* meaning "fatherland"). As shown in FIGURE 16.1, geographic isolation can occur when a subset of individuals from a larger population colonizes a new area of habitat that is physically separated from that larger population. For example, a single large population of field mice might be split into two smaller populations as geographic barriers change over time. For example, a river might change course and divide a

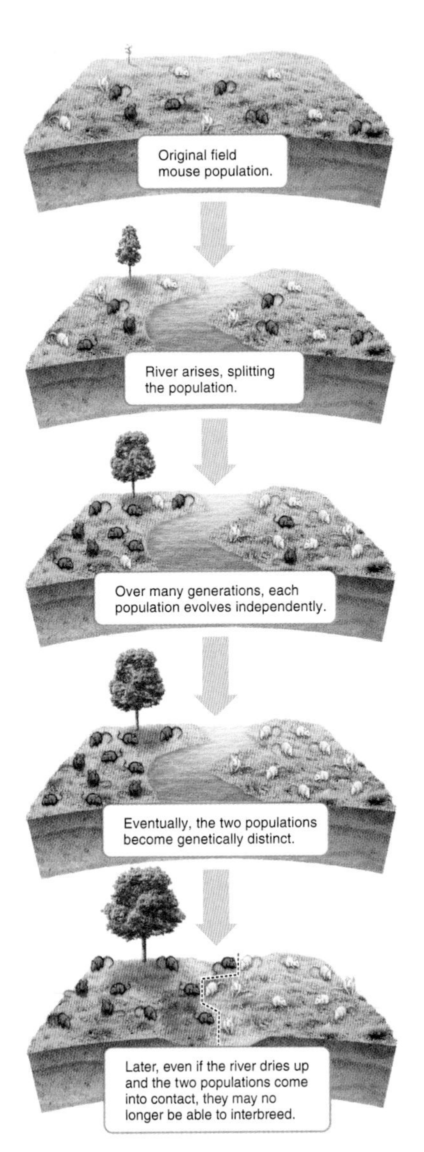

FIGURE 16.1 **Allopatric speciation.** Geographic barriers can split populations. Natural selection may favor different traits in the environment of each isolated population, resulting in different adaptations. Over time, the two populations may become so genetically distinct that they are no longer capable of interbreeding.

large prairie into two halves, a large lake might split into two smaller lakes, or a new mountain range could rise. In such cases, the genetic composition of the isolated populations might diverge over time, either because of random processes or because natural selection favors different adaptations on each side of the barrier.

If the two separated habitats differ in environmental conditions, such as temperature, precipitation, or the occurrence of predators, natural selection will favor different phenotypes in each of the habitats. If individuals cannot move between the populations, then over time the two geographically isolated populations will continue to become more and more genetically distinct. Eventually, the two populations will be separated not only by geographic isolation but also by **reproductive isolation,** which means the two populations of a species have evolved separately to the point that they can no longer interbreed and produce viable offspring. At this point, the two populations will have become distinct species.

Allopatric speciation is thought to be responsible for the diversity of the group of birds known as Darwin's finches. When Charles Darwin visited the Galápagos Islands, located just west of Ecuador, he noted a large variety of finch species, each of which seemed to live in different habitats or to eat different foods. Research on these birds has demonstrated that they all share a common ancestor that colonized the islands from the mainland long ago. FIGURE 16.2 is a phylogenetic tree for these finches. Over a few million years, as Darwin discovered, the finches that were geographically isolated on different islands became genetically distinct and eventually became reproductively isolated.

Sympatric Speciation

Allopatric speciation is thought to be the most common way in which evolution generates new species. However, it is not the only way. **Sympatric speciation** is the evolution of one species into two species without geographic isolation. It usually happens through a process known as polyploidy. Most organisms are diploid: They have two sets of chromosomes. In polyploidy, the number of chromosomes increases to three, four, or even six sets. Such increases can occur during the division of reproductive cells, either accidentally in nature or as a result of deliberate human actions. Plant breeders, for example, have found several ways to interrupt the normal cell division process. Polyploid organisms include some species of snails and salamanders, 15 percent of

Reproductive isolation The result of two populations within a species evolving separately to the point that they can no longer interbreed and produce viable offspring.

Sympatric speciation The evolution of one species into two, without geographic isolation.

Teaching Tip: Journal Prompt

Ask students to describe the process of sympatric speciation. *Sympatric speciation is the evolution of one species into two without geographic isolation. It usually occurs with polyploidy. Polyploid organisms have more sets of chromosomes than the typical two (diploid). A polyploid generally cannot breed with its diploid ancestors. Species that are reproductively isolated from each other are considered distinct species.*

Teaching Tip: Review

Reproduce the flow chart without the terms in italics. Ask students to complete the chart to compare and contrast allopatric and sympatric speciation.

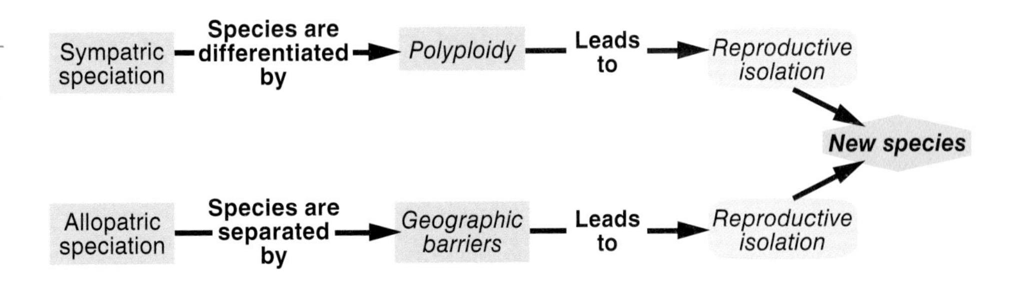

Teaching Tip: Warm-up

Ask students to identify two factors that make it difficult for a population to survive environmental change. *The population has a slow generation time and cannot evolve rapidly enough. The population has low genetic diversity on which natural selection can act.*

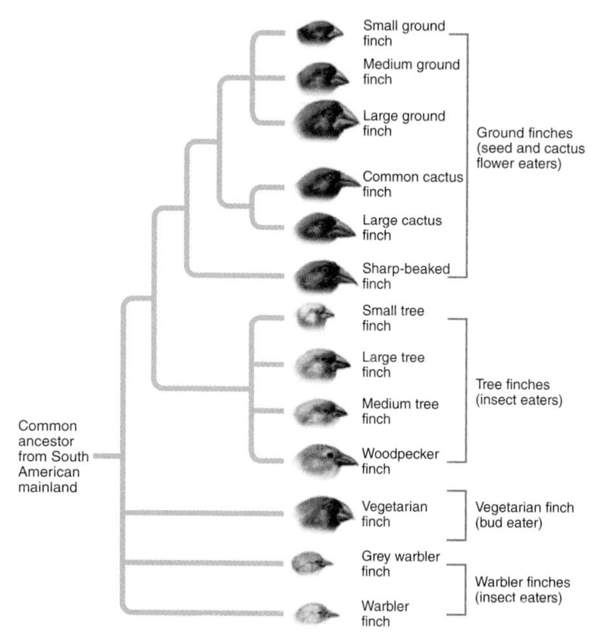

FIGURE 16.2 **Allopatric speciation of Darwin's finches.** In the Galápagos Islands, allopatric speciation has led to a large variety of finch species, all descended from a single species that colonized the islands from the South American mainland.

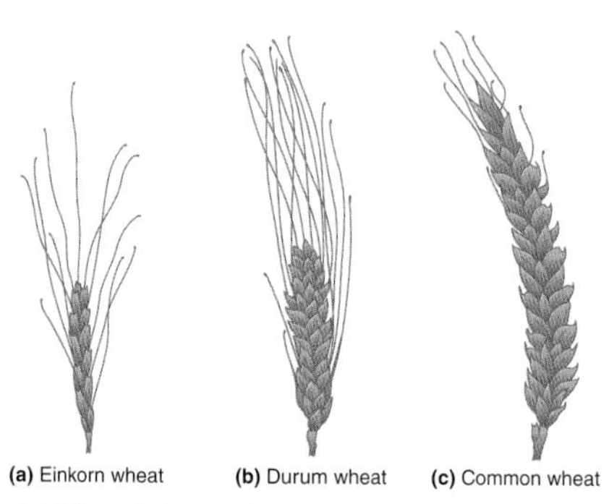

FIGURE 16.3 **Sympatric speciation.** Flowering plants such as wheat commonly form new species through the process of polyploidy, an increase in the number of sets of chromosomes beyond the normal two sets. (a) The ancestral einkorn wheat (*Triticum boeoticum*) has two sets of chromosomes and produces small seeds. (b) Durum wheat (*Triticum durum*), which is used to make pasta, was bred to have four sets of chromosomes and produces medium-sized seeds. (c) Common wheat (*Triticum aestivum*), which is used mostly for bread, was bred to have six sets of chromosomes and produces the largest seeds.
(After http://zr.molbiol.ru/poaceae_znachenije.html)

all flowering plant species, and a wide variety of agricultural crops such as bananas, strawberries, and wheat. As FIGURE 16.3 shows for wheat, polyploidy often results in larger plants and larger fruits.

The key feature of polyploid organisms is that once they become polyploid, they generally cannot interbreed with their diploid ancestors. At the instant polyploidy occurs, the polyploid and diploid organisms are reproductively isolated from each other and are therefore distinct species, even though they may continue to live in the same place.

The pace of evolution depends on several factors

How long does evolution take? A significant change in a species' genotype and phenotype, such as an adaptation to a completely different food source, can take anywhere from hundreds to millions of years. In this section, we will consider examples of rapid evolution by natural selection and very rapid evolution by artificial selection.

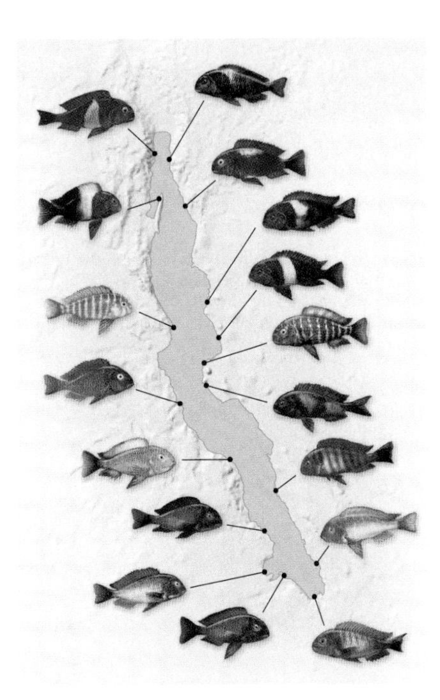

FIGURE 16.4 **Rapid evolution.** The cichlid fishes of Lake Tanganyika have evolved approximately 200 distinct and colorful species in the relatively short period since the lake formed in eastern Africa. The location where each species can be found in the lake is indicated by a corresponding black dot on the map. *(After http://www.uni-graz.at/~sefck/)*

Rapid Evolution by Natural Selection

Sometimes evolution can occur rapidly, as in the case of the cichlid fishes of Lake Tanganyika, one of the African Great Lakes. You can see a variety of these species in FIGURE 16.4. Evidence indicates that the roughly 200 different species of cichlids in the lake evolved from a single ancestral species over a period of several million years. During this period, some cichlid species specialized to become insect eaters and others to become fish eaters, while still others evolved to eat invertebrates such as snails and clams.

Although the cichlids of Lake Tanganyika evolved quickly in evolutionary terms, the pupfishes of the Death Valley region of California and Nevada evolved even more rapidly. In the 20,000 to 30,000 years since the large lakes of the region were reduced to isolated springs, several species of pupfish have evolved.

The ability of a species to survive an environmental change depends greatly on how quickly it evolves the adaptations needed to thrive and reproduce under the new conditions. If a species cannot adapt quickly enough, it will go extinct. This can happen when the rate of environmental change is faster than the rate at which evolution can respond. Slow rates of evolution can occur when a population has long generation times or when a population contains low genetic variation on which natural selection can act.

Very Rapid Evolution by Artificial Selection

The pace of evolution by artificial selection can be incredibly fast. Such rapid evolution is occurring in many species of commercially harvested fish, including the Atlantic cod (*Gadus morhua*). Intensive fishing over several decades has targeted the largest adults, selectively removing most of those individuals from the population and, therefore, also removing the genes that produce large adults. Because larger fish tend to reach sexual maturity later, the genes that code for a later onset of sexual maturity have also been removed. As a result, after just a few decades of intensive fishing, the Atlantic cod population has evolved to reach reproductive maturity at a smaller size and a younger age. This evolution of shorter generation times also means that the cod may be able to evolve even faster in the future.

Evolution occurs even more rapidly in populations of *genetically modified organisms*. Using genetic engineering techniques, scientists can now copy genes from a species with some desirable trait, such as rapid growth or disease resistance. Scientists can insert these genes into other species of plants, animals, or microbes to produce **genetically modified organisms (GMO).** When those organisms reproduce, they pass on the inserted genes to their offspring. For example, scientists have found that a soil bacterium (*Bacillus thuringiensis*) naturally produces an insecticide as a defense against being consumed by insects in the soil. Plant breeders have identified the bacterial genes that are responsible for making the insecticide, copied those genes, and inserted them into the genomes of crop plants. Such crops can now naturally produce their own insecticide, which makes them less attractive to insect herbivores. Common examples include Bt-corn and Bt-cotton, so named because they contain genes from soil bacterium. As you might guess, inserting genes into an organism is a much faster way to produce desired traits than traditional plant and animal breeding, which can only select from the naturally available variation in a population.

Genetically modified organism (GMO) An organism produced by copying genes from a species with a desirable trait and inserting them into another species.

Teaching Tip: Discussion Starter

Ask students: Why is the pace of evolution so rapid in populations of genetically modified organisms? *Inserting genes into an organism is a much faster way to produce desired traits than traditional plant and animal breeding, which can only select from the naturally available variations in a population.*

Teaching Tip: Activity

Howard Hughes Medical Institute (HHMI) has a series that is particularly relevant to the topics introduced in Chapter 5. The set of resources comprises the study of various species of anole lizards found in the Caribbean islands. Begin by showing students the short animation called "Anole Lizards: An Example of Speciation." If more time is available, the longer film clip provides a thorough overview of concepts taught in Chapter 5. To teach more about the practice of science, such as measurement, collection of data, and graphing, use the lab resource entitled "Lizard Evolution Virtual Lab."

The link to these resources can be found by clicking on the link buttons in the TE-book, opening the TRFD, or logging onto the book's companion website bcs.whfreeman.com/friedlandapes2e (teacher login required).

Chapter 5 Web Resources

Teaching Tip: Debate the Issue

Genetically Modified Organisms

Have students research the subject of genetically modified organisms in preparation for a debate to answer the question:

Do the potential benefits of genetically modified organisms outweigh the unknown consequences and potential dangers?

Students should be prepared to discuss these topics in their debate:

- the procedures used to engineer organisms
- pros and cons of genetically modified organisms
- potential and current applications of genetic engineering
- environmental concerns
- environmental benefits
- human health concerns
- human health benefits
- social, religious, and economic concerns

In class, select groups to debate the issue. Allow each side 2 minutes to prepare, 3 minutes to present, and 1 minute for rebuttal. Have the remainder of the class take notes, ask questions, and decide which side has the most persuasive argument.

module

16

REVIEW

In this module, we learned that species evolve through the mechanisms of allopatric and sympatric speciation. We also saw that there are a number of factors that can affect the pace of evolution including how quickly the environment changes, how much genetic variation exists in the population, the size of the population, and the generation time of the species. In the next module, we will examine how the evolution of species affects where they live and how they make their living.

Answers to Module 16 AP® Review Questions

1. d
2. a
3. b
4. c

Module 16 AP® Review Questions

1. Which of the following contribute to allopatric speciation?
   I. Genetic convergence over time
   II. Geographic isolation
   III. Reproductive isolation

   (a) I and II only
   (b) I and III only
   (c) II only
   (d) II and III only
   (e) I, II, and III

2. Which of the following is often a cause of sympatric speciation?
   (a) Polyploidy
   (b) Geographic separation
   (c) Macroevolution
   (d) Artificial selection
   (e) Genetic modification

3. Which of the following would cause the most-rapid evolution?
   (a) Environmental change
   (b) Artificial selection
   (c) Geographic isolation
   (d) Recombination
   (e) Natural selection

4. The Bt-cotton is an example of
   (a) evolution through geographic isolation.
   (b) evolution through polyploidy.
   (c) evolution through genetic modification.
   (d) evolution through natural selection.
   (e) macroevolution.

module

17

Evolution of Niches and Species Distributions

Because evolution alters the traits that species express and because different traits perform well in some environments but not in others, the evolution of species affects where species are able to live on Earth. As environments have changed over millions of years, species have responded by altering their

distributions. This response also suggests that environmental changes in the future will continue to affect the distributions of species. Species that can move easily will likely persist whereas species that cannot adjust or move will likely go extinct.

Learning Objectives

After reading this module you should be able to

- explain the difference between a fundamental and a realized niche.
- describe how environmental change can alter species distributions.
- discuss how environmental change can cause species extinctions.

Every species has a niche

Every species has an optimal environment in which it performs particularly well. All species have a **range of tolerance,** or limits to the abiotic conditions they can tolerate, such as extremes of temperature, humidity, salinity, and pH. FIGURE 17.1 illustrates this concept using one environmental factor—temperature. As conditions move further away from the ideal, individuals may be able to survive, and perhaps even to grow, but not be able to reproduce. As conditions continue to move away from the ideal, individuals can only survive. If conditions move beyond the range of tolerance, individuals will die. Because the combination of abiotic conditions in a particular environment fundamentally determines whether a species can persist there, the suite of abiotic conditions under which a species can survive, grow, and reproduce is the **fundamental niche** of the species.

The fundamental niche establishes the abiotic limits for the persistence of a species. However, biotic factors can further limit the locations where a species can live. Common biotic limitations include the presence of competitors, predators, and diseases. For example, even if abiotic conditions are favorable for a plant species in a particular location, other plant species may be better competitors for water and soil nutrients. Those competitors might prevent the species from growing in that environment. Similarly, even if a small rodent can tolerate the temperature and humidity of a tropical forest, a deadly rodent disease might prevent the species from persisting in the forest. Therefore, biotic factors further narrow the fundamental niche that a species actually uses. The range of abiotic and biotic conditions under which a species actually lives is called its **realized niche.** Once we determine what contributes to the realized niche of a species, we have a better understanding of the **distribution** of the species, or the areas of the world in which the species lives.

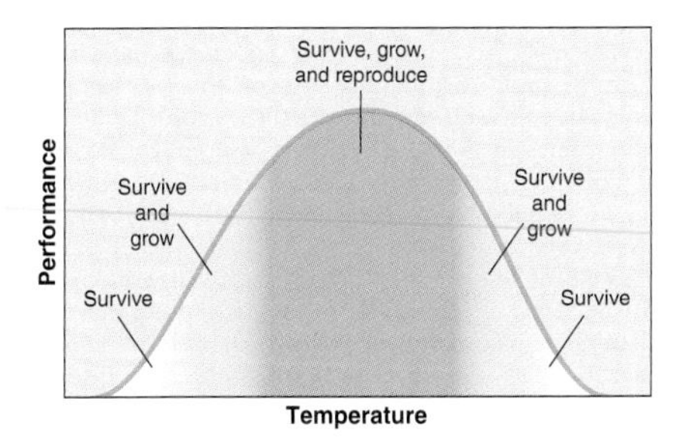

FIGURE 17.1 **Range of tolerance.** All species have an ideal range of abiotic conditions, such as temperature, under which their members can survive, grow, and reproduce. Under more extreme conditions, their ability to perform these essential functions declines.

Range of tolerance The limits to the abiotic conditions that a species can tolerate.

Fundamental niche The suite of abiotic conditions under which a species can survive, grow, and reproduce.

Realized niche The range of abiotic and biotic conditions under which a species actually lives.

Distribution Areas of the world in which a species lives.

Teaching Tip: Warm-up

Ask students to differentiate between fundamental niche and realized niche. *A fundamental niche is the suite of abiotic conditions under which a species will survive, grow, and reproduce. The realized niche is a subset of the fundamental niche because it includes both the abiotic and biotic conditions under which a species will survive, grow, and reproduce.*

▲ TEACHING with FIGURES

Show students Figure 17.1 and ask the following:

- At the extreme temperature ranges, which function(s) are organisms capable of performing? *At the extreme temperature ranges, organisms are only likely to survive. Organisms will not grow or reproduce until the conditions change to more optimal temperatures.*
- Will a population continue to persist at extreme temperatures temperatures? *If the temperatures remain at these extremes, then the population is not likely to persist since organisms will be unlikely to reproduce.*

Teaching Tip: Warm-up

Ask students: What are the major differences between niche specialists and niche generalists? *A specialist is a species that lives in a specific habitat or feeds on a small group of species. A generalist can live in a broad range of abiotic and biotic conditions. Generalists fare well under changing conditions while specialists are more vulnerable to extinction if conditions change.*

Teaching Tip: Journal Prompt

Ask students: How does environmental change determine species distribution? When does it lead to extinction? *Because species are adapted to a particular environment, changes in these conditions will lead to changes in species distribution. Species may go extinct if they cannot adapt to the changes quickly enough, if there is not a more favorable environment close by where they can move, or if biotic factors in the more favorable region prohibit the species from thriving.*

(a)

(b)

FIGURE 17.2 **Generalists and specialists.** (a) Some organisms, such as this meadow spittlebug, are niche generalists that have broad diets and wide habitat preferences. (b) Other organisms, such as this skeletonizing leaf beetle, are niche specialists with narrow diets and highly specific habitat preferences. *(a: Ray Wilson/Alamy; b: © Bo Zaremba 2009)*

When we examine the realized niches of species in nature, we see that some species, known as **niche generalists,** can live under a very wide range of abiotic or biotic conditions. For example, some insects, such as the meadow spittlebug (*Philaenus spumarius*) (FIGURE 17.2a), feed on numerous plant species and can live in different habitats. Other species, known as **niche specialists,** are specialized to live under a very narrow range of conditions or feed on a small group of species. For example, the skeletonizing leaf beetle (*Trirhabda virgata*) (Figure 17.2b) feeds on only a single species or genus of plant. Niche specialists can persist quite well when environmental conditions remain relatively constant, but they are more vulnerable to extinction if conditions change because the loss of a favored habitat or food source leaves them with few alternatives for survival. In contrast, niche generalists should fare better under changing conditions because they have a number of alternative habitats and food sources available.

Environmental change can alter the distribution of species

Because species are adapted to particular environmental conditions, we would expect that changes in environmental conditions would alter the distribution of species on Earth. For example, scientists have found evidence for the relationship between environmental change and species distribution in layers of sediments that have accumulated over time at the bottom of modern lakes. Each sediment layer contains pollen from plants that lived in the region when the sediments were deposited. In some cases, this pollen record goes back thousands of years.

Niche generalist A species that can live under a wide range of abiotic or biotic conditions.

Niche specialist A species that is specialized to live in a specific habitat or to feed on a small group of species.

In much of northern North America, lakes formed 12,000 years ago at the end of the last ice age when temperatures warmed and the glaciers slowly retreated to the north. The retreating glaciers left behind a great deal of barren land, which was quickly colonized by plants, including trees. Some of the pollen produced by these trees fell into lakes and was buried in the lake sediments. Scientists can measure the ages of these sediment layers with carbon dating (see Chapter 2). Furthermore, because each tree species has uniquely shaped pollen, it is possible to determine when a particular tree species arrived near a particular lake and how the entire community of plant species changed over time. FIGURE 17.3 shows pollen records for three tree species in North America. These pollen records suggest that changes in climatic conditions after the ice age produced substantial changes in the distributions of plants over time. Pine trees, spruce trees, and birch trees all moved north with the retreat of the glaciers. As the plants moved north, the animals followed.

Because past changes in environmental conditions have led to changes in species distributions, it is reasonable to ask whether current and future changes in environmental conditions might also cause changes in species distributions. For example, as the global climate warms, some areas of the world are expected to receive less precipitation, while other areas are predicted to receive more. If our predictions of future environmental

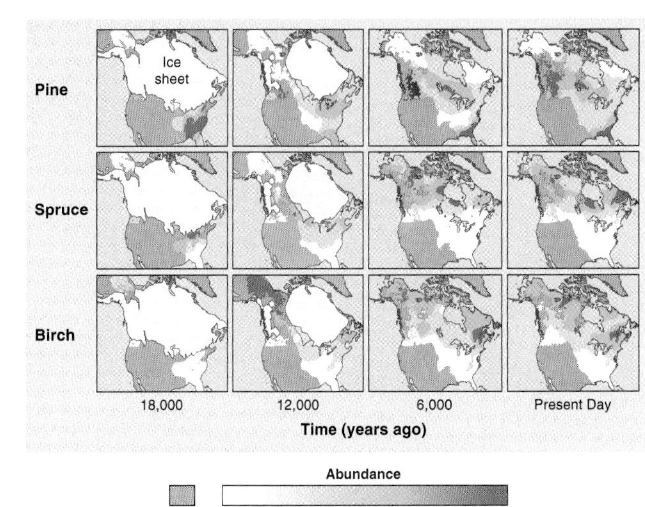

FIGURE 17.3 Changes in tree species distributions over time. Pollen recovered from lake sediments indicates that plant species moved north as temperatures warmed following the retreat of the glaciers, beginning about 12,000 years ago. Areas shown in color or white were sampled for pollen, whereas areas shown in gray were not sampled. *(Data from http://veimages.gsfc.nasa.gov//3453/boreal_model.gif)*

changes are correct, and if we have a good understanding of the niche requirements of many species, we should be able to predict how the distribution of species will change in the future. North American trees, for example, are expected to have more northerly distributions with future increases in global temperatures. As FIGURE 17.4 shows, the loblolly pine (*Pinus taeda*) is expected to move from its far southern distribution and become common throughout the eastern half of the United States.

Species vary in their ability to move physically across the landscape as the environment changes. Some species are highly mobile at particular life stages: Adult birds and wind-dispersed seeds, for example, move across the landscape easily. Other organisms, such as the desert tortoise (*Gopherus agassizii*), are slow movers. Furthermore, the movements of many species may be impeded by obstacles built by humans, including roads and dams. It remains unclear how species that face these challenges will shift their distributions as global climate change occurs.

Environmental change can cause species extinctions

If environmental conditions change, species that cannot adapt to the changes or move to more favorable environments will eventually go extinct. The average life span of a species appears to be only between about 1 million to 10 million years. In fact, 99 percent of the species that have ever lived on Earth are now extinct.

There are several reasons why species might go extinct. First, there may be no favorable environment

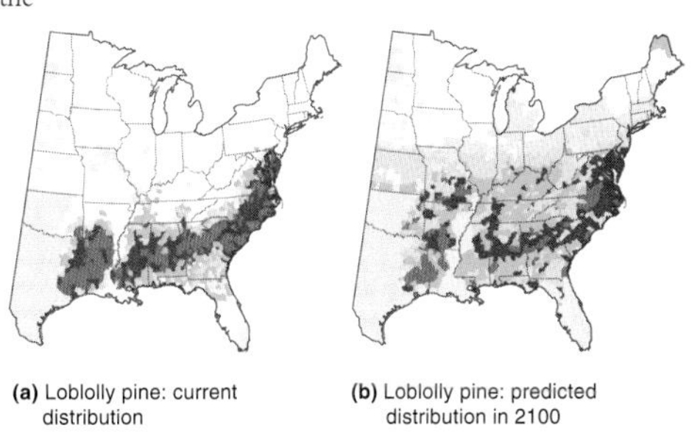

(a) Loblolly pine: current distribution

(b) Loblolly pine: predicted distribution in 2100

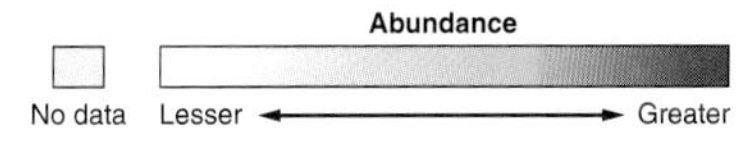

FIGURE 17.4 Predicting future species distributions. Based on our knowledge of the niche requirements of the loblolly pine tree, we can predict how the distributions might change as a result of future changes in environmental conditions. *(After http://www.fs.fed.us/ne/delaware/atlas/web_atlas.html#)*

◀ TEACHING with FIGURES

Ask students to observe the changes in tree distribution over time in Figure 17.3. Then ask the following:

- How do you think our current climate change might affect species distribution? *As the climate warms, some species may relocate to northern regions for cooler temperatures.*
- How might scientists benefit from understanding past changes in environmental conditions and their effect on species distribution? *By understanding the relationships between environmental conditions and species distribution, scientists can predict future species distribution patterns. This may help conservationists in their efforts to protect species and their habitats.*

Teaching Tip: Discussion Starter

Ask students: Why do we see a continuous pattern of extinction events throughout the history of Earth? *Because the Earth is constantly changing and has historically experienced dramatic shifts in climate, species that cannot adapt quickly enough to these changing conditions will go extinct.*

close enough to which they can move. For example, on the southwest coast of Australia is a small area of land that experiences a current climate that supports a woodland/shrubland biome (see Figure 12.9 on page 128). This biome is surrounded by a large desert biome to the north and east and an ocean to the south and west. If scientists are right in predicting that the coming century will bring a hotter and drier climate to southwestern Australia, then the unique species of plants that live in this small biome will have nowhere to go where they can survive.

Even if there is an alternative favorable environment to which a species can move, it may already be occupied by other species against which the moving populations cannot successfully compete. For example, the predicted northern movement of the loblolly pine, shown in Figure 17.4b, might not happen if another pine tree species in the northern United States is a better competitor and prevents the loblolly pine from surviving in that area. Finally, an environmental change may occur so rapidly that the species does not have time to evolve new adaptations.

FIGURE 17.5 **Fossils.** Fossils, such as this fish discovered in Fossil Butte National Monument in Wyoming, are a record of evolution. *(David R. Frazier)*

The Fossil Record

Much of what we know about the evolution of life is based on fossils, which are the remains of organisms that have been preserved in rock. Most dead organisms decompose rapidly; the elements they contain are recycled and nothing of the organism is preserved. Occasionally, though, organic material is buried and protected from decomposition by mud or other sediments. That material may eventually become fossilized, which means it has been hardened into rocklike material as it was buried under successive layers of sediment (FIGURE 17.5). When these layers are uncovered, they reveal a record of at least some of the organisms that existed at the time the sediments were deposited. Because of the way layers of sediment are deposited on top of one another over time, the oldest fossilized organisms are found in the deepest layers of the fossil record. We can therefore use the fossil record to determine when different species existed on Earth.

Mass extinction A large extinction of species in a relatively short period of time.

The Five Global Mass Extinctions

Throughout Earth's history, individual species have evolved and gone extinct at random intervals. The fossil record has revealed five periods of global **mass extinction,** in which large numbers of species went extinct over relatively short periods of time. The times of these mass extinctions are shown in FIGURE 17.6. Note that because species are not always easy to discriminate in the fossil record, scientists count the number of genera, rather than species, that once roamed Earth but are now extinct.

The greatest mass extinction on record took place 251 million years ago when roughly 90 percent of marine species and 70 percent of land vertebrates went extinct. The cause of this mass extinction is not known.

A better-known mass extinction occurred at the end of the Cretaceous period (65 million years ago), when roughly one-half of Earth's species, including the dinosaurs, went extinct. The cause of this mass extinction has been the subject of great debate, but there is now a near-consensus that a large meteorite struck Earth and produced a dust cloud that circled the planet and blocked incoming solar radiation. This resulted in an almost complete halt to photosynthesis, and thus an almost total lack of food at the bottom of the food chain. Among the few species that survived was a small squirrel-sized primate that was the ancestor of humans.

Many scientists view extinctions as the ultimate result of change in the environment. Environmental scientists can learn about the potential effects of both large and small environmental changes by studying historic environmental changes and applying the lessons learned to help predict the effects of the environmental changes that are taking place on Earth today.

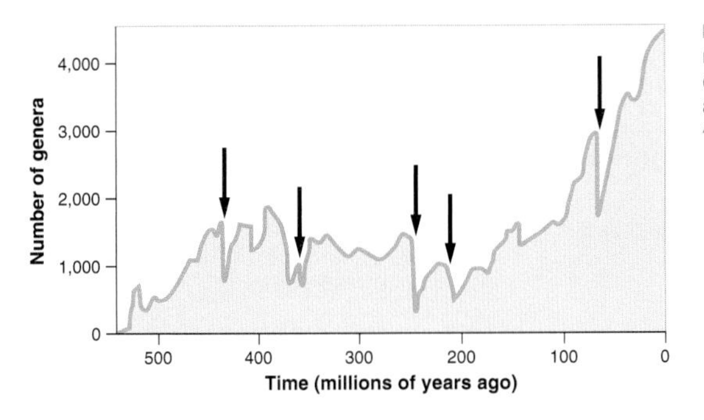

FIGURE 17.6 **Mass extinctions.** Five global mass extinction events have occurred since the evolution of complex life roughly 500 million years ago. *(Data from GreenSpirit, http://www.greenspirit.org.uk/resources/TimeLines.jpg)*

The Sixth Mass Extinction

During the last 2 decades, scientists have reached a consensus that we are currently experiencing a sixth global mass extinction of a magnitude within the range of the previous five mass extinctions. Estimates of extinction rates vary widely, ranging from 2 percent to as many as 25 percent of species going extinct by 2020. However, in contrast to some previous mass extinctions, there is agreement among scientists that the current mass extinction has human causes. These wide-ranging causes include habitat destruction, overharvesting, introductions of invasive species, climate change, and emerging diseases. We will examine all of these factors in detail in Chapters 18 and 19.

Because much of the current environmental change caused by human activities is both dramatic and sudden, environmental scientists contend that many species may not be able to move or adapt in time to avoid extinction.

The recovery of biodiversity from earlier mass extinctions took about 10 million years, an unthinkably long time from a human perspective. Recovery from the present mass extinction could take just as long—500,000 human generations. Much of the current debate among environmental scientists and government officials centers on the true magnitude of this crisis and on the costs of reducing the human impact on extinction rates.

◀ **TEACHING with FIGURES**

Ask students to use Figure 17.6 to determine the approximate length of time that has passed since each extinction event. *First: 420 million years ago; second: 350 million years ago; third: 240 million years ago; fourth: 200 million years ago; fifth: 60 million years ago.*

Teaching Tip: Warm-up

Rapid environmental change is one major reason species go extinct. Ask students: What can you conclude about each of the five mass extinctions and the sixth extinction event occurring now? *Large numbers of species probably went extinct because they were unable to adapt to rapid changes in the environment.*

module

17

REVIEW

In this module, we learned that species differ in their tolerance to abiotic conditions and that the range of abiotic and biotic conditions that allow survival, growth, and reproduction all help to determine where a species can live. Past changes in environmental conditions have altered species distributions and caused extinctions. Future environmental changes are expected to have similar effects.

Module 17 AP® Review Questions

1. The abiotic conditions under which a species can survive and reproduce is called its
   (a) range of tolerance.
   (b) realized niche.
   (c) range of persistence.
   (d) environmental distribution.
   (e) fundamental niche.

2. Which would be expected for a niche specialist?
   (a) A large distribution
   (b) Many food sources
   (c) A narrow fundamental niche
   (d) Adaptability to changing conditions
   (e) No difference between its realized and fundamental niche

Answers to Module 17 AP® Review Questions

1. e
2. c
3. d
4. c
5. e

Teaching Tip: Journal Prompt

Ask students: What distinguishes the sixth mass extinction from the previous five events? *The sixth mass extinction is probably caused by human activity including habitat destruction, overharvesting, climate change, introduction of invasive species, and emergent diseases.*

Teaching Tip: Activity

After teaching about the sixth mass extinction, assign each student an animal from the International Union for Conservation of Nature (IUCN) endangered species list. Students should create a poster describing the animal, including a picture, information on where the animal lives, what it eats, and an explanation of why it is on the endangered species list. The student must then list at least five actions that could help protect this animal. Depending on the animal, the preservation list might include refusing to buy the animal in the pet trade, donating money to conserving this animal's habitat, planting trees, etc.

The link to the IUCN endangered species list resource can be found by clicking on the link buttons in the TE-book, opening the TRFD, or logging onto the book's companion website bcs.whfreeman.com/friedlandapes2e (teacher login required).

Chapter 5 Web Resources

3. Which of the following will NOT affect the distribution of a species?
   (a) A localized disease
   (b) Climate change
   (c) A change in the distribution of the species' food
   (d) A new food source outside its range of tolerance
   (e) The introduction of a predator species

4. How many mass extinction events have there been in Earth's history?
   (a) 2
   (b) 4
   (c) 5
   (d) 7
   (e) 12

5. Which is NOT a significant cause of the current mass extinction event?
   (a) Invasive species
   (b) Climate change
   (c) Overharvesting
   (d) Habitat destruction
   (e) Extreme weather events

working toward  **sustainability**

Protecting the Oceans When They Cannot Be Bought

For over 50 years, The Nature Conservancy (TNC) has protected biodiversity by using a simple strategy: Buy it. The Conservancy uses grants and donations to purchase privately owned natural areas or to buy development rights to those areas. TNC owns over 0.8 million hectares (2 million acres) of land and has protected over 46 million hectares (115 million acres) of land by buying development rights. As a nonprofit, nongovernmental organization, TNC has great flexibility to use innovative conservation and restoration techniques on natural areas in its possession.

TNC focuses its efforts on areas containing rare species or that have high biodiversity, including the Florida Keys in southern Florida and Santa Cruz Island in California. Recently, it has set its sights on the oceans, including coastal marine ecosystems. Coastal ecosystems have experienced steep declines in the populations of many fish and shellfish, including oysters, clams, and mussels, due to a combination of overharvesting and pollution. By preserving these coastal ecosystems, TNC hopes to create reserves that will serve as breeding grounds for declining populations of overharvested species. In this way, protecting a relatively small area of ocean will benefit much larger unprotected areas, and even benefit the very industries that have led to the population declines.

Shellfish are particularly valuable in many coastal ecosystems because they are filter feeders: They remove tiny organisms, including algae, from large quantities of water, cleaning the water in the process. However, shellfish worldwide have been harvested unsustainably, leading to a cascade of effects throughout many coastal regions. For example, oyster populations in the Chesapeake Bay were once sufficient to filter the water of the entire bay in 3 to 6 days. Now there are so few oysters that it would take a year for them to filter the same amount of water. As a result, the bay has become much murkier, and excessive algae have led to lowered oxygen levels that make the bay less hospitable to fish.

Conserving marine ecosystems is particularly challenging because private ownership is rare. State and

Buying the oceans. Because the ocean floor cannot be privately owned, The Nature Conservancy has implemented a plan to lease the harvesting rights to imperiled areas and then either not harvest shellfish in the area or harvest in a sustainable way. *(Arkady Chubykin/iStockphoto.com)*

federal governments generally do not sell areas of the ocean. Instead, they have allowed industries to lease the harvesting or exploitation rights to marine resources such as oil, shellfish, and physical space for marinas and aquaculture. So how can a conservation group protect coastal ecosystems if it cannot buy an area of the ocean? The Nature Conservancy's strategy is to purchase harvesting and exploitation rights and use them as a conservation tool. In some cases, TNC will not harvest any shellfish in order to allow the populations to rebound. In many cases, the leases require at least some harvesting, and TNC hopes to demonstrate sustainable management practices that will serve as an example of how shellfish harvests can be conducted while restoring the shellfish beds.

In 2002, TNC acquired the rights to 4,650 ha (11,500 acres) of oyster beds in New York's Great South Bay, along the southern shore of Long Island. These rights, which were donated to TNC by the Blue Fields Oyster Company, were valued at $2 million. TNC developed restoration strategies and in 2010 reported that populations of shellfish were starting to show signs of recovery. After the oyster populations have rebounded, TNC hopes to engage in sustainable harvesting over part of this area and conduct research in the rest of it. TNC has similar projects under way off the coasts of Virginia, North Carolina, and Washington State. In California, TNC has purchased trawling permits and, by allowing them to go unused, has secured a no-trawl area the size of Connecticut. By 2009, TNC had accumulated the rights to 10,000 ha (25,000 acres) of marine fisheries along the coasts of the United States. In the future, it hopes to lease these permits to other harvesters of fish and shellfish that will use sustainable practices.

TNC has recently expanded its efforts in California, where it buys not only fishing leases but even fishing boats. It then hires the former fishermen to fish in a more sustainable manner. For example, TNC has its fisherman post any information about areas of the ocean where they accidentally catch protected species, which helps other fishing boats avoid these areas. One result is that the amount of unintended species caught, known as "bycatch," has declined from between 15 to 20 percent to only 1 percent. In making these changes, TNC has found effective ways to continue the tradition of fishing that provides jobs to the local communities while working to ensure that the fish populations persist long into the future.

Critical Thinking Questions

1. What are the advantages of working to increase fish populations by leasing areas of the ocean and banning fishing in those areas versus leasing areas and working with local commercial fishing boats to change fishing practices?
2. What are some of the challenges in getting commercial fishing operations to change their practices?

References

The Nature Conservancy. 2002. *Leasing and Restoration of Submerged Lands: Strategies for Community-Based, Watershed-Scale Conservation.*

The Nature Conservancy. 2006. *Annual Report.* http://www.nature.org/aboutus/annualreport/file/annualreport2008.pdf.

Partnership Preserves Livelihoods and Fish Stocks. *New York Times,* November 27, 2011. http://www.nytimes.com/2011/11/28/science/earth/nature conservancy-partners-with-california-fishermen.html?_r=0.

chapter 5 REVIEW

In this chapter, we have learned about the biodiversity of Earth, how this biodiversity came to be, and how environmental changes can cause it to decline. The estimated number of species on Earth varies widely, but many scientists agree on approximately 10 million. In any given location, we can quantify the diversity of species in terms of both species richness and species evenness. The diversity that exists came about through the process of evolution. We can view patterns of evolution by placing species on a phylogeny. Evolution occurs when there are changes in the genetic composition of a population. This can happen through artificial selection, natural selection, or random processes. The species that evolve through these processes each have a niche that helps to determine their geographic distributions. Historic changes in the environment have altered species distributions and caused many species to go extinct; current and future environmental changes are expected to have similar effects.

Suggested Answers to Critical Thinking Questions

1. A ban on fishing is guaranteed to reduce the harvesting pressure on the fish populations, and is thus more likely to result in an increase in fish populations. Working with local fishing communities will require more time and training to ensure that they are following sustainable practices.

2. Changing to more sustainable fishing practices will almost always decrease the income of those engaged in fishing operations due to the decreased catch, at least in the short term. While people who fish for a living may recognize the potential for long-term benefit, many may need that money to stay financially stable and will not be willing to risk economic hardship for longer-term gains.

Exploring the Literature

Coyne, J. A. 2009. *Why Evolution Is True.* Penguin.

Evolution: A Journey into Where We're From and Where We're Going. http://www.pbs.org/wgbh/evolution/.

Magurran, A. E. 2003. *Measuring Biological Diversity.* Wiley-Blackwell.

University of California Museum of Paleontology. *Understanding Evolution: Your One-Stop Source for Information on Evolution.* http://evolution.berkeley.edu/.

Zimmer, C., and D. Emlen. 2012. *Evolution: Making Sense of Life.* Roberts Publishers.

Download a printable version of this list by clicking on the link buttons in the TE-book, opening the TRFD, or logging onto the book's companion website bcs.whfreeman.com/friedlandapes2e (teacher login required).

 **Exploring the Literature 5**

Teaching Tip: Measuring Your Impact

The True Cost of a Green Lawn
Measuring Your Impact 5 asks students to consider a potential option to improve lawn biodiversity and consider the economic value of doing so. Download this resource by clicking on the link buttons in the TE-book, opening the TRFD, or logging onto the book's companion website bcs.whfreeman.com/friedlandapes2e (teacher login required).

 **Measuring Your Impact 5:** The True Cost of a Green Lawn

AP® Exam Tip

The characteristics of biodiverse ecosystems and human and natural impacts on those systems are covered in the 2013 AP® Exam, Question 4. To answer this question, students must

- describe the characteristics of biodiverse ecosystems
- explain human and natural impacts on those systems
- describe two ecological benefits that greater biodiversity provides

Key Terms

Species richness
Species evenness
Phylogeny
Evolution
Microevolution
Macroevolution
Gene
Genotype
Phenotype
Mutation
Recombination
Evolution by artificial selection
Evolution by natural selection
Fitness
Adaptation
Gene flow
Genetic drift
Bottleneck effect
Extinction
Founder effect
Geographic isolation
Allopatric speciation
Reproductive isolation
Sympatric speciation
Genetically modified organism
Range of tolerance
Fundamental niche
Realized niche
Distribution
Niche generalist
Niche specialist
Mass extinction

Learning Objectives Revisited

Module 14 The Biodiversity of Earth

- **Understand how we estimate the number of species living on Earth.**

  Scientists have estimated the number of species on Earth by collecting samples of diverse groups of organisms, determining the proportion of all known species, and then extrapolating these numbers to other groups to estimate the total number of species.

- **Quantify biodiversity.**

  Biodiversity, can be quantified using a variety of measurements including species richness, species evenness, or both. Such measurements provide scientists with a baseline they can use to determine how much an ecosystem has been affected by a natural or anthropogenic disturbance.

- **Describe patterns of relatedness among species using a phylogeny.**

  Patterns of relatedness are depicted as phylogenies. Phylogenies indicate how species are related to one another and the likely steps in evolution that gave rise to current species.

Module 15 How Evolution Creates Biodiversity

- **Identify the processes that cause genetic diversity.**

  Every individual has a genotype that, in combination with the environment, determines its phenotype. In a population, genetic diversity is produced by the processes of mutation and recombination.

- **Explain how evolution can occur through artificial selection.**

  Evolution by artificial selection occurs when humans select individuals with a particular phenotypic goal in mind. Such selection has produced various breeds of domesticated animals and numerous varieties of crops. It has also produced harmful outcomes, including selection for pesticide-resistant pests and drug-resistant bacteria.

- **Explain how evolution can occur through natural selection.**

  Evolution by natural selection occurs when individuals vary in traits that can be passed on to the next generation and this variation in traits causes different abilities to survive and reproduce in the wild. Natural selection does not target particular traits, but simply favors any trait changes that result in higher survival or reproduction.

- **Explain how evolution can occur through random processes.**

  Because evolution is defined as a change in the genetic composition of a population, evolution can also occur when there are mutations within a population or gene flow into or out of a population. It can also occur due to the processes of genetic drift, the bottleneck effect, or the founder effect.

Module 16 Speciation and the Pace of Evolution

- **Explain the processes of allopatric and sympatric speciation.**

  Allopatric speciation occurs when a portion of a population experiences geographic isolation from the rest of the population. The composition of isolated populations diverges over time due to random processes or natural selection. Sympatric speciation occurs when one species separates into two species without any geographic isolation. The production of polyploidy individuals is a common mechanism of sympatric speciation.

- **Understand the factors that affect the pace of evolution.**
  Evolution can produce adapted populations more easily when environmental changes are slow rather than rapid. The populations are also more likely to adapt when they have high genetic variation, as is often found in large populations. Should there be a beneficial mutation arise, however, the mutation can spread through small populations more rapidly than large populations. Finally, evolution can occur more rapidly in populations that have shorter generation times.

Module 17 Evolution of Niches and Species Distributions

- **Explain the difference between a fundamental and a realized niche.**
  Every species has a range of tolerance to the abiotic conditions of the environment. The conditions under which a species can survive, grow, and reproduce is known as its fundamental niche. The portion of the fundamental niche that a species actually occupies due to biotic interactions, including predation, competition, and disease, is known as the realized niche. Some species are niche generalists whereas other species are niche specialists. The niche a species occupies determines its distribution.
- **Describe how environmental change can alter species distributions.**
  Because a species niche represents the environmental conditions under which a species can live, environmental change can cause a change in the distribution of a species.
- **Discuss how environmental change can cause species extinctions.**
  When environmental changes are too rapid and too extensive to permit evolutionary changes, or if a species is unable to move to more hospitable environments, a species will not be able to persist and will go extinct.

Chapter 5 AP® Environmental Science Practice Exam

Section 1: Multiple-Choice Questions

Choose the best answer for questions 1–12.

1. Which is NOT a measure of biodiversity?
   (a) Economic diversity
   (b) Ecosystem diversity
   (c) Genetic diversity
   (d) Species diversity
   (e) Species richness

2. Which statement describes an example of artificial selection?
   (a) Cichlids have diversified into nearly 200 species in Lake Tanganyika.
   (b) Thoroughbred racehorses have been bred for speed.
   (c) Whales have evolved tails that help propel them through water.
   (d) Darwin's finches have beaks adapted to eating different foods.
   (e) Ostriches have lost the ability to fly.

3. The following table represents the number of individuals of different species that were counted in three forest communities. Which statement best interprets these data?

Species	Community A	Community B	Community C
Deer	95	20	10
Rabbit	1	20	10
Squirrel	1	20	10
Mouse	1	20	10
Chipmunk	1	20	10
Skunk			10
Opossum			10
Elk			10
Raccoon			10
Porcupine			10

   (a) Community A has greater species evenness than Community B.
   (b) Community A has greater species richness than Community B.
   (c) Community B has greater species evenness than Community C.
   (d) Community C has greater species richness than Community A.
   (e) Community A has greater species evenness than Community C.

Additional Multiple-Choice AP® Practice Exam

Download an additional Multiple-Choice AP® Environmental Science Practice Exam for Chapter 5 by clicking on the link buttons in the TE-book, opening the TRFD, or logging onto the book's companion website bcs.whfreeman.com/friedlandapes2e (teacher login required).

Multiple-Choice AP® Practice Exam 5

Answers to Chapter 5 AP® Exam Multiple-Choice Questions

1. a
2. d
3. b
4. a
5. c
6. a
7. d
8. e
9. e
10. b
11. c
12. e

Answers to Chapter 5 AP® Exam

Download the full answers to the Chapter 5 AP® Practice Exam by clicking on the link buttons in the TE-book, opening the TRFD, or logging onto the book's companion website bcs.whfreeman.com/friedlandapes2e (teacher login required).

Answers to Chapter 5 AP® Exam

4. The yellow perch (*Perca flavescens*) is a fish that breeds in spring. A single female can produce up to 40,000 eggs at one time. This species is an example of which of the key ideas of Darwin's theory of evolution by natural selection?
   (a) Individuals produce an excess of offspring.
   (b) Humans select for predetermined traits.
   (c) Individuals vary in their phenotypes.
   (d) Phenotypic differences in individuals can be inherited.
   (e) Different phenotypes have different abilities to survive and reproduce.

5. In 2002, Peter and B. Rosemary Grant studied a population of Darwin's finches on one of the Galápagos Islands that feeds on seeds of various sizes. After a drought that caused only large seeds to be available to the birds, they found that natural selection favored those birds that had larger beaks and bodies. Once the rains returned and smaller seeds became much more abundant, however, natural selection favored those birds that had smaller beaks and bodies. Which process is the best interpretation of this scenario?
   (a) Genetic drift
   (b) Founder effect
   (c) Microevolution
   (d) Macroevolution
   (e) Bottleneck effect

6. When a population of monkeys migrates to a new habitat across a river and encounters another population of the same species, what evolutionary effect may occur as a result?
   (a) Gene flow
   (b) The founder effect
   (c) Genetic drift
   (d) The bottleneck effect
   (e) Recombination

7. The northern elephant seal (*Mirounga angustirostris*) was once hunted to near-extinction. Only 20 animals remained alive in 1890. Then, after the species was protected from hunting, its population grew to nearly 30,000 animals. However, this large population possesses very low genetic variation. Which process is the best interpretation of this scenario?
   (a) Evolution by natural selection
   (b) Evolution by artificial selection
   (c) Evolution by the founder effect
   (d) Evolution by the bottleneck effect
   (e) Evolution by genetic drift

8. Which statement is NOT correct?
   (a) Most speciation is thought to occur through allopatric speciation.
   (b) Polyploidy is an example of sympatric speciation.
   (c) Speciation can be caused by either natural selection or random processes.
   (d) Geographic isolation can eventually lead to reproductive isolation.
   (e) Speciation cannot occur without geographic isolation.

9. Which allows more-rapid evolution?
   (a) Long generation times
   (b) Rapid environmental change
   (c) Large population sizes
   (d) Low genetic variation
   (e) High genetic variation

10. Which conditions do NOT define the fundamental niche of a species?
    (a) Humidity
    (b) Predators
    (c) Temperature
    (d) Salinity
    (e) pH

11. Some scientists estimate that the current global extinction rate is about 30,000 species per year. If there are currently 10,000,000 species on Earth, how long will it take to destroy all of Earth's biodiversity?
    (a) Less than 100 years
    (b) Between 100 and 300 years
    (c) Between 300 and 500 years
    (d) Between 500 and 700 years
    (e) Between 700 and 1,000 years

12. Global climate change can cause the extinction of species due to all of the following EXCEPT
    (a) the inability of individuals to move.
    (b) the absence of a favorable environment nearby.
    (c) the presence of another species occupying a similar niche in neighboring areas.
    (d) the rapid rate of environment change.
    (e) the presence of high genetic variation.

Section 2: Free-Response Questions

Write your answer to each part clearly. Support your answers with relevant information and examples. Where calculations are required, show your work.

1. Look at the photograph below and answer the following questions.

(Radius Images/Alamy)

   (a) Explain how this human impact on a forest ecosystem might affect the ability of some species to move to more suitable habitats as Earth's climate changes. (2 points)
   (b) Propose and explain one alternative plan that could have preserved this forest ecosystem. (2 points)
   (c) Distinguish between the terms microevolution and macroevolution. Explain how the organisms in the forest on the left could evolve into species different from those in the forest on the right. (6 points)

2. Read the following article, which appears courtesy of The University of Texas Health Science Center at San Antonio, and answer the questions that follow.

Drug-Resistant E. coli and Klebsiella Bacteria Found in Hospital Samples and Elsewhere in U.S.

A research team from The University of Texas Health Science Center at San Antonio, examining bacterial isolates obtained in hospital and non-hospital clinical settings between 2000 and 2006, has identified drug-resistant strains of *E. coli* and *Klebsiella* bacteria in more than 50 blood, urine and respiratory samples. These resistant strains, which resemble bacteria reported in Latin America, Asia, and Europe, were thought to be rare in the United States.

"This antibiotic resistance problem is likely to become widespread," said paper co-author Jan Evans Patterson, M.D., professor of medicine, infectious diseases, and pathology at the UT Health Science Center. "It affects the way we will treat infections in the future. In the past, we were concerned with antibiotic resistance in the hospital primarily, but in this review many of the strains we detected were from the community. This tells us antibiotic resistance is spreading in the community as well, and will affect how we choose antibiotics for outpatient infections."

If the trend continues, it may become difficult to select appropriate antibiotic therapy for urinary tract infections, for example. "The trend over the last decade has been to treat urinary infections empirically, to pick the drug that has worked," said James Jorgensen, Ph.D., professor of pathology, medicine, microbiology, and clinical laboratory sciences at the Health Science Center. "Now it is important for physicians to culture the patient's urine to be sure they have selected the right antibiotic. The top three drugs that are often prescribed may not be effective with these resistant bacteria."

   (a) Explain how drug-resistant strains of bacteria could evolve in a hospital. (4 points)
   (b) According to the article, what are scientists now concerned about that they were not concerned about in the past? (2 points)
   (c) Explain how new drugs could be viewed as restricting the fundamental niche of a particular bacterial species. (2 points)
   (d) Propose two possible solutions to the current problem of drug-resistant bacteria. (1 point each)

Answers to Chapter 5 AP® Exam Free-Response Questions

1. (a) Roads can act as barriers that prevent some species from dispersing as the climate changes. Cars can injure or kill species that attempt to cross the road, and the lack of tree cover can make animals crossing the road more vulnerable to predators.
(b) Possible alternative plans:

- Route the road around forested area rather than through it.
- Elevate the road over the forest.
- Tunnel the road under the forest.
- Purchase the land or development rights and build the road somewhere else.

(c) Microevolution: genetic changes that alter the phenotype of a population or species over time. Macroevolution: genetic changes that give rise to new species, genera, families, or orders over time. Steps to speciation:

- Larger populations are subdivided into smaller populations with different allele frequencies.
- Two separate environments cause selection for different traits.
- Over time the isolated populations become more and more genetically different until they can no longer interbreed and produce fertile offspring even if the barrier is removed. At this point, the two populations will have achieved reproductive isolation. Their evolution into two distinct species has occurred through the process of allopatric speciation.

2. (a) Because there is considerable genetic variation in most populations, it is possible that a few of the bacteria possess mutations that make them resistant to the antibiotic. Because the antibiotic will kill all bacteria that are not resistant, exposure to the antibiotic will select for resistant bacteria. Over time, the population will become dominated by the resistant genotypes of bacteria.
(b) Doctors will no longer be able to simply prescribe a single effective antibiotic because it will not be effective against bacteria that have evolved resistance to the drug. Instead, doctors will have to determine whether a patient possesses drug-resistant strains of the bacteria and, if the patient does carry a resistant strain, prescribe specific drugs for which resistance has not yet evolved.
(c) Because drugs are chemicals, they represent part of the abiotic environment in which the bacteria live. When the bacteria cannot live in an antibiotic environment, their fundamental niche—the range of environments in which the bacteria can survive, grow, and reproduce—is restricted. The more people that take an antibiotic that kills the bacteria, the fewer people that can be infected by the bacteria.
(d) Doctors could prescribe antibiotics only to people who really need the drugs so that fewer people would serve as selective environments for evolving drug resistance. Doctors could prescribe a mixture of different antibiotics, which would make it harder for a bacterial population to evolve resistance.

Answers to Unit 2 AP® Exam Multiple-Choice Questions

1. b
2. a
3. d
4. c
5. c
6. a
7. b
8. a
9. d
10. c
11. a
12. d
13. e
14. e
15. c
16. b
17. e
18. a
19. e
20. e
21. e
22. d
23. c

Answers to Unit 2 AP® Exam

Download the full answers to the Unit 2 AP® Practice Exam by clicking on the link buttons in the TE-book, opening the TRFD, or logging onto the book's companion website bcs.whfreeman.com /friedlandapes2e (teacher login required).

Answers to Unit 2 AP® Exam

Unit 2 AP® Environmental Science Practice Exam

Section 1: Multiple-Choice Questions

Choose the best answer for questions 1–20.

1. Ecosystem boundaries are best defined as
   (a) borders that prevent migration or dispersal.
   (b) spatial divisions in abiotic and biotic conditions.
   (c) vague areas that often lack characteristic attributes.
   (d) the border of a lake or stream.
   (e) borders encompassing large areas of land or water.

Question 2 refers to the following diagram:

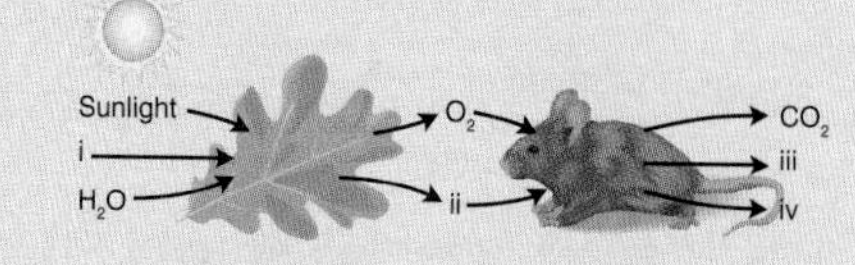

2. By ingesting leaf tissue, herbivores utilize the products of photosynthesis for cellular respiration. The diagram above shows the flow of inputs and outputs for these two processes. Labels i, ii, iii, and iv refer to
   (a) CO_2, $C_6H_{12}O_6$, energy, and water.
   (b) CO_2, CO_2, energy, and excreted waste.
   (c) nutrients, $C_6H_{12}O_6$, O_2, and water.
   (d) CO_2, $C_6H_6O_6$, energy, and O_2.
   (e) heat, organic tissue, energy, and O_2.

3. To estimate gross primary production, researchers can use all of the following variable combinations EXCEPT
   (a) net primary production and plant respiration.
   (b) standing crop, plant respiration, consumer biomass, and mass of consumer waste.
   (c) CO_2 taken up by plant in sunlight and CO_2 produced by plants in the dark.
   (d) CO_2 released by plants in sunlight and CO_2 produced by plants in the dark.
   (e) the amount of photosynthesis occurring over time.

4. Which explains why the distribution of biomass in ecosystems is often structured like a pyramid?
   I. The second law of thermodynamics
   II. Low ecological efficiency in natural ecosystems
   III. Photosynthesis is more efficient than cellular respiration

   (a) I only
   (b) II only
   (c) I and II
   (d) I and III
   (e) I, II, and III

5. Which process is NOT part of the carbon cycle?
   (a) Growth of algae in the open ocean
   (b) Combustion of fossil fuels
   (c) Weathering of mineral rock
   (d) Decomposition of organic material
   (e) Cellular respiration

6. Production of synthetic fertilizers is an example of ______ and one process that enables plants to use fertilizer is ________.
   (a) nitrogen fixation/nitrification
   (b) nitrification/nitrogen fixation
   (c) nitrification/assimilation
   (d) denitrification/nitrification
   (e) mineralization/leaching

7. Phosphorus is often limiting in aquatic environments because
   (a) phosphorus is in higher demand by aquatic organisms relative to all other minerals.
   (b) phosphorus is not easily leached from soils and rapidly precipitates in water.
   (c) weathering of rock is an extremely slow process.
   (d) human activity regularly absorbs phosphorus from aquatic environments.
   (e) phosphorus rarely changes form as it is cycled from land to water.

8. Intertidal communities are frequently disturbed by storms that generate large waves. Following a period of intermediate wave disturbance, a group of researchers examined the species richness of north and south Pacific intertidal communities. Four days after the disturbance, they found that northern communities had returned to their original state whereas southern communities were still recovering. This result suggests that northern intertidal communities
   (a) have greater resilience.
   (b) have greater resistance.
   (c) follow predictions of the intermediate disturbance hypothesis.
   (d) have greater species richness.
   (e) experience more frequent disturbance.

9. Which statement does NOT describe a function of ozone gas on Earth?
   (a) Ozone absorbs ultraviolet-B radiation from the Sun.
   (b) Ozone absorbs ultraviolet-C radiation from the Sun.
   (c) Ozone absorbs X-ray radiation from the Sun.
   (d) Ozone causes the upper stratosphere to become warmer than the lower stratosphere.
   (e) Ozone prevents DNA damage.

10. If Earth's axis of rotation were tilted at 45 degrees instead of 23.5 degrees, how many hours of daylight would the Northern Hemisphere receive during the March and September equinoxes?
    (a) 0 hours
    (b) 6 hours
    (c) 12 hours
    (d) 18 hours
    (e) 24 hours

11. Which order of processes results in the occurrence of hot and dry deserts at 30°S and 30°N latitudes?
    (a) Adiabatic cooling, condensation, latent heat release, adiabatic heating
    (b) Adiabatic heating, condensation, latent heat release, adiabatic cooling
    (c) Latent heat release, adiabatic cooling, air displacement, condensation
    (d) Condensation, adiabatic heating, adiabatic cooling, latent heat release
    (e) Air displacement, condensation, adiabatic heating, adiabatic cooling

12. Lush vegetation on the western slopes of the Colorado Rockies may be attributed to
    I. trade winds generated by the Coriolis effect.
    II. a rain shadow.
    III. adiabatic cooling.

    (a) I only
    (b) III only
    (c) I and II
    (d) I and III
    (e) I, II, and III

13. The mixing of surface water and deep water in the oceans that occurs because of differences in salinity is known as
    (a) upwelling.
    (b) a gyre.
    (c) the El Niño–Southern Oscillation.
    (d) an isocline.
    (e) thermohaline circulation.

14. Which of the following is NOT a cause of the El Niño–Southern Oscillation?
    (a) The position of Australia with respect to South America
    (b) A reversal of trade wind direction
    (c) Movement of warm surface water from east to west
    (d) A buildup of warm water along the coast of South America
    (e) Upwelling of water along the coast of South America

Questions 15 and 16 refer to the following graph:

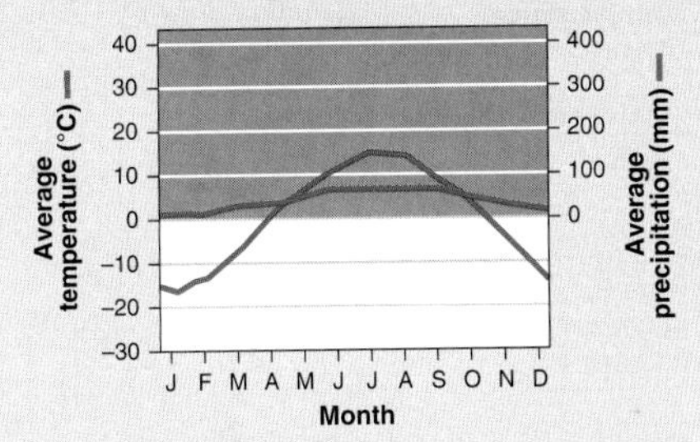

15. For the region represented by the graph, when is the growing season?
    (a) January to December
    (b) January to July
    (c) April to October
    (d) November to March
    (e) July to December

16. What biome is this graph most likely to represent?
    (a) Temperate rainforest
    (b) Boreal forest
    (c) Temperate seasonal forest
    (d) Tropical seasonal forest
    (e) Cold desert

17. Which list contains terms for lake classification, from systems with the lowest primary productivity to systems with the highest primary productivity?
    (a) Eutrophic, mesotrophic, oligotrophic
    (b) Oligotrophic, eutrophic, mesotrophic
    (c) Mesotrophic, eutrophic, oligotrophic
    (d) Mesotrophic, oligotrophic, eutrophic
    (e) Oligotrophic, mesotrophic, eutrophic

18. Iron is a limiting nutrient for algae in the open ocean. After researchers released 5,000 kg of iron into the ocean, they notice an algal bloom in the ________ zone.
    (a) photic
    (b) profundal
    (c) aphotic
    (d) intertidal
    (e) benthic

Unit 2 AP® Environmental Science Practice Exam 181

Question 19 refers to the following table comparing forest communities in different regions of North America.

Total percent statewide cover

Species	Connecticut (CT)	Pennsylvania (PA)	Georgia (GA)
Red maple	25	40	20
Black oak	30	20	50
White pine	15	10	10
Eastern hemlock	20	10	0
Black cherry	20	20	30

19. Researchers collected data on tree abundance for a group of forest communities in the three states shown in the table above. Which statement best describes the forest communities within each state?
    (a) CT has the highest species richness; PA has the highest species evenness.
    (b) PA has the highest species richness; CT has the highest species evenness.
    (c) CT and PA have equal species richness; GA has the highest species evenness.
    (d) GA and PA have equal species richness; GA has the highest species evenness.
    (e) CT and PA have equal species richness; CT has the highest species evenness.

20. Humans have developed varieties of grape vines that produce grapes that make fine wines if grown under particular environmental conditions. Therefore, the flavor of grapes on a given vine depends on
    I. genes.
    II. environments.
    III. artificial selection.

    (a) I only
    (b) II only
    (c) III only
    (d) I and II
    (e) I, II, and III

21. Which could cause a dramatic decline in genetic diversity?
    I. Mutation
    II. A bottleneck effect
    III. A founder effect

    (a) I only
    (b) II only
    (c) III only
    (d) I and II
    (e) II and III

22. Which is an example of sympatric speciation?
    (a) A stream diversion divides a single turtle population into two isolated populations that evolve into two species.
    (b) Forest fragmentation isolates two species of mice.
    (c) Two bird species from separate forests evolve reproductive isolation.
    (d) Two fish species in a single lake evolve from a single ancestor that once lived in the lake.
    (e) Artificial selection by humans for certain breeds of dog.

23. Which of the following is the best definition of a realized niche?
    (a) The range of abiotic and biotic conditions under which a species can live
    (b) The total amount of area in which a species can live
    (c) The range of abiotic and biotic conditions under which a species actually lives
    (d) The total amount of area in which a species actually lives
    (e) The range of abiotic and biotic conditions in which a species can live but its predators cannot live

182 UNIT 2 ■ AP® Environmental Science Practice Exam

Section 2: Free-Response Questions

Write your answer to each part clearly. Support your answers with relevant information and examples. Where calculations are required, show your work.

1. In temperate forests, average daily temperatures during winter often drop to 0° C or lower. During this time, many organisms burrow into the soil and remain in a dormant stage until the spring thaw. Although these organisms are able to withstand very cold temperatures, they also rely on the insulating capacity of snow, which prevents soil from experiencing temperature extremes. However, during winters when snowfall is less abundant, soil is exposed to freezing temperatures that kill many of the bacteria and fungi residing in the very top layers of soil. When bacteria and fungi die, nutrients and other organic material are easily leached out of cell membranes and these materials enter the soil.
   (a) What are two consequences of reduced winter snowfall on the nitrogen cycle in temperate forests? (4 points)
   (b) What is one consequence of reduced winter snowfall on the carbon cycle in temperate forests? (2 points)
   (c) Describe two ways in which stream and pond food webs might be affected by reduced snowfall during the winter. (2 points)
   (d) Exposure of the soil to extreme temperatures can also kill larger organisms, such as frogs and burrowing insects. For example, wood frogs (*Lithobates sylvaticus*) that burrow too close to the soil surface are likely to die whereas frogs that burrow deeper in the soil are likely to survive. Describe one way in which reduced winter snowfall could alter the local population genetics of wood frogs. (2 points)

2. During El Niño–Southern Oscillation (ENSO) years, the buildup of warm water off the coast of South America alters the circulation of the Hadley cell. One consequence of this altered circulation is a dip in the subtropical jet stream that carries storms from west to east across North America. Subsequently, southern California experiences greater precipitation during El Niño years.
   (a) Draw a diagram that shows both Hadley and Ferrell cells, noting latitudinal locations, air direction and areas of condensation, adiabatic heating, and adiabatic cooling. (4 points)
   (b) Why would the western slopes of the southern Colorado Rockies experience greater precipitation than the eastern slopes? (2 points)
   (c) A researcher wants to test the hypothesis that gross primary productivity (GPP) on the western slopes of the Rockies is greater during ENSO years.
      (i) Write an equation that defines gross primary productivity. (1 point)
      (ii) What characteristics of local plants should the researcher measure to test the hypothesis? Provide justification for your responses. (2 points)
      (iii) Provide a null hypothesis for this study. (1 point)

Answers to the Unit 2 Free-Response Questions can be found in the Answer Appendix on page ANS-3.

Teaching Tip: Unit 2 Additional Free-Response Question

An additional practice free-response question covering concepts in Unit 2 is available. This question presents a short research report on the *Anolis* lizards. To answer this question, students must

- define natural selection and the founder effect
- explain divergent evolutionary trends between mainland and island populations based on different conditions
- describe potential risks associated with small populations

Download this resource by clicking on the link buttons in the TE-book, opening the TRFD, or logging on to the book's companion website bcs.whfreeman.com /friedlandapes2e (teacher login required).

Unit 2 Additional Free-Response Question

Teaching Tip: Engage

This Science Applied article introduces students to biodiversity hotspots—isolated areas that contain a high proportion of all the species found on earth. Students will learn about the characteristics that define a hotspot. They will discover areas that have been identified as hotspots around the world. Students must also consider the costs as well as the benefits of conserving these areas. This article provides an opportunity for students to extend the information learned in Chapter 5 to biodiversity conservation methods.

Jenny Pate/age fotostock

2

science**applied**

How Should We Prioritize the Protection of Species Diversity?

As a result of human activities, we have seen a widespread decline in biodiversity across the globe. Many people agree that we should try to slow or even stop this loss. But how do we do this? While ideally we might want to preserve all biodiversity, in reality preserving biodiversity requires compromises. For example, in order to preserve the biodiversity of an area, we might have to set aside land that would otherwise be used for housing developments, shopping malls, or strip mines. If we cannot preserve all biodiversity, how do we decide which species receive our attention? (FIGURE SA2.1)

FIGURE SA2.1 **The California tiger salamander (*Ambystoma californiense*).** This salamander is endemic to the biodiversity hotspot of California and is threatened with extinction due to habitat destruction and the introduction of non-native predators. *(James Gerholdt/Getty Images)*

In 1988, Oxford University professor Norman Myers noted that much of the world's biodiversity is concentrated in areas that make up a relatively small fraction of the globe. Part of the reason for this uneven pattern of biodiversity is that so many species are *endemic species*. **Endemic species** are species that live in a very small area of the world and nowhere else, often in isolated locations such as the Hawaiian Islands. Because they are home to so many endemic species, these isolated areas end up containing a high proportion of all the species found on Earth. Myers called these areas **biodiversity hotspots.**

Endemic species A species that lives in a very small area of the world and nowhere else.

Biodiversity hotspot An area that contains a high proportion of all the species found on Earth.

Scientists originally identified 10 biodiversity hotspots, including Madagascar, western Ecuador, and the Philippines. Myers argued that these 10 areas were in need of immediate conservation attention because human activities there could have disproportionately large negative effects on the world's biodiversity. A year later, the group Conservation International adopted Myers's concept of biodiversity hotspots to guide its conservation priorities. As of 2013, Conservation International had identified the 34 biodiversity hotspots shown in FIGURE SA2.2. Although these hotspots collectively represent only 2.3 percent of the world's land area, more than 50 percent of all

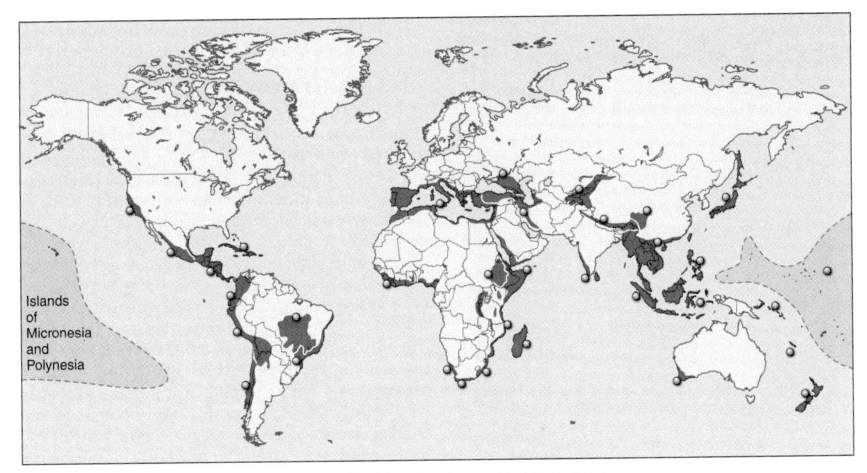

FIGURE SA2.2 **Biodiversity hotspots.** Conservation International has identified 34 biodiversity hotspots that have at least 1,500 endemic plant species and a loss of at least 70 percent of all vegetation. *(Data from Conservation International at http://www.conservation.org/where/priority_areas/hotspots/Documents/CI_Biodiversity-Hotspots_2013_Map.pdf)*

plant species and 42 percent of all vertebrate species are confined to these areas. As a result of this categorization, major conservation organizations have adjusted their funding priorities and are spending hundreds of millions of dollars to conserve these areas. What does environmental science tell us about the hotspot approach to conserving biodiversity?

What makes a hotspot hot?

Since Norman Myers initiated the idea of biodiversity hotspots, scientists have debated which factors should be considered most important when deciding where to focus conservation efforts. For example, most scientists agree that species richness is an important factor. There are more than 1,300 bird species in the small nation of Ecuador—more than twice the number of bird species living in the United States and Canada. For this reason, protecting a habitat in Ecuador has the potential to save many more bird species than protecting the same amount of habitat in the United States or Canada. From this point of view, the choice to protect areas with a lot of species makes sense.

Identifying biodiversity hotspots is challenging, however, because scientists have not yet discovered and identified all the species on Earth. Because the distribution of plants is typically much better known than that of animals, the most practical way to identify hotspots has been to locate areas containing high numbers of endemic plant species. It is reasonable to expect that areas with high plant diversity will contain high animal diversity as well.

Conservation International considers two criteria when determining whether an area qualifies as a hotspot. First, the area must contain at least 1,500 endemic plant species. By conserving plant diversity, the hope is that we will simultaneously conserve animal diversity, especially for those groups, such as insects, that are poorly cataloged. Second, the area must have lost more than 70 percent of the vegetation that contains those endemic plant species. In this way, high-diversity areas with a high level of habitat loss receive the highest conservation priority. High-diversity areas that are not being degraded receive lower conservation priority.

What else can make a hotspot hot?

The number of endemic species in an area is undoubtedly important in identifying biodiversity hotspots, but other scientists have argued that this criterion alone is not enough. They suggest that we also consider the total number of species in an area or the number of species currently threatened with extinction in an area. Would all three approaches identify similar regions of conservation priority? A recent analysis of birds suggests they would not. When scientists identified bird diversity hotspots using each of the three criteria—endemic species, total species richness, and threatened species—their results, shown in **TABLE SA2.1**, identified very different areas. Scientists using the three

Science Applied 2 How Should We Prioritize the Protection of Species Diversity? 185

TABLE SA2.1 Biodiversity hotspots for birds, identified by three criteria

Rank	Total number of species	Number of endemic species	Number of threatened species
1	Andes	Andes	Andes
2	Amazon Basin	New Guinea and Bismarck Archipelago	Amazon Basin
3	Western Great Rift Valley	Panama and Costa Rica Highlands	Guiana Highlands
4	Eastern Great Rift Valley	Caribbean	Himalayas
5	Himalayas	Lesser Sunda Islands	Atlantic Coastal Forest, Brazil

Source: Data from C.D. Orme et al., Global hotspots of species richness are not congruent with endemism or threat, *Nature* 436 (2005): 1016–1019.

criteria to identify hotspots of mammal diversity reached the same conclusion.

As we can see, some areas of the world that have high species richness do not contain high numbers of endemic species. The Amazon, for example, has a high number of bird species, but not a particularly high number of endemic bird species compared with other more-isolated regions of the world, such as the islands of the Caribbean. Similarly, areas with high numbers of threatened or endangered species do not always have high numbers of endemic species. These findings highlight the critical problem of deciding whether conservation efforts should be focused on areas containing the greatest number of species, areas containing the greatest number of threatened species, or areas containing the greatest number of endemic species. All three approaches are reasonable.

In addition to considering species diversity, some scientists have argued that we must also consider the size of the human population in diverse areas. For example, we might expect that natural areas containing more people face a greater probability of being affected by human activities. Furthermore, if we wish to project into the future, we must consider not only the size of the human population today, but also the expected size of the human population several decades from now. Scientists have found that many hotspots for endemic species have human population densities that are well above the world's average. Whereas the world has an average human population density of 42 people per square kilometer, the average hotspot has a human population of 73 people per square kilometer. Such places may be at a higher risk of degradation from human activities. This risk should be considered when determining priority areas for conservation, and it should motivate us to promote development that does not come at the cost of species diversity.

What are the costs and benefits of conserving biodiversity hotspots?

Conservation efforts that focus on regions with large numbers of species place a clear priority on preserving the largest number of species possible. However, people making these efforts do not explicitly consider the likelihood of succeeding in this goal, nor do they necessarily consider the costs associated with the effort. For example, there may be many ways of helping a species persist in an area, including buying habitat, entering into agreements with landowners not to develop their land, or removing threats such as invasive species. In the case of the California tiger salamander, for example, while eliminating invasive predators may not be feasible, it is possible to protect salamander habitat. In 2011, the U.S. Fish and Wildlife Service agreed to designate more than 20,000 ha (50,000 acres) of habitat as being critical for the salamander's persistence and in 2012 the agency agreed to develop a plan to help populations of the salamander to recover. Each option will have a different impact on the number of species that will be helped, and each option will have different costs of implementation. Given the limited funds that are available for protecting species, it is certainly worth comparing the expected costs and benefits of different options, both within and among biodiversity hotspots. In this way, we have the potential to maximize the return on our conservation investment.

What about biodiversity coldspots?

The concept of biodiversity hotspots assumes that our primary goal is to protect the maximum number of species. That goal is admirable, but it could come at the cost of many important ecosystems that do not fall within hotspots. Yellowstone National Park, for example, has a relatively low diversity of species, yet it is one of the few places in the United States that contains remnant populations of large mammals, including wolves, grizzly bears, and bison. Does this mean that places such as Yellowstone National Park should receive decreased conservation attention?

Biodiversity coldspots also provide ecosystem services that humans value at least as much as species diversity. For example, wetlands in the United States are incredibly important for flood control, water purification, wildlife habitat, and recreation. Many wetlands, however, have relatively low plant diversity and, as a result, would not be identified as biodiversity hotspots. It is true that increased species richness leads to improved ecosystem services, but only as we move from very low species richness to moderate species richness. Moving from moderate species richness to high species richness generally does not further

186 Science Applied 2 How Should We Prioritize the Protection of Species Diversity?

improve the functioning of an ecosystem. Since very high species diversity is not expected to provide any substantial improvement in ecosystem function, protecting more and more species produces diminishing returns in terms of protecting ecosystem services. Hence, if our primary goal is to preserve the functioning of the ecosystems that improve our lives, we do not necessarily need to preserve every species in those ecosystems.

How can we reach a resolution?

During the past 2 decades, it has become clear that scientists and policy makers need to set priorities for the conservation of biodiversity. No single criterion may be agreed upon by everyone. However, it is important to appreciate the bias of each approach and to consider the possible unintended consequences of favoring some geographic regions over others. Our investment in conservation cannot be viewed as an all-or-nothing choice of some areas over others. Instead, our decisions must take into account the costs and benefits of alternative conservation strategies and incorporate current and future threats to both species diversity and ecosystem function. In this way, we can strike a balance between our desire to preserve Earth's species diversity and our desire to protect the functioning of Earth's ecosystems.

Questions

1. If you were in the role of a decision maker, how might you balance your desire to protect biodiversity around the world with the high cost of preserving biodiversity in any given location?
2. When identifying hotspots around the world, why is it important to simultaneously consider the biodiversity of different areas and the future human population size in each of these locations?
3. If you had to make a choice between protecting biodiversity hotspots and biodiversity coldspots, how would you make this decision?

Free-Response Question

Florida Reef is the only living coral barrier reef in the continental United States. Like other coral reefs on the planet, Florida Reef is home to many species of coral, fish, and other aquatic life forms. Also like other coral reefs, its health is at risk due to such factors as coral bleaching. However, Conservation International does not list this as a biodiversity hotspot.

(a) List two reasons why Florida Reef might not be internationally recognized as a hotspot. (2 points)
(b) List and describe two ecosystem services provided by Florida Reef. (2 points)
(c) What is coral bleaching? (2 points)
(d) Researchers recently found that the disturbance generated by hurricanes is important for maintaining the diversity of coral reef ecosystems. However, massive hurricanes often put many coral reef species at risk of local extinction. Generate a hypothesis to explain this pattern. (4 points)

References

Bacchetta, G., et al. 2012. A new method to set conservation priorities in biodiversity hotspots. *Plant Biosystems* 146:638–648.

Joppa, L. N., et al. 2011. Biodiversity hotspots house most undiscovered plant species. *PNAS* 108:13171–13176.

Kareiva, P., and M. Marvier. 2003. Conserving biodiversity coldspots. *American Scientist* 91:344–351.

Myers, N., et al. 2000. Biodiversity hotspots for conservation priorities. *Nature* 403:853–858.

Orme, C. D., et al. 2005. Global hotspots of species richness are not congruent with endemism or threat. *Nature* 436:1016–1019.

Key Terms

Endemic species
Biodiversity hotspots

The answer to the Science Applied 2 Free-Response Question can be found in the Answer Appendix on page ANS-4.

Answer to Science Applied 2 Free-Response Question

Download the full answer to the Science Applied 2 Free-Response Question by clicking on the link buttons in the TE-book, opening the TRFD, or logging onto the book's companion website bcs.whfreeman .com/friedlandapes2e (teacher login required).

Answer to Science Applied 2 Free-Response Question

Suggested Answers to Science Applied Questions

1. While it would be ideal to be able to protect biodiversity everywhere in the world, some areas may not be feasible or cost-effective to protect. With limited funds it would make sense to focus on those areas with the highest diversity and on areas where protection can be done inexpensively.

2. Areas with high human population are much more likely to be negatively affected by human activities, and thus are at much greater risk for species extinction.

3. Answers will vary. One possible answer might be to select biodiversity hotspots because they have more of an impact on global biodiversity. Students might also select biodiversity coldspots because they provide valuable ecosystem services.

chapter

6 Population and Community Ecology

PD Overview

This chapter examines population characteristics, factors that drive populations to change, and different reproductive strategies. It looks at community relationships, how members of a community share resources, and how ecosystems evolve over time. The chapter presents both exponential growth and logistic growth and their relationship to carrying capacity. Limiting resources typically prevent exponential growth. Environmental pressures often cause population growth to stabilize over time to a sustainable level. We call this level carrying capacity. The logistic growth model, an S-shaped curve, depicts the relationship between population growth and carrying capacity. Chapter 6 is an ideal place to reinforce graphing tools. Students should be able to graph the growth models, survivorship curves, and predator–prey relationships. In addition, students should have a firm understanding of population distribution, growth models, *r*-selected and *K*-selected species, outcomes of competition, and primary versus secondary succession.

Module 18: The Abundance and Distributions of Populations

This module begins with a broad examination of the different levels of complexity in nature and then looks specifically at the population level. Characteristics of populations include abundance and distribution. Density-dependent factors influence population growth and density-independent factors affect populations of all sizes equally.

Module 19: Population Growth Models

Models help explain relationships in the natural world and predict how changes in different factors can affect these relationships in the future. Population ecologists use population growth models that incorporate density-dependent and density-independent factors to explain and predict changes in population size. Exponential growth models assume unlimited resources. Logistic growth models incorporate a carrying capacity and the associated density-dependent factors that occur in natural populations. Growth models can also help explain how a population might overshoot its carrying capacity and experience a die-off or oscillate because of predator–prey interactions. Populations can be characterized as either *K*-selected or *r*-selected and have unique types of survivorship curves. A metapopulation model looks at multiple, interconnected populations. These models are of interest to environmental scientists for many reasons. From a practical standpoint, they can help people who wish to protect endangered populations, manage a commercially harvested species, and control an insect pest.

Module 20: Community Ecology

Community ecology is the study of interactions among species. Numerous factors affect population distribution patterns across the planet. Two conditions determine whether a species will persist in a location. First, if a species is able to disperse to a location, its ability to persist there depends on its fundamental niche, which is the range of abiotic conditions that it can tolerate. The survival of a species in a habitat is also determined by the set of interactions that it has with other species. Some interactions such as competition, predation, parasitism, and herbivory negatively affects one or both species. Other interactions such as mutualisms and commensalisms have neutral or positive effects on both species. Keystone species have particularly large effects on a community.

Module 21: Community Succession

Even without human activity, natural communities do not stay the same forever. Change in the species composition of communities over time is a perpetual process in nature. Ecological succession—the predictable replacement of one group of species by another group of species over time—occurs in virtually every community. Depending on the community type, ecological succession can occur over time spans varying from decades to centuries. This module considers succession in both terrestrial and aquatic environments. In terrestrial communities, an area that begins with no soil undergoes primary succession whereas an area that begins with soil undergoes secondary succession. In aquatic communities, lakes and ponds can slowly fill in with sediments over thousands of years and eventually become terrestrial habitats. The species richness of a community is affected by the latitude of a habitat, the length of time that the habitat has been present, the size of the habitat, and the distance from the habitat to other habitats that serve as sources of species.

Alignment to AP® Environmental Science Course Description

AP® Outline	Module	
II. The Living World (10–15%)		
C. Ecosystem Diversity	Module 20	Community Ecology
	Module 21	Community Succession
III. Population (10–15%)		
A. Population Biology Concepts	Module 18	The Abundance and Distribution of Populations
	Module 19	Population Growth Models

Chapter Learning Objectives

After completing this chapter students will be able to

- explain how nature exists at several levels of complexity.
- discuss the characteristics of populations.
- contrast the effects of density-dependent and density-independent factors on population growth.
- explain the exponential growth model of populations, which produces a J-shaped curve.
- describe how the logistic growth model incorporates a carrying capacity and produces an S-shaped curve.
- compare the reproductive strategies and survivorship curves of different species.
- explain the dynamics that occur in metapopulations.
- identify species interactions that cause negative effects on one or both species.
- discuss species interactions that cause neutral or positive effects on both species.
- explain the role of keystone species.
- explain the process of primary succession.
- explain the process of secondary succession.
- explain the process of aquatic succession.
- describe the factors that determine the species richness of a community.

Chapter 6 Pacing Guide

This pacing guide is based on a schedule with 120 sessions of 50 minutes each before the AP® Exam. If you have a different number of sessions before the exam, you can modify the pacing to suit your needs. If you have additional time, consider incorporating quizzes, released AP® Environmental Science free-response and multiple-choice questions, or additional activities.

Module	Standard Schedule Days	Block Schedule Days
Module 18	2	1
Module 19	3	1½
Module 20	1	½
Module 21	1	½
Assessment	1	½

Chapter 6 Resources

The link to these resources can be found by clicking on the link buttons in the Teacher's e-Book (TE-book), opening the Teacher's Resource Flash Drive (TRFD), or logging onto the book's companion website bcs.whfreeman.com/friedlandapes2e (teacher login required).

- **PowerPoint Presentation 6**
- **Optimized Art PowerPoint and JPEG Files 6**
- **Do the Math Videos**
- **Measuring Your Impact 6:** The Living Planet Index
- **Activity 6.1:** Observing Duckweed Population Growth
- **Activity 6.2:** Traits of *K*-selected and *r*-selected Species
- **Activity 6.3:** Simulating the Theory of Island Biogeography
- **Lab 6.1:** Determining Population Size
- **Chapter 6 Web Resources**
- **Exploring the Literature 6**
- **Answers to Chapter 6 AP® Exam**
- **Multiple-Choice AP® Practice Exam 6**

Free-Response Questions from Previous AP® Environmental Science Exams

Free-response questions from prior AP® Environmental Science Exams are available on the AP® website, https://apstudent.collegeboard.org/apcourse/ap-environmental-science/exam-practice. Students should be able to answer all of the questions listed below with material learned in this chapter. When a question requires students to understand material from multiple chapters, the question will be listed in the last chapter required to complete the entire question. Questions marked with an asterisk (*) are from exams with released multiple-choice questions. You may want to save these questions until the end of the year so you can give your students a complete released exam for practice. Questions marked with double asterisks (**) require math to calculate a problem. Look for references to these questions throughout the chapter.

Year	Question	Content
2010	2 **	• Distribution and abundance of a population

PD Chapter 6 Overview

Watch the video overview of Chapter 6 (for teachers) by clicking on the link buttons in the TE-book, opening the TRFD, or logging onto the book's companion website bcs.whfreeman.com /friedlandapes2e (teacher login required).

TRM PowerPoint Presentation 6

Download the PowerPoint presentation for Chapter 6 by clicking on the link buttons in the TE-book, opening the TRFD, or logging onto the book's companion website bcs.whfreeman .com/friedlandapes2e (teacher login required).

A former New England farm is now a forest. *(Joseph Leuchter-Mindel/Alamy Images)*

chapter

Population and Community Ecology

New England Forests Come Full Circle

When the Pilgrims arrived in Massachusetts in 1620, they found immense areas of undisturbed temperate seasonal forest containing a variety of tree species, including sugar maple (*Acer saccharum*), American beech (*Fagus grandifolia*), white pine (*Pinus strobus*), and eastern hemlock (*Tsuga canadensis*). Over the next 200 years, settlers cut down most of the trees to clear land for farming and housing. This deforestation peaked in the 1800s, at which point up to 80 percent of all New England forests had been cleared. Between 1850 and 1950, however, many people abandoned their New England farms to take jobs in the growing textile industry. Others moved to the Midwest, where farmland was more fertile and considerably less expensive.

What happened to the former farmland is a testament to the resilience of the forest ecosystem. The transformation began shortly after the farmers left. Seeds of grasses and wildflowers were carried to the abandoned fields by birds or blown there by the wind. Within a year, the fields were carpeted with a large variety of plant species. Eventually, a single group of plants—the goldenrods—came to dominate the fields by growing taller and outcompeting other species of plants for sunlight. The other species remained in the fields, but they were not very abundant. Nevertheless, the dominance of the goldenrods was short-lived.

> The old stone walls that are so common in the New England countryside are the only evidence that this forest was once farmland.

Goldenrods and other wildflowers play an important part in old-field communities because they support a diverse group of plant-eating insects. Some of these herbivorous insects are generalists that feed on a

189

Teaching Tip: Chapter Opening Case

The chapter opening case, "New England Forests Come Full Circle," illustrates the complexity of community interactions and provides a historical example of ecological recovery and secondary succession. The case follows the New England ecosystem through a series of changes including deforestation and conversion to croplands, field abandonment, and the reestablishment of grasses and trees. This case exemplifies the multifaceted interactions among species that drive the ever-changing ecosystem composition.

Teaching Tip: Warm-up

After students read the chapter opening case, divide the class into four groups and ask them to brainstorm on the following questions for 2 minutes. Have each group share their ideas with the class.

- What does it mean for an ecosystem to be highly resilient? *Resilience is the rate at which an ecosystem returns to its original state after a disturbance. An ecosystem is highly resilient if it returns to its original state relatively rapidly.*
- Based on the opening case, to what degree is the New England forest resilient? *The forest is resilient because the forest is once again composed of many tree species after a relatively short amount of time (100 years). However, in ecosystems that are even more productive, this reestablishment might have happened faster.*
- Identify and describe some abiotic factors that might affect the forest's resilience. *Succession will occur more rapidly where high rainfall and warm temperatures promote increased growth. Windy conditions may facilitate greater seed dispersal and increased biodiversity. Nutrient-rich soils also contribute to increased productivity.*
- Identify and describe three biotic influences that changed the New England forest ecosystem. *Humans arrived and cut down most of the forest for farming and grazing land. Humans later abandoned farming in the area. Goldenrods outcompeted the other plant species and became dominant. White pines cast so much shade over the goldenrods that the goldenrod population declined. Humans began harvesting white pine lumber, which allowed beech and sugar maple trees to populate the forest.*

wide range of plant species, while others specialize on only a small number of plant species. The number of individuals of each insect species varies from year to year, and occasionally some species experience very large population increases, or outbreaks. One such species is a leaf beetle (*Microrhopala vittata*) that specializes in eating goldenrods. Periodic outbreaks of this species in the abandoned fields of New England dramatically reduced goldenrod populations. With fewer goldenrods, other plant species could compete and prosper.

The complex interactions among populations of goldenrods, insects, and other species created an ever-changing ecosystem. For example, as the leaf beetle population increased in the community, so did the populations of predators and parasites that fed on them. As these predators and parasites reduced the population of leaf beetles, the goldenrod population began to rebound. As the goldenrods surged, they once again caused other plant species to decline in numbers.

Over time, tree seeds arrived and tree seedlings began to grow, which changed the species composition of the old fields once again. One species in particular, the fast-growing white pine, eventually came to dominate. The pine trees cast so much shade that the goldenrods and other sunlight-loving plant species could not survive.

White pines dominated the old-field communities until humans began harvesting them for lumber in the 1900s. Just as the reduction of goldenrod populations made room for other plant species, logging of the white pines made room for broadleaf tree species. Two of these broadleaf species, American beech and sugar maple, are dominant in New England forests today. The New England fields that had been abandoned earlier were slowly transformed into communities that resemble the original forests of centuries ago, with a mix of pines, hemlocks, and broadleaf trees. The old stone walls that are so common in the New England countryside are the only evidence that this forest was once farmland.

The story of the New England forests shows us that populations can increase or decrease dramatically over time. It also illustrates how species interactions within a community can alter species abundance. Finally, it demonstrates how human activity can alter the distribution and diversity of species within an ecosystem.

Sources:
W. P. Carson and R. B. Root, Herbivory and plant species coexistence: Community regulation by an outbreaking phytophagous insect, *Ecological Monographs* 70 (2000): 73–99; T. Wessels, *Reading the Forested Landscape* (Countryman Press, 1997).

A New England forest is a wonderful reminder of the intricate complexity of the natural world. As we saw in this account, there are clear patterns in the distribution and abundance of species over space and time. Understanding the factors that generate these patterns can help us find ways to preserve global biodiversity. These factors include the ways in which populations increase and decrease in size and how species interact in ecological communities. In this chapter, we will examine the factors that help determine the abundance and distribution of populations. We will then look at interactions among species that live within ecological communities and how these interactions further determine whether a species can persist in a particular location on Earth. Finally, we will examine how ecological communities change over time.

module

18

The Abundance and Distribution of Populations

Complexity in the natural world ranges from a single individual to all of the biotic and abiotic components of Earth. In this module, we will begin by broadly examining the different levels of complexity in nature and then narrow our focus to the population level. We will consider the characteristics of populations and then discuss factors that determine how populations increase and decrease over time.

Learning Objectives

After reading this module you should be able to

- explain how nature exists at several levels of complexity.
- discuss the characteristics of populations.
- contrast the effects of density-dependent and density-independent factors on population growth.

Teaching Tip: Journal Prompt

Ask students to describe each level of complexity and identify what the scientist studies at each level.

- ***Individual:*** *One organism of a species. Scientists might study characteristics of this type of species.*
- ***Population:*** *All members of the same species in a given area. Scientists might study factors that cause the population numbers to increase or decrease. They might also look at changes within the population.*
- ***Community:*** *All the biotic factors within a given area. Scientists might study species interactions.*
- ***Ecosystems:*** *All the biotic and abiotic factors within an area. Scientists might study flows of energy and matter.*
- ***Biosphere:*** *All of Earth's ecosystems. Scientists might study the movement of air, water, and heat around the globe.*

Nature exists at several levels of complexity

As FIGURE 18.1 shows, the environment around us exists at a series of increasingly complex levels: individuals, populations, communities, ecosystems, and the biosphere. The simplest level is the individual—a single organism. As we saw in Chapter 5, natural selection operates at the level of the individual because it is the individual that must survive and reproduce.

The second level of complexity is a *population*. A **population** is composed of all individuals that belong to the same species and live in a given area at a particular time. Evolution occurs at the level of the population. Scientists who study populations are also interested in the factors that cause the number of individuals to increase or decrease. As we saw in Chapter 1, the boundaries of a population are rarely clear and may be set arbitrarily by scientists. For example, depending on what we want to learn, we might study the entire population of white-tailed deer in North America, or we might focus on the deer that live within a single state, or even within a single forest.

The third level of complexity is the *community*. A **community** incorporates all of the populations of organisms within a given area. Like those of a population, the boundaries of a community may be defined by the state or federal agency responsible for managing it. Scientists who study communities are generally interested in how species interact with one another. In Chapter 4, we saw that terrestrial communities can be grouped into biomes that contain plants with similar

Population The individuals that belong to the same species and live in a given area at a particular time.

Community All of the populations of organisms within a given area.

Teaching Tip: Video

Population Ecology: The Texas Mosquito Mystery

This short video illustrates the importance of population ecology by relating concepts in this chapter to a real problem in Dallas, Texas. In 2012, 12 people died and 300 were infected by the West Nile Virus carried by mosquitoes. This was a significant increase in disease transmission compared to the prior year. The video communicates the importance of understanding population density, distribution patterns, and limiting factors when trying to understand how this sudden increase occurred in one year. Students will be introduced to the concepts presented later in the chapter and this story can act as a point of reference when teaching concepts such as carrying capacity and density-dependent and density-independent factors.

After students view the video, ask the following questions:

- What are two factors that allowed the population of mosquitoes to grow at exponential rates in the summer of 2012? *Space and temperature.*
- Why might this be important for population ecologists to know? *This information can help scientists anticipate the potential for large populations of disease-carrying mosquitoes.*
- How might residents of Dallas benefit from knowing mosquito population trends? *Municipalities can plan to use pesticides only as needed when population numbers are very high. Residents can plan to stay indoors to protect themselves.*

The link to this video may be found by clicking on the link buttons in the TE-book, opening the TRFD, or logging onto the book's companion website bcs.whfreeman.com/friedlandapes2e (teacher login required).

TRM Chapter 6 Web Resources

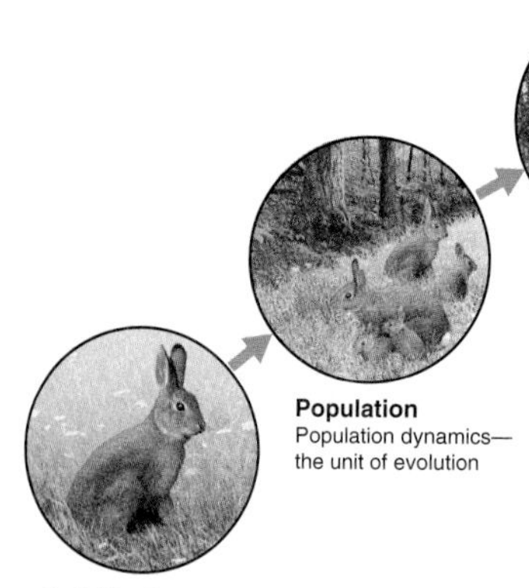

FIGURE 18.1 Levels of complexity. Environmental scientists study nature at several different levels of complexity, ranging from the individual organism to the biosphere. At each level, scientists focus on different processes.

growth forms. However, the actual composition of tree species varies from community to community. While the temperate seasonal forests of the eastern United States and Europe experience similar patterns of temperature and precipitation, they contain different tree species.

Communities exist within an ecosystem, which consists of all of the biotic and abiotic components in a particular location. Ecosystem ecologists study flows of energy and matter, such as the cycling of nutrients through the system.

The largest and most complex system environmental scientists study is the biosphere, which incorporates all of Earth's ecosystems. Scientists who study the biosphere are interested in the movement of air, water, and heat around the globe.

Populations have distinctive characteristics

When we study nature at the level of the population, one of the first things that becomes apparent is that populations are dynamic—that is, they are constantly changing. Individuals in the population die, new individuals are produced, and individuals can move from one population to another. The study of factors that cause populations to increase or decrease is the science of **population ecology.**

There are many circumstances in which scientists find it useful to identify the factors that influence population size over time. For example, in the case of endangered species such as the California condor (*Gymnogyps californianus*), knowing the factors that affect its population size has helped us to implement measures that improve its survival and reproduction. Similarly, knowing the factors that influence the population size of a pest species can help us control it. For instance, population ecologists are currently studying the emerald ash borer (*Agrilus planipennis*), an insect from Asia that was accidentally introduced to the American Midwest and is causing the widespread death of ash trees. Once we understand the population ecology of this destructive insect, we can begin to explore and develop strategies to control or eradicate it.

To understand how populations change over time, we must first examine the basic characteristics of populations. These characteristics are *size, density, distribution, sex ratio,* and *age structure.*

Population ecology The study of factors that cause populations to increase or decrease.

Population size (*N*) The total number of individuals within a defined area at a given time.

Population Size

Population size (*N*) is the total number of individuals within a defined area at a given time. For example, the California condor once ranged throughout California and the southwestern United States. Over the past 2 centuries, however, a combination of poaching, poisoning, and accidents (such as flying into electric power lines) greatly reduced the population's size. By 1987, only 22 birds remained in the wild. Scientists realized that the species was nearing extinction and decided to capture all the wild birds and start a captive breeding program in zoos. As a result of these captive breeding programs and other conservation efforts, the condor population size had increased to more than 437 birds by 2014.

Teaching Tip: Engage

Ask students: Why is it important to understand the factors that influence population size? *Understanding factors that influence population size can help conservation efforts to save endangered species. Understanding populations can also help us develop strategies to control pest species.*

Lab
Determining Population Size

This lab gives students the opportunity to use the quadrant method to study communities and to calculate the density and frequency of species in a particular community.

Download this resource by clicking on the link buttons in the TE-book, opening the TRFD, or logging onto the book's companion website bcs.whfreeman.com/friedlandapes2e (teacher login required).

 **Lab 6.1** Determining Population Size

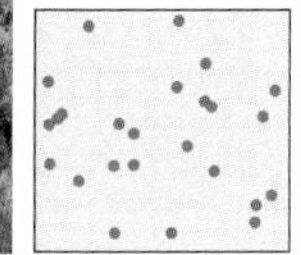

(a) Random distribution

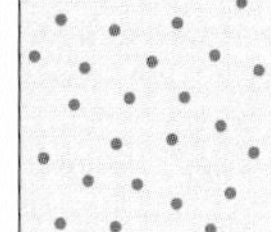

(b) Uniform distribution

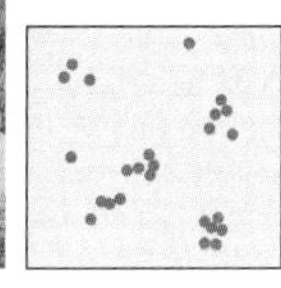

(c) Clumped distribution

FIGURE 18.2 **Population distributions.** Populations in nature distribute themselves in three ways. (a) Many of the tree species in this New England forest are randomly distributed, with no apparent pattern in the locations of individuals. (b) Territorial nesting birds, such as these Australasian gannets (Morus serrator), exhibit a uniform distribution, in which all individuals maintain a similar distance from one another. (c) Many pairs of eyes are better than one at detecting approaching predators. The clumped distribution of these meerkats (*Suricata suricatta*) provides them with extra protection. *(a: David R. Frazier Photolibrary, Inc./Science Source; b: Michael Thompson/Earth Scenes/Animals Animals; c: Clem Haagner/ARDEA)*

Population Density

Population density is the number of individuals per unit area (or volume, in the case of aquatic organisms) at a given time. Knowing a population's density, in addition to its size, can help scientists estimate whether a species is rare or abundant. For example, the density of coyotes (*Canis latrans*) in some parts of Texas might be only 1 per square kilometer, but in other parts of the state it might be as high as 12 per square kilometer. Scientists also study population density to determine whether a population in a particular location is so dense that it might outstrip its food supply.

Population density can be a particularly useful measure for wildlife managers who must set hunting or fishing limits on a species. For example, managers may divide the entire population of an animal species that is hunted or fished into management zones. Management zones may be designated political areas, such as counties, or areas with natural boundaries, such as the major water bodies in a state. Wildlife managers might offer more hunting or fishing permits for zones with a high-density population and fewer permits for zones with a low-density population.

Population Distribution

In addition to population size and density, population ecologists are interested in how a population occupies space. **Population distribution** is a description of how individuals are distributed with respect to one another. FIGURE 18.2 shows three types of population distributions. In some populations, such as a population of trees in a natural forest, the distribution of individuals is *random* (Figure 18.2a). In other words, there is no pattern to the locations where the individual trees grow.

In other populations, such as a population of trees in a plantation, the distribution of individuals is uniform, or evenly spaced (Figure 18.2b). Uniform distributions are common among territorial animals, such as nesting birds that defend areas of similar sizes around their nests. Uniform distributions are also observed among plants that produce toxic chemicals to prevent other plants of the same species from growing close to them.

In still other populations, the distribution of individuals is *clumped* (Figure 18.2c). Clumped distributions, which are common among schooling fish, flocking birds, and herding mammals, are often observed when living in large groups provides enhanced feeding opportunities or protection from predators.

Population Sex Ratio

The **sex ratio** of a population is the ratio of males to females. In most sexually reproducing species, the sex ratio is usually close to 50:50, although sex ratios can be far from equal in some species. In fig wasps, for example, there may be as many as 20 females for every male. Because the number of offspring produced is

Population density The number of individuals per unit area at a given time.

Population distribution A description of how individuals are distributed with respect to one another.

Sex ratio The ratio of males to females in a population.

Teaching Tip: Concept Map

Scientists work to understand how a population might change over time. Population characteristics help predict these changes. Ask each student to create a concept map that includes the boxes shown below. Divide the class into five groups. Assign each group a population characteristic to describe. They should also explain the significance of the characteristic. Ask each group to present its characteristic to the class.

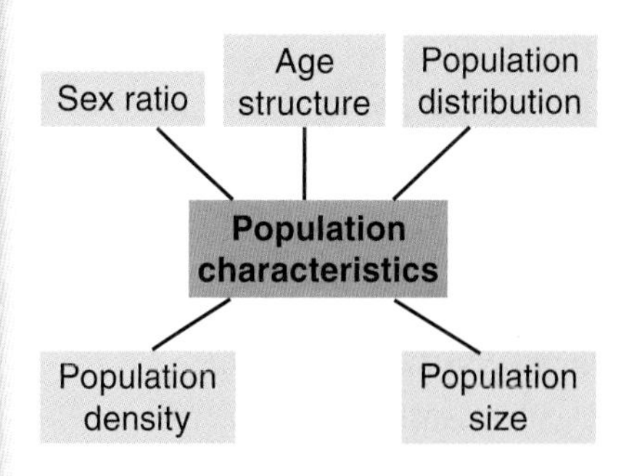

Population size: *Describes the number of individuals within a population. Knowing population size is essential for conservation efforts.*
Population density: *Describes the number of individuals per unit area. Knowing population density allows scientists to describe populations as rare or abundant.*
Population distribution: *Describes how individuals are distributed with respect to one another. Distribution patterns include random, clumped, and uniform. Knowing a population's distribution is essential to conservation and the management of pest species.*
Sex ratio: *Describes the ratio of males to females. Because the number of offspring produced is primarily a function of how many females there are in a population, knowing the sex ratio helps scientists estimate the number offspring to be expected in the next generation.*
Age structure: *Describes how many members of the population fit into a particular age category. Knowing the age structure helps ecologists predict how rapidly a population will grow.*

Teaching Tip: Journal Prompt

Ask students to describe population distribution patterns and give examples.

- ***Random:*** *Seed dispersal is a random process. For example, wildflowers will grow where they are distributed and optimal conditions exist.*
- ***Uniform:*** *Territorial animals such as nesting Australasian gannets will maintain equal distance from each other.*
- ***Clumped:*** *Herding animals such as elephants move in groups to enhance feeding opportunities and increase protection from predators.*

Teaching Tip: Warm-up

Ask students the following questions:

- What is a limiting resource? *A limiting resource is a resource that a population cannot live without—for example, water—and which occurs in quantities lower than the population would need if it were to increase in size.*
- What is carrying capacity? *Carrying capacity is the maximum number of individuals a particular ecosystem can support.*
- How does a limiting resource influence carrying capacity? *Carrying capacity is determined by limiting resources such as availability of water, nutrients, or light.*

Teaching Tip: Remember

Students struggle to differentiate between density-dependent and density-independent factors. The easiest way to help them remember is to correlate density-independent factors with disasters such as droughts, tsunamis, or forest fires. Ask students to consider how a factor affects a single organism. Is the factor more likely to cause death if the organism has more neighbors? If so, it is density-dependent. For example, food availability is density-dependent. If only one deer lives in a large area of the forest, there will be plenty of food. If the forest is crowded with deer, there may not be enough food. When a factor is equally likely to cause death whether the organism is alone or with others, it is density-independent. For example, a forest fire is likely to cause the death of a deer regardless of how many deer are present. Density-dependent factors increase in intensity with higher population numbers while density-independent factors do not.

primarily a function of how many females there are in the population, knowing a population's sex ratio helps scientists estimate the number of offspring a population will produce in the next generation.

Population Age Structure

Many populations are composed of individuals of varying ages. A population's **age structure** describes how many individuals fit into particular age categories. Knowing a population's age structure helps ecologists predict how rapidly a population can grow. For instance, a population with a large proportion of old individuals no longer capable of reproducing, or with a large proportion of individuals too young to reproduce, will produce far fewer offspring than a population that has a large proportion of individuals of reproductive age.

Population size is affected by density-dependent and density-independent factors

Factors that influence population size can be classified as *density dependent* or *density independent*. We will look at each type in turn.

Density-dependent Factors

In 1932, Russian biologist Georgii Gause conducted a set of experiments that demonstrated that food supply can control population growth. Gause monitored population growth in two species of *Paramecium* (a type of single-celled aquatic organism) living under ideal conditions in test tubes. Each day he added a constant amount of food. As the graph in FIGURE 18.3a shows, both species of *Paramecium* initially experienced rapid population growth, but over time the rate of growth began to slow. Eventually, the two population sizes reached a plateau and remained there for the rest of the experiment.

Gause suspected that *Paramecium* population growth was limited by food supply. To test this hypothesis, he conducted a second experiment in which he doubled the amount of food he added to the test tubes. Again, both species of *Paramecium* experienced rapid population growth early in the experiment, and the rate of growth, again, slowed over time. However, as Figure 18.3b shows, their maximum population sizes were approximately double those observed in the first experiment.

Gause's results confirmed that food is a *limiting resource* for *Paramecium*. A **limiting resource** is a resource that a population cannot live without and that occurs in quantities lower than the population would require to increase in size. If a limiting resource decreases, so does the size of a population that depends on it. For terrestrial plant populations, water and nutrients such as nitrogen and phosphorus are common limiting resources. For animal populations, food, water, and nest sites are common limiting resources.

Factors such as food that influence an individual's probability of survival and its amount of reproduction in a manner that depends on the size of the population are called **density-dependent factors.** For example, because a smaller population requires less total food, food scarcity will have little or no effect on the survival and reproduction of individuals in a small population but it will have large negative effects on a large population.

In Gause's *Paramecium* experiments, population growth slowed as population size increased because there was a limit to how many individuals the food supply could sustain. This limit to the number of individuals that can exist in a population is called the **carrying capacity** of the environment and is denoted as ***K*.** Knowing the carrying capacity for a species and its limiting resource helps us predict how many individuals an environment can sustain.

Density-independent Factors

Whereas density-dependent factors affect an individual's probability of survival and amount of reproduction differently depending on whether the population is large or small, **density-independent factors** have the same effect on an individual's probability of survival and amount of reproduction regardless of the population size. A tornado, for example, can uproot and kill a large number of trees in an area. However, a given tree's probability of being killed does not depend on whether it resides in a forest with a high or low density of other trees. Other density-independent factors include hurricanes, floods, fires, volcanic eruptions, and other climatic events. An individual's likelihood of mortality increases during such an event regardless of whether the population happens to be at a low or high density.

Age structure A description of how many individuals fit into particular age categories in a population.

Limiting resource A resource that a population cannot live without and that occurs in quantities lower than the population would require to increase in size.

Density-dependent factor A factor that influences an individual's probability of survival and reproduction in a manner that depends on the size of the population.

Carrying capacity (*K*) The limit of how many individuals in a population the environment can sustain.

Density-independent factor A factor that has the same effect on an individual's probability of survival and the amount of reproduction at any population size.

Teaching Tip: Warm-up

Ask students: How do density-independent and density-dependent factors influence population size? *Density-dependent factors intensify the effect on an organism as population numbers increase, while density-independent factors do not.*

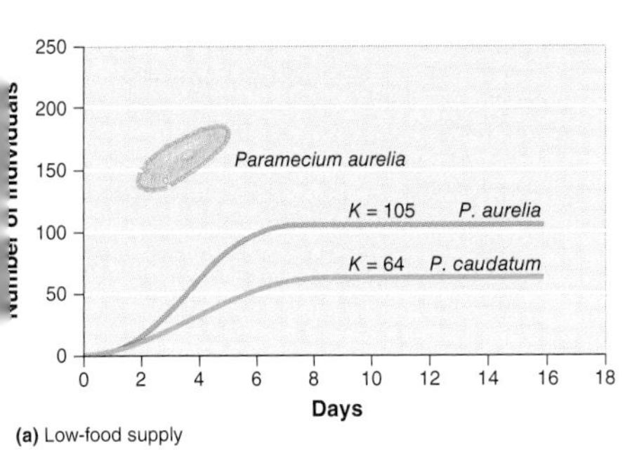

(a) Low-food supply

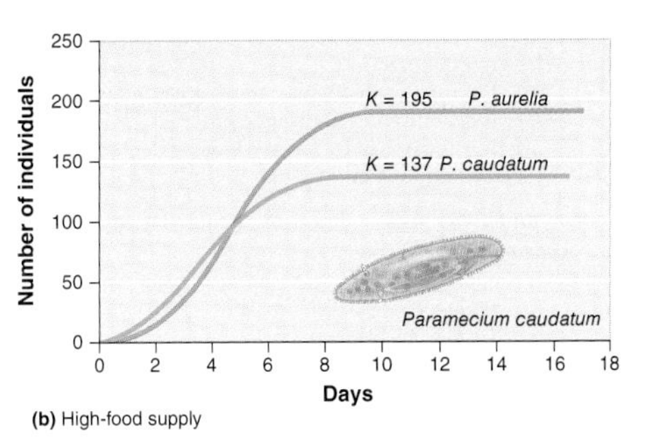

(b) High-food supply

FIGURE 18.3 **Gause's experiments.** (a) Under low-food conditions, the population sizes of two species of *Paramecium* initially increased rapidly, but then leveled off as their food supply became limiting. (b) When twice as much food was provided, both species attained population sizes that were nearly twice as large, but they again leveled off. *(Data from Gause, 1932)*

Bird populations are often regulated by density-independent factors. For example, in the United Kingdom, a particularly cold winter can freeze the surfaces of ponds, making amphibians and fish inaccessible to wading birds such as herons. With their food supply no longer available, herons would have an increased risk of starving to death, regardless of whether the heron population is at a low or a high density.

module

18

REVIEW

In this module, we learned that nature exists at a series of different levels of complexity, which include individuals, populations, communities, and ecosystems. We then examined the level of the population and observed that populations possess a number of characteristics that can be used to describe them, including their abundance and distribution. Finally, we discussed how density-dependent factors can regulate populations more strongly as populations grow whereas density-independent factors can regulate populations at any population size. In the next module, we will see how scientists use mathematical models of populations to obtain insights into how populations change in abundance over time.

Module 18 AP® Review Questions

1. Which is the correct order of ecological levels from basic to complex?
   (a) Individual, population, ecosystem, biosphere, community
   (b) Individual, community, ecosystem, population, biosphere
   (c) Individual, population, community, ecosystem, biosphere
   (d) Ecosystem, biosphere, community, population, individual
   (e) Individual, population, community, biosphere, ecosystem

2. Population distribution is
   (a) often clumped in response to predation.
   (b) used by wildlife managers when regulating hunting and fishing.
   (c) measured relative to other species.
   (d) uniform in most tree species.
   (e) important when estimating the number of offspring expected.

Practicing Science: Using Scientific Method

After students evaluate the data from Gause's experiments in Figure 18.3, ask them to write a conclusion statement to include the effect of different amounts of food given to the paramecium. Tell them to include the independent and dependent variables in their statement and to note whether the availability of food is a density-dependent or density-independent factor. *As food availability increases, numbers of paramecium increase as well. Food availability is an important determinant of survival and reproduction. Availability of food is a density-dependent factor because the amount of food available to an individual depends on the population size.*

Answers to Module 18 AP® Review Questions

1. c
2. a
3. c
4. d
5. a

3. Which is true about a population's carrying capacity?
(a) It is denoted as C.
(b) It is usually used when studying ecosystems.
(c) It depends on a limiting resource.
(d) It is controlled by density-independent factors.
(e) The population of a species cannot exceed it.

4. Which population is typically regulated by density-independent factors?
(a) Algae
(b) Predators
(c) *Paramecium* bacteria
(d) Birds
(e) Trees

5. Which does NOT have a significant effect on the number of offspring produced by a population?
(a) Population distribution
(b) Sex ratio
(c) Age structure
(d) Population size
(e) Carrying capacity

module 19

Population Growth Models

Scientists often use models to help them explain how things work and to predict how things might change in the future. Population ecologists use growth models that incorporate density-dependent and density-independent factors to explain and predict changes in population size. These models are important tools for population ecologists, whether they are protecting an endangered condor population, managing a commercially harvested fish species, or controlling an insect pest. In this module we will look at several growth models and other tools for understanding changes in population size.

Learning Objectives

After reading this module you should be able to

- explain the exponential growth model of populations, which produces a J-shaped curve.
- describe how the logistic growth model incorporates a carrying capacity and produces an S-shaped curve.
- compare the reproductive strategies and survivorship curves of different species.
- explain the dynamics that occur in metapopulations.

more **math** practice:

Graphing

Most students need a thorough review, if not primary instruction, on creating and interpreting graphs. The following simple problem is a good way for students to practice doing calculations and plotting data in a graph. The resulting graph shows exponential growth.

Present the data in the table below without the answers in the "interest gained" column and have the students make the calculation requested in the problem after the table. Then instruct them to build a graph plotting years on the *x* axis and total dollars on the *y* axis. Note that this problem requires the use of a calculator.

Growth of $1,000 investment at 10 percent per year

Year	Investment ($)	Interest gained at 10%($)
0	1,000.00	*100.00*
1	1,100.00	*110.00*
2	1,210.00	*121.00*
3	1,331.00	*133.10*
4	1,464.10	*146.41*
5	1,610.51	*161.05*
6	1,771.56	*177.16*
7	1,948.72	*194.87*
8	2,143.59	*214.36*
9	2,357.95	*235.79*
10	2,593.74	*259.37*
11	2,853.12	*285.31*
12	3,138.43	*313.84*
13	3,452.27	*345.23*
14	3,797.50	*379.75*
15	4,177.25	*417.72*
16	4,594.97	*459.50*
17	5,054.47	*505.45*
18	5,559.92	*555.99*
19	6,115.91	*611.59*
20	6,727.50	*672.75*
21	7,400.25	*740.02*
22	8,140.27	*814.03*
23	8,954.30	*895.43*
24	9,849.73	*984.97*
25	10,834.71	*1,083.47*
26	11,918.18	*1,191.82*
27	13,109.99	*1,311.00*
28	14,420.99	*1,442.10*
29	15,863.09	*1,586.31*
30	17,449.40	*1,744.94*

Problem:
You decide to invest $1,000 in a savings account. Your investment will grow at a rate of 10 percent each year. Assuming that you reinvest the interest each year, how much money will you have in 30 years? *The total investment over the 30-year period would be $17,449.40.*

[Note: Please remind your students that a 10 percent return on investment is very high. Currently most banks will give you a savings account with an annual percentage of 1.5 to 3.0 percent.]

Exponential growth at 10 percent return

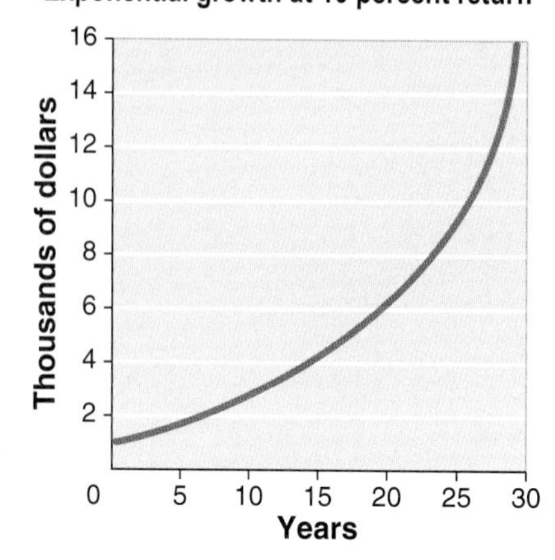

The exponential growth model describes populations that continuously increase

Population growth models are mathematical equations that can be used to predict population size at any moment in time. In this section, we will examine one commonly used growth model.

As we saw in Gause's experiments, a population can initially grow very rapidly when its growth is not limited by scarce resources. We can define **population growth rate** as the number of offspring an individual can produce in a given time period, minus the deaths of the individual or its offspring during that same period. Under ideal conditions, with unlimited resources available, every population has a particular maximum potential for growth, which is called the **intrinsic growth rate** and denoted as ***r***. When food is abundant, individuals have a tremendous ability to reproduce. For example, domesticated hogs (*Sus domestica*) can have litters of 10 piglets, and American bullfrogs (*Rana catesbeiana*) can lay up to 20,000 eggs. Under these ideal conditions, the probability of an individual surviving also increases. Together, a high number of births and a low number of deaths produce a high population growth rate. When conditions are less than ideal due to limited resources, a population's growth rate will be lower than its intrinsic growth rate because individuals will produce fewer offspring (or forgo breeding entirely) and the number of deaths will increase.

If we know the intrinsic growth rate of a population (r) and the number of reproducing individuals that are currently in the population (N_0), we can estimate the population's future size (N_t) after some period of time (t) has passed. The formula that allows us to estimate future population is known as the **exponential growth model**

$$N_t = N_0 e^{rt}$$

where e is the base of the natural logarithms (the e^x key on your calculator, or 2.72) and t is time. This equation tells us that, under ideal conditions, the future size of the population (N_t) depends on the current size of the population (N_0), the intrinsic growth rate of the population (r), and the amount of time (t) over which the population grows.

When populations are not limited by resources, growth can be very rapid because more births occur with each step in time. When graphed, the exponential growth model produces a **J-shaped curve**, as shown in FIGURE 19.1. The J-shape of the curve represents the change in a growing population over time. At first the population is so small that it cannot increase rapidly because there are few individuals present to reproduce. As the population increases, there are more reproducing individuals, and so the growth rate increases.

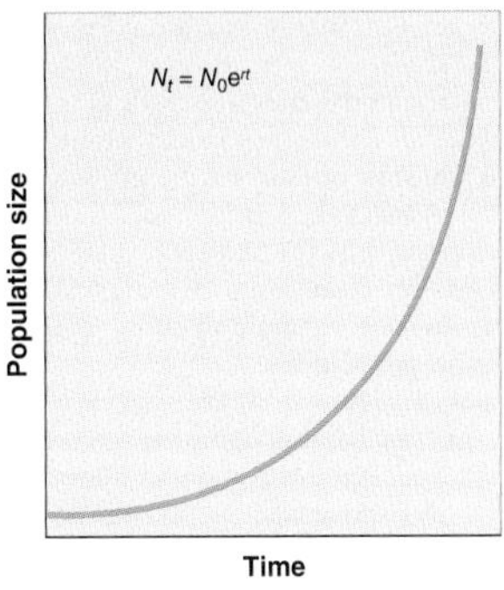

FIGURE 19.1 **The exponential growth model.** When populations are not limited by resources, their growth can be very rapid. More births occur with each step in time, creating a J-shaped growth curve.

One way to think about exponential growth in a population is to compare it to the growth of a bank account where N_0 is the account balance and r is the interest rate. Let's say you put \$1,000 in a bank account at an annual interest rate of 5 percent. After 1 year the new balance in the account would be:

$$N_t = \$1{,}000 \times e^{(0.05 \times 1)}$$

$$N_t = \$1{,}051.27$$

In the second year, the balance in the account would be:

$$N_t = \$1{,}000 \times e^{(0.05 \times 2)}$$

$$N_t = \$1{,}105.17$$

In the tenth year, the account would grow to a balance of \$1,648.72. Moving forward to the twentieth year,

Population growth models Mathematical equations that can be used to predict population size at any moment in time.

Population growth rate The number of offspring an individual can produce in a given time period, minus the deaths of the individual or its offspring during the same period.

Intrinsic growth rate (*r*) The maximum potential for growth of a population under ideal conditions with unlimited resources.

Exponential growth model ($N_t = N_0e^{rt}$) A growth model that estimates a population's future size (N_t) after a period of time (t), based on the intrinsic growth rate (r) and the number of reproducing individuals currently in the population (N_0).

J-shaped curve The curve of the exponential growth model when graphed.

Teaching Tip: Warm-up

Ask students to explain why the exponential growth model is not a realistic model for most populations. *Populations rarely experience ideal conditions in which resources are unlimited. They experience both density-dependent and density-independent factors. Therefore populations do not typically continue to grow indefinitely.*

TEACHING with FIGURES ▶

Ask students: At which points on Figure 19.2 would you expect growth to be the slowest and why? *Growth rate is slowest at the beginning and at the end of the time scale. In the beginning, growth is slow because there are low numbers of reproducing organisms. Growth slows toward the end because resources become scarce.*

Teaching Tip: Activity

Observing Duckweed Population Growth
This duckweed population activity is useful to help students gain a better understanding of carrying capacity. This activity also gives students a chance to practice their graphing skills. After completing this activity, students should produce a graph that looks similar to Figure 19.2.

Download this activity by clicking on the link buttons in the TE-book, opening the TRFD, or logging onto the book's companion website bcs.whfreeman.com /friedlandapes2e (teacher login required).

Activity 6.1 Observing Duckweed Population Growth

the same 5 percent interest rate would produce a balance of $2,718.28.

Applying an annual rate of growth to an increasing amount, whether money in a bank account or a population of organisms, produces rapid growth over time. Exponential growth is density independent because the value will grow by the same percentage every year. "Do the Math: Calculating Exponential Growth" gives a step-by-step example to show how this principle works in populations.

The exponential growth model is an excellent starting point for understanding population growth. Indeed, there is solid evidence that real populations—even small ones—can grow exponentially, at least initially. However, no population can experience exponential growth indefinitely. In Gause's experiments with *Paramecium*, the two populations initially grew exponentially until they approached the carrying capacity of their test-tube environment, at which point their growth slowed and eventually leveled off to reflect the amount of food that was added daily. We now turn to another model that gives a more complete view of population growth.

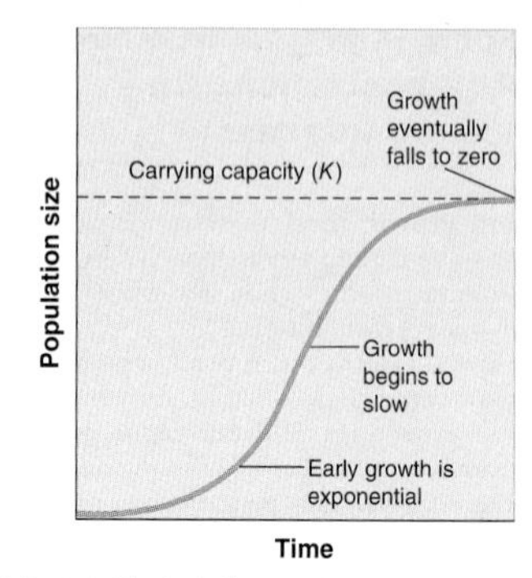

FIGURE 19.2 **The logistic growth model.** A small population initially experiences exponential growth. As the population becomes larger, however, resources become scarcer, and the growth rate slows. When the population size reaches the carrying capacity of the environment, growth stops. As a result, the pattern of population growth follows an S-shaped curve.

The logistic growth model describes populations that experience a carrying capacity

While the exponential growth model describes a continuously increasing population that grows at a fixed rate, populations do not experience exponential growth indefinitely. For this reason, ecologists have modified the exponential growth model to incorporate environmental limits on population growth, including limiting resources. The **logistic growth model** describes a population whose growth is initially exponential, but slows as the population approaches the carrying capacity of the environment (*K*). As we can see in FIGURE 19.2, if a population starts out small, its growth can be very rapid. As the population size nears about one-half of the carrying capacity, however, population growth begins to slow. As the population size approaches the carrying capacity, the population stops growing. When graphed, the logistic growth model produces an **S-shaped curve.** We observed this pattern in Gause's *Paramecium* experiments: At the carrying capacity, the populations stopped growing and remained at a constant size (see Figure 18.3).

The logistic growth model is used to predict the growth of populations that are subject to density-dependent constraints as the population grows, such as increased competition for food, water, or nest sites. Because density-independent factors such as hurricanes and floods are inherently unpredictable, the logistic growth model does not account for them.

One of the assumptions of the logistic growth model is that the number of offspring produced depends on the current population size and the carrying capacity of the environment. However, many species of mammals mate during the fall or winter, and the number of offspring that develop depends on the food supply at the time of mating. Because these offspring are not actually born until the following spring, there is a risk that food availability will not match the new population size. If there is less food available in the spring than needed to feed the offspring, the population will experience an **overshoot** by becoming larger than the environment's carrying capacity. As a result of this overshoot, there will not be enough food for all the individuals in the population, and the population will experience a **die-off,** which is a rapid population decline due to death.

The reindeer (*Rangifer tarandus*) population on St. Paul Island in Alaska is a good example of this pattern. As you can see in FIGURE 19.3, a small population of 25 reindeer that was introduced to the island in 1910 grew exponentially until it reached more than 2,000

Logistic growth model A growth model that describes a population whose growth is initially exponential, but slows as the population approaches the carrying capacity of the environment.

S-shaped curve The shape of the logistic growth model when graphed.

Overshoot When a population becomes larger than the environment's carrying capacity.

Die-off A rapid decline in a population due to death.

Teaching Tip: Warm-up

Ask students: Which factors are accounted for when using a logistic growth model to determine the carrying capacity of a population: density-dependent factors, density-independent factors, or both? *The logistic growth model accounts only for density-dependent factors because density-independent factors are unpredictable and therefore are not included in the model.*

AP® Exam Tip

Distribution and abundance of a population is covered in the 2010 AP® Exam, Question 2. In order to answer this question, students must

- analyze termite microclimate data from a tropical rainforest
- calculate methane emitted by termites in a tropical rainforest
- explain why termite density increases after a tropical rainforest is cleared
- describe one additional way that tropical rainforest destruction contributes to climate change

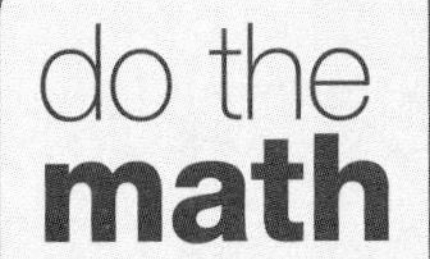

Calculating Exponential Growth

Consider a population of rabbits that has an initial population size of 10 individuals ($N_0 = 10$). Let's assume that the intrinsic rate of growth for a rabbit is $r = 0.5$ (or 50 percent), which means that each rabbit produces a net increase of 0.5 rabbits each year. With this information, we can predict the size of the rabbit population 2 years from now:

$$N_t = N_0 e^{rt}$$
$$N_t = 10 \times e^{0.5 \times 2}$$
$$N_t = 10 \times e^1$$
$$N_t = 10 \times (2.72)^1$$
$$N_t = 10 \times 2.72$$
$$N_t = 27 \text{ rabbits}$$

We can then ask how large the rabbit population will be after 4 years:

$$N_t = 10 \times e^{0.5 \times 4}$$
$$N_t = 10 \times e^2$$
$$N_t = 10 \times (2.72)^2$$
$$N_t = 10 \times 7.4$$
$$N_t = 74 \text{ rabbits}$$

We can also project the size of the rabbit population 10 years from now:

$$N_t = 10 \times e^{0.5 \times 10}$$
$$N_t = 10 \times e^5$$
$$N_t = 10 \times (2.72)^5$$
$$N_t = 10 \times 148.9$$
$$N_t = 1{,}489 \text{ rabbits}$$

Your Turn Now assume that the intrinsic rate of growth is 1.0 for rabbits. Calculate the predicted size of the rabbit population after 1, 5, and 10 years. Create a graph that shows the growth curves for an intrinsic rate of growth at 0.5, as calculated above, and an intrinsic rate of growth at 1.0. (Note that you will need to use your calculator to complete this problem.)

individuals in 1938. After 1938, the population crashed to only 8 animals, most likely because the reindeer ran out of food.

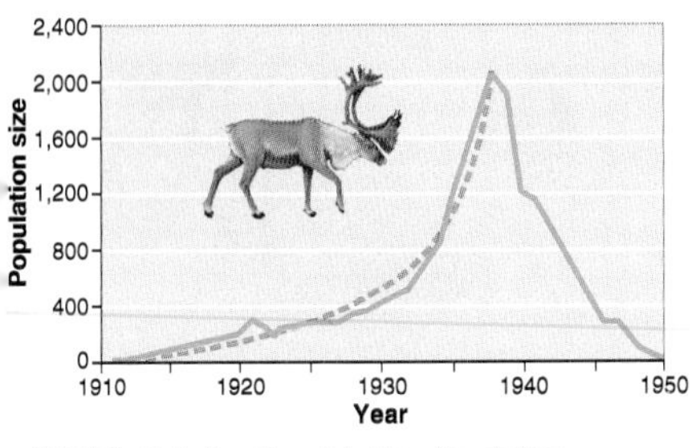

FIGURE 19.3 **Growth and decline of a reindeer population.** Humans introduced 25 reindeer to St. Paul Island, Alaska, in 1910. The population initially experienced rapid growth (blue line) that approximated a J-shaped exponential growth curve (orange line). In 1938, the population crashed, probably because the animals exhausted the food supply. *(Data from V. B. Scheffer, "The rise and fall of a reindeer herd," Scientific Monthly (1951): 356–362.)*

Such die-offs can take a population well below the carrying capacity of the environment. In subsequent cycles of reproduction, the population may grow large again. FIGURE 19.4 illustrates the recurring cycle of overshoots and die-offs that causes populations to oscillate around the carrying capacity. In many cases, these oscillations decline over time and approach the carrying capacity.

So far we have considered only how populations are limited by resources such as food, water, and the

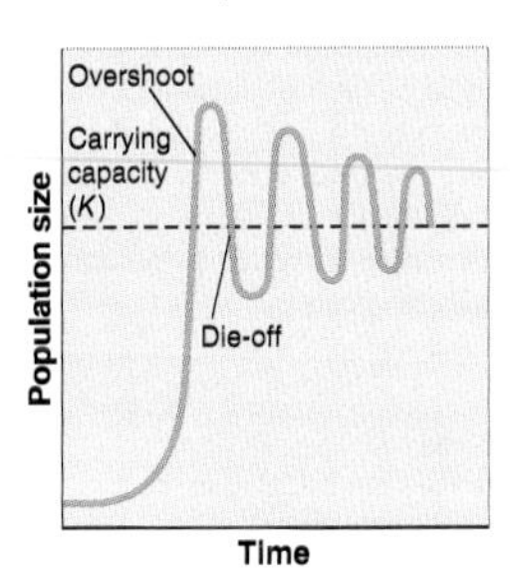

FIGURE 19.4 **Population oscillations.** Some populations experience recurring cycles of overshoots and die-offs that lead to a pattern of oscillations around the carrying capacity of their environment.

▲ TEACHING with FIGURES

Show students Figure 19.3 and ask the following questions:

- What was the approximate population size in 1938? *The population size in 1938 was approximately 2,000 reindeer.*
- What happened to the population after 1938? *After 1938 the population declined to zero in a 10-year time frame, probably because the reindeer exhausted the food supply.*
- What might conservationists do to ensure reindeer will thrive in an area? *Conservationists could manage reindeer density to ensure the food supply is not overeaten so that the population can continue to thrive at a sustainable level.*

Answer to Your Turn

Calculate the predicted sizes using $r = 1.0$ the same way we calculated for $r = 0.5$. Using the exponential growth equation, $N_t = N_0 e^{rt}$, we have $N_0 = 10$ rabbits, $r = 1.0$ year, and $t = 1$, 5, and 10 years.

$t = 1$: $N_t = N_0 e^{rt} = 10 e^{1.0 \times 1} = 10 e^1 = 10 \times 2.718 = 27$ rabbits

$t = 5$: $N_t = N_0 e^{rt} = 10 e^{1.0 \times 5} = 10 e^5 = 10 \times 148.4 = 1{,}484$ rabbits

$t = 10$: $N_t = N_0 e^{rt} = 10 e^{1.0 \times 10} = 10 e^{10} = 10 \times 22{,}026.5 = 220{,}265$ rabbits

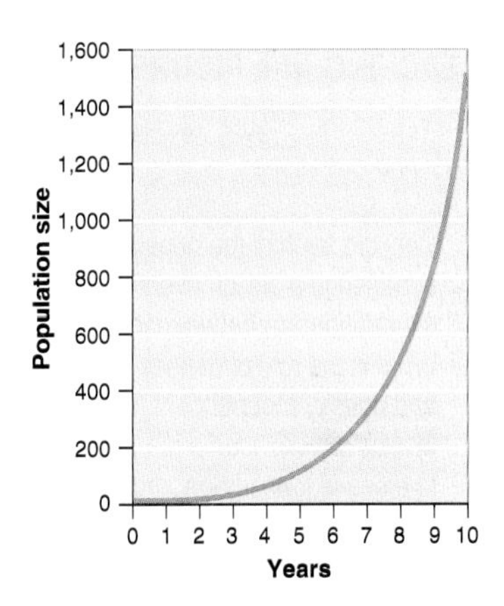

more math practice: Graphing Growth Models

Have students sketch the following on a graph with "Time" on the x axis and "Population size" on the y axis. Tell students to use different colors to represent each trend and key the graph. Instruct students to follow these steps:

- Draw a dashed line that represents carrying capacity.
- Draw a line that represents intrinsic growth rate.
- Draw a line that represents logistic growth.
- Draw a line that shows how a reproductive lag time can lead to a series of overshoot and die-off events.

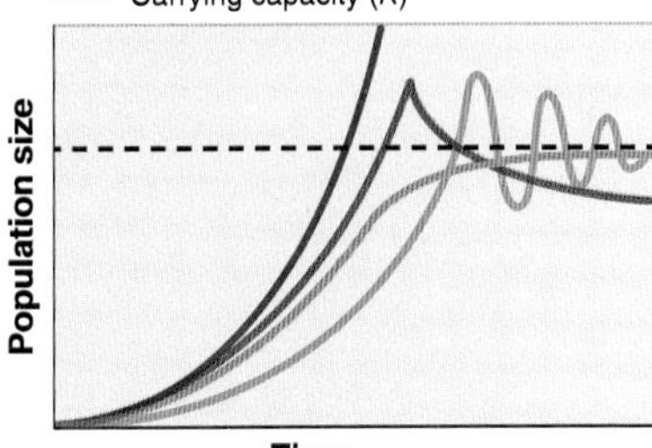

TEACHING with FIGURES ▶

Show students Figure 19.5 and ask them to use the graph to summarize the effect of predator–prey relationships on population sizes. *Prey populations can increase with an increase in available food. As prey populations increase, the food supply increases for the predator. The availability of food allows the predator populations to increase. As predator numbers increase, prey populations decrease. Without the prey as food, predator populations decrease and the cycle starts again.*

TEACHING with FIGURES ▶

Show students Figure 19.6 and ask the following questions:

- Why did each of the populations experience a die-off event and what is the relationship between the two events? What is the role of density-dependent and density-independent factors? *The wolves experienced a die-off event due to disease (density-dependent factor). The moose population soared because predators were absent. As the moose population increased, its food source became scarce (density-dependent) and the population crashed.*
- What is the role of predators in an ecosystem? *Predators maintain balance among all members of the community. If one population dominates the ecosystem, it is likely a resource will be used at unsustainable levels.*

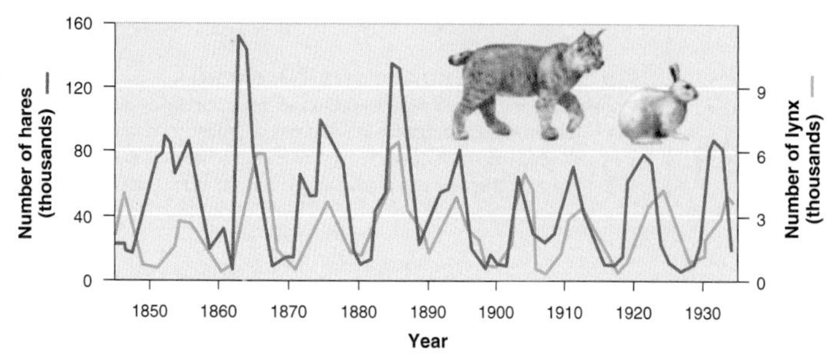

FIGURE 19.5 **Population oscillations in lynx and hares.** Both lynx and hares exhibit repeated oscillations of abundance, with the lynx population peaking 1 to 2 years after the hare population. When hares are not abundant, there is plenty of food, which allows the hare population to increase. As the hare population increases, there are more hares for lynx to eat, so then the lynx population increases. As the hare population becomes very abundant, they start to run out of food and the hare population dies off. As hares become less abundant, the lynx population subsequently dies off. With less predation and more food once again available, the hare population increases again, and the cycle repeats. *(Data from Hudson's Bay Company)*

availability of nest sites. Predation may play an important additional role in limiting population growth. A classic example is the relationship between snowshoe hares (*Lepus americanus*) and lynx (*Lynx canadensis*) that prey on them in North America. Trapping records from the Hudson's Bay Company, which purchased hare and lynx pelts for nearly 90 years in Canada, indicate that the populations of both species cycle over time. FIGURE 19.5 shows how this interaction works. The lynx population peaks 1 or 2 years after the hare population peaks. As the hare population increases, it provides more prey for the lynx, and thus the lynx population begins to grow. As the hare population reaches a peak, food for hares becomes scarce, and the hare population dies off. The decline in hares leads to a subsequent decline in the lynx population. Because the low lynx numbers reduce predation, the hare population increases again.

A similar case of predator control of prey populations has been observed on Isle Royale, Michigan, an island in Lake Superior. Wolves (*Canis lupus*) and moose (*Alces alces*) have coexisted on Isle Royale for several decades. Starting in 1981, the wolf population declined sharply, probably as a result of a deadly canine virus. As wolf predation on moose declined, the moose population grew rapidly until it ran out of food and experienced a large die-off that began in 1995. FIGURE 19.6 shows the changes in both populations over a period of more than 50 years.

K-selected species A species with a low intrinsic growth rate that causes the population to increase slowly until it reaches carrying capacity.

Species have different reproductive strategies and distinct survivorship curves

Population size most commonly increases through reproduction. Population ecologists have identified a range of reproductive strategies in nature.

K-selected Species

***K*-selected species** are species that have a low intrinsic growth rate that causes the population to increase

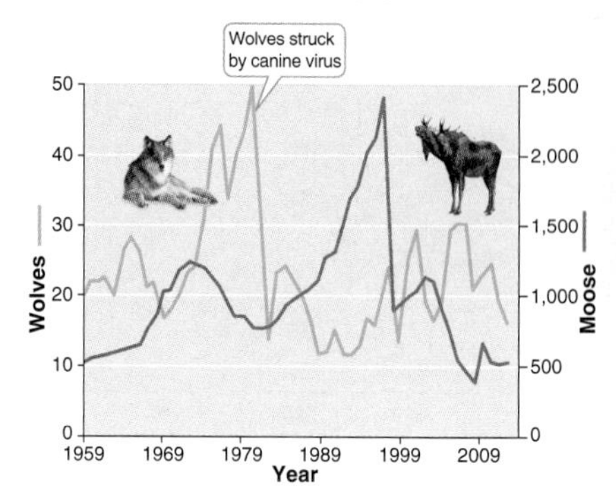

FIGURE 19.6 **Predator control of prey populations.** As the population of wolves on Isle Royale succumbed to a canine virus, their moose prey experienced a dramatic population increase. *(Data from J. A. Vucetich and R. O. Peterson, Ecological Studies of Wolves on Isle Royale: Annual Report 2007–2008, School of Forest Resources and Environmental Science, Michigan Technological University.)*

slowly until it reaches the carrying capacity of the environment. As a result, the abundance of *K*-selected species is determined by the carrying capacity, and their population fluctuations are small. The name "*K*-selected species" refers to the fact that these populations commonly exist close to their carrying capacity, denoted as *K* in population models.

K-selected species have certain traits in common. For instance, *K*-selected animals are typically large organisms that reach reproductive maturity relatively late, produce a few, large offspring, and provide substantial parental care. Elephants, for example, do not become reproductively mature until they are 13 years old, breed only once every 2 to 4 years, and produce only one calf at a time. Large mammals and most birds are *K*-selected species. For environmental scientists interested in biodiversity management or protection, the slow growth of *K*-selected species poses a challenge; in practical terms, it means that an endangered *K*-selected species cannot respond quickly to efforts to save it from extinction.

r-selected Species

At the opposite end of the spectrum from *K*-selected species, ***r*-selected species** have a high intrinsic growth rate that often leads to population overshoots and die-offs. Such populations reproduce often and produce large numbers of offspring. The name *r*-selected species refers to the fact that the intrinsic growth rate is designated as *r* in population models. In contrast to *K*-selected species, populations of *r*-selected species do not typically remain near their carrying capacity, but instead exhibit rapid population growth that is often followed by overshoots and die-offs. Among animals, *r*-selected species tend to be small organisms that reach reproductive maturity relatively early, reproduce frequently, produce many small offspring, and provide little or no parental care. House mice (*Mus musculus*), for example, become reproductively mature at 6 weeks of age, can breed every 5 weeks, and produce up to a dozen offspring at a time. Other *r*-selected organisms include small fishes, many insect species, and weedy plant species. Many organisms that humans consider to be pests, such as cockroaches, dandelions, and rats, are *r*-selected species.

TABLE 19.1 summarizes the traits of *K*-selected and *r*-selected species. These two categories represent opposite ends of a wide spectrum of reproductive strategies. Most species fall somewhere in between these two extremes, and many exhibit combinations of traits from the two extremes. For example, tuna and redwood trees are both long-lived species that take a long time to reach reproductive maturity. Once they do, however, they produce millions of small offspring that receive no parental care.

TABLE 19.1 Traits of *K*-selected and *r*-selected species

Trait	*K*-selected species	*r*-selected species
Life span	Long	Short
Time to reproductive maturity	Long	Short
Number of reproductive events	Few	Many
Number of offspring	Few	Many
Size of offspring	Large	Small
Parental care	Present	Absent
Population growth rate	Slow	Fast
Population regulation	Density dependent	Density independent
Population dynamics	Stable, near carrying capacity	Highly variable

In addition to different reproductive strategies, species have distinct patterns of survival over the life span of individuals. These patterns can be plotted on a graph as **survivorship curves,** as shown in FIGURE 19.7. There are three basic types of survivorship curves. A **type I survivorship curve** has high survival throughout most of the life span, but then individuals start to die in large numbers as they approach old age. Species with type I curves include *K*-selected species such as elephants, whales, and humans. In contrast, a **type II survivorship curve** has a relatively constant decline in survivorship throughout most of the life span. Species with a type II curve include corals and squirrels. A **type III survivorship curve** has low survivorship early in life with few individuals reaching adulthood. Species with type III curves include *r*-selected species such as mosquitoes and dandelions.

***r*-selected species** A species that has a high intrinsic growth rate, which often leads to population overshoots and die-offs.

Survivorship curve A graph that represents the distinct patterns of species survival as a function of age.

Type I survivorship curve A pattern of survival over time in which there is high survival throughout most of the life span, but then individuals start to die in large numbers as they approach old age.

Type II survivorship curve A pattern of survival over time in which there is a relatively constant decline in survivorship throughout most of the life span.

Type III survivorship curve A pattern of survival over time in which there is low survivorship early in life with few individuals reaching adulthood.

Teaching Tip: Activity

Traits of K-selected and r-selected Species
This card-sort activity can be used as a quick assessment to ensure that students understand the different characteristics exhibited by *K*-selected and *r*-selected species.

Download this activity by clicking on the link buttons in the TE-book, opening the TRFD, or logging onto the book's companion website bcs.whfreeman.com/friedlandapes2e (teacher login required).

Activity 6.2 Traits of *K*-selected and *r*-selected Species

TEACHING with FIGURES ▶

Ask students to compare type I and type III species in Figure 19.7. They should be able to conclude that most type I species die in old age while type III organisms die early. Explain that type II species experience constant mortality rate (uniform death rates) at all ages.

Teaching Tip: Discussion Starter

Ask students: How are humans contributing to a greater prevalence of metapopulations? *Humans build roads and neighborhoods, log forests, and drain wetlands. All these practices fragment large areas of land and inadvertently create metapopulations.*

Teaching Tip: Journal Prompt

Ask students: Why is a metapopulation more likely to persist than a small, completely isolated population? Offer at least two reasons. *Individuals can move between subpopulations that make up a metapopulation. This movement introduces new genetic material, which increases diversity and fitness. Because a metapopulation consists of numerous, somewhat isolated populations, disease may be less likely to destroy an entire population; a disease might wipe out one population but other populations can persist. Single isolated populations are more vulnerable to extinction from catastrophic events such as a fire or a particularly harsh winter.*

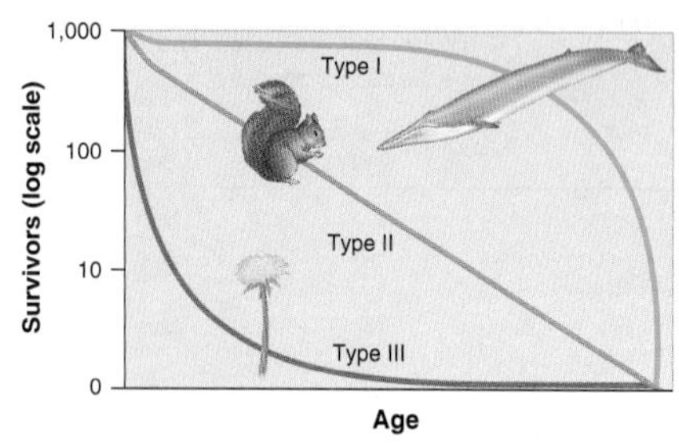

FIGURE 19.7 **Survivorship curves.** Different species have distinct patterns of survivorship over the life span. Species range from exhibiting excellent survivorship until old age (type I curve) to exhibiting a relatively constant decline in survivorship over time (type II curve) to having very low rates of survivorship early in life (type III curve). *K*-selected species tend to exhibit type I curves, whereas *r*-selected species tend to exhibit type III curves.

Interconnected populations form metapopulations

Cougars (*Puma concolor*)—also called mountain lions or pumas—once lived throughout North America but, because of habitat destruction and overhunting, they are now found primarily in the remote mountain ranges of the western United States. In New Mexico, cougar populations are distributed in patches of mountainous habitat scattered across the desert landscape. These mountain habitats allow the cats to avoid human activities and provide them with reliable sources of water and of prey such as mule deer (*Odocoileus hemionus*).

Because areas of desert separate the cougar's mountain habitats, we can consider the cougars of each mountain range to be a distinct population. Each population has its own dynamics based on local abiotic conditions and prey availability: Large mountain ranges support large cougar populations and smaller mountain ranges support smaller cougar populations. As FIGURE 19.8 shows, cougars sometimes move between mountain ranges, often using strips of natural habitat that connect the separated populations. Strips of natural habitat that connect populations are known as **corridors** and provide some connectedness among the populations. A group of spatially distinct populations that are connected by occasional movements of individuals between them is called a **metapopulation.**

The connectedness among the populations within a metapopulation is an important part of each population's overall persistence. Small populations are more likely than large ones to go extinct. As we have seen, small populations contain relatively little genetic variation and therefore may not be able to adapt to changing environmental conditions. Small populations can also experience *inbreeding depression*. **Inbreeding depression** occurs when individuals with similar genotypes— typically relatives—breed with each other and produce offspring that have an impaired ability to survive and reproduce. This impaired ability occurs when each parent carries one copy of a harmful mutation in his or her genome. When the parents breed, some of their offspring receive two copies of the harmful mutation and, as a result, have poor chances of survival and successful reproduction.

Small populations are also more vulnerable than large populations to catastrophes such as particularly harsh winters that drive down their populations to critically low numbers. In a metapopulation, occasional immigrants from larger nearby populations can add to the size of a small population and introduce new genetic diversity, both of which help reduce the risk of extinction.

Metapopulations can also provide a species with some protection against threats such as diseases. A disease could cause a population living in a single large habitat patch to go extinct. But if a population living in an isolated habitat patch is part of a much larger metapopulation, then, while a disease could wipe out that isolated population, immigrants from other populations could later recolonize the patch and help the species to persist.

Because many habitats are naturally patchy across the landscape, many species are part of metapopulations. For instance, numerous species of butterflies specialize on plants with patchy distributions. Some amphibians live in isolated wetlands, but occasionally disperse to other wetlands. The number of species that exist as metapopulations is growing because human activities have fragmented habitats, dividing single large populations into several smaller populations. Identifying and managing metapopulations is thus an increasingly important part of protecting biodiversity.

Corridor Strips of natural habitat that connect populations.

Metapopulation A group of spatially distinct populations that are connected by occasional movements of individuals between them.

Inbreeding depression When individuals with similar genotypes—typically relatives—breed with each other and produce offspring that have an impaired ability to survive and reproduce.

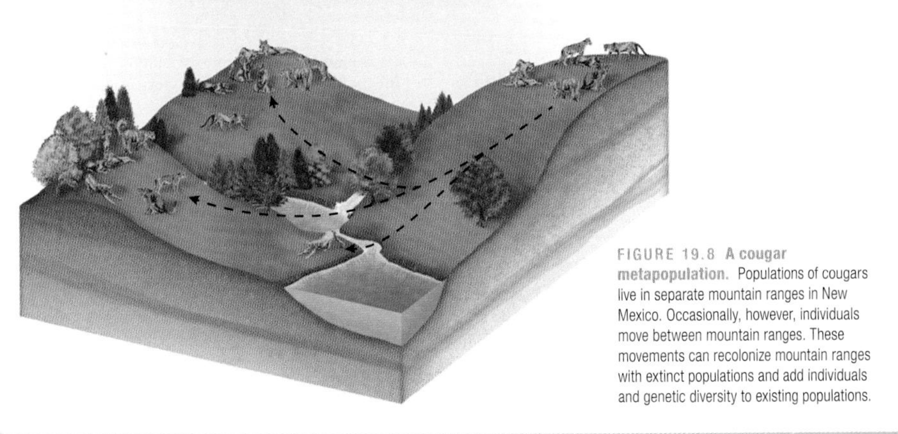

FIGURE 19.8 A cougar metapopulation. Populations of cougars live in separate mountain ranges in New Mexico. Occasionally, however, individuals move between mountain ranges. These movements can recolonize mountain ranges with extinct populations and add individuals and genetic diversity to existing populations.

module

19

REVIEW

In this module, we examined population growth models, which help us understand population increases and decreases. Exponential growth models are the simplest because they assume unlimited resources. Logistic growth models are more realistic because they incorporate a carrying capacity and the associated density-dependent factors that occur in natural populations. By varying assumptions of growth models, we see that a population can overshoot its carrying capacity and experience a die-off. Populations can also oscillate due to predator-prey interactions. Populations can be characterized as either *K*-selected or *r*-selected and have unique types of survivorship curves. Finally, we have seen that a species can exist as a metapopulation composed of multiple, interconnected populations. In the next module, we will move from the population level to the community level and examine how species interactions help to determine which species can persist in natural communities.

Module 19 AP® Review Questions

1. The intrinsic growth rate of a population
   (a) occurs at the population's carrying capacity.
   (b) depends on the limiting resources of the population.
   (c) increases as the population size increases.
   (d) only occurs under ideal conditions.
   (e) decreases as the population size increases.

2. Population growth using the exponential growth model
   (a) increases at a constant rate.
   (b) applies to most populations only after a long time.
   (c) has an increasing intrinsic growth rate.
   (d) represents ideal conditions that rarely occur in natural populations.
   (e) incorporates the carrying capacity of the population.

3. An *r*-selected species characteristically has
   (a) a type I survivorship curve.
   (b) few offspring.
   (c) a population near carrying capacity.
   (d) significant parental care.
   (e) a fast population growth rate.

4. Which is true of a population overshoot?
   (a) It occurs when reproduction quickly responds to changes in food supply.
   (b) It is followed by a die-off.
   (c) It is most likely to be experienced by *K*-selected species.
   (d) It rarely occurs in species with type III survivorship curves.
   (e) It occurs when a species stops growing after reaching the carrying capacity.

Answers to Module 19 AP® Review Questions

1. d
2. d
3. e
4. b
5. a

AP® Exam Tip

Many free-response questions require students to "design an experiment." Most students don't know where to begin this process. Breaking down how points might be awarded makes the questions seem less overwhelming. Tell students that the hypothesis is always worth a point. Remind them that the independent and dependent variables must be included in a hypothesis statement. For example, if two different species of yeast are grown together (independent variable), the growth rate (dependent variable) of both species will decline. Ask students to identify an experimental group and a control group in their experiment. Ask them to define a reasonable time frame over which the experiment should take place. Students are allowed to make up reasonable data that could be collected from the experiment. And finally, ask students to state that experiments must be repeated to be considered valid. If students can remember to incorporate these ideas into an experimental design, they can always earn points on this type of question.

5. Inbreeding depression
   (a) rarely occurs in highly connected metapopulations.
   (b) results from the creation of corridors between populations.
   (c) results in increased rates of reproduction.
   (d) occurs in species that are experiencing overshoot.
   (e) rarely occurs in *K*-selected species.

module 20

Community Ecology

We have explored the factors that determine population size, as well as ways to model or predict how populations will grow or decline. However, the size of a population tells us nothing about what determines the distribution of populations across the planet.

There are two factors that determine whether a species will persist in a location. As we learned in Chapter 5, if a species is able to disperse to an area, its ability to persist there depends on the fundamental niche of that species, which is the range of abiotic conditions that it can tolerate. The survival of a species in a habitat is also determined by the set of interactions that it has with other species. In this module we will explore these relationships.

Learning Objectives

After reading this module you should be able to

- identify species interactions that cause negative effects on one or both species.
- discuss species interactions that cause neutral or positive effects on both species.
- explain the role of keystone species.

Some species interactions cause negative effects on one or both of the species

The science of **community ecology** is the study of interactions among species. Some interactions have negative effects on one or both species. Other interactions involve neutral or positive effects on the two species. As we will see, throughout the world species have **symbiotic relationships,** which means that the two species are living in close association with each other.

Some of the best-known species interactions are those that cause one or both species to be negatively affected by the other species. These interactions include *competition, predation, parasitism,* and *herbivory*.

Competition

In 1934, two years after his experiments on logistic growth in populations of *Paramecium*, Georgii Gause studied how different *Paramecium* species affected each other's population growth. FIGURE 20.1 shows the results of his experiments. When the two species—*P. caudatum* and *P. aurelia*—were grown in separate laboratory cultures, each species thrived and reached a relatively high population size within 10 days. However, when the two species were grown together, *P. aurelia* continued to thrive but *P. caudatum* declined to extinction. **Competition,** the struggle of individuals to obtain a shared limiting resource, caused *P. aurelia* to do better. Gause's observations, combined with additional experiments with other organisms, led researchers to formulate the **competitive exclusion principle,** which states that two species competing for the same limiting resource cannot coexist. Under a given set of environmental conditions, when two species have the same realized niche, one species will perform better and will drive the other species to extinction.

We see competition at work throughout nature. For example, many plant species influence the distribution and abundance of other plant species. Goldenrods are dominant in the old fields of New England because of their superior competitive ability: They can grow taller than other wildflowers and obtain more of the available sunlight. Similarly, the wild oat plant (*Avena fatua*) can outcompete crop plants on the Great Plains of North America because its seeds ripen earlier, permitting oat seedlings to start growing before the other species.

Competition for a limiting resource can lead to **resource partitioning,** in which two species divide a resource based on differences in their behavior or morphology. In evolutionary terms, when competition reduces the ability of individuals to survive and reproduce, natural selection will favor individuals that overlap less with other species in the resources they use. FIGURE 20.2 shows how this process works. Let's imagine two species of birds that eat seeds of different sizes. In species 1, represented by blue, some individuals eat small seeds and others eat medium seeds. In species 2, represented by yellow, some individuals eat medium seeds and others eat large seeds. As a result, some

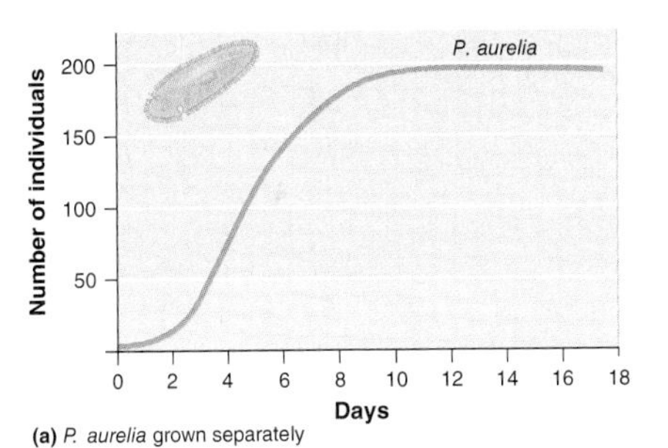

(a) *P. aurelia* grown separately

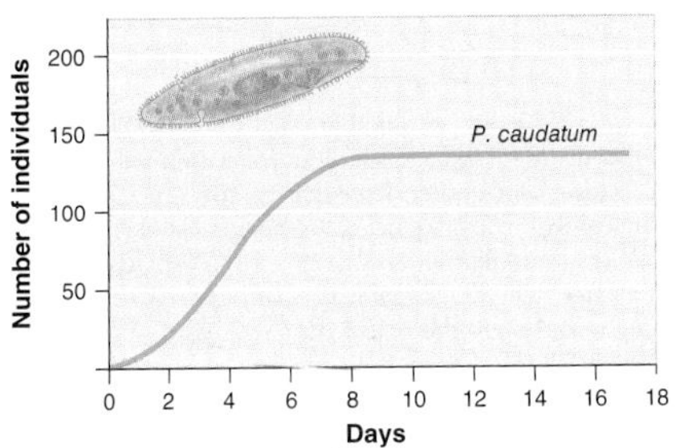

(b) *P. caudatum* grown separately

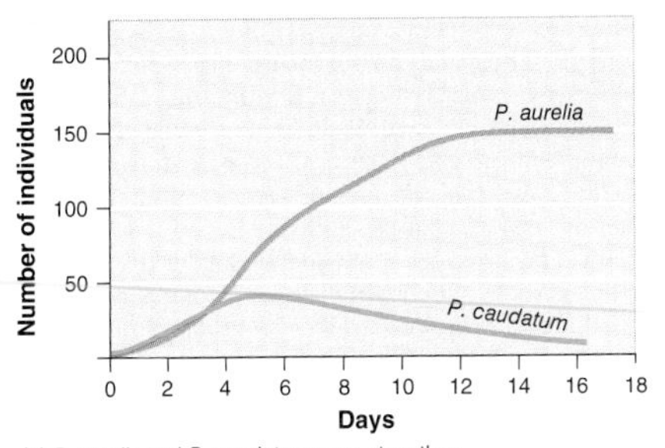

(c) *P. aurelia* and *P. caudatum* grown together

FIGURE 20.1 **Competition for a limiting resource.** When Gause grew two species of Paramecium separately, both achieved large population sizes. However, when the two species were grown together, *P. aurelia* continued to grow well, while *P. caudatum* declined to extinction. These experiments demonstrated that two species competing for the same limiting resource cannot coexist. *(Data from Gause 1934)*

Community ecology The study of interactions between species.

Symbiotic relationship The relationship between two species that live in close association with each other.

Competition The struggle of individuals to obtain a shared limiting resource.

Competitive exclusion principle The principle stating that two species competing for the same limiting resource cannot coexist.

Resource partitioning When two species divide a resource based on differences in their behavior or morphology.

Practicing Science: Analyze and Predict

Prior to discussing Gause's experiments, or showing them Figure 20.1, have students graph the following data:

	Grown alone		Grown together	
Time (days)	*P. aurelia*	*P. caudatum*	*P. aurelia*	*P. caudatum*
0	2	2	2	2
1	2	5	3	4
2	3	18	4	19
3	31	24	27	17
4	92	27	88	11
5	174	42	169	14
6	210	67	207	6
7	324	79	291	7
8	357	82	326	3
9	381	87	352	0
10	395	92	351	0

Students should analyze their graph, determine the carrying capacity for each set of data, and create a concluding statement describing the interactions of these two populations. *When each species is grown separately, each population shows exponential growth initially and then levels off. When both populations are grown together, P. aurelia once again experiences rapid population growth whereas P. caudatum experiences a lack of food due to competition and after a small initial increase the population ultimately declines to zero. Food is a limiting resource for these populations and it can be categorized as a density-dependent factor because the effect of food on the population changes as numbers of organisms increase.*

▲ TEACHING with FIGURES

Ask students to draw a conclusion about limiting resources and competition based on the data in Figure 20.1. *When greater numbers of organisms are present, competition for resources is greater. One species will be at a disadvantage and die off over time in the presence of the more dominant species. Two species do not typically share the same niche.*

Teaching Tip: Journal Prompt

Ask students to compare and contrast the competitive exclusion principle and resource partitioning. *According to the competitive exclusion principle, one species will outcompete another and the less fit species will be pushed to extinction. Resource partitioning occurs when species evolve to share resources so that competition is reduced and both species can thrive.*

TEACHING with FIGURES ▶

Ask students to compare the two graphs in Figure 20.2. Explain the difference between parts (a) and (b). Why does natural selection favor individuals who eat larger seeds? Predict what might happen if Species 1 evolved to prefer larger seeds. *Part (a) shows that Species 1 eats small- and medium-sized seeds. Species 2 eats medium- and large-sized seeds. However, Species 1 is a stronger competitor and will outcompete Species 2 for medium-sized seeds. Members of Species 2 that eat larger seeds do better because they do not have to compete with Species 1. The members of Species 2 that eat large seeds will gain more energy and therefore reproduce more than any members of Species 2 that hunt for medium-size seeds. Assuming that this behavior is inherited, this will result in more individuals eating large-size seeds in the next generation of Species 2.*

Part (b) shows that over many generations, individuals that eat large seeds are successful in reproduction and the population has shifted its niche. Resource partitioning has occurred and competition between the two species in reduced. If Species 1 began to eat larger seeds, it would again compete with Species 2 and Species 2 might die off because it is not as strong a competitor.

Teaching Tip: Activity

Divide students into three groups. Have each group describe one of the three categories of partitioning listed below. Ask each group to share their findings with the rest of the class. The description should include potential benefits for the populations involved. In all three examples, the benefit should be the same, and students should begin to understand that by occupying different niches and reducing competition, greater numbers of species thrive within the community.

- **Temporal Resource Partitioning:** *Different species use a resource at different times to alleviate competition and therefore coexist. An example would be a herd of elephants that use the same watering hole as a herd of okapi. Using the watering hole at different times of the day allows each animal the space needed to obtain water. Therefore, both populations can thrive.*
- **Spatial Resource Partitioning:** *Different species use the same resource by using different parts or areas. An example would be different species of fish feeding at different depths of the lake. By sharing the same resource but eliminating the competition, all species can thrive.*
- **Morphological Resource Partitioning:** *This occurs when two competing species evolve different morphologies that enable them to use the same resource in different ways. An example would be two different species of weasels. Differences in skull and tooth size allow each species to specialize on different sizes of prey. By reducing competition for food sources, both species can thrive.*

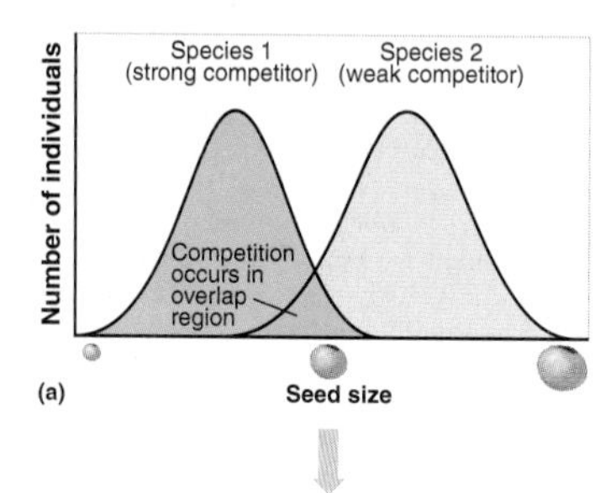

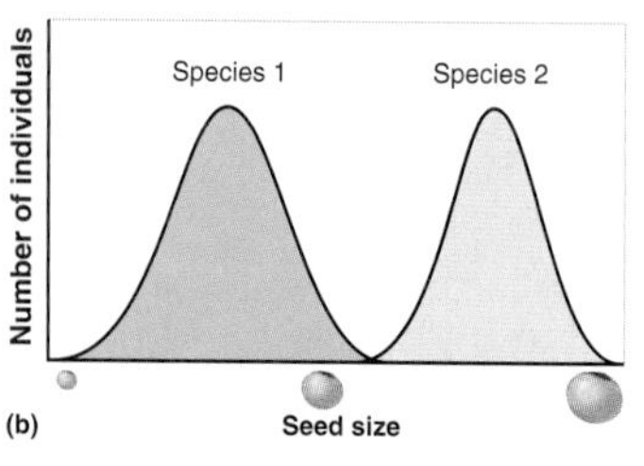

FIGURE 20.2 **The evolution of resource partitioning.** (a) When two species overlap in their use of a limiting resource, selection favors those individuals of each species whose use of the resource overlaps the least with that of the other species. (b) Over many generations, the two species can evolve to reduce their overlap and thereby partition their use of the limiting resource.

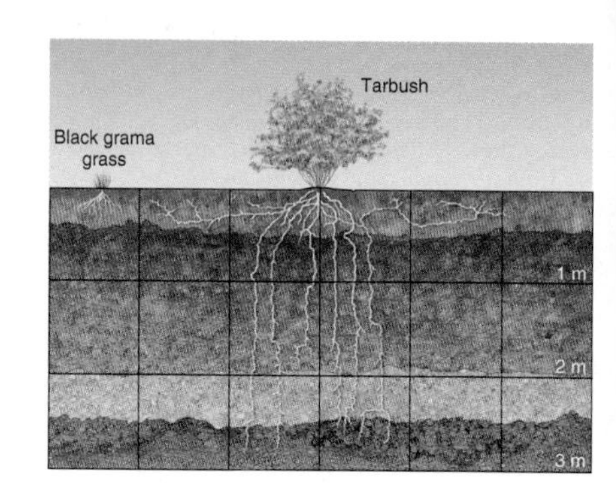

FIGURE 20.3 **Spatial resource partitioning.** Species that face competition for a shared limiting resource may evolve to partition the resource and reduce competition. Among desert plants, tarbush grows very deep roots to reach deep sources of water, whereas grama grass has an extensive shallow root system that allows it to take up water rapidly during rare rain events.

individuals of both species compete for medium seeds. This overlap is represented by green. If species 1 is the better competitor for medium seeds, then individuals of species 2 that compete for medium seeds will have poor survival and reproduction. After several generations, species 2 will evolve to contain fewer individuals that feed on medium seeds. This process of resource partitioning reduces competition between the two species.

Species in nature reduce resource overlap in several ways. One strategy is temporal resource partitioning, a process in which two species utilize the same resource but at different times. For example, wolves and coyotes that live in the same territory are active at different times of day to reduce the overlap in the times that they are hunting. Similarly, some plants reduce competition for pollinators by flowering at different times of the year.

If two species reduce competition by using different habitats, they are exhibiting spatial resource partitioning. As shown in FIGURE 20.3, desert plant species have evolved a variety of different root systems that reduce competition for water and soil nutrients; some have very deep roots while others have shallow roots. For example, we see that black grama grass (*Bouteloua eriopoda*) has shallow roots that extend over a large area to capture rainwater, whereas tarbush (*Flourensia cernua*) sends roots deep into the ground to tap deep sources of water.

A third method of reducing resource overlap is by morphological resource partitioning, which is the evolution of differences in body size or shape. As we saw in Chapter 5, Charles Darwin observed morphological resource partitioning among the finches of the Galápagos Islands (see Figure 16.2 on page 166). The 14 species of finches on the islands descended from a single finch species that colonized the islands from mainland South America. This ancestral finch species would have consumed seeds on the ground. Over time, however, the finches have evolved into 14 species with uniquely shaped beaks that allow each species to eat different foods. The beaks of some species are well suited for crushing seeds, whereas the beaks of others are better suited for catching insects. This morphological resource partitioning reduces competition among the finch species.

Predation

One of the most dramatic species interactions occurs between predators and their prey. **Predation** is an interaction in which one animal typically kills and consumes another animal. Predators include African lions (*Panthera leo*) that eat gazelles and great horned owls (*Bubo virginianus*) that eat small rodents. Organisms

Predation An interaction in which one animal typically kills and consumes another animal.

(a)

(b)

(c)

(d)

FIGURE 20.4 **Prey defenses.** Predation has favored the evolution of fascinating antipredator defenses. (a) The camouflage of this stone flounder (*Kareius bicoloratus*) makes it difficult for predators to see it. (b) Sharp spines protect this porcupine (*Erethizon dorsatum*) from predators. (c) The poison dart frog (*Epipedobates bilinguis*) has a toxic skin. (d) This nontoxic frog (Allobates zaparo) mimics the appearance of the poison dart frog. *(a: Stephen Frink Collection/Alamy; b: B. von Hoffman/ClassicStock/The Image Works; c: Morley Read/Alamy; d: Luis Louro/Alamy)*

of all sizes may be predators, and their effects on their prey vary widely.

One specialized type of predator is the group known as **parasitoids,** which are animals that lay eggs inside other organisms—referred to as their host. When the eggs hatch, the parasitoid larvae slowly consume the host from the inside out, eventually leading to the host's death. Parasitoids include certain species of wasps and flies.

As we learned from the lynx-hare interactions in Canada and the wolf-moose interactions on Isle Royale, predators can play an important role in controlling the abundance of prey. In Sweden, red foxes (*Vulpes vulpes*) prey on several species of hares and grouse. In the 1970s and 1980s, this fox population was reduced by a disease called mange. With fewer foxes around, the foxes' prey increased in abundance: Populations of grouse doubled in size, and populations of hares increased sixfold. The predators in all of these examples play a critical role in their ecosystems: that of regulating prey populations.

To avoid being eaten or harmed by predators, many prey species have evolved defenses. These defenses may be behavioral, morphological, or chemical, or may simply mimic another species' defense. FIGURE 20.4 shows several examples of antipredator defenses. Animal prey commonly use behavioral defenses, such as hiding and reduced movement, so as to attract less attention from predators. Other prey species have evolved impressive morphological defenses, including camouflage to help them hide from predators and spines to help deter predators. For example, flounders are a group of fish that lay on the ocean floor with a body color that blends in with their surroundings. Many plants, for example, have evolved spines that deter herbivores from grazing on their leaves and fruits. Similarly, many animals, such as porcupines, stingrays, and puffer fish, have spines that deter predator attack.

Chemical defenses are another common mechanism of protection from predators. Several species of insects, frogs, and plants emit chemicals that are toxic or distasteful to their predators. For example, the poison dart frog shown in Figure 20.4 produces a toxin on its skin that is toxic to many species of predators. Many toxic prey are also brightly colored, and predators learn to recognize and avoid consuming them. In some cases, prey have not evolved toxic defenses of their own, but instead have evolved to mimic the physical characteristics of other prey species that do possess chemical defenses. By mimicking toxic species, these nontoxic species can fool predators into not attacking them.

Parasitism

Parasitism is an interaction in which one organism lives on or in another organism—referred to as the host. Because parasites typically consume only a small fraction of their host, a single parasite rarely causes the death of its host. Parasites include tapeworms that live in the intestines of animals as well as the protists that live in the bloodstream of animals and cause malaria.

Parasites that cause disease in their host are called **pathogens.** Pathogens include viruses, bacteria, fungi,

Parasitoid A specialized type of predator that lays eggs inside other organisms—referred to as its host.

Parasitism An interaction in which one organism lives on or in another organism.

Pathogen A parasite that causes disease in its host.

Teaching Tip: Discussion Starter

Ask students to evaluate the images shown in Figure 20.4. Discuss the various types of defense mechanisms listed below and name other examples of species that exhibit these defense mechanisms.

- **Chemical:** *Several species of plants, frogs, and insects emit chemicals that are toxic or distasteful to their predators. An example would be the walkingstick insect that can spray chemical toxins on approaching predators.*
- **Morphological**: *Organisms have developed physical features such as spines or camouflage to protect themselves from predators. An example would be the leopard's spotted coat that helps it blend into the African landscape.*
- **Behavioral**: *Animals have developed behaviors to attract less attention, such as reduced movement and hiding. An example would be a jackrabbit remaining motionless as a predator passes by. The rabbit's lack of movement increases the chances it will remain camouflaged in the grasses.*
- **Mimicry:** *Species that do not have their own chemical defenses have evolved to mimic other prey species that do have their own defenses. An example would be a nontoxic frog that has the same appearance as a poison dart frog. The frog's predators will be fooled by its appearance and avoid attacking it.*

Teaching Tip: Journal Prompt

Ask students to explain what might happen to forest vegetation if predators of herbivores are eradicated. State the importance of predators in an ecosystem. *Vegetation will be consumed and herbivores will exhaust their food source. The herbivore population will decrease. Predators in a population balance numbers of organisms within an ecosystem and create balance.*

Teaching Tip: Activity

Have students create a table to compare and contrast predation and parasitism. (Answers are provided in italics.)

	Definition	Example	Similarity
Predator	*Predators kill their prey and consume most of what they kill.*	*African lions eat gazelles.*	*In both relationships, one organism benefits while the other is harmed. One organism uses the other as its source of energy.*
Parasite	*Parasites live in or on the organism but only consume a fraction of their host. Parasites rarely kill their hosts.*	*Tapeworms live in the intestines of animals.*	

FIGURE 20.5 **Herbivory.** When fences are erected to exclude herbivory by deer, plant growth typically increases dramatically, especially for species that deer prefer to eat. *(Jean-Louis Martin, RGIS/ CEFE-CNRS)*

protists, and wormlike organisms called helminths. These organisms cause many of the most well-known diseases in the world, ranging from the common cold to some forms of cancer.

Herbivory

Herbivory is an interaction in which an animal consumes a producer. They typically eat only a portion of a producer without killing it. The gazelles of the African plains are well-known herbivores, as are the various species of deer, rabbits, goats, and sheep. In aquatic systems, herbivores include sea urchins in the ocean and tadpoles in ponds that consume various species of algae. When herbivores become abundant, they can have dramatic effects on producers. One way to assess the effect of herbivores is by excluding herbivores from an area and seeing how the producers respond. For example, when deer are excluded from an area by putting up a high fence, scientists commonly observe an extraordinary increase in plants, especially in those species of plants that deer prefer to consume (FIGURE 20.5). As in the case of predators and prey, many species of producers have evolved defenses against herbivores that include sharp spines and distasteful chemicals.

Herbivory An interaction in which an animal consumes a producer.

Mutualism An interaction between two species that increases the chances of survival or reproduction for both species.

Other species interactions cause neutral or positive effects on one or both species

Some species interactions can be quite beneficial to the participants. In this section, we will example two such positive interactions: *mutualisms* and *commensalisms*.

Mutualisms

One of the most ecologically important interactions is the relationship between plants and their pollinators, which include birds, bats, and insects. The plants depend on the pollinators for their reproduction, and the pollinators depend on the plants for food. In some cases, one pollinator species might visit many species of plants, while in others, a range of pollinator species may visit many plant species. In still other cases, one animal species pollinates only one plant species, and that plant is pollinated only by that animal species. For example, there are about 900 species of fig trees, and almost every one is pollinated by a particular species of fig wasp.

When two interacting species benefit each other by increasing both species' chances of survival or reproduction, we call it a **mutualism.** Each species in a mutualistic interaction is ultimately assisting the other species in order to benefit itself. If the benefit is too small, the interaction will no longer be worth the cost of helping the other species. Under such conditions, natural selection will favor individuals that no longer engage in the mutualistic interaction.

One well-studied example of plant-animal mutualism is the interaction between acacia trees and several species of *Pseudomyrmex* ants in Central America. The acacia tree supplies the ants with food and shelter, and the ants protect the tree from herbivores and competitors. The ants eat nectar produced by the tree in special structures called nectaries, and they live in the tree's large thorns, which have soft cores that the ants can easily hollow out (FIGURE 20.6). The ants protect the tree by attacking anything that falls on it—for example, by stinging intruding insects that might eat the acacia's leaves and destroying vines that might shade the tree's branches. Through natural selection over generations of close association, the ants and the trees have both evolved traits that make this mutualism work. The ants have evolved several behavioral adaptations: living in the hollowed-out thorns of the tree, extensive patrolling of branches and leaves, and attacking all foreign animals and plants regardless of whether they are useful to the ants as food. For its part, the acacia tree has evolved specialized thorns, year-round leaf production, and modified leaves and nectaries that provide food for the ants.

Two other well-known mutualistic relationships are those found in coral reefs and in lichens. As we saw in Chapter 4, algae live within the tiny coral animals that

Teaching Tip: Warm-up

Ask students to give an example of a mutualistic relationship. *The acacia tree provides ants with food and shelter and ants provide the tree with protection from herbivores and other competitors.*

FIGURE 20.6 **Mutualism.** Acacia trees and Pseudomyrmex ants have each evolved adaptations that enhance their mutualistic interaction. The trees' thorns serve as nest sites for the ants (inset) and provide them with food in the form of nectar produced in specialized nectaries (arrow). In return, the ants protect the trees from herbivores and competing plants. *(Oxford Scientific/Getty Images; inset: Alex Wild Photography)*

Teaching Tip: Warm-up

Ask students to describe the composition of lichens. What type of symbiotic relationship do lichens exhibit? *Lichens are composed of fungus and alga. This mutualistic relationship provides carbohydrates to the fungus from the alga and in return the alga receives nutrients from the fungus.*

build the coral reefs. Their relationship is critical: If the algae die, the coral will die as well. Many of the lichens that grow on the surfaces of rocks and trees are made up of an alga and a fungus (and often bacteria) living in close association. The fungus provides many of the nutrients the alga needs, and the alga provides carbohydrates for the fungus via photosynthesis (FIGURE 20.7).

Commensalisms

Sometimes species can interact in a way that benefits one species but has no effect on the other species. For example, tree branches can serve as perch locations and nest sites for birds. The birds benefit from the presence of the tree because it provides them perches to search for food and a place to raise their offspring. However, the survival and reproduction of the tree is not affected by the presence of the bird. Species interactions in which one species benefits but the other is neither harmed nor helped are called **commensalisms.** Commensalisms are common in nature. For example, many species of fish use coral reefs as hiding places to avoid predators. The fish receive a major benefit from the presence of the coral, but the species of coral that construct the reefs are typically neither harmed nor benefited by the presence of the fish.

FIGURE 20.7 **Lichens.** Lichens, such as this species named "British soldiers" (Cladonia cristatella), are composed of a fungus and an alga that are tightly linked together in a mutualistic interaction. The fungus provides nutrients to the alga, and the alga provides carbohydrates for the fungus via photosynthesis. *(Ed Reschke/Peter Arnold Inc./Getty Images)*

Interactions among species are important in determining which species can live in a community. TABLE 20.1 summarizes these interactions and the effects they have on each of the interacting species, whether positive (+), negative (–), or neutral (0). Competition for a limiting resource has a negative effect on both of the competing species. In contrast, predation, parasitism, and herbivory each has a positive effect on the consumer, but a negative effect on the organism being consumed. Mutualism has positive effects on both interacting species. Commensalism has a positive effect on one species and no effect on the other species.

TABLE 20.1 Interactions between species and their effects

Type of interaction	Species 1	Species 2
Competition	–	–
Predation	+	–
Parasitism	+	–
Herbivory	+	–
Mutualism	+	+
Commensalism	+	0

Commensalism A relationship between species in which one species benefits and the other species is neither harmed nor helped.

Teaching Tip: Journal Prompt

Have students create and complete a table like the one below to compare interactions between species. (Answers are provided in italics.)

Interaction	Effect species 1	Effect species 2	Definition	Example
Competition	*Negative*	*Negative*	*The struggle between two individuals to obtain a limiting resource*	*Two plants competing for sunlight*
Predation	*Positive*	*Negative*	*An interaction in which one animal typically kills and consumes another animal*	*Great horned owls eat small rodents*
Parasitism	*Positive*	*Negative*	*An interaction in which one organism lives in or on another organism*	*A tapeworm lives in the intestine of an animal*
Mutualism	*Positive*	*Positive*	*The interaction of two species that increases the chances of survival and reproduction for both species*	*The relationship between plants and their pollinators such as hummingbirds and a honeysuckle vine*
Commensalism	*Positive*	*Neutral*	*One species benefits and the other species is neither harmed nor helped*	*Fish using a coral reef to hide from predators*

Teaching Tip: Warm-up

Ask students to describe the various ways in which species interact with each other. *Species that live in close association with each other have a symbiotic relationship. Examples of symbiotic relationships include mutualism, commensalism, and parasitism. Organisms can also compete with one another for resources. Predation occurs when one animal kills and consumes another animal.*

Teaching Tip: Activity

Using the food web of Yellowstone National Park shown below, create cards with pictures and a description of which organism consumes another. Pass out the cards so each student has one. Follow the steps listed below.

- Ask students to organize themselves into a food web using yarn to make connections between organisms.
- To model the effect of a keystone species, start by asking those who are prey of the gray wolf to sit. Then ask: What will happen if the gray wolf is removed from the ecosystem? *For the herbivores, a limiting factor to population growth will be removed and they will grow above carrying capacity. The potential is high for their food source to be exhausted.*
- Next, ask the producers to sit because they have been overeaten by the abundant herbivores. The primary consumers who are attached to the herbivores must also sit.
- Ask the secondary consumers who are attached to the primary consumers to sit.
- Continue to tertiary and quaternary consumers.
- When the food web literally collapses, students see the effect on the community very visually. Make sure to point out that the keystone species does not exist in high numbers and its population makes up very little biomass relative to the entire food web. Even with such small numbers, its absence causes a disproportionately large effect on the ecosystem.

After you have completed the demonstration, ask the following questions:

- What is the significance of the gray wolf in Yellowstone National Park? *The gray wolf acts as a predator and demonstrates predator-mediated competition. A predator like the gray wolf keeps other populations from becoming dominant and essentially maintains balance in the community.*
- What is another keystone species in this food web? *The beaver acts as an ecosystem engineer building habitats for many other species in the Yellowstone community.*

Gray wolf, Red-tailed hawk, Bald eagle, Bighorn sheep, Grizzly bear, Coyote, Osprey, Bison, Cutthroat trout, Deer mouse, Pronghorn, Beaver, Elk, Mayfly, Wheatgrass, Pond lily, Lodgepole pine, Quaking aspen, Green algae, Gray willow

Data from: http://visityellowstonenationalparkyall.weebly.com/yellowstones-wildlife.html

FIGURE 20.8 **Beavers.** Beavers are an important species to the community because of the role they play in creating new pond and wetland habitat. For a species that is not particularly abundant, the beaver has a strong influence on the presence of other organisms in the community. *(Eric and David Hosking/Corbis; inset: Thomas & Pat Leeson/Science Source)*

Keystone species have large effects on communities

Some species that are not abundant can still have very large effects on a community. The beaver (*Castor canadensis*) is a prime example. Although they make up only a small percentage of the total biomass of the North American forest, beavers play a critical role in the forest community. They build dams that convert narrow streams into large ponds, thereby creating new habitat for pond-adapted plants and animals (FIGURE 20.8). These ponds also flood many hectares of forest, causing the trees to die and creating habitat for animals that rely on dead trees. Several species of woodpeckers and some species of ducks make their nests in cavities that are carved into the dead trees.

A **keystone species** plays a far more important role in its community than its relative abundance might suggest. The name *keystone species* is a metaphor that comes from architecture. As FIGURE 20.9 shows, in a stone arch, the keystone is the single center stone that supports all the other stones. Without the keystone, the arch would collapse. Typically, the most abundant species or the major energy producers, while vital to the health of a community, are not keystone species. Keystone species typically exist in low numbers. They may be predators, sources of food, mutualistic species, or providers of some other essential service.

The role of keystone predators was well demonstrated in a classic experiment conducted in intertidal communities off the coast of Washington State, as shown in FIGURE 20.10. These intertidal communities include mobile animal species such as sea stars and snails as well as dozens of other species that make their living attached to the rocky substratum, including mussels, barnacles, and algae. When sea stars (*Pisaster ochraceus*) are present, they prey on mussels (*Mytilus californianus*). This predation continually clears spaces where other species can attach to the rocks. When researchers removed the sea stars from the community, the mussels were no longer subject to predation by sea stars, and they outcompeted the other species in the community. The mussels became numerically dominant, while 25 other species declined in abundance. Thus the predatory sea stars, while not particularly numerous, played a key role in reducing the abundance of a superior competitor—the mussel—and allowing inferior competitors to persist.

The ability of predators to alter the outcome of competition is common in nature. We observed it in the opening of this chapter, where we described the effects of outbreaks of the leaf beetle that specializes on goldenrods in the old fields of New England. An outbreak of these herbivorous beetles reduces populations of goldenrods, which are superior competitors, and increases the abundance of many wildflower species that are inferior competitors.

Some species are considered keystone species because of the importance of their mutualistic interactions with other species. For example, most animal pollinators are abundant or provide a service that can be duplicated by other species. Some communities, however, rely on relatively rare pollinator species, which are therefore keystone species. On many South Pacific islands, a species of bat known as the flying fox (*Pteropus vampyrus*) is the only pollinator and seed disperser for hundreds of

Keystone species A species that plays a far more important role in its community than its relative abundance might suggest.

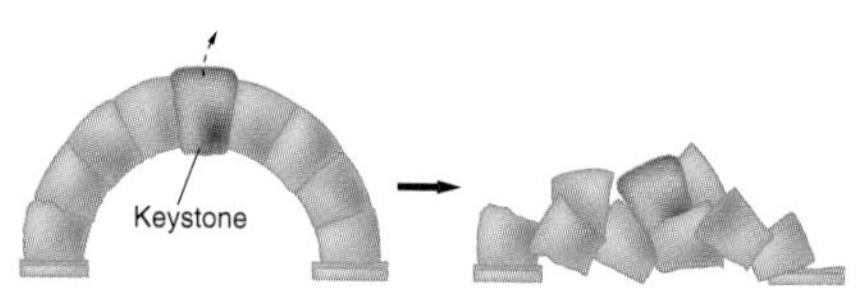

FIGURE 20.9 **Keystone.** Keystone species get their name from the keystone of an arch. Without the keystone in place, the arch would fall apart.

FIGURE 20.10 **Keystone predators.** Sea stars are keystone predators in their rocky intertidal communities in Washington State. When sea stars are present, they consume mussels, which are strong competitors for space. This predation creates open spaces that inferior competitors can colonize. As a result, the diversity of species is high. In the absence of sea stars, the mussels dominate the surfaces of the intertidal rocks, and the diversity of species declines dramatically. *(Data from T. T. Paine, "Intertidal community structure: Experimental studies on the relationship between a dominant competitor and its principal predator," Oecologia 15 (1974): 93–120. Photos by (top) blueeyes/Shutterstock; (bottom) Alex L. Fradkin/Stockbyte)*

tropical plant species. Flying foxes have been hunted for food to near-extinction. Without the pollinating and seed-dispersing functions of the flying foxes, many plant species may become extinct, dramatically changing these island communities.

Mycorrhizal fungi are another group of mutualists that serve as keystone species. These fungi are found on and in the roots of many plant species, where they increase the plants' ability to extract nutrients from the soil. They play a critical role in the growth of plant species, which in turn provide habitat and resources for other members of a forest or field community.

Finally, a keystone species that creates or maintains habitat for other species is known as an **ecosystem engineer.** We have already seen that beavers are ecosystem engineers. Alligators play a similar role in their communities by digging deep "gator holes" in summer. These holes serve as critical sources of water for many other animals during dry months.

Ecosystem engineer A keystone species that creates or maintains habitat for other species.

module

20 REVIEW

In this module, we saw that interactions among species affect whether a species can persist. Some interactions such as competition, predation, parasitism, and herbivory have negative effects on one or both species. Other interactions such as mutualisms and commensalisms have neutral or positive effects on both species. Keystone species have particularly large effects on a community; they can dramatically alter the habitat, such as when beavers construct dams. In the next module, we will examine how species in communities change over time as a result of changes in the biotic and abiotic environment.

Teaching Tip: Journal Prompt

Ask students to explain how the following exemplify keystone roles within their community:

- **Predator-mediated competition:** *By reducing the abundance of superior competitor species, predator-mediated competition is often essential to the survival of many other populations within a community.*
- **Mutualistic interactions:** *Mutualistic relationships such as pollinators and flowers are essential to the reproduction of most plant species. Since plants are a foundation of the food chain, pollinators must provide this ecosystem service or the other populations will collapse.*
- **Ecosystem engineers:** *Ecosystem engineers often create habitats for many other species.*

Teaching Tip: Discussion Starter

Ask students: How does an ecologist identify a keystone species? *Answers will vary. Scientists would be looking for one organism whose presence maintains a balance among competitors. For example, sea stars are predators of mussels. In one experiment, when researchers removed the sea star, mussels became numerically dominant while 25 other species declined in abundance. The loss of the predatory sea star, which was not particularly numerous, had a disproportionately large effect on the surrounding community.* Note: This is a good time to talk about challenges and limitations in environmental research.

Answers to Module 20 AP® Review Questions

1. b
2. d
3. b
4. d
5. a

Module 20 AP® Review Questions

1. Resource partitioning
   (a) occurs in mutualisms.
   (b) can occur through morphological differences between competing species.
   (c) can cause the extinction of a competing species.
   (d) is not the result of behavioral changes.
   (e) does not occur among competing predators.

2. The interaction between bees and sunflowers is an example of
   (a) predation.
   (b) herbivory.
   (c) parasitism.
   (d) mutualism.
   (e) commensalism.

3. Pathogens are a type of
   (a) mutualist.
   (b) parasite.
   (c) predator.
   (d) herbivore.
   (e) commensalist.

4. Which of the following is NOT typical of a keystone species?
   (a) It can shape and maintain habitats for other species.
   (b) It can limit a dominant competitive species.
   (c) It can have a relatively low abundance.
   (d) It is at the top of the food chain.
   (e) It is rarely a primary producer.

5. Which interaction harms both species involved?
   (a) Competition
   (b) Predation
   (c) Parasitism
   (d) Mutualism
   (e) Commensalism

module 21

Community Succession

Even without human activity, natural communities do not stay the same forever. Change in the species composition of communities over time is a perpetual process in nature. In this module we will look at how both terrestrial and aquatic communities change over time.

Learning Objectives

After reading this module you should be able to

- explain the process of primary succession.
- explain the process of secondary succession.
- explain the process of aquatic succession.
- describe the factors that determine the species richness of a community.

Primary succession starts with no soil

Virtually every community experiences **ecological succession,** which is the predictable replacement of one group of species by another group of species over time. Depending on the community type, ecological succession can occur over time spans varying from decades to centuries.

Some terrestrial communities begin with bare rock and no soil. For example, a community may begin forming on newly exposed rock left behind after a glacial retreat, newly cooled lava after a volcanic eruption, or an abandoned parking lot. When succession begins with bare rock and no soil, we call it **primary succession.**

FIGURE 21.1 shows the process of primary succession in a temperate forest biome in New England. The bare rock can be colonized by organisms such as algae, lichens, and mosses—organisms that can survive with little or no soil. As these early-successional species grow, they excrete acids that allow them to take up nutrients directly from the rock. The resulting chemical alteration of the rock also makes it more susceptible to erosion. When the algae, lichens, and mosses die, they become the organic matter that mixes with minerals eroded from the rock to create new soil.

Over time, soil develops on the bare surface and it becomes a hospitable environment for plants with deep root systems. Mid-successional plants such as grasses and wildflowers are easily dispersed to such areas. These species are typically well adapted to exploiting open, sunny areas and are able to survive in the young, nutrient-poor soil. The lives and deaths of these mid-successional species gradually improve the quality of the soil by increasing its ability to retain nutrients and water. As a result, new species colonize the area and outcompete the mid-successional species.

The type of community that eventually develops is determined by the temperature and rainfall of the region. In the United States, succession produces forest communities in the East, grassland communities in the Midwest, and shrubland communities in the Southwest. In some areas, the number of species increases as succession proceeds. In others, late-successional communities have fewer species than early-successional communities.

Secondary succession starts with soil

Secondary succession occurs in areas that have been disturbed but have not lost their soil. Secondary succession follows an event, such as a forest fire or

Ecological succession The predictable replacement of one group of species by another group of species over time.

Primary succession Ecological succession occurring on surfaces that are initially devoid of soil.

Secondary succession The succession of plant life that occurs in areas that have been disturbed but have not lost their soil.

Exposed rocks
Lichens and mosses
Annual wildflowers
Perennial wildflowers and grasses
Shrubs
Aspen, cherry, and young pine forest
Beech and maple broadleaf forest
Time

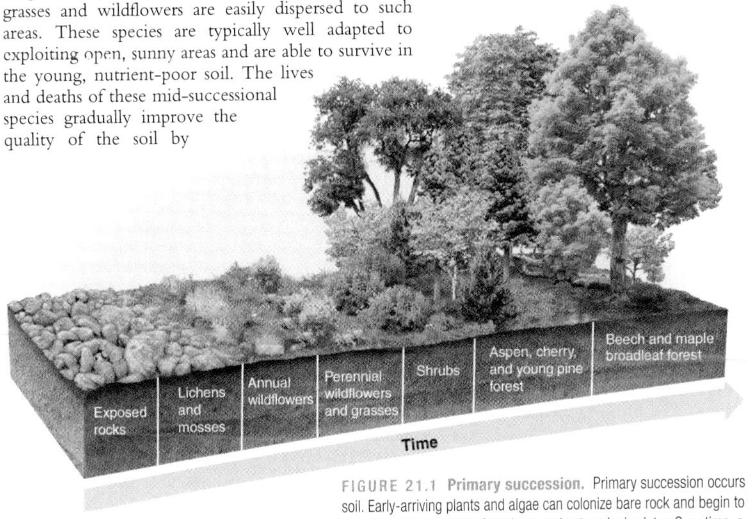

FIGURE 21.1 **Primary succession.** Primary succession occurs in areas devoid of soil. Early-arriving plants and algae can colonize bare rock and begin to form soil, making the site more hospitable for other species to colonize later. Over time, a series of distinct communities develops. In this illustration, representing an area in New England, bare rock is initially colonized by lichens and mosses and later by grasses, shrubs, and trees.

COMMON MISCONCEPTIONS
Primary and secondary succession

Students sometimes assume that primary succession happens first and secondary succession follows. To dispel this common misconception, ask the following questions:

- What events would lead to primary succession? *Glacial retreat and volcanic eruptions that create a new island are two examples of events that might lead to primary succession.*
- What events might lead to secondary succession? *Secondary succession happens in areas that have experienced a disturbance that leaves bare soil. For example, a forest fire or agricultural activity might lead to secondary succession.*
- Which succession process is longer? *The process of primary succession often takes hundreds of years, while secondary succession can occur over a decade.*

Teaching Tip: Warm-up

Ask students to differentiate between early successional species in primary succession and pioneer species. *Early successional species such as algae, mosses, and lichens are necessary in soil formation as they secrete acids that break down rock. Pioneer species are characterized by their ability to colonize areas quickly and to grow in full sunlight. These include species such as aspen trees and cherry trees.*

▲ TEACHING with FIGURES

Ask students to identify the differences between Figure 21.1 and Figure 21.2 (see page 214). *Students should see that in primary succession no soil is present, while in secondary succession the ecosystem has been disturbed and soil is already available. The first two events in primary succession (bare rock/lichens and mosses) are not included in secondary succession. Allow students to draw this conclusion so that they will remember the critical difference between primary and secondary succession.*

Teaching Tip: Activity

Have students create cards that include the events of primary and secondary succession. Ask students to arrange the sequence of events on their desks and walk around to check their understanding.

Teaching Tip: Journal Prompt

Ask students to explain how dispersal ability contributes to succession. *Species whose seeds are carried by wind or are eaten and deposited by birds or other animals are more likely to colonize remote areas or new ecosystems.*

Teaching Tip: Warm-up

Ask students to describe why climax communities may be an unrealistic description of late succession. *Late succession is never final as ecosystems are constantly undergoing change. For example, fire, outbreaks of pests, and floods can return an ecosystem to an earlier stage.*

FIGURE 21.2 **Secondary succession.** Secondary succession occurs where soil is present, but all plants have been removed. Early-arriving plants set these areas on a path of secondary succession. Secondary succession in a New England forest begins with grasses and wildflowers, which are later replaced by trees.

hurricane, that removes vegetation but leaves the soil mostly intact. Secondary succession also occurs on abandoned agricultural fields, such as the New England farms we discussed at the beginning of the chapter.

FIGURE 21.2 shows the process of secondary succession, again for a typical forest in New England. It usually begins with rapid colonization by plants that can easily disperse to the disturbed area. The first to arrive are typically such plants as grasses and wildflowers that have light, wind-borne seeds. As in primary succession, these species are eventually replaced by species that are better competitors for sunlight, water, and soil nutrients. In regions that receive sufficient rainfall, trees replace grasses and flowers. The first tree species to colonize the area are those that can disperse easily and grow rapidly. For example, the seeds of aspen trees are carried on the wind, and cherry tree seeds are carried by birds that consume the fruit and excrete the seeds on the ground. Trees such as aspen and cherry are often called **pioneer species** because of their ability to colonize new areas rapidly and grow well in full sunshine. As the pioneer trees increase in number and grow larger, however, they cause an increased amount of shade on the ground. Because pioneer trees need full sunshine to grow, new seedlings of these trees cannot persist. In contrast, species that are more shade tolerant, including many species of beech and maple, survive and grow well in the shade of the pioneer tree species. These shade-tolerant species grow up through the pioneer canopy and eventually outcompete the pioneer trees and dominate the forest community.

Pioneer species A species that can colonize new areas rapidly and grow well in full sunshine.

The process of secondary succession happens on many spatial scales, from small-scale disturbances such as a single treefall to huge areas cleared by natural disasters such as hurricanes or wildfires. The process is similar in both cases. For example, if a large tree falls and creates an opening for light to reach the forest floor, the seedlings of trees that grow rapidly in full sunlight outcompete other seedlings, and secondary succession begins again.

Historically, ecologists have described succession as ending with a *climax stage*. In forests, for example, ecologists considered the oldest forests to be *climax forests*. However, it is now recognized that, because natural disturbances such as fire, wind, and outbreaks of insect herbivores are a regular part of most communities, a late-successional stage is typically not final because at any moment it can be reset to an earlier stage.

Succession occurs in a variety of aquatic ecosystems

Disturbances also create opportunities for succession in aquatic environments. For example, the rocky intertidal zone along the Pacific coast of North America is exposed to the air during low tide and lies under water during high tide. From time to time, major storms turn over rocks or clear their surfaces of living things. These bare rocks can then be colonized through the process of primary succession. Diatoms and short-lived species of red and green algae are usually the first to arrive. Not long afterward, the rocks are colonized by barnacles and several species of long-lived red algae. This mid-successional stage develops rather quickly—in about 15 months. Finally, if the rock is not disturbed again for 3 years or more, the area may become dominated by several species of attached barnacles and mussels.

FIGURE 21.3 shows the pattern of succession in freshwater lakes. A glacier may carve out a lake basin, scouring it of sediments and vegetation. Over time, algae and aquatic plants colonize the lake. The growth of plants and algae and the erosion of surrounding rock and soil slowly fill the basin with sediment and organic matter, making it increasingly shallow. This process, which may take hundreds or even thousands of years, will eventually fill in the lake completely and make it a terrestrial habitat.

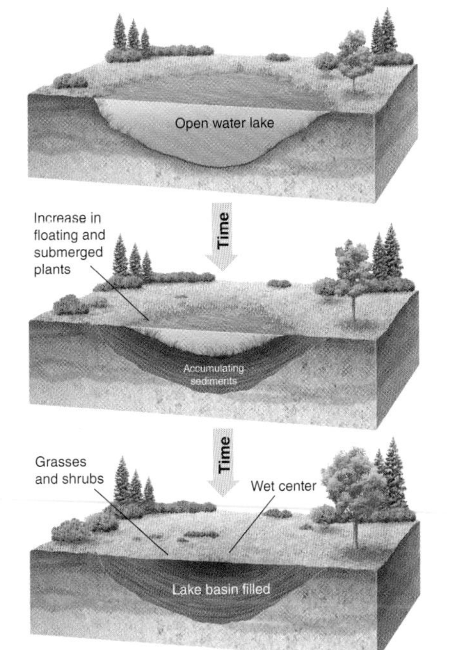

FIGURE 21.3 **Succession in lakes.** Over a time span of hundreds to thousands of years, lakes are filled with sediments and slowly become terrestrial habitats.

The species richness of a community is influenced by many factors

As we have seen, species are not distributed evenly on Earth. They are organized into biomes by global climate patterns and into communities whose composition changes regularly as species interact. In a given region within a biome, the number and types of species present are determined by three basic processes: colonization of the area by new species, speciation within the area, and losses from the area by extinction. The relative importance of these processes varies from region to region and is influenced by four factors: *latitude, time, habitat size,* and *distance from other communities*.

Latitude

As we move from the equator toward the North or South Pole, the number of species declines. For example, the southern latitudes of the United States support more than 12,000 species of plants, whereas a similar-sized area in northern Canada supports only about 1,700 species. This latitudinal pattern is also observed among birds, reptiles, amphibians, and insects. For more than a century, scientists have sought to understand the reasons for this pattern, yet those reasons remain unclear.

Time

Patterns of species richness are also regulated by time. The longer a habitat exists, the more colonization, speciation, and extinction can occur in that habitat. For example, Lake Baikal in Siberia—more than 25 million years old—is one of the oldest lakes in the world. Its benthic zone is home to over 580 species of invertebrates. By contrast, only 4 species of invertebrates inhabit the benthic zone of the Great Slave Lake in northern Canada—a lake that is similar in size and latitude to Lake Baikal, but is only a few tens of thousands of years old. This difference suggests that older communities have had more opportunities for speciation.

Teaching Tip: Journal Prompt

Ask students to describe how each of the factors below affects the species richness of a community.

- **Latitude:** *As we move from the poles toward the equator, species richness increases.*
- **Time:** *The longer a habitat exists, the more colonization, speciation, and extinction can occur. If the rate of speciation and the rate of colonization exceed the rate of extinction, then a habitat present for more time will have more species.*
- **Habitat size:** *The larger the habitat, the more species, niches, and potential for dispersion exists, leading to greater numbers of species.*
- **Distance from other communities:** *Many species can disperse short distances, but few can disperse long distances. If a new island is closer to a colonized land mass, then it will have a greater number of species than an island of the same size that is farther away.*

Teaching Tip: Discussion Starter

Ask students to describe how succession happens in aquatic environments. *Aquatic succession may occur along a coastline. If a storm turns over a rock it can be colonized by algae and then by barnacles and mussels. In a freshwater lake, erosion can fill the lake basin with sediment. Aquatic plants can thrive and fill the basin with organic matter over time. As the basin fills it can then develop into a terrestrial habitat.*

Practicing Science: Analyze

Ask students to examine Figure 21.4 and write a summary statement about the relationship between the size of an island and species richness. Students should also give an explanation for this trend. *As habitat size increases, numbers of species increase as well. Larger habitats can support more species. Dispersal species are more likely to find larger islands. Finally, larger islands contain a greater range of environmental conditions that provide more niches for greater numbers of species.*

Teaching Tip: Activity

Simulating the Theory of Island Biogeography

This activity is designed to visually demonstrate the theory of island biogeography for students. Students will consider the two factors that are the basis of island biogeography. They will model the impact of distance and habitat size on species richness and ease of colonization. Students will also consider how effectively organisms can disperse to new areas.

Download this activity by clicking on the link buttons in the TE-book, opening the TRFD, or logging onto the book's companion website bcs.whfreeman.com/friedlandapes2e (teacher login required).

 **Activity 6.3**: Simulating the Theory of Island Biogeography

Teaching Tip: Journal Prompt

Ask students: How can the concepts of island biogeography be applied to conservation efforts? *Island biogeography tells us that larger areas are more likely to sustain a greater diversity of species. When setting aside a natural habitat for conservation, larger areas should help maintain diversity. Limiting distance between protected areas will enable colonists to move more easily between the areas.*

Habitat size and distance from a source of species

The final two factors that influence species richness are the size of the habitat and the distance of that habitat from a source of colonizing species. These factors are the basis for the **theory of island biogeography,** which demonstrates the dual importance of habitat size and distance from a mainland in determining species richness.

Larger habitats typically contain more species. In FIGURE 21.4, for example, we can see that larger islands of reed habitat in Hungary contain a greater number of bird species than smaller islands. There are three reasons for this pattern. First, dispersing species are more likely to find larger habitats than smaller habitats, particularly when those habitats are islands. Second, at any given latitude, larger habitats can support more species than smaller habitats. Larger habitats are capable of supporting larger populations of any given species, and larger populations are less prone to extinction. Third, larger habitats often contain a wider range of environmental conditions, which in turn provide more niches that support a larger number of species. A wider range of environmental conditions also provides greater opportunities for speciation over time.

The distance between a habitat and a source of colonizing species is the second factor that affects the species richness of communities. For example, oceanic islands that are more distant from continents generally have fewer species than islands that are closer to continents. Distance matters because, while many species can disperse short distances, only a few can disperse long distances. In other words, if two islands are the same size and contain the same resources, the nearer island should accumulate more species than the farther island because it has a higher rate of immigration by new species.

Theory of island biogeography A theory that demonstrates the dual importance of habitat size and distance in determining species richness.

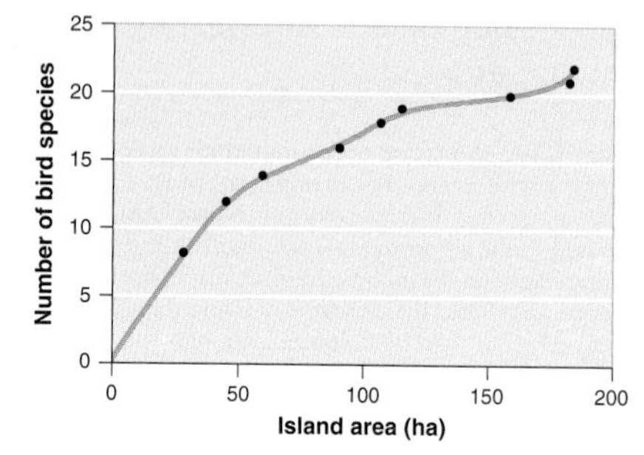

FIGURE 21.4 **Habitat size and species richness.** Species richness increases as the size of the habitat increases. In this example, researchers counted the number of bird species that inhabited reed islands in Lake Velence, Hungary. As island area increased, the number of bird species initially rose quickly and then began to slow. *(Data from A. Baldi and T. Kisbenedek, "Bird species numbers in an archipelago of reeds at Lake Velence, Hungary," Global Ecology and Biogeography 9 (2000): 451–461.)*

Conservation and island biogeography

The effects of colonization, speciation, and extinction on the species richness of communities have important implications for conservation. The theory of island biogeography was originally applied to oceanic islands, but it has since been applied to "habitat islands" within a continent, such as the "islands" of protected habitat represented by national parks. These habitat islands are often surrounded by less hospitable habitats that have been dramatically altered by human activities. If we wish to set aside natural habitat for a given species or a group of species, we need to consider both the size of the protected area and the distance between the protected area and other areas that could provide colonists.

module 21 REVIEW

In this module, we learned that communities change over time in a process known as succession. In terrestrial communities, an area that begins with no soil undergoes primary succession whereas an area that begins with soil undergoes secondary succession. In aquatic communities, lakes and ponds can slowly fill in with sediments over thousands of years to eventually become terrestrial habitats. We also learned that the species richness of a community is affected by the latitude of a habitat, the length of time that the habitat has been present, the size of the habitat, and the distance from the habitat to other habitats that serve as sources of species.

Module 21 AP® Review Questions

1. Which is true of primary succession?
   (a) It starts with bare soil.
   (b) As it progresses the number of species initially decreases.
   (c) It occurs after forest fires.
   (d) It begins with colonization by algae, lichens, and mosses.
   (e) It results in an ecosystem of grasses and small plants.

2. The process of succession in lakes
   (a) results in a terrestrial ecosystem.
   (b) occurs rarely because disturbances are rare.
   (c) depends primarily on the fish species within the lake.
   (d) progresses fastest in very deep lakes.
   (e) has no climax species.

3. Which tree species is a pioneer species in North American forests?
   (a) Beech
   (b) Fir
   (c) Aspen
   (d) Maple
   (e) Oak

4. Which factor does NOT affect species richness?
   (a) Latitude
   (b) Survivorship curves
   (c) Time
   (d) Habitat size
   (e) Distance from other communities

5. Which is NOT a factor in the theory of island biogeography?
   (a) Dispersing species are more likely to find a large habitat.
   (b) Islands can also be isolated areas within a continent.
   (c) Large populations are less likely to go extinct.
   (d) Larger habitats contain a larger range of environmental conditions.
   (e) Islands farther from the continent have more species due to increased speciation.

Answers to Module 21 AP® Review Questions

1. d
2. a
3. c
4. b
5. e

working toward  sustainability

Bringing Back the Black-footed Ferret

Throughout the western United States, the Great Plains were once covered with large congregations of prairie dogs. Spanning several different species, these prairie dog "towns" consisted of networks of underground tunnels. They were great attractions for out-of-town tourists, but the local ranchers who had to live with the prairie dogs were not nearly as fond of the little rodents. Prairie dogs are effective herbivores that consume a great deal of plant biomass, including agricultural crops. The ranchers viewed prairie dogs as competitors for their crops and sought a variety of ways, such as poisoning, to eradicate them. An unintended consequence of the ranchers' victory against the prairie dogs was the near-extinction of another species, the black-footed ferret (*Mustela nigripes*).

The black-footed ferret is a member of the weasel family and the only species of ferret native to North America. It lives in burrows and preys on prairie dogs. In fact, a single ferret can consume 125 to 150 prairie dogs in a year. Although prairie dogs had been the ferrets' main source of food for millions of years, things started to change when settlers began plowing the Great Plains for agriculture. This destroyed many prairie dog towns, while other prairie dog towns were poisoned. Together, the plowing and poisoning have reduced the prairie dog population by 98 percent.

As we would expect in a density-dependent scenario, the reduction in the prairie dog population reduced the carrying capacity of the environment for the black-footed ferret. The poisoning campaign poisoned ferrets as well as prairie dogs. Because both small and large populations of ferrets were poisoned, and a large proportion of ferrets died in both cases, the poisoning had a density-independent effect on the ferret population.

In 1967, the black-footed ferret was officially listed as an endangered species, and the last known population died out in 1974. People feared that the black-footed ferret was extinct. In 1981, however, a small population of 130 ferrets was discovered in Wyoming. Conservation efforts began, but a highly lethal disease known as canine distemper passed through the population, reducing the entire species to a mere 18 animals.

Teaching Tip: Beyond the Classroom

Divide students into groups and ask them to read "Working Toward Sustainability: Bringing Back the Black-footed Ferret." Using the article as an example, ask students to design a conservation plan for an endangered animal. Ask them to include management practices that take into consideration density-dependent and density-independent factors as well as community interactions.

As part of a collaboration between federal biologists, private landowners, and several zoos, all 18 ferrets were immediately brought into captivity in the hope that a captive breeding program might be able to restore their numbers. The black-footed ferret is a *K*-selected species; it breeds once a year and has 3 to 4 offspring, for which the parents provide a great deal of parental care. As a result, population increases could not occur rapidly. Nevertheless, the captive population grew to 120 animals by 1989.

At this point, the captive breeding program was considered successful enough to allow reintroduction of the species into the wild. However, the biologists first had to consider exactly where they wanted to reintroduce the ferrets. They understood the risks of introducing all the animals into a single site because another round of disease could kill the entire population. To manage the reintroduction process, the biologists decided to make use of the metapopulation concept. They reintroduced the ferrets in several places across the Great Plains, choosing locations with healthy prairie dog populations that had a low risk of extermination.

From 1991 to 2013, black-footed ferrets were reintroduced at 19 sites, from northern Mexico all the way up to southern Canada. These efforts are clearly paying off. Today, more than 1,000 ferrets live in the wild, and hundreds more are part of the ongoing captive breeding program. But the rescue effort is not over. During years of high rainfall, outbreaks of plague, a bacterial disease carried by fleas, kill many ferrets. To combat plague, biologists have been dusting prairie dog towns with an insecticide powder that kills fleas. They are also working on a new vaccine against plague. Their goal is to achieve a wild population of 1,500 ferrets, with more than 30 animals at each of 10 reintroduction sites. If the biologists meet this goal, the ferret can be moved from the endangered species list to the less-perilous classification of "threatened."

Black-footed ferret. Once critically endangered, populations of the black-footed ferret are rebounding as a result of collaborative conservation efforts. *(D. Robert & Lorri Franz/Corbis)*

Critical Thinking Questions

1. How can humans balance the interests of ranchers who want to control prairie dog populations with the interests of conservationists who want to prevent the extinction of the black-footed ferret?
2. How has the metapopulation concept helped conservationists manage the recovery of the black-footed ferret?

References

Black-footed Ferret Recovery Implementation Team. *The Black-footed Ferret Recovery Program.* http://www.blackfootedferret.org.

Robbins, J. 2008. Efforts on 2 fronts to save a population of ferrets. *New York Times,* July 15.

Suggested Answers to Critical Thinking Questions

1. Maintaining healthy populations of black-footed ferrets can help control the prairie dog population through predation. If the ranchers stop using poison and other methods that kill both ferrets and prairie dogs, they should be able to prevent large prairie dog populations from forming while allowing continued survival of the ferrets.

2. The conservationists have created a metapopulation by introducing the ferrets to several new areas and trying to establish multiple subpopulations. This metapopulation will be more resistant to extinction, especially extinction due to a single cause, since even if one subpopulation dies out the others are likely to survive.

chapter 6 REVIEW

Community ecology examines how species interactions help to determine the species that are present in a community. Different characteristics of populations affect their abundance and distribution including density-dependent and density-independent factors. Population growth models help us understand how populations increase and decrease over time. Species also have distinctive reproductive strategies and growth curves, which affect population size and characteristics. At the community level, major types of species interactions include competition, predation, parasitism, mutualism, and commensalism. Communities experience ecological succession. The species richness of a given community depends on latitude, elapsed time, habitat size, and habitat distance to other sources of species.

Teaching Tip: Measuring Your Impact

The Living Planet Index

Measuring Your Impact 6 offers students an opportunity to practice graphing skills and evaluate biodiversity trends. Download this item by clicking on the link buttons in the TE-book, opening the TRFD, or logging onto the book's companion website bcs.whfreeman.com/friedlandapes2e (teacher login required).

Measuring Your Impact 6:
The Living Planet Index

Key Terms

Population
Community
Population ecology
Population size (N)
Population density
Population distribution
Sex ratio
Age structure
Limiting resource
Density-dependent factor
Carrying capacity (K)
Density-independent factor
Population growth model
Population growth rate
Intrinsic growth rate (r)
Exponential growth model ($N_t = N_0e^{rt}$)
J-shaped curve
Logistic growth model
S-shaped curve
Overshoot
Die-off
K-selected species
r-selected species
Survivorship curve
Type I survivorship curve
Type II survivorship curve
Type III survivorship curve
Corridor
Metapopulation
Inbreeding depression
Community ecology
Symbiotic relationship
Competition
Competitive exclusion principle
Resource partitioning
Predation
Parasitoid
Parasitism
Pathogen
Herbivory
Mutualism
Commensalism
Keystone species
Ecosystem engineer
Ecological succession
Primary succession
Secondary succession
Pioneer species
Theory of island biogeography

Learning Objectives Revisited

Module 18 The Abundance and Distribution of Populations

- **Explain how nature exists at several levels of complexity.**

  Nature exists at several levels of complexity: individuals, populations, communities, ecosystems, and the biosphere.

- **Discuss the characteristics of populations.**

  Populations can have distinct population sizes, densities, distributions, sex ratios, and age structures.

- **Contrast the effects of density-dependent and density-independent factors on population growth.**

  Density-dependent factors influence an individual's probability of survival and reproduction in a manner that is related to the size of the population. Density-independent factors have the same effect on an individual's probability of survival and reproduction in populations of any size.

Module 19 Population Growth Models

- **Explain the exponential growth model of populations, which produces a J-shaped curve.**

  The exponential growth model describes rapid growth under ideal conditions when resources are not limited. The J-shaped curve occurs because the population initially grows slowly when few individuals are present to reproduce but then grows rapidly as the number of reproducing individuals increases.

- **Describe how the logistic growth model incorporates a carrying capacity and produces an S-shaped curve.**

  The logistic growth model incorporates density-dependent factors that allow rapid initial growth but then cause population growth to slow down as populations approach their carrying capacity. When we allow lag times between when resources change in abundance and when populations produce new offspring, we can observe population overshoots and die-offs.

- **Compare the reproductive strategies and survivorship curves of different species.**

  Organisms have a range of reproductive patterns. At the extremes are *r*-selected species, which experience rapid population growth rates, and *K*-selected species, which experience high survivorship and slow population growth rates. Patterns of survivorship over the life span can be graphically represented as type I, II, and III survivorship curves.

- **Explain the dynamics that occur in metapopulations.**

  Metapopulations are groups of spatially distinct populations that are connected by occasional movements of individuals. These movements reduce the probability of any of the populations going extinct.

Exploring the Literature

Hanski, I., and O. E. Gaggiotti. 2004. *Ecology, Genetics, and Evolution of Metapopulations.* Elsevier.
Pfennig, D. W., and K. S. Pfennig. 2012. *Evolution's Wedge: Competition and the Origins of Diversity.* University of California Press.
Ricklefs, R., and R. A. Relyea. 2015. *Ecology: The Economy of Nature.* 7th ed. W. H. Freeman.
Rockwood, L. L. 2006. *Introduction to Population Ecology.* Blackwell.
U.S. Department of the Interior, National Park Service. *Exploring Nature: Preserving Biodiversity.* http://www.nature.nps.gov/biodiversity/index.cfm.

Download a printable version of this list by clicking on the link buttons in the TE-book, opening the TRFD, or logging onto the book's companion website bcs.whfreeman.com/friedlandapes2e (teacher login required).

Exploring the Literature 6

Module 20 Community Ecology

- **Identify species interactions that cause negative effects on one or both species.**

  Competition is an interaction between two species that share a limiting resource. Over time, competition for a resource can cause natural selection to favor those individuals that have reduced overlap in resource use and this can lead to spatial, temporal, or morphological resource partitioning. Predation is an interaction in which animals partially or entirely consume another animal. Predators can affect the abundance of prey populations and cause the evolution of anti-predator defenses in prey populations. Parasitism is an interaction in which one organism lives on or in another organism. Those parasites that can cause disease in their hosts are known as pathogens. Herbivory is an interaction in which animals consume producers. In some cases, herbivores can have dramatic effects on plant and algal communities by removing the most palatable species.

- **Discuss species interactions that cause neutral or positive effects on both species.**

  Mutualisms are interactions that benefit two interacting species by increasing the chances of survival or reproduction for both. One of the most common mutualisms is the interaction between flowering plants and their pollinators. A second well-known mutualism is between acacia trees and the ants that defend the trees in exchange for food and a place to live. Commensalisms are interactions in which one species benefits but the other is neither harmed nor helped. Examples include birds perching on trees and marine fish using coral reefs for protection from predators.

- **Explain the role of keystone species.**

  Keystone species play a role in the community that is far more important than its relative abundance might suggest. Common examples include predators that alter the outcome of competition in intertidal communities and beavers that create large ponds by constructing dams on streams.

Module 21 Community Succession

- **Explain the process of primary succession.**

  Primary succession occurs on surfaces that are initially devoid of soil, such as bare rock that is exposed after the retreat of glaciers or the cooled lava from a volcanic eruption. Over time, plants and animals arrive at the site and modify the environment, making it more favorable for other species to arrive and persist.

- **Explain the process of secondary succession.**

  Secondary succession occurs in areas that have been disturbed but have not lost their soil. A common example is the bare soil left behind when farmers stop planting crops in a field. Over time, plants and animals colonize the site, alter the environmental conditions, and favor the persistence of other species.

- **Explain the process of aquatic succession.**

  Lakes and ponds experience sedimentation over long periods of time and this slowly fills in the basin. Over thousands of years, the lakes and ponds can be slowly converted into terrestrial habitats.

- **Describe the factors that determine the species richness of a community.**

  The species richness of a community is typically higher at latitudes that are closer to the equator. Richness is also higher in older sites where evolution has been producing new species for longer periods of time. Finally, more species exist in larger habitats and habitats that are closer to sources of new species, as is the case for oceanic islands that are located close to continents.

Section 1: Multiple-Choice Questions

Choose the best answer for questions 1–12.

1. Which is NOT an example of a density-independent factor?
(a) Drought
(b) Competition
(c) Forest fire
(d) Hurricane
(e) Flood

2. As the size of a white-tailed deer population increases,
(a) the carrying capacity of the environment for white-tailed deer will be reduced.
(b) a volcanic eruption will have a greater proportional effect than it would on a smaller population.
(c) the effect of limiting resources will decrease.
(d) the number of gray wolves, a natural predator of white-tailed deer, will increase.
(e) white-tailed deer are more likely to become extinct.

3. The following graph shows population growth of Canada geese in Ohio between 1955 and 2002.

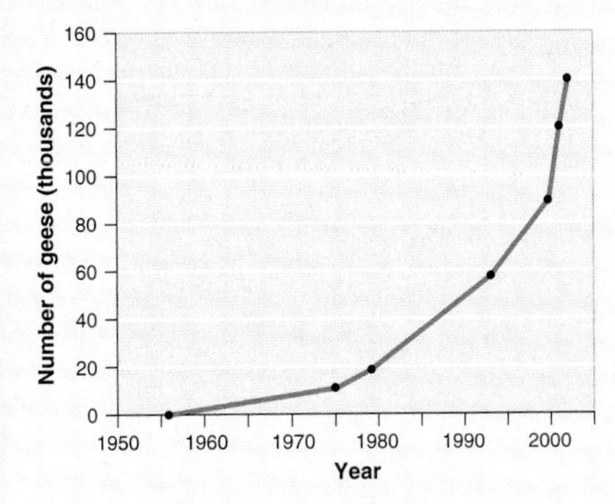

This graph can best be described as:
(a) an exponential growth curve.
(b) a logistic growth curve.
(c) a stochastic growth curve.
(d) oscillation between overshoot and die-off.
(e) approaching the carrying capacity.

4. Which is NOT a true statement based on the logistic growth model?
(a) Population growth is limited by density-dependent factors.
(b) A population will initially increase exponentially and then level off as it approaches the carrying capacity of the environment.
(c) Future population growth cannot be predicted mathematically.
(d) Population growth slows as the number of individuals approaches the carrying capacity.
(e) A graph of population growth produces an S-shaped growth curve over time.

5. Which characteristics are typical of *r*-selected species?
I. They produce many offspring in a short period of time.
II. They have very low survivorship early in life.
III. They take a long time to reach reproductive maturity.

(a) I only
(b) II only
(c) III only
(d) I and II
(e) II and III

6. A high intrinsic growth rate would most likely be characteristic of
(a) a *K*-selected species such as elephants.
(b) an *r*-selected species such as the American bullfrog.
(c) a *K*-selected species that lives close to its carrying capacity.
(d) a species that is near extinction.
(e) a species with a low reproductive rate that takes a long time to reach reproductive maturity.

Answers to Chapter 6 AP® Exam Multiple-Choice Questions

1. b
2. d
3. a
4. c
5. d
6. b
7. e
8. a
9. c
10. d
11. d
12. b

Additional Multiple-Choice AP® Practice Exam

To download an additional Multiple-Choice AP® Environmental Science Practice Exam for Chapter 6, click on the link buttons in the Teacher's e-Book (TE-book), open the Teacher's Resource CD (TRCD), or log onto the book's companion website www.whfreeman.com/APES2e (teacher login required).

Multiple-Choice AP® Practice Exam 6

Answers to Chapter 6 AP® Exam

Download the full answers to the Chapter 6 AP® Practice Exam by clicking on the link buttons in the TE-book, opening the TRFD, or logging onto the book's companion website bcs.whfreeman.com/friedlandapes2e (teacher login required).

Answers to Chapter 6 AP® Exam

7. The following graph presents three survivorship curves.

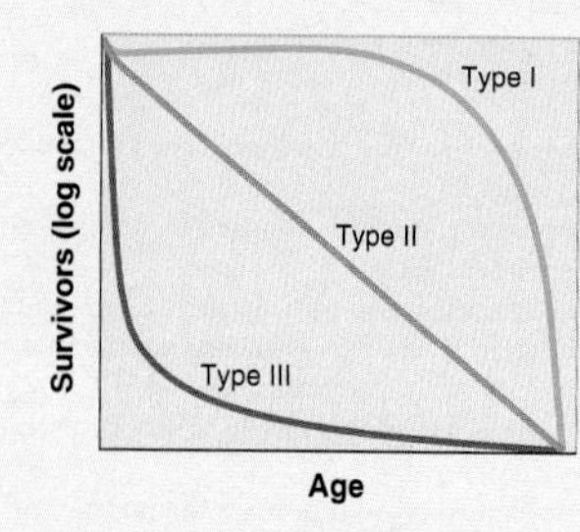

Choose the description that best describes the graph.
(a) Type I could represent the house mouse, which is a typical *r*-selected species.
(b) Type II could be the average of *r*-selected and *K*-selected species in a specific area.
(c) Type III could represent elephants, which are typical K-selected species.
(d) Type I could represent an oak tree species that experiences very low survivorship early and late in life.
(e) Type II could represent a coral species that has a constant decline in survivorship throughout its life.

8. In the coniferous forests of Oregon, eight species of woodpeckers coexist. Four species select their nesting sites based on tree diameter, while the fifth species nests only in fir trees that have been dead for at least 10 years. The sixth species also nests in fir trees, but only in live or recently dead trees. The two remaining species nest in pine trees, but each selects trees of different sizes. This pattern is an example of
(a) resource partitioning.
(b) commensalism.
(c) predation.
(d) predator-mediated competition.
(e) a keystone species.

9. Which statement about ecological succession is correct?
(a) Secondary succession begins in a community lacking soil.
(b) Primary succession occurs over a shorter time span than secondary succession.
(c) Succession is influenced by competition for limiting resources such as available soil, moisture, and nutrients.
(d) In forest succession, less shade-tolerant trees replace more shade-tolerant trees.
(e) Forest fires and hurricanes lead to primary succession because a soil base still exists.

10. Which sequence of secondary succession would be likely to occur in abandoned farmland in the eastern United States?
(a) Bare soil, lichens, mosses, grasses, deciduous trees
(b) Bare rock, lichens, mosses, grasses, shrubs, mixed shade- and sunlight-tolerant trees
(c) Bare soil, grasses and wildflowers, shrubs, shade-tolerant trees
(d) Bare soil, grasses and wildflowers, shrubs, sunlight-tolerant trees, shade-tolerant trees
(e) Bare rock, grasses and wildflowers, lichens, mosses, shrubs, coniferous trees

11. The theory of island biogeography suggests that species richness is affected by which of the following factors?
I. Island distance from mainland
II. How the island is formed
III. Island size

(a) I only
(b) II only
(c) III only
(d) I and III
(e) II and III

12. Which combination of factors would result in the highest biodiversity?
(a) A small island, close to a continent
(b) A large island, close to a continent
(c) A medium island, distant from a continent
(d) A small island, distant from a continent
(e) A large island, distant from a continent

Section 2: Free-Response Questions

Write your answer to each part clearly. Support your answers with relevant information and examples. Where calculations are required, show your work.

1. The California Department of Fish and Wildlife is developing a plan to connect mountain "habitat islands" that are separated by open areas of flat, arid land in the deserts of southeastern California. These mountain areas are habitats for desert bighorn sheep (*Ovis canadensis*), which move extensively among the islands through habitat corridors. The habitat corridors provide opportunities for recolonization, seasonal migration, and maintenance of genetic variation among the metapopulation of desert bighorn sheep.
   (a) Explain what is meant by a metapopulation and how it relates to the desert bighorn sheep. (1 point)
   (b) Identify two density-dependent factors and one density-independent factor that could affect the populations of desert bighorn sheep. (3 points)
   (c) Explain the consequences to the desert bighorn sheep population if the plan to connect the mountain habitat islands is not implemented. (2 points)
   (d) Explain how the theory of island biogeography applies to the mountainous areas of southeastern California. (4 points)

2. Read the following information, which was posted in the great apes exhibit of the Fremont Zoo, and answer the questions that follow.

   The western lowland gorilla (*Gorilla gorilla*) lives in the moist tropical rainforests of western Africa. The gorillas' primary diet consists of fruits, leaves, foliage, and sometimes ants and termites. Occasionally they venture onto farms and feed on crops. Their only natural enemy is the leopard, the only animal other than humans that can successfully kill an adult gorilla. However, disease, particularly the Ebola virus, is threatening to decimate large populations of these gorillas in Congo.

   The western lowland gorilla has a life span of about 40 years and produces one offspring every 4 years. Males usually mate when they are 15 years old. Females reach sexual maturity at about 8 years of age, but rarely mate before they are 10 years old.

   (a) Based on the information above, is the western lowland gorilla an *r*-selected or a *K*-selected species? Provide evidence to support your answer. (3 points)
   (b) Identify and explain four community interactions that involve the western lowland gorilla. (4 points)
   (c) Explain what is meant by secondary succession and describe how it may be initiated in a tropical forest. (3 points)

Answers to Chapter 6 AP® Exam Free-Response Questions

1. (a) A metapopulation is a group of separate, isolated populations that have individuals occasionally disperse between the habitat islands, allowing recolonization of islands that might go extinct.
(b) Density-dependent factors:

- scarcity of food resources
- disease
- crowding
- increase in predators
- competition for mates

Density-independent factors:

- drought
- fires
- storms
- mudslides
- freezes

(c) If the plan to connect the mountain islands is not implemented, the following could occur:

- Because the smaller isolated populations are more prone to extinction, they could not be recolonized without dispersal from neighboring islands.
- Dispersal from neighboring islands could also bring additional genetic diversity, which could help an isolated population persist.

(d) The theory of island biogeography applies to the mountainous areas of southeastern California in the following ways:

- The larger the area of the mountain "island" containing more different types of habitats, the greater the number of species that will be present.
- The larger the area of the mountain "island," the larger the populations of given species; larger populations are less prone to extinction.
- The more connected the mountain "islands" are, the greater the chance for dispersal or migration of species, resulting in higher rates of immigration of new species.

2. (a) The Western lowland gorilla is a *K*-selected species.
Evidence:

- Long time to reach sexual maturity—15 years for males and 8 to10 years for females
- Low number of offspring—one every 4 years

(b) Community interactions and explanations:

- Competition exists between gorillas and humans for crops.
- The Ebola virus is a parasite living in the gorilla.
- Gorillas are herbivores when they feed on fruit, leaves, and foliage.
- Gorillas are true predators when they feed on ants and termites.
- Leopards are true predators that feed on gorillas as their prey.
- Gorillas are omnivores because they eat vegetation as well as ants and termites.

(c) Secondary succession is succession that occurs on an area of land that is devoid of vegetation but has not experienced complete destruction of the soil. In a moist tropical area, the secondary communities are eventually composed of trees.
Possible events that would lead to secondary succession in the tropical forest:

- abandoning farmland
- cutting down and removing trees for the lumber
- forest fires

chapter

The Human Population

PD Overview

This chapter covers human population growth, including the social, economic, and environmental factors that determine growth. This chapter also addresses the relationship between resource consumption and population growth. By the time they reach the end of this chapter, students should know how to calculate population growth rates and interpret age structure diagrams. They should understand the phases of the demographic transition model and be able to correlate economic development to family size. Other essential topics in this chapter are the relationship between education and total fertility rates and the environmental impacts of both poverty and affluence.

Module 22: Human Population Numbers

Although it may seem surprising, environmental scientists do not know the human carrying capacity of Earth. However, they are able to discuss the factors that contribute to the carrying capacity of humans on Earth and what drives human population growth. It is also possible to look at patterns of past human population growth to gain an understanding of what might occur in future decades.

Module 23: Economic Development, Consumption, and Sustainability

This module describes demographic transitions, the effect of these transitions on the environment, and the relationship between economic development and sustainability. It examines the changes human populations undergo as environmental and social conditions change over time. Some of these changes occur with specific patterns that can be described and explained. For example, slower population growth is typical of many wealthier countries. Countries that have gone through a period of industrialization also typically increase their consumption of resources, with adverse impacts on the environment. At the same time, as a country gains wealth, it also gains the resources to address environmental issues.

Alignment to AP® Environmental Science Course Description

AP® Outline	Module	
III. Population (10–15%)		
B. Human Population	Module 22	Human Population Numbers
	Module 23	Economic Consumption, Development, and Sustainability

Chapter Learning Objectives

After completing this chapter students will be able to

- **explain factors that may potentially limit the carrying capacity of humans on Earth.**
- **describe the drivers of human population growth.**
- **read and interpret an age structure diagram.**
- **describe how demographic transition follows economic development.**
- **explain how relationships among population size, economic development, and resource consumption influence the environment.**
- **describe why sustainable development is a common but elusive goal.**

Chapter 7 Pacing Guide

This pacing guide is based on a schedule with 120 sessions of 50 minutes each before the AP® Exam. If you have a different number of sessions before the exam, you can modify the pacing to suit your needs. If you have additional time, consider incorporating quizzes, released AP® Environmental Science free-response and multiple-choice questions, or additional activities.

Module	Standard Schedule Days	Block Schedule Days
Module 22	3	1½
Module 23	2	1
Assessment	1	½

Chapter 7 Resources

The link to these resources can be found by clicking on the link buttons in the Teacher's e-Book (TE-book), opening the Teacher's Resource Flash Drive (TRFD), or logging onto the book's companion website bcs.whfreeman.com/friedlandapes2e (teacher login required).

- **PowerPoint Presentation 7**
- **Optimized Art PowerPoint and JPEG Files 7**
- **Do the Math Videos**
- **More Math Practice 7.1:** Population Calculations
- **Measuring Your Impact 7:** National Footprints
- **Activity 7.1:** Be a Demographer
- **Activity 7.2:** Demographic Transition Model
- **Lab 7.1:** World Population
- **Chapter 7 Web Resources**
- **Exploring the Literature 7**
- **Answers to Chapter 7 AP® Exam**
- **Multiple-Choice AP® Practice Exam 7**
- **Answers to Unit 3 AP® Exam**
- **Unit 3 Additional Free-Response Question**
- **Answer to Science Applied 3 Free-Response Question**

Free-Response Questions from Previous AP® Environmental Science Exams

Free-response questions from prior AP® Environmental Science Exams are available on the AP® course website, https://apstudent.collegeboard.org/apcourse/ap-environmental-science/exam-practice. Students should be able to answer all of the questions listed below with material learned in this and previous chapters. When a question requires students to understand material from multiple chapters, the question will be listed in the last chapter required to complete the entire question. Questions marked with an asterisk (*) are from exams with released multiple-choice questions. You may want to save these questions until the end of the year so you can give your students a complete released exam for practice. Questions marked with double asterisks (**) require math to calculate a problem. Look for references to these questions throughout the chapter.

Year	Question	Content
2000	4	• Age structure diagrams
2003*	2**	• Demographic transition model
2008*	4**	• Worldwide trends in fertility

PD **Chapter 7 Overview**

Watch the video overview of Chapter 7 (for teachers) by clicking on the link buttons in the TE-book, opening the TRFD, or logging onto the book's companion website bcs.whfreeman.com/friedlandapes2e (teacher login required).

TRM **PowerPoint Presentation 7**

Download the PowerPoint presentation for Chapter 7 by clicking on the link buttons in the TE-book, opening the TRFD, or logging onto the book's companion website bcs.whfreeman.com/friedlandapes2e (teacher login required).

There are more than 215 million children under the age of 15 in China today.
(Pierre Montavon/Strates/Panos Pictures)

chapter

The Human Population

The Environmental Implications of China's Growing Population

Human population size, affluence, and resource consumption all have interrelated impacts on the environment. The example of China is striking. With more than 1.3 billion people—almost 20 percent of the human population on Earth—China is the world's most populous nation. Because of its rapid economic development, it may one day become the world's largest economy. Once-scarce consumer goods such as automobiles and refrigerators are becoming increasingly commonplace in China. Although the United States, with 315 million people, has historically been the world's largest consumer of resources and the greatest producer of many pollutants, China is rapidly surpassing the United States in both consumption and pollution. It is already the largest emitter of carbon dioxide and sulfur dioxide, and it consumes one-third of commercial fish and seafood. The Chinese are facing considerable environmental challenges as their affluence increases.

The Chinese are facing considerable environmental challenges as their affluence increases.

Sustainably managing the presence of humans on Earth requires addressing both population growth and resource consumption. China has already taken dramatic steps to limit its population growth. For many decades, China has had a "one-child" policy. Couples that restricted themselves to a single child were rewarded financially, while those with three or more children faced sanctions, such as a 10 percent salary reduction. Chinese officials have used numerous tools—many controversial—to meet population targets, including abortions, sterilizations, and the designation of certain pregnancies as "illegal."

China is one of only a few countries where government-mandated population control measures have contributed to reducing population growth. After decades of having one of the world's highest fertility rates, China now has a fertility rate of 1.5 births per woman. In addition to the

225

Teaching Tip: Chapter Opening Case

The chapter opening case, "The Environmental Implications of China's Growing Population," provides an engaging introductory discussion about the topics addressed in Chapter 7. Students will get a first glimpse at China's large population numbers and the country's growing affluence. The one-child policy is introduced in this case. Students can deliberate China's controversial strategies for population control. The case also discusses how China's growing affluence affects resource use and pollution. This case provides an opportunity to introduce differences in resource use between developed and developing nations as well as environmental impacts as consumption increases across the world.

Teaching Tip: Discussion Starter

China's large population places pressure on resource use. Ask students: Which do you believe plays a bigger role in moving toward sustainability: managing population size or economic development? *In reality both are important when considering sustainability.*

Use this question to determine misconceptions and to introduce some of the ideas presented in Chapter 7. For example, affluence (increased economic prosperity) can cause an increased use of resources but affluence also contributes to a nation's ability to make environmental improvements and increase its efficiency in the use of resources. Many students already understand that a larger population creates greater pressure on available resources, but they might not realize that a large number of people living in poverty can lead to local environmental degradation.

Teaching Tip: Video

Dare to Educate Afghan Girls

Students often struggle to understand how decisions about the environment might be related to culture and politics. After reading the opening case and having a discussion about China's one-child policy, use the TED talk "Dare to Educate Afghan Girls" to introduce another culture very different from that of the United States. Ask students why they believe education is important for society. As students broaden their awareness of the world around them, they are more likely to understand how social attitudes can affect attitudes toward the environment.

Find a link to this video by clicking on the link buttons in the TE-book, opening the TRFD, or logging onto the book's companion website bcs.whfreeman.com/friedlandapes2e (teacher login required).

Chapter 7 Web Resources

Teaching Tip: Debate the Issue

China's One-Child Policy
Have students research China's one-child policy in preparation for a debate to answer the question: Was China's one-child policy a successful population control measure?

Students should be prepared to discuss these topics in their debate:

- details of government restrictions and efforts to implement the campaigns
- strengths and weaknesses of the program
- outcomes after implementation
- social concerns
- recent changes to the policy

In class, select groups to debate the issue. Allow each side 2 minutes to prepare, 3 minutes to present, and 1 minute for rebuttal. Have the remainder of the class take notes, ask questions, and decide which side has the most persuasive argument. Because this particular debate has the potential to become a highly emotional discussion, ensure the debate remains focused on the policy by setting some ground rules, including the following:

- The debate is not whether abortion is right or wrong or whether it should be legal.
- The debate topic is not capitalism versus socialism.
- Debate points should be factual statements, not opinions.

Explain that a debate is different from an argument. One side presents its factual evidence while the other listens. Then, the other side may respond with its factual evidence. It is not a discussion between the teams. Enforce this rule by deducting points for any interruptions or whispering between team members. Students who are not respectful of their peers or who direct personal comments at others should be disqualified from participation.

one-child policy, economic prosperity has also contributed to slowing population growth. If China's current population dynamics continue, its population will reach a maximum by 2030 and begin declining shortly after that.

Population decline is only one part of the picture, however. Even if China's population were to stop growing today, the country's resource consumption would continue to increase as standards of living improve. Greater numbers of Chinese people are purchasing cars, home appliances, and other material goods that are common in Western nations. All of these products require resources to produce and use. Manufacturing a refrigerator requires mining and processing raw materials such as steel and copper, producing plastic from oil, and using large quantities of electricity. Having a refrigerator in the home increases daily electricity demand. All of these processes generate carbon dioxide, air and water pollution, and other waste products.

A look at a typical Chinese city street is evidence of the country's growing affluence. Between 1985 and 2002, China's population increased 30 percent, but the number of motor vehicles used in China grew by over 500 percent, from 3 million to 20 million. Today, China has over 50 million private cars on the road, and by 2020 the total will be 140 million, according to Chinese government estimates. China is already the second largest consumer of petroleum (after the United States), and concentrations of urban air pollutants, such as carbon monoxide and photochemical smog, are on the rise. In 2013, 7 of the 10 most polluted cities in the world were in China, according to an Asian Development Bank study.

There is some good news, however. China already has higher fuel efficiency standards for cars than does the United States, and it is quickly becoming a leader in the manufacturing of renewable energy technologies.

China's influence on the environment is dramatic because of its size and its increasing industrial activity. However, increasing industrial activity is occurring in many other parts of the world as human populations and consumption increase. What will be the environmental impact of humans in 2020, and what can we do to reduce it?

Sources:
H. Kan, B. Chen, and C. Hong, Health impact of outdoor air pollution in China: Current knowledge and future research needs, *Environmental Health Perspectives* 117 (2009): A187; Asian Development Bank, *Toward an Environmentally Sustainable Future: Country Environmental Analysis of the People's Republic of China,* August 2012 Report.

Human population growth and associated resource consumption have large impacts on the environment. In 2014, Earth's human population was 7.2 billion and increasing by 235,000 people per day. At the same time, people in many parts of the world are using more resources than ever before. Almost all aspects of environmental science are affected by the numbers of people on Earth and their activities. This chapter describes how human populations grow, what limits human population numbers, and the relative impacts of human population size and human consumption behavior.

COMMON MISCONCEPTIONS
Developed nations and environmental degradation

Students sometimes believe that developed nations such as the United States have less environmental impact than developing nations. Often students focus on population growth and the subsequent environmental degradation that occurs in less developed nations. Remind them that people living in developed nations tend to use more resources and those resources sometimes have greater impacts than resources used in developing countries.

Teaching Tip: Discussion Starter

Nova, televised by the Public Broadcasting Service, offers a "Global Trends Quiz" you can use to introduce the topics covered in this chapter.

Find a link to this resource by clicking on the link buttons in the TE-book, opening the TRFD, or logging onto the book's companion website bcs.whfreeman.com/friedlandapes2e (teacher login required).

Chapter 7 Web Resources

module

22

Human Population Numbers

Although it may seem surprising, environmental scientists do not know the human carrying capacity of Earth. However, we are able to discuss the factors that contribute to the carrying capacity of humans on Earth and what drives human population growth. We can also look at patterns of past human population growth to gain an understanding of what might occur in future decades.

Learning Objectives

After reading this module you should be able to

- explain factors that may potentially limit the carrying capacity of humans on Earth.
- describe the drivers of human population growth.
- read and interpret an age structure diagram.

Scientists disagree on Earth's carrying capacity

Every 5 days, the global human population increases by more than a million lives: 1.9 million infants are born and 800,000 people die. The human population has not always grown at this rate, however. As FIGURE 22.1 shows, until a few hundred years ago the human population was relatively stable: Deaths and births occurred in roughly equal numbers. This situation changed about 400 years ago, when agricultural output increased and sanitation began to improve. Better living conditions caused death rates to fall, but birth rates remained relatively high. This was the beginning of a period of rapid population growth that has brought us to the current human population of 7.2 billion people.

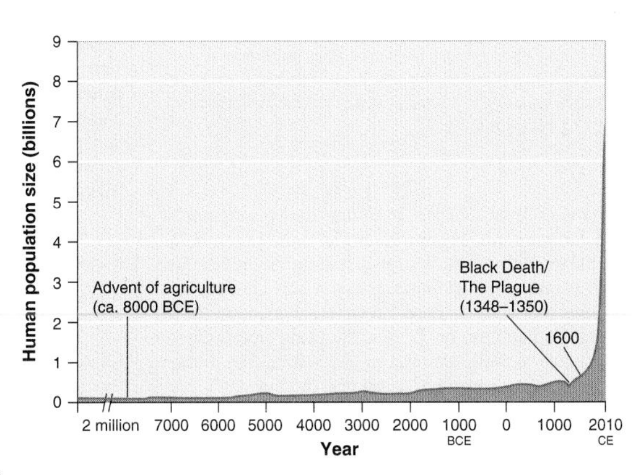

FIGURE 22.1 **Human population growth.** The global human population has grown more rapidly in the last 400 years than at any other time in history.

Teaching Tip: Engage

Ask students: What are potential limits to human population growth? Are these limits density-independent or density-dependent? *Density-dependent limits to human population growth might include fresh water supplies and food. Density-independent limits to human population growth might include catastrophic flooding or an earthquake.*

◀ TEACHING with FIGURES

Ask students to examine human population growth in Figure 22.1. Then ask them to determine the approximate year the human population began to experience exponential growth. *According to the graph, exponential growth begins to occur approximately 400 years ago in the year 1600* CE.

Teaching Tip: Engage

View the interactive “Human Numbers Through Time” on PBS.org. The visual begins 2,000 years ago and projects global population numbers through 2050. As greater numbers of dots fill the map students can see the dramatic growth of the human population.

Find a link to this resource by clicking on the link buttons in the TE-book, opening the TRFD, or logging onto the book's companion website bcs.whfreeman.com/friedlandapes2e (teacher login required).

Chapter 7 Web Resources

TEACHING with FIGURES ▶

Ask students to examine Figure 22.2 and describe the difference between the two graphs. Ask them to identify how humans might differ from other organisms in terms of carrying capacity. *Figure 22.2a shows the human population exceeding available food supply, resulting in a significant food deficit. Figure 22.2b models increases in food supply with significant improvements in agriculture, which would allow the human population to enjoy a surplus of food. Humans can alter Earth's carrying capacity by employing innovation and technology.*

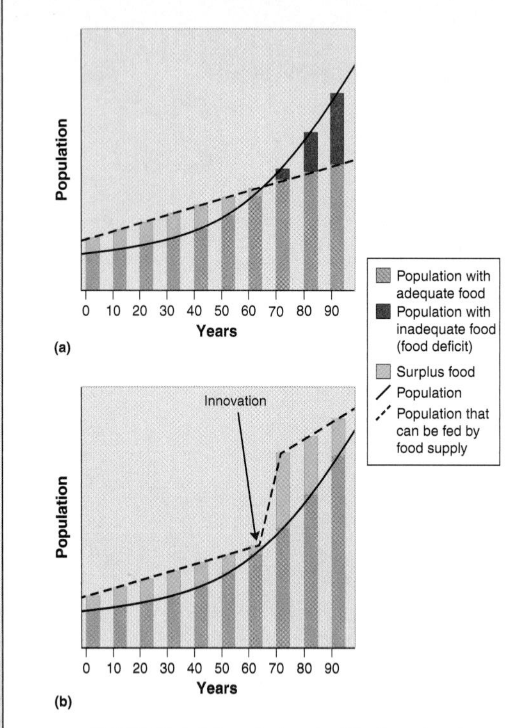

FIGURE 22.2 **A theoretical model of food supply and population size.** (a) In a theoretical 100-year period *without* significant improvements in agricultural technology, the human population grows exponentially, while the food supply grows linearly. Consequently, a food surplus is followed by a food deficit. (b) In a theoretical 100-year period *with* significant improvements in agricultural technology, the food supply increases suddenly. Consequently, there is a continuing food surplus.

As we saw in Chapter 6, under ideal conditions all populations grow exponentially. In most cases, exponential growth slows or stops when an environmental limit is reached. The limiting factor can be a scarcity of resources such as food or water or an increase in predators, parasites, or diseases. Limiting factors determine the carrying capacity of a habitat. Are human populations constrained by limiting factors?

Environmental scientists have differing opinions on Earth's carrying capacity for humans. Some scientists believe we have already outgrown, or eventually will outgrow, the available supply of food, water, timber, fuel, and other resources on which humans rely. One of the first proponents of the notion that the human population could exceed Earth's carrying capacity was English clergyman and professor Thomas Malthus. In 1798, Malthus observed that the human population was growing exponentially while the food supply we rely on was growing linearly. In other words, the food supply increases by a fixed amount each year, while the human population increases in proportion to its own increasing size. Malthus concluded that the human population size would eventually exceed the food supply. FIGURE 22.2a shows this projection graphically.

A number of environmental scientists today subscribe to Malthus's view that humans will eventually reach the carrying capacity of Earth, after which the rate of population growth will decline. Other scientists do not believe that Earth has a fixed carrying capacity for humans. They argue that the growing population of humans provides an increasing supply of intellect that leads to increasing amounts of innovation. Humans can alter Earth's carrying capacity by employing creativity—one of the fundamental ways in which humans differ from most other species on Earth.

For example, in the past whenever the food supply seemed small enough to limit the human population, major technological advances increased food production. This progression began thousands of years ago. The development of arrows made hunting more efficient, which allowed hunters to feed a larger number of people. Early farmers increased crop yields with hand plows and later with oxen- or horse-driven plows. More recently, mechanical harvesters made farming even more efficient. Each of these inventions increased the planet's carrying capacity for humans, as Figure 22.2b shows.

The ability of humans to innovate in the face of challenges has led some scientists to expect that we will continue to make technological advances indefinitely. This expectation is reasonable, but questions remain. Based on our history, should we assume that humans will continue to find ways to feed a growing population? Are there other limits to human population growth? And how do we know if we have exceeded Earth's carrying capacity?

Many factors drive human population growth

As we know from our study of biological populations, a variety of factors influence the growth, reproduction, and success of plant and animal species. Many of these same factors influence human populations as well. Population size, birth and death rates, fertility, life expectancy, and migration are factors that influence population size in countries. In order to understand the impact of the human population on the environment,

Teaching Tip: Engage

Students are usually fascinated by population clocks. Using a population clock can help students see that the world population is climbing rapidly. Visualizing this growth will motivate student interest in population calculations.

Find a link to a United States Bureau of the Census population clock by clicking on the link buttons in the TE-book, opening the TRFD, or logging onto the book's companion website bcs.whfreeman.com/friedlandapes2e (teacher login required).

TRM Chapter 7 Web Resources

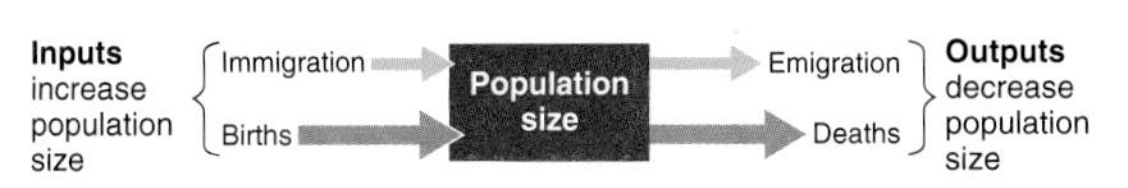

FIGURE 22.3 **The human population as a system.** We can think of the human population as a system, with births and immigration as inputs and deaths and emigration as outputs.

we must first understand what drives human population growth. The study of human populations and population trends is called **demography,** and scientists in this field are called **demographers.** By analyzing specific data such as changes in population size, fertility, life expectancy, and migration, demographers can offer insights—some of them surprising—into how and why human populations change and what can be done to influence rates of change.

Changes in Population Size

We can view the human population as a system with inputs and outputs, like all biological systems. If there are more births than deaths, the inputs are greater than the outputs, and the system expands. For most of human history, total births slightly outnumbered total deaths, resulting in very slow population growth. If the reverse had been true, the human population would have decreased and would have eventually become extinct.

When demographers look at population trends in individual countries, they take into account inputs and outputs. As FIGURE 22.3 shows, inputs include both births and **immigration,** which is the movement of people into a country or region from another country or region. Outputs, or decreases, include deaths and **emigration,** which is the movement of people out of a country or region. When inputs to the population are greater than outputs, the growth rate is positive. Conversely, if outputs are greater than inputs, the growth rate is negative.

Demographers use specific measurements to determine yearly birth and death rates. The **crude birth rate (CBR)** is the number of births per 1,000 individuals per year. The **crude death rate (CDR)** is the number of deaths per 1,000 individuals per year. Worldwide, there were 20 births and 8 deaths per 1,000 people in 2014. We do not factor in migration for the global population because, even though people move from place to place, they do not leave Earth. Thus, in 2014, the global population increased by 12 people per 1,000 people. This rate can be expressed mathematically as a percentage:

$$\text{Global population growth rate} = \frac{[CBR - CDR]}{10} = \frac{[20-8]}{10} = 1.2\%$$

We divide by 10 to arrive at the percentage because the birth and death rates are expressed per 1,000 people.

To calculate the population growth rate for a single nation, we take immigration and emigration into account:

$$\text{National population growth rate} = \frac{[(CBR + \text{immigration}) - (CDR + \text{emigration})]}{10}$$

If we know the growth rate of a population and assume that growth rate is constant, we can calculate the number of years it takes for a population to double, which is known as its **doubling time.** As a population grows rapidly, the doubling time gives us a better sense of the magnitude of the change than the growth rate alone. Because growth rates may change in future years, we can never determine a country's doubling time with certainty. Therefore, we say that a population will double in a certain number of years if the growth rate remains constant.

The doubling time can be approximated mathematically using a formula called the rule of 70:

$$\text{Doubling time (years)} = \frac{70}{\text{growth rate}}$$

Therefore, a population growing at 2 percent per year will double every 35 years:

$$\frac{70}{2} = 35 \text{ years}$$

Note that this is true of any population growing at 2 percent per year, regardless of the size of that population. At a 2 percent growth rate, a population of 50,000 people will increase by 50,000 in 35 years, and a population of 50 million people will increase by 50 million in 35 years.

Demography The study of human populations and population trends.

Demographer A scientist in the field of demography.

Immigration The movement of people into a country or region, from another country or region.

Emigration The movement of people out of a country or region.

Crude birth rate (CBR) The number of births per 1,000 individuals per year.

Crude death rate (CDR) The number of deaths per 1,000 individuals per year.

Doubling time The number of years it takes a population to double.

Teaching Tip: Warm-up

Ask students the following questions to help them understand how populations are dynamic:

- What factors increase the size of a population? *Births and immigration increase the size of a population.*
- What factors decrease the size of a population? *Deaths and emigration decrease the size of a population.*
- Write a formula to describe these changes. *$(B + I) - (D + E) =$ change in population size.*

more **math** practice

The Rule of 70

If the population of rabbits in an ecosystem grows at a rate of approximately 4 percent per year, the number of years required for the rabbit population to double is closest to

(a) 4 years.
(b) 8 years.
(c) 12 years.
(d) 17 years.
(e) 25 years.

Doubling time (years) $= 70 \div$ growth rate

$70 \div 4 = 17.5$ years

The closest answer to 17.5 would be "d" 17 years.

more **math** practice

The Mathematical Basis for the Rule of 70

If students want to know where the rule of 70 originated, you can tell them that it comes from the equation

$$N_t = N_0\, e^{(rt)}$$

where N_t is population at some time in the future and N_0 is the population now, r is the growth rate expressed as a decimal and t is the time in years. When you set $N_t = N_0 \times 2$ (double) and solve for t, you get $rt = \ln 2$ and

$$t = \frac{r}{0.69}$$

AP® Exam Tip

Remember, the rule of 70 is an approximation; the actual rule is 69.3. However the AP® exam will most likely ask for an approximate value, which means the student should know how to work with the rule of 70.

Teaching Tip: Activity

Be a Demographer
To compare demographic data between diverse countries, use Activity 7.1, "Be a Demographer." This activity will help students build graphing and analysis skills. It will also increase familiarity with vocabulary used in demographic language.

Find a link to this resource by clicking on the link buttons in the TE-book, opening the TRFD, or logging onto the book's companion website bcs.whfreeman.com/friedlandapes2e (teacher login required).

Activity 7.1 Be a Demographer

Teaching Tip: Engage

Ask students: What is the difference between replacement-level fertility and total fertility rate (TFR)? *Total fertility rate is an estimate of the average number of children a woman in a population will bear in her lifetime. Replacement level fertility is the total fertility rate required to offset the average number of deaths in a population in order to maintain current population size.*

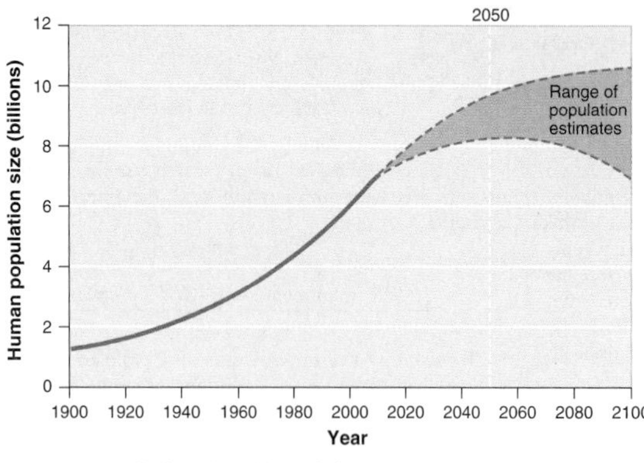

FIGURE 22.4 **Projected world population growth.** Demographers project that the global human population will be between 8.1 billion and 9.6 billion by 2050. By 2100, it is projected to be between 6.8 billion and 10.5 billion. The dashed lines represent estimated values. *(After Millennium Ecosystem Assessment, 2005)*

As we saw in Figure 22.1, Earth's population has doubled several times since 1600. It is almost certain, however, that Earth's population will not double again. FIGURE 22.4 shows the current projections through the year 2100. Most demographers believe that the human population will be somewhere between 8.1 billion and 9.6 billion in 2050 and will stabilize between 6.8 billion and 10.5 billion by roughly 2100.

Fertility

To understand more about the role births play in population growth, demographers look at the **total fertility rate (TFR),** an estimate of the average number of children that each woman in a population will bear throughout her childbearing years (between the onset of puberty and menopause). For example, in the United States in 2014, the TFR was 1.9, meaning that, on average, each woman of childbearing age gave birth to just under 2 children. Note that, unlike crude birth rate and crude death rate, TFR is not calculated per 1,000 people. Instead, it is a measure of births per woman.

To gauge changes in population size, demographers also calculate **replacement-level fertility,** the TFR required to offset the average number of deaths in a population so that the current population size remains stable. Typically, replacement-level fertility is just over 2 children. Two children replace the parents who conceive them when the parents die. Replacement level fertility is higher than 2, however, because it also must account for children who die before they are able to have children or people who otherwise do not have children. As we shall see, the rate of death among children depends on a country's economic status.

In **developed countries**—countries with relatively high levels of industrialization and income—we typically see a replacement-level fertility of about 2. In **developing countries**—those with relatively low levels of industrialization and incomes of less than $3 per person per day—a TFR of greater than 2.1 is needed to achieve replacement-level fertility. Replacement level fertility is higher in developing countries because mortality among young people tends to be higher.

In a country where TFR is equal to replacement-level fertility, and where immigration and emigration are equal, the country's population is stable. A country with a TFR of less than 2.1 and no net increase from immigration is likely to experience a population decrease because that country's TFR is below replacement-level fertility. In contrast, a developed country with a TFR of more than 2.1 and no net decrease from emigration is likely to experience population growth because that country's TFR is above replacement-level fertility.

Total fertility rate (TFR) An estimate of the average number of children that each woman in a population will bear throughout her childbearing years.

Replacement-level fertility The total fertility rate required to offset the average number of deaths in a population in order to maintain the current population size.

Developed country A country with relatively high levels of industrialization and income.

Developing country A country with relatively low levels of industrialization and income.

Life expectancy The average number of years that an infant born in a particular year in a particular country can be expected to live, given the current average life span and death rate in that country.

Life Expectancy

To understand more about the outputs in a human population system, demographers study the human life span. **Life expectancy** is the average number of years that an infant born in a particular year in a particular country can be expected to live, given the current average life span and death rate in that country. Life expectancy is generally higher in countries with better health care. A high life expectancy also tends to be a good predictor of high resource consumption rates and environmental impacts. FIGURE 22.5 shows life expectancies around the world.

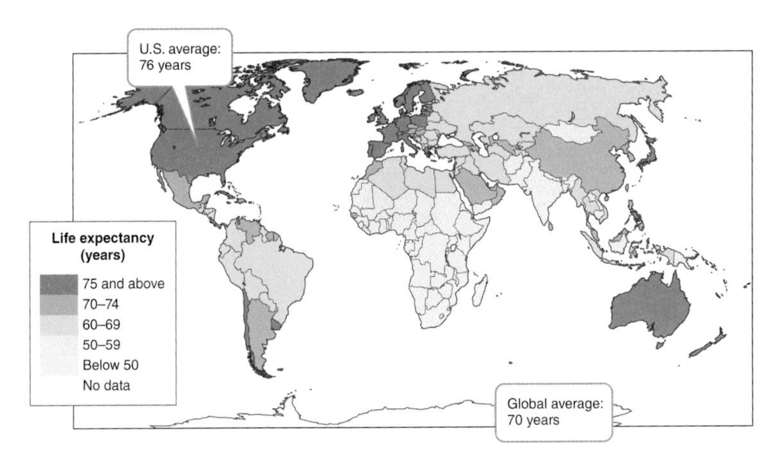

FIGURE 22.5 **Average life expectancies around the world.** Life expectancy varies significantly by continent and in some cases by country. *(Data from CountryWatch, http://www.countrywatch.com/facts/facts_default.aspx?type=image&img=LEAG; data from http://www.worldlifeexpectancy.com/world-life-expectancy-map)*

Life expectancy is often reported in three different ways: for the overall population of a country, for males only, and for females only. For example, in 2014, global life expectancy was 70 years overall, 68 years for men, and 72 years for women. In the United States, life expectancy was 76 years overall, 73 years for men, and 81 years for women. In general, human males have higher death rates than human females, leading to a shorter life expectancy for men. In addition to biological factors, men have historically tended to face greater dangers in the workplace, made more hazardous lifestyle choices, and been more likely to die in wars. Cultures have changed over time, however, and as more and more women enter the workforce and the armed forces, the life expectancy gap between men and women will probably decrease.

Infant and Child Mortality

The availability of health care, access to good nutrition, and exposure to pollutants are all factors in life expectancy, *infant mortality*, and *child mortality*. The **infant mortality** rate is defined as the number of deaths of children under 1 year of age per 1,000 live births. The **child mortality** rate is defined as the number of deaths of children under age 5 per 1,000 live births. FIGURE 22.6 shows infant mortality rates around the world.

If a country's life expectancy is relatively high and its infant mortality rate is relatively low, it is likely that the country has a high level of available health care, an adequate food supply, potable drinking water, good sanitation, and a moderate level of pollution. Conversely, if its life expectancy is relatively low and its infant mortality rate is relatively high, it is likely that the country's population does not have sufficient health care or sanitation and that potable drinking water and food are in limited supply. Pollution and exposure to other environmental hazards may also be high. In 2014, the global infant mortality rate was 41. In the United States, the infant mortality rate was 5.9. In other developed countries, such as Sweden (2.6) and France (3.3), the infant mortality rate was even lower. Availability of prenatal care is an important predictor of the infant mortality rate. For example, the infant mortality rate is 63 in Liberia and 40 in Bolivia, both countries where many women do not have good access to prenatal care.

Sometimes, life expectancy and infant mortality in a given sector of a country's population differ widely from life expectancy and infant mortality in the country as a whole. In this case, even when the overall numbers seem to indicate a high level of health care throughout the country, the reality may be starkly different for a portion of its population. For example, whereas the infant mortality rate for the U.S. population as a whole is 6.0, it is 12.4 for African Americans, 8.5 for Native Americans, and 5.3 for Caucasians. This variation in infant mortality rates is probably related to socioeconomic status and varying degrees of access to adequate nutrition and health care. These differences are often issues of environmental justice, a topic we discuss in more detail in Chapter 20.

Infant mortality The number of deaths of children under 1 year of age per 1,000 live births.

Child mortality The number of deaths of children under age 5 per 1,000 live births.

◀ TEACHING with FIGURES

Show students Figures 22.5 and 22.6 (see page 232.). Ask students the following questions:

- Which countries or regions of the world experience long life expectancies and low infant mortality rates? *The United States, Europe, and Australia experience long life expectancies and have low infant mortality rates.*
- Which regions have short life expectancies and high infant mortality rates? *Most of sub-Saharan Africa experiences short life expectancies and high infant mortality rates.*
- Name the common characteristics shared among nations that enjoy long life expectancies and low infant mortality rates. *Countries with long life expectancies and low infant mortality rates have a high level of health care, adequate food supplies, potable drinking water, good education, and low levels of pollution.*

Teaching Tip: Journal Prompt

Infant mortality rates can vary widely, even within a given country. Ask students: What are some reasons for this variation? *Variation in infant mortality rates is likely due to differences in socioeconomic status, race, ethnicity, and access to adequate nutrition and health care.*

Teaching Tip: Beyond the Classroom

In the online *Demography Lab*, students capture and record birth and death rates from around the country at different periods and consider factors that could have influenced these rates such as advances in medicine and environmental protection. Students will also evaluate future population trends, taking into account socioeconomic factors, social stability, and disease.

Find a link to this resource by clicking on the link buttons in the TE-book, opening the TRFD, or logging onto the book's companion website bcs.whfreeman.com /friedlandapes2e (teacher login required)

Chapter 7 Web Resources

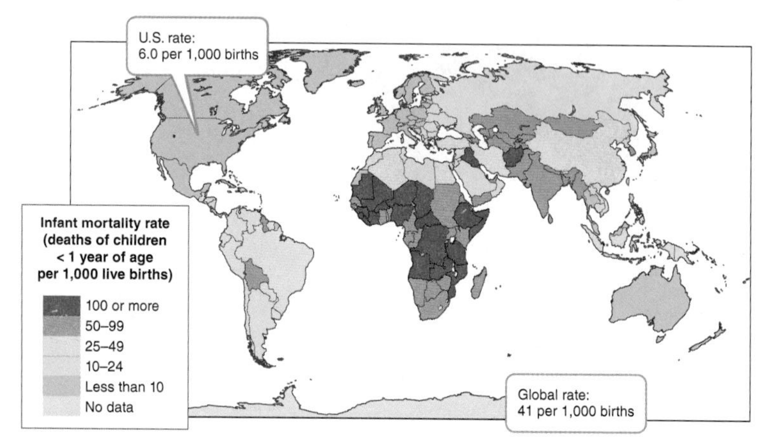

FIGURE 22.6 **Infant mortality around the world.** Infant mortality rates are lower in developed countries and higher in developing countries. *(Data from http://www.mapsofworld.com/infant-mortality-rate-map.htm#bot)*

Aging and Disease

Even with a high life expectancy and a low infant mortality rate, a country may have a high crude death rate, in part because it has a large number of older individuals. The United States, for example, has a higher standard of living than Mexico, which is consistent with the higher life expectancy and lower infant mortality rate in the United States. At the same time, the United States has a much higher CDR, at 8 deaths per 1,000 people on average, than Mexico, which has 5 deaths per 1,000 people on average. This higher CDR results from the much larger elderly population in the United States, with 13 percent of its population aged 65 years or older, compared with the 6 percent of the population aged 65 or older in Mexico.

Disease is an important regulator of human populations. According to the World Health Organization, infectious diseases—those caused by microbes that are transmissible from one person to another—are the second biggest killer worldwide after heart disease. In the past, tuberculosis and malaria were two of the infectious diseases responsible for the greatest number of human deaths. Today, the human immunodeficiency virus (HIV), which causes acquired immune deficiency syndrome (AIDS), is responsible for more deaths annually than either tuberculosis or malaria. Between 1990 and 2012, AIDS-related illnesses killed more than 28 million adults and children. Because HIV disproportionately infects people aged 15 to 49—the most productive years in a person's life span—HIV has had a more disruptive effect on society than other illnesses that affect the very young and the very old.

HIV has a significant effect on infant mortality, child mortality, population growth, and life expectancy. In Lesotho, in southern Africa, where 23 percent of the adult population is infected with HIV, life expectancy fell from 63 years in 1995 to 40 years in 2009. It has since been increasing and is now approximately 48.

As FIGURE 22.7 shows, approximately 34 million people were living with HIV in 2011, 22 million of

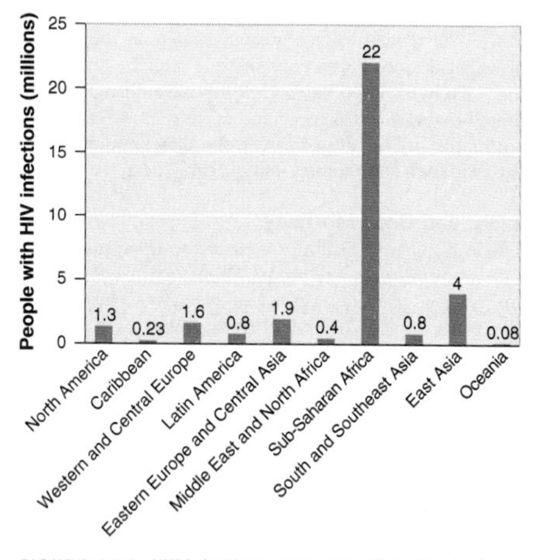

FIGURE 22.7 **HIV infection worldwide.** Worldwide, about 33 million people are living with HIV, two-thirds of them in sub-Saharan Africa. *(Data from World Health Organization, UNAIDS)*

AP® Exam Tip

Trends in worldwide fertility are covered in the 2008 AP® Exam, Question 4. To answer this question, students must

- create a graph of TFR from population data
- identify causes of the trends in worldwide TFR shown in the graph they create
- discuss economic and societal factors that explain the differences in TFR of a developed and a developing nation
- describe human activities associated with population growth that have affected biodiversity on Earth

Because this question is written in a straightforward manner and the scoring guidelines are clear, this free-response question provides an excellent opportunity for peer review. Teach students to use the scoring guidelines and to evaluate the work of others as well as their own work. Evaluating free-response questions helps students to better understand how to answer exam questions effectively.

Teaching Tip: Warm-up

Ask students to explain how a country that exhibits long life expectancies can still have a high crude death rate. *If a country has a large elderly population, the crude death rate can be high even if individuals live to older ages.*

▲ TEACHING with FIGURES

Explain to students that disease is a limiting factor in human population growth. HIV/AIDS has had a disproportionately large effect on society because it affects people in their most productive years. Show students Figure 22.7 and ask them to name the region of the world that is most affected by this disease. *Sub-Saharan Africa is most affected by this disease. Of the 34 million people living with HIV worldwide, 22 million are located in sub-Saharan Africa.*

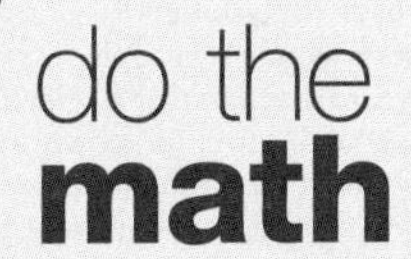

Calculating Population Growth

In 2012, New Zealand had a population of 4.3 million people, a TFR of 2.1, and a net migration rate of 2 per 1,000. How many people will New Zealand gain in the following year as a result of immigration? If the TFR stays the same for the next century, and the net migration rate stays the same as well, when will the population of New Zealand double?

$$\text{Net migration rate} = \frac{\text{number of immigrants/year}}{\text{number of people in the population}}$$

A TFR of 2.1 for a developed country suggests that the country is at replacement-level fertility and, therefore, the population is stable. The migration rate suggests that

$$\frac{2}{1{,}000} = \frac{x}{4{,}300}$$

and therefore that

$$x = 8{,}600 \text{ people/year}$$

So, 8,600 people are added to New Zealand each year by migration. If there is no growth due to biological replacement, then the rate of increase is

$$\frac{8{,}600 \text{ people/year}}{4{,}300{,}000 \text{ people}} = 0.2\%/year$$

Your Turn How many years will it take for the New Zealand population to double if the population increases due to migration only? Recall that to calculate doubling time, we use the rule of 70:

$$\text{Doubling time (years)} = \frac{70}{\text{growth rate}}$$

them in sub-Saharan Africa. The annual number of deaths due to AIDS reached a peak of 2.1 million in 2005 but has decreased since then. We will talk more about AIDS and other infectious diseases in Chapter 17.

Migration

Regardless of its birth and death rates, a country may experience population growth, stability, or decline as a result of migration. **Net migration rate** is the difference between immigration and emigration in a given year per 1,000 people in a country. A positive net migration rate means there is more immigration than emigration, and a negative net migration rate means the opposite. For example, approximately 1 million people immigrate to the United States each year, and only a small number emigrate. With a U.S. population of 315 million, these rates are equal to 3.2 immigrants per 1,000 people.

A country with a relatively low CBR but a high immigration rate may still experience population growth. For example, the United States has a TFR of 1.9, but it has a high net migration. As a result, the U.S. population will probably increase by 30 percent by 2050. Canada has a net migration rate of 0.7 per 1,000 and a TFR of 1.6, which is well below replacement-level fertility. Therefore, both the United States and Canada will experience net population growth over the next few decades, but the growth will come from immigration rather than from births originating within the existing population. "Do the Math: Calculating Population Growth" shows how this can happen.

In countries with a negative net migration rate and a low TFR, the population actually decreases over time. Very few countries fit this model. One is the country of Georgia, in western Asia. It has a growth rate of 0.2 percent, a TFR of 1.7, and a net migration rate of −5 per 1,000. Georgia is projected to have a 20 percent population decrease by 2050.

Net migration rate The difference between immigration and emigration in a given year per 1,000 people in a country.

Answer to Your Turn

The growth rate is 0.2 percent per year.

$$\text{Doubling time} = \frac{70}{0.2} = 350 \text{ years}$$

more **math** practice

Calculating Population Growth

More Math Practice 7.1 offers students practice with calculating population growth statistics. Download this resource by clicking on the link buttons in the TE-book, opening the TRFD, or logging onto the book's companion website bcs.whfreeman.com/friedlandapes2e (teacher login required).

 **More Math Practice 7.1**
Population Calculations

AP® Exam Tip

Students are expected to know the approximate population sizes of the world and the three most populous countries, China, India, and the United States. This information is essential to solving math problems on previous exams, but it is not always given in the problem. Be sure to look up current statistics because population numbers will change. GeoHive is a good website to quickly review current statistics.

Find a link to this resource by clicking on the link buttons in the TE-book, opening the TRFD, or logging onto the book's companion website bcs.whfreeman.com /friedlandapes2e (teacher login required).

 **Chapter 7 Web Resources**

Teaching Tip: Review

Ask students the following questions:

- Explain the difference between a positive net migration rate and negative net migration rate. *A positive net migration rate means there is more immigration than emigration while negative net migration means the opposite.*
- How might a country with low CBR still experience population growth? *High immigration rates maylead to population growth.*
- What would you expect to happen to population size in a country that has negative net immigration and low TFR? *The population size would decrease.*

Teaching Tip: Engage

Some countries experience both high levels of emigration and a high CBR. Fiji, for example, has a growth rate of 1.7 percent, a TFR of 2.6, and a net migration rate of −8 per 1,000. With a population of 800,000, Fiji loses 6,400 people to emigration every year. However, the increase in population due to more births than deaths exceeds the losses from emigration. Thus, Fiji's population is still projected to increase by 12 percent over the next 40 years.

Teaching Tip: Video

Population Pyramids: Powerful Predictors of the Future

Show this short video from TED-Ed to give a good visual explanation of age structure diagrams and how they are used.

Find a link to this resource by clicking on the link buttons in the TE-book, opening the TRFD, or logging onto the book's companion website bcs.whfreeman.com/friedlandapes2e (teacher login required).

 **Chapter 7 Web Resources**

Teaching Tip: Warm-up

Ask students the following questions:

- What information does an age structure diagram include? *An age structure diagram shows the distribution of age groups within a population and is divided according to the number of males and females.*
- Why is age structure of a population important to demographers? *Demographers use data on age to predict how rapidly a population is growing and the size it will be in the future.*

We have noted that although the movement of people around the world does not affect the total number of people on the planet, migration is still an important issue in environmental science. The movement of people displaced because of disease, natural disasters, environmental problems, or conflict can create crowded, unsanitary conditions, and shortages of food and water. In some cases people are moved into refugee camps where they have little opportunity to improve their conditions through employment or emigration. All of these situations can easily become humanitarian and environmental health issues. The movement of people from developing countries to developed countries tends to increase the ecological footprint of those people because, over time, immigrants typically adopt the lifestyle and consumption habits of their new country. A person who migrates from Mexico to the United States, for example, is likely to use more resources as a U.S. resident than as a resident of Mexico because the United States typically has a more affluent lifestyle.

Age structure diagrams describe how populations are distributed across age ranges

Demographers use data on age to predict how rapidly a population will increase and what its size will be in the future. The age structure of a population describes how its members are distributed across age ranges, usually in 5-year increments. **Age structure diagrams,** examples of which are shown in FIGURE 22.8, are visual representations of the number of individuals within specific age groups for a country, typically expressed for males and females. Each horizontal bar of the diagram represents a 5-year age group. The total area of all the bars in the diagram represents the size of the whole population.

Age structure diagram A visual representation of the number of individuals within specific age groups for a country, typically expressed for males and females.

Population pyramid An age structure diagram that is widest at the bottom and smallest at the top, typical of developing countries.

Population momentum Continued population growth after growth reduction measures have been implemented.

While every nation has a unique age structure, we can group countries very broadly into three categories. A country with many more younger people than older people has an age structure diagram that is widest at the bottom and narrowest at the top, as shown in Figure 22.8a. This type of age structure diagram, called a **population pyramid,** is typical of developing countries, such as Venezuela and India. The wide base of the graph compared with the levels above it indicates that the population will grow because a large number of females aged 0 to 15 have yet to bear children. Even if each one of these future potential mothers has only two children, the population will grow simply because there are increasing numbers of women able to give birth.

The population pyramid can also be used to illustrate how long a time it takes for changes to affect a growing population. **Population momentum** is continued population growth after growth reduction measures have been implemented. It occurs because there are relatively large numbers of individuals at reproductive maturity in the population. Population momentum has been compared with the momentum of a long, heavy freight train, which takes longer to stop than a shorter, lighter freight train. It is the reason why a population keeps on growing after birth control policies or voluntary birth reductions have begun to lower the CBR of a country. Eventually, over several generations, those actions will bring the population to a more stable growth rate but the momentum of all the individuals who have recently reached child-bearing age will carry the population forward for a number of years.

A country with little difference between the number of individuals in younger age groups and in older age groups has an age structure diagram that looks more like a column from age 0 through age 50, as shown in Figure 22.8b. If a country has few individuals in the younger age classes, we can deduce that it has slow population growth or is approaching no growth at all. The United States, Canada, Australia, Sweden, and many other developed countries have this type of age structure diagram. A number of developing countries that have recently lowered their growth rates should begin to show this pattern within the next 10 to 15 years.

A country with a greater number of older people than younger people has an age structure diagram that resembles an inverted pyramid. Such a country has a total fertility rate below 2.1 and a decreasing number of females within each younger age range. Such a population will continue to shrink. Italy, Germany, Russia, and a few other developed countries display this pattern, seen in Figures 22.8c and 22.8d. China is in the very early stages of showing this pattern.

Teaching Tip: Journal Prompt

Ask students to describe the phenomenon of population momentum and to identify a country that has recently experienced population momentum. *Population momentum is the time it takes for a change such as a decline in TFR to affect a growing population. Because there will be large numbers of individuals at reproductive maturity, the population will continue to increase even after growth reduction measures have been implemented. India is one example. Even though India has a TFR of 2.5, just a little over replacement level fertility, its population will continue to grow substantially because more than 30% of the population is under age 15 and will reach reproductive maturity in the coming years. Thus we describe India as having population momentum.*

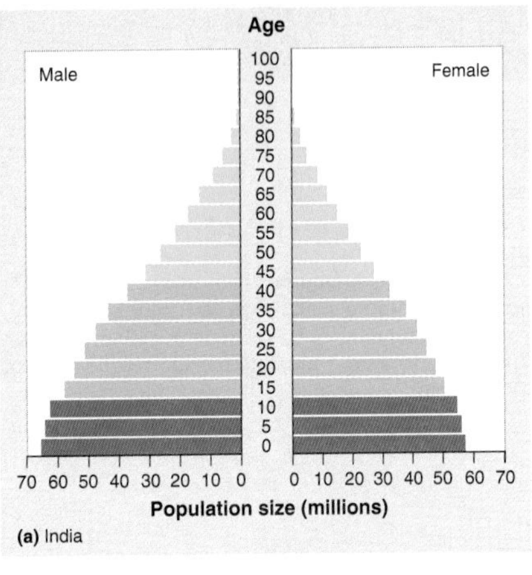

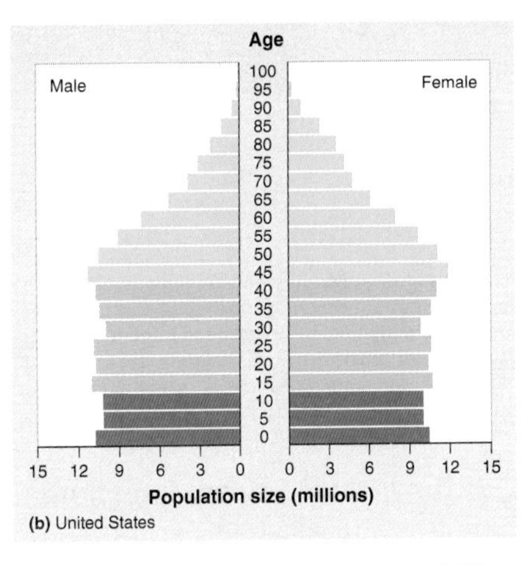

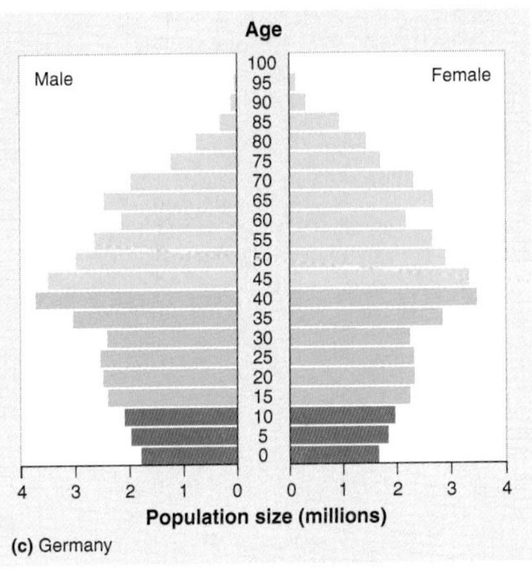

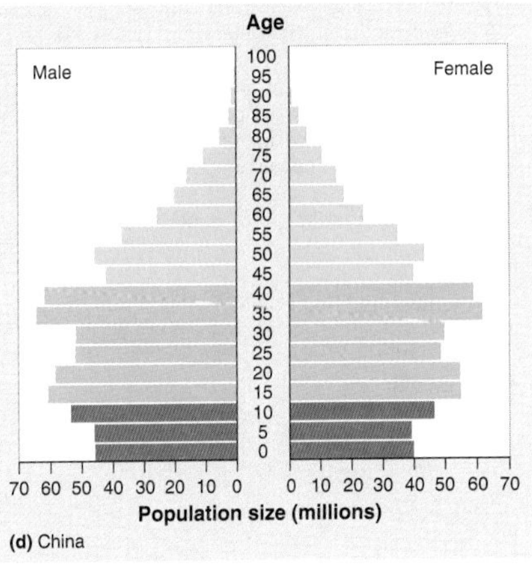

FIGURE 22.8 **Age structure diagrams in 2010.** The horizontal axis of the age structure diagram shows the population size in millions for males and females in each 5-year age group shown on the vertical axis. (a) A population pyramid illustrates a rapidly growing population. (b) A column-shaped age structure diagram indicates population stability. (c) In some developed countries, the population is declining. (d) China's population control measures will eventually lead to a population decline. *(Data from http://www.census.gov/ipc/www/idb/pyramids.html)*

lab
World Population

This lab teaches students how to research population data, and how to read and create a population diagram.

Download this resource by clicking on the link buttons in the TE-book, opening the TRFD, or logging onto the book's companion website bcs.whfreeman.com/friedlandapes2e (teacher login required).

Lab 7.1 World Population

◀ TEACHING with FIGURES

Ask students to describe the shape and trends of the three broad categories of age structure diagrams.

- ***Population Pyramid:*** *A broad base with many individuals approaching reproductive maturity. This age structure shape, such as the one shown in Figure 22.8a, indicates potential for rapid growth.*
- ***Even Distribution of Ages:*** *Age structure diagrams with little differences between numbers of individuals in age groups are shaped more like a column, as shown in Figure 22.8b. This type of population will experience slow growth or zero population growth.*
- ***Inverted Pyramid****: A country with more older people than younger people will look like an upside down pyramid as seen in Figure 22.8c. These populations will experience a decline in population numbers.*

AP® Exam Tip

Age structure diagrams are covered in the 2000 AP® Exam, Question 4. To answer this question, students must

- read and compare population growth among three age structure diagrams
- project likely infant mortality rates in two countries represented
- describe changes in birth and death rates for countries experiencing a transition from a preindustrial to an industrial economy
- describe an incentive a government could offer to reduce birth rates and a possible drawback of this incentive

Released free-response questions are an excellent resource that can be used to assess how well students understand the material. Furthermore, using released free-response questions frequently throughout the year will help familiarize students with the test format and prepare them for the exam in May.

module

22

REVIEW

In this module we have seen that scientists disagree about the human carrying capacity of Earth. While some believe that we have exceeded human carrying capacity on Earth, others suggest that human ingenuity and innovation will allow an almost infinite increase in the carrying capacity for human existence. Many factors drive human population growth, including crude birth rate, crude death rate, and fertility rates. The age structure diagram is an important tool for evaluating the age composition of populations.

Module 22 AP® Review Questions

1. Thomas Malthus introduced
   (a) the study of human demographics.
   (b) the hypothesis that humans could exceed Earth's carrying capacity.
   (c) the concept of variation in replacement-level fertility due to differences in industrialization.
   (d) the age structure diagram.
   (e) the concept of population momentum.

2. Doubling time can be approximated mathematically by
   (a) $\frac{70\%}{\text{growth rate}}$.
   (b) $\frac{10\%}{\text{growth rate}}$.
   (c) $2\% \times$ growth rate.
   (d) $\frac{CBR - CDR}{10}$.
   (e) $\frac{20}{CBR - CDR}$.

3. Replacement-level fertility is
   (a) just under 2 children in the United States.
   (b) the number of births per 1,000 women.
   (c) always greater than 2.
   (d) the base level of 1 child to replace the mother.
   (e) higher in developed countries than developing countries.

4. A country has a net immigration rate of 3 per 1,000, a CBR of 9 per 1,000, and a CDR of 11 per 1,000. What is the growth rate of this country?
   (a) 0.1%
   (b) 0.2%
   (c) 0.5%
   (d) 1%
   (e) 2%

5. An age structure diagram with roughly similar numbers of individuals in each group is representative of which country?
   (a) India
   (b) Mexico
   (c) Sweden
   (d) Venezuela
   (e) Russia

Answers to Module 22 AP® Review Questions

1. b
2. a
3. c
4. a
5. c

module

23

Economic Development, Consumption, and Sustainability

Human populations undergo change as a variety of natural and societal conditions in those populations change over time. Some of these changes occur with specific patterns that can be described and explained. In this module we will look at demographic transitions, the effect of these transitions on the environment, and the relationship between economic development and sustainability.

Learning Objectives

After reading this module you should be able to

- describe how demographic transition follows economic development.
- explain how relationships among population size, economic development, and resource consumption influence the environment.
- describe why sustainable development is a common but elusive goal.

Demographic transition follows economic development

As we saw in Module 22, populations in different countries change over time and the population of Earth has also fluctuated. Demographers are interested in understanding the reasons behind fluctuations in population growth in the past and whether they apply to contemporary or future demographic issues. We will begin this module by looking at an important theory of how populations change. We will also look at family planning because it is an important component of demographic transition.

The Theory of Demographic Transition

Historically, nations that have gone through similar processes of economic development have experienced similar patterns of population growth. Scientists who studied the population growth patterns of European countries in the early 1900s described a four-phase process they referred to as a demographic transition. The **theory of demographic transition** says that as

Theory of demographic transition The theory that as a country moves from a subsistence economy to industrialization and increased affluence it undergoes a predictable shift in population growth.

AP® Exam Tip

The 2003 AP® Exam, Question 2, provides a challenging graphing and calculation problem using population data. To answer this question, students must

- use population data to plot crude birth and death rates
- calculate the annual growth rate and birth rate
- consider factors that account for a decrease in death rates and determine reasons that the birth rate remains high

This question assesses understanding of the demographic transition model. Students can earn one point for scaling and labeling the axis correctly. Although this may seem like a simple task, many students scale the axis according to the data given rather than using equal increments. In the year this question was given, many students did not even attempt to answer it. When working with students on this question, remind them that getting easy points could be enough to exceed the average score.

To answer this question, students must understand the definition of "crude." Numbers are given as crude rates and also described as per 1,000. Part (d) requires students to understand and apply the rule of 70. Students should calculate two doubling times from 1895 to 1951 and will therefore need to double the population size twice. The difficulty of this free-response question makes it a good problem to teach in parts.

Teaching Tip: Activity

Use Activity 7.2, "The Demographic Transition Model," to teach trends in birth rate, death rate, and population size. This activity allows students to process concepts of the demographic transition model to include socioeconomic characteristics countries experience as they move through the demographic transition model.

Find a link to this resource by clicking on the link buttons in the TE-book, opening the TRFD, or logging onto the book's companion website bcs.whfreeman.com /friedlandapes2e (teacher login required).

Activity 7.2 The Demographic Transition Model

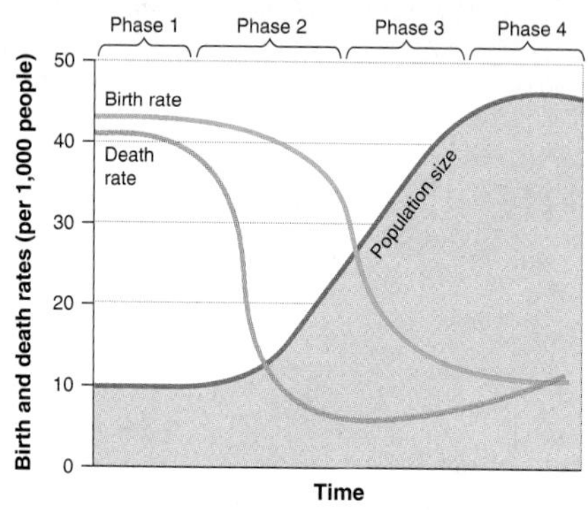

FIGURE 23.1 **Demographic transition.** The theory of demographic transition models the way that birth, death, and growth rates for a nation change with economic development. Phase 1 is a preindustrial period characterized by high birth rates and high death rates. In phase 2, as the society begins to industrialize, death rates drop rapidly, but birth rates do not change. Population growth is greatest at this point. In phase 3, birth rates decline for a variety of reasons. In phase 4, the population stops growing and sometimes begins to decline as birth rates drop below death rates.

a country moves from a subsistence economy to industrialization and increased affluence, it undergoes a predictable shift in population growth. The four phases of a demographic transition are shown in FIGURE 23.1.

The theory of demographic transition, while helpful as a learning tool, does not adequately describe the population growth patterns of some developing countries either today or during the last quarter-century. Both birth and death rates have declined rapidly in a number of developing countries because of a variety of factors that are not yet entirely understood. In some developing countries governments have taken measures to improve health care and sanitation and promote birth control, in spite of the country's poverty.

Despite the limitations of the theory of demographic transition, it is worth examining in more detail because it allows us to understand the way some countries influence the environment as they undergo growth and development.

Phase 1: Slow Population Growth

Phase 1 represents a population that is nearly at steady state. The size of the population will not change very quickly because high birth rates and high death rates offset one another. In other words, CBR equals CDR. This pattern is typical of countries before they begin to modernize. In these countries, life expectancy for adults is relatively short due to difficult and often dangerous working conditions. The infant mortality rate is also high because of disease, lack of health care, and poor sanitation. In a subsistence economy, where most people are farmers, having numerous children is an asset. Children can do jobs such as collecting firewood, tending crops, watching livestock, and caring for younger siblings. With no social security system, parents also count on having many children to care for them when they become old.

Western Europe and the United States were in phase 1 before the Industrial Revolution, which began in the late eighteenth century. Today, crude birth rates exceed crude death rates in almost every country, so even the poorest nations have moved beyond phase 1. However, an increase in crude death rates due to war, famine, and diseases such as AIDS has pushed some countries back in the direction of phase 1. For example, a decade ago, Lesotho (CBR = 26, CDR = 28 in 2005) moved back into phase 1 as a result of its high death rate.

Phase 2: Rapid Population Growth

In phase 2, death rates decline while birth rates remain high and, as a result, the population grows rapidly. As a country modernizes, better sanitation, clean drinking water, increased access to food and goods, and access to health care, including childhood vaccinations, all reduce the infant mortality rate and CDR. However, the CBR does not markedly decline. Couples continue to have large families because it takes at least one generation, if not more, for people to notice the decline in infant mortality and adjust to it. This is another example of population momentum. It also takes time to implement educational systems and birth control measures.

A phase 2 country is in a state of imbalance: Births outnumber deaths. India is in phase 2 today. The United States population exhibited a phase 2 population pyramid in the early twentieth century when there were high birth rates, low death rates, and a large total fertility rate.

Phase 3: Stable Population Growth

A country enters phase 3 as its economy and educational system improve. In general, as family income increases, people have fewer children, as FIGURE 23.2 shows. As a result, the CBR begins to fall. Phase 3 is typical of many developed countries, including the United States and Canada.

Why do people produce fewer children as their income increases? As societies transition from subsistence farming to more complex economic specializations, having large numbers of children may become a financial burden rather than an economic benefit. Relative **affluence**, more time spent pursuing education, and the

Affluence The state of having plentiful wealth including the possession of money, goods, or property.

Teaching Tip: Journal Prompt

Ask students to complete the following chart of demographic transition. (Answers are provided in italics.)

	Phase I Preindustrial	Phase II Transitional	Phase III Industrial	Phase IV Postindustrial
CBR and CDR patterns	*CBR and CDR both are high*	*CBR remains high, CDR begins to fall*	*CBR begins to fall, CDR continues to decrease*	*CBR is below the CDR*
Socioeconomic trends	• *Subsistence economy* • *Lack of medical care, sanitation, clean water*	• *Increased access to food, health care, and sanitation* • *Lack of birth control, education*	• *Increased affluence and education* • *Declining birth rates*	• *High affluence and economic development* • *More elderly* • *Possible increase of birth rates*
Impact on population growth rate	*Zero population growth*	*High population growth*	*Stable population growth*	*Negative population growth*

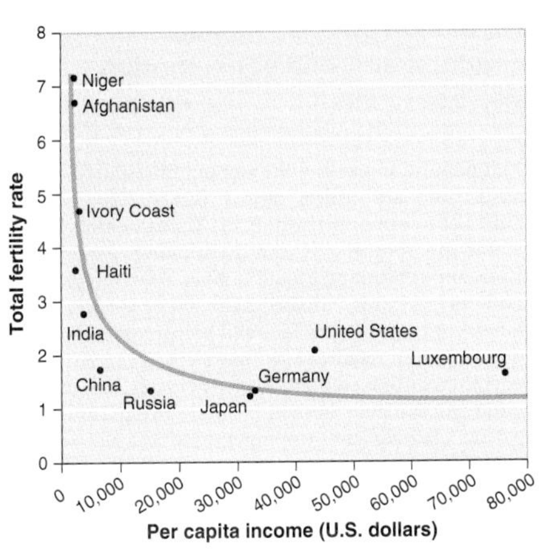

FIGURE 23.2 **Total fertility rate and per capita income.** Wealthier nations tend to have lower total fertility rates. *(Data from http://www.gapminder.org)*

availability of birth control increase the likelihood that people will choose to have smaller families. However, it is important to note that cultural, societal, and religious norms may also play a role in birth rates.

As birth rates and death rates decrease in phase 3, the system returns to a steady state. Population growth levels off during this phase, and population size does not change very quickly, because low birth rates and low death rates cancel each other out.

Phase 4: Declining Population Growth

Phase 4 is characterized by declining population size and often by a relatively high level of affluence and economic development. Japan, the United Kingdom, Germany, Russia, and Italy are phase 4 countries, with the CBR well below the CDR.

The declining population in phase 4 means fewer young people and a higher proportion of elderly people (FIGURE 23.3). This demographic shift can have important social and economic effects. With fewer people in the labor force and more people retired or working part-time, the ratio of dependent elderly to wage earners increases, and the costs of pension programs and social security services will increase the tax burden on each wage earner. There may be a shortage of health care workers to care for an aging population. Governments may encourage immigration as a source of additional workers. In some countries, such as Japan, the government provides economic incentives to encourage families to have more children in order to offset the demographic shift.

Recent studies on demographic shifts in highly developed countries suggest that the TFR actually increases after reaching a low point between 1.2 and 1.5. The reasons for this increase are unclear, but it appears that when a population becomes affluent and well educated, it becomes somewhat easier for women to raise children, and they choose to do so in slightly greater numbers. Such a pattern is occurring in Norway, Italy, the United States, and other developed nations.

FIGURE 23.3 **Elderly populations.** Some countries have very large elderly populations. Here, men and women gather at a senior residence home in San Diego, California. *(Stockbroker/Alamy Images)*

Family Planning

We have already observed that as family income increases, people tend to have fewer children. In fact, there is a link between higher levels of education and affluence among females, in particular, and lower birth rates. As the educational levels of women increase and women enter the workplace, fertility generally decreases. Even in developed countries where the TFR has increased slightly after hitting its apparent low point, women have fewer children than those in developing countries. Educated and working women tend to have fewer children than other women, and many delay having children because of the demands of school and work. Having a first child at an older age means that a woman is likely to have fewer children in her lifetime.

Women with more education and income also tend to have more access to information about methods of birth control, they are more likely to interact with their partners as equals, and they may choose to practice *family planning* with or without the consent of their partners. **Family planning** is regulation of the number or spacing of offspring through the use

Family planning The practice of regulating the number or spacing of offspring through the use of birth control.

◀ **TEACHING with FIGURES**

Ask students to evaluate Figure 23.2. What does the graph tell us about the relationship between income and fertility, and what are some probable reasons for this trend? *People with higher incomes tend to have lower total fertility rates. Probable reasons for this trend include access to medical care and birth control. Education becomes a priority and marriage may be delayed, resulting in fewer births per woman. Children in more affluent countries are more of a financial burden than an asset, and therefore families will have fewer children.*

Teaching Tip: Journal Prompt

Kenya, Thailand, and China have all implemented effective family planning campaigns. Ask students to list some reasons for the effective reduction in total fertility rates. What are some possible ethical concerns with government-led family planning campaigns? *All three countries made family planning accessible. Efforts were made to educate the public about benefits of smaller family size and how to use birth control. Possible ethical concerns may include the disregard for religious and cultural values; human rights violations could include forced sterilizations and infanticide of females. If a couple are permitted only one child, they may choose to abort a female fetus to ensure they have a male child to carry on the family name and traditions. For this reason, restricting families to one child has led to a disproportionate population ratio of males to females. Only children may incur the sole burden of caring for their parents and possibly their grandparents.*

Teaching Tip: Engage

Ask students to brainstorm the benefits of family planning as well as possible reasons some people might not use a family planning strategy.

Teaching Tip: Warm-up

Ask students: In what ways beyond population control might family planning be beneficial? *Sexually transmitted diseases will decrease with increased condom use. Women may be able to delay starting a family, which can open the door to higher education. With fewer numbers of children, families are more able to provide more education, medical care, and food. Levels of poverty decrease.*

TEACHING with FIGURES ▶

Show students Figure 23.4 and ask: What can you conclude about the effect of education among females? *Availability of secondary education for females in developing nations is associated with lower total fertility rates.*

Teaching Tip: Activity

Ask students to visit the Global Footprint Network and complete the interactive calculator. Students should be prepared to discuss how many Earths it takes to support their lifestyle and ways they can reduce their footprint. For example, a student might eat less meat, take public transportation, and carpool. Since students often have limited control over how and where they live at this stage in their lives, ask them how they plan to live in the future. For example, would they be open to living in a small apartment downtown where they can walk to work? Would they be willing to reduce their meat consumption?

Find a link to this resource by clicking on the link buttons in the TE-book, opening the TRFD, or logging onto the book's companion website bcs.whfreeman.com /friedlandapes2e (teacher login required).

Chapter 7 Web Resources

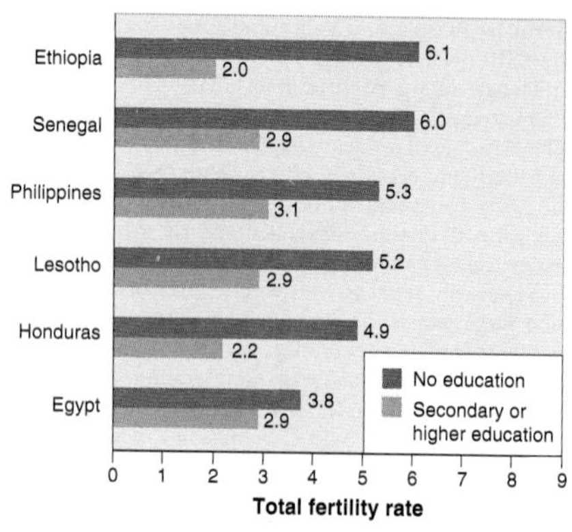

FIGURE 23.4 Total fertility rates for educated and uneducated women in six countries. Fertility is strongly related to female education in many developing countries. *(Data from Population Reference Bureau, 2007 World Population Data Sheet, http://www.prb.org/pdf07/07WPDS_Eng.pdf)*

of birth control. When women have the option to use family planning, crude birth rates tend to drop. FIGURE 23.4 shows how female education levels correlate with crude birth rates. In Ethiopia, for instance, women with a secondary school education or higher have a TFR of 2.0, whereas the TFR among uneducated women is 6.1.

There have been many examples of effective family planning campaigns in the last few decades. In the 1980s, Kenya had one of the highest population growth rates in the world, and its TFR was almost 8. By 1990, its TFR was about 4—one-half of the previous rate. Kenya's government achieved these dramatic results by implementing an active family planning campaign. The campaign, which began in the 1970s, encouraged smaller families. Advertising directed toward both men and women emphasized that overpopulation led to both unemployment and harm to the natural environment. The ads also promoted condom use.

Thailand also successfully used family planning campaigns to lower its growth rate and TFR. Beginning in 1971, national population policy encouraged married couples to use birth control. Contraceptive use increased from 15 to 70 percent, and within 15 years the population growth rate fell from 3.2 to 1.6 percent. Today, Thailand's growth rate is 0.6 percent, among the lowest in Southeast Asia. Some of the credit for this hugely successful reduction in the growth rate is given to a creative and charismatic government official who gained a great deal of attention, in part by handing out condoms in public places.

Population size, economic development, and consumption interact to influence the environment

Both population size and the amount of resources each person uses are critical factors that determine the impact of humans on Earth. Every human exacts a toll on the environment by eating, drinking, generating waste, and consuming products. Even relatively simple foods such as beans and rice require energy, water, and mineral resources to produce and prepare. Raising and preparing meat requires even more resources: Animals require crops to feed them as well as water and energy resources. Building homes, manufacturing cars, and making clothing and consumer products all require energy, water, wood, steel, and other resources. These and many other human activities contribute to environmental degradation.

Both population and economic development contribute to the consumption of resources and to human impact on the environment. In this section we will examine how relationships among population size, economic development, and resource consumption influence the environment. We will then look more closely at local, urban, and global impacts.

Resource Use

In Module 22 we described developing countries as those with relatively low levels of industrialization and incomes of less than $3 per person per day. In contrast, developed nations have relatively high levels of both industrialization and income. Of Earth's 7.2 billion human inhabitants, roughly 6.0 billion live in developing countries, and 1.2 billion live in developed countries. As FIGURE 23.5 shows, 9 of the 12 most populous nations on Earth are developing countries. FIGURE 23.6 charts the relationship between economic development and population growth rate for developing nations. Populations in developing parts of the world have continued to grow relatively rapidly, at an average rate of 1.4 percent per year. At the same time, populations in the developed world have almost leveled off, with an average growth rate of 0.1 percent per year. Impoverished countries are increasing their populations more rapidly than are affluent countries.

Differences in resource use are striking in terms of how population and wealth affect the environment. Calculating the per capita ecological footprint for a country provides a way to measure the effect of affluence—the state of having plentiful wealth that includes the possession of money, goods, or property—and consumption on the planet. Although affluence tends to be associated with higher consumption, it is possible to be affluent without having a large ecological footprint. FIGURE 23.7 shows some examples of

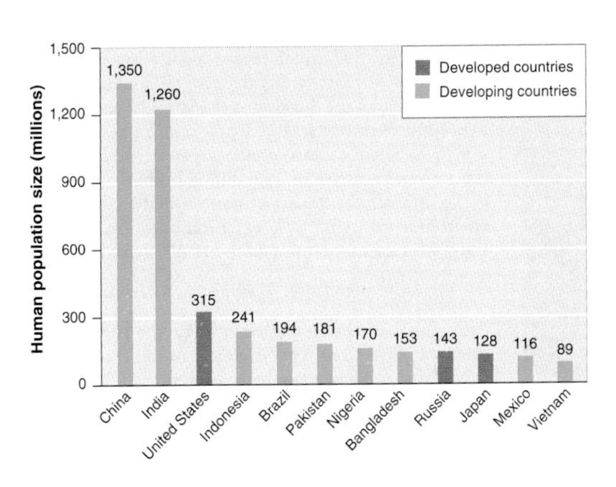

FIGURE 23.5 **The 12 most populous countries in the world.** China and India are by far the largest nations in the world. Only 3 of the 12 most populous countries are developed nations. *(Data from Population Reference Bureau, 2012 data)*

ecological footprints for selected countries. The world average ecological footprint is 2.7 ha (6.7 acres) per capita. The United States has the largest ecological footprint of any nation, at 9.0 ha (22 acres) per capita. China's footprint is 1.8 ha (4.5 acres) per capita. Haiti, the poorest country in the Western Hemisphere, has a footprint of 0.5 ha (1.2 acres) per capita. In other words, a person living in the United States has more than 5 times the environmental impact of a person living in China and 18 times that of a person living in Haiti.

In addition to looking at data on a per capita basis, it is useful to examine the footprints of entire countries. We can do this by multiplying the per capita ecological footprint of a country by the number of people in the country. We find that the United States has a footprint of 2,810 million hectares (6,944 million acres). China's footprint is 2,790 million hectares (6,894 million acres), and Ivory Coast (in western Africa) has a footprint of 18.6 million hectares (46.0 million acres). Looked at another way, the United States, with only one-fourth the population of China, has an ecological

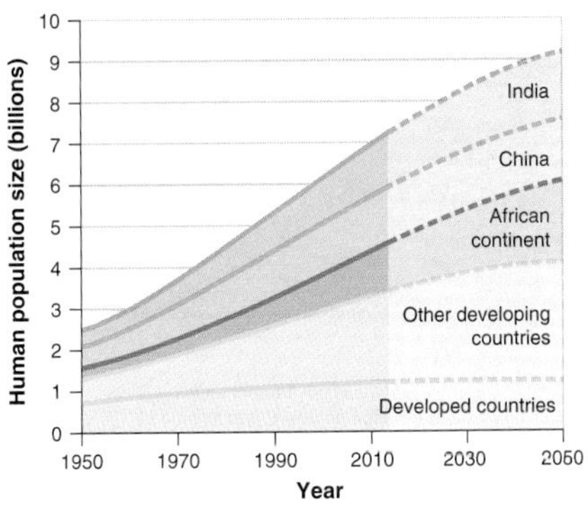

FIGURE 23.6 **Population growth past and future.** Population growth in developed countries has mostly ceased, while that in developing countries is slowing, but is expected to continue beyond 2050. *(Data from United Nations Population Division)*

◀ TEACHING with FIGURES

Show students Figure 23.6 and ask: How does population growth in developed nations compare to population growth in developing nations? *Growth rates in developed nations have leveled off. Population growth rates in developing nations still continue to increase.*

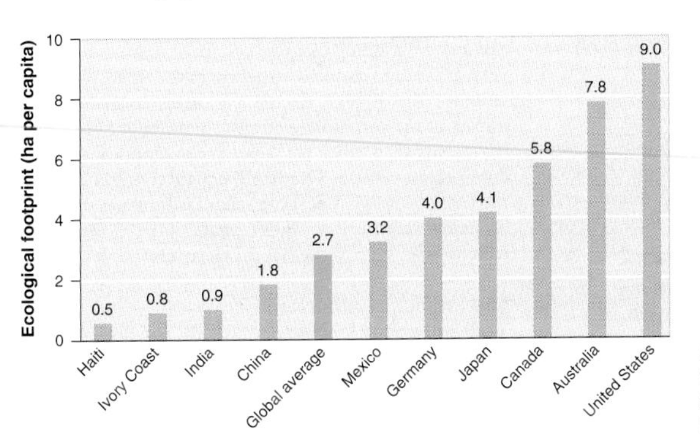

FIGURE 23.7 **Per capita ecological footprints.** Many countries exceed the global average footprint of 2.7 ha per capita. *(Data from Global Footprint Network, 2009 Data Sheet)*

▲ TEACHING with FIGURES

Ask students to compare their results from the Global Footprint Network (see the "Footprint Calculator" activity on page 240) to the data in Figure 23.7. How do they compare to the U.S. average? How do they compare to the global average? How do the two most populous countries, India and China, compare to the United States? Which country has the greatest impact on Earth? *Students may answer that China and India have a greater impact than the United States because they outnumber the U.S. population 4:1. But the argument can also be made that the United States has a greater impact because its ecological footprint per individual is far greater than that of an individual in either China or India.*

Teaching Tip: Journal Prompt

Ask students: What are the three factors in the IPAT equation? Describe how each factor impacts the environment.

- ***Population:*** *Two people consume twice as much as one; more people will have a greater impact on Earth.*
- ***Affluence:*** *More wealth leads to greater consumption levels.*
- ***Technology:*** *Destructive technology such as equipment that uses CFCs can degrade the environment.*

Teaching Tip: Journal Prompt

GDP is the most common measure of a country's wealth. Ask students the following questions about GDP:

- Name the four economic activities that make up a country's GDP. *GDP includes consumer spending, investments, government spending, and exports minus imports.*
- How is GDP correlated to environmental impact? *Increased wealth often correlates with higher pollution levels. However, as GDP continues to increase, a turning point may be reached where a country can afford technologies that will prevent pollution and is able to implement environmental regulations that improve environmental quality.*

(a)

(b)

FIGURE 23.8 **Material possessions.** Most families in developing countries have few possessions compared with their counterparts in developed countries. In each of these photos, members of a family are shown outside their homes with all their possessions. (a) A family in a rural region of Tibet. (b) An individual in an urban part of China, where more development has occurred. *(Huang Qingjun [http://www.huangqingjun.com/])*

footprint comparable to that of China because of its high levels of consumption. However, China's rapid development, as described at the beginning of this chapter, means that its ecological footprint is likely to continue to grow in the coming years.

The IPAT Equation

The total environmental impact of 7.2 billion people is hard to appreciate and even more difficult to quantify. Some people consume large amounts of resources and have a negative impact on environmental systems, while others live much more lightly on the land (FIGURE 23.8). Living lightly can be intentional, as when people in the developed world make an effort to live "green," or sustainably. But it can also be unintentional, as when poverty prevents people from acquiring material possessions or building homes.

To estimate the impact of human lifestyles on Earth, environmental scientists Barry Commoner, Paul Ehrlich, and John Holdren developed the **IPAT equation**

$$\text{impact} = \text{population} \times \text{affluence} \times \text{technology}$$

Although it is written mathematically, the IPAT equation is a conceptual representation of the three major factors that influence environmental impact. Impact in this context is the overall environmental effect of a human population multiplied by affluence, multiplied by technology. It is useful to look at each of these factors individually.

Population has a straightforward effect on impact. All else being equal, two people consume twice as much as one. Therefore, when we compare two countries with similar economic circumstances, the one with more people is likely to have a larger impact on the environment.

Affluence is created by economic opportunity and does not have as simple a relationship to impact as population does. One person in a developed country can have a greater impact than two or more people in a developing country. A family of four in the United States that owns two large sport utility vehicles and lives in a spacious home with a lawn and swimming pool uses a much larger share of Earth's resources than a Bangladeshi family of four living in a two-room apartment and traveling on bicycles and buses. The more affluent a society or individual is, the higher the environmental impact.

The effect of technology is even more complicated. Technology can both degrade the environment and create solutions to minimize our impact on the environment. For example, the manufacturing of chlorofluorocarbons (CFCs) resulted in safe and effective refrigeration and air conditioning that was beneficial to human health, yet CFCs led to ozone destruction in the stratosphere. In contrast, the hybrid electric car helps to reduce the impact of the automobile on the environment because it has greater fuel efficiency than a conventional internal combustion vehicle. The IPAT equation originally used the term *technology*, but some scientists now use the term *destructive technology* to differentiate it from beneficial technologies such as the hybrid electric car.

The Impact of Affluence

To help gauge a country's wealth and its potential impact on the environment, environmental scientists often turn to the most commonly used measure of a nation's wealth. **Gross domestic product (GDP)** is the value of all products and services produced in one

IPAT equation An equation used to estimate the impact of the human lifestyle on the environment: Impact = population × affluence × technology.

Gross domestic product (GDP) A measure of the value of all products and services produced in one year in one country.

Teaching Tip: Warm-up

Some environmental scientists argue that increasing the GDP of developing nations is the best way to save the environment. Ask students to explain the reasoning behind this argument. After you have discussed the point, ask students if they agree or if they see a flaw in the argument. After you have discussed the point, ask students if they agree or if they see a flaw in the argument. *Rising income generally correlates with falling birth rates. Reducing population size will reduce the overall environmental impact. Wealthier countries can afford to make environmental improvements and so increase their efficiency of resource use.*

year in one country. GDP is made up of four types of economic activity: consumer spending, investments, government spending, and exports minus imports. A country's per capita GDP often correlates with its pollution levels. At very low levels of per capita GDP, industrial activity is too low to produce much pollution; the country uses very little fossil fuel and generates relatively little waste. Many developing countries fit this pattern.

As GDP increases, a nation begins to be able to afford to burn fossil fuels, especially coal, which, although relatively inexpensive, emits a substantial amount of pollution. The country may also rely on rudimentary, inefficient equipment that emits large amounts of pollutants. It is at this point in its development that a country emits pollution at the highest levels. The United States fit this pattern during the twentieth century. China, which is going through a similar rapid industrialization, currently relies on coal as its primary energy source. Many people view this shift as a trade off that occurs as a country's GDP increases: Breathing dirty air poses a different risk to human health than poverty. Although it is not always a more desirable risk, many people would choose dirty air over poverty.

As a nation's GDP increases further, it may reach a turning point. It can afford to purchase equipment that burns fossil fuels more efficiently and cleanly, which helps to reduce the amounts and types of pollution generated. People may also be willing to expend resources and support government efforts to regulate polluting industries. Wealthier societies are also able to afford better policing and enforcement mechanisms that ensure environmental regulations are being followed. Western European countries and the United States fit this scenario today. We will explore this turning point in more depth in Chapter 20.

Some environmental scientists argue that increasing the GDP of developing nations is the best way to save the environment, for at least two reasons. First, as we have seen, rising income generally correlates with falling birth rates, and a reduced population size should lead to a reduction in environmental impact. Second, wealthier countries can afford to make environmental improvements and increase their efficiency of resource use.

Local versus Global Impacts, and Urban Impacts

Impacts on the environment may occur locally—within the borders of a region, city, or country—or they may be global in scale. The scale of impact depends on the nature of the economy and the degree to which the society has developed. For example, a person may create a local impact by using products created within the country's borders. That same person may create a global impact by using materials that are imported. In addition to local and global impacts, some impacts are specific to people who live in urban environments.

FIGURE 23.9 **Deforested land in Brazil.** Clearing forests and grasslands for agriculture can lead to erosion, soil degradation, and habitat loss. This photo shows erosion caused by clear-cutting in Brazil. *(Publiphoto/Science Source)*

Local versus Global

In general, highly localized impacts are typical of rural, agriculturally based societies. Most of the materials consumed in developing countries are produced locally. While this may benefit the local economy, it can lead to regional overuse of resources and environmental degradation. Chapter 3 described deforestation in Haiti as an example of such overuse.

Two commonly overused local resources are the land itself and woody biomass from trees and other plants. A growing population requires increasing amounts of food. In developing countries that do not import their food, local demand for agricultural land increases with population size. To put more land into cultivation, farmers may convert forests or natural grasslands into cropland.

Brazil provides an example of land overuse in a developing nation. The United Nations Food and Agriculture Organization estimates that in Brazil, approximately 3 million hectares (7.4 million acres) were cleared per year during the peak years between 2000 and 2005. Some of this land was used for small-scale agriculture and some for industrial production of soybeans, sugarcane, and corn. The local environmental impacts of converting land to agricultural use include erosion, soil degradation, and habitat loss (FIGURE 23.9). Agriculture has global as well as local impacts, whether it occurs in developed or developing countries. Conversion of land to agriculture reduces the total amount of atmospheric carbon dioxide uptake by plants, which affects the global carbon cycle.

Teaching Tip: Beyond the Classroom

The *World in Balance* website offers a glimpse of average families across the world with their possessions in pictures. Each picture is accompanied by a brief family description and the country's statistical data. Have students view these images and read about families from Mali, the United States, Japan, China, and India.

Find a link to this resource by clicking on the link buttons in the TE-book, opening the TRFD, or logging onto the book's companion website bcs.whfreeman.com /friedlandapes2e (teacher login required).

Chapter 7 Web Resources

Teaching Tip: Journal Prompt

Ask students the following questions:

- Distinguish between local and global impacts. Give examples of each type of impact. *Local impacts are typically more common in less developed nations where societies are agriculturally based. Local impacts include the overuse of resources within a region, such as overuse of land that results in deforestation and local water pollution. Global impacts are more common in affluent or urban societies because they use imported fossil fuels, food, and other resources.*
- Explain how a local impact can become a global impact. *A local impact can become a global impact when a local action affects a global common good. For example, deforestation in Brazil is a local impact because it leads to soil erosion and habitat loss. It becomes a global impact because forest removal means losing a carbon sink. Increased levels of atmospheric carbon lead to global climate change.*

Teaching Tip: Beyond the Classroom

Have students brainstorm ways to cut their consumption. Challenge students to then keep a journal for 2 weeks that records ways they have reduced consumption, thereby reducing their impact on the planet.

Teaching Tip: Warm-up

Ask students to discuss both environmental concerns as well as benefits associated with living in an urban area. *Urban areas produce greater amounts of solid waste, pollution, and carbon dioxide; however, they tend to have smaller ecological footprints because of greater access to public transportation and services. In less developed countries, affluent urban areas have safe drinking water, sanitation, and waste disposal. However, in less affluent urban areas of developing nations, the very poor often construct shantytowns that include unsafe dwellings that lack basic amenities such as water and sanitation.*

Teaching Tip: Journal Prompt

Ask students: What is the difference between economic development and sustainable development? *Economic development focuses solely on meeting human needs and increasing prosperity. Sustainable development focuses on meeting the essential needs of people without compromising the ability of future generations to meet their needs.*

In addition, an increase in the use of fertilizers made from fossil fuels increases the release of greenhouse gases into the atmosphere.

Global impacts are more common in affluent or urban societies because they tend to specialize production in the industrial and high technology sectors. For example, more than half the ecological footprint of the United States comes from its use of fossil fuels, of which approximately 50 percent are imported. China's ecological footprint has more than doubled since 1970, and its ratio of local to global impact has shifted. Most of its ecological footprint was previously driven by demand for food, fiber crops such as cotton, hemp, and flax, and woody biomass. However, in recent years, China's demand for fossil fuels has increased dramatically.

Families in suburban areas of developed countries such as the United States consume far fewer local resources than rural families in developing countries, but they have a much greater impact on the global environment. In general, populations with large global impacts tend to deplete more environmental resources. Much of the impact comes from consumption of imported energy sources such as oil and other imported resources such as food. When people are affluent, they are more likely to purchase imported bananas, fish, and coffee from other countries, drive long distances in automobiles that were manufactured in factories hundreds or thousands of miles away, and live in homes surrounded by lawns that require large quantities of water, fertilizer, and pesticides.

Urban Impacts

Urban populations represent one-half of the human population but consume three-fourths of Earth's resources. While definitions vary by country, an **urban area,** according to the U.S. Census Bureau, contains more than 386 people per square kilometer (1,000 people per square mile). New York City is the most densely populated city in the United States, with 10,400 people per square kilometer (27,000 people per square mile). Mumbai, India, is the most densely populated city in the world, with 23,000 people per square kilometer (60,000 people per square mile).

More than 75 percent of people in developed countries live in urban areas, as FIGURE 23.10 shows, and that number is expected to increase slightly over approximately the next 20 years. In developing countries, 44 percent of people live in urban areas, but that number is increasing more rapidly than in developed countries and will probably reach 56 percent by 2030. **TABLE 23.1** shows that, of the 20 largest cities in the world, 16 are in developing countries. Worldwide, almost 5 billion people are expected to live in urban areas by 2030.

Urban area An area that contains more than 385 people per square kilometer (1,000 people per square mile).

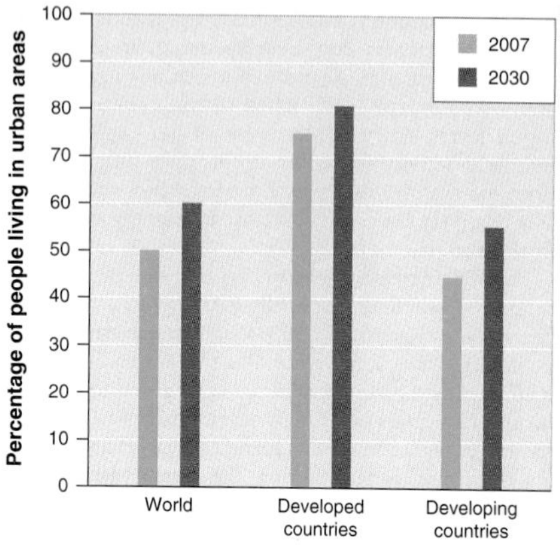

FIGURE 23.10 **Urban growth.** More than one-half of the world's population will live in urban settings by 2030. *(Data from United Nations Population Fund)*

Urban living in both developed and developing countries presents environmental challenges. Most developed countries employ city planning to some degree. As urban areas expand, experts design and install public transportation facilities, water and sewer lines, and other municipal services. In addition, while urban areas produce greater amounts of solid waste, pollution, and carbon dioxide emissions than suburban or rural areas, they tend to have smaller per capita ecological footprints. There are many reasons for this difference, including greater access to public transportation and services that are nearby, such as shopping.

In developing countries, the relatively affluent portions of urban areas have safe drinking water, sewage treatment systems, and systems for disposal of household solid waste to minimize their impact on the surrounding environment. However, many less-affluent urban residents have no access to these services. Rapid urbanization in the developing world often results in an influx of the very poor who often cannot afford permanent housing and instead construct temporary shelters with whatever materials they find available, including mud, cardboard, or plastic. Whether they are squatter settlements, shantytowns, or slums, these overcrowded and underserved living situations are a common fact of life in

TABLE 23.1 The largest 20 urban areas in the world

Rank	City, Country	Population (millions)
1	Tokyo, Japan	37.2
2	Delhi, India	22.7
3	Mexico City, Mexico	20.4
4	New York-Newark, United States	20.4
5	Shanghai, China	20.2
6	São Paulo, Brazil	19.9
7	Mumbai, India	19.7
8	Beijing, China	15.6
9	Dhaka, Bangladesh	15.4
10	Kolkata, India	14.4
11	Karachi, Pakistan	13.9
12	Buenos Aires, Argentina	13.5
13	Los Angeles-Long Beach-Santa Ana, United States	13.4
14	Rio de Janeiro, Brazil	12.0
15	Manila, Philippines	11.9
16	Moscow, Russia	11.6
17	Osaka-Kobe, Japan	11.5
18	Istanbul, Turkey	11.3
19	Lagos, Nigeria	11.2
20	Cairo, Egypt	11.2

Source: United Nations Population Division.
Note: Data are from 2011 and contain the areas defined by the United Nations as "urban agglomerations." Other agencies agglomerate urban areas differently and obtain slightly different results.

most cities in the developing world. The United Nations organization UN-HABITAT estimates that 1 billion people live in squatter settlements and other similar areas throughout the world. Most residents live in housing structures without flooring, safe walls and ceilings, or such basic amenities as water, sanitation, or health care (FIGURE 23.11).

Sustainable development is a common, if elusive, goal

We have seen that economic development and advancement improve human well-being, but it also has a strong influence on the environmental impact of a society. While many people believe that we cannot have both economic development and environmental protection, a growing number of social and natural scientists maintain that, in fact, sustainable economic development is possible.

As we saw in Chapter 1, sustainable development goes beyond economic development to meet the essential needs of people in the present without compromising the ability of future generations to meet their needs. In other words, sustainable development strives to improve standards of living—which involves greater expenditures of energy and

FIGURE 23.11 **Shantytowns.** A squatter settlement on the outskirts of Lima, Peru. Such shantytowns expand outward from the urban center. *(Reuters/Mariana Bazo/Landov)*

Teaching Tip: Discussion Starter

The Millennium Ecosystem Assessment project reports the following: "Human actions are depleting Earth's natural capital, putting such strain on the environment that the ability of the planet's ecosystems to sustain future generations can no longer be taken for granted." Have students predict some potential challenges to achieving global sustainability and discuss possible solutions to overcoming the challenges we face.

resources—without causing additional environmental harm. Many cities in Scandinavia have achieved something close to sustainable development. For example, Övertorneå, a Swedish town of 5,600 people, recently revived the local economy by focusing its economic activities on renewable energy, free public transportation, organic agriculture, and land preservation. The city still has an ecological footprint, of course, but it is much smaller, and thus much more sustainable, than it was previously.

How can sustainable development be achieved? There are no simple answers to that question, nor is there a single path that all people must follow. The Millennium Ecosystem Assessment project, completed in 2005, offers some insights. This project's reports constitute a global analysis of the effects of the human population on ecosystem services such as clean water, forest products, and natural resources. They are also a blueprint for sustainable development. The reports, prepared at the request of the United Nations, concluded that human demand for food, water, lumber, fiber, and fuel has led to a large and irreversible loss of biodiversity. The Millennium Ecosystem Assessment drew several other conclusions:

- Ecosystem sustainability will be threatened if the human population continues along its current path of resource consumption around the globe.
- The continued alterations to ecosystems that have improved human well-being (greater access to food, clean water, suitable housing) will also exacerbate poverty for some populations.
- If we establish sustainable practices, we may be able to improve the standard of living for a large number of people.

The project's reports state that "human actions are depleting Earth's natural capital, putting such strain on the environment that the ability of the planet's ecosystems to sustain future generations can no longer be taken for granted." They further suggest that sustainability, as well as sustainable development, will be achieved only with a broader and accelerated understanding of the connections between human systems and natural systems. This means that governments, nongovernmental organizations, and communities of people will have to work together to raise standards of living while understanding the impacts of those improvements on the local, regional, and global environments.

module 23 REVIEW

In this module we examined the theory of the demographic transition, which states that in some countries, the process of industrialization causes a reduction in the population growth rate. Slower population growth is typical of many countries that have industrialized and become wealthier. Some nations that have slowed in population growth have not industrialized. Those countries that have industrialized typically increase their consumption of resources, with adverse impacts on the environment. Impacts can occur locally and globally. Some locations are beginning to take on the challenge of fostering industrialization and economic development in a sustainable manner.

Module 23 AP® Review Questions

1. According to the theory of demographic transition, countries move through growth phases in which order?
   (a) Stable growth, rapid growth, slow growth, declining growth
   (b) Rapid growth, slow growth, stable growth, declining growth
   (c) Slow growth, declining growth, rapid growth, stable growth
   (d) Slow growth, rapid growth, stable growth, declining growth
   (e) Stable growth, rapid growth, declining growth, slow growth
2. Which of the following have contributed to increased family planning worldwide?
   I. Women's education
   II. Increased income
   III. Advertising campaigns

Answers to Module 23 AP® Review Questions

1. d
2. e
3. b
4. c
5. d

(a) I and II only
(b) I and III only
(c) II and III only
(d) I and III only
(e) I, II, and III

3. Which of the following is NOT a factor in gross domestic product?
   (a) Consumer spending
   (b) Cost of environmental damage
   (c) Government spending
   (d) Imports and exports
   (e) Investments

4. The IPAT equation was developed to estimate
   (a) the affluence of a country's population.
   (b) the rate of demographic transition.
   (c) the impact of human lifestyles on Earth.
   (d) the gross domestic product of a country.
   (e) the effect of urbanization on resource use.

5. Approximately what percentage of the population of developed countries lives in urban areas?
   (a) 20%
   (b) 44%
   (c) 50%
   (d) 75%
   (e) 86%

working toward  sustainability

Gender Equity and Population Control in Kerala

India is the world's second most populous country and will probably overtake China as the most populous before 2030. Since the 1950s, India has tried various methods of population control, steadily reducing its growth rate to the current 1.5 percent. Although this growth rate would suggest below-replacement-level fertility, India's population will continue to increase for some time because of its large number of young people. The Population Reference Bureau estimates that India's population will not stabilize until after 2050.

In the 1960s, India attempted to enforce nationwide population control through sterilization, but massive protests against this coercive approach led to changes in policy. Since the 1970s, India has emphasized family planning and reproductive health, although each Indian state chooses its own approach to population stabilization. Some states have had tremendous success in lowering growth rates. Kerala, in southwestern India, is one of those states.

The state of Kerala is about twice the size of Connecticut. But Connecticut has a population of roughly 3.5 million people while Kerala has a population of more than 30 million. Kerala's population density is about 820 people per square kilometer (2,100 people per square mile), almost 3 times that of India as a whole, and 25 times that of the United States. Until 1971, Kerala's population was growing even faster than the population in the rest of India, in large part because the state government had implemented an effective health care system that decreased infant and adult mortality rates.

Since 1971, Kerala's birth rate, like its mortality rate, has fallen to levels similar to those in North America and other industrialized countries. The current total

Computer students in Kerala, India. Literacy rates similar to those in developed countries may be a factor in the population stabilization that has occurred in the state of Kerala, India. *(Owen Franken/Corbis)*

Suggested Answers to Critical Thinking Questions

1. Sterilization, family planning, and reproductive health.

2. Kerala used the "three Es": education, employment, and equality. By increasing the education and equality of men and women, the birth rate has decreased to the levels of many developed nations.

3. The ratio of old to young people is closer to 1 at replacement level fertility, which can increase the burden that caring for the elderly places on their children and grandchildren.

4. The matriarchal tradition in Kerala made the model much easier to implement. In many other places there may be resistance to the changes that allowed Kerala to reduce its growth rate.

Exploring the Literature

Cohen, J. E. 2003. Human population: The next half century. *Science* 304:1172–1175.

Dodds, W. K. 2008. *Humanity's Footprint.* Columbia University Press.

Ellis, Erle C. 2013. Overpopulation Is Not the Problem. *New York Times*, September 13, p. A19. http://www.nytimes.com/2013/09/14/opinion/overpopulation-is-not-the-problem.html?

Global Footprint Network. http://www.footprintnetwork.org.

Sabin, P. 2013. *The Bet: Paul Ehrlick, Julian Simon, and Our Gamble over Earth's Future.* Yale University Press.

Special Population Issue. 2011. *Science* 333: (6042): 489–660,

Weeks, J. R. 2011. *Population: An Introduction to Concepts and Issues.* 11th ed. Cengage.

Download a printable version of this list by clicking on the link buttons in the TE-book, opening the TRFD, or logging onto the book's companion website bcs.whfreeman.com/friedlandapes2e (teacher login required).

TRM **Exploring the Literature 7**

fertility rate in Kerala is about 2.0—lower than the U.S. TFR of 2.1 and much lower than the TFR of 2.7 for India as a whole. As a result, Kerala's population has stabilized. How did the state's population stabilize despite its poverty?

A special combination of social and cultural factors seems to be responsible for Kerala's sustainable population growth. Kerala has emphasized "the three Es" in its approach: education, employment, and equality. In addition to its accessible health care system, Kerala has good schools that support a literacy rate of over 90 percent, the highest in India. Furthermore, unlike rates in the rest of India, male and female literacy rates in Kerala are almost identical. In other parts of India and in most other developing countries, women attend school, on average, only half as long as men do. As we have learned, higher education levels for women lead to greater female empowerment and increased use of family planning. In Kerala, 63 percent of women use contraceptives, compared with 48 percent in the rest of India, and many women delay childbirth and join the workforce before deciding to start families. Kerala has a strong matriarchal tradition in which women are highly valued and their education is encouraged. Education and empowerment allow these women to be in a better position to make decisions regarding family size.

The World Bank estimates that if the rest of the developing world had followed Kerala's lead 30 years ago and equalized education for men and women, current TFR throughout the developing world would be close to replacement-level fertility. Evidence for this view can be found in the reduced fertility rates of other relatively poor countries where women and men have equal status, such as Sri Lanka and Cuba. In fact, gender equity was made a cornerstone of an international program to control population growth at the 1994 Third International Conference on Population and Development in Cairo, Egypt.

Critical Thinking Questions

1. Name some typical ways countries have tried to lower birth rates.
2. What are some other innovative ways to lower birth rates?
3. Can achieving replacement-level fertility sometimes cause difficulties? If so, describe some problems that might occur.
4. What is preventing the model used to promote slower population growth in Kerala from being used elsewhere in the world?

References

Franke, R. W., and B. H. Chasin. 2005. Kerala: Radical reform as development in an Indian state. In *The Anthropology of Development and Globalization: From Classical Political Economy to Contemporary Neoliberalism,* ed. M. Edelman and A. Haugerud. Blackwell.

Pulsipher, L. M., and A. Pulsipher. 2011. *World Regional Geography.* 5th ed. W. H. Freeman.

chapter 7 REVIEW

Throughout this chapter, we have examined the ways in which the human population changes in response to natural biological factors and in response to other factors that are specifically human in nature. Together, these factors determine the carrying capacity of Earth, the total number of human beings that the world can support. People in the developed world are currently exhibiting very different behavior from those in the developing world. In the developed world, population growth has almost stopped while consumption of resources is still quite high. In the developing world, growth rates, while slowing, are still high, and consumption of resources is still rather low. As the world becomes more urban, and the impacts of development and consumption from larger numbers of people increase, the challenge is to achieve sustainable growth while also promoting the improvement of living conditions in the developing world.

Key Terms

Demography
Demographer
Immigration
Emigration
Crude birth rate (CBR)
Crude death rate (CDR)
Doubling time
Total fertility rate (TFR)
Replacement-level fertility
Developed country
Developing country
Life expectancy
Infant mortality
Child mortality
Net migration rate
Age structure diagram
Population pyramid
Population momentum
Theory of demographic transition
Family planning
Affluence
IPAT equation
Gross domestic product (GDP)
Urban area

Learning Objectives Revisited

Module 22 Human Population Numbers

- **Explain factors that may potentially limit the carrying capacity of humans on Earth.**

  Scientists disagree about the size of Earth's carrying capacity for humans. Some scientists believe we have already exceeded that carrying capacity. Others believe that innovative approaches and new technologies will allow the human population to continue to grow beyond the environmental limits currently imposed by factors such as the supply of food, water, and natural resources.

- **Describe the drivers of human population growth.**

  The human population is currently 7.2 billion people, and it is growing at a rate of about a million people every 5 days. If we think of the human population—as a whole or in individual countries—as a system, there are more inputs—births and immigration—than outputs—deaths and emigration. To understand changes in population size, demographers measure crude birth rate, crude death rate, total fertility rate, replacement-level fertility, life expectancy, infant and child mortality, age structure, and net migration rate.

- **Read and interpret an age structure diagram**

  Age structure diagrams are visual representations of age structures for males and females. Each horizontal bar in the diagram represents an age group. The total area of all the bars in the diagram represents the size of the whole population. A country with more younger people than older people has an age structure diagram that is widest at the bottom and narrowest at the top. This is often called a population pyramid and is typical of developing countries. Developed countries tend to have closer to an even age distribution and their age structure diagrams are more vertical or column-like.

Module 23 Economic Development, Consumption, and Sustainability

- **Describe how demographic transition follows economic development.**

  A number of countries have undergone a demographic transition as their economies have modernized. Economic development generally leads to increased affluence, increased education, less need for children to help their families generate subsistence income, and increased family planning. These factors have reduced the average size of families in developed countries, which leads to slower population growth. Eventually, population size may even decline.

- **Explain how relationships among population size, economic development, and resource consumption influence the environment.**

  Most population growth today is occurring in developing countries. Only one-fifth of the global population lives in developed countries, but those countries consume more than half the world's energy and resources. The IPAT equation states that the environmental impact of a population is a result of population size, affluence, and technology. A relatively small population can have a high environmental impact if its affluence leads to high consumption and extensive use of destructive technology. However, an affluent nation can more easily take measures to reduce its environmental impact through the use of technology that counters pollution and increases the efficiency of resource use.

- **Explain why sustainable development is a common but elusive goal.**

  Sustainable development attempts to raise standards of living without increasing environmental impact. The Millennium Ecosystem Assessment is a blueprint for sustainable development.

Teaching Tip: Measuring Your Impact 7

National Footprints

Measuring Your Impact 7 offers your students an opportunity to evaluate national ecological footprints and to compare these data with actual availability of land. In addition, students will consider how many planets would be necessary if the entire world population lived the way the U.S. population does.

Download this resource by clicking on the link buttons in the TE-book, opening the TRFD, or logging onto the book's companion website bcs.whfreeman.com /friedlandapes2e (teacher login required).

Measuring Your Impact 7: National Footprints

Teaching Tip: Activity

The Science Applied article on page 255, "How Can We Manage Overabundant Deer Populations?", provides an opportunity for students to tie together concepts presented in Chapters 6 and 7. Often students think of the human influence causing declines in animal populations. This article asks students to consider human influence as a cause of animal population explosions. Use this article to jump-start a class discussion. Ask students if they know of any local examples of human-induced population growth. Then use the free-response question at the end of Science Applied to assess student learning and provide another opportunity to practice writing.

Answers to Chapter 7 AP® Exam Multiple-Choice Questions

1. a
2. b
3. e
4. e
5. c
6. e
7. c
8. c
9. a
10. d
11. d
12. d

Chapter 7 AP® Environmental Science Practice Exam

Section 1: Multiple-Choice Questions

Choose the best answer for questions 1–12.

1. Which of the following does NOT support the theory that humans can devise ways to expand their carrying capacity on Earth?
   (a) The development of CFCs for use in refrigeration
   (b) The development of hydraulic fracturing to reach natural gas reserves
   (c) The use of arrows for hunting animals
   (d) The use of horse-driven plows
   (e) The use of waste methane from landfills for electricity generation

2. A metropolitan region of 100,000 people has 2,000 births, 500 deaths, 200 emigrants, and 100 immigrants over a 1-year period. Its population growth rate is
   (a) 1.2 percent.
   (b) 1.4 percent.
   (c) 1.6 percent.
   (d) 1.8 percent.
   (e) 2.0 percent.

3. Which of the following pairs of indicators best reflects the availability of health care in a country?
   (a) Crude death rate and growth rate
   (b) Crude death rate and crude birth rate
   (c) Growth rate and life expectancy
   (d) Infant mortality rate and crude death rate
   (e) Infant mortality rate and life expectancy

4. In 2013, the population of Earth was about _____ billion, with about _____ billion living in China.
   (a) 6.1/1.1
   (b) 6.1/1.2
   (c) 6.3/1.4
   (d) 7.1/1.2
   (e) 7.1/1.4

Use the following age structure diagram to answer question 5.

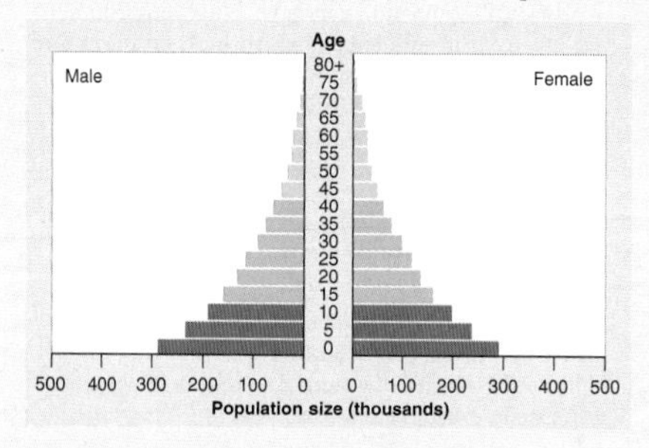

5. A country with an age structure diagram like the one shown is most likely experiencing
   (a) a high life expectancy.
   (b) slow population growth.
   (c) a short doubling time.
   (d) a low infant mortality rate.
   (e) replacement-level fertility.

6. Which statement about total fertility rate is correct?
   (a) TFR is equal to the crude birth rate minus the crude death rate.
   (b) TFR is the average number of children each woman must have to replace the current population.
   (c) TFR is generally higher in developed countries than in developing countries.
   (d) TFR is equal to the growth rate of a country.
   (e) TFR is the average number of children each woman will give birth to during her childbearing years.

7. Even if a country reduces its birth rate and maintains replacement-level fertility, its population will still continue to grow for several decades because of
   (a) lower death rates.
   (b) increased income.
   (c) population momentum.
   (d) better health care.
   (e) increased life expectancy.

8. At current growth rates, which country will probably be the most populous in the world after 2050?
   (a) China
   (b) Brazil
   (c) India
   (d) Indonesia
   (e) United States

9. Which characteristics are typical of developed countries?
   I. High technology use
   II. Low GDP
   III. Small-scale sustainable agriculture

   (a) I only
   (b) II only
   (c) I and III only
   (d) II and III only
   (e) I, II, and III

10. What percentage of the world's population lives in developing countries?
   (a) 34
   (b) 50
   (c) 66
   (d) 82
   (e) 98

Additional Multiple-Choice AP® Practice Exam

Download an additional Multiple-Choice AP® Environmental Science Practice Exam for Chapter 7, by clicking on the link buttons in the TE-book, opening the TRFD, or logging onto the book's companion website bcs.whfreeman.com /friedlandapes2e (teacher login required).

Multiple-Choice AP® Practice Exam 7

Answers to Chapter 7 AP® Exam

Download the full answers to the Chapter 7 AP® Practice Exam by clicking on the link buttons in the TE-book, opening the TRFD, or logging onto the book's companion website bcs.whfreeman.com /friedlandapes2e (teacher login required).

Answers to Chapter 7 AP® Exam

11. Which country best exemplifies phase 4 of a demographic transition?
    (a) Argentina
    (b) China
    (c) India
    (d) Japan
    (e) Mexico

12. As Brazil has become more developed and industrialized, its population growth has stabilized. At the same time, the use of technology and raw materials has increased to meet the demands of a wealthier and more prosperous population. This increased consumption is predicted by
    (a) Thomas Malthus.
    (b) the Millennium Ecosystem Assessment.
    (c) the theory of demographic transition.
    (d) the IPAT equation.
    (e) population momentum.

Section 2: Free-Response Questions

Write your answer to each part clearly. Support your answers with relevant information and examples. Where calculations are required, show your work.

1. Answer the following questions about the theory of demographic transition.
    (a) Draw a fully labeled diagram that shows how birth and death rates change as a country undergoes the four phases of a demographic transition. (3 points)
    (b) For each of the phases labeled in your diagram, explain the changes occurring in each phase and describe what is causing them. (3 points)
    (c) Describe a strategy that a government might implement to slow its population growth that could be utilized by a country undergoing a demographic transition. Explain how your proposed strategy would work, and describe one potential drawback to its implementation. (4 points)

2. Look at the age structure diagrams for country A and country B below and answer the following questions.

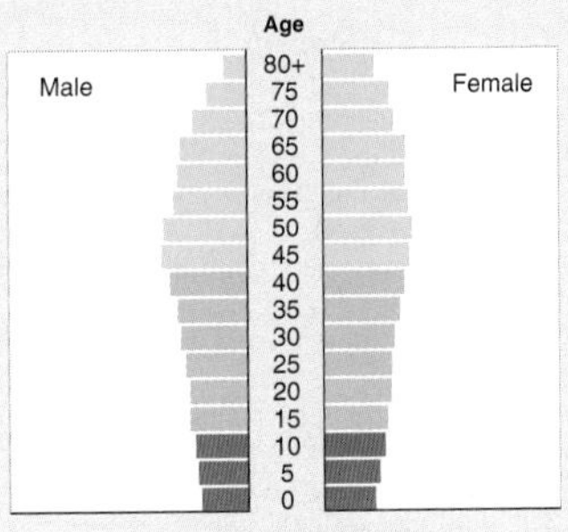

Population size (country A)

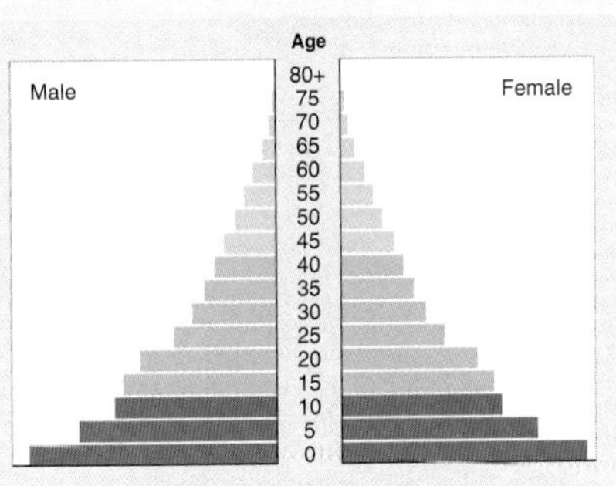

Population size (country B)

    (a) What observations and educated predictions can you make about the following characteristics of country A?
        (i) The age structure of its population (1 point)
        (ii) The total fertility rate of the country (1 point)
        (iii) The life expectancy of the population (1 point)
        (iv) The growth rate and doubling time of the population (2 points)
    (b) Describe one socioeconomic feature of country A. (2 points)
    (c) Explain how country B differs from country A in terms of
        (i) the age structure of its population. (1 point)
        (ii) its infant mortality rate. (1 point)
        (iii) its rate of population growth. (1 point)

Answers to the Chapter 7 Free-Response Questions can be found in the Answer Appendix on page ANS-4.

Answers to Unit 3 AP® Exam Multiple-Choice Questions

1. a
2. c
3. b
4. e
5. e
6. d
7. a
8. d
9. e
10. c
11. d
12. b
13. d
14. b
15. a
16. c
17. e
18. d
19. c
20. a

Unit 3 AP® Environmental Science Practice Exam

Section 1: Multiple-Choice Questions

Choose the best answer for questions 1–20.

1. Which of the following is NOT true regarding limiting resources?
   (a) A limiting resource is an important density-independent factor.
   (b) A limiting resource is common to all members of a population.
   (c) Foraging territory is an example of a limiting resource.
   (d) The number of reproducing individuals in a population is not a limiting resource.
   (e) A population at carrying capacity has consumed all limiting resources.

2. Which of the following is most likely to result in a random population distribution?
   (a) Clumping of individuals around limiting resources
   (b) Competition among individuals
   (c) Density-independent factors
   (d) Variable carrying capacities among populations within a metapopulation
   (e) An age structure biased toward younger individuals

3. The common reed (*Phragmites australis*) is a perennial grass found throughout temperate and tropical regions of the world. Researchers recently discovered that many North American populations of the reed are composed entirely of genetically identical clones. Which is most likely true of these populations?
   (a) The age structure is unlikely to vary among populations.
   (b) Evolution by natural selection occurs very slowly in the absence of seed dispersal among populations.
   (c) Density-dependent factors will be more important than density-independent factors in regulating population densities.
   (d) Density-dependent factors will be less important than density-independent factors in regulating population densities.
   (e) Individual populations will not reach carrying capacity.

4. Species with a type III survivorship curve are likely to exhibit
   I. fast population growth rate.
   II. large amounts of parental care.
   III. density-independent population regulation.
   (a) I only
   (b) II only
   (c) III only
   (d) I and II
   (e) I and III

5. In an experimental grassland plot, researchers planted 20 seeds of a wildflower species that exhibits exponential growth with an intrinsic growth rate of 0.60. Approximate the maximum population size after 10 years.
   (a) 20×0.60
   (b) $20 \div 0.60$
   (c) $e^{0.60 \times 10}$
   (d) $e^{0.60 \times 11}$
   (e) $20 \times e^{0.60 \times 10}$

6. Which of the following is NOT a likely consequence of resource niche partitioning?
   (a) Predator-prey interactions
   (b) Clumped species distributions
   (c) Morphological variation within species
   (d) Competition within a species
   (e) Coexistence among species

7. Mycorrhizal fungi are commonly found attached to plant roots. The fungi are more efficient than plants at extracting nutrients from the soil. In exchange for nutrients, plants provide the fungi with sugars made through photosynthesis. However, when nutrients in the soil are abundant, researchers have found that plant growth increases if mycorrhizal fungi are removed. This phenomenon suggests that
   (a) mycorrhizae have switched from being mutualists to parasites.
   (b) mycorrhizae have switched from being mutualists to predators.
   (c) mycorrhizae have switched from being mutualists to commensalists.
   (d) mycorrhizae have ceased to act as a keystone species.
   (e) mycorrhizae are not able to partition resources as well as plants.

Answers to Unit 3 AP® Exam

Download the full answers to the Unit 3 AP® Practice Exam by clicking on the link buttons in the TE-book, opening the TRFD, or logging onto the book's companion website bcs.whfreeman.com/friedlandapes2e (teacher login required).

Answers to Unit 3 AP® Exam

8. Given two islands of the same size, the theory of island biogeography suggests that the __________ island should have a higher __________.
   (a) farther/extinction rate
   (b) closer/speciation rate
   (c) farther/species diversity
   (d) closer/species richness
   (e) closer/population size

9. Ecological succession
   (a) begins with primary succession, followed by secondary succession, and ends with a climax stage.
   (b) begins with either primary or secondary succession and ends with a climax stage.
   (c) cannot occur if there are no species initially present at a site.
   (d) starts with bare soil and ends with a climax forest.
   (e) does not end with a climax stage.

10. The theory of island biogeography does not apply to
   (a) protected wildlife areas.
   (b) islands very far away from the mainland.
   (c) metapopulations.
   (d) mountaintop communities.
   (e) bacterial communities.

11. Which may lead to an increase in the carrying capacity of the global human population?
   I. Niche partitioning
   II. Technological advancement
   III. Reduction in per-capita consumption rate

   (a) I only
   (b) II only
   (c) I and II
   (d) II and III
   (e) I, II, and III

12. In Vietnam, there were 16 births and 6 deaths out of every 100 people in the year 2011. Using the rule of 70, estimate the doubling time for Vietnam's population.
   (a) 0.7 years
   (b) 7 years
   (c) 0.14 years
   (d) 1.4 years
   (e) 14 years

13. Which is NOT likely to describe a population in a developing country?
   (a) Replacement level fertility is higher than 2.1.
   (b) Infant mortality is 10 percent or more.
   (c) Immigration is relatively low or nonexistent.
   (d) The age structure will be shaped like an inverted pyramid.
   (e) Per capita ecological footprint is much lower than in developed countries.

14. World Bank records indicate that in 2010, the total fertility rate in Puerto Rico was 1.6 and population size was 4,000,000. If net migration rate was −1.0, then the best estimate of the population in 2012 is
   (a) 2,500,000.
   (b) 3,700,000.
   (c) 4,000,000.
   (d) 4,500,000.
   (e) 4,600,000.

15. A developing country with a total fertility rate of 4.0 and a net migration rate of −3.5 is likely to have an age structure
   (a) shaped like a pyramid.
   (b) shaped like an inverted pyramid.
   (c) shaped like a column.
   (d) that is variable depending on the demographic.
   (e) shaped like a diamond.

16. Which is NOT a factor that determines population growth under the theory of demographic transition?
   (a) Improved sanitation and health care leads to rapid population growth.
   (b) Population momentum holds birth rates stable while death rates decline.
   (c) Density independent factors slow economic growth.
   (d) Education reduces a population growth rate.
   (e) A burgeoning proportion of elderly people slows advancement.

17. Increasing the gross domestic product of a nation can be associated with
   I. an increase in pollution.
   II. a decrease in pollution.
   III. an increase in population size.

   (a) I only
   (b) II only
   (c) III only
   (d) I and II
   (e) I, II, and III

Teaching Tip: Unit 3 Additional Free-Response Question

An additional free-response question is available for you to test concepts from Unit 3 with your students. The question presents a short discussion about the history of the Galapagos tortoise. To answer this question, students must

- discuss whether the Galapagos tortoise is a keystone species, an ecosystem engineer, or both
- describe how the Galapagos tortoise might contribute to the ecological succession of plants on the islands
- explain why differences might have developed in the tortoises on different islands in the Galapagos
- use the theory of island biogeography to predict colonization and extinction rates of tortoises on islands of varying sizes

Download this resource by clicking on the link buttons in the TE-book, opening the TRFD, or logging onto the book's companion website bcs .whfreeman.com/friedlandapes2e (teacher login required).

Unit 3 Additional Free-Response Question

Answers to Unit 3 AP® Exam Free-Response Questions

1. (a) For any parasite, an important limiting resource is the number of hosts available. Another limiting resource is the number of eggs that a female host can produce, since maternal inheritance is the primary way *Wolbachia* persist in a population from generation to generation.
(b) Inside the host, the population trajectory of *Wolbachia* is most likely to follow a logistic growth model. This includes an initial stage of exponential growth, followed by slowing growth as carrying capacity is reached. Alternatively, *Wolbachia* populations may overshoot carrying capacity and experience a subsequent die-off. If the die-off was not very severe, *Wolbachia* populations will likely begin to increase again and population size will oscillate around carrying capacity. This latter scenario may be particularly harmful to the host, as an overshoot would consume more resources than the host can supply to *Wolbachia* without host death.
(c) When *Wolbachia* first invades a host population, it is likely to experience rapid population growth due to the large availability of hosts. As all individuals in a population become infected, the growth rate of the *Wolbachia* population will decrease because all available hosts are infected (i.e., carrying capacity is reached). This will resemble a logistic population growth model. In addition, as the number of infected female hosts increase, the carrying capacity of *Wolbachia* populations will likely decrease. This occurs because *Wolbachia* consumes host resources used for reproduction and infected females produce fewer eggs, as stated in the problem. If fewer host eggs are produced, then fewer host individuals will be present in the next generation. This will reduce the carrying capacity of *Wolbachia*.
(d) Given that *Wolbachia* infection reduces the number of host offspring and *Wolbachia* carrying capacity, natural selection would select for traits that reverse this trend. One possibility is that there will be selection pressure for *Wolbachia* to preferentially infect male eggs and reduce the infection rates of female hosts. A second possibility is that there will be selection pressure for *Wolbachia* to alter the sex ratio of host populations by selectively killing male offspring. This will free female host resources for the production of more female offspring. A third possibility is that there will be selection pressure for *Wolbachia* populations to limit their population growth within female hosts such that a minimum number of eggs are always produced to ensure the production of hosts for future generations of *Wolbachia*.
(e) If nearby host populations are available for infection, a *Wolbachia* population in one host population may use up all host resources in that population and migrate to another host population. The presence of alternative pools of host resources may allow for *Wolbachia* overshoots since die-offs are not likely to occur when more resources are available. The presence of nearby host populations may reduce the selection pressures discussed in the answer to question (d).

Answers continue in the Answer Appendix on page ANS-5.

Questions 18 and 19 refer to the following table:

Country	Ecological footprint (km^2)	Average monthly salary	Number of coal burning plants	Population growth rate (%)
Argentina (119)	1,000,000	1,100	1	0.98
Colombia (108)	800,000	700	4	1.10
Philippines (64)	1,100,000	300	9	1.84
Brazil (133)	5,500,000	800	7	0.83
Panama (89)	100,000	800	1	1.38

Sources: Global Energy Observatory, *http://www.globalenergyobservatory.org*; United Nations International Labour Organization, *http://www.ilo.org*; World Bank, *http://data.worldbank.org*.

18. Based solely on the IPAT equation, _____ is likely to have the greatest population size and _____ is likely to have smallest population size.
(a) Argentina/Colombia
(b) Colombia/Philippines
(c) Philippines/Panama
(d) Brazil/Panama
(e) Brazil/Argentina

19. Which country is most likely to be in the second phase of demographic transition?
(a) Argentina
(b) Colombia
(c) Philippines
(d) Brazil
(e) Panama

20. If everyone moved to urban areas, it is likely that
(a) the global ecological footprint would decrease.
(b) the global ecological footprint would increase.
(c) the global ecological footprint would remain the same.
(d) global sustainability would be achieved.
(e) global population increase would cease.

Section 2: Free-Response Questions

Write your answer to each part clearly. Support your answers with relevant information and examples. Where calculations are required, show your work.

1. *Wolbachia* is one of the world's most common parasitic microbes. It infects an enormous number of species, including many insects. Estimates indicate that as many as 70 percent of all insect species are infected with *Wolbachia*, including mosquitoes, spiders, and mites. The bacteria reproduce inside a host and are transmitted directly from mother to offspring or when infected males mate with uninfected females. After infection, *Wolbachia* consume host resources, which leaves the host with fewer resources for reproduction. As a result, infected females often have fewer offspring.
(a) What is a possible limiting resource for *Wolbachia*, aside from the amount of resources within the host? (2 points)
(b) Describe the population trajectory of *Wolbachia* inside of a host. (2 points)
(c) Describe the population trajectory of *Wolbachia* in a population of hosts. (2 points)
(d) Based on your answer for the previous question, what is one possible way that natural selection is altering the interaction between *Wolbachia* and its hosts? (2 points)
(e) Discuss how the presence of nearby host populations might alter the spread of *Wolbachia*. (2 points)

2. China's one-child policy has drastically reduced total fertility rate to 1.4, and has led to a reduction in population growth rate to 0.50 percent in 2012. One problem associated with this is the development of an inverted population pyramid. Care for the elderly requires a substantial amount of energy, resources, and manpower. Chinese elderly primarily rely on their families for health care.
(a) Given China's one-child policy, provide two examples of how China's growing elder population will affect future population growth. (3 points)
(b) Describe one potential benefit and one potential problem associated with the emigration of elderly individuals from rural to urban areas. (2 points)
(c) The total population size of China was 1,400,000,000 in 2012, with a crude birth rate of 12 births per 1,000 individuals. The number of individuals that emigrated from China was 700,000 and the number of individuals that immigrated to China was 200,000. Calculate the crude death rate. (3 points)
(d) What would the total fertility rate be if all families in China obeyed the one-child policy rule? (1 point)

3

Harry Hu/Shutterstock

science applied

How Can We Manage Overabundant Animal Populations?

The growing human population has caused population declines in many animal species, while, at the same time, other species have flourished as the result of human alterations of the landscape. Increases in the populations of these species have had many consequences. How have these increases happened, and what, if anything, should we do about them?

What causes animal population explosions?

The white-tailed deer is a high-profile example of dramatic animal population growth in the United States. Before European colonists arrived, its densities were approximately 4 to 6 deer per square kilometer (10 to 15 deer per square mile). In many areas today, however, deer densities are up to 20 times higher. In New York State, for example, unregulated hunting and the clearing of preferred deer habitats for farming had nearly eliminated the deer population by 1900, but the population has since grown exponentially to 1 million deer.

This population growth can be explained by processes we discussed in Chapters 6 and 7. For example, white-tailed deer prefer to live in early-successional fields containing wildflowers and shrubs. Abandoned farms that have undergone secondary succession during the twentieth century, such as the old fields of New England that we described in Chapter 6, are prime deer habitat. Such increases in habitat and food supply have increased the carrying capacity of the environment for deer. With a higher carrying capacity, density-dependent controls on the population are weaker, and reproduction increases.

In addition to providing deer with habitat and food, humans have eliminated many of the deer's major predators, including wolves and cougars. During much of the twentieth century, regulated hunting held deer populations in check, taking the place of natural predators that had largely been eliminated. Over the last few decades, however, the number of deer hunters has steadily declined, allowing the deer population to increase. In short, a higher carrying capacity and a lower mortality by both predators and hunters have allowed deer populations to grow exponentially and exceed their historic carrying capacity (FIGURE SA3.1).

The white-tailed deer is not the only species whose populations are exploding. Animal overpopulation is a global issue. Many large herbivorous mammals have benefited from anthropogenic alterations of the environment, including predator extinctions, reduced

FIGURE SA3.1 **Deer overpopulation.** As humans have provided increased food and habitat for white-tailed deer and simultaneously removed their major predators, deer populations have increased to the point that they exceed the historic carrying capacity. *(Gary Griffen/Earth Scenes/Animals Animals)*

Science Applied 3 How Can We Manage Overabundant Animal Populations? 255

FIGURE SA3.2 **Overabundant kangaroos.** Abnormally high populations of kangaroos consume all of their natural food supply and consequently move into human-dominated landscapes, where they come into conflict with humans. *(David Wall/Alamy)*

hunting, increased habitat, and increased food supplies provided by agriculture and by people who feed them. The sika deer (*Cervus nippon*) of Nara, Japan, for example, lives in an area that has been protected as a natural monument since World War II. Many factors have contributed to a population explosion of this species. Human visitors to the area feed the deer, hunting is restricted, and Japan's native wolves—natural predators of the deer—are extinct. In Australia, eastern gray kangaroos (*Macropus giganteus*), which normally consume wild grasses, are also consuming agricultural crops and the grass on golf courses. (FIGURE SA3.2). Their natural predator, the Tasmanian wolf (*Thylacinus cynocephalus*), is now extinct. In South Africa, the population of African elephants (*Loxodonta africana*) doubled within 10 years of being granted legal protection from hunting.

Large herbivorous mammals are not the only animals that experience overpopulation as the result of human activity. Florida's 500,000 monk parakeets (*Myiopsitta monachus*) are the offspring of individuals that have escaped from the pet trade since the 1960s. Other former pets, including dogs and cats, quickly establish feral populations when released into the wild. Populations of Canada geese (*Branta canadensis*) have grown along with the number of agricultural fields and golf courses on which they can feed.

What effects do overabundant species have on communities?

Populations that exceed their historic carrying capacity can cause widespread problems. Large populations of herbivore species such as deer, elephants, and kangaroos can overgraze the plants in their community, leaving little food for other herbivorous species. In Australia, grazing by the overabundant kangaroos removes many of the plants needed by other herbivores, including several species of endangered animals that depend on these plants for food and habitat.

These large animal populations also have direct effects on humans. Herbivorous mammals consume agricultural crops, landscape plants, and valuable wild plant species. In the United States, for example, the deer population so greatly exceeds the land's natural carrying capacity that it is now difficult to regenerate several valuable forest tree species, including northern red oak (*Quercus rubra*) and sugar maple. Deer also consume suburban landscape plants and damage $60 million worth of agricultural crops each year.

Wildlife overabundance increases wildlife-human encounters, such as deadly collisions of vehicles with deer or kangaroos. More frequent contact with wildlife also increases the risk of humans contracting diseases that the animals carry. White-tailed deer, for example, carry ticks that can be infected by the bacterium that causes Lyme disease. According to the Centers for Disease Control and Prevention, nearly 100,000 people in the United States contracted Lyme disease in 1992, a number that climbed to more than 300,000 by 2013.

How can we control overabundant species?

Controlling the population of any species involves issues that are ecological, political, moral, and ethical. From an ecological perspective, it is clear how to control a wildlife population: reduce the available food and habitat to lower the carrying capacity, compensate for the missing predators by killing individuals in the population, or slow the population's ability to reproduce. On a local scale, it can be helpful to discourage people from feeding wildlife. On a larger scale, however, reducing food and habitat is often the least viable option because it would typically require restricting the access of large herbivores to millions of hectares of land.

In 2009, the Australian government ordered the army to shoot 6,000 kangaroos as a way to reduce the population of 9,000 kangaroos living on a large military compound. The objective was to reduce the amount of grazing done by the kangaroo herd and therefore protect rare plants and the insects that live on those plants. Culling the kangaroo population on the military compound was a controversial decision that was opposed by animal rights activists and by many Australians who consider the kangaroo a national symbol that should be protected. The activists obtained a court injunction to stop the culling of the species, but after a delay, the kangaroo cull proceeded. Throughout the world, even when all parties agree that the overabundance of a species poses major problems for humans and for the environment, culling a herd through shooting is an increasingly unpopular strategy.

256 Science Applied 3 How Can We Manage Overabundant Animal Populations?

FIGURE SA3.3 **Administering contraceptives to wild animals.** In South Africa, a wildlife biologist shoots a dart containing contraceptives at a female elephant. The goal is to reduce the overabundant elephant population and reduce the destructive effects of these elephants on the habitat. *(Gialuigi Guercia/AFP/Getty Images)*

A strategy that has more public support is the use of birth control. This approach is common in cities, where many feral cats are trapped, neutered, and returned to the wild. Researchers are also investigating ways to give animals contraceptive drugs in order to reduce their populations over time. The drugs being developed differ in their modes of action, ranging from chemicals that prevent animals from making critical reproductive hormones to chemicals that prevent sperm from fertilizing an egg. Contraceptive strategies are currently being tested on a variety of species, including deer, kangaroos, elephants, and monk parakeets (FIGURE SA3.3).

Reducing animal overpopulation with contraceptives has widespread public appeal because it eliminates the need to kill animals. Getting the contraceptive drugs into the animals, however, presents substantial challenges. Drugs can be administered to some animals via dart guns. For others, such as monk parakeets, researchers are working on ways to administer medications in food. Some contraceptive drugs last for a single breeding season and require a new dose each season, which can be difficult and expensive. The cost of giving an animal a single dose of a contraceptive drug ranges from $100 to $500 per animal.

The issue of overabundant wildlife highlights a number of other environmental issues. The underlying causes of animal overpopulation are typically associated with human population increases. These human-caused wildlife population explosions can, in turn, have substantial negative effects on humans and their environment. Although there are several potential solutions to the problem of animal overpopulation, successful solutions must take into account the public's affection for wildlife, its opposition to killing animals, and its frequent lack of knowledge about the negative effects of overabundant species.

Questions

1. If humans are responsible for environmental changes that cause increases in the populations of native species, under what conditions should we work to reverse these population increases?
2. How might the options to control overabundant populations differ in rural versus urban areas?

Free-Response Question

(a) List three density-dependent factors and three density-independent factors that can influence the population size of white-tailed deer. (2 points)

(b) Describe two ways that deer populations at low abundance can accelerate the process of ecological succession in a forest toward a climax stage. (4 points)

(c) Consider a region of temperate forest that contains 100,000 deer with a sex ratio of 3 females to 1 male, an infant mortality rate of 10, a crude birth rate of 40, and a crude death rate of 20. Only 10 percent of the females are able to reproduce.

(i) Calculate population growth rate for white-tailed deer in this region. (2 points)

(ii) Calculate the number of deer that hunters would need to shoot in order to keep the deer population stable. (2 points)

References

Broache, A. 2005. Oh deer! *Smithsonian,* October.

New York Times. 2004. Putting nature on the pill. August 31. (http://www.nytimes.com/2004/08/31/science/putting-nature-on-the-pill.html)

New York Times. 2008. Australia stirs controversy over a plan to cull 400 kangaroos. March 14. (http://www.nytimes.com/2008/03/14/world/asia/14australia.html)

New York Times. 2013. A kinder, gentler way to thin the deer herd. July 5. (http://www.nytimes.com/2013/07/06/nyregion/providing-birth-control-to-deer-in-an-overrun-village.html)

Suggested Answers to Science Applied Questions

1. We should start investigating how to reverse an increase in population caused by human actions when that increase in population starts to have negative effects on other species or on the ecosystem.

2. While hunting or trapping is usually a viable option for controlling populations in rural areas, in urban areas it poses too great a risk to humans.

The answer to the Science Applied 3 Free-Response Question can be found in the Answer Appendix on page ANS-6.

Answer to Science Applied 3 Free-Response Question

Download the full answer to the Science Applied 3 Free-Response question by clicking on the link buttons in the TE-book, opening the TRFD, or logging onto the book's companion website bcs.whfreeman.com/friedlandapes2e (teacher login required).

Answer to Science Applied 3 Free-Response Question

chapter

Earth Systems

PD Overview

Chapter 8 covers mineral resources, geologic processes, weathering, and soil science. Plate boundaries increase seismic activity and drive the formation of volcanoes and mountain ranges. The rock cycle includes the processes of weathering and erosion. The rock cycle also distributes resources on the surface of Earth. Soil, which is essential for life on Earth, has various physical, chemical, and biological properties. The chapter concludes with a comprehensive discussion of mining processes that remove valuable mineral resources and often harm the environment. Recent legislation has sought to minimize damage, and ongoing reclamation efforts to repair damage from mining have seen some success.

Module 24: Mineral Resources and Geology

Almost all of the mineral resources on Earth accumulated when the planet formed 4.6 billion years ago. But Earth is a dynamic planet. Earth's geologic processes form and break down rocks and minerals, drive volcanic eruptions and earthquakes, determine the distribution of scarce mineral resources, and create the soil in which plants grow. This module examines the distribution of Earth's mineral resources and some of the geological processes that continue to affect this distribution.

Module 25: Weathering and Soil Science

Soil is a combination of geologic and organic material that forms a dynamic membrane over much of the surface of Earth. A variety of processes that occur in soil connect the overlying biology with the underlying geology. This module describes how the weathering of rocks leads to the formation of soil and the development of specific soil horizons. It examines the physical, chemical, and biological processes that take place in soils. The module concludes with a discussion of mining and its consequences.

Alignment to AP® Environmental Science Course Description

AP® Outline	Module	
I. Earth Systems and Resources (10–15%)		
A. Earth Science Concepts	Module 24	Mineral Resources and Geology
D. Soil and Soil Dynamics	Module 25	Weathering and Soil Science
IV. Land and Water Use (10–15%)		
E. Mining	Module 25	Weathering and Soil Science

Chapter Learning Objectives

After completing this chapter students will be able to

- describe the formation of Earth and the distribution of critical elements on Earth.
- define the theory of plate tectonics and discuss its relevance to the study of the environment.
- describe the rock cycle and discuss its importance in environmental science.
- understand how weathering and erosion occur and how they contribute to element cycling and soil formation.
- explain how soil forms and describe its characteristics.
- describe how humans extract elements and minerals and the social and environmental consequences of these activities.

Chapter 8 Pacing Guide

This pacing guide is based on a schedule with 120 sessions of 50 minutes each before the AP® Exam. If you have a different number of sessions before the exam, you can modify the pacing to suit your needs. If you have additional time, consider incorporating quizzes, released AP® Environmental Science free-response and multiple-choice questions, or additional activities.

Module	Standard Schedule Days	Block Schedule Days
Module 24	3	1½
Module 25	2	1
Assessment	1	½

Chapter 8 Resources

The link to these resources can be found by clicking on the link buttons in the Teacher's e-Book (TE-book), opening the Teacher's Resource Flash Drive (TRFD), or logging onto the book's companion website bcs.whfreeman.com/friedlandapes2e (teacher login required).

- **PowerPoint Presentation 8**
- **Optimized Art PowerPoint and JPEG Files 8**
- **Do the Math Videos**
- **Measuring Your Impact 8**: What Is the Impact of Your Diet on Soil Dynamics?
- **Activity 8.1**: Geologic Timescale
- **Lab 8.1**: Soil Texture
- **Chapter 8 Web Resources**
- **Exploring the Literature 8**
- **Answers to Chapter 8 AP® Exam**
- **Multiple-Choice AP® Practice Exam 8**

Free-Response Questions from Previous AP® Environmental Science Exams

Free-response questions from prior AP® Environmental Science Exams are available on the AP® course website, https://apstudent.collegeboard.org/apcourse/ap-environmental-science/exam-practice. Students should be able to answer all of the questions listed below with material learned in this and previous chapters. When a question requires students to understand material from multiple chapters, the question will be listed in the last chapter required to complete the entire question. Questions marked with an asterisk (*) are from exams with released multiple-choice questions. You may want to save these questions until the end of the year so you can give your students a complete released exam for practice. Questions marked with double asterisks (**) require math to calculate a problem. Look for references to these questions throughout the chapter.

Year	Question	Content
2004	4	• Soil testing • Soil conservation • Inorganic fertilizers • Humus-rich soil conditions
2014	3	• Subduction zones • Volcanic activity • Primary succession and soil formation from volcanic rock • Tsunamis and plate movement • Transform faults

PD Chapter 8 Overview

Watch the video overview of Chapter 8 (for teachers) by clicking on the link buttons in the TE-book, opening the TRFD, or logging onto the book's companion website bcs.whfreeman.com/friedlandapes2e (teacher login required).

TRM PowerPoint Presentation 8

Download the PowerPoint presentation for Chapter 8 by clicking on the link buttons in in the TE-book, opening the TRFD, or logging onto the book's companion website bcs.whfreeman.com/friedlandapes2e (teacher login required).

Lithium is a vital component of environmentally friendly hybrid-electric cars but mining lithium has adverse environmental consequences. This lithium mine is in Bolivia. *(Robin Hammond/Panos Pictures)*

chapter

Earth Systems

Module **24** Mineral Resources and Geology

Module **25** Weathering and Soil Science

Are Hybrid Electric Vehicles as Environmentally Friendly as We Think?

Many people in the environmental science community believe that hybrid electric vehicles (HEV) and all-electric vehicles are some of the most exciting innovations of the last decade. Cars that run on electric power or on a combination of electricity and gasoline are much more efficient in their use of fuel than similarly sized internal combustion (IC) automobiles. Some of these cars use no gasoline at all, while others are able to run as much as twice the distance as a conventional IC car on the same amount of gasoline.

Although HEV and all-electric vehicles reduce our consumption of liquid fossil fuels, they do come with environmental trade offs. The construction of HEV vehicles uses scarce metals, including neodymium, lithium, and lanthanum. Neodymium is needed to form the magnets used in the electric motors, and lithium and lanthanum are used in the compact high-performance batteries the vehicles require. At present, there appears to be enough lanthanum available in the world to meet the demand of the Toyota Motor Corporation, which has manufactured more than 3 million Prius HEV vehicles. Toyota obtains its lanthanum from China. There are also supplies of lanthanum in various geologic deposits in California, Australia, Bolivia, Canada, and elsewhere, but most of these deposits have not yet been developed for mining. Until this happens, some scientists believe that the production of HEVs and all-electric vehicles will eventually be limited by the availability of lanthanum.

Although HEV and all-electric vehicles reduce our consumption of liquid fossil fuels, they do come with environmental trade offs.

In addition to the scarcity of metals needed to make HEV and all-electric vehicles, we have to consider how we acquire these metals. Wherever mining occurs, it has a number of environmental consequences. Material extraction leaves a landscape fragmented by holes, and road construction necessary for access to and from the mining site further alters the habitat. Erosion and water contamination are also common results of mining.

A typical Toyota Prius HEV uses approximately 1 kg (2.2 pounds) of

Teaching Tip: Chapter Opening Case

The chapter opening case, "Are Hybrid Electric Vehicles as Environmentally Friendly as We Think?", asks students to consider the environmental impacts of hybrid electric vehicles and all-electric vehicles. Because these vehicles allow us to reduce our use of fossil fuels, many believe they are the answer to our growing transportation needs. However, these vehicles also have environmental impacts. For example, there is a limited supply of metals needed to manufacture batteries, and extracting these metals through mining creates acid mine drainage and fragments habitats.

Teaching Tip: Discussion Starter

Before reading the opening case, ask students which type of vehicle they believe is most environmentally friendly: electric vehicles or those with internal combustion engines. Have them give two reasons to support their answer.

Teaching Tip: Journal Prompt

As students read the opening case, they are likely to discover some environmental issues that are new to them. This is a good time to have them create a chart to compare an electric vehicle and a gas vehicle. Ask them to list the pros and cons of each type of vehicle and decide which has the bigger environmental footprint. For example, as we learn from the case, the scarce resources of neodymium and lanthanum must be mined, with many negative consequences. After listing negative impacts for an electric vehicle, they should list all of the negative impacts for traditional gasoline-powered vehicles, such as potential spills from ocean drilling. They should go on to list the benefits of each; for example, electric vehicles produce less air pollution when in use, while gas vehicles use an established infrastructure that allows for easy fueling. A comparison chart will help students make informed decisions when they purchase a vehicle in the future.

neodymium and 10 kg (22 pounds) of lanthanum. Mining these elements involves pumping acids into deep boreholes to dissolve the surrounding rock and then removing the acids and resulting mineral slurry. Lithium is extracted from certain rocks, and lithium carbonate is extracted from brine pools and mineral springs adjacent to or under salt flats. Both extraction procedures are types of surface mining, which can have severe environmental impacts. The holes, open pits, and ground disturbance created by mining these minerals provide the opportunity for air and water to react with other minerals in the rock, such as sulfur, to form an acidic slurry. As this acid mine drainage flows over the land or underground toward rivers and streams, it dissolves metals and other elements. As a result, water near surface mining operations is highly acidic—sometimes with a pH of 2.5 or lower. It may also contain harmful levels of dissolved metals and minerals.

Current HEV technology does represent a step forward in our search to reduce the use of fossil fuels. However, to make good decisions about the use of resources, we must have a full understanding of the costs and benefits.

Sources:
M. Armand and J.-M. Tarascon, Building better batteries, *Nature* 451 (2008): 652–657; Anonymous, Rare earth metals may trigger trade wars, *www.discovery.com*, February 11, 2013.

As we saw in our account of the hybrid-electric car, decisions about resource use are not always simple. Why are some of Earth's mineral resources so limited? Why do certain elements occur in some locations but not in others? What processes create minerals and other Earth materials, and what are the consequences of extracting them? Understanding the answers to these questions helps environmental scientists make informed decisions about the environmental and economic costs and benefits of resource use.

In this chapter we will explore the subjects of resources, geology, and soil science. We will look inside Earth to explore its structure, its formation, and the ongoing processes that affect the composition and availability of elements and minerals on our planet. We will then turn to the surface of Earth to explore how the rock cycle distributes these resources. With this foundation, we will examine the formation of soil, which is essential for so many biological activities on Earth. Finally, we will look at the problems we face in extracting resources from Earth—an issue of great concern to environmental scientists and others.

Teaching Tip: Activity

Use the Howard Hughes Medical Institute interactive entitled "Deep History of Life on Earth" to expand student learning. This interactive "click and learn" format allows students to explore the record of life on Earth. Short written excerpts and video clips provide interest along this interactive timeline.

Find the link to this resource by clicking on the link buttons in the TE-book, opening the TRFD, or logging onto the book's companion website bcs.whfreeman.com/friedlandapes2e (teacher login required).

TRM **Chapter 8 Web Resources**

module

24

Mineral Resources and Geology

Almost all of the mineral resources on Earth accumulated when the planet formed 4.6 billion years ago. But Earth is a dynamic planet. Earth's geologic processes form and break down rocks and minerals, drive volcanic eruptions and earthquakes, determine the distribution of scarce mineral resources, and create the soil in which plants grow. In this module we will examine the distribution of Earth's mineral resources and some of the geological processes that continue to affect this distribution.

Learning Objectives

After reading this module you should be able to

- describe the formation of Earth and the distribution of critical elements on Earth.
- define the theory of plate tectonics and discuss its relevance to the study of the environment.
- describe the rock cycle and discuss its importance in environmental science.

The availability of Earth's resources was determined when the planet was formed

Earth's history is measured using the geologic time scale, shown in FIGURE 24.1. It's hard to believe that events that took place 4.6 billion years ago continue to have such an influence on humans and their interactions with Earth. The distribution of chemicals, minerals, and ores around the world is in part a function of the processes that occurred during the formation of Earth.

The Formation and Structure of Earth

Nearly all the elements found on Earth today are as old as the planet itself. FIGURE 24.2 illustrates how Earth formed roughly 4.6 billion years ago from cosmic dust in the solar system. The early Earth was a hot, molten sphere. For a period of time, additional debris from the formation of the Sun bombarded Earth. As this molten material slowly cooled, the elements within it separated into layers according to their mass. Heavier elements such as iron sank toward Earth's center, and lighter elements such as silica floated toward its surface. Some gaseous elements left the solid planet and became part of Earth's atmosphere. Although asteroids occasionally strike Earth today, the bombardment phase of planet formation has largely ceased and the elemental composition of Earth has stabilized. In other words, the elements and minerals that were present when the planet formed—and which are distributed unevenly around the globe—are all that we have.

Some minerals such as silicon dioxide—the primary component of sand and glass—are readily available

Teaching Tip: Warm-up

Ask students the following questions:

- Describe how elements accumulated on Earth. *Earth formed from cosmic dust in the solar system. For a period of time, additional debris bombarded Earth. The molten Earth began to cool and elements separated by mass.*
- The formation of Earth has largely ceased and Earth is considered a closed system; what does this mean for our available resources? *The elements that were present when Earth formed are all we have today. Our available resources are limited.*

Teaching Tip: Activity

Geologic Timescale
Activity 8.1 asks students to construct a timeline for the appearance of organisms on the geologic timescale. This activity provides students with a perspective on the vast timescale in which various species evolved on Earth.

Download this resource by clicking on the link buttons in the TE-book, opening the TRFD, or logging onto the book's companion website bcs.whfreeman.com/friedlandapes2e (teacher login required).

Activity 8.1: Geologic Timescale

TEACHING with FIGURES ▶

Show students Figure 24.1 and ask them to identify the approximate number of years since the following events occurred (answers are provided in italics):

- **Origin of Earth:** *4.5 billion years ago*
- **Oldest evidence of life:** *3.5–3.8 billion years ago*
- **Oxygen builds up in the atmosphere:** *2.25–2.75 billion years ago*
- **First vertebrates:** *490–543 million years ago*
- **First land plants:** *443 million years ago*
- **Largest mass extinction:** *240 million years ago*
- **First dinosaurs:** *206 million years ago*
- **First humans:** *1–1.8 million years ago*

Teaching Tip: Journal Prompt

Ask students to identify Earth's layers and their composition.

- ***Core:*** *Solid inner core/liquid outer core; nickel and iron*
- ***Mantle:*** *Molten rock (magma) circulates in convection currents*
- ***Asthenosphere:*** *Outer part of the mantle; semi-molten rock*
- ***Lithosphere:*** *Outermost layer of Earth including the mantle and crust*

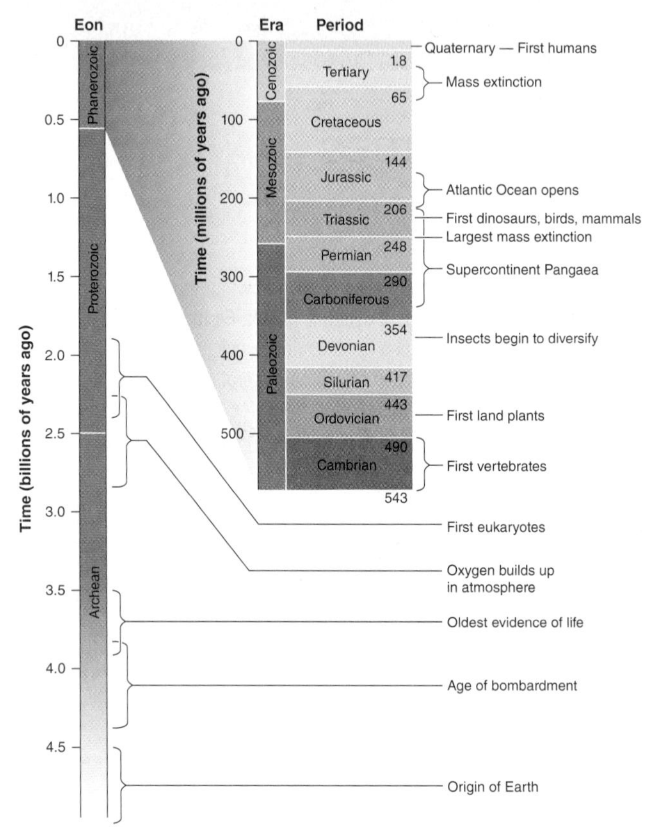

FIGURE 24.1 **The geologic time scale.** Time since the origin of Earth through the present is divided into three eons. The most recent eon, the Phanerozoic, is broken down into three eras. The Phanerozoic can also be divided into 11 periods, the most recent of which is the Quaternary.

worldwide on beaches and in shallow marine and glacial deposits. Others such as diamonds—which are formed from carbon that has been subjected to intense pressure—are found in relatively few isolated locations. Over the course of human history, this uneven geographic distribution has driven many economic and political conflicts.

Because Earth's elements settled into place according to their mass, the planet is characterized by distinct vertical zonation. If we could slice into Earth as shown in FIGURE 24.3 we would see concentric layers composed of various materials. The innermost zone is the planet's **core,** over 3,000 km (1,860 miles) below Earth's surface. The core is a dense mass largely made of nickel and some iron. The inner core is solid, and the outer core is liquid. Above the core is the **mantle,** containing molten rock, or **magma,** that slowly circulates in convection cells, much as the atmosphere

Core The innermost zone of Earth's interior, composed mostly of iron and nickel. It includes a liquid outer layer and a solid inner layer.

Mantle The layer of Earth above the core, containing magma.

Magma Molten rock.

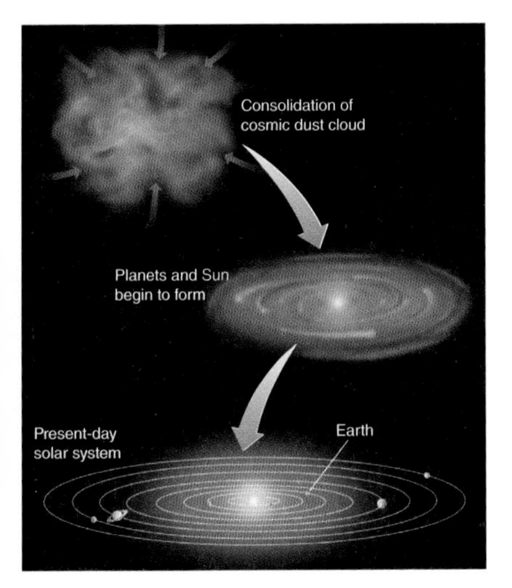

FIGURE 24.2 **Formation of Earth and the solar system.** The processes that formed Earth 4.6 billion years ago determined the distribution and abundance of elements and minerals today.

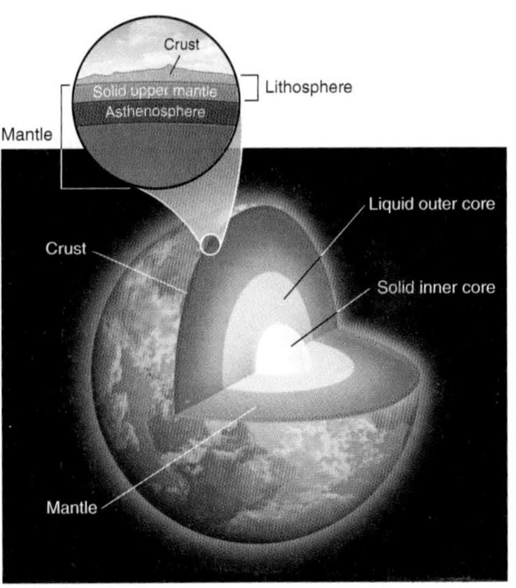

(a) Earth's vertical zonation

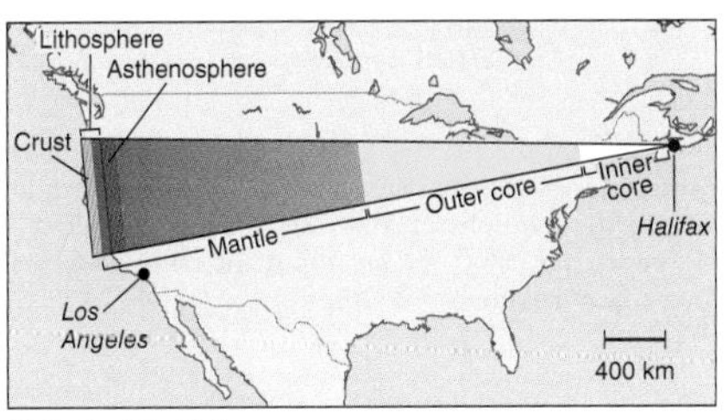

(b) Scale of Earth's layers

FIGURE 24.3 **Earth's layers.** (a) Earth is composed of concentric layers. (b) If we were to slice a wedge from Earth, it would cover the width of the United States.

does. The **asthenosphere,** located in the outer part of the mantle, is composed of semi-molten, ductile (flexible) rock. The brittle outermost layer of the planet, called the **lithosphere** (from the Greek word *lithos,* which means "rock"), is approximately 100 km (60 miles) thick. It includes the solid upper mantle as well as the **crust,** the chemically distinct outermost layer of the lithosphere. It is important to recognize that these regions overlap: The lowest part of the lithosphere is also the uppermost portion of the mantle.

The lithosphere is made up of several large and numerous smaller plates, which overlie the convection cells within the atmosphere. Over the crust lies the thin layer of soil that allows life to exist on the planet. The crust and overlying soil provide most of the chemical elements that make up life.

Because Earth contains only a finite supply of mineral resources, we will not be able to extract resources from the planet indefinitely. In addition, once we have mined the deposits of resources that are most easily obtained, we must use more energy to extract the remaining resources. Both of these realities provide an incentive for us to minimize our use of mineral resources and to reuse and recycle them whenever possible.

Hot Spots

One of the critical consequences of Earth's formation and elemental composition is that the planet remains very hot at its center. The high temperature of Earth's outer core and mantle is thought to be the result of the radioactive decay of various isotopes of elements such as potassium, uranium, and thorium, which release heat. The heat causes plumes of hot magma to well

Asthenosphere The layer of Earth located in the outer part of the mantle, composed of semi-molten rock.

Lithosphere The outermost layer of Earth, including the mantle and crust.

Crust In geology, the chemically distinct outermost layer of the lithosphere.

Teaching Tip: Warm-up

Ask students: What is the reason for the high temperatures of Earth's core and mantle? *High temperatures in Earth's core and mantle are the result of radioactive decay of various isotopes.*

Teaching Tip: Journal Prompt

Ask students: What is a hot spot? *A hot spot is a place where molten material from Earth's mantle reaches the lithosphere.*

COMMON MISCONCEPTIONS
Plate tectonics is ongoing

Students tend to think that the process of plate tectonics is something that occurred millions of years ago and is not happening today. But Earth's geological systems are dynamic; plates are constantly shifting and islands are forming. There are many examples such as the earthquake in Haiti and the activity of Mount St. Helens.

Teaching Tip: Journal Prompt

Ask students the following questions:

- What is the theory of plate tectonics? *The theory of plate tectonics states that Earth's lithosphere is divided into plates, most of which are in constant motion and have existed in several different configurations over time.*
- What is the evidence that supports the theory of plate tectonics? *Evidence of plate tectonics includes observations of identical rock formations on both sides of the Atlantic Ocean and fossil evidence of the same species on continents that are separated by oceans.*

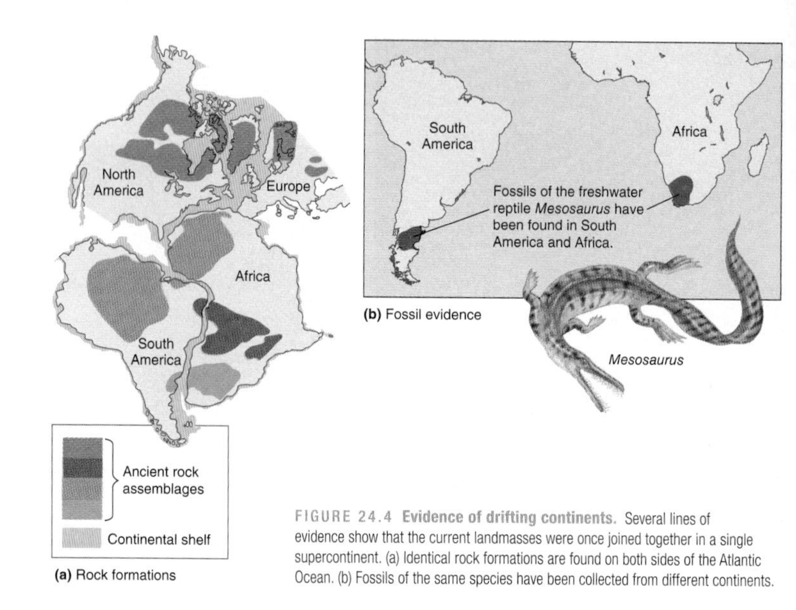

FIGURE 24.4 Evidence of drifting continents. Several lines of evidence show that the current landmasses were once joined together in a single supercontinent. (a) Identical rock formations are found on both sides of the Atlantic Ocean. (b) Fossils of the same species have been collected from different continents.

upward from the mantle. These plumes produce **hot spots:** places where molten material from the mantle reaches the lithosphere. As we shall see in the following pages, hot spots are an important component of the surface dynamics of Earth.

The theory of plate tectonics describes the movement of the lithosphere

Prior to the 1900s, scientists believed that the major features of Earth—such as continents and oceans—were fixed in place. In 1912, a German meteorologist named Alfred Wegener published a revolutionary hypothesis proposing that the world's continents had once been joined in a single landmass, which he called "Pangaea." His evidence included observations of identical rock formations on both sides of the Atlantic Ocean, as shown in FIGURE 24.4a. The positions of these formations suggested that a single supercontinent may have broken up into separate landmasses. Fossil evidence also suggested that a single large continent existed in the past. Today, we can find fossils of the same species on different continents that are separated by oceans; Figure 24.4b shows one example.

Many resisted the idea that Earth's lithosphere could move laterally. However, following the publication of Wegener's hypothesis, scientists found additional evidence that Earth's landmasses had existed in several different configurations over time. This led to the theory of **plate tectonics,** which states that Earth's lithosphere is divided into plates, most of which are in constant motion. The **tectonic cycle** is the sum of the processes that build up and break down the lithosphere.

The theory of plate tectonics is often called a unifying theory in geology and earth sciences because it relates to so many different aspects of the earth sciences.

Hot spot In geology, a place where molten material from Earth's mantle reaches the lithosphere.

Plate tectonics The theory that the lithosphere of Earth is divided into plates, most of which are in constant motion.

Tectonic cycle The sum of the processes that build up and break down the lithosphere.

Plate Movement

We now know that the lithosphere consists of a number of plates. Oceanic plates lie primarily beneath the oceans, whereas continental plates lie beneath landmasses. The crust of oceanic plates is dense and rich in iron, while the crust of continental plates generally contains more silicon dioxide, which is much less dense than iron. The continental plates are therefore lighter and

Teaching Tip: Warm-up

Ask students: How do the properties of oceanic plates differ from continental plates? *The crust of oceanic plates is dense and rich in iron. The crust of continental plates contains more silicon dioxide and is much lighter. Typically, continental plates will rise above the oceanic plates.*

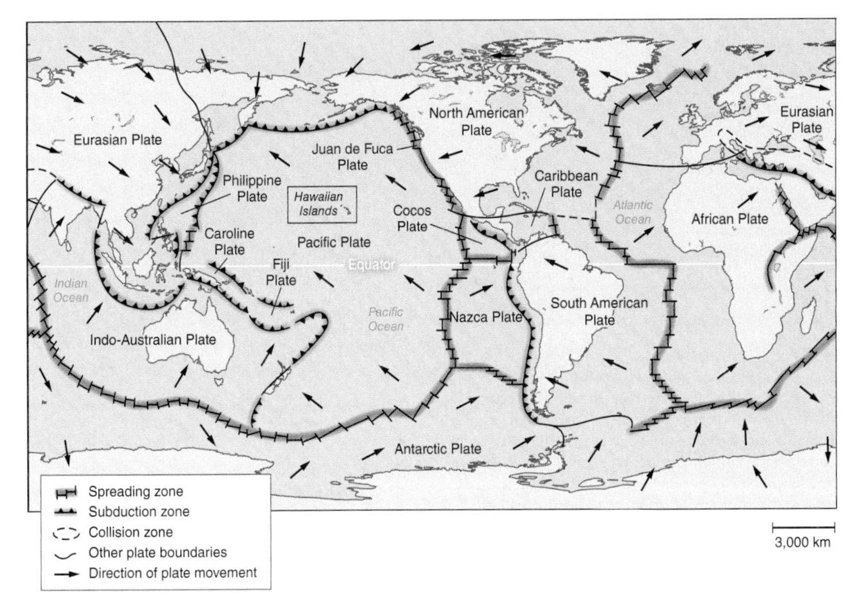

FIGURE 24.5 **Tectonic plates.** Earth is covered with tectonic plates, most of which are in constant motion. The arrows indicate the direction of plate movement. New lithosphere is added at spreading zones and older lithosphere is recycled into the mantle at subduction zones.

typically rise above the oceanic plates. FIGURE 24.5 identifies Earth's major plates.

Oceanic and continental plates "float" on top of the denser material beneath them. Their slow movements are driven by convection cells in Earth's mantle. The heat from Earth's core creates these convection cells, which are similar to those in the atmosphere. (See Figure 10.6 on page 115.) Mantle convection drives continuous change: the creation and renewal of Earth materials in some locations of the lithosphere and destruction and removal of Earth material in other locations. As oceanic plates move apart, rising magma forms new oceanic crust on the seafloor at the boundaries between those plates. This process, called seafloor spreading, is shown in the center of FIGURE 24.6. Where oceanic plates meet continental plates, old oceanic crust is pulled

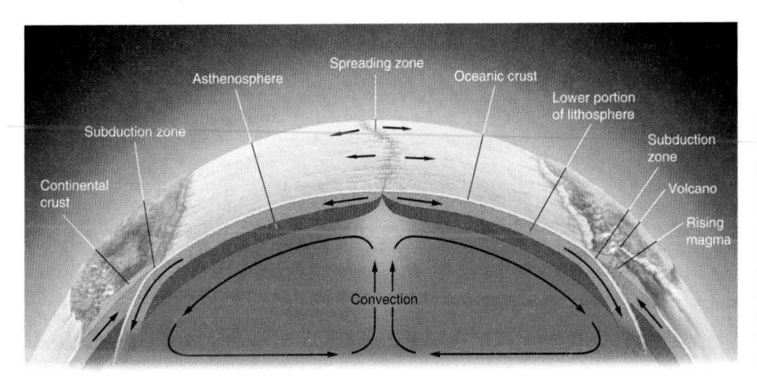

FIGURE 24.6 **Convection and plate movement.** Convection in the mantle causes oceanic plates to spread apart as new rock rises to the surface at spreading zones. Where oceanic and continental plate margins come together, older oceanic crust is subducted.

◀ TEACHING with FIGURES

Ask students to look at Figure 24.5 and name locations where the following occur:

- **Formation of oceanic crust:** *Occurs primarily in spreading zones (highlighted red); an example location would be through the middle of the Atlantic Ocean between the South American Plate and the African Plate.*
- **Volcanic activity:** *Occurs primarily at subduction zones (highlighted blue); an example location would be along the west coast of South America between the Nazca Plate and the South American Plate.*
- **Formation of mountain ranges:** *Occurs at convergent boundaries between continental plates (highlighted blue); an example location would be the Himalayas.*
- **Increased seismic activity:** *Occurs at transform faults when plates slide past each other; an example location would be between the Pacific Plate and the North American Plate along the West Coast of the United States.*

Teaching Tip: Video

The Pangaea Pop-up
This video clip from TED-Ed introduces the basics of tectonic plate movement, consequences of plate movement, and types of plate boundaries.

Find the link to this resource by clicking on the link buttons in the TE-book, opening the TRFD, or logging onto the book's companion website bcs.whfreeman.com/friedlandapes2e (teacher login required).

 **Chapter 8 Web Resources**

TEACHING with FIGURES ▶

Use Figure 24.6 to explain what we see in Figure 24.7—how the Hawaiian Islands formed: A plume of magma (a column of lava) lies at a fixed position. As the Pacific Plate moves over the plume, a volcano forms, creating a steady succession of volcanic islands.

Teaching Tip: Connections

Ask students to describe the impact on species as continents drift over Earth's surface. *Climates and geographic barriers change. Species will either adapt or go extinct. In some regions where individuals of the same species became separated by geographic barriers, they evolved into different species over time (allopatric speciation).*

Teaching Tip: Concept Map

There are three zones of plate contact. Have each student create a concept map that includes the titles listed below. Describe each type of boundary and give a likely outcome resulting from the identified type of plate contact.

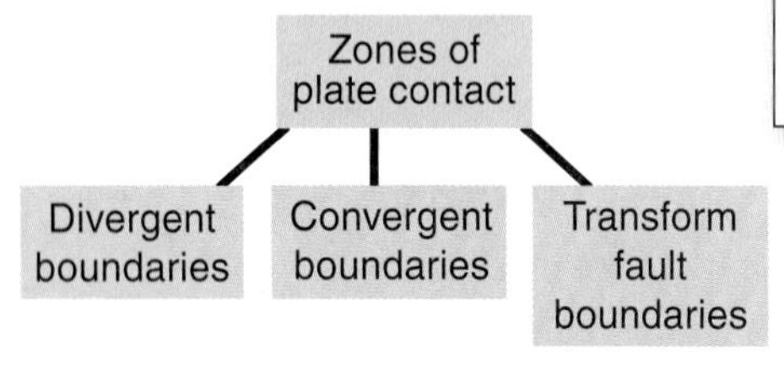

- **Divergent boundaries:** *Plates move away from each other. Magma from the mantle reaches Earth's surface and pushes the plates outward while forming new rock on the ocean floor (seafloor spreading).*
- **Convergent boundaries:** *Plates move toward one another and collide. At convergent boundaries, volcanoes are likely to form. If two continental plates meet, then mountain ranges may be formed.*
- **Transform fault boundaries:** *Plates move sideways past each other. Earthquakes are likely to occur along these boundaries.*

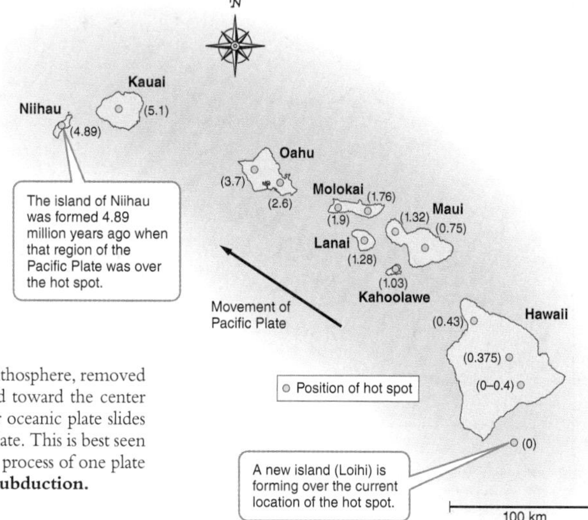

FIGURE 24.7 Plate movement over a hot spot. The Hawaiian Islands were formed by volcanic eruptions as the Pacific Plate traveled over a geologic hot spot. The chain of inactive volcanoes to the northwest of Hawaii shows that those locations used to be over the hot spot. Numbers indicate how long ago each area was located over the hot spot (in millions of years).

downward beneath the continental lithosphere, removed from the ocean bottom, and pushed toward the center of Earth. In this process, the heavier oceanic plate slides underneath the lighter continental plate. This is best seen on the left side of Figure. 24.6. This process of one plate passing under another is known as **subduction.**

Consequences of Plate Movement

Because the plates move, continents on those plates slowly drift over the surface of Earth. As the continents have drifted, their climates have changed and geographic barriers were formed or removed, and as a result, species evolved and adapted, or slowly or rapidly went extinct. In some places, as the plates moved a continent that straddled two plates broke apart, creating two separate smaller continents or islands in different climatic regions. As you may recall from our discussion of allopatric speciation in Chapter 5, species that become separated can take different evolutionary paths and over time evolve into two or more separate species. The fossil record tells us how species adapted to the changes that took place over geologic time. Climate scientists and ecologists can use this information to anticipate how species will adapt to the relatively rapid climate changes happening on Earth today.

Although the rate of plate movement is too slow for us to notice, geologic activity provides vivid evidence that the plates are in motion. As a plate moves over a geologic hot spot, heat from the rising mantle plume melts the crust, forming a **volcano**: a vent in Earth's surface that emits ash, gases, and molten lava. Volcanoes are a natural source of atmospheric carbon dioxide, particulates, and metals. Over time, as the plate moves past the hot spot, it can leave behind a trail of extinct volcanic islands, each with the same chemical composition. The Hawaiian Islands, shown in FIGURE 24.7, are an excellent example of this pattern.

Subduction The process of one crustal plate passing under another.

Volcano A vent in the surface of Earth that emits ash, gases, or molten lava.

Divergent plate boundary An area beneath the ocean where tectonic plates move away from each other.

Seafloor spreading The formation of new ocean crust as a result of magma pushing upward and outward from Earth's mantle to the surface.

Types of Plate Contact

Many other geologic events occur at the zones of contact that result from the movements of plates relative to one another. These zones of plate contact can be classified into three types: *divergent plate boundaries, convergent plate boundaries,* and *transform fault boundaries.*

Beneath the oceans, plates move away from each other at **divergent plate boundaries,** as illustrated in FIGURE 24.8a. At these boundaries, oceanic plates move apart as if on a giant conveyer belt. As magma from the mantle reaches Earth's surface and pushes upward and outward, new rock is formed, a phenomenon, called **seafloor spreading.** Seafloor spreading creates new lithosphere and brings important elements such as copper, lead, and silver to the surface of Earth. However, this new rock typically lies under the deep ocean. Over tens to hundreds of millions of years, as the tectonic cycle continues, some of that material forms new land that contains these valuable resources.

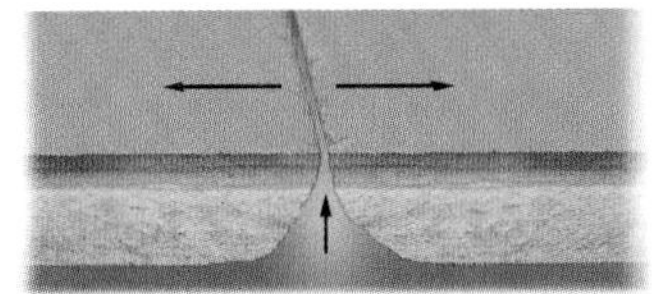

(a) Divergent plate boundary

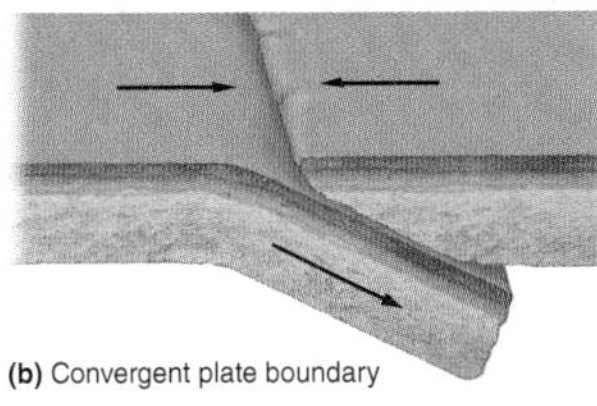

(b) Convergent plate boundary

(c) Transform fault boundary

FIGURE 24.8 **Types of plate boundaries.** (a) At divergent plate boundaries, plates move apart. (b) At convergent plate boundaries, plates collide. (c) At transform fault boundaries, plates slide past each other.

Clearly, if tectonic plates are diverging in one place, and if the surface of Earth has a finite area, the plates must be moving together somewhere else. **Convergent plate boundaries** form where plates move toward one another and collide, as shown in Figure 24.8b. The plates generate a great deal of pressure as they push against one another.

When plates move sideways past each other, the result is a **transform fault boundary,** shown in Figure 24.8c. A **fault** is a fracture in rock across which there is movement. Where this occurs, it is said that there is a high level of **seismic activity,** which is the frequency and intensity of earthquakes experienced over time. A **fault zone** is a large expanse of rock where a fault has occurred. Fault zones—also called areas of high seismic activity—form in the brittle upper lithosphere where two plates meet or slide past one another. In these large expanses of rock where movement has occurred, the rock near the plate margins becomes fractured and deformed from the immense pressures exerted by plate movement. We'll discuss faults and earthquakes in more detail in the next section.

If two continental plates meet, both plate margins may be lifted, forming a mid-continental mountain range such as the Himalayas in Asia (FIGURE 24.9a). As shown in Figure 24.9b, when both plates are composed of material of equal density, one does not get subducted under the other. Instead, they are forced into one another, and the force of the plate movement pushes material upward in one of the processes that leads to the formation of mountains.

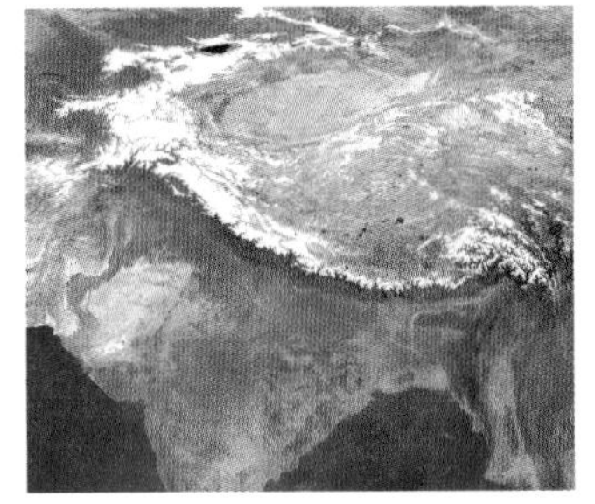

(a) The Himalayas from space

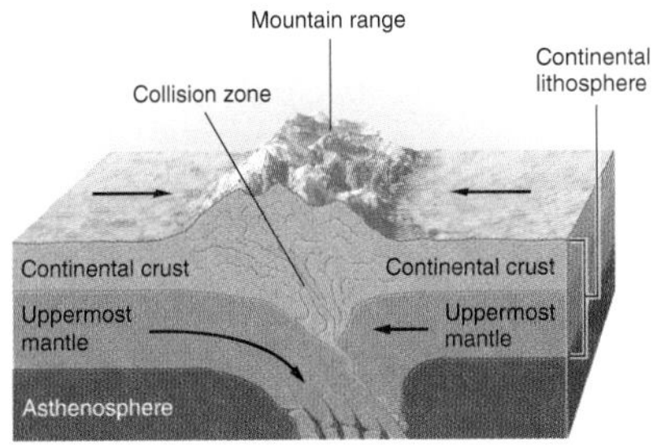

(b) Formation of the Himalayas

FIGURE 24.9 **Collision of two continental plates.** (a) A satellite image of the Himalayas, which include the highest mountains on Earth. (b) The Himalayan mountain range was formed when the collision of two continental plates forced the margins of both plates upward. *(European Space Agency)*

Most plates and continents move at about the same rate as your fingernails grow: roughly 36 mm, or 1.4

Convergent plate boundary An area where plates move toward one another and collide.

Transform fault boundary An area where tectonic plates move sideways past each other.

Fault A fracture in rock caused by a movement of Earth's crust.

Seismic activity The frequency and intensity of earthquakes experienced over time.

Fault zone A large expanse of rock where a fault has occurred.

Practicing Science: Demo

Use sandwich cookies to demonstrate the three types of plate boundaries. This should engage kinesthetic learners. The bottom cookie represents the solid inner core. The filling represents the mantle. The top cookie will be broken in half and represents two plates floating over the asthenosphere. To demonstrate convergent boundaries, pull the two cookie halves (plates) apart. To demonstrate transform fault boundaries, slide the plates past each other. To demonstrate convergent boundaries slide one plate under the other. In a high school setting this activity is a quick way to assess student understanding. Ask students to demonstrate and explain the three types of plate movement to a partner or lab group.

Teaching Tip: Concept Map

To explore the concept of convergent boundaries, have each student create a concept map to include the events listed in the blue boxes below. Ask students to determine the most likely outcome of the event. (Answers provided in italics.)

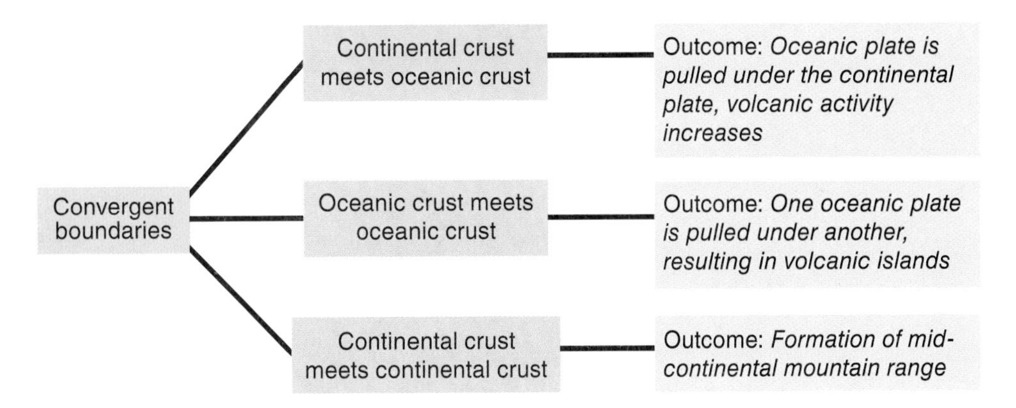

do the **math**

Answer to Your Turn

Time = distance ÷ rate

$$91.44 \text{ m} = 91.44 \times 1{,}000 \text{ mm} = 91{,}440 \text{ mm}$$

$$\frac{91{,}440}{20 \text{ mm/year}} = 4{,}572 \text{ years}$$

Teaching Tip: Warm-up

Ask students: At which type of plate boundary is an earthquake most likely to occur? *An earthquake is most likely to occur at a transform fault boundary.*

do the math

Plate Movement

If two cities lie on different tectonic plates, and those plates are moving so that the cities are approaching each other, how many years will it take for the two cities to be situated adjacent to each other?

Los Angeles is 630 km (380 miles) southeast of San Francisco. The plate under Los Angeles is moving northward at about 36 mm per year relative to the plate under San Francisco. Given this average rate of plate movement, how long will it take for Los Angeles to be located next to San Francisco?

The distance traveled is 630 km, and the net distance moved is 36 mm per year. We can use this formula to determine the answer

time = distance ÷ rate

$$630 \text{ km} = 630{,}000 \text{ m} = 630{,}000{,}000 \text{ mm}$$

$$\frac{630{,}000{,}000 \text{ mm}}{36 \text{ mm/year}} = 17{,}500{,}000 \text{ years}$$

We could also put these dimensional relationships together as follows and then cancel units that occur in both the numerator and denominator. We are left with an answer in numbers of years

$$630 \text{ km}\left(\frac{1{,}000 \text{ m}}{1 \text{ km}}\right) \times \left(\frac{1{,}000 \text{ mm}}{1 \text{ m}}\right) \times \left(\frac{1 \text{ year}}{36 \text{ mm}}\right) = 17{,}500{,}000 \text{ years}$$

It will take about 18 million years for Los Angeles to be located alongside San Francisco.

Your Turn How long will it take for a plate that moves at 20 mm per year to travel the distance of one football field? Note that a football field is 91.44 m (100 yards) long.

inches, per year. While this movement is far too slow to notice on a daily basis, the two plates underlying the Atlantic Ocean have spread apart and come together twice over the past 500 million years, causing Europe and Africa to collide with North America and South America and separate from them again. "Do the Math: Plate Movement" shows how we can calculate the time it takes for plates to move.

Faults, Earthquakes, and Volcanoes

Although the plates are always in motion, their movement, while generally slow, is not necessarily smooth. Imagine rubbing two rough and jagged rocks past each other. The rocks would resist that movement and get stuck together. The rock along a fault is also jagged and thus resists movement, but the mounting pressure eventually overcomes the resistance and the plates give way, slipping quickly. This is an **earthquake,** which is a sudden movement of Earth's crust caused by the release of potential energy along a fault, causing vibration or movement at the surface. Earthquakes occur when the rocks of the lithosphere rupture unexpectedly along a fault. The plates can move up to several meters in just a few seconds. Earthquakes are common in fault zones, which are areas of seismic activity. FIGURE 24.10 shows one such area, the San Andreas Fault in California, which is a transform fault. The **epicenter** of an earthquake is the exact point on the surface of Earth directly above the location where the rock ruptures, also shown in Figure 24.10.

Earthquakes are a direct result of the movement of plates and their contact with each other. Volcanic eruptions happen when molten magma beneath the crust is released to the atmosphere. Sometimes the two events are observed together, most often along plate boundaries where tectonic activity is high. FIGURE 24.11 shows one example, in which earthquake locations and volcanoes form a circle of tectonic activity, called the "Ring of Fire," around the Pacific Ocean.

Earthquake The sudden movement of Earth's crust caused by a release of potential energy along a geologic fault and usually causing a vibration or trembling at Earth's surface.

Epicenter The exact point on the surface of Earth directly above the location where rock ruptures during an earthquake.

Richter scale A scale that measures the largest ground movement that occurs during an earthquake.

Teaching Tip: Journal Prompt

Ask students: How does an earthquake occur? *As plates move past each other, movement may be restricted and pressure builds. Mounting pressure eventually overcomes resistance and the plates slip past each other quickly. The potential energy released causes a vibration or movement at the Earth's surface.*

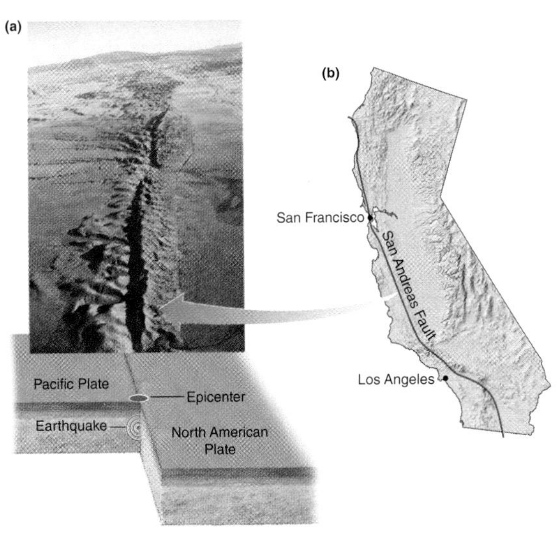

FIGURE 24.10 **A fault zone.** Many areas of seismic activity, including the San Andreas Fault in California, are characterized by a transform fault. The epicenter of an earthquake is the point on the surface of Earth directly above the location where the rock ruptures. *(age fotostock/SuperStock)*

The Environmental and Human Toll of Earthquakes and Volcanoes

Plate movements, volcanic eruptions, seafloor spreading, and other tectonic processes bring molten rock from deep beneath Earth's crust to the surface, and subduction sends surface crust deep into the mantle. This tectonic cycle of surfacing and sinking is a continuous Earth process. When humans live in close proximity to areas of seismic or volcanic activity, however, the results can be dramatic and devastating.

Earthquakes occur many times a day throughout the world, but most are so small that humans do not feel them. The magnitude of an earthquake is reported on the **Richter scale,** a measure of the largest ground movement that occurs during an earthquake. The Richter scale, like the pH scale described in Chapter 2, is logarithmic. On a logarithmic scale, a value increases by a factor of 10 for each unit increase. Thus a magnitude 7.0 earthquake, which causes serious damage, is 10 times greater than a magnitude 6.0

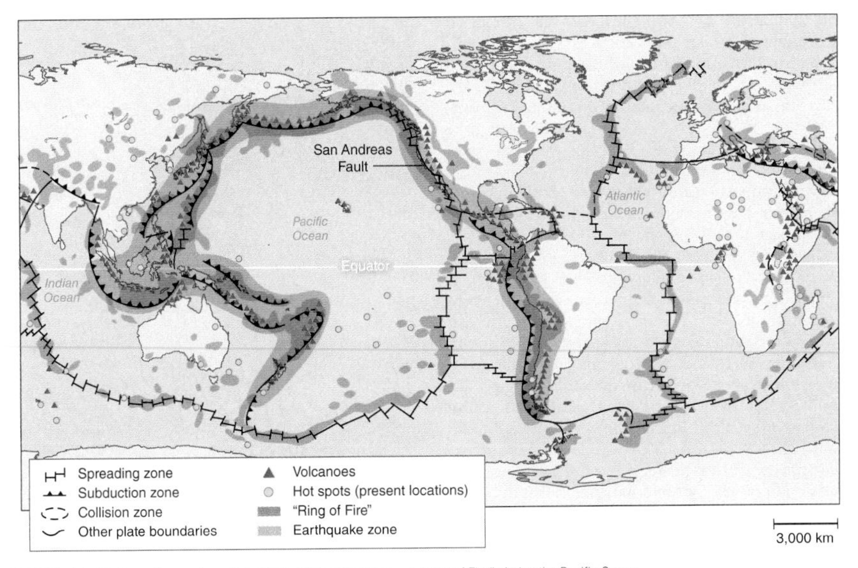

FIGURE 24.11 **Locations of earthquakes and volcanoes.** A "Ring of Fire" circles the Pacific Ocean along plate boundaries. Other zones of seismic and volcanic activity, including hot spots, are also shown on this map.

more **math** practice: Understanding the Richter Scale

The Richter scale is a measure of the largest ground movement that occurs during an earthquake. Like the pH scale, it is on a logarithmic scale. Students often do not know what it means when a scale is explained as a "power of 10." To help students understand that each change in magnitude is a power of 10, explain that for every place you move along the scale, you add that number of zeros to the base of 1. For example the difference between 4 and 5 on the Richter scale is one place, therefore one zero is added to the base of 1, making a "5" ten times stronger than a "4." The difference between 2 and 6 on the Richter scale is four places. Therefore four zeros are added to the base of 1, making a "6" 10,000 times stronger than a "2."

After providing the explanation above, ask students: How much stronger is an earthquake that registers 9.0 on the Richter scale than an earthquake that registers 6.0? *It is 1,000 times stronger.*

◀ TEACHING with FIGURES

Ask students to examine Figure 24.11 and answer the following questions:

- What is the "Ring of Fire"? *The Ring of Fire circles the Pacific Ocean along plate boundaries. Tectonic activity is so high along these boundaries that they include hotspots, seismic activity, and volcanoes.*
- Which characteristics distinguish the Ring of Fire from other geologically active areas around the globe? *In the Ring of Fire, earthquakes and volcanoes may be observed together.*

Teaching Tip: Journal Prompt

Earthquakes can cause environmental destruction such as habitat loss. Ask students to describe how humans are affected by earthquakes. *Earthquakes can lead to collapsed structures and buildings, contaminated water supplies, ruptured dams that provide drinking water and hydroelectric power, and human death. Nuclear power plants built in areas of seismic activity are designed to withstand significant movement and are programmed to shut down if movement exceeds a certain threshold. However, they have the potential for nuclear disaster if an earthquake is severe, like the 2011 earthquake in Japan.*

FIGURE 24.12 Earthquake damage in Haiti. The 2010 earthquake in Haiti killed more than 200,000 people and destroyed most of the structures in the capital, Port-au-Prince. *(Dominic Nahr/Magnum Photos)*

earthquake and 1,000 (10^3) times greater than a magnitude 4.0 earthquake, which only some people can feel or notice. Worldwide, there may be as many as 800,000 small earthquakes of magnitude 2.0 or less per year, but an earthquake of magnitude 8.0 occurs approximately once every year.

Even a moderate amount of Earth movement can be disastrous. Moderate earthquakes (defined as magnitudes 5.0 to 5.9) lead to collapsed structures and buildings, fires, contaminated water supplies, ruptured dams, and deaths. Loss of life is more often a result of the proximity of large population centers to the epicenter than of the magnitude of the earthquake itself. The quality of building construction in the affected area is also an important factor in the amount of damage that occurs. In 2008, a magnitude 7.9 earthquake in the southwestern region of Sichuan Province, China, killed more than 69,000 people. The epicenter was near a populated area where many buildings were probably not built to withstand a large earthquake. In 2010, a magnitude 7.0 earthquake in Haiti killed more than 200,000 people. Many of the victims were trapped under collapsed buildings (FIGURE 24.12).

Extra safety precautions are needed when dangerous materials are used in areas of seismic activity. Nuclear power plants are designed to withstand significant ground movement and are programmed to shut down if movement above a certain threshold occurs. The World Nuclear Association estimates that 20 percent of nuclear power plants operate in areas of significant seismic activity. Between 2004 and 2009, in four separate incidents, nuclear power plants in Japan shut down operation because of ground movement that exceeded the threshold. In 2011, seismic activity there led to a devastating major earthquake and tsunami that caused the second-worst nuclear power plant accident ever to occur, as we will see in Chapter 13.

Volcanoes, when active, can be equally disruptive and harmful to human life. Active volcanoes are not distributed randomly over Earth's surface; 85 percent of them occur along plate boundaries. As we have seen, volcanoes can also occur over hot spots. Depending on the type of volcano, an eruption may eject cinders, ash, dust, rock, or lava into the air (FIGURE 24.13). Volcanoes

FIGURE 24.13 A volcanic eruption. This 2001 explosive eruption from the Etna volcano in Italy threatened several nearby towns. *(Mario Cipollini/Aurora)*

AP® Exam Tip

When answering free-response questions, students must pay particular attention to the key words *environmental, economic, social,* and *ecological.* Teach students to differentiate among these words throughout the year so that when they see them on the exam, their answers will be automatic. The exercise below allows students to practice answering questions accurately.

Ask students to describe an ecological, an environmental, a social, and an economic impact associated with plate movement.

Ecological: *Impacts on communities (biotic); volcanic activity can destroy plant life as lava flows through established ecosystems.*

Environmental: *Impacts on ecosystems (biotic and abiotic); volcanic activity spews particulate matter and sulfur dioxide into the air as it erupts, decreasing air quality for animals. (Remind students not to mention people when asked about environmental impacts.)*

Social: *Impacts on people; earthquakes can be disastrous in highly populated areas. Many people have been killed because of collapsed structures.*

Economic: *Impacts on people and nations; earthquakes can cause major structural damage to buildings and interrupt both transportation and communication networks vital to business. Repair or rebuilding can be very expensive. Entire economies can be disrupted. (Advise students that money or jobs should be included in the language of the answer.)*

(a)

(b)

(c)

FIGURE 24.14 **Some common minerals.** (a) Pyrite (FeS_2), also called "fool's gold." (b) Graphite, a form of carbon (C). (c) Halite, or table salt (NaCl). *(a–b: John Grotzinger/Ramón Rivera-Moret/Harvard Mineralogical Museum; c: The Natural History Museum/Alamy)*

can result in loss of life, habitat destruction and alteration, reduction in air quality, and many other environmental consequences.

The world gained a new awareness of the impact of volcanoes when eruptions from a volcano in Iceland disrupted air travel to and from Europe in April 2010. Ash from the eruption entered the atmosphere in a large cloud and prevailing winds spread it over a wide area. The ash contained small particles of silicon dioxide, which have the potential to damage airplane engines. Air travel was suspended in many parts of Europe, and millions of travelers were stranded in what may have been the greatest travel disruption ever to have been caused by a volcano.

The rock cycle recycles scarce minerals and elements

The second part of the geologic cycle is the **rock cycle**, which refers to the constant formation and destruction of rock. The rock cycle is the slowest of all of Earth's cycles. Environmental scientists are most often concerned with the part of the rock cycle that occurs at or near Earth's surface.

Rock, the substance of the lithosphere, is composed of one or more minerals (FIGURE 24.14). Minerals are solid chemical substances with uniform (often crystalline) structures that form under specific temperatures and pressures. They are usually compounds, but may be composed of a single element such as silver or gold.

Formation of Rocks and Minerals

FIGURE 24.15 shows the processes of the rock cycle. Rock forms when magma from Earth's interior reaches the surface, cools, and hardens. Once at Earth's surface, rock masses are broken up, moved, and deposited in new locations by processes such as weathering and erosion. New rock may be formed from the deposited material. Eventually, the rock is subducted into the mantle, where it melts and becomes magma again. The rock cycle slowly but continuously breaks down rock and forms new rock.

While magma is the original source of all rock, there are three major ways in which the rocks we see at Earth's surface can form: directly from molten magma; by compression of sediments; and by exposure of rocks and other Earth materials to high temperatures and pressures. These three modes of formation lead to three distinct rock types: *igneous, sedimentary,* and *metamorphic.*

Igneous Rocks

Igneous rocks form directly from magma. They are classified by their chemical composition as basaltic or granitic, and by their mode of formation as intrusive or extrusive.

Basaltic rock is dark-colored rock that contains minerals with high concentrations of iron, magnesium, and calcium. It is the dominant rock type in the crust of oceanic plates. Granitic rock is lighter-colored rock made up of the minerals feldspar, mica, and quartz, and contains elements such as silicon, aluminum, potassium, and calcium. It is the dominant rock type in the crust of continental plates. When granitic rock breaks down due to weathering, it forms sand. Soils that develop from granitic rock tend to be more permeable

Rock cycle The geologic cycle governing the constant formation, alteration, and destruction of rock material that results from tectonics, weathering, and erosion, among other processes.

Igneous rock Rock formed directly from magma.

Teaching Tip: Concept Map

Have students create a concept map that identifies the three major types of rock and notes the way they are formed. (Answers are provided in italics.)

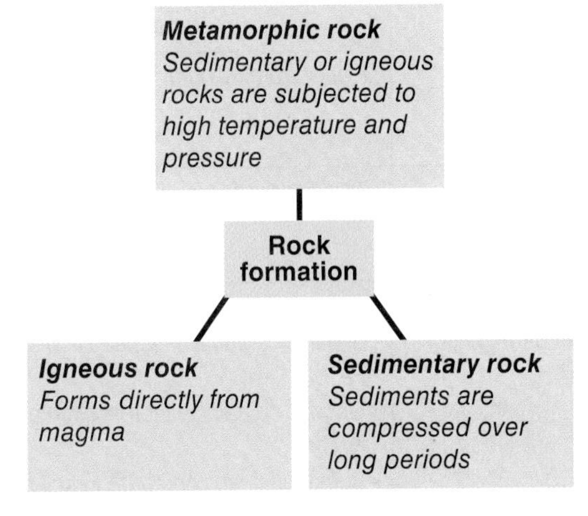

Teaching Tip: Journal Prompt

Ask students to describe the relationship among rock, minerals, compounds, and elements. *Rock is the substance of the Earth's lithosphere that is composed of one or more minerals. Minerals are uniform substances, usually compounds of two or more elements that are chemically combined. They may also be composed of a single element such as silver or gold.*

Teaching Tip: Warm-up

Ask students: What is the role of weathering and erosion in the formation of sedimentary rock? *Rock masses are broken down by weathering and erosion. Particles of rock settle and accumulate over time. These smaller particles of rock, which include sand and gravel, are compressed over time to form sedimentary rock.*

Teaching Tip: Journal Prompt

Have students create their own drawing of the rock cycle using Figure 24.15.

AP® Exam Tip

Students are expected to know six cycles: carbon, nitrogen, phosphorus, sulfur, water, and rock. By the time they start this chapter they should have covered all but the rock cycle. Even though the other five cycles are not in this chapter, students should be reviewing and building on these concepts all year. This is an excellent point in the course to do a thorough review of all biogeochemical cycles.

Teaching Tip: Journal Prompt

Have students complete the table below as a way to study the different types of rock. (Answers are provided in italics.)

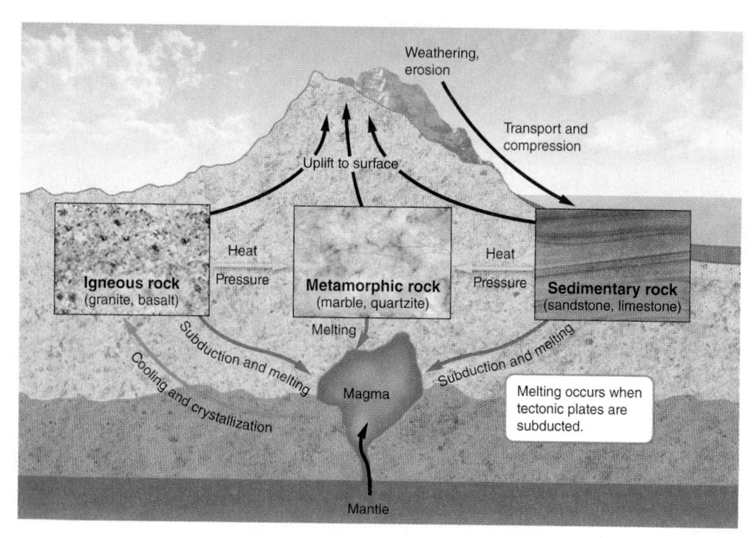

FIGURE 24.15 **The rock cycle.** The rock cycle slowly but continuously forms new rock and breaks down old rock. Three types of rock are created in the rock cycle: Igneous rock is formed from magma; sedimentary rock is formed by the compression of sedimentary materials; and metamorphic rock is created when rocks are subjected to high temperatures and pressures.

than those that develop from basaltic rock, but both types of rock can form fertile soil.

Intrusive igneous rocks form within Earth as magma rises up and cools in place underground. **Extrusive igneous rocks** form when magma cools above Earth's surface, such as when it is ejected from a volcano or released by seafloor spreading. Extrusive rocks cool rapidly, so their minerals have little time to expand into large individual crystals. The result is fine-grained, smooth types of rock such as obsidian. Both extrusive and intrusive rocks can be either granitic or basaltic in composition.

The formation of igneous rock often brings to the surface rare elements and metals that humans find economically valuable, such as the lanthanum described at the beginning of this chapter. When rock cools, it is subject to stresses that cause it to break. Cracks that occur when rock cools, known as **fractures,** can occur in any kind of rock. Water from the surface of Earth running through fractures may dissolve valuable metals, which may precipitate out in the fractures to form concentrated deposits called *veins*. These deposits are important sources of gold- and silver-bearing ores as well as rare metals such as tantalum, which is used to manufacture electronic components of cell phones.

Intrusive igneous rock Igneous rock that forms when magma rises up and cools in a place underground.

Extrusive igneous rock Rock that forms when magma cools above the surface of Earth.

Fracture In geology, a crack that occurs in rock as it cools.

Sedimentary rock Rock that forms when sediments such as muds, sands, or gravels are compressed by overlying sediments.

Metamorphic rock Rock that forms when sedimentary rock, igneous rock, or other metamorphic rock is subjected to high temperature and pressure.

Sedimentary Rocks

Sedimentary rocks form when sediments such as muds, sands, or gravels are compressed by overlying sediments. Sedimentary rock formation occurs over long periods when environments such as sand dunes, mudflats, lake beds, or areas prone to landslides are buried and the overlying materials create pressure on the materials below. The resulting rocks may be uniform in composition, such as sandstones and mudstones that formed from ancient oceanic or lake environments. Alternatively, they may be highly heterogeneous, such as conglomerate rocks formed from mixed cobbles, gravels, and sands.

Sedimentary rocks hold the fossil record that provides a window into our past. When layers of sediment containing plant or animal remains are compressed

	Igneous	Sedimentary	Metamorphic
Formation	*Directly from magma*	*Sediments are compressed over long periods*	*Sedimentary, igneous, or other metamorphic rocks are subjected to high temperatures and pressure*
Types	*Basaltic or granitic*	*May be uniform in composition or heterogeneous*	*Slate, marble, and anthracite (coal)*
Unique characteristics	• *Dominant rock type in continental plates* • *Basaltic contains minerals such as Ca, Fe, Mg* • *Granitic contains minerals such as Si, Al, K, Ca* • *Both types can contribute to soil fertility*	*Sedimentary rock holds the fossil record*	*Structurally strong and visually attractive; historically used as building materials*

over eons, those organic materials may be preserved, as described in Chapter 5.

Metamorphic Rocks

Metamorphic rocks form when sedimentary rocks, igneous rocks, or other metamorphic rocks are subjected to high temperatures and pressures. The pressures that form metamorphic rock cause profound physical and chemical changes in the rock. These pressures can be exerted by overlying rock layers or by tectonic processes such as continental collisions, which cause extreme horizontal pressure and distortion. Metamorphic rocks include stones such as slate and marble as well as anthracite, a type of coal. Metamorphic rocks have long been important as building materials in human civilizations because they are structurally strong and visually attractive.

module

24 REVIEW

In this module, we looked at the formation of Earth and the distribution of minerals. After Earth formed, heavier elements sank to the core, which accounts for the distribution and abundance of elements we see at the surface of Earth. The plates that overlay Earth and move at different rates further distribute elements. Plates are in constant motion around the globe and the resulting seismic activity contributes to various environmental hazards and has led to the creation of different landforms such as mountain ranges. Minerals and rocks of various compositions result from the chemical composition of the material that forms them, and the geologic conditions under which they form. When rocks and minerals break down as a result of various environmental conditions, they release chemical elements and the precursors of soils. In the next module we will examine the conditions under which rocks break down and how that contributes to the variety of soils that form around the world.

Module 24 AP® Review Questions

1. Which layer of Earth is composed primarily of iron and nickel?
   (a) The core
   (b) The crust
   (c) The asthenophere
   (d) The mantle
   (e) The lithosphere

2. Subduction
   (a) is the reason similar fossils appear on both sides of the Atlantic.
   (b) is the result of a hot spot moving near a plate boundary.
   (c) occurs when one plate passes under another.
   (d) occurs when oceanic plates diverge and form volcanoes.
   (e) is the process in transform boundaries that results in earthquakes.

3. The Hawaiian islands were formed
   (a) at a divergent plate boundary.
   (b) at a hot spot.
   (c) at a convergent plate boundary.
   (d) at a transform fault.
   (e) at a mid-ocean ridge.

4. How far will a plate travel in 60,000 years if it moves at net rate of 25 mm/yr?
   (a) 24 m
   (b) 1,500 m
   (c) 3,000 m
   (d) 4,800 m
   (e) 12,000 m

5. Which rock is formed at high temperatures and pressures?
   (a) Extrusive igneous
   (b) Intrusive igneous
   (c) Basaltic
   (d) Sedimentary
   (e) Metamorphic

6. Earthquake epicenters are often at
   (a) divergent boundaries.
   (b) convergent boundaries.
   (c) transform boundaries.
   (d) hot spots.
   (e) subduction zones.

Answers to Module 24 AP® Review Questions

1. a
2. c
3. b
4. b
5. e
6. c

AP® Exam Tip

Plate movement activity along the Ring of Fire is covered in the 2014 AP® Exam, Question 3. To answer this question, students must

- describe a subduction zone and how this leads to volcanic activity
- describe primary succession and soil formation from volcanic rock
- explain how tsunamis form as a result of plate movement
- describe transform fault boundaries as an earthquake occurs and after the earthquake has occurred

When students answer this free-response question, remind them to include temporal language in their description. For example, students lost a point if they said that earthquakes occur at transform boundaries when "plates move past each other," but did not note the time frame of movement as "rapidly" or "suddenly". This is an excellent opportunity to point out that descriptions must be thorough. One word can make all the difference in earning a point.

module

25 Weathering and Soil Science

Soil is a combination of geologic and organic material that forms a dynamic membrane over much of the surface of Earth. A variety of processes that occur in soil connect the overlying biology with the underlying geology. In this module we will explore the weathering of rocks that leads to the formation of soil and the development of specific soil horizons. We will discuss physical, chemical, and biological processes that take place in soils. Finally, we will examine human activities that degrade soils, including the process of mining.

Learning Objectives

After reading this module, you should be able to

- understand how weathering and erosion occur and how they contribute to element cycling and soil formation.
- explain how soil forms and describe its characteristics.
- describe how humans extract elements and minerals and the social and environmental consequences of these activities.

The processes of weathering and erosion contribute to the recycling of elements

We have seen that rock forms beneath Earth's surface under intense heat, pressure, or both heat and pressure. When rock is exposed at Earth's surface, it begins to break down through the processes of weathering and erosion. These processes are components of the rock cycle, returning chemical elements and rock fragments to the crust by depositing them as sediments through the hydrologic cycle. This physical breakdown and chemical alteration of rock begins the cycle all over again, as shown in Figure 24.15. Without this part of the rock cycle, elements would never be recycled and the precursors of soils would not be present.

Physical weathering The mechanical breakdown of rocks and minerals.

Weathering

Weathering occurs when rock is exposed to air, water, certain chemical compounds, or biological agents such as plant roots, lichens, and burrowing animals. There are two major categories of weathering—*physical* and *chemical*—that work in combination to degrade rocks.

Physical weathering is the mechanical breakdown of rocks and minerals, shown in FIGURE 25.1. Physical

(a)

(b)

FIGURE 25.1 **Physical weathering.** (a) Water can work its way into cracks in rock, where it can wash away loose material. When the water freezes and expands, it can widen the cracks. (b) Growing plant roots can force rock sections apart.

weathering can be caused by water, wind, or variations in temperature such as seasonal freeze-thaw cycles. When water works its way into cracks or fissures in rock, it can remove loose material and widen the cracks, as illustrated in Figure 25.1a. When water freezes in the cracks, the water expands, and the pressure of its expansion can force rock to break. Different responses to temperature can cause two minerals within a rock to expand and contract differently, which also results in splitting or cracking. Coarse-grained rock formed by slow cooling or metamorphism tends to weather more quickly than fine-grained rock formed by rapid cooling or metamorphism.

Biological agents can also cause physical weathering. Plant roots can work their way into small cracks in rocks and pry them apart, as illustrated in Figure 25.1b. Burrowing animals may also contribute to the breakdown of rock material, although their contributions are usually minor. However it occurs, physical weathering exposes more surface area and makes rock more vulnerable to further degradation. By producing more surface area for weathering processes to act on, physical weathering increases the rate of *chemical weathering*. **Chemical weathering** is the breakdown of rocks and minerals by chemical reactions, the dissolving of chemical elements from rocks, or both these processes. It releases essential nutrients from rocks, making them available for use by plants and other organisms.

Chemical weathering occurs most rapidly on newly exposed minerals, known as primary minerals. It alters primary minerals to form secondary minerals and the ionic forms of their constituent chemical elements. For example, when feldspar—a mineral found in granitic rock—is exposed to natural acids in rain, it forms clay particles and releases ions such as potassium, an essential nutrient for plants. Lichens can break down rock in a similar way by producing weak acids. Their effects can commonly be seen on soft gravestones and masonry. Rocks that contain compounds that dissolve easily, such as calcium carbonate, tend to weather quickly. Rocks that contain compounds that do not dissolve readily are often the most resistant to chemical weathering. In examining element cycles in Chapter 3, we noted that weathering of rocks is an important part of the phosphorus cycle.

Recall from Chapter 2 that solutions can be basic or acidic. Depending on the starting chemical composition of rock and the pH of the water that comes in contact with it, hundreds of different chemical reactions can take place. For example, as we saw in Chapter 2, carbon dioxide in the atmosphere dissolves in water vapor to create a weak acid, called carbonic acid. When waters containing carbonic acid flow into geologic regions that are rich in limestone, they dissolve the limestone (which is composed of calcium carbonate) and create spectacular cave systems (FIGURE 25.2).

Some chemical weathering is the result of human activities. For example, sulfur emitted into the atmosphere from fossil fuel combustion combines with oxygen to form sulfur dioxide. That sulfur dioxide reacts with water vapor in the atmosphere to form sulfuric acid, which then causes *acid precipitation*. **Acid precipitation,** also called **acid rain,** is precipitation high in sulfuric acid and nitric acid from reactions between water vapor and sulfur and nitrogen oxides in the atmosphere. Acid precipitation is responsible for the rapid degradation of certain old statues and gravestones and other limestone and marble structures. When acid precipitation falls on soil, it can promote

Chemical weathering The breakdown of rocks and minerals by chemical reactions, the dissolving of chemical elements from rocks, or both.

Acid precipitation Precipitation high in sulfuric acid and nitric acid from reactions between water vapor and sulfur and nitrogen oxides in the atmosphere. *Also known as* **Acid rain.**

Teaching Tip: Concept Map

Ask students to complete the concept map below to distinguish between chemical and physical weathering. They should give examples of each type of weathering and identify the environmental benefits weathering provides. (Answers are provided in italics.)

Weathering

Physical	**Environmental benefits**	**Chemical**
Definition: *Mechanical breakdown of rocks and minerals*	*Formation of soil precursors*	Definition: *Breakdown of rocks and minerals by chemical reactions*
Example: *Freeze-thaw cycles of water physically break apart rock*	*Reduction of atmospheric carbon dioxide*	Examples: *Acid rain can chemically break down rocks, lichens secrete acids that chemically weather rock*
	Contribution of additional elements to an ecosystem	
	Renewal of soil fertility	

Teaching Tip: Warm-up

Ask students to distinguish between weathering and erosion. *Weathering describes the breakdown of rock while erosion describes the movement of rock fragments.*

Teaching Tip: Discussion Starter

Ask students to describe the two mechanisms that drive erosion. *The first mechanism is wind, water, and ice that move soil and other materials. The second mechanism is living organisms that burrow under the soil.*

Teaching Tip: Journal Prompt

Ask students to explain how humans have increased rates of erosion. *Poor land use practices such as deforestation, overgrazing, unmanaged construction activity, and road building can create and accelerate erosion.*

FIGURE 25.2 **Chemical weathering.** Water that contains carbonic acid wears away limestone, sometimes forming spectacular caves. *(Mauritius/SuperStock)*

FIGURE 25.3 **Erosion.** Some erosion, such as the erosion that created these formations in the Badlands of South Dakota, occurs naturally as a result of the effects of water, glaciers, or wind. The Badlands are the result of the erosion of softer sedimentary rock types, such as shales and clays. Harder rocks, including many types of metamorphic and igneous rocks, are more resistant to erosion. *(welcomia/Shutterstock)*

chemical weathering of certain minerals in the soil, releasing elements that may then be taken up by plants or leached from the soil into groundwater and streams.

Chemical weathering, due to either natural processes or acid precipitation, can contribute additional elements to an ecosystem. Knowing the rate of weathering helps researchers assess how rapidly soil fertility can be renewed in an ecosystem. In addition, because the chemical reactions involved in the weathering of certain granitic rocks consume carbon dioxide from the atmosphere, weathering can actually reduce atmospheric carbon dioxide concentrations.

Erosion

We have seen that physical and chemical weathering results in the breakdown and chemical alteration of rock. **Erosion** is the physical removal of rock fragments (sediment, soil, rock, and other particles) from a landscape or ecosystem. Erosion is usually the result of two mechanisms. In one, wind, water, and ice move soil and other materials by downslope creep under the force of gravity. In the other, living organisms, such as animals that burrow under the soil, cause erosion. After eroded material has traveled a certain distance from its source, it accumulates. Deposition is the accumulation or depositing of eroded material such as sediment, rock fragments, or soil.

Erosion The physical removal of rock fragments from a landscape or ecosystem.

Erosion is a natural process: Streams, glaciers, and wind-borne sediments continually carve, grind, and scour rock surfaces (FIGURE 25.3). In many places, however, human land use contributes substantially to the rate of erosion. Poor land use practices such as deforestation, overgrazing, unmanaged construction activity, and road building can create and accelerate erosion problems. Furthermore, erosion usually leads to deposition of the eroded material somewhere else, which may cause additional environmental problems. We discuss human-caused erosion further in Chapters 10 and 11.

Soil links the rock cycle and the biosphere

Soil has a number of functions that benefit organisms and ecosystems. As you can see in FIGURE 25.4, soil is a medium for plant growth. It also serves as the primary filter of water as water moves from the atmosphere into rivers, streams, and groundwater. Soil contributes greatly to biodiversity by providing habitat for a wide variety of living organisms—from bacteria, algae, and fungi to insects and other animals. Soil and the organisms within it filter chemical compounds deposited by air pollution and by household sewage systems; some of these materials remain in the soil and some are released to the atmosphere or into groundwater.

FIGURE 25.4 **Ecosystem services provided by soil.** Soil serves as a medium for plant growth, as a habitat for other organisms, and as a recycling system for organic wastes. Soil also helps to filter and purify water.

In this section we will look at the formation and properties of soil.

The Formation of Soil

In order to appreciate the role of soil in ecosystems, we need to understand how and why soil forms and what happens to soil when humans alter it. It takes hundreds to thousands of years for soil to form. Soil is the result of physical and chemical weathering of rocks and the gradual accumulation of detritus from the biosphere. We can determine the specific properties of a soil if we know its parent rock type, the amount of time it has been forming, and its associated biotic and abiotic components.

FIGURE 25.5 shows the stages of soil development from rock to mature soil. The processes that form soil work in two directions simultaneously. The breakdown of rocks and primary minerals by weathering provides the raw material for soil from below. The deposition of organic matter from organisms and their wastes contributes to soil formation from above. What we normally think of as "soil" is a mix of these mineral and organic components. A poorly developed (young) soil has substantially less organic matter and fewer nutrients than a more developed (mature) soil. Very old soils may also be nutrient poor because over time plants remove many essential nutrients and water leaches away others. Five factors simultaneously determine the properties of soils: *parent material,* climate, topography, organisms, and time.

Parent Material

A soil's **parent material** is the underlying rock material from which a soil's inorganic components are derived. Different soil types arise from different parent materials. For example, a quartz sand (made up of silicon dioxide) parent material will give rise to a soil that is nutrient poor, such as those along the Atlantic coast of the United States. By contrast, a soil that has calcium carbonate as its parent material will contain an abundant supply of calcium, have a high pH, and may also support high agricultural productivity. Such soils are found in the area surrounding Lake Champlain in Vermont and northern New York, as well as in many other locations.

Parent material The rock material from which the inorganic components of a soil are derived.

◀ TEACHING with FIGURES

Ask students to identify the roles of soil using Figure 25.4 as a guide. *Soil is a medium for plant growth. It filters water, recycles nutrients, and provides a habitat for organisms.*

Teaching Tip: Concept Map

To review factors that determine soil properties, have each student create a concept map that includes the titles listed below. Describe each factor and how it contributes to soil properties.

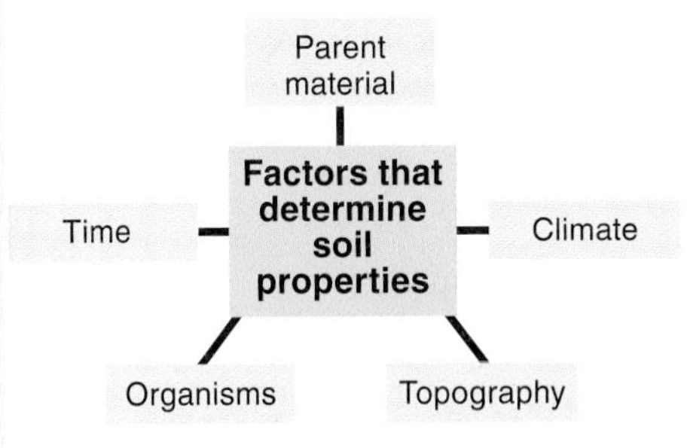

Parent material: *This is the rock underlying the soil. Parent material gives rise to inorganic components of soil. Different soil types arise from different parent material.*

Climate: *Soils do not develop if temperatures are below freezing because decomposition is limited. In the tropics soil formation can be accelerated due to rapid weathering and decomposition. Climate can have an indirect effect on soil formation because it affects the type of vegetation that develops and therefore the type of detritus in soil.*

Topography: *This is the surface slope and arrangement of a landscape. Soils that form on a slope are constantly subjected to erosion while soils that form at the bottom of steep slopes accumulate material.*

Organisms: *Organisms influence soil formation. Plants remove nutrients from the soil and secrete acids that promote chemical weathering. Animals burrow and effectively distribute organic and mineral matter. Organisms act as nutrient and mineral recyclers.*

Time: *The amount of time soil has to develop affects a variety of characteristics. Continual inputs of organic matter from highly productive communities can develop into deep and fertile soils after hundreds of thousands of years. Soils exposed to large amounts of water and that have less productive communities may develop into relatively infertile soils.*

TEACHING with FIGURES ▶

Ask students to examine Figure 25.5 and answer the following questions:

- Describe the development of soils from parent rock to mature soil. *Parent rock is weathered and fragments including minerals move upward. Organic material accumulates as plants and other organisms die.*
- Explain how nutrient values of soil change with age. *As soil continues to mature, greater amounts of organic material accumulate in the soil and nutrient values increase. Very young soil may be nutrient poor due to low organic matter while very old soils may be poor because plants remove many essential nutrients.*

Teaching Tip: Journal Prompt

Ask students to describe how soil development occurs in two directions simultaneously. *The breakdown of rocks and primary minerals by weathering provides the raw material for soil from below. The deposition of organic matter from organisms and their wastes contributes to soil formation from above.*

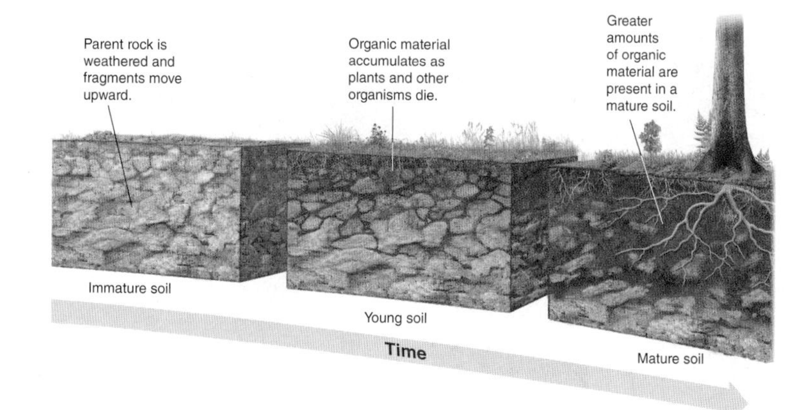

FIGURE 25.5 **Soil formation.** Soil is a mixture of organic and inorganic matter. The breakdown of rock and primary minerals from the parent material provides the inorganic matter. The organic matter comes from organisms and their wastes.

Climate

Climate influences soil formation in a number of ways. Soils do not develop well when temperatures are below freezing because decomposition of organic matter and water movement are both extremely slow in frozen or nearly frozen soils. Therefore, soils at high latitudes of the Northern Hemisphere are composed largely of organic material in an undecomposed state, as we saw in Chapter 4. In contrast, soil development in the humid tropics is accelerated by rapid weathering of rock and soil minerals, leaching of nutrients, and decomposition of organic detritus. Climate also has an indirect effect on soil formation because it affects the type of vegetation that develops, and therefore the type of detritus left after the vegetation dies.

Topography

Topography—the surface slope and arrangement of a landscape—is another factor in soil formation. Soils that form on steep slopes are constantly subjected to erosion and, on occasion, more drastic mass movements of material as happens in landslides. In contrast, soils that form at the bottoms of steep slopes may continually accumulate material from higher elevations and become quite deep.

Organisms

Many organisms influence soil formation. Plants remove nutrients from soil and excrete organic acids that speed chemical weathering. Animals that tunnel or burrow—for example, earthworms, gophers, and voles—mix the soil, uniformly distributing organic and mineral matter. Collectively, soil organisms act as recyclers of organic matter. In the process of using dead organisms and wastes as an energy source, soil organisms break down organic detritus and release mineral nutrients and other materials that benefit plants.

Human activity has dramatic effects on soils. For centuries, the use and overuse of land for agriculture, forestry, and other human activities has led to significant **soil degradation**: the loss of some or all of the ability of soils to support plant growth. One of the major causes of soil degradation is soil erosion, which occurs when topsoil is disturbed—for example, by plowing—or when vegetation is removed. As we saw in Chapter 3, these activities lead to erosion by water or wind (FIGURE 25.6).

Soil degradation The loss of some or all of a soil's ability to support plant growth.

FIGURE 25.6 **Erosion from human activity.** Erosion in this cornfield in Tennessee is obvious after a brief rainstorm. *(Tim McCabe/USDA Natural Resources Conservation Service)*

278 CHAPTER 8 ■ Earth Systems

Teaching Tip: Warm-up

Ask students: How have humans contributed to soil degradation? *The use and overuse of land for agriculture, forestry, and other human activities have led to soil degradation. Soil erosion occurs when topsoil is disturbed or when vegetation is disturbed. Soil that is compacted by machines, humans, and livestock has less vegetation and is therefore more subject to erosion. Intensive agriculture, including irrigation, depletes nutrients, and pesticides pollute the soil.*

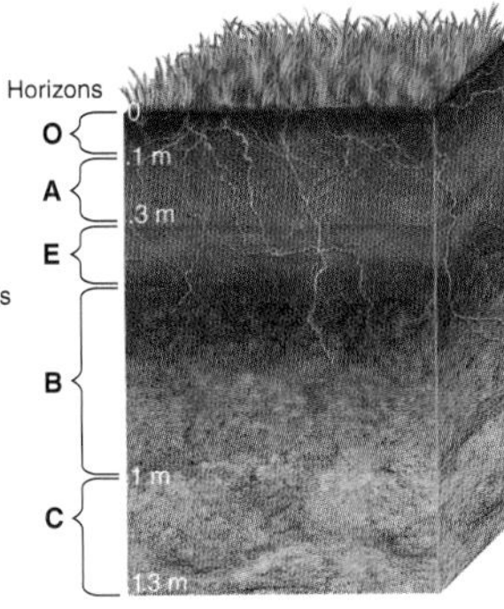

O horizon: Organic matter in various stages of decomposition
A horizon (topsoil): Zone of overlying organic material mixed with underlying mineral material
E horizon: Zone of leaching of metals and nutrients; occurs in some soils beneath either the O horizon or the A horizon.
B horizon (subsoil): Zone of accumulation of metals and nutrients
C horizon (subsoil): Least-weathered portion of the soil profile, similar to the parent material
Horizons
O
A
E
B
C
.1 m
.3 m
1 m
1.3 m

FIGURE 25.7 **Soil horizons.** All soils have horizons, or layers, which vary depending on soil-forming factors such as climate, organisms, and parent material. Most soils have either an O or A horizon and usually not both. Some soils that have an O horizon also have an E horizon.

While topsoil loss can happen rapidly—in as little as a single growing season—it takes centuries for the lost topsoil to be replaced. Compaction of soil by machines, humans, and livestock can alter its properties and reduce its ability to retain moisture. Compaction and drying of soil can, in turn, reduce the amount of vegetation that grows in the soil and thereby increase erosion. Intensive agricultural use and irrigation can deplete soil nutrients, and the application of agricultural pesticides can pollute the soil. In Chapters 11 and 14 we will return to the ways in which human activity affects the soil.

Time

The final factor that determines the properties of a soil is the amount of time during which the soil has developed. As soils age, they develop a variety of characteristics. The grassland soils that support much of the food crop and livestock feed production in the United States are relatively old soils. Because they have had continual inputs of organic matter for hundreds of thousands of years from the grassland and prairie vegetation growing above them, they have become deep and fertile. Other soils that are equally old, but with less productive communities above them and perhaps greater quantities of water moving through them, can become relatively infertile.

Soil Horizons

As soils form, they develop characteristic **horizons,** which are horizontal layers with distinct physical features such as color or texture, shown in FIGURE 25.7. The specific composition of those horizons depends largely on climate, vegetation, and parent material. At the surface of many soils is a layer known as the **O horizon,** composed of organic detritus such as leaves, needles, twigs, and even animal bodies, all in various stages of decomposition. The O horizon is most pronounced in forest soils and is also found in some grasslands. Organic matter is sometimes called humus (pronounced "hu-mus," but often mispronounced by beginning students as hummus, the delicious food made from chickpeas, olive oil, and garlic). In actuality, only the most decomposed organic matter at the lowest part of the organic horizon is truly humus.

In a soil that is mixed, either naturally or by human agricultural practices, the top layer is the **A horizon,** also known as **topsoil,** a zone of organic material and minerals that have been mixed together. In some acidic soils, an **E horizon**—a zone of leaching, or eluviation—forms under the O horizon or, less often, the A horizon. When an E horizon is present, iron, aluminum, and dissolved organic acids from the overlying horizons are transported through and removed from the E horizon and then deposited in the B horizon, where they accumulate. When an E horizon is present, it always occurs above the *B horizon*. The **B horizon,** commonly known as subsoil, is composed primarily of mineral material with very little organic matter. If nutrients are present in the soil, they will be in the B horizon. The **C horizon**—the least weathered soil horizon—occurs beneath the B horizon and is similar to the parent material.

Horizon A horizontal layer in a soil defined by distinctive physical features such as texture and color.

O horizon The organic horizon at the surface of many soils, composed of organic detritus in various stages of decomposition.

A horizon Frequently the top layer of soil, a zone of organic material and minerals that have been mixed together. *Also known as* **Topsoil.**

E horizon A zone of leaching, or eluviation, found in some acidic soils under the O horizon or, less often, the A horizon.

B horizon A soil horizon composed primarily of mineral material with very little organic matter.

C horizon The least-weathered soil horizon, which always occurs beneath the B horizon and is similar to the parent material.

Teaching Tip: Journal Prompt

Have students draw their own version of Figure 25.7 and describe each soil horizon.

- ***O horizon***: *Composed of organic detritus or decomposed organic matter*
- ***E horizon***: *A zone of leaching or eluviation*
- ***B horizon***: *Subsoil, composed primarily of mineral material with very little organic material*
- ***C horizon***: *Least weathered soil horizon, similar to the parent material*

Teaching Tip: Warm-up

Ask students: What is humus and where is it found in the soil profile? *Humus is the most decomposed portion of organic matter. It is found in the lowest part of the organic horizon.*

TEACHING with FIGURES ▶

Using Figure 25.8 (b) as a guide, ask students to rank the soil particles in size from largest to smallest.
Sand → Silt → Clay

Lab
Soil Texture

In this lab, each group gets three "unknown" samples of soil. The students must test the permeability, texture, bulk density, and chemical content of each sample. They then determine the most appropriate type of fertilizer for that soil. This lab provides a good opportunity to discuss sustainable farming practices such as using leguminous crops for added nitrogen content.

Download this resource by clicking on the link buttons in the TE-book, opening the TRFD, or logging onto the book's companion website bcs.whfreeman.com/friedlandapes2e (teacher login required).

TRM **Lab 8.1** Soil Texture

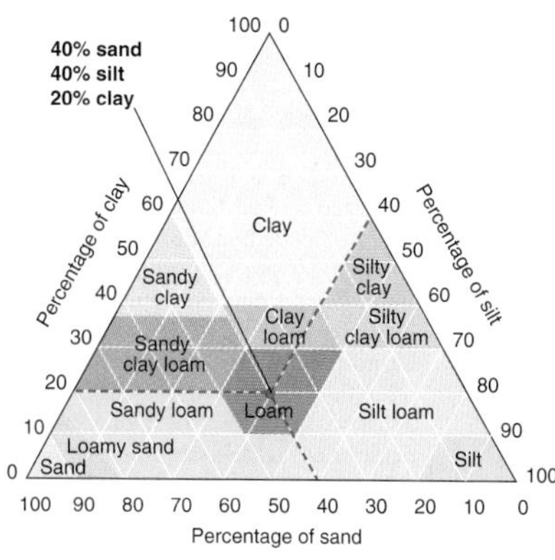

(a) Soil texture chart

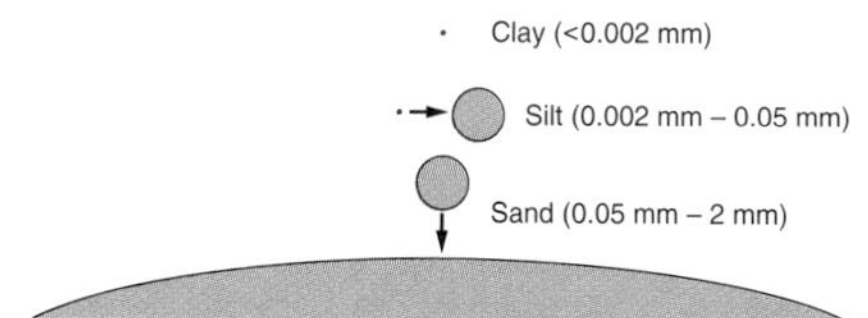

(b) Relative soil particle sizes (magnified approximately 100 times)

FIGURE 25.8 **Soil properties.** (a) Soils consist of a mixture of clay, silt, and sand. The relative proportions of these particles determine the texture of the soil. (b) The relative sizes of sand, silt, and clay.

Properties of Soil

Soils with different properties serve different functions for humans. For example, some soil types are good for growing crops and others are more suited for building a housing development. Therefore, to understand and classify soil types, we need to understand the physical, chemical, and biological properties of soils.

Physical Properties of Soil

The physical properties of soil refer to features related to physical characteristics such as size and weight. Sand, silt, and clay are mineral particles of different sizes. The texture of a soil is determined by the percentages of sand, silt, and clay it contains. FIGURE 25.8a plots those percentages on a triangle-shaped diagram that allows us to identify and compare soil types. Each location on the diagram has three determinants: the percentages of sand, of silt, and of clay. A point in the middle of the "loam" category (approximately at the "a" in "loam" in Figure 25.8) represents a soil that contains 40 percent sand, 40 percent silt, and 20 percent clay. We can determine this by following the lines from that point to the scales on each of the three sides of the triangle. If you are not certain which line to follow from a given point, always follow the line that leads to the lower value. For example, if you want to determine the percentage of sand in a sample represented by the red dot in Figure 25.8, you might follow the line out to either 60 percent sand or 40 percent sand; however, since you should always follow the line to the lower value, in this case the percentage of sand is 40 percent. The sum of sand, silt, and clay will always be 100 percent. Conversely, in the laboratory, a soil scientist can determine the percentage of each component in a soil sample and then plot the results. The name for the soil, for example "silty clay loam," follows from the percentage of the components in the soil.

The permeability of soil—how quickly soil drains—depends on its texture, shown in FIGURE 25.9. Sand particles—the largest of the three components—pack together loosely. Water can move easily between the particles, making sand quick to drain and quick to dry out. Soils with a high proportion of sand are also easy for roots to penetrate, making sandy soil somewhat advantageous for growing plants such as carrots and potatoes. Clay particles—the smallest of the three components—pack together much more tightly than sand particles. As a result, there is less pore space in a soil dominated by clay, and water and roots cannot easily move through it. Silt particles are intermediate both in size and in their ability to drain or retain water.

FIGURE 25.9 **Soil permeability.** The permeability of soil depends on its texture. Sand, with its large, loosely packed particles, drains quickly. Clay drains much more slowly.

Teaching Tip: Journal Prompt

Ask students to complete the graphic organizer to the right to distinguish the properties of soil. (Answers are provided in italics.)

Physical properties	Chemical properties	Biological properties
Texture *Percentages of sand, silt, and clay within a soil sample*	**Cation exchange** *Nutrient holding capacity*	**Microorganisms (bacteria, fungi, and protozoans)** *Consume dead plant material; some bacteria fix nitrogen*
Permeability *How quickly soil drains; depends on texture*	**Base saturation** *Proportion of soil bases to soil acids expressed as a percentage; soil acids are usually detrimental while bases promote growth*	**Detritivores (earthworms, snails, and slugs)** *Consume dead plant material; earthworms responsible for abundant humus formation*

The best agricultural soil is a mixture of sand, silt, and clay. This mixture promotes balanced water drainage and retention. In natural ecosystems, however, various herbaceous and woody plants have adapted to growing in wet, intermediate, and dry environments, so there are plants that thrive in soils of virtually all textures.

Soil texture can have a strong influence on how the physical environment responds to environmental pollution. For example, the ground water of western Long Island in New York State has been contaminated over the years by toxic chemicals discharged from local industries. One major reason for the contamination is that Long Island is dominated by sandy soils that readily allow surface water to drain into the groundwater. While soil usually serves as a filter that removes pollutants from the water moving through it, sandy soils are so permeable that pollutants move through them quickly and therefore are not filtered effectively.

Clay is particularly useful where a potential contaminant needs to be contained. Many modern landfills are lined with clay, which helps keep the contaminants in solid waste from leaching into the soil and groundwater beneath the landfill.

Chemical Properties of Soil

Chemical properties are also important in determining how a soil functions. Clay particles contribute the most to the chemical properties of a soil because of their ability to attract positively charged mineral ions, referred to as cations. Because clay particles have a negative electrical charge, cations are adsorbed—held on the surface—by the particles. The cations can be subsequently released from the particles and used as nutrients by plants.

The ability of a particular soil to adsorb and release cations is called its **cation exchange capacity (CEC),** sometimes referred to as the nutrient holding capacity. The overall CEC of a soil is a function of the amount and types of clay particles present. Soils with high CECs have the potential to provide essential cations to plants and therefore are desirable for agriculture. If a soil is more than 20 percent clay, however, its water retention becomes too great for most crops as well as many other types of plants. In such waterlogged soils, plant roots are deprived of oxygen. Thus there is a trade off between CEC and permeability.

The relationship between soil bases and soil acids is another important soil chemical property. Calcium, magnesium, potassium, and sodium are collectively called soil bases because they can neutralize or counteract soil acids such as aluminum and hydrogen. Soil acids are generally detrimental to plant nutrition, while soil bases tend to promote plant growth. With the exception of sodium, all the soil bases are essential for plant nutrition. **Base saturation** is the proportion of soil bases to soil acids, expressed as a percentage.

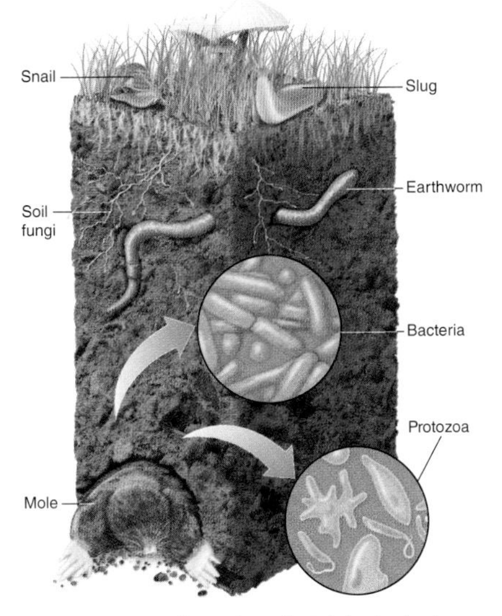

FIGURE 25.10 **Soil organisms.** Bacteria, fungi, and protozoans account for 80 to 90 percent of soil organisms. Also present are snails, slugs, insects, earthworms, and rodents.

Because of the way they affect nutrient availability to plants, CEC and base saturation are important determinants of overall ecosystem productivity. If a soil has a high CEC, it can retain and release plant nutrients. If it has a relatively high base saturation, its clay particles will hold important plant nutrients such as calcium, magnesium, and potassium. A soil with both high CEC and high base saturation is likely to support high productivity.

Biological Properties of Soil

As we have seen, a diverse group of organisms populates the soil. FIGURE 25.10 shows a representative sample. Three groups of organisms account for 80 to 90 percent of the biological activity in soils: fungi, bacteria, and protozoans (a diverse group of single-celled organisms). Rodents and earthworms contribute to soil mixing and the breakdown of large organic materials into smaller pieces. Earthworms are responsible for abundant humus formation in soils. Some soil organisms, such as snails

Cation exchange capacity (CEC) The ability of a particular soil to absorb and release cations.

Base saturation The proportion of soil bases to soil acids, expressed as a percentage.

Teaching Tip: Journal Prompt

Ask students to answer the following questions and explain their answers:

- Identify the type of soil texture that is ideal for agriculture. *The best agricultural soil is a mixture of sand, silt, and clay (loam). This mixture promotes balanced water retention and drainage.*
- Which type of soil is best for containment of pollution? *Soils high in clay content do not allow water to drain readily and therefore keep contaminants from leaching into surrounding soil and groundwater.*
- What is the ideal combination of chemical properties to support high productivity in a soil? *High CEC and moderate base saturation will promote high productivity. High CEC means the soil has high nutrient holding capacity. Nutrients are essential for plant growth. Moderate base saturation indicates that elements such as calcium, magnesium, and potassium are available for plant growth as well.*

TEACHING with FIGURES ▶

Ask students to use Figure 25.11 to identify the crustal abundance of the top four elements.

- ***Oxygen:*** *46 percent*
- ***Silicon:*** *28 percent*
- ***Aluminum:*** *8 percent*
- ***Iron:*** *6 percent*

and slugs, are herbivores that eat plant roots as well as the aboveground parts of plants. However, the majority of soil organisms are detritivores, which consume dead plant and animal tissues and recycle the nutrients they contain. Some soil bacteria also fix nitrogen, which, as we saw in Chapter 3, is essential for plant growth.

The distribution of mineral resources on Earth has social and environmental consequences

The tectonic cycle, the rock cycle, and soil formation and erosion all influence the distribution of rocks and minerals on Earth. These resources, along with fossil fuels, exist in finite quantities, but are vital to modern human life. In this section we will discuss some important nonfuel mineral resources and how humans obtain them; we will discuss fuel resources in Chapters 12 and 13. Some of these resources are abundant, whereas others are rare and extremely valuable.

Abundance of Ores and Metals

As we saw at the beginning of this chapter, early Earth cooled and differentiated into distinct vertical zones. Heavy elements sank toward the core, and lighter elements rose toward the crust. **Crustal abundance** is the average concentration of an element in Earth's crust. Looking at FIGURE 25.11, we can see that four elements—oxygen, silicon, aluminum, and iron—constitute over 88 percent of the crust. However, the chemical composition of the crust is highly variable from one location to another.

Environmental scientists and geologists study the distribution and types of mineral resources around the planet in order to locate them and to manage their extraction or conservation. **Ores** are concentrated accumulations of minerals from which economically valuable materials can be extracted. Ores are typically characterized by the presence of valuable metals, but accumulations of other valuable materials, such as salt or sand, can also be considered

Crustal abundance The average concentration of an element in Earth's crust.

Ore A concentrated accumulation of minerals from which economically valuable materials can be extracted.

Metal An element with properties that allow it to conduct electricity and heat energy, and to perform other important functions.

Reserve In resource management, the known quantity of a resource that can be economically recovered.

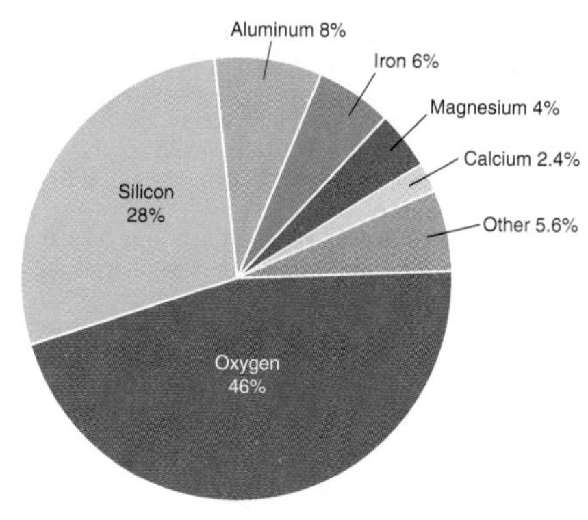

FIGURE 25.11 **Elemental composition of Earth's crust.** Oxygen is the most abundant element in the crust. Silicon, aluminum, and iron are the next three most abundant elements.

ores. **Metals** are elements with properties that allow them to conduct electricity and heat energy and to perform other important functions. Copper, nickel, and aluminum are common examples of metals. They exist in varying concentrations in rock, usually in association with elements such as sulfur, oxygen, and silicon. Some metals, such as gold, exist naturally in a pure form.

Ores are formed by a variety of geologic processes. Some ores form when magma comes into contact with water, heating the water and creating a solution from which metals precipitate, while others form after the deposition of igneous rock. Some ores occur in relatively small areas of high concentration, such as veins, and others, called disseminated deposits, occur in much larger areas of rock, although often in lower concentrations. Still other ores, such as copper, can be deposited both throughout a large area and in veins. Nonmetallic mineral resources, such as clay, sand, salt, limestone, and phosphate, typically occur in concentrated deposits. These deposits occur as a result of their chemical or physical separation from other materials by water, in conjunction with the tectonic and rock cycles. Some ores, such as bauxite—the ore in which aluminum is most commonly found—are formed by intense chemical weathering in tropical regions.

The global supply of mineral resources is difficult to quantify. Because private companies hold the rights to extract certain mineral resources, information about the exact quantities of resources is not always available to the public. The publicly known estimate of how much of a particular resource is available is based on its **reserve**: the known quantity of the resource that can be economically recovered.

TABLE 25.1 Approximate Supplies of Metal Reserves Remaining

Metal	Global reserves remaining (years)	U.S. reserves remaining (years)
Iron (Fe)	120	40
Aluminum (Al)	330	2
Copper (Cu)	65	40
Lead (Pb)	20	40
Zinc (Zn)	30	25
Gold (Au)	30	20
Nickel (Ni)	75	0
Cobalt (Co)	50	0
Manganese (Mn)	70	0
Chromium (Cr)	85	0

Sources: Data from S. Marshak, *Earth: Portrait of a Planet*, 3rd ed. (W. W. Norton, 2007); U.S. Geological Survey Mineral Commodity Summaries 2013, http://minerals.er.usgs.gov/minerals/pubs/mcs/.

TABLE 25.1 lists the estimated number of years of remaining supplies of some of the most important metal resources commonly used in the United States, assuming that rates of use do not change. Some important metals, such as tantalum, have never existed in the United States. The United States has used up all of its reserves of some other metals, such as nickel, and must now import those metals from other countries.

Mining Techniques

Mineral resources are extracted from Earth by mining the ore and separating any other minerals, elements, or residual rock away from the sought-after element or mineral. As illustrated in FIGURE 25.12, two kinds of mining take place on land: surface mining and *subsurface mining*. Each method has different benefits and costs in terms of environmental, human, and social perspectives.

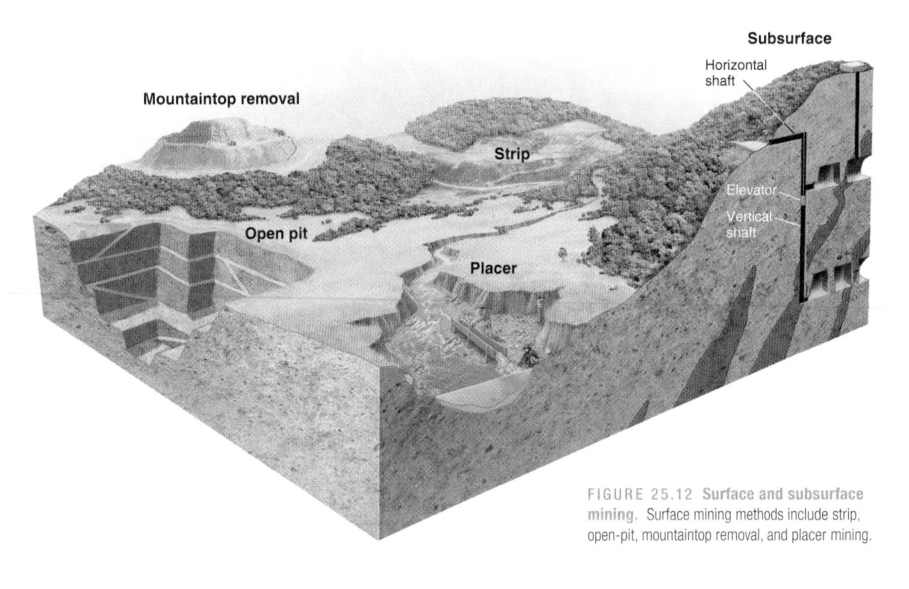

FIGURE 25.12 Surface and subsurface mining. Surface mining methods include strip, open-pit, mountaintop removal, and placer mining.

Teaching Tip: Discussion Starter

Ask students to evaluate Table 25.1 and predict potential economic consequences the United States might face because of an unequal distribution of metal reserves. *Economic consequences could include higher prices of rarer metals for U.S. residents. Nickel, cobalt, manganese, and chromium reserves are limited worldwide, which means they will be expensive to import. Lead is more abundant in the United States than elsewhere. It could be exported for economic gain.*

Teaching Tip: Journal Prompt

Ask students to identify and describe the various types of mining operations and their uses.

- ***Strip mining**: Removal of strips of soil and rock to expose the underlying ore. Used when ore is relatively close to Earth's surface and runs parallel to it.*
- ***Open-pit mining**: Mining that creates a large visible pit or open hole in the ground. Used when the resource is close to the surface but extends beneath the surface both horizontally and vertically.*
- ***Mountaintop removal**: Removal of the entire top of a mountain with explosives.*
- ***Placer mining**: The process of looking for metals and precious stones in river sediments.*
- ***Subsurface mining**: Digging a tunnel to access a resource that is more than 100 m below Earth's surface.*

Teaching Tip: Video

How Do They Do It—Coal Mining
This video clip from the Science Channel introduces the process of coal mining and gives students a visual perspective of the mining process.

Find a link to this resource by clicking on the link buttons in the TE-book, opening the TRFD, or logging onto the book's companion website bcs.whfreeman.com/friedlandapes2e (teacher login required).

 **Chapter 8 Web Resources**

Surface Mining

A variety of surface mining techniques can be used to remove a mineral or ore deposit that is close to the surface of Earth. **Strip mining,** or the removal of "strips" of soil and rock to expose the underlying ore, is used when the ore is relatively close to Earth's surface and runs parallel to it, which is often the case for deposits of sedimentary materials such as coal and sand. In these situations, miners remove a large volume of material, extract the resource, and return the unwanted waste material, called **mining spoils** or **tailings,** to the hole created during the mining. A variety of strategies can be used to restore the affected area to something close to its original condition.

Open-pit mining, a mining technique that creates a large visible pit or hole in the ground, is used when the resource is close to the surface but extends beneath the surface both horizontally and vertically. Copper mines are usually open-pit mines. One of the largest open-pit mines in the world is the Kennecott Bingham Canyon mine near Salt Lake City, Utah. This copper mine is 4.4 km (2.7 miles) across and 1.1 km (0.7 miles) deep.

In **mountaintop removal,** miners remove the entire top of a mountain with explosives. Large earth-moving equipment removes the resource and deposits the tailings in lower-elevation regions nearby, often in or near rivers and streams.

Placer mining is the process of looking for metals and precious stones in river sediments. Miners use the river water to separate heavier items, such as diamonds, tantalum, and gold, from lighter items, such as sand and mud. The prospectors in the California gold rush in the mid 1800s were placer miners, and the technique is still used today.

Subsurface Mining

When the desired resource is more than 100 m (328 feet) below Earth's surface, miners must turn to **subsurface mining,** which is mining that occurs below the surface of Earth. Typically, a subsurface mine begins with a horizontal tunnel dug into the side of a mountain or other feature containing the resource. From this horizontal tunnel, vertical shafts are drilled, and elevators are used to bring miners down to the resource and back to the surface. The deepest mines on Earth are up to 3.5 km (2.2 miles) deep. Coal, diamonds, and gold are some of the resources removed by subsurface mining.

Strip mining The removal of strips of soil and rock to expose ore.

Mining spoils Unwanted waste material created during mining. *Also known as* **Tailings.**

Open-pit mining A mining technique that uses a large visible pit or hole in the ground.

Mountaintop removal A mining technique in which the entire top of a mountain is removed with explosives.

Placer mining The process of looking for minerals, metals, and precious stones in river sediments.

Subsurface mining Mining techniques used when the desired resource is more than 100 m (328 feet) below the surface of Earth.

The Environment and Safety

The extraction of mineral resources from Earth's crust has a variety of environmental impacts on water, soil, biodiversity, and other areas. In addition, mineral resource extraction can have human health consequences that affect the miners directly as well as other individuals who are affected by the mining process.

Mining and the Environment

As you can see in **TABLE 25.2**, all forms of mining affect the environment. Mining almost always requires the construction of roads, which can result in soil erosion, damage to waterways, and habitat fragmentation. In addition, all types of mining produce tailings, the residue that is left behind after the desired metal or ore is removed, and some tailings contaminate land and water with acids and metals.

In mountaintop removal, the mining spoils are typically deposited in the adjacent valleys, sometimes blocking or changing the flow of rivers. Mountaintop removal is used primarily in coal mining and is safer for workers than subsurface mining. In environmental terms, mining companies do sometimes make efforts to restore the mountain to its original shape. However, there is considerable disagreement about whether these reclamation efforts are effective. Damage to streams and nearby groundwater during mountaintop removal cannot be completely rectified by the reclamation process.

Placer mining can also contaminate large portions of rivers, and the areas adjacent to the rivers, with sediment and chemicals. In certain parts of the world, the toxic metal mercury is used in placer mining of gold and silver. Mercury is a highly *volatile* metal; that is, it moves easily among air, soil, and water. Mercury is harmful to plants and animals and can damage the central nervous system in humans; children are especially sensitive to its effects.

The environmental impacts of subsurface mining may be less apparent than the visible scars left behind by surface mining. One of these impacts is acid mine drainage. To keep underground mines from flooding, pumps must continually remove the water, which can

Teaching Tip: Journal Prompt

Ask students to compare the consequences of surface mining and subsurface mining. How has mining legislation tried to reduce the negative effects of mining? *Subsurface mining produces much less dust than surface mining. Both processes can lead to contamination of water as it percolates through tailings. With surface mining, most of the soil removed from the mining site can be replaced if reclamation occurs. Significantly more habitat destruction occurs with surface mining than with subsurface mining. Surface mining is less hazardous to human health than subsurface mining; humans are more likely to contract respiratory disease or die in mining accidents with subsurface mines.*

The Surface Mining Control and Reclamation Act of 1977 regulates surface mining of coal and the surface effects of subsurface coal. The law mandates the land be minimally disturbed and reclaimed after mining.

TABLE 25.2 Types of Mining Operations and Their Effects

Type of operation	Effects on air	Effects on water	Effects on soil	Effects on biodiversity	Effects on humans
Surface mining	Significant dust from earth-moving equipment	Contamination of water that percolates through tailings	Most soil removed from site; may be replaced if reclamation occurs	Habitat alteration and destruction over the surface areas that are mined	Minimal in the mining process, but air quality and water quality can be adversely affected near the mining operation
Subsurface mining	Minimal dust at the mining site, but emissions from fossil fuels used to power mining equipment can be significant	Acid mine drainage as well as contamination of water that percolates through tailings		Road construction to mines fragments habitat	Occupational hazards in mine; possibility of death or chronic respiratory diseases such as black lung disease

have an extremely low pH. Drainage of this water lowers the pH of nearby soils and streams and can cause damage to the ecosystem.

Mining Safety and Legislation

Subsurface mining is a dangerous occupation. Hazards to miners include accidental burial, explosions, and fires. In addition, the inhalation of gases and particles over long periods can lead to a number of occupational respiratory diseases, including black lung disease and asbestosis, a form of lung cancer. In the United States, between 1900 and 2006, more than 11,000 coal miners died in underground coal mine explosions and fires. A much larger number died from respiratory diseases. Today, there are relatively few deaths per year in coal mines in the United States, in part because of improved work safety standards and in part because there is much less subsurface mining. In other countries, especially China, mining accidents remain fairly common.

As human populations grow and developing nations continue to industrialize, the demand for mineral resources continues to increase. But as the most easily mined mineral resources are depleted, extraction efforts become more expensive and environmentally destructive. The ores that are easiest to reach and least expensive to remove are always recovered first. When these sources are exhausted, mining companies must turn to deposits that are more difficult to reach. These extraction efforts result in greater amounts of mining spoils and more of the environmental problems we have already noted. Learning to use and reuse limited mineral resources more efficiently will help protect the environment as well as human health and safety.

Governments have sought to regulate the mining process for many years. Early mining legislation was primarily focused on promoting economic development, but later legislation became concerned with worker safety as well as environmental protection. The effectiveness of these mining laws has varied.

Congress passed the Mining Law of 1872 to regulate the mining of silver, copper, and gold ores as well as fuels, including natural gas and oil, on federal lands. This law, also known as the General Mining Act, allowed individuals and companies to recover ores or fuels from federal lands. The law was written primarily to encourage development and settlement in the western United States and, as a result, it contains very few provisions for environmental protection.

The Surface Mining Control and Reclamation Act of 1977 regulates surface mining of coal and the surface effects of subsurface coal mining. The act mandates that land be minimally disturbed during the mining process and reclaimed after mining is completed. Mining legislation does not regulate all of the mining practices that can have harmful effects on air, water, and land. In later chapters we will learn about other U.S. legislation that does, to some extent, address these issues, including the Clean Air Act, Clean Water Act, and Superfund Act.

Teaching Tip: Debate the Issue

Do the economic and social benefits of mountaintop removal outweigh the environmental impacts of this extraction method?

To discuss this question, students should research and be prepared to discuss these topics in their debate:

- the procedures used in mountaintop removal
- economic benefits of domestically produced coal
- the growing demand for electrical energy and energy supplied by coal
- pros and cons compared to other mining methods
- human health concerns
- environmental concerns including biodiversity, air, soil, and water quality
- potential for reclamation of a mining area

In class, select groups to debate the issue. Allow each side 2 minutes to prepare, 3 minutes to present, and 1 minute for rebuttal. Have the remainder of the class take notes, ask questions, and decide which side has the most persuasive argument.

module

25

REVIEW

In this module, we saw that rocks and minerals undergo physical and chemical weathering and become products that are precursors for soil. Weathered materials are subject to erosion, which is a natural process that can be enhanced by human activity. Erosion also influences the precursors to soil. Soil forms from geologic material as well as biological material. Soil properties result from physical, chemical, and biological processes and are influenced by five soil forming factors. Concentrated accumulations of elements and minerals in and below soils that are economically valuable are called ores. When ores are extracted, a variety of consequences affect humans and the environment.

Answers to Module 25 AP® Review Questions

1. d
2. e
3. d
4. a
5. e
6. e

Module 25 AP® Review Questions

1. Acid precipitation directly causes
   I. erosion.
   II. physical weathering.
   III. chemical weathering.

   (a) I only
   (b) I and II only
   (c) I and III only
   (d) III only
   (e) I, II, and III

2. What are the five primary soil formation factors?
   (a) Altitude, climate, parent material, latitude, organisms
   (b) Climate, parent material, pH, organisms, topography
   (c) Parent material, topography, organisms, time, latitude
   (d) Parent material, climate, pH, latitude, altitude
   (e) Climate, parent material, topography, organisms, time

3. Which is the correct order of soil horizons starting at the surface?
   (a) A, B, C, E, O
   (b) O, A, B, C, E
   (c) O, E, A, B, C
   (d) O, A, E, B, C
   (e) E, A, B, O, C

4. What type of soil would be best for a man-made pond, where the goal is to have as little leakage of water as possible?
   (a) Mostly clay
   (b) Mostly silt
   (c) Mostly sand
   (d) Equal amounts of sand, silt, and clay
   (e) Equal amounts of silt and sand

5. Which of the following, if added to soil, would lower the base saturation?
   (a) Sodium
   (b) Potassium
   (c) Magnesium
   (d) Calcium
   (e) Aluminum

6. Tailings are
   (a) minerals found in metamorphic rock.
   (b) magma ore resulting from seafloor spreading.
   (c) the remaining supply of metals on Earth.
   (d) the nutrients that leach downward in soil.
   (e) the waste material from mining.

working toward sustainability

Mine Reclamation and Biodiversity

One of the environmental impacts of surface mining is the amount of land surface it disturbs. Once a mining operation is finished, the mining company may try to restore it to its original condition. In the United States, the Surface Mining Control and Reclamation Act of 1977 requires coal mining companies to restore the lands they have mined. Regulations also require other types of mining operations to do some level of restoration.

A disturbed ecosystem can return to a state similar to its original condition only if the original physical, chemical, and biological properties of the land are re-created. Therefore, reclamation after mining involves several steps. First, the mining company must fill in the hole or depression it has created in the landscape. Second, the fill material must be shaped to follow the original contours of the land that existed before the mining began.

The mining company usually scrapes off the topsoil that was on the land and puts it aside at the beginning of the mining operation. This topsoil must be returned and spread over the landscape after mining is completed. Finally, the land must be replanted. In order to re-create the communities of organisms that inhabited the area before mining, the vegetation planted on the site must be native to the area and foster the process of natural succession. Properly completed reclamation makes the soils physically stable so that erosion does not occur and water infiltration and retention can proceed as they did before mining. The materials used in the reclamation must be relatively free of metals, acids, and other compounds that could potentially leach into nearby bodies of water.

Many former mining areas have not been reclaimed properly. However, there are an increasing number of reclamation efforts that have achieved conditions that equal those that existed prior to the mining operation. The Trapper Mine in Craig, Colorado, and other mines like it illustrate reclamation success stories.

The Trapper Mine produces about 2 million tons of coal per year that is sold to a nearby electricity generation plant. Although all coal mining operations are required to save excavated rock and topsoil, managers at the Trapper Mine have stated that they are meticulous about saving all the topsoil they remove. The rock they save is used to fill the lower portions of excavated holes. Workers then install drainage pipes and other devices to ensure proper drainage of water. The topsoil that has been set aside is spread over the top of the restored ground and contoured as it was before mining. Trapper Mine reclamation staff then replant the site with a variety of native species of grasses and shrubs, including the native sagebrush commonly found in the high plateaus of northwestern Colorado.

Government officials, and even the Colorado branch of the Sierra Club, have expressed approval of the Trapper Mine reclamation process. But

(a)

(b)

Mining and reclamation. At a strip mine operation in Colorado, not far from the Trapper Mine: (a) active mining underway; (b) restored native habitat after reclamation. *(Courtesy of Seneca Coal Company)*

Suggested Answers to Critical Thinking Questions

1. Measures and conditions to consider would include increases in the populations of native species, the levels of toxic metals in surface water, and refilling holes created by mining.

2. Heavy metal contamination of both surface water and groundwater as well as erosion of topsoils are problems that could occur if a mining site was not reclaimed.

Teaching Tip: Measuring Your Impact

What Is the Impact of Your Diet on Soil Dynamics?

Measuring Your Impact 8 asks students to consider the effect of their diet on soil dynamics. Students will keep a daily record of what they eat and drink over the course of a week. Students will consider the data given to assess resource use and potential environmental impacts. Students will evaluate methods to minimize their impact.

Download this resource by clicking on the link buttons in the TE-book, opening the TRFD, or logging onto the book's companion website bcs.whfreeman.com/friedlandapes2e (teacher login required).

TRM Measuring Your Impact 8:
What Is the Impact of Your Diet on Soil Dynamics?

perhaps the strongest evidence of its success is the wildlife that now inhabits the formerly mined areas. The Columbian sharp-tailed grouse (*Tympanuchus phasianellus columbianus*), a threatened bird species, has had higher annual survival and fertility rates on the reclaimed mine land than it has on native habitat in other parts of Colorado. Populations of elk, mule deer, and antelope have all increased on reclaimed mine property.

Reclamation issues must be addressed for each particular area where mining occurs. As we have seen, subsurface and surface water runoff in certain locations can contain high concentrations of toxic metals and acids. In other situations, although certain native species may increase in abundance after reclamation, other native species may decrease. Nevertheless, with supervision, skill, and enough money to pay for the proper reclamation techniques, a reclaimed mining area can become a satisfactory or even an improved habitat for many species.

Critical Thinking Questions

1. If you were on a committee asked to determine if a mine reclamation project had been successful, what three measures or conditions would you want to consider in your evaluation?
2. Identify two environmental problems that might occur in the area where you live if a mining site was not reclaimed.

References

Department of the Interior, Office of Surface Mining Reclamation and Enforcement. http://www.osmre.gov/.

Raabe, S. 2002. Nature's Comeback: Trapper Mine reclamation attracts wildlife, wins praise. *Denver Post*, November 12, p. C01.

chapter 8 REVIEW

In this chapter, we have examined how geologic processes such as plate tectonics, earthquakes, and volcanism have led to the differential distribution of elements and minerals on the surface of Earth. These geologic processes, which have occurred at different rates over long periods of time, have led to the formation of different rocks and minerals on or near the surface of Earth. Rocks and minerals have undergone weathering at different rates and have been eroded and deposited elsewhere on Earth. This has been one of the contributors to soil formation. Soils are a membrane that covers much of the land surface on Earth and these soils contain a mixture of geologic material from below and organic material from plants and animals from above. We have also examined how the actions involved in removal of valuable mineral resources from on top and below the surface of Earth have affected a number of environmental processes.

Key Terms

Core
Mantle
Magma
Asthenosphere
Lithosphere
Crust
Hot spot
Plate tectonics
Tectonic cycle
Subduction
Volcano
Divergent plate boundary
Seafloor spreading
Convergent plate boundary
Transform fault boundary
Fault
Seismic activity
Fault zone
Earthquake
Epicenter
Richter scale
Rock cycle
Igneous rock
Intrusive igneous rock
Extrusive igneous rock
Fracture
Sedimentary rock
Metamorphic rock
Physical weathering
Chemical weathering
Acid precipitation
Acid rain
Erosion
Parent material
Soil degradation
Horizon

O horizon
A horizon
Topsoil
E horizon
B horizon
C horizon
Cation exchange capacity (CEC)
Base saturation
Crustal abundance
Ore
Metal
Reserve
Strip mining
Mining spoils
Tailings
Open-pit mining
Mountaintop removal
Placer mining
Subsurface mining

Learning Objectives Revisited

Module 24 Mineral Resources and Geology

- **Describe the formation of Earth and the distribution of critical elements on Earth.**

  Earth formed from cosmic dust in the solar system. As it cooled, heavier elements, such as iron, sank toward the core, while lighter elements, such as silica, floated toward the surface. These processes have led to an uneven distribution of elements and minerals throughout the planet.

- **Define the theory of plate tectonics and discuss its relevance to the study of the environment.**

  Earth is overlain by a series of plates that move at rates of a few millimeters per year. Plates can move away from each other, move toward each other, or slide past each other. One plate can be subducted under another. These tectonic processes create mountains, earthquakes, and volcanoes.

- **Describe the rock cycle and discuss its importance in environmental science.**

  Rocks are made up of minerals, which are formed from the various chemical elements in Earth's crust. The processes of the rock cycle lead to the formation, breakdown, and recycling of rocks.

Module 25 Weathering and Soil Science

- **Understand how weathering and erosion occur and how they contribute to element cycling and soil formation.**

  Physical weathering is the mechanical breakdown of rocks and minerals while chemical weathering is a result of chemical reactions. Both occur as a result of natural processes and can be accelerated by human activities. Erosion is the physical removal of rock fragments and weathering products that are subsequently deposited elsewhere.

- **Explain how soil forms and describe its characteristics.**

  Soil forms as the result of physical and chemical weathering of rocks and the gradual accumulation of organic detritus from the biosphere. The factors that determine soil properties are parent material, climate, topography, soil organisms, and time. The relative abundances of sand, silt, and clay in a soil determine its texture.

- **Describe how humans extract elements and minerals and the social and environmental consequences of these activities.**

  Concentrated accumulations of minerals from which economically valuable materials can be extracted are called ores. Ores are removed by surface or subsurface mining operations. Surface mining generally results in greater environmental impacts, whereas subsurface mining is more dangerous to miners. With the exception of coal mining, legislation directly related to mining does not address most environmental considerations.

Exploring the Literature

Brady, N.C., and R. R. Weil. 2007. *The Nature and Properties of Soils*. 14th ed. Prentice Hall.

Christopherson, R. W. 2012. *Geosystems: An Introduction to Physical Geography*. 8th ed. Prentice Hall.

Grotzinger, J., and T. H. Jordan. 2014. *Understanding Earth*. 7th ed. W. H. Freeman.

McPhee, J. 1982. *Basin and Range*. Farrar, Straus and Giroux.

U.S. Geological Survey. http://www.usgs.gov/.

Download a printable version of this list by clicking on the link buttons in the TE-book, opening the TRFD, or logging onto the book's companion website bcs.whfreeman.com/friedlandapes2e (teacher login required).

Exploring the Literature 8

Answers to Chapter 8 AP® Exam Multiple-Choice Questions

1. d
2. c
3. c
4. e
5. e
6. d
7. c
8. b
9. e
10. e
11. c
12. a
13. c

Additional Multiple-Choice AP® Practice Exam

Download an additional Multiple-Choice AP® Environmental Science Practice Exam for Chapter 8 by clicking on the link buttons in the TE-book, opening the TRFD, or logging onto the book's companion website bcs.whfreeman.com/friedlandapes2e (teacher login required).

Multiple-Choice AP® Practice Exam 8

Answers to Chapter 8 AP® Exam

Download the full answers to the Chapter 8 AP® Practice Exam by clicking on the link buttons in the TE-book, opening the TRFD, or logging onto the book's companion website bcs.whfreeman.com/friedlandapes2e (teacher login required).

Answers to Chapter 8 AP® Exam

Chapter 8 AP® Environmental Science Practice Exam

Section 1: Multiple-Choice Questions

Choose the best answer for questions 1–13.

1. As Earth slowly cooled
   (a) lighter elements sank to the core and heavier elements moved to the surface.
   (b) lighter elements mixed with heavier elements and sank to the core.
   (c) lighter elements mixed with heavier elements and moved to the surface.
   (d) lighter elements moved to the surface and heavier elements sank to the core.
   (e) lighter and heavier elements dispersed evenly from the core to the surface.

2. The correct vertical zonation of Earth above the core is
   (a) asthenosphere-mantle-soil-lithosphere.
   (b) asthenosphere-lithosphere-mantle-soil.
   (c) mantle-asthenosphere-lithosphere-soil.
   (d) mantle-lithosphere-soil-asthenosphere.
   (e) soil-mantle-lithosphere-asthenosphere.

3. Evidence for the theory of plate tectonics includes
   I. deposits of copper ore around the globe.
   II. identical rock formations on both sides of the Atlantic.
   III. fossils of the same species on distant continents.

   (a) I only
   (b) I and III only
   (c) II and III only
   (d) I and II only
   (e) I, II, and III

4. Which type of mining is usually most directly harmful to miners?
   (a) Mountaintop removal
   (b) Open-pit mining
   (c) Placer mining
   (d) Strip mining
   (e) Subsurface mining

5. Measured on the Richter scale, an earthquake with a magnitude of 7.0 is ____ times greater than an earthquake with a magnitude of 2.0.
   (a) 10
   (b) 100
   (c) 1,000
   (d) 10,000
   (e) 100,000

For questions 6 to 9, select from the following choices:
(a) Seismic activity center
(b) Divergent plate boundary
(c) Convergent plate boundary
(d) Transform fault boundary
(e) Epicenter

6. At which type of boundary do tectonic plates move sideways past each other?

7. At which type of boundary does subduction occur?

8. At which type of boundary does seafloor spreading occur?

9. Which term refers to the point on Earth's surface directly above an earthquake?

10. Fossil records are found in
    (a) intrusive rock.
    (b) extrusive rock.
    (c) igneous rock.
    (d) metamorphic rock.
    (e) sedimentary rock.

11. Where would you expect to find extrusive igneous rock?
    (a) Along a transform fault
    (b) In the Himalayas
    (c) Along the ocean floor
    (d) On the coast of a continent
    (e) At a continental subduction zone

12. Which of the following statements about soil is NOT correct?
    (a) Soil is fairly static and does not change.
    (b) Soil is the medium for plant growth.
    (c) Soil is a primary filter of water.
    (d) A wide variety of organisms live in soil.
    (e) Soil plays an important role in biogeochemical cycles.

13. The soil horizon commonly known as subsoil is the
    (a) A horizon.
    (b) O horizon.
    (c) B horizon.
    (d) C horizon.
    (e) E horizon.

Section 2: Free-Response Questions

Write your answer to each part clearly. Support your answers with relevant information and examples. Where calculations are required, show your work.

1. The rock cycle plays an important role in the recycling of Earth's limited amounts of mineral resources.
   (a) The mineral composition of rock depends on how it is formed. Name the *three* distinct rock types, explain how each rock type is formed, and give one specific example of a rock of each type. (6 points)
   (b) Explain either the physical *or* the chemical weathering process that leads to the breakdown of rocks. (2 points)
   (c) Describe the natural processes that can lead to the formation of soil, and discuss how human activities can accelerate the loss of soil. (2 points)

2. Read the following articles that appeared on the website of the organization "Eatwild," at http://www.eatwild.com/environment.html, and answer the questions that follow.

Grass Farming Benefits the Environment

Grazing better for the soil than growing grain
Six Minnesota pasture-based ranchers asked researchers to compare the health of their soil with soil from neighboring farms that produced corn, soybean, oats, or hay. At the end of four years of monitoring, researchers concluded that the carefully managed grazed land had

- 53% greater soil stability
- 131% more earthworms
- Substantially more organic matter
- Less nitrate pollution of groundwater
- Improved stream quality
- Better habitat for grassland birds and other wildlife

Depending on the way that cattle are managed, they can either devastate a landscape or greatly improve the health of the soil. ["Managed Grazing as an Alternative Manure Management Strategy," Jay Dorsey, Jodi Dansingburg, Richard Ness, USDA-ARS, Land Stewardship Project.]

Pasture reduces topsoil erosion by 93 percent
Currently, the United States is losing three billion tons of nutrient-rich topsoil each year. Growing corn and soy for animal feed using conventional methods causes a significant amount of this soil loss. Compared with row crops, pasture reduces soil loss by as much as 93 percent. [Ontario Ministry of Agriculture and Food, Robert P. Stone and Neil Moore, Fact Sheet 95-089.]

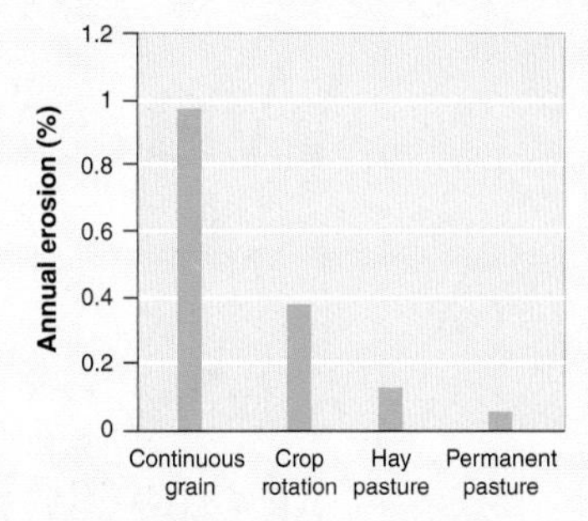

(a) Explain the consequences of the observations in the studies described and comment on the long-term sustainability of each agricultural practice.
   (i) What roles do earthworms play in maintaining soil stability? (1 point)
   (ii) How does the presence of organic matter benefit the soil? (1 point)
   (iii) Suggest why there is less nitrate contamination of groundwater from the permanent pastures. (1 point)
   (iv) In what ways could the stream quality have improved? (1 point)

(b) Discuss one negative consequence of grazing cattle on pastureland instead of feeding them grain. (3 points)

(c) Describe one potential negative effect on rivers and streams of grazing cattle on pastureland. (3 points)

Answers to Chapter 8 AP® Exam Free-Response Questions

1. (a) Igneous rocks are formed directly from liquid magma and are classified by their chemical composition. When the magma is high in iron, basalt is formed. When the magma is high in silica, granite is formed. Intrusive igneous rocks can be either granitic or basaltic and are formed as the magma rises up through Earth and cools in place. Extrusive igneous rocks are formed when magma is ejected or released, and cool very rapidly above Earth's surface.

Sedimentary rocks form from the compression and cementing together of earth materials such as mud, sand, or gravel that have been under pressure for a long time. This occurs when sand dunes, mudflats, or lakebeds are buried and the overlying materials create pressure on the materials below. Sandstone is an example of a sedimentary rock.

Metamorphic rock is formed when sedimentary rock, igneous rock, or even another older metamorphic rock is subjected to heat and pressure, causing profound chemical or physical change. Metamorphic rock makes up a large part of Earth's crust and an example of such rock is gneiss. (Slate, marble, schist, and quartzite could also be given as examples.)

(b) Both physical and chemical weathering are important processes that act to complete the rock cycle and to begin it again by returning the chemical elements and rock fragments back into the crust via the hydrological cycle, or to sediments. This process recycles elements.

Physical weathering: The breakdown of rocks and minerals occurs through water, wind, variations in temperature, and other mechanical agents. Water can get into cracks in the rock and widen them; it can also freeze in the cracks and split the rock. Temperature differentials cause different minerals in a rock to expand and contract at varying rates and split the rock. Plant roots can split rocks apart and burrowing animals may also play a minor role in the process. More surface area is exposed and further degradation can take place.

Chemical weathering: The breakdown of rock material occurs by chemical reactions or the dissolution of chemical elements from the rock. Unweathered, fresh minerals are known as primary minerals. During the chemical weathering process, these become secondary minerals, sediment, and chemical elements. Depending on the original chemical composition of the rock and the chemistry of the water and acids that come into contact with the mineral, hundreds of different chemical reactions can take place.

(c) Erosion is the process of physically removing rock fragments or solids such as sediment or soil particles. It occurs naturally through transport by wind, water, or ice, by the downslope creep of soil under the force of gravity, or by living organisms such as burrowing animals. Human activities can rapidly accelerate the rate of natural erosion, leading to significant environmental problems due to the enhanced erosion rate. In many cases the underlying cause is human land use. This can be poor land use practices such as deforestation, overgrazing, construction activity, road and highway building, and surface mining operations.

Answers continue in the Answer Appendix on page ANS-6.

chapter

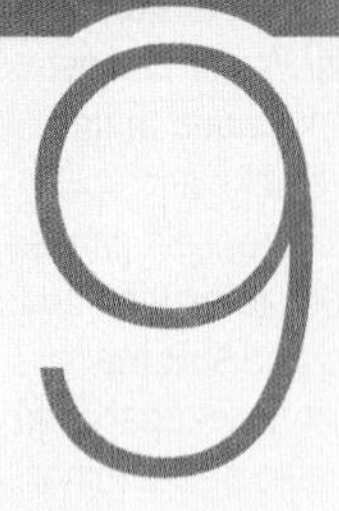

Water Resources

Overview

Chapter 9 demonstrates that fresh water is a critical, scarce resource. The major sources of fresh water are groundwater, surface water, and atmospheric water. When groundwater is extracted faster than it is replaced, the result can be dry water wells and saltwater intrusion. Humans alter the availability of water by constructing levees, dikes, and dams to constrain water movement. Aqueducts transport water. Salt water is desalinized through distillation or reverse osmosis. Developed countries typically have a larger water footprint than developing countries. Most water is used for agriculture, followed by industrial and household use. Recent changes in each of these sectors have improved water use and reuse efficiency and conservation.

Module 26: The Availability of Water

This module examines the major sources of fresh water on Earth. Fresh water comprises a tiny fraction of all water on Earth and most of this fresh water is locked up in the form of ice and glaciers. One fourth of fresh water resides underground as either confined or unconfined aquifers. In some parts of the world the rate of withdrawal for human use is faster than the rate of aquifer recharge. Surface waters such as streams, rivers, wetlands, and lakes are also important sources of fresh water, and wetlands can be important for flood control. Climatic differences in the availability of precipitation can interact with human land use to cause dramatic floods and droughts in different parts of the world.

Module 27: Human Alteration of Water Availability

Humans have found a number of ways to live with variations in water availability. We use levees and dikes to block the flow of water and to protect human settlements. We use dams to store water. Water can be diverted from rivers and lakes and transported to distant locations through the use of aqueducts. Desalination technologies make more fresh water available. This module considers the costs and benefits of each of these water distribution methods.

Module 28: Human Use of Water Now and in the Future

This module examines how humans use water. Approximately 70 percent of freshwater use goes to agriculture, 20 percent is used in manufacturing, and the remaining 10 percent goes to households. The module considers these uses in detail. It explores how water ownership and conservation will determine the availability of water in the future. The future availability of water depends on agreements regarding water ownership and whether steps will be taken to reduce the amount of water humans use.

Alignment to AP® Environmental Science Course Description

AP® Outline	Module	
I. Earth Systems and Resources (10–15%)		
C. Global Water Resources and Use	Module 26	The Availability of Water
	Module 27	Human Alteration of Water Availability
	Module 28	Human Use of Water Now and in the Future

Chapter Learning Objectives

After completing this chapter students will be able to

- describe the major sources of groundwater.
- identify some of the largest sources of fresh surface water.
- explain the effects of unusually high and low amounts of precipitation.
- compare and contrast the roles of levees and dikes.
- explain the benefits and costs of building dams.
- explain the benefits and costs of building aqueducts.
- describe the processes used to convert salt water to fresh water.
- compare and contrast the four methods of agricultural irrigation.
- describe the major industrial and household uses of water.
- discuss how water ownership and water conservation are important in determining future water availability.

Chapter 9 Pacing Guide

This pacing guide is based on a schedule with 120 sessions of 50 minutes each before the AP® Exam. If you have a different number of sessions before the exam, you can modify the pacing to suit your needs. If you have additional time, consider incorporating quizzes, released AP® Environmental Science free-response and multiple-choice questions, or additional activities.

Module	Standard Schedule Days	Block Schedule Days
Module 26	2	1
Module 27	1	½
Module 28	3	1½
Assessment	1	½

Chapter 9 Resources

The link to these resources can be found by clicking on the link buttons in the Teacher's e-Book (TE-book), opening the Teacher's Resource Flash Drive (TRFD), or logging onto the book's companion website bcs.whfreeman.com/friedlandapes2e (teacher login required).

- **PowerPoint Presentation 9**
- **Optimized Art PowerPoint and JPEG Files 9**
- **Do the Math Videos**
- **Measuring Your Impact 9:** Saving Water
- **Lab 9.1:** EcoBottle
- **Chapter 9 Web Resources**
- **Exploring the Literature 9**
- **Answers to Chapter 9 AP® Exam**
- **Multiple-Choice AP® Practice Exam 9**
- **Answers to Unit 4 AP® Exam**
- **Unit 4 Additional Practice Free-Response Question**
- **Answer to Science Applied 4 Free-Response Question**

Free-Response Questions from Previous AP® Environmental Science Exams

Free-response questions from prior AP® Environmental Science Exams are available on the AP® course website, https://apstudent.collegeboard.org/apcourse/ap-environmental-science/exam-practice. Students should be able to answer all of the questions listed below with material learned in this and previous chapters. When a question requires students to understand material from multiple chapters, the question will be listed in the last chapter required to complete the entire question. Questions marked with an asterisk (*) are from exams with released multiple-choice questions. You may want to save these questions until the end of the year so you can give your students a complete released exam for practice. Questions marked with double asterisks (**) require math to calculate a problem. Look for references to these questions throughout the chapter.

Year	Question	Content
2002	2	• Issues associated with water diversion from rivers
2003	3	• Human impacts on coastal estuaries and wetlands

PD **Chapter 9 Overview**

Watch the video overview of Chapter 9 (for teachers) by clicking on the link buttons in the TE-book, opening the TRFD, or logging onto the book's companion website bcs.whfreeman.com/friedlandapes2e (teacher login required).

TRM **PowerPoint Presentation 9**

Download the PowerPoint presentation for Chapter 9 by clicking on the link buttons in the TE-book, opening the TRFD, or logging onto the book's companion website bcs.whfreeman.com/friedlandapes2e (teacher login required).

The beautiful Klamath River flows from southern Oregon and down through northern California. *(Kevin Schafer/Alamy)*

chapter 9

Water Resources

Dams and Salmon on the Klamath River

The Klamath River is a beautiful stretch of water that runs for 400 km (250 miles) from southern Oregon through northern California, where it empties into the ocean. The Klamath River was once a spectacular habitat for salmon and was the third largest salmon fishery on the West Coast. During the past 100 years, however, the Klamath River has seen many changes. It has been dammed for electricity, and farmers have settled the land and diverted water to irrigate their crops. The salmon population has been greatly reduced, and local Native American tribes and the commercial fishing industry have suffered as a result.

The story of the Klamath River began thousands of years ago, when six tribes of Native Americans were the sole inhabitants of the region. For more than 300 generations, these tribes relied on salmon as an important part of their diet. Historically, salmon eggs hatched in the upstream portion of the river. Young fish migrated down the river into the ocean, where they spent several years growing into mature fish. Once mature, they migrated back up the river, returning to lay eggs in the same place where they had hatched. Several species of salmon were abundant in the river. An estimated 800,000 mature chinook salmon (*Oncorhynchus tshawytscha*)—which can grow as large as 45 kg (100 pounds)—migrated up the river each year.

> The Klamath River was once a spectacular habitat for salmon.

This migratory behavior worked well for thousands of years, until the region began to change during the twentieth century. In the early 1900s, the U.S. Bureau of Reclamation decided to build canals to drain two large lakes that were the source of the river. The exposed lake bottoms were turned into farmland, and the water that drained into the canals provided a constant source of irrigation for this new farmland. In addition, four hydroelectric dams were built on the river between 1908 and 1962. These dams generated enough electricity to power 70,000 homes. Because hydroelectric dams generate less pollution than other sources of electricity, these dams provided a substantial benefit to the region. But their presence imposed a barrier to the natural migration of salmon. In addition, the formation of

Teaching Tip: Chapter Opening Case

The chapter opening case, "Dams and Salmon on the Klamath River," provides an excellent opportunity to introduce the importance of water resources to human interests and to ecosystem stability. This case describes the historic use of the Klamath River, which extends from Oregon to California. It has been used over the years as a salmon fishery, a source of irrigation, and a generator of hydroelectric power. For 300 generations it has provided salmon, a traditional food source for Native Americans. Water diversion, overuse, and climate change have harmed the river ecosystem in a variety of ways. In 2009, conservationists, government agencies, the hydroelectric company, and local farmers reached an agreement to ensure that the Klamath River would continue to flow and provide a habitat for salmon populations. An increase in conflicts over water resources around the world makes this case a good model for resolving conflicts in the future.

Teaching Tip: Discussion Starter

National Geographic features a photo gallery entitled "Reuniting a River." Use these pictures and descriptions to help students visualize the concerns presented in the opening case.

The link to this website may be found by clicking on the link buttons in the TE-book, opening the TRFD, or logging onto the book's companion website bcs.whfreeman.com/friedlandapes2e (teacher login required).

Chapter 9 Web Resources

Teaching Tip: Journal Prompt

In addition to the destruction of fish habitat examined in the opening case, there are many other environmental problems associated with dams and water diversion. However, water diversion has also provided substantial benefits to humans. Have students do a side-by-side comparison of the benefits and costs of water diversion projects presented in this opening case so they gain a better understanding of some of the key issues in water resource use. (Suggested answers are provided in italics.)

Benefits	Costs
Water diversion provides water for agricultural irrigation, household use, and industry.	*Habitat alteration occurs as dams are constructed, causing upstream areas to flood and downstream habitats to go dry.*
Lakes created by dams provide recreational areas.	*Warmer temperatures occur in flooded areas behind the dams, which alter abiotic conditions affecting the ability of species, such as the salmon, to thrive.*
Hydroelectric power is generated without adding to air pollution.	*Overuse and overwithdrawal result in lack of water flow.*

AP® Exam Tip

The opening case is a good way to introduce other dam projects that have had large environmental impacts, such as China's Three Gorges Dam, the James Bay project in northern Quebec, and the damming of the Colorado River. These large water diversion projects are important to the study of environmental science. Questions about them have appeared on released AP® exams and several free-response questions. Make sure students are familiar with the major dams and the pros and cons of damming rivers.

large pools behind the dams caused a rise in water temperature that is not favorable to salmon.

As the decades passed, the multiple uses of the Klamath River began to come into conflict. In addition, a gradual warming of temperatures in the region has caused less snow to fall in the mountains. Although less snow means less water for the river, agricultural use of the river water continued, which reduced the river flow to low levels that created an additional threat to the salmon. In 1997, the coho salmon (*Oncorhynchus kisutch*) was given protection under the U.S. Endangered Species Act. To protect the salmon, water withdrawals could not reduce the river's flow below the minimal level in which the salmon could thrive. That restriction prevented farmers from receiving enough water to irrigate their crops and threatened their livelihoods.

The conflict among the various interest groups intensified. Native Americans, the hydroelectric company, farmers, the commercial fishing industry, conservationists, and multiple government agencies all debated regional water management policy from different points of view.

In 2002, the migrating salmon experienced a massive die-off in the Klamath River. This catastrophe was caused by unusually warm water, the growth of toxic algae in the dammed waters, and diseases. The sight of 30,000 large, dead chinook salmon brought worldwide attention to the river. Nearly 30 interest groups began meeting in order to find a solution to the river's complex problems.

In 2009, the parties reached a decision. The hydroelectric company, which realized it would soon have to spend $460 million on dam renovations, agreed to remove the four dams. The farmers agreed to conserve water by planting crops that required less water and updating their irrigation technology. These changes are expected to improve river flow, thereby improving conditions for the salmon and those who depend on them. The agreement among the parties was signed in 2010 and the U.S. Interior Department supported the proposal in 2013. Once the plan is approved by Congress, the dams are scheduled to be removed by 2020 at a cost of $450 million. This endeavor represents the largest dam removal project in history. When completed, salmon will have 420 miles of uninterrupted river for the first time in a century.

Conflicts over the use of water resources are found throughout the world as human populations grow and place ever-increasing demands on limited water resources. The compromise reached on the Klamath has the potential to serve as a model for resolving such conflicts in the future.

Sources:
R. Rymer, Reuniting a river, *National Geographic*, December 2008, http://ngm.nationalgeographic.com/2008/12/klamath-river/rymer-text/1; C. Sullivan, Landmark agreement to remove four Klamath River dams, *New York Times*, September 30, 2009, http://www.nytimes.com/gwire/2009/09/30/30greenwire-landmark-agreement-to-remove-4-klamath-river-d-72992.html; L. Zuckerman, Interior Department recommends removal of dams on Klamath River to aid salmon, Reuters, April 4, 2013, http://www.reuters.com/article/2013/04/05/us-usa-environment-dams-idUSBRE93402020130405.

Water is a critical resource for all organisms but it can be challenging to obtain. As we saw in Chapter 4, climate variation causes some regions of the world to possess abundant supplies of water, whereas other regions have very little water. This not only affects the plants and animals that live in different regions of the world but also affects growing human populations that have limited water availability. In this chapter, we will examine the many sources of fresh water, which is the type of water that humans can drink. We will then consider how humans have developed ways of altering the availability of water on Earth. We will also examine how humans use water and the prospects for the availability of water in the future.

Teaching Tip: Video

Where We Get Our Fresh Water
This video clip from TED-Ed introduces the topic of distribution of our freshwater on Earth and how it is used.

The link to this video may be found by clicking the link buttons in the TE-book, opening the TRFD, or logging onto the book's companion website bcs.whfreeman.com/friedlandapes2e (teacher login required).

Chapter 9 Web Resources

module

26 The Availability of Water

A very small amount of Earth's aboveground water is found in the atmosphere and in the form of water bodies such as streams, rivers, wetlands, and lakes. In this module, we will examine the major sources of fresh water on Earth. We will also consider the effects of unusually high and low amounts of precipitation.

Learning Objectives

After reading this module you should be able to

- describe the major sources of groundwater.
- identify some of the largest sources of fresh surface water.
- explain the effects of unusually high and low amounts of precipitation.

Groundwater can be extracted for human use

Nearly 70 percent of Earth's surface is covered by water. As FIGURE 26.1 shows, this water is found in five main repositories, not including the water vapor and precipitation contained in the atmosphere.

The vast majority of Earth's water—more than 97 percent— is found in the oceans as salt water. The remainder—less than 3 percent—is fresh water, the type of water that can be consumed by humans. Of this small percentage of Earth's water that is fresh water, nearly three-fourths is aboveground, though mostly in the form of ice and glaciers, and therefore generally not available for human consumption.

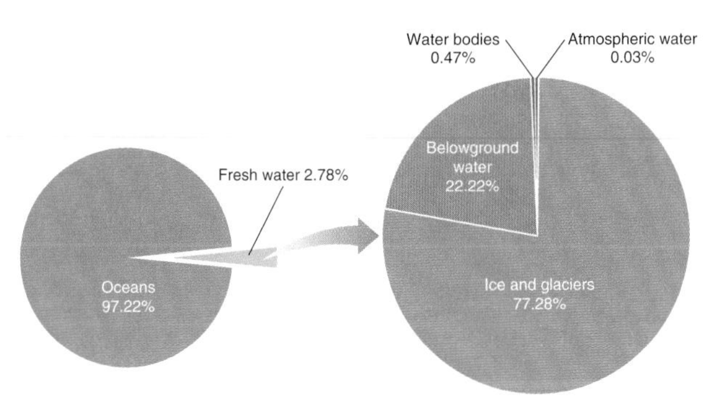

FIGURE 26.1 Distribution of water on Earth. Fresh water represents less than 3 percent of all water on Earth, and only about three-fourths of that fresh water is surface water. Most of that surface water is frozen as ice and in glaciers. Therefore, less than 1 percent of all water on the planet is accessible for use by humans. *(Data from R. W. Christopherson,* Geosystems, *7th ed. Pearson/Prentice Hall, 2009.)*

◀ TEACHING with FIGURES

Ask students to first examine Figure 26.1 and then answer the following questions:

- What are the primary repositories of fresh water on Earth? *Belowground water, ice and glaciers, water bodies, and atmospheric water.*
- Which repository is the largest? *The largest repository is held in ice and glaciers.*
- Of all the water on Earth, what is the approximate percentage of fresh water? *Approximately 3 percent of water on Earth is fresh.*
- What percentage of fresh water is surface water? *About 75 percent of fresh water is surface water.*
- What percentage of water is accessible for human use? *Less than 1 percent of all water on Earth is accessible by humans.*

Teaching Tip: Discussion Starter

Ask students: What is the difference between a confined and an unconfined aquifer? How do their recharge rates differ? *Water can easily flow in and out of an unconfined aquifer because it is covered by soil. A confined aquifer is surrounded by a layer of impermeable rock or clay that impedes water flow. Unconfined aquifers can recharge quickly as precipitation falls over the area and flows through the soil. Confined aquifers can only recharge if there is an opening in the impermeable layer of rock that allows water to flow in. It may take a confined aquifer 10,000 to 20,000 years to recharge.*

Teaching Tip: Journal Prompt

Ask students to explain the similarities and differences between an artesian well and a spring. *Artesian wells are similar to natural springs because, in both cases, groundwater bubbles up against the forces of gravity to reach the surface. Unlike natural springs, though, artesian wells are drilled. In modern wells, spring water is pumped to the surface. Artesian wells are drilled into confined aquifers where sufficient pressure forces water to rise up to the surface.*

Only one-fourth resides underground in the form of groundwater.

Groundwater exists in the multitude of small spaces found within permeable layers of rock and sediment called **aquifers.** FIGURE 26.2 shows how different types of aquifers are situated. Water can easily flow in and out of an **unconfined aquifer,** which is made of porous rock covered by soil. In contrast, **confined aquifers** are surrounded by a layer of impermeable rock or clay, which impedes water flow to or from the aquifer. The uppermost level at which the groundwater in a given area fully saturates the rock or soil is called the **water table.**

The process by which water from precipitation percolates through the soil and works its way into the groundwater is known as **groundwater recharge.** If water falls on land that contains a confined aquifer, however, it cannot penetrate the impermeable layer of rock. A confined aquifer can only be recharged if the impermeable layer of rock has a surface opening that can serve as a recharge area.

Aquifers serve as important sources of fresh water for many organisms. Plant roots can access the groundwater in an aquifer and draw down the water table. Water from some aquifers naturally percolates up to the ground surface as **springs** (FIGURE 26.3). Springs serve as a natural source of water for freshwater aquatic biomes, and they can be used directly by humans as sources of drinking water. Humans discovered centuries ago that water can also be obtained from aquifers by digging a well—essentially a hole in the ground. Most modern wells are very deep and water is pumped to the surface against the force of gravity. In some confined aquifers, however, the water is under tremendous pressure from the impermeable layer of rock that surrounds it. Drilling a hole into a confined aquifer releases the pressure, which allows the water to burst out of the aquifer and rise up in the well, as shown in Figure 26.2. A well created by drilling a hole into a confined aquifer is called an **artesian well.** If the pressure is sufficiently great, the water can rise all the way up to the ground surface, in which case no pump is required to extract the water from the ground.

The age of water in aquifers varies, as does the rate at which aquifers are recharged. The water in an unconfined aquifer may originate from water that fell to the ground last year or even last week. This direct and rapid connection with the surface is one of the reasons water from unconfined aquifers is much more likely to be contaminated with chemicals released by human activities. Confined aquifers, on the other hand, are generally recharged very slowly, perhaps

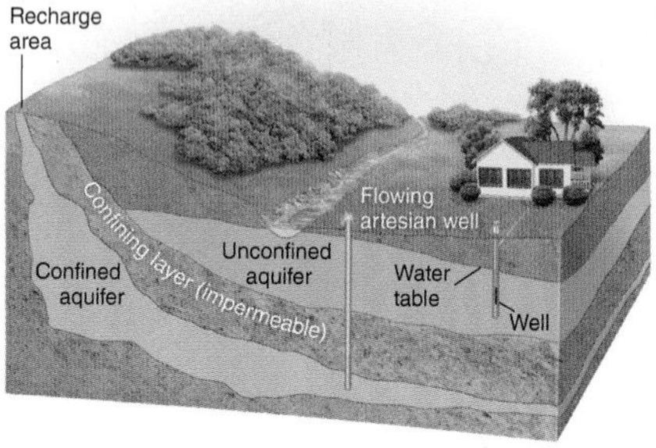

FIGURE 26.2 **Aquifers.** Aquifers are sources of usable groundwater. Unconfined aquifers are rapidly recharged by water that percolates downward from the land surface. Confined aquifers are capped by an impermeable layer of rock or clay, which can cause water pressure to build up underground. Artesian wells are formed when a well is drilled into a confined aquifer and the natural pressure causes water to rise toward the ground surface.

Aquifer A permeable layer of rock and sediment that contains groundwater.

Unconfined aquifer An aquifer made of porous rock covered by soil out of which water can easily flow.

Confined aquifer An aquifer surrounded by a layer of impermeable rock or clay that impedes water flow.

Water table The uppermost level at which the water in a given area fully saturates rock or soil.

Groundwater recharge A process by which water percolates through the soil and works its way into an aquifer.

Spring A natural source of water formed when water from an aquifer percolates up to the ground surface.

Artesian well A well created by drilling a hole into a confined aquifer.

FIGURE 26.3 **Natural spring.** When an aquifer has an opening at the land surface, the water can flow out to form a spring. Springs, such as this one in New Mexico, can be important sources of water for organisms and serve as the initial water source for many streams and rivers. (*Jerry Moorman/iStockphoto.com*)

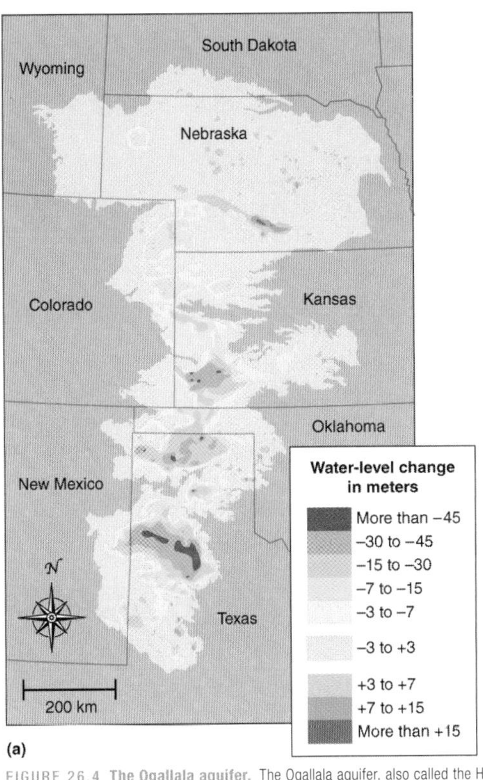

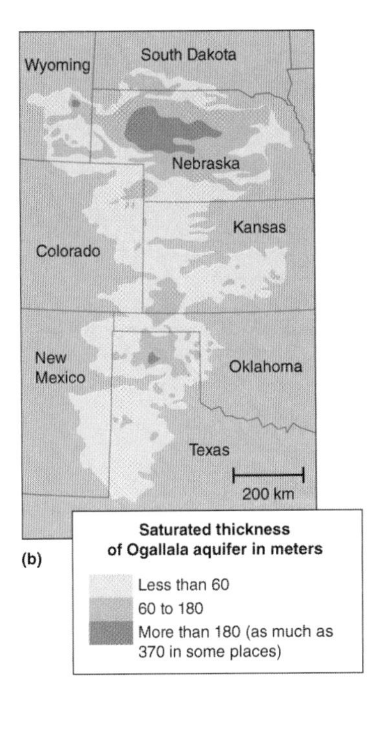

FIGURE 26.4 The Ogallala aquifer. The Ogallala aquifer, also called the High Plains aquifer, is the largest in the United States, with a surface area of about 450,000 km^2 (174,000 miles^2). (a) The change in water level from 1950 to 2005, mostly due to withdrawals for irrigation that have exceeded the aquifer's rate of recharge. (b) The current thickness of the aquifer. *(Data from http://pubs.usgs.gov/fs/2004/3097/ and http://ne.water.usgs.gov/ogw/hpwlms/)*

over 10,000 to 20,000 years. For this reason, water from a confined aquifer is usually much older and less likely to be contaminated by anthropogenic chemicals than is water from an unconfined aquifer.

Large-scale use of water from a confined aquifer is unsustainable because the withdrawal of water is not balanced by recharge. The largest aquifer in the United States is the massive Ogallala aquifer in the Great Plains. Large amounts of water have been withdrawn from this aquifer for household, agricultural, and industrial uses. Unfortunately the slow rate of recharge is not keeping pace with the fast rate of water withdrawal, as FIGURE 26.4 shows. For this reason, the Great Plains region could run out of water during this century.

FIGURE 26.5 shows what happens when more water is withdrawn from an aquifer than enters the aquifer. As the water table drops farther from the ground surface, springs that once bubbled up to the surface no longer emerge, and spring-fed streams dry up. In addition, some shallow wells no longer reach the water table. When water is rapidly withdrawn from a well it can create an area of that contains no ground water around the well, which is known as a **cone of depression.** In other words, rapid pumping of a deep well can cause adjacent, shallower wells to go dry.

Cone of depression An area lacking groundwater due to rapid withdrawal by a well.

Teaching Tip: Engage

Ask students: Which type of aquifer is most likely to be contaminated by anthropogenic chemicals? *Unconfined aquifers are most likely to be contaminated because they hold relatively new water that is in closer contact with the surface where pollution occurs.*

◀ TEACHING with FIGURES

Show students Figure 26.4 and ask: Where has the greatest withdrawal of water occurred according to the data given in the map? *Northwest Texas (the Panhandle) and through western Oklahoma and Kansas*

Teaching Tip: Journal Prompt

Ask students to explain how a cone of depression is formed. *When water is withdrawn rapidly from an aquifer it can remove enough groundwater to cause adjacent shallow wells to run dry.*

Teaching Tip: Warm-up

Ask students to explain why large-scale use of water from a confined aquifer is unsustainable. *It is unsustainable because the withdrawal of water is not balanced by recharge.*

Lab
EcoBottle

This lab offers students an opportunity to build and observe a land and water ecosystem. Students will observe abiotic and biotic interactions between the two systems. Students will also have the opportunity to observe and review carbon, nitrogen, and water cycles during this investigation.

Download the lab by clicking on the link buttons in the TE-book, opening the TRFD, or logging onto the book's companion website bcs.whfreeman .com/friedlandapes2e (teacher login required).

 **Lab 9.1** EcoBottle

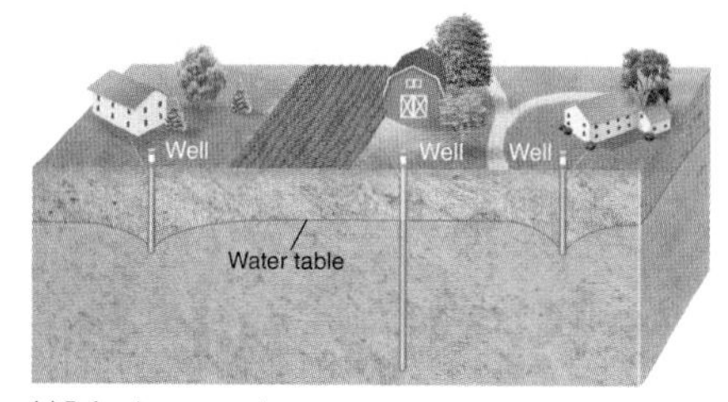

(a) Before heavy pumping

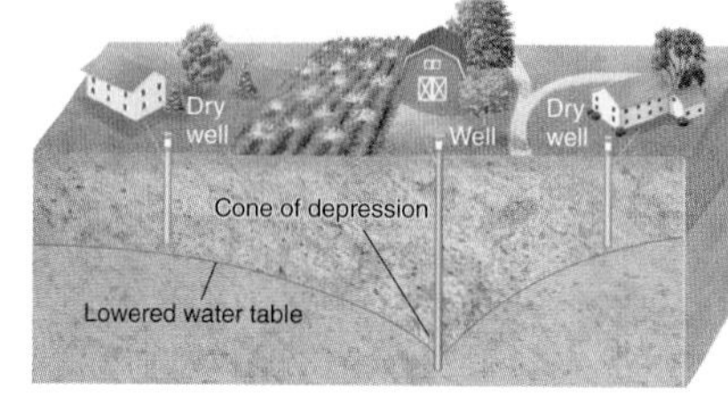

(b) After heavy pumping

FIGURE 26.5 **Cone of depression.** (a) When a deep well is not heavily pumped, the recharge of the water table keeps up with the pumping. (b) In contrast, when a deep well pumps water from an aquifer more rapidly than it can be recharged, it can form a cone of depression in the water table and cause nearby shallow wells to go dry.

In some situations, water quality is compromised when fresh water is pumped out of wells faster than the aquifer can be recharged. FIGURE 26.6 shows how this can happen when wells are drilled near coastlines. Some coastal regions have an abundance of fresh water underground. This groundwater exerts downward pressure that prevents the adjacent ocean water from flowing into aquifers and mixing with the groundwater. As humans drill thousands of wells along these coastlines, however, rapid pumping draws down the water table, reduces the depth of the groundwater, and thereby lessens that pressure. The adjacent salt water is then able to infiltrate the area of rapid pumping, making the water in the wells salty. An infiltration of salt water in an area where groundwater pressure has been reduced from extensive drilling of wells is called **saltwater intrusion** and it is a common problem in coastal areas.

Surface water is the collection of aquatic biomes

Surface water is the fresh water that exists aboveground and includes streams, rivers, ponds, lakes, and wetlands. As mentioned in Chapter 4, these waters typically represent freshwater biomes that perform a number of important functions.

The world's three largest rivers, as measured by the volume of water they carry, are the Amazon in South America, the Congo in Africa, and the Yangtze in China. Early human civilizations typically settled along major rivers not only because these waterways served as important means of transportation but also because the land surrounding rivers is often highly fertile. Most streams and rivers naturally overflow their banks

Saltwater intrusion An infiltration of salt water in an area where groundwater pressure has been reduced from extensive drilling of wells.

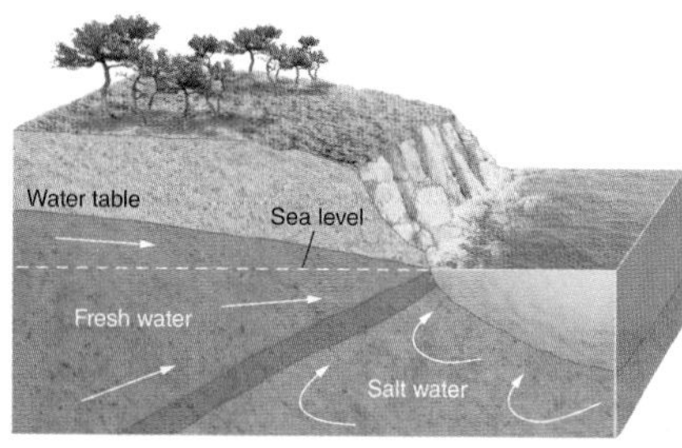

(a)

(b)

FIGURE 26.6 **Saltwater intrusion.** (a) When there are few wells along a coastline, the water table remains high and the resulting pressure prevents salt water from intruding. (b) Rapid pumping of wells drilled in aquifers along a coastline can lower the water table. Lowering the water table reduces water pressure in the aquifer, allowing the nearby salt water to move into the aquifer and contaminate the well water with salt.

Teaching Tip: Concept Map

Have students complete the following concept map to review surface water biomes. (Answers are provided in italics.)

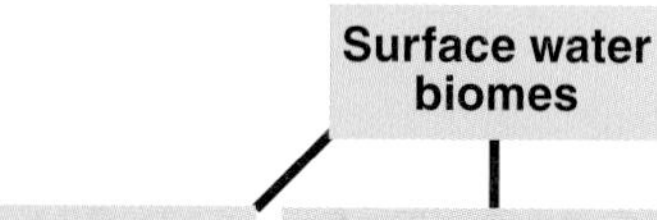

Streams and rivers
Flowing fresh water that may originate from underground springs, runoff from rain, or melting snow

Lakes and ponds
Standing water that forms in depressions filled by precipitation and runoff

Wetlands
Land submerged in water for most of the year

Teaching Tip: Journal Prompt

Ask students to describe how saltwater intrusion occurs. *Saltwater intrusion can occur near coastlines when groundwater is pumped out too rapidly, reducing the pressure that keeps saltwater out. This allows saltwater to flow into groundwater systems, making the well water salty.*

during periods of spring snowmelt or heavy rainfall. This excess water spreads onto the land adjacent to the river, called the **floodplain,** where it deposits nutrient-rich sediment that improves the fertility of the soil. The ancient Egyptians built a large and prosperous civilization by using the Nile River for commerce and its fertile floodplains for agriculture. The use of rivers as a means of transport continues today. Many large cities in North America, including Pittsburgh, New Orleans, and St. Louis, developed on the banks of major rivers.

Ponds and lakes typically form in depressions in the landscape that are filled by precipitation, runoff that is not absorbed by the surrounding landscape, and groundwater that flows into the depression. As you can see in FIGURE 26.7, we can compare the size of Earth's lakes according to surface area, depth, or volume. Some lakes, such as Lake Victoria in Africa, have a large surface area but are not especially deep. Other lakes, such as Lake Baikal in Asia, do not have a large surface area, but are extremely deep and therefore have a large volume.

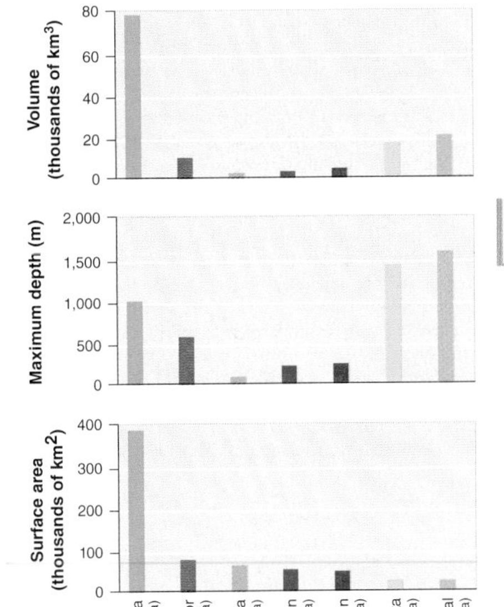

FIGURE 26.7 **The world's largest lakes.** Measurements of surface area, depth, and volume are shown for each lake. *(Data from http://wldb.ilec.or.jp/LakeDB2/)*

Lakes can be created by a variety of processes, including tectonic activity and glaciation. When tectonic activity causes areas of land to rise up, it can isolate a region of the ocean. Because lakes formed in this way were once part of the ocean, they have higher salinities than other lakes. The Caspian Sea in western Asia is an excellent example of a salty lake that was formed by tectonic uplift. Tectonic activity can also cause land to split open into long, deep fissures that subsequently fill with water. Some of the world's largest lakes have formed this way, including Lake Victoria and Lake Tanganyika in Africa and Lake Baikal in Asia. Another way that lakes are formed is through the movement of glaciers. Over thousands of years, glacial movement scrapes large depressions in the land that subsequently fill with water. This process is thought to have played a major role in the creation of the Great Lakes in North America.

As we saw in Chapter 4, freshwater wetlands play an important role in water distribution and regulation. During periods of heavy rainfall that might otherwise lead to flooding, these freshwater wetlands—as well as salt marshes and mangrove swamps—can absorb and store the excess water and release it slowly, thereby reducing the likelihood of a flood.

Wetlands are measured by surface area. There is some debate about which freshwater wetland is the world's largest, but most researchers consider the Pantanal in South America to be the largest wetland and the Florida Everglades to be the second largest.

Atmospheric water produces precipitation

Although the atmosphere contains only a very small percentage of the water on Earth, that atmospheric water is essential to global water distribution. People living in arid regions rely heavily on precipitation, in the form of rain and snow, for their water needs. In certain regions, such as East Africa, annual rainfall follows a fairly regular pattern, with April, May, October, and November normally rainy months. However, even areas with predictable rainfall patterns can experience unexpected droughts. In the fall of 2000, the United Nations Food and Agriculture Organization reported that a succession of droughts had put 16 million people at risk of starvation in eastern Africa. In 2004, the worst drought in more than a decade adversely affected many parts of southern Africa, destroying crops, killing cattle, and causing millions of people to go hungry.

Floodplain The land adjacent to a river.

Teaching Tip: Journal Prompt

Ask students to explain how lakes form. *When tectonic activity causes land to rise up, regions of the ocean can be isolated, creating a sea or salt lake. Tectonic activity can cause the land to split and fill with water. In addition, when glaciers move, they scrape large depressions out of the land that can fill with water.*

AP® Exam Tip

Human effects on coastal wetlands are covered in the 2003 AP® Exam, Question 3. To answer this question, students must

- discuss two reasons that the temperature or salinity of an estuary varies
- explain the ecological and economical significance of coastal wetlands
- list three ways in which humans have harmed coastal wetlands
- suggest an environmental policy to address one of the ways in which humans have harmed coastal wetlands

▲ TEACHING with FIGURES

Ask students to study Figure 26.7 and answer the following questions:

- Which lake has the greatest depth? *Lake Baikal has the greatest depth.*
- Which lake has the largest surface area? *The Caspian Sea has the largest surface area.*
- Which lake holds the greatest volume of water? *The Caspian Sea holds the greatest volume of water.*

Teaching Tip: Warm-up

Ask students to describe the importance of atmospheric water. *Atmospheric water is essential to global water distribution because it is the source of water in precipitation.*

Teaching Tip: Journal Prompt

Ask students to list some human and ecological impacts associated with droughts. *Droughts can destroy crops, kill cattle, and cause millions of people to go hungry. With prolonged droughts, soil fertility can be affected, erosion is increased, and hardened land becomes impermeable to rainfall.*

FIGURE 26.8 **The Dust Bowl.** This photo shows a farmhouse in Stratford, Texas, just prior to being hit by a massive dust storm in 1935. Poor agricultural practices combined with prolonged droughts can produce severe dust storms that carry away topsoil and rob the land of its fertility. *(NOAA/Science Source)*

In addition to the direct losses of human lives, livestock, and crops that droughts cause, they also have long-term effects on soil. Because the cycling of many nutrients important to ecosystem productivity, such as nitrogen and phosphorus, depends on the movement of water, droughts affect nutrient cycling and thus soil fertility. Furthermore, prolonged droughts can dry out the soil to such an extent that the fertile upper layer—the topsoil—blows away in the wind. As a result, the land may become useless for agriculture for decades or longer. Finally, soil that has become severely parched may harden and become impermeable. When the rains finally arrive, the water runs off over the hardened land surface rather than soaking in, and this causes the topsoil to erode.

Human activities can contribute to the negative effects of droughts. In the midwestern United States in the 1920s and 1930s, for example, the amount of conversion from native grasslands to wheat fields increased dramatically. Because wheat fields are more susceptible to soil erosion by wind than are native grasslands, this conversion produced many more hectares of land susceptible to erosion. In the early 1930s, a severe and prolonged drought caused a decade of crop failures. With few crop plants remaining alive to hold the soil in place, the drought led to massive dust storms as winds picked up the parched topsoil and blew it away (FIGURE 26.8). Indeed, one dust storm, on April 14, 1935, was so severe that the Sun was entirely blocked out in the southern Great Plains, leading to the name "Black Sunday." Topsoil from these dust storms traveled as far as Washington, D.C., and earned the southern Great Plains the nickname of "the Dust Bowl." The decade of dust storms and the resulting losses of topsoil led to large-scale human migration out of the region and impacted the economy of the entire nation. Today, improved farming practices have reduced the susceptibility of soil to erosion by wind, making major dust storms in the Great Plains and elsewhere much less likely.

Flooding occurs when water input exceeds the ability of an area to absorb that water. Many drought-prone areas of the world rarely experience high amounts of rainfall, but when it does happen, severe flooding can occur. California, for example, is a relatively dry region of the United States, but occasionally it does experience heavy rainstorms. Because there is no system in place to capture this water, the state on occasion has serious flooding problems.

Human activities can also worsen the risk of flooding. In healthy natural areas, porous soil and wetlands soak up excess rainwater. However, in areas where the soil has been baked hard by drought, water that comes from heavy rainfalls cannot soak into the ground. This is also true in urban and suburban areas with large areas of **impermeable surfaces**—pavement or buildings that do not allow water to penetrate the soil. Instead, storm waters run off over impermeable soil or paved surfaces into storm sewers or nearby streams. Excess water that the ground does not absorb fills streams and rivers, which can overflow their banks and may flood lowland areas. Like droughts, floods can lead to crop and property damage as well as losses of animal and human lives.

Impermeable surface Pavement or buildings that do not allow water penetration.

Teaching Tip: Engage

Ask students to explain how human activities worsen the effects of droughts and floods. *Conversion of grasslands to croplands can increase erosion by wind, which can lead to massive dust storms and loss of topsoil. Impermeable surfaces in urban and suburban areas cause runoff and flooding in low-lying areas.*

module

26

REVIEW

In this module, we learned that fresh water makes up a tiny fraction of all water on Earth and that most of this fresh water is locked up in the form of ice and glaciers. Groundwater is a type of available freshwater and it can be found as either confined or unconfined aquifers that may experience fast or slow recharge. In some parts of the world the rate of water withdrawal for human use is faster than the rate of aquifer recharge. Surface waters such as streams, rivers, wetlands, and lakes are important sources of freshwater, and wetlands can be important for flood control. Finally, we learned that the climatic differences in the availability of precipitation can interact with human land use to cause dramatic floods and droughts in different parts of the world. In the next module, we will examine how humans have worked to alter the natural distribution of water in ways that better suit human needs.

Module 26 AP® Review Questions

1. Which is one of the top three largest rivers by volume?
   (a) The Nile
   (b) The Congo
   (c) The Mississippi
   (d) The Yellow River
   (e) The Volga

2. When deep wells are heavily pumped, one result can be
   (a) decreased groundwater recharge.
   (b) spring formation.
   (c) a cone of depression.
   (d) increased groundwater recharge.
   (e) a rise in the water table.

3. Which is true of floodplains?
   (a) They have increased fertility.
   (b) Humans are unable to use them.
   (c) They occur near ponds and lakes.
   (d) They are the result of glaciation.
   (e) They occur in areas with saltwater intrusions.

4. Groundwater recharge
   (a) is the result of precipitation.
   (b) sometimes occurs as a spring.
   (c) occurs rapidly in confined aquifers.
   (d) can cause saltwater intrusions.
   (e) does not occur in unconfined aquifers.

Answers to Module 26 AP® Review Questions

1. b
2. c
3. a
4. a

Teaching Tip: Warm-up

Ask students how levees, dikes, dams, and aqueducts differ from one another and the primary purpose of each.

- **Levee:** *An enlarged bank built to prevent rivers from flooding*
- **Dike:** *A structure built to prevent ocean waters from flooding adjacent land*
- **Dam:** *A barrier that runs across a river or stream to control the flow of water*
- **Aqueduct:** *A canal or ditch used to carry water from one location to another*

Teaching Tip: Journal Prompt

Ask students to describe some of the negative consequences of building levees along river banks.
Fertility of the surrounding lands is reduced because floodwaters, which typically distribute sediment and nutrients that add to the fertility of floodplains, no longer flow onto the land. Instead, sediments and nutrients are deposited in the ocean. Levees prevent flooding at one location but may force more water downstream and increase flooding there. When floodwaters become too high they can overtop or erode a hole in the levee, resulting in massive flooding of surrounding areas.

module 27

Human Alteration of Water Availability

We have seen that the availability of water around the world can be unpredictable. In the Klamath River, for example, years of low precipitation caused an insufficient water supply for human needs. As a result, humans have learned to live with variations in water availability in several ways. We can channel the flow of flood waters with *levees* and *dikes,* block the flow of rivers with *dams* to store water, divert water from rivers and lakes and transport it to distant locations, and even obtain fresh water by removing the salt from salt water. In this module we will look at each of these water distribution methods and examine their costs and benefits.

Learning Objectives

After reading this module you should be able to

- compare and contrast the roles of levees and dikes.
- explain the benefits and costs of building dams.
- explain the benefits and costs of building aqueducts.
- describe the processes used to convert salt water into fresh water.

Levees and dikes are built to prevent flooding

Before the intervention of humans, most rivers periodically overflowed their banks. These occasional overflows of nutrient-rich water made their floodplains particularly fertile. While throughout history humans have taken advantage of these fertile floodplains for agriculture, in more recent times, they have sought ways to prevent flooding so floodplain land could be developed for residential and commercial use. One way to prevent flooding is by constructing a **levee,** an enlarged bank built up on each side of the river. The Mississippi River has the largest system of levees in the world. This river is enclosed by more than 2,400 km (1,500 miles) of levees that offer flood protection to more than 6 million hectares (15 million acres) of floodplains.

Levee An enlarged bank built up on each side of a river.

The use of levees has produced several major challenges. First, the fertility of these lands is reduced because natural floodwaters no longer add fertility to floodplains by depositing sediments there. Second, because the sediments do not leave the river, they are carried farther downstream and settle out where the river enters the ocean. Third, levees may prevent flooding at one location, but by doing so they force floodwater farther downstream where it can cause even worse flooding. Finally, the building of levees encourages development in floodplains, although these areas will still occasionally flood. This practice raises the question of whether development should be allowed in areas where flooding remains a high risk.

When floodwaters become too high, levees can either collapse due to the tremendous pressure of the water, or water can come over the top and quickly erode a large hole in the levee. Both events result in massive flooding (FIGURE 27.1). One of the most famous levee failures occurred in New Orleans in 2005 as the result of Hurricane Katrina. The levees could not contain the storm surge and heavy rainfall associated with the hurricane. The water overtopped nearly 50 levees, quickly washing out the banks of dirt and flooding many of the neighborhoods that the levees were built to protect. The hurricane and the devastating floodwaters caused into 1,800 deaths and over $80 billion in damage to homes and businesses.

Dikes are enlarged banks built up to prevent ocean waters from flooding adjacent land. Functionally, dikes are similar to levees in that they are built to keep water from flooding onto the land. Dikes are common in northern Europe, where large areas of farmland lie below sea level. Perhaps the best known dikes are those of the Netherlands, where dikes have been used for nearly 2,000 years. Approximately 27 percent of the land in the Netherlands is below sea level. The dikes, combined with pumps that move any intruding water back out to the ocean, have allowed the country to inhabit and farm areas that would otherwise be under water. The original pumps were powered by windmills, which is why we often associate the Netherlands with wind energy. Today, however, the water is pumped with electric and diesel pumps.

Dams are built to restrict the flow of streams and rivers

A **dam** is a barrier that runs across a river or stream to control the flow of water. The water body created by damming a river or stream is called a **reservoir.** Dams hold water for a wide variety of purposes, including human consumption, generation of electricity, flood control, and recreation. In 2013, there were 845,000 dams in the world and more than 84,000 dams in the United States. The largest reservoir system in the United States is along the Missouri River. With six dams and reservoirs in Montana, North Dakota, South Dakota, and Nebraska, this system stores nearly 90 trillion liters (24 trillion gallons) of water.

For centuries, humans have used dams to do work, from turning waterwheels that operated grain mills to powering modern turbines that generate electricity in hydroelectric plants. Although hydroelectric dams (discussed further in Chapter 13) are some of the largest dams in the world, they represent only 3 percent of all dams in the United States. In contrast, 18 percent of U.S. dams were built with the primary purpose of flood control—to reduce or prevent flooding farther down the river. Another 38 percent of U.S. dams were constructed primarily for recreation. Dams have also been built to create scenic lakes for housing developments.

FIGURE 27.1 **Levees.** Levees are built to prevent rivers from flowing over their banks and onto the floodplain. In 2008, this levee, just north of St. Louis, Missouri, collapsed and allowed the floodwaters to spread over the surrounding fields. *(Anthony Souffle/MCT/Landov)*

Dams provide substantial benefits to humans, but they come with financial, societal, and environmental costs. The world's largest dam, the Three Gorges Dam built across the Yangtze River in China (FIGURE 27.2), illustrates these costs and benefits well. This massive structure is 2 km (1.3 miles) wide and 185 m (610 feet) high, and has created a reservoir 660 km (410 miles) long behind it. The Three Gorges Dam, completed in 2006, took 13 years to construct. The reservoir required another 2 years to fill with water.

The benefits of the Three Gorges Dam include the large amount of hydroelectric power it generates, which reduces the extraction and use of fossil fuels. In addition, the dam helps prevent the seasonal flooding that damaged downstream cities and villages and killed more than 1 million people over the past 100 years.

The costs of building dams, both to people and the environment, are substantial. Building a dam, like other large-scale construction, uses large amounts of energy and materials. Projects on such a large scale may displace many people. For example, the Three Gorges Dam and the reservoir it created flooded 13 cities, 140 towns, and 1,350 villages; more than 1.3 million people were forced to relocate.

Dike A structure built to prevent ocean waters from flooding adjacent land.

Dam A barrier that runs across a river or stream to control the flow of water.

Reservoir The water body created by a damming a river or stream.

Teaching Tip: Warm-up

Ask students to identify the primary purposes for which dams are constructed. *Dams are constructed to hold water for human consumption, to generate electricity, to aid in flood control, and to create recreational areas.*

Teaching Tip: Journal Prompt

Ask students to complete the chart to evaluate the pros and cons of dams. (Answers are provided in italics.)

Pros	Cons
Provides hydroelectric power, which reduces the need for extraction and use of fossil fuels	*Large amounts of energy and resources needed for construction*
Prevents seasonal flooding	*Construction may require people to relocate*
Provides reservoir for recreation	*Interruption of natural water flow disturbs ecosystems both upstream and downstream*
Provides a stored water source for human use	*Interferes with fish migration, and the associated equipment may kill fish*

Teaching Tip: Warm-up

Ask students how we can address the interruption of fish migration caused by dams. *Fish ladders (see Figure 27.3) have been constructed alongside dams, allowing fish to migrate upstream and reach their breeding grounds.*

Teaching Tip: Journal Prompt

Ask students to complete the chart to evaluate the pros and cons of aqueducts. (Answers are provided in italics.)

Pros	Cons
Large cities depend on aqueducts to meet daily water needs	*Older aqueducts can lose as much as 55% of the water they carry due to evaporation*
Aqueducts can distribute water to areas that are used for agriculture	*Construction is expensive and fragments habitats*
	Water diversion can lead to decreased flow of rivers and drain surface lakes

FIGURE 27.2 **Dams.** The photo shows world's largest dam, the Three Gorges Dam on the Yangtze River in China. Dams serve a wide variety of purposes, including flood control and electricity generation. *(AP Photo/Xinhua, Du Huaju)*

FIGURE 27.3 **Fish ladder.** Because dams are an impediment to fish such as salmon that migrate upstream to breed, fish ladders like this one on the Rogue River in Oregon have been designed to allow the fish to get around the dam and continue on their traditional path of migration. *(Kevin Schafer/Alamy)*

One of the most common environmental problems associated with dams is the interruption of the natural flow of water to which many organisms are adapted. For migrating fish such as salmon, dams represent an insurmountable obstacle to breeding. The loss of these fish can have a cascading effect on other organisms that depend on them, such as bears that feed on migrating salmon. To alleviate this problem, **fish ladders,** which are built like a set of stairs with water flowing over them, have been added to some dams. Migrating fish can swim up the fish ladders and reach their traditional breeding grounds (FIGURE 27.3). Fish are also impacted when water is released from a dam to generate electricity. Passing millions of gallons of water through a spinning turbine kills millions of fish each year.

Using dams to prevent seasonal flooding has other consequences for the ecology of an area. Seasonal flooding represents a natural disturbance that scours out pools and shorelines, and this disturbance favors colonization by certain plants and animals. Some of the species that are highly dependent on such disturbances are very rare. In recent years, managers have begun to experiment with releasing large amounts of water from reservoirs to simulate seasonal flooding; the results have been very promising. In some areas, dams have been removed where they are no longer needed, as described at the opening of this chapter. In these cases, the natural flow of water has been restored and much of the natural ecology has been quickly restored.

Fish ladder A stair-like structure that allows migrating fish to get around a dam.

Aqueduct A canal or ditch used to carry water from one location to another.

Aqueducts carry water from one location to another

While dams are designed to hold water back and store it, we often also need to move water from one place to another. **Aqueducts** are canals or ditches used to carry water from one location to another. Typically, aqueducts remove water from a lake or river to a place where it is needed. Although aqueducts are famously associated with the Roman Empire, their use dates back to at least the seventh century BCE in Greece. The earliest aqueducts were made of limestone, but modern aqueducts include concrete canals and pressurized steel pipes laid above or under the ground. These structures are more efficient water carriers than the ditches and stone-lined canals of historic times. Older aqueducts, many of which are still in use, can lose as much as 55 percent of the water they carry through leakage or evaporation. These losses can be particularly troublesome in such arid regions as Israel and Jordan where water is already scarce.

In the United States, two of the country's largest cities, New York and Los Angeles, depend on aqueducts to meet their daily water needs. The Catskill Aqueduct brings clean, fresh water over 200 km (120 miles) from the streams and lakes of the Catskill Mountains to New York City. The Colorado River Aqueduct is a canal that carries water 400 km (250 miles) from the Colorado River to Los Angeles (FIGURE 27.4).

Using aqueducts to transport water supplies comes with costs and benefits. Although bringing water to cities from pristine areas may ensure its cleanliness, the construction of aqueducts is expensive and disturbs

FIGURE 27.4 **Aqueducts.** Aqueducts deliver water from places where it is abundant to places where it is needed. This section of the Colorado River Aqueduct, which diverts water from the Colorado River to Los Angeles, passes through the Mojave Desert. *(Mark Hanauer/Corbis)*

natural habitats. Aboveground aqueducts can fragment an environment. Even if an aqueduct is buried as an underground pipeline, a great deal of disruption occurs during construction. Often water is diverted from a natural river where it has flowed for millennia. Some major rivers in the United States, including the Colorado River and the Rio Grande, now lose so much water in multiple locations that during certain periods the rivers go dry before they reach the ocean.

Some water diversion projects have international impacts. India and Bangladesh, for example, share nearly 250 rivers that originate in the Himalayas of India and flow into Bangladesh. In 2009, India proposed a large-scale water diversion project, involving more than 50 rivers, that has caused Bangladesh some major concerns. By diverting water from these rivers, India stands to gain 170 billion liters (55 billion gallons) of water each year for agricultural and household use. However, Bangladesh is situated downstream of these river diversions and Bangladeshi officials worry that the project will end up dramatically reducing the flow of river water into their country and affect local fish populations and river navigability for commerce. Even farther downstream, the project may reduce the flow of fresh water into estuaries, allowing ocean water to move farther up into the river. This influx of seawater raises the salinity of the estuary and harms aquatic life that is not adapted to such salty water.

At the same time that India is diverting water that would naturally flow into Bangladesh, neighboring China has begun the construction of dams on the Yarlung-Zangbo River, which flows from China into India. The first dam is of modest size and is scheduled to be completed in 2015; larger dams are scheduled to follow. The Chinese dams are primarily for the generation of electricity, but there is concern that the water behind the dams could later be diverted to agricultural use in China, which could result in substantially less water flowing through India and Bangladesh. These conflicts over water rights have led many people to call for treaties among the countries in the region to reach a mutually agreeable decision regarding water use.

The most infamous river diversion project happened in the 1950s, when the Soviet Union diverted two rivers that fed the Aral Sea in Central Asia. Diversion of the rivers dramatically decreased freshwater input into the Aral Sea, increased the salinity of the remaining lake water, and destroyed the local fish populations. Although the Aral Sea was originally the fourth largest lake in the world in terms of surface area, the diversion project reduced its surface area by more than 60 percent, as shown in FIGURE 27.5. The withdrawals of water caused the Aral Sea to split into two parts: the North Aral Sea and the South Aral Sea. Dust storms have eroded soil, salt, and pesticide residues from the dry lake bed and carried particles of these substances into the atmosphere. The reduction in the size of the lake has also affected the local climate. Without the moderating effect of a large body of

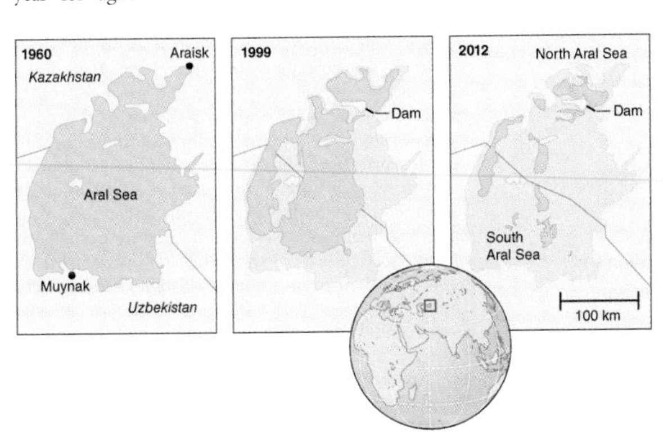

FIGURE 27.5 **Consequences of river diversion.** Diverting river water can have devastating impacts downstream. The Aral Sea, on the border of Kazakhstan and Uzbekistan, was once the world's fourth largest lake. Since the two rivers that fed the lake were diverted, its surface area has declined by 60 percent and the lake has split into two parts: the North and South Aral seas.

Teaching Tip: Journal Prompt

Ask students to describe the series of events that occurred as a result of water diversions from the Aral Sea during the Soviet period.

1. *Diversion of two rivers away from the Aral Sea decreased freshwater input.*
2. *Salinity of the water increased.*
3. *Local fish populations were destroyed.*
4. *The surface area of the lake was reduced by 60 percent.*
5. *The sea was split into two parts.*
6. *Dust storms eroded dried lakebed soil, salt, and pesticide residues, and carried particles of these substances into the atmosphere.*
7. *The climate moderating effect of the lake was reduced.*
8. *Summers have become hotter and winters have become colder in the region.*

AP® Exam Tip

The Colorado River and environmental issues associated with water diversion are covered in the 2002 AP® Exam, Question 2. To answer this question, students must

- list two environmental issues associated with water diversion
- make an argument in favor of water diversion for urban use, for agricultural use, or for maintaining a natural ecosystem
- identify another large-scale water diversion project and identify two environmental concerns associated with that project

Teaching Tip: Warm-up

Ask students: What is the benefit of desalinizing water? *Desalination converts saltwater to freshwater that can be used for drinking and agriculture. The process makes fresh water available in areas where it is scarce.*

Teaching Tip: Video

A Country with No Water
This TED talk by Fahad Al-Attiya acquaints students with the need for water in the Middle East. In the energy-rich, water-poor country of Qatar, desalination proves to be a solution to its water needs.

The link to this video may be found by clicking on the link buttons in the TE-book, opening the TRFD, or logging onto the book's companion website bcs.whfreeman.com/friedlandapes2e (teacher login required).

Chapter 9 Web Resources

Teaching Tip: Warm-up

Ask students: What are some drawbacks of desalination? *A great deal of energy is required to desalinate water. Brine has a very high salt concentration, which can contaminate the soil and potentially harm plant and animal life. If brine is dumped into a bay, coastal area, or open ocean, marine life can be harmed. Desalination systems require a large investment to build, maintain, and repair.*

water on climate, summers in the region are now hotter and winters are colder. While efforts are being made to save the North Aral Sea, saving the South Aral Sea has been deemed too expensive and all efforts have been abandoned. The drying of the South Aral Sea continues today: The shallower eastern half of the South Aral Sea is predicted to be completely dry in 10 years.

Desalination converts salt water into fresh water

Today, some water-poor countries are able to obtain fresh water by removing the salt from salt water, a process called **desalination,** or **desalinization** (FIGURE 27.6). The salt water usually comes from the ocean, but it can also come from salty inland lakes. Currently, the countries of the Middle East produce 50 percent of the world's desalinated water. During the past decade, a number of technological advances have been made and this has greatly reduced the cost of desalination.

The two most common desalination technologies are *distillation* and *reverse osmosis.* **Distillation** (Figure 27.6a) is a process of desalination in which water is boiled and the resulting steam is captured and condensed to yield pure water. As the water is converted into steam, salt is left behind. A great deal of energy is required to boil the water and then condense it, so distillation can be a monetarily and environmentally expensive process.

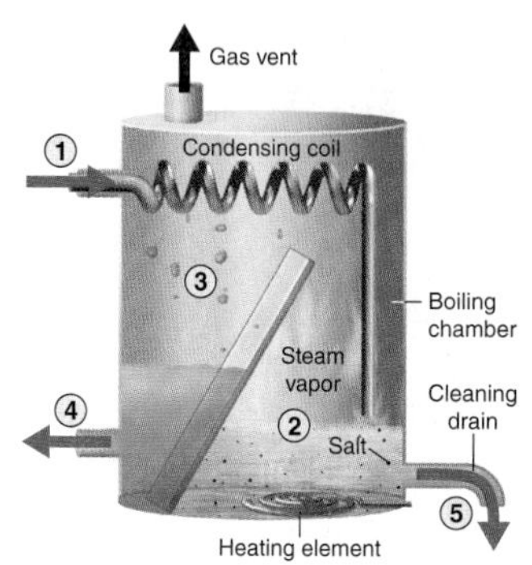

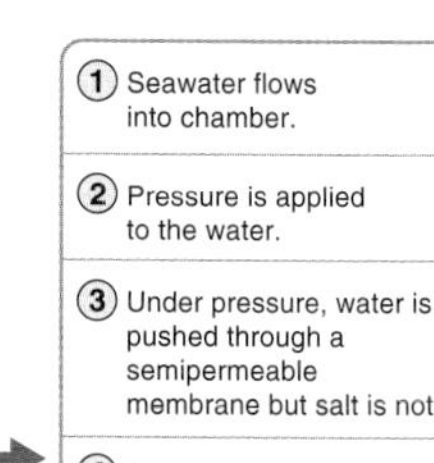

FIGURE 27.6 **Desalination technologies.** Salt water can be converted into fresh water in one of two ways. (a) Distillation uses heat to convert pure water into steam that is later condensed, leaving the salt behind. (b) Reverse osmosis uses pressure to force pure water through a semipermeable membrane, leaving the salt behind.

Reverse osmosis (Figure 27.6b) is a process of desalination in which water is forced through a thin semipermeable membrane at high pressure. Water can pass through the membrane, but salt cannot. Reverse osmosis is a newer technology that is more efficient and often less costly than distillation. However, the liquid that remains, called brine, has a very high salt concentration. It cannot be deposited on land because its high salt content would contaminate the soil, potentially harming plant and animal life. If dumped into a bay or coastal area, it can harm fish and other aquatic life. Typically, brine is returned to the open ocean, though its high salt concentrations can still cause harm to ocean life in areas where it is dumped.

Ultimately, all water management systems require a large investment to build, maintain, and repair. This means that many water-poor countries are not able to implement large-scale methods of water management to provide for their needs. Water availability varies greatly around the world, as FIGURE 27.7 shows. Middle Eastern and North African countries represent about 5≈percent of the world's population, yet they have less than 1 percent of the fresh water available for drinking. The United Nations estimates that nearly 1.2 billion people live in regions that have a scarcity of water.

Desalination The process of removing the salt from salt water. *Also known as* **Desalinization.**

Distillation A process of desalination in which water is boiled and the resulting steam is captured and condensed to yield pure water.

Reverse osmosis A process of desalination in which water is forced through a thin semipermeable membrane at high pressure.

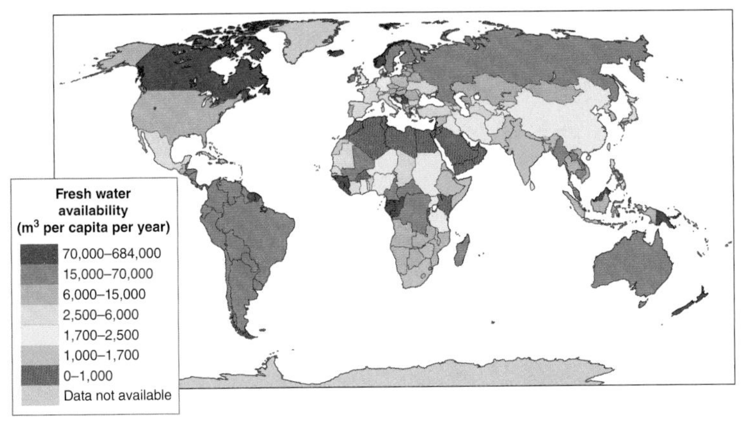

FIGURE 27.7 **Water availability per capita.** The amount of water available per person varies tremendously around the world. North Africa and the Middle East are the regions with the lowest amounts of available fresh water.

◀ TEACHING with FIGURES

Ask students to study Figure 27.7 and determine the most water-poor regions of the world. *The Middle East and North Africa have the lowest amounts of water available per capita.*

module 27

REVIEW

In this module, we have seen that humans have developed numerous methods of altering the distribution of water including the use of levees, dikes, dams, and aqueducts. We also learned that desalination technologies can be used to convert undrinkable salt water into drinkable fresh water. In the next module, we will examine how humans consume all of this water for agriculture, industry, and home use.

Module 27 AP® Review Questions

1. Which is NOT true of levees?
   (a) They limit the benefits of floodplains.
   (b) They are very similar in function to dikes.
   (c) They can lead to increased flooding downstream.
   (d) They discourage the use of floodplains.
   (e) They increase sediment deposition in the ocean.
2. Which is NOT a primary use of dams?
   (a) Recreation
   (b) Electricity generation
   (c) Flood control
   (d) Habitat restoration
   (e) Agricultural uses
3. Which cannot reduce the effects of flooding?
   (a) Levees
   (b) Dams
   (c) Freshwater wetlands
   (d) Impermeable surfaces
   (e) Dikes
4. Water diversion has significantly reduced the size of
   (a) Lake Erie.
   (b) the Aral Sea.
   (c) the Nile.
   (d) Lake Baikal.
   (e) the Baltic Sea.
5. Which is true of desalination?
   (a) The technique of distillation requires less energy than reverse osmosis.
   (b) Desalination is primarily used in North Africa.
   (c) The technique of reverse osmosis leaves brine.
   (d) Desalination plants are common in developing nations.
   (e) Brine can be dumped onto land without affecting plant and animal life.

Answers to Module 27 AP® Review Questions

1. d
2. d
3. d
4. b
5. c

Teaching Tip: Video

Fresh Water Scarcity: An Introduction to the Problem
This video clip from TED-Ed introduces the idea that there is not currently enough water in the world to meet human needs and offers potential solutions to global water shortages.

The link to this video may be found by clicking on the link buttons in the TE-book, opening the TRFD, or logging onto the book's companion website bcs.whfreeman.com/friedlandapes2e (teacher login required).

 **Chapter 9 Web Resources**

Teaching Tip: Engage

Have students calculate their water footprint using the interactive "Water Footprint Calculator" on the *National Geographic* website. Students will have the opportunity to consider how to reduce their water usage and to make a pledge. Please be aware that the student must enter an email or phone number to make the pledge. (This part of the interactive is not necessary to complete the activity and it should not be a required component of the assignment.)

The link to this website may be found by clicking on the link buttons in the TE-book, opening the TRFD, or logging onto the book's companion website bcs.whfreeman.com/friedlandapes2e (teacher login required).

Chapter 9 Web Resources

module 28

Human Use of Water Now and in the Future

A human can survive without food for 3 weeks or more, but cannot survive without water for more than a few days. Water is also essential for producing food. In fact, 70 percent of the world's freshwater consumption is used for agriculture. The remaining 30 percent is split between industrial and household uses, the proportion of which varies from country to country. On average, experts estimate that about 20 percent of the world's freshwater use is for industry and about 10 percent is for household uses. In this module, we will examine the major uses of water by humans in the areas of agriculture, industry, and households. We will then investigate how water ownership and conservation will determine the availability of water in the future.

Learning Objectives

After reading this module you should be able to

- compare and contrast the four methods of agricultural irrigation.
- describe the major industrial and household uses of water.
- discuss how water ownership and water conservation are important in determining future water availability.

Water is used for agriculture

As we have seen with so many other resources, the per capita daily use of water varies dramatically among the nations of the world. FIGURE 28.1 shows total daily per capita use of fresh water for a number of countries, which is known as the **water footprint** of a nation. This water use reflects the total water use by a country for agriculture, industry, and residences divided by the population of that country. This allows us to compare water use among nations. For example, a person living in the United States, Spain, or Canada uses about three times more water than a person living in Kenya or China.

Water footprint The total daily per capita use of fresh water.

As we have just noted, the largest use of water worldwide is for agriculture. During the last 50 years, as agricultural output has grown along with the human population, the amount of water used for irrigation throughout the world has more than doubled. Indeed, producing a metric ton of grain (1,000 kg, or 2,200 pounds) requires more than 1 million liters of water (264,000 gallons). Together, India, China, the United States, and Pakistan account for more than half the irrigated land in the world. In the United States, approximately one-third of all freshwater use is for irrigation. Raising livestock for meat also requires vast quantities of water. For example, producing 1 kg (2.2 pounds) of beef in the United States requires about 11 times more

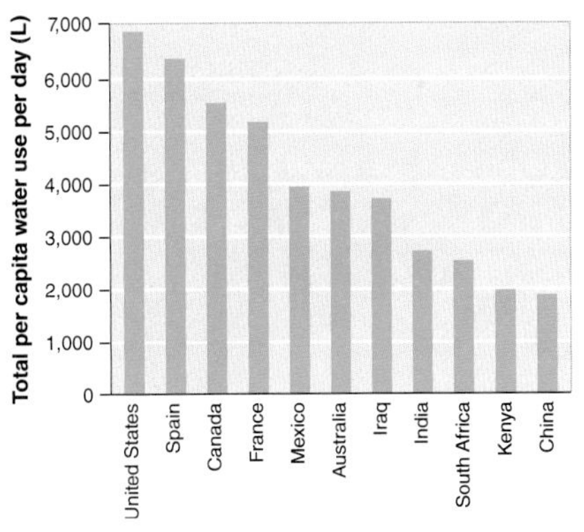

FIGURE 28.1 **Total per capita water use per day.** The total water use per person for agriculture, industry, and households varies tremendously by country.

water than producing 1 kg of wheat. In this section, we will consider some of the technological advances for irrigating crops that make efficient use of water.

Irrigation

Since agriculture is the greatest consumer of fresh water throughout the world, agriculture poses a large potential for conserving water. One way to conserve water in agriculture is by changing irrigation practices. There are four major techniques for irrigating crops: furrow irrigation, flood irrigation, spray irrigation, and drip irrigation (FIGURE 28.2).

The oldest technique is furrow irrigation (Figure 28.2a), which is easy and inexpensive. The farmer digs trenches, or furrows, along the crop rows and fills them with water, which seeps into the ground and provides moisture to plant roots. Furrow irrigation is about 65 percent efficient; 65 percent of the water is accessible to the plants and the other 35 percent either runs off the field or evaporates.

(a) Furrow irrigation

(b) Flood irrigation

(c) Spray irrigation

(d) Drip irrigation

FIGURE 28.2 **Irrigation techniques.** Several techniques are used for irrigating agricultural crops, each with its own set of costs and benefits. *(a: Jenny E. Ross/Corbis; b: Jeff Vanuga/NRCS/USDA; c: Peter Walton Photography/Getty Images; d: Lynn Betts/NRCS/USDA)*

◀ TEACHING with FIGURES

Ask students to examine Figure 28.1 and determine which country uses the most water each day on a per capita basis and which country uses the least. *The United States uses the most water per capita while China uses the least.*

Teaching Tip: Journal Prompt

Ask students to describe the different irrigation methods and comment on their efficiency.

- **Furrow irrigation:** *Trenches are dug along crop rows and filled with water that seeps into the ground and waters the roots of plants. Approximately 35 percent of this water runs off or is lost to evaporation.*
- **Flood irrigation:** *This involves flooding an entire field and letting water soak in evenly. About 20 percent of this water is lost to runoff or evaporation.*
- **Spray irrigation:** *Nozzles spray water across the field like giant lawn sprinklers. This is more efficient than furrow or flood irrigation, but it is more expensive and requires a fair amount of energy to pump water into the apparatus.*
- **Drip irrigation:** *This is the most efficient method of irrigation. Drip irrigation uses a hose that slowly drips water and that is either laid on the ground near the roots or buried beneath the soil.*

Teaching Tip: Warm-up

Ask students: In which applications do humans use the most water? *Humans use the most water for agriculture, industry, and household use.*

Teaching Tip: Warm-up

Ask students the following questions:

- What is hydroponic agriculture? *Hydroponic agriculture is the cultivation of crop plants under greenhouse conditions with roots immersed in a nutrient-rich solution rather than in soil.*
- What are the benefits associated with hydroponic agriculture? *Water that plants do not use can be recirculated. This method uses 95 percent less water than traditional methods of agriculture. Because crops grow under ideal conditions, production is higher than traditional agriculture. It uses little or no pesticides, and crops can be grown all year.*

Teaching Tip: Review

Ask students: Of all water used in the United States, how much goes toward generation of electricity? *Half of the water used in the United States goes toward the generation of electricity.*

Flood irrigation (Figure 28.2b) involves flooding an entire field with water and letting the water soak in evenly. This technique is generally more disruptive to plant growth than furrow irrigation, but is also slightly more efficient, ranging from 70 to 80 percent efficiency. In spray irrigation (Figure 28.2c), water is pumped from a well into an apparatus that contains a series of spray nozzles that spray water across the field, like giant lawn sprinklers. The advantage of spray irrigation is that it is 75 to 95 percent efficient, but it is also more expensive than furrow or flood irrigation and uses a fair amount of energy.

The most efficient method of irrigation, drip irrigation (Figure 28.2d), uses a slowly dripping hose that is either laid on the ground or buried beneath the soil. Drip irrigation using buried hoses is over 95 percent efficient. It has the added benefit of reducing weed growth because the surface soil remains dry, which discourages weed germination. Drip irrigation systems are particularly useful in fields containing perennial crops such as orchard trees, where the hoses do not have to be moved each year in order to plow the field.

Efficient irrigation technology benefits the environment by reducing both water consumption and the amount of energy needed to deliver the water. Many new technologies are being developed to more carefully control when plants are irrigated. As with all human activities, the costs and benefits of each irrigation technique need to be weighed to determine the best solution for each situation.

Hydroponic Agriculture

For some crops, *hydroponic agriculture* is an alternative to traditional irrigation. **Hydroponic agriculture** is the cultivation of crop plants under greenhouse conditions with their roots immersed in a nutrient-rich solution, but with no soil (FIGURE 28.3). Water not taken up by the plants can be reused, so this method uses up to 95 percent less water than traditional irrigation techniques. Hydroponic operations can produce more crops per hectare than traditional farms. They can also grow crops under ideal conditions, grow the crops during every season of the year, and often grow the crops with little or no use of pesticides. Hydroponic agriculture is growing in popularity, especially for vegetables; hydroponic tomatoes, for example, have won awards for their flavor. Although the cost of hydroponic agriculture is higher, consumers are willing to pay more for hydroponically grown vegetables and a

FIGURE 28.3 **Hydroponic agriculture.** These greenhouse tomatoes are growing with their roots immersed in a solution of water and nutrients. By recycling the water, hydroponic operations can use up to 95 percent less water than traditional farms. *(Kerry Sherck/Aurora Photos)*

number of businesses have been successful. One company in Georgia now grows hydroponic vegetables in a huge facility that covers 129 ha (318 acres).

Hydroponic agriculture The cultivation of plants in greenhouse conditions by immersing roots in a nutrient-rich solution.

Water is also used for industrial processes and households

While most water is used for agriculture, the remainder is used for industrial purposes and households.

Industrial Water Use

Water is required for many industrial processes, such as generating electricity, cooling machinery, and refining metals and paper. In the United States, approximately one-half of all water used goes toward generating electricity. It is important to distinguish between water that is withdrawn from and then returned to its source and water that is withdrawn and consumed. For example, water that passes through a turbine at a hydroelectric dam is withdrawn from a reservoir to generate electricity, but that water is not consumed. Rather, it passes from the reservoir through the turbine and back to the river flowing away from the dam. That is not to say that this process has no effect on the environment; recall our discussion of the ecological impacts of damming rivers earlier in this chapter.

Some processes that generate electricity do consume water. This means that some of the water used is not returned to the source from which it was removed, but instead enters the atmosphere as water

Teaching Tip: Discussion Starter

Ask students: What are some major industrial uses for water? *Generating electricity, cooling machinery, refining metals, and making paper.*

FIGURE 28.4 **Water consumption in nuclear power plants.** When nuclear reactors, such as this one in Germany, heat water to make the steam that turns electrical turbines, the steam must be cooled back down. During the cooling of the steam, a great deal of water vapor is released to the atmosphere. *(Hans F. Meier/iStockphoto.com)*

vapor. Thermoelectric power plants, including the many plants that generate heat using coal or nuclear reactors, are large consumers of water. These plants use heat to convert water into steam that is used to turn turbines. The steam needs to be cooled and condensed before it can be returned to its source. In many plants, this cooling is accomplished by using massive cooling towers. If you have ever seen a nuclear power plant, even from a distance, you may have seen the large plumes of water vapor rising up for thousands of meters from the cooling towers. The towers allow much of the steam from the plant to condense and cool into liquid water, but a large fraction of it is lost to the atmosphere. This water vapor represents the water that is consumed by nuclear reactors (FIGURE 28.4).

Industrial processes such as refining metals and making paper also require large amounts of water. Copper, used extensively for electrical wiring, requires 440 L (116 gallons) of water per kilogram to refine. Aluminum, used in products as diverse as automobiles and aluminum foil for cooking, requires 410 L (108 gallons) per kilogram. Steel, used to manufacture home appliances, cars, buildings, and other products, requires 260 L (68 gallons) per kilogram. When we use paper, we indirectly use water. A kilogram of paper requires 125 L (33 gallons) of water to manufacture. As we will see later in this chapter, there are opportunities to reduce the use of water during these processes.

Household Water Use

According to the U.S. Geological Survey, household use accounts for approximately 10 percent of all water used in the United States. The quantity of water used in households depends on the types of infrastructure available. Households in less developed countries generally do not have the appliances and bathroom fixtures that are common in more developed countries. As a result of such disparities, per capita household water use varies dramatically among nations. For example, on average, an individual in the United States uses 595 L (157 gallons) per day, whereas an average individual in Kenya uses only 41 L (11 gallons) per day. FIGURE 28.5 shows per capita daily household water use for the United States and 10 other countries.

Indoor use of water is quite similar across the United States since households across the country are typically equipped with bathrooms, washing machines, and cooking appliances. FIGURE 28.6 shows the fraction of indoor water use that goes to each of these functions. Of all household water that is used indoors, 41 percent is used for flushing toilets, 33 percent for bathing, 21 percent for laundry, and 5 percent for cooking and drinking.

Outdoor water use—for watering lawns, washing cars, and filling swimming pools—varies tremendously across the United States by region. In California, for example, the typical household uses about six times more water outdoors than does the typical Pennsylvania family.

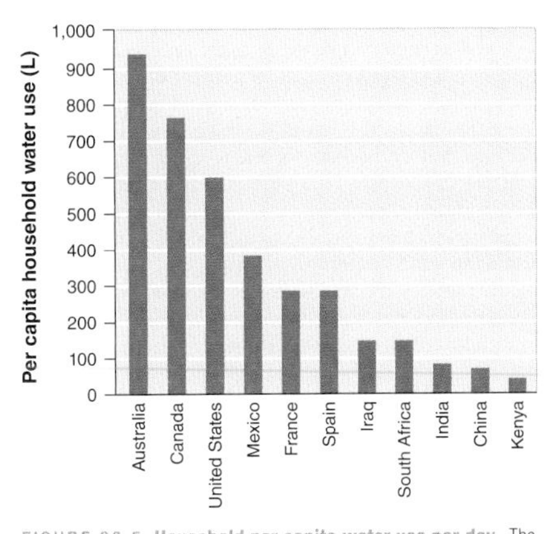

FIGURE 28.5 **Household per capita water use per day.** The amount of household water use per capita is different from that of total water use, shown in Figure 28.1. *(A. K. Chapagain and A. Y. Hoekstra, Water Footprints of Nations, Vol. 1, Main Report, UNESCO-IHE. Research Report Series #16, 2004.)*

Teaching Tip: Journal Prompt

Ask students to explain how water is lost when generating electricity in thermoelectric power plants. *Thermoelectric plants use heat to convert water into steam that turns a turbine, which generates the electricity. The steam must be cooled before it is returned to its source. Cooling takes place in massive cooling towers. These towers allow much of the steam to be condensed into water but a large fraction of it is lost to the atmosphere.*

◀ TEACHING with FIGURES

Have students examine Figure 28.5 and answer the following questions:

- Which three countries have the highest per capita daily use of household water? *Australia, Canada, and the United States consume the most water for household use on a daily per capita basis.*
- Which three countries consume the least amount of water for household use per day? *Kenya, China, and India consume the least water for household use on a daily per capita basis.*

TEACHING with FIGURES ▶

Show students Figure 28.6 and ask: Which three household activities have the greatest impact on water consumption in the United States? *Toilet flushing, bathing, and laundry have the greatest impact on water consumption in the United States.*

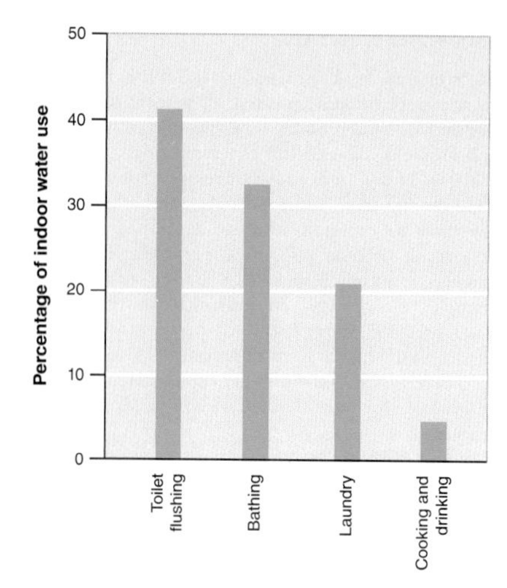

FIGURE 28.6 **Indoor household water use.** Most water used indoors is used in the bathroom. *(Data from U.S. Environmental Protection Agency, 2003, http://esa21.kennesaw.edu/activities/water-use-overview-epa.pdf)*

Although drinking water represents a relatively small percentage of household water use, it is particularly important. If you live in a developed country, you may not have given much thought to the availability and safety of drinking water. However, more than 1 billion people—nearly 15 percent of the world's population—lack access to clean drinking water. Every year, 1.8 million people die of diarrheal diseases related to contaminated water, and 90 percent of those people are children under 5 years of age. This means that 5,000 people in the world die each day in large part because they do not have access to clean water. As we will see in Chapter 14, developed countries typically have modern sanitation systems, stronger environmental laws, and the technologies to remove harmful contaminants and waterborne pathogens from public water supplies. Poor conditions in many developing countries do not provide safeguards that ensure the availability of clean drinking water.

The future availability of water depends on water ownership and water conservation

The future of water availability will depend on many things, including how we resolve issues of water ownership, how we improve water conservation, and—as world population grows—how we develop new water-saving technologies.

Water Ownership

Water is an essential resource, but who actually owns it? This is a rather complex question. In a particular area, such as the Klamath River region described at the beginning of the chapter, it is clear that multiple interest groups can claim a right to use and consume the water. However, having a right to use the water is not the same as owning it. For example, regional and national governments often set priorities for water distribution, but of course they have no control over whether or not a particular year will bring an abundance of rain and snow. In California, the state government promises specific amounts of water for cities, suburbs, farmers, and fish, but these promises can exceed the actual amount of water that is available in many years.

Throughout the world, the issues of water rights and ownership have created many conflicts. Earlier in this chapter we discussed India's plan to divert water from rivers that flow from the Himalayas into Bangladesh (FIGURE 28.7). Since the water originates in India, does India own the water? Does Bangladesh have any legitimate claim to some of this water that its people have relied on for millennia? In the water-poor Middle East, water rights have long contributed to political tensions. In both the 1967 Arab-Israeli War and the 1980s Iran-Iraq War, disputes over water use added to the conflict. Water experts predict that as populations in these arid regions continue to grow, conflicts over water will increase.

FIGURE 28.7 **A river in the Himalayas.** The government of India is proposing to divert water from some of the rivers flowing out of the Himalayas to provide more water for its citizens. However, rivers, such as this one flowing down from Mera Peak in Nepal, subsequently flow into Bangladesh, raising interesting questions regarding who owns the water. *(David Pickford/Robert Harding World Imagery/Alamy)*

Teaching Tip: Journal Prompt

Ask students the following questions:

- What are the factors that limit access to clean water in many developing nations? *Developing countries typically lack modern sanitation systems, strong environmental laws, and technologies to remove harmful contaminants and pathogens from public water supplies.*
- How does the lack of access to clean water affect people living in developing nations? *Many people suffer from diarrheal diseases and many die each year from dehydration.*

Teaching Tip: Journal Prompt

Ask students the following questions:

- Why is it more difficult to determine ownership for water than for many other resources? *Water moves through various locations, and the abundance of rainfall and future availability of water are unpredictable.*
- What is one possible method to resolve water-ownership conflicts? *One solution proposed by economists is to allow all interested parties to compete for water and let market forces determine its price.*

Teaching Tip: Warm-up

Ask students to identify the factors that will determine water availability in the future. *Water availability will be determined by how we resolve issues of water ownership, how we improve water conservation, and how we develop new water-saving technologies.*

One solution that has been proposed by economists is to allow all interested parties to openly compete for water and let market forces determine its price. In this way, they argue, water could be owned, but the true value of water would be realized and paid for. In 1981, a free-market system of water distribution was initiated in Chile. While water is public property, Chile allows individuals and corporations to buy and sell the water. The idea behind this approach is that having to buy water will encourage more efficient use. While market forces can be useful in determining the appropriate distribution of water among competing needs, government oversight helps to ensure that the needs of the people and the environment are balanced with the needs of agriculture and industry. As we will see in the Science Applied section that follows this chapter ("Is There a Way to Resolve the California Water Wars?"), competing interests are now attempting to implement this more-balanced approach.

Water Conservation

Ultimately, there is a finite amount of water that we all must share. In some regions, such as much of the northeastern United States, water is abundant. In regions where water is scarce, such as the southwestern United States, water conservation becomes more of a necessity.

In recent years, many countries have begun to find ways to use water more efficiently through technological improvements in water fixtures, faucets, and washing machines. In 1994, new federal standards were issued for toilets and showerheads in the United States. For example, a toilet manufactured before 1994 typically uses 27 L (7 gallons) per flush, but toilets manufactured after January 1994 must use 6 L (1.6 gallons) or less per flush, representing a 78 percent reduction in water use. Australia and some countries in Europe and Asia have moved to dual-flush toilets. First invented by an Australian company in 1980, this type of toilet allows the user to push one button for a normal 6 L flush to remove solid waste, but another button produces a much more efficient 3 L (0.8 gallon) flush to remove liquid waste. These dual-flush toilets are increasingly used in the United States. Consumers also have embraced improved efficiencies in showerheads and washing machines. For example, a 10-minute shower with an older showerhead might use 150 L (40 gallons) of water, but revised federal standards for new reduced-flow showerheads call for a 10-minute shower to use no more than 95 L (25 gallons)—a 37 percent reduction in water use. Washing machines are not subject to the new federal standards, but newer, more efficient front-loading machines are now available to consumers, although these more efficient models cost nearly twice as much to purchase. "Do the Math: Selecting the Best Washing Machine" looks at the costs and benefits of these newer, more energy-efficient washing machines.

In countries where a percentage of the population can be considered wealthy, household water uses can include watering lawns and filling swimming pools, both of which require large amounts of water. In some regions of the United States, homeowners have been encouraged, or even required, to plant vegetation that is appropriate to the local habitat. For example, the city of Las Vegas, Nevada, paid homeowners to remove water-intensive turf grass from their lawns and replace it with more water-efficient native landscaping. Changing the vegetation in this manner can result in a savings of 2,000 L of water per square meter (520 gallons per 10 square feet) of lawn per year (FIGURE 28.8).

One of the best ways to reduce industrial water consumption is by producing more efficient manufacturing equipment. In the United States, businesses and

FIGURE 28.8 **Landscaping in the desert.** By using plants that are adapted to a desert environment, homeowners in Arizona can greatly reduce the need for irrigation, resulting in a considerable reduction in water use compared with that required for growing grass. *(Jan Paul Schrage/iStockphoto)*

Teaching Tip: Debate the Issue

Competing for Water

Have students read Science Applied 4, "Is There a Way to Resolve the California Water Wars?", on page 324. In this article students will learn about the historical events that led to California's water wars, the impact on California farmers if the government should raise water prices, and the competing interests involved in this conflict. Use this article as a starting point for students to debate the question: "Would water supplies be managed better if people were allowed to compete for water and market forces determined the price?"

Students should be prepared to discuss these topics in their debate:

- water availability for agricultural needs, manufacturing, and households
- pros and cons of using free-market forces to determine water value
- needs of the environment
- social justice concerns
- consideration of continuous water flow through various states and nations
- a plan to resolve conflict between interest groups

In class, select groups to debate the issue. Allow each side 2 minutes to prepare, 3 minutes to present, and 1 minute for rebuttal. Have the remainder of the class take notes, ask questions, and decide which side has the most persuasive argument.

Teaching Tip: Journal Prompt

Ask students to list some ways in which people conserve water.

- *Technological improvements have led to water fixtures, toilets, washing machines, and faucets that use substantially less water.*
- *Homeowners in low-water areas replace plants that use a lot of water, such as grass, with those that do not.*
- *Newer manufacturing equipment is more efficient and reduces industrial water consumption.*
- *People collect rainwater.*
- *Wastewater may be used for irrigation after completing the water treatment process.*

Teaching Tip: Beyond the Classroom

"The Wealth of Water" is a case study prepared by the National Center for Case Study Teaching in Science (NCCSTS). This case study challenges students to understand limited water resources, appreciate access to clean water, consider opposing views on water privatization, and analyze current data related to water consumption.

The link to this resource may be found by clicking on the link buttons in the TE-book, opening the TRFD, or logging onto the book's companion website bcs.whfreeman.com/friedlandapes2e (teacher login required).

TRM **Chapter 9 Web Resources**

do the **math**

Answer to Your Turn

Annual savings: 30,000 L/year × $1.00/1,000 L = $30/year

Time to pay for: ($900 − $300) ÷ $30/year = 20 years

do the math

Selecting the Best Washing Machine

Suppose you move into a new house and need to purchase a washing machine. You've heard that the new front-loading machines use less water, but that they are more expensive. A traditional washing machine costs $300, whereas the water-efficient washing machine costs $900. So how do you choose?

If the average family washes 300 loads of laundry per year, and traditional washing machines use 100 L more water per load than water-efficient machines, how many liters would an efficient washing machine save per year?

300 loads/year × 100 L/load = 30,000 L/year

If the average cost of water in the United States is $0.50 for every 1,000 L, how much money would you save annually by using the more efficient washing machine?

30,000 L/year × $0.50/1,000 L = $15/year

Given your calculated annual savings, how many years would it take for your water savings to pay for the more expensive, high-efficiency washer?

($900 − $300) ÷ $15/year = 40 years

Your Turn If the average cost of water in the United States doubled to $1.00 for every 1,000 L, how much money would you save annually by using the more efficient washing machine and how many years would it take to pay for the more expensive, high-efficiency washer?

factories have achieved more sustainable water use in the last 15 years, mainly through the introduction of equipment that either uses less water or reuses water. For example, industries that need water for cooling machinery have switched from once-through water systems—systems that bring in water for cooling and then pump the heated water back into the environment—to recirculating water systems.

Some simple ways to conserve water can be used throughout the world. For example, the impervious surfaces of buildings represent a potential water-collecting surface. A gutter system can collect rainwater and channel it into rain barrels or, for greater capacity, a large underground water tank. As we will see in Chapter 14, some countries are now using wastewater for irrigation after sending it through a sewage treatment process. There is also great interest in developing monitoring technologies that will detect leaks in the pipes that distribute water over large areas.

The world's growing population and the associated expansion of irrigated agriculture have increased global water withdrawals more than fivefold in the last hundred years. Global water use is expected to continue to grow along with the human population through the early part of this century. However, as FIGURE 28.9 shows, despite the growing population in the United States, water withdrawals have leveled

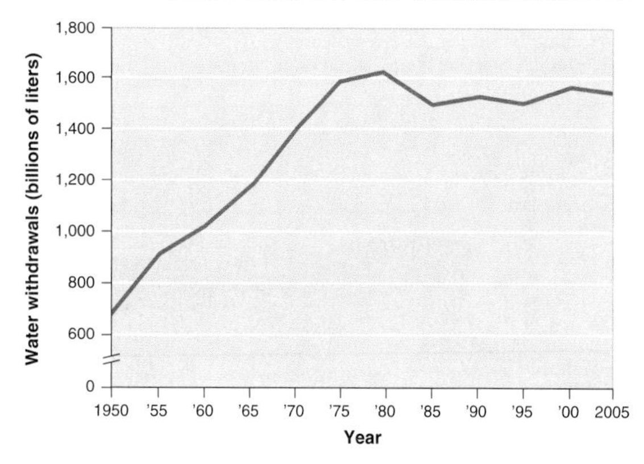

FIGURE 28.9 **Water withdrawals in the United States from 1950 to 2005.** Although the U.S. population continues to increase, the country's water use has leveled off due to the use of water-efficient technologies. *(After U.S. Geological Survey, 2009)*

more **math** practice: Calculating Energy Use

The Draper family wanted to find ways to live more sustainably. They analyzed their water and energy usage. Each person in the family showers twice a day with an average of 6 minutes per shower. The shower has a flow rate of 5.0 gallons per minute. Their standard hot water heater raises the water temperature to 130° F, which requires 0.2 kWh per gallon at a cost of $0.10/kWh.

Calculate the total amount of water the Draper family uses for showering per year.

(6 min/shower) × (5.0 gallons/minute) × (2 showers per day per person) × (4 people) × (365 days/year) = 87,600 gallons/year used for showers

Calculate the annual cost of the electricity needed to heat the water the Draper family uses for showers. Assume that 2.5 gallons per minute of hot water is being used.

(Hint: The other 2.5 gallons of water used in the 5 gallons/minute is cold water.)

First, find the total amount of hot water used in gallons. 87,600 gallons/year ÷ 2 = 43,800 gallons of hot water per year

Then, find the cost per year to heat 43,800 gallons. (43,800 gallons/year) × (0.20 kWh/gallon) × ($0.10/kWh) = $876 per year

▲ TEACHING with FIGURES

Show students Figure 28.9 and ask: What is the overall trend for water usage in the United States? *Water usage peaked in 1980 and then declined slightly in 1985. Since that time, water usage has leveled off.*

off since they peaked in 1980. This is largely a result of greater efficiency in the use of water for agricultural irrigation, electricity generation, and household appliances. Reductions in water use are projected to continue until at least 2020. Of course, the continued development of more efficient technologies would allow additional reductions in water use.

module

28

REVIEW

In this module, we learned that about 70 percent of water that is consumed goes to agriculture, 20 percent goes to industry, and 10 percent goes to households. Much of the agricultural use of water is for irrigation, which includes low-efficiency furrow and flood irrigation, moderate-efficiency spray irrigation, and high-efficiency drip irrigation. An alternative method of growing crops with highly efficient water use is hydroponic agriculture. Nearly half of the water used in industry is for generating electricity. Some of this water is returned to its source and some is released to the atmosphere as water vapor. Industries that manufacture paper and refine metals also use large amounts of water. Among households, those in developed countries often have more water-consuming appliances, swimming pools, and irrigated lawns, and therefore use more water per person than households in developing countries. The future availability of water depends on countries reaching agreements on water ownership and taking steps to reduce the amount of water used in agriculture, industry, and households.

Module 28 AP® Review Questions

1. In the United States, most water is used for
   (a) agriculture.
   (b) households.
   (c) industry.
   (d) transportation.
   (e) hydroelectric plants.

2. Which irrigation method is the most efficient?
   (a) Furrow irrigation
   (b) Spray irrigation
   (c) Flood irrigation
   (d) Drip irrigation
   (e) Condensate irrigation

3. A water-efficient washer costs $400 more than a regular washer and saves 100 L of water for each load. If water costs $5 per 1,000 L and you wash 100 loads each year, how many years will it take you to recoup the extra cost of the machine?
   (a) 10
   (b) 5
   (c) 4
   (d) 8
   (e) 1

4. The primary use of industrial water in the United States is
   (a) mining.
   (b) paper products.
   (c) electricity generation.
   (d) construction.
   (e) cooling and heating.

5. The issue of water ownership is complicated by
   I. the irregularity of precipitation.
   II. the distance rivers can travel.
   III. increased demand for water.

   (a) I and II
   (b) I and III
   (c) II and III
   (d) III only
   (e) I, II, and III

Answers to Module 28 AP® Review Questions

1. a
2. d
3. d
4. c
5. e

Teaching Tip: Beyond the Classroom

After students read "Working Toward Sustainability: Is the Water in Your Toilet Too Clean?" have them brainstorm ways to educate others and to develop meaningful conservation practices within their communities. For example, students might present strategies for conserving water at home to younger students at a local elementary school. Students could also design and build a project to collect, store, and recycle water. They might consider creating a garden that uses drip irrigation.

Suggested Answers to Critical Thinking Questions

1. While bacteria from gray water used to prepare food could cause disease, gray water can safely be used to irrigate lawns and trees. State governments can follow the lead of Arizona lawmakers and create guidelines for safe usage.

2. Answers will vary. Students might suggest highlighting the potential cost savings from reduced water use as well as the reduced load on the town's wastewater treatment system. In addition, they might note the low risk of disease or contamination associated with gray water use as long as it is properly handled.

working toward  **sustainability**

Is the Water in Your Toilet Too Clean?

In certain parts of the world, such as the United States, sanitation regulations impose such high standards on household wastewater that we classify relatively clean water from bathtubs and washing machines as contaminated. This water must then be treated as sewage. On the other hand, we use clean, drinkable water to flush our toilets and water our lawns. Can we combine these two observations to come up with a way to save water? One idea that is gaining popularity throughout the developed world is to reuse some of the relatively clean water from bathtubs and washing machines that we normally discard as sewage.

This idea has led creative homeowners and plumbers to distinguish between two categories of wastewater in the home: *gray water* and *contaminated water*. **Gray water** is defined as the wastewater from baths, showers, bathroom sinks, and washing machines. Although no one would want to drink it, gray water is perfectly suitable for watering lawns, washing cars, and flushing toilets. In contrast, **contaminated water** is defined as the wastewater from toilets, kitchen sinks, and dishwashers; such water contains a good deal of waste and contaminants and should therefore be disposed of as sewage.

Around the world, there is a growing number of commercial and homemade systems used for storing gray water to flush toilets and water lawns or gardens. For example, a Turkish inventor has designed a household system allowing the homeowner to pipe wastewater from the washing machine to a storage tank that dispenses this gray water into the toilet bowl with each flush.

Many cities in Australia have considered the use of gray water as a way to reduce both the consumption of fresh water and the volume of contaminated water that is sent to sewage treatment plants. The city of Sydney estimates that 70 percent of the water used in the greater metropolitan area is used in households, and that perhaps 60 percent of that water becomes gray water. The Sydney Water Corporation, provider of water services to metropolitan Sydney, estimates that the use of gray water for outdoor purposes could save up to 50,000 L (13,000 gallons) per household per year.

Unfortunately, many local and state regulations in the United States and around the world do not allow households to use gray water. Some localities allow the use of gray water only if it is treated, filtered, or delivered to lawns and gardens through underground drip irrigation systems to avoid potential bacterial contamination. Arizona, a state in the arid Southwest, has some of the least restrictive regulations. As long as a number of guidelines are followed, homeowners are permitted to reuse gray water. In 2010, in the face of a severe water shortage, California reversed earlier restrictions on gray water use and agreed to allow gray water to be used for irrigating lawns and trees. Given that the typical household in the United States produces 227,000 L (60,000 gallons) of gray water annually, using gray water for irrigation presents a major opportunity for water conservation.

Critical Thinking Questions

1. Given that gray water can contain more bacteria than clean water, how should we balance a desire to irrigate with clean water against the need to conserve limited water supplies?
2. After reading the two references listed below, describe how you could formulate an argument to convince a town council to permit the use of gray water for watering lawns.

References

California Gray Water Guide. http://www.thegreywaterguide.com/california.html

Grey Water Action. http://greywateraction.org/content/about-greywater-reuse

Reusing gray water. A Turkish inventor has designed a washing machine that pipes the relatively clean water left over after washing a load of laundry, termed gray water, to a toilet, where it can be reused for flushing. Such technologies can reduce the amount of drinkable water used and the volume of water going into sewage treatment plants. *(Sevin Coskun)*

Gray water Wastewater from baths, showers, bathrooms, and washing machines.

Contaminated water Wastewater from toilets, kitchen sinks, and dishwashers.

chapter 9

REVIEW

In this chapter, we have learned that fresh water is a critical, scarce resource. The major sources of fresh water are groundwater, surface water, and atmospheric water. When groundwater is extracted faster than it is replaced, we can experience dry water wells and saltwater intrusion. Humans alter the availability of water by constructing levees, dikes, and dams to constrain water movement. We transport water from one place to another using aqueducts and we are able to convert salt water into fresh water through distillation or reverse osmosis. The use of water among countries can be determined by calculating the water footprint of a country; developed countries typically have a larger water footprint than developing countries. Most water is used for agriculture, followed by industrial and household use. Recent changes in each of these sectors have improved water use and reuse efficiency and conservation.

Key Terms

Aquifer
Unconfined aquifer
Confined aquifer
Water table
Groundwater recharge
Spring
Artesian well
Cone of depression
Saltwater intrusion
Floodplain
Impermeable surface
Levee
Dike
Dam
Reservoir
Fish ladder
Aqueduct
Desalination
Desalinization
Distillation
Reverse osmosis
Water footprint
Hydroponic agriculture
Gray water
Contaminated water

Learning Objectives Revisited

Module 26 The Availability of water

- **Describe the major sources of groundwater.**
  Groundwater can exist as either confined or unconfined aquifers. Unconfined aquifers can be rapidly recharged, but they are susceptible to contamination from chemicals released by human activities. Confined aquifers are typically recharged slowly and are less likely to be contaminated by anthropogenic chemicals. Large withdrawals of water from an aquifer can cause cones of depression that result in wells going dry. In coastal areas, large withdrawals for fresh water can cause saltwater intrusion.

- **Identify some of the largest sources of fresh surface water.**
  Sources of fresh surface water include streams, rivers, wetlands, and lakes. Streams and rivers have traditionally overflowed their banks and deposited nutrient-rich sediments along floodplains. Rivers have also been an important resource for transportation. Lakes are formed through a variety of processes including tectonic activity and glaciation.

- **Explain the effects of unusually high and low amounts of precipitation.**
  The abundance of precipitation varies with climate around the world and this affects the availability of water for ecosystems and humans. Periods of drought can interact with human activities to alter the cycling of nutrients, cause the topsoil to be blown away, and cause the soil surface to become impermeable to future precipitation.

Module 27 Human Alteration of Water Availability

- **Compare and contrast the roles of levees and dikes.**
  Levees and dikes both use barriers to prevent flood water, but levees are built along rivers to prevent rising rivers from spilling over onto the floodplain whereas dikes are built near the ocean to prevent ocean waters from flooding the adjacent land.

Teaching Tip: Measuring Your Impact

Saving Water
Measuring Your Impact 9 asks students to calculate the impact of installing low-flow shower heads and toilets that use less water. Students will not only consider the amount of water conserved but also consider the economic benefits of conserving water.

The link to this resource may be found by clicking on the link buttons in the TE-book, opening the TRFD, or logging onto the book's companion website bcs.whfreeman.com/friedlandapes2e (teacher login required).

Measuring Your Impact 9: Saving Water

Exploring the Literature

Ghaitidak, D. M., and K. D. Yadav. 2013. Characteristics and treatment of greywater—A review. *Environmental Science and Pollution Research* 20:2795–2809.

Gleick, P. H. 2009. *The World's Water 2008–2009.* Island Press.

Kemp, P.S., and J. R. O'Hanley. 2010. Procedures for evaluating and prioritizing the removal of fish passage barriers: A synthesis. *Fisheries Ecology and Management* 17:297–322.

Postel, S., and B. Richter. 2004. *Rivers for Life: Managing Water for People and Nature.* Island Press.

Water Conservation Tips. *National Geographic.* http://environment.nationalgeographic.com/environment/freshwater/water-conservation-tips/

Download a printable version of this list by clicking on the link buttons in the TE-book, opening the TRFD, or logging onto the book's companion website bcs.whfreeman.com/friedlandapes2e (teacher login required).

Exploring the Literature 9

Answers to Chapter 9 AP® Exam

Download the full answers to the Chapter 9 AP® Practice Exam by clicking on the link buttons in the TE-book, opening the TRFD, or logging onto the book's companion website bcs.whfreeman.com/friedlandapes2e (teacher login required).

Answers to Chapter 9 AP® Exam

Answers to Chapter 9 AP® Exam Multiple-Choice Questions

1. a
2. b
3. c
4. e
5. a
6. a
7. b
8. e
9. c
10. a
11. d

- **Explain the benefits and costs of building dams.**
  Dams can provide multiple benefits including electric generation, flood control, and recreation. However, dams require a large amount of resources to construct, they often displace people, and they have ecological effects including interference with fish migration and prevention of seasonal flooding that some species require to persist.
- **Explain the benefits and costs of building aqueducts.**
  Aqueducts are effective at moving water from places that have water to places that need water. Aqueducts built as open canals can lose a large fraction of water through leaks and evaporation. Newer aqueducts are commonly built as pipelines to reduce water loss, but they still require a large amount of resources to construct and their construction can have negative impacts on the habitat during the construction process.
- **Describe the processes used to convert salt water into fresh water.**
  The process of converting salt water into fresh water is known as desalination or desalinization. It can be accomplished either through distillation, which boils water and then condenses the salt-free water vapor, or through reverse osmosis, which pushes salt water against a membrane through which water can pass but salt cannot.

Module 28 Water Use by Humans Now and in the Future

- **Compare and contrast the four methods of agricultural irrigation.**
  Furrow irrigation involves flooding the furrows between plant rows; it is the easiest and least expensive method, but it also is the least efficient. Flood irrigation involves flooding an entire field with water; it is only slightly more efficient than furrow irrigation. Spray irrigation involves distributing water through spray nozzles that resemble giant lawn sprinklers; it can be 75 to 95 percent efficient. Drip irrigation involves the use of a hose that slowly drips water on or below the ground surface; its efficiency is over 95 percent.
- **Describe the major industrial and household uses of water.**
  About half of the water used for industry is used for generating electricity. Of the remainder, major users include refiners of paper and metal. Of all indoor water use in U.S. households, we use 41 percent for flushing toilets, 33 percent for bathing, 21 percent for laundry, and 5 percent for cooking and drinking. Outdoor water use varies tremendously by region as a result of differences in climate and in the use of water for swimming pools and watering lawns.
- **Discuss how water ownership and water conservation are important in determining future water availability.**
  Water ownership is a complicated issue because multiple groups can have a right to use the water but it is often unclear who actually owns the water. Among nations, water rights are often determined through treaties. In some regions of the world, everyone competes for water by purchasing it. The demand from different users helps to determine the price while government oversight can help ensure that there is a proper balance between the needs of industry, agriculture, households, and the environment.

Chapter 9 AP® Environmental Science Practice Exam

Section 1: Multiple-Choice Questions

Choose the best answer for questions 1–11.

1. What percentage of Earth's water is fresh water?
   (a) 3 percent
   (b) 10 percent
   (c) 50 percent
   (d) 90 percent
   (e) 97 percent

2. Which of the following contrasts between confined and unconfined aquifers is correct?
   (a) Confined aquifers are more rapidly recharged.
   (b) Only confined aquifers can produce artesian wells.
   (c) Only unconfined aquifers are overlain by a layer of impermeable rock.
   (d) Only unconfined aquifers have a water table above them.
   (e) Only unconfined aquifers can be drilled for wells to extract water.

Additional Multiple-Choice AP® Practice Exam

Download an additional Multiple-Choice AP® Environmental Science Practice Exam for Chapter 9 by clicking on the link buttons in the TE-book, opening the TRFD, or logging onto the book's companion website bcs.whfreeman.com/friedlandapes2e (teacher login required).

Multiple-Choice AP® Practice Exam 9

3. Which of the following statements about surface waters is NOT correct?
   (a) Historically, most rivers regularly spilled over their banks.
   (b) The largest river in the world is the Amazon River.
   (c) Levees are used to make reservoirs.
   (d) Dikes are human-made structures that keep ocean water from moving inland.
   (e) Wetlands play an important role in reducing the likelihood of flooding.

4. Which of the following statements about dams is NOT correct?
   (a) Dams are used to reduce the risk of flooding.
   (b) Dams can cause increased water temperatures.
   (c) The water held back by a dam is called a reservoir.
   (d) Fish ladders allow migrating fish to move past dams.
   (e) Most dams are built to generate electricity.

5. Which statement about aqueducts is correct?
   (a) Aqueducts designed as open canals can lose a lot of water through evaporation.
   (b) Aqueducts are a modern invention.
   (c) Aqueducts do not divert water from lakes.
   (d) Aqueducts do not affect the amount of water remaining in rivers.
   (e) Aqueducts move water from locations where the demand for water is high.

6. Which of the following statements about desalination is correct?
   (a) Distillation requires more energy than reverse osmosis.
   (b) The brine left over from desalination is not harmful when returned to the ocean.
   (c) Large-scale desalination of water is affordable to all nations.
   (d) Desalination of ocean water is not yet a feasible endeavor.
   (e) Most desalination occurs in North America.

7. Which of the following lists of household water uses is in the correct order, from highest to lowest?
   (a) Toilet, laundry, cooking and drinking, bathing
   (b) Toilet, bathing, laundry, cooking and drinking
   (c) Laundry, bathing, toilet, cooking and drinking
   (d) Bathing, toilet, laundry, cooking and drinking
   (e) Bathing, toilet, cooking and drinking, laundry

8. Which of the following statements about the industrial use of water is NOT correct?
   (a) It is used to refine metals.
   (b) It is used to create steam.
   (c) It is important in generating electricity.
   (d) It plays a role in making paper products.
   (e) Its use is becoming less efficient.

9. Which of the following lists of agricultural irrigation techniques is in the correct order, from least efficient to most efficient?
   (a) Drip irrigation, furrow irrigation, flood irrigation, spray irrigation
   (b) Spray irrigation, furrow irrigation, flood irrigation, drip irrigation
   (c) Furrow irrigation, flood irrigation, spray irrigation, drip irrigation
   (d) Furrow irrigation, spray irrigation, drip irrigation, flood irrigation
   (e) Furrow irrigation, flood irrigation, drip irrigation, spray irrigation

10. Which of the following is NOT a water conservation technique?
   (a) Flood irrigation
   (b) Reduced-flow showerheads
   (c) Recycling of industrial water
   (d) Dual-flush toilets
   (e) Front-loading washing machines

11. Drought conditions
   (a) can result in increased nutrient cycling.
   (b) decrease the severity of flooding.
   (c) are rarely affected by human activities.
   (d) cause increased erosion.
   (e) are most common in northern latitudes.

Section 2: Free-Response Questions

Write your answer to each part clearly. Support your answers with relevant information and examples. Where calculations are required, show your work.

1. An important source of fresh water is groundwater. Answer the following questions about human use of groundwater.
   (a) Using a cross section of the ground, draw the water table, an unconfined aquifer, and a confined aquifer. (4 points)
   (b) Explain the factors that create an artesian well. (2 points)
   (c) Describe the concerns that scientists have regarding the pumping of water from confined aquifers. (2 points)
   (d) Explain how cones of depression near the coasts of continents can lead to saltwater intrusion of water wells. (2 points)

2. Answer the following questions about the world's approximately 845,000 dams.
   (a) What are the benefits that dams provide to humans? (4 points)
   (b) What are the costs of dams to human society? (2 points)
   (c) What are the negative impacts of dams on the environment? (2 points)
   (d) If dams have both costs and benefits, how should we decide when and where a dam should be built? (2 points)

Answers to Chapter 9 AP® Exam Free-Response Questions

1. (a) See Figure 26.2 .
(b) Groundwater collects when precipitation moves through cracks in the Earth (see Figure 26.2) until it can go no further. That is, the water hits some impervious surface and stops. A confined aquifer is completely surrounded by impervious surfaces, so it is trapped underground at high pressures. If a shaft is drilled into the confined aquifer, or occurs naturally, this pressure is released, the water rushes out, and an artesian well is formed.
(c) Water residing in a confined aquifer is recharged very slowly, so it is quite easy to withdraw water at a faster rate than it can be replenished. This has been observed in the Ogallala aquifer located in the Great Plains. In many places within that aquifer, it is no longer economically viable to use the remaining water for irrigation because it costs too much to drill a well deep enough to reach it.
(d) When the water in a well is pumped faster than the groundwater can be recharged, a cone of depression is formed in the water table. In coastal areas, lowering the water table by overpumping wells reduces the pressure in the aquifer, allowing the adjacent saltwater to move into the aquifer and contaminate the water table with salt water.

2. (a) Dams serve a wide variety of purposes, including holding water for consumption, generating power using a renewable resource, controlling floodwaters, and recreation.
(b) There are substantial financial costs incurred in building a dam. There are also costs to communities, because the construction of dams often requires the relocation of communities where the reservoir will be located.
(c) Dams interrupt migration routes for fish. Dams interrupt the natural flow of rivers and streams, including seasonal flooding that brings nutrients to terrestrial ecosystems and natural scouring that creates favorable habitats for many species.
(d) This question should be analyzed on a case-by-case basis. For example, the Chinese have decided that their energy and flood control needs along the Yangtze River demand the construction of a dam. Over time, however, these needs may change and the balance between benefits and costs may change. Changing assessments of costs and benefits over time is one of the reasons why some dams in the United States are now being removed.

Answers to Unit 4 AP® Exam Multiple-Choice Questions

1. d
2. a
3. e
4. c
5. b
6. c
7. a
8. a
9. c
10. e
11. a
12. b
13. a
14. e
15. d
16. b
17. e
18. c
19. c
20. c

Answers to Unit 4 AP® Exam

Download the full answers to the Unit 4 AP® Exam by clicking on the link buttons in the TE-book, opening the TRFD, or logging onto the book's companion website bcs.whfreeman.com /friedlandapes2e (teacher login required).

Answers to Unit 4 AP® Exam

Unit 4 AP® Environmental Science Practice Exam

Section 1: Multiple-Choice Questions

Choose the best answer for questions 1–20.

1. Which of the following is the order of layers in Earth's structure, from innermost to outermost?
   (a) Core, asthenosphere, lithosphere, mantle
   (b) Core, lithosphere, asthenosphere, mantle
   (c) Core, asthenosphere, mantle, lithosphere
   (d) Core, mantle, asthenosphere, lithosphere
   (e) Core, mantle, lithosphere, asthenosphere

2. Which of the following does NOT provide evidence for the theory of plate tectonics?
   (a) Plumes of magma from the mantle that reach the crust
   (b) Similar rock formations on both sides of the Atlantic Ocean
   (c) Discovery of similar fossils in Nigeria and Brazil
   (d) Fault zones that extend for hundreds of miles across a landscape
   (e) Mountain ranges along the northern boundaries of India

3. An earthquake occurs as a direct result of
   (a) volcanic eruption.
   (b) conversion of kinetic energy to tectonic plate movement.
   (c) subduction.
   (d) magma convection in the mantle.
   (e) potential energy release along a fault.

4. The dominant type of rock in the Hawaiian Islands is most likely
   (a) metamorphic.
   (b) sedimentary.
   (c) basaltic.
   (d) granitic.
   (e) intrusive igneous.

Question 5 refers to the following diagram:

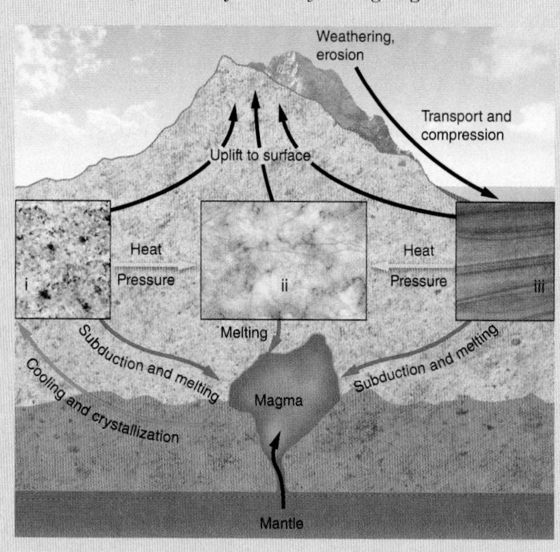

5. The types of rock represented by i, ii, and iii are
   (a) igneous, sedimentary, and metamorphic.
   (b) granite, marble, and limestone.
   (c) sedimentary, marble, and limestone.
   (d) basalt, marble, and quartz.
   (e) metamorphic, igneous, and sedimentary.

6. Which of the following statements are true?
   I. Physical weathering of rocks can increase rates of chemical weathering.
   II. High rates of physical weathering can lead to higher primary production.
   III. Weathering by acid precipitation occurs faster than physical weathering.
   (a) I only
   (b) II only
   (c) I and II
   (d) I and III
   (e) I, II, and III

Teaching Tip: Unit 4 Additional Free-Response Question

An additional practice free-response question covering concepts from Unit 4 is available. The question presents a short discussion of invasive earthworms and leaf litter. To answer this question, students must

- discuss the soil horizons affected by the activity of the invasive earthworm
- suggest how the earthworms might alter the cation exchange capacity of forest soils
- describe how earthworms are likely to affect the movement of groundwater and nutrients through the soil
- explain why earthworms might be beneficial in agricultural fields where soil is compacted and plant density is high, but detrimental in forests were soil is relatively loose and plant density is low
- hypothesize why earthworms are altering cycles in their invaded environment

Download this resource by clicking on the link buttons in the TE-book, opening the TRFD, or logging on to the book's companion website bcs.whfreeman.com/friedlandapes2e (teacher login required).

Unit 4 Additional Free-Response Question

7. Which of the following areas is likely to have soil with the greatest accumulation of minerals available for plant growth?
   (a) The bottom of an old lake bed
   (b) A grassland in Oklahoma
   (c) The top of Mount Whitney in the Colorado Rockies
   (d) A boreal peatland
   (e) The eastern slopes of the northern Colorado Rockies

8. The ________ soil horizon is the most dominant feature of boreal forests, whereas the ________ soil horizon is the most dominant feature in agricultural lands.
   (a) O/A
   (b) O/E
   (c) E/A
   (d) B/C
   (e) B/A

Question 9 refers to the following diagram:

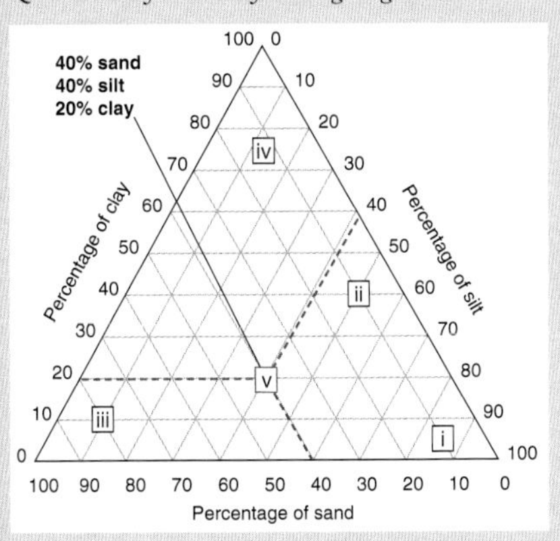

9. Arrange the labels i, ii, iii, iv, and v in order from the soil that holds the most amount of water to the soil that holds the least amount of water.
   (a) v, ii, iv, i, iii
   (b) iv, iii, i, v, ii
   (c) iv, v, i, ii, iii
   (d) ii, iii, v, i, iv
   (e) iv, iii, i, ii, v

10. Which of the following best describes the differences between the cation exchange capacity (CEC) and base saturation of a soil?
   (a) CEC is the proportion of nutrients to clay particles; base saturation is the proportion of soil bases to soil acids.
   (b) CEC is the proportion of nutrients to clay particles; base saturation is the proportion of soil acids to bases.
   (c) CEC is an index of potential plant growth; base saturation is the proportion of beneficial to detrimental soil minerals that are accessible to plants.
   (d) CEC is the ability of a soil to retain beneficial nutrients for plant production; base saturation is the proportion of beneficial minerals to detrimental minerals in soil.
   (e) CEC is the ability of a soil to adsorb and release nutrients; base saturation is the proportion of soil bases to soil acids.

11. Several types of minerals are mined from locations around subduction zones. As subduction occurs, hot magma is pushed upward into the subsurface. The heat of magma causes groundwater to rise and cool. As the groundwater moves, it extracts minerals from the subsurface and deposits them near the surface in veins that flow in all directions. The most efficient way of mining for these minerals is
   (a) open-pit mining.
   (b) mountaintop removal.
   (c) placer mining.
   (d) subsurface mining.
   (e) strip mining.

12. The Surface Mining Control and Reclamation Act of 1977
   (a) protects workers from the hazards of subsurface and surface mining.
   (b) mandates that the land be minimally disturbed during the mining process.
   (c) requires that mined land be fully restored to its original condition.
   (d) mandates the type of minerals than can be safely mined.
   (e) regulates all mining practices than can impact the environment.

13. A cone of depression in an unconfined aquifer can be caused by
   I. excessive pumping of water from a well.
   II. low rainfall.
   III. saltwater intrusion.

   (a) I only
   (b) II only
   (c) III only
   (d) I and II
   (e) II and III

14. Which of the following is NOT a likely consequence of severe drought?
    (a) Reduced rates of nutrient cycling
    (b) Loss of topsoil
    (c) Severe erosion of soil during subsequent rains
    (d) Starvation
    (e) Slower recharge of confined aquifers

15. All of the following may be consequences of dam construction except
    (a) reduced upstream migration of fish.
    (b) reduced downstream migration of fish.
    (c) less frequent and less severe inundation of downstream floodplains.
    (d) higher risk of levee failure downstream of the dam.
    (e) lower fertility of floodplains downstream.

16. Desalination by reverse osmosis
    (a) is less efficient than distillation.
    (b) produces a brine that is saltier than seawater.
    (c) has less effect on local wildlife than other methods of desalination.
    (d) is primarily a chemical process.
    (e) can be accomplished without any harmful by-products.

17. Which of the following factors is NOT considered when determining a nation's water footprint?
    I. The daily per capita use of fresh water
    II. The amount of water used to produce exported crops
    III. The daily per capita amount of precipitation

    (a) I only
    (b) II only
    (c) III only
    (d) I and II
    (e) II and III

18. Common methods of irrigation, in order from least efficient to most efficient, include
    (a) flood, furrow, drip, spray, and hydroponic irrigation.
    (b) furrow, drip, spray, and flood irrigation.
    (c) furrow, spray, drip, and hydroponic irrigation.
    (d) flood, furrow, and spray irrigation.
    (e) hydroponic, flood, and drip irrigation.

19. Which of the following is NOT a likely consequence of the free-market system of water distribution?
    (a) The free-market system will encourage more efficient use of water.
    (b) The free-market system will encourage illegal diversion of water resources.
    (c) The price of water for homeowners will go down.
    (d) The free-market system will encourage the development of desalination technologies.
    (e) Farmers will begin planting crops that are more water efficient.

20. What is one possible method to increase water use efficiency?
    (a) Discouraging free-market distribution of water
    (b) Replacing aging aqueducts
    (c) Replacing furrow irrigation systems with drip irrigation systems
    (d) Monetarily subsidizing the water use of farmers
    (e) Damming rivers to regulate water resources

Section 2: Free-Response Questions

Write your answer to each part clearly. Support your answers with relevant information and examples. Where calculations are required, show your work.

1. Mountaintop mining is the process of removing 50–200 vertical meters of mountaintop to extract coal from underlying seams. Following the removal of topsoil, material overlying coal seams is removed with explosives. This process leaves behind a tremendous amount of soil, rock, and coal tailings. The tailings are often pushed into adjacent valleys that contain headwater streams and floodplains. Much of the coal tailings become oxidized, and can release sulfuric acid after a rainfall. When coal extraction is complete, the mountains are not rebuilt. However, mining companies are required to contour the landscape so that rainfall and snowmelt drain from the mountain along similar paths as in the pre-mining landscape. Companies are also required to add topsoil to the mountain and seed the soil to facilitate plant growth.
   (a) Topsoil is often added by large tractors that compact the soil. How does soil compaction alter the flow of water down the mountain? (2 points)
   (b) List two ways in which soil compaction may alter plant growth on the mountain following all reclamation activities. (2 points)
   (c) List two environmental consequences that are likely to result from filling valleys with mountaintop tailings, and two consequences that might arise downstream of the mining site. (4 points)
   (d) What other coal mining practice(s) could be used to remove the coal from mountains that would have less of an environmental impact than mountaintop mining? (2 points)

2. Predicting earthquakes could save many lives and prevent costly damage. Currently, we are only able to predict what areas are likely to experience earthquakes but not when an actual event will occur.
   (a) Describe how an earthquake occurs. In your answer, describe the location of the epicenter. (2 points)
   (b) We may be able to predict earthquakes by measuring the electric charge of minerals in the ground. When pressure is applied to many crystals, they emit a small electric charge that increases with pressure, a phenomenon is known as "piezoelectricity." Describe how we might use piezoelectricity to predict the future occurrence of an earthquake. (4 points)
   (c) The amount of ground movement is directly proportional to the magnitude of the earthquake. The amount of ground movement produced by an earthquake with a magnitude of 4.0 on the Richter scale is equal to 10 millimeters. What is the amount of ground movement that would occur during an earthquake of magnitude 6.0 on the Richter scale? (2 points)
   (d) If an area of land experiences an earthquake of magnitude 6.0 on Richter scale every 40 years, how long will it take for that area to move 10 meters? (2 points)

Answers to Unit 4 AP® Exam Free-Response Questions

1. (a) The rate of groundwater flow is determined by several factors, including the composition of the soil and the amount of space within the soil, which defines the permeability of the soil. For example, soil that is composed mostly of clay does not allow water to pass through, whereas soil that is composed mostly of sand tends to release water very quickly. This is because clay has very little space between particles, whereas sand has relatively more space between particles. By compacting the soil, tractors reduce the space between soil particles and reduce the permeability of the soil. As a result, rainwater will likely flow on top of the soil instead of penetrating through the soil surface. This will increase erosion of the topsoil. In addition, the water that does penetrate the soil will not flow down the mountain.

(b) Mining and soil compaction can alter plant growth on the mountain in several ways.

- Soil compaction can decrease the permeability of the soil to water and make it more difficult for plant roots to extract water from the soil. This will increase wilting rates and decrease plant productivity.
- Soil compaction can increase the amount of water flow on the soil surface. This causes erosion and may wash the seeds away before they germinate. This can also wash away valuable nutrients.
- Soil compaction can also compact the seeds within the soil and prevent germination.

(c) Filling the valleys with the tailings from mountaintop mining can have many environmental consequences within the valley, as well as downstream of the valley. Within the valley, the environment is likely to experience

- loss of stream habitat that carries water, nutrients, pollutants, and debris out of the forest
- loss of diversity that results from loss of terrestrial and aquatic habitats
- altered soil structure that reduces plant productivity and diversity
- increased water retention due to greater amounts of available soil to absorb rainfall and mountain runoff
- increased pollutants resulting from the oxidization of coal tailings

Downstream of the valley, the environment is likely to experience

- high amounts of acid mine drainage that results from the oxidization of coal tailings in the valley
- changes in hydrology (Depending on the reclamation activities, downstream areas may experience greater amounts of runoff relative to pre-mining activities. If soil runoff is directed away from the valleys that were filled in, downstream areas may experience greater amounts of stream flow than in the past. If soil runoff remains directed at filled valleys, downstream areas may experience lower amounts of stream flow than in the past.)
- greater sediment load in streams due to increased erosion from the mountain
- loss of biodiversity from pollutants and changes in flow regimes

(d) Both strip mining and subsurface mining might provide more environmentally friendly alternatives to mountaintop removal. Strip mining would be useful if the seams of coal ran in single directions and if the coal seams were relatively close to the surface. However, many coal seams are buried deep in the parent material and run in several directions. For this reason, subsurface mining may be an alternative practice. Unfortunately, subsurface mining is more costly than mountaintop removal and far more dangerous to the miners. Both strip mining and subsurface mining would produce fewer tailings than mountaintop removal and would likely have less of an impact on the environment.

Answers continue in the Answer Appendix on page ANS-7.

Inga Spence/Getty Images

4

science**applied**

Is There a Way to Resolve the California Water Wars?

Pick up any tomato, broccoli spear, or carrot in the grocery store and chances are it came from California. California's $37 billion agricultural industry is more than twice the size of the agricultural industry of any other state, and it accounts for nearly one-half of all the fruits, nuts, and vegetables grown in the United States. California's temperatures are ideal for many crops, and its mild climate permits a long growing season. However, water is a scarce resource and most of California's growing regions lack the water that is necessary for farming. To remedy this situation, the state has developed an extensive irrigation system that moves water from areas where it is abundant to areas where it is scarce, as you can see in FIGURE SA4.1.

Although agriculture in California has been economically successful, it has increased competition among the state's residents for water. Nearly two-thirds of California's population lives in the southern part of the state, where residential and business demand for water has long exceeded the locally available supply. Los Angeles, for example, has been importing water from other areas of the state for more than 100 years. Demand for water has continued to grow with the state's population, which has been increasing by about 1 percent per year.

Wildlife also depends on a readily available supply of fresh water. In many regions of California, diverting water away from streams and rivers threatens the habitats of fish and other aquatic organisms, including several endangered species. Salmon, for instance, require cold, clean water to lay their eggs. In the delta region where the San Joaquin and Sacramento rivers come together just east of San Francisco, so much water has been pumped out for use in the southern part of the state that the salmon in the region's rivers have experienced dramatic declines.

What historical factors have led to California's water wars?

At 144 trillion liters (38 trillion gallons) of fresh water per day, California's water use is more than twice that of any other state. The vast majority of that water is used in the Central Valley region, which produces more than 80 percent of California's farm income. In 1935, the U.S. Department of Reclamation began an immense endeavor, called the Central Valley Project, to divert 8.6 trillion liters (2.3 trillion gallons) of water per year from rivers and lakes throughout the state for both agricultural and municipal uses.

While the project made agriculture possible in many of southern California's desert regions, it also locked into place long-term contracts that have subsidized several decades of cheap water for farmers, who pay rates that are only 5 to 10 percent of those paid by southern California towns and cities. Interestingly, northern California also has farms, but much of its abundant water is sent to the farms and municipalities of southern California.

In 2005, the federal government renewed nearly 200 large water supply contracts with farmers. These contracts were for 25 years, with an option available to extend them another 25 years, at prices so low that the payments did not even cover the cost of the electricity used to transport the water. Organizations ranging

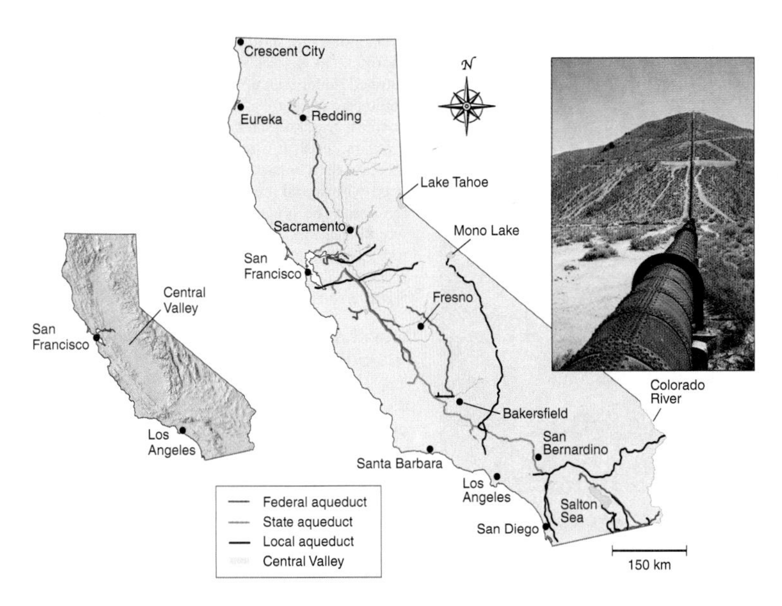

FIGURE SA4.1 **Water distribution in California.** To supply water to the agricultural areas in the Central Valley and to supply drinking water to municipalities, the state has developed an elaborate system of aqueducts. These aqueducts consist of both open canals and pipes laid above or below the ground. *(Jim West/The Image Works)*

from the Cato Institute to the Natural Resources Defense Council protested this decision. The controversy raised the question of whether farmers should continue to be given cheap access to water to protect the existing agricultural industry or be required to pay the same price for water as everyone else.

How might farmers respond to increased water prices?

What would happen if the federal government raised the price of water for California farmers? If water were more expensive, farmers would face four choices: raise the prices of their crops, switch to less water-intensive crops, use more efficient irrigation methods, or sell their farms. Each of these options would have different effects on the state's environment and economy.

One obvious way to offset higher water prices would be for farmers to charge more for the produce they grow. For example, if California broccoli growers had to pay the same price for water as municipalities paid, that cost would add about 13 cents to the price of broccoli per pound—an increase of less than 10 percent. This action would not change the environmental threat of water depletion, but would have a small negative effect on the demand for broccoli and on the consumer's wallet.

Other farmers faced with higher water prices might decide to switch to crops that require less water. For example, alfalfa, which uses 25 percent of California's irrigation water and has a low resale value compared with other crops, would no longer be a profitable crop. A farm using 296 million liters (78 million gallons) of water can produce about $60,000 worth of alfalfa per year, according to the Natural Resources Defense Council. If farmers had to purchase water at the unsubsidized rate, that water would cost them $168,000; if the alfalfa continued to sell for only $60,000, the farmers would lose more than $100,000 a year growing alfalfa. Tomatoes, broccoli, and potatoes use about half as much water per acre as alfalfa and have a higher resale value (FIGURE SA4.2). Just by replacing alfalfa with these crops, California would reduce its statewide water use by 20 percent. This action would have a positive environmental impact, since less overall water would be needed for agriculture in the state.

Investment in new irrigation technology that uses less water would also reduce a farmer's costs. As we saw earlier in this chapter, the four major irrigation techniques differ in cost and efficiency. If farmers had to purchase water at the unsubsidized rate, converting

FIGURE SA4.2 **Irrigating broccoli.** Crops such as potatoes, tomatoes, and broccoli use less water and have a higher resale value than crops such as alfalfa. Growing such crops would reduce the water needed for agriculture in California. *(Ed Young/age fotostock)*

to more-efficient irrigation techniques could rapidly pay for itself since the initial investment in equipment would be offset by savings on water. As long as the cost of water continues to be subsidized by the government, farmers have little incentive to upgrade their irrigation technology.

If water prices rise, not all farmers will be able to afford the new costs. This recognition brings us to the final option: Some farmers could be driven out of business and forced to sell their land. This outcome would probably reduce the amount of food grown in the region. Whether conversion from farming to other uses would reduce water demand would depend on how the land was used. For example, if it became a housing development in which residents practiced water conservation, then the residents would use much less water per acre than the amount used for agriculture. However, if the residents chose to plant and water large lawns and were not careful about their use of water indoors, they could use even more water than was used for agriculture.

Because charging more for water could force some farmers out of business, many politicians are reluctant to support this change. In response, some economists argue that farmers who possess long-term subsidized water contracts should be allowed to sell their water allotment to the highest bidder, something that is not allowed under the current law. In this scenario, a farmer could buy water from the government at the contracted low cost and then sell that water to a city at a substantial profit rather than using it on the farm. Farmers would increase their income, cities would get the water they need without taking additional water from natural systems, and the government agencies that sell the water would get the same amount of money that they do today. Many critics argue, however, that the government should not be giving preferential water rates to a few users who can then make a profit on the water.

How have the water wars played out in the political arena?

The California water wars have developed because of the large number of competing interests involved, including northern California farmers and municipalities that are losing water, southern California farmers and municipalities that need more water, endangered species that are declining due to the diversion of water, and beautiful natural areas that have experienced reduced water capacity, seen increased salinity, or gone completely dry.

Competition for California's water has been marked by political maneuvering and numerous lawsuits. In the 1960s, for example, there was a large push to build dams that would create reservoirs to collect water. Today there are about 1,200 dams in the state. Because of the money and time needed to construct new dams and the environmental impact of dams, many Californians do not consider more dams to be a viable solution to today's water problems.

The past few years have witnessed a number of lawsuits that have produced very interesting court decisions. For example, the Natural Resources Defense Council won a lawsuit regarding the 2005 water contracts for farmers in the eastern San Joaquin Valley, who use water diverted from the San Joaquin River. The judge ruled that the contracts were illegal because they did not consider how diverting the water might affect endangered species that rely on the river.

In 2007, a federal court was asked to consider how the large, powerful pumps used to divert water from the San Joaquin River and Sacramento delta region affected local fish populations—particularly the delta smelt (*Hypomesus transpacificus*), a federally protected fish whose population in the delta region was declining. The court ruled that pumping from the delta region—the source of much of the state's fresh water—must be reduced. Because this decision came during a multi-year drought, its impact on southern California was even larger than it otherwise would have been. A number of California congressmen accused the state and federal governments of caring more about fish than people.

Answer to Science Applied 4 Free-Response Question

Download the full answer to the Science Applied 4 Free-Response Question by clicking on the link buttons in the TE-book, opening the TRFD, or logging onto the book's companion website bcs .whfreeman.com/friedlandapes2e (teacher login required).

Answer to Science Applied 4 Free-Response Question

The problem of water scarcity in California rose to a crisis level in June 2008 when Governor Arnold Schwarzenegger declared that the state was officially in a drought. As a result of a lack of rain and a reduced amount of snow in the Sierra Nevada, the spring of 2008 was the driest in 88 years. The California legislature passed a bill to cut water use by 20 percent. But the legislature put most of the responsibility for reducing water use on residential users, despite the fact that agriculture uses a much larger portion of the state's water. Some municipalities started rationing water, and a number of municipalities began enforcing a 2001 state law that requires new building developments to show plans that ensure a 20-year water supply.

As the drought continued into 2009, additional ideas to promote water conservation were proposed. In June 2009, for example, the Los Angeles Department of Water and Power announced that customers who replaced turfgrass with drought-resistant plants or mulch would be credited $11 per square meter.

In November 2009, the California legislature passed a series of bills to address the water crisis. The $40 billion package called for the restoration of the San Joaquin and Sacramento delta ecosystem as well as the building of new dams and a series of water conservation efforts to reduce water demand. A few days later, the governor signed the bill into law.

What can be done?

Given the scarcity of water in California, it stands to reason that its sustainable use is an important topic with many environmental, economic, and even political implications. Economists generally argue that all users should be charged the same amount for a resource so that only those who value it most will pay to use it. Doing so would provide increased motivation for more efficient water use by farms, residences, and businesses. Increased water costs could be helpful in reducing the demand for water, but we cannot forget the simultaneous need for all users to better conserve water.

Scenarios similar to the California water wars are being played out in many other parts of the world. Competition for water exists not only among farmers and municipalities, but also among neighboring countries. Indeed, disagreements over water have turned into real wars between nations. If the competing interests in California can devise creative solutions to the state's water problems, those solutions can serve as a model for the rest of the world so that we can continue to have economic prosperity without severely harming the environment.

Questions

1. What are the challenges of balancing the need for water in agriculture with the need for water to protect endangered species?
2. How might a California farmer who receives subsidized costs of water argue that the current water distribution rules are beneficial not only to farmers but also to society?

Free-Response Question

The Los Angeles aqueduct spans 375 km (233 miles) from the Owens Valley through the Mojave Desert to the San Fernando Valley. The aqueduct collects snowmelt from the mountains and is capable of carrying 13.7 m^3 of water per second. The water flows the entire distance by gravity. In the years after the aqueduct was completed in 1913, the area in and around Owens Valley has become increasingly dry and farmers have suffered a tremendous loss of productivity.

(a) Surveys taken from the Los Angeles Department of Water and Power claim that the water table in the Owens Valley has not changed, yet residents in the Owens Valley claim that wells are running dry. Why might this be? (3 points)

(b) Aside from moving out of the area, list three possible ways that farmers may mitigate the loss of water supply and continue farming. (3 points)

(c) Los Angeles is situated along the shore of the Pacific Ocean. Describe two methods of desalinization that Los Angeles could use to obtain fresh water. List two costs associated with each method. (2 points)

(d) Considering the position of Owens Valley on the windward side of the Sierra Nevada mountain range, how might water diversion through the aqueduct affect air quality in the Owens Valley? (2 points)

References

Martin, G. 2005. The California water wars: Water flowing to farms, not fish. *San Francisco Chronicle*, October 23.

Steinhauer, J. 2008. Governor declares drought in California. *New York Times*, June 5.

Obegi, D. 2011. California water wars: It isn't fish vs. farmers. *Los Angeles Times*, August 10.

Suggested Answers to Science Applied Questions

1. Most people are interested in protecting endangered species. However, when the protection of endangered species appears to interfere with food production for humans, many people will argue that the food should be more important and they will be less likely to support efforts to protect the endangered species.

2. A California farmer could argue that the subsidized cost of water benefits society by reducing the cost of food, particularly fresh vegetables, making them more affordable for people with low incomes.

Answer to Science Applied Free-Response Question

(a) Pumping water from a well can remove groundwater from the area around the well faster than it can be replenished. This results in a cone of depression around the well; the well can run dry while the overall level of the water table has not decreased. Members of the Owens Valley community are continually pumping water from wells for personal needs, so it is likely that there are cones of depression around many wells. If the Los Angeles Department of Water and Power doesn't measure water table depth near private wells, they will not detect these cones of depression. Instead, they will report that the water table depth has not changed.

(b) Farmers in the Owens Valley may mitigate the loss of water supply in several ways:

- Farmers can plant less water-intensive crops, such as alfalfa. This option will not work if the water supply is severely diminished.
- Farmers can switch to raising cattle, which is less water intensive than raising crops but requires importing feed from outside the community.
- Farmers can raise the price of their crops while raising fewer crops. This is not likely to be a viable option because they will still have to compete with lower crop prices throughout the region.
- Farmers can switch to methods of agriculture that require less water, such as hydroponics or drip irrigation.

(c) Two methods of desalinization that Los Angeles could use are distillation and reverse osmosis. Distillation is the process of boiling salt water. Freshwater is converted to steam, which is then condensed and collected; salt is left behind and disposed of. Reverse osmosis is the process of forcing salt water through a thin semipermeable membrane at high pressure. Water passes through the membrane but salt does not. The liquid that remains, known as brine, is then disposed of. There are several costs associated with these processes. For distillation, a tremendous amount of energy is required to boil the water. For reverse osmosis, energy is required to pressurize the salt water. Both processes require a substantial amount of energy to dispose of the salt or brine. Disposal of salt and brine also comes at an environmental cost because dumping locations become extremely salty, which can kill species that are not tolerant of salty environments. In addition, energy must be used to pump the salt water from the ocean to the desalinization plant, and then from the plant to residents, businesses, and farmers.

(d) Water diversion from the Owens Valley has led to a much more arid environment with very dry topsoil. Because the Owens Valley is located on the windward side of the Colorado Rockies, wind easily blows particles from the soil surface. This has created massive dust storms that severely reduce air quality.

chapter

10 Land, Public and Private

PD Overview

Chapter 10 explores how our use of land affects the environment and what we can do to minimize negative impacts. It examines land use and management in both private and public sectors, as well as sustainable land use practices. The chapter also covers urban sprawl and the problems it creates. Finally, it describes smart growth as a solution to urban sprawl.

Module 29: Land Use Concepts and Classification

Humans value land for what it can provide: food, shelter, resources, and intrinsic beauty. However, human use of the land for these purposes can create problems, some of which can be prevented or addressed by regulations. This module examines how human land use affects the environment. It covers land use concepts and classification both abroad and within the United States.

Module 30: Land Management Practices

Whether resources are conserved or over-exploited depends on a number of factors related to economics, human behavior, and specific management practices. Proper resource management also depends on specific aspects of the type of land involved as well as the agency and the guidelines under which that agency operates. This module examines land management practices in rangelands and forests, with a focus on timber management practices and fire management. It also considers trends in residential land use and the causes and consequences of urban sprawl.

Alignment to AP® Environmental Science Course Description

AP® Outline	Module	
IV. Land and Water Use (10–15%)		
A. Agriculture	Module 29	Land Use Concepts and Classification
	Module 30	Land Management Practices
B. Forestry	Module 29	Land Use Concepts and Classification
	Module 30	Land Management Practices
C. Rangelands	Module 29	Land Use Concepts and Classification
	Module 30	Land Management Practices

Chapter Learning Objectives

After completing this chapter students will be able to

- **explain how human land use affects the environment.**
- **describe the various categories of public land used globally and in the United States.**
- **explain specific land management practices for rangelands and forests.**
- **describe contemporary problems in residential land use and some potential solutions.**

Chapter 10 Pacing Guide

This pacing guide is based on a schedule with 120 sessions of 50 minutes each before the AP® Exam. If you have a different number of sessions before the exam, you can modify the pacing to suit your needs. If you have additional time, consider incorporating quizzes, released AP® Environmental Science free-response and multiple-choice questions, or additional activities.

Module	Standard Schedule Days	Block Schedule Days
Module 29	2	1
Module 30	2	1
Assessment	1	½

Chapter 10 Resources

The link to these items can be found by clicking on the link buttons in the Teacher's e-Book (TE-book), opening the Teacher's Resource Flash Drive (TRFD), or logging onto the book's companion website bcs.whfreeman.com/friedlandapes2e (teacher login required).

- **PowerPoint Presentation 10**
- **Optimized Art PowerPoint and JPEG Files 10**
- **Do the Math Videos**
- **More Math Practice 10.1:** Ratios
- **Measuring Your Impact 10:** The Costs of Commuting
- **Chapter 10 Web Resources**
- **Exploring the Literature 10**
- **Answers to Chapter 10 AP® Exam**
- **Multiple-Choice AP® Practice Exam 10**

Free-Response Questions from Previous AP® Environmental Science Exams

Free-response questions from prior AP® Environmental Science Exams are available on the AP® course website, https://apstudent.collegeboard.org/apcourse/ap-environmental-science/exam-practice. Students should be able to answer all of the questions listed below with material learned in this and previous chapters. When a question requires students to understand material from multiple chapters, the question will be listed in the last chapter required to complete the entire question. Questions marked with an asterisk (*) are from exams with released multiple-choice questions. You may want to save these questions until the end of the year so you can give your students a complete released exam for practice. Questions marked with double asterisks (**) require math to calculate a problem.

Year	Question	Content
2008*	3	• Negative impact of fire suppression policies • Healthy Forest Initiative

PD Chapter 10 Overview

Watch the video overview of Chapter 10 (for teachers) by clicking on the link buttons in in the TE-book, opening the TRFD, or logging onto the book's companion website bcs .whfreeman.com/friedlandapes2e (teacher login required).

TRM PowerPoint Presentation 10

Download the PowerPoint presentation for Chapter 10 by clicking on the link buttons in the TE-book, opening the TRFD, or logging onto the book's companion website bcs.whfreeman.com /friedlandapes2e (teacher login required).

The Headwaters Forest Reserve in California contains the largest remaining intact forest of ancient redwood trees. (Paul Rogers/MCT/Landov)

chapter 10

Land, Public and Private

Who Owns a Tree? Julia Butterfly Hill versus Maxxam

For most of its history, northern California's Pacific Lumber Company was a leader in environmental stewardship. It pioneered the practice of selectively cutting some trees on its land while leaving nearby trees intact. Unlike the usual practice of clear-cutting—clearing all the trees from an area of land—selective cutting allows the roots of the remaining trees to retain water and hold soil in place. In addition, the surviving trees foster regrowth that more closely mimics natural succession after a disturbance. Pacific Lumber developed a 100-year sustainable logging plan that ensured relatively healthy forests as well as job security for the loggers and their families.

In 1986, however, Pacific Lumber was purchased by new owners and renamed Maxxam. To increase revenue from the company's land holdings, the new management clear-cut hundreds of thousands of acres of redwood forests. Because the new harvesting method removed vegetation and tree roots, Maxxam's practices left the forests susceptible to soil erosion and landslides. On December 31, 1996, an immense landslide began on a steeply sloped Maxxam site above the town of Stafford, California, destroying a number of homes and drastically altering the natural environment.

Julia Butterfly Hill proved indomitable. In her most famous tree sit, she occupied Luna, a 55-m (180-foot) redwood tree that was nearly a thousand years old.

The adjacent uncut land, also owned by Maxxam, was home to the largest remaining intact forest of ancient redwood trees as well as to a number of endangered species. Environmental activist Julia Butterfly Hill was appalled by Maxxam's forest management practices, and in response she joined a "tree sit" on Maxxam's land organized by the direct action group Earth First!

The ultimate goal of a tree sit is to remain in a tree long enough to save it from logging, although many such actions merely delay the inevitable. Still, activists support tree sits because they gain time for further legal proceedings and increase media awareness. Julia Butterfly Hill proved indomitable. In her most famous tree sit, she occupied Luna, a 55-m (180-foot) redwood tree that was nearly a thousand years old, and resolved to

Teaching Tip: Chapter Opening Case

The chapter opening case, "Who Owns a Tree? Julia Butterfly Hill versus Maxxam," introduces students to the practice of timber harvesting. It discusses two types of timber harvesting practices: clear-cutting and selective cutting. It describes an event that occurred on privately owned land in a redwood forest in Stafford, California. A company planned to clear-cut the forest, but activist Julia Butterfly Hill took action to protect a single tree. This case provides a good opportunity to talk about some of the conflicts that occur with human use of land and land management practices.

Teaching Tip: Beyond the Classroom

Ask students to research more details of Julia Butterfly Hill's story and write a short report. What are the arguments in favor of and against a tree sit on private land? Students can share and discuss their findings with the class.

Teaching Tip: Journal Prompt

Ask students to discuss the following questions:

- Why is it important to preserve redwood trees? *Answers will vary but might include: redwoods provide a home for many species; redwoods keep soil intact and prevent landslides; trees play an important role in the carbon cycle and the hydrologic cycle.*
- What type of climatic data do trees tell us? *Tree rings help scientists understand past changes in climate. Thicker rings indicate more favorable growing conditions while thinner rings indicate less favorable conditions.*

remain there until Maxxam agreed to spare the tree. She spent 2 years in the tree without coming down once, staying through gale-force winds, two winters, intimidation tactics, the death of another activist, and the removal of most of the trees in the stand that she occupied. Julia's efforts were rewarded when Maxxam agreed to protect Luna, and a 61-m (200-foot) buffer around it, in perpetuity. In a separate, complex deal that included many stipulations about logging practices, Maxxam and another company also agreed to sell the U.S. government 3,035 ha (7,500 acres) of ancient forest for $480 million. That property has now become the Headwaters Forest Reserve.

Hill's actions cost Maxxam millions of dollars in delays and lost revenue and brought unsustainable logging practices to the attention of many people. A large debt load forced Maxxam to file for bankruptcy protection in January 2007, and its depleted assets were reorganized and transferred to a new company in 2008.

Maxxam's logging practices appeared to place profit before sustainability. The new company, though, appears to have implemented more sustainable practices with the trees that remain on its holdings.

The conflict over logging practices raises many questions. Do citizens of the United States have the right to influence what activities occur on private lands? What if the land is public? Was Julia Butterfly Hill a hero or a villain? Some see her as having made a personal sacrifice to save an ancient tree and to bring the issue of unsustainable logging practices to the public's attention. Others see her as a lawbreaker who trespassed on private property in an effort to prevent a legal activity.

Sources:
J. B. Hill, *The Legacy of Luna: The Story of a Tree, a Woman, and the Struggle to Save the Redwoods* (HarperCollins, 2000); H. Sims, The scion, *North Coast Journal*, July 20, 2006.

As this opening account about ownership and management of land containing ancient redwood trees illustrates, issues involving land use are complex, and there are no easy solutions to many of the conflicts that arise. In this chapter we will explore how our use of land affects the environment, and how specific land-management practices can be implemented to minimize negative impacts. We will look at land use and management on both public and private land, and we will explore the topic of sustainable land use practices. We will also look at how land use has changed with shifting and growing populations.

module

29

Land Use Concepts and Classification

Humans value land for what it can provide: food, shelter, resources, and intrinsic beauty. There are also various classifications of land that are made for a variety of scientific and political purposes. However, human use of the land for these purposes can create problems that are prevented or addressed by regulations. In this module we will look at how human land use affects the environment and examine land use concepts and classification.

Learning Objectives

After reading this module, you should be able to

- explain how human land use affects the environment.
- describe the various categories of public land used globally and in the United States.

Teaching Tip: Concept Map

Start a concept map with the words "Land use" in the center like the one shown below. Ask students to give examples of different ways humans use land.

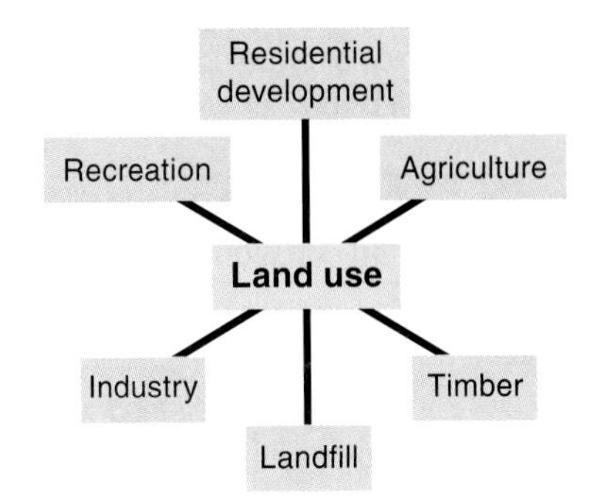

Human land use affects the environment in many ways

Agriculture, housing, recreation, industry, mining, and waste disposal are all land uses that benefit humans. But, as we have seen in previous chapters, these activities also have negative consequences. Extensive logging may lead to mudslides, and deforestation of large areas contributes to climate change and many other environmental problems. Change to the landscape is the single largest cause of species extinctions today. Paving over land surfaces reroutes water runoff, and these paved surfaces also absorb heat from the sun and reradiate it, creating urban "heat islands." Overuse of farmland can lead to soil degradation and water pollution (FIGURE 29.1).

Every human use of land alters it in some way. Furthermore, individual activities on any parcel of land can have wide-ranging effects on other land. For this reason, communities around the world use a number of methods, including laws, to influence or regulate both private and public land use.

As we saw at the beginning of this chapter, people do not always agree on land use and management priorities. Do we save a beautiful, ancient stand of trees, or do we harvest the trees in order to gain benefits in the form of jobs, profits, building materials, and economic development? Such conflicts can arise on both public and private land. To understand the issues involved in the use of land, we will look at three concepts: the *tragedy of the commons*, *externalities*, and *maximum sustainable yield*.

The Tragedy of the Commons

In certain societies, land was viewed as a common resource: Anyone could use land for foraging, growing crops, felling trees, hunting, or mining. But as populations increased, such common lands tended to become degraded—overgrazed, overharvested, and deforested. In 1968, ecologist Garrett Hardin brought the issue of overuse of common resources to the attention of the broader scientific community when he described the **tragedy of the commons**: the tendency of a shared,

Tragedy of the commons The tendency of a shared, limited resource to become depleted because people act from self-interest for short-term gain.

Teaching Tip: Warm-up

Ask students the following questions:

- What is the tragedy of the commons? *The tragedy of the commons is the tendency of a shared, limited resource to become depleted because people act from self-interest for short-term gain.*
- What is an externality? *An externality is a cost or benefit of a good or service that is not included in the purchase price of that good or service.*

Teaching Tip: Activity

Read the Dr. Seuss story *The Lorax*, which describes a tragedy of the commons situation. As they hear the story, have your students fill in a chart with a list of the abused common areas on the left side and descriptions of the resulting problems on the right, like the sample shown below.

Common area	Environmental problem and consequences
Truffula trees	*Deforestation destroys the Brown Bar-ba-loots' habitat, and their food source (Truffula fruit) is gone*
Air	*The air is polluted by factories; Swomee-Swans have to fly away*
Pond	*Pond is polluted with chemicals; Humming-fish can't hum because their gills are all gummed*

After students hear the story, note that for an economy to be successful, business is necessary, and that sustainable business practice does not degrade resources. Then discuss the following questions:

- How could the Once-ler have created a more sustainable business? *Answers will vary. Students might suggest that the Once-ler could collect Truffula seeds and start a planting initiative to replace trees he cuts down. The Once-ler could install a water treatment plant to clean the water that comes from the factory before it reenters the lake.*
- What protections do we have in place to prevent a tragedy of the commons like the one in the story? *Answers will vary but could include conservation associations to protect public land and federal legislation such as the Clean Water Act and the Clean Air Act.*

(a)

(b)

(c)

(d)

FIGURE 29.1 **Some negative consequences of land use by humans.** (a) A mudslide caused by logging and poor land management. (b) Habitat conversion and land degradation due to shifting agriculture. (c) Logging and other habitat alteration can adversely affect many species, including the spotted owl. (d) Extensive paving of land due to the growth of cities reduces the amount of land available for vegetation and water infiltration. *(a: David McNew/Getty Images; b: Ryan Pyle/Corbis; c: DLILLC/Corbis; d: Richard Cummins/SuperStock)*

limited resource to become depleted because people act from self-interest for short-term gain. Hardin observed that when many individuals share a common resource without agreement on or regulation of its use, it is likely to become overused very quickly.

For example, imagine a communal pasture on which many farmers graze their sheep, like the one shown in FIGURE 29.2. At first, no single farmer appears to have too many sheep. But because an individual farmer gains from raising as many sheep as possible, each farmer may

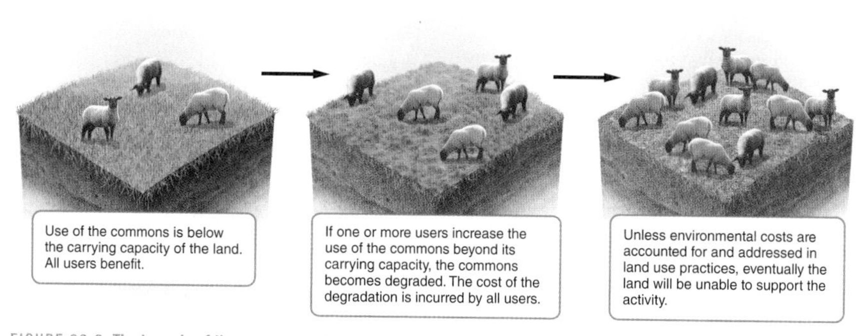

FIGURE 29.2 **The tragedy of the commons.** If the use of common land is not regulated in some way—by the users or by a government agency—the land can easily be degraded to the point at which it can no longer support that use.

Teaching Tip: Engage

Ask students about the common areas in your school. Do people keep them clean? Do students leave trash in the cafeteria or in other areas? What are other examples of common areas or resources? *Other examples of common areas or resources might include streams, lakes and ponds, oceans, playground areas, air, public grazing lands, parking lots, and beaches.*

be tempted to add sheep to the pasture. However, if the total number of sheep owned by all the farmers continues to grow, the number of sheep will soon exceed the carrying capacity of the land. The sheep will overgraze the common pasture to the point at which plants will not have a chance to recover. The common land will be degraded and the sheep will no longer have an adequate source of nourishment. Over a longer period, the entire community will suffer. When the farmers make decisions that benefit only their own short-term gain and do not consider the common good, everybody loses.

The tragedy of the commons applies not only to agriculture, but to any publicly available resource that is not regulated, including land, air, and water. For example, the use of global fisheries as commons has led to the overexploitation and rapid decline of many commercially harvested fish species, and has upset the balance of entire marine ecosystems.

Externalities

The tragedy of the commons is the result of an economic phenomenon called a *negative externality*. More generally, an **externality** is a cost or benefit of a good or service that is not included in the purchase price of that good or service. For example, if a bakery moves into the building next to you and you wake up every morning to the delicious smell of freshly baked bread, you are benefiting from a positive externality. On the other hand, if the bakers arrive at three in the morning and make so much noise that they interrupt your sleep, making you less productive at your job later that day, you are suffering from a negative externality.

In environmental science, we are especially concerned with negative externalities because they so often lead to serious environmental damage for which no one is held legally or financially responsible. For example, if one farmer grazes too many sheep in a common pasture, his action will ultimately result in more total harm than total benefit. But, as long as the land continues to support grazing, the individual farmer will not have to pay for the harm he is causing; ultimately, this cost is *externalized* to the other farmers. If the farmer responsible for the extra sheep had to bear the cost of his overuse of the land, he would not graze the extra sheep on the commons; the cost of doing so would exceed the benefit. From this example, we can see that in order to calculate the true cost of using a resource, we must always include the externalized cost. In other words, we must account for any potential harm that comes from the use of that resource.

Some economists maintain that private ownership can prevent the tragedy of the commons. After all, a landowner is much less likely to overgraze his own land than he would common land. Regulation is another approach. For example, a local government could prevent overuse of a common pasture by passing an ordinance that permits only a certain number of sheep to graze there.

Challenging the idea that government regulation is necessary, the late Professor Elinor Ostrom (1933–2012) of Indiana University showed that many commonly held resources can be managed effectively at the community level or by user institutions. Ostrom's work, for which she was awarded the 2009 Nobel Prize in Economics, has shown that self-regulation by resource users can prevent the tragedy of the commons.

Maximum Sustainable Yield

When we want to obtain the maximum amount of a resource, we need to know how much of a given plant or animal population can be harvested without harming the resource as a whole.

Imagine a situation in which deer hunting in a public forest is unregulated, with each hunter free to harvest as many deer as possible. As a result of unlimited hunting, the deer population could be depleted to the point of endangerment. This, in turn, would disrupt the functioning of the forest ecosystem. On the other hand, if hunting were prohibited entirely, the deer herd might grow so large that there would not be enough food in the forests and fields to support the herd. In extreme cases, such as that of the reindeer of St. Paul Island (see Figure 19.3 on page 156), the population could grow unchecked until it crashed due to starvation.

Some intermediate amount of hunting will leave enough adult deer to reproduce at a rate that will maintain the population, but not leave so many that there is too much competition for food. This intermediate harvest is called the *maximum sustainable yield*. Specifically, the **maximum sustainable yield (MSY)** of a renewable resource is the maximum amount that can be harvested without compromising the future availability of that resource. In other words, it is the maximum harvest that will be adequately replaced by population growth.

MSY varies case by case. A reasonable starting point is to assume that population growth is the fastest at about one-half the carrying capacity of the environment, as shown on the S-shaped curve in FIGURE 29.3.

Looking at the graph, we can see that at a small population size, the growth curve is shallow and growth is relatively slow. As the population increases in size, the slope of the curve is steeper, indicating a faster growth rate. As the population size approaches the carrying capacity, the growth rate slows. The MSY is the amount of harvest that keeps the resource population at about one-half the carrying capacity of the environment.

Externality The cost or benefit of a good or service that is not included in the purchase price of that good or service.

Maximum sustainable yield (MSY) The maximum amount of a renewable resource that can be harvested without compromising the future availability of that resource.

Teaching Tip: Warm-up

Ask students: What is maximum sustainable yield? *The maximum sustainable yield (MSY) of a renewable resource is the maximum amount that can be harvested without compromising the future availability of the resource. In other words, it is the maximum harvest that will be adequately replaced by population growth.*

more **math** practice:
Graphing Maximum Sustainable Yield (MSY)

Help your students create a maximum sustainable yield graph according to the following steps:

1. Label the *x* and *y* axis. Tell students that one variable is population size and the other is time (years). See if they know which variable goes on which axis. Remember that time is almost always on the *x* axis on a line graph. The population size should go on the *y* axis.
2. Draw an S-shaped curve starting at the origin and continuing to the upper right corner of the graph. If students have trouble drawing an S-shaped curve, review the logistic growth model in Chapter 6, Figure 19.2, page 198.
3. Label the carrying capacity with a dotted line. This line should run perpendicular to the *y* axis and be at the top of the graph.
4. Finally, label three different locations on the S-shaped curve: two areas of slow growth and one area of fast growth. Once students have finished drawing the graph, have them compare their graph to Figure 29.3.

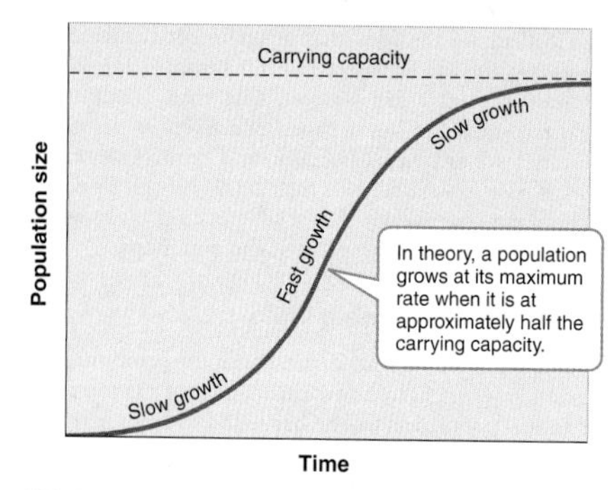

FIGURE 29.3 **Maximum sustainable yield.** Every population has a point at which a maximum number of individuals can be harvested sustainably. That point is often reached when the population size is about one-half the carrying capacity.

Forest trees, like animal populations, have a maximum sustainable yield. Loggers may remove a particular percentage of the trees at a site in order to allow a certain amount of light to penetrate to the forest floor and reach younger trees. If they cut too many trees, however, an excess of sunlight will penetrate and dry the forest soil. This drying may create conditions inhospitable to tree germination and growth, thus inhibiting adequate regeneration of the forest.

In theory, harvesting the maximum sustainable yield should permit an indefinite use without depletion of the resource. In reality, though, it is very difficult to calculate MSY with certainty because in a natural ecosystem, it is difficult to obtain necessary information such as birth rates, death rates, and the carrying capacity of the system. Once an MSY calculation is made, we cannot know if a yield is truly sustainable until months or years later when we can evaluate the effect of the harvest on reproduction. By that time, if the harvest rate has been too great, it is too late to prevent harm to the population.

Public lands are classified according to their use

All countries have public lands, which they manage for a variety of purposes, including environmental protection. The 2003 United Nations List of Protected Areas—the most recent global study of protected areas—includes almost 1.7 billion hectares (4.2 billion acres) of land in a variety of categories. Given that Earth's total land area is about 15 billion hectares (37 billion acres), this means that approximately 11 percent, or one-ninth, of the land area of Earth is protected in one way or another. These areas, as well as protected marine areas, are shown on the map in FIGURE 29.4. Let's look at both international and national categories of land protection.

International Categories of Public Lands

The 2003 United Nations List of Protected Areas classifies protected public lands into six categories according to how they are used: national parks, managed resource protected areas, habitat/species management areas, strict nature reserves and wilderness areas, protected landscapes and seascapes, and national monuments.

National Parks

There are roughly 3,400 national parks in the world, covering more than 400 million hectares (1 billion acres). This means that national parks make up about 2.7 percent of Earth's land area. National parks are managed for scientific, educational, and recreational use, and sometimes for their beauty or unique landforms. In most cases, they are not used for the extraction of resources such as timber or ore. Some of the most famous national parks in the world are found in Africa and include Amboseli National Park in Kenya and Kruger National Park in South Africa. Parks like these generally exist to protect animal species such as elephants, rhinoceroses, and lions, as well as areas of great natural beauty. They also generate tourism, which can provide much revenue. On the negative side, in order to create and maintain national parks, governments have sometimes evicted and excluded indigenous human populations from the land. For example, in the winter of 2009, a new round of evictions from the Mau Forest in the Rift Valley of Kenya led to the displacement of 20,000 families. A Kenyan parliamentary report estimated that thousands of the evictees had not been resettled by 2012. Such programs continue to generate controversy in Kenya and other countries.

Managed Resource Protected Areas

This classification allows for the sustained use of biological, mineral, and recreational resources. In most countries, these areas are managed for multiple uses. There are approximately 4,100 such sites in the world, encompassing more than 440 million hectares (1.1 billion acres). In the United States, national forests are one example of this kind of area.

Habitat/Species Management Areas

These areas are actively managed to maintain biological communities, for example through provision of fire prevention or predator control. There are approximately 27,600 such sites, covering more than 300 million hectares (740 million acres). Karelia, in Northwest Russia and bordering areas of Finland, has one of the highest proportions of protected areas in Europe: 5 percent of its total area. Of this total, more than one-half consists of habitat

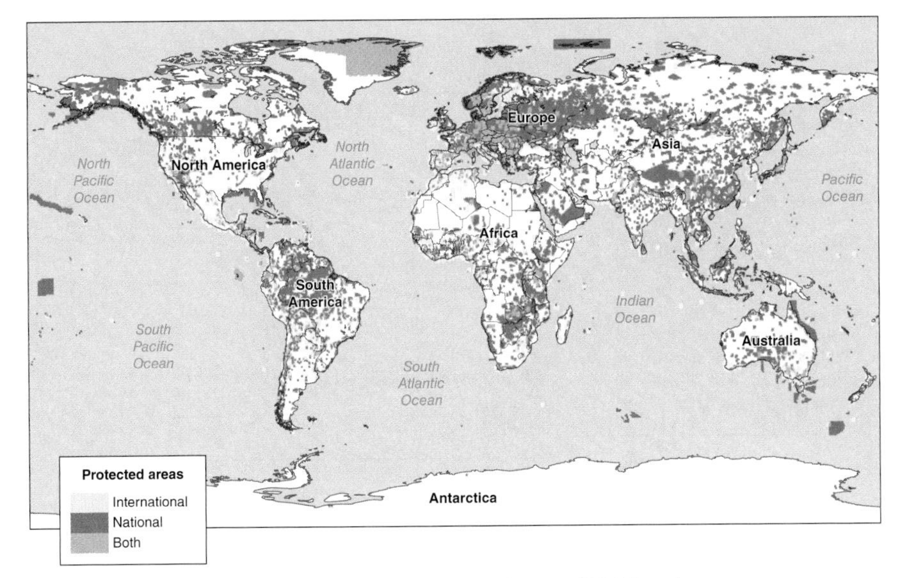

FIGURE 29.4 **Protected land and marine areas of the world.** Protected areas are distributed around the globe. *(Data from http://protectedplanet.net/#2_43.5_-72.25_0)*

or species management areas that are actively managed for hunting and conservation.

Strict Nature Reserves and Wilderness Areas.

These areas are established to protect species and ecosystems. There are approximately 6,000 such sites worldwide, covering more than 200 million hectares (490 million acres). The Chang Tang Reserve on the Tibetan Plateau in China was set aside to protect a number of species—including the declining population of wild yak—from hunting, habitat destruction, and hybridization with domesticated animals.

Protected Landscapes and Seascapes

These areas combine the nondestructive use of natural resources with opportunities for tourism and recreation. Orchards, villages, beaches, and other such areas make up the 6,500 such sites worldwide, which cover more than 100 million hectares (250 million acres). Among these protected areas is the Batanes Protected Landscape and Seascape in the northernmost islands of the Philippines, home to several endemic plant and animal species as well as important marine habitats.

National Monuments

National monuments are set aside to protect unique sites of special natural or cultural interest. There are almost 20,000 national monuments and landmarks in the world, covering nearly 28 million hectares (69 million acres). Most of these are established to protect historical landmarks, such as the Arc de Triomphe in Paris, France.

Public Lands in the United States

In the United States, publicly held land may be owned by federal, state, or local governments. Of the nation's land area, 42 percent is publicly held—a larger percentage than in any other nation. As you can see in FIGURE 29.5, the federal government is by far the largest single landowner in the United States: It owns 240 million hectares (600 million acres), or roughly 25 percent of the country. Most of this land—55 percent—is in the 11 western continental states, and an additional 37 percent is in Alaska. Less than 10 percent of federal land is located in the Midwest and on the East Coast.

Public Land Classifications

Public lands in the United States include rangelands, national forests, national parks, national wildlife refuges, and wilderness areas. Since the founding of the nation, many different individuals and groups have expressed interest in using these public lands. However, most environmental policies, laws, and management plans have been based, at least partially, on the *resource conservation*

Teaching Tip: Warm-up

Ask students to name the six categories that the United Nations List of Protected Areas uses to classify protected public lands, and give an example of each.

1. ***National Park:*** *Kruger National Park, South Africa*
2. ***Managed Resource Protected Area:*** *Great Smoky Mountains National Park, United States*
3. ***Habitat/Species Management Area:*** *Karelia, Russia*
4. ***Strict Nature Reserves and Wilderness Area:*** *The Chang Tang Reserve, China*
5. ***Protected Landscape and Seascape:*** *Batanes Protected Landscape and Seascape, Philippines*
6. ***National Monument:*** *Arc de Triomphe, Paris, France*

TEACHING with FIGURES ▶

Show students Figure 29.5 and ask the following questions:

- Do you observe any patterns in the location of federal lands? *The western states have the most federal lands.*
- Why do most of the federal lands in the United States occur in the west? *A large portion of the land in the west is dry and arid, which has prevented large cities from developing in states like New Mexico, Utah, Wyoming, Idaho, and Montana. Also, many Native American reservations are located in this area, as well as land that has been set aside for mining.*

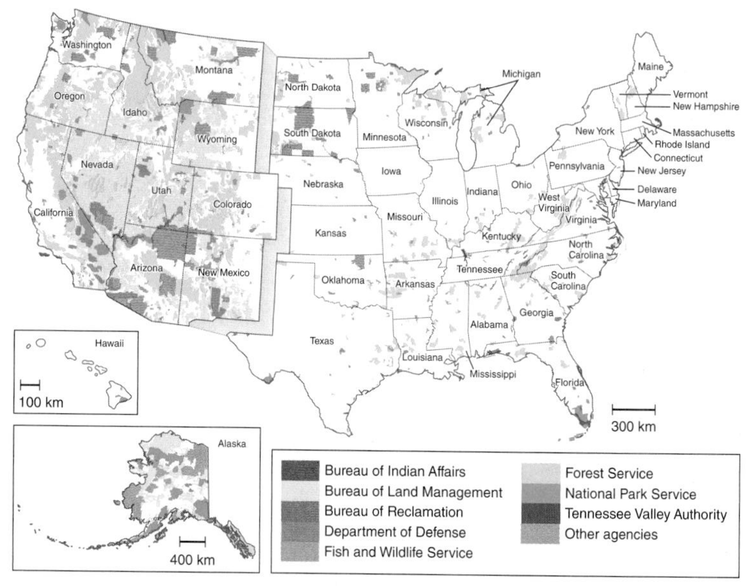

FIGURE 29.5 **Federal lands in the United States.** Approximately 42 percent of the land in the United States is publicly owned, with 25 percent of the nation's land owned by the federal government. *(After http://nationalatlas.gov)*

ethic, which calls for policy makers to consider the resource or monetary value of nature. The **resource conservation ethic** states that people should maximize resource use based on the greatest good for everyone. In conservation and land use terms, it has meant that areas are preserved and managed for economic, scientific, recreational, and aesthetic purposes.

Of course, many of these purposes are in conflict. In order to manage competing interests, the U.S. government has, for decades, adopted the principle of *multiple use* in managing its public resources. Some public lands are in fact classified as **multiple-use lands,** and may be used for recreation, grazing, timber harvesting, and mineral extraction. Others are designated as protected lands in order to maintain a watershed, preserve wildlife and fish populations, or maintain sites of scenic, scientific, and historical value.

Resource conservation ethic The belief that people should maximize use of resources, based on the greatest good for everyone.

Multiple-use lands A U.S. classification used to designate lands that may be used for recreation, grazing, timber harvesting, and mineral extraction.

Land Use and Federal Agencies

As shown in FIGURE 29.6, land in the United States, both public and private, is used for many purposes. These uses can be divided into a number of categories.

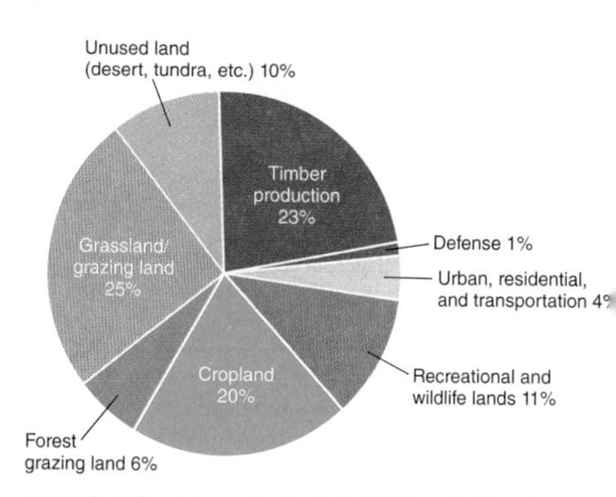

FIGURE 29.6 **Land use in the United States.** Public and private land in the United States is used for many purposes. *(Data from R. N. Lubowski et al., Major Land Uses in the United States, Economic Research Service, USDA, 2002.)*

▲ TEACHING with FIGURES

Show students Figure 29.6 and ask: What are the main uses of public and private lands in the United States? *The main uses of public and private lands are grassland/grazing land, 25 percent; timber production, 23 percent; cropland, 20 percent; recreational and wildlife lands, 11 percent; unused land, desert, and tundra, 10 percent; forest grazing land, 6 percent; urban, residential, and transportation, 4 percent; and defense, 1 percent.*

The probable use of public land determines how it is classified and which federal agency will manage it. More than 95 percent of all federal lands are managed by four federal agencies: the Bureau of Land Management (BLM), the United States Forest Service (USFS), the National Park Service (NPS), and the Fish and Wildlife Service (FWS). BLM, USFS, and NPS lands are typically classified as multiple-use lands because most, and sometimes all, public uses are allowed on them.

Although individual tracts may differ, the following are typical divisions of public land uses:

- BLM lands: grazing, mining, timber harvesting, and recreation
- USFS lands: timber harvesting, grazing, and recreation
- NPS lands: recreation and conservation
- FWS lands: wildlife conservation, hunting, and recreation

module 29

REVIEW

In this module, we have seen that the tragedy of the commons describes the tendency of shared, common resources to be over-exploited. Externalities are costs or benefits not included in the price of a good or service, and the tragedy of the commons can be the result of negative externalities. Maximum sustainable yield is a management technique that allows for the greatest amount of a resource that can be harvested from land without compromising the sustainability of that land and the resource. Public lands are classified and designated for various uses differently in different countries. In the United States, public lands fall under the jurisdiction of numerous federal agencies and can be managed for multiple uses. In the next module we will look in more detail at land management practices.

Module 29 AP® Review Questions

1. The tragedy of the commons can be prevented by
   (a) externalities.
   (b) fisheries in international waters.
   (c) the use of harvest permits.
   (d) the use of public land for grazing.
   (e) multiple-use land management.

2. Which is NOT an example of an externality?
   (a) The use of national forests by private companies
   (b) The use of the atmosphere for air pollution
   (c) The use of forested watersheds for aquifer purification
   (d) Roads maintained with a tax on gasoline
   (e) The use of oceans for stormwater drainage

3. The maximum sustainable yield of a population usually occurs
   (a) at the maximum population size.
   (b) at the minimum growth rate.
   (c) at half the carrying capacity.
   (d) at approximately 90 percent of the carrying capacity.
   (e) at twice the maximum growth rate.

4. Which type of protected land might allow limited mining?
   (a) A nature reserve
   (b) A national monument
   (c) A habitat management area
   (d) A managed resource protected area
   (e) A protected landscape

5. Which federal agencies typically classify their federal lands as multiple-use?
   I. The Bureau of Land Management
   II. The United States Forest Service
   III. The Fish and Wildlife Service

   (a) I only
   (b) I and II only
   (c) I and III only
   (d) II and III only
   (e) I, II, and III

Answers to Module 29 AP® Review Questions

1. c
2. d
3. c
4. d
5. e

Teaching Tip: Warm-up

Ask students the following questions:

- Why was the Taylor Grazing Act of 1934 created? *The Taylor Grazing Act of 1934 was created to halt overgrazing.*
- What were the goals of the Taylor Grazing Act of 1934? *It converted federal rangelands from a commons to a permit-based grazing system. The goal of a permit-based system is to limit the number of animals grazing in a particular area and thereby avoid a tragedy of the commons situation.*
- Have the goals of this legislation been achieved? *Critics maintain that the low cost of the permits continues to encourage overgrazing. In 2006, the federal government spent seven times more money managing its rangelands than it received in permit fees. In effect, grazing is subsidized with federal funds.*

module 30

Land Management Practices

We have seen that resources can be managed or they can be over-exploited, depending on a number of factors related to economics, human behavior, and specific management practices. Proper resource management is also dependent on specific aspects of the type of land involved as well as the agency and the guidelines under which it operates. In this module, we will examine land management practices in rangelands and forests, with a focus on timber management practices and fire management. We will also examine trends in residential land use and the causes and consequences of urban sprawl.

Learning Objectives

After reading this module, you should be able to

- explain specific land management practices for rangelands and forests.
- describe contemporary problems in residential land use and some potential solutions.

Land management practices vary according to land use

Now that we have a basic picture of how public land is classified and of the relationship between public land use and management agencies in the United States, let's turn to some of the specific issues involved in managing different types of public lands. Note that many of the management issues we discuss here apply to private lands as well.

Rangeland A dry open grassland.

Rangelands

Rangelands are dry, open grasslands primarily used for the grazing of cattle, which is the most common use of land in the United States. Rangelands are semiarid ecosystems and are therefore particularly susceptible to fires and other environmental disturbances. When humans overuse rangelands, those rangelands can easily lose biodiversity.

Like most human activities, livestock grazing has mixed environmental effects. One environmental benefit of grazing is that ungulates—hoofed animals such as cattle and sheep—can be raised on lands that are too dry to farm. In addition, grazing these animals uses less fossil fuel energy than raising them in feedlots. However, improperly managed livestock can damage stream banks and pollute surface waters. Grazing too many animals can quickly denude a region of vegetation

FIGURE 30.1 **Overgrazed rangeland.** Overgrazing can rapidly strip an area of vegetation. *(Rod Planck/Science Source)*

(FIGURE 30.1), which leaves the land exposed to wind erosion and makes it difficult for soils to absorb and retain water when it rains.

Many environmental scientists argue that rangeland ecosystems are too fragile for multiple uses. Certain environmental organizations have suggested that as much as 55 percent of U.S. rangeland soils are in poor or very poor condition, due in large part to overgrazing. However, the BLM, which manages most public rangelands in the United States, has maintained that the percentage of poor or very poor soils is not nearly that high. Reconciling this difference is a challenge because of the many factors that influence how soil condition is determined.

The Taylor Grazing Act of 1934 was passed to halt overgrazing. It converted federal rangelands from a commons into a permit-based grazing system. The goal of a permit-based system is to limit the number of animals grazing in a particular area and thereby avoid a tragedy of the commons situation. However, critics maintain that the low cost of the permits continues to encourage overgrazing. In 2009, the latest year for which data are available, the federal government spent seven times more money managing its rangelands than it received in permit fees. Thus, in effect, grazing is subsidized with federal funds.

The BLM focuses on mitigating the damage caused by grazing and considers "rangeland health" when it sets grazing guidelines. For example, state and regional rangeland managers are required to ensure healthy watersheds, maintain ecological processes such as nutrient cycles and energy flow, preserve water quality, maintain or restore habitats, and protect endangered species. However, because these managers are not given detailed guidance, and the BLM regulations do not require the involvement of environmental scientists, the managers have wide latitude to set their own guidelines and standards. As a result, BLM regulations are not consistently successful in preserving vulnerable rangeland ecosystems.

Forests

Forests are dominated by trees and other woody vegetation. Approximately 73 percent of the forests used for commercial timber operations in the United States are privately owned. Many national forests were originally established to ensure a steady and reliable source of timber, and commercial logging companies are currently allowed to harvest U.S. national forests, usually in exchange for a royalty that is a percentage of their revenues. The federal government typically spends more money managing the timber program and building and maintaining logging roads than it receives from these royalties, essentially subsidizing the timber program as it does grazing on rangelands.

Timber Harvest Practices

The two most common ways in which trees are harvested for timber production are *clear-cutting* and *selective cutting,* both of which are illustrated in FIGURE 30.2. **Clear-cutting** involves removing all, or almost all, the trees within an area (Figure 30.2a); it is the easiest harvesting method and, in most cases, the most economical. When a stand, or cluster, of trees has been clear-cut, foresters often replant or reseed the area. Often the entire area will be replanted at the same time, so all the resulting trees will be the same age. Because they are exposed to full sunlight, clear-cut tracts of land are ideal for fast-growing tree species that achieve their maximum growth rates with large amounts of direct sunlight. Because other species may not be so successful in these conditions, there can be an overall reduction in biodiversity. However, if a commercially valuable tree species constitutes only a small fraction of a stand of trees, it may not be economically efficient to clear-cut the entire stand. This is particularly true in many tropical forests, where economically valuable species constitute only a small percentage of the trees and are mixed in with many other less valuable species.

Clear-cutting, especially on slopes, increases wind and water erosion, which causes the loss of soil and

Forest Land dominated by trees and other woody vegetation and sometimes used for commercial logging.

Clear-cutting A method of harvesting trees that involves removing all or almost all of the trees within an area.

more **math** practice:

Ratios

Several free-response questions over the years have asked students to use ratios to solve math problems. More Math Practice 10.1 provides several problems for students to practice ratios.

Download this worksheet by clicking on the link buttons in in the TE-book, opening the TRFD, or logging onto the book's companion website bcs.whfreeman.com/friedlandapes2e (teacher login required).

 **More Math Practice 10.1:** Ratios

AP® Exam Tip

Questions about tree-harvesting practices have appeared on both the multiple-choice and free-response sections of the AP® exam. Make sure your students understand tree harvest techniques and the environmental costs and benefits of each method. For example, clear-cutting leaves the forest open to erosion; regrowth after clear-cutting will consist entirely of trees that are the same age; and because there are no large trees for shade, increased sunlight can raise the temperature of nearby rivers and streams. The benefit of this type of forestry is that it is less expensive and allows trees and other plants with high sunlight requirements to flourish as regrowth.

Teaching Tip: Journal Prompt

Ask students to answer the following questions:

- Define clear-cutting and selective cutting. *Clear-cutting involves removing all or almost all the trees within an area. Selective cutting removes single trees or relatively small numbers of trees from among many in a forest.*
- Does either clear-cutting or selective cutting provide more economic value? *Typically, clear-cutting a forest is more profitable because it takes less time and labor to harvest the trees. Selective cutting takes more time because the machinery has to work around remaining trees.*
- Which timber harvesting practice affects the environment more and why? *Clear-cutting, especially on slopes, increases wind and water erosion, which causes the loss of soil and nutrients. Erosion also adds silt and sediment to nearby streams, which harms aquatic populations. In addition, the denuded slopes are prone to dangerous landslides. Clear-cutting also increases the amount of sunlight that reaches rivers and streams, which raises water temperatures and can adversely affect certain aquatic species.*
- Which timber harvesting practice reduces biodiversity the most? *Clear-cutting reduces biodiversity the most because organisms that use the trees for food and shelter can no longer survive in the area. With selective cutting, not all trees are removed and species are able to remain in the area. Typically, bird populations will decrease after clear-cutting but some insect populations might increase due to large amounts of scrap wood piles left to rot.*

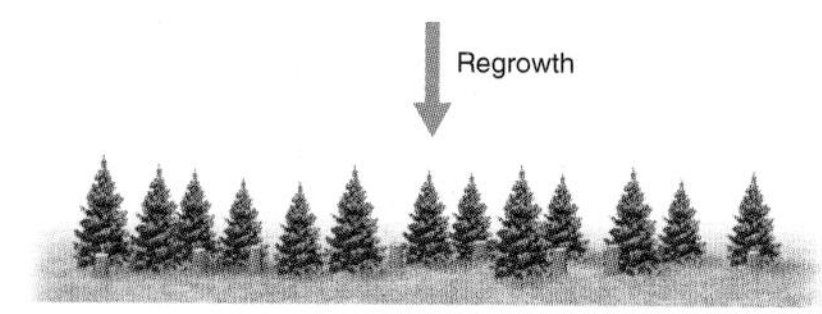

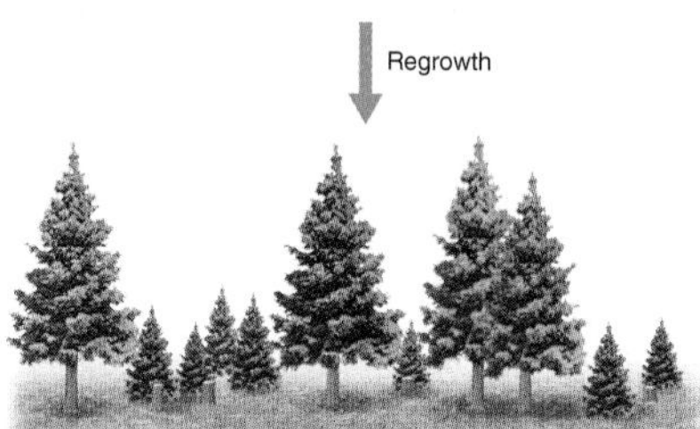

FIGURE 30.2 **Timber harvest practices.** (a) Clear-cutting removes most, if not all, trees from an area and is often coupled with replanting. The resulting trees are then all the same age. (b) In selective cutting, single trees or small numbers of trees are harvested. The resulting forest consists of trees of varying ages.

FIGURE 30.3 **Clear-cut forest near Eureka, California.** A clear-cut on a steep slope increases the likelihood of erosion and delayed regeneration of vegetation. *(Peter Menzel/Science Source)*

nutrients (FIGURE 30.3). Erosion also adds silt and sediment to nearby streams, which harms aquatic populations. In addition, the denuded slopes are prone to dangerous mudslides like the Stafford, California, mudslide described at the beginning of this chapter. Clear-cutting also increases the amount of sunlight that reaches rivers and streams, which raises water temperatures that can adversely affect certain aquatic species. Even the replanting process can have negative environmental consequences. Timber companies often use fire or herbicides to remove bushy vegetation before a clear-cut is replanted. These practices reduce soil quality, and herbicides may contaminate water that runs off into streams and rivers. Many environmental scientists identify clear-cutting as a cause of habitat alteration and destruction as well as forest fragmentation. These effects decrease biodiversity and sometimes lower the aesthetic value of the affected forest.

Selective cutting (Figure 30.2b) removes single trees or a relatively small number of trees from the larger forest. This method creates many small openings in a stand where trees can reseed or young trees can be planted, so the regenerated stand contains trees of different ages. Because seedlings and young trees must grow next to larger, older trees, selective cutting produces optimum growth only among shade-tolerant tree species.

The environmental impact of selective cutting is less extensive than that of clear-cutting. However, many of the negative environmental impacts associated with logging remain the same. For example, whether a company uses clear-cutting or selective cutting, it will need to construct logging roads to carry equipment and workers into the area that will be harvested. These roads fragment the forest habitat, which affects species diversity, and compact the soil, which causes nutrient loss and reductions in water infiltration.

Selective cutting The method of harvesting trees that involves the removal of single trees or a relatively small number of trees from among many in a forest.

Ecologically sustainable forestry An approach to removing trees from forests in ways that do not unduly affect the viability of other trees.

FIGURE 30.4 **Sustainable forestry.** Logging without the use of fossil fuels, as is done in this forest in Oregon, further enhances the sustainability of a forestry project. *(Michael Thompson/Earth Scenes/Animals Animals)*

A third approach to logging—**ecologically sustainable forestry**—removes trees from the forest in ways that do not unduly affect the viability of other, noncommercial tree species. This approach has a goal of maintaining both plants and animals in as close to a natural state as possible. Some loggers have even returned to using animals such as horses to pull logs in order to reduce soil compaction, although the costs of such methods make it difficult to compete economically with mechanized logging practices. (FIGURE 30.4)

Logging, Deforestation, and Reforestation

Approximately 30 percent of all commercial timber in the world is produced in the United States and Canada. Compared with South America and Africa, forest losses in these two major timber-producing countries have been relatively small over the last several decades. Still, timber production presents important ecological challenges in these two countries.

Throughout this chapter we have seen examples of the conflicts over land use created by competing interests and values. Perhaps nowhere is this conflict so clear as in the case of logging. For example, while timber production has always been a part of the mission of the United States Forest Service, maintaining biodiversity is an equally important goal. It would seem that we can't have both.

All logging disrupts habitat and usually has an effect, either negative or positive, on plant and animal species. One such species is the marbled murrelet (*Brachyramphus marmoratus*). This bird spends most of its life along the coastal waters of the Pacific Northwest, but it nests in coastal redwood forests. With the intensive logging of these forests, the marbled murrelet has become endangered. The tree sit described at the beginning of this chapter brought this species, among others, to the public's attention.

Logging often replaces complex forest ecosystems with **tree plantations,** which are large areas typically planted with a single rapidly growing tree species. These same-aged stands can be easily clear-cut for commercial purposes, such as pulp and wood, and then replanted. Because of this cycle of planting and harvesting, tree plantations never develop into mature, ecologically diverse forests. If too many planting and harvesting cycles occur, the soil may become depleted of important nutrients such as calcium.

Since 1982, federal regulations have required the USFS to provide appropriate habitat for plant and animal communities while at the same time meeting multiple-use goals. However, because these regulations fail to specify how biodiversity protection should be achieved or how the results should be quantified, the USFS has had to choose its own approach to biodiversity management. Critics charge that the USFS is not adequately protecting biodiversity and forest ecosystems, but the USFS maintains that it is doing the best it can at meeting many different objectives.

Fire Management

In many ecosystems, fire is a natural process that is important for nutrient cycling and regeneration. As discussed briefly in Chapter 3, when fires periodically move through an ecosystem, they liberate nutrients tied up in dead biomass. In addition, areas where vegetation is killed by fire provide openings for early-successional species.

Humans have followed a variety of management policies with respect to fire. For many years, managers of forest ecosystems, including the USFS, worked to suppress fires. This strategy led to the accumulation of large quantities of dead biomass on the forest floor, which built up until a large fire became inevitable. One method for reducing the accumulation of dead biomass is a **prescribed burn,** in which a fire is deliberately set under controlled conditions. Prescribed burns help reduce the risk of uncontrolled natural fires as well as providing some of the other benefits of fire. More recently, forest managers have allowed certain fires that have occurred naturally to burn. This policy appears to have been accepted in many parts of the United States, as long as human life and property are not threatened.

Probably the best-known forest fires in the United States are those that occurred in Yellowstone National Park in the summer of 1988, the driest year on record

Tree plantation A large area typically planted with a single rapidly growing tree species.

Prescribed burn A fire deliberately set under controlled conditions in order to reduce the accumulation of dead biomass on a forest floor.

AP® Exam Tip

Fire management and healthy forests are covered in the 2008 AP® Exam, Question 3. To answer this question, students must

- identify two characteristics of forests that develop when fires are suppressed and explain why fire suppression increases the risk of forest fires
- describe the positive and negative effects of the Healthy Forests Initiative
- describe ecosystem services that forests provide
- identify another biome that is naturally maintained by fire and explain the role of fire in this biome

Teaching Tip: Video

Prescribed Burns in National Parks
This brief video describes the importance of conducting prescribed burns. The video also shows students that prescribed burns are important for recycling nutrients back into the ecosystem to enhance future growth of native species.

Find a link to this resource by clicking on the link buttons in in the TE-book, opening the TRFD, or logging onto the book's companion website bcs .whfreeman.com/friedlandapes2e (teacher login required).

Chapter 10 Web Resources

(a)

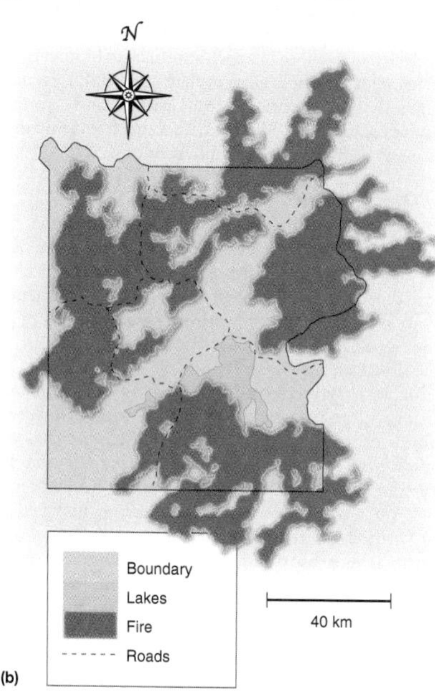

(b)

FIGURE 30.5 **Yellowstone fires of 1988.** As can be seen from the map, extensive areas of the park were burned in this exceptionally hot and dry year. *(Jonathan Blair/Corbis)*

at the park where a combination of human activity and lightning set off multiple fires (FIGURE 30.5). Over 25,000 people participated in fighting the fires. While this effort saved human lives and property, the firefighters had little impact on the fires. When the fires were out, more than one-third of Yellowstone National Park had burned. Initially, many people were outraged that the NPS and others had "allowed" the park to burn. However, within a few years, it became clear that the fires had created new, nutrient-rich habitat for early-successional plant species that attracted elk and other herbivores. Ultimately, researchers and forest managers concluded that the Yellowstone fires of 1988 provided many benefits to the Yellowstone ecosystem. Today, a typical visitor viewing a portion of the park that burned in 1988 probably wouldn't even know that there had been a major fire 25 years earlier.

National Parks

As we have already noted, many national parks were established to preserve scenic views and unusual landforms. Today, national parks are managed for scientific, educational, aesthetic, and recreational use. Since Yellowstone National Park was founded in 1872, 57 additional national parks have been established in the United States. Today the NPS manages a total of 400 national parks and other areas, such as historical parks and national monuments, and the list continues to grow.

The Goals of National Park Management

Management of national parks, like that of national forests, is based on the multiple-use principle. Unlike national forests, U.S. national parks were set aside specifically to protect ecosystems. In establishing Yellowstone National Park, Congress mandated the Department of the Interior to regulate the park in a manner consistent with the preservation of timber resources, mineral resources, and "natural curiosities." However, it did not require a management process based on ecological principles.

It was not until the 1960s that ecology became a focus of national park management. In 1963, an advisory board on wildlife management presented the Leopold Report to Secretary of the Interior Stewart Udall. This report established the guiding principles of national park management that are followed today. It proposed that the primary purpose of NPS should be to maintain the parks in the same biotic condition in which they were first found by European settlers. To this end, the authors believed that NPS should focus its efforts on conservation and on protection of wildlife species and their habitats. The report claimed that human activity had severely affected normal ecological processes in the parks and that active intervention was required to achieve the goal of a return to a more "natural" state.

Today, NPS applies knowledge gained from environmental science to maintain biodiversity and ecosystem function in all national parks. As we saw in the example of the Yellowstone fires, NPS policies continue to be controversial. Each national park adapts U.S. policy to its specific needs. In parks with high levels of endemic biodiversity, for example, management focuses on conserving endemic species. The Channel Islands National Park off the coast of southern California, for instance, is an important breeding ground for many seabirds, including a pelican species rarely found elsewhere in the western United States. Because of this, the Channel Islands National Park is managed primarily to conserve its biodiversity. Other national parks balance biodiversity protection with recreational use.

National Parks and Human Activities

Reducing the impact of human activities both outside and inside park borders is the primary challenge of most national parks throughout the world. Air and water pollution from distant sources can reduce biodiversity, recreational value, and economic opportunities. Ongoing development adjacent to park boundaries can be particularly problematic. In many locations, national parks have become islands of biodiversity amid increasing human development, but even the protected status of the parks cannot defend them completely from invasive species and other problems, such as pollution. These external threats require large-scale evaluation, planning, and management that extend beyond the borders of individual parks.

National parks are also victims of their popularity. Although the park system was established in part to make areas of great beauty accessible to people, human overuse can harm the very environment that attracts visitors. For example, all-terrain vehicles (ATVs) are a major cause of air and noise pollution in national parks, as well as a direct cause of habitat destruction (FIGURE 30.6). Today, many parks strictly limit or even ban the use of ATVs. Still, park managers grapple with how best to determine appropriate limits on human activity. In many cases, there is no easy answer to the trade off between short-term recreational uses and long term protection of biodiversity.

FIGURE 30.6 **ATV-caused damage near Olympic National Park in Washington State.** Although today many national parks limit ATV use, the conflict between those who want to use the parks for this form of recreation and those who wish to protect biodiversity remains. *(Transtock/SuperStock)*

Wildlife Refuges and Wilderness Areas

National wildlife refuges are the only federal public lands managed for the primary purpose of protecting wildlife. The Fish and Wildlife Service manages more than 450 national wildlife refuges and 28 waterfowl production areas on 34.4 million hectares (85 million acres) of publicly owned land.

National wilderness areas are set aside with the intent of preserving large tracts of intact ecosystems or landscapes. Sometimes only a portion of an ecosystem is included. Wilderness areas are created from other public lands, usually national forests or rangelands, and are managed by the same federal agency that managed them prior to their designation as wilderness areas.

These areas allow only limited human use and are designated as roadless. Although logging, road building, and mining are banned in national wilderness areas, roads that existed before the designation sometimes remain in use, and activities such as mining that were previously permitted on the land are allowed to continue. More than 38.5 million hectares (95 million acres) of federal land, 60 percent of which is in Alaska, are classified as wilderness.

Federal Regulation of Land Use

Government regulation can influence the use of private as well as public lands. The 1969 **National Environmental Policy Act (NEPA)** mandates an environmental assessment of all projects involving federal money or federal permits. Along with other major laws of the 1960s and 1970s, such as the Clean Air Act, the Clean Water Act, and the *Endangered Species Act*, NEPA creates an environmental regulatory process designed to ensure protection of the nation's resources.

Before a project can begin, NEPA rules require the project's developers to file an *environmental impact statement*. An **environmental impact statement (EIS)** typically outlines the scope and purpose of the project,

National wildlife refuge A federal public land managed for the primary purpose of protecting wildlife.

National wilderness area An area set aside with the intent of preserving a large tract of intact ecosystem or a landscape.

National Environmental Policy Act (NEPA) A 1969 U.S. federal act that mandates an environmental assessment of all projects involving federal money or federal permits.

Environmental impact statement (EIS) A document outlining the scope and purpose of a development project, describing the environmental context, suggesting alternative approaches to the project, and analyzing the environmental impact of each alternative.

Teaching Tip: Warm-up

Ask students: What is the significance of the national wilderness area designation for parts of federally owned lands? *The national wilderness areas are set aside with the intent of preserving large tracts of intact ecosystems or landscapes. Sometimes only a portion of an ecosystem is included. Wilderness areas are created from other public lands (usually national forests or rangelands) and are managed by the same federal agency that managed them prior to their designation as a wilderness.*

Teaching Tip: Warm-up

Ask students: What is NEPA, and what is an environmental impact statement (EIS)? *NEPA stands for National Environmental Policy Act (NEPA), which mandates an environmental assessment of all projects involving federal money or federal permits. Before a project can begin, NEPA rules require the project's developers to file an EIS. An EIS typically outlines the scope and purpose of the project, describes the environmental context, suggests alternative approaches to the project, and analyzes the environmental impact of each alternative.*

Teaching Tip: Beyond the Classroom

Ask students to use the questions below to write a short report on one of the 59 U.S. national parks.

- Describe the park. How large is it? What is its geography? Does it include any waterways?
- What year was this national park created?
- Describe the climate in the park. Does it fall within one biome or more than one?
- Describe the plants and animals. Is it a breeding ground for one or more species?
- Choose one species of plant or animal found in the national park and describe its role in the park's ecosystem.

Teaching Tip: Journal Prompt

Ask students to describe the difference between suburban and exurban areas and give examples from their personal experience. *Suburban areas surround metropolitan centers and have low population densities compared with urban areas. Exurban areas are similar to suburban areas, but are unconnected to any central city or densely populated area.*

Teaching Tip: Warm-up

Ask students: What is urban sprawl and what are the four main causes of it in the United States? *Urban sprawl is urbanized areas that spread into rural areas and remove clear boundaries between the two. Automobiles and highway construction, living costs, urban blight, and government policies are four causes of urban sprawl.*

describes the environmental context, suggests alternative approaches to the project, and analyzes the environmental impact of each alternative. NEPA does not require that developers proceed in the way that will have the least environmental impact. However, in some situations, NEPA rules may stipulate that building permits or government funds be withheld until the developer submits an **environmental mitigation plan** stating how it will address the project's environmental impact. In addition, preparation of the EIS sometimes uncovers the presence of endangered species in the area under consideration. When this occurs, managers must apply the protection measures of the **Endangered Species Act,** a 1973 law designed to protect species from extinction.

Members of the public are entitled to comment on the environmental assessment and decision makers are required to respond. And, although developers are not obligated to act in accordance with public wishes, in practice public concern often improves the project's outcome. For this reason, attending informational sessions and providing input is a good way for concerned citizens to learn more about local land use decisions and to help reduce the environmental impact of land development.

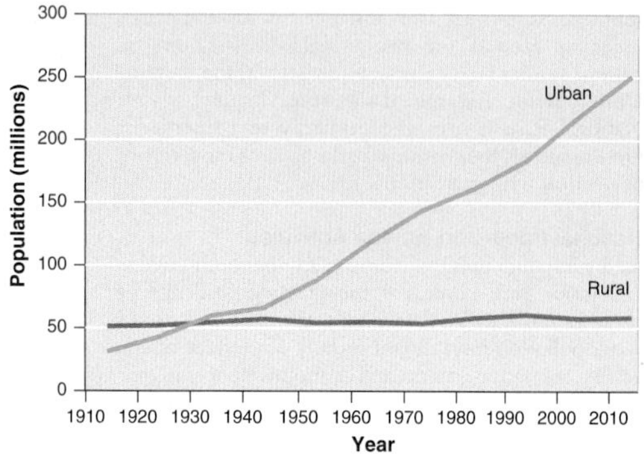

FIGURE 30.7 Distribution of urban and rural populations in the United States between 1910 and 2012. This graph shows a dramatic shift in the population from rural to urban areas. *(Data from http://www.census.gov/geo/reference/ua/urban-rural-2012.html)*

Residential land use is expanding

Roughly a century ago, large numbers of people in the United States began to move from rural areas to large cities. The last 50 years have seen an increased movement from those cities to the surrounding areas, which has resulted in the growth of two classes of communities: *suburban* and *exurban*. **Suburban** areas surround metropolitan centers and have low population densities compared with urban areas. **Exurban** areas are similar to suburban areas, but are not connected to any central city or densely populated area. Since 1950, more than 90 percent of the population growth in metropolitan areas has occurred in suburbs, and two out of three people now live in suburban or exurban communities.

The U.S. census bureau currently categorizes a population as urban if the area contains more than 2,500 people. FIGURE 30.7 shows U.S. urban and rural populations between 1910 and 2012. The rural population in the United States has been roughly the same since 1910, and a century later it makes up less than a fifth of the total U.S. population. Although this population shift has brought with it a new set of environmental problems associated with *urban sprawl,* attempts to find creative solutions have been increasingly successful.

Environmental mitigation plan A plan that outlines how a developer will address concerns raised by a project's impact on the environment.

Endangered Species Act A 1973 U.S. act designed to protect species from extinction.

Suburb An area surrounding a metropolitan center, with a comparatively low population density.

Exurb An area similar to a suburb, but unconnected to any central city or densely populated area.

Urban sprawl Urbanized areas that spread into rural areas, removing clear boundaries between the two.

Causes and Consequences of Urban Sprawl

If you have ever been to a strip mall, you are familiar with the phenomenon known as **urban sprawl**—urbanized areas that spread into rural areas and remove clear boundaries between the two. The landscape in these areas is characterized by clusters of housing, retail shops, and office parks, which are separated by miles of road. Large feeder roads and parking lots that separate "big box" and other retail stores from the road discourage pedestrian traffic.

Urban sprawl has had a dramatic environmental impact. Dependence on the automobile causes suburban residents to drive more than twice as much as people who live in cities. Between 1950 and 2000, the number of vehicle miles traveled per person in U.S. suburban areas tripled. Because suburban house lots tend to be significantly larger than their urban counterparts, suburban communities also use more than twice as much land per person as urban communities. Urban sprawl tends to occur at the edge of a city, often replacing farmland and increasing the distance between farms and consumers. In its most recent survey, the U.S. Department of Agriculture estimated that between 1982 and 2007, U.S. farmland was being converted to residential uses at a rate of 405,000 ha (1 million acres) per year. There are four main causes of urban sprawl in the United States: automobiles and highway construction, living costs, urban blight, and government policies.

Automobiles and Highway Construction

Before automobiles and highway systems existed, transportation into and out of cities was difficult: Horses were slow, roads were bad, and trolley services rarely went far beyond the city limits. In those days, if you wanted to take advantage of the many amenities of city life, such as job opportunities, cultural institutions, shopping, and social activities, you had to live within a few miles of the city center.

The advent of the automobile, and the subsequent development of the interstate highway system in the 1950s and 1960s, changed everything. Today we think nothing of working in the city during the day and commuting home to the suburbs at night. And if you live in the suburbs but want to get into the city on a Saturday night for dinner, a concert, or a movie, that's no problem either. With rapid, comfortable transportation between urban and suburban areas, it became possible to work or play in the city and then return to a large home in a quiet neighborhood. For the first time in history, people could enjoy the best of both worlds.

Living Costs

Land in the suburbs, because it is readily available, is relatively inexpensive when compared with land in the city. For the cost of a tiny one-bedroom condominium in a desirable section of a city, you may be able to purchase a five-bedroom house with a big yard in the suburbs. In addition, because suburban governments often provide fewer public services than cities, tax rates in the suburbs are likely to be lower. For these reasons, moving to the suburbs allows some people to enjoy a higher standard of living than they could afford in the city.

Although single-family homes in the suburbs are likely to be less expensive than desirable housing in the city, they are still out of reach for many people. And because single-family homes are usually the only housing option available in the suburbs, those with lower incomes are excluded. Most suburban communities have little, if any, low-income or "affordable" housing. Furthermore, in many suburban locations, it is difficult to commute to the city without a car, which compounds the difficulties for lower-income individuals. Even when public transportation is available, commuting costs can be high.

Urban Blight

We have seen that urban sprawl encourages population shifts to the suburbs. As people leave the city, city revenue from sources such as property, sales, and service taxes begins to shrink. However, the cost of maintaining urban services, including public transportation, police and fire protection, and social services, does not decrease. Faced with declining tax receipts, cities are forced to reduce services, raise tax rates, or both. As services decline, crime rates may increase, either because police resources are stretched thin or because conditions for lower-income residents decline even further. Infrastructure also deteriorates, leading to a decline in the quality of the built environment. These problems, combined with higher taxes, have made cities less attractive places to live in previous decades, and those who can afford to move away have often been likely to do so.

In addition, as the population shifts to the suburbs, jobs and services follow. Suburbanization has spawned suburban office parks, which have led to an increase in suburb-to-suburb commuting. Commuting patterns develop around cities rather than into and out of them, and these new traffic patterns make it more difficult to provide public transportation to the spreading region. As wealthy and middle-income people leave cities, urban retail stores lose customers and are forced to close, giving people even fewer reasons to go to the city to shop. This further decreases the customer base for the remaining stores. This cascade of effects leads to the positive feedback loop shown in FIGURE 30.8. This loop creates **urban blight**—the degradation of the built and social environments of the city that often accompanies and accelerates migration to the suburbs.

Historically, urban blight has contributed to racial segregation. In the 1950s and 1960s, when migration to the suburbs began in earnest, those leaving the cities for the suburbs were predominantly middle- or upper-income Caucasians. This so-called "white flight" resulted in highly concentrated minority populations in city centers and almost entirely white populations in the suburbs. Over time, the disparity of opportunity increased because tax revenues in suburban communities often allowed for better schools. In a number of U.S. cities, some of this segregation has begun to decline as Caucasians return to the cities and more minorities, after having accumulated wealth, move out to the suburbs. In addition, many young professionals are making the decision to work and live in cities. Nevertheless, racial disparities remain.

Urban blight The degradation of the built and social environments of the city that often accompanies and accelerates migration to the suburbs.

Teaching Tip: Measuring Your Impact

The Cost of Commuting

Measuring Your Impact 10 offers students the opportunity to calculate the effect of commuting to work from the suburbs versus walking or taking public transportation. Download this resource by clicking on the link buttons in in the TE-book, opening the TRFD, or logging onto the book's companion website bcs.whfreeman.com /friedlandapes2e (teacher login required).

Measuring Your Impact 10: The Cost of Commuting

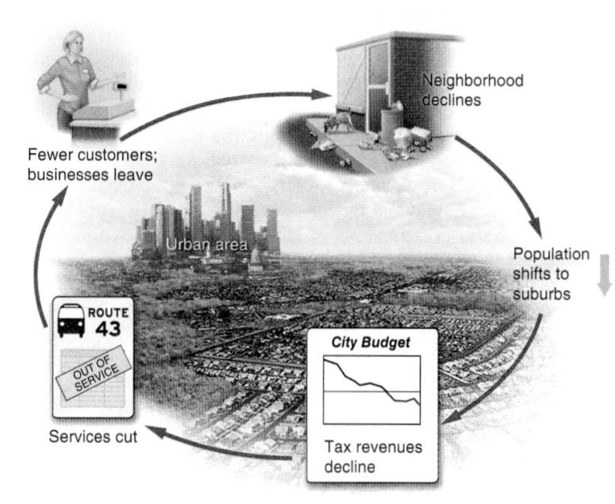

FIGURE 30.8 **Urban blight.** As people move away from a city to suburbs and exurbs, the city often deteriorates, which causes yet more people to leave. This cycle is an example of a positive feedback system. The green arrow indicates the starting point of the cycle.

Teaching Tip: Activity

After reading and discussing the four main causes of urban sprawl in the United States, ask students to describe the steps that lead to urban blight and draw a positive feedback loop that depicts these steps.

- ***Step 1:*** *The population shifts to the suburbs, taking with it revenue such as property taxes, sales taxes on goods, and service taxes on water and gas.*
- ***Step 2:*** *The city's budget begins a slow decline, leaving less money for maintaining urban services.*
- ***Step 3:*** *There is less money to operate and maintain public transportation, the police department, the fire department, education, and other social services.*
- ***Step 4:*** *As services decline, crime rates may increase because of fewer police resources. Infrastructure also deteriorates, leading to a decline in the quality of the built environment. There are fewer customers for the urban businesses, so many businesses close or relocate.*
- ***Step 5:*** *Neighborhoods decline, and residents who can afford to move away.*

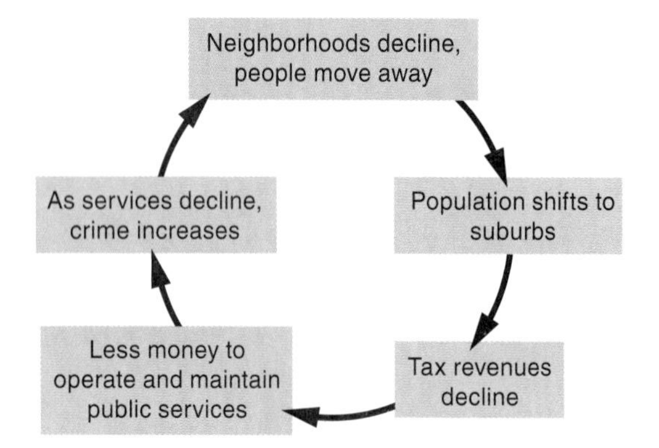

Government Policies

Urban sprawl has also been enhanced by federal and local laws and policies, including the Highway Trust Fund, zoning laws, and subsidized mortgages.

The **Highway Trust Fund,** begun by the Highway Revenue Act of 1956 and funded by a federal gasoline tax, pays for the construction and maintenance of roads and highways. We have already seen that highways allow people to live farther from where they work. FIGURE 30.9 shows the resulting positive feedback loop. More highways mean more driving and more gasoline purchases, which lead to more gasoline tax receipts that pay for more roads, and so on. As people move farther away from their jobs, traffic congestion increases, and roads are expanded. But the new, even larger roads encourage even more people to live farther away from work. This cycle exemplifies a phenomenon known as **induced demand,** in which an increase in the supply of a good causes demand to grow.

Governments may use *zoning* to address issues of traffic congestion, urban sprawl, and urban blight. **Zoning** is a planning tool developed in the 1920s and used to separate industry and business from residential neighborhoods, thereby creating quieter, safer communities. Governments that use zoning can classify land areas into "zones" in which certain land uses are restricted. For instance, zoning ordinances might prohibit developers from building a factory or a strip mall in a residential area or a multi-dwelling apartment building in a single-home neighborhood. Nearly all metropolitan governments across the United States have adopted zoning.

Zoning often regulates much more than land use. The number of parking spaces a building must have, how far from the street a building must be placed, or even the size and location of a home's driveway are among the development features that zoning may stipulate. Zoning laws have been helpful in addressing issues of safety and sometimes in minimizing environmental damage caused by new construction. One negative aspect of zoning, however, is that it generally prohibits suburban neighborhoods from developing a traditional "Main Street" with shops, apartments, houses, and businesses clustered together. Many communities are now attempting to incorporate **multi-use zoning,** which allows retail and high-density residential development to coexist in the same area. However, most zoning in the United States continues to promote automobile-dependent development.

Highway Trust Fund A U.S. federal fund that pays for the construction and maintenance of roads and highways.

Induced demand The phenomenon in which an increase in the supply of a good causes demand to grow.

Zoning A planning tool used to separate industry and business from residential neighborhoods.

Multi-use zoning A zoning classification that allows retail and high-density residential development to coexist in the same area.

Teaching Tip: Discussion Starter

Ask students: What is zoning and how can it help reduce urban sprawl? *Zoning is a planning tool developed in the 1920s to separate industry and business from residential neighborhoods with a goal of creating quieter, safer communities. Governments that use zoning can classify land areas into "zones" in which certain land uses are restricted.*

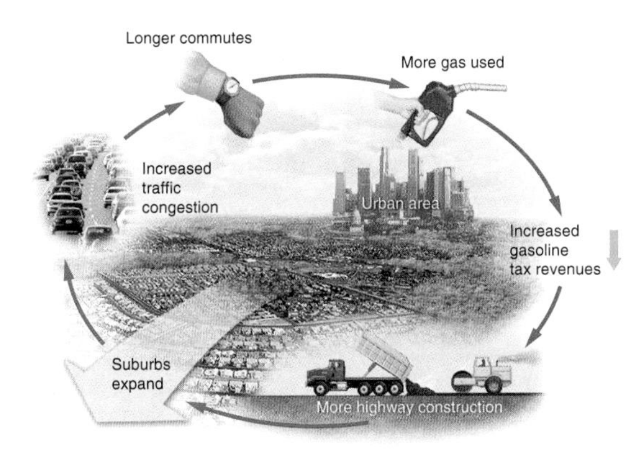

FIGURE 30.9 **Induced demand as a cause of traffic congestion and urban sprawl.** The use of gasoline tax money to build highways leads to the development of suburbs and traffic congestion, at which point yet more money is spent on highways to alleviate the congestion. The green arrow indicates the starting point of the cycle.

The federal government has also played a large part in encouraging the growth of suburbs through the Federal Housing Administration (FHA). Congress established the FHA during the Great Depression of the 1930s, in part to jump-start the economy by creating more demand for new housing. Through the FHA, people could apply for federally subsidized mortgages that offered low interest rates, which allowed many people who otherwise could not afford a house to purchase one. However, the FHA financed mortgages only for homes in financially "low risk" areas that were almost always the newly built, low-density suburbs. At the end of World War II, the GI Bill extended generous credit terms to war veterans, and the suburban housing boom continued. Suburbs became even more attractive as cars became more affordable and mass-produced housing allowed developers to sell suburban homes at affordable prices.

Smart Growth

People are beginning to recognize and address the problems of urban sprawl. One approach, **smart growth,** focuses on strategies that encourage the development of sustainable, healthy communities. The Environmental Protection Agency lists 10 basic principles of smart growth:

1. *Create mixed land uses.* Smart growth mixes residential, retail, educational, recreational, and business land uses. Mixed-use development allows people to walk or bicycle to various destinations and encourages pedestrians to be in a neighborhood at all times of the day, increasing safety and interpersonal interactions.
2. *Create a range of housing opportunities and choices.* By providing housing for people of all income levels, smart growth counters the concentration of poverty in failing urban neighborhoods. Mixed housing also allows more people to find jobs near where they live, improves schools, and generates strong support for neighborhood transit stops, commercial centers, and other services.
3. *Create walkable neighborhoods.* Walkable neighborhoods are created by mixing land uses, reducing the speed of traffic, encouraging businesses to build stores directly up to the sidewalk, and placing parking lots behind buildings. In neighborhoods that encourage walking, people use their cars less, reducing fossil fuel use and traffic congestion and providing health benefits. Communities with more pedestrians tend to see more interaction among neighbors because people stop to talk with each other. This, in turn, creates opportunities for civic engagement. When people interact, the environment usually benefits.

Smart growth A set of principles for community planning that focuses on strategies to encourage the development of sustainable, healthy communities.

Teaching Tip: Activity

Divide your students into 10 groups and assign one of the Environmental Protection Agency's 10 basic principles of smart growth (listed on pages 347–349) to each group. The groups should prepare a 3- to 5-minute presentation on the assigned principle. This can be done in class or as a homework assignment.

FIGURE 30.10 **The French Quarter of New Orleans.** One principle of smart growth is to foster communities with a strong sense of place. The French Quarter of New Orleans, Louisiana, is known for its architecture, food, and, especially, music. *(Design Pics Inc./Alamy)*

4. *Encourage community and stakeholder collaboration in development decisions.* There is no single "right" way to build a neighborhood; residents and **stakeholders**—people with an interest in a particular place or issue—need to work together to determine how their neighborhoods will appear and be structured.
5. *Take advantage of compact building design.* Smart growth incorporates multistory buildings and parking garages—as opposed to sprawling parking lots—to reduce a neighborhood's environmental footprint and protect more open space. Ideally, shops, cafés, and small businesses should be easily accessible to pedestrian traffic on the ground floor near sidewalks, with two or three stories of apartments and offices above.
6. *Foster distinctive, attractive communities with a strong sense of place.* A **sense of place** is the feeling that an area has a distinct and meaningful character. Many cities have such unique neighborhoods. For example, the French Quarter of New Orleans (FIGURE 30.10) has a sense of place that adds to the quality of life for people living there. Smart growth attempts to foster this sense of place through development that fits into the neighborhood.
7. *Preserve open space, farmland, natural beauty, and critical environmental areas.* Working farmland is a source of fresh local produce and other goods. Open space provides opportunities for recreation and enjoyment. Land protected from development through restricted growth also provides habitats for a variety of species.
8. *Provide a variety of transportation choices.* **Transit-oriented development (TOD)** attempts to focus dense residential and retail development around public transportation stops, giving people convenient alternatives to driving. Bicycle racks and safe roads, pleasant sidewalks for walking, frequent bus service, and light rail service can all aid in this goal (FIGURE 30.11). Car-sharing networks such as Zipcar can provide easy access to a fleet of rental automobiles. This increasingly popular

Stakeholder A person or organization with an interest in a particular place or issue.

Sense of place The feeling that an area has a distinct and meaningful character.

Transit-oriented development (TOD) Development that attempts to focus dense residential and retail development around stops for public transportation, a component of smart growth.

Infill Development that fills in vacant lots within existing communities.

Urban growth boundary A restriction on development outside a designated area.

FIGURE 30.11 **Light rail system.** Light rail is one option in transit-oriented development (TOD), which strives to develop communities with a denser mix of retail and residential components around a convenient public transportation system. *(AP Photo/Steve C. Wilson)*

service reduces the need for private car ownership where public transportation is not available.

9. *Strengthen and direct development toward existing communities.* Development strategies can help to reinvigorate urban neighborhoods that are caught in a vicious cycle of depopulation and blight and can protect rural lands from sprawl. Development that fills in vacant lots within existing communities, rather than expanding into new land outside the city, is known as **infill.** Some cities, such as Portland, Oregon, have had success with **urban growth boundaries,** which place restrictions on development outside a designated area.

10. *Make development decisions predictable, fair, and cost-effective.* Suburban developments throughout the country often look the same because standardized designs allow developers to move rapidly through the permitting process. A streamlined approval process that encourages smart growth could increase the number of individualized plans rather than promoting cookie-cutter development.

Of course, no individual new development or neighborhood plan is likely to incorporate all of these ideals, but as a guide to thinking about how to build communities, the smart growth concept has been quite successful.

Smart growth can have important environmental benefits. For example, compact development reduces the amount of impervious surface, reducing runoff and flooding downstream. A 2000 study found that smart growth in New Jersey would reduce water pollution by 40 percent compared with the more common, dispersed growth pattern. By mixing uses and providing transportation options, smart growth also reduces fossil fuel consumption. A 2005 study in Seattle found that residents of neighborhoods incorporating just a few of the techniques to make non-auto travel more convenient traveled 26 percent fewer vehicle miles than residents of more dispersed, less-connected neighborhoods. A 2012 San Francisco State University study found that similar smart growth practices led to a 20 percent reduction in miles traveled in 18 urban areas across the United States.

module

30 REVIEW

In this module, we saw that land management varies by land type. Rangelands are managed to maintain ecological processes such as nutrient cycles and to preserve water quality. Forests may be clear-cut or selectively cut. Clear-cutting may have greater environmental consequences than selective cutting. Tree plantations are sometimes established as replacement forest after clear-cutting. Management of forests may include the use of fire. Residential land use is expanding and urban and suburban sprawl are major environmental problems. Smart growth is a response to the problems of sprawl and includes many innovative strategies that are being implemented around the United States.

Module 30 AP® Review Questions

1. Which of the following is NOT a problem in rangeland management?
   (a) Fires
   (b) Overgrazing
   (c) Poor soil conditions
   (d) The high cost of grazing permits
   (e) Inconsistent management standards and guidelines

2. Which is used to reduce the impact of timber harvesting?
   (a) Tree plantations
   (b) Selective cutting
   (c) Prescribed burns
   (d) Mechanized logging
   (e) Clear-cutting

3. The use of environmental impact statements was mandated by
   (a) the Endangered Species Act of 1973.
   (b) the Leopold Report in 1963.
   (c) the Clean Air Act of 1963.
   (d) the National Environmental Policy Act of 1969.
   (e) the Clean Water Act of 1972.

Answers to Module 30 AP® Review Questions

1. d
2. b
3. d
4. b
5. d

Teaching Tip: Debate the Issue

Should the government have the right to take a resident's property?
To prepare for the debate, have students research the following questions:

- How does eminent domain work?
- What are the economic costs and benefits of eminent domain?
- What are the environmental costs and benefits of eminent domain?
- What are the social costs and benefits of eminent domain?

In class, select groups to debate the issue. Allow each side 2 minutes to prepare, 3 minutes to present, and 1 minute for rebuttal. Have the remainder of the class take notes, ask questions, and decide which side has the most persuasive argument.

4. Which of the following is NOT a cause of urban sprawl?
   (a) Automobile and highway construction
   (b) Mixed land use
   (c) Living costs
   (d) Urban blight
   (e) Government policies

5. The creation of walkable neighborhoods is
   (a) a result of urban sprawl.
   (b) one goal of the Federal Housing Administration.
   (c) discouraged by multi-use zoning.
   (d) a principal of smart growth.
   (e) a common feature of suburbs.

working toward  **sustainability**

What Are the Ingredients for a Successful Neighborhood?

The Dudley Street area of Roxbury and North Dorchester in Boston was once a prime example of urban blight. In the 1980s, years of urban decay and the loss of large numbers of primarily Caucasian families to the suburbs had left 21 percent of the Dudley Street neighborhood vacant—amounting to 1,300 abandoned parcels. Almost 30 percent of neighborhood residents had an income below the federal poverty level, making the neighborhood one of the poorest in Boston. Fires were a particular problem; in some cases, arsonists attempted to gain insurance money on homes that could not be sold. The residents felt that they were being ignored by City Hall. Some suggested that this was because 96 percent of the neighborhood's residents were members of ethnic minority groups.

In 1984, residents banded together to turn their neighborhood around. They formed the Dudley Street Neighborhood Initiative (DSNI), designed to allow the residents to move toward a common vision for a sustainable neighborhood. Participants chose a large board of directors—34 members—so they would hear a diversity of perspectives and, in coming to consensus, ensure that decisions had broad support. The DSNI also obtained something no other neighborhood organization had: the power of *eminent domain*. **Eminent domain** allows a government to acquire property at fair market value even if the owner does not wish to sell it. It is frequently used to acquire land for highway projects, but also has been used recently, and controversially, in urban redevelopment.

By 1987, DSNI had worked with community members to develop a comprehensive revitalization plan, which has been periodically updated since then. The main goals of the DSNI plan are

- to rehabilitate existing housing.
- to construct homes that are affordable according to criteria set by residents.
- to assemble parcels of vacant land for redevelopment, using the power of eminent domain if necessary.

Eminent domain A principle that grants government the power to acquire a property at fair market value even if the owner does not wish to sell it.

The Dudley Street neighborhood. This neighborhood in Boston, Massachusetts, was once a symbol of urban decay. Today it is a thriving urban community that has adopted many of the principles of smart growth. *(George Rizer/The Boston Globe via Getty Images)*

- to plan environmentally sound, affordable development that is physically attractive.
- to convert some vacant properties into safe play areas, gardens, and facilities that the entire community can enjoy.
- to run a full summer camp program for area children.
- to develop strong public and private partnerships to ensure the economic vitality of the neighborhood.
- to increase both the economic and political power of residents.

DSNI has had many successes. Its first major action was to force the city to remove trash, appliances, and abandoned cars that littered the streets and vacant lots. The city also cleaned up two illegal dumps in the area. Residents planted community gardens to grow produce and flowers. DSNI successfully reduced drug dealing in the neighborhood park, although keeping the drug dealers out remains a constant struggle. And, perhaps most significantly, its work led to the construction of 300 new homes on formerly vacant lots. New residents help to add vitality to the neighborhood, reversing the cycle of depopulation and business closure.

It is evident that the individuals involved in DSNI have taken many of the principles of smart growth to heart. The community mixes residential with retail development, and residents live within walking distance of a grocery store, ethnic markets, and other amenities. Moreover, since the founding of DSNI, two small manufacturing businesses—a furniture maker and an electronics company—have moved into the neighborhood, providing additional jobs within walking distance for residents.

The Dudley Street neighborhood still has relatively low per capita income, and its development choices have not been without controversy. However, it serves as one example of hundreds of neighborhoods that have begun to turn the positive feedback loop of urban decay into one of urban renewal and hope.

Critical Thinking Questions

1. What are some of the features of a community that will make it sustainable?
2. How has the Dudley Street neighborhood engaged in "smart growth"?
3. List and describe three ideas for "smart growth" in your own community.

References

Bernfield, F. K., J. Terris, and N. Vorsanger. 2003. *Solving Sprawl: Models of Smart Growth in Communities Across America*. Island Press.

Dudley Street Neighborhood Initiative. http://www.dsni.org/

Suggested Answers to Critical Thinking Questions

1. Features of a community that make it sustainable include having a nearby area for residents to work, having a good education system, and having a mix of land uses that allow for different community activities and opportunities.

2. DSNI has attracted businesses and created a more walkable community, created and restored natural/garden areas in previously vacant lots, and developed more housing options for new residents.

3. Possible answers may include:
- creating paths and lanes for the use of bicycles and encouraging public transportation options
- making vacant lots into community gardens or parks
- encourage a mix of housing options as well as new businesses

chapter 10 REVIEW

In this chapter we have looked at issues involving human use of land. The tragedy of the commons suggests that common-use resources may be overexploited. Maximum sustainable yield is a management technique that should allow for the greatest amount of harvest from a resource. Public lands are classified in a variety of ways around the world and in the United States. Rangeland and forest land are managed differently. In forests, clear-cutting and selective cutting are two ways to manage tree removal. Fire is also a tool used to manage forests. An increasingly greater percentage of the U.S. population lives in or near cities, a fact that has created a number of environmental problems. Some of these problems are being addressed by the principals of smart growth.

Exploring the Literature

Farr, D. 2007. *Sustainable Urbanism: Urban Design with Nature.* John Wiley & Sons.

Flint, A. 2006. *This Land: The Battle over Sprawl and the Future of America.* Johns Hopkins University Press.

McCann, B. 2013. *Completing Our Streets: The Transition to Safe and Inclusive Transportation Networks.* Island Press.

Randolph, J. 2011. *Environmental Land Use Planning and Management,* 2nd ed. Island Press.

U.S. Environmental Protection Agency Smart Growth. http://www.epa.gov/dced/index.htm.

Download a printable version of this list by clicking on the link buttons in the TE-book, opening the TRFD, or logging onto the book's companion website bcs.whfreeman.com/friedlandapes2e (teacher login required).

 **Exploring the Literature 10**

Key Terms

Tragedy of the commons
Externality
Maximum sustainable yield (MSY)
Resource conservation ethic
Multiple-use lands
Rangeland
Forest
Clear-cutting
Selective cutting
Ecologically sustainable forestry
Tree plantation
Prescribed burn
National wildlife refuge
National wilderness area
National Environmental Policy Act (NEPA)
Environmental impact statement (EIS)
Environmental mitigation plan
Endangered Species Act
Suburb
Exurb
Urban sprawl
Urban blight
Highway Trust Fund
Induced demand
Zoning
Multi-use zoning
Smart growth
Stakeholder
Sense of place
Transit-oriented development (TOD)
Infill
Urban growth boundary
Eminent domain

Learning Objectives Revisited

Module  29 Land Use Concepts and Classification

- **Explain how human land use affects the environment.**

  Individuals have no incentive to conserve common resources when they do not bear the cost of using those resources. A cost or benefit not included in the price of a good is an externality. The lack of incentive to conserve common resources leads to the overuse of these resources, which may be degraded if their use is not regulated. The maximum sustainable yield (MSY) is the largest amount of a renewable resource that can be harvested indefinitely. Harvesting at the MSY keeps the resource population at about one-half the carrying capacity of the environment.

- **Describe the various categories of public land used globally and in the United States.**

  In the United States, public land is managed for multiple uses, including grazing, timber harvesting, recreation, and wildlife conservation. The Bureau of Land Management manages rangeland, which is used for grazing. The United States Forest Service manages national forests, which are used for timber harvesting, recreation, and other uses.

Module 30 Land Management Practices

- **Explain specific land management practices for rangelands and forests.**

  Rangelands are grazed by cattle and sheep and need to be managed to prevent overgrazing. Timber can be harvested by clear-cutting or selective cutting, both of which have environmental impacts, or by ecologically sustainable forestry methods.

- **Describe contemporary problems in residential land use and some potential solutions.**

  Causes of urban sprawl include the development of the automobile, construction of highways, less expensive land at the urban fringe, and urban blight. Government institutions and policies, such as the federal Highway Trust Fund, zoning, and subsidized mortgages, have also contributed to the problem. The result of urban sprawl is automobile dependence, traffic congestion, and social isolation, including less involvement in community affairs. Smart growth is one possible response to urban sprawl. It advocates more compact, mixed-use development that encourages people to walk, bicycle, or use public transportation. Smart growth consumes less land and offers numerous other environmental benefits.

Chapter 10 AP® Environmental Science Practice Exam

Section 1: Multiple-Choice Questions

Choose the best answer for questions 1–12.

1. Which of the following is NOT an example of the tragedy of the commons?
   (a) Overgrazing by sheep on community-owned pastures
   (b) Depletion of fish stocks in international waters
   (c) Automobile congestion in Yellowstone National Park
   (d) Depletion of soil minerals by farmers on private land
   (e) Tropical deforestation due to clearing public land for agriculture and then moving on to another location

Use the following graph to answer Question 2.

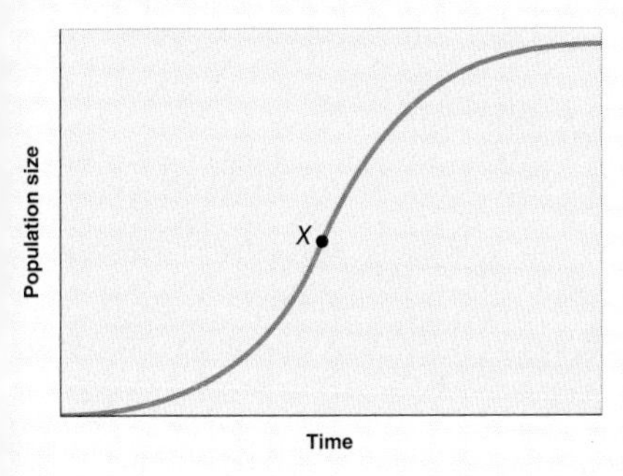

2. This graph shows the population growth of the common pheasant, one of the world's most hunted birds. On the graph, *X* represents
   (a) carrying capacity.
   (b) maximum sustainable yield.
   (c) resource depletion.
   (d) endangered species designation.
   (e) population overshoot.

3. Under the provisions of the National Environmental Policy Act (NEPA), which of the following would require the preparation of an environmental impact statement (EIS)?
   (a) Construction of a house on privately owned land
   (b) Paving of a parking lot for a local business
   (c) Expansion of an interstate highway
   (d) Planting of trees in front of City Hall
   (e) Revision of local zoning ordinances

4. Federally owned land in the United States can best be described as
   (a) 25 percent of all land, with the majority of it in the west.
   (b) 42 percent of all land, with the majority of it in the east.
   (c) 28 percent of all land, with most of it in Texas.
   (d) 20 percent of all land, with 10 percent of it in the west.
   (e) 35 percent of all land, with the majority of it in the east.

5. Which type of protected area is least likely to support tourism and recreation?
   (a) Protected landscapes and seascapes
   (b) National parks
   (c) Managed resource protected areas
   (d) National monuments
   (e) Habitat/species management areas

6. The four major public land management agencies in the United States operate under the principle of multiple use. Which of the following uses is common to lands managed by all four agencies?
   (a) Hunting
   (b) Mining
   (c) Grazing
   (d) Timber harvesting
   (e) Recreation

7. For many years, forest fires were suppressed to protect lives and property. This policy has led to
   (a) a buildup of dead biomass that can fuel larger fires.
   (b) many forest species being able to live without having their habitats destroyed.
   (c) increased solar radiation in most ecosystems.
   (d) soil erosion on steep slopes.
   (e) economic instability.

8. When we purchase an item, we are charged for the labor and supply costs of producing that item. However, we are not charged for the costs of any environmental damage that occurred in manufacturing that item. Those costs are known as
   (a) externalities.
   (b) environmental mitigation.
   (c) infill.
   (d) marginal costs.
   (e) induced demand.

Answers to Chapter 10 AP® Exam Multiple-Choice Questions

1. d
2. b
3. c
4. a
5. e
6. e
7. a
8. a
9. c
10. e
11. d
12. b

Answers to Chapter 10 AP® Exam

Download the full answers to the Chapter 10 AP® Practice Exam by clicking on the link buttons in the TE-book, opening the TRFD, or logging onto the book's companion website bcs.whfreeman.com /friedlandapes2e (teacher login required).

Answers to Chapter 10 AP® Exam

Additional Multiple-Choice AP® Practice Exam

To download an additional Multiple-Choice AP® Environmental Science Practice Exam for Chapter 10, click on the link buttons in the TE-book, open the TRFD, or log onto the book's companion website bcs.whfreeman.com /friedlandapes2e (teacher login required).

Multiple-Choice AP® Practice Exam 10

9. Which of the following is NOT an environmental consequence of clear-cutting?
   (a) Increased soil erosion and sedimentation in nearby streams
   (b) Decreased biodiversity due to habitat fragmentation
   (c) Increased fish populations due to the influx of nutrients into streams
   (d) Decreased tree species diversity due to the loss of shade-tolerant species
   (e) Stands of same-aged trees

10. Which of the following are environmental impacts of urban sprawl?
   I. Greater reliance on the automobile and increased fossil fuel consumption
   II. Increased consumption of land for housing and highway construction
   III. Loss of valuable farmlands

   (a) I only
   (b) II only
   (c) I and II only
   (d) II and III only
   (e) I, II, and III

11. Which of the following has contributed to urban sprawl over the past 50 years?
   (a) Migration of people from suburban areas to rural areas
   (b) Increased availability of public transportation
   (c) Lower property taxes in urban areas
   (d) Use of the federal gasoline tax to construct and maintain highways
   (e) Improved infrastructure and reduced crime rates in urban areas

12. Which of the following is NOT an environmental benefit of smart growth?
   (a) Reduced flooding
   (b) Increased impervious surfaces
   (c) Reduced fossil fuel consumption
   (d) Increased open space
   (e) Decreased water pollution

Section 2: Free-Response Questions

Write your answer to each part clearly. Support your answers with relevant information and examples. Where calculations are required, show your work.

1. The property pictured below is the Farm Barn at Shelburne Farms, a National Historic Landmark, nonprofit environmental education center, and 1,400-acre working farm on the shores of Lake Champlain in Vermont. However, for the sake of this exercise, let's assume that the property pictured below belongs to the federal government.

(Farrell Grehan/Corbis)

   (a) Identify and explain which of the four public land management agencies would be involved in managing this public land. (2 points)
   (b) Applying any three of the basic principles of smart growth, explain how the private land surrounding this federally owned property might be developed to minimize environmental impacts. (4 points)
   (c) Define *environmental impact statement* and describe one condition under which an EIS might be required for the use of either the privately owned or the federally owned lands associated with this tract. (4 points)

2. The town of Fremont met recently to discuss the pros and cons of protecting prairie dogs. Prairie dogs are burrowing rodents the size of rabbits that live in colonies underground in grasslands and prairies. Their numbers have been greatly reduced over the last few decades. Dr. Masser, a local biologist, pointed out that prairie dogs are an important part of the prairie food web, as they are prey for many birds and mammals. Without federal protection from both the Bureau of Land Management and the U.S. Fish and Wildlife Service, they could become extinct in a few years.
Dr. Masser also explained that two of the five species of prairie dogs are already listed as either threatened or endangered. Local ranchers disagreed with Dr. Masser. Mr. Smith, a Fremont area rancher, stated that he will continue to poison or shoot the prairie dogs because they destroy the grasses that are needed by his livestock, and he encouraged the BLM to do the same on public lands.
   (a) Explain the tragedy of the commons in general terms. Using the information you just read about the prairie dog and any other relevant information, incorporate the town of Fremont's discussion into your explanation. (4 points)
   (b) Identify and discuss one argument in favor of preserving western grasslands as habitat for prairie dogs and one argument in favor of maintaining those grasslands for the grazing of livestock. (3 points)
   (c) Identify one action that the Bureau of Land Management and one action that the U.S. Fish and Wildlife Service could take to resolve this land use conflict. (3 points)

Answers to the Chapter 10 Free-Response Questions can be found in the Answer Appendix on page ANS-7.

chapter

11 Feeding the World

PD Overview

This chapter explores how agriculture has developed to meet the needs of a growing population. Technological advances such as synthetic fertilizers, pesticides, and genetically modified crops have helped humans grow increasing amounts of food. However, these advances do not come without some environmental costs. Recent alternatives to industrial farming methods include sustainable agricultural techniques such as no-till, intercropping, crop rotation, contour plowing, and integrated pest management. In addition to advances in agriculture, there have also been changes in the way we raise animals and harvest fish; concentrated animal feeding operations (CAFOs) and aquaculture are relatively new methods that have helped increase our food supply.

Module 31: Human Nutritional Needs

Human nutritional requirements are met to varying degrees throughout the world. Since agriculture first developed more than 10,000 years ago, farmers have been able to increase agricultural output through improved technology and more efficient use of resources. While these improvements have been beneficial to the overall output of food resources, they have not necessarily been beneficial to human well-being. In this module, students will look at basic human nutritional needs, how those needs are met, and some reasons why they are not always met.

Module 32: Modern Large-Scale Farming Methods

Modern agricultural methods are widespread throughout the developed world. They involve efficient, large-scale practices that produce the greatest amount of food at the lowest economic price, but they do not necessarily cause the least environmental impact. This module explores modern large-scale farming methods, their benefits, and their costs.

Module 33: Alternatives to Industrial Farming Methods

As problems with industrial agriculture become more apparent, alternatives are gaining more attention. Some of the newer techniques have actually been used in the developing world for thousands of years. This module examines some of these alternative farming practices. Students will also look at the large-scale farming of meat and fish.

Alignment to AP® Environmental Science Course Description

AP® Outline	Module	
IV. Land and Water Use (10–15%)		
A. Agriculture	Module 31	Human Nutritional Needs
	Module 32	Modern Large-Scale Farming Methods
	Module 33	Alternatives to Industrial Farming Methods
F. Fishing	Module 33	Alternatives to Industrial Farming Methods

Chapter Learning Objectives

After completing this chapter students will be able to

- describe human nutritional requirements.
- explain why nutritional requirements are not being met in various parts of the world.
- describe modern, large-scale agricultural methods.
- explain the benefits and consequences of genetically modified organisms.
- discuss the large-scale raising of meat and fish.
- describe alternatives to conventional farming methods.
- explain alternative techniques used in farming animals and in fishing and aquaculture.

Chapter 11 Pacing Guide

This pacing guide is based on a schedule with 120 sessions of 50 minutes each before the AP® Exam. If you have a different number of sessions before the exam, you can modify the pacing to suit your needs. If you have additional time, consider incorporating quizzes, released AP® Environmental Science free-response and multiple-choice questions, or additional activities.

Module	Standard Schedule Days	Block Schedule Days
Module 31	1	½
Module 32	1	½
Module 33	2	1
Assessment	1	½

Chapter 11 Resources

The link to these resources can be found by clicking on the link buttons in the Teacher's e-Book (TE-book), opening the Teacher's Resource Flash Drive (TRFD), or logging onto the book's companion website bcs.whfreeman.com/friedlandapes2e (teacher login required).

- **PowerPoint Presentation 11**
- **Optimized Art PowerPoint and JPEG Files 11**
- **Do the Math Videos**
- **Measuring Your Impact 11:** The Ecological Footprint of Food Consumption
- **Chapter 11 Web Resources**
- **Exploring the Literature 11**
- **Answers to Chapter 11 AP® Exam**
- **Multiple-Choice AP® Practice Exam 11**
- **Answers to Unit 5 AP® Exam**
- **Unit 5 Additional Free-Response Question**
- **Answer to Science Applied 5 Free-Response Question**

Free-Response Questions from Previous AP® Environmental Science Exams

Free-response questions from prior AP® Environmental Science Exams are available on the AP® course website: https://apstudent.collegeboard.org/apcourse/ap-environmental-science/exam-practice. Students should be able to answer all of the questions listed below with material learned in this chapter. When a question requires students to understand material from multiple chapters, the question will be listed in the last chapter required to complete the entire question. Questions marked with an asterisk (*) are from exams with released multiple-choice questions. You may want to save these questions until the end of the year so you can give your students a complete released exam for practice. Questions marked with double asterisks (**) require math to calculate a problem. Look for references to these questions throughout the chapter.

Year	Question	Content
1999	4	• Pros and cons of pesticide use • Alternatives to pesticides
2004	4	• Soil testing • Synthetic fertilizers • Soil rich in humus
2005	2**	• Global meat production • Efficiency of grain production compared to meat production • Problems with meat production
2006	4	• Commercial fishing industry • Decline in fish harvest • International fishing law • Tragedy of the commons and commons management
2009	4**	• Genetically modified crops • Agricultural practices to maintain or improve soil quality
2011	4**	• Land needed to feed the human population • Soil properties and quality
2012	3	• Use of pesticides • Integrated pest management • Agricultural practices that increase crop yields

PD **Chapter 11 Overview**

Watch the video overview of Chapter 11 (for teachers) by clicking on the link buttons in the TE-book, opening the TRFD, or logging onto the book's companion website bcs.whfreeman.com /friedlandapes2e (teacher login required).

TRM **PowerPoint Presentation 11**

Download the PowerPoint presentation for Chapter 11 by clicking on the link buttons in the TE-book, opening the TRFD, or logging onto the book's companion website bcs.whfreeman.com /friedlandapes2e (teacher login required).

Joel Salatin's Polyface Farm has become a showcase for agricultural techniques that are consistent with ecological principles.
(Michael Wickes/The Image Works)

chapter 11

Feeding the World

A Farm Where Animals Do Most of the Work

It is common for a farmer growing juicy tomatoes or sweet apples to become a local hero, but how often does a farmer become known throughout an entire country?

Joel Salatin of Swoope, Virginia, has been featured in a best-selling book and two major motion pictures. He is the author of numerous popular books, including *Holy Cows and Hog Heaven: The Food Buyer's Guide to Farm Friendly Food* and *Everything I Want to Do Is Illegal: War Stories from the Local Food Front.* His farming has drawn the attention of authors, filmmakers, and many others because he raises vegetables and livestock in a sustainable, organic way. Salatin understands that farming, done properly, has to replicate many of the processes that exist in nature.

Salatin calls himself a "grass farmer" because grass is at the foundation of the trophic feeding pyramid. On his Polyface Farm, he grows grass that is harvested, dried by the Sun, and turned into hay, which he feeds to cattle. The cattle are processed into beef and consumed by humans. But that is only one piece of the intricate food chain that Salatin has created on his farm—all without chemical pesticides or synthetic fertilizers.

> Salatin understands that farming, done properly, has to replicate many of the processes that exist in nature.

As cattle graze a field, they leave behind large piles of manure that become the energy source for large numbers of grubs and fly larvae. Before the larvae can hatch into flies and cause a problem on the farm, Salatin brings in his "sanitation crew"—not people spraying pesticides, but chickens. As the chickens eat the larvae, they spread the nitrogen-rich cow manure and deposit some of their own. They continue to lay eggs, which are tasty and nutritious. Salatin explains that his method reproduces the process in nature in which birds typically follow herbivores to take advantage of their waste.

Salatin uses other animals to take care of the buildup of cow manure during winter when his cattle remain in a barn. Some farmers push that manure

Teaching Tip: Chapter Opening Case

The chapter opening case, "A Farm Where Animals Do Most of the Work," points out that using cow and chicken manure is a good way to recycle nutrients back into the soil. Since nitrogen is one of the major components of fertilizer, this would be a good time to review the nitrogen cycle. Have your students draw the nitrogen cycle starting with atmospheric nitrogen. Make sure to discuss how consumers (cows) eat producers (grass) in the process of assimilation.

Teaching Tip: Activity

After reading the chapter opening case, have students research organic farming in your community or state. Ask the class to create interview questions to ask an organic farmer. Either invite an organic farmer to be a guest speaker or take a field trip to a farm. Sample interview questions might include:

- How do you define "organic agriculture"?
- What sustainable practices does your farm use?
- Why is it important for you to be organic?
- Do you sell your produce directly to consumers? If so, are your prices comparable to those found in grocery stores?
- What are the disadvantages of organic farming?

out the doorway and end up with huge manure piles in the spring but Salatin continuously layers straw, which consists of dried stalks of plants, and wood chips on top of the manure. He allows the floor of his barn to rise higher and higher during winter. Occasionally he throws down some corn. During the winter, the manure-straw-corn layers decompose and produce heat, which allows the cattle to use less energy and their food to stay warm. In spring, when the cattle go out to pasture, he brings pigs into the barn. Pigs have an excellent sense of smell, and they dig up the layers of manure and straw with their snouts to get at the fermented corn that they find especially delectable. Within a few weeks, the pigs provide Salatin with a barn full of thoroughly mixed compost that required no machinery to make: His animals have mixed the manure and straw for him as they gained nutrients and calories.

These are just two examples of the ways one farmer uses natural processes and knowledge of plants, animals, and the natural world to grow food for human consumption. Joel Salatin and other farmers interested in sustainable farming are finding ways to feed the human population by working with, not against, the natural world.

Sources:
M. Pollan, *The Omnivore's Dilemma*, Penguin, 2007; R. Kenner (director /producer) and E. Schlosser (producer), *Food Inc.*, Magnolia Home Entertainment, 2009.

Most agricultural methods, especially modern ones practiced in the developed countries of the world, tend to fight the basic laws of ecology to provide large amounts of food for humans. These modern methods also use substantial quantities of fossil fuel in every step of the process, from the use of machinery in plowing and harvesting crops to the production of fertilizers and pesticides to the transportation and processing of the food. However, there are ways to grow food efficiently without using as much fossil fuel, manufactured fertilizers, and pesticides as currently used. In this chapter, we will explore the current state of human nutrition around the world, and examine conventional, large-scale methods of agriculture. We will also look at alternatives to conventional agriculture including sustainable and organic agriculture, which have the potential to reduce adverse impacts of agriculture on the environment.

module 31

Human Nutritional Needs

In both the developed and developing worlds, human nutritional requirements are being met to varying degrees. Since agriculture first started to be practiced more than 10,000 years ago, farmers have been able to increase agricultural output through improved technology and more efficient use of resources. While these improvements have been beneficial to the overall output of food resources, they have not necessarily been beneficial to human well-being. In this module, we will look at basic human nutritional needs, how those needs are met, and the reasons why they are not always met.

Learning Objectives

After reading this module, you should be able to

- describe human nutritional requirements.
- explain why nutritional requirements are not being met in various parts of the world.

Human nutritional requirements are not always satisfied

Advancements in agricultural methods are believed to have greatly improved the human diet over the last 10,000 years. In particular, tremendous gains in agricultural productivity and food distribution were made in the twentieth century. But despite these advances, many people throughout the world do not receive adequate nutrition. As FIGURE 31.1 shows, over 800 million people worldwide lack access to adequate amounts of food. Although that number has been declining for decades, it is still way above target numbers set by various world agencies. Currently, as many as 24,000 people starve to death each day—8.8 million people each year. In this section we will explore human nutritional requirements.

Chronic hunger, or **undernutrition,** means not consuming enough calories to be healthy. Food calories are converted into usable energy by the human body. Not receiving enough food calories leads to an energy deficit. An average person needs approximately

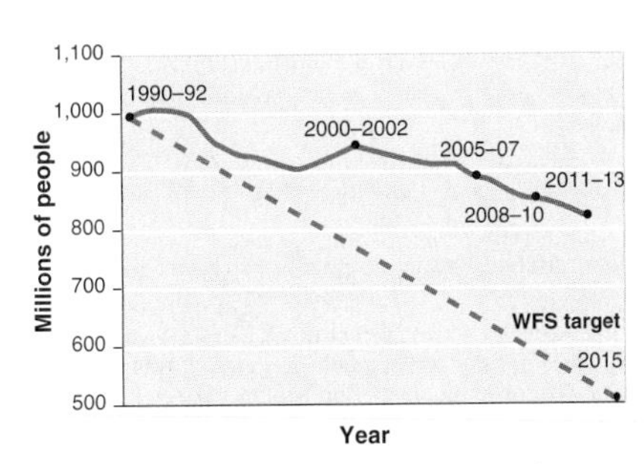

FIGURE 31.1 **Global undernutrition.** The number of undernourished people in the world has declined from 1990 through the present (solid line), but is still greater than World Food Summit targets (dashed line). *(Data from FAO, IFAD, and WFP. 2013. The State of Food Insecurity in the World 2013. The multiple dimensions of food security. Rome, Food and Agriculture Organization.)*

Undernutrition The condition in which not enough calories are ingested to maintain health.

◀ TEACHING with FIGURES

Show students Figure 31.1. Ask them to use the graph to answer the following questions:

- How many undernourished people were there in the world from 2011 to 2013? *The number of undernourished people ranged between 860 million in 2011 and 820 million in 2013.*
- How do the 2013 data compare to the World Food Summit target? *The World Food Summit target for 2013 was 550 million undernourished people. Actual data for that year show approximately 820 million people were undernourished. So the difference would be 820 million − 550 million = 270 million people above the target number.*
- How much has global undernutrition dropped since 1990? *Estimates show that in 1990 1 billion people were undernourished. In 2013 the number of undernourished people was 820 million. So global undernutrition dropped by approximately 180 million people in 23 years.*

Teaching Tip: Warm-up

Ask students: What are undernutrition and malnutrition? *Chronic hunger, or undernutrition, occurs when a person does not consume enough calories to be healthy. Food calories are converted into usable energy by the human body. Malnutrition occurs when a person's diet lacks the correct balance of proteins, carbohydrates, vitamins, and minerals.*

Teaching Tip: Warm-up

Ask students the following questions:

- What is food insecurity? *Food insecurity refers to the condition in which people do not have adequate access to food.*
- What are some of the possible reasons for food insecurity in the world today? *Possible reasons include weather conditions such as drought, or social and political upheaval.*

Teaching Tip: Activity

Divide the class into groups of two. Give each group 20 index cards and the vocabulary terms in the list below. Have each team label one side of each card with the numbers 1 through 20. On the blank sides of cards 1 through 10, students should write one of the terms using a new card for each term. Instruct students to shuffle the order of the terms; they should not duplicate the order of the list. On the blank sides of cards numbered 11 through 20, students should write the definition or some other information that connects the term to the definition of the terms on cards 1 through 10, making sure to keep an answer key of which numbered term card matches which numbered definition card. The groups should lay out their cards number side up in two rows—one with terms and one with definitions. Have the groups swap places and answer keys.

The first player in each pair then chooses one card from each row. Read the cards to see if they are a match. If they are, that person keeps the cards and takes another turn. If they are not a match, the second member of the team has a turn. Each group continues playing until all the cards are gone. The person with the most cards is the winner.

- undernutrition
- malnourished
- food security
- food insecurity
- famine
- anemia
- overnutrition
- meat
- poverty
- grains

2,200 kilocalories per day, though this amount varies with gender, age, and weight. A long-term food deficit of only 100 to 400 kilocalories per day—in other words, an intake of 100 to 400 kilocalories less than one's daily need—deprives a person of the energy needed to perform daily activities and makes the person more susceptible to disease. In children, undernutrition can lead to improper brain development and lower IQ.

In addition, a person lacking sufficient food calories is probably lacking sufficient protein and other nutrients. According to the World Health Organization (WHO), 3 billion people—nearly one-half of the world's population—are **malnourished;** that is, regardless of the number of calories they consume, their diets lack the correct balance of proteins, carbohydrates, vitamins, and minerals. These people experience malnutrition.

The Food and Agriculture Organization of the United Nations (FAO) defines **food security** as the condition in which people have access to sufficient, safe, and nutritious food that meets their dietary needs for an active and healthy life. Access refers to the economic, social, and physical availability of food. **Food insecurity** refers to the condition in which people do not have adequate access to food.

Famine is a condition in which food insecurity is so extreme that large numbers of deaths occur in a given area over a relatively short period. Actual definitions of famine vary widely depending on the agency using the term. One relief agency defines a famine as an event in which there are more than 5 deaths per day per 10,000 people due to lack of food. According to this definition, there is an annual mortality rate of 18 percent during a famine. Famines are often the result of crop failures, sometimes due to drought, although famines can have social and political causes.

Malnourished Having a diet that lacks the correct balance of proteins, carbohydrates, vitamins, and minerals.

Food security A condition in which people have access to sufficient, safe, and nutritious food that meets their dietary needs for an active and healthy life.

Food insecurity A condition in which people do not have adequate access to food.

Famine The condition in which food insecurity is so extreme that large numbers of deaths occur in a given area over a relatively short period.

Anemia A deficiency of iron.

Overnutrition Ingestion of too many calories and a lack of balance of foods and nutrients.

Meat Livestock or poultry consumed as food.

Even when people have access to sufficient food, a deficit in just one essential vitamin or mineral can have drastic consequences. The WHO estimates that more than 250,000 children per year become blind due to a vitamin A deficiency. Iron deficiency, known as **anemia,** is the most widespread nutritional deficiency in the world. The WHO estimates that there are 3 billion anemic people in the world—many in developing countries, but also in developed countries. Lack of dietary iron is the major cause of anemia, but there are other causes, including malaria, AIDS, and parasite infestations. An increase in the ingestion of iron-rich foods, such as certain grains, herbs, vegetables, and meats, can reduce anemia.

In the last few decades, one other form of malnutrition has been increasing. **Overnutrition,** the ingestion of too many calories combined with a lack of proper balance in foods and nutrients, causes a person to become both overweight and malnourished. The WHO estimates that there are over 1 billion people in the world who are overweight, and that roughly 300 million of those people are *obese,* meaning they are more than 20 percent above their ideal weight. Overnutrition is a type of malnutrition that puts people at risk for a variety of diseases, including type 2 diabetes, hypertension, heart disease, and stroke. While overnutrition is common in developed countries such as the United States, it can also coexist with malnutrition in developing countries. Childhood obesity is a related condition that has started to occur in greater numbers. Overnutrition is often a function of the availability and affordability of certain kinds of foods. For example, overnutrition in the United States has been attributed in part to the easy availability and low cost of processed foods containing ingredients such as high-fructose corn syrup, which are often high in calories but lack other nutritional content.

Humans eat a variety of foods, but grains—the seedlike fruits of corn (maize), rice, wheat, and rye, among others—make up the largest component of the human diet. Worldwide, there are roughly 50,000 edible plant species, but just three of them—corn, rice, and wheat—constitute 60 percent of human energy intake. **Meat,** the second largest component of the human diet, is usually defined as livestock (beef, veal, pork, and lamb) and poultry (chicken, turkey, and duck) consumed as food. Globally, we produce over 300 kg (660 pounds) of grain per person per year, followed by meat at 43 kg (95 pounds) per person per year, and fish at about 21 kg (46 pounds) per person per year. As income increases with economic growth, people tend to add more meat to their diet. As FIGURE 31.2 shows, meat consumption has been increasing globally, although it has decreased slightly in the United States in recent years. Nevertheless, the United States is still the second highest per capita consumer of meat in the world, after Luxembourg.

AP® Exam Tip

Meat production is covered in the 2005 AP® Exam, Question 2. To answer this question, students must

- calculate the per capita meat production in 1950 and 2000 using provided data
- calculate the change in global per capita meat production during the 50-year period in a percentage of the 1950 value
- discuss why production of grain is more efficient than production of meat, considering both land use and energy use
- identify advantages and disadvantages of a human diet that contains little meat

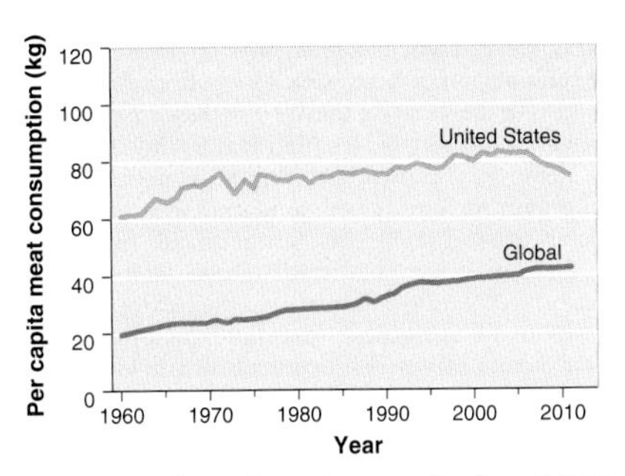

FIGURE 31.2 Per capita meat consumption through 2012. Per capita meat consumption has begun to decrease in the United States. It has been steadily increasing worldwide. *(Data from www.fao.org and http://www.earth-policy.org/data_highlights/2012/highlights25.)*

Undernutrition and malnutrition occur primarily because of poverty

Currently, the world's farmers grow enough grain to feed at least 8 billion people, which would appear to be more than enough for the world's population of 7.1 billion. And grain is only a little more than one-half of the food we produce, in terms of caloric content or biomass. Why, then, do we find that undernutrition and malnutrition are common in so many countries of the world? There are many ways to examine this question, and the answers are complex.

The primary reason for undernutrition and malnutrition is poverty: the lack of resources that allows a person to have access to food. According to many food experts, starvation on a global scale is the result of unequal distribution of food rather than absolute scarcity of food. In other words, the food exists but not everyone has access to it. This may mean that people cannot afford to buy the food they need, which is a problem that cannot be solved just by producing more grain.

In addition, political and economic factors play an important role. For example, refugee populations that have fled their homes due to war or natural disasters may not have access to food that they grew and stored but had to leave behind when they became refugees. Many times in modern history the lack of an adequate food supply has led to political unrest because people without the means to feed themselves or their families may resort to crime or violence in an attempt to improve their situation. More recently, a rise in food prices in 2008 led to food riots in Haiti, Egypt, Ivory Coast, Cameroon, Yemen, and elsewhere. Poor governance and political unrest can lead to inadequate food supplies as well, so they can be both causes and effects of undernutrition and malnutrition.

Food researchers have also observed that large amounts of agricultural resources are diverted to feed livestock and poultry rather than people. In fact, roughly 40 percent of the grain grown in the world is used to feed livestock. In the United States, the two largest agricultural crops, corn and soybeans, are grown more for animal feed than for people (although more corn is used for ethanol than for either livestock or people). When these foods are fed to livestock, the low efficiency of energy transfer causes much of the energy they contain to be lost from the system, as we saw in Chapter 3. Ultimately, perhaps only 10 to 15 percent of the calories in grain or soybeans fed to cattle are converted into calories in beef. If people ate producers, such as grains and soybeans, rather than primary consumers, such as cattle, it is possible that more food would be available for people.

FIGURE 31.3 shows global grain production since 1950. A large number of factors influence grain production, including the amount of land under cultivation, global weather and precipitation patterns, world prices for grain, and the productivity of the land on which

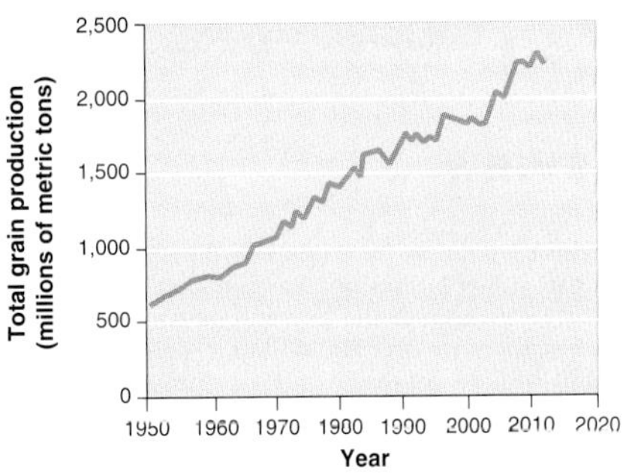

(a) Total grain production

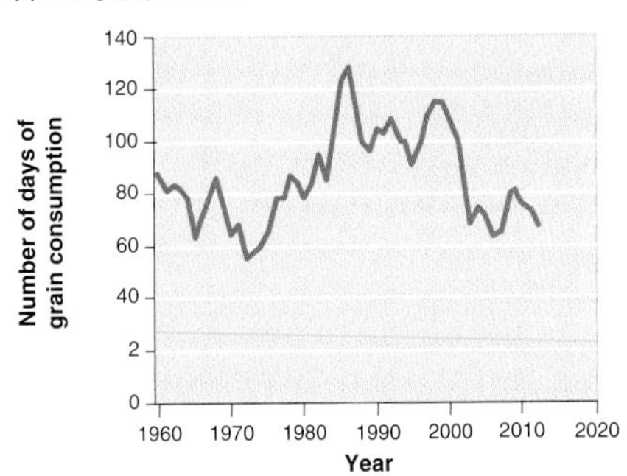

(b) World grain supply as days of grain consumption

FIGURE 31.3 Global grain production, 1950–2012. (a) Global grain production grew rapidly from 1950 through the mid-1980s. Growth has continued since then, but per capita growth has slowed. (b) World grain supply (days of supply for everyone in the world) has been declining. *(Data from http://www.earth-policy.org/indicators/C54)*

more **math** practice: Calculating a Change in Percentage

Numerous free-response questions in past exams have asked students to extract data from a graph and then use the data to calculate a change in percentage. Remind students that to calculate a change in percentage, they can use the rule of thumb: "part" divided by "whole."

Provide students with the data table and ask them to plot a graph with or without a computer.

Year	U.S. per capita meat consumption (kg)	Global per capita meat consumption (kg)
1960	60	20
1965	66	22
1970	74	22.5
1975	71	23
1980	75	26
1985	76	29
1990	76	32
1995	78	38
2000	80	39.5
2005	82	40
2010	76	42
2012	75	42

Per Capita Meat Consumption, Global and U.S. 1960–2012.

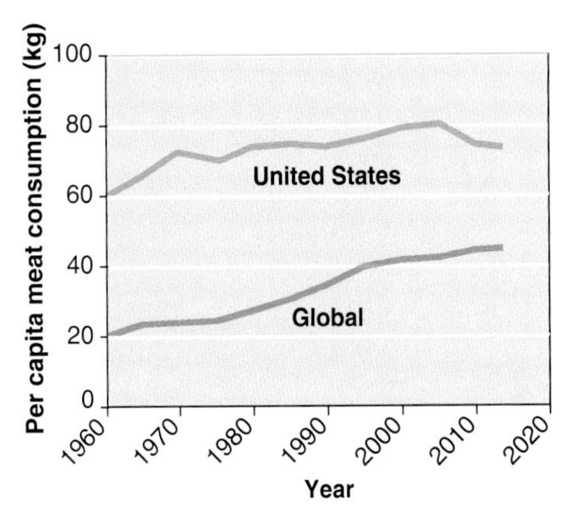

When students have completed their graphs, ask them to compare the global per capita meat consumption trends with those of the United States. *Global meat consumption has slowly increased over the 52-year period. Meat consumption in the United States increased at different rates over the 52-year period and has recently fallen from its highest level in 2005.*

Next, ask students to calculate the percentage change in global meat consumption and in U.S. meat consumption from 1960 to 2012.

Global 1960 = 20 kg per person per year
Global 2012 = 42 kg per person per year
42 kg − 20 kg = 22 kg more per person per year in 2012
22 kg ÷ 20 kg = 1.1
1.1 × 100 = 110%

There was a 110 percent change in global meat consumption over the 52-year period.

United States 1960 = 60 kg per person per year
United States 2012 = 75 kg per person per year
75 kg − 60 kg = 15 kg
15 kg ÷ 60 kg = 0.25
0.25 × 100 = 25%

There was a 25 percent increase in U.S. meat consumption over the 52-year period.

After you have reviewed the calculations, point out that meat consumption more than doubled from 1960 to 2012 (20 kg to 42 kg) and there was never a decrease in percentage consumption. In contrast, meat consumption in the United States had several fluctuations over the 52-year period; for example, there was a decrease in meat consumption from 1970 to 1975 and again from 2005 to 2012.

grain is being grown. As Figure 31.3a shows, grain production has been steadily increasing, though there have been some years within the last decade with no increase in production. Number of days of grain consumption in reserve, as Figure 31.3b shows, has varied considerably in the last three decades but is roughly similar to numbers of days of reserve available in 1980. There is not a perfectly clear answer as to why global per capita grain production has not continued to increase. Some environmental scientists suggest that we have finally reached the limit of our ability to supply the human population with food while others blame some of the political and social factors that we mentioned above.

In Chapter 7 we learned that the human population is expected to be between 8 and 10 billion by 2050. To feed everyone in the future, we will have to put more land into agricultural production, improve crop yields, reduce the consumption of meat, harvest more from the world's fisheries, or use some combination of these strategies.

Experts disagree on whether it is feasible to greatly expand food production. Optimists point to new techniques for crop development as well as unused land in tropical rainforests and grasslands that could be farmed, while pessimists argue that climate change, decreasing biodiversity, dwindling supplies of water for irrigation, diminishing topsoil, and other factors may make it difficult to produce more crops. To understand what we can do to improve food security in all nations of the world and reduce the environmental consequences of growing food, we will need to examine the farming methods we use today and the trade offs involved in using them.

module 31 REVIEW

In this module, we have seen that the advent of agriculture 10,000 years ago has brought about many positive changes to the human population. Nevertheless, there are over 800 million people around the world who do not receive adequate amounts of food each day. Nutritional requirements are met by meat consumption to a greater extent in the developed world than in the developing world. Undernutrition, overnutrition, and malnutrition remain issues of concern around the globe.

Answers to Module 31 AP® Review Questions

1. e
2. b
3. c
4. a
5. a

Module 31 AP® Review Questions

1. Approximately how many people are malnourished worldwide?
   (a) 10 percent
   (b) 18 percent
   (c) 25 percent
   (d) 30 percent
   (e) 45 percent

2. Undernutrition means
   (a) lacking the proper balance of proteins, vitamins, and minerals.
   (b) not consuming a sufficient number of calories.
   (c) lacking a specific nutrient such as vitamin A or iron.
   (d) a lack of food such that deaths begin to occur.
   (e) the ingestion of too many poor quality calories.

3. Anemia, the most widespread nutritional deficiency, is the result of insufficient
   (a) vitamin A.
   (b) protein.
   (c) iron.
   (d) calories.
   (e) vitamin D.

4. The primary reason for malnutrition is
   (a) poverty.
   (b) political unrest.
   (c) insufficient food production.
   (d) diversion of food to livestock.
   (e) excessive food waste in developed nations.

5. Which nutritional trend is true?
   (a) Three species of plants contribute over 60 percent of human energy intake.
   (b) Meat consumption is decreasing globally.
   (c) Growing more grain is the only way to feed the world's population.
   (d) Overnutrition has remained constant over the last century worldwide.
   (e) Grain production has increased rapidly in recent years.

module

32

Modern Large-Scale Farming Methods

Modern agricultural methods are widespread throughout the developed world. They involve efficient, large-scale practices that produce the greatest amount of food at the lowest economic price, but they do not necessarily result in the least environmental impact. It is important to understand these methods, their benefits, and their costs before examining alternatives.

Learning Objectives

After reading this module you should be able to

- describe modern, large-scale agricultural methods.
- explain the benefits and consequences of genetically modified organisms.
- discuss the large-scale raising of meat and fish.

Modern industrial farming methods have transformed agriculture

The advent of agriculture roughly 10,000 years ago enabled people to move beyond a subsistence level of existence. At the same time, it did have some negative consequences. Deliberate cultivation of food also initiated a level of environmental degradation never before experienced on Earth. The abundance of food supplied by agriculture is also one factor that has led to the exponential growth of the human population. An increase in food production leads to a positive feedback loop: With more food come more people, who require even more food production.

In the twentieth century, farming became more mechanized, and the use of fossil fuel energy in food production increased. These changes have led to increasing food output as well as a variety of environmental impacts. **Industrial agriculture,** or **agribusiness,** applies the techniques of the Industrial Revolution—mechanization and standardization—to the production of food. Today's modern agribusinesses are quite different from the small family farms that dominated agriculture only a few decades ago. In this section we will look at the ways in which modern industrial agriculture is dependent on energy and we will describe the activities those additional energy inputs allow.

The Energy Subsidy in Agriculture

The activities associated with growing, harvesting, processing, and preparing food require a great deal of energy input beyond solar energy. The fossil fuel energy and human energy input per calorie of food produced is called the **energy subsidy.** In other words, if we use

Industrial agriculture Agriculture that applies the techniques of mechanization and standardization. *Also known as* **agribusiness.**

Energy subsidy The fossil fuel energy and human energy input per calorie of food produced.

Teaching Tip: Warm-up

Ask students to explain the energy subsidy in agriculture. *The energy subsidy is the fossil fuel energy and human energy input per calorie of food produced. In other words, if we use 5 calories of energy to produce food, and we receive 1 calorie of energy when we eat the food, then the food has an energy subsidy of 5.*

5 calories of human and fossil fuel energy to produce food, and we receive 1 calorie of energy when we eat that food, then the food has an energy subsidy of 5. We can think of this in another way if we use mass of inputs and outputs as a substitute measure for energy. If it takes 20 kg (44 pounds) of grain to feed to cattle to produce 1 kg (2.2 pounds) of beef, the energy subsidy is 20. If it takes 2.8 kg (6 pounds) of grain to feed to chickens to produce 1 kg (2.2 pounds) of chicken meat, the energy subsidy is 2.8, which is considerably smaller than the energy subsidy for beef. The comparison is not perfect, because the energy content of 1 kg of beef is greater than that of 1 kg of grain. Nevertheless, it should help you to understand the relationship between energy inputs and outputs in producing food.

As FIGURE 32.1 shows, traditional small-scale agriculture requires a relatively small energy subsidy: It uses few energy inputs per calorie of food produced. By contrast, food writer and journalism professor Michael Pollan estimates that if you eat the average modern U.S. diet, which contains foods mostly produced by modern agricultural methods, there is a 10-calorie energy input for every calorie you eat. Most of that energy input is fossil fuel energy. Therefore, food choices are also energy choices.

"Do the Math: Land Needed for Food" shows you how to calculate the different land requirements for obtaining sufficient calories from corn and from beef.

Most of the energy subsidies in modern agriculture are in the form of fossil fuels, which are used to produce fertilizers and pesticides, to operate tractors, to pump water for irrigation, and to harvest food and prepare it for transport. Other energy subsidies take place off the farm. For example, the average food item travels 2,000 km (1,240 miles) from the farm to your plate, so we often spend far more energy on transporting food than we get from the food itself. The Department of Energy reports that in the United States, 17 percent of total commercial energy use goes into growing, processing, transporting, and cooking food. Those of us eating a supermarket diet in the developed world are highly dependent on fossil fuel for our food; the modern agricultural system would not work without it.

The Green Revolution

How did we get to our current levels of energy use for food production? In the twentieth century, the agricultural system was dramatically transformed from a system of small farms relying mainly on human labor with relatively low fossil fuel inputs to a system of large industrial operations with fewer people but much more machinery. This shift in farming methods, known as the **Green Revolution,** involved new management techniques and mechanization as well as the triad of fertilization, irrigation, and improved crop varieties. These changes increased food production dramatically, and farmers were able to feed many more people.

The Green Revolution began with the work of crop scientists, particularly the American scientist Norman Borlaug (1914–2009), who won the Nobel Peace Prize for his contribution to increasing the world food supply. Through intensive breeding, Borlaug and other agricultural researchers developed strains of wheat that were disease resistant and produced higher yields. They also used fertilizers and irrigation to improve yields. In the 1940s, researchers brought these techniques to developing nations such as Mexico and the Philippines to help increase their agricultural output and feed their growing populations. From the 1950s through the 1970s, many countries, particularly those in the developing world, underwent similar shifts in the way they farmed.

Green Revolution A shift in agricultural practices in the twentieth century that included new management techniques, mechanization, fertilization, irrigation, and improved crop varieties, and that resulted in increased food output.

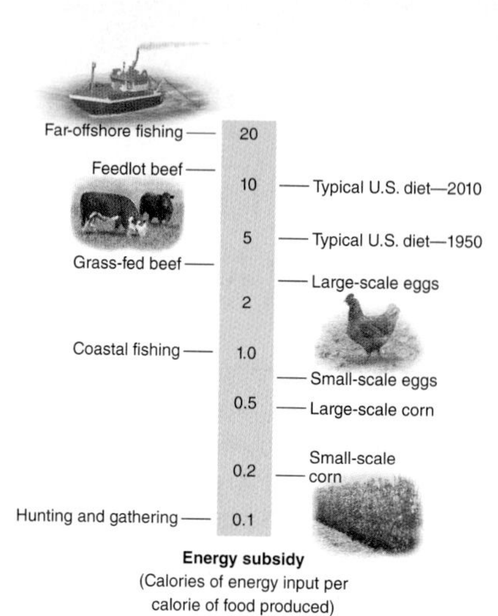

FIGURE 32.1 **Energy subsidies for various methods of food production and diets.** Energy input per calorie of food obtained is greater for modern agricultural practices than for traditional agriculture. Energy inputs for hunting and gathering and for small-scale food production are mostly in the form of human energy, whereas fossil fuel energy is the primary energy subsidy for large-scale modern food production. All values are approximate, and for any given method there is a large range of values.

Teaching Tip: Engage

Ask students: What are the major features of the Green Revolution? *The major features of the Green Revolution are new management techniques, new mechanization, improved methods of fertilization and irrigation, and improved crop varieties.*

Teaching Tip: Beyond the Classroom

The Green Revolution involved new management techniques and mechanization as well as the triad of fertilization, irrigation, and improved crop varieties. Ask groups of students to research factors that contributed to the Green Revolution and to prepare a class presentation. Possible topics include the following:

- How did Norman Borlaug contribute to the Green Revolution?
- What is a combine harvester and how did it lead to a change in the size of farms?
- When were inorganic or synthetic fertilizers created? How have they affected the Green Revolution?
- What is monocropping and how has it changed the way farmers grow crops?
- How have pesticides contributed to the Green Revolution? How do pesticides help a farmer maintain profitability?
- When did farmers begin to plant crops that have been genetically modified? How have genetically modified crops helped farmers increase their yields?

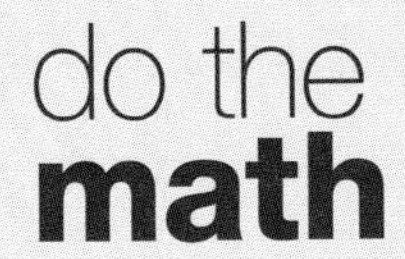

Land Needed for Food

We have seen that raising beef requires more resources than growing corn. Let's look at some of the actual numbers.

On farms in the midwestern United States, a hectare of land yields roughly 370 bushels of corn (equivalent to 150 bushels per acre). A bushel consists of 1,250 ears of corn, and each ear typically contains 80 kilocalories. Assume that a person eats only corn and requires 2,000 kilocalories per day. Although this assumption is not very realistic, it allows an approximation of how much land it would take to feed that person.

The person's food requirement is

$$2{,}000 \text{ kilocalories/day} \times 365 \text{ days/year} = 730{,}000 \text{ kilocalories/year}$$

A hectare of corn produces

$$370 \text{ bushels/hectare} \times 1{,}250 \text{ ears/bushel} \times 80 \text{ kilocalories/year} = 37{,}000{,}000 \text{ kilocalories/hectare}$$

$$730{,}000 \text{ kilocalories/year} \div 37{,}000{,}000 \text{ kilocalories/hectare} = 0.02 \text{ ha } (0.05 \text{ acres})$$ of land to feed one person for a year

Thus, one person eating only corn can obtain sufficient calories in a year from 0.02 ha (0.05 acres) of land. What if that person ate only beef? We have seen that it takes 20 kg of grain to produce 1 kg of beef. So it would take 20 times as much land, or 0.4 ha (1 acre), to feed a person who ate only beef.

What happens if we extend this analysis to a global scale? If Earth has about 1.5 billion ha (3.7 billion acres) of land suitable for growing food, is there sufficient land to feed all 6.8 billion inhabitants of the planet if they all eat a diet of only beef?

$$6.8 \text{ billion people} \times 0.4 \text{ ha/person} = 2{,}720{,}000{,}000 \text{ ha}$$

So 2.72 billion ha (6.72 billion acres) would be needed, and the answer is no.

Your Turn How many people eating a beef-only diet can Earth support?

The upward trend in world grain production since 1950 (see Figure 31.3a) is the result of the Green Revolution. From the mid-1960s through the mid-1980s, world grain production doubled. By the 1990s, there were at least 18 organizations around the world promoting and developing Green Revolution techniques, including mechanization, irrigation, use of fertilizer, *monocropping*, and use of pesticides. However, the Green Revolution has also had negative environmental impacts. Let's examine some Green Revolution practices in more detail.

Mechanization

Farming involves many kinds of work. Fields must be plowed, planted, irrigated, weeded, protected from pests, harvested, and prepared for the next season. Even after harvesting, crops must be dried, sorted, cleaned, and prepared for market in almost as many different ways as there are different crops. Machines do not necessarily do this work better than humans or animals, but it can be economically advantageous to replace humans or animals with machinery, particularly if fossil fuels are abundant, fuel prices are relatively low, and labor prices are relatively high. In developed countries, where wages are relatively high, less than 5 percent of the workforce works in agriculture. In developing countries, where wages tend to be much lower, 40 to 75 percent of the working population is employed in agriculture.

Since the advent of mechanization, large farms producing staple crops such as beans or corn have generally been more profitable than small farms. Size matters because of **economies of scale,** which means that the average costs of production fall as output increases. For example, a new combine harvester—a machine that harvests the crop and separates out the grain or seed for transport—costs between $150,000 and $400,000. This large up-front expenditure is a good investment for a large farm, where the cost of the machine is justified by the profits on the increased production that comes from using it. A small farm, however, does not

Economies of scale The observation that average costs of production fall as output increases.

Answer to Your Turn

From "Do the Math," we see that 6.8 billion people on a beef-only diet require 2.72 billion ha of land. We are also told that we have 1.5 billion ha of land available. Therefore the easiest way to approach the problem is to set up a proportion.

$$\frac{6.8 \text{ billion people}}{2.72 \text{ billion ha}} = \frac{x \text{ billion people}}{1.5 \text{ billion ha}}$$

$$x = \frac{(1.5 \text{ billion ha}) \times (6.8 \text{ billion people})}{2.72 \text{ billion ha}}$$

$$= 3.75 \text{ billion people}$$

Therefore 3.75 billion people eating a beef-only diet could be supported on 1.5 billion ha of land.

Teaching Tip: Engage

Ask students the following questions:

- What are organic and synthetic fertilizers? How are they different? *Organic fertilizers are composed of organic matter from plants and animals. They are typically made up of animal manure and plant material that has been allowed to decompose. Synthetic fertilizers are produced commercially. Nitrogen fertilizers are often produced by combusting natural gas, which allows nitrogen from the atmosphere to be fixed and captured.*
- How are the two types of fertilizers applied to the soil? *Organic fertilizer is frequently generated at the same farm or close to the farm where it is being used. It must be collected, composted, and then spread. Synthetic fertilizers come in a liquid or solid form. They are purchased and transported and are typically much easier to spread.*
- How do the costs of organic and synthetic fertilizers compare? *Typically, synthetic fertilizers are much more expensive then organic fertilizers. Many large farms purchase organic fertilizers from local sources such as farms that raise chickens or cows.*
- What are environmental consequences of using organic or synthetic fertilizers? *Organic fertilizers increase the water retention ability of the soil. The manufacture of synthetic fertilizers requires fossil fuel. Use of synthetic fertilizers is more likely to lead to runoff and subsequent problems in local waterways.*

have enough land to be able to recoup the cost of such expensive equipment. Because of economies of scale, profits tend to increase with size, and large agricultural operations generally outcompete small ones. As a result, between 1950 and 2000, the average farm size in Iowa more than doubled, from about 70 ha (173 acres) to over 140 ha (346 acres).

Mechanization also means that single-crop farms are generally more efficient than farms that grow many crops. Because mechanized crop planting and picking require specialized, expensive equipment specific to each crop, planting and harvesting only a single type of crop reduces equipment costs.

Irrigation

Irrigation systems like those described in Chapter 9 can increase crop growth rates or even enable crops to grow where they could not otherwise be grown (FIGURE 32.2). For example, irrigation has transformed approximately 400,000 ha (1 million acres) of former desert in the Imperial Valley of southeastern California into a major producer of fruits and vegetables. In other situations, irrigation can allow productive land to become extremely productive land. One estimate suggests that the 16 percent of the world's agricultural land that is irrigated produces 40 percent of the world's food.

While irrigation has many benefits, including more efficient use of water in some places where it is scarce, it can also have a number of negative consequences over time. As we saw in Chapter 9, it can deplete groundwater, draw down aquifers, and cause saltwater intrusion into freshwater wells. It can also contribute to soil degradation through *waterlogging* and *salinization*, shown in FIGURE 32.3. **Waterlogging,** a form of soil degradation that occurs when soil remains under water for prolonged periods, impairs root growth because roots cannot get oxygen. **Salinization** occurs when the small amounts of salts in irrigation water become highly concentrated on the soil surface through evaporation. These salts can eventually reach toxic levels and impede plant growth.

Waterlogging A form of soil degradation that occurs when soil remains under water for prolonged periods.

Salinization A form of soil degradation that occurs when the small amount of salts in irrigation water becomes highly concentrated on the soil surface through evaporation.

Organic fertilizer Fertilizer composed of organic matter from plants and animals.

Synthetic fertilizer Fertilizer produced commercially, normally with the use of fossil fuels. *Also known as* **inorganic fertilizer.**

FIGURE 32.2 **Irrigation circles.** The green circles in this aerial photograph from Oregon are obvious evidence of irrigation. *(Doug Wilson/ARS/USDA)*

Fertilizers

Agriculture removes organic matter and nutrients from soil, and if these materials are not replenished, they can be quickly depleted. Industrial agriculture, because it keeps soil in constant production, requires large amounts of fertilizers to replace lost organic matter and nutrients. Fertilizers contain essential nutrients for plants—primarily nitrogen, phosphorus, and potassium—and they foster plant growth where one or more of these nutrients is lacking. There are two types of fertilizers used in agriculture: *organic* and *synthetic*.

As their name suggests, **organic fertilizers** are composed of organic matter from plants and animals. They are typically made up of animal manure that has been allowed to decompose. Traditional farmers often spread animal manure and crop wastes onto fields to return some of the nutrients that were removed from those fields when crops were harvested.

Synthetic fertilizers, or **inorganic fertilizers,** are produced commercially, normally with the use of fossil fuels. Nitrogen fertilizers are often produced by combusting natural gas, which allows nitrogen from

AP® Exam Tip

Soil tests, synthetic fertilizers, soil conservation, and soil rich in humus are covered in the 2004 AP® exam, Question 4. To answer this question, students must

- identify and describe a chemical and physical soil test, and explain how the soil test results would help a soils scientist to make recommendations for agricultural purposes
- give one advantage and one disadvantage of using synthetic fertilizers
- name two soil conservation practices that would help to decrease soil erosion
- name a biome that is characterized by soil that is rich in humus; explain how humus originated in the soils of this biome, and why humus improves plant growth

Teaching Tip: Journal Prompt

Ask students to describe the pros and cons of using organic fertilizer. *Pros: Organic fertilizers do not require fossil fuels for production; they help increase the water retention ability of the soil; and they are much less expensive than synthetic fertilizers. Cons: It is not possible to obtain the exact amount of needed nutrients as easily; organic fertilizers are much more difficult to spread.*

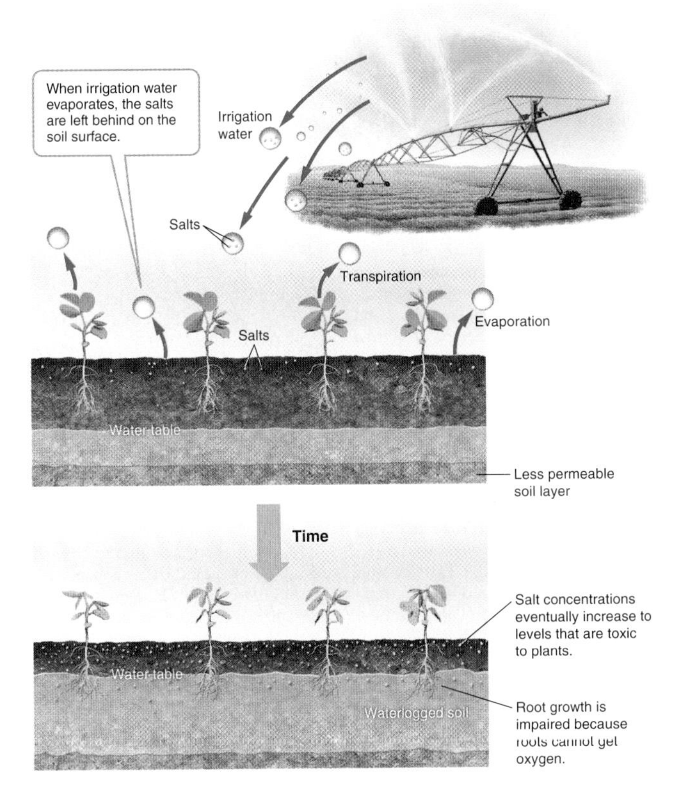

FIGURE 32.3 Irrigation-induced salinization and waterlogging. Over time, irrigation can degrade soil by leaving a layer of highly concentrated salts at the soil surface and waterlogged soil below.

the atmosphere to be fixed and captured in fertilizer. Fertilizers produced in this way are highly concentrated, and their widespread use has increased crop yields tremendously since the Green Revolution began. Synthetic fertilizers have many advantages over organic fertilizers. They are designed for easy application, their nutrient content can be targeted to the needs of a particular crop or soil, and plants can easily absorb them, even in poor soils. Worldwide, synthetic fertilizer use increased from 20 million metric tons in 1960 to nearly 200 million metric tons in 2011. It can be argued that without synthetic fertilizers, we could not feed all the people in the world.

Despite these advantages, synthetic fertilizers can have several adverse effects on the environment. The process of manufacturing synthetic fertilizers uses large quantities of fossil fuel energy. Producing nitrogen fertilizer is an especially energy-intensive process. Whereas synthetic fertilizers are more readily available for plant uptake than are organic fertilizers, they are also more likely to be carried by runoff into adjacent waterways and aquifers. In surface waters, this nutrient runoff can cause algae and other organisms to proliferate. After these organisms die, they decompose and reduce oxygen levels in the water, a process we will learn more about when we look at water pollution in Chapter 14. Finally, synthetic fertilizers do not add organic matter to the soil, as organic fertilizers do. As we saw in Chapter 8, organic matter contributes many beneficial properties to soil, including increased cation exchange capacity and water retention abilities.

AP® Exam Tip

The amount of land required to feed the human population is covered in the 2011 AP® Exam, Question 4. To answer this question, students must

- plot time-series data on the amount of land needed to feed the world's population from 1900 through 2060
- determine the year in which humans are likely to run out of arable land for agriculture
- identify two physical or chemical properties of soils and the role these properties play in soil quality
- describe two strategies to reduce the amount of land humans require for agriculture
- describe how salinization occurs in soil and how it can be remediated

The United States uses somewhat less fertilizer and consequently experiences less nutrient runoff than other nations with similar agricultural output. Still, large amounts of nitrogen and other nutrients run into waterways in intensively farmed regions such as California's Central Valley, farming regions along the East Coast, and the Mississippi River watershed (FIGURE 32.4).

Monocropping

Both the mechanization of agriculture and the use of synthetic fertilizers encourage large plantings of a single species or variety, a practice known as **monocropping** (FIGURE 32.5). Monocropping is the dominant agricultural practice in the United States, where wheat and cotton are frequently grown in monocrops of 405 ha (1,000 acres) or more.

Monocropping has greatly improved agricultural productivity. This technique allows large expanses of land to be planted and then harvested, all at the same time. With the use of large machinery, the harvest can be done easily and efficiently. If fertilizer or pesticide treatments are required, those treatments can also be applied uniformly over large fields, which, because they are planted with the same crop, have the same pesticide or nutritional needs.

Despite the benefits of increased efficiency and productivity, monocropping can lead to environmental degradation. First, soil erosion can become a problem. Because fields that are monocropped are readied for planting or harvesting all at once, soil will be exposed over many hectares at the same time. On a 405-ha (1,000-acre) field that has not yet been planted, wind can blow for over 1.6 km (1 mile) without encountering anything but bare soil. Under these circumstances, the wind can gain enough speed to carry dry soil away from the field. Certain farmland in the United States loses an average of 1 metric ton of topsoil per hectare (2.5 metric tons per acre) per year to wind erosion. As we saw in Chapter 8, this topsoil contains important nutrients and its loss can reduce productivity.

FIGURE 32.4 **Fertilizer runoff after fertilization.** The proximity of this drainage ditch to an agricultural field may lead to runoff of fertilizer during a heavy rainstorm. *(Lynn Betts/USDA Natural Resources Conservation Service)*

Monocropping also makes crops more vulnerable to attack by pests. Large expanses of a single plant species represent a vast food supply for any pests that specialize on that particular plant. Such pests will establish themselves in the monocrop and reproduce rapidly. Their populations may experience exponential growth similar to what we saw in Georgii Gause's *Paramecium* populations that were supplied with unlimited food (see Figure 18.3 on page 195). Natural predators may not be able to respond rapidly to the exploding pest population. Many predators of crop pests, such as ladybugs and parasitic wasps, are attracted to the pests that feed on monocrops. But these predators also rely on non-crop plants for habitat, which monocropping removes. Therefore, predators that might otherwise control the pest population are largely absent.

Monocropping An agricultural method that utilizes large plantings of a single species or variety.

Pesticide A substance, either natural or synthetic, that kills or controls organisms that people consider pests.

Insecticide A pesticide that targets species of insects and other invertebrates that consume crops.

Herbicide A pesticide that targets plant species that compete with crops.

Broad-spectrum pesticide A pesticide that kills many different types of pest.

Pesticides

Pesticides are substances, either natural or synthetic, that kill or control organisms that people consider pests. The use of pesticides has become routine and widespread in modern industrial agriculture. In the United States, over 227 million kilograms (500 million pounds) of pesticides are applied to food crops, cotton, and fruit trees. The United States accounts for about one-third of worldwide pesticide use.

Insecticides target species of insects and other invertebrates that consume crops, and **herbicides** target plant species that compete with crops. Some pesticides are **broad-spectrum pesticides,** meaning that they

Teaching Tip: Warm-up

Ask students to compare broad-spectrum pesticides and selective pesticides. How are they different? How are they similar? *Broad-spectrum pesticides kill many different types of pests. Selective pesticides focus on a narrower range of organisms. Both broad-spectrum and selective pesticides increase crop yields by killing pests that would otherwise eat the crops. Broad-spectrum pesticides can kill many different varieties of pests; however, they often kill beneficial insects as well as harmful ones. Selective pesticides focus on one type of organism and are less likely to kill beneficial species.*

kill many different types of pest, and some are **selective pesticides** that target a narrow range of organisms. The broad-spectrum insecticide dimethoate, for example, kills almost any insect or mite—relatives of ticks and spiders—while the more selective acequinocyl kills only mites.

The application of pesticides is a rapid and relatively easy response to an infestation of pests on an agricultural crop. In many cases, a single application can significantly reduce a pest population. By preventing crop damage, pesticides allow greater crop yields on less land, thereby reducing the area disturbed by agriculture and making agriculture more efficient.

But the application of pesticides, like many other industrial agricultural practices, presents some environmental problems. For example, pesticides often injure or kill more than their intended targets. Some pesticides, such as dichlorodiphenyltrichloroethane, also known as DDT, are **persistent pesticides,** meaning that they remain in the environment for a long time. In 1972, DDT was banned in the United States, in part because it was found to accumulate in the fatty tissues of animals, such as eagles and pelicans. The increasing concentrations of DDT in these animals caused them to lay eggs with thin shells that easily cracked during incubation by the parents. We will discuss this process in more detail in Chapter 17.

Other pesticides, such as the herbicide glyphosate, known by the trade name Roundup, are **nonpersistent pesticides,** meaning that they break down relatively rapidly, usually in weeks to months. Nonpersistent pesticides have fewer long-term effects, but because they must be applied more often, their overall environmental impact is not always lower than that of persistent pesticides.

Another disadvantage of pesticide use is that pest populations may evolve resistance to pesticides over time, as described in Chapter 5. Pest populations are usually large and thus contain significant genetic diversity. In those vast gene pools, there are usually a few individuals that are not as susceptible to a pesticide as others, and those individuals may survive an initial application of the pesticide. Surviving individuals are said to be **pesticide resistant** to that pesticide. If the pesticide is successful in reducing the pest population, the next generation will contain a larger fraction of pesticide resistant individuals in the population. As time goes by, resistant individuals will make up a larger and larger portion of the population. Often the resistance becomes more effective, which makes the pesticide significantly less useful. At this point, crop scientists and farmers must search for a new pesticide. The cycle of pesticide development and pest resistance, illustrated in FIGURE 32.6, is known as the **pesticide treadmill.** The pesticide treadmill is an example of a positive feedback system.

Pesticides can cause even wider environmental effects. They may kill organisms that benefit farmers, such as predatory insects that eat crop pests, pollinator insects that pollinate crop plants, and plants that fix nitrogen and improve soil fertility. Furthermore, chemical pesticides, like fertilizers, can run off into surrounding surface waters and pollute groundwater, a problem we will look at in detail in Chapter 14. The toxicity of pesticides to farmworkers has been well documented. The risk to humans who ingest food that has been treated with pesticides is a subject of debate, which will be covered in Chapter 17.

FIGURE 32.5 **Monocropping.** This large field in Oregon contains only one crop species, soybeans. There are both advantages and disadvantages to monocropping. *(Scott Sinklier/Alamy Images)*

Selective pesticide A pesticide that targets a narrow range of organisms.

Persistent pesticide A pesticide that remains in the environment for a long time.

Nonpersistent pesticide A pesticide that breaks down rapidly, usually in weeks or months.

Pesticide resistance A trait possessed by certain individuals that are exposed to a pesticide and survive.

Pesticide treadmill A cycle of pesticide development, followed by pest resistance, followed by new pesticide development.

AP® Exam Tip

Pesticide use is covered in the 1999 AP® Exam, Question 4. To answer this question, students must

- evaluate four viewpoints about the use of pesticides and write a scientific response to the point each makes
- identify a pest that is harmful to crops or human health and name a way other than the use of pesticides by which the pest could be controlled

Teaching Tip: Warm-up

Ask students the following questions:

- What is a genetically modified organism? *A GMO is an organism that has been produced by isolating a gene from one organism and transferring it into the genetic material of another, often very different, organism.*
- What are the major benefits of GMOs? *GMOs can offer greater yields, improved nutritional benefits in some crops, reduced use of pesticides, and, often, higher profits.*
- What are the disadvantages of using GMOs? *There is some concern over the safety of GMOs. In addition, some people fear that genetically modified crop plants might crossbreed with wild or non-GMO plants that are similar and that the modified genetic sequence will spread and reduce worldwide plant diversity. The spread of modified genomes might alter or eliminate natural plant varieties.*

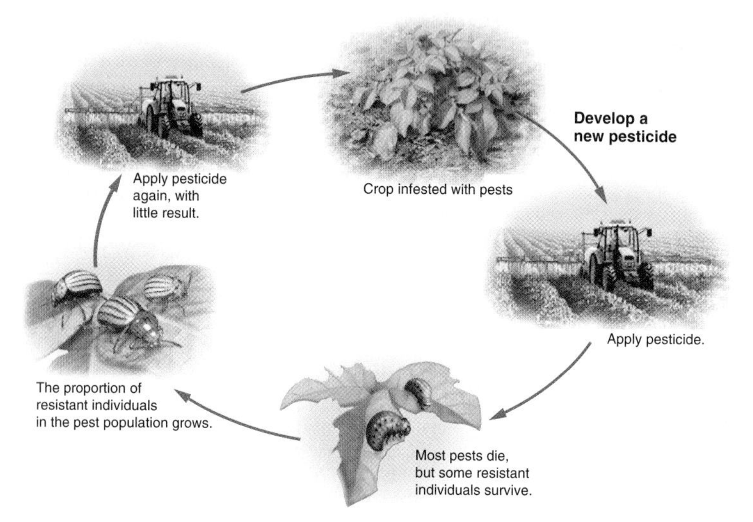

FIGURE 32.6 **The pesticide treadmill.** Over time, pest populations evolve resistance to pesticides, which requires farmers to use higher doses or to develop new pesticides.

Genetic engineering is revolutionizing agriculture

As noted in Chapter 5, humans have modified plants and animals by artificial selection for thousands of years. The modern techniques of genetic engineering, however, go far beyond traditional practices. Scientists today can isolate a specific gene from one organism and transfer it into the genetic material of another, often very different, organism to produce a genetically modified organism, or GMO. By manipulating specific genes, agricultural scientists can rapidly produce organisms with desirable traits that may be impossible to develop with traditional breeding techniques. Genetically modified organisms present both benefits and drawbacks.

The Benefits of Genetic Engineering

Genetically modified crops and livestock offer the possibility of greater yield and food quality, reductions in pesticide use, and higher profits for the agribusinesses that use them. They are also seen as a way to help reduce world hunger by increasing food production and reducing losses to pests and varying environmental conditions.

Increased Crop Yield and Quantity

Genetic engineering can increase food production in several ways. It can create strains of organisms that are resistant to pests and harsh environmental conditions such as drought or high salinity. In addition, agricultural scientists have begun to engineer plants that produce essential nutrients for humans. For example, they have inserted a gene for the production of vitamin A into rice plants, creating new seeds known as golden rice (FIGURE 32.7). Although golden rice is still an experimental product, some scientists hope that it will help reduce the incidence of blindness resulting from vitamin A deficiency. Crop plants, animals, and bacteria have also been modified to produce pharmaceuticals and other compounds, a process that can make these products far less expensive to manufacture. A number of projects are under way to create genetically modified animals for food production, including a salmon that grows to its full size of 3.6 kg (8 pounds) in 18 months—half the growing time of an unmodified fish.

Potential Changes in Pesticide Use

Genetic engineering for resistance to pests could reduce the need for pesticides. Corn, for example, is subject to attacks from the bollworm, European corn borer (*Ostrinia nubilalis*), and lepidopteran (butterfly and moth) larvae. *Bacillus thuringiensis* is a natural soil bacterium that produces a toxin that can kill lepidopterans. The insecticidal gene of this bacterium, known as Bt, has been inserted into the genetic material of corn plants, resulting in a genetically modified plant that produces a natural insecticide in its leaves. By

FIGURE 32.7 **White rice and golden rice.** Crop scientists have inserted a gene that synthesizes a precursor to vitamin A in white rice. The resulting genetically modified rice is called golden rice. *(Erik De Castro/Reuters/Corbis)*

2009, 63 percent of the land area planted with corn in the United States was planted with Bt corn. Growers of Bt corn have been able to reduce the amount of synthetic pesticide used on their corn crops.

A similar technique has been used to create crop plants that are resistant to the herbicide Roundup. The "Roundup Ready" gene allows growers to spray the herbicide on their fields to control the growth of weeds without harming the crop plants. These Bt genes are now widely used in corn, soybean, and cotton plants. The success of no-till agriculture, which we will discuss later in this chapter, rests largely on the use of herbicides and the introduction of herbicide-resistant crops.

Increased Profits

Because pesticides can be a significant expense on any farm, genetically modified crops have the potential to reduce expenses. And because GMO crops often produce greater yields, there is also the potential for an increase in revenues. Both of these changes can lead to higher incomes for farmers, lower food prices for consumers, or both.

Concerns about Genetically Modified Organisms

Industrial agriculture relies more heavily on genetically modified crops each year. In 2012, 88 percent of the corn, 93 percent of the soybeans, and 94 percent of the cotton planted in the United States came from genetically modified seeds. However, many European countries, as well as a number of people in the United States, question the safety of GMOs. Genetically modified crops and livestock are the source of considerable controversy, and concerns have been raised about their safety for human consumption and their effects on biodiversity. Regulation of GMOs is also an issue, both in the United States and abroad.

Safety for Human Consumption

Some people are concerned that the ingestion of genetically modified foods may be harmful to humans, although so far there is little evidence to support these concerns. However, researchers are studying the possibility that GMOs may cause allergic reactions when people eat a food containing genes transferred from another food to which they are allergic.

Effects on Biodiversity

There is some concern that if genetically modified crop plants are able to breed with their wild relatives—as many domesticated crop plants are—the newly added genes will spread to the wild plants. The spread of such genes might then alter or eliminate natural plant varieties. Examples of GMOs crossing with wild relatives do exist. Because of these concerns, attempts have been made to introduce buffer zones around genetically modified crops.

The use of genetically modified seeds is contributing to a loss of genetic diversity among food crops. As with any reduction in biodiversity, we cannot know what beneficial genetic traits might be lost. For example, researchers recently discovered that one variety of sorghum, a cereal grass grown for its sweet juice extract, has a natural genetic variation that gives it resistance to a pest called the greenbug. This particular variety has since been used extensively to confer greenbug resistance on the U.S. sorghum crop. If growers had been growing only one or two genetically modified varieties of sorghum, this naturally resistant variety might have been eliminated before its beneficial trait was discovered.

Regulation of Genetically Modified Organisms

Currently, there are no regulations in the United States that mandate the labeling of genetically modified foods. The European Union, in contrast, allows very few genetically modified foods. In France, Germany, and Italy, almost all GMOs are banned. Those opposed to labeling argue that labeling of foods containing GMOs might suggest to consumers that there is something wrong with GMOs. They also argue that such labeling would be too difficult because small amounts of GMO materials are found throughout the U.S. agricultural system. Those who want to avoid consuming GMOs can safely purchase organic food; the federal definition of "organic" excludes genetically modified foods. (See "Science Applied 5: How Do We Define Organic Food?" on page 392).

The U.S. government has not yet approved any genetically modified animals for market, but there are currently applications for a number of genetically modified animals, most notably salmon, under consideration by the Food and Drug Administration.

Teaching Tip: Debate the Issue

Do the benefits of genetically modified organisms outweigh the concerns? After reading about genetically modified organisms on pages 370–371, ask students to research the subject in preparation for a debate. Students should be prepared to discuss these topics:

- the procedures used to create genetically modified organisms
- potential and current applications of genetic engineering
- effects on biodiversity
- environmental concerns
- environmental benefits
- human health concerns
- human health benefits

In class, select groups to debate the issue. Allow each side 2 minutes to prepare, 3 minutes to present, and 1 minute for rebuttal. Have the remainder of the class take notes, ask questions, and decide which side has the most persuasive argument.

AP® Exam Tip

Genetically modified organisms are covered on the 2009 AP® Exam, Question 4. To answer this question, students must

- calculate the increase in the area of land used for growing genetically modified crops based on historical data provided in a graph
- calculate the annual rate of increase in land area used for growing genetically modified crops
- describe one environmental advantage and one environmental disadvantage of using genetically modified crops
- describe two viable agricultural practices that farmers use to help maintain or improve soil conditions

Modern agribusiness includes farming meat and fish

We have seen that in order to remain economically viable and feed large numbers of people, modern agriculture in the United States has had to become larger and more mechanized. While many of the goals described in the preceding sections have remained in place as agricultural activities moved to a larger scale, some have been abandoned. In particular, as agriculture developed to supply meat and poultry to large numbers of people at a low cost, the primary goal became faster growth of animals.

FIGURE 32.8 **Cattle in a concentrated animal feeding operation in Texas.** CAFOs are large indoor or outdoor structures that allocate a very small amount of space to each animal. *(Michael Dwyer/Alamy)*

High-Density Animal Farming

In 2011, according to the U.S. Department of Agriculture, roughly 150 million animals were slaughtered for beef, pork, and lamb, along with billions of chickens, turkeys, and ducks. Many of these animals were raised in feed-lots, or **concentrated animal feeding operations (CAFOs),** which are large indoor or outdoor structures designed for maximum output (FIGURE 32.8).This type of high-density animal farming is used for beef cattle, dairy cows, hogs, and poultry, all of which are confined or allowed very little room for movement during all or part of their life cycle. A CAFO may contain as many as 2,500 hogs or 55,000 turkeys in a single building. By keeping animals confined, farmers minimize land costs, improve feeding efficiency, and increase the fraction of food energy that goes into the production of animal body mass. The animals are given antibiotics and nutrient supplements to reduce the risk of adverse health effects and diseases, which would normally be high in such highly concentrated animal populations.

High-density animal farming has many environmental and health consequences. There is evidence that antibiotics given to confined animals are contributing to an increase in antibiotic-resistant strains of microorganisms that can affect humans. Waste disposal is another serious problem. An average CAFO produces over 2,000 tons of manure annually, or about as much as a town of 5,000 people would produce. The waste is usually used to fertilize nearby agricultural fields, but if over-applied, it can cause the same nutrient runoff problems as synthetic fertilizer. Sometimes animal wastes are stored in lagoons adjacent to feedlots but, during heavy rainstorms, runoff from these lagoons can contaminate nearby waterways. Animal wastes have also been dumped, either inadvertently or intentionally, into natural waters. The U.S. Environmental Protection Agency has concluded that chicken, hog, and cattle waste has caused pollution along 56,000 km (35,000 miles) of rivers in 22 states and has caused some degree of groundwater contamination in 17 states.

Concentrated animal feeding operation (CAFO) A large indoor or outdoor structure designed for maximum output.

Fishery A commercially harvestable population of fish within a particular ecological region.

Harvesting Fish and Shellfish

Fish is the third major source of food for humans, after grain and meat. In many coastal areas, particularly in Asia and Africa, fish accounts for nearly all animal protein that some people consume. As FIGURE 32.9 shows, the global production of fish has increased more than 30 percent since 1980.This increase masks two divergent trends: a rapid increase in farmed fish production and a decrease in wild fish caught in the world's oceans.

A **fishery** is a commercially harvestable population of fish within a particular ecological region. The tragedy of the commons, which we learned about in Chapter 10, is particularly applicable to ocean fisheries. Since most fish do not live their entire lives within national borders, if one individual fisher, or even an entire country, limits its catch, others are likely to make up the difference because those others will not limit their catch. No country has an incentive to protect fish stocks or to attempt to replenish them because fish in the ocean do not belong to any one individual or nation. As a result, competition for fish has led to a precipitous decline in fish populations.

In 2003, the journal *Nature* reported a dramatic decline in the number of large predatory ocean fish caught over the past 50 years, even though both the number of fishing vessels and the amount of time spent fishing had increased over that period. Fishers were

AP® Exam Tip

The commercial fishing industry is covered in the 2006 AP® Exam, Question 4. To answer this question, students must

- calculate the greatest decline in fish harvest over a 5-year period based on data presented in a graph
- discuss two commercial fishing practices and how these practices hurt the sustainability of the local waters.
- identify another commons affected by human activity and suggest some practical management strategies

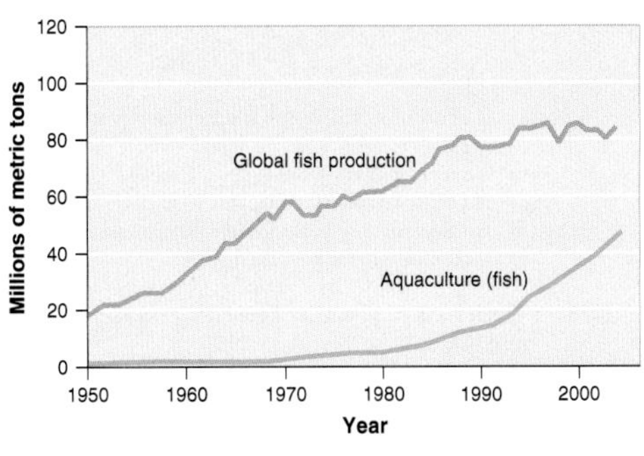

FIGURE 32.9 **Global fish production.** Global fish production has increased by more than 30 percent since 1980, primarily as a result of the large increase in aquaculture. The graph shows data for aquaculture-raised fish (blue) and global fish production (orange), which includes both wild-caught fish and aquaculture-raised fish. *(Data from K. M. Brander, PNAS 104 (2007): 19709–19714)*

working harder, but catching fewer fish. A study in 2006 found that 30 percent of fisheries worldwide had experienced a 90 percent decline in fish populations. The decline of a fish population by 90 percent or more is referred to as **fishery collapse.** Many studies on this subject focus on large predatory fish. However, a 2011 study in the *Proceedings of the National Academy of Sciences* noted an unexpected increase in fishery collapse among smaller species as well.

Ocean harvests used to be limited by the difficulty of finding fish in a vast ocean as well as by lack of capacity of small boats and nets, but this is no longer the case. Current fishing methods make it easy to catch large numbers of fish. Factory ships can stay at sea for months at a time, processing and freezing their harvest without having to return to port. Most marine fish are now caught either by large nets pulled behind one of these ships or by very long fishing lines bearing hundreds or even thousands of baited hooks. Fishers in pursuit of high-value species such as tuna use spotter planes and sonar to locate schools. They encircle schools with nets that can capture up to 3,000 adult tuna at a time—almost a million pounds of fish. Fish species that live on or close to the ocean bottom, as well as many shellfish, are caught in dragnets, which are weighted so that they can be pulled across the ocean floor.

Large-scale, high-tech fishing can adversely affect both target and nontarget species. Dragnets can damage ocean-bottom habitats by scouring them of coral, sea sponges, and plants. Many commercially important fish are keystone species, so a decline or loss of their populations can have cascading effects on other marine species. Intensive fishing leads to the loss of juvenile fish of the target species as well as to the loss of noncommercial species that are accidentally caught by nets and lines. This unintentional catch of nontarget species, referred to as **bycatch,** has significantly reduced populations of fish species such as sharks and has endangered other organisms such as sea turtles. Some countries now require the use of technology that minimizes bycatch or harm to endangered species.

Fishery collapse The decline of a fish population by 90 percent or more.

Bycatch The unintentional catch of nontarget species while fishing.

module 32 REVIEW

In this module, we have seen that modern, large-scale farming methods have transformed agriculture. Large amounts of fossil fuel energy and fertilizer are common inputs to most agriculture in the developed world. The Green Revolution introduced a number of changes to agriculture in the 1950s and 1960s that continue to result in increased agricultural yield today. Genetic engineering has also changed agriculture, but it remains controversial. Animals are typically raised in concentrated feeding areas that allow for maximum profit but can result in a variety of environmental concerns. Fishery production is also dependent on modern mechanized systems that have resulted in increased yield at the expense of the wild populations.

Answers to Module 32 AP® Review Questions

1. c
2. d
3. d
4. e
5. e

Module 32 AP® Review Questions

1. If 12,000,000 kilocalories of chicken can be produced per hectare, how much land is needed to provide someone with 2,000 kilocalories/day for a year?
   (a) 0.01 ha
   (b) 0.02 ha
   (c) 0.06 ha
   (d) 0.2 ha
   (e) 0.4 ha

2. The Green Revolution
   (a) began in the mid-nineteenth century.
   (b) discouraged the mechanization of agriculture.
   (c) decreased the energy subsidy of most food.
   (d) encouraged the use of monocropping.
   (e) pertained to leafy green plants only.

3. Broad-spectrum pesticides
   (a) do not cause increased pesticide resistance.
   (b) are almost always nonpersistent.
   (c) are banned in the United States.
   (d) are likely to kill beneficial insects.
   (e) are only produced synthetically.

4. Bycatch
   (a) is a common problem with increased pesticide use.
   (b) is a management technique in CAFOs.
   (c) is a cause of fishery collapse.
   (d) is a side effect of some genetic engineering.
   (e) is common in large-scale fishing.

5. Which is NOT a benefit of genetically modified organisms?
   (a) Decreased pesticide use
   (b) Increased profits
   (c) Increased resistance to extreme weather
   (d) Increased crop yield
   (e) Increased genetic diversity

module

33

Alternatives to Industrial Farming Methods

As problems with industrial agriculture become more apparent, alternatives are gaining more attention. Some of the newer techniques have actually been used in the developing world for thousands of years. In this module, we will examine some of these alternative farming practices. We will also look at the large-scale farming of meat and fish.

Learning Objectives

After reading this module, you should be able to

- describe alternatives to conventional farming methods.
- explain alternative techniques used in farming animals and in fishing and aquaculture.

Alternatives to industrial farming methods are gaining more attention

Industrial agriculture has been so successful and widespread that it has come to be known as conventional agriculture. However, in situations in which the cost of labor is not the most important consideration, traditional farming techniques may also be economically successful. Small-scale farming is common in the developing world where labor is less expensive than are machinery and fossil fuels. In these countries, there are still many farmers who grow crops on small plots of land. Traditional farming methods that differ from those of industrial agriculture include *shifting agriculture* and *nomadic grazing,* which are not always sustainable, and more sustainable methods such as *intercropping* and *agricultural forestry*.

Shifting Agriculture and Nomadic Grazing

In locations with a moderately warm climate and relatively nutrient-poor soils, such as the rainforests of Central and South America, a large percentage of the nutrients is contained within the vegetation. In these places, farmers sometimes use **shifting agriculture,** in which the land is cleared and used for only a few years until the soil is depleted of nutrients. This traditional method of agriculture uses a technique sometimes called "slash-and-burn," in which existing trees and vegetation are cut down, placed in piles, and burned (FIGURE 33.1).The resulting ash is rich in potassium, calcium, and magnesium, which makes the soil more fertile, although these nutrients are usually depleted quickly. If the deforestation occurs in an area of heavy rainfall, nutrients may be washed away, along with some of the soil, which further reduces the availability of nutrients. After a few years, the farmer usually moves on to another plot and repeats the process.

If a plot is used for a few years and then abandoned for a number of decades, over time the soil may recover its organic content and nutrient supplies and the vegetation may have a chance to regrow. However, population pressures may cause the land to be used too frequently to allow for its full recovery. In that case, soil productivity can decrease rapidly, leaving the land suitable only for animal grazing. In addition, the burning process oxidizes carbon, meaning that it converts it into the oxide compounds carbon monoxide (CO) and carbon dioxide (CO_2). In this way, carbon from the vegetation and the soil is released into the atmosphere and ultimately contributes to higher atmospheric CO_2 concentrations.

In semiarid environments, dry, nutrient-poor soils can be easily degraded by agriculture to the point at which they are no longer viable for any production at all. Irrigation can cause salinization, and topsoil is eroded away because the shallow roots of annual crops fail to hold the soil in place. The transformation of arable,

FIGURE 33.1 **Shifting agriculture.** This forest has been cleared for agriculture. Clearing land often involves burning it, which may make nutrients and soils vulnerable to erosion. *(Pete Oxford/DanitaDelimont.com)*

productive land to desert or unproductive land due to climate change or destructive land use is known as **desertification.** The world map in FIGURE 33.2 shows the parts of the world that are most vulnerable to desertification. Today, desertification is occurring most rapidly in Africa, where the Sahara is expanding at a rate of up to 50 km (31 miles) per year. Unsustainable farming practices in northern China are also leading to rapid desertification.

The only sustainable way for people to use soil types with very low productivity is **nomadic grazing,** in which they move herds of animals, often over long distances, to seasonally productive feeding grounds. If grazing animals move from region to region without lingering in any one place for too long, the vegetation can usually regenerate.

Shifting agriculture and nomadic grazing worked well under the conditions in which they were first developed—namely, low human population densities

Shifting agriculture An agricultural method in which land is cleared and used for a few years until the soil is depleted of nutrients.

Desertification The transformation of arable, productive land to desert or unproductive land due to climate change or destructive land use.

Nomadic grazing The feeding of herds of animals by moving them to seasonally productive feeding grounds, often over long distances.

Teaching Tip: Discussion Starter

Ask students: How does shifting agriculture often lead to desertification? *Shifting agriculture involves clearing land and using it for only a few years until the soil is depleted of nutrients. In semiarid environments, dry, nutrient-poor soils can be easily degraded by agriculture to the point at which those soils are no longer viable for any production at all.*

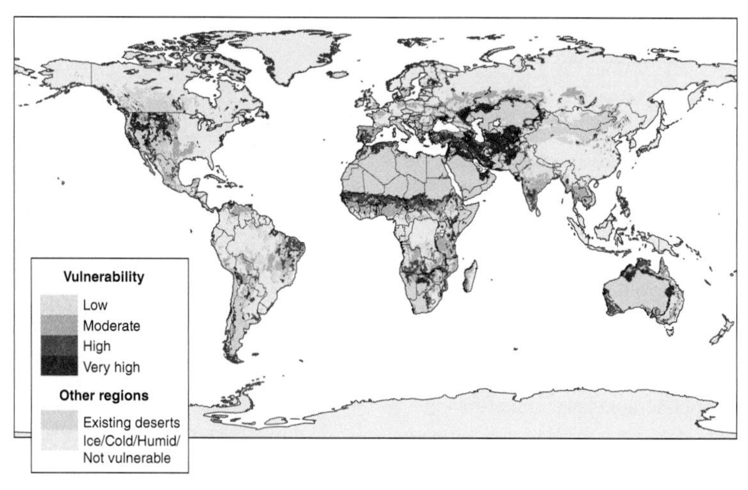

FIGURE 33.2 **Vulnerability to desertification.** Certain regions of the world are much more vulnerable to desertification than others. *(Data from www.fao.org)*

and subsistence farming—but as populations increase, both these forms of traditional agriculture become less sustainable. Sometimes the relocation of people for political or other reasons can cause traditionally sustainable agricultural techniques to become unsustainable. For example, in the early part of the twentieth century, many subsistence farmers in Central America were relocated away from rich floodplains to mountainous areas where the plowing methods that had worked in the flat areas caused severe erosion in the mountains and could not be sustained.

Sustainable Agriculture

Is it possible to produce enough food to feed the world's population without destroying the land, polluting the environment, or reducing biodiversity? **Sustainable agriculture** fulfills the need for food and fiber while enhancing the quality of the soil, minimizing the use of nonrenewable resources, and allowing economic viability for the farmer. It emphasizes the ability to continue agriculture on a given piece of land indefinitely through conservation and soil improvement. Sustainable agriculture often requires more labor than industrial agriculture, which makes it more expensive in places where labor costs are high. But practitioners of sustainable agriculture consider the improved long-term productivity of the land to be worth this extra cost.

Many of the practices used in sustainable agriculture are traditional farming methods (FIGURE 33.3). Subsistence farmers in India, Kenya, and Thailand typically use animal and plant wastes as fertilizer because they cannot obtain or afford synthetic fertilizers. Such traditional farmers may also practice **intercropping** (Figure 33.3a), in which two or more crop species are planted in the same field at the same time to promote a synergistic interaction between them. For example, corn, which requires a great deal of nitrogen, can be planted along with peas, a nitrogen-fixing crop. **Crop rotation** achieves the same effect by rotating the crop species in a field from season to season. For example, peas can be planted in a field for one year, leaving excess nitrogen in the soil to nourish the corn crop that is planted there in the following year.

Sustainable agriculture Agriculture that fulfills the need for food and fiber while enhancing the quality of the soil, minimizing the use of nonrenewable resources, and allowing economic viability for the farmer.

Intercropping An agricultural method in which two or more crop species are planted in the same field at the same time to promote a synergistic interaction.

Crop rotation An agricultural technique in which crop species in a field are rotated from season to season.

Teaching Tip: Video

How to Intercrop in a Vegetable Garden: Berry Gardening, Fertilizers, and Vegetables
Use this video on intercropping to help your students understand how different types of crops can be grown together to maximize space and crop yields.

The link to this video may be found by clicking on the link buttons in the TE-book, opening the TRFD, or logging onto the book's companion website bcs.whfreeman.com/friedlandapes2e (teacher login required).

TRM **Chapter 11 Web Resources**

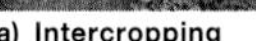

(a) Intercropping

(c) Contour plowing

(b) Agroforestry

FIGURE 33.3 **Sustainable farming methods.** A variety of farming methods can be used to improve agricultural yield and retain soil and nutrients. *(a: Inga Spence/Getty Images; b: Chris R. Sharp/Science Source; c: Tim McCabe/USDA Natural Resources Conservation Service)*

Intercropping trees with vegetables—a practice that is sometimes called **agroforestry**—allows vegetation of different heights, including trees, to act as windbreaks and to catch soil that might otherwise be blown away, greatly reducing erosion (Figure 33.3b). The trees not only protect the vegetable crops and the soil, but also provide fruit and firewood.

Alternative methods of land preparation and use can also help to conserve soil and prevent erosion. For instance, **contour plowing**—plowing and harvesting parallel to the topographic contours of the land—helps prevent erosion by water while still allowing for the practical advantages of plowing (Figure 33.3c). Some farmers plant an autumn crop, such as winter wheat, that will sprout before frost sets in, so that the land does not remain bare between regular plantings.

No-Till Agriculture

Perennial plants live for multiple years, and there is usually no need to disturb the soil. In contrast, **annual plants,** such as wheat and corn, live only one season and must be replanted each year. Conventional

Agroforestry An agricultural technique in which trees and vegetables are intercropped.

Contour plowing An agricultural technique in which plowing and harvesting are done parallel to the topographic contours of the land.

Perennial plant A plant that lives for multiple years.

Annual plant A plant that lives only one season.

Teaching Tip: Activity

Ask students to create a table that describes and compares sustainable farming practices. (Answers are provided in italics.)

Sustainable farming practice	Definition	Advantage	Disadvantage
Intercropping	*An agricultural method in which two or more crop species are planted in the same field and promote a synergistic interaction*	*Can reduce the amount of fertilizer required*	*Requires scientific expertise on plant interactions; can be more labor intensive*
Crop rotation	*An agricultural technique in which crop species in a field are rotated from season to season*	*Can reduce the amount of fertilizer required*	*Requires scientific expertise on plant interactions; can be more expensive*
Agroforestry	*An agricultural technique in which trees and vegetables are intercropped*	*Reduces erosion by wind*	*Land management is more complex; can be more labor intensive and costly*
Contour plowing	*An agricultural technique in which plowing and harvesting are done parallel to the topographic contours of the land*	*Conserves soil and helps to prevent erosion by water*	*Land management is more complex; can be more labor intensive and costly*
No-till agriculture	*An agricultural method in which farmers do not turn the soil between seasons so that topsoil erosion is reduced*	*Reduces topsoil erosion and reduces emissions of* CO_2	*Farmers may need to use herbicides before and after planting crops*

TEACHING with FIGURES ▶

Show students Figure 33.4 and ask the following questions:

- What are some of the causes of soil degradation? *Soil degradation is caused by overgrazing of livestock, deforestation due to clearing for farmland, and mismanagement of agricultural activities.*
- What areas of the world have experienced the worst soil degradation? *Central Africa, South Africa, Indonesia, Mexico, and Western Europe are areas that have experienced the worst soil degradation.*

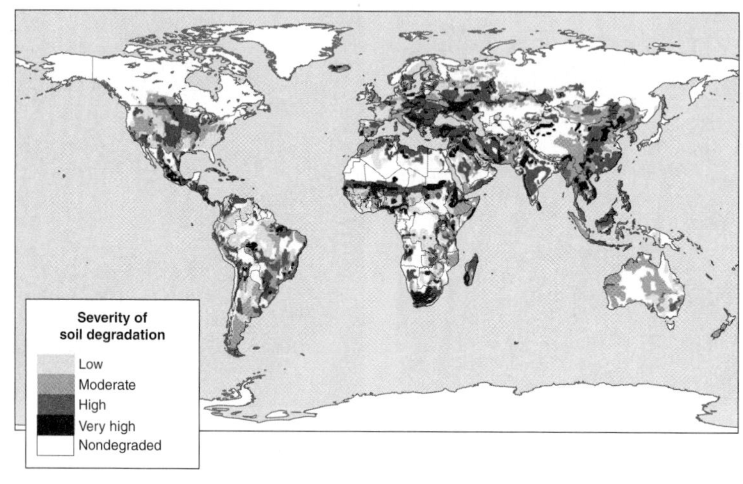

FIGURE 33.4 **Global distribution of soil degradation.** Soil degradation is a global problem caused by overgrazing and deforestation as well as agricultural mismanagement. *(Data from United Nations Environment Programme)*

agriculture of annual plants therefore relies on plowing and tilling, processes that physically turn the soil upside down and push crop residues under the topsoil, thereby killing weeds and insect pupae. Critics argue that plowing and tilling have negative effects on soils. As we learned in Chapter 8, soils may take hundreds or even thousands of years to develop as organic matter accumulates and soil horizons form. Every time soil is plowed or tilled, soil particles that were attached to other soil particles or to plant roots are disturbed and broken apart and become more susceptible to erosion. In addition, repeated plowing increases the exposure of organic matter deep in the soil to oxygen. This exposure leads to oxidation of organic matter, a reduction in the organic matter content of the soil, and an increase in atmospheric CO_2 concentrations. Tilling, in addition to irrigation and overproduction, has led to severe soil degradation in many parts of the world. The world map in FIGURE 33.4 indicates areas of severe soil degradation.

No-till agriculture is an agricultural method in which farmers do not turn the soil between seasons and is used as a means of reducing topsoil erosion.

No-till agriculture An agricultural method in which farmers do not turn the soil between seasons as a means of reducing topsoil erosion.

No-till agriculture is designed to avoid the soil degradation that comes with conventional agricultural techniques. Farmers using this method leave crop residues in the field between seasons (FIGURE 33.5). The intact roots hold the soil in place, reducing both wind and water erosion, and the undisturbed soil is able to

FIGURE 33.5 **No-till agriculture.** Rows of soybeans emerge between the residues of a corn crop left over from the previous season. *(Lynn Betts/USDA Natural Resources Conservation Service)*

regenerate natural soil horizons. No-till agriculture also reduces emissions of CO_2 because the intact soil undergoes less oxidation. In many cases, however, in order for no-till agriculture to be successful, farmers must apply herbicides to the fields before, and sometimes after, planting so that weeds do not compete with the crops. Therefore, the downside of no-till methods is an increase in the use of herbicides.

Integrated Pest Management

Another alternative agricultural practice, **integrated pest management (IPM),** uses a variety of techniques designed to minimize pesticide inputs. These techniques include crop rotation and intercropping, the use of pest-resistant crop varieties, the creation of habitats for predators of pests, and limited use of pesticides.

Crop rotation and the use of pest-resistant crop varieties prevent pest infestations. Crop rotation can foil insect pests that are specific to one crop and that may have laid eggs in the soil. It can also hinder crop-specific diseases that may survive on infected plant material from the previous season. Intercropping, as stated earlier, also makes it harder for specialized pests that succeed best with only one crop present to establish themselves. Farmers can also provide habitat for species that prey on crop pests (FIGURE 33.6). Agroforestry encourages the presence of insect-eating birds (although birds can also damage some crops). Many herbs and flowers attract beneficial insects.

Although IPM practitioners do use pesticides, they limit applications through very careful observation. Farmers regularly inspect their crops for the presence of insect pests and other potential crop hazards to catch them early so they can be treated using natural controls or smaller doses of pesticides than would be needed at later stages of an infestation. These more-targeted methods of pest control can result in significant savings on pesticides as well as improved yields. FIGURE 33.7—a case study from IPM training in Indonesia—shows the difference IPM can make. Farmers who learned how to determine whether a pesticide application was warranted were able to cut their pesticide applications, and their expenditures on pesticides, in half. Yields also improved after farmers learned IPM methods.

When farmers take time to inspect their fields carefully, as required by IPM methods, they often notice other crop needs, and this additional attention improves overall crop management. The trade off for these benefits is that farmers must be trained in IPM methods and must spend more time inspecting their crops. But once farmers are trained, the extra income and reduced costs associated with IPM often outweigh the extra time they must spend in the field. IPM has been especially successful in many parts of the developing world where the high-input industrial farming model is not viable because labor costs are low and farmers lack financial resources.

FIGURE 33.6 **Beneficial insect habitat.** Practitioners of integrated pest management often provide habitat for insects that prey on crop pests. This wasp is laying eggs in a caterpillar, which it has paralyzed. *(Science Source)*

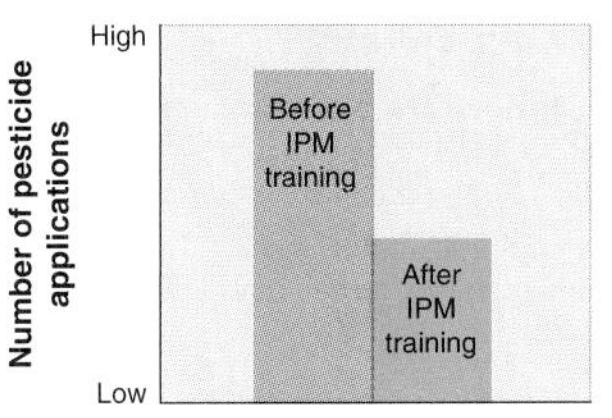

(a) Pesticide use

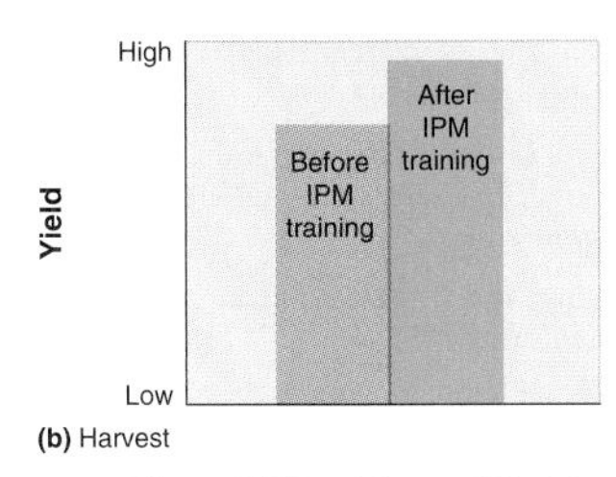

(b) Harvest

FIGURE 33.7 **Effects of IPM training.** (a) IPM training of farmers in Indonesia led to a significant reduction in pesticide applications. (b) Yield improvements also occurred after the training because of the additional attention the farmers gave to their crops. *(Data from www.fao.org)*

Integrated pest management (IPM) An agricultural practice that uses a variety of techniques designed to minimize pesticide inputs.

◀ TEACHING with FIGURES

Show students Figure 33.7 and ask the following questions:

- What is IPM training? *IPM stands for integrated pest management. IPM uses a variety of techniques designed to minimize pesticide inputs. During IPM training, farmers learn how to reduce pesticide use. Techniques include crop rotation, intercropping, and the creation of habitat for beneficial species that feed on pests.*
- What can we learn from looking at Figure 33.7a and Figure 33.7b? *After IPM training, farmers used less pesticide and their crop yields increased.*
- What are the benefits of IPM? *Farmers spend less money on pesticides and have a higher crop yield. There is less pesticide runoff in local waterways. Nontarget species are not subject to extermination from pesticides intended for pest species.*

Teaching Tip: Beyond the Classroom

Beneficial insects perform valuable services such as pollination or pest control. Ask students to research and write a short report on a beneficial insect. Examples of these beneficial insects to research include:

- ladybug
- dragonfly
- lacewing
- praying mantis
- minute pirate bug
- earwig
- big-eyed bug
- assassin bug
- damsel bug
- mealybug
- ladybird
- soldier beetle
- hoverfly
- tachinid fly
- ichneumon wasp
- trichogramma wasp

AP® Exam Tip

Pesticide use and integrated pest management are covered in the 2012 AP® Exam, Question 3. To answer this question, students must

- design a laboratory experiment to determine whether a new pesticide is harmful to minnows
- describe experimental results that would cause the student to reject his or her hypothesis
- explain integrated pest management and name an environmental benefit of using IPM
- describe two agricultural practices to increase crop yields

Teaching Tip: Activity

Use Science Applied 5 on page 392, “How Do We Define Organic Food?” to introduce your students to organic farming. Conduct a pretest and a post-test with these questions:

- How did the organic food movement begin?
- What does it mean to be organic?
- Does organic food mean family farms?
- What doesn’t the organic label mean?

For the pretest, divide the class into small groups and give them 10 minutes to answer all the questions without using any textbooks or the Internet. After collecting the answers, ask them to read the Science Applied feature. Have them take another 10 minutes to answer the same questions. Ask groups to share their answers with the rest of the class. When the groups are finished presenting their post-test answers, have them reevaluate pretest responses.

Organic Agriculture

Organic agriculture is the production of crops without the use of synthetic pesticides or fertilizers. Organic agriculture follows several basic principles:

- Use ecological principles and work with natural systems rather than dominating those systems.
- Keep as much organic matter and as many nutrients in the soil and on the farm as possible.
- Avoid the use of synthetic fertilizers and synthetic pesticides.
- Maintain the soil by increasing soil mass, biological activity, and beneficial chemical properties.
- Reduce the adverse environmental effects of agriculture.

In the developed world, organic farming has increased in popularity over the past 3 decades. The U.S. Organic Foods Production Act (OFPA) was enacted as part of the 1990 farm bill to establish uniform national standards for the production and handling of foods labeled organic. “Science Applied 5: How Do We Define Organic Food?” on page 392 discusses organic food labeling in more detail.

For a long time, all organic farms were small, but today this is not necessarily the case. However, because most organic farmers plant diverse crops and encourage beneficial insects, it is common for them to keep their farms relatively small. Organic farmers also must manage the soil carefully because if they lose soil nutrients and see a decline in the health of their crops, they have fewer options than conventional farmers. These practices usually increase labor costs significantly. However, farmers can recoup extra labor costs by selling their harvest at a premium price to consumers who prefer to buy organic food (FIGURE 33.8).

Organic agriculture is not without some adverse environmental consequences. Because organic farmers typically do not use herbicides, they are less likely than conventional farmers to be able to use no-till methods successfully. And alternative pest control methods are not always environmentally friendly. For example, in order to keep crops such as carrots free of weeds, organic farmers may treat the soil with a flame fueled by propane before planting. While this technique protects carrots without the use of herbicides, it does use propane, a fossil fuel.

Organic agriculture Production of crops without the use of synthetic pesticides or fertilizers.

FIGURE 33.8 **Organic farming.** Many of the consumers at this farmers’ market are willing to pay higher prices for organically grown food. *(Stephen Matera/DanitaDelimont.com)*

Alternative techniques for farming animals and fish are becoming more popular

Animals and fish are in general more energy intensive to raise and prepare for eating than vegetable crops, but there are methods for animal and fish production that have a lower environmental impact. This section covers some of those processes.

Not all meat comes from CAFOs. Free-range chicken and beef are becoming increasingly popular in the United States. Some people find it more ethically acceptable to eat a chicken or cow that has wandered free than one that has spent its entire life confined in a small space. Free-range meat, if properly produced, is more likely to be sustainable than meat produced in CAFOs. Because these free-range animals are not as likely to spread disease as those that are kept in close quarters, the use of antibiotics and other medications can be reduced or eliminated. The animals graze or feed on the natural productivity of the land, with little or no supplemental feeding, so less fossil fuel goes into the raising of free-range meat. Finally, manure and urine are dispersed over the range area and are naturally processed by detritivores and decomposers in the soil. As a result, there is no need to treat and dispose of massive quantities of manure. On the negative side, free-range operations use more land than CAFOs do, and the cost of meat produced using these techniques is usually significantly higher.

More Sustainable Fishing

In the interest of creating and supporting sustainable fisheries, many countries around the world have developed fishery management plans, often in cooperation with one another. International cooperation is particularly important because fish migrate across national borders, some marine ecosystems span national borders, and many of the world's most important fisheries lie in international waters.

The northwestern Atlantic fisheries, for example, comprise several continental shelf ecosystems that stretch from the northeastern United States to southeastern Canada. Historically, these fisheries were among the most productive in the world. However, overfishing by international fleets of factory ships led to a catastrophic depletion of fish stocks, particularly of cod and pollock, by the early 1990s, as FIGURE 33.9 shows. The fisheries were forced to close because of the depleted stocks, and the Canadian and U.S. governments imposed a moratorium on bottom fishing in the area. The majority of the Atlantic Canadian fisheries are still closed today. Most fisheries in the United States are now open, at least for part of the year.

In response to the fishery collapse, and in order to restore the depleted stocks and manage the ecosystem as a whole, the U.S. Congress passed the Sustainable Fisheries Act in 1996. This act shifted fisheries management from a focus on economic sustainability to an approach that increasingly stressed conservation and the sustainability of species. The act calls for the protection of critical marine habitat, which is important for both commercial fish species and nontarget species. For many commercial species considered to be in danger, such as cod, a sustainable fishery means that no fishing will be permitted until populations recover.

One successful fishery management plan was developed in Alaska, where the commercial salmon fishery declined rapidly between 1940 and 1970. Managers first tried to increase salmon populations by limiting the fishing season, but by 1970, when the season was restricted to less than a week, so many fishers participated that populations continued to drop.

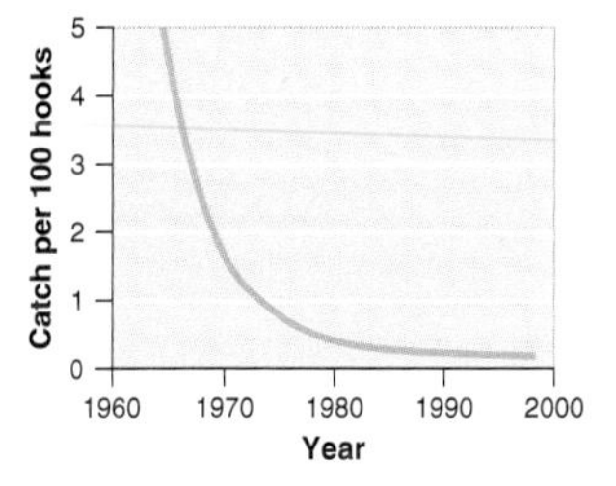

FIGURE 33.9 **Fishery collapse in the northwestern Atlantic Ocean.** Most Canadian fisheries have still not opened since this collapse. U.S. fisheries are open for at least part of the year.

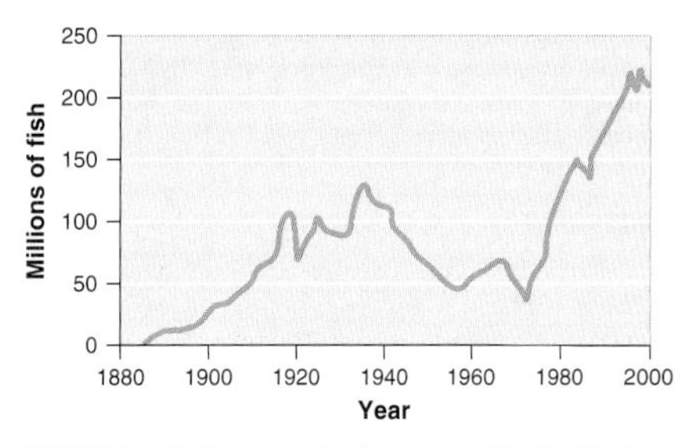

FIGURE 33.10 **Commercial salmon harvest in the Alaska fishery.** After a peak harvest in 1940, overfishing led to a decline in the number of fish caught. In 1973, fishery managers introduced a system of individual transferable quotas. By 1980, the fishery had rebounded.

In 1973, fishery managers introduced a system of **individual transferable quotas (ITQs),** a fishery management program in which individual fishers are given a total allowable catch of fish for a season that they can either catch themselves or sell to others. Before the start of each salmon season, fishery managers establish a total allowable catch and distribute or sell quotas to individual fishers or fishing companies, favoring those with long-term histories in the fishery. Fishers with ITQs have a secure right to catch their quota so they have no need to spend money on bigger boats and better equipment in order to outcompete others. If fishers cannot catch enough salmon to remain economically viable, they can sell all or part of their quota to another fisher. FIGURE 33.10 shows the results: Since the beginning of the ITQ program, the salmon population and harvest have increased—at times very rapidly—and costs to fishers have been reduced.

In Alaska, ITQs are sold primarily to small family-run fishing operations. In New Zealand, however, ITQs have been used effectively to control overfishing by large fishing companies. The ITQ system is being used successfully in many other fisheries around the world.

Not all fisheries are declining, but it is often difficult for consumers to know which fish are being overharvested and which are not. To help consumers choose more sustainable fish, the Environmental Defense Fund and other organizations have compiled lists of popular

Individual transferable quota (ITQ) A fishery management program in which individual fishers are given a total allowable catch of fish in a season that they can either catch or sell.

Teaching Tip: Beyond the Classroom

Divide students into groups that will create a presentation on alternative techniques for farming and harvesting fish and marine organisms. Assign one of the following industrial fishing techniques to each group:

- trawl nets or trawler fishing
- drift-net fishing
- purse-seine fishing
- longline fishing
- fish farming in cages
- deep-sea aquaculture cage fishing

You may offer the following guidelines for the presentation:

- Describe the technique. Describe the size of the net or lines and where the fish are caught or farmed.
- Describe the target fish harvested with this technique. Address whether commercial fishing practices have contributed to the depletion of this fish in the past.
- Describe one international regulation or U.S. law that applies to the harvesting of marine organisms in general and the type of organism this technique specifically harvests.
- Explain how human activities affect ocean resources such as fish and suggest one practical method for managing ocean resources more sustainably.

Teaching Tip: Debate the Issue

Should the U.S. government subsidize aquaculture?

Have your students research fish farming in the ocean. In the debate, students should be prepared to answer the following questions:

- What is fish farming? How does it work?
- What are the environmental pros and cons of fish farming in the ocean?
- Do you think fish farming in the ocean is a good idea?
- How does the environmental impact of fish farming in the ocean compare to traditional methods of commercial fishing?

In class, select groups to debate the issue. Allow each side 2 minutes to prepare, 3 minutes to present, and 1 minute for rebuttal. Have the remainder of the class take notes, ask questions, and decide which side has the most persuasive argument.

food fish, dividing them into three categories, depending on how sustainable their stocks are. "Best" choices include wild Alaskan salmon and farmed rainbow trout. "Worst" choices include shark and Chilean sea bass.

Aquaculture

The demand for fish has increased even as wild fish catches have been falling. In response, many scientists, government officials, and entrepreneurs have been developing ways to increase the production of seafood through **aquaculture**: the farming of aquatic organisms such as fish, shellfish, and seaweeds. Aquaculture involves constructing an aquatic ecosystem by stocking the organisms, feeding them, and protecting them from diseases and predators. It usually requires keeping the organisms in enclosures (FIGURE 33.11), and it may require providing them with food and antibiotics. Almost all of the catfish and trout eaten in the United States, as well as half of the shrimp and salmon, are produced by aquaculture.

Proponents of aquaculture believe it can alleviate some of the human-caused pressure on overexploited fisheries while providing much-needed protein for the more than 1 billion undernourished people in the world. Aquaculture also has the potential to boost the economies of many developing countries.

Critics of aquaculture, though, point out that it can create many environmental problems. In a typical aquaculture facility, clean water is pumped in at one end of a pond or marine enclosure, and wastewater containing feces, uneaten food, and antibiotics is pumped back into the river or ocean at the other end. The wastewater may also contain bacteria, viruses, and pests that thrive in the high-density habitat of aquaculture facilities and this can infect wild fish and shellfish populations outside the facility. In addition, fish that escape from aquaculture facilities may harm wild fish populations by competing with them, interbreeding with them, or spreading diseases and parasites. Overall, however, aquaculture has many promising characteristics as a means of sustainable food production.

Aquaculture Farming aquatic organisms such as fish, shellfish, and seaweeds.

FIGURE 33.11 A salmon farming operation in Chile. Uneaten food and waste released from salmon farms can cause significant nutrient input into natural marine ecosystems. *(Morten Andersen/Corbis)*

Teaching Tip: Review

Ask students the following questions:

- What are some of the ways in which modern agribusiness produces meat and fish? *Modern agribusiness produces large amounts of meat in areas known as CAFOs (concentrated animal feeding operations). CAFOs are large indoor or outdoor structures designed for maximum output. For fish, large agribusinesses harvest fish in the ocean using large dragnets or trawling nets. Fish are also raised in aquaculture tanks or cages where fish are actually farm raised.*
- How does the concept of the tragedy of the commons apply to fisheries? *The ocean is a common area. If it is overfished, there can be a fishery collapse. The decline of a fish population by 90 percent or more is referred to as fishery collapse.*
- What is aquaculture? *Aquaculture is the farming of aquatic organisms such as fish, shellfish, and seaweeds. Aquaculture involves the construction of an aquatic ecosystem by stocking the organisms, feeding them, and protecting them from diseases and predators.*

module 33

REVIEW

Alternatives to modern, industrial agriculture feature a number of characteristics that tend to emulate natural processes. Intercropping of different plants is similar to multi-species associations of plants that exist in nature. Crop rotation is a practice that simulates the natural shifts in species that occur. Integrated pest management uses a variety of techniques to control pests with a minimum of pesticide use. Organic agriculture is the production of crops without the use of synthetic pesticides or fertilizers.

Module 33 AP® Review Questions

1. Which of the agricultural methods is most similar to intercropping?
   (a) Contour plowing
   (b) Agroforestry
   (c) No-till agriculture
   (d) Integrated pest management
   (e) Nomadic grazing

2. Which agricultural method is used to prevent erosion?
   (a) Intercropping
   (b) Integrated pest management
   (c) Contour plowing
   (d) CAFOs
   (e) Annual plants

3. Which is NOT a principle of organic farming?
   (a) Avoiding synthetic fertilizers
   (b) Keeping as much organic matter as possible on the farm
   (c) Increasing soil mass and biological activity
   (d) Avoiding the use of fossil fuels
   (e) Working with natural systems

4. Which is a part of integrated pest management?
   (a) No-till
   (b) Increased application of pesticides
   (c) Contour plowing
   (d) Desertification
   (e) Crop rotation

5. Individual transferable quotas
   (a) are used in organic pest management.
   (b) are a part of sustainable fishing.
   (c) are an alternative method used for CAFOs.
   (d) are used to increase yields in communal farms.
   (e) are used in aquaculture.

Answers to Module 33 AP® Review Questions

1. b
2. c
3. d
4. e
5. b

working toward  sustainability

The Prospect of Perennial Crops

Whenever soil is plowed and prepared for another growing season, soil erosion is a problem. One long-term approach to reducing agriculture-related erosion is being explored by plant geneticist Wes Jackson, president of the Land Institute in Salina, Kansas. Jackson maintains that the single most beneficial new development in agriculture would be the development of food crops that do not need to be replanted every year.

Annual plants, such as wheat and corn, live only one season and must be replanted each year, which causes enormous disruption to the soil. In contrast, perennial plants do not require plowing. Perennials have a longer growing season than annuals: They can continue producing roots and storing energy even after harvest, and they often emerge earlier in spring than annuals. In addition, they can rely on root systems that were established in previous years, so they can allocate more resources in the current year to the production of stems, fruits, and seeds. All of these characteristics make them more productive than annuals.

Researchers at the Land Institute are exploring a variety of ways to grow crops in systems that mimic the natural world. Using a combination of conventional selective breeding and technology, they are attempting to convert annual species such as wheat, sorghum, sunflowers, and corn into perennials. At the same time, they are collecting wild perennials such as wheatgrass, then domesticating them and selecting for higher seed yield, size, and quality. Through these efforts, researchers hope to assemble communities of perennial plants, animals, fungi, and microorganisms that will be stable, productive, and resistant to insect pests and diseases.

The ultimate goal of the Land Institute is to develop sustainable crops that will produce sufficient amounts of food for harvest by humans, reduce soil erosion, and reduce or eliminate the need for synthetic fertilizers, pesticides, and irrigation. The researchers have already managed to double the size of seeds of a species of wheatgrass and increase its seed production by 20 percent without losing other important qualities of the plant. They have also had some success in crossing specially treated domesticated sorghum with perennial sorghum varieties to produce progeny that are both fertile and perennial.

Critics contend that it may take many decades to develop perennial seed crops that provide usable produce. Jackson doesn't seem overly concerned about this. According to Jackson, "If you're working on something you can finish in your lifetime, you're not thinking big enough."

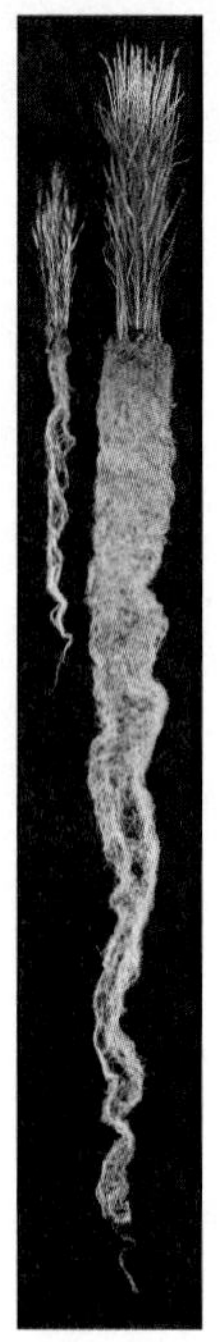

Annuals and perennials.
The root system of annual wheat (*left*) is much smaller than that of intermediate wheatgrass, a perennial (*right*). *(The Land Institute, Salinas, Kansas)*

Critical Thinking Questions

1. What are some of the disadvantages of annual crops?
2. What are the difficulties in developing perennial crops?

Suggested Answers to Critical Thinking Questions

1. Annual crops cause increased erosion and require labor to replant each year.

2. Many of the crops we grow do not have perennial variations, and thus new varieties must be developed that are both productive and still maintain the benefits of being perennial.

References

Glover, J.D., et al. 2010. Increased food and ecosystem security via perennial grains. *Science* 328:1638–1639.

The Land Institute: http://www.landinstitute.org.

chapter 11 REVIEW

In this chapter, we examined human nutritional requirements and contemporary problems of undernutrition and overnutrition. Modern industrial agriculture in the developed world uses large amounts of fossil fuel inputs and relies on other practices such as irrigation, fertilization, and pesticide application. Genetically modified organisms are also a component of modern agriculture in many parts of the world. Raising meat and fish have specific environmental impacts related to concentrated feeding areas for meat and removal of large numbers of fish from the oceans. Alternatives to conventional modern agriculture exist and often make use of natural patterns observed in nature such as crop rotation and interplanting different species. Organic agriculture is an agricultural practice that does not use chemical pesticides or fertilizers. A variety of sustainable organic and other alternative agricultural practices are being utilized in a variety of locations around the world.

Key Terms

Undernutrition
Malnourished
Food security
Food insecurity
Famine
Anemia
Overnutrition
Meat
Industrial agriculture
Agribusiness
Energy subsidy
Green Revolution
Economies of scale
Waterlogging
Salinization
Organic fertilizer
Synthetic fertilizer
Inorganic fertilizer
Monocropping
Pesticides
Insecticide
Herbicide
Broad-spectrum pesticide
Selective pesticide
Persistent pesticide
Nonpersistent pesticide
Pesticide resistance
Pesticide treadmill
Concentrated animal feeding operation (CAFO)
Fishery
Fishery collapse
Bycatch
Shifting agriculture
Desertification
Nomadic grazing
Sustainable agriculture
Intercropping
Crop rotation
Agroforestry
Contour plowing
Perennial plant
Annual plant
No-till agriculture
Integrated pest management (IPM)
Organic agriculture
Individual transferable quota (ITQ)
Aquaculture

Learning Objectives Revisited

Module 31 Human Nutritional Needs

- **Describe human nutritional requirements.**
  Humans require both a certain number of calories each day and a balance of proteins and other nutrients. Undernutrition is a condition in which not enough food calories are ingested each day. Malnutrition is a condition in which the diet lacks the correct balance of proteins, carbohydrates, vitamins, and minerals.
- **Explain why nutritional requirements are not being met in various parts of the world.**
  Malnutrition occurs due to poverty and political situations that prohibit the efficient and equitable distribution of food. In addition, some grains that could be fed to humans are used to feed livestock and are not available for direct human consumption.

Module 32 Modern Large-Scale Farming Methods

- **Describe modern, large-scale agricultural methods.**
  In the twentieth century, a variety of modern agricultural innovations including the Green Revolution transformed agriculture from a system of small farms relying mainly on human labor into a system of large industrial operations that use machinery run by fossil fuels. This shift has resulted in larger farms and monocropping. Irrigation can increase crop yields dramatically, but it can also draw down aquifers and lead to soil degradation through waterlogging and salinization. Fertilizers also increase crop yields dramatically, but can run off into surface waters and cause damage to ecosystems. The negative environmental consequences of pesticides include loss of beneficial nontarget organisms, human health problems, and surface water contamination. When pests become resistant to pesticides over time, a pesticide treadmill can develop.
- **Explain the benefits and consequences of genetically modified organisms and large-scale raising of meat and fish.**
  Agricultural scientists are using genetic engineering to produce genetically modified organisms with desirable traits. GMOs can increase yields and reduce the use of pesticides. Concerns about GMOs include safety for human consumption and their effects on biodiversity. Concentrated animal feeding operations and aquaculture facilities enable efficient animal growth and economically inexpensive food production. However, the disposal of concentrated animal waste presents a problem and the environment can be harmed by runoff containing antibiotics and other waste products.

Teaching Tip: Measuring Your Impact

The Ecological Footprint of Food Consumption
Measuring Your Impact 11 asks students to examine data on the amount of different foods consumed by a typical person in the United States and the ecological footprint of producing these items. Students are then asked to calculate the ecological footprint for the total amount of each food item consumed in the United States and to compare this footprint with that of others in the world.

The link to this resource may be found by clicking on the link buttons in the TE-book, opening the TRFD, or logging onto the book's companion website bcs.whfreeman.com/friedlandapes2e (teacher login required).

Measuring Your Impact 11:
The Ecological Footprint of Food Consumption

Exploring the Literature

Butterly, J. R., and J. Shepherd. 2010. *Hunger: The Biology and Politics of Starvation.* Dartmouth College Press.

Ikerd, J. 2008. *Crisis and Opportunity: Sustainability in American Agriculture.* Bison Books.

Pollan, Michael. 2007. *The Omnivore's Dilemma.* Penguin.

Seafood Watch, Monterey Bay Aquarium: http://www.montereybayaquarium.org/cr/seafoodwatch.aspx.

The State of Food Insecurity in the World. 2009. United Nations Food and Agriculture Organization.

U.S. Department of Agriculture: http://www.usda.gov/wps/portal/usdahome.

U.S. Department of Agriculture, National Agricultural Statistics Service: http://www.nass.usda.gov.

Download a printable version of this list by clicking on the link buttons in the TE-book, opening the TRFD, or logging onto the book's companion website bcs.whfreeman.com/friedlandapes2e (teacher login required).

TRM **Exploring the Literature 11**

Answers to Chapter 11 AP® Exam

Download the full answers to the Chapter 11 AP® Exam by clicking on the link buttons in the TE-book, opening the TRFD, or logging onto the book's companion website bcs.whfreeman.com/friedlandapes2e (teacher login required).

 **Answers to Chapter 11 AP® Exam**

Answers to Chapter 11 AP® Exam Multiple-Choice Questions

1. b
2. e
3. d
4. c
5. e
6. a
7. c
8. b
9. d
10. d
11. a
12. b
13. e
14. c
15. a
16. e

Module **33** Alternatives to Industrial Farming Methods

- **Describe alternatives to conventional farming methods.**

Traditional farming techniques such as intercropping, crop rotation, agroforestry, and contour plowing can sometimes improve agricultural yields and conserve soil and other resources. No-till agriculture is another way to reduce soil erosion and degradation. Integrated pest management reduces the use of pesticides, thus saving money and reducing environmental damage. IPM requires more labor, however, and practitioners must be trained to identify potential hazards to their crops. Organic agriculture focuses on maintaining the soil and avoids the use of synthetic fertilizers and pesticides. This approach often results in more labor-intensive and smaller farms, where many alternative agricultural techniques must be applied.

- **Explain alternative techniques used in farming animals and in fishing and aquaculture.**

Free-range animals can be raised with much lower environmental cost than those raised in feedlots.

Wild-caught fish, when caught in appropriate numbers, can be managed sustainably. Some fish species can be raised on fish farms with minimal environmental harm.

Chapter 11 AP® Environmental Science Practice Exam

Section 1: Multiple-Choice Questions

Choose the best answer for questions 1–16.

1. Which of the following does NOT explain the rise of the modern farming system?
   (a) The cost of labor varies from country to country.
   (b) Small farms are usually more profitable than large farms.
   (c) Irrigation contributes to greater crop yields.
   (d) Fertilizers improve crop yields and are easy to apply.
   (e) Mechanization facilitates monocropping and improves profits.

2. Irrigation can result in which of the following environmental problems?
   I. Reduction of evaporation rates
   II. Accumulation of salts in soil
   III. Waterlogging of soil and plant roots

   (a) I only
   (b) II only
   (c) III only
   (d) I and II
   (e) II and III

3. The use of synthetic fertilizers increases crop yields, but also
   (a) destroys the nitrifying bacteria in the soil.
   (b) increases fish populations in nearby streams.
   (c) decreases phosphorus concentrations in the atmosphere.
   (d) increases nutrient runoff into bordering surface waters.
   (e) slows the release of organic nutrients from compost.

4. Which of the following statements best describes the pesticide treadmill?
   (a) Broad-spectrum pesticides degrade into selective pesticides, thereby killing a wide range of insect pests over a long period.
   (b) Pesticides accumulate in the fatty tissues of consumers and increase in concentration as they move up the food chain.
   (c) Some pest populations evolve resistance to pesticides, which become less effective over time so that new pesticides must be developed.
   (d) Beneficial insects and natural predators are killed at a faster rate than the pest insects.
   (e) Testing of the toxicity of pesticides to humans cannot keep pace with the discovery and production of new pesticides.

5. In which of the following ways did the Green Revolution increase food production?
   I. The development of disease-resistant and high-yielding crop plants
   II. Monocropping and the widespread use of machinery
   III. The application of fertilizers and the use of irrigation techniques

   (a) I only
   (b) II only
   (c) III only
   (d) I and III
   (e) I, II, and III

Additional Multiple-Choice AP® Practice Exam

Download an additional Multiple-Choice AP® Environmental Science Practice Exam for Chapter 11 by clicking on the link buttons in the TE-book, opening the TRFD, or logging onto the book's companion website bcs.whfreeman.com/friedlandapes2e (teacher login required).

 **Multiple-Choice AP® Practice Exam 11**

6. Which of the following is NOT a traditional farming technique that is used in sustainable agriculture?
   (a) Nomadic herding
   (b) Intercropping
   (c) Crop rotation
   (d) Agroforestry
   (e) Contour plowing

7. Which of the following is an environmental advantage of no-till agriculture?
   (a) The use of herbicides improves the stability of the soil.
   (b) Migratory bird populations are reduced.
   (c) The undisturbed soil is less susceptible to erosion.
   (d) The crop residues reduce the soil profile.
   (e) The concentration of CO_2 in the fields is increased.

8. Which of the following practices is NOT a part of integrated pest management?
   (a) Crop rotation
   (b) Elimination of pesticides
   (c) Use of pest-resistant crops
   (d) Introduction of predators
   (e) Frequent inspection of crops

9. Farmers who practice organic agriculture have less of an impact on the environment than farmers who practice industrial agriculture because they
   (a) use no-till agriculture exclusively.
   (b) import soil to maintain soil fertility.
   (c) maintain large farms with a single crop.
   (d) avoid pesticides and synthetic fertilizers.
   (e) have lower labor costs.

10. Critics of using genetically modified organisms as food crops warn of which of the following dangers?
    I. Introduction of new allergens into the food supply
    II. Loss of genetic diversity in food crops
    III. Decreases in food production worldwide

    (a) I only
    (b) II only
    (c) III only
    (d) I and II
    (e) I and III

11. Concentrated animal feeding operations (CAFOs) can best be described as
    (a) facilities where a large number of animals are housed and fed in a confined space.
    (b) a method of producing more meat at a higher cost.
    (c) a means of producing great quantities of manure to fertilize fields organically.
    (d) an experimental plan to test the effectiveness of antibiotics.
    (e) the storing and compacting of grain for use as a nutrient supplement for cattle.

12. Which of the following is NOT an environmental or health problem that has been associated with CAFOs?
    (a) The increase of antibiotic-resistant bacteria potentially harmful to humans
    (b) The overgrazing of large tracts of land
    (c) The runoff of animal wastes into natural waters
    (d) The production of huge quantities of manure, creating a waste disposal problem
    (e) The use of grain as feed, reducing the food supply available to humans

Use the following graphs to answer Question 13.

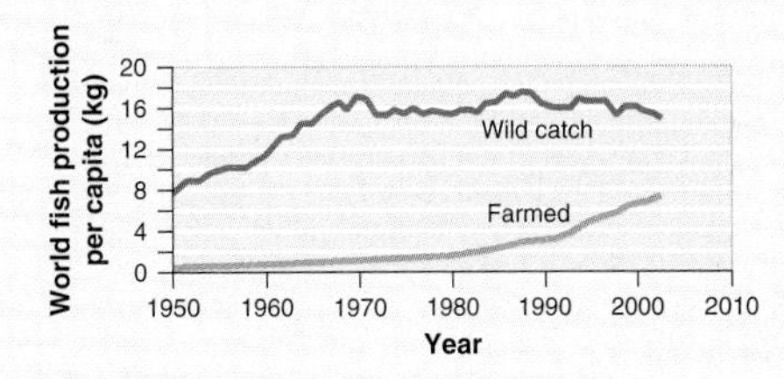

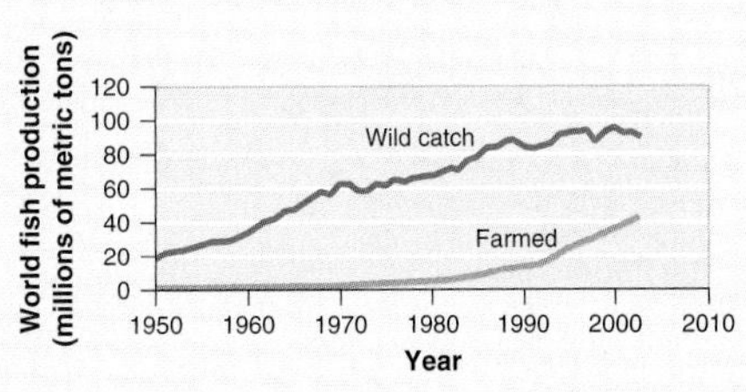

13. The data presented in the graphs, compiled by the FAO, describe world fish production from 1950 to 2003. Based on these data, which of the following statements best describes the global trend in fish production?
    (a) The U.S. Sustainable Fisheries Act of 1996 has successfully banned the catch of threatened fish species.
    (b) The use of individual transferrable quotas (ITQs) has led to the overfishing of wild species.
    (c) Fish production through aquaculture exceeds fish production in the ocean.
    (d) As the human population continues to increase, the per capita wild catch also continues to increase.
    (e) As the human population continues to increase, per capita farmed fish production also continues to increase.

Chapter 11 AP® Environmental Science Practice Exam 387

Answers to Chapter 11 AP® Exam Free-Response Questions

1. (a) 200 L (consumed)/cow/day + 120 L (washing)/cow/day = 320 L (total)/cow/day 1,000 cows × 320 L/cow/day = 320,000 L water required
(b) 1,000 cows × 55 kg/cow/day = 55,000 kg = 55 metric tons/day of manure

2. (a) 15.9% (very likely) + 31.9% (somewhat likely) = 47.8% probably would purchase GM foods 38.9% (not too likely) + 7.1% (not at all likely) = 46.0% probably would not purchase GM foods
(b) Benefits identified must be related to health. Possible benefits might include: Food could be more nutritious, as genes that produce certain vitamins can be inserted into crops (as in the case of Vitamin A in rice), which would reduce malnourishment and the resulting diseases. Food production could be increased, as food crops can be made resistant to drought, pests, or salt. This would reduce undernourishment and the resulting diseases. Crops could be grown without the use of pesticides. The insecticidal gene Bt has been inserted into corn, eliminating the use of pesticides and their toxic effects on humans.
(c) Dangers identified must be related to safety. Possible dangers might include: GM foods could cause allergic reactions if people were to eat foods that contain genes transferred from other foods to which they are allergic. The use of GM foods could cause a loss of genetic diversity in food crops, increasing the threat of pest or disease outbreaks, which would, in turn, reduce the food supply and increase malnourishment and undernourishment. GMOs modified for drug production could turn up in the food supply, causing unintentional exposure to pharmaceuticals and resulting in adverse drug reactions in the consumer.
(d) Three federal agencies are currently responsible for the regulation of GMOs:

- U.S. Department of Agriculture
- U.S. Food and Drug Administration
- U.S. Environmental Protection Agency

14. Which of the following is an environmental effect of bycatch?
(a) Juvenile fish are too small and slip through the nets, thereby ensuring the next generation of commercial fish.
(b) Predators caught in the nets feed on the commercially important fish also caught, thereby reducing the total catch.
(c) Nontarget fish populations have declined.
(d) Cod populations have increased due to the removal of competitor species.
(e) Endangered species, such as sea turtles, are being marketed through the ITQ system.

15. It has been projected that aquaculture could supply about one-half the global demand for seafood by 2025. However, future production through aquaculture could be limited because
(a) concentrated waste from aquaculture facilities contaminates rivers and oceans.
(b) raising fish in a protected environment could lead to a fish population overshoot.
(c) the economies of developing countries would be negatively affected.
(d) ocean fishing operations would go out of business.
(e) the ITQ system would prohibit their expansion.

16. What is the primary benefit of perennial plants?
(a) Increased crop yields
(b) Decreased pesticide use
(c) Increased nitrogen fixing
(d) Easy integration of crop rotation
(e) Decreased erosion

Section 2: Free-Response Questions

Write your answer to each part clearly. Support your answers with relevant information and examples. Where calculations are required, show your work.

1. Maintaining dairy cattle in a CAFO requires large quantities of water and produces vast quantities of wastewater and manure. The increasing number of CAFO dairies in eastern New Mexico and west Texas is contributing to significant groundwater contamination and the depletion of the Ogallala aquifer. Consider a CAFO with 1,000 dairy cows and answer the following questions.
(a) According to the U.S. Department of Agriculture, the average dairy cow can consume up to 200 L of water daily. An additional 120 L per cow per day is required to wash the milking equipment and milking area. How many liters of water are required to operate this dairy daily? (2 points)
(b) According to the U.S. Environmental Protection Agency, the average dairy cow produces 55 kg of wet manure daily. How many metric tons would this dairy produce each day (1,000 kg = 1 metric ton)? (2 points)

2. The table below shows the results of a survey of 113 people conducted in Trinidad, West Indies. It appeared in a 2006 issue of the *British Food Journal*.
(a) Considering the "very" and "somewhat" responses as one group and the "not too" and "not at all" responses as another group, what percentage of the survey respondents would probably purchase GM foods? What percentage would not? (1 point)
(b) Referring to the second survey question, identify and explain two possible benefits of the inclusion of GM foods in the diet that might convince people to purchase these products. (4 points)
(c) Referring to the third survey question, identify and explain two possible dangers of the use of GM foods that might convince people not to purchase these products. (4 points)
(d) Identify one U.S. federal agency that is responsible for the regulation of GMOs. (1 point)

Questions	Very likely (%)	Somewhat likely (%)	Not too likely (%)	Not at all likely (%)
Would you be willing to purchase these foods (corn oil, potatoes, tomatoes, rice) which are bought regularly if they were genetically modified?	15.9	31.9	38.9	7.1
Would you purchase GM foods if they were considered healthier than conventionally produced foods?	30.1	34.5	25.6	9.7
Would you be willing to purchase GM foods if they were safe?	15.9	31.9	38.9	7.1

Unit 5 AP® Environmental Science Practice Exam

Section 1: Multiple Choice Questions

Choose the best answer for questions 1–20.

1. Which is an example of a positive externality?
   (a) Individual farmers exploit a common grazing area.
   (b) Deforestation causes the loss of valuable ecosystem services.
   (c) A change in local zoning causes a rise in property value.
   (d) Selective logging occurs in a nationally managed forest.
   (e) Industrial activities cause an increase in air pollution.

2. Which of the following are possible methods to prevent the tragedy of the commons?
   I. Private ownership of land
   II. Government regulation of public land
   III. Self-regulation by communities and stakeholders

   (a) I only
   (b) II only
   (c) III only
   (d) II and III
   (e) I, II, and III

3. Determination of a maximum sustainable yield of plant harvest is most likely to be useful for which classification of land?
   (a) National park
   (b) Protected seascape
   (c) Nature reserve
   (d) Habitat management area
   (e) National monument

4. Which of the following is NOT a consideration that guides the regulation of public land use in the United States?
   (a) Revenue from tourism
   (b) Profits to private companies
   (c) Resource conservation ethic
   (d) Value of ecosystem services
   (e) Economic benefits resulting from land use by private companies

5. The Taylor Grazing Act of 1934
   (a) has resulted in a positive net revenue for the federal government.
   (b) has prevented a tragedy of the commons situation for rangelands.
   (c) converted federal rangelands from a permit-based to common-use grazing system.
   (d) subsidizes grazing activities with federal funds.
   (e) requires the development of an environmental mitigation plan.

6. For several decades, managers of a forest in West Virginia have selectively logged oak trees. After logging, they reseeded cleared areas with shade-tolerant tree species. However, recently the survival of seedlings in cleared areas has been greatly reduced. Which of the following activities is likely to increase seedling survival?
   (a) Prescribed burns around logged areas
   (b) Clear-cutting the forest
   (c) Planting shade-intolerant seeds
   (d) Attracting a greater diversity of bird species to the forest
   (e) Stopping all logging practices

7. Which of the following is NOT required by the National Environmental Policy Act of 1969?
   (a) An environmental impact statement that analyzes the impact of the project and all possible alternatives
   (b) A plan for construction that has as little environmental impact as possible
   (c) Protection of all endangered species
   (d) An environmental assessment of all projects involving federal money
   (e) Response to public concerns and comments on development plans

8. Which of the following is likely to have the greatest environmental impact?
   (a) Population movement into suburbs
   (b) Condensation of a population into cities
   (c) Population movement into exurbs
   (d) Development of rapid transit between cities and suburbs
   (e) Zoning an urban park for business development

9. What would be one method to reduce the rate of urban blight in a city?
   (a) Increase the taxes of individuals living in cities
   (b) Subsidize failing businesses within the city
   (c) Reduce police resources in low-income areas
   (d) Increase public transportation within the city
   (e) Reduce funding for highway construction

10. The Environmental Protection Agency's principles of smart growth are designed to alleviate the problems associated with
    (a) urban blight.
    (b) urban sprawl.
    (c) improper zoning.
    (d) city planning.
    (e) "Main Street" development.

Answers to Unit 5 AP® Exam Multiple-Choice Questions

1. c
2. e
3. d
4. b
5. b
6. a
7. b
8. c
9. e
10. b
11. b
12. e
13. c
14. d
15. e
16. b
17. e
18. c
19. b
20. e

Answers to Unit 5 AP® Exam

Download the full answers to the Unit 5 AP® Exam by clicking on the link buttons in the TE-book, opening the TRFD, or logging onto the book's companion website bcs.whfreeman.com/friedlandapes2e (teacher login required).

Answers to Unit 5 AP® Exam

11. Which of the following factors does NOT increase the rate of malnutrition in a country?
(a) Increased use of grain used to produce ethanol
(b) The number of people with anemia
(c) Political and economic factors
(d) Increased availability of inexpensive food
(e) Drought and other natural disasters

12. On average, 1 kg of corn contains 3,000 kilocalories, and a single hectare can produce 8,000 kg of corn per year. Approximately how many humans eating 2,000 calories per day can a single hectare of corn feed in a year?
(a) 5
(b) 12
(c) 21
(d) 25
(e) 33

13. A single ear of corn contains 80 kilocalories. To produce a single ear of corn requires 40 kilocalories of fuel. What is the energy subsidy of corn production?
(a) 0.05
(b) 0.1
(c) 0.5
(d) 1.0
(e) 5.0

14. Due to the economies of scale, it is economically beneficial for
(a) an owner of a small farm to purchase industrial farming equipment.
(b) farm owners to seek the best purchasing price for their produce.
(c) an owner of a large farm to lease sections of his farm to neighbors.
(d) an owner of a small farm to share farming machinery with neighboring farms.
(e) farmers to switch from the production of small crops (e.g., peas) to large crops (e.g., corn).

15. Monocropping generally requires more __________ than intercropping.
I. Fertilization
II. Irrigation
III. Pesticide application

(a) I only
(b) II only
(c) III only
(d) I and II
(e) I and III

16. Which of the following is NOT true regarding the creation of genetically modified organisms (GMOs) for use as crop species?
(a) Use of GMO crops has reduced pesticide use.
(b) Use of GMO crops has eliminated the pesticide treadmill.
(c) Use of GMO crops has increased livestock production.
(d) Use of GMO crops can be combined with the practices of intercropping.
(e) Use of GMO crops has reduced the prices of produce.

17. Which of the following factors has NOT contributed to the collapse of ocean fisheries?
(a) Tragedy of the commons in open waters
(b) Increase in offshore aquaculture development
(c) Use of large dragnets
(d) Bycatch harvest
(e) Use of international transferable quotas

18. No-till agriculture and contour plowing are methods that typically result in a decrease in
(a) pesticide use.
(b) energy subsidies.
(c) loss of topsoil.
(d) ecological footprints.
(e) pest outbreaks.

19. For produce to be labeled as organic in the United States, the food must have been grown
I. without the use of any pesticide.
II. without the use of synthetic fertilizer.
III. on small, sustainable farms.

(a) I only
(b) II only
(c) III only
(d) I and II
(e) I, II, and III

20. Which of the following parameters are most likely to determine the ecological footprint of a concentrated free-range livestock ranch?
(a) Number of cattle, method of waste disposal, amount of land allocated for cattle
(b) Number of cattle, method of tilling crop lands, amount of land allocated for cattle
(c) Diversity of livestock, type of grain fed to cattle, method of waste disposal
(d) Diversity of livestock, public interest in the ranch, amount of land allocated for cattle
(e) Diversity of livestock, number of cattle, amount of land allocated for cattle

Section 2: Free-Response Questions

Write your answer to each part clearly. Support your answers with relevant information and examples. Where calculations are required, show your work.

1. Detroit, once a bustling city similar to Los Angeles, now resembles a ghost town. For over 3 decades, the city has experienced the departure of many Caucasian residents, along with racial and ethnic turmoil. In 2009, two of the city's major employers, Chrysler and General Motors, filed for bankruptcy and eliminated a large number of jobs. The population of the city has declined by 90 percent from what it once was. Many residential areas are now demolished and urban businesses have been abandoned.
   (a) Describe two principles related to the strategy of smart growth that might be applied to reinvigorate the city of Detroit. (4 points)
   (b) Describe the positive feedback loop of urban blight. (2 points)
   (c) There has been a recent proposal to convert several abandoned residential neighborhoods into urban parks. Describe one positive externality and one negative externality of this proposal. (2 points)
   (d) Detroit is located halfway between Lake Erie and Lake Huron. Some officials have suggested that allowing increased fishing activities could attract people to the area. What are four parameters that state officials would need to measure in order to determine the maximum sustainable yield of Great Lakes fisheries? (2 points)

2. Bt-corn, a genetically modified plant, is now widely used in the United States because it greatly reduces the need to spray insecticides on croplands. More recently, scientists have begun inserting Bt genes into other plant types, such as rice, canola, and cotton. Consumers are wary of the potential effects these genetically modified organisms may have on human health. However, many environmental scientists are more concerned with the potential effects of the plants on the ecology of land surrounding farms. Consider this recent report from a study examining the Bt-crops:

   We observed slower decomposition of leaf litter and stems from Bt-positive transgenic strains of rice, tobacco, canola, cotton, and potato relative to decomposition of litter from Bt-negative strains. The amount of carbon dioxide released by soil bacteria and fungi surrounding Bt-positive litter was also less, indicating less microbial activity and slower microbial growth. These results may be caused by the production of toxic enzymes by the foliage of Bt-positive crops that deters herbivory while the plant is living, but leaches into the soil as the plant decomposes.
   (a) Why is the decomposition of plant litter important on farmlands? (2 points)
   (b) Discuss how and why the use of Bt-positive crop strains might influence future crop production on a farm. (2 points)
   (c) How might the use of Bt-positive crops influence the evolution of pest species? (2 points)
   (d) On average, one kilogram of potato contains 850 kilocalories. Suppose that growing a kilogram of nontransgenic potatoes on a conventional farm requires 1,100 kilocalories of fuel to plant the potatoes, 700 kilocalories of fuel to spray pesticides on the potatoes, and 200 kilocalories to harvest the potatoes. If a farmer switches to planting Bt-potatoes, then the energy investment in spraying pesticides is no longer needed. What is the difference in energy subsidies between nontransgenic and transgenic potatoes? (4 points)

Answers to the Unit 5 Free-Response Questions can be found in the Answer Appendix on page ANS-8.

Teaching Tip: Unit 5 Additional Free-Response Question

An additional practice free-response question covering concepts from Unit 5 is available. To answer this question students must

- describe the tragedy of the commons and name the 1930 act designed to prevent the tragedy
- name the four federal agencies responsible for managing public land
- describe the difference in designated use for wildlife refuges, national parks, and rangelands
- create a graph that shows cattle population size over time under two conditions—unrestricted grazing with no communal regulation (tragedy of the commons) and communal regulation through limited grazing

Unit 5 Additional Free-Response Question

(Andersen Ross/Spacca/age fotostock)

5

science**applied**

How Do We Define Organic Food?

If you've ever spent time roaming through the aisles of a grocery store, you have undoubtedly seen food labeled "organic." You may even have purchased organic food, despite the fact that it is generally more expensive than conventionally grown food. Some people prefer organic food because they believe it is healthier and tastes better. Others buy it because they think organic food is safer to eat. Still others believe organic food is produced in a way that is healthier for the environment and safer for farmworkers. Are these beliefs accurate? What exactly does the organic food label mean?

How did the organic food movement begin?

The notion of organic food emerged in the 1940s, when farmers first started using synthetic pesticides in agriculture. Three decades later, interest in organic food increased as concerns about environmental contaminants became more widespread. Organic food enthusiasts wanted food that was free of chemicals, including pesticides. At the same time, the popularity of food grown locally on small farms was increasing. The push for organic produce in the 1970s was part of a counterculture movement against the farming practices of large-scale commercial agriculture. Organic food was perceived as the food of people rebelling against the mainstream culture.

An organic farmer is fundamentally concerned with the health of the soil. He or she will work hard to ensure that the organic matter content, base saturation, and cation exchange capacity of the soil will increase each year. Actions that will degrade the soil or promote erosion are carefully avoided (FIGURE SA5.1).

FIGURE SA5.1 An organic farm. This mixed-crop organic farm in Australia is a certified organic producer. *(redbrickstock.com/Alamy)*

Today, organic food is much more common and appeals to people from all walks of life. Indeed, organic farming has increased rapidly during the past decade. According to the U.S. Department of Agriculture, the number of acres of organic farmland increased more than fivefold from 1995 to 2011. The states with the most land devoted to organic farmland are California, Wyoming, Oregon, Texas, and New Mexico. In 2011, there were about 2.2 million ha (5.4 million acres) of organic cropland and pastureland in the United States. But although organic food has been the fastest-growing category of food in the nation, it still represents less than 1 percent of all cropland and pastureland and only about 3 percent of all food sales.

FIGURE SA5.2 **The USDA organic seal.** Food that meets USDA organic certification standards can carry the USDA organic seal. *(U.S. Department of Agriculture)*

What does it mean to be organic?

The rise in popularity of organic products has brought increased pressure to define what exactly it means to be organic. Without government guidelines, anyone could claim to be selling organic food. In 1990, the U.S. Department of Agriculture (USDA) began to develop guidelines for organic food. Its goal was to set national organic certification standards that would assure consumers that the food was produced and processed with minimal use of chemicals. Food that met these standards could carry the USDA Organic seal (FIGURE SA5.2).

In 1995, the USDA determined that "Organic agriculture is an ecological production management system that promotes and enhances biodiversity, biological cycles, and soil biological activity. It is based on minimal use of off-farm inputs and on management practices that restore, maintain, and enhance ecological harmony." What does this statement mean in practice? First, it means that to be considered organic, food must be produced and processed with minimal use of pesticides. It also means that organic meat, eggs, and dairy products come from animals that have not been given antibiotics or growth hormones. Finally, it means that food labeled organic cannot have been fertilized with synthetic fertilizers or sewage sludge. Rather than relying on chemicals that come from outside the farm, organic farmers must use nonchemical methods of pest control and fertilization, as discussed in Chapter 11.

Pest control is perhaps one of the greatest challenges in organic farming. Agricultural pests include weeds that compete with the crops, insects that eat the crops, and fungal pathogens that can kill the crops. Organic farmers are required to try to control pests without synthetic pesticides. Such efforts can include mechanically destroying weeds and using plant-derived insecticides or the release of insect enemies to combat the pests. If these tactics fail, organic farmers are permitted to use a synthetic pesticide from a relatively small list of approved chemicals.

Having standards for organic food is effective only if those standards are enforced. One of the important functions of the USDA's National Organic Program is to conduct inspections of organic food producers and processors to ensure that organic food meets the national standards. Since 1990, the government has allowed inspections to be conducted by independent companies that are hired by producers and processors to inspect organic farms and factories to ensure that the organic standards are being met.

As part of this effort, the certifying companies were required to conduct spot checks. In 2010, the USDA announced that it had failed to enforce the spot check requirement and that, as a result, many of the largest independent certifiers had been conducting spot checks on organic food only if they suspected a problem. When problems were not suspected, producers could operate for years without any spot checks. In fact, some growers had been selling nonorganic food under the USDA Organic seal and obtaining a higher price for their products. In response, the federal government announced in 2010 that the funding and staff of the National Organic Program would nearly double to ensure that the certifying companies were conducting the required spot checks.

Does organic food mean family farms?

Most organic food production was originally conducted on small family farms. As we saw in Chapter 11, organic farming methods can require more labor, time, and money than conventional methods, so the price of organic food can be significantly higher. As farmers have learned, however, many consumers are willing to pay a premium for what they believe to be a premium product. This consumer behavior has attracted the attention of large factory farms owned by corporations, which have begun to conduct large-scale organic agriculture.

Although organic farms have a history of being small farms, there is no inherent reason why large farms cannot grow organic food (FIGURE SA5.3). An increasing number of corporations have decided to enter the business of organic food production. One of the most significant events to drive this movement occurred in 2006 when Wal-Mart, the largest grocery retailer in the United States, announced that it would begin selling a substantial number of organic foods in its stores at prices only slightly higher than those of nonorganic foods. Wal-Mart's expansion into this market has brought organic food to a much larger number of consumers. As a result, demand for organic food has increased, and more agribusinesses

Science Applied 5 How Do We Define Organic Food? 393

FIGURE SA5.3 **Large-scale organic farming.** Although organic farms traditionally have been small-scale family farms, some organic farms, such as this organic dairy farm in California, operate at much larger scales. *(Inga Spence/Alamy)*

are producing it. Increased production of organic food by larger, more efficient farms has driven down the cost of organic food.

As more factory farms entered the organic food business, there was increasing pressure to permit the use of some nonorganic and synthetic chemicals during the production stage, when the food from the field is turned into a form that can be sold in the grocery store. Proponents of the rule change, including agribusinesses, argued that many processed foods, such as frozen meals, could not be made in an organic version unless the rules about additives were changed. Opponents argued that if processed foods could not be made without additives, then processed foods should not be certified as organic. In 2002, the USDA designated several nonorganic and synthetic additives as permissible in organic processed foods. Permitted nonorganic additives include citric acid, pectin, and cornstarch. Synthetic chemicals permitted in the processing of organic foods include alcohol (used for disinfecting machinery) and aspirin (used for medical treatment of animals). The number of nonorganic additives has increased from 77 additives in 2002 to 250 additives in 2013.

What doesn't the organic label mean?

The USDA definition of "organic" does not match what many people think or hope "organic" should mean. For example, many people associate organic food with small family farms, but as we have seen, organic food can be grown on farms of all sizes. The USDA organic seal therefore tells us nothing about the size of the farm that grew the food. Nor does the seal tell us anything about the safety of the food, including whether it contains fewer dangerous food-related pathogens than nonorganic food. Nor does it indicate whether eggs, meat, and dairy products came from animals that were treated humanely or were allowed to roam freely outdoors (known as "free range"). It tells us nothing about where the food was grown, or about labor conditions, such as compensation for the farmworkers who produced it. Furthermore, there are currently no standards for certifying nonagricultural products, such as seafood or cotton, as organic. Although you may see these products labeled "organic," there is currently no agreed-upon definition for such products. Similarly, companies sometimes label their products "natural," a word that seems to imply "organic," but in fact has no standard definition.

In response to public demand, farmers sometimes use food labels that go beyond the federal requirements for organic certification. They may include information about humane treatment of animals, the conditions and wages of the farmworkers, and whether genetically modified organisms were used. Such detailed labels have seen growing popularity.

Although the USDA organic certification has a somewhat narrow definition, it does offer a number of important benefits. As reported in a 2012 study by researchers at Stanford University, there is no evidence that organic food is healthier for humans to eat than nonorganic food. However, the certification program at least provides a standardized set of requirements so that informed consumers understand what they are buying. Given the reduction in synthetic pesticide use, the program also allows consumers to support farming practices that are much better for soils and ecosystems and better for farmworkers—who face decreased risks from applying pesticides—than conventional farming practices.

Questions

1. Why do many people think that organic food comes from small farms?
2. Why might large agricultural companies lobby the federal government to permit many more synthetic chemicals to be used in processing organic food?

Suggested Answers to Science Applied Questions

1. The organic food movement started with small farms. Although there are now many large organic farms, organic farming is more time intensive and costly, which people associate with smaller operations.

2. Chemicals make it easier to process and store food. As the demand for organic food has grown, companies are trying to find ways to meet production demands.

Answer to Science Applied 5 Free-Response Question

Download the full answer to the Science Applied 5 Free-Response question by clicking on the link buttons in the TE-book, opening the TRFD, or logging onto the book's companion website bcs.whfreeman.com/friedlandapes2e (teacher login required).

Answer to Science Applied 5 Free-Response Question

Free-Response Question

Write your answer to each part clearly. Support your answers with relevant information and examples. Where calculations are required, show your work.

For many people, organic farming involves far more than the reduction of pesticide and fertilizer use. Organic farming is also a method of caring for the land in a manner that will benefit landscape ecology and sustain crop production for many years.

(a) Succession is the ecological process by which plants replace each other on a landscape through competition and facilitation. How can the principle of succession be used to generate a sustainable organic farm? (3 points)

(b) Describe three methods of reducing soil erosion. (3 points)

(c) What is integrated pest management, and how might it be applied to organic farming? (2 points)

(d) Before a farm can become certified as organic, it must undergo a 3-year transition period. During this period, no prohibited fertilizers or pesticides can be used on the land. Due to the high cost of farming without the aid of synthetic chemicals, transitioning to organic farming represents a financial risk. Describe two ways that this risk can be reduced. (2 points)

References

Chang, K. 2012. Organic food vs. conventional food. *New York Times,* September 4. http://well.blogs.nytimes .com/2012/09/04/organic-food-vs-conventional-food/.

Strom, S. 2012. Has organic been oversized? *New York Times,* July 7. http://www.nytimes.com/2012/07/08/business /organic-food-purists-worry-about-big-companies-influence.html.

U.S. Department of Agriculture. *What is organic production?* http://www.nal.usda.gov/afsic/pubs/ofp/ofp.shtml.

Answer to Science Applied Free-Response Question

1. (a) Like any ecosystem, a farm is a collection of interacting species that compete with or rely on each other. During succession, some plants compete with each other for nutrients, whereas other plants facilitate the arrival and persistence of new species. In addition, birds and other mobile organisms spread plant seeds and encourage the germination of plants. Farmers can use these interactions to their advantage by minimizing competing interactions and encouraging facilitating interactions. One example is intercropping plant species that facilitate each other, such as planting a nitrogen-fixing legume next to a nutrient-demanding plant. Another example may be to plant trees that provide minimal shade but have roots that aerate the soil. Trees also encourage visits by birds and other insect predators that can reduce pest abundance.

(b) Methods of reducing soil erosion include:

- intercropping trees with vegetables (i.e., agroforestry) to generate windbreaks that reduce wind erosion
- contour plowing, which is plowing and harvesting parallel to the topographic contours of the land
- no-till agriculture, which leaves crop residues in the field between seasons; intact roots hold the soil in place, and reduce wind and water erosion
- using alternative methods of irrigation, such as drip irrigation, that minimize the flow of water across the land
- planting crops for each season so that the soil is rarely bare

(c) Integrated pest management uses a variety of techniques designed to minimize pesticide inputs. These techniques include crop rotation and intercropping, the use of pest-resistant crop varieties, the creation of habitats for predators of pests, and limited use of pesticides. Crop rotation deters insects that are specific to one crop and hinders diseases that may survive on infected plant material from the previous season. Intercropping makes it harder for specialized pests that succeed in establishing themselves best where there is only one crop present. Agroforestry can create habitat for insect-eating birds (although birds can also damage some crops), and many herbs and flowers attract beneficial insects. Use of pest-resistant crop varieties and limited use of pesticides increase crop yields and decrease the expense of pesticide spraying. All of these techniques may be applied to the practice of organic farming, as long as any pesticide used is not synthetic.

(d) Even if a farm cannot yet label its food organic, it can use the following strategies:

- Farmers can sell food at local markets and talk to customers about their farming practices.
- Farmers can still use labels such as "naturally grown" to demonstrate that their crops are produced using more environmentally friendly methods.
- Farmers can use the 3-year period to perfect farming practices that will ensure maximum yield during the first year of organic certification.
- Farmers can use their land to raise free-range livestock that is likely to sell at a premium price to meat markets.

chapter

12 Nonrenewable Energy Resources

PD Overview

Chapter 12 explores the use of fossil fuels and nuclear energy resources, and how to project future supplies of these resources. This chapter offers an opportunity for students to focus on the advantages and disadvantages of the nonrenewable energy resources. For example, coal is abundant, energy-dense, easy to obtain, and relatively inexpensive. However, coal has impurities that are released into the atmosphere when it is burned, it leaves behind large deposits of ash, and it contributes to the increasing atmospheric concentrations of carbon dioxide, sulfur dioxide, and particulates. An understanding of energy is vital to the student's ability to understand environmental science because energy use impacts so many areas: air pollution, water pollution, food and agriculture, waste disposal, health, and climate change among others. This chapter on nonrenewable energy resources and the following chapter on renewable energy resources follow up on energy basics presented in Chapter 2.

Module 34: Patterns of Energy Use

This module examines patterns of energy use worldwide and in the United States. It demonstrates that evaluating energy efficiency helps us determine the best application for different energy sources. Finally, because electricity accounts for such a large percentage of our overall energy use, the module takes a close look at the ways in which electricity is generated.

Module 35: Fossil Fuel Resources

Fossil fuels provide most of the commercial energy in the world. They are also responsible for a good deal of the pollution that occurs, a topic that will be discussed in Chapter 15. This module describes the major fossil fuels—coal, petroleum, and natural gas—and discusses the advantages and disadvantages of each. It also considers the future use of fossil fuels.

Module 36: Nuclear Energy Resources

The combustion of fossil fuels releases large quantities of CO_2 into the atmosphere. Because nuclear energy has relatively low emissions of CO_2, it has received increasing interest as an energy source. This module explores how nuclear energy works and examines its advantages and disadvantages.

Alignment to AP® Environmental Science Course Description

AP® Outline	Module
V. Energy Resources and Consumption (10–15%)	
A. Energy Concepts	Module 34 Patterns of Energy Use
	Module 35 Fossil Fuel Resources
	Module 36 Nuclear Energy Resources
B. Energy Consumption	Module 34 Patterns of Energy Use
C. Fossil Fuel Resources and Use	Module 35 Fossil Fuel Resources
D. Nuclear Energy	Module 36 Nuclear Energy Resources

Chapter Learning Objectives

After completing this chapter students will be able to

- describe the use of nonrenewable energy in the world and in the United States.
- explain why different forms of energy are best suited for certain purposes.
- understand the primary ways that electricity is generated in the United States.
- discuss the uses of coal and its consequences.
- discuss the uses of oil and its consequences.
- discuss the uses of natural gas and its consequences.
- discuss the uses of oil sands and liquefied coal and its consequences.
- describe future prospects for fossil fuel use.
- describe how nuclear energy is used to generate electricity.
- discuss the advantages and disadvantages of using nuclear fuels to generate electricity.

Chapter 12 Pacing Guide

This pacing guide is based on a schedule with 120 sessions of 50 minutes each before the AP® Exam. If you have a different number of sessions before the exam, you can modify the pacing to suit your needs. If you have additional time, consider incorporating quizzes, released AP® Environmental Science free-response and multiple-choice questions, or additional activities.

Module	Standard Schedule Days	Block Schedule Days
Module 34	1	½
Module 35	3	1½
Module 36	2	1
Assessment	1	½

Chapter 12 Resources

The link to these resources can be found by clicking on the link buttons in the Teacher's e-Book (TE-book), opening the Teacher's Resource Flash Drive (TRFD), or logging onto the book's companion website bcs.whfreeman.com/friedlandapes2e (teacher login required).

- **PowerPoint Presentation 12**
- **Optimized Art PowerPoint and JPEG Files 12**
- **Do the Math Videos**
- **Measuring Your Impact 12:** Choosing a Car: Conventional or Hybrid?
- **Lab 12.1:** Coal Investigations
- **Exploring the Literature 12**
- **Answers to Chapter 12 AP® Exam**
- **Multiple-Choice AP® Practice Exam 12**

Free-Response Questions from Previous AP® Environmental Science Exams

Free-response questions from prior AP® Environmental Science Exams are available on the AP® course website: https://apstudent.collegeboard.org/apcourse/ap-environmental-science/exam-practice. Students should be able to answer all of the questions listed below with material learned in this and previous chapters. When a question requires students to understand material from multiple chapters, the question will be listed in the last chapter required to complete the entire question. Questions marked with an asterisk (*) are from exams with released multiple-choice questions. You may want to save these questions until the end of the year so you can give your students a complete released exam for practice. Questions marked with double asterisks (**) require math to calculate a problem. Look for references to these questions throughout the chapter.

Year	Question	Content
1998*	2	• The function of parts of a nuclear power plant • Environmental problems associated with nuclear power • The future of nuclear power use in the United States and the environmental implications
2005	3	• Reclamation of land formerly used for mining coal • Restoration of arid land • Environmental impacts of coal extraction and use • The future of coal consumption in the United States
2005	4 **	• Environmental costs of oil extraction from ANWR • Characteristics of a tundra biome • Use of oil in the United States and ways to use it more efficiently
2007	2 **	• Calculation of household water and electricity usage • Calculation of cost savings in using an energy-efficient hot water heater • Measures to reduce household energy and water use
2012	1	• Water-related environmental problems of hydraulic fracturing • The benefits of natural gas compared to coal • The economic benefits of fracking • Environmental problems associated with nuclear power
2014	1	• Alternative viewpoints on safety and impact of nuclear power • Environmental problems associated with construction of a nuclear power plant • Environmental impacts and ecological consequences of the pollution generated by a nuclear power plant • Reduction of electricity use

PD Chapter 12 Overview

Watch the video overview of Chapter 12 (for teachers) by clicking on the link buttons in the TE-book, opening the TRFD, or logging onto the book's companion website bcs.whfreeman.com /friedlandapes2e (teacher login required).

TRM PowerPoint Presentation 12

Download the PowerPoint presentation for Chapter 12 by clicking on the link buttons in the TE-book, opening the TRFD, or logging onto the book's companion website bcs.whfreeman.com /friedlandapes2e (teacher login required).

This massive structure is an oil refinery in Antwerp, Belgium. *(Eberhard Streichan/Corbis)*

chapter 12

Nonrenewable Energy Resources

All Energy Use Has Consequences

The modern world is dependent on fossil and nuclear fuels for energy. Many of the benefits of our modern society—health care, comfortable living conditions, easy travel, abundant food—rely on readily accessible and relatively affordable fossil and nuclear fuels. But the costs to society are high.

For example, on April 20, 2010, an explosion and fire occurred at the British Petroleum Deepwater Horizon oil rig in the Gulf of Mexico. Oil gushed from the rig until it was capped 87 days later, on July 15. The BP accident killed 11 workers on the drilling platform, injured 17 others, and released more than 780 million liters (206 million gallons) of oil into the gulf. Scientists from the Center for Biological Diversity later estimated that at least 6,000 sea turtles, 26,000 marine mammals, and more than 82,000 birds were killed as a result of the spill. The oil spread through the Gulf of Mexico and washed up on the shores of Louisiana, Mississippi, Alabama, and Florida.

The modern world is dependent on fossil and nuclear fuels for energy but the costs to society are high.

Before the Deepwater Horizon incident, the largest spill in U.S. waters had been the March 1989 Exxon Valdez accident, when the Valdez, a supertanker carrying 200 million liters (53 million gallons) of oil, crashed into a reef in Prince William Sound, Alaska. Roughly 42 million liters (11 million gallons) of oil spilled into the sound, much of which washed up on shore, contaminating the coastline and killing perhaps half a million birds and thousands of marine mammals. The number of dead animals was so much greater in the Exxon Valdez accident because the spill occurred in a relatively enclosed sound rather than in open waters.

Teaching Tip: Chapter Opening Case

The chapter opening case, "All Energy Use Has Consequences," discusses some of the risks associated with use of nonrenewable fossil fuels. It mentions both the *Exxon Valdez* spill in 1989 and the BP *Deepwater Horizon* oil rig blowout in 2010, as well as the more recent events at the Fukushima nuclear power plant in Japan. Countries of the developed world consume more energy (EJ) per capita than undeveloped countries (see Figure 34.2 on page 400). The technological progress associated with increased energy consumption provides many benefits, but our dependence on fossil fuels for energy also has many long-term costs.

Teaching Tip: Beyond the Classroom

Ask students to research and write a short report on a major oil spill or coal mining disaster. Possible topics include:

- Santa Barbara blowout (1969)
- *Exxon Valdez* oil spill (1989)
- BP oil refinery explosion (2005)
- BP *Deepwater Horizon* oil well blowout (2010)
- West Virginia coal mining accident (2010)

Students should address the following in their report:

- Where and when did the event occur?
- How did it happen?
- What were the environmental implications?
- What were the economic implications?
- Did any laws or policies change as a result of the event?

Teaching Tip: Warm-up

Have students read the chapter opening case, which discusses some of the unintended consequences of using fossil fuels. To start a discussion about our use of fossil fuels, ask the following questions:

- What is a fossil fuel? What are some examples of fossil fuels? *A fossil fuel is a fuel derived from biological material that became fossilized millions of years ago. Some examples include coal, oil, and natural gas.*
- Why are fossil fuels classified as a nonrenewable energy source? *A fossil fuel cannot be easily replenished once it is used up. Fossil fuels take hundreds of thousands of years to form.*
- What are some daily benefits you experience from the use of fossil fuels? *Answers will vary but could include comfortable living conditions (lighting, heating and cooling, hot water), food preparation, access to a variety of foods, travel, use of information technology, entertainment (movies, television, video games), and products made from petroleum.*

Oil spills remain an ongoing environmental problem in the modern world. Some spills are caused by leaks or explosions at wells, while others occur when oil is being transported by pipeline or tanker. Accidents can happen even after oil is extracted and transported to a refinery. In 2005, 15 workers died in an explosion at a British Petroleum oil refinery in Texas. And the hazards do not end with production. After the oil is refined into gasoline, jet fuel, or diesel fuel, it is used to run vehicles and heat buildings. This combustion process emits pollutants, which cause a number of environmental and human health problems.

Other fossil fuels pose similar risks. In the early 1900s, there were hundreds of accidental coal-mining deaths each year in the United States. While safety measures have improved since then, mining is still a dangerous business. In April 2010, the worst U.S. coal mine explosion in 40 years killed 29 miners in West Virginia. Long after they leave the mines, tens of thousands of coal miners develop black lung disease and other respiratory ailments that lead to disability or death.

Natural gas is often considered to be "clean" because its combustion produces lower amounts of particulates, sulfur dioxide, and carbon dioxide than does oil or coal. However, combustion still results in emission of carbon dioxide, which is the major greenhouse gas produced by human activity. The production of natural gas also has negative consequences. "Thumper trucks," which generate seismic vibrations to identify natural gas deposits underground, can disturb soil and alter groundwater flow, causing some areas to flood and some wells to go dry. Drilling for natural gas can also contaminate drinking water. Construction of natural gas pipelines is disruptive to the environment and is often opposed by those who live in the affected communities. And a growing number of studies suggest that an unknown but possibly substantial percentage of natural gas is lost during extraction and transport.

Nuclear energy is another *nonrenewable* energy resource. Unfortunately, it has contributed to a few energy catastrophes as well. Most recently, a 2011 earthquake off the coast of Japan resulted in a tsunami that damaged the nuclear reactors at the Fukushima nuclear power plant, leading to the release of radioactive gases. Even when operating properly, a nuclear power plant generates radioactive nuclear waste. Storage of this waste is currently an unresolved environmental challenge.

Sources:
S. McGraw, *The End of Country* (Random House, 2011); L. Margonelli, *Oil on the Brain* (Broadway Books, 2008).

We use energy in all aspects of our daily lives: heating and cooling, cooking, lighting, communications, and travel. In these activities, humans convert energy resources such as natural gas and oil into useful forms of energy such as motion, heat, and electricity, with varying degrees of efficiency and environmental effects. Each energy choice we make has both positive and negative consequences. In a society like the United States, where each person averages 10,000 watts of energy use continuously—24 hours per day, 365 days per year—there are a lot of consequences to understand and evaluate.

In this chapter, we will look at the supplies of fossil fuel and nuclear fuel that we currently use. Chapter 13 addresses the renewable energy resources that we use to some extent now and that many people expect we will use even more in the future.

Teaching Tip: Beyond the Classroom

Many students have questions about how large oil spills are cleaned up. Have students research the difference between chemical dispersants and chemical coagulants. Questions to research include:

- What are chemical dispersants? *Chemical dispersants are a mixture of chemicals that break down oil into smaller drops that mix more easily with water.*
- What are chemical coagulants? *Coagulants are chemicals used to change a liquid by thickening it into a gel or solid mass.*
- How do these substances clean up oil? *Chemical dispersants do not remove oil from the water but allow it to move below the surface. Coagulants cause the oil to solidify so it is easier to collect.*

After you discuss the difference between dispersants and coagulants, ask students the following questions:

- What are the pros of using genetically engineered microorganisms to clean up oil spills? *Microorganisms can be very efficient at consuming oil in both fresh and salt water.*
- What are the cons of using genetically engineered microorganisms to clean up oil spills? *We can't know how other living things in the water will react to genetically engineered microorganisms. These organisms could harm plankton or disrupt the food chain.*

module

34

Patterns of Energy Use

In this module we begin our study of nonrenewable energy sources by looking at patterns of energy use throughout the world and in the United States. We will see how evaluating energy efficiency can help us determine the best application for different energy sources. Finally, because electricity accounts for such a large percentage of our overall energy use, we will examine the ways in which electricity is generated.

Learning Objectives

After reading this module, you should be able to

- describe the use of nonrenewable energy in the world and in the United States.
- explain why different forms of energy are best suited for certain purposes.
- understand the primary ways that electricity is generated in the United States.

Nonrenewable energy is used worldwide and in the United States

Fossil fuels are fuels derived from biological material that became fossilized millions of years ago. Fuels from this source provide most of the energy used in both developed and developing countries. The vast majority of the fossil fuels we use—coal, oil, and natural gas—come from deposits of organic matter that were formed 50 million to 350 million years ago. As we saw in Chapter 3 (see Figure 7.2 on page 83), when organisms die, decomposers break down most of the dead biomass aerobically, and it quickly reenters the food web. However, in an anaerobic environment—for example in places such as swamps, river deltas, and the ocean floor—a large amount of detritus may build up quickly. Under these conditions, decomposers cannot break down all of the detritus. As this material is buried under succeeding layers of sediment and exposed to heat and pressure, the organic compounds within it are chemically transformed into high-energy solid, liquid, and gaseous components that are easily combusted. Because fossil fuel cannot be replenished once it is used up, it is known as a **nonrenewable energy resource. Nuclear fuel,** derived from radioactive materials that give off energy, is another major source of nonrenewable energy on which we depend. The supplies of these energy types are finite.

Every country in the world uses energy at different rates and relies on different energy resources. Factors that determine the rate at which energy is used include the resources that are available and affordable. In the past few decades, people have also begun to consider environmental impacts in some energy-use decisions.

Fossil fuel A fuel derived from biological material that became fossilized millions of years ago.

Nonrenewable energy resource An energy source with a finite supply, primarily the fossil fuels and nuclear fuels.

Nuclear fuel Fuel derived from radioactive materials that give off energy.

more math practice

Scientific Notation

Scientific notation is a method of writing numbers that makes it easier for us to handle very large or very small numbers. For example, instead of writing 256,000, we write 2.56×10^5. We can think of 2.56×10^5 as the product of two numbers: 2.56 (the digit term) and 10^5 (the exponential term). Here are some additional examples of numbers written in scientific notation.

$$10{,}000 = 1 \times 10^4$$
$$100 = 1 \times 10^2$$
$$1 = 1 \times 10^0$$
$$34{,}267 = 3.4267 \times 10^4$$
$$488 = 4.88 \times 10^2$$
$$0.053 = 5.3 \times 10^{-2}$$

As you can see, the exponent of 10 is the number of places the decimal point must be shifted to write the number in long form. A positive exponent shows that the decimal point is shifted to the right. A negative exponent shows that the decimal point is shifted to the left. Here are some pointers for teaching students how to work with scientific notation.

Addition and Subtraction: All numbers are first converted to the same power of 10, and then the digit terms are added or subtracted.

$$(4.215 \times 10^{-2}) + (3.2 \times 10^{-4})$$
$$= (4.215 \times 10^{-2}) + (0.032 \times 10^{-2})$$
$$= 4.247 \times 10^{-2}$$

Multiplication: The digit terms are multiplied but the exponents are added. The resulting answer is standardized to leave only one nonzero digit to the left of the decimal.

$$(3.4 \times 10^6) \times (4.2 \times 10^3)$$
$$= (3.4) \times (4.2) \times 10^{(6+3)}$$
$$= 14.28 \times 10^9 = 1.4 \times 10^{10}$$

Division: The digit terms are divided but the exponents are subtracted. The resulting answer (quotient) is standardized (if necessary) to leave only one nonzero digit to the left of the decimal.

$$(6.4 \times 10^6) \div (8.9 \times 10^2)$$
$$= (6.4) \div (8.9) \times 10^{(6-2)}$$
$$= 0.719 \times 10^4 = 7.2 \times 10^3$$

more math practice

Unit Prefixes

Remind students that a prefix in front of a unit of measure indicates a multiple or fraction of that unit. For example, the prefix *kilo* (k) means 1,000, or 10^3, so a kilogram (kg) is equal to 1,000 g. The prefix *milli* (m) means 1/1000 or 10^{-3}, so a milligram (mg) is equal to 0.001 g. The prefix *mega* (M) means 1,000,000 or 10^6. In this chapter students may encounter prefixes for units of energy and power, for example:

$$1 \text{ gigajoule} = 1 \text{ GJ} = 1{,}000{,}000{,}000 \text{ J} = 10^9 \text{ J}$$
$$1 \text{ kilowatt} = 1 \text{ kW} = 1{,}000 \text{ watts} = 10^3 \text{ watts}$$

COMMON MISCONCEPTIONS
Units of energy and units of power

Units of energy and power can often be difficult for students to understand. Review the difference between energy and power and make sure students know the units used for each.

Energy is the ability to do work or transfer heat. Below are some common units of energy that might appear on the exam. See Table 5.1 on page 44 for common usages.

- **joule (J):** The amount of energy used when a 1-watt electrical device is turned on for 1 second.
- **British thermal unit (Btu):** The amount of energy required to heat 1 pound of water by 1 degree Fahrenheit.
- **calorie:** The amount of energy it takes to heat 1 gram of water by 1 degree Celsius.
- **kilowatt hour (kWh):** The amount of energy expended by using 1 kilowatt of electricity for 1 hour.

Power is the rate at which work is done. The units for power describe the amount of work done over time. Below are some common units of power that may appear on the exam.

- **watt (W):** 1 watt is equal to 1 joule per second (1 J/s).
- **horsepower (hp):** One horsepower is equal to 746 watts.

COMMON MISCONCEPTIONS
Calories and calories

Students often confuse calories and Calories. Remind them that calorie is a basic scientific unit for energy. A food calorie, like that listed on the packaging of a product, is written as Calorie (uppercase C) and is equal to 1,000 calories, or 1 kilocalorie (kcal).

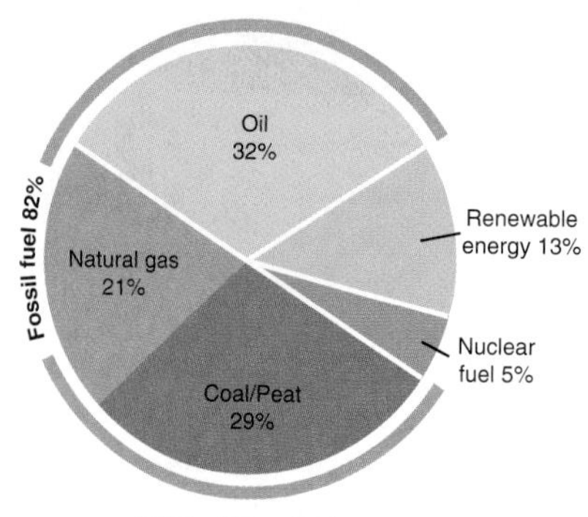

FIGURE 34.1 **Worldwide annual energy consumption, by resource, in 2011.** Oil, coal and peat, and natural gas are the major sources of energy for the world. *(Data from the International Energy Agency, 2013)*

In this section we will look at patterns of energy use worldwide and in the United States.

To talk about quantities of energy used, it is helpful to use specific measures. Recall from Chapter 2 that the basic unit of energy is the joule (J). One gigajoule (GJ) is 1 billion (1×10^9) joules, or about as much energy as is contained in 30 L (8 gallons) of gasoline. One exajoule (EJ) is 1 billion (1×10^9) gigajoules. In some figures, we also present the quad, a unit of energy used only by the U.S. government to report energy consumption. The quad is 1 quadrillion, or 1×10^{15}, British thermal units, or Btu. One quad is equal to 1.055 EJ.

Worldwide Patterns of Energy Use

As FIGURE 34.1 shows, in 2011 total world energy consumption was approximately 550 EJ per year. This number amounts to roughly 75 GJ per person per year. Oil, coal, and natural gas were the three largest energy sources. Peat, a precursor to coal, is sometimes combined with coal for reporting purposes in certain countries, mostly in the developing world.

Energy use is not evenly distributed throughout the world. FIGURE 34.2 shows that energy consumption in the United States was 325 GJ per person per year, almost 5 times greater than the world average. In fact, although only 20 percent of the world's population lives in developed countries, those people use 70 percent of the world's energy each year. Note that of the countries shown in Figure 34.2, China has the greatest total energy consumption, whereas Canada has the greatest per capita energy consumption. At 0.12 EJ per year, Tanzania has the lowest annual energy consumption of the countries shown; annual per capita energy consumption in Tanzania is less than 3 GJ per person per year.

There are a variety of reasons for the patterns we see in Figure 34.2. In developed countries and in urban areas of some developing countries, individuals are likely to use fossil fuels such as coal, oil, and natural gas—either directly or indirectly through the use of electricity that is generated by burning those fuels. However, people living in rural areas of developing countries primarily still use such fuels as wood, charcoal, or animal waste. These differences lead us to distinguish between *commercial* and *subsistence* energy sources. **Commercial energy sources** are those that are bought and sold, such as coal, oil, and natural gas, although sometimes wood, charcoal, and animal waste are also sold commercially. **Subsistence energy sources** are those gathered by individuals for their own immediate needs and include straw, sticks, and animal dung. There is much greater use of subsistence energy sources in the developing world, especially in rural areas.

Changes in energy demand generally reflect the level of industrialization in a country or region. As energy demand increases, societies change the types of fuels they use. Today, we see the same patterns of changing energy use in developing countries that have been observed historically in the United States. For example, as more people own automobiles, demand for gasoline and diesel fuel increases. As industries develop and factories are built, demand for electricity

Commercial energy source An energy source that is bought and sold.

Subsistence energy source An energy source gathered by individuals for their own immediate needs.

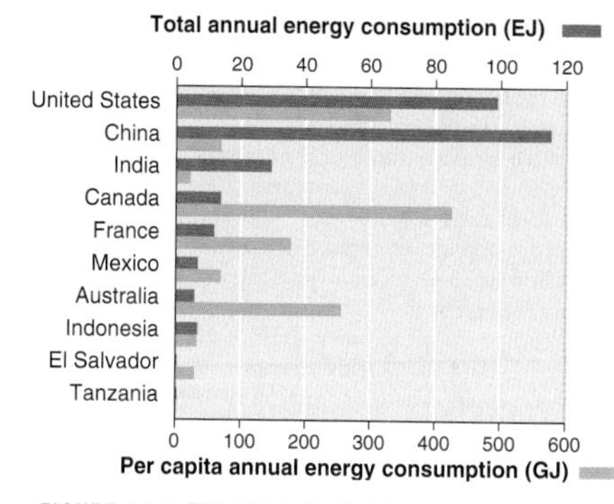

FIGURE 34.2 **Global variation in total annual energy consumption and per capita energy consumption.** The 10 countries shown are among the largest and the smallest energy users in the world. *(Data from the U.S. Department of Energy, Energy Information Administration, 2012)*

more **math** practice
Conversions

You can convert from one unit of energy to another:

1 calorie = 4.184 J
1 Btu = 1,055 J
1 kWh = 3,600,000 J = 3.6 Megajoules (MJ)
1 hp = 746W

Here are two problems to try with your class:

1. A runner burns 600 Calories in track practice. How many joules (J) did the runner burn?

600 Calories × (1,000 calories/Calorie) × (4.184 J/calorie) = 2,510,400 J

2. A lawnmower has 5 hp. How many watts would that equal?
5 hp × (746 W/hp) = 3,730 W

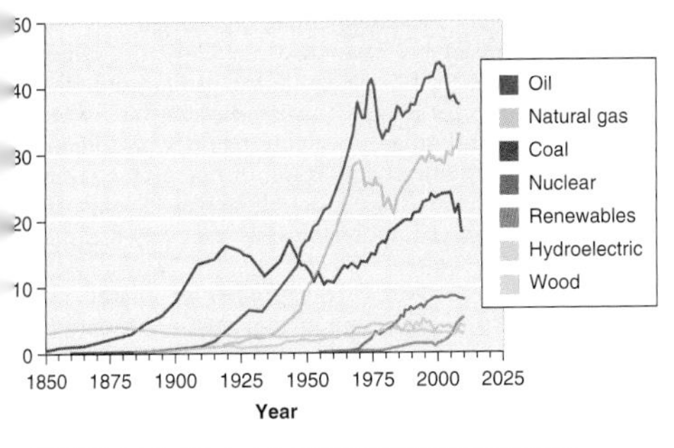

FIGURE 34.3 Energy consumption in the United States from 1850 through 2012. Wood and then coal once dominated our energy supply. Today a mix of three fossil fuels accounts for most of our energy use. The recent increase in natural gas and decrease in oil and coal is quite evident. *(After U.S. Department of Energy, Energy Information Administration, 2013)*

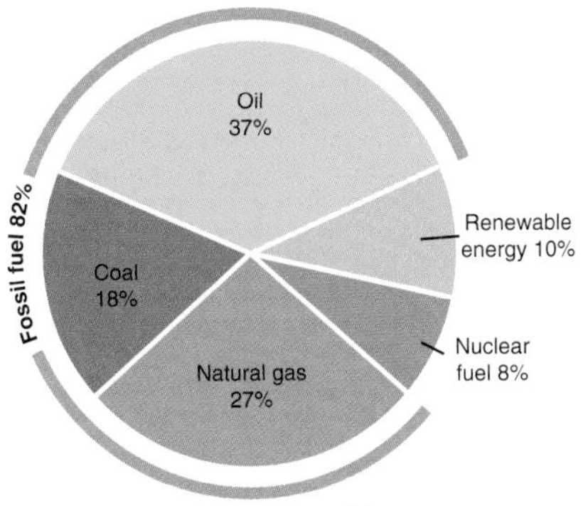

(a) Total = 100 exajoules (95 quads) per year

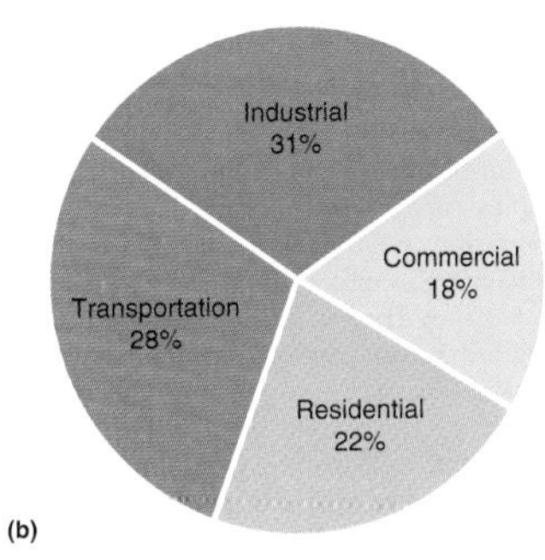

(b)

FIGURE 34.4 United States annual energy consumption by resource and end use in 2012. These graphs show energy consumption and end use in the United States. (a) United States annual energy consumption by fuel type in 2012. (b) United States end use energy sectors in 2012. Commercial includes businesses and schools. *(Data from U.S. Department of Energy, Energy Information Administration, 2013)*

and nuclear fuel increases. Although worldwide energy use varies considerably, the United States is a particularly large energy consumer.

Patterns of Energy Use in the United States

FIGURE 34.3 shows the history of energy use in the United States. Wood was the predominant energy source until about 1875, when coal came into wider use. Starting in the early 1900s, oil and natural gas joined coal as the primary sources of energy. By 1950, electricity generated by nuclear energy became part of the mix, and hydroelectricity became more prominent. The 1970s saw a decline of oil and a resurgence of coal. These changes were the result of political, economic, and environmental factors that will continue to shape energy use into the future. Today, the three resources that supply the majority of the energy used in the United States—in order of importance—are oil, natural gas, and coal. The recent increase in the consumption of natural gas evident in Figure 34.3 is a result of its increase in availability and decrease in price due to hydraulic fracturing (fracking).

Energy use in the United States today is the result of the inputs and outputs of an enormous system. The boundaries of the system are political and technological as well as physical. For example, oil inputs enter the U.S. energy system from both domestic production and imports from other countries. Hydroelectric energy comes from water that flows within the physical boundaries of the country, as well as from neighboring Canada, but it is not an energy input until we move it into a technological system, such as a hydroelectric dam. One major output from the system is work—the end use of the energy, such as in transportation, residential, commercial, and industrial. The other major output is waste: heat, CO_2, and other pollutants that are released as energy is converted and entropy increases.

As you can see in FIGURE 34.4, U.S. energy consumption in 2012 was approximately 100 EJ (1.0×10^{18} J) per year. The energy mix of U.S. consumption is 82 percent fossil fuel, 8 percent nuclear fuel, and 10 percent renewable energy resources. The United States produces 85 percent of the energy it needs, while the remainder—roughly 15 percent—comes from other countries, primarily in the form of petroleum imports. The percentage of energy imports has been decreasing as domestic production of natural gas and oil increases; these trends are expected to continue. Industry uses the most energy, followed by the transportation sector.

Energy use varies regionally and seasonally. In the midwestern and southeastern states, coal is the primary

◀ TEACHING with FIGURES

Show students Figure 34.3 and ask the following questions:

- How will oil consumption change in 10 years? In 50 years? *In 10 years oil consumption will probably still be increasing. However, if the supply is slowed or disrupted and oil becomes more expensive, use may decline. In 50 years, oil consumption might begin to decline, unless we find new technology to extract oil from deeper parts of Earth's crust.*
- Compare the use of oil over time to the use of natural gas. *The use of natural gas has risen more in recent years than the use of oil, although more oil is used than natural gas.*

Teaching Tip: Discussion Starter

Ask students the following questions:

- Why was wood the predominant source of energy used until 1875? *Wood was the main fuel used by the machines that existed before the internal combustion engine came into widespread use in the 1870s.*
- Is wood used today as an energy source in the United States? If so, where? *Wood is still used today to heat homes in rural areas in the United States. Urban areas have less access to wood, and the cost increases when wood, which is heavy, has to be shipped.*

more **math** practice: Graphing

Have students use the following data tables to create two pie charts and then have them answer the questions that follow.

Distribution of annual energy consumption worldwide by resource

Oil	32%
Natural gas	21%
Coal	29%
Renewable energy	13%
Nuclear energy	5%

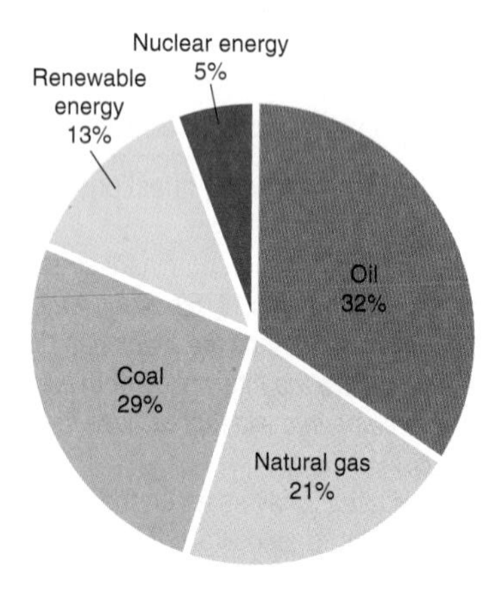

Distribution of annual energy consumption in the United States by resource

Oil	37%
Natural gas	27%
Coal	18%
Renewable energy	10%
Nuclear energy	8%

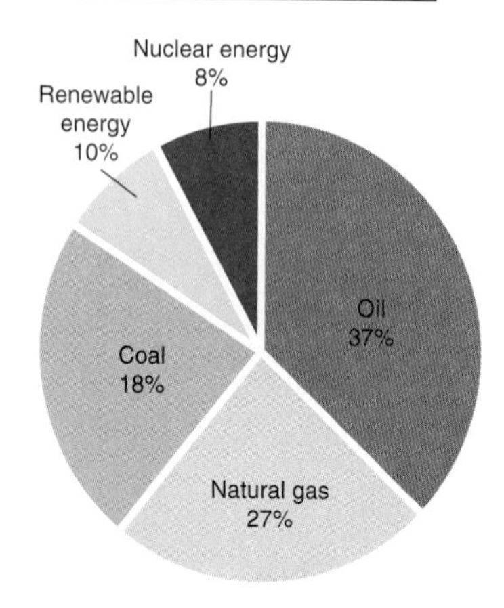

fuel burned for electricity generation. The western and northeastern states generate electricity using a mix of nuclear fuels, natural gas, and hydroelectric dams. Highly populated areas tend to use less coal, which creates more air pollution than any other fuel. Regional use of fuel varies according to the climate and the season. Northern areas consume more oil and natural gas during the winter months to meet the demand for heating while southern areas consume more electricity in the summer months to meet the demand for air conditioning.

Expanded domestic production of natural gas and oil has led to a rapidly changing energy portfolio in the United States and elsewhere. However, the type of energy used for a particular application is a function of many factors, including its characteristics. Factors to be considered include the ease with which the fuel can be transported and the amount of energy a given mass of the fuel contains.

Different energy forms are best suited for specific purposes

The best form of energy to use depends on the particular purpose for which it is needed. For example, for transportation, we usually prefer gasoline or diesel fuel—liquid energy sources that are relatively compact, meaning that they have a high energy-to-mass ratio. Imagine running your car on coal or firewood: To travel the same distance as a conventional car on one tank of gasoline, you would have to carry around a much larger volume of material. Gasoline, the current fuel of choice for personal transportation, gets you a lot farther on a much smaller volume.

Energy-to-mass ratio is not the only consideration for fuel choice, however. A wood or coal fire starts relatively slowly and if used in an automobile, would not allow it to accelerate quickly. Gasoline and diesel are also ideal fuels for vehicles because they can provide energy quickly and can be shut off quickly. Unfortunately, compared with other energy sources, such as natural gas or hydroelectricity, gasoline produces large amounts of air pollution per joule of energy released. In addition, unlike coal or wood, gasoline requires a good deal of refining—chemical processing—to produce. So there are many factors to consider when determining the most suitable energy source for a particular application.

Quantifying Energy Efficiency

Although all conventional nonrenewable energy sources have environmental impacts, an understanding of energy efficiency can help us make better energy use decisions. Energy efficiency refers to the efficiency of the process we use to obtain the fuel and the efficiency of the process that converts it into the work that is needed.

In Chapter 2 we discussed energy efficiency as well as energy quality, a measure of the ease with which stored energy can be converted into useful work. The second law of thermodynamics dictates that when energy is transformed, its ability to do work diminishes because some energy is lost during each conversion. In addition to these losses, there is an expenditure of energy involved in obtaining almost every fuel that we use.

FIGURE 34.5 outlines the process of energy use from extraction of a resource to electricity generation and disposal of waste products from the power plant. The red arrows indicate that there are many opportunities for energy loss, each of which reduces energy efficiency. You may recall that the efficiency of converting coal into electricity is approximately 35 percent (see Figure 5.6, on page 50). In other words, about two-thirds of the energy that enters a coal-burning electricity generation plant ends up as waste heat or other undesired outputs. If we included the energy used to extract the coal as well as the energy used to build the coal extraction machinery, to construct the power plant, and to remove and dispose of the waste material from the power plant, the efficiency of the process would be even lower. These other energy inputs will be discussed in greater detail in Chapter 16.

Every energy source, from coal to oil to wind, requires an expenditure of energy to obtain. The most direct way to account for the energy required to produce a fuel, or energy source, is by calculating the energy return on energy investment (EROEI), the amount of energy we get out of an energy source for every unit of energy expended on its production. EROEI is calculated as follows:

$$\text{EROEI} = \frac{\text{Energy obtained from the fuel}}{\text{Energy invested to obtain the fuel}}$$

For example, in order to obtain 100 J of coal from a surface coal mine, 5 J of energy is expended. Therefore,

$$100\text{ J EROEI} = \frac{100\text{ J}}{5\text{ J}} = 20$$

As you might expect, a larger value for EROEI suggests a more efficient and more desirable process. "Science Applied 6: Should Corn Become Fuel?" following Chapter 13 calculates the EROEI for ethanol, a fuel made from corn.

Finding the Right Energy Source for the Job

When deciding between two energy sources for a given job, it is essential to consider the overall system efficiency. Sometimes the trade offs are not immediately apparent. The home hot water heater is an excellent illustration of this principle.

402 CHAPTER 12 ■ Nonrenewable Energy Resources

When students have created their pie charts, ask the following questions:

- Compare the world's use of oil to that of the United States. How do you explain the difference?
  In the United States, 37 percent of all energy use comes from oil. In the world, 32 percent of all energy use comes from oil. In the United States, oil is less expensive and more readily obtained than it is in other countries. Also, in the United States, overall wealth is higher than many other countries so people can afford to buy more expensive fuels.
- Compare the world's use of renewable energy sources to that of the United States. Which types of renewable energy are increasing in usage most rapidly? *World usage of renewable energy sources as a percentage of total energy use is approximately 3 percent higher than that of the United States. The United States has been slower to research and invest in renewables. Many European and Asian countries have promoted renewables by offering subsidies. Use of solar power, wind energy, and hydroelectric power have all increased in other counties more rapidly than in the United States. In addition, fossil fuels are less expensive in the United States so there is less incentive for people to explore other fuel sources.*

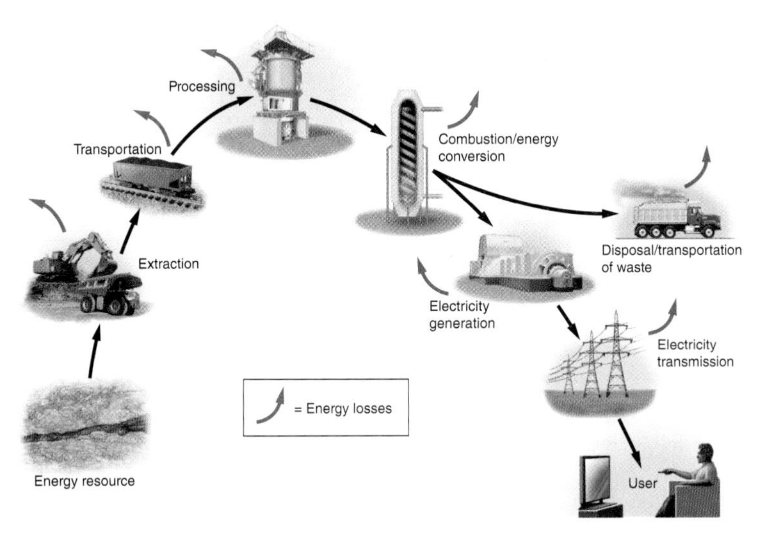

FIGURE 34.5 **Inefficiencies in energy extraction and use.** Coal provides an example of inefficiencies in energy extraction and use. Energy is lost at each stage of the process, from extraction, processing, and transport of the fuel to the disposal of waste products.

Electric hot water heaters are often described as being highly efficient. Even though it is very difficult to convert an energy supply entirely to its intended purpose, converting electricity into hot water in a water heater comes very close. That's because heat, the waste product that usually makes an energy conversion system less efficient, is actually the intended product of the conversion.

If the conversion from electricity to heat occurs inside the tank of water, which is usually the case with electric hot water heaters, very little energy is lost, and we can say the efficiency is 99 percent. By contrast, a typical natural gas water heater, which transfers energy to water with a flame below the tank and vents waste heat and by-products of combustion to the outside, has an efficiency of about 80 percent. The overall efficiency is actually lower, though, because we need to include the energy expended to extract, process, and deliver natural gas to the home. However, if a coal-fired power plant is the source of the electricity that fuels the electric water heater, we have to consider that conversion of fossil fuel into electricity is only about 35 percent efficient. This means that even though an electric water heater has a higher direct efficiency than a natural gas water heater, the overall efficiency of the electric water heating system is lower—35 percent for the electric water heater compared with something close to 80 percent for the gas water heater. There may be many situations in which it is a better choice to heat water with electricity rather than with natural gas, but from an environmental perspective, it is important to look at the overall system efficiency when considering the pros and cons of an energy choice.

Efficiency and Transportation

Because nearly 30 percent of energy use in the United States is for transportation, energy efficiency in transportation is particularly important. The movement of people and goods occurs primarily by means of vehicles that are fueled by petroleum products, such as gasoline and diesel fuel, and by electricity. These vehicles contribute to air pollution and greenhouse gas emissions. However, some modes of transportation are more efficient than others.

As you might expect, public transportation—train or bus travel—is much more efficient than traveling by car, especially when there is only one person in the car. And public ground transportation is usually more efficient than air travel. **TABLE 34.1** shows the efficiencies of different modes of transportation. Note that the energy values report only energy consumed, in megajoules (MJ, 10^6 J), per passenger-kilometer traveled and do not include the embodied energy used to build the different

- In the last 10 years in the United States, which nonrenewable energy source has had the largest percentage change? Why do you think this is the case? *In the United States, natural gas has had the largest percentage change in the last 10 years because of fracking.*
- Why has nuclear energy production for the past 40 years remained about the same in the United States? *U.S. nuclear power plants are getting old. As older plants go offline, we have not been replacing them. Many people believe nuclear energy is more risky than other forms of energy production. In addition, storage of radioactive waste is problematic.*

Teaching Tip: Connections

When discussing different energy sources, it is important to consider the energy return on energy investment (EROEI). This is the most direct way to compare the energy required to produce a fuel and how much energy we get out of that fuel after it is produced.

Coal

To obtain 100 J of energy from coal that is extracted using surface mining, 5 J of energy is expended. Therefore

$$\text{Surface mine EROEI} = 100\text{ J} \div 5\text{ J} = 20$$

To obtain 100 J of energy from coal that is extracted using subsurface mining, 10 J of energy is expended. Therefore

$$\text{Subsurface mine EROEI} = 100\text{ J} \div 10\text{ J} = 10$$

Ask students: What explains the difference in EROEI for the two extraction methods? *A larger value for EROEI suggests a more efficient process. Surface mining uses less energy to extract coal and is therefore more efficient than subsurface mining.*

Hydroelectric Power

To obtain 100 J of energy from hydroelectric power, 1.25 J of energy is expended. Therefore

$$\text{Hydroelectric EROEI} = 100\text{ J} \div 1.25\text{ J} = 80$$

Ask students: Why do you think the EROEI for hydroelectric power is around 80? *Most energy involved in generating hydroelectric power is expended when the dam is first constructed. After that, very little energy is required to obtain energy.*

Ethanol

Note that ethanol will be discussed in Chapter 13. You might wish to mention that some of the energy inputs to produce ethanol include planting the crop, running farm machinery, producing and using fertilizers, harvesting the crop, drying and shipping the harvested crop to an ethanol plant, and removing by-products of corn or soybeans. All of these inputs use energy and decrease the EROEI.

To obtain 100 J of energy from ethanol, 78 J of energy is expended. Therefore

$$\text{Ethanol EROEI} = 100\text{ J} \div 78\text{ J} = 1.3$$

TABLE 34.1 Energy expended for different modes of transportation in the United States	
Mode	MJ per passenger-kilometer
Air	2.1
Passenger car (driver alone)	3.6
Motorcycle	1.1
Train (Amtrak)	1.1
Bus	1.7

vehicles. Trains and motorcycles are the most energy-efficient modes of transportation shown. If a car contained four passengers, we would divide the value in Table 34.1 for a lone driver by four, and thus the car would then be the most efficient means of transportation. However, cars traveling with four riders are relatively rare in the United States; single-occupant vehicles are the most common means of transportation. "Do the Math: Efficiency of Travel" shows you how to calculate the efficiencies of different modes of transportation.

Transportation efficiency calculations do not take into account convenience, comfort, or style. Many people in the developed world are quite particular about how they get from place to place and want the independence of a personal vehicle. They also tend to have strong feelings about what type of personal vehicle is most desirable. In the United States, light trucks—a category that usually includes sport-utility vehicles (SUVs), minivans, and pickup trucks—account for roughly one-half of total automobile sales, while hybrid electric vehicles account for roughly 3 percent of total sales.

Light trucks are comparatively heavy vehicles and so generally have mileage ratings of less than 8.5 km per liter, or 20 miles per gallon (mpg). Because they are exempt from certain vehicle emission standards, they emit more of certain air pollutants per liter of fuel combusted than passenger cars. Smaller cars with standard internal combustion engines can travel up to 19 km per liter (45 mpg) on the highway. Hybrid passenger cars, which use a gasoline engine, electric motors, and special braking systems, obtain closer to 21 km per liter (50 mpg). Electric cars and plug-in hybrid electric cars, which are becoming more common in the United States, obtain even better fuel efficiency.

Despite the availability of fuel-efficient vehicles, many people drive vehicles that yield relatively low fuel efficiencies. As FIGURE 34.6 shows, the overall fuel efficiency of the U.S. personal vehicle fleet declined from 1985 through 2005 as people chose light trucks and SUVs over cars. Only in the last 8 years have vehicle choice changed and vehicle efficiency slowly increased. Recently, legislation was passed to increase the average fuel efficiency of new cars and light trucks sold each year so as to deliver a combined fleet average of 15 km per liter (35 mpg) by 2016.

do the math

Efficiency of Travel

Imagine that you had unlimited time and you needed to get from Washington, D.C., to Cleveland, Ohio. The distance is roughly 600 km (370 miles). For each mode of transportation, calculate how many megajoules of energy you would use.

Using the data in Table 34.1, we can determine the following:

Air 2.1 MJ/passenger-kilometer × 600 km/trip = 1,260 MJ/passenger-trip

Car 3.6 MJ/passenger-kilometer × 600 km/trip = 2,160 MJ/passenger-trip

Train 1.1 MJ/passenger-kilometer × 600 km/trip = 660 MJ/passenger-trip

Bus 1.7 MJ/passenger-kilometer × 600 km/trip = 1,020 MJ/passenger-trip

If a gallon of gasoline contains 120 MJ, how many gallons of gasoline does it take to make the trip by car?

$$\frac{2{,}160 \text{ MJ/passenger-trip}}{120 \text{ MJ/gallon}} = 18 \text{ gallons/passenger-trip}$$

Examining total joules expended makes it clear that the train is the most energy-efficient means of travel. Driving alone in a car is the least energy-efficient.

Your Turn If you could carpool with three other people who needed to make the same trip, what would the energy expenditure be for each person?

do the math

Answer to Your Turn

Assume the total energy required by the car does not increase with the increased number of passengers.

3.6 MJ/passenger-kilometer ÷ 4 people
= 0.9 MJ/passenger-kilometer

0.9 MJ/passenger-kilometer
× 600 km/trip
= 540 MJ/passenger-trip

Teaching Tip: Activity

To investigate how vehicle efficiency has changed, divide your class into groups and have each group research the average miles per gallon (mpg) of one of the following vehicles:

- Nissan Leaf
- Chevrolet Volt
- Toyota Prius
- smart fortwo
- Tesla Model S
- Harley-Davidson LiveWire
- Mitsubishi i-MiEV

Ask students to compare the information they discover with statistics for a comparable, less efficient vehicle.

Have the groups share their results with the class.

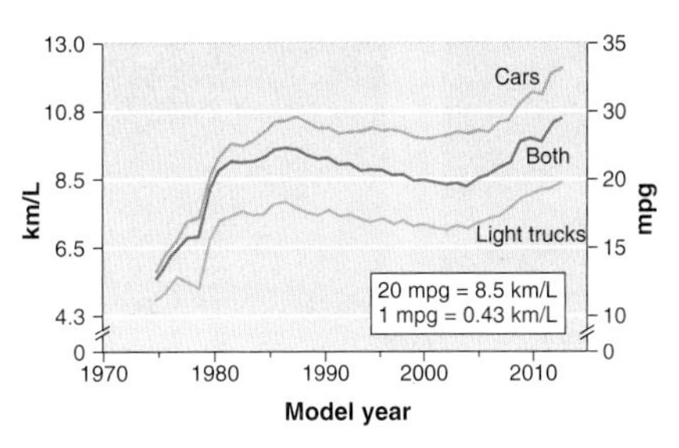

FIGURE 34.6 **Overall fuel efficiency of U.S. automobiles from 1975 through 2013.** As more buyers moved from cars to light trucks (a category that includes pickup trucks, minivans, and SUVs) for their personal vehicles, the fuel economy of the total U.S. fleet declined. Only recently has it begun to increase. *(After U.S. Environmental Protection Agency, 2013)*

Energy efficiency is an important consideration when making fuel and technology choices, but it is not the only factor we must consider. Determining the best fuel for the job is not always easy, and it involves trade offs among convenience, ease of use, safety, cost, and pollution.

Electricity accounts for 40 percent of our energy use

Because electricity can be generated from many different sources, including fossil fuels, wind, water, and the Sun, it is a form of energy in its own category. Coal, oil, and natural gas are primary sources of energy. Electricity is a secondary source of energy, meaning that we obtain it from the conversion of a primary source. As a secondary source, electricity is an **energy carrier**—something that can move and deliver energy in a convenient, usable form to end users.

Approximately 40 percent of the energy consumed in the United States is used to generate electricity. But because of conversion losses during the electricity generation process, of that 40 percent, only 13 percent of this energy is available for end uses. In this section we will look at some of the basic concepts and issues related to generating electricity from fossil fuels. Chapter 13 discusses electricity generation from renewable energy sources.

The Process of Electricity Generation

Electricity is produced by conversion of primary sources of energy such as coal, natural gas, or wind. Electricity is clean at the point of use; no pollutants are emitted in your home when you use a light bulb or computer. When electricity is produced by combustion of fossil fuels, however, pollutants are released at the location of its production. And, as we have seen, the transfer of energy from a fuel to electricity is only about 35 percent efficient. Therefore, although electricity is highly convenient, from the standpoint of efficiency of the overall system and the total amount of pollution released, it is more desirable to transfer heat directly to a home with wood or oil combustion, for example, than via electricity generated from the same materials. The energy source that entails the fewest conversions from its original form to the end use is likely to be the most efficient.

Many types of fossil fuels, as well as nuclear fuels, can be used to generate electricity. Regardless of which fuel is used, all thermal power plants work in the same basic way—they convert the potential energy of a fuel into electricity.

FIGURE 34.7 illustrates the major features of a typical coal-burning power plant. Fuel—in this case, coal—is delivered to a boiler, where it is burned. The burning fuel transfers energy to water, which becomes steam. The kinetic energy contained within the steam is transferred to the blades of a **turbine,** a device with blades that can be turned by water, steam, or wind. As the energy in the steam turns the turbine, the shaft in the center of the turbine turns the generator, which generates electricity. The electricity that is generated is then transported along a network of interconnected transmission lines known as the **electrical grid,** which connects power plants together and links them with end users of electricity. Once the electricity is on the grid, it is distributed to homes, businesses, factories, and other consumers of electricity, where it may be converted into heat energy for cooking, kinetic energy in motors, or radiant energy in lights or used to operate electronic and electrical devices. After the steam passes through the turbine, it is condensed back into water. Sometimes the water is cooled in a cooling tower or discharged into a nearby body of water. As we saw in Chapter 9, once-through use of water for thermal electricity generation is responsible for about half the water consumption in the United States.

Energy carrier Something that can move and deliver energy in a convenient, usable form to end users.

Turbine A device with blades that can be turned by water, wind, steam, or exhaust gas from combustion that turns a generator in an electricity-producing plant.

Electrical grid A network of interconnected transmission lines that joins power plants together and links them with end users of electricity.

◀ TEACHING with FIGURES

Show students Figure 34.6 and ask: Will there be a trend to purchase gas-efficient compact cars or will Americans continue to purchase larger, less fuel-efficient vehicles? *If prices rise and gas is more expensive for an extended period, overall efficiency of U.S. automobiles that are purchased will increase. People will begin to buy smaller, more compact cars that achieve a higher mpg. If the price per gallon falls, people will not have an incentive to economize on fuel use and may continue to purchase vehicles that have a lower mpg.*

AP® Exam Tip

Review kinetic and potential energy before discussing how coal or other fossil fuels produce electricity. This is a good time to review the first and second laws of thermodynamics. Use these concepts to describe what happens in a coal-fired electricity generation plant. In such plants, the potential energy in coal is converted to the kinetic energy of water that is converted into steam, which, in turn, spins a turbine. The second law of thermodynamics tells us that when energy is transformed, the quantity of energy remains the same, but its ability to do work diminishes.

AP® Exam Tip

Household electricity and water usage is covered on the 2007 AP® Exam, Question 2. To answer this question, students must

- use provided data to calculate the amount of water a family uses and the yearly cost of electricity for the family to shower
- calculate the time needed to recoup the investment in an energy-efficient hot water heater
- describe how a family can reduce overall water use
- list measures to reduce total energy use

Teaching Tip: Discussion Starter

Ask students: What is the basic process by which the energy in a fuel is converted into electricity? *Electricity is a secondary source of energy, meaning that we obtain it from the conversion of a primary source such as coal, oil, or natural gas. By the nature of being a secondary source, electricity is an energy carrier—something that can move and deliver energy in a convenient, usable form to end users.*

TEACHING with FIGURES ▶

Students often struggle to understand how electricity is generated. Have students work in small groups to examine Figure 34.7. Ask each group to create a table like the one shown below, listing each feature from the diagram. Have each group write a brief description of what occurs at the stage of the process or part of the plant that is labeled. (Answers provided in italics.)

Feature	Description of what occurs
Air	*Needed for flame*
Coal	*Fuel burned to make steam*
Chimney stack	*Removes smoke from the boiler and releases it into the atmosphere*
Exhaust gases	*CO_2, NO_x, So_x, and particulate matter produced when coal is burned*
Pulverizer	*Breaks up large solid pieces of coal into a dust that can be easily burned*
Boiler	*Device in which water is heated to make steam*
Ash	*Residue from burned coal that must be removed and taken for proper disposal*
Water	*Changed to steam that turns a turbine*
Steam	*Contains kinetic energy that is transferred to the blades of the turbine*
Turbine	*Device with a center shaft that turns the generator*
Condenser	*Fixture in which steam is changed back to liquid form (water)*
Generator	*Device connected to turbine that creates electrical current*
Electrical grid	*A network of interconnected transmission lines that join power plants together and link them with end users of electricity*
Transmission line	*Wires that carry the energy from the power plant throughout the electric grid*
Cooling tower or body of water	*Where water is cooled and discharged*

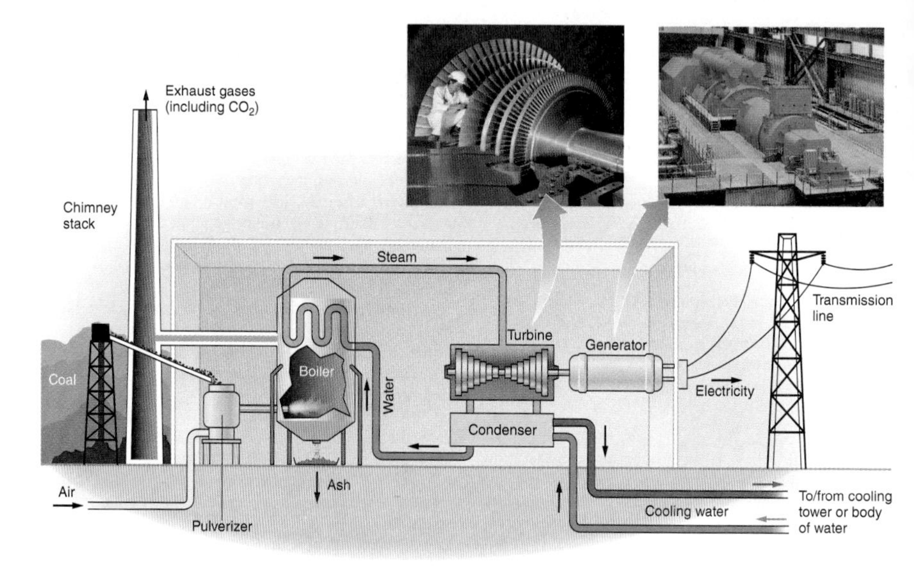

FIGURE 34.7 A coal-fired electricity generation plant. Energy from coal combustion converts water into steam, which turns a turbine. The turbine turns a generator, which produces electricity. *(left: Peter Bowater /Alamy; right: Peter Bowater/Science Source)*

Efficiency of Electricity Generation

Whereas a typical coal-burning power plant has an efficiency of about 35 percent, newer coal-burning power plants may have slightly higher efficiencies. Power plants using other fossil fuels can be even more efficient. An improvement in gas combustion technology has led to the **combined cycle** natural gas–fired power plant, which uses both exhaust gases and steam turbines to generate electricity. Natural gas is combusted, and the combustion products turn a gas turbine. In addition, the waste heat from this process boils water, which turns a conventional steam turbine. For this reason, a combined cycle plant can achieve efficiencies of up to 60 percent.

Capacity

A typical power plant in the United States might have a **capacity**—its maximum electrical output—of 500 megawatts (MW). This means that when the plant is operating, it generates 500 MW of electricity. If the plant operated for one day, it would generate 500 MW × 24 hours = 12,000 megawatt-hours (MWh). Most home electricity is measured in kilowatt-hours (kWh). So the typical power plant we've just described would generate 12,000,000 kWh in a day. If it operated for 365 days per year, it would generate 365 times that daily amount.

Most power plants, however, do not operate every day of the year. They must be shut down for periods of time to allow for maintenance, refueling, or repairs. Therefore, it is useful to measure the amount of time a plant actually operates in a year. This number—the fraction of time a plant is operating in a year—is known as its **capacity factor.** Most thermal power plants have capacity factors of 0.9 or greater. As we will see in Chapter 13, power plants using some forms of renewable energy, such as wind, may have a capacity factor of only about 0.25. "Do the Math: Calculating Energy Supply" shows you how to calculate the amount of energy a power plant can supply.

When it is time to start up a power plant, both nuclear and coal-fired plants may take a number of hours, or even a full day, to come up to full generating capacity. Because of the time it takes for them to become operational, electric companies tend to keep

Combined cycle A power plant that uses both exhaust gases and steam turbines to generate electricity.

Capacity In reference to an electricity-generating plant, the maximum electrical output.

Capacity factor The fraction of time a power plant operates in a year.

Teaching Tip: Discussion Starter

Ask students the following questions:

- What is the difference between energy efficiency and energy quality? *Energy efficiency is the ratio of the amount of work done to the total amount of energy introduced to the system. Energy quality is the ease with which an energy source can be used for work.*
- How do we determine the overall efficiency of energy use in a system? *We can use the equation EROEI. This helps people understand which energy type is most efficient and theoretically most desirable.*

Teaching Tip: Beyond the Classroom

Ask students to investigate a power plant in your state. What energy source does it use? When was it built? How many megawatt-hours does it typically produce in a year?

Calculating Energy Supply

According to the U.S. Department of Energy, a typical home in the United States uses approximately 900 kWh of electricity per month. On an annual basis, this is

$$900 \text{ kWh/month} \times 12 \text{ months/year} = 10{,}800 \text{ kWh/year}$$

How many homes can a 500 MW power plant with a 0.9 capacity factor support? Begin by determining how much electricity the plant can provide per month:

$$500 \text{ MW} \times 24 \text{ hours/day} \times 30 \text{ days/month} \times 0.9 = 324{,}000 \text{ MWh/month}$$

1 MWh equals 1,000 kWh, so to convert MWh per month into kWh per month, we multiply by 1,000:

$$324{,}000 \text{ MWh/month} \times 1{,}000 \text{ kWh/MWh} = 324{,}000{,}000 \text{ kWh/month}$$

So

$$\frac{324{,}000{,}000 \text{ kWh/month}}{900 \text{ kWh/month/home}} = 360{,}000 \text{ homes}$$

On average, a 500 MW power plant can supply roughly 360,000 homes with electricity.

Your Turn During summer months, in hot regions of the United States, some homes run air conditioners continuously. How many homes can the same power plant support if average electricity usage increases to 1,200 kWh/month during summer months?

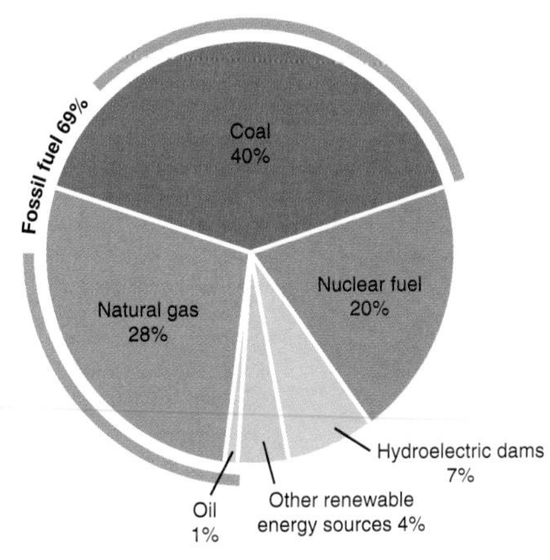

FIGURE 34.8 **Fuels used for electricity generation in the United States in 2012.** Coal is the fuel most commonly used for electricity generation. However, the electricity fuel mix in the United States is changing rapidly due to the increased availability and decreasing price of natural gas. *(Data from U.S. Department of Energy, Energy Information Administration, 2013)*

nuclear and coal-fired plants running at all times. As demand for electricity changes during the day or week, plants that are more easily powered up, such as those that use natural gas, oil, water, or wood, are used.

Cogeneration

The use of a fuel to generate electricity and produce heat is known as **cogeneration,** also called **combined heat and power.** Cogeneration is a method employed by certain users of steam for obtaining greater efficiencies. If steam used for industrial purposes or to heat buildings is diverted to turn a turbine first, the user will achieve greater overall efficiency than by generating heat and electricity separately. Cogeneration efficiencies can be as high as 90 percent, whereas steam heating alone might be 75 percent efficient, and electricity generation alone might be 35 percent efficient.

There are over 17,000 power plants in the United States. In 2012, they generated approximately 3.7 billion MWh. FIGURE 34.8 shows the fuels that were used to

Cogeneration The use of a fuel to generate electricity and produce heat. *Also known as* **combined heat and power.**

Answer to Your Turn

$$\frac{324{,}000{,}000 \text{ kWh/month}}{1{,}200 \text{ kWh/month/home}} = 270{,}000 \text{ homes}$$

more **math** practice:

Cost to Run an Appliance

To practice working with units of energy, have students solve the following problem.

A family in a small apartment uses the following electrical devices:

- A fixture with two 60-W incandescent light bulbs is on for 3 hours each day.
- An HD television that uses 200 W is on for 2 hours each day.
- A ceiling fan that uses 90 W is on while the family watches television.

The cost per kWh on the family's electricity bill is $0.105 per kWh. Calculate the yearly number of kWh used by each item and the cost to run it. Then calculate the yearly total number of kWh used and the total cost of the electricity.

Lamp

Total kWh per year:

$$\frac{(2) \times (60\ W) \times (3\ hours) \times (365\ days)}{1{,}000} = 131.4\ kWh$$

Total cost per year:

$$(131.4\ kWh) \times (\$0.105/kWh) = \$13.80 \text{ per year}$$

Television

Total kWh per year:

$$\frac{(200\ W) \times (2\ hours) \times (365\ days)}{1{,}000} = 146.0\ kWh$$

Total cost per year:

$$(146\ kWh) \times (\$0.105/kWh) = \$15.33 \text{ per year}$$

Ceiling Fan

Total kWh per year:

$$\frac{(90\ W) \times (2\ hours) \times (365\ days)}{1{,}000} = 65.7 \text{ kWh per year}$$

Total cost per year:

$$(65.7\ kWh) \times (\$0.105/kWh) = \$6.90 \text{ per year}$$

Yearly total kWh used by family:

$131.4 + 146.0 + 65.7 = 343.1$

Yearly total cost of electricity used by family:

$\$13.80 + \$15.33 + \$6.90 = \$36.03$ *per year*

generate this electricity. As we can see, coal-fired power plants are the backbone of electricity generation in the United States, responsible for 40 percent of all electricity produced. Natural gas, at 28 percent, and nuclear energy, at 20 percent, account for most of the remainder of the generating capacity. Together, these three sources, plus a very small amount of oil, account for 89 percent of electricity generation in the United States. Water and other renewable energy resources such as wind and solar energy account for the remainder of the electricity generated in the country. As we discussed earlier, expanded domestic production of natural gas has increased the fraction of electricity generated from natural gas while the fraction of electricity generated from coal has decreased. In 2005, coal accounted for 50 percent of electricity generation.

module 34

REVIEW

In this module, we have seen that fossil fuels are the most common fuels used in the United States and in other countries throughout the world. In the developed world, per capita energy use is many times higher than it is in the developing world. Inefficiencies during energy extraction and use account for losses of usable energy throughout the energy cycle. The fuels that have been consumed in the United States have varied over time and according to the end use. Oil and natural gas together make up 64 percent of overall energy use in the United States. Different fuels are suitable for different tasks, depending on a variety of factors including their energy content and whether they are liquid or solid. Coal and natural gas account for two-thirds of electricity generation. In the next module, we will consider the major fossil fuel resources used around the world as well as in the United States.

Answers to Module 34 AP® Review Questions

1. d
2. c
3. c
4. c
5. e

Module 34 AP® Review Questions

1. Which energy source does not originate from the Sun?
   (a) Coal
   (b) Solar
   (c) Oil
   (d) Nuclear
   (e) Natural gas

2. Traveling alone in a car uses 3.6 MJ of energy per kilometer. If 4 people go on a trip of 400 miles, what is the MJ used per person?
   (a) 200 MJ
   (b) 320 MJ
   (c) 580 MJ
   (d) 860 MJ
   (e) 1440 MJ

3. The major source of energy in the United States is
   (a) natural gas.
   (b) coal.
   (c) oil.
   (d) nuclear.
   (e) renewables.

4. Which is an example of a secondary energy source?
   (a) Solar
   (b) Coal
   (c) Electricity
   (d) Heat
   (e) Nuclear

5. Cogeneration is
   (a) the use of two or more energy sources to generate electricity.
   (b) the use of two separate turbines to generate electricity.
   (c) a method of electricity generation that includes renewable energy.
   (d) a method of increasing a power plant's capacity factor.
   (e) the use of a fuel to generate heat and electricity.

module

35 Fossil Fuel Resources

We have seen that fossil fuels provide most of the commercial energy in the world. These fuels are also responsible for a good deal of the pollution that occurs, a topic that will be discussed in Chapter 15. Because fossil fuels are such an important energy source, environmental scientists must study them closely. In this module, we will examine the major fossil fuels—coal, petroleum, and natural gas—and discuss their advantages and disadvantages. We will also consider the future use of fossil fuels.

Learning Objectives

After reading this module, you should be able to

- discuss the uses of coal and its consequences.
- discuss the uses of petroleum and its consequences.
- discuss the uses of natural gas and its consequences.
- discuss the uses of oil sands and liquefied coal and their consequences.
- describe future prospects for fossil fuel use.

Coal is the most abundant and dirtiest of the fossil fuels

Coal is a solid fuel formed primarily from the remains of trees, ferns, and other plant materials that were preserved 280 million to 360 million years ago. Coal has been the "work horse" of fossil fuels in the United States and in many other parts of the world. It is abundant in many areas and often is relatively easy to extract, handle, and process. We have seen that coal is the fuel most commonly used for electricity generation in the United States. There are three types of coal, ranked from lesser to greater age, exposure to pressure, and energy content; they are lignite, bituminous, and anthracite. A precursor to coal, called peat, is made up of partly decomposed organic material, including mosses. The formation of coal takes hundreds of millions of years. FIGURE 35.1 represents factors involved in the formation of the different types of coal. Starting with an organic material such as peat, increasing time and pressure produce successively denser coal with more carbon molecules, and more potential energy, per kilogram.

The largest coal reserves are found in the United States, Russia, China, and India. The countries that are currently producing the greatest amounts of coal are China, the United States, India, and Australia.

Coal A solid fuel formed primarily from the remains of trees, ferns, and other plant materials preserved 280 million to 360 million years ago.

lab

Coal Investigations

In this lab, students will measure and calculate the percentage of water found in the four different types of coal.

Download this resource by clicking on the link buttons in the TE-book, opening the TRFD, or logging onto the book's companion website bcs.whfreeman.com/friedlandapes2e (teacher login required).

 **Lab 12.1** Coal Investigations

TEACHING with FIGURES ▶

Figure 35.1 illustrates how coal is formed. Point out that all coal starts from living vegetation, and make sure students understand how long this process takes. This is also a good place to introduce the four basic types of coal: peat, lignite, bituminous, and anthracite.

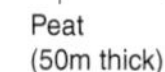
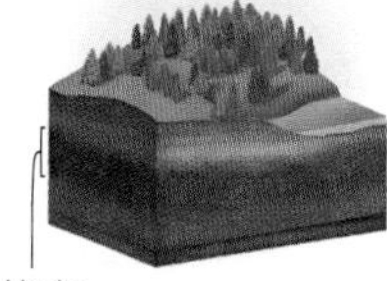

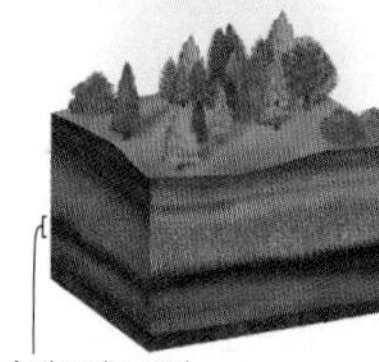

Peat (50m thick)
Lignite (10m thick)
Bituminous coal (5m thick)
Anthracite coal (5m thick)
0 years
Millions of years
Hundreds of millions of years
280 to 360 million years
Ancient forests cover much of land surface. The vegetation dies and is buried under anaerobic conditions, forming peat (partially decomposed organic matter).
As layers of peat become buried deeper below the surface of Earth, they are compressed into lignite.
Lignite layers are buried even deeper. The increased pressure compresses lignite into soft, bituminous coal.
Deeper burial, years of increased pressure, tectonic activity, and heat transform bituminous coal into harder anthracite coal.

FIGURE 35.1 **The coal formation process.** Peat is the raw material from which coal is formed. Over millions of years and under increasing pressure due to burial under more and more layers of rock and sediment, various types of coal are formed.

Advantages of Coal

Because it is energy-dense and plentiful, coal is used to generate electricity and in industrial processes such as making steel. In many parts of the world, coal reserves are relatively easy to exploit by surface mining. The technological demands of surface mining are relatively small and the economic costs are low. Once coal is extracted from the ground, it is relatively easy to handle and needs little refining before it is burned. It can be transported to power plants and factories by train, barge, or truck. All of these factors make coal a relatively easy fuel to use, regardless of technological development or infrastructure.

Disadvantages of Coal

Although coal is a relatively inexpensive fossil fuel, its use does have several disadvantages. As we saw in Chapter 8, the environmental consequences of the tailings from surface mining are significant. As surface coal is used up and becomes harder to find, however, subsurface mining becomes necessary. With subsurface mining, the technological demands and costs increase, as do the consequences for human health.

Coal contains a number of impurities, including sulfur, that are released into the atmosphere when the coal is burned. The sulfur content of coal typically ranges from 0.4 to 4 percent by weight. Lignite and anthracite have relatively low sulfur contents, whereas the sulfur content for bituminous coal is often much higher. Trace metals such as mercury, lead, and arsenic are also found in coal. Combustion of coal results in the release of these impurities, which leads to an increase of sulfur dioxide and other air pollutants, such as particulates, in the atmosphere, as we will see in Chapter 15. Compounds that are not released into the atmosphere remain behind in the resulting ash.

In order to reduce the chemical compounds released into the air, coal companies wash their coal in a variety of organic compounds. In some cases, these compounds have not been widely tested for toxicity or their effect on humans and ecosystems. In January 2014, a chemical storage tank containing one such coal cleaning compound leaked in West Virginia. Residents nearby immediately noticed a sweet odor and ultimately over 300,000 residents were advised not to drink their water for a number of days. Unfortunately, chemical spills are not the only accidents related to coal combustion. The residual ash from coal combustion can also be a problem.

According to the U.S. Department of Energy, there are 1,450 coal mines in the United States and they produced roughly 1 billion metric tons of coal in 2012. Most of that coal is burned in the United States, with anywhere from 3 to 20 percent of it remaining behind as ash. Large deposits of this ash are often stored near coal-burning power plants. One such ash deposit—a mixture of ash and water—was kept in a holding pond at a power plant near Knoxville, Tennessee. In December 2008, the retaining wall that contained the ash gave way and spilled 4.1 billion liters (1.1 billion gallons) of ash (FIGURE 35.2). Three houses were destroyed by the flow of muddy ash, which covered over 121 ha (300 acres) of land. The event was the largest of its kind in U.S. history. It took 5 years and over $1 billion to clean up the bulk of the ash.

AP® Exam Tip

Land reclamation after mining is covered in the 2005 AP® Exam, Question 3. To answer this question, students must

- name the steps needed to restore land that once was used for coal extraction
- explain why it is more difficult to restore arid land
- list the difficulties of successful reclamation when sulfur remains in former mines
- name two environmental impacts of coal
- explain why coal consumption in the United States is likely to increase in the future

FIGURE 35.2 **Results of a coal ash spill.** In 2008, this Tennessee home was buried in ash when an ash holding pond at a nearby power plant gave way. *(J. Miles Cary/Knoxville News Sentinel/ AP Images)*

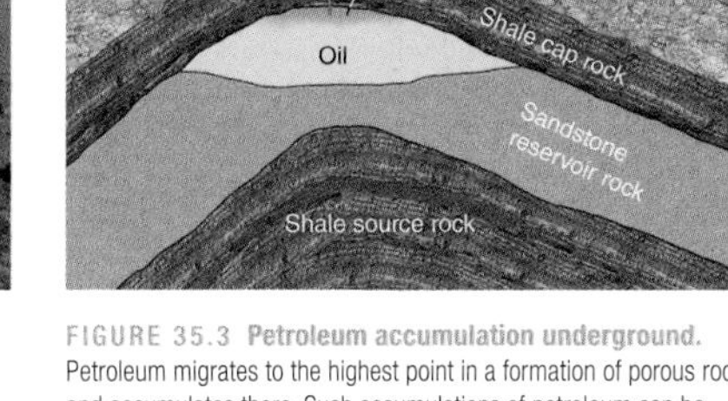

FIGURE 35.3 **Petroleum accumulation underground.** Petroleum migrates to the highest point in a formation of porous rock and accumulates there. Such accumulations of petroleum can be removed by drilling a well.

Finally, coal is a significant source of air pollution. Coal is 60 to 80 percent carbon. When it is burned, most of that carbon is converted into CO_2 and energy is released in the process. Coal produces far more CO_2 per unit of energy released than either oil or natural gas and it contributes to the increasing atmospheric concentrations of CO_2.

Petroleum is cleaner than coal

Petroleum, another widely used fossil fuel, is a fluid mixture of hydrocarbons, water, and sulfur that occurs in underground deposits. While coal is ideal for stationary combustion applications such as those in power plants and industry, the fluid nature of petroleum products such as oil and gasoline makes them more suitable for mobile combustion applications, such as in vehicles.

Petroleum is formed from the remains of ocean-dwelling phytoplankton (microscopic algae) that died 50 million to 150 million years ago. Deposits of phytoplankton are found in locations where porous sedimentary rocks, such as sandstone, are capped by nonporous rocks. Petroleum forms over millions of years and fills the pore spaces in the rock. Geologic events related to the tectonic cycle we discussed in Chapter 8 may deform the rock layers so that they form a dome. The petroleum is less dense than the rock, so over time, it migrates upward toward the highest point in the porous rock, where it is trapped by the nonporous rock, as FIGURE 35.3 shows. In certain locations, petroleum flows out under pressure the way water flows from an artesian well, as described in Chapter 9. But usually petroleum producers must drill wells into a deposit and extract the petroleum with pumps. After extraction, the petroleum must be transported to a petroleum refinery, by pipeline if the well is on land, or by supertanker if the well is underwater.

Petroleum contains natural gas. As we can see from Figure 35.3, some of this gas separates out naturally. If you have ever seen a burning flame in photographs of oil wells, it is a gas flare created when oil workers flare, or burn off, the natural gas, which is done under controlled conditions to prevent an explosion. As we will see in the next section, some of the gas is also extracted for use as fuel.

Liquid petroleum that is removed from the ground is known as **crude oil.** The U.S. Department of Energy refers to oil, crude oil, and petroleum as equivalent substances, and we will do the same in this chapter. Crude oil can be further refined into a variety of compounds. These compounds, including tar and asphalt, gasoline, diesel, and kerosene, are distinguished by the temperature at which they boil and can therefore be separated by heating the petroleum. This process takes place in an oil refinery, a large factory of the kind shown in this chapter's opening photograph. The refining process is complex and dangerous, and requires a major financial

Petroleum A fossil fuel that occurs in underground deposits, composed of a liquid mixture of hydrocarbons, water, and sulfur.

Crude oil Liquid petroleum removed from the ground.

investment. There are roughly 150 oil refineries in the United States; some of the larger ones can refine over 80 million liters (21 million gallons) per day. Oil production and sales are measured in barrels of oil; one barrel equals 160 L (42 gallons).

As we saw in Figure 34.4a, the United States uses more petroleum than any other fuel—roughly 3.1 billion liters (816 million gallons) of petroleum products per day. Gasoline accounts for roughly one-half of that amount. While the primary use of petroleum products is for transportation, petroleum is also the raw material for petrochemicals, such as plastics, and for lubricants, pharmaceuticals, and cleaning solvents. Worldwide petroleum consumption is almost 14 billion liters (3.7 billion gallons) per day, of which the United States is responsible for about 22 percent.

The top petroleum-producing countries are Saudi Arabia, Russia, the United States, Iran, China, Canada, and Mexico, in that order. These seven countries account for roughly one-half of worldwide oil production.

FIGURE 35.4 **View of an oil spill.** Oil from the BP Deepwater Horizon blowout is visible on the surface waters of the Gulf of Mexico, near the Louisiana coast, in 2010. *(Peter van Agtmael/Magnum Photos)*

Advantages of Petroleum

Because petroleum is a liquid, it is extremely convenient to transport and use. It is relatively energy-dense and is cleaner-burning than coal. For these reasons, it is an ideal fuel for mobile combustion engines such as those found in automobiles, trucks, and airplanes. Because it is a fossil fuel, it releases CO_2 when burned, although for every joule of energy released, oil produces only about 85 percent as much CO_2 as coal.

Disadvantages of Petroleum

Oil, like coal, contains sulfur and trace metals such as mercury, lead, and arsenic, which are released into the atmosphere when it is burned. Some sulfur can be removed during the refining process, so it is possible, though more expensive, to obtain low-sulfur oil.

As we have seen, oil must be extracted from under the ground or beneath the ocean. Whenever oil is extracted and transported, there is the potential for oil to leak from the wellhead or to be spilled from a pipeline or tanker. Some oil naturally escapes from the rock in which it was stored and seeps into water or out onto land. However, commercial oil extraction has greatly increased the number of leakage and spillage events and the amount of oil that has been lost to land and water around the world. As noted at the beginning of the chapter, the largest oil spill in the United States until 2010 was the *Exxon Valdez* oil tanker accident in 1989. More recently, the blowout of the BP Deepwater Horizon oil well, drilled 81 km (50 miles) off the coast of Louisiana in 1,524 m (5,000 feet) of water, led to a spill of well over 780 million liters (206 million gallons) of oil (FIGURE 35.4).

Larger oil spills have occurred elsewhere in the world. For example, during the 1991 Persian Gulf War, approximately 912 million liters (240 million gallons) of petroleum were spilled when wellheads were deliberately sabotaged or destroyed by the Iraqi army in Kuwait.

Oil is spilled into the natural environment in various ways. A 2003 National Academy of Sciences study found that oil extraction and transportation were responsible for a relatively small fraction of the oil spilled into marine waters worldwide. It found that roughly 85 percent of the oil entering marine waterways came from runoff from land and rivers, airplanes, and small boats and personal watercraft, including both deliberate and accidental releases of waste oil.

In the United States, debates continue over the trade off between domestic oil extraction and the consequences for habitat and species living near oil wells or pipelines. For example, when a 1,300-km (800-mile) pipeline was constructed to transport oil overland from the North Slope of Alaska to tankers that would carry it south to the contiguous United States, wildlife biologists predicted that the pipeline might melt permafrost and interfere with the calving grounds of caribou. Scientists continue to monitor the pipeline, but so far have come to no conclusions about its environmental impact. The transportation of petroleum products by means other than pipeline can have significant consequences as well. There have been a number of serious railway accidents in recent years as a result of increased domestic drilling for oil in the United States. In an accident in Quebec, Canada, near the Maine border, a railroad train carrying oil derailed and exploded in a small town in the early morning hours, killing 47 people and destroying many buildings.

The debate about the environmental effects of land-based oil extraction has continued with the proposal to allow oil exploration in the Arctic National Wildlife Refuge (ANWR), a 7.7 million hectare (19 million acre) tract of land in northeastern Alaska, shown in FIGURE 35.5. Proponents of exploration suggest that ANWR might yield 95 billion liters (25 billion gallons) to 1.4 trillion liters (378 billion gallons) of oil and substantial quantities of natural gas. Opponents maintain that opening ANWR to exploration and petroleum extraction will harm pristine habitat for many species as well as adversely affect people living in the area.

Humans, as well as wildlife, have been harmed by oil extraction. In Nigeria and many other developing countries, oil fields are adjacent to villages. Thick, gelatinous crude oil covers the ground where people walk, sometimes in bare feet. Oil flaring—the burning off of excess natural gas—takes place close to homes. Concerns about the effects of oil extraction on health, human rights, and environmental justice have led to violent political protests against oil companies in Nigeria and elsewhere.

Natural gas is the cleanest of the fossil fuels

We have already mentioned natural gas in connection with petroleum, since it exists as a component of petroleum in the ground as well as in gaseous deposits found separately from petroleum. Natural gas is 80 to 95 percent methane (CH_4) and 5 to 20 percent ethane, propane, and butane. Because natural gas is lighter than oil, it lies above oil in petroleum deposits (see Figure 35.3). Natural gas is generally extracted in association with petroleum; only recently has exploration specifically for natural gas been conducted.

The two largest uses of natural gas in the United States are for electricity generation and industrial processes. Natural gas is also used to manufacture nitrogen fertilizer, and it is used in homes as an efficient fuel for cooking, heating, and operating clothes dryers and water heaters. Compressed natural gas can be used as a fuel for vehicles, but because it must be transported by pipeline, it is not accessible in all parts of the United States and is therefore unlikely to become an important fuel for cars. Liquefied petroleum gas (LPG)—which is similar to natural gas, but in a liquid form—is a slightly less energy-dense substitute. LPG can be transported via train or truck and stored at the point of use in tanks. This fuel is available practically everywhere in the United States and is used in place of natural gas and for portable barbecue grills and heaters. Overall, natural gas and LPG supply 27 percent of the energy used in the United States.

Advantages of Natural Gas

Because of the extensive natural gas pipeline system in many parts of the United States, roughly one-half of homes use natural gas for heating. Compared with

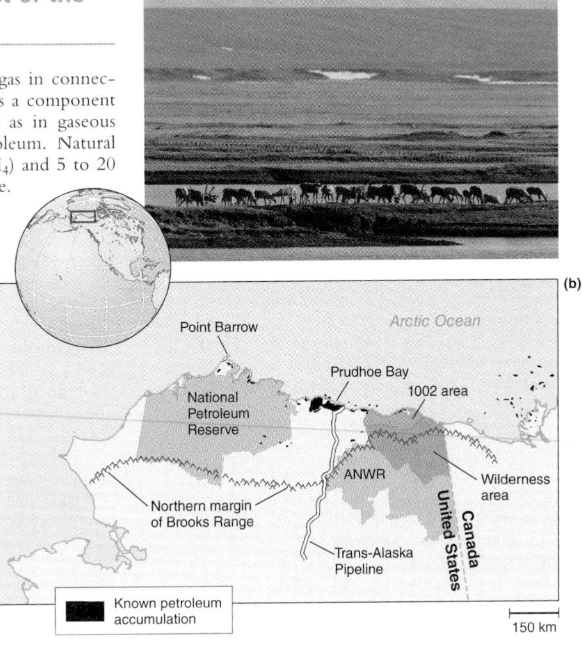

FIGURE 35.5 **The Arctic National Wildlife Refuge (ANWR).** (a) Caribou are among the many species that live in ANWR. (b) This map shows ANWR and adjacent areas on the North Slope of Alaska. The area in orange (the coastal plain, also called the "Ten-O-Two" area) is under consideration for petroleum exploration and extraction. *(David Shaw/Alamy)*

Teaching Tip: Debate the Issue

Should the U.S. government allow drilling for oil in the Arctic National Wildlife Refuge?

Students should research and be prepared to discuss these questions in their debate:

- How large is the Arctic National Wildlife Refuge (ANWR), and what type of biome is it?
- What are the economic benefits of opening ANWR to oil exploration?
- What are the environmental concerns with opening ANWR to oil exploration?
- What are the social concerns of opening ANWR to oil exploration?

In class, select groups to debate the issue. Allow each side 2 minutes to prepare, 3 minutes to present, and 1 minute for rebuttal. Have the remainder of the class take notes, ask questions, and decide which side has the most persuasive argument.

AP® Exam Tip

The effects of oil extraction on fragile tundra environments, specifically in the ANWR, is covered in the 2005 AP® Exam, Question 4. To answer this question, students must

- use provided data to calculate how long recoverable oil in ANWR could supply the total U.S. demand for oil
- describe two characteristics of a tundra biome and what factors make this fragile biome susceptible to human impact
- discuss how extracting oil from ANWR would affect the environment
- describe the use of oil in the United States and ways to use it more efficiently

Teaching Tip: Connections

This is a good place to review hydraulic fracturing, or fracking, which was first discussed in the Chapter 1 opening case. Ask students the following questions:

- What is the process of hydraulic fracturing? *The process entails drilling vertically and then horizontally. Explosive charges are used to enlarge existing fissures in the rock formation. Then water, sand, and chemicals are used to liberate the natural gas.*
- What are the environmental consequences of fracking? *Fracking often leads to groundwater contamination from the drilling and the natural gas wells that are created.*
- What are the benefits of fracking? *The greatest benefit is economic. Fracking offers a domestic fossil fuel that we typically have to get from other countries. Another benefit is that natural gas is cleaner than other fossil fuels.*
- What is one risk of fracking? *Risks include contaminating the aquifer, which is very difficult to clean once contaminated.*

coal and oil, natural gas contains fewer impurities and therefore emits almost no sulfur dioxide or particulates during combustion. And for every joule of energy released during combustion, natural gas emits only 60 percent as much CO_2 as coal. So natural gas is the cleanest of the fossil fuels and as long as it can be supplied by pipeline, it is a very convenient and desirable fossil fuel. In some locations where natural gas pipelines are not present, LPG is used although it is slightly less convenient.

Disadvantages of Natural Gas

While natural gas when combusted releases the least carbon dioxide of all the fossil fuels, unburned natural gas—methane—that escapes into the atmosphere is itself a potent greenhouse gas that is 25 times more efficient at absorbing infrared energy than CO_2. Natural gas that leaks after extraction is a suspected contributor to the steep rise in atmospheric methane concentrations that was observed in the 1990s. While natural gas is referred to as the "clean" fossil fuel, extraction and use still lead to environmental problems (FIGURE 35.6).

The process of exploring for natural gas involves the "thumper trucks" already mentioned. As described in Chapter 1, the process of hydraulic fracturing (fracking) releases natural gas from host rock. Fracking uses large amounts of chemicals and because those chemicals do not need to be named, their effects are unknown. Fracking also uses large quantities of water that can become contaminated and must be disposed of afterward. There can be groundwater contamination resulting from the drilling of natural gas wells. As hydraulic fracturing has grown more widespread, a number of scientists and gas extraction experts have attempted to quantify exactly how much natural gas escapes during the extraction and transportation process. Estimating the amount of escaped natural gas has become a controversial subject due to the large uncertainties in the natural gas extraction process, measurement difficulties, and industrial secrecy. Presently, estimates of escaped gas range from 2 to 9 percent of the total amount of gas extracted.

Oil sands Slow-flowing, viscous deposits of bitumen mixed with sand, water, and clay.

Bitumen A degraded petroleum that forms when petroleum migrates to the surface of Earth and is modified by bacteria.

FIGURE 35.6 **Natural gas field in Wyoming.** Even though natural gas is relatively clean compared with other fossil fuels, its extraction impacts large amounts of land. *(Joel Sartore/National Geographic/Getty Images)*

Oil sands and liquefied coal are also fossil fuels

Other types of fossil fuel deposits contain a great deal of energy, but are not readily available. When we consider using these energy resources, it is important to determine the energy return on energy investment. Two of the less readily available fossil fuels are oil sands and liquid coal.

Oil Sands

Not all petroleum is easily extractable in conventional oil wells. **Oil sands** are slow-flowing, viscous deposits of *bitumen* mixed with sand, water, and clay. **Bitumen** is a degraded type of petroleum that forms when a petroleum deposit is not capped with nonporous rock. The petroleum migrates close to the surface, where bacteria metabolize some of the light hydrocarbons while others evaporate. The remaining mix no longer flows at ambient temperatures and pressures, but it can be extracted by surface mining.

Although oil sand exploitation promises to extend our petroleum supply, it could have serious negative environmental impacts. The mining of bitumen is much more energy-intensive than conventional drilling for crude oil. As we saw in Chapter 8, surface mining creates large open pits. Extraction of the bitumen from the other material contaminates roughly 2 to 3 L of water for every liter of bitumen obtained, and many oil sands are located in areas where water is

Teaching Tip: Video

Shale Gas Drilling: Pros and Cons
This episode of the television program *60 Minutes* explores different viewpoints on extracting natural gas from shale.

The link to this video may be found by clicking the link buttons in the TE-book, opening the TRFD, or logging onto the book's companion website bcs.whfreeman.com/friedlandapes2e (teacher login required).

Chapter 12 Web Resources

AP® Exam Tip

Hydraulic fracturing and natural gas are covered in the 2012 AP® Exam, Question 1. The question also asks about nuclear power, which is covered later in this chapter. To answer this question, students must

- identify water-related environmental problems of hydraulic fracturing
- describe why natural gas is better for the environment than coal
- explain economic benefits of using hydraulic fracturing to obtain natural gas from shale
- describe the environmental problems associated with nuclear power

not an abundant resource. In addition, because oil sands require so much energy before they arrive at the refinery, the overall system efficiency is lower, and the resulting CO_2 release greater, than for conventional oil production.

Liquid Coal

Whereas the availability of petroleum may become much more limited in this century as supplies diminish, the United States and China both have immense coal reserves. The technology to convert solid coal into a liquid fuel—a process known as **CTL,** short for "coal to liquid"—has been available for decades. CTL was widely used by the German military during World War II and has been used by other countries since then. But producing liquefied coal is relatively expensive and has many of the same drawbacks found in the exploitation of oil sands. In terms of total energy content, there is over 1,000 times more energy in the world's coal reserves than in the world's petroleum reserves. Because there is so much coal in the United States, CTL has the potential to eliminate U.S. dependence on foreign oil. On the other hand, the U.S. Environmental Protection Agency estimates that total greenhouse gas emissions from liquefied coal are more than twice those from conventionally produced oil, and as we have seen, the environmental impacts of coal mining can be severe. For these reasons, liquid coal is not highly promising as an important future energy source.

Fossil fuels are a finite resource

Although we know that the supply of fossil fuels is finite, there is some discussion within the environmental science community on whether or not that matters. Recall our discussion in Chapter 7 about human creativity. Many people believe that we will apply human ingenuity to develop new energy sources. In the meantime, total energy use has leveled off, and energy use per person has decreased slightly. In addition, energy use per unit of gross domestic product (GDP), known as **energy intensity,** has been steadily decreasing, as FIGURE 35.7 shows. In other words, we are using energy more efficiently, but because the human population has grown so much and we are doing more things that use energy, our overall energy use has not decreased. For example, think about how many electronic devices you have, then ask your parents or grandparents how many electronic devices they had when they were high school students. We will begin this section with a survey of ideas about the

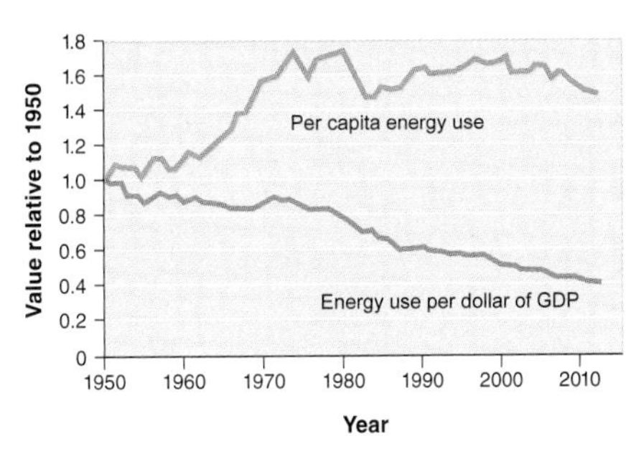

FIGURE 35.7 **U.S. energy use per capita and energy intensity.** Our energy use per capita was level and has been dropping in recent years. Our energy intensity, or energy use per dollar of GDP, has been decreasing steadily since 1980. However, because of the increasing U.S. population, total energy use of the nation has been roughly constant between 2000 and 2012. *(Data from U.S. Department of Energy, Energy Information Administration, 2013)*

availability of fuel and its use over time, and then we will look at possible patterns of fossil fuel energy use in the future.

The Hubbert Curve

We have seen that fossil fuels, our primary source of energy, are finite. Because fossil fuels take millions of years to form, they are nonrenewable resources, at least on a human timescale. By definition, then, the use of fossil fuels is not sustainable because there is no way to limit our consumption to the rate at which these fuels are being formed. Experts have debated whether our economy will at some point be limited by the availability of this energy resource, although in recent years, concerns have shifted away from the actual supply of fossil fuels to the consequences of fossil fuel combustion, particularly the release of CO_2 and its contribution to global warming. Many environmental scientists believe that these consequences will manifest themselves in adverse ways long before we run out of fossil fuels. However, there is still a group of energy analysts deeply concerned about "the end of oil."

In 1969, M. King Hubbert, a geophysicist and oil company employee, published a graph showing a

CTL (coal to liquid) The process of converting solid coal into liquid fuel.

Energy intensity The energy use per unit of gross domestic product.

Teaching Tip: Discussion Starter

Ask students: How is petroleum formed and why does it need to be refined? *Petroleum is formed from the remains of ocean-dwelling phytoplankton (microscopic algae) that died 50 million to 150 million years ago. The oil in this form, known as crude oil, must be further refined to be usable. The compounds into which crude oil is refined include tar and asphalt, gasoline, diesel, and kerosene. These are distinguished by the temperature at which they boil.*

Teaching Tip: Warm-up

Ask students: What is the relationship between energy intensity and energy use per capita? *Energy intensity is the energy use per unit of gross domestic product. Energy use per capita is the amount of energy each individual uses in a country. The energy use per capita in the United States has been level, while energy intensity (energy use per dollar of GDP) has been decreasing in recent years. Because the population of the United States has been increasing, overall energy use continues to remain the same or increase.*

Teaching Tip: Warm-up

Ask students to describe the Hubbert curve. *The Hubbert curve projects the point at which world oil production would reach a maximum and the point at which the world would exhaust all oil supplies.*

Teaching Tip: Discussion Starter

Ask students: Should we be concerned about future supplies of fossil fuels? *If current global use patterns continue, we could run out of conventional oil supplies in less than 50 years. Natural gas supplies will last slightly longer. Coal supplies will last for at least 200 years, and probably much longer. While these predictions assume that we will continue our current use patterns, advances in technology, a shift to nonfossil fuels, or changes in social choices and population patterns could alter them.*

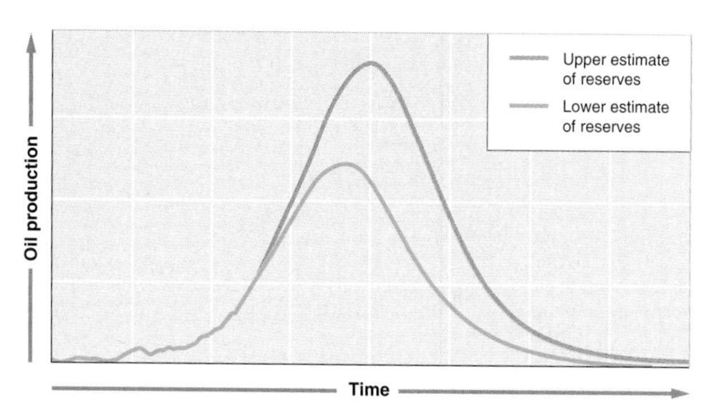

FIGURE 35.8 **A generalized version of the Hubbert curve.** Whether an upper estimate or a lower estimate of total petroleum reserves is used, the date by which petroleum reserves will be depleted does not change substantially. *(After M. K. Hubbert, "The energy resources of the Earth," in* Energy and Power, *A Scientific American Book [W. H. Freeman, 1971].)*

bell-shaped curve representing oil use. Shown in FIGURE 35.8, the **Hubbert curve** is a bell-shaped curve representing oil use and projecting both when world oil production will reach a maximum and when the oil will finally be depleted. Hubbert used two estimates of total world petroleum reserves: an upper estimate and a lower estimate. He found that the total reserves did not greatly influence the time it would take to use up all of the oil in known reserves. He predicted that oil extraction and use would increase steadily until roughly half the supply had been used up, a point known as **peak oil.** Hubbert predicted that at peak oil, extraction and use would begin to decline. Some oil experts believe we have already reached peak oil, while others maintain that we may reach it very soon. Back in 1969, Hubbert predicted that 80 percent of the world's total oil supply would be used up in roughly 60 years.

Although there have been discoveries of large oil fields since Hubbert did his work, the conclusion he drew from his model still holds. Regardless of the exact amount of the total reserves, the total number of years we use petroleum will fall within a relatively narrow time window. When we identify a fuel source, we tend to use it until we come upon a better fuel source. As a number of energy experts are fond of saying, "We did not move on from the Stone Age because we ran out of stones." In a similar vein, many people believe that ingenuity and technological advances in the renewable energy sector will one day render oil much less desirable. A growing number of energy analysts maintain that wondering about how much oil remains and whether or not we have reached peak oil are no longer meaningful questions to ask. As the undesirability of fossil fuels becomes greater they ask, "What fuel will we use next?"

Hubbert curve A bell-shaped curve representing oil use and projecting both when world oil production will reach a maximum and when the world will run out of oil.

Peak oil The point at which half the total known oil supply is used up.

The Future of Fossil Fuel Use

If current global use patterns continue, and no significant additional oil supplies are discovered, we may run out of conventional oil supplies in less than 50 years. Natural gas supplies will last longer. Coal supplies will last for at least 200 years, and probably much longer. While these predictions assume that we will continue our current use patterns, advances in technology, a shift to nonfossil fuels, or changes in social choices and population patterns could alter them.

As people have come to accept the theory that anthropogenic increases in atmospheric greenhouse gas concentrations have caused global climate change, a large number of researchers have suggested that we should turn our attention to how we might transition from fossil fuels before their use causes further problems.

Concerns about scarcity, environmental impacts (especially the influence of CO_2 on global climate change), and rising costs of fossil fuels present many opportunities. Rising oil prices create a powerful incentive to invest in alternative energy resources and conservation. As we will see in Chapter 13, people have begun to explore the possibility of a new infrastructure to deliver wind energy, hydroelectricity, and solar energy. On the other hand, the higher price of oil makes formerly unproductive mining and extraction methods cost-competitive, which may encourage greater total production of fossil fuels.

Teaching Tip: Activity

Have your students complete the table below to compare the advantages and disadvantages of nonrenewable energy sources: oil, coal, natural gas, and nuclear. (Suggested answers are provided in italics.)

	Oil	Coal	Natural gas	Nuclear
Advantages	• *Ideal for mobile combustion, quick ignition, and turning off capability* • *Cleaner burning than coal*	• *Energy-dense* • *Abundant in the United States* • *No refining necessary* • *Easy and safe to transport* • *Economic backbone of some small towns*	• *Combined cycle power plants can have efficiencies up to 60%* • *Efficient for cooking, home heating, etc.* • *Fewer impurities than coal or oil*	• *Emits no CO_2 once plant is operating* • *Offers independence from fossil fuels* • *High energy density* • *Ample supply*
Disadvantages	• *Significant refining required* • *Oil spills cause significant environmental damage* • *Significant dust and emissions from combustion* • *Limited supply*	• *Mining is dangerous* • *Mining dramatically alters natural landscapes* • *Coal-fired plants are slow to reach full operating capacity* • *Combustion of coal is a large contributing factor to acid rain*	• *Risk of leaks and explosions* • *Twenty-five times more effective as a greenhouse gas then CO_2* • *Transportation by pipelines limits availability*	• *Lacks public support* • *Plants are expensive to build because of legal challenges* • *Meltdown could be catastrophic* • *Possible target for terrorist attacks*

module

35

REVIEW

In this module, we have seen that fossil fuels have significant advantages as well as significant disadvantages. Coal is an abundant fuel that requires minimal processing and is easily transported. However, it leads to significant human health problems when mined below ground, while surface mining causes environmental damage. Coal releases more carbon dioxide per unit of energy released than do other fossil fuels. Petroleum is less abundant than coal but requires more processing. The possibility of spills is another disadvantage of petroleum. Natural gas is easily transported via pipeline and releases less carbon dioxide per unit of energy released than other fossil fuels. Its use has increased substantially in recent years for electricity generation and other purposes, mostly because of increased availability and decreased price due to fracking. In the last 50 years, estimates of fossil fuel availability and the rate at which it will be depleted have been replaced by concerns about the pollution from fossil fuels and questions about future fossil fuel energy use patterns. As we will see in the next module, there is one other conventional, nonrenewable fuel that offers a very attractive feature: It does not contribute significantly to the addition of greenhouse gases to the atmosphere. That fuel is uranium.

Module 35 AP® Review Questions

1. Which type of coal has the highest energy density?
   (a) Lignite
   (b) Peat
   (c) Bituminous
   (d) Anthracite
   (e) Subbituminous

2. What makes petroleum convenient to use as fuel for transportation?
   I. High energy density
   II. Clean burning
   III. Its liquid state

   (a) I only
   (b) I and III only
   (c) II and III only
   (d) III only
   (e) I, II, and III

3. Natural gas is primarily
   (a) ethane.
   (b) propane.
   (c) butane.
   (d) methane.
   (e) kerosene.

4. Bitumen is
   (a) a form of liquid coal.
   (b) a degraded type of petroleum.
   (c) a by-product of natural gas extraction.
   (d) a fast-forming fossil fuel.
   (e) a petroleum product used for plastic production.

5. The Hubbert curve predicts that
   (a) peak oil will occur once half of the supply is used up.
   (b) finding additional reserves could greatly increase the time to peak oil.
   (c) natural gas will be cheaper than oil by the early twenty-first century.
   (d) the cost of oil is independent of the world's oil supply.
   (e) coal will run out faster than oil.

Answers to Module 35 AP® Review Questions

1. d
2. e
3. d
4. b
5. a

COMMON MISCONCEPTIONS
Fusion and fission

Students often confuse the terms fusion and fission. Fission is a nuclear reaction in which a neutron strikes a relatively large atomic nucleus, which then splits into two or more parts. This releases a great deal of energy and more neutrons. Fusion is the joining of two smaller nuclei to form a larger nucleus, which is accompanied by a release of energy. Fusion is a natural process that occurs in stars at extremely high temperatures. Point out to your students that at present, we are only able to use *fission* as a source of energy here on Earth, not *fusion*. Scientists are trying to find ways for sustained fusion reactions to occur at lower temperatures so they might be used as a source of energy on Earth, but so far this has proved to be extremely difficult.

AP® Exam Tip

Nuclear energy is covered in the 2014 AP® Exam, Question 1. To answer this question students must

- take a position on statements regarding the safety and impact of nuclear power
- describe the waste generated by nuclear power
- explain the environmental problems associated with the construction of a nuclear power plant
- describe the environmental impacts and ecological consequences of the pollution generated by a nuclear power plant
- suggest ways to reduce the use of electricity

module 36

Nuclear Energy Resources

Because the combustion of fossil fuels releases large quantities of CO_2 into the atmosphere, nuclear energy has received increasing interest. Concerns about nuclear energy include radioactivity, the proliferation of radioactive fuels that could be used in weapons, and the potential for accidents. Recently, however, nuclear energy has received positive attention, even from self-proclaimed environmentalists, because of its relatively low emissions of CO_2. In this module we will examine how nuclear energy works and the advantages and disadvantages of nuclear energy.

Learning Objectives

After reading this module, you should be able to

- describe how nuclear energy is used to generate electricity.
- discuss the advantages and disadvantages of using nuclear fuels to generate electricity.

Nuclear reactors use fission to generate electricity

Electricity generation from nuclear energy uses the same basic process as electricity generation from fossil fuels: Steam turns a turbine that turns a generator that generates electricity. The difference is that a nuclear power plant uses a radioactive isotope, uranium-235 (^{235}U), as its fuel source. We presented the concepts of isotopes, radioactive decay, and half-lives in Chapter 2. Radioactive decay occurs when a parent radioactive isotope emits alpha or beta particles or gamma rays. Here, we need to introduce one more concept before being able to fully describe a nuclear power plant. The naturally occurring isotope ^{235}U, as well as other radioactive isotopes, undergoes a process called *fission*.

Fission A nuclear reaction in which a neutron strikes a relatively large atomic nucleus, which then splits into two or more parts, releasing additional neutrons and energy in the form of heat.

Fuel rod A cylindrical tube that encloses nuclear fuel within a nuclear reactor.

Fission, depicted in FIGURE 36.1, is a nuclear reaction in which a neutron strikes a relatively large atomic nucleus, which then splits into two or more parts. This process releases additional neutrons and energy in the form of heat. The additional neutrons can, in turn, promote additional fission reactions, which leads to a chain reaction of nuclear fission that gives off an immense amount of heat energy. In a nuclear power plant, that heat energy is used to produce steam, just as in any other thermal power plant. However, 1 g of ^{235}U contains 2 million to 3 million times the energy of 1 g of coal.

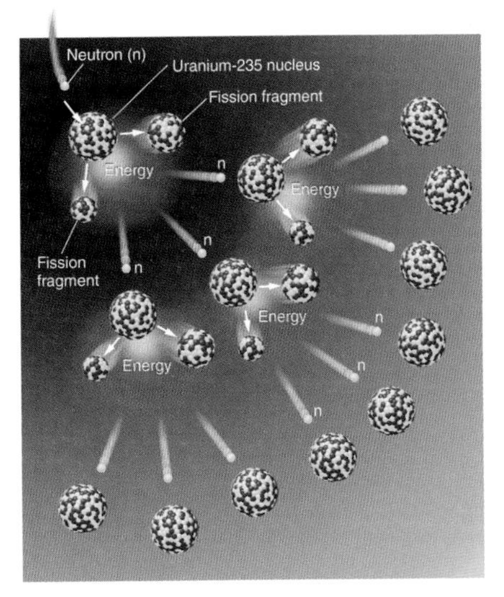

FIGURE 36.1 **Nuclear fission.** Energy is released when a neutron strikes a large atomic nucleus, which then splits into two or more parts.

Uranium-235 is one of the more easily fissionable isotopes, which makes it ideal for use in a nuclear reactor. A neutron colliding with ^{235}U splits the uranium atom into smaller atoms, such as barium and krypton, and three neutrons. The reaction is as follows:

$$1 \text{ neutron} + {}^{235}U \rightarrow {}^{142}Ba + {}^{91}Kr + 3 \text{ neutrons in motion (kinetic energy)}$$

Many other radioactive daughter products are released as well. A properly designed nuclear reactor will harness the kinetic energy from the three neutrons in motion to produce a self-sustaining chain reaction of nuclear fission. The by-products of the nuclear reaction include radioactive waste that remains hazardous for many half-lives—that is, hundreds of thousands of years or longer.

FIGURE 36.2 shows how a nuclear reactor works. The containment structure encloses the nuclear fuel—which is contained within cylindrical tubes called **fuel rods**—and the steam generator. Uranium fuel is processed into pellets, which are then put into the fuel rods. A typical nuclear reactor might contain hundreds of bundles of fuel rods in the reactor core within the containment structure.

Heat from nuclear fission is used to heat water within the containment structure, which circulates in a loop. This loop passes close to another loop of water,

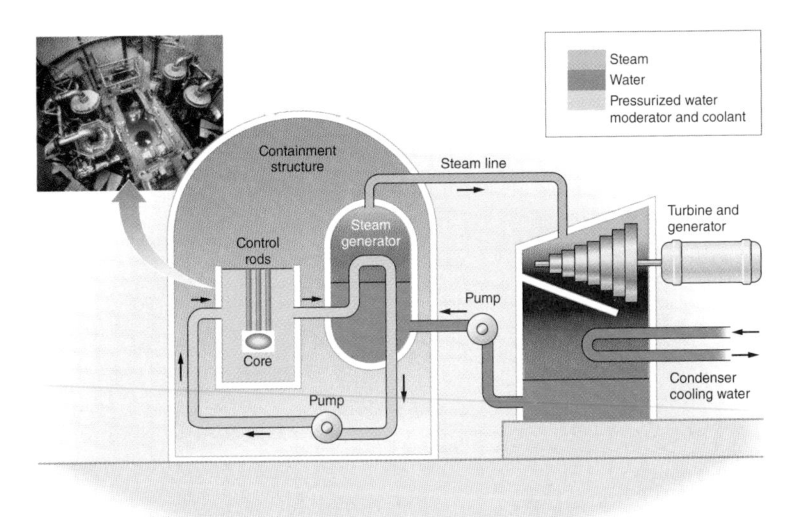

FIGURE 36.2 **A nuclear reactor.** This schematic shows the basic features of the light-water reactor, the type of reactor found in the United States. *(U.S. Department of Energy/Science Source)*

Teaching Tip: Beyond the Classroom

Have students work in groups to research and write a report on a major nuclear accident. Some major nuclear power plant accidents include Three Mile Island in Pennsylvania, Chernobyl in the Ukraine, and Fukushima in Japan. Students should address the following in their report:

- Where and when did the accident occur?
- How did it happen?
- What were the environmental implications?
- What were the economic implications?
- Did any laws or policies change as a result of the event?

▲ TEACHING with FIGURES

Have students study Figure 36.2 and complete the table below. (Suggested answers are provided in italics.)

Major parts of nuclear power plant	Function
Control rods	*A cylindrical device inserted between the fuel rods in a nuclear reactor to absorb excess neutrons and slow or stop the fission reaction*
Fuel rods	*A cylindrical tube that encloses nuclear fuel within a nuclear reactor*
Core	*Location of the fission reaction*
Containment structure	*Structure that contains the core with thick concrete walls to protect against radiation leaks*
Steam generator	*Area within the containment structure where liquid water is changed to gaseous steam*
Steam line	*Pipe that transports heated steam to turn turbine*
Turbine and generator	*Turbine: device with blades propelled by steam to turn a generator* *Generator: device that produces electricity*
Cooling tower	*Structure in which hot water is cooled*

Teaching Tip: Debate the Issue

Should U.S. policy make it easier to construct new nuclear reactors for generating electricity?

The use of nuclear energy for power has been debated for many decades. The United States has not built a new nuclear power plant facility since 1978, so almost all U.S. facilities are between 35 to 50 years old. To prepare a debate on the question of U.S. nuclear policy, ask students to research the following questions:

- What are the risks and benefits of nuclear power?
- Are other countries in the world using more nuclear power compared to the United States?
- Are older plants at greater risk of accidents or more likely to cause environmental damage?
- What are the issues to consider in shutting down old plants?
- What are the environmental, economic, and social ramifications of building new plants compared to maintaining old plants?

In class, select groups to debate the issue. Allow each side 2 minutes to prepare, 3 minutes to present, and 1 minute for rebuttal. Have the remainder of the class take notes, ask questions, and decide which side has the most persuasive argument.

AP® Exam Tip

Nuclear power plants and the environmental implication of using or not using nuclear power are covered in the 1998 AP® Exam, Question 2. To answer this question, students must

- identify and describe the function of three parts of a nuclear power plant shown in a diagram
- describe and discuss two environmental problems associated with nuclear power
- discuss the future of nuclear power use in the United States and the environmental implications

and heat is transferred from one loop to the other. In the process, steam is produced, which turns a turbine, which turns a generator, just as in most other thermal power plants. The nuclear power plant shown in Figure 36.2 is a light-water reactor, the only type of reactor used in the United States and the most common type used elsewhere in the world.

A nuclear power plant is designed to harness heat energy from fission to make steam. But the plant must be able to slow the fission reaction to allow collisions to take place at the appropriate speed. To do this, nuclear reactors contain a moderator, such as water, to slow down the neutrons so that they can effectively trigger the next chain reaction. Because there is also a risk that the reaction will run out of control, nuclear reactors contain **control rods,** cylindrical devices that can be inserted between the fuel rods to absorb excess neutrons, thus slowing or stopping the fission reaction. This is done routinely during the operation of the plant because nuclear fuel rods left uncontrolled will quickly become too hot and melt—an event called a meltdown—or cause a fire, either of which could lead to a catastrophic nuclear accident. Control rods are also inserted when the plant is being shut down during an emergency or for maintenance and repairs.

Depending on the ore, it may take up to 900 kg (2,000 pounds) of uranium ore to produce 3 kg (6.6 pounds) of nuclear fuel. In order to obtain uranium, miners remove large amounts of the host rock, extract and concentrate the uranium, and leave the remaining material in slag piles. Australia, the western United States, and parts of Canada have large commercial uranium mining operations for nuclear fuel. As we saw in Chapter 8, the mining of any material requires fossil fuel energy and results in mine tailings. This is also true for uranium, although, as we have noted, a much smaller volume and mass of uranium is needed to generate a similar quantity of electricity than would be the case with coal.

Nuclear power plants rely on ^{235}U as their fuel. However, most uranium ores contain as much as 99 percent ^{238}U, another isotope of uranium that occurs with ^{235}U but does not fission as easily. Therefore, when uranium ore is mined, it must be chemically enriched—a process to increase its concentration—in ^{235}U to be useful as a fuel. Typically, suitable nuclear fuel contains 3 percent or greater ^{235}U.

Control rod A cylindrical device inserted between the fuel rods in a nuclear reactor to absorb excess neutrons and slow or stop the fission reaction.

Nuclear energy has advantages and disadvantages

Nuclear power plants do not produce air pollution during their operation, so proponents of nuclear energy consider it "clean" energy. In countries with limited fossil fuel resources, nuclear energy is one way to achieve independence from imported oil. Nuclear energy generates 70 percent of electricity in France, and it is widely used in Lithuania, Germany, Spain, the United Kingdom, Japan, China, and South Korea, as well as in other countries.

As we saw in Figure 34.8, 20 percent of the electricity generated in the United States comes from nuclear energy. Early proponents of nuclear energy in the 1950s and 1960s claimed that it would be "too cheap to meter," meaning that it would be so inexpensive that there would be no point in trying to figure out how much each customer used. However, construction of new nuclear power plants became more and more expensive in the United States, in part because public protests, legal battles, and other delays increased the cost of construction. Public protests arose because of concerns that a nuclear accident would release radioactivity into the surrounding air and water. Other concerns included uncertainty about appropriate locations for radioactive waste disposal and fear that radioactive waste could fall into the hands of individuals seeking to make a nuclear weapon. By the 1980s, it had become prohibitively expensive—both monetarily and politically—to attempt the construction of new nuclear power plants.

There are currently 104 nuclear power plants in the United States—the same number as 2 decades ago. However, the relatively low CO_2 emissions associated with nuclear energy has caused a resurgence of interest in constructing additional nuclear power plants. There are certainly CO_2 emissions related to mining, processing, and transporting nuclear fuel as well as constructing the actual nuclear power plant. However, these emissions are perhaps 10 percent or less of those related to generating an equivalent amount of electricity from coal. Two major issues of environmental concern remain, however: the possibility of accidents and the disposal of radioactive waste. We will also consider nuclear fusion and summarize the advantages and disadvantages of nuclear power relative to other energy sources.

Nuclear Accidents

Two accidents contributed to the global protests against nuclear energy in the 1980s and 1990s. On March 28, 1979, at the Three Mile Island nuclear power plant in Pennsylvania, operators did not notice that a cooling water valve had been closed the previous day. This oversight led to a lack of cooling water around the reactor core, which overheated and suffered a partial meltdown. The reactor core was severely damaged, and a large part of the containment structure became highly radioactive with an unknown amount of radiation released to the outside environment. A few thousand schoolchildren and pregnant women were evacuated from the area surrounding the plant by order of the governor of Pennsylvania, although an estimated 200,000 other people chose to evacuate as well.

In the days following the accident, residents of the area experienced a great deal of anxiety and fear, especially as reports of a potentially explosive gas bubble in the containment structure were evaluated. Although there has been no documented increase in local health problems as a result of this accident, concerns remain and several investigators maintain that infant mortality rates and cancer rates increased in the following years. This nuclear reactor, one of two at the Three Mile Island nuclear facility, has not been used since the accident.

The Three Mile Island event, compounded by the coincidental release of the film *The China Syndrome* about safety violations and a near-catastrophic accident at a nuclear power plant, led to widespread public fear and anger not only in Pennsylvania but elsewhere as well.

A much more serious accident occurred on April 26, 1986, at a nuclear power plant in Chernobyl, Ukraine. The accident occurred during a test of the plant when, in violation of safety regulations, operators deliberately disconnected emergency cooling systems and removed the control rods. With no control rods and no coolant, the nuclear reactions continued without control and the plant overheated. These "runaway" reactions led to an explosion and fire that damaged the plant beyond use. At the time of the accident, 31 plant workers and firefighters died from acute radiation exposure and burns, and many more died later of causes related to exposure.

After the accident, winds blew radiation from the plant across much of Europe, where it contaminated crops and milk from cows grazing on contaminated grass. More than a hundred thousand people were evacuated from the area around Chernobyl. Estimates of health effects vary widely, in part because of the paucity of information provided by the Soviet government, but a U.S. National Academy of Sciences panel estimated that 4,000 additional cancer deaths (over and above the average number of expected deaths) would occur over the next 50 years among people who lived near the plant or worked on the cleanup. There have been approximately 5,000 cases of thyroid cancer, most of them nonfatal, among children who were younger than 18 at the time of the accident and lived near the Chernobyl plant. Thyroid cancer may be caused by the absorption of radioactive iodine, one of the radioactive elements emitted during the accident.

An accident considered to be almost equivalent in severity to Chernobyl occurred in Japan on March 11, 2011. A magnitude 9.0 earthquake off the country's coast generated a tsunami that was 15 m high. This caused flooding in northeastern Japan, where the Fukushima nuclear power plant is located. The resulting structural damage to the plant and interruption of the regional electrical supply led to fires, hydrogen gas explosions, and the release of radioactive gases from nuclear reactors within the containment structure. Ultimately, four out of a total of six reactors were damaged beyond repair, radioactive gases were released into the surrounding environment, and over 100,000 people were evacuated from their homes. Approximately 20,000 people, including 3 workers at the Fukushima plant, were killed by the earthquake and tsunami. However, at present, there have been no deaths attributed to the release of radioactivity from the reactors.

Radioactive Waste

When nuclear fuel is no longer useful in a power plant, yet continues to emit radioactivity, it is considered **radioactive waste.** Long after nuclear fuel can produce enough heat to be useful in a power plant, it continues to emit radioactivity. Because radioactivity can be extremely damaging to living organisms, radioactive materials—the nuclear waste—must be stored in special, highly secure locations.

The use of nuclear fuels produces three kinds of radioactive waste: high-level waste in the form of used fuel rods; low-level waste in the form of contaminated protective clothing, tools, rags, and other items used in routine plant maintenance; and uranium mine tailings, the residue left after uranium ore is mined and enriched. Disposal of all three types is regulated by the government, but because it has the greatest potential impact on the environment, we will focus here on high-level radioactive waste.

After a period of time, nuclear fuel rods become spent, meaning they are not sufficiently radioactive to generate electricity efficiently. As we discussed in Chapter 2, each radioactive isotope has a specific half-life: the time it takes for one-half of its radioactive

Radioactive waste Nuclear fuel that can no longer produce enough heat to be useful in a power plant but continues to emit radioactivity.

more **math** practice: Radioactive Decay

Have students practice working with the concept of half-lives and radioactive decay by completing the table below. The table shows what happens as 2,000 g of a substance with a half-life of 1,000 years decays over time. Leave the last four rows (8, 9, 10, and 11) blank for students to complete. (Answers are provided in italics.) You might want to work together as a class to fill in the values for the first eight half-lives. For simplicity, have students round values for percent and mass to two decimal places. Here is an example for row 8:

Years: Calculate the number of years that would pass in eight half-lives.

Since each half-life is 1,000 years, add 1,000 to the number of years it took for seven half-lives to pass:

$$7{,}000\text{ years} + 1{,}000\text{ years} = 8{,}000\text{ years}$$

Fraction remaining: Determine the fraction remaining after eight half-lives.

Start with the fraction remaining at seven half-lives $= \frac{1}{128}$

Divide this fraction by 2 to determine the fraction remaining at eight half-lives:

$$\frac{1}{128} \div 2 = \frac{1}{256}$$

Percentage remaining: Convert the fraction remaining at eight half-lives to a percentage.

To convert a fraction to a percentage, divide the numerator by the denominator and multiply by 100:

$$\frac{1}{256} \times 100 = 0.39\%$$

Mass remaining: Determine the mass of the substance remaining after eight half-lives.

Start with the mass remaining at seven half-lives: 15.63 g

Divide the mass remaining after seven half-lives by 2:

$$15.63\text{ g} \div 2 = 7.82\text{ g}$$

Number of half-lives	Years	Fraction remaining	Percentage remaining	Mass (g) remaining
0	0	1	100.00%	2,000 g
1	1,000	$\frac{1}{2}$	50.00%	1,000 g
2	2,000	$\frac{1}{4}$	25.00%	500 g
3	3,000	$\frac{1}{8}$	12.50%	250 g
4	4,000	$\frac{1}{16}$	6.25%	125 g
5	5,000	$\frac{1}{32}$	3.13%	62.5 g
6	6,000	$\frac{1}{64}$	1.56%	31.25 g
7	7,000	$\frac{1}{128}$	0.78%	15.63 g
8	*8,000*	$\frac{1}{256}$	*0.39%*	*7.82 g*
9	*9,000*	$\frac{1}{512}$	*0.20%*	*3.91 g*
10	*10,000*	$\frac{1}{1024}$	*0.10%*	*1.96 g*
11	*11,000*	$\frac{1}{2048}$	*0.05%*	*0.98 g*

Answer to Your Turn

First find the number of half-lives in 1,325 years.

$$1{,}325 \text{ years} \times \frac{1 \text{ half-life}}{265 \text{ years}} = 5 \text{ half-lives}$$

Half of the remaining material decays after every half-life; therefore, start with the original amount and multiply by one-half for every half-life.

$180 \text{ g} \times ½ = 90 \text{ g}$	1 half-life
$90 \text{ g} \times ½ = 45 \text{ g}$	2 half-lives
$45 \text{ g} \times ½ = 22.5 \text{ g}$	3 half-lives
$22.5 \text{ g} \times ½ = 11.25 \text{ g}$	4 half-lives
$11.25 \text{ g} \times ½ = 5.625 \text{ g}$	5 half-lives

Teaching Tip: Journal Prompt

To review nuclear energy, ask students the following questions:

- How does a nuclear reactor work? *Inside a containment structure, uranium fuel is processed into pellets, which are then put into the fuel rods. A typical nuclear reactor might contain hundreds of bundles of fuel rods in the center. Within the containment structure, heat from nuclear fission is used to heat water, which circulates in a loop. This loop passes close to another loop of water, and heat is transferred from one loop to the other. Finally, steam is produced, which turns a turbine. The turbine powers a generator, which generates electricity.*
- What makes nuclear energy a desirable energy option? *Nuclear energy produces almost no harmful air emissions. It is a very dense energy source.*
- What are two major concerns about nuclear energy? *Spent fuel rods are radioactive for hundreds of thousands of years. A meltdown could be catastrophic to nearby areas. Nuclear energy plants can be a target for terrorist attacks.*

do the math

Calculating Half-Lives

Strontium-90 is a radioactive waste product from nuclear reactors. It has a half-life of 29 years. How many years will it take for a quantity of strontium-90 to decay to $\frac{1}{16}$ of its original mass?

It will take 29 years to decay to $\frac{1}{2}$ its original mass; another 29 years to $\frac{1}{4}$; another 29 years to $\frac{1}{8}$; and another 29 years to $\frac{1}{16}$:

$$29 + 29 + 29 + 29 = 116 \text{ years}$$

Your Turn You have 180 g of a radioactive substance. It has a half-life of 265 years. After 1,325 years, what mass remains?

atoms to decay. Uranium-235 has a half-life of 704,000,000 years, which means that 704 million years from today, a sample of ^{235}U will be only half as radioactive as it is today. In another 704 million years it will lose another half of its radioactivity, so it will be one-fourth as radioactive as it is today.

Radiation can be measured with a variety of units. A **becquerel (Bq)** measures the rate at which a sample of radioactive material decays; 1 Bq is equal to the decay of one atom per second. A **curie,** another unit of measure for radiation, is 37 billion decays per second. If a material has a radioactivity level of 100 curies and has a half-life of 50 years, the radioactivity level in 200 years will be

$$\frac{200 \text{ years}}{50 \text{ years/half-life}} = 4 \text{ half-lives}$$

100 curies → 50 curies (one half-life) → 25 curies (two half-lives) → 12.5 curies (three half-lives) → 6.3 curies (four half-lives)

You can gain more experience in working with these calculations in "Do the Math: Calculating Half-Lives."

Because spent fuel rods remain a threat to human health for 10 or more half-lives, they must be stored until they are no longer dangerous. At present, nuclear power plants are required to store spent fuel rods at the plant itself. Initially, all fuel rods were stored in pools of water at least 6 m (20 feet) deep. The water acts as a shield from radiation emitted by the rods. Currently, more than 100 sites around the country are storing spent fuel rods. However, some facilities have run out of pool storage space and are now storing them in lead-lined dry containers on land (FIGURE 36.3). Eventually, all of this material will need to be moved to a permanent radioactive waste disposal facility.

Disposing of radioactive waste is a challenge. This waste cannot be incinerated, safely destroyed using chemicals, shot into space, dumped on the ocean floor, or buried in an ocean trench because all of these options involve the potential for large amounts of radioactivity to enter the oceans or atmosphere. Therefore, at present, the only solution is to store it safely somewhere on Earth indefinitely. The physical nature of the storage site must ensure that the waste will not leach into the groundwater or otherwise escape into the environment. It must be far from human habitation in case of any accidents and secure against terrorist attack. In addition, the waste has to be transported to the storage site in a way that minimizes the risk of accidents or theft by terrorists.

In 1978, the U.S. Department of Energy began examining a site at Yucca Mountain, Nevada, 145 km (90 miles) northwest of Las Vegas, as a possible long-term repository for the country's spent nuclear fuel. The Yucca Mountain proposal has generated enormous protest and controversy. In 2006, the Department of Energy released a report confirming the soundness of the research supporting the Yucca Mountain site. However, a few years later, the Yucca Mountain project was canceled.

Nuclear energy raises a unique question of sustainability. It is a source of electricity that releases much less CO_2 than coal, and none during the actual generation of electricity. But some argue that the use of a fuel that consistently generates large quantities of high-level radioactive waste is not a sustainable practice. Many

Becquerel (Bq) Unit that measures the rate at which a sample of radioactive material decays; 1 Bq = decay of 1 atom or nucleus per second.

Curie A unit of measure for radiation; 1 curie = 37 billion decays per second.

FIGURE 36.3 **Storage of nuclear waste.** Nuclear waste is often stored in metal containers at the power plant where it was produced. *(AP Photo/Holtec International)*

critics of nuclear energy maintain that there will never be a place safe enough to store high-level radioactive waste.

Fusion Power

As we have seen, nuclear fission occurs when the nuclei of radioactive atoms are broken apart into smaller, lighter nuclei. **Nuclear fusion,** the reaction that powers the Sun and other stars, occurs when lighter nuclei are forced together to produce heavier nuclei. In the process, a great deal of heat is generated. The nuclear fusion reaction that is most promising for electricity generation is that of two hydrogen isotopes fusing together into a helium atom. As the reaction occurs, a small amount of mass is lost and an immense amount of energy is liberated.

Nuclear fusion seems to promise a seemingly unlimited source of energy that requires only hydrogen as an input and produces relatively small amounts of radioactive waste. Unfortunately, creating fusion on Earth requires a reactor that will heat material to temperatures 10 times those in the core of the Sun. These temperatures make containment extremely difficult. So far, the most promising techniques have involved suspending superhot material in a magnetic field, but the amount of energy required is greater than the energy output. Most experts believe that it will be several decades, or perhaps longer, before the promise of fusion power can be realized.

Summary of Nuclear Power and Comparison with Other Fuels

At present, our reliance on nuclear energy in the United States is subject to speculation and projections, but it is hard to know whether the positive aspects of nuclear energy will come to outweigh the risks of accidents and containment of radioactive waste. Indeed, in 2006, applications for new nuclear power plants in the United States were filed with the Nuclear Regulatory Commission for the first time in more than 2 decades. In 2012, licenses were approved for two nuclear power plants. It does appear that up to five new nuclear power plants may come online in the United States in the next decade.

TABLE 36.1 summarizes the major benefits and consequences of the conventional fuels we have discussed in this chapter.

Nuclear fusion A reaction that occurs when lighter nuclei are forced together to produce heavier nuclei.

Teaching Tip: Discussion Starter

Ask students to describe the promising aspects and the problems of nuclear fusion as a potential energy source on Earth. *The nuclear fusion reaction that is most promising for electricity generation is that of two hydrogen isotopes fusing together into a helium atom. As the reaction occurs, a small amount of mass is lost and an immense amount of energy is liberated. In this way, a large amount of energy can be obtained from easily obtainable hydrogen. To conduct this type of fusion on Earth would require a reactor that heats material to temperatures 10 times greater than that of the materials in the core of the Sun. Such high temperatures would be extremely difficult to contain.*

Teaching Tip: Remember

Students can use Table 36.1 to review nonrenewable energy. First ask students to read the table. Then read the 10 nonrenewable energy facts listed below. Ask students to identify the energy type that corresponds to the fact. You can do this either as a written review or as an oral exercise.

- Which nonrenewable energy source is known as the highest emitter of pollutants and greenhouse gas? *Coal*
- Which nonrenewable energy resource has the highest energy return on energy investment (EROEI)? *Coal*
- Which nonrenewable energy source emits no CO_2 during plant operation? *Nuclear*
- Which nonrenewable energy source is rarely used to generate electricity? *Oil*
- Which nonrenewable energy source will probably be much less available in the next 40 to 60 years? *Oil*
- Which nonrenewable energy source has many different uses including heating/cooling homes, cooking, and power production? *Natural gas*
- Which nonrenewable energy source would help a country become energy independent yet generate a highly radioactive waste product? *Nuclear*
- Which nonrenewable energy source is the most abundant in the United States and could potentially supply the United States with energy for 200 years? *Coal*
- Which nonrenewable energy source would have the highest upfront costs to become operational? *Nuclear*
- Which nonrenewable energy source has the highest electricity cost per kWh? *Nuclear*

TABLE 36.1 Comparison of nonrenewable energy fuels

Energy Type	Advantages	Disadvantages	Pollutant and greenhouse gas emissions	Electricity (cents/kWh)	Energy return on energy investment*
Oil/gasoline	• Ideal for mobile combustion (high energy/mass ratio) • Quick ignition/turn-off capability • Cleaner burning than coal	• Significant refining required • Oil spill potential effect on habitats near drilling sites • Significant dust and emissions from fossil fuels used to power earth-moving equipment • Human rights/ environmental justice issues in developing countries that export oil • Will probably be much less available in the next 40 years or so	• Second highest emitter of CO_2 among fossil fuels • Hydrocarbons • Hydrogen sulfide	• Relatively little electricity is generated from oil	4.0 (gasoline) 5.7 (diesel)
Coal	• Energy-dense and abundant—U.S. resources will last at least 200 years • No refining necessary • Easy, safe to transport • Economic backbone of some small towns	• Mining practices frequently risk human lives and dramatically alter natural landscapes • Coal power plants are slow to reach full operating capacity • A large contributing factor to acid rain in the United States	• Highest emitter of CO_2 among energy sources • Sulfur • Trace amounts of toxic metals such as mercury	5 cents/kWh	14
Natural Gas	• Cogeneration power plants can have efficiencies up to 60% • Efficient for cooking, home heating, etc. • Fewer impurities than coal or oil	• Risk of leaks/ explosions • Twenty-five times more effective as a greenhouse gas than CO_2 • Not available everywhere because it is transported by pipelines	• Methane • Hydrocarbons • Hydrogen sulfide	6–8 cents/kWh	8
Nuclear Energy	• Emits no CO_2 once plant is operational • Offers independence from imported oil • High energy density, ample supply	• Very unpopular; generates protests • Plants are very expensive to build because of legal challenges • Meltdown could be catastrophic • Possible target for terrorist attacks	• Radioactive waste is dangerous for hundreds of thousands of years • No long-term plan currently in place to manage radioactive waste • No air pollution during production	12–15 cents/kWh	8

*Estimates vary widely.

module

36

REVIEW

In this module, we have seen that the use of nuclear fuels for generating electricity has significant advantages and disadvantages. Nuclear energy is a relatively clean means of electricity generation, although fossil fuels are used in constructing nuclear power plants and mining and processing the uranium fuels. The possibility of accidents during plant operation and the difficulty of radioactive nuclear waste disposal are major environmental hazards of nuclear energy used for the generation of electricity. Neither of these issues has been satisfactorily resolved at present. Throughout the nuclear electricity generation cycle, carbon dioxide emissions are much less than during electricity generation from fossil fuels.

Module 36 AP® Review Questions

1. One g of U-235 produces approximately how much more energy than 1 g of coal?
   (a) 3,000 times
   (b) 19,000 times
   (c) 80,000 times
   (d) 400,000 times
   (e) 2,000,000 times

2. Control rods slow nuclear reactions by
   (a) reducing the amount of fuel available.
   (b) absorbing the heat produced.
   (c) increasing the rate heat transfers to the water.
   (d) absorbing excess neutrons.
   (e) adding additional barium to the reaction.

3. The process of fusion
   (a) splits atoms.
   (b) requires extremely high temperatures.
   (c) uses plutonium instead of uranium.
   (d) requires several radioactive elements.
   (e) does not produce radioactive waste.

4. Increased interest in nuclear power is due to
   (a) low energy costs.
   (b) lack of significant accidents.
   (c) low carbon dioxide emissions.
   (d) new solutions for waste disposal.
   (e) decreased energy independence.

5. For a sample of Ti-44 with a half-life of 63 years, how long until $\frac{1}{16}$ of the original amount is left?
   (a) 63 years
   (b) 126 years
   (c) 189 years
   (d) 252 years
   (e) 315 years

Answers to Module 36 AP® Review Questions

1. e
2. d
3. b
4. c
5. d

Teaching Tip: Beyond the Classroom

Take a field trip to a house or building that has been certified by Leadership in Energy and Environmental Design (LEED). Alternatively, have students research a LEED building. Construction that is certified by LEED has been rated according to a set of best practices and standards created by the United States Green Building Council for green building construction and maintenance. Have students generate a set of questions about the building before the tour. Suggestions and possible answers include the following:

- What type of LEED certification has this building been awarded? *There are three types of LEED certifications in order from lowest to highest: silver, gold, and platinum.*
- What four design standards has the building met in order to achieve certification? *Answers will vary.*
- Does the building have special insulation to reduce heating and cooling costs? How does it work? *Answers will vary.*
- What type of water conservation practices are in use in this building? *Answers will vary; however, LEED buildings must have dual flush toilets, rain barrels, and motion detecting water faucets.*
- In what way does the building use renewable energy resources? *Answers will vary.*

working toward sustainability

Meet TED: The Energy Detective

It is generally accepted that peer pressure influences behavior. When hotels placed small signs in bathrooms telling their guests that *other* guests were reusing their towels rather than insisting on receiving fresh clean towels each day, the need for towel laundering dropped drastically. Many studies have shown that when people receive feedback about their electricity consumption, they will often respond just as the hotel guests did. A number of electric companies have experimented with mailings that show homeowners how much electricity they use in comparison with neighbors in similar-size homes. However, if homeowners are using less electricity than their neighbors, they may not be inclined to reduce their use further.

For decades, environmentalists have dreamed about a magical device that could allow homeowners to receive an instantaneous reading of actual electricity use in the home. Today, environmental scientists have produced such devices. One example is The Energy Detective (TED), which provides a readout on a small device that can sit on the kitchen table or be viewed on a laptop computer. And there are many other similar devices and software packages available, including some that allow you to view your home electricity use on a laptop over the Internet.

TED and other such electricity monitoring devices provide an instantaneous readout of electricity use in your home. Suppose you are getting ready to head out for the evening. The refrigerator is plugged in, but is quiet, meaning that the compressor is not running at that exact moment. Before you turn out the last light and leave, you glance at TED and see that your house is drawing 500 watts. Wait a minute, that's not right! Then you remember that you left your Xbox 360 game console and 42-inch flat screen TV on in the other room. You run over and turn them both off. TED now reads that only 45 watts are being used in your home. But if everything is turned off, why is your house using 45 watts? It's probably due to the phantom load—unnecessary standby electricity—drawn by battery chargers, instant-on features on televisions, computers in sleep mode, and other electrical devices that are on even when you think they are off. Some TED owners have gone around their homes installing power strips that allow them to truly turn an appliance off and have seen their phantom loads drop as a result.

Since we know that all electricity use has environmental implications, a reduction in electricity use by any means is beneficial. The reductions that come from simple changes in behavior are some of the easiest to achieve.

Critical Thinking Questions

1. How does the energy meter described here compare with typical energy meters?
2. What are some major contributors to phantom load in a home and what prevents manufacturers from making electrical items that don't draw a phantom load?

References

Ayres, I., S. Raseman, and A. Shih. 2009. *Evidence from Two Large Field Experiments That Peer Comparison Feedback Can Reduce Residential Energy Usage.* Working Paper 15386, National Bureau of Economic Research.

Schor, J. 2004. *Born to Buy.* Scribner.

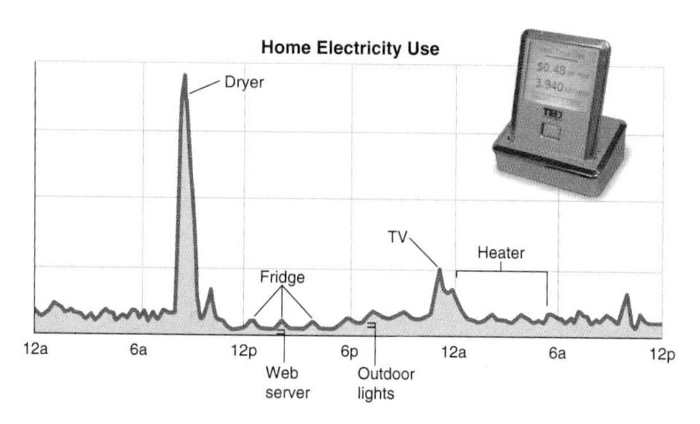

The Energy Detective. TED allows a user to instantaneously monitor electricity use in a home. *(Energy, Inc.)*

Suggested Answers to Critical Thinking Questions

1. A typical electricity meter shows the total amount of electricity used. If it is read once per month, the resident learns how much electricity was used in the past month. The TED meter provides an instantaneous measure of electricity usage. So if someone left an appliance or lights on when they were not being used, the TED meter would provide instantaneous feedback, which might lead someone to change his or her behavior. The person might wonder what was causing the peak in the TED usage graph and look around the house and possibly turn off or reduce the items using the electricity.

2. A phantom load or standby demand is electricity used by an item that allows it to have an "instant on" feature. Televisions, sound systems, and power supplies for phones and computers often have a phantom load. If you absolutely need a device to turn on instantly, it probably needs to be on "standby" at all times and thus will have a phantom load. But there are many devices that, when plugged in and not turned on, will draw zero electricity. Manufacturers don't pay for electricity—the consumer does. So manufacturers don't care if their appliances have a phantom load. The worst offender is the cable box, which is usually provided by the cable company. Some cable boxes use 20 W whether they are turned on or turned off. Video game consoles are another device that often has a large phantom load.

chapter
12 REVIEW

In this chapter, we have examined the nonrenewable fossil fuels that are used around the world and in the United States. We observed that energy use varies widely in different countries. Fossil fuel resources are used for different purposes and electricity draws 40 percent of U.S. energy demand. In the United States, oil, natural gas, and coal, in that order, meet 82 percent of the nation's energy requirements. Each fossil fuel has its own advantages and disadvantages. Nuclear fuels, which generate 20 percent of the electricity in the United States, have their own set of advantages and disadvantages. One reasonable conclusion is that all energy choices have adverse consequences.

Key Terms

Fossil fuel
Nonrenewable energy resource
Nuclear fuel
Commercial energy source
Subsistence energy source
Energy carrier
Turbine
Electrical grid
Combined cycle
Capacity
Capacity factor
Cogeneration
Combined heat and power
Coal
Petroleum
Crude oil
Oil sands
Bitumen
CTL (coal to liquid)
Energy intensity
Hubbert curve
Peak oil
Fission
Fuel rod
Control rod
Radioactive waste
Becquerel (Bq)
Curie
Nuclear fusion

Learning Objectives Revisited

Module 34 Patterns of Energy Use

- **Describe the use of nonrenewable energy in the world and in the United States.**

  Energy use has changed over time along with the level of industrial development in a country. The United States and the rest of the developed world have moved from a heavy reliance on wood and coal to other fossil fuels and nuclear energy. The developing world still relies largely on wood, charcoal, and animal waste.

- **Explain why different forms of energy are best suited for certain purposes.**

  Each source of energy is best suited for certain activities, and less well suited for others. Energy efficiency is an important consideration in determining the environmental impacts of energy use. In general, the energy source that entails the fewest conversions from its original form to its end use is likely to be the most efficient.

- **Understand the primary ways that electricity is generated in the United States.**

  Electricity generation plants convert the chemical energy of fuel into electricity. Coal, oil, natural gas, and nuclear fuels are the energy sources most commonly used for generating electricity. The electrical grid is a network of interconnected transmission lines that tie power plants together and link them with end users of electricity.

Module 35 Fossil Fuel Resources

- **Discuss the uses of coal and its consequences.**

  Coal is an energy-dense fossil fuel that is a major energy source in the generation of electricity. Coal combustion, however, is a major source of air pollution and greenhouse gas emissions.

- **Discuss the uses of petroleum and its consequences.**

  Petroleum includes both crude oil and natural gas. The United States uses more petroleum than any other

Teaching Tip: Measuring Your Impact

Choosing a Car: Conventional or Hybrid? Measuring Your Impact 12 asks students to determine the pros and cons of choosing a hybrid or a conventional vehicle. The link to this resource may be found by clicking on the link buttons in the TE-book, opening the TRFD, or logging onto the book's companion website bcs.whfreeman.com/friedlandapes2e (teacher login required).

 **Measuring Your Impact 12:** Choosing a Car: Conventional or Hybrid?

Exploring the Literature

Hinrichs, R.A., and M.H. Kleinbach. 2012. *Energy: Its Use and the Environment.* 5th ed. Cengage.

Nersesian, R. L. 2010. *Energy for the 21st Century.* 2nd ed. M. E. Sharpe.

Ristinen, R. A., and J. J. Kraushaar. 2006. *Energy and the Environment.* 2nd ed. John Wiley & Sons.

Toossi, R. 2008. *Energy and the Environment.* 2nd ed. Verve Publishers.

U.S. Department of Energy, Energy Information Administration. http://www.eia.doe.gov.

Wolfson, R. 2011. *Energy, Environment and Climate.* 2nd ed. W. W. Norton.

Download a printable version of this list by clicking on the link buttons in the TE-book, opening the TRFD, or logging onto the book's companion website bcs.whfreeman.com/friedlandapes2e (teacher login required).

 **Exploring the Literature 12**

Additional Multiple-Choice AP® Practice Exam

Download an additional Multiple-Choice AP® Environmental Science Practice Exam for Chapter 12 by clicking on the link buttons in the TE-book, opening the TRFD, or logging onto the book's companion website bcs.whfreeman.com/friedlandapes2e (teacher login required).

Multiple-Choice AP® Practice Exam 12

Answers to Chapter 12 AP® Exam Multiple-Choice Questions

1. d
2. d
3. a
4. d
5. e
6. b
7. a
8. e
9. c
10. b
11. c

Answers to Chapter 12 AP® Exam

Download the full answers to the Chapter 12 AP® Exam by clicking on the link buttons in the TE-book, opening the TRFD, or logging onto the book's companion website bcs.whfreeman.com/friedlandapes2e (teacher login required).

Answers to Chapter 12 AP® Exam

fuel, primarily for transportation. Petroleum produces air pollution as well as greenhouse gas emissions. Oil spills are a major hazard to organisms and habitat.

- **Discuss the uses of natural gas and its consequences.**
  Natural gas is a relatively clean fossil fuel. The presence of natural gas pipelines makes natural gas a convenient fuel for electricity generation, home heating, and manufacturing processes as well as fertilizer production. Hydraulic fracturing has greatly increased the availability of natural gas in the United States.

- **Discuss the uses of oil sands and liquefied coal and their consequences.**
  Oil sands and liquefied coals have the potential to become more important in the U.S. energy portfolio but both require relatively large energy inputs to be obtained and processed into a usable form.

- **Describe future prospects for fossil fuel use.**
  Most observers believe that oil production will begin to decline sometime in the next few decades. The transition away from oil will have important environmental consequences, depending upon how quickly it occurs and whether we move to renewable energy resources or alternative fossil fuels.

Module 36 Nuclear Energy Resources

- **Describe how nuclear energy is used to generate electricity.**
  In nuclear fission, a neutron strikes a relatively large atom such as uranium, and two or more smaller parts split off, releasing a great deal of energy. This energy is used to convert water into steam, which turns a turbine, which then turns a generator. A small amount of nuclear fuel can release a great deal of energy and generate a large quantity of electricity.

- **Discuss the advantages and disadvantages of using nuclear fuels to generate electricity.**
  All forms of energy have advantages and disadvantages. Considering only emissions during electricity generation, nuclear energy is relatively clean. However, the construction of nuclear power plants and the mining of uranium both use fossil fuels. The major environmental hazards of electricity generation from nuclear energy are the potential for accidents during plant operation and the challenges of radioactive waste disposal.

Chapter 12 AP® Environmental Science Practice Exam

Section 1: Multiple-Choice Questions

Choose the best answer for questions 1–11.

1. Which of the following is NOT a nonrenewable energy resource?
   (a) Oil
   (b) Coal
   (c) Natural gas
   (d) Wind
   (e) Nuclear fuels

2. The fact that global transfer of energy from fuels to electricity is about 35 percent efficient is mostly a consequence of
   (a) the Hubbert curve.
   (b) the law of conservation of matter.
   (c) the first law of thermodynamics.
   (d) the second law of thermodynamics.
   (e) the law of limiting factors.

3. Which of the following is the most fuel-efficient mode of transportation in terms of joules per passenger-kilometer?
   (a) Train
   (b) Bus
   (c) Airplane
   (d) Car with one passenger
   (e) Car with three passengers

4. Which of the following is NOT associated with the surface extraction of coal?
   (a) Low death rates among miners
   (b) Land subsidence and collapse
   (c) Large piles of tailings
   (d) Underground tunnels and shafts
   (e) Acid runoff into streams

5. Which of the following statements regarding petroleum is correct?
   I. It is formed from the decay of woody plants.
   II. It contains natural gas as well as oil.
   III. It migrates through pore spaces in rocks.
   (a) I, II, and III
   (b) I and III
   (c) II only
   (d) I and II
   (e) II and III

428 CHAPTER 12 ■ Nonrenewable Energy Resources

6. Which is a disadvantage of natural gas?
   (a) High sulfur emissions
   (b) Groundwater contamination
   (c) High carbon dioxide emissions
   (d) Low EROEI due to extraction processes
   (e) Significant waste disposal

7. Nuclear power plants produce electricity using energy from the radioactive decay of
   (a) uranium-235.
   (b) uranium-238.
   (c) uranium-239.
   (d) plutonium-235.
   (e) plutonium-238.

8. Currently, most high-level radioactive waste from nuclear reactors in the United States is
   (a) stored in deep ocean trenches.
   (b) buried in Yucca Mountain.
   (c) reprocessed into new fuel pellets.
   (d) chemically modified into safe materials.
   (e) stored at the power plant that produced it.

9. A radioactive isotope has a half-life of 40 years and a radioactivity level of 4 curies. How many years will it take for the radioactivity level to become 0.25 curies?
   (a) 80
   (b) 120
   (c) 160
   (d) 200
   (e) 240

10. Which of the following energy sources is responsible for the largest fraction of electricity generation in the United States?
   (a) Natural gas
   (b) Coal
   (c) Uranium
   (d) Oil
   (e) Wood

11. In 1969, M. King Hubbert published a graph known as the Hubbert curve. This graph shows
   (a) the amount of nuclear fuel available in North America.
   (b) the amount of nuclear fuel available in the world.
   (c) the point at which world oil production will reach a maximum and the point at which we will run out of oil.
   (d) the point at which world oil production will increase.
   (e) the coal reserves found in the United States, China, and Russia.

Section 2: Free-Response Questions

Write your answer to each part clearly. Support your answers with relevant information and examples. Where calculations are required, show your work.

1. Many college students have a mini fridge in their dorm room. A standard mini fridge costs roughly $100, uses about 100 watts of electricity, and can be expected to last for 5 years. The refrigerator is plugged into an electrical socket 24 hours a day, but is usually running only about 12 hours a day. Assume that electricity costs $0.10/kWh.
   (a) Calculate the lifetime monetary cost of owning and operating the refrigerator. (2 points)
   (b) Assume that the electricity used to power the refrigerator comes from a coal-burning power plant. One metric ton of coal contains 29.3 GJ (8,140 kWh) of energy. Because of the inefficiency of electricity generation and transmission, only one-third of the energy in coal reaches the refrigerator. How many tons of coal are used to power the refrigerator during its lifetime? (2 points)
   (c) Assume that 15 percent of the mass of the coal burned in the power plant ends up as coal ash, a potentially toxic mixture that contains mercury and arsenic. How many tons of coal ash are produced as a result of the refrigerator's electricity use over its lifetime? (2 points)
   (d) What externalities does your answer from part (a) not include? Describe one social and one environmental cost associated with using this appliance. (2 points)
   (e) Describe two ways a college student could reduce the electricity use associated with having a mini fridge in his or her dorm room. (2 points)

2. A number of U.S. electric companies have filed applications with the Nuclear Regulatory Commission for permits to build new nuclear power plants to meet future electricity demands.
   (a) Explain the process by which electricity is generated by a nuclear power plant. (2 points)
   (b) Describe the two nuclear accidents that occurred in 1979 and 1986, respectively, that led to widespread concern about the safety of nuclear power plants. (2 points)
   (c) Discuss the environmental benefits of generating electricity from nuclear energy rather than coal. (2 points)
   (d) Describe the three types of radioactive waste produced by nuclear power plants and explain the threats they pose to humans. (2 points)
   (e) Discuss the problems associated with the disposal of radioactive waste and outline the U.S. Department of Energy's proposal for its long-term storage. (2 points)

Answers to Chapter 12 AP® Exam Free-Response Questions

1. (a) (100 W × $0.10/kWh × 12 hrs/day × 1 kW/1,000 W × 365 days/year × 5 years) + $100 initial cost = $319 lifetime cost
(b) 2,190 kWh × 3 [efficiency factor] = 6,570 kWh lifetime energy use
6,570 kWh ÷ 8,141 kWh/ton = 0.81 tons of coal
(c) 0.81 tons × 0.15 = 0.12 tons (or 240 pounds)
(d) Externalities would include all costs not paid by the company or the consumer in the extraction, transportation, and end uses of the coal. Answers will vary, but may include mention of respiratory illness, habitat destruction, and climate change due to increased emissions of carbon dioxide.
(e) Answers will vary, but may include mention of buying an Energy Star certified appliance or sharing a refrigerator with roommates.

Answers continue in the Answer Appendix on page ANS-9.

chapter

13 Achieving Energy Sustainability

PD Overview

This chapter looks at one of the most significant topics in the AP® environmental science course: renewable energy. The chapter also addresses the issue of energy conservation and efficiency. Major renewable energy sources considered include biomass, solar, hydroelectric, geothermal, wind, and fuel cell technology. As you go through the chapter, make sure students are confident that they understand the advantages and disadvantages of each renewable energy source.

Module 37: Conservation, Efficiency, and Renewable Energy

When considering energy use, energy conservation and energy efficiency are crucial factors to consider. The discussion of renewable energy begins with a look at conservation and efficiency. The module then explores the range of renewable energy resources that are available.

Module 38: Biomass and Water

This module presents two important renewable energy sources: biomass and water, which are currently major sources of renewable energy in both the developed and the developing worlds.

Module 39: Solar, Wind, Geothermal, and Hydrogen

This module explores energy from the Sun and wind, which represent the fastest growing forms of energy development throughout the world.

Module 40: Planning Our Energy Future

Although renewable energy is a more sustainable energy choice than nonrenewable energy, all forms of energy have an impact on the environment. Because all energy choices have environmental consequences, minimizing energy use through conservation and efficiency is the best approach to energy use.

Alignment to AP® Environmental Science Course Description

AP® Outline	Module	
V. Energy Resources and Consumption (10–15%)		
F. Energy Conservation	Module 37	Conservation, Efficiency, and Renewable Energy
	Module 40	Planning Our Energy Future
G. Renewable energy	Module 38	Biomass and Water
	Module 39	Solar, Wind, Geothermal, and Hydrogen

Chapter Learning Objectives

After completing this chapter students will be able to

- describe strategies to conserve energy and increase energy efficiency.
- explain differences among the various renewable energy resources.
- describe the various forms of biomass.
- explain how energy is harnessed from water.
- list the different forms of solar energy and their application.
- describe how wind energy is harnessed and its contemporary uses.
- discuss the methods of harnessing the internal energy from Earth.
- explain the advantages and disadvantages of energy from hydrogen.
- discuss the environmental and economic options we must assess in planning our energy future.
- consider the challenges of a renewable energy strategy.

Chapter 13 Pacing Guide

This pacing guide is based on a schedule with 120 sessions of 50 minutes each before the AP® Exam. If you have a different number of sessions before the exam, you can modify the pacing to suit your needs. If you have additional time, consider incorporating quizzes, released AP® Environmental Science free-response and multiple-choice questions, or additional activities.

Module	Standard Schedule Days	Block Schedule Days
Module 37	1	½
Module 38	2	1
Module 39	3	1½
Module 40	1	½
Assessment	1	½

Chapter 13 Resources

The link to these resources can be found by clicking on the link buttons in the Teacher's e-Book (TE-book), opening the Teacher's Resource Flash Drive (TRFD), or logging onto the book's companion website bcs.whfreeman.com/friedlandapes2e (teacher login required).

- **PowerPoint Presentation 13**
- **Optimized Art PowerPoint and JPEG Files 13**
- **Do the Math Videos**
- **Measuring Your Impact 13:** Choosing a Light Bulb
- **Lab 13.1:** Solar Energy
- **Chapter 13 Web Resources**
- **Exploring the Literature 13**
- **Answers to Chapter 13 AP® Exam**
- **Multiple-Choice AP® Practice Exam 13**
- **Answers to Unit 6 AP® Exam**
- **Unit 6 Additional Free-Response Question**
- **Answer to Science Applied 6 Free-Response Question**

Free-Response Questions from Previous AP® Environmental Science Exams

Free-response questions from prior AP® Environmental Science Exams are available on the AP® course website: https://apstudent.collegeboard.org/apcourse/ap-environmental-science/exam-practice.

Students should be able to answer all of the questions listed below with material learned in this and previous chapters. When a question requires students to understand material from multiple chapters, the question will be listed in the last chapter required to complete the entire question. Questions marked with an asterisk (*) are from exams with released multiple-choice questions. You may want to save these questions until the end of the year so you can give your students a complete released exam for practice. Questions marked with double asterisks (**) require math to calculate a problem. Look for references to these questions throughout the chapter.

Year	Question	Content
1999	2	• Renewable versus nonrenewable energy sources • Total resource use per capita in developed and developing countries • Sustainable resource use • Policies that help and hinder sustainable resource use
2001	1**	• The amount and cost of natural gas needed to heat a house • Advantages and disadvantages of heating with wood
2004	2**	• Costs and benefits of wind-generated electricity
2006	1	• Environmental benefit and cost of using photovoltaic system • Promotion of passive solar design
2008*	1**	• Biodiesel from microalgae and soybeans • Problems associated with producing fuel from corn
2009	2**	• Use of methane to produce electricity
2011	3	• Geothermal energy • Per capita emissions of CO_2 versus total emissions • Production of fuel from either sugarcane or tar sands
2012	2**	• Alternative energy sources to heat a school • Calculation of forest needed to offset CO_2 emissions
2013	2**	• Battery electric vehicles versus internal combustion engine vehicles

PD Chapter 13 Overview

Watch the video overview of Chapter 13 (for teachers) by clicking on the link buttons in the TE-book, opening the TRFD, or logging onto the book's companion website bcs.whfreeman.com /friedlandapes2e (teacher login required).

TRM PowerPoint Presentation 13

Download the PowerPoint presentation for Chapter 13 by clicking on the link buttons in the TE-book, opening the TRFD, or logging onto the book's companion website bcs.whfreeman.com /friedlandapes2e (teacher login required).

William Kamkwamba's cousin climbs one of the wind turbines that William built.
(Lucas Oleniuk/Toronto Star via Getty Images)

chapter 13

Achieving Energy Sustainability

Energy from the Wind

In a small village in the African nation of Malawi, a 14-year-old boy named William Kamkwamba and his family did not have enough to eat. A famine gripped his country, and his family could not pay the tax required in order to send him to school. School taxes are common in many parts of Africa, and a child whose parents cannot pay the tax cannot attend school. So instead of attending school, he spent his days in a public library funded by the U.S. government, trying to teach himself. In the library, he studied one book over and over: a textbook titled *Using Energy*. The cover of the book featured a series of windmills. Although William had never seen a windmill, within months he was building his own from abandoned bicycles and old parts he found in scrap heaps. William used the fundamentals of physics he learned from *Using Energy* and his inherent skills at tinkering and fixing things. He did not have any teachers or mentors but he did rely on assistance from some of his friends. He worked hard and made many attempts to construct something that in his world was seemingly impossible. At first his neighbors thought he was mentally disturbed or was even practicing magic. But when he was able to illuminate a small light bulb at the top of what they had called his "junk" tower, people rushed from great distances to see it, and he became a local

Although William had never seen a windmill, within months he was building his own from abandoned bicycles and old parts he found in scrap heaps.

Teaching Tip: Chapter Opening Case

Have students read the chapter opening case, "Energy from the Wind," which discusses how William Kamkwamba, a young man in Malawi, built a windmill out of local resources. Many people in his village thought he was either mentally unbalanced or practicing magic. However, when his windmill first generated enough electricity to light up a light bulb, they began to think differently.

Teaching Tip: Beyond the Classroom

As a summer assignment or during a school break, have students read the *New York Times* bestseller *The Boy Who Harnessed the Wind* by William Kamkwamba, the subject of the chapter opening case. After reading this book, students can write a one-page reaction to the challenges Kamkwamba faced. Students can also discuss the effects of bringing electricity to the village. How did Kamkwamba's windmill change the lives of his family and other villagers?

Teaching Tip: Beyond the Classroom

Ask students to research wind farm projects in your state or region. Ask them to produce a table that includes information about the capacity in gigawatts and the actual or expected number of gigawatt-hours each installation will produce per year. Then ask them to identify the largest capacity wind farm project in the United States and compare the local farms to that one.

hero. William had generated electricity without any conventional fuel and far from the nearest power plant. Because there were no visible inputs like fuel and no waste piles or pollution outputs, in many ways it did seem like magic. William used the electricity he generated from wind to light his house and charge cell phones, and eventually to irrigate his family's crops.

The use of windmills, also known as wind turbines, to generate electricity is growing in both the developing and developed worlds. The mechanics of how to build a windmill are widely discussed in online sources including YouTube, *Wikipedia*, and "how to build it" instructional videos. A few years after he built his first windmill with the help of only one book, and after learning about the vast resources available over the Internet that he did not have access to previously, William exclaimed to Jon Stewart on *The Daily Show*, "Where was this Internet when I needed it?" William recently graduated from Dartmouth College, where he majored in environmental studies, and has cowritten a book, *The Boy Who Harnessed the Wind*. It has sold thousands of copies and has been adopted as summer reading in high schools and colleges around the United States and elsewhere in the world.

Sources:
W. Kamkwamba and B. Mealer, *The Boy Who Harnessed the Wind* (Harper, 2009);
R. Wolfson, *Energy, Environment and Climate*, 2nd ed. (Norton, 2012).

Throughout this book we have discussed sustainability as the foundation of the environmental health of our planet. Sustainability is particularly important to consider with respect to energy because energy is a resource humans cannot live without and energy use often has many consequences for the environment. Currently, only a very small fraction of the energy we use, particularly in the developed world, comes from renewable resources. While developing renewable energy resources is an important step, achieving energy sustainability will require us to rely as much, or more, on reducing the amount of energy that we use.

In Chapter 12 we discussed the finite nature of traditional energy resources such as fossil fuels and the environmental consequences of their use. In this chapter we outline the components of a sustainable energy strategy, beginning with ways to reduce our use of energy through conservation and increased efficiency. We define renewable forms of energy and discuss an important carbon-based energy resource, biomass, as well as energy that is obtained from flowing water. We then describe innovations in obtaining energy from non-carbon-based resources such as the Sun, wind, internal heat from Earth, and hydrogen. We conclude the chapter with a discussion of our energy future.

Teaching Tip: Review

Chapter 13 is a good time to review the first and second laws of thermodynamics. Students should write a simple definition of each law and give two examples. (Sample answers are provided in italics.)

	First law of thermodynamics	Second law of thermodynamics
Definition	*Energy is neither created nor destroyed, but can change from one form to another.*	*Whenever energy is converted from one form into another, some energy is lost.*
Example	*The potential energy contained in firewood never goes away, but it is transformed into heat energy permeating a room when the wood is burned in a fireplace.*	*When using an incandescent light bulb, 5 percent of the energy entering the lightbulb is converted to useful light energy and the remaining 95 percent is lost to the environment as heat.*

module

37

Conservation, Efficiency, and Renewable Energy

In any discussion of energy use—whether renewable or nonrenewable—energy conservation and increased energy efficiency rank among the most crucial factors to consider. We begin our discussion of renewable energy with a look at conservation and efficiency. We will then explore the range of renewable energy resources that are available.

Learning Objectives

After reading this module you should be able to

- describe strategies to conserve energy and increase energy efficiency.
- explain differences among the various renewable energy resources.

We can use less energy through conservation and increased efficiency

A truly sustainable approach to energy use must incorporate both *energy conservation* and energy efficiency. Conservation and efficiency efforts save energy that can then be used later, just as you might save money in a bank account to use later when the need arises. In this sense, conservation and efficiency are sustainable energy "sources."

Energy conservation and energy efficiency are the least expensive and most environmentally sound options for maximizing our energy resources. In many cases, they are also the easiest approaches to implement because they require fairly simple changes to existing systems rather than a switch to a completely new technology. In this section we will examine ways to achieve both objectives.

Energy Conservation and Efficiency

Energy conservation means finding and implementing ways to use less energy. As we saw in Chapters 2 and 12, increasing energy efficiency means obtaining the same work from a smaller amount of energy. Energy conservation and energy efficiency are closely linked. One can conserve energy by not using an electrical appliance; doing so results in less energy consumption. But one can also conserve energy by using a more efficient appliance—one that does the same work but uses less energy.

Conservation

FIGURE 37.1 lists some of the ways that an individual might conserve energy, including lowering the household thermostat during cold months, consolidating errands in order to drive fewer miles, or turning off a computer when it is not being used. On a larger scale, a government might implement energy conservation measures that encourage or even require individuals to adopt strategies or habits that use less energy. One such top-down approach is to improve the availability of public transportation. Governments can also facilitate energy conservation by taxing electricity, oil, and

Energy conservation Finding and implementing ways to use less energy.

Teaching Tip: Journal Prompt

Ask students to take the role of a mayor or congressional representative with a mandate to reduce energy consumption. They should describe ways in which government policy can encourage people to reduce energy consumption. Examples might include improving public transportation, instituting a tax, or providing a subsidy. For example, the city might build a subway system or institute a carpooling initiative. Students should describe the feasibility of the initiative and be ready to read their journal entry to the class.

AP® Exam Tip

Numerous past free-response questions have asked students to suggest ways they or others might practice conservation in their daily lives.

Teaching Tip: Activity

Ask students to create and fill in a table like the one on the right. Ask them to consider ways of reducing energy use in each of the three categories. (Suggested answers are provided in italics.)

Home	Transportation	Electronic Devices
Replace single-pane windows with double- or triple-pane windows	*Ride a bike to school or work*	*Reduce phantom drain*
Turn down thermostat	*Carpool with coworkers*	*Use electronic devices less*
Weatherize anywhere cold air or hot air is entering home	*Take mass transit*	*Buy Energy Star or the most efficient model when replacing devices*

Home

- Weatherize (insulate, seal gaps).
- Turn thermostat down in winter, up in summer.
- Reduce use of hot water (do laundry in cold water/take shorter showers).
- Replace incandescent bulbs with compact fluorescents or LEDs.

Transportation

- Walk or ride a bike.
- Take public transportation.
- Carpool.
- Consolidate trips.

Electrical and electronic devices

- Buy Energy Star devices and appliances.
- Unplug when possible or use a power strip.
- Use a laptop rather than a desktop computer.

FIGURE 37.1 **Reducing energy use.** There are many ways individuals can reduce their energy use in and outside the home. *(top: Imagenet/Shutterstock; middle: Lawrence Manning/Corbis; bottom: W. H. Freeman & Company/Worth Publishers)*

natural gas, since higher taxes discourage use. Alternatively, governments might offer rebates or tax credits for retrofitting a home or business so it will operate on less energy. Some electric companies bill customers with a **tiered rate system** in which customers pay a low rate for the first increment of electricity they use and pay higher rates as use goes up. All of these practices encourage people to reduce the amount of energy they use.

As we saw in Chapter 12, the demand for energy varies with time of day, season, and weather. When electricity-generating plants are unable to handle the demand during high-use periods, brownouts or blackouts may occur. To avoid this problem, electric companies must be able to provide enough energy to satisfy **peak demand,** the greatest quantity of energy used at any one time. Peak demand may be several times the overall average demand, which means that substantially more energy must be available than is needed under average conditions. To meet peak demand for electricity, electric companies often keep backup sources of energy available—primarily fossil fuel–fired generators.

Therefore, an important aspect of energy conservation is the reduction of peak demand, which would make it less likely that electric companies will have to build excess generating capacity that is used only sporadically. One way of reducing peak demand is to set up a variable price structure under which customers pay less to use electricity when demand is lowest (typically in the middle of the night and on weekends) and more when demand is highest. This approach helps even out the use of electricity, which both reduces the burden on the generating capacity of the utility and rewards the electric consumer at the same time.

The second law of thermodynamics tells us that whenever energy is converted from one form into another, some energy is lost as unusable heat. In a typical thermal fossil fuel or nuclear power plant, only about one-third of the energy consumed goes to its intended purpose; the rest is lost during energy conversions. We need to consider these losses in order to fully account for all energy conservation savings. So, the amount of energy we save is the sum of both the energy we did not use together with the energy that would have been lost in converting that energy into the form in which we would have used it. For example, if we can reduce our electricity use by 100 kWh, we may actually be conserving 300 kWh of an energy resource such as coal, since we save both the 100 kWh that we decide not to use and the 200 kWh that would have been lost during the conversion process to make the 100 kWh available to us.

Efficiency

Modern changes in electric lighting are a good example of how steadily increasing energy efficiency results in overall energy conservation. Compact fluorescent light bulbs use one-fourth as much energy to provide the same amount of light as incandescent bulbs. LED (light-emitting diode) light bulbs are even more efficient; they use one-sixth as much energy as incandescent bulbs. Over time, the widespread adoption of these efficient bulbs will result in substantially less energy used to provide lighting.

Another way in which consumers can increase energy efficiency is by switching to products that meet the efficiency standards of the Energy Star program set by the U.S. Environmental Protection Agency. For example, an Energy Star air conditioner may use 0.2 kilowatt-hours (kWh, or 200 watt-hours) less electricity per hour than a non–Energy Star unit. In terms of cost, a single consumer may save only 2 to 5 cents per hour by switching to an Energy Star unit. However, if 100,000 households in a city switched to Energy Star air conditioners, the city would reduce its energy use by 20 MW, or 4 percent of the output of a typical power plant. "Do the Math: Energy Star" on page 436 shows you how to calculate Energy Star savings.

Tiered rate system A billing system used by some electric companies in which customers pay higher rates as their use goes up.

Peak demand The greatest quantity of energy used at any one time.

more **math** practice:

Graphing

Have students research the grid energy demands for your state. If you can't find the daily energy demand of your state use the data in the table provided below. Have students create a line graph depicting energy demands for a 24-hour period. Place hours on the *x* axis starting at midnight and finishing at midnight 24 hours later. Place amount of energy in gigawatts (GW) on the *y* axis. (See sample graph below.)

After the graph is complete, ask students the following questions. (Answers are provided in italics for the sample data and graph.)

- What is the 4-hour range where peak usage occurs? *Peak usage occurs between 2 PM to 6 PM.*
- When is energy use at its lowest? *The energy demand is 11 GW at 4 AM. It is the lowest power demanded in a 24-hour period.*
- What happens if the power company doesn't meet the peak power demands? *If the power company doesn't meet peak power demands, there could be a brownout or a blackout.*

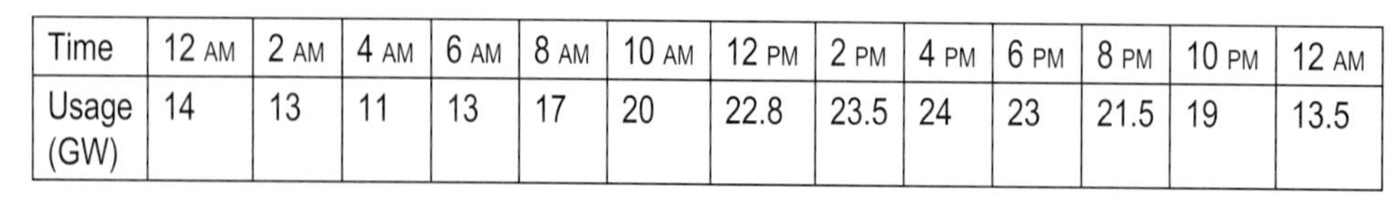

Time	12 AM	2 AM	4 AM	6 AM	8 AM	10 AM	12 PM	2 PM	4 PM	6 PM	8 PM	10 PM	12 AM
Usage (GW)	14	13	11	13	17	20	22.8	23.5	24	23	21.5	19	13.5

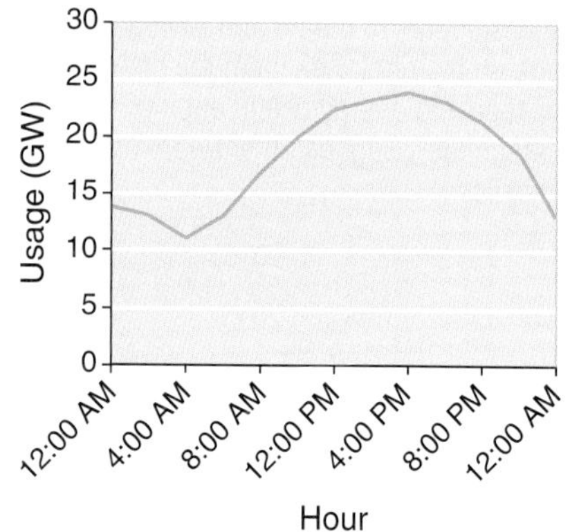

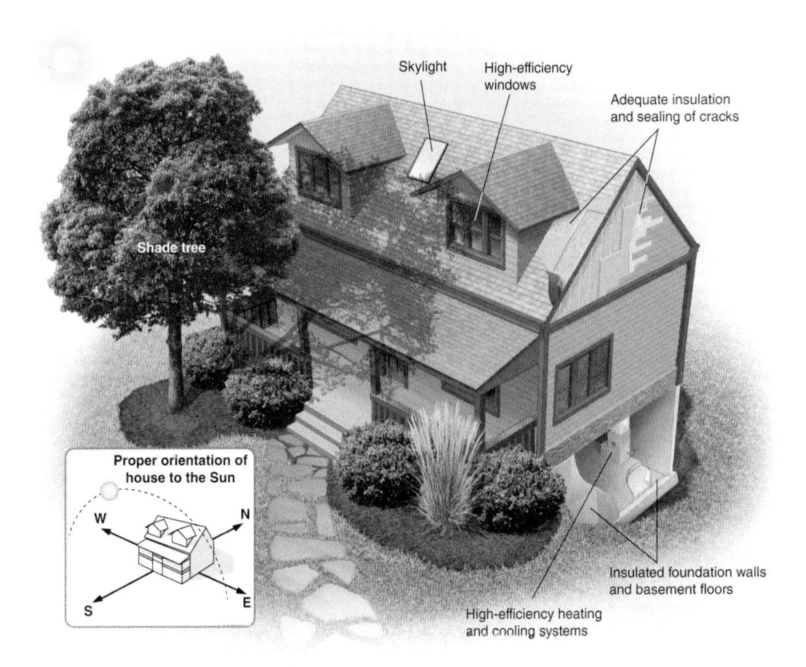

FIGURE 37.2 **An energy-efficient home.** A sustainable building design incorporates proper solar orientation and landscaping as well as insulated windows, walls, and floors. In the Northern Hemisphere, a southern exposure allows the house to receive more direct rays from the Sun in winter when the path of the Sun is in the southern sky.

Sustainable Design

Sustainable design can improve the efficiency of the buildings and communities in which we live and work. FIGURE 37.2 shows some key features of sustainable design applied to a single-family dwelling. Insulating foundation walls and basement floors, orienting a house properly in relation to the Sun, and planting shade trees in warm climates are all appropriate design features. As we saw in Chapter 10, good community planning also conserves energy. Building houses close to where residents work reduces reliance on fossil fuels used for transportation, which in turn reduces the amount of pollution and carbon dioxide released into the atmosphere.

Buildings consume a great deal of energy for cooling, heating, and lighting. Many sustainable building strategies rely on **passive solar design,** a construction technique designed to take advantage of solar radiation without the use of active technology. FIGURE 37.3 illustrates key features of passive solar design. Passive solar design stabilizes indoor temperatures without the need for pumps or other mechanical devices. For example, in the Northern Hemisphere, constructing a house with windows along a south-facing wall allows the Sun's rays to penetrate and warm the house, especially in winter when the Sun is more prominent in the southern sky. Double-paned windows insulate while still

Passive solar design Construction designed to take advantage of solar radiation without active technology.

Type and brand	Incandescent bulb Philips	Compact fluorescent EcoSmart	LED Cree
Wattage	*60 watts*	*17 watts*	*9.5 watts*
Estimated life expectancy	*0.9 years*	*9.1 years*	*22.8 years*
Estimated yearly energy cost based on 3 hours per day	*$7.23*	*$1.69*	*$1.14*
Cost of individual bulb	*$0.80*	*$1.50*	*$7.00–$10.00*
Brightness (lumens)	*860 lumens*	*800 lumens*	*800 lumens*

Teaching Tip: Discussion Starter

Ask students: How can building design contribute to energy conservation and efficiency? *A sustainable building design considers a home as a system and attempts to minimize energy inputs and losses from that system. It incorporates proper solar orientation and landscaping as well as insulated windows, walls, and floors.*

Practicing Science: Demo

In this demonstration, students compare incandescent, compact fluorescent, and LED light bulbs. Find light bulbs that can work in the same light fixture. You will also need an electricity usage monitor that plugs into a wall outlet and provides a socket for your lamp.

For each bulb, measure and record the wattage. Have students note the cost and life expectancy of the bulb, which should be listed on the package. Finally, use a thermometer to record the temperature of the bulb, holding the thermometer 3 cm away from the bulb for 5 minutes.
Record the data in a table like the one below. (Sample answers are provided in italics.)

Ask students to answer the following questions about the three different types of light bulbs:

- Which light bulb uses the most wattage? Does the temperature of the bulbs tell us anything about the efficiency of the light bulb used? *Incandescent light bulbs use the most watts. They will also have the highest temperature reading on the thermometer because the filament that produces the light also produces wasted heat energy. You should notice that LED light bulbs do not produce much heat and use only about 9 watts of electricity to generate the same amount of light as a 60-watt incandescent bulb.*
- Considering the purchase price of the light bulb, its life expectancy, and the cost of electricity used to run the bulb during its lifetime, do you think LED bulbs save consumers money? Explain your answer. *Although an LED bulb costs more to purchase than other types of bulbs, it uses less electricity over its lifetime and provides the same amount of light. Even when you include the purchase price of the LED bulb, the consumer ends up spending less money over the lifetime of the LED bulb.*

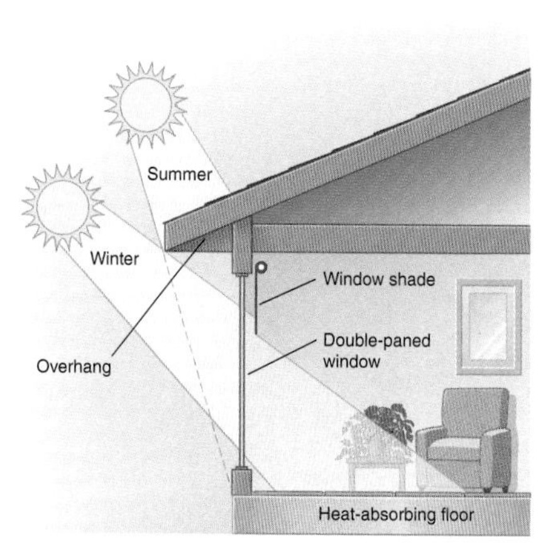

FIGURE 37.3 **Passive solar design.** Passive solar design uses solar radiation to maintain indoor temperature. Roof overhangs make use of seasonal changes in the Sun's position to reduce energy demand for heating and cooling. In winter, when the Sun is low in the sky, it shines directly into the window and heats the house. In summer, when the Sun is higher in the sky, the overhang blocks incoming sunlight and the room stays cool. High-efficiency windows and building materials with high thermal inertia are also components of passive solar design.

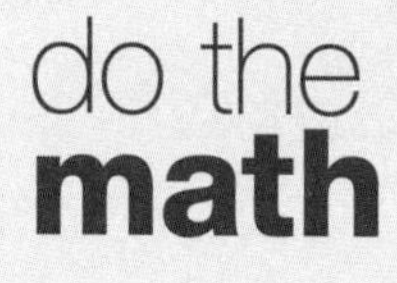

do the math

Energy Star

Thinking clearly about energy efficiency and energy conservation can save you a lot of money in some surprising ways. If you are saving money on your electric bill, you are also saving energy and reducing the emission of pollutants. Consider the purchase of an air conditioner. Suppose you have a choice: an Energy Star unit for \$300 or a standard unit for \$200. The two units have the same cooling capacity but the Energy Star unit costs 5 cents per hour less to run. If you buy the Energy Star unit and run it 12 hours per day for 6 months of the year, how long does it take to recover the \$100 extra cost?

You would save

$$\$0.05/\text{hour} \times 12 \text{ hours/day} = \$0.60/\text{day}$$

Six months is about 180 days, so in the first year you would save

$$\$0.60/\text{day} \times 180 \text{ days} = \$108$$

Spending the extra \$100 for the Energy Star unit actually saves you \$8 in just 1 year of use. In 3 years of use, the savings will more than pay for the entire initial cost of the unit ($3 \times \$108 = \$324$), and after that you pay only for the operating costs.

Your Turn You are about to invest in a 66-inch flat screen TV. These TVs come in both Energy Star and non–Energy Star models. The cost of electricity is \$0.15 per kilowatt-hour, and you expect to watch TV an average of 4 hours per day.

1. The non–Energy Star model uses 0.5 kW (half a kilowatt). How much will it cost you per year for electricity to run this model?
2. If the Energy Star model uses only 40 percent of the amount of electricity used by the non–Energy Star model, how much money would you save on your electric bill over 5 years by buying the efficient model?

 do the **math**

Answer to Your Turn

The cost of electricity is \$0.15 per kWh and we wish to run the TV for 4 hours per day. Calculate the cost to run the non-Energy Star compliant TV first.

1. Non-Energy Star TV

This TV uses 0.5 kW. Therefore:

$$0.5 \text{ kW} \times \left(\frac{4 \text{ hours}}{\text{day}}\right) \times \left(\frac{365 \text{ days}}{\text{year}}\right) \times \left(\frac{\$0.15}{\text{kWh}}\right) = \$109.50/\text{year}$$

2. Energy Star TV

This TV uses 40 percent of the energy used by the first TV, meaning we save 60 percent of the energy. If we save 60 percent of the energy, we save 60 percent of the cost. Therefore, in one year:

$$0.60 \times \left(\frac{\$109.50}{\text{year}}\right) = \$65.70/\text{year}$$

So over 5 years we would save:

$$5 \text{ years} \times \$65.70/\text{year} = \$328.50 \text{ savings}$$

FIGURE 37.4 **The California Academy of Sciences.** The sustainable design of this San Francisco research institution maximizes the use of natural light and ventilation. The building generates much of its own electricity with solar panels on its roof and captures water in its rooftop garden. *(Nancy Hoyt Belcher/Alamy)*

allowing incoming solar radiation to warm the house. Carefully placed windows also allow the maximum amount of light into a building and reduce the need for artificial lighting. Dark materials on the roof or exterior walls of a building absorb more solar energy than light-colored materials, further warming the structure. Conversely, using light-colored materials on a roof reflects heat away from the building, which keeps it cooler. In summer, when the Sun is high in the sky for much of the day, an overhanging roof helps block out sunlight during the hottest period, which makes the indoor temperature cooler and reduces the need for ventilation fans or air conditioning. Window shades can also reduce solar energy entering the house.

To reduce demand for heating at night and for cooling during the day, builders can use construction materials that have high *thermal mass*. **Thermal mass** is a property of a building material that allows it to retain heat or cold. Materials with high thermal mass stay hot once they have been heated and cool once they have been cooled. Stone and concrete have high thermal mass, whereas wood and glass do not; think of how a cement sidewalk stays warm longer than a wooden boardwalk after a hot day. A south-facing room with stone walls and a stone floor will heat up on sunny winter days and retain that heat long after the Sun has set.

Although building a house into the side of a hill or roofing a building with soil and plants are less-common approaches, these measures also provide insulation and reduce the need for both heating and cooling. While "green roofs"—roofs with soil and growing plants—are somewhat unusual in the United States, many European cities, such as Berlin, have them on new or rebuilt structures. They are especially common on high-rise buildings in downtown areas that have little natural plant cover. These green roofs cool and shade the buildings and the surrounding environment. And the addition of plants to an urban environment also improves overall air quality.

The use of recycled building materials is another method of energy conservation. Recycling reduces the need for new construction materials, which reduces the amount of energy required to produce the components of the building. For example, many buildings now use recycled denim insulation in the walls and ceilings, and fly ash (a byproduct recovered from coal-fired power plants) in the foundation.

Many homes constructed today incorporate some or all of these sustainable design strategies, but it is possible to achieve energy efficiency even in very large buildings. The building that houses the California Academy of Sciences in Golden Gate Park in San Francisco is a showcase for several of these sustainable design techniques (FIGURE 37.4). This structure, which incorporates a combination of passive solar design, radiant heating, solar panels, and skylights, actually uses

Thermal mass A property of a building material that allows it to maintain heat or cold.

Practicing Science: Demo

In this demonstration, students use different materials to see which has the highest thermal mass. Materials might include red brick, gray stone, glass, or wood. Make sure students handle hot materials with gloves and follow all appropriate safety precautions.

Thermal mass is a property that allows a building material to retain heat or cold. Materials with high thermal mass stay hot once they have been heated and cool once they have been cooled. Stone and concrete have high thermal mass, whereas wood and glass have low thermal mass.

Choose several materials and put them in constant sunlight for 30 minutes. You can do this near a large window inside the classroom or you can set up the experiment outside. After 30 minutes the temperature will be greater. Place a thermometer on the materials and record the temperatures. Block the sunlight by closing the blinds to the window or putting up an umbrella if you are outside. Wait for 10 minutes and record the temperatures again. Calculate the change. Ask students to predict which material will store the most heat over 30 minutes. Measure the temperatures once more after 30 minutes of blocked sunlight and check this reading against student predictions.

30 percent less energy than the amount permitted under national building energy requirements. Natural light fills 90 percent of the office space and many of the public areas. Windows, blinds, and skylights open as needed to allow air to circulate, capturing the ocean breezes and ventilating the building. Recycled denim insulation in the walls and a soil-covered rooftop garden provide insulation that reduces heating and cooling costs. As an added benefit, the living green roof grows native plants and captures 13.6 million liters (3.6 million gallons) of rainwater per year, which is then used to recharge groundwater stores.

In addition to these passive techniques, the designers of this building incorporated active technologies that further reduce its use of energy. An efficient radiant heating system carries warm water through tubes embedded in the concrete floor, using a fraction of the energy required by a standard forced-air heating system. To produce some of the electricity used in the building directly, the designers added 60,000 photovoltaic solar cells to the roof. These solar panels convert energy from the Sun into 213,000 kWh of electricity per year and reduce greenhouse gas emissions by about 200 metric tons per year.

The California Academy of Sciences took an innovative approach to meeting its energy needs through a combination of energy efficiency and use of renewable energy resources. However, many, if not all, of these approaches will have to become commonplace if we are going to use energy in a sustainable way.

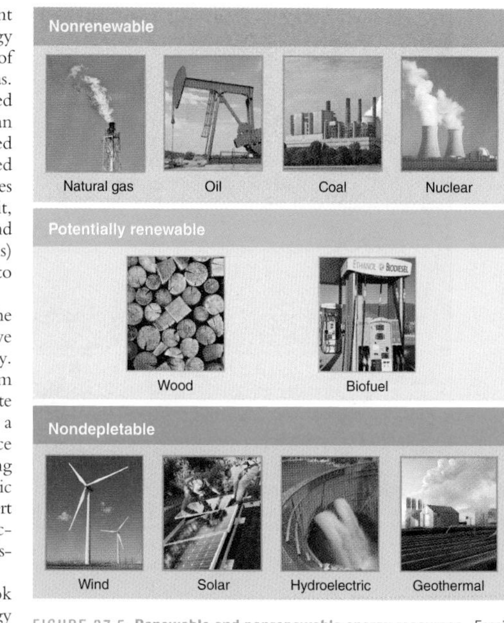

FIGURE 37.5 Renewable and nonrenewable energy resources. Fossil fuels and nuclear fuels are nonrenewable energy resources. Renewable energy resources include potentially renewable energy sources such as biomass, which is renewable as long as humans do not use it faster than it can be replenished, and nondepletable energy sources, such as solar radiation and wind. *(top row: Ian Hamilton/iStockphoto.com, Sandra Nicol/iStockphoto.com, Michael Utech/iStockphoto.com, Hans F. Meier/iStockphoto.com; middle row: Bernd Lang/iStockphoto.com, Inga Spence/Science Source; bottom row: acilo/iStockphoto.com, Don Mason/Blend Images/Corbis, amana images inc./Alamy, Rhoberazzi/iStockphoto.com)*

Renewable energy is either potentially renewable or nondepletable

As fossil fuels become less available and more expensive, what will take their place? Probably it will be a mix of energy efficiency strategies, energy conservation, and new energy sources. In the rest of this chapter we will explore our renewable energy options.

In Chapter 12 we learned that conventional energy resources, such as petroleum, natural gas, coal, and uranium ore, are nonrenewable. From a systems analysis perspective, fossil fuels constitute an energy reservoir we are depleting much faster than it can ever be replenished. Similarly, we have a finite amount of uranium ore available to use as fuel in nuclear reactors.

In contrast, some other sources of energy can be regenerated rapidly. Biomass energy resources are **potentially renewable** because those resources can be regenerated indefinitely as long as we do not consume them more quickly than they can be replenished. There are still other energy resources that cannot be depleted no matter how much we use them. Solar, wind, geothermal, hydroelectric, and tidal energy are essentially **nondepletable** in the span of human time; no matter how much we use there will always be more. The amount of a nondepletable resource available tomorrow does not depend on how much we use today. In this book we refer to potentially renewable and nondepletable energy resources together as **renewable** energy resources. FIGURE 37.5 illustrates the categories of energy resources.

Potentially renewable An energy source that can be regenerated indefinitely as long as it is not overharvested.

Nondepletable An energy source that cannot be used up.

Renewable In energy management, an energy source that is either potentially renewable or nondepletable.

Many renewable energy resources have been used by humans for thousands of years. In fact, before humans began using fossil fuels, the only available energy sources were wood and plants, animal manure, and fish or animal oils. Today, in parts of the developing world where there is little access to fossil fuels, people still rely on local biomass energy sources such as manure and wood for cooking and heating—sometimes to such an extent that they overuse the resource. For example, according to the U.S. Energy Information Administration, biomass is currently the source of 86 percent of the energy consumed in sub-Saharan Africa (excluding South Africa) and much of it is not harvested sustainably.

As FIGURE 37.6 shows, renewable energy resources account for approximately 13 percent of the energy used worldwide, most of which is in the form of biomass. In the United States, which depends heavily on fossil fuels, renewable energy resources provide only about 7 percent of the energy used. That 7 percent, shown in detail in FIGURE 37.7, comes primarily from biomass and hydroelectricity.

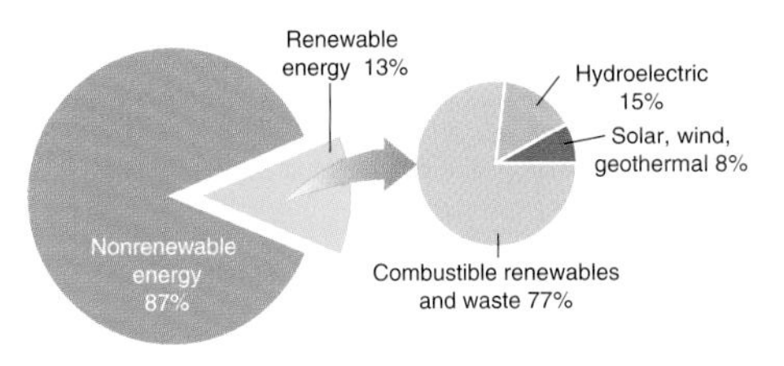

FIGURE 37.6 **Global energy use.** Renewable energy resources provide about 13 percent of energy worldwide. *(Data from International Energy Agency for the 2011 calendar year, World Energy Outlook, 2013)*

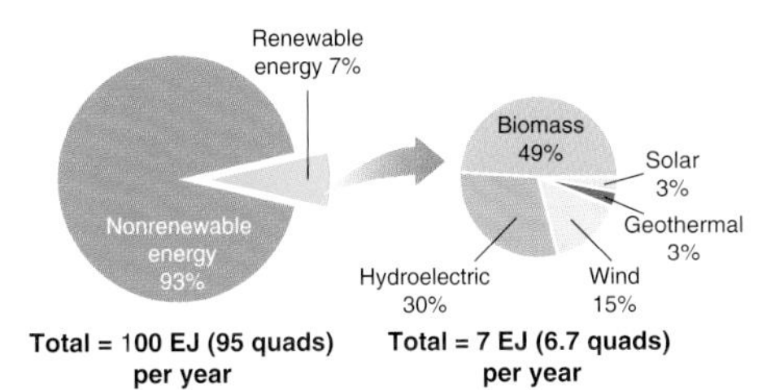

FIGURE 37.7 **Energy use in the United States.** Only 7 percent of the energy used in the United States comes from renewable energy resources. *(Data from U.S. Department of Energy for the 2012 calendar year, EIA, 2013)*

module

37 REVIEW

In this module, we have seen that to achieve energy sustainability we should begin with conserving energy and increasing energy efficiency. Energy conservation refers to finding ways to use less energy. Increasing energy efficiency means achieving the same amount of work from a smaller quantity of energy. Sustainable design of buildings and communities can decrease energy use. We also looked at different categories of renewable energy resources. Potentially renewable energy sources can regenerate if we do not consume them more quickly than they can be replaced. Nondepletable energy resources such as the wind cannot be depleted no matter how much we use. In the next module, we examine two important renewable resources, biomass and water that is used for generating electricity.

Module 37 AP® Review Questions

1. Which is an energy efficiency improvement?
   (a) Riding a bike
   (b) Adjusting the thermostat in a building
   (c) Using a power strip
   (d) Using cold water instead of hot water
   (e) Replacing incandescent bulbs with LEDs

Answers to Module 37 AP® Review Questions

1. e
2. e
3. c
4. d
5. b

2. Peak demand
   (a) decreases electricity-generating capacity.
   (b) is caused by higher electricity prices.
   (c) occurs only in early morning.
   (d) can be managed using a tiered rate system.
   (e) can cause brownouts.

3. If an Energy Star refrigerator costs 2 cents less per hour to run, and if it runs for 16 hours a day, how much will it save in a year?
   (a) $44
   (b) $82
   (c) $117
   (d) $174
   (e) $236

4. Which is NOT used in sustainable design in North America?
   (a) Recycled building materials
   (b) Double-paned windows
   (c) An overhanging roof
   (d) Windows on the northern wall
   (e) Building materials with high thermal mass

5. Which energy source is NOT nondepletable?
   (a) Wind
   (b) Biomass
   (c) Tidal
   (d) Solar
   (e) Hydroelectric

module 38

Biomass and Water

As we discussed in Chapter 12, the Sun is the ultimate source of fossil fuels. Fossil fuels are created from dead plants and animals that are buried deep in sediments and that are slowly transformed into petroleum or coal. Most types of renewable energy are also derived from the Sun and cycles driven by the Sun, including solar, wind, and hydroelectric energy as well as plant biomass such as wood. In this module, we present two important renewable energy sources: biomass and water.

Learning Objectives

After reading this module you should be able to

- describe the various forms of biomass.
- explain how energy is harnessed from water.

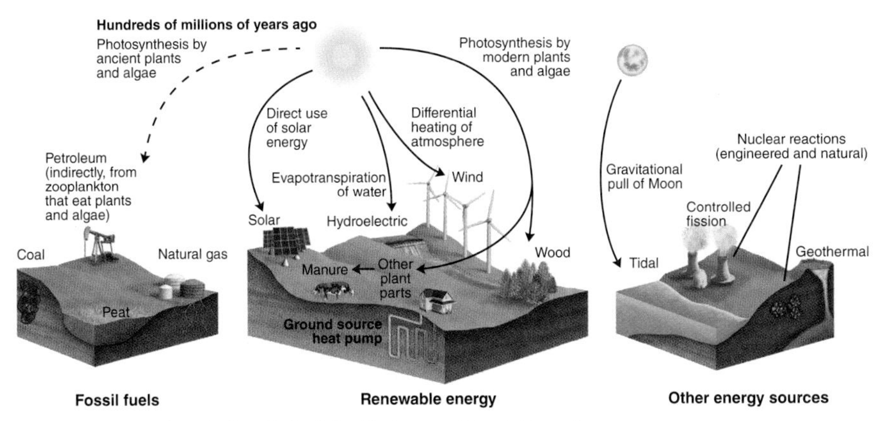

FIGURE 38.1 **Energy from the Sun.** The Sun is the ultimate source of almost all types of energy.

Biomass is energy from the Sun

As FIGURE 38.1 shows, all fossil fuel and most renewable energy sources ultimately come from the Sun. Biomass energy resources encompass a large class of fuel types that include wood and charcoal, animal products and manure, plant remains, and municipal solid waste (MSW), as well as liquid fuels such as ethanol and biodiesel. Many forms of biomass used directly as fuel, such as wood and manure, are readily available all over the world. Because these materials are inexpensive and abundant, they account for more than 10 percent of world energy consumption (see Figure 37.6), with a much higher percentage in many developing countries. Biomass can also be processed or refined into liquid fuels such as ethanol and biodiesel, known collectively as **biofuels.** These fuels are used in more limited quantities due to the technological demands associated with their use. For example, it is easier to burn a log in a fire than it is to develop the technology to produce a compound such as ethanol.

Biomass—including ethanol and biodiesel—accounts for roughly one-half of the renewable energy and approximately 3.5 percent of all the energy consumed in the United States today (see Figure 37.7 on page 439). However, the mix of biomass used in the United States differs from that found in the developing world. Two-thirds of the biomass energy used in the United States comes from wood, with the remaining one-third divided evenly between MSW and biofuels, while in the developing world, a larger percentage of biomass energy comes from wood and animal manure.

Modern Carbon versus Fossil Carbon

Like fossil fuels, biomass contains a great deal of carbon, and burning it releases that carbon into the atmosphere. Given the fact that both fossil fuels and biomass raise atmospheric carbon concentrations, is it really better for the environment to replace fossil fuels with biomass? The answer depends on how the material is harvested and processed and on how the land is treated during and after harvest. It also depends on how long the carbon has been stored. The carbon found in plants growing today was in the atmosphere in the form of carbon dioxide until recently when it was incorporated into the bodies of the plants through photosynthesis. Depending on the type of plant it comes from, the carbon in biomass fuels may have been captured through photosynthesis as recently as a few months ago, as in the case of a corn plant, or perhaps up to several hundred years ago, as in the case of wood from a large tree. We call the carbon in biomass **modern carbon,** in contrast to the carbon in fossil fuels, which we call **fossil carbon.**

Biofuel Liquid fuel created from processed or refined biomass.

Modern carbon Carbon in biomass that was recently in the atmosphere.

Fossil carbon Carbon in fossil fuels.

◀ TEACHING with FIGURES

Ask students to work in groups of two and answer the following questions about Figure 38.1:

- How are fossil fuels related to the Sun? *All fossil fuels are created by decomposition of ancient plants and animals. Plants require sunlight to grow and animals require plants for food, either directly or indirectly. Therefore, energy from the Sun is the source of the energy in fossil fuels.*
- How are wind, hydroelectric, and solar energy related to the Sun? *Wind is the result of the unequal heating of Earth and the atmosphere. Hydroelectric energy is indirectly related to the Sun through the water cycle in which the Sun evaporates water. Finally solar energy is energy harnessed from sunlight.*
- What are the sources of tidal, nuclear, and geothermal energy? *The gravitational pull of the moon creates tides. Nuclear energy is produced by uranium, which is a radioactive element mined from Earth's crust. Geothermal energy uses the heat from radioactive elements including uranium from Earth's interior.*

Teaching Tip: Journal Prompt

Ask students to compare solid biomass and liquid biomass with the following questions:

- What are the different types of solid biomass and liquid biomass? *Solid biomass includes wood, charcoal, and manure. Liquid biomass includes ethanol and biodiesel.*
- How are the different types of biomass used? *Solid biomass is used primarily for heating homes and cooking food. Liquid biomass is used as a substitute for petroleum. It is often mixed with gasoline.*
- Where are the different types of biomass used? *Solid biomass is mostly used in developing countries because wood and manure are less expensive than other fuels. Liquid biomass is used more in developed countries where people can afford it.*
- How are liquid biomass and solid biomass similar? *Both solid and liquid biomass are derived from the Sun. They both are considered modern carbon inputs.*
- Describe the environmental effects of using solid biomass and the effects of using liquid biomass. *Solid biomass such as wood and manure produces large amounts of particulate matter, carbon dioxide, and nitrous oxides. Liquid biomass burns cleaner than gasoline.*

AP® Exam Tip

The use of animal manure as a fuel is covered in the 2009 AP® Exam, Question 2. To answer this question, students must

- summarize an article that outlines the steps by which methane is used to produce electricity in a methane digester
- discuss environmental benefits of a methane digester
- calculate the amount of electricity a farmer can generate through cow manure
- calculate how much the farmer would save by producing electricity in a methane digester and the number of years it would take to recover the installation costs
- calculate the minimum number of cows the farmer would need to produce 800,000 kWh of electricity from methane in a year

Unlike modern carbon, fossil carbon has been buried for millions of years. Fossil carbon is carbon that was out of circulation until humans discovered it and began to use it in increasing quantities. The burning of fossil fuels results in a rapid increase in atmospheric CO_2 concentrations because we are unlocking or releasing stored carbon that was last in the atmosphere millions of years ago. In theory, the burning of biomass (modern carbon) should not result in a net increase in atmospheric CO_2 concentrations because we are returning the carbon to the atmosphere, where it had been until recently. And, if we allow vegetation to grow back in areas where biomass was recently harvested, that new vegetation will take up an amount of CO_2 more or less equal to the amount we released earlier by burning the biomass. Over time, the net change in atmospheric CO_2 concentrations should be zero. An activity that does not change atmospheric CO_2 concentrations is referred to as **carbon neutral.**

Whether using biomass is truly carbon neutral, however, is an important question that is currently being discussed by scientists and policy makers. Sometimes the use of modern carbon ends up releasing CO_2 into the atmosphere that would otherwise have remained in the soil or in fossil fuels.

Solid Biomass: Wood, Charcoal, and Manure

Throughout the world, 2 billion to 3 billion people rely on wood for heating or cooking. In the United States, approximately 3 million homes use wood as the primary heating fuel, and more than 20 million homes use wood for energy at least some of the time. In addition, the pulp and paper industries, power plants, and other industries use wood waste and by-products for energy. In theory, cutting trees for fuel is sustainable if forest growth keeps up with forest use. Unfortunately, many forests, like those in Haiti, are cut intensively, allowing little chance for regrowth.

Removing more timber than is replaced by growth, or **net removal** of forest, is an unsustainable practice that will eventually lead to deforestation. This net removal of forest together with the burning of wood results in a net increase in atmospheric CO_2: The CO_2 released from the burned wood is not balanced by photosynthetic carbon fixation that would occur in new tree growth. Harvesting the forest may also release carbon from the soil that would otherwise have remained buried deep in the A and B horizons. Although the mechanism is not entirely clear, it appears that some of this carbon release may occur because logging equipment disturbs the soil.

Tree removal can be sustainable if we allow time for forests to regrow. In addition, in some heavily forested areas, extracting individual trees of abundant species and opening up spaces in the canopy will allow other plants to grow and increase habitat diversity, which may even increase total photosynthesis. More often, however, tree removal has the potential to cause soil erosion, to increase water temperatures in nearby rivers and streams, and to fragment forest habitats when logging roads divide them. Tree removal may also harm species that are dependent on old-growth forest habitat.

Many people in the developing world use wood to make charcoal, which is a superior fuel for many reasons. Charcoal is lighter than wood and contains approximately twice as much energy per unit of weight. A charcoal fire produces much less smoke than wood and does not need to be tended constantly, as does a wood fire. Although it is more expensive than wood, charcoal is a fuel of choice in urban areas of the developing world and for families who can afford it. However, harvesters who clear an area of land for charcoal production often leave it almost completely devoid of trees (FIGURE 38.2).

In regions where wood is scarce, such as parts of Africa and India, people often use dried animal manure as a fuel for indoor heating and cooking. Burning manure can be beneficial because it removes harmful microorganisms from surrounding areas, which reduces the risk of disease transmission. However, burning manure also releases particulate matter and other pollutants into the air that cause a variety of respiratory illnesses, from emphysema to cancer. The problem is exacerbated when the manure is burned indoors in poorly ventilated rooms, a common situation in many developing countries. The World Health Organization estimates that indoor air pollution is responsible for nearly 2 million deaths annually. Chapters 15 and 17 cover indoor air pollution and human health in more detail.

Whether indoors or out, burning biomass fuels produces a variety of air pollutants, including particulate matter, carbon monoxide (CO), and nitrogen oxides (NO_x), which are important components of air pollution (FIGURE 38.3).

Biofuels: Ethanol and Biodiesel

The liquid biofuels—ethanol and biodiesel—can be used as substitutes for gasoline and diesel, respectively. **Ethanol** is an alcohol made by converting starches and sugars from plant material into alcohol and carbon

Carbon neutral An activity that does not change atmospheric CO_2 concentrations.

Net removal The process of removing more than is replaced by growth, typically used when referring to carbon.

Ethanol Alcohol made by converting starches and sugars from plant material into alcohol and CO_2.

AP® Exam Tip

The use of wood as an energy source and nonrenewable energy sources are covered in the 2001 AP® Exam, Question 1. To answer this question, students must

- calculate the number of cubic feet of natural gas used to heat a house over the course of a winter, and the cost, using data provided
- describe actions residents can take to conserve energy and lower heating costs
- discuss the environmental advantages and disadvantages of using wood as an energy source for home heating

Teaching Tip: Journal Prompt

Many students don't understand the difference between biofuel and biodiesel. Ask students to describe the difference between gasohol, E-85, and B-20. *Gasohol is 10 percent ethanol and 90 percent gasoline. Gasohol is currently used in the United States for cars and light trucks. E-85 is 85 percent ethanol and 15 percent gasoline. E-85 is available in the United States but is typically only found in midwestern states. B-20 is a mixture of 80 percent petroleum diesel and 20 percent biodiesel. B-20 is available at some gas stations scattered around the United States and can be used in any diesel engine.*

(a)

(b)

FIGURE 38.2 **Charcoal as fuel in the developing world.** (a) Many people in developing countries rely on charcoal for cooking and heating. This photo shows a charcoal market in the Philippines. (b) Charcoal production can strip the land of all trees *(a: Jay Directo/AFP/Getty Images; b: Eduardo Martino/Panos Pictures)*

FIGURE 38.3 **Particulate emissions from burning biomass fuels.** Burning biomass fuels contributes to air pollution. This photo shows smog and decreased visibility in Montreal, Canada, caused by emissions of particulate matter due to the extensive use of woodstoves. *(Andrew Lank/iStockphoto.com)*

dioxide. More than 90 percent of the ethanol produced in the United States comes from corn and corn by-products, although ethanol can also be produced from sugarcane, wood chips, crop waste, or switchgrass. **Biodiesel,** a diesel substitute produced by extracting and chemically altering oil from plants, is a substitute for regular petroleum diesel. It is usually produced by extracting oil from algae and plants such as soybean and palm. In the United States, many policy makers are encouraging the production of ethanol and bio-diesel as a way to reduce the need to import foreign oil while also supporting U.S. farmers and declining rural economies.

Ethanol

The United States is the world leader in ethanol production, manufacturing more than 50 billion liters (13.3 billion gallons) in 2012. Brazil, the world's second

Biodiesel A diesel substitute produced by extracting and chemically altering oil from plants.

Teaching Tip: Activity

For a closer look at the issues involved in using food crops for fuel, have students read "Science Applied 6: Should Corn Become Fuel?" on page 476. Use the questions on page 479 for class discussion.

AP® Exam Tip

Biodiesel is covered on the 2008 AP® Exam, Question 1. To answer this question, students must

- use provided data to calculate the amount of microalgae or soybeans needed to produce 1,000 gallons of oil in a year
- explain the advantage of microalgae as a source for biodiesel over other crops
- compare the effect on atmospheric CO_2 concentrations from burning biodiesel and fossil fuels
- describe two benefits not related to the atmosphere of increasing biodiesel use over the next 50 years
- discuss two economic or social problems of using corn for fuel

AP® Exam Tip

The use of wood as an energy source and forest conservation are covered on the 2012 AP® Exam, Question 2. To answer this question, students must

- describe a way to reduce a school's carbon footprint, including environmental benefits and environmental drawbacks
- explain the ecological benefits of an intact forest system
- using provided data, calculate the mass of carbon accumulated in a forest in 1 year, the mass of carbon emitted by a school from fuel consumption in 1 year, the amount of forest reserve needed to offset the carbon emissions of the school, and how much money the school would need to raise to preserve the amount of forest it needs to reduce its carbon footprint

Teaching Tip: Debate the Issue

Flex-Fuel Vehicles
Have students research flex-fuel vehicles in preparation for a debate to answer the question: Should the government encourage people to purchase flex-fuel vehicles?

Students should be prepared to discuss these questions in their debate:

- What is a flex-fuel car and how is it different from a regular vehicle?
- What are the environmental pros or cons to using flex-fuel vehicles?
- What is E-85 and how might it help the United States become energy independent?
- Would using more corn for ethanol result in a net environmental benefit?

In class, select groups to debate the issue. Allow each side 2 minutes to prepare, 3 minutes to present, and 1 minute for rebuttal. Have the remainder of the class take notes, ask questions, and decide which side has the most persuasive argument.

Teaching Tip: Video

Chemists Patent New Formula for Cleaner Cheaper Diesel Fuel
This short film describes diesel fuel and a new process for creating cleaner, cheaper diesel fuel.

The link to this video may be found by clicking the link buttons in the TE-book, opening the TRFD, or logging onto the book's companion website bcs.whfreeman.com/friedlandapes2e (teacher login required).

Chapter 13 Web Resources

largest ethanol producer, is making biofuels a major part of its sustainable energy strategy. Brazil manufactures ethanol from sugarcane, which is easily grown in its tropical climate. Unlike corn, which must be replanted every year, sugarcane is replanted every 6 years and is sometimes harvested by hand, factors that reduce the amount of fossil fuel energy needed.

Ethanol is usually mixed with gasoline, most commonly at a ratio of one part ethanol to nine parts gasoline. The result is gasohol, a fuel that is 10 percent ethanol. Gasohol has a higher oxygen content than gasoline alone and produces less of some air pollutants when combusted. In certain parts of the midwestern United States, especially in corn-growing states, a fuel called E-85 (85 percent ethanol, 15 percent gasoline) is available. **Flex-fuel vehicles** can run on either gasoline or E-85. However, a study by General Motors several years ago revealed that most of the owners of the 7 million flex-fuel vehicles in use at that time did not know that their cars could run on E-85.

Proponents of ethanol claim that it is a more environmentally friendly fuel than gasoline, although opponents dispute that claim. Ethanol does have disadvantages. The carbon bonds in alcohol have a lower energy content than those in gasoline, which means that a 90 percent gasoline/10 percent ethanol mix reduces gas mileage by 2 to 3 percent when compared with 100 percent gasoline fuel. As a result, a vehicle needs more gasohol to go the same distance it could go on gasoline alone. Furthermore, growing corn to produce ethanol uses a significant amount of fossil fuel energy, as well as land that can otherwise be devoted to growing food. Ethanol production has led to concern among economic analysts that this periodically contributes to short-term food shortages. Furthermore, some scientists argue that using ethanol actually creates a net increase in atmospheric CO_2 concentrations. The benefits and drawbacks of using ethanol as a fuel are discussed in more detail in "Science Applied 6: Should Corn Become Fuel?" that follows this chapter.

Research is under way to find viable alternatives to corn as sources for U.S. ethanol production. Switchgrass is one possibility. It is a perennial crop, which means that farmers can harvest it without replanting, minimizing soil disturbance and erosion. Furthermore, switchgrass does not require as much fossil fuel input as corn to produce. However, crops such as corn and sugarcane produce ethanol more readily due to their high sugar content because sugars are readily and rapidly converted into ethanol. In contrast, switchgrass and other alternative materials, such as wood chips, are composed primarily of cellulose—the material that constitutes plant cell walls—which must be broken down into sugars before it can be used in ethanol production. Scientists have not yet developed an efficient breakdown process for large-scale ethanol production from switchgrass, although such a process would increase the energy and carbon advantages of ethanol.

Flex-fuel vehicle A vehicle that runs on either gasoline or a gasoline/ethanol mixture.

Biodiesel

Biodiesel is a direct substitute for petroleum-based diesel fuel. It is usually more expensive than petroleum diesel, although the difference varies depending on market conditions and the price of petroleum. Biodiesel is typically diluted to "B-20," a mixture of 80 percent petroleum diesel and 20 percent biodiesel, and is available at some gas stations scattered around the United States. It can be used in any diesel engine without modification.

Because biodiesel tends to solidify into a gel at low temperatures, higher concentrations of biodiesel work effectively only in modified engines. However, with a kit sold commercially, a skilled individual or automobile mechanic can modify any diesel vehicle to run on 100 percent straight vegetable oil (SVO), typically obtained as a waste product from restaurants and filtered for use as fuel. Groups of students in the United States have driven buses around the country almost exclusively on SVO, and some municipalities, such as Portland, Maine, have community-based SVO recycling facilities (FIGURE 38.4). Although there is unlikely to be enough waste vegetable oil to significantly reduce fossil fuel consumption, SVO is nevertheless a potential transition fuel that may temporarily reduce our use of petroleum.

In the United States, most biodiesel comes from soybean oil or processed vegetable oil. However, scientists are working on ways to produce large quantities of biodiesel directly from wood or other forms of

FIGURE 38.4 **An SVO recycling facility.** Used cooking oil collected from 800 restaurants in Maine and Massachusetts is stored in vats and converted into biodiesel fuel. Shown here are chief operating officer of Maine Standard Biofuels, Matt Pemberton, and founder and co-owner, Jarmin Kaltsas. *(Gabe Souza/Staff Photographer Portland Press Herald)*

Teaching Tip: Discussion Starter

Ask students: Why is biodiesel a better option for the environment than conventional diesel fuel? What are some of the concerns about biodiesel? *Emissions of carbon monoxide from combustion of biodiesel are lower than those from conventional petroleum diesel. Since it contains modern carbon rather than fossil carbon, biodiesel should be carbon neutral, although, as with ethanol, some critics question whether biodiesel is truly carbon neutral. For example, producing biodiesel from soybeans requires less fossil fuel input per liter of fuel than producing ethanol from corn, but because soybeans require more cropland, they may actually transfer more carbon from the soil to the atmosphere. In contrast, producing biodiesel from wood waste or algae may require very little or no cropland and a minimal amount of other land.*

cellulose—especially waste wood from logging and sawmills. In addition, some species of algae appear to have great potential for producing biodiesel. Algae are photosynthetic microorganisms that can be grown almost anywhere and, of all biodiesel options, produce the greatest yield of fuel per hectare of land area per year and utilize the least amount of energy and fertilizer per quantity of fuel. One study reported that algae produce 15 to 300 times more fuel per area used than did conventional crops. Algae can be grown on marginal lands, in brackish water, on rooftops, and in other places that are not traditionally thought of as agricultural space.

Emissions of carbon monoxide from combustion of biodiesel are lower than those from petroleum diesel. Since it contains modern carbon rather than fossil carbon, biodiesel should be carbon neutral, although, as with ethanol, some critics question whether biodiesel is truly carbon neutral. For instance, producing biodiesel from soybeans requires less fossil fuel input per liter of fuel than producing ethanol from corn, but soybeans require more cropland, and so they may actually transfer more carbon from the soil to the atmosphere. In contrast, producing biodiesel from wood waste or algae may require very little or no cropland and a minimal amount of other land.

The kinetic energy of water can generate electricity

Hydroelectricity is electricity generated by the kinetic energy of moving water. It is the second most commonly used form of renewable energy in the United States and in the world, and it is the form most widely used for electricity generation. As we saw in Chapter 12 (Figure 34.8 on page 407), hydroelectricity accounts for approximately 7 percent of the electricity generated in the United States. More than one-half of that hydroelectricity is generated in three states: Washington, California, and Oregon. Worldwide, nearly 20 percent of all electricity comes from hydroelectric power plants, with China the leading producer, followed by Brazil, Canada, and the United States. In this section we will look at ways in which hydroelectricity is generated, and we will consider whether hydroelectricity is sustainable.

Methods of Generating Hydroelectricity

Moving water, either falling over a vertical distance or flowing with a river or tide, contains kinetic energy. A hydroelectric power plant captures this kinetic energy and uses it to turn a turbine in the same way that the kinetic energy of steam turns a turbine in a coal-fired power plant. The turbine, in turn, transforms the kinetic energy of water or steam into electricity, which is then exported to the electrical grid via transmission lines.

The amount of electricity that can be generated at any particular hydroelectric power plant depends on the flow rate, the vertical distance the water falls, or both. Where falling water is the source of the energy, the amount of electricity that can be generated depends on the vertical distance the water falls; the greater the distance, the more potential energy the water has, and the more electricity it can generate (see Figure 5.2 on page 45). The amount of electricity a hydroelectric power plant can generate also depends on the flow rate: the amount of water that flows past a certain point per unit of time. The higher the flow rate, the more kinetic energy is present, and the more electricity can be generated.

Run-of-the-River Systems

In **run-of-the-river** hydroelectricity generation, water is retained behind a low dam and runs through a channel before returning to the river. Run-of-the-river systems do not store water in a reservoir. These systems have several advantages that reduce their environmental impact: Relatively little flooding occurs upstream, and seasonal changes in river flow are not disrupted. However, run-of-the-river systems are generally small and, because they rely on natural water flows, electricity generation can be intermittent. Heavy spring runoff from rains or snowmelt cannot be stored, and the system cannot generate any electricity in hot, dry periods when the flow of water is low.

Water Impoundment Systems

Storing water in a reservoir behind a dam is known as **water impoundment.** FIGURE 38.5 illustrates the various features of a water impoundment system. By managing the opening and closing of the gates, the dam operators control the flow rate of the water that turns the turbine—and in turn the generator—and thereby influence the amount of electricity produced.

Water impoundment is the most common method of hydroelectricity generation because it usually allows for the generation of electricity on demand. The largest hydroelectric water impoundment dam in the United States is the Grand Coulee Dam in Washington State, which generates 6,800 MW at peak capacity.

Hydroelectricity Electricity generated by the kinetic energy of moving water.

Run-of-the-river Hydroelectricity generation in which water is retained behind a low dam or no dam.

Water impoundment The storage of water in a reservoir behind a dam.

Teaching Tip: Review

Ask students to answer the following questions as you review biomass:

- What are the major forms of biomass energy? *Biomass energy resources encompass a large class of fuel types that include wood and charcoal, animal products and manure, plant remains and municipal solid waste (MSW), as well as liquid fuels such as ethanol and biodiesel.*
- Why is energy from modern carbon potentially carbon neutral? *The carbon in biomass fuels may have been captured through photosynthesis as recently as a few months ago, in the case of a corn plant, or up to several hundred years ago, in the case of wood from a large tree. In theory, the burning of biomass should not result in a net increase in atmospheric CO_2 concentrations because the carbon it releases has recently been in the atmosphere and presumably will be taken up by the plants that have replaced the corn or trees that were removed.*
- Why is it important to find abundant sources of biomass energy? What are the advantages and disadvantages of different forms of biomass energy? *Biodiesel combustion emits less carbon monoxide than petroleum diesel combustion. Also, because biofuels contain modern carbon, biodiesel should be carbon neutral. Advantages: Plants can be grown in many different climates on Earth, and biomass is potentially renewable, can reduce dependence on fossil fuels, and is possibly more environmentally friendly than fossil fuels. Disadvantages: Biomass causes the loss of agricultural land, higher food costs, lower gas mileage, and a possible net increase in greenhouse gas emissions.*

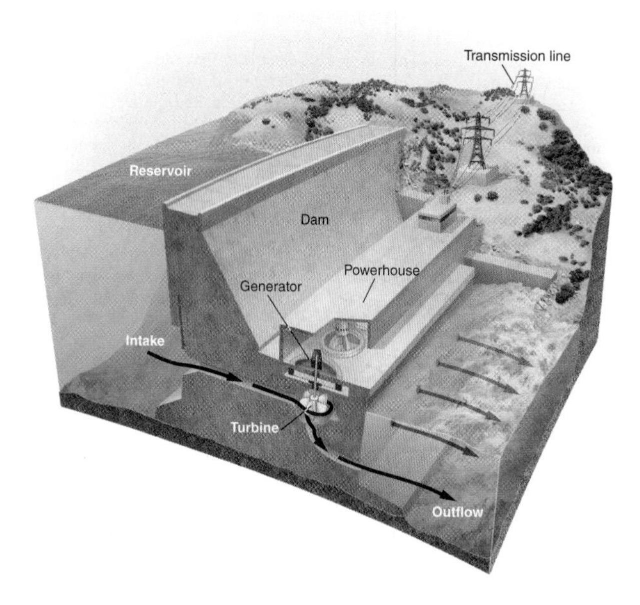

FIGURE 38.5 A water impoundment hydroelectric dam. Water impoundment, a common method of hydroelectricity generation, allows for electricity generation on demand. Dam operators control the rate of water flow by opening and closing the gates. This determines the amount of electricity generated. The arrows indicate the path of water flow.

The Three Gorges Dam on the Yangtze River in China (see Figure 27.2, page 304) is the largest dam in the world. It has a capacity of 18,000 MW and can generate almost 85 billion kilowatt-hours per year, approximately 11 percent of China's total electricity demand.

Tidal Systems

Tidal energy also comes from the movement of water, although the movement in this case is driven by the gravitational pull of the Moon. Tidal energy systems use gates and turbines similar to those used in run-of-the-river and water impoundment systems to capture the kinetic energy of water flowing through estuaries, rivers, and bays and convert this energy into electricity.

Although tidal power plants are operating in many parts of the world, including France, Korea, and Canada, tidal energy does not have the potential to become a major energy source. In many places the difference in water level between high and low tides is not great enough to provide sufficient kinetic energy to generate a large amount of electricity. In addition, to transfer the energy generated, transmission lines must be constructed on or near a coastline or estuary. This infrastructure may have a disruptive effect on coastal, shoreline, and marine ecology as well as on tourism that relies on the aesthetics of a coastal region.

Tidal energy Energy that comes from the movement of water driven by the gravitational pull of the Moon.

Hydroelectricity and Sustainability

Major hydroelectric dam projects have brought renewable energy to large numbers of rural residents in many countries, including the United States, Canada, India, China, Brazil, and Egypt. Although hydroelectric dams are expensive to build, once built, they require a minimal amount of fossil fuel for operation. In general, the benefits of water impoundment hydroelectric systems are great: They generate large quantities of electricity without creating air pollution, waste products, or CO_2 emissions. Electricity from hydroelectric power plants is usually less expensive for the consumer than electricity generated using nuclear fuels or natural gas. In the United States, the price of hydroelectricity ranges from 5 cents to 11 cents per kilowatt-hour.

In addition, the reservoir behind a hydroelectric dam can provide recreational and economic opportunities as well as downstream flood control for flood-prone areas. For example, Lake Powell, the reservoir

Teaching Tip: Beyond the Classroom

Ask students to research a tidal energy power plant. There are presently only a few locations, including Roosevelt Island Tidal Energy Project (RITE) in New York City, Sihwa Lake Tidal Power Station in South Korea, and the Tidal Lagoon Project in Swansea, Wales. Some questions for students to consider in their report include:

- Where is the plant located?
- When was the plant constructed?
- How much energy does it provide?
- What are the environmental benefits and environmental costs of the project?

impounded by the hydroelectric Glen Canyon Dam, draws more than 2.5 million visitors to the Glen Canyon National Recreation Area each year and generates more than $400 million annually for the local and regional economies in Arizona and Utah (FIGURE 38.6). In China, the Three Gorges Dam provides flood control and protection to many millions of people.

Water impoundment, however, does have negative environmental consequences. In order to form an impoundment, a free-flowing river must be held back. The resulting reservoir may flood hundreds or thousands of hectares of prime agricultural land or canyons with great aesthetic or archeological value. It may also force people to relocate. As we saw in Chapter 9, the construction of the Three Gorges Dam displaced more than 1.3 million people and submerged ancient cultural and archaeological sites as well as large areas of farmland. Impounding a river in this way may also make it unsuitable for organisms or recreational activities that depend on a free-flowing river. Large reservoirs of standing water hold more heat and contain less oxygen than do free-flowing rivers, thereby affecting which species can survive in the waters. Certain human parasites also become more abundant in impounded waters in tropical regions.

By regulating water flow and flooding, dams also alter the dynamics of the river ecosystem downstream. Some rivers, for example, have sandbars created during periods of low flow that follow periods of flooding. Some plant species, such as cottonwood trees, cannot reproduce in the absence of these sandbars. The life cycles of certain aquatic species, such as salmon, certain trout species, and freshwater clams and mussels, also depend on seasonal variations in water flow. Impoundment systems disrupt these life cycles by controlling the flow of water so it is consistently plentiful for hydroelectricity generation.

It is possible to address some of these problems. For example, as we saw in Chapter 9, the installation of a fish ladder (see Figure 27.3 on page 304) may allow fish to travel upstream around a dam. Such solutions are not always optimal, however; some fish species fail to utilize them and some predators learn to monitor the fish ladders for their prey.

Other environmental consequences of water impoundment systems include the release of greenhouse gases to the atmosphere, both during dam construction and after filling the reservoir. Production of cement—a major component of dams—is responsible for approximately 5 percent of global anthropogenic CO_2 emissions to the atmosphere. Once the dam is completed, the impounded water usually covers forests or grasslands. The dead plants and organic materials in the flooded soils decompose anaerobically and release methane, a potent greenhouse gas. Some researchers assert that in tropical regions, the methane released from a hydroelectric water impoundment contributes more to climate warming than a coal-fired power plant with about the same generating capacity.

The accumulation of sediments in reservoirs has negative consequences not only for the environment but also for the electricity-generating capacity of hydroelectric dams. A fast-moving river carries sediments that settle out when the river feeds into the reservoir. The accumulation of these sediments on the

FIGURE 38.6 A recreation area created by water impoundment. Lake Powell at the Glen Canyon National Recreation Area was created by water impoundment at the hydroelectric Glen Canyon Dam in northern Arizona. *(Sal Maimone/SuperStock)*

Teaching Tip: Review

To review hydroelectric energy, ask students the following questions:

- How can water be used as a renewable energy resource? *Moving water, either falling over a vertical distance or flowing with a river or tide, contains kinetic energy. A hydroelectric power plant captures this kinetic energy and uses it to turn a turbine, just as the kinetic energy of steam turns a turbine in a coal-fired power plant.*
- What are the types of hydroelectricity? *There are two main types of hydroelectric power. The first type is called a run-of-the-river system. Water is retained behind the dam and runs through a channel before returning to the river. The second is called a water impoundment system. Water impoundment is the most common system. By controlling the opening and closing of the gates, the dam operators determine the flow rate of water and therefore the amount of electricity generated.*
- What are the trade offs associated with using hydroelectricity compared with biomass energy? *Hydroelectric power is a renewable resource while biomass is potentially renewable. Hydroelectric power has high initial construction costs while biomass does not. Environmentally, hydroelectric power produces almost no emissions while liquid biomass releases carbon dioxide and methane and solid biomass releases carbon monoxide, particulate matter, and nitrogen oxides.*

(a)

(b)

FIGURE 38.7 **Dam removal and environmental restoration.** In recent years, a number of dams have been dismantled due to environmental concerns or heavy siltation. (a) This photo shows Oregon's Marmot Dam, on the Sandy River, before removal. (b) The same stretch of the Sandy River after dam removal in 2007 shows how the natural landscape has been restored. *(Portland General Electric)*

bottom of the reservoir is known as **siltation.** Over time, as the reservoir fills up with sediments, the amount of water that can be impounded, and thus the generating capacity and life span of the dam, is reduced. This process may take hundreds of years or only decades, depending on the geology of the area. The only way to reverse the siltation process is by dredging, removal of the sediment, usually with machinery that runs on fossil fuels.

Because of either environmental concerns or heavy siltation, a number of hydroelectric dams are being dismantled, as we saw at the beginning of Chapter 9. In 1999, the Edwards Dam was removed from the Kennebec River in Maine. More than a decade later, native fishes such as bass and alewives have returned to the waters and are flourishing. In 2007, the Marmot Dam on Oregon's Sandy River was removed using explosives. The restored river now hosts migrating salmon and steelhead trout (*Oncorhynchus mykiss*) for the first time since 1912 (FIGURE 38.7).

Siltation The accumulation of sediments, primarily silt, on the bottom of a reservoir.

Teaching Tip: Journal Prompt

Ask students to review the definition of siltation and answer the following questions:

- What are the negative consequences of the accumulation of sediment in reservoirs? *As a reservoir fills up with sediment, less water can be impounded. This reduces the generating capacity and the lifespan of a dam. A dam prevents nutrient-rich sediment from overflowing the river and fertilizing the low-lying agricultural areas.*
- How can siltation be remediated or fixed? What are some of the environmental disadvantages of these techniques? *Removal is the only way to reverse excess siltation. This is done by dredging silt from behind the dam or using large equipment to dig it out. Both are labor intensive, use fossil fuels, and are expensive.*

module 38

REVIEW

In this module, we have seen that biomass and water are two important renewable energy sources. Biomass is a modern source of carbon, which was formed between a few years ago and hundreds of years ago, as opposed to fossil fuels, which contain carbon formed millions of years ago. Solid biomass includes wood, charcoal, and animal manure, all of which are relatively low-quality sources of energy and release a fair amount of particulates and other pollutants when burned. Liquid fuels include ethanol, an alcohol derived from plant material such as corn, and biodiesel, produced from vegetable oils such as soybean. Algae is another source of oil for biodiesel. Hydroelectricity is generated from the energy in water. The largest hydroelectric projects come from impounding water behind a large dam and releasing it periodically when electricity is needed. The impounded water behind a dam can promote recreational and economic opportunities but can also have numerous impacts on the environment. Water availability in a region may be somewhat variable. In the next module we will examine two continuous, nondepletable sources of renewable energy: the Sun and wind.

Module 38 AP® Review Questions

1. Which is NOT a form of biomass?
   (a) Coal
   (b) Charcoal
   (c) Manure
   (d) Municipal solid waste
   (e) Ethanol

2. A hydroelectric power plant's rate of electricity generation depends on
   I. The flow rate of the water
   II. The vertical distance the water falls
   III. The amount of water behind the dam

   (a) I only
   (b) I and II only
   (c) II only
   (d) II and III only
   (e) I and III only

3. Which is true of solid biofuels?
   (a) Charcoal is the primary replacement when wood is scarce.
   (b) Indoor air pollution from them results in millions of deaths annually.
   (c) Switchgrass is a newly developed replacement for wood.
   (d) They are carbon neutral due to the net removal of forests.
   (e) The primary use of manure is as a heat source in developed nations.

4. Cellulosic ethanol is produced from
   (a) corn.
   (b) beets.
   (c) sugarcane.
   (d) switchgrass.
   (e) soy.

5. The most common method of hydroelectric generation is
   (a) run-of-the-river.
   (b) tidal.
   (c) water impoundment.
   (d) gorge dams.
   (e) siltation.

Answers to Module 38 AP® Review Questions

1. a
2. b
3. b
4. d
5. c

module

39

Solar, Wind, Geothermal, and Hydrogen

After biomass and water, the most important forms of renewable energy come from the Sun and wind. These nondepletable sources of renewable energy represent the fastest growing forms of energy development throughout the world.

Learning Objectives

After reading this module, you should be able to

- list the different forms of solar energy and their application.
- describe how wind energy is harnessed and its contemporary uses.
- discuss the methods of harnessing the internal energy from Earth.
- explain the advantages and disadvantages of energy from hydrogen.

TEACHING with FIGURES ▶

Show students Figure 39.1 and ask the following questions:

- What do the colors on the map indicate? *The colors indicate the amount of solar energy available to a flat photovoltaic solar panel in kilowatt-hours per square meter per day, averaged over a year.*
- What is the kilowatt hours per square meter per day at each of the following locations?

    Houston, TX: *5.0–5.5 kWh/m²/day*
    New York City: *4.5–5.0 kWh/m²/day*
    Phoenix, AZ: *6.0–6.5 kWh/m²/day*
    Cleveland, OH: *4.0–4.5 kWh/m²/day*

- What is the kilowatt hours per square meter per day at your location? (*Answers will vary.*)
- If you were the governor of the state of Arizona, what type of incentive would you recommend to increase the number of photovoltaic cells used in residential homes? *Answers will vary. For example, the state could offer a tax deduction for those who install a photovoltaic system of 2 kW or more in their homes. The state could also offer a subsidy for those who purchase a photovoltaic system.*

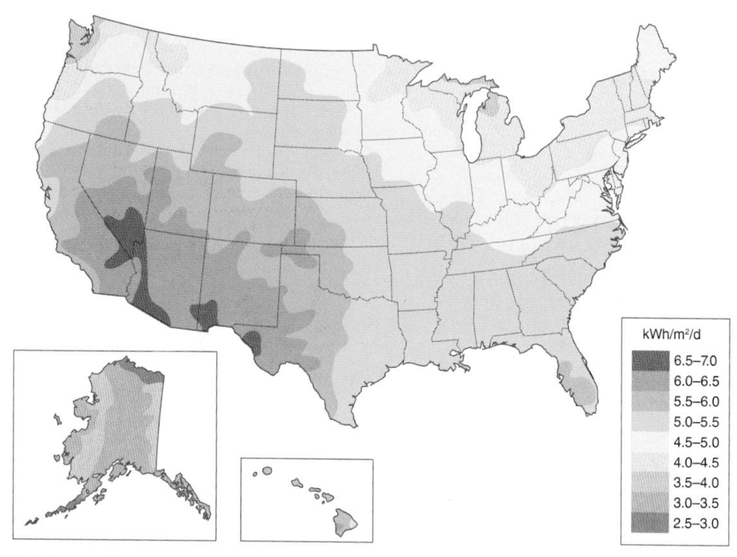

FIGURE 39.1 **Geographic variation in solar radiation in the United States.** This map shows the amount of solar energy available to a flat photovoltaic solar panel in kilowatt-hours per square meter per day, averaged over a year. *(Data from National Renewable Energy Laboratory, U.S. Department of Energy)*

The energy of the Sun can be captured directly

In addition to driving the natural cycles of water and air movement that we can tap as energy resources, the Sun also provides energy directly. Every day, Earth is bathed in solar radiation, an almost limitless source of energy. The amount of solar energy available in a particular place varies with amount of cloudiness, time of day, and season. The average amount of solar energy available varies geographically. As FIGURE 39.1 shows, average daily solar radiation in the continental United States ranges from 3 kWh of energy per square meter in the Pacific Northwest to almost 7 kWh per square meter in parts of the Southwest.

Passive Solar Heating

We have already seen several applications of passive solar heating, including positioning windows on south-facing walls to admit solar radiation in winter, covering buildings with dark roofing material in order to absorb the maximum amount of heat, and building homes into the side of a hill. None of these strategies relies on intermediate pumps or technology to supply heat. Solar ovens are another practical application of passive solar heating. For instance, a simple "box cooker" concentrates sunlight as it strikes a reflector on the top of the oven. Inside the box, the solar energy is absorbed by a dark base and a cooking pot and is converted into heat energy. The heat is distributed throughout the box by reflective material lining the interior walls and is kept from escaping by a glass top. On sunny days, such box cookers can maintain temperatures of 175°C (350°F), heat several liters of water to boiling in under an hour, or cook traditional dishes of rice, beans, or chicken in 2 to 5 hours.

Solar ovens have both environmental and social benefits. The use of solar ovens in place of firewood reduces deforestation and, in areas unsafe for travel, having a solar oven means not having to leave the relative safety of home to seek firewood. For example, over 10,000 solar ovens have been distributed in refugee camps in the Darfur region of western Sudan in Africa, where leaving the camps to find cooking fuel would put women at risk of attack (FIGURE 39.2).

Active solar energy Energy captured from sunlight with advanced technologies.

Active Solar Energy Technologies

In contrast to passive solar design, **active solar energy** technologies capture the energy of sunlight with the

FIGURE 39.2 **Solar cookers.** Residents of this refugee camp in Chad use solar cookers to conserve firewood and reduce travel outside the camp. *(Tim Dirven/Panos Pictures)*

use of technologies that include small-scale solar water heating systems, photovoltaic solar cells, and large-scale concentrating solar thermal systems for electricity generation.

Solar Water Heating Systems

Solar water heating applications range from providing domestic hot water and heating swimming pools to a variety of heating purposes for business and home. In the United States, heating swimming pools is the most common application of solar water heating, and it is also the one that pays for itself the most quickly.

A household solar water heating system, like the domestic hot water system shown in FIGURE 39.3, allows heat energy from the Sun to be transferred directly to water or another liquid, which is then circulated to a hot water heating system. The circulation of the liquid is driven either by a pump (in active systems) or by natural convection (in passive systems). In both cases, cold liquid is heated as it moves through a solar collector mounted on the roof or wall of a building or situated on the ground.

The simplest solar water heating systems pump cold water directly to the collector to be heated; the heated water then flows back to an insulated storage tank. In areas that are sunny but experience temperatures below freezing, the water is kept in the storage tank and a "working" liquid containing nontoxic antifreeze circulates in pipes between the storage tank and the solar collector. The nonfreezing circulating liquid is heated by the Sun in the solar collector, then returned to the storage tank where it flows through a heat exchanger that transfers its heat to the water. The energy needed to run the pump is usually much less than the energy gained from using the system, especially if the pump runs on electricity from the Sun. Solar water heating systems typically include a backup energy source, such as an electric heating element or a connection to a fossil fuel–based central heating system, so that hot water is available even when it is cloudy or very cold.

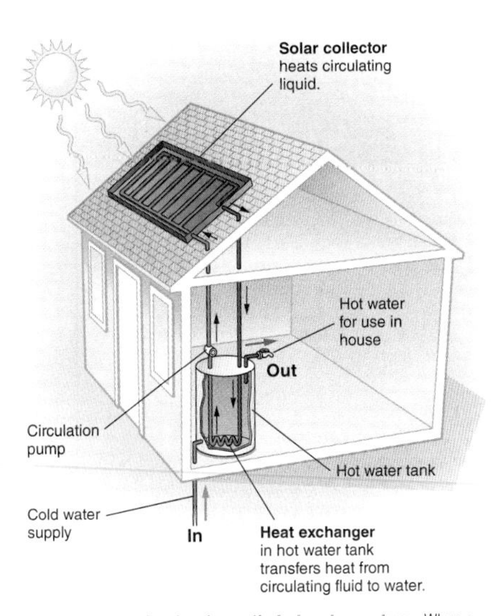

FIGURE 39.3 **A solar domestic hot water system.** When a solar hot water system is used to heat a house, a nonfreezing liquid is circulated by an electric pump through a closed loop of pipes. This circulating liquid moves from a water storage tank to a solar collector on the roof, where it is heated, and then sent back to the tank, where a heat exchanger transfers the heat to water.

Teaching Tip: Discussion Starter

Ask students the following questions to compare and contrast passive solar heating and active solar energy technologies:

- What equipment do passive solar heating and active solar energy systems use? *Passive systems need no extra equipment. Passive systems include positioning windows on south-facing walls to admit solar radiation in winter and covering buildings with dark roofing material in order to absorb the maximum amount of heat. Active systems for homes typically involve a nonfreezing liquid that moves from a water storage tank to a solar collector on the roof, where it is heated and then returns to the tank, where a heat exchanger transfers the heat to water.*
- What is the primary use of a passive solar heating system? What is the primary use of an active solar heating system? *A passive solar heating system is primarily used to heat homes. An active solar heating system is used for heating homes and can be used for heating water.*
- What is the relative cost of passive and active solar heating systems? *The cost of a passive system can be relatively low. The house would need to add double- or triple-pane windows, and the windows must be positioned to face south. The cost of an active system is higher because of the piping, heat exchanger, and solar collector.*

AP® Exam Tip

Photovoltaic systems are covered on the 2006 AP® Exam, Question 1. To answer this question, students must

- describe environmental costs and benefits of photovoltaic systems
- write an argument in favor of a solar-power system described in an accompanying article, including an analysis of the pros and cons of that system
- create and describe two possible ways that government or industry could promote photovoltaic systems for homeowners
- discuss two methods homeowners could use to incorporate passive solar design or systems into the energy plan for their homes and how these methods would reduce their energy costs

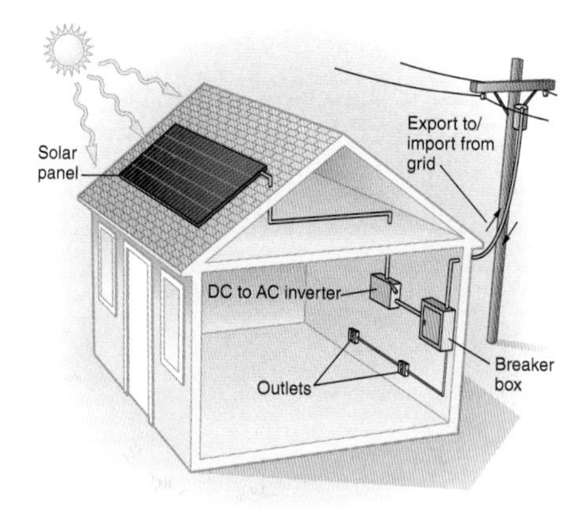

(a)

(b)

FIGURE 39.4 **Photovoltaic solar energy.** (a) In this domestic photovoltaic system, photovoltaic solar panels convert sunlight into direct current (DC). An inverter converts DC into alternating current (AC), which supplies electricity to the house. Any electricity not used in the house is exported to the electrical grid. (b) Photovoltaic panels on the roof of this house in California provide 4,200 kWh of electricity per year—nearly all of the electricity this family uses. *(Inga Spence/Alamy)*

Photovoltaic Systems

In contrast to solar water heating systems, **photovoltaic solar cells** capture energy from the Sun as light, not heat, and convert it directly into electricity. FIGURE 39.4a shows how a photovoltaic system, also referred to as PV, delivers electricity to a house. A photovoltaic solar cell makes use of the fact that certain semiconductors—very thin, ultraclean layers of material—generate a low-voltage electric current when they are exposed to direct sunlight (Figure 39.4b). The low-voltage direct current is usually converted into higher-voltage alternating current for use in homes or businesses. Typically, photovoltaic solar cells are 12 to 20 percent efficient in converting the energy of sunlight into electricity.

Electricity produced by photovoltaic systems can be used in several ways. Solar panels—arrays of photovoltaic solar cells—on a roof can be used to supply electricity to appliances or lights directly, or they can be used to charge batteries. The vast majority of photovoltaic systems are tied to the electrical grid, meaning that any extra electricity generated and not needed is sent to the electric utility, which buys it or gives the customer credit toward the cost of future electricity use. Homes that are "off the grid" may rely on photovoltaic solar cells as their only source of electricity, using batteries to store the electricity until it is needed. Photovoltaic solar cells have other uses in locations far from the grid where a small amount of electricity is needed on a regular basis. For example, small photovoltaic solar cells charge the batteries that keep highway emergency telephones working. In several U.S. cities, photovoltaic solar cells provide electricity for streetside trash compactors and for new "smart" parking meter systems that have replaced aging coin-operated parking meters.

Photovoltaic solar cell A system of capturing energy from sunlight and converting it directly into electricity.

Concentrating Solar Thermal Electricity Generation

Concentrating solar thermal (CST) systems are a large-scale application of solar energy to electricity generation. CST systems use lenses or mirrors and tracking systems to focus the sunlight falling on a large area into a small beam, in the same way you might use a magnifying glass to focus energy from the Sun and perhaps burn a hole in a piece of paper. In this case, however, the heat of the concentrated beam is used to evaporate water and produce steam that turns a turbine to generate electricity. CST power plants operate much like conventional thermal power plants; the only difference is that the energy to produce the steam comes directly from the Sun, rather than from fossil fuels. The arrays of lenses and mirrors required are large, so CST power plants are best constructed in desert areas where there is consistent sunshine and plenty of open space (FIGURE 39.5).

Although CST systems have existed for 10 years or more, they are now becoming more common. In the United States, several plants are under development in California and in the Southwest. One 35 MW plant being planned in California calls for reflectors to cover 65 ha (160 acres) of land. These plants, though, have drawbacks that include the large amount of land required and their inability to generate electricity at night.

Teaching Tip: Activity

Many students have a hard time differentiating types of solar energy. Ask students to create the following table and work with a partner to fill it in. (Answers are provided in italics.)

	Description	Advantages	Disadvantages
Passive design	*Captures the sunlight to heat a residential home using existing windows and insulation*	• *Cheapest solar design to install* • *Can help cut down heating costs*	• *Windows must face south*
Active design	*Typically involves a nonfreezing liquid circulated by an electric pump through a closed loop of pipes from a water storage tank to a solar collector on the roof, where it is heated and sent back to the tank, where a heat exchanger transfers the heat to water*	• *Relatively inexpensive to install* • *After initial investment, no cost to harvest energy* • *Reduces costs for home heating and hot water*	• *Manufacturing materials requires high input of metals and water* • *Geographically limited* • *High initial cost*
Photovoltaic cells	*Capture energy from the Sun as light, not heat, and convert it directly into electricity*	• *Nondepletable resource* • *After initial investment, no cost to harvest energy* • *Can connect to main grid so owner can sell excess energy*	• *Manufacturing materials require high input of metals and water* • *No plan in place to recycle solar panels* • *Geographically limited* • *High initial cost* • *Storage batteries required for off—grid system*
Solar thermal power plants (CST)	*Large-scale application of solar energy to electricity generation; use lenses or mirrors and tracking systems to focus the sunlight falling on a large area into a small beam*	• *Can produce energy with no fossil fuels* • *After initial investment, no cost to harvest energy*	• *Requires a large amount of land* • *Unable to generate electricity at night*

FIGURE 39.5 **A concentrating solar thermal power plant.** Mirrors and reflectors concentrate the energy of the Sun onto a "power tower," which uses the sunlight to heat water and make steam for electricity generation. *(Lowell Georgia/Science Source)*

Benefits and Drawbacks of Active Solar Energy Systems

Active solar energy systems offer many benefits such as generating hot water or electricity without producing CO_2 or polluting the air or water during operation. In addition, photovoltaic solar cells and CST power plants can produce electricity when it is needed most: on hot, sunny days when demand for electricity is high, primarily for air conditioning. By producing electricity during peak demand hours, these systems can help reduce the need to build new fossil fuel power plants.

In many areas, small-scale solar energy systems are economically feasible. For a new home located miles away from the grid, installing a photovoltaic system may be much less expensive than running electrical transmission lines to the home site. When a house is near the grid, the initial cost of a photovoltaic system may take 5 to 20 years for payback; once the initial cost is paid back, however, the electricity it generates is almost free.

Despite these advantages, a number of drawbacks have inhibited the growth of solar energy use in the United States. Photovoltaic solar panels are expensive to manufacture and install. Although the technology is changing rapidly as industrial engineers and scientists seek better, cheaper photovoltaic materials and systems, the initial cost to install a photovoltaic system can be daunting and the payback period is a long one. In parts of the country where the average daily solar radiation is low, the payback period can be even longer. Some countries, such as Germany, have made solar energy a part of their sustainable energy agenda by subsidizing their solar industry. In the United States, recent tax breaks, rebates, and funding packages instituted by various states and the federal government have made solar electricity and water heating more affordable for consumers and businesses.

The use of photovoltaic solar cells has environmental as well as financial costs. Manufacturing photovoltaic solar cells requires a great deal of energy and water and involves a variety of toxic metals and industrial chemicals that can be released into the environment during the manufacturing process, although newer types of these solar cells may reduce reliance on toxic materials. For systems that use batteries for energy storage, there are environmental costs associated with manufacturing, disposing of, or recycling the batteries, as well as energy losses during charging, storage, and recovery of electricity in batteries. The end-of-life reclamation and recycling of photovoltaic solar cells is another potential source of environmental contamination, particularly if the cells are not recycled properly. However, solar energy advocates, and even most critics, agree that the energy expended to manufacture photovoltaic solar cells is usually recovered within a few years of their operation, and that if the life span of photovoltaic solar cells can be increased to between 30 and 50 years, they will be a very promising source of renewable energy.

Wind energy is the most rapidly growing source of electricity

The wind is another important source of nondepletable, renewable energy. **Wind energy** is energy generated from the kinetic energy of moving air. As discussed in Chapter 4, winds are the result of the unequal heating of the surface of Earth by the Sun. Warmer air rises and cooler, denser air sinks, creating circulation patterns similar to those in a pot of boiling water. Ultimately, the Sun is the source of all winds—it is solar radiation and ground surface heating that drive air circulation.

Before the electrical grid reached rural areas of the United States in the 1920s, windmills dotted the landscape. Today, wind energy is the fastest-growing major source of electricity in the world. As FIGURE 39.6 shows, global installed wind energy capacity has risen from less than 10 gigawatts in 1996 to more than 300 gigawatts today. FIGURE 39.7 shows installed wind

Wind energy Energy generated from the kinetic energy of moving air.

lab
Solar Energy

This lab allows students to experiment with solar cells and to compare energy output with different variables in order to learn how passive solar systems work.

Download this lab by clicking the link buttons in the TE-book, opening the TRFD, or logging onto the book's companion website bcs.whfreeman.com/friedlandapes2e (teacher login required).

 **Lab 13.1** Solar Energy

Teaching Tip: Review

Ask students the following questions about solar energy:

- In what ways do humans capture solar energy for their use? *Humans capture solar energy in four main ways: passive energy systems, active energy systems, photovoltaic energy systems, and solar thermal power plants. Some humans also capture energy from the Sun to cook food in a solar cooker.*
- Why are active solar energy systems not feasible everywhere? *The average amount of solar energy available varies geographically. In the continental United States, average daily solar radiation ranges from 3 kWh of energy per square meter in the Pacific Northwest to almost 7 kWh per square meter in parts of the Southwest.*

Teaching Tip: Video

How Solar Panels Work
Use this short video clip from Science 360 to help students understand how solar panels work.

The link to this video may be found by clicking the link buttons in the TE-book, opening the TRFD, or logging onto the book's companion website bcs.whfreeman.com/friedlandapes2e (teacher login required).

 **Chapter 13 Web Resources**

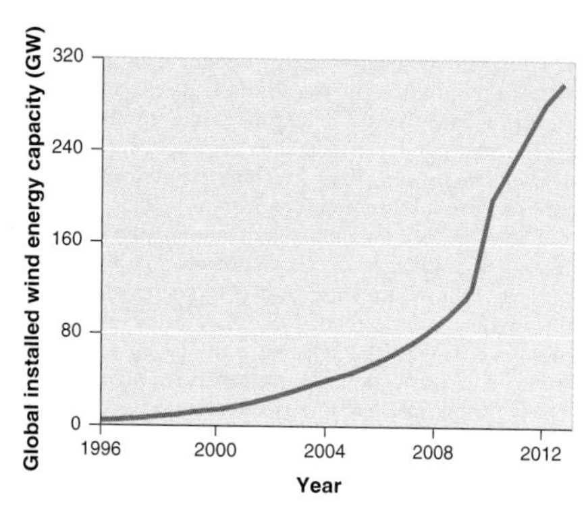

FIGURE 39.6 **Global growth of installed wind energy capacity.** Worldwide, installed wind energy capacity is now almost 300 gigawatts (GW). *(Data from Global Wind Energy Council)*

energy generating capacity and the percentage of electricity generated by wind for a number of countries. China has the largest wind energy generating capacity in the world, followed by the United States, Germany, Spain, and India.

Despite its large generating capacity, the United States obtains less than 6 percent of its electricity from wind. The largest amounts are generated in California and Texas, although more than half of U.S. states produce at least some wind-generated electricity. Denmark, a country of 5.5 million people, generates about 26 percent of its electricity from wind and hopes to increase this figure to 50 percent by 2020. Although the United States currently obtains only a small percent of its electricity from wind, it is the fastest growing source of electricity in the country.

Generating Electricity from Wind

A **wind turbine** converts the kinetic energy of moving air into electricity in much the same way that a hydroelectric turbine harnesses the kinetic energy of moving water. As you can see in FIGURE 39.8, wind turns the blades of the wind turbine and the blades transfer energy to the gear box that in turn transfers energy to the generator that generates electricity. A modern wind turbine, like the one shown, may sit on a tower as tall as 100 m (330 feet) and have blades 40 to 75 m (130–250 feet) long. Under average wind conditions, a wind turbine on land might produce electricity 25 percent of the time. While it is spinning, it might generate between 2,000 and 3,000 kW (2–3 MW), and in a year it might produce more than 4.4 million kilowatt-hours of electricity, enough to supply more than 400 homes. Offshore wind conditions are even more desirable for electricity generation, and turbines can be made even larger in an offshore environment.

Wind turbines on land are typically installed in rural locations, away from buildings and population centers. However, they must also be close to electrical transmission lines with enough capacity to transport the electricity they generate to users. For these

Wind turbine A turbine that converts wind energy into electricity.

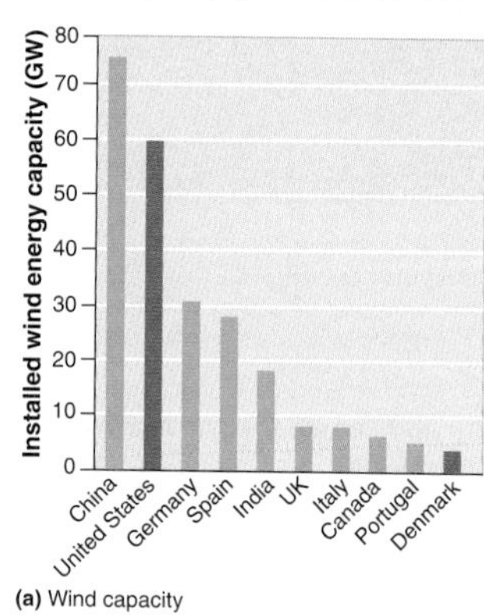

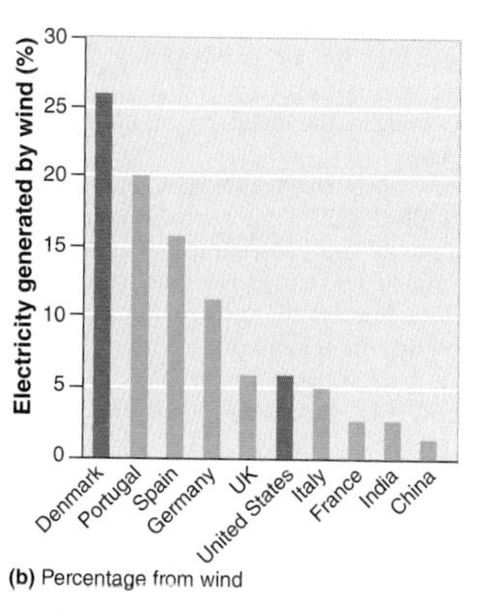

FIGURE 39.7 **Installed wind energy capacity by country.** (a) The United States generates more electricity from wind energy than any other country. (b) However, some relatively small countries, such as Denmark, generate a much higher percentage of their electricity from wind. *(After Global Wind Energy Council, 2013)*

AP® Exam Tip

The subject of converting from coal-generated electricity to wind-generated electricity is covered in the 2004 AP® Exam, Question 2. To answer this question, students must

- calculate the kWh produced by a coal-burning power plant based on data provided
- determine the amount of kWh a community needs per year based on usage data provided
- explain why the amount of kWh generated by the coal-burning plant and the amount used are not the same
- calculate the cost of the wind power the community would need to meet its current needs over a number of years based on data provided
- describe two environmental benefits and two environmental costs of switching to wind-generated electricity

Teaching Tip: Video

Energy 101: Wind Turbines—2014 Update
This brief video from the Department of Energy describes how wind turbines work and also shows the different types of locations that can be used to harvest wind.

The link to this video may be found by clicking the link buttons in the TE-book, opening the TRFD, or logging onto the book's companion website bcs.whfreeman.com/friedlandapes2e (teacher login required).

TEACHING with FIGURES ▲

Show students Figure 39.7a and 39.7b. Ask them to explain why China has a high wind energy capacity (GW) but a low percentage of electricity generation by wind. *China produces about 77 gigawatts of wind power per year, which is the largest amount in the world. However, since China has such a large population it uses a large quantity of electricity. Approximately 2 percent of the energy it uses comes from wind and 98 percent comes from other sources, primarily fossil fuels and hydroelectric.*

TRM Chapter 13 Web Resources

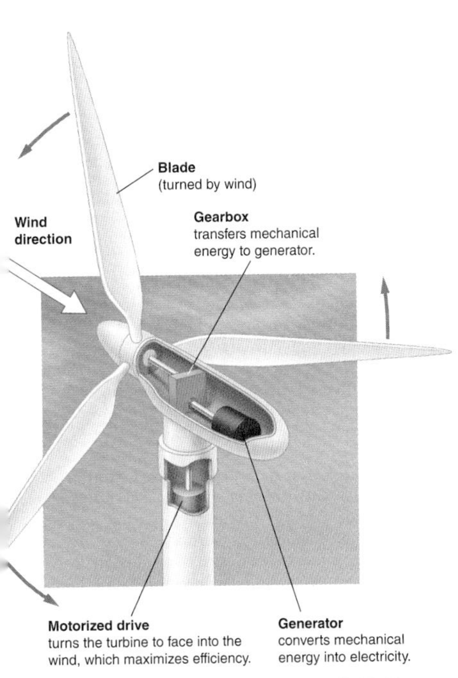

FIGURE 39.8 **Generating electricity with a wind turbine.** The wind turns the blade, which is connected to the generator, which generates electricity.

FIGURE 39.9 **Offshore wind parks.** Capacity factors at near-offshore locations like this one in Denmark are generally higher than on land. *(Max Mudie/Alamy)*

reasons, as well as for political and regulatory reasons and to facilitate servicing the equipment, the usual practice is to group wind turbines into wind farms or wind parks.

The number of wind farms is increasing in the United States and around the world. Wind farms are often placed on land in places where the wind blows up to 25 percent of the time. However, near-offshore coastal locations are even more desirable because there the wind blows up to 35 percent of the time. Offshore wind parks, which are clusters of wind turbines, are often located in the ocean within a few miles of the coastline (FIGURE 39.9). Such parks are operating in Denmark, the Netherlands, the United Kingdom, Sweden, and elsewhere. A proposed project located off Cape Cod, Massachusetts, in Nantucket Sound may become the first offshore wind farm in the United States. The project would feature 130 wind turbines with the potential to produce up to 420 MW of electricity, or up to 75 percent of the electricity used by Cape Cod and the nearby islands.

A Nondepletable Resource

Wind energy offers many advantages over other energy resources. Like sunlight, wind is a nondepletable, clean, and free energy resource; the amount available tomorrow does not depend on how much we use today. Furthermore, once a wind turbine has been manufactured and installed, the only significant energy input comes from the wind. The only substantial fossil fuel input required, once the turbines are installed, is the fuel workers need to travel to the wind farm to maintain the equipment. Thus, wind-generated electricity produces no pollution and no greenhouse gases. Finally, unlike hydroelectric, CST, and conventional thermal power plants, wind farms can share the land with other uses. For example, wind turbines on land may share the area with grazing cattle.

Wind-generated electricity does have some disadvantages, however. Currently, most off-grid residential wind energy systems rely on batteries to store electricity, but as we have discussed, batteries are expensive to

Teaching Tip: Beyond the Classroom

Ask students to research a proposed wind-farm project and write a short report. The following questions can guide their research:

- Where is the proposed wind farm going to be built? Is it on land or offshore?
- What will be the amount of kWh produced when the project is complete?
- How many turbines will the farm include and what will they cost?
- Are any local groups against the construction of the wind farm? Describe their objections.

Teaching Tip: Review

Ask students the following questions to review wind energy:

- Where are wind farms typically located and why? *Wind farms on land are typically installed in rural locations, away from buildings and population centers. However, they must also be close to electrical transmission lines that have enough capacity to transport the electricity the wind farms generate to users.*
- How is wind used to generate electricity? *A wind turbine converts the kinetic energy of moving air into electricity. A modern wind turbine sits on a tower as tall as 100 m, where the blades are positioned. While the blades are spinning, a gearbox transfers mechanical energy to the generator, which then produces electricity.*
- What are the similarities and differences between near-offshore and land-based wind farms? *Offshore wind conditions are more desirable for the generation of electricity, and larger turbines can be used in an offshore environment. Land-based wind turbines are often installed in rural locations, away from buildings and population centers. Both need to be near electrical transmission lines to move the electricity from where it is produced to where it is used by the consumer.*

AP® Exam Tip

Geothermal energy as well as other renewable and nonrenewable sources of energy are covered on the 2011 AP® Exam, Question 3. To answer this question, students must

- explain how energy is generated from a geothermal source
- describe the difference between per capita emissions of CO_2 and total emissions of CO_2
- name at least one problem in addition to CO_2 emissions that China faces with the energy sources it uses
- explain the advantages and disadvantages of using either sugarcane or tar sands as an energy source, and consider the sustainability of that source

produce and hard to dispose of or recycle. In addition, birds and bats may be killed by collisions with wind turbine blades. According to the National Academy of Sciences, as many as 40,000 birds may be killed by wind turbine blades in the United States each year—approximately four deaths per turbine. Bat deaths are not as well quantified. New turbine designs and location of wind farms away from migration paths have reduced these deaths to some extent, along with some of the noise and aesthetic disadvantages.

There has been resistance to wind farms in some regions of the United States. For example, the proposal that we mentioned to place a number of windmills in the waters off the coast of Cape Cod, Massachusetts, has languished as numerous hearings, protests, and court decisions have slowed development. In Vermont, a state often considered to be very environmentally friendly, more and more towns and individuals have argued against the installation of commercial wind projects on or near ridgelines, citing habitat fragmentation and alteration, noise, and aesthetics, among other reasons. Other states have slowed wind development by resisting the construction of electrical transmission lines, which also fragment habitat but are needed to move renewable electricity through forested areas.

Earth's internal heat is a source of nondepletable energy

Unlike most forms of renewable energy, *geothermal energy* does not come from the Sun. **Geothermal energy** is heat that comes from the natural radioactive decay of elements deep within Earth. As we saw in Chapter 8, convection currents in Earth's mantle bring hot magma toward the surface of Earth. Wherever magma comes close enough to groundwater, that groundwater is heated. The pressure of the hot groundwater sometimes drives it to the surface, where it visibly manifests itself as geysers and hot springs, like those in Yellowstone National Park. Where hot groundwater does not naturally rise to the surface, humans may be able to reach it by drilling.

Many countries obtain clean, renewable energy from geothermal resources. The United States, China, and Iceland, all of which have substantial geothermal resources, are the largest geothermal energy producers.

Geothermal energy Heat energy that comes from the natural radioactive decay of elements deep within Earth.

Ground source heat pump A technology that transfers heat from the ground to a building.

Harvesting Geothermal Energy

Geothermal energy can be used directly as a source of heat. Hot groundwater can be piped directly into household radiators to heat a home. In other cases, heat exchangers can collect heat by circulating cool liquid underground, where it is heated, and then returned to the surface. Iceland, a small nation with vast geothermal resources, heats 87 percent of its homes this way.

Geothermal energy can also be used to generate electricity. The electricity-generating process is much the same as that in a conventional thermal power plant although, in this case, the steam to run the turbine comes from water evaporated by Earth's internal heat instead of by burning fossil fuels.

The heat released by decaying radioactive elements deep within Earth is essentially nondepletable in the span of human time. However, the groundwater that so often carries that heat to Earth's surface can be depleted. As we learned in Chapter 9, groundwater, if used sustainably, is a renewable resource. Unfortunately, long periods of harvesting groundwater from a site may deplete it to the point at which the site is no longer a viable source of geothermal energy. Returning the water to the ground to be reheated is one way to use geothermal energy sustainably.

Iceland currently produces about 25 percent of its electricity using geothermal energy. In the United States, geothermal energy accounts for about 2 percent of the renewable energy used. Geothermal power plants are currently in operation in many states including California, Nevada, New Mexico, Oregon, Hawaii, and Utah. Geothermal energy has less growth potential than wind or solar energy because it is not easily accessible everywhere. Hazardous gases and steam may also escape from geothermal power plants, another drawback of geothermal energy.

Ground Source Heat Pumps

Another approach to tapping Earth's thermal resources is the use of **ground source heat pumps,** which take transfer heat from the ground to a building by taking advantage of the high thermal mass of the ground. Earth's temperature about 3 m (10 feet) underground remains fairly constant year-round, at 10°C to 15°C (50°F–60°F), because the ground retains the Sun's heat more effectively than does the ambient air. We can take advantage of this fact to heat and cool residential and commercial buildings. Although the heat tapped by ground source heat pumps is often referred to informally as "geothermal," it comes not from geothermal energy but from solar energy.

FIGURE 39.10 shows how a ground source heat pump transfers heat from the ground to a house. In contrast to the geothermal systems just described,

Teaching Tip: Video

Energy 101: Geothermal Energy
This animated video produced by the Department of Energy describes how geothermal energy works. It reviews three different types of geothermal energy and shows students how each is different.

Find the link to this video by clicking the link buttons in the TE-book, opening the TRFD, or logging onto the book's companion website bcs .whfreeman.com/friedlandapes2e (teacher login required).

Chapter 13 Web Resources

Teaching Tip: Video

Energy 101: Geothermal Heat Pumps
This animated video produced by the Department of Energy describes how ground source heat pumps or geothermal heat pumps work. It reviews how homeowners can reduce their heating and cooling bills by installing such a system.

Find the link to this video by clicking the link buttons in the TE-book, opening the TRFD, or logging onto the book's companion website bcs .whfreeman.com/friedlandapes2e (teacher login required).

Chapter 13 Web Resources

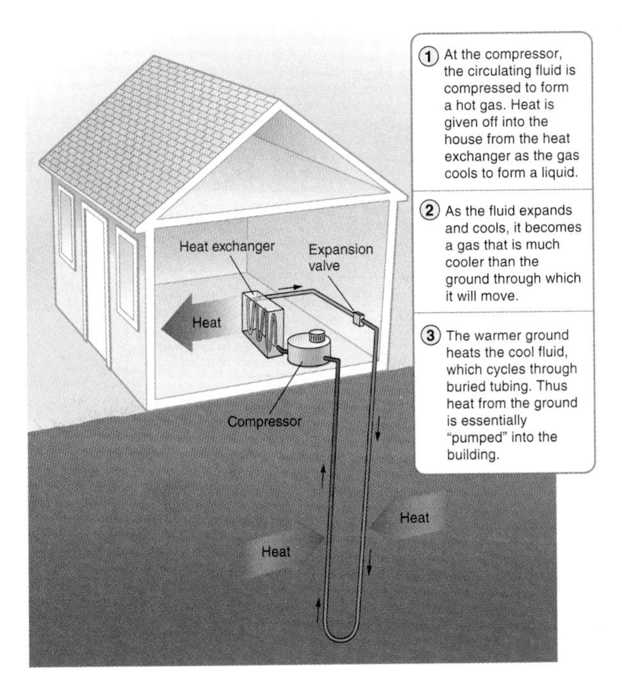

1 At the compressor, the circulating fluid is compressed to form a hot gas. Heat is given off into the house from the heat exchanger as the gas cools to form a liquid.
2 As the fluid expands and cools, it becomes a gas that is much cooler than the ground through which it will move.
3 The warmer ground heats the cool fluid, which cycles through buried tubing. Thus heat from the ground is essentially "pumped" into the building.
Heat exchanger
Expansion valve
Heat
Compressor
Heat
Heat

FIGURE 39.10 **Heating and cooling with a ground source heat pump.** By exchanging heat with the ground, a ground source heat pump can heat and cool a building using 30 to 70 percent less energy than traditional furnaces and air conditioners.

ground source heat pumps do not remove steam or hot water from the ground. In much the same way that a solar water heating system works, a ground source heat pump cycles fluid through pipes buried underground. In winter, this fluid absorbs heat from underground. The slightly warmed fluid is compressed in the heat pump to increase its temperature even more, and the heat is distributed throughout the house. The fluid is then allowed to expand, which causes it to cool and run through the cycle again, picking up more heat from the ground. In summer, when the underground temperature is lower than the ambient air temperature, the fluid is cooled underground and then pulls heat from the house as it circulates, resulting in a cooler house as heat is transferred underground.

Ground source heat pumps can be installed anywhere in the world, regardless of whether there is geothermal energy accessible in the vicinity. The operation of the pump requires some energy, but in most cases the system uses 30 to 70 percent less energy to heat and cool a building than a standard furnace or air conditioner.

Hydrogen fuel cells have many potential applications

We end our coverage of sustainable energy types with one additional energy technology that has received a great deal of attention for many years: hydrogen fuel cells.

The Basic Process in a Fuel Cell

A **fuel cell** is an electrical-chemical device that converts fuel, such as hydrogen, into an electrical current. A fuel cell operates much like a common battery, but with one key difference. In a battery, electricity is generated by a reaction between two chemical reactants, such as nickel and cadmium. This reaction happens in a closed container to which no additional materials can be added; eventually the reactants are

Fuel cell An electrical-chemical device that converts fuel, such as hydrogen, into an electrical current.

Teaching Tip: Video

Energy 101: Fuel Cell Technology
This animated video produced by the Department of Energy describes many possible applications of hydrogen fuel cell technology. It reviews how hydrogen is obtained and how it can revolutionize the way we travel.

The link to this video may be found by clicking the link buttons in the TE-book, opening the TRFD, or logging onto the book's companion website bcs .whfreeman.com/friedlandapes2e (teacher login required).

 **Chapter 13 Web Resources**

Teaching Tip: Review

Ask students: What is geothermal energy, and what is its source? *Geothermal energy is heat that comes from the natural radioactive decay of elements deep within Earth. Groundwater is heated wherever magma comes close enough. The pressure of the hot groundwater sometimes drives it to the surface as geysers or hot springs, like those in Yellowstone National Park. In some locations, humans may be able to reach the hot magma by drilling.*

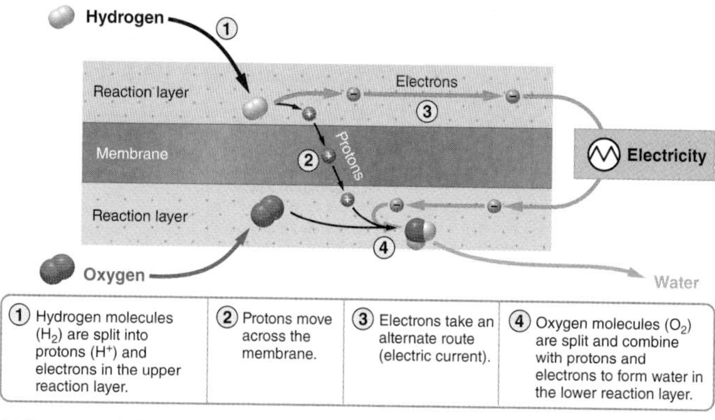

(a) One common fuel cell design

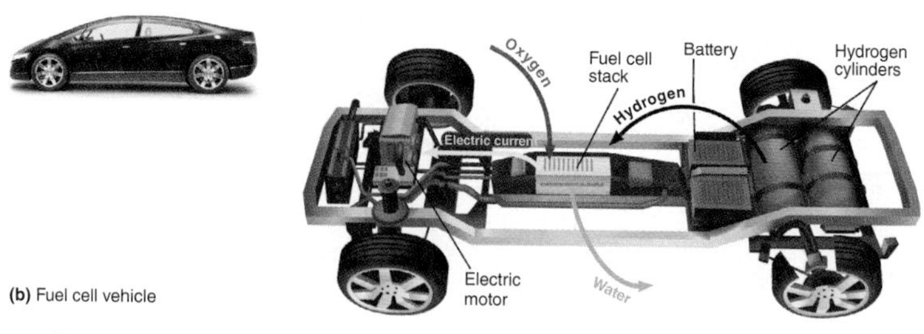

(b) Fuel cell vehicle

FIGURE 39.11 **Power from a hydrogen fuel cell.** (a) Hydrogen gas enters the cell from an external source. Protons from the hydrogen molecules pass through a membrane, while electrons flow around it, producing an electric current. Water is the only waste product of the reaction. (b) In a fuel cell vehicle, hydrogen is the fuel that reacts with oxygen to provide electricity to run the motor.

Teaching Tip: Warm-up

Ask students: Before hydrogen fuel cells can become a viable source of energy for transportation, three problems must be investigated. What are they?

- *How can we obtain hydrogen without expending more fossil fuel energy than we would save by using it?*
- *How can hydrogen suppliers safely deliver hydrogen to consumers?*
- *How can hydrogen be stored?*

used up and the battery goes dead. In a fuel cell, however, the reactants are added continuously to the cell, so the cell produces electricity for as long as it continues to receive fuel.

FIGURE 39.11 shows how hydrogen functions as one of the reactants in a hydrogen fuel cell. Electricity is generated by the reaction of hydrogen with oxygen, which forms water:

$$2H_2 + O_2 \rightarrow \text{energy} + 2H_2O$$

Although there are many types of hydrogen fuel cells, the basic process forces protons from hydrogen gas through a membrane, while the electrons take a different pathway. The movement of protons in one direction and electrons in another direction generates an electric current.

Using a hydrogen fuel cell to generate electricity requires a supply of hydrogen. Supplying hydrogen is a challenge, however, because free hydrogen gas is relatively rare in nature and because the gas is explosive. Hydrogen tends to bond with other molecules, forming compounds such as water (H_2O) or natural gas (CH_4). Producing hydrogen gas requires separating it from these compounds using either heat or electricity. Currently, most commercially available hydrogen is produced by an energy-intensive process of burning natural gas in order to extract its hydrogen; carbon dioxide is a waste product of this combustion. In an alternative process, known as **electrolysis,** an electric current is applied to water to "split" it into hydrogen and oxygen. Energy scientists are looking for other ways to obtain hydrogen; for example, under certain conditions, some photosynthetic algae and bacteria,

Electrolysis The application of an electric current to water molecules to split them into hydrogen and oxygen.

using sunlight as their energy source, can give off hydrogen gas.

Although it may seem counterintuitive to use electricity to create electricity, the advantage of hydrogen is that it can act as an energy carrier. Renewable energy sources such as wind and the Sun cannot produce electricity constantly, but the electricity they produce can be used to generate hydrogen, which can be stored until it is needed. Thus, if we could generate electricity for electrolysis using a clean, nondepletable energy resource such as wind or solar energy, hydrogen could potentially be a sustainable energy carrier.

The Viability of Hydrogen

Some policy makers consider hydrogen fuel cells to be the future of energy and the solution to many of the world's energy problems. Hydrogen fuel cells are 80 percent efficient in converting the potential energy of hydrogen and oxygen into electricity, with water as their only by-product. In contrast, thermal fossil fuel power plants are only 35 to 50 percent efficient, and they produce a wide range of pollutants as by-products. However, there are some who believe that hydrogen fuel cells will not provide a solution to our energy problems.

Despite the many advantages of hydrogen as a fuel, it also has a number of disadvantages. First, scientists must learn how to obtain hydrogen without expending more fossil fuel energy than its use would save. This means that the energy for the hydrogen generation process must come from a renewable resource such as wind or solar energy rather than fossil fuels. Second, suppliers will need a distribution network to safely deliver hydrogen to consumers—something similar to our current system of gasoline delivery trucks and gasoline stations. Hydrogen can be stored as a liquid or as a gas, although each storage medium has its limitations. In a fuel cell vehicle, hydrogen would probably be stored in the form of a gas in a large tank under very high pressure. Vehicles would have to be redesigned with fuel tanks much larger than current gasoline tanks to achieve an equivalent travel distance per tank. There is also the risk of a tank rupture, in which case the hydrogen might catch fire or explode.

Given these obstacles, why is hydrogen even considered a viable energy alternative? Ultimately, hydrogen-fueled vehicles could be a sustainable means of transportation because a hydrogen-fueled car would use an electric motor. Electric motors are more efficient than internal combustion engines: While an internal combustion engine converts about 20 percent of the fuel's energy into the motion of the drive train, an electric motor can convert 60 percent of its energy into motion. So if we generated electricity from hydrogen at 80 percent efficiency, and used an electric motor to convert that electricity into vehicular motion at 60 percent efficiency, we would have a vehicle that is much more efficient than one with an internal combustion engine. Thus, even if we obtained hydrogen by burning natural gas, the total amount of energy used to move an electric vehicle using hydrogen might still be substantially less than the total amount needed to move a car fueled by gasoline. Using solar or wind energy to produce the hydrogen would lower the environmental cost even more, and the energy supply would be renewable. In those circumstances, an automobile could be fueled by a truly renewable source of energy that is both carbon neutral and pollution free.

module

39 REVIEW

In this module, we have seen that the Sun and wind provide viable sources of renewable energy in many locations. Solar energy can be used both passively, such as locating a building in a particular direction, as well as actively, such as using photovoltaic cells. Wind energy is harnessed through the use of a wind turbine, which converts the energy of moving air into electricity. Wind is the fastest growing form of new electricity generation in the world. Harnessing geothermal energy also is a good source of energy in certain locations. Hydrogen has great potential that has not yet been realized. We can use all of these energy forms in appropriate locations under the proper circumstances, which is the focus of the next module.

Teaching Tip: Journal Prompt

Ask students the following questions:

- How do we obtain hydrogen for use in fuel cells? *Supplying hydrogen is a challenge because free hydrogen gas is relatively rare in nature and because it is explosive. Hydrogen tends to bond with other molecules, forming compounds such as water (H_2O) or natural gas (CH_4). Producing hydrogen gas requires separating it from these compounds with heat or electricity. Currently, most commercially available hydrogen is produced by an energy-intensive process of burning natural gas in order to extract its hydrogen. Another process known as electrolysis uses electrical current to split water and separate hydrogen from oxygen.*
- How does a fuel cell work? *Electricity is generated by the reaction of hydrogen with oxygen, which forms water:*

  $$2\,H_2 + O_2 \rightarrow \text{energy} + H_2O$$

  The basic process forces protons from hydrogen gas through a membrane, while the electrons take a different pathway. The movement of protons in one direction and electrons in another direction generates an electric current.
- Why is hydrogen useful as an alternative to fossil fuels? *The advantage of hydrogen is that it can act as an energy carrier. Renewable energy sources such as wind and the Sun cannot produce electricity constantly, but the electricity they produce can be used to generate hydrogen, which can be stored until it is needed. Thus if we could generate electricity for electrolysis using a clean nondepletable energy resource such as wind or solar energy, hydrogen could potentially be a sustainable energy carrier.*

Answers to Module 39 AP® Review Questions

1. e
2. e
3. b
4. d
5. c

AP® Exam Tip

A comparison of battery electric vehicles (BEVs) and internal combustion engine vehicles (ICE) is covered on the 2013 Exam, Question 2. To answer this question, students must

- identify three strategies the federal government might use to encourage the use of BEVs
- calculate the cost of gasoline per mile used by an ICE, using data provided
- calculate the cost to charge a battery on a BEV and the cost of electricity per mile, using data provided
- calculate and compare the CO_2 release of each vehicle from provided data
- describe two economic impacts of having more BEVs on the road

Module 39 AP® Review Questions

1. Which is an application of passive solar technology?
   (a) Concentrating solar thermal
   (b) Photovoltaic cells
   (c) Heat pumps
   (d) Solar water heating
   (e) Solar ovens

2. On average, what percentage of time does a land-based wind turbine generate electricity?
   (a) 70 percent
   (b) 60 percent
   (c) 45 percent
   (d) 30 percent
   (e) 25 percent

3. All of the following are true about geothermal energy except
   (a) it is only available in limited areas.
   (b) it cannot be used for cooling.
   (c) it can be locally depleted due to heavy use.
   (d) ground source heat pumps require an additional source of energy.
   (e) there is a risk of hazardous gas release.

4. A hydrogen fuel cell is most similar to
   (a) an engine.
   (b) a photovoltaic cell.
   (c) a source of coal.
   (d) a battery.
   (e) a ground source heat pump.

5. Which is NOT a benefit of solar energy systems?
   (a) They typically produce electricity during peak demand.
   (b) They require very little maintenance.
   (c) They produce electricity cheaply.
   (d) They do not produce pollution while generating electricity.
   (e) They can easily be used for small-scale applications.

module 40

Planning Our Energy Future

Although renewable energy is a more sustainable energy choice than nonrenewable energy, using any form of energy has an impact on the environment. Biomass, for instance, is a renewable resource only if it is used sustainably. Overharvesting wood leads to deforestation and degradation of the land, as we saw in the description of Haiti in Chapter 3. Wind turbines can kill birds and bats, and hydroelectric turbines kill millions of fish. Manufacturing photovoltaic solar panels requires heavy metals and a great deal of water. Because all energy choices have environmental consequences, minimizing energy use through conservation and efficiency is the best approach to energy use. After we achieve that, we must make energy choices wisely, depending on a variety of environmental, economic, and convenience factors.

Learning Objectives

After reading this module, you should be able to

- discuss the environmental and economic options we must assess in planning our energy future.
- consider the challenges of a renewable energy strategy.

Our energy future depends on efficiency, conservation, and the development of renewable and nonrenewable energy resources

Each of the renewable energy resources we have discussed in this chapter has unique advantages. None of these resources, however, is a perfect solution to our energy needs. **TABLE 40.1** lists some of the advantages and limitations of each. In short, no single energy resource that we are currently aware of can replace nonrenewable energy resources in a way that is completely renewable, nonpolluting, and free of impacts on the environment. A sustainable energy strategy, therefore, must combine energy efficiency, energy conservation, and the development of renewable and nonrenewable energy resources, taking into account the costs, benefits, and limitations of each. Convenience and reliability are also important factors. Finally, logistical considerations, such as where an energy source is located and how we transport the energy from that source to users, are also important. This is particularly important with the generation of electricity from renewable sources in remote regions, which requires an electrical transmission grid to get it to users.

A renewable energy strategy presents many challenges

Energy expert Amory Lovins suggests that innovation and technological advances, not the depletion of a resource, have provided the driving force for moving from one energy technology to the next. Extending this concept to the present, one can argue that we will develop new energy technologies before we run out of the fuels on which we currently depend.

Despite their tremendous potential, however, renewable energy resources are unlikely to replace fossil fuels completely in the immediate future unless nations commit to supporting their development and use through direct funding and financial incentives such as tax cuts and consumer rebates. In fact, the U.S. Department of Energy predicts that fossil fuel consumption will continue to increase in the United States well into the middle of the twenty-first century. In spite of their extremely rapid growth, wind and solar energy still account for far less than 1 percent of all the energy produced in the United States. Government funding or other sources of capital are needed to support research to overcome the current limitations of many renewable energy resources. One limitation that is already evident relates to the transmission of renewable electricity over the electrical distribution network. Other limitations to consider are energy cost and storage.

Improving the Electrical Grid

An increased reliance on renewable energy means that energy will be obtained in many locations and will need to be delivered to other locations. Delivery can be particularly problematic when electricity for an urban area is generated at a remote location. The electrical distribution system—the grid that we described in Chapter 12—was not originally designed for this purpose. So in addition to investing in new energy sources, the United States will have to upgrade its existing electrical infrastructure—its power plants, storage capacity, and distribution networks. Approximately 40 percent of the energy used in the United States is used to generate electricity. The U.S. electricity distribution system is outdated and subject to overloads and outages, which cost the U.S. economy over $100 billion per year. There are regions of the country that cannot supply enough generating capacity to meet local needs, while in other locations the electrical infrastructure cannot accommodate all the electricity that is generated. Furthermore, the current system requires that electricity be moved long distances from power plants to consumers. Approximately 5 to 10 percent of the electricity generated is lost as it is transported along electrical transmission lines, and the greater the distance, the more that is lost. While the storage capacity of batteries improves each year, batteries are not a sustainable solution for this problem of energy loss.

Teaching Tip: Journal Prompt

Ask students to explain why approximately 5 to 10 percent of the electricity generated in the United States is lost as it is transported along electrical transmission lines. *The current electric grid is outdated and inefficient. In some situations electricity must be transmitted long distances. The greater the distance that the electricity must move across the transmission lines, the more energy that is lost or wasted. Creating a smarter grid would help electric companies become more efficient and would save energy and money for the consumer.*

Teaching Tip: Beyond the Classroom

Ask students, or groups of students, to create a poster based on a single renewable energy source. Make sure each of the resources listed in Table 40.1 is covered. The poster should include a picture and answer the following questions:

- How does this technology work?
- How can this energy source be used?
- What are some examples of its current use? Can you find any local examples?
- What are the apparent environmental impacts associated with it?
- Are there hidden environmental or social costs?
- Is this technology widely used today? Why or why not?

Call on each student or group to present at the beginning of the relevant class so the presentations form the first few minutes of your class discussion on each particular type of renewable energy.

TABLE 40.1 Comparison of renewable energy resources

Energy resource	Advantages	Disadvantages
Liquid biofuels	• Potentially renewable • Can reduce our dependence on fossil fuels • Reduce trade deficit • Possibly more environmentally friendly than fossil fuels	• Loss of agricultural land • Higher food costs • Lower gas mileage • Possible net increase in greenhouse gas emissions
Solid biomass	• Potentially renewable • Eliminates waste from environment • Available to everyone • Minimal technology required	• Deforestation • Erosion • Indoor and outdoor air pollution • Possible net increase in greenhouse gas emissions
Photovoltaic solar cells	• Nondepletable resource • After initial investment, no cost to harvest energy	• Manufacturing materials requires high input of metals and water • No plan in place to recycle solar panels • Geographically limited • High initial costs • Storage batteries required for off-grid systems
Solar water heating systems	• Nondepletable resource • After initial investment, no cost to harvest energy	• Manufacturing materials requires high input of metals and water • After initial investment, no cost to harvest energy and water • No plan in place to recycle solar panels • Geographically limited • High initial costs
Hydroelectricity	• Nondepletable resource • Low cost to run • Flood control • Recreation	• Limited amount can be installed in any given area • High construction costs • Threats to river ecosystems • Loss of habitat, agricultural land, and cultural heritage; displacement of people • Siltation
Tidal energy	• Nondepletable resource • After initial investment, no cost to harvest energy	• Potential disruptive effect on some marine organisms • Geographically limited
Geothermal energy	• Nondepletable resource • After initial investment, no cost to harvest energy • Can be installed anywhere (ground source heat pump)	• Emits hazardous gases and steam • Geographically limited
Wind energy	• Nondepletable resource • After initial investment, no cost to harvest energy • Low up-front cost	• Turbine noise • Deaths of birds and bats • Geographically limited to windy areas near transmission lines • Aesthetically displeasing to some • Storage batteries required for off-grid systems
Hydrogen fuel cell	• Efficient • Zero Pollution	• Producing hydrogen is an energy-intensive process • Lack of distribution network • Hydrogen storage challenges

Many people are focusing their attention on improving the electrical grid to make it as efficient as possible at moving electricity from one location to another, thereby reducing the need for storage capacity.

An energy economy based on nondepletable energy sources requires reliable electricity storage and affordable—or at least effective and efficient—distribution networks. U.S. energy scientists maintain that because we currently do not have a cost-effective, reliable means of storing energy, we should not depend on intermittent sources such as wind and solar energy for more than about 20 percent of our total electricity production since it could lead to risky instability in the grid.

Smart grid An efficient, self-regulating electricity distribution network that accepts any source of electricity and distributes it automatically to end users.

One solution currently in development may be the **smart grid,** an efficient, self-regulating electricity distribution network that accepts any source of electricity and distributes it automatically to end users.

Teaching Tip: Review

Write the major renewable energy sources on the board: hydroelectric, solar, geothermal, wind, biomass, tidal. Ask students to write down the advantages and disadvantages of each. Choose one student for each resource to share his or her answers with the rest of the class. Correct student answers as needed.

ssions (pollutants and enhouse gases)	Electricity cost ($/kWh)	Energy return on energy investment
and methane		1.3 (from corn) 8 (from sugar cane)
Carbon monoxide Particulate matter Nitrogen oxides Possible toxic metals from MSW Danger of indoor air pollutants		
None during operation Some pollution generated during manufacturing of panels	0.2	8
None during operation Some pollution generated during manufacturing of panels		
Methane from decaying flooded vegetation	.05–.11	12
None during operation		15
None during operation	.05–.30	8
None during operation	.04–.06	18
None during operation		8

A smart grid uses computer programs and the Internet to tell electricity generators when electricity is needed and electricity users when there is excess capacity on the grid. In this way, it coordinates energy use with energy availability. In late 2009, President Obama announced a plan to invest $3.4 billion in smart grid technology. Since that time, industry contributions have brought the total investment to almost $8 billion, which has been used to fund roughly 100 smart grid projects around the country.

How does a smart grid work? FIGURE 40.1 shows one example. With "smart" appliances plugged into a smart grid, at bedtime a consumer could set an appliance such as a dishwasher to operate before he wakes up the following morning. A computer on the dishwasher would be programmed to run it anytime between midnight and 5:00 AM, depending on when there is a surplus of electricity. The dishwasher's computer would query the smart grid and determine the optimal time, in terms of electricity availability, to turn on the appliance. The smart grid could also help manage electricity demand so that peak loads do not become too great. We cannot control the timing of all electricity demand, but by improving consumer awareness of electricity abundance and shortages, using smart appliances, and setting variable pricing for electricity, we can make electricity use much more regular, and thus more sustainable.

Our current electrical infrastructure relies on a system of large energy producers—regional electricity generation plants. When one plant goes off-line or shuts down, the reduction in available generating capacity puts greater demands on the rest of the system. Some energy experts maintain that a better system would consist of a large number of small-scale electricity generation "parks" that rely on a mix of fossil fuel and renewable energy sources. These experts maintain that a system of decentralized energy parks would be the least expensive and most reliable electrical infrastructure to meet our future needs. Small, local energy parks would save money and energy by transporting electricity a shorter distance. Such decentralized generators would also be less likely to suffer breakdowns or sabotage. Since each small energy park might serve only a few thousand people, widespread outages would be much less likely.

Addressing Energy Cost and Storage

The major impediments to widespread use of wind, solar, and tidal energy—the forms of renewable energy with the least environmental impact—are cost and the limitations of energy storage technology. Fortunately, the cost of renewable energy has been falling. For example, in some markets wind energy is now cost-competitive with natural gas and coal. Throughout this book we have seen that the efficiency of production improves with technological advances and experience. In general, as we produce more of something, and get experience from making it, we learn to produce it less expensively. Production processes become dramatically more efficient, more companies enter the market, and developing new technologies has a clear payoff. For the consumer, this technological advancement also has the benefit of lowering prices: For electricity generation from solar, wind, and natural gas, we have seen that

Teaching Tip: Journal Prompt

Ask students: What factors must be considered for a successful sustainable energy strategy for the United States? How achievable do you think such a strategy would be in the next decade? Explain your answer. *Answers will vary.*

AP® Exam Tip

The difference between renewable and nonrenewable resources is covered on the 1999 AP® Exam, Question 3. To answer this question, students must

- describe the difference between renewable and nonrenewable resources and provide an example of each
- contrast total per capita resource use in developed and developing nations
- explain the meaning of sustainable resource use and provide an example
- discuss one policy that helps with the goal of sustainable resource use and one policy that hinders it

Teaching Tip: Discussion Starter

Ask students to describe a scenario that uses smart grid technology and how that technology would reduce a homeowner's electric bill. *With "smart" appliances plugged into a smart grid, a consumer can set an appliance such as a dishwasher to operate during the night when a surplus of electricity exists on the power grid. The dishwasher's computer would query the smart grid and determine the optimal time, in terms of electricity availability, to turn on the appliance. The smart grid could also help manage electricity demand so that peak loads do not become too great.*

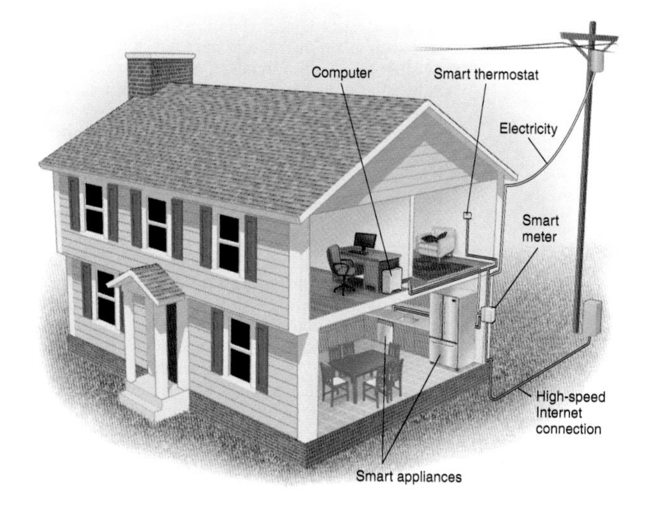

FIGURE 40.1 Using a smart grid. A smart grid optimizes the use of energy in a home by continuously coordinating energy use with energy availability.

costs tend to decline in a fairly regular way as installed capacity grows.

What are the implications of this relationship between experience and efficiency? In general, any technology that has been in widespread use has an advantage over a newer technology because it is familiar and because the less expensive something is, the more people will buy it, leading to further reductions in its price. State and federal subsidies and tax incentives also help to lower the price of a technology. Tax credits and rebates have been instrumental in reducing the cost of solar and wind energy systems for consumers.

Similarly, in time, researchers will develop solutions to the problem of creating efficient energy storage systems, which might reduce the need to transport electricity over long distances. One very simple and effective approach is by using the excess capacity of off-peak hours to pump water uphill to a reservoir. Then, during hours of peak demand, operators can release the water through a turbine to generate the necessary electricity— cleanly and efficiently. Research into battery technology and hydrogen fuel cell technology continues.

Progress on these and other technologies may accelerate with government intervention, taxes on industries that emit carbon dioxide, or a market in which consumers are willing to pay more for technologies with minimal environmental impacts. In the immediate future, we are more likely to move toward a sustainable energy mix if nonrenewable energy becomes more expensive. Consumers have shown more willingness to convert in large numbers to renewable energy sources, or to engage in further energy conservation, when fossil fuel prices increase. We have already seen instances of this shift in behavior. In 2008, energy conservation increased when oil prices rose rapidly to almost $150 per barrel and gasoline in most of the United States cost more than $4 per gallon. People used public transportation more often, drove more fuel-efficient vehicles, and carpooled more than they did before the price spike.

Other ways to spur conservation are initiatives that regulate the energy mix itself—for example, by encouraging that a certain fraction of electricity be generated using renewable energy sources. One such initiative is the Regional Greenhouse Gas Initiative (RGGI), whereby nine eastern states have committed to reducing greenhouse gas emissions from electricity generation plants in this decade.

module

40

REVIEW

In this module, we have seen that conservation and efficiency need to be considered simultaneously with a wide variety of renewable energy sources. Environmental, economic, and convenience considerations must be evaluated when comparing renewable energy options against each other and with nonrenewable fuels. The path we take in our energy future will depend on the potential for a given energy source to meet the needs of the country, along with its availability and issues of reliability, storage, and accessibility. Challenges include improving the electrical grid and the adoption of smart grid technologies to address some of the challenges that are specific to renewable and sometimes intermittent energy supplies.

Module 40 AP® Review Questions

1. Which of the following is NOT a disadvantage of liquid biofuels?
   (a) They are associated with lower gas mileage.
   (b) They create more carbon monoxide than fossil fuels.
   (c) They can contribute to a loss of agricultural land.
   (d) They can increase food costs.
   (e) They can lead to an increase in greenhouse gas emissions.

2. Which aspects of renewable energy electricity generation require updating the electricity transmission grid?
   I. Electricity generators are located in numerous, remote locations.
   II. There is a need to transport electricity long distances.
   III. There are storage problems due to the unpredictable nature of some renewables.

   (a) I and II only
   (b) I and III only
   (c) II and III only
   (d) III only
   (e) I, II, and III

3. The smart grid does NOT
   (a) use the Internet to coordinate energy use and energy availability.
   (b) reduce the variability in electricity demand.
   (c) have the potential to provide a cheap way to store electricity.
   (d) increase the need for variable pricing of electricity.
   (e) allow for the use of hydrogen fuel cells as a part of the electricity grid.

4. Which renewable energy source has become cost-competitive with fossil fuels?
   (a) Tidal
   (b) Geothermal
   (c) Wind
   (d) Solar photovoltaic
   (e) Concentrated solar

5. Which will NOT increase adoption of renewable technologies?
   (a) Increased cost of fossil fuels
   (b) A carbon dioxide emissions tax
   (c) Cheaper energy storage
   (d) Decreased government subsidies
   (e) Increased production efficiency

Answers to Module 40 AP® Review Questions

1. b
2. e
3. c
4. c
5. d

Building an Alternative Energy Society in Iceland

The people of Iceland use more energy per capita than the people of any other nation, including the United States. However, the energy Icelanders use is almost all in the form of local, renewable resources that do not pollute or contribute greenhouse gases to the environment.

This isolated European island nation has had to learn to be self-sufficient in energy or suffer the high cost of importing fuel. When the Vikings first came to Iceland over a thousand years ago, they relied on biomass, in the form of birch wood and peat, for fuel. The resulting deforestation, and the slow regrowth of forests in Iceland's cold temperatures and limited sunlight, restricted human population growth and economic development for the next thousand years. With the beginning of the Industrial Revolution in the eighteenth century, Iceland began to supplement its biomass fuel with imported coal, but the expense of importing coal also limited its economic growth.

In the late nineteenth and early twentieth centuries, Iceland began to look to its own resources for energy. It began by tapping its abundant freshwater resources to generate hydroelectricity, which became the country's major energy source for residential and industrial use. This transition led to the general electrification and economic modernization of the country and greatly reduced its dependence on imported fossil fuels. Iceland did not stop with hydroelectricity, however, but sought ways to utilize its other major renewable energy source: the thousands of geysers and hot springs on this volcanic island that would provide ready access to geothermal energy.

Geothermal energy is now the primary energy source for home heating in Iceland, and geothermal and hydroelectric resources provide energy for nearly all electricity generation in the country. Even so, the potential of these resources remains relatively underutilized. Iceland has harvested only about 17 percent of its hydroelectric potential, and there is even more potential in its geothermal resources.

Despite its commitment to renewable energy, Iceland is still dependent on imported fossil fuels to run its cars, trucks, buses, and fishing vessels. In a further push for energy independence, Iceland is now developing new technologies that could harness another vast and clean source of energy, hydrogen. In 2000, in partnership with the automobile company DaimlerChrysler, the European energy giant Norsk Hydro, and the oil company Royal Dutch Shell, Iceland embarked on an ambitious project that could wean the country from fossil fuels by 2050. The plan is to use its sustainably generated electricity to split water and obtain hydrogen, then use that hydrogen as fuel.

In April 2003, Iceland opened one of the world's first filling stations for hydrogen-fueled vehicles. In August of that year, hydrogen-fueled buses for public transportation were introduced in Reykjavik, the capital of Iceland. Hydrogen-fueled rental cars are also available. Although the global banking crisis of 2008 changed near-term projections significantly, Iceland could have the world's first hydrogen-fueled ocean fishing fleet by 2030, and could become a net exporter of clean hydrogen energy. Iceland is an example of an ongoing experiment that takes advantage of one's resources to develop clean, sustainable energy sources.

Critical Thinking Questions

1. Can the knowledge and experience gained in Iceland be applied to many other parts of the world? Why or why not?
2. What is the most up-to-date information on hydrogen fuel cell use in automobiles in the United States? How many hydrogen fueling stations are there and where are they?

References

Blanchette, S. 2008. A hydrogen economy and its impact on the world as we know it. *Energy Policy* 36:522–530.

Maack, M. H., and J. B. Skulason. 2006. Implementing the hydrogen economy. *Journal of Cleaner Production* 14:52–64.

A hydrogen-fueled bus in Reykjavik, Iceland. All 80 buses in the Reykjavik fleet were converted to hydrogen fuel cells in 2003. *(Martin Bond/Science Source)*

Suggested Answers to Critical Thinking Questions

1. While some of the knowledge and advances developed can be applied to other regions, the abundance of hydroelectricity and geothermal energy is unique to Iceland.

2. This will vary based on when the question is answered.

chapter 13

REVIEW

In this chapter, we have examined the role of conservation as well as increased energy efficiency in reducing the demand for energy. We have described the different categories of renewable energy and examined the two most prominent renewable energy sources: biomass and energy from flowing and standing water. Biomass energy contains modern carbon and can be obtained from wood, charcoal, and animal wastes. Energy can be harnessed from both standing water and free flowing water, typically to generate electricity. Solar energy can be harnessed both passively and actively. The most prominent active collection of solar energy comes from photovoltaic cells that convert sunlight into electricity. Wind energy is harnessed directly and a wind turbine is very similar to the turbines used to generate electricity from fossil fuels. Geothermal energy from Earth can be used in specific locations. Hydrogen is a fuel that has much promise but is not likely to be used widely anytime soon. Each renewable energy resource has its advantages and disadvantages and these can be considered from both environmental and economic perspectives.

Key Terms

Energy conservation
Tiered rate system
Peak demand
Passive solar design
Thermal mass
Potentially renewable
Nondepletable
Renewable
Biofuel
Modern carbon
Fossil carbon
Carbon neutral
Net removal
Ethanol
Biodiesel
Flex-fuel vehicle
Hydroelectricity
Run-of-the-river
Water impoundment
Tidal energy
Siltation
Active solar energy
Photovoltaic solar cell
Wind energy
Wind turbine
Geothermal energy
Ground source heat pump
Fuel cell
Electrolysis
Smart grid

Learning Objectives Revisited

Module 37 Conservation, Efficiency, and Renewable Energy

- **Describe strategies to conserve energy and increase energy efficiency.**

  Turning down the thermostat and driving fewer miles are examples of steps individuals can take to conserve energy. Buying appliances that use less energy and switching to compact fluorescent light bulbs are examples of steps individuals can take to increase energy efficiency. Buildings that are carefully designed for energy efficiency can save both energy resources and money. Reducing the demand for energy can be an equally effective or a more effective means of achieving energy sustainability than developing additional sources of energy.

- **Explain differences among the various renewable energy resources.**

  Renewable energy resources include nondepletable energy resources, such as the Sun, wind, and moving water, and potentially renewable energy resources, such as biomass. Potentially renewable energy resources will be available to us as long as we use them sustainably.

Module 38 Biomass and Water

- **Describe the various forms of biomass.**

  Biomass is one of the most common sources of energy in the developing world, but biomass energy is also used in developed countries. In theory, biomass energy is carbon neutral; that is, the carbon

Teaching Tip: Measuring Your Impact

Choosing a Light Bulb
Measuring Your Impact 13 asks students to calculate the costs of using incandescent and fluorescent light bulbs. The link to this resource may be found by clicking on the link buttons in the TE-book, opening the TRFD, or logging onto the book's companion website bcs.whfreeman.com/friedlandapes2e (teacher login required).

Measuring Your Impact 13: Choosing a Light Bulb

Exploring the Literature

Friedland, A. J., and K. T. Gillingham. 2010. Carbon accounting a tricky business (letter). *Science* 327:410–411.

International Energy Agency: http://www.iea.org.

Kruger, P. 2006. *Alternative Energy Resources.* John Wiley & Sons.

Randolph, J., and G. M. Masters. 2008. *Energy for Sustainability.* Island Press.

Smith, T. P. 2014. Fish, tides and turbines. *Natural History* 121:32–39.

U.S. Department of Energy, National Renewable Energy Laboratory: http://www.nrel.gov.

U.S. Energy Information Administration, Independent Statistics and Analysis: http://www.eia.doe.gov.

Wolfson, R. 2011. *Energy, Environment, and Climate.* 2nd ed. W. W. Norton.

Download a printable version of this list by clicking on the link buttons in the TE-book, opening the TRFD, or logging onto the book's companion website bcs.whfreeman.com/friedlandapes2e (teacher login required).

Exploring the Literature 13

produced by combustion of biomass should not add to atmospheric carbon concentrations because it comes from modern, rather than fossil, carbon sources. Wood is a potentially renewable resource because, if harvests are managed correctly, it can be a continuous source of biomass energy. Ethanol and biodiesel have the potential to supply large amounts of renewable energy, but growing and processing these fuels makes demands on land and energy resources.

- **Explain how energy is harnessed from water.**

  Most hydroelectric systems use the energy of water impounded behind a dam to generate electricity. Run-of-the-river hydroelectric systems impound little or no water and have fewer environmental impacts, although they often produce less electricity.

Module 39 Solar, Wind, Geothermal, and Hydrogen

- **List the different forms of solar energy and their application.**

  Passive solar energy takes advantage of relatively inexpensive strategies such as the direction windows are facing in a building. Active solar technologies use technology to obtain heat or electrical energy from the Sun and have high initial costs but can potentially supply relatively large amounts of energy. Active solar applications can be small, such as those that fit on a rooftop or in a field, or they can be extremely large, on an industrial scale.

- **Describe how wind energy is harnessed and its contemporary uses.**

  Wind turbines can be located on land or in the near-offshore environment. Frequently, a number of wind turbines are grouped together in wind farms. Wind is the most rapidly growing source of renewable electricity. It is a clean, nondepletable energy resource, but objections to wind farms are increasing because of aesthetics, sound, and hazards the turbines pose to birds and bats.

- **Discuss the methods of harnessing the internal energy from Earth.**

  Geothermal energy from underground can heat buildings directly or can generate electricity. However, geothermal power plants must be located in places where geothermal energy is accessible.

- **Explain the advantages and disadvantages of energy from hydrogen.**

  The only waste product from a hydrogen fuel cell is water, but obtaining hydrogen gas for use in fuel cells is an energy-intensive process. If hydrogen could be obtained using renewable energy sources, it could become a truly renewable source of energy.

Module 40 Our Energy Future

- **Discuss the environmental and economic options we must assess in planning our energy future.**

  Although many scenarios have been predicted for the world's energy future, conserving energy, increasing energy efficiency, relying on renewable energy sources to a greater extent, and improving energy distribution and storage through new technologies will all be necessary to achieve energy sustainability. Despite these clearly stated goals, fossil fuel consumption continues and it does not appear that its use will decrease any time soon.

- **Consider the challenges of a renewable energy strategy.**

  Improving the electrical grid in the United States is vital if we are to increase reliance on renewable forms of electricity. However, because the grid is so widespread, expanding and maintaining its geographic spread and electrical capacity are expensive. Even with an expanded grid, there are numerous obstacles to increasing renewable electricity generation. The high economic cost—at least initially—for renewable forms of electricity is a challenge. Also, the difficulty and economic cost of storing electricity have no easy solutions at present.

468 Chapter 13 Achieving Energy Sustainability

Chapter 13 AP® Environmental Science Practice Exam

Section 1: Multiple-Choice Questions

Choose the best answer for questions 1–13.

1. Which of the following is NOT an example of a potentially renewable or nondepletable energy source?
   (a) Hydroelectricity
   (b) Solar energy
   (c) Nuclear energy
   (d) Wind energy
   (e) Geothermal energy

2. Renewable energy sources are best described as
   (a) those that are the most cost-effective and support the largest job market.
   (b) those that are, or can be, perpetually available.
   (c) those that are dependent on increasing public demand and decreasing supply.
   (d) those that are being depleted at a faster rate than they are being replenished.
   (e) those that are reusable, and therefore eliminate waste energy released into the environment.

3. An energy-efficient building might include all of the following except
   (a) building materials with low thermal mass.
   (b) a green roof.
   (c) southern exposure with large double-paned windows.
   (d) reused or recycled construction materials.
   (e) photovoltaic solar cells as a source of electricity.

4. Which source of energy is NOT (ultimately) solar-based?
   (a) Wind
   (b) Biomass
   (c) Tides
   (d) Coal
   (e) Hydroelectricity

5. Which of the following demonstrate(s) the use of passive solar energy?
   I. A south-facing room with stone walls and floors
   II. Photovoltaic solar cells for the generation of electricity
   III. A solar oven

   (a) I only
   (b) II only
   (c) III only
   (d) I and III
   (e) II and III

6. A study of small wind turbines in the Netherlands tested the energy output of several models. The results for two models are shown in the graphs below. Which statement can be inferred from these data?
   (a) As wind speed increases, energy output decreases.
   (b) The annual energy output of model 1 can exceed 6,000 kWh.
   (c) As energy output surpasses 50 kWh per week, noise pollution increases.
   (d) Model 2 is likely to cause more bird and bat deaths.
   (e) Model 1 is more cost-effective.

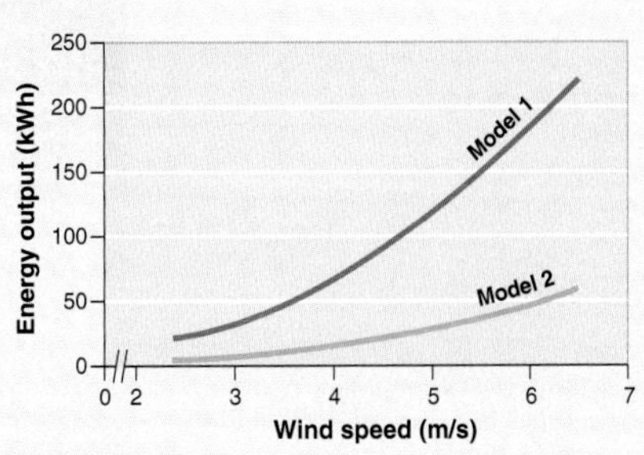

(a) Weekly energy output

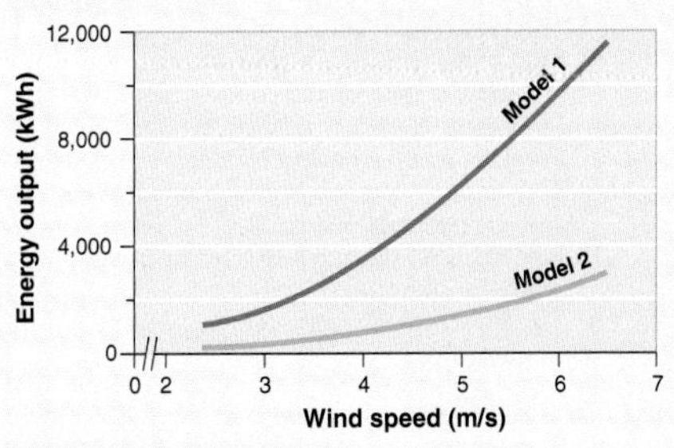

(b) Annual energy output

7. The primary sources of renewable energy in the United States are
   (a) solar and wind energy.
   (b) hydroelectricity and tidal energy.
   (c) biomass and hydroelectricity.
   (d) geothermal and tidal energy.
   (e) wind and geothermal energy.

Answers to Chapter 13 AP® Exam Multiple-Choice Questions

1. c
2. b
3. a
4. c
5. d
6. b
7. c
8. d
9. d
10. c
11. d
12. c
13. e

Answers to Chapter 13 AP® Exam

Download the full answers to the Chapter 13 AP® Practice Exam by clicking on the link buttons in the TE-book, opening the TRFD, or logging onto the book's companion website bcs.whfreeman.com/friedlandapes2e (teacher login required).

Answers to Chapter 13 AP® Exam

Additional Multiple-Choice AP® Practice Exam

Download an additional Multiple-Choice AP® Environmental Science Practice Exam for Chapter 13 by clicking on the link buttons in the TE-book, opening the TRFD, or logging onto the book's companion website bcs.whfreeman.com/friedlandapes2e (teacher login required).

Multiple-Choice AP® Practice Exam 13

8. The environmental impacts of cutting down a forest to obtain wood as fuel for heating and cooking could include
   I. deforestation and subsequent soil erosion.
   II. release of particulate matter into the air.
   III. a large net rise in atmospheric concentrations of sulfur dioxide.

   (a) I only
   (b) II only
   (c) III only
   (d) I and II
   (e) II and III

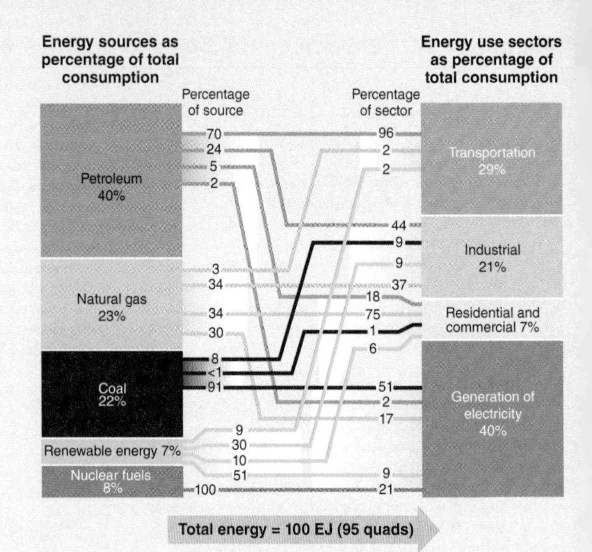

Use the diagram above, which represents annual U.S. energy consumption by source and sector for 2007, to respond to questions 9 and 10.

9. Which statement best describes the sources of energy in U.S. energy consumption patterns?
   (a) Most of the renewable energy is used in the industrial, residential, and commercial sectors.
   (b) Most of the electricity generated in the United States comes from nuclear energy.
   (c) The industrial sector is heavily dependent on coal and renewable energy.
   (d) Fossil fuels continue to be the major energy source for all sectors.
   (e) The transportation sector uses the greatest amount of energy.

10. Which best describes U.S. energy use?
    (a) Transportation is the largest end use of energy in the United States.
    (b) Transportation is fueled mainly by coal.
    (c) Electricity generation is the largest end use of energy in the United States.
    (d) Electricity generation is powered mainly by nuclear energy.
    (e) Industry is the largest end use of energy in the United States.

11. Which statement best describes the role of renewable energy in the United States?
    (a) It is the dominant source of energy.
    (b) It is the largest contributor of greenhouse gases.
    (c) It is a large contributor to the transportation sector.
    (d) Its largest contribution is to the electricity generation sector.
    (e) It is never sustainable.

12. What is the fuel source for a flex-fuel vehicle?
    (a) Electricity
    (b) Biodiesel
    (c) E-85
    (d) Solar
    (e) Cellulose

13. In order to best achieve energy sustainability, humans must consider which of the following strategies?
    I. Building large, centralized power plants
    II. Improving energy efficiency
    III. Developing new energy technologies

    (a) I only
    (b) II only
    (c) III only
    (d) I and II
    (e) II and III

470 Chapter 13 Achieving Energy Sustainability

Section 2: Free-Response Questions

Write your answer to each part clearly. Support your answers with relevant information and examples. Where calculations are required, show your work.

1. Hydroelectricity provides about 7 percent of the electricity generated in the United States.
   (a) Explain how a hydroelectric power plant converts energy stored in water into electricity. (2 points)
   (b) Identify two factors that determine the amount of electricity that can be generated by an individual hydroelectric power plant. (2 points)
   (c) Describe the two main types of land-based hydroelectric power plants. (2 points)
   (d) Describe two economic advantages and two environmental disadvantages of hydroelectricity. (4 points)

2. The following table shows the amounts of electricity generated by photovoltaic solar cells and by wind in the United States from 2002 to 2007 (in thousands of megawatt-hours):

Energy Source	Year					
	2002	2003	2004	2005	2006	2007
PV Solar Cells	555	534	575	550	508	606
Wind	10,354	11,187	14,143	17,810	26,589	32,143

   (a) Describe the trend in electricity generation by photovoltaic solar cells from 2002 to 2007. Calculate the approximate percentage change between 2002 and 2007. (2 points)
   (b) Describe the trend in electricity generation by wind from 2002 to 2007. Calculate the approximate percentage change between 2002 and 2007. (2 points)
   (c) Identify and explain any difference between the two trends you described in (a) and (b). (2 points)
   (d) A homeowner wants to install either photovoltaic solar cells or wind turbines to provide electricity for her home in Nevada, which gets both ample sunlight and wind. Provide two arguments in favor of installing one of these technologies, and explain two reasons for not choosing the other. (2 points)
   (e) Would the installation of either PV solar cells or wind turbines be considered an application of energy conservation or of energy efficiency? Explain. (2 points)

Answers to Chapter 13 AP® Exam Free-Response Questions

1. (a) As water falls over a vertical distance, the potential energy stored in it is converted to kinetic energy. A hydroelectric power plant captures this kinetic energy and uses it to turn a turbine, just as the kinetic energy of steam turns a turbine in a coal-fired electricity generating plant. The turbine transforms the kinetic energy into electrical energy.
(b) The amount of electricity that can be generated at a particular power plant depends on two factors. First, it depends on the vertical distance through which the water falls. The greater the distance that the water falls, the more potential energy it has and the more electricity that is generated. Second, the flow rate, or amount of water that flows past a certain point per unit of time, affects the amount of electricity the power plant can generate. The higher the flow rate, the more kinetic energy is present, and the more electricity is generated.
(c) There are two main types of hydroelectric generation on land: run-of-the-river and water impoundment. In run-of-the-river hydroelectric generation, dams are usually constructed across a river. Water flows over the dam or through a channel and then returns to the river without being stored in a reservoir. In run-of-the-river hydroelectric power, relatively little flooding occurs upstream and the seasonal changes in river flow are not disrupted. This greatly alleviates the environmental impact of the project. However, it also significantly limits the generating potential and reliability of the energy production, because heavy spring runoff from rains (and in some regions, snowmelt) cannot be stored. Also, in hot, dry periods such as during the summer, the flow of water is low and energy is not produced.
Water impoundment is the storage of water in a reservoir behind a dam. It is the most common type of hydro power because it usually allows for electricity generation on demand, rather than only during periods of heavy water flow. As with run-of-the-river, electricity is generated by water flowing through a turbine, but the flow rate of water and subsequent energy generation is controlled by the gates of the dam. Water impoundment can produce more power than run-of-the-river generation.
(d) Economic advantages of hydroelectricity:

- The operating expenses of the dam are not subject to fluctuations in energy costs.
- Revenue from recreational use of the lake behind the dam can help the local economy.
- In the absence of drought, hydropower is often very low cost.

Environmental disadvantages of hydroelectricity:

- It can cause a disruption in the life cycle of certain species.
- It alters the ecological systems both above and below the dam, often displacing people or flooding ancient cultural or archaeological sites.

Answers continue in the Answer Appendix on page ANS-10.

Answers to Unit 6 AP® Exam Multiple-Choice Questions

1. c
2. b
3. b
4. c
5. b
6. d
7. d
8. e
9. a
10. d
11. b
12. a
13. f
14. e
15. a
16. c
17. b
18. c
19. e
20. a

Unit 6 AP® Environmental Science Practice Exam

Section 1: Multiple-Choice Questions

Choose the best answer for questions 1–20.

1. Within a developing nation, an increase in the use of subsistence energy sources would most likely be caused by
   (a) a decrease in the availability of straw, sticks, animal dung, and other local sources of fuel.
   (b) an increase in the availability of oil.
   (c) an increase in the cost of oil.
   (d) the loss of forested land.
   (e) a reduction in waste energy generated by subsistence energy sources.

Questions 2 and 3 refer to following table:

Energy type	Energy return on energy investment	MJ per kilogram of fuel
Biodiesel	1	40
Coal	80	24
Ethanol from corn	1	30
Ethanol from sugarcane	5	30
Natural gas	10	54

2. Which fuel is most likely the least expensive fuel to produce?
   (a) Biodiesel
   (b) Coal
   (c) Ethanol from corn
   (d) Ethanol from sugarcane
   (e) Natural gas

3. How many kilograms of fuel will be consumed if two people travel 200 km in a single car that uses biodiesel fuel, assuming that the car expends 3 MJ per passenger-kilometer?
   (a) 7.5 kg
   (b) 15 kg
   (c) 22 kg
   (d) 66 kg
   (e) 100 kg

4. ________ is likely to reduce the efficiency of a power plant.
   (a) Increasing the capacity factor of the plant
   (b) Using anthracite coal instead of bituminous coal
   (c) Shutting off a turbine driven by exhaust gases
   (d) Pumping cold water from a nearby stream into the condenser
   (e) Releasing ash residues into the environment

5. Which of the following is a benefit of using petroleum instead of coal for energy?
   I. Petroleum releases about 15 percent more CO_2 than coal.
   II. Petroleum is easier to transport than coal.
   III. Mining and transport of petroleum is less harmful to the environment.

   (a) I only
   (b) II only
   (c) III only
   (d) I and II
   (e) II and III

6. Which of the following is true regarding natural gas?
   (a) Extraction and combustion of natural gas have less effect on global warming than extraction and combustion of coal.
   (b) Contamination of water during the extraction process is of little concern.
   (c) Liquefied petroleum gas is a slightly more energy-dense than natural gas.
   (d) Pipelines are the primary means of transporting natural gas.
   (e) Exploration for natural gas requires less energy than exploration for other energy sources.

7. In 1969, M. King Hubbert calculated lower and upper estimates for the volume of total world petroleum reserves. Regardless of which estimate was used, he predicted that 80 percent of the total reserves would be used up in approximately 60 years. Which of the following likely explains why his predictions were the same for both estimates?
   I. The lower and upper estimates were not different enough to cause a major significant shift in his predictions.
   II. Per capita energy use will increase with available energy.
   III. Availability and use of petroleum is inversely correlated with cost.

   (a) I only
   (b) II only
   (c) III only
   (d) II and III
   (e) I, II, and III

Answers to Unit 6 AP® Exam

Download the full answers to the Unit 6 AP® Practice Exam by clicking on the link buttons in the TE-book, opening the TRFD, or logging onto the book's companion website bcs.whfreeman .com/friedlandapes2e (teacher login required).

Answers to Unit 6 AP® Exam

8. In a nuclear power plant, control rods are used to
   (a) control the placement of fuel rods.
   (b) increase the efficiency of nuclear reactions.
   (c) transfer heat energy from the fuel rods into water.
   (d) increase the capacity factor of the plant.
   (e) absorb excess neutrons emitted by fuel rods.

9. One particularly large nuclear power plant produces about 400 kilocuries of krypton per year. Krypton has a half-life of 10 years. How radioactive will the krypton waste generated at the power plant in a single year be after 30 years?
   (a) 50 kilocuries
   (b) 100 kilocuries
   (c) 200 kilocuries
   (d) 800 kilocuries
   (e) 7,500 Becquerels

10. Which of the following reduces the capacity factor of nuclear power plants?
    (a) The need for long-term storage of radioactive waste
    (b) Government regulation of low-level radioactive waste
    (c) Competition with plants that use coal to produce energy
    (d) The tendency for fuel rods to overheat and risk a meltdown
    (e) The relative supplies of ^{235}U and ^{238}U isotopes

11. Which of the following is a renewable energy source that does not originate from solar radiation?
    (a) Biomass
    (b) Geothermal
    (c) Nuclear
    (d) Wind
    (e) Ground heat

12. A homeowner wants to save money and energy by heating her home using a passive solar design. She could consider
    (a) building a green roof.
    (b) using only Energy Star appliances.
    (c) placing photovoltaic cells on the roof.
    (d) installing a ground source heat pump.
    (e) storing solar energy in a battery.

13. Suppose you own a diesel car and the current price of diesel gasoline is $4.50 per gallon. You pay $1,500 for parts that allow the engine to run on straight vegetable oil (SVO), which costs $1.50 per gallon. The car gets 50 miles per gallon when running on either type of fuel. After how many miles will you recoup the cost of engine conversion?
    (a) 1,000
    (b) 1,500
    (c) 5,000
    (d) 10,000
    (e) 15,000
    (f) 20,000

14. Relative to burning fossil carbon, the use of modern carbon for energy
    (a) rarely contributes to the removal of vegetation.
    (b) does not use energy that originates from the Sun.
    (c) is always subsidized by government programs.
    (d) is cheaper and more efficient.
    (e) is more likely to be carbon-neutral.

15. Which of the following consequences of water impoundment for hydroelectric energy production is most likely to release greenhouse gases?
    (a) Flooding of forests and grasslands
    (b) Siltation
    (c) The generation of energy by use of turbines
    (d) Transfer of energy to power lines
    (e) Use of excess energy from the hydroelectric plant to pump water into the impoundment

Unit 6 AP® Environmental Science Practice Exam 473

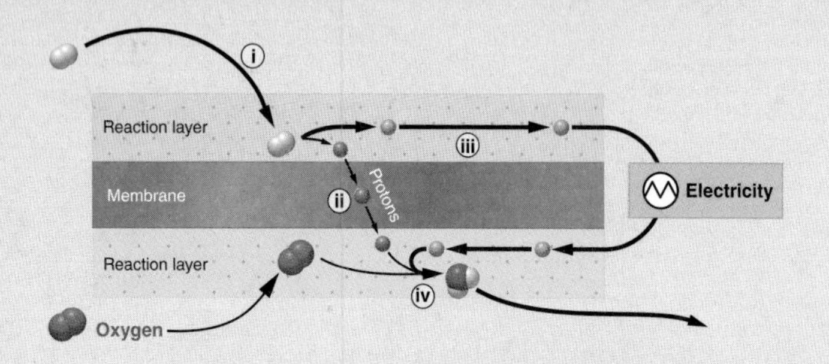

Question 16 refers to the diagram above:

16. In this depiction of a hydrogen powered fuel cell, components i, ii, iii, and iv refer to
    (a) hydrogen, protons, neutrons, water.
    (b) hydrogen, protons, hydrogen ions, water.
    (c) hydrogen, protons, electrons, water.
    (d) hydrogen, electrons, protons, water.
    (e) hydrogen, electrons, protons, hydrogen ions.

17. The combined use of concentrated solar thermal (CST) and fossil fuel energy generation occurs most efficiently when
    (a) CST and fossil fuel plants are built near each other.
    (b) CST energy is used during peak hours whereas fossil fuels are used during off-peak hours.
    (c) CST and fossil fuel energy are equally distributed to the power grid.
    (d) CST plants are built in desert areas where there is consistent sunshine and plenty of open space.
    (e) CST contributes to more energy production when the price of fossil fuels increases.

18. Which of the following sources of electricity employ the use of turbines?
    (a) Photovoltaic cells, windmills, run-of-the-river hydroelectric plants
    (b) Photovoltaic cells, windmills, water impoundment hydroelectric dams
    (c) Concentrated solar thermal plants, nuclear plants, windmills
    (d) Concentrated solar thermal plants, water impoundment hydroelectric dams, fuel cells
    (e) Nuclear plants, fossil fuel plants, photovoltaic cells.

19. Which of the following is NOT likely to increase the efficiency of energy use and production in the United States?
    (a) Government subsidies for active solar construction designs
    (b) Increasing the cost of fossil fuels
    (c) Replacement of large power plants with several smaller plants
    (d) Use of smart grid systems and smart appliances
    (e) Increasing the production of batteries

20. A smart grid system
    I. regulates electrical energy usage according to electrical energy availability.
    II. requires more energy-efficient electricity production.
    III. will reduce the cost of energy production.

    (a) I only
    (b) II only
    (c) III only
    (d) I and II
    (e) I, II, and III

Section 2: Free-Response Questions

Write your answer to each part clearly. Support your answers with relevant information and examples. Where calculations are required, show your work.

1. Although hybrid and electric vehicles release less carbon dioxide than fully gas-powered vehicles, there remains much debate over whether hybrid vehicles are better for the environment and whether they are truly more fuel-efficient.
   (a) Provide three reasons why hybrid and electric vehicles might be less energy-efficient than gas-powered cars. (3 points)
   (b) How might the lifetime of a vehicle alter its total ecological footprint? (2 points)
   (c) Suppose a new all-electric vehicle costs $40,000 and requires 1.5 MJ per passenger-mile. You trade in your old gas-powered vehicle, which required 4 MJ per passenger-mile, for $2,000.
      (i) Supposing that a gallon of gasoline contains 20 MJ and costs $4.00, how many miles must you drive before your purchase becomes cost effective? (3 points)
      (ii) How might the use of "smart grid" technology reduce the environmental footprint of an electric vehicle? (2 points)

2. The process of growing algae for use as a biofuel has become an attractive and exciting prospect for the energy needs of our future. However, there is still much debate regarding the ultimate environmental benefits of algae production at the scale needed to significantly replace other forms of biodiesel.
   (a) Why is biofuel produced by algae considered to be carbon neutral? (2 points)
   (b) Many have questioned the ultimate sustainability of algae biofuel production based on physical principles.
      (i) What is the second law of thermodynamics? (1 point)
      (ii) What does the second law of thermodynamics suggest about the efficiency of algal biofuel production over many decades? (1 point)
   (c) What are two similarities of algae biofuel production to the production of corn ethanol, and what are two differences? (2 points)
   (d) Where may be the best location to place an algal-biofuel generating plant that would maximize the efficiency of fuel production? (2 points)
   (e) Many researchers are exploring how to genetically engineer strains of algae that are fast-growing and lipid-rich in order to maximize the amount of fuel created in relation to production area. List two ways in which this could potentially harm the environment and suggest two ways to mitigate harmful effects. (2 points)

Answers to the Unit 6 Free-Response Questions can be found in the Answer Appendix on page ANS-10.

Teaching Tip: Unit 6 Additional Free-Response Question

An additional free-response question is available for you to test concepts from Unit 6 with your students. This free-response question asks students to consider the long-term benefits of wind energy. To answer this question, students must

- describe how the sun and rotation of the Earth combine to produce the energy needed to run a wind turbine
- explain what determines the capacity factor of a windmill
- use provided data to calculate the number of wind turbines needed to generate enough electricity for the United States
- list three factors related to the construction and installation of wind turbines that might offset the carbon benefits of the turbines

Download this resource by clicking on the link buttons in the TE-book, opening the TRFD, or logging onto the book's companion website bcs.whfreeman.com /friedlandapes2e (teacher login required).

Unit 6 Additional Free-Response Question

Fedor Selivanov/Shutterstock

6

scienceapplied

Should Corn Become Fuel?

Corn-based ethanol is big business—so big, in fact, that to offset demand for petroleum, U.S. policy has required an increase in annual ethanol production from 34 billion liters (9 billion gallons) in 2008 to 136 billion liters (36 billion gallons) by 2022.

Ethanol proponents maintain that substituting ethanol for gasoline decreases air pollution, greenhouse gas emissions, and our dependence on foreign oil. Opponents counter that growing corn and converting it into ethanol uses more energy than we recover when we burn the ethanol for fuel. Perhaps more importantly, those energy inputs release additional carbon into the atmosphere. Are the policies that encourage corn ethanol production firmly grounded in science, or are they simply a concession to politically powerful farming interests?

Does ethanol reduce air pollution?

Ethanol (C_2H_6O) and gasoline (a mixture of several compounds, including heptane, C_7H_{16}) are both hydrocarbons. Under ideal conditions, in the presence of enough oxygen, burning hydrocarbons produces only water and carbon dioxide. In reality, however, gasoline-only vehicles always produce some carbon monoxide (CO) because there can be insufficient oxygen present at the time of combustion. Carbon monoxide has direct effects on human health and also contributes to the formation of photochemical smog (see Chapter 15 for more on CO and air pollution). Modern car engines regulate the fuel/oxygen mix to maximize the combustion of gasoline and minimize CO production. However, these regulatory systems are not fully operational until a car has warmed up to its normal operating temperature. As a result, whenever you start your engine, you release some CO. Many older cars do not have these regulators at all.

Because ethanol is an **oxygenated fuel**—a fuel with oxygen as part of the molecule—adding ethanol to the fuel mix of a car should ensure that more oxygen is present and that combustion is more complete, which would reduce the production of CO. However, an ethanol/gasoline blend evaporates more readily than pure gasoline, and in the atmosphere, this evaporated fuel produces photochemical smog in the same way that CO does. In other words, the fuel may burn more cleanly, but the extra evaporation may counteract some of this benefit.

Oxygenated fuel A fuel with oxygen as part of the molecule.

Does ethanol reduce greenhouse gas emissions?

Biofuels are modern carbon, not fossil carbon. In theory, burning biofuels should not introduce additional carbon into the atmospheric reservoir because the carbon captured in growing the crops and the carbon released in burning the fuel should cancel each other out. However, agriculture in the United States depends heavily on inputs of fossil fuels to grow and process crops. Does the argument that ethanol use is carbon neutral hold up when we examine the entire ethanol production cycle from farm to filling station?

To analyze this question further, we need to examine the energy return on energy investment (EROEI) for ethanol, or how much energy we get out of ethanol for every unit of energy we put in.

Scientists at the U.S. Department of Agriculture have analyzed this problem, examining the energy it takes to grow corn and convert it into ethanol (the inputs) and the return on this energy investment (the outputs). Energy to run farm machinery, to produce chemicals (especially nitrogen fertilizer), and to dry the corn are inputs. Additional energy is required to transport the corn, convert it into ethanol, and ship it. Ethanol is the primary output, but in the course of growing and processing the corn, several by-products are produced, including distiller's grains, corn gluten, and corn oil. Each by-product would have required energy to produce if it had been manufactured independently of the ethanol manufacturing process, and so energy "credit" is assigned to these by-products. As FIGURE SA6.1 shows, there is a slight gain of usable energy when corn is converted into ethanol: For every unit of fossil fuel energy we put in, we produce about 1.3 units of ethanol. A land area of 0.4 ha (1 acre) produces enough net energy to displace about 236 L (62 gallons) of gasoline.

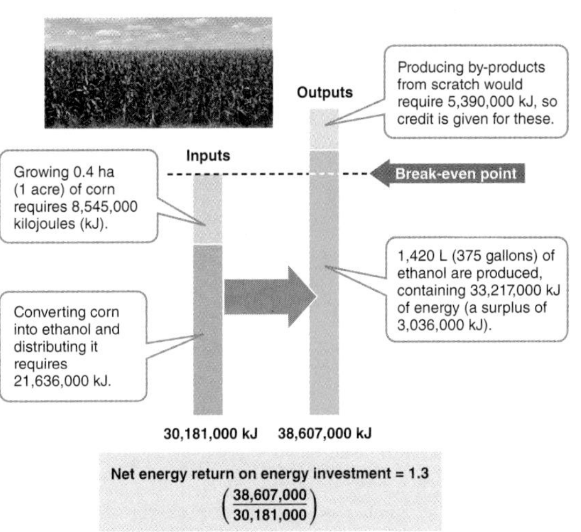

FIGURE SA6.1 **Energy required to produce ethanol.** An analysis of the energy costs of growing and converting 0.4 ha (1 acre) of corn into ethanol shows a slight gain of usable energy when corn is converted into ethanol. *(Lee Wilcox)*

Our analysis indicates that replacing part of our gasoline needs with ethanol will displace some greenhouse gases. There will be a climate benefit, but perhaps not to the extent that one would imagine because we rely so heavily on fossil fuels to support agriculture. Moreover, in the United States, the ethanol production process currently uses more coal than natural gas. Because coal emits nearly twice as much CO_2 per joule of energy as natural gas (see Chapter 12), producing the ethanol may reverse many of the benefits of replacing the fossil carbon in gasoline with the modern carbon in ethanol. Quite possibly, producing ethanol with coal releases as much carbon into the atmosphere as simply burning gasoline in the first place.

Finally, we have to take into account the corn-growing process itself. Various aspects of corn production, such as plowing and tilling, may release additional CO_2 into the atmosphere from organic matter that otherwise would have remained undisturbed in the A and B horizons of the soil. Furthermore, greater demand for corn will increase pressure to convert land that is forest, grassland, or pasture into cropland. There is increasing evidence from recent studies that these conversions result in a net transfer of carbon from the soil to the atmosphere and lead to additional increases in atmospheric CO_2 concentrations.

Does ethanol reduce our dependence on gasoline?

If there is a positive energy return on energy investment, then using ethanol should reduce the amount of gasoline we use and therefore the amount of foreign oil we must import. However, if in our attempt to use less gasoline we use more coal to produce ethanol, we might increase the release of greenhouse gases, so the overall benefit is questionable. Furthermore, if we create greater demand for a crop that until now has been primarily used as a food source, there are other implications as well.

Replacing a significant percentage of U.S. gasoline consumption with corn-based ethanol would involve the large-scale conversion of cropland from food to fuel production. Even if we converted every acre of potential cropland to ethanol production, we could not produce enough ethanol to displace more than 20 percent of U.S. annual gasoline consumption. Furthermore, if all cropland in the U.S. were devoted to ethanol production, all agricultural products destined for the dinner table would have to be imported from other countries. Clearly this is not a practical solution.

What seems more likely is that we will be able to replace some smaller fraction of gasoline consumption with biofuels. Lester Brown of the Earth Policy Institute points out, however, that the 10 bushels of corn that it takes to produce enough ethanol to fill a 95-L (25-gallon) SUV fuel tank contain the number of calories needed to feed a person for about a year. He argues that higher ethanol demand would increase corn and other grain prices and would thus make it harder for lower-income people around the world to afford food (FIGURE SA6.2).

FIGURE SA6.2 **Rising food prices.** Increased demand for corn-based ethanol has accounted for about 10 to 15 percent of the rise in food prices in recent years, according to the Congressional Budget Office. Here activists dressed in corn costumes protest the use of food for fuel in front of the European parliament in Brussels. *(AP Photo/ Yves Logghe)*

Indeed, in the summer of 2007, corn prices in the United States rose to $4 per bushel, roughly double the price in prior years, primarily because of the increased demand for ethanol. Since then, prices have stayed above $3 per bushel and at times have gone above $7 per bushel. People in numerous countries have had difficulty obtaining food because of these higher grain prices. In a number of years, most notably 2008 and 2011, there were food riots in Afghanistan, Algeria, Bangladesh, Egypt, Haiti, and Tunisia, among other places around the globe.

Our analysis indicates that corn-based ethanol has the potential to displace a small amount of total U.S. gasoline consumption, but that increasing corn ethanol consumption to the levels suggested by some policy makers may require troublesome trade offs between driving vehicles and feeding the world.

Are there alternatives to corn ethanol?

Stimulating demand for corn ethanol may spur the development of another ethanol technology—**cellulosic ethanol,** an ethanol derived from cellulose. Cellulose is the material that makes up the cell walls of plants: Grasses, trees, and plant stalks are made primarily of cellulose. If we were able to produce large quantities of ethanol from cellulose, we could replace fossil fuels with fuel made from a number of sources. Ethanol could be manufactured from fast-growing grasses such as switchgrass or *Miscanthus* (FIGURE SA6.3), tree species that require minimal energy input, as well as from many waste products such as discarded paper and agricultural waste. It is also possible that algae could be used as the primary material for ethanol.

Producing cellulosic ethanol requires breaking cellulose into its component sugars before distillation. This is a difficult and expensive task because the bonds between the sugar molecules are very strong. One method of breaking down cellulose is to mix it with enzymes that sever these bonds. In 2007, the first commercial cellulosic ethanol plant was built in Iowa. At the moment, however, cellulosic ethanol is more expensive to produce than corn ethanol.

How much land would it take to produce significant amounts of cellulosic ethanol? Some scientists suggest that the impact of extensive cellulosic ethanol production would be very large, while others have calculated that, with foreseeable technological improvements, we could replace all of our current gasoline consumption without large increases in land under cultivation or significant losses in food production. Because the technology is so new, it is not yet clear who is correct. There will still be other considerations, such as the impact on biodiversity whenever land is dedicated to growing biofuels.

The good news is that many of the raw materials for cellulosic ethanol are perennial crops such as grasses. These crops do not require the high energy, fertilizer, and water inputs commonly used to grow annual plants such as corn. Furthermore, the land used to grow grass would not need to be plowed every year. Fertilizers and pesticides would also be unnecessary, eliminating the large inputs of energy needed to produce and apply them.

Algae may be an even more attractive raw material for cellulosic ethanol because its production would not need to utilize land that could otherwise be used for growing food crops or serving as a repository for carbon.

FIGURE SA6.3 **A potential source of cellulosic ethanol.** *Miscanthus*, a fast-growing tall grass, may be a source of ethanol in the future. *(Frank Dohleman)*

Cellulosic ethanol An ethanol derived from cellulose, the cell wall material in plants.

478 Science Applied 6 Should Corn Become Fuel?

Many forms of ethanol are now being produced or are very close to being produced commercially. Production methods that use biomass grown with fewer fossil fuel inputs at a lower level of intensity on agricultural land, or without the use of land at all, are the most desirable from virtually all environmental perspectives.

What's the bottom line?

Adding ethanol to gasoline reduces carbon monoxide formation and may reduce photochemical smog if evaporation can be controlled. But when we look at the entire "life cycle" of ethanol, from farm to tank, it's not clear if ethanol will actually reduce greenhouse gas emissions. Because its EROEI is fairly low, and because the amount of available agricultural land is limited, corn ethanol will not replace more than a small fraction of gasoline consumption. Cellulosic ethanol shows the potential to have a significant effect on fossil fuel use, at least in part because of lower energy inputs required to obtain the raw material and convert it into a fuel. Corn should not be used for fuel to any great extent, but it may serve as a stepping-stone to other biofuels.

Questions

1. Legislation requiring that ethanol be added to gasoline was passed to improve air quality and reduce dependence on foreign oil. Do you think the latter goal is achieved by adding ethanol to gasoline?
2. What are some of the other environmental impacts of using corn-based ethanol that are not illustrated by energy calculations shown here?

Free-Response Question

Write your answer to each part clearly. Support your answers with relevant information and examples. Where calculations are required, show your work.

The production of ethanol from corn is much like the production of any alcoholic beverage. Corn is harvested, ground, and cooked to form a "mash." Yeast are added to the mash, which use the sugar to grow and reproduce. As they grow, yeast respire CO_2 and produce ethanol as a waste product (i.e., through fermentation). At this point, the mixture now resembles a "corn wine." Since ethanol has a lower boiling point than water, the fermented product can be boiled to evaporate the ethanol, which is a process known as distillation. Evaporated ethanol is then collected through condensation.

(a) Identify six energy inputs in the process of producing corn ethanol. (2 points)
(b) Using the answers provided in question (a), write an equation that would calculate the energy return on energy investment (EROEI) for ethanol production. (2 points)
(c) The leftover mash consists of water and corn. Ethanol producers will often dry this mash, collect the corn, and sell it as livestock feed. Why might this livestock feed be less beneficial for livestock than unfermented corn? (2 points)
(d) If the price of food increases around the world as a result of corn ethanol production, poorer communities may revert to subsistence energy production. Explain how this reversion could impact the total environmental benefit of ethanol production. (2 points)
(e) Cellulosic ethanol production is similar to corn ethanol production, except producers make use of enzymes that increase the sugar content of the mash, which in turn increases the energy available for yeast to grow. Describe two benefits and two disadvantages of cellulosic ethanol production relative to corn ethanol production.

References

R.F. Service. 2013. Battle for the Barrel. *Science* 339: 1374–1379.

Searchinger, T. D., et al. 2009. Fixing a critical climate accounting error. *Science* 326:527–528.

Key Terms

Oxygenated fuel
Cellulosic ethanol

Suggested Answers to Science Applied Questions

1. Yes, it will reduce the amount of foreign oil required, although not by very much. The other side effects mean that there are probably much more effective ways to reduce the dependence on foreign oil.

2. Those calculations do not show other environmental impacts of growing corn, such as the conversion of the land to growing corn instead of whatever its prior use was along with the damage caused by herbicides and pesticides used to grow the corn.

Answer to Science Applied Free-Response Question

Download the full answer to the Science Applied 6 Free-Response Question by clicking on the link buttons in the TE-book, opening the TRFD, or logging onto the book's companion website bcs .whfreeman.com/friedlandapes2e (teacher login required).

Answer to Science Applied 6 Free-Response Question

Answer to Science Applied Free-Response Question

(a) Energy inputs related to corn ethanol production include:

- planting the corn
- fertilizing and irrigating the corn
- harvesting the corn
- transporting the corn to the ethanol plant
- grinding the corn
- cooking the mash
- boiling the fermented mash
- condensing the ethanol vapors

(b) Answers for (a) may vary, the formula for EROEI remains the same:

$$\text{EROEI} = \frac{\text{Energy obtained from the fuel}}{\text{Energy invested to obtain the fuel}}$$

In place of "Energy invested to obtain the fuel," the student should write a summation of all the answers provided for (a).

(c) Livestock feed that is produced as a result of drying the leftover corn in the mash is less beneficial for cattle than unfermented corn because much of the sugar in the corn has been converted to ethanol and CO_2. Hence, the cattle receive less energy from the leftover mash of fermented corn than they would from unfermented corn.

(d) Subsistence energy sources are those gathered by individuals for their own immediate needs. These sources include straw, sticks, and animal dung. Often, the use of these sources requires harvesting large amounts of plant material and burning that material in highly inefficient ways. This results in loss of plant life and release of greenhouse gases. Due to the inefficiency of subsistence energy use, the amount of greenhouse gases released per kilogram of subsistence energy may be more than the amount release per kilogram of corn ethanol. Hence, large-scale production of corn ethanol might have a net-negative effect on the environment.

(e) Benefits of cellulosic energy production include a higher yield of ethanol per kilogram of plant material, the need for less land devoted to crop production, and a wider diversity of existing plants that can generate ethanol.

The primary disadvantage of cellulosic ethanol production relative to corn ethanol production is the need to invest energy in producing enzymes that can break down cellulose, which may be costly and energy-intensive to manufacture. In addition, the fermentation process is likely to require longer periods of time, owing to the time needed for enzymatic breakdown of cellulose.

chapter

14 Water Pollution

PD Overview

Chapter 14 introduces students to many different ways that water can be compromised. All organisms, including humans, require water to live, but increases in human population combined with industrialization have led to the contamination of water supplies. This chapter explores the main sources of water pollution: wastewater from humans and livestock, heavy metals from industry, oil pollution from mining and transportation, and nonchemical pollution such as thermal pollution and sediment. This chapter also describes ways in which water pollution can be remediated and discusses U.S. federal laws designed to protect waterways.

Module 41: Wastewater from Humans and Livestock

Wastewater is produced by livestock operations and human activities, including human sewage from toilets and gray water from bathing and washing clothes and dishes. This module reviews how wastewater pollution affects natural water bodies and the technologies that have been developed to treat wastewater and minimize its effects.

Module 42: Heavy Metals and Other Chemicals

Some compounds such as nitrogen and phosphorus cause environmental problems by overfertilizing the water. Other inorganic compounds, including heavy metals (lead, arsenic, and mercury), acids, and synthetic compounds (pesticides, pharmaceuticals, and hormones), can directly harm humans and other organisms. This module examines how these chemicals enter water bodies and the effects that they can have on ecosystems and humans.

Module 43: Oil Pollution

Chapter 12 noted that the pollution of Earth's oceans and shorelines from crude oil and other petroleum products is an ongoing problem. Petroleum products are highly toxic to many marine organisms, including birds, mammals, and fish, as well as to algae and microorganisms that form the base of the aquatic food chain. Oil is a persistent substance that can spread below and across the surface of the water for hundreds of kilometers and leave shorelines with a thick, viscous covering that is extremely difficult to remove. This module examines the many sources of oil pollution and then describes some of the ways currently used to clean up oil and to reduce its harmful effects.

Module 44: Nonchemical Water Pollution

When students think about water pollution, they most commonly envision scenes of dirty water contaminated with toxic chemicals. Though such scenarios certainly receive a great deal of public attention, there are other, less familiar types of water pollution. This module examines four types of nonchemical water pollution: solid waste, sediment, heat, and noise.

Module 45: Water Pollution Laws

A country's water quality improves when its citizens demand it and the country is affluent enough to afford measures to clean up pollution, and to take steps to prevent it in the future. This module looks at U.S. laws that protect water from pollution and ensure safe drinking water. It also examines how laws in developing nations are changing to address water pollution.

Alignment to AP® Environmental Science Course Description

AP® Outline	Module	
VI. Pollution (25–30%)		
A. Pollution Types	Module 41	Wastewater from Humans and Livestock
	Module 42	Heavy Metals and Other Chemicals
	Module 43	Oil Pollution
	Module 44	Nonchemical Water Pollution
B. Impacts on the Environment and Human Health	Module 42	Heavy Metals and Other Chemicals
	Module 43	Oil Pollution
	Module 44	Nonchemical Water Pollution
C. Economic Impacts	Module 45	Water Pollution Laws

Chapter Learning Objectives

After completing this chapter students will be able to

- discuss the three major problems caused by wastewater pollution.
- explain the modern technologies used to treat wastewater.
- explain the sources of heavy metals and their effect on organisms.
- discuss the sources and effects of acid deposition and acid mine drainage.
- explain how synthetic organic compounds can affect aquatic organisms.
- identify the major sources of oil pollution.
- explain some of the current methods to remediate oil pollution.
- identify the major sources of solid waste pollution.
- explain the harmful effect of sediment pollution.
- discuss the sources and consequences of thermal pollution.
- understand the causes of noise pollution.
- explain how the Clean Water Act protects against water pollution.
- discuss the goals of the Safe Drinking Water Act.
- understand how water pollution legislation is changing in developing countries.

Chapter 14 Pacing Guide

This pacing guide is based on a schedule with 120 sessions of 50 minutes each before the AP® Exam. If you have a different number of sessions before the exam, you can modify the pacing to suit your needs. If you have additional time, consider incorporating quizzes, released AP® Environmental Science free-response and multiple-choice questions, or additional activities.

Module	Standard Schedule Days	Block Schedule Days
Module 41	2	1
Module 42	2	1
Module 43	1	½
Module 44	1	½
Module 45	1	½
Assessment	1	½

Chapter 14 Resources

The link to these resources can be found by clicking on the link buttons in the Teacher's e-Book (TE-book), opening the Teacher's Resource Flash Drive (TRFD), or logging onto the book's companion website bcs.whfreeman.com/friedlandapes2e (teacher login required).

- PowerPoint Presentation 14
- Optimized Art PowerPoint and JPEG Files 14
- Do the Math Videos
- **Measuring Your Impact 14:** Gaining Access to Safe Water and Proper Sanitation
- **Lab 14.1:** Waste and Its Effects on Atmospheric Carbon Dioxide
- Chapter 14 Web Resources
- Exploring the Literature 14
- Answers to Chapter 14 AP® Exam
- Multiple-Choice AP® Practice Exam 14

Free-Response Questions from Previous AP® Environmental Science Exams

Free-response questions from prior AP® Environmental Science Exams are available on the AP® course website: https://apstudent.collegeboard.org/apcourse/ap-environmental-science/exam-practice. Students should be able to answer all of the questions listed below with material learned in this and previous chapters. When a question requires students to understand material from multiple chapters, the question will be listed in the last chapter required to complete the entire question. Questions marked with an asterisk (*) are from exams with released multiple-choice questions. You may want to save these questions until the end of the year so you can give your students a complete released exam for practice. Questions marked with double asterisks (**) require math calculations. Look for references to these questions throughout the chapter.

Year	Question	Content
2001	4	• Water contamination from a farm's animal waste • The Clean Water Act
2007	1	• Treatment and disinfection of primary and secondary sewage • Problems in disposing of solid waste from sewage treatment • U.S. federal laws regulating quality of treated sewage released into waterways
2012	4	• Wetland ecoservices • Food webs in wetlands • Relationship between wastewater treatment and wetland processes
2013*	1	• The effect of human activities on sediment flow in Gulf Coast ecosystems • Environmental and economic impact of fertilizer runoff • Strategies for reducing fertilizer use that contributes to the runoff problem
2014	2**	• Pollutants in storm water • Calculation of stormwater volume • Problems with stormwater runoff

PD **Chapter 14 Overview**

Watch the video overview of Chapter 14 (for teachers) by clicking on the link buttons in the TE-book, opening the TRFD, or logging onto the book's companion website bcs.whfreeman.com /friedlandapes2e (teacher login required).

TRM **PowerPoint Presentation 14**

Download the PowerPoint presentation for Chapter 14 by clicking on the link buttons in the TE-book, opening the TRFD, or logging onto the book's companion website bcs.whfreeman.com /friedlandapes2e (teacher login required).

The Chesapeake Bay receives water from a very large geographic area that carries sediments, nutrients, and chemicals from a variety of urban, suburban, and agricultural areas. *(Peter Essick/Getty Images)*

chapter 14

Water Pollution

The Chesapeake Bay

The Chesapeake Bay is the largest estuary in the United States and it faces some of the biggest environmental challenges. Located between the states of Virginia and Maryland, the Chesapeake Bay receives fresh water from numerous rivers and streams that mixes with the salt water of the ocean to produce an extremely productive estuary. Many of the rivers and streams that dump into the bay travel long distances and drain water from a large watershed of urban, suburban, and agricultural areas. Indeed, the watershed that supplies the bay extends from northern Virginia all the way up into central New York State. One of the consequences of receiving water from such a large watershed is that the water coming into the bay contains an abundance of nutrients, sediments, and chemicals.

Among the major nutrients that enter the bay each year are 272 million kg (600 million pounds) of nitrogen and 14 million kg (30 million pounds) of phosphorus. The nitrogen and phosphorus come from three major sources. The first source is water discharged from sewage treatment facilities; it carries high amounts of nutrients from human waste. The second source is animal waste produced by concentrated animal feeding operations; these operations generate large amounts of manure that can make their way into nearby streams and rivers. The third source is fertilizer that is spread on both agricultural fields and suburban lawns; much of this fertilizer leaches out of the soil and into local streams that eventually flow into the Chesapeake Bay. When these nutrients reach the bay, the algae in the bay experience an explosive population growth that is known as an algal bloom.

> Anthropogenic chemicals appeared to be responsible for 82 percent of male smallmouth bass developing into hermaphrodites.

Increased sediments are also an issue in the Chesapeake Bay. Sediments are soils washed away from fields and forests as well as soils washed away from the banks of streams and the ocean shoreline. According to current estimates, 8.2 billion kilograms (18.7 billion pounds) of sediments come into the bay each year. The tiniest soil particles stay suspended in the water, make the water cloudy, and prevent sunlight from reaching the grasses that historically have been abundant in the bay. These grasses are important because they serve as a habitat for fish and blue crabs (*Callinectes sapidus*).

Teaching Tip: Chapter Opening Case

The chapter opening case, "The Chesapeake Bay," can set the stage for a field trip. If possible, take your class to a local stream, creek, or river to conduct field tests of the water, or visit a sewage treatment plant. If field trips are not an option, you can collect water from a local river or stream and bring it to the classroom for the students to test. Common water tests include measuring the dissolved oxygen, carbon dioxide, phosphorus, and nitrite, as well as the temperature and pH. Also, there are many online sewage treatment plant tours.

Teaching Tip: Journal Prompt

Ask students the following questions:

- What is an estuary? *An estuary is an ocean inlet where salt water and fresh water meet. Multiple rivers may feed into an estuary, as is the case with the Chesapeake Bay. An estuary is subject to changing tides.*
- What are two major nutrients that enter the Chesapeake Bay and how do these nutrients harm the bay? *The two major nutrients entering the bay are nitrogen and phosphorus. They cause algae to grow in excessive amounts and, when the algae die, decomposition can reduce the levels of oxygen in the water.*
- How can sediment harm the water quality of the Chesapeake Bay? *A buildup of sediment can cause the water to become cloudy and prevent sunlight from reaching the aquatic grasses that historically have been abundant in the bay. Reduced sunlight inhibits photosynthesis.*

Over the past several decades, the blue crab population has experienced substantial declines.

Inputs of anthropogenic chemicals, including pesticides and pharmaceutical drugs, are also prevalent. Pesticides arrive in the bay by direct application over water, by running off the surface of the land when it rains, or by being carried by the wind immediately after application. Pharmaceuticals pass through the human body and enter sewage treatment plants that eventually discharge into the streams and rivers that feed the bay. In 2009, the U.S. Fish and Wildlife Service announced that estrogen from sewage in the water was the most likely reason that 23 percent of male largemouth bass and 82 percent of male smallmouth bass developed into hermaphrodites; their male sex organs grow eggs, a function that normally only occurs in the sex organs of female fish. This impact not only concerns fish, but also is a concern for humans who share many similarities in their reproductive systems and may consume water contaminated with estrogens.

Because the Chesapeake Bay watershed is so large, cleaning it up has required a sustained and monumental effort. In 2000, the states surrounding the Chesapeake Bay formed a partnership along with multiple federal departments to develop the Chesapeake Bay Action Plan. This plan outlines a series of goals to reduce the impacts of nutrients, sediments, and chemicals coming into the bay. An assessment in 2010 concluded that many of the Action Plan's goals were being met, including a reduction in nitrogen, an increase in water clarity, and an increase in the number of blue crabs. Overall, the improved water quality combined with limits in the blue crab harvest allowed the blue crab population to increase dramatically. In 2012, the blue crab population in the bay reached its highest density in 20 years. In 2013, the total population size declined, but biologists were encouraged to discover that the remaining population had a higher proportion of reproductively mature females, which suggested that the future of the blue crab population looked bright. Much more work still needs to be done, such as resolving the problem of anthropogenic chemicals that cause hermaphroditic smallmouth bass.

The story of the Chesapeake Bay serves as an excellent example of the wide variety of pollutants that can impact aquatic ecosystems and of effective and substantial efforts that can only be made when all parties work together toward a common goal.

Sources:
Chesapeake Bay Program, http://www.chesapeakebay.net/; U.S. Fish and Wildlife Service, News Release, April 22, 2009, http://www.fws.gov/ChesapeakeBay/pdf/IntersexPR.pdf.

Water is a key resource for life on Earth. All organisms, including humans, require water to live, but increases in human populations combined with industrialization have led to the contamination of water supplies. **Water pollution** is generally defined as the contamination of streams, rivers, lakes, oceans, or groundwater with substances produced through human activities. As we will see, water pollution occurs from a wide range of substances from various sources. The prevalence and impact of pollutants in water also varies. Although this chapter focuses on the contaminants that exist in water and their effects on aquatic organisms and humans, because there are many ecological connections between aquatic and terrestrial ecosystems, water pollution can also affect terrestrial organisms.

The broad range of pollutants that can be found in water includes human and animal waste, inorganic substances, organic compounds, synthetic organic compounds, and nonchemical pollutants. In this chapter we will consider the origin of each category of pollutant, its negative effects on humans and the environment, and what can be done to reduce the pollutant.

Water pollution The contamination of streams, rivers, lakes, oceans, or groundwater with substances produced through human activities.

module

41

Wastewater from Humans and Livestock

A major type of water pollution is **wastewater,** which is the water produced by livestock operations and human activities, including human sewage from toilets and gray water from bathing and washing of clothes and dishes. In this module, we will review the effects that wastewater pollution can have on natural water bodies and then discuss the technologies that have been developed to treat wastewater and minimize its effects.

Learning Objectives

After reading this module you should be able to

- discuss the three major problems caused by wastewater pollution.
- explain the modern technologies used to treat wastewater.

Wastewater from humans and livestock poses multiple problems

Regardless of the specific contaminant, pollution can come from either *point sources* or *nonpoint sources* (FIGURE 41.1). A **point source** is a distinct location from which pollution is directly produced. Examples of point sources include a particular factory that pumps its waste directly into a nearby stream or a sewage treatment plant that discharges its wastewater from a pipe into the ocean. A **nonpoint source** is a more diffuse area that produces pollution, such as an entire farming region, a suburban community with many lawns and septic systems, or storm runoff from parking lots. The distinction between the two sources can help us control pollutant inputs to waterways. For example, if a municipality can determine one or two point sources for most of its water pollution, it can target those specific sources to reduce its pollution output. Controlling pollution from nonpoint sources is more challenging because it represents a negative externality. As we saw in Chapter 10, it is difficult to address negative externalities since they are not typically reflected in the cost of making the products that generate them.

Environmental scientists are concerned about wastewater as a pollutant for three major reasons. First, wastewater dumped into bodies of water naturally undergoes decomposition by bacteria, which creates a large demand for oxygen in the water. Second, the nutrients that are released from wastewater decomposition can make the water more fertile. Third, wastewater can carry a wide variety of disease-causing organisms.

Wastewater Water produced by livestock operations and human activities, including human sewage from toilets and gray water from bathing and washing of clothes and dishes.

Point source A distinct location from which pollution is directly produced.

Nonpoint source A diffuse area that produces pollution.

Teaching Tip: Discussion Starter

Ask students: What are point and nonpoint sources of pollution? How do they differ? *A point source is a distinct location from which pollution is directly produced, such as a pipe that discharges pollutants into a nearby stream or river. A nonpoint source is a more diffuse area that produces pollution, such as an entire farming region, a suburban community with many lawns and septic systems, or storm runoff from parking lots.*

Teaching Tip: Connections

As described in the chapter opening case, the Chesapeake Bay has been affected by both point and nonpoint sources of pollution. Ask students to make a poster that shows the two types of pollution sources. They can search the internet for examples of point source pollution and nonpoint source pollution and either print out pictures or draw their own. Ask students to arrange the pictures on poster board with point source pollution on one side and nonpoint source pollution on the other. At the bottom of the poster, have your students define point source and nonpoint source.

Teaching Tip: Engage

Ask students: Why are environmental scientists concerned about wastewater as a pollutant? *Environmental scientists are concerned about wastewater as a pollutant for three major reasons. First, wastewater dumped into bodies of water naturally undergoes decomposition by bacteria, which creates a large demand for oxygen in the water. Second, the nutrients that are released from wastewater decomposition can make the water more fertile. Finally, wastewater can carry a wide variety of disease-causing organisms.*

Teaching Tip: Activity

Divide the class into groups of two. Give each group 20 index cards and the vocabulary terms in the list below. Have each team label one side of each card with the numbers 1 through 20. On the blank sides of cards 1 through 10, students should write one of the key terms below, using a new card for each term. Instruct students not to duplicate the order of the list when they create cards 1 through 10. On the blank sides of cards numbered 11 through 20, students should write the definition or some other information connected to the terms on cards 1 through 10, making sure to keep an answer key of which numbered term card matches which numbered definition card. The groups should then lay out their cards number side up in two rows—one with terms (1–10) and one with definitions (11–20). Have the groups swap places and answer keys so each group is in front of a set of cards they did not make.

The first player in each pair then chooses one card from each row. The player reads the two cards to see if they are a match. If they are, that player keeps the cards and takes another turn. If the cards are not a match, the player returns the cards to the correct row and the second member of the team has a turn. Each group continues playing until all the cards are gone. The person with the most cards is the winner.

- wastewater
- biochemical oxygen demand (BOD)
- dead zone
- eutrophication
- cultural eutrophication
- indicator species
- fecal coliform bacteria
- septic system
- septic tank
- sludge

(a)

(b)

FIGURE 41.1 **Two types of pollution sources.** Pollution can enter water bodies either from (a) point sources, as in sewage pipes, or from (b) nonpoint sources, as in rainwater that runs off hundreds of square kilometers of agricultural fields and into streams. *(a: age fotostock/SuperStock; b: The Irish Image Collection/SuperStock)*

Oxygen Demand

When organic matter enters a body of water, microbes that are decomposers feed on it. Because these microbes require oxygen to decompose the waste, the more waste that enters the water, the more the microbes grow and the more oxygen they demand. The amount of oxygen a quantity of water uses over a period of time at a specific temperature is its **biochemical oxygen demand (BOD).** Lower BOD values indicate that a water body is less polluted by wastewater, and higher BOD values indicate that a water body is more polluted by wastewater. If we were to test the BOD of natural waters over a 5-day period in a liter of water, we might find a BOD of 5 to 20 mg of oxygen coming from the decomposition of leaves, twigs, and perhaps a few dead organisms. In contrast, wastewater might have a BOD of 200 mg of oxygen.

Biochemical oxygen demand (BOD) The amount of oxygen a quantity of water uses over a period of time at specific temperatures.

Dead zone In a body of water, an area with extremely low oxygen concentration and very little life.

Eutrophication A phenomenon in which a body of water becomes rich in nutrients.

Cultural eutrophication An increase in fertility in a body of water, the result of anthropogenic inputs of nutrients.

When microbial decomposition uses a large amount of oxygen in a body of water, the amount of oxygen remaining for other organisms can be very low. Low oxygen concentrations are lethal to many organisms including fish. Low oxygen can also be lethal to organisms that cannot move, such as many plants and shellfish. Some areas, known as **dead zones,** become so depleted of oxygen that little life survives. Dead zones can be self-perpetuating; organisms that die from lack of oxygen decompose and cause microbes to use even more oxygen.

Nutrient Release

When wastewater decomposes it releases nitrogen and phosphorus. As you may recall from Chapter 3, nitrogen and phosphorus are generally the two most important elements for limiting the abundance of producers in aquatic ecosystems. The decomposition of wastewater releases nitrogen and phosphorus, which in turn provide an abundance of fertility to a water body, a phenomenon known as **eutrophication.** As noted at the beginning of this chapter, the Chesapeake Bay experiences eutrophication from wastewater decomposition as well as from nutrients leached from fertilized agricultural lands during precipitation. When a body of water experiences an increase in fertility due to anthropogenic inputs of nutrients, it is called **cultural eutrophication.**

Teaching Tip: Discussion Starter

Ask students the following questions:

- How does high BOD influence water quality? *Higher BOD values indicate that a water body is more polluted by wastewater. If we were to test the BOD of a liter of natural water over a 5-day period, we might find a BOD of 5 to 20 mg of oxygen, the result of decomposition of leaves, twigs, and perhaps a few dead organisms. In contrast, wastewater might have a BOD of 200 mg of oxygen.*
- What is a dead zone? How do nitrogen and phosphorus contribute to dead zones? *Dead zones are areas that become so depleted of oxygen that little life survives. Dead zones can be self-perpetuating; organisms that die from lack of oxygen decompose and cause microbes to use even more oxygen. Nitrogen and phosphorus can provide an abundance of fertility to a water body, a phenomenon known as eutrophication.*

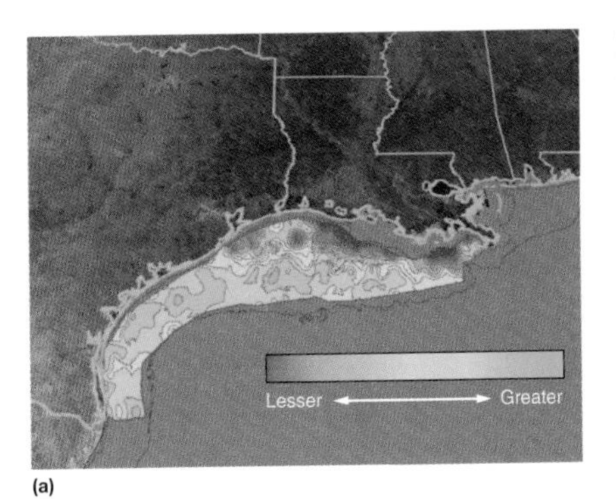

(a)

(b)

FIGURE 41.2 **Dead zone.** When raw sewage is dumped directly into bodies of water, subsequent decomposition by microbes can consume nearly all of the oxygen in the water and cause a dead zone to develop. Shown here are (a) oxygen concentrations in Gulf Coast waters and (b) a massive fish die-off that occurred as a result of low oxygen conditions in Lake Trafford, Florida. *(a: NASA/Goddard Space Flight Center Scientific Visualization Studio; b: Steven David Miller/Nature Picture Library)*

Eutrophication initially causes a rapid growth of algae, known as an algal bloom. As we discussed in Chapter 3, this enormous amount of algae eventually dies and microbes then rapidly begin digesting the dead algae; the increase in microbes consumes most of the oxygen in the water (see Figure 7.5 on page 88). In short, the release of nutrients from wastewater initiates a chain of events that eventually leads to a lack of oxygen and the creation of dead zones once again. One of the most impressive dead zones in the world occurs where the Mississippi River dumps into the Gulf of Mexico, shown in FIGURE 41.2a. The Mississippi River receives water from 41 percent of the land of the continental United States. Each summer there is an influx of wastewater and fertilizer that causes large algal blooms followed by substantial decreases in oxygenated water and massive die-offs of fish (Figure 41.2b). While some dead zones arise naturally, most are caused by human activities. In 1910, 4 dead zones were known around the world. This number increased to 87 dead zones in the 1980s and more than 400 dead zones by 2008, which is the most recent estimate.

Disease-Causing Organisms

For centuries, humans have faced the challenge of keeping wastewater from contaminating drinking water. This can be difficult because throughout the world many people routinely use the same water source for drinking, bathing, washing, and disposing of sewage (FIGURE 41.3). Wastewater can carry a variety of pathogens including viruses, bacteria, and protists. Pathogens in wastewater are responsible for a number of diseases that can be contracted by humans or other organisms coming in contact with or ingesting the water. Such pathogens cause cholera,

FIGURE 41.3 **Washing clothes in the Tuo River of China.** Using water for bathing, washing, and disposing of sewage without contaminating sources of drinking water is a long-standing challenge. *(Bamboosil/age footstock)*

Teaching Tip: Video

Sustainability: Water—Nutrient Loading in Lake Erie

This video from the National Science Foundation looks at Lake Erie, a vital source of drinking water for 11 million people. Researchers discuss how farming practices and severe weather can increase the amount of fertilizer-derived nutrients in the water, which harms water quality and is a threat to the lake's ecosystem as well as to public health.

The link to this video may be found by clicking the link buttons in the TE-book, opening the TRFD, or logging onto the book's companion website bcs.whfreeman.com/friedlandapes2e (teacher login required).

 **Chapter 14 Web Resources**

Teaching Tip: Video

Sustainability: Water—Baltimore's Urban Streams

This video from the National Science Foundation describes the travel times of pollutants in the urban streams in and around the city of Baltimore, Maryland. The video discusses the urban water cycle and how municipalities can better prevent pollutants from contaminating the greater watershed.

The link to this video may be found by clicking the link buttons in the TE-book, opening the TRFD, or logging onto the book's companion website bcs.whfreeman.com/friedlandapes2e (teacher login required).

 **Chapter 14 Web Resources**

Teaching Tip: Discussion Starter

Ask students: What problems are associated with sewage? *Sewage contains waterborne pathogens that cause disease. Many deaths worldwide can be attributed to lack of safe drinking water and exposure to raw sewage.*

FIGURE 41.4 Cholera is prevalent in raw sewage. Children who play in water contaminated by raw sewage, such as this girl in Cambodia, face a high risk of contracting the cholera pathogen. *(imagebroker/Alamy)*

typhoid fever, various types of stomach flu, and diarrhea.

Worldwide, the major waterborne diseases are cholera and hepatitis. Cholera, which claims thousands of lives annually in developing countries (FIGURE 41.4), is not common in the United States. However, hepatitis A is appearing more frequently in the United States, usually originating in restaurants that lack adequate sanitation practices. The bacterium *Cryptosporidium* has caused a number of outbreaks of gastrointestinal illness in this country. Large-scale disease outbreaks from municipal water systems are relatively rare in the United States, but they do occasionally occur. They are relatively common in many regions of the developing world.

The World Health Organization estimates that 1.1 billion people, nearly one-sixth of the world's population, do not have access to sufficient supplies of safe drinking water. In addition, half of the 3.1 million annual deaths from diarrheal diseases and malaria could be prevented if all people had safe drinking water, proper sanitation, and proper hygiene. Approximately 42 percent of the world's population lacks access to proper sanitation and over half of these people live in China and India. In sub-Saharan Africa, only 36 percent of the population has access to proper sanitation.

Indicator species A species that indicates whether or not disease-causing pathogens are likely to be present.

Fecal coliform bacteria A group of generally harmless microorganisms in human intestines that can serve as an indicator species for potentially harmful microorganisms associated with contaminated sewage.

Because of the risk that pathogens pose, we need the ability to easily test for the presence of pathogens in our drinking water. It is not feasible to test for all of the many different pathogens that can exist in drinking water. Instead, scientists have settled on using an **indicator species,** an organism that indicates whether or not disease-causing pathogens are likely to be present. The best indicators for water that is potentially harmful are **fecal coliform bacteria,** a group of generally harmless microorganisms that live in the intestines of human beings and other animals. One of the most common species of fecal coliform bacteria is *Escherichia coli,* abbreviated *E. coli.* Most strains of *E. coli* live naturally in humans and are not harmful, although there are strains that can be deadly to people who are very young, very old, or possess weak immune systems. Given that *E. coli* is commonly found in human intestines, detecting *E. coli* in a body of water is a reliable indicator that human waste has entered the water. This does not necessarily mean that the water is harmful to drink, but the presence of *E. coli* does indicate an increased risk that wastewater pathogens are in the water.

Public water supplies, such as drinking water sources and swimming pools, are routinely tested for *E. coli.* Homeowners with a single-family well might test their water less frequently or not at all. Public health authorities recommend declaring water unsuitable for human consumption if any bacteria are present. For safe swimming and fishing, the acceptable level of *E. coli* is higher; for example, swimming at a public beach or in a river is considered safe as long as the fecal coliform bacteria levels are less than 500 to 10,000 colonies per 100 megaliters of water. A pool, beach, or campground with contaminated water likely would be posted with a sign such as "The Department of Health has closed this water supply because of the presence of fecal coliform bacteria."

We have technologies to treat wastewater

Because proper sanitation is so important, we need ways of treating wastewater to reduce the risk of waterborne pathogens. Humans have devised a number of ways to handle wastewater. The various solutions all have the same basic approach: Bacteria are used to break down the organic matter into carbon dioxide and inorganic compounds, which include nitrogen and phosphorus, and the harmful pathogens are outcompeted by nonharmful pathogens. In this section we will look at the two most widespread systems for treating human sewage: *septic systems* and *sewage treatment plants.* We will also describe manure lagoons, which are used to treat waste from large livestock operations.

Septic Systems

In rural areas, houses often treat their own sewage with a **septic system,** which is a relatively small and simple sewage treatment system that is made up of a *septic tank* and a *leach field*. Septic systems are often used for homes in rural areas. As shown in FIGURE 41.5, the **septic tank** is a large container that receives wastewater from the house. Having a capacity of 1,900 to 4,700 L (500–1,250 gallons), the septic tank is buried underground adjacent to the house. Wastewater from the house flows into the tank at one end and leaves the tank at the other end. After the tank has been operating for some time, three layers develop. Anything that floats will rise to the top of the tank and form a scum layer. Anything heavier than water, including many pathogens, will sink to the bottom of the tank and form the **sludge** layer, which we define as the solid waste material from wastewater. In the middle is a fairly clear water layer called **septage.** The septage contains large quantities of bacteria and may also contain some pathogens and inorganic nutrients such as nitrogen and phosphorus.

Through gravity, septage moves out of the septic tank into several underground pipes laid out across a lawn below the surface. The combination of pipes and lawn makes up the **leach field.** The pipes contain small perforations so the water can slowly seep out and spread across the leach field. The septage that seeps out of the pipes is slowly absorbed and filtered by the surrounding soil. The harmful pathogens in the septage can be outcompeted by other microorganisms in the septic tank and therefore diminish in abundance, or can be degraded by soil microorganisms after the septage seeps out into the leach field. The organic matter found in the septage is broken down into carbon dioxide and inorganic nutrients. Eventually, the water and nutrients seeping out into the leach field are taken up by plants or enter a nearby stream or aquifer.

There are significant environmental advantages to septic systems. Because most septic systems rely on gravity—water from the house flows downhill to the septic tank, and water from the septic tank flows downhill to the leach field—no electricity is needed to run a septic system. However, sludge from the septic tank must be pumped out periodically (every 5 to 10 years) and taken to a sewage treatment plant.

Sewage Treatment Plants

Although household septic systems work well for rural areas in which each house has sufficient land for a leach field, they are not feasible for more developed areas with greater population densities and little open land. In developed countries, municipalities use centralized sewage treatment plants that receive the wastewater from hundreds or thousands of households via a network of underground pipes. In a traditional sewage treatment plant, wastewater is handled using a primary treatment followed by a secondary treatment.

As shown in FIGURE 41.6, the primary treatment allows the solid waste material to settle into a sludge layer. The remaining wastewater undergoes a secondary treatment that uses bacteria to break down 85 to 90 percent of the organic matter and convert it into carbon dioxide and inorganic nutrients such as nitrogen and phosphorus. The secondary treatment typically aerates the water to add oxygen; this promotes the growth of aerobic bacteria, which emit less offensive odors than anaerobic bacteria. The treated water sits for several days to allow particles to settle out. Settled particles are added to the sludge from the primary treatment. The remaining water is disinfected, using chlorine, ozone, or ultraviolet light to kill any

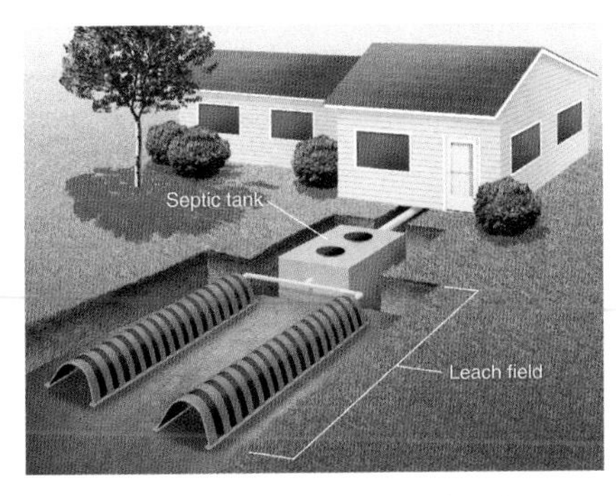

FIGURE 41.5 **A septic system.** Wastewater from a house is held in a large septic tank where solids settle to the bottom and bacteria break down the sewage. The liquid moves through a pipe at the top of the tank and passes through perforated pipes that distribute the water through a leach field.

Septic system A relatively small and simple sewage treatment system, made up of a septic tank and a leach field, often used for homes in rural areas.

Septic tank A large container that receives wastewater from a house as part of a septic system.

Sludge Solid waste material from wastewater.

Septage A layer of fairly clear water found in the middle of a septic tank.

Leach field A component of a septic system, made up of underground pipes laid out below the surface of the ground.

AP® Exam Tip

Sewage treatment is covered in the 2007 AP® Exam, Question 1. To answer this question, students must

- identify one component of sewage that is removed by primary treatment and one that is removed by secondary treatment
- describe how primary and secondary pollutants are removed from sewage
- explain how sewage treatment plants create a problem with solid waste
- describe environmental problems associated with disposal of solid waste from sewage
- name one pollutant that disinfection of sewage targets and a common method of disinfection
- name a U.S. federal law that requires monitoring the quality of treated sewage that is released into waterways

Teaching Tip: Activity

Have your students take a few minutes to create a table like the one below to compare and contrast septic systems and sewage treatment plants. (Answers are provided in italics.)

	Location	Application	Parts/stages	Environmental advantages	Environmental disadvantages
Septic system	*Rural areas*	*Smaller amounts of waste such as a single residence*	• *Tank* • *Leach field*	• *No electricity needed*	• *Manages small amount of waste* • *Tank must be pumped out every 5–10 years*
Sewage treatment plant	*Urban or suburban areas*	*Large amounts of waste from many households and businesses*	• *Primary* • *Secondary* • *Tertiary*	• *Manages waste from urban areas* • *Some sludge can be used as fertilizer*	• *Release of inorganic nutrients into waterways* • *Sludge must be removed*

AP® Exam Tip

Problems with storm water runoff are covered on the 2014 AP® Exam, Question 2. To answer this question, students must

- identify two pollutants in stormwater runoff that degrade surface water
- use provided data to calculate the volume of rainwater runoff in a shopping center
- use provided data to calculate the volume of untreated water that bypasses a local sewage treatment plant in a storm event
- describe two ways that stormwater runoff can be reduced
- discuss one environmental problem other than water runoff created by paved areas

AP® Exam Tip

Wastewater treatment and its similarity to the ecosystem services that wetlands provide is covered on the 2012 AP® Exam, Question 4. To answer this question, students must

- describe two characteristics scientists use to define wetlands
- complete a food web of a wetland ecosystem showing energy flow
- explain why it takes many hectares of wetlands to support a pair of eagles
- describe two economic benefits of wetlands
- describe a human activity that degrades wetlands
- explain the similarity in terms of water quality functions provided by wetlands and by wastewater treatment plants, with specific reference to primary and secondary treatment

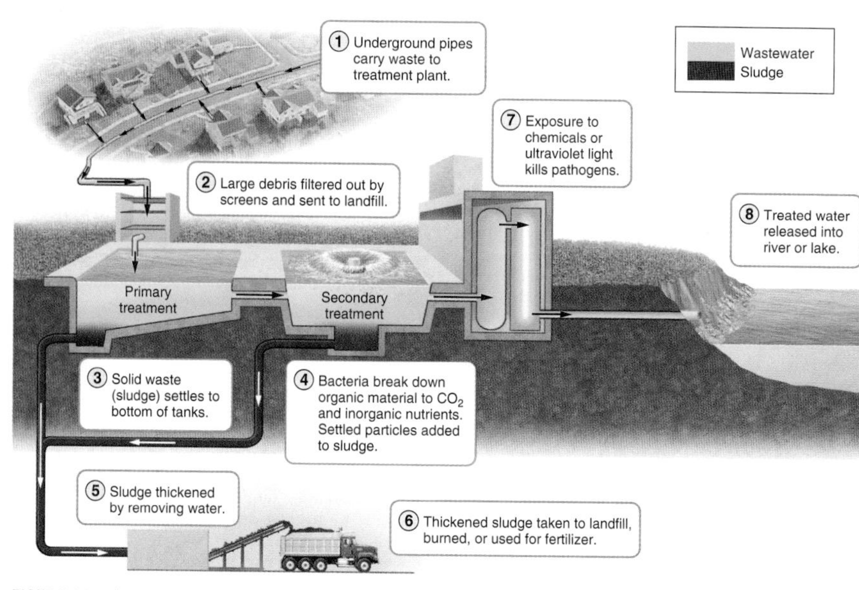

FIGURE 41.6 **A sewage treatment plant.** In large municipalities, great volumes of wastewater are handled by separating the sludge from the water and then using bacteria to break down both components.

remaining pathogens. The disinfected water is then released into a nearby river or lake, where it is once again part of the global water cycle.

The combined sludge from the primary and secondary treatments must be taken away from the sewage treatment plant. To reduce the volume of material prior to transport and help remove many of the pathogens, sludge is typically exposed to bacteria that can digest it. After digestion, most of the water is removed from the sludge, which reduces its volume and weight. This final form of sludge can be placed into a landfill, burned, or converted into fertilizer pellets for agricultural fields, lawns, and gardens.

Sewage treatment plants are very effective at breaking down the organic matter into carbon dioxide and inorganic nutrients. Unfortunately, these inorganic nutrients can still have undesirable effects on the waterways into which they are released. Although nitrogen and phosphorus are important nutrients for increasing primary productivity, when high concentrations of these nutrients are released into bodies of water from sewage treatment plants, they fertilize the water, which can lead to large increases in the abundance of algae and aquatic plants. In response to this problem, large sewage treatment plants are now developing tertiary treatments that remove nitrogen and phosphorus from the wastewater. The ultimate goal is to release wastewater that is similar in quality to the water body that is receiving it.

Legal Sewage Dumping

Sewage treatment plants are critical to human health because they remove a great deal of harmful organic matter and associated pathogens that cause human illness. It might surprise you to know that even in the most developed countries, raw sewage can sometimes be directly pumped into rivers and lakes. Sewage treatment plants are typically built to handle wastewater from local households and industries. However, many older sewage treatment plants also receive water from stormwater drainage systems. During periods of heavy rain, the combined volume of storm water and wastewater overwhelms the capacity of the plants. When this happens, the treatment plants are allowed to bypass their normal treatment protocol and pump vast amounts of water directly into an adjacent body of water.

How big a problem is this? According to the U.S. Environmental Protection Agency, overflows of raw sewage occur approximately 40,000 times per year in the United States. In Indianapolis, for example, more

Teaching Tip: Journal Prompt

Ask students: What is the role of bacteria in the treatment of human and animal waste? *In both septic systems and wastewater treatment systems, bacteria help break down the organic matter and convert it to carbon dioxide and inorganic nutrients such as nitrogen and phosphorus.*

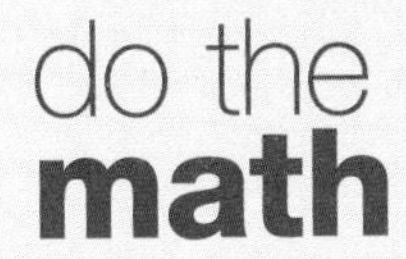

Building a Manure Lagoon

Concentrated animal feeding operations typically use manure lagoons to hold the manure produced by the cattle that are being held. If an individual animal produces 50 L of manure each day and the average concentrated animal feeding operation holds 800 cattle on any given day, how much manure is produced each day?

Daily manure production = 50 L/animal × 800 animals = 40,000 L

Your Turn

1. If a manure lagoon needs to hold 30 days' worth of manure production for 900 cattle, what is the minimum capacity of the lagoon a farmer would need?
2. After the manure has broken down, the manure must be spread onto farm fields. A modern manure spreader can hold 40,000 L of liquid manure. How many trips will it take for the manure spreader to remove the 30 days' worth of manure that is held in the manure lagoon?

than 3.8 billion liters (1 billion gallons) of raw sewage are dumped into surrounding water bodies each year during periods of high rain. Around the country, such incidents result in the contamination of drinking water, beaches, fish, and shellfish, and can lead to human illness. Between 1.8 million and 3.5 million illnesses are associated with swimming in sewage-contaminated water in the United States each year, and 500,000 illnesses are linked to drinking sewage-contaminated water. Illnesses caused by eating contaminated shellfish are estimated to cost $2.5 million to $22 million annually. The solution to this problem is straightforward though expensive. Municipalities facing sewage overflow need to modernize their sewage treatment systems to prevent the influx of storm water from overwhelming their capacity to treat human wastewater. Unfortunately, the cost of such work can be considerable.

FIGURE 41.7 **Feedlot and manure lagoon.** Large-scale agricultural operations, such as this one in Georgia, produce great amounts of waste that can be held in lagoons until pumped out and removed. *(Jeff Vanuga/NRCS/USDA)*

Animal Feed Lots and Manure Lagoons

Although the impact of human waste on the environment and human health is generally understood, people are less familiar with the problem of animal waste. On a small scale, animal manure can contaminate waters when farm animals are allowed access to streams for food and water because, inevitably, they will defecate in the stream. As you may recall from Chapter 11, large-scale manure issues arise with concentrated animal feeding operations that raise thousands of cattle, hogs, and poultry. The manure from these operations contains digested food and often also contains a variety of hormones and antibiotics that farmers use to improve the growth and health of their animals.

Farms that raise thousands of animals in a single location often use a *manure lagoon* to dispose of the tremendous amount of manure produced (FIGURE 41.7). A **manure lagoon** is a large, human-made pond lined with rubber to prevent the manure from leaking into the groundwater. After bacteria has broken down the manure—the same process that occurs in sewage treatment plants—the manure can be spread onto farm fields

Manure lagoon Human-made pond lined with rubber built to handle large quantities of manure produced by livestock.

Answer to Your Turn

1. Daily manure production = 50 L/animal × 900 animals = 45,000 L

Minimum lagoon capacity = 45,000 L × 30 days = 1,350,000 L

2. Number of trips = 1,350,000 L × (1 trip ÷ 40,000 L) = 33.75 trips

Round to 34 trips.

AP® Exam Tip

Contamination of water from animal waste is covered in the 2001 AP® Exam, Question 4. To answer this question, students must

- use provided data and a map to assess whether a farm's animal waste is contaminating a local waterway
- describe two tests that could be used to monitor water quality in various locations and suggest expected results at various locations along the waterway
- discuss changes to the local environment that could happen as a result of waste being discharged into the waterway
- explain which provisions of the Clean Water Act apply to the situation described in the problem

lab

Waste and Its Effects on Carbon Dioxide

In this lab, students have the opportunity to evaluate the effect of the presence of pollutants such as sewage and agricultural runoff on atmospheric carbon dioxide.

Download the lab by clicking on the link buttons in the TE-book, opening the TRFD, or logging onto the book's companion website bcs.whfreeman.com/friedlandapes2e (teacher login required).

 **Lab 14.1** Waste and Its Effects on Carbon Dioxide

Answers to Module 41 AP® Review Questions

1. b
2. c
3. a
4. c
5. e

to serve as a fertilizer. One major risk of manure lagoons is the possibility of developing a leak in the liner that would allow the waste to seep into and contaminate the underlying groundwater. Another risk is a possible overflow into adjacent water bodies. Like human wastewater, overflow of animal waste into rivers can lead to disease outbreaks in humans and wildlife. Finally, the application of manure as fertilizer can create runoff that moves into nearby water bodies. "Do the Math: Building a Manure Lagoon" shows you how to calculate the appropriate size for a manure lagoon.

module 41

REVIEW

In this module, we learned that wastewater from livestock and humans can cause harmful effects on the water bodies that receive it. Some of the most important effects are the high oxygen demand associated with the decomposition of the organic matter that can cause dead zones, the production of inorganic nutrients such as nitrogen and phosphorus that can cause algal blooms, and the introduction of disease-causing organisms. Fortunately, humans have invented several technologies to handle wastewater including septic tanks, sewage treatment plants, and manure lagoons. These technologies rely in part on decomposition of organic matter and help reduce the presence of pathogens. In some cases, they also reduce the amount of inorganic nutrients before the wastewater is dumped into a natural water body. In the next module, we will examine how a number of heavy metals and other chemicals affect aquatic ecosystems.

Module 41 AP® Review Questions

1. Which is a point source of pollution?
   (a) Farm runoff
   (b) A sewage treatment plant
   (c) Storm runoff
   (d) A suburban community
   (e) A harbor

2. Eutrophication
   (a) is the result of excess carbon.
   (b) causes low BOD.
   (c) can cause dead zones.
   (d) is rarely caused by human activities.
   (e) only occurs in fresh water.

3. Fecal coliform bacteria
   (a) is used as an indicator of water quality.
   (b) is the cause of many waterborne diseases.
   (c) is another name for *Cryptosporidium*.
   (d) usually causes diarrhea.
   (e) is rarely found in septage.

4. Leach fields are
   (a) used to remove sludge.
   (b) part of tertiary water treatment.
   (c) used to filter septage.
   (d) a result of sewage dumping.
   (e) often used with manure lagoons.

5. A manure lagoon is being built for a dairy with 700 cows, each of which produces 40 L of manure each day. How large must the lagoon be to hold 30 days' worth of manure?
   (a) 21,000 L
   (b) 28,000 L
   (c) 120,000 L
   (d) 560,000 L
   (e) 840,000 L

module

42

Heavy Metals and Other Chemicals

We have seen that some compounds such as nitrogen and phosphorus cause environmental problems by overfertilizing the water. Other inorganic compounds, including heavy metals (lead, arsenic, and mercury), acids, and synthetic compounds (pesticides, pharmaceuticals, and hormones), can directly harm humans and other organisms. In this module, we will examine how each of these chemicals enters water bodies and the effects that they can have on ecosystems and humans.

Learning Objectives

After reading this module you should be able to

- explain the sources of heavy metals and their effect on organisms.
- discuss the sources and effects of acid deposition and acid mine drainage.
- explain how synthetic organic compounds can affect aquatic organisms.

Heavy metals are highly toxic to organisms

Heavy metals are a group of chemicals that can pose serious health threats to humans and other organisms. In this section, we will focus on three heavy metals that are of particular concern: lead, arsenic, and mercury.

Lead

Lead is rarely found in natural sources of drinking water, but it contaminates water that passes through lead-lined pipes and other materials that contain lead such as brass fittings and solder used to fasten pipes together. Fetuses and infants are the most sensitive to lead, and exposure can damage the brain, nervous system, and kidneys. A series of federal guidelines for building construction implemented during the past 3 decades now requires the installation of lead-free pipes, pipe fittings, and pipe solders. These changes and the increased use of water filtration systems in many homes are gradually reducing the problem of lead in drinking water, although it remains a concern in older houses and apartments.

Arsenic

Arsenic is a compound that occurs naturally in Earth's crust and can dissolve into groundwater. As a result, naturally occurring arsenic in rocks can lead to high concentrations of arsenic in groundwater and drinking water. Human activity also contributes to higher arsenic concentrations in groundwater. For example, mining breaks up rocks deep underground, and industrial uses of arsenic for items such as wood preservatives can add to the amount of arsenic found in drinking water. Fortunately, arsenic can be removed from water via fine membrane filtration, distillation, and reverse osmosis (see Figure 27.6 on page 306).

Teaching Tip: Journal Prompt

Ask students to explain why lead in the U.S. water supply has become less of a risk than previously. *Building materials such as lead fittings and solder used to fasten pipes are no longer used in new construction. Many old buildings are being refitted to eliminate these materials.*

Teaching Tip: Activity

Ask students to choose a partner and take 15 minutes to fill out the heavy metal organizational table below. (Answers are provided in italics.) After the tables are completed, invite groups to share their results and discuss.

Metal	Source	Removal	Health effect
Lead	• *Lead-lined pipes in plumbing fixtures, brass fittings; solder used to fasten pipes, paint, some toys* • *Mostly found in older buildings and apartments*	• *Replacement of old pipes and paint* • *Federal guidelines for building construction materials have banned lead* • *Increased use of water filtration systems* • *Laws prohibit the import of toys made with lead paint*	• *Fetuses and infants most sensitive; can cause brain, nervous system, and kidney damage*
Arsenic	• *Found naturally in Earth's crust; can dissolve in groundwater* • *Mining practices can contribute to its presence*	• *Can be removed from water via fine membrane filtration, distillation, and reverse osmosis*	• *Can cause cancer of the skin, lung, kidney, and bladder*
Mercury	• *Released into atmosphere by burning of coal and incineration of garbage, medical, and dental supplies* • *Once, airborne, it can fall into rivers, lakes, and streams*	• *Reduce use of coal*	• *Increases in toxicity as it moves up food chain* • *Can damage central nervous system, which can impair coordination* • *Especially dangerous to fetuses and young children*

Teaching Tip: Warm-up

Ask students to describe the primary dangers associated with the presence of heavy metals in water. *Some compounds such as nitrogen and phosphorus cause environmental problems by overfertilizing the water. Other inorganic compounds, including heavy metals (lead, arsenic, and mercury), can directly harm humans and other organisms.*

TEACHING with FIGURES ▶

Show students Figure 42.1 and then ask the following questions:

- Which states have the highest level of arsenic concentrations in well water? *States with high levels of arsenic in well water include California, Nevada, Arizona, Texas, and Idaho.*
- In what region of the United States is a concentration of arsenic above 3 µg/L very unlikely? *A concentration of arsenic above 3 µg/L is very unlikely in the Southeast.*

FIGURE 42.1 indicates where high concentrations of arsenic have been found in well water throughout the United States. As you can see, concentrations are highest in the Midwest and West. Arsenic in drinking water is associated with cancers of the skin, lungs, kidneys, and bladder. These illnesses can take 10 years or more after exposure to develop. Even very low concentrations of arsenic, such as those found in many wells around the world and measured in parts per billion (ppb), can cause severe health problems. In fact, though cancers can develop with less than 50 ppb of arsenic in drinking water, 50 ppb was set as the upper "safe" limit for U.S. drinking water from 1942 to 1999. In 1999, the EPA lowered the upper limit to 10 ppb, a compromise reached after much debate between environmental groups pushing for 5 ppb and water, mining, and wood preservative industries arguing that even the 10 ppb limit would be too expensive to implement. In 2001, the EPA delayed the implementation of the 10 ppb standard until more data could be evaluated. Later that year, the U.S. National Academy of Sciences recommended a standard of 5 ppb and the EPA decided to implement a standard of 10 ppb. As a result, after substantial investment required to improve many water treatment facilities, people in the United States will now have lower amounts of cancer-causing arsenic in their drinking water.

The problem of arsenic in drinking water is a worldwide issue. During the 1980s and 1990s, for example, water engineers in regions of Bangladesh and eastern India drilled millions of very deep wells in an attempt to find sources of groundwater that were not contaminated by local pollution. While this had the desired effect of obtaining less-polluted water, officials were not aware that this deeper groundwater was contaminated by naturally occurring arsenic from the rocks deep underground. Currently, 140 million people in this region drink arsenic-contaminated water and thousands of individuals have been diagnosed with arsenic poisoning. Current efforts to solve this problem include plans to collect uncontaminated rainwater as a source of drinking water and research to develop inexpensive filters that can remove arsenic from the well water.

Mercury

Mercury is another naturally occurring heavy metal found in increased concentrations in water as a result of human activities. FIGURE 42.2 shows mercury releases from different regions of the world as the result of activities such as burning coal (see Chapter 12). Among regions of the world, 6 percent of human-produced mercury comes from North America and more than half comes from Asia, including 28 percent from China.

Approximately two-thirds of all mercury produced by human activities comes from the burning of fossil fuels, especially coal. Other important sources of mercury include the incineration of garbage, hazardous waste, and medical and dental supplies. One of the less well-known sources of mercury comes from the raw materials that go into the manufacturing of cement for construction. The limestone used to make cement can contain mercury, and this mercury is released during a heating process that is needed when making cement. Moreover, the source of heat for cement manufacturing is often coal that also releases mercury when burned.

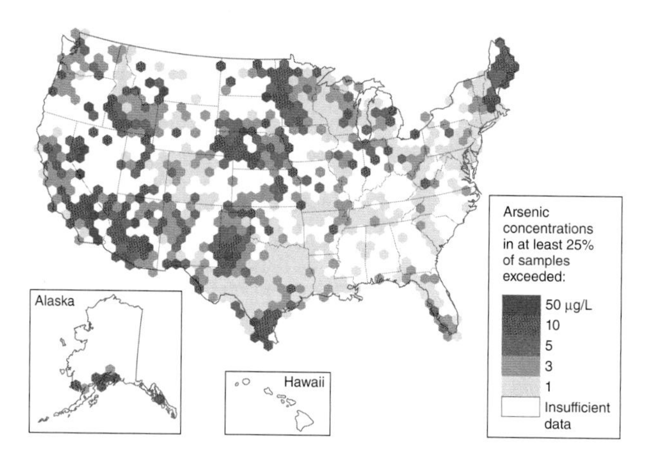

FIGURE 42.1 Arsenic in U.S. well water. The highest concentrations of arsenic are generally found in the upper Midwest and the West. *(Source: http://water.usgs.gov/nawqa/trace/pubs/geo_v46n11/fig3.html)*

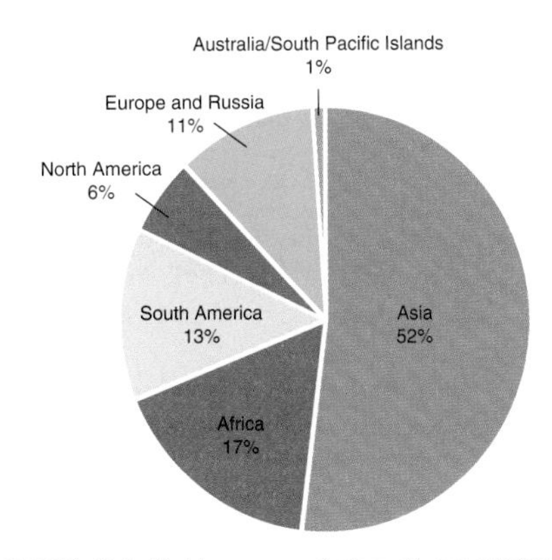

FIGURE 42.2 **World mercury emissions.** Mercury emissions from human activities vary greatly among regions of the world. *(Data from AMAP/UNEP, 2013. Technical Background Report for the Global Mercury Assessment 2013. Arctic Monitoring and Assessment Programme, Oslo, Norway/UNEP Chemicals Branch, Geneva, Switzerland. vi + 263 pp.)*

Petroleum exploration is a source of both mercury and lead pollution. Each petroleum well produces roughly 681,374 L (180,000 gallons) of contaminated wastewater and mud over its lifetime. This water is usually dumped at the drilling site and, depending on the soils and topography, can either run into nearby waterways or infiltrate the soil and contaminate the underlying groundwater.

The mercury emitted by these activities eventually finds its way into water. Inorganic mercury (Hg) is not particularly harmful, but its release into the environment can be hazardous because of a chemical transformation it undergoes. In wetlands and lakes, bacteria convert inorganic mercury into methylmercury, which is highly toxic to humans. Methylmercury damages the central nervous system, particularly in young children and in the developing embryos of pregnant women. The result impairs coordination and the senses of touch, taste, and sight.

Human exposure to methylmercury occurs mostly through eating fish and shellfish. Methylmercury can move up the food chain in aquatic ecosystems, which results in the top consumers containing the highest concentrations of mercury in their bodies. Given that oceans are contaminated with mercury and that tuna are top predators, it is not surprising that these fish contain high concentrations of mercury. In 2008, a reporter for the *New York Times* purchased tuna sushi from 20 locations in New York City and, after analysis by a private laboratory, found that, at most restaurants, a diet of six pieces of sushi per week would exceed the EPA standard for human consumption of mercury. In 2009, the EPA reported that the concentration of mercury in the tuna of the North Pacific Ocean had increased by 30 percent since 1990. If China's plans to build more electric-generating plants that burn coal go forward, the EPA estimates that these concentrations in tuna will increase by another 50 percent by 2050.

What can be done about mercury pollution? In 2013 the United States announced a new agreement with more than 140 other countries to reduce global mercury pollution. In a step toward this goal, the EPA has proposed that cement-manufacturing plants reduce mercury emissions by 92 percent. New EPA rules are also being written to reduce mercury emissions from other major sources, including coal-burning power plants.

Acid deposition and acid mine drainage affect terrestrial and aquatic ecosystems

About 40 years ago, people throughout the northeastern United States, northern Europe, China, and Russia began to notice that the forests, soils, lakes, and streams were becoming more and more acidic. As a consequence, some trees were killed and some bodies of water became too acidic to sustain fish. After much debate, it became clear that the source of the lower pH of the water was the presence of very tall smokestacks of industrial plants that were burning coal and releasing sulfur dioxide and nitrogen dioxide into the air. These tall smokestacks kept the emissions away from local residents, but sent the chemicals into the atmosphere where they were converted into sulfuric acid and nitric acid that returned to Earth hundreds of kilometers away. Acids deposited on Earth as rain and snow or as gases and particles that attach into the surfaces of plants, soil, and water are known as **acid deposition.** As described in Chapter 8, wet-acid deposition, also known as acid precipitation or acid rain, occurs in the form of rain and snow, whereas dry-acid deposition occurs as gases and particles that attach to the surfaces of plants, soil, and water. Acid deposition reduces the pH of water bodies from 5.5 or 6 to below 5, which can be lethal to many aquatic organisms, leaving these water bodies devoid of many species.

To combat the problem of acids being released into the atmosphere, many coal-burning facilities have installed coal scrubbers. Coal scrubbers pass the hot

Acid deposition Acids deposited on Earth as rain and snow or as gases and particles that attach to the surfaces of plants, soil, and water.

◀ TEACHING with FIGURES

Show students Figure 42.2 and ask the following questions:

- Which region of the world emits the largest amount of mercury? *Asia is the largest emitter of mercury, at 52 percent.*
- Why do you think China has the highest mercury emissions in the world? *China is currently going through a period of industrialization and is a heavy user of coal. Other sources of mercury such as incinerating garbage and making cement are associated with large populations and economic development.*

Teaching Tip: Discussion Starter

Ask students the following questions:

- What two gases are largely responsible for acid deposition? *The two gases primarily responsible for acid deposition are sulfur dioxide and nitrogen dioxide.*
- What technological advance has helped combat the problem of acids being released into the atmosphere? How does it work? *To combat the problem of acids being released into the atmosphere, many coal-burning facilities have installed coal scrubbers. Coal scrubbers pass the hot gases produced during combustion through a limestone slurry. The limestone reacts with the acidic gases and removes them before the hot gases leave the smokestack.*

Teaching Tip: Video

Commonly Used Pesticide Turns Honeybees into "Picky Eaters"
This video describes how a small dose of a commonly used crop pesticide interferes with honeybee activity, and raises the question of what pesticides should be applied to bee-pollinated crops. It also examines the recent declines in honeybee colonies.

The link to this video may be found by clicking the link buttons in the TE-book, opening the TRFD, or logging onto the book's companion website bcs.whfreeman.com/friedlandapes2e (teacher login required).

Chapter 14 Web Resources

gases produced during combustion through a limestone slurry. The limestone reacts with the acidic gases and removes them before the hot gases leave the smokestack. We will look at the issues raised by acid deposition in greater detail in Chapter 15 on air pollution.

Low pH in water bodies also occurs when very acidic water comes from below ground. This problem begins with the development of underground mines that, once abandoned, flood with groundwater. The combination of water and air allows pyrite, a type of rock, to break down and produce iron and hydrogen ions. The increase in hydrogen ions produces acidic water with a low pH. Water in the mine containing these ions can find its way up to the surface in the form of springs that feed into streams. As we discussed in Chapter 8, a similar effect occurs during mountaintop mining operations in which the tops of mountains are removed and the soil is dumped into stream valleys. The streams that are fed by springs from these mines are infiltrated with water that can have a pH close to zero. Low-pH water can cause many other harmful metal ions to become soluble, including zinc, copper, aluminum, and manganese. Although some water bodies such as bogs are naturally acidic and contain species that are adapted to acidic conditions, the combination of very low pH and toxic metals can produce environments that are too harsh for most organisms to survive. Because much of the dissolved iron precipitates out of solution as the low-pH water of the mine mixes with the high-pH water of the stream, these streams often have a striking red or yellow color (FIGURE 42.3).

Unfortunately, many mining companies responsible for making streams uninhabitable for fish and other organisms are no longer in business. As a result, they are often not held accountable for the environmental damage they caused. This presents a major challenge for state and local governments that bear the cost of dealing with the problem of acidic water from mines and the toxic metals that are produced. Researchers are currently investigating a number of strategies to counteract the acidity of these streams. One possible approach is to pass stream water through a limestone treatment facility that raises the pH of the water and removes toxic metals to levels tolerable to stream organisms.

Synthetic organic compounds are human-produced chemicals

Synthetic, or human-made, compounds can enter the water supply either from industrial point sources where they are manufactured or from nonpoint sources when they are applied over very large areas. These organic (carbon-containing) compounds include pesticides, pharmaceuticals, military compounds, and industrial compounds. Synthetic organic compounds have a variety of effects on organisms. They can be toxic, cause genetic defects, and, in the case of compounds that resemble animal hormones, interfere with growth and sexual development.

Pesticides and Inert Ingredients

Pesticides serve an important role in helping to control pest organisms that pose a threat to crop production and human health (FIGURE 42.4). Although natural pesticides such as arsenic have been used for centuries, the first generation of synthetic pesticides was developed during World War II. As we discussed in

FIGURE 42.3 **Acid mine drainage.** The low pH of water emerging from abandoned mines mixes with stream water to lower the pH of streams, causing iron to precipitate out of solution and form a rusty red oxidized iron. This problem occurs around the world, including this stream in Colline Metallifere, Italy. *(Elbardamu/Alamy)*

FIGURE 42.4 **Applying pesticides.** Pesticides provide benefits to humans, but they also can have unexpected effects on humans and other nonpest organisms that are not fully understood and have not been adequately investigated. These airplanes are spraying insecticides over a lake to help control mosquito populations. *(Nathan Benn/Alamy)*

Chapter 11, these chemicals proved to be very effective in killing a variety of undesired plants (herbicides), fungi (fungicides), and insects (insecticides). In the decades that have followed, however, environmental scientists have identified a number of concerns about the unintended effects of pesticides.

Most pesticides do not target particular species of organisms, but generally kill a wide variety of related organisms. For example, an insecticide that is sprayed to kill mosquitoes is typically lethal to many other species of invertebrates, including insects that might be desirable as predators of the mosquito. Some pesticides are lethal to unrelated species. For example, researchers have recently discovered that the insecticide endosulfan, a chemical designed to kill insects, is highly lethal to amphibians even at very low concentrations. Even a pesticide that is not directly lethal to a species can indirectly affect organisms by altering the species composition of the community.

Synthetic pesticides are generally designed specifically to target particular aspects of a pest's physiology. However, they can also alter physiological functions of other species. For example, most insecticides target the nervous system of a particular pest, yet they can have unintended impacts on other pests as well as on many nonpest species. The insecticide DDT (dichlorodiphenyltrichloroethane) is a good example. While DDT was designed to alter nerve transmissions in insects, the chemical can move up an aquatic food chain all the way to eagles that consume fish. Eagles that consumed DDT-contaminated fish produced eggs with thinner shells that would prematurely break during incubation. Despite the fact that DDT was designed to kill mosquitoes, its unintended impact was the thinning of bird eggshells. After the United States banned the spraying of DDT in 1972, the bald eagle and other birds of prey increased in numbers. However, DDT is still manufactured in developed nations and sprayed in developing countries as a preferred way to control the mosquitoes that carry the deadly malaria parasite.

Another concern about pesticides is the effect of inert ingredients added to commercial formulations. Inert ingredients are additives that make a pesticide more effective—for example they allow it to dissolve in water for spraying or to penetrate inside a pest species. Although the term "inert" may suggest that these chemicals are harmless, this is often not true. The popular herbicide Roundup, for example, is composed of a chemical that is highly effective at killing plants but has difficulty getting past the waxy outer layer of leaves without the help of an added inert ingredient. Since inert ingredients are legally classified as trade secrets and most are not required to be tested for safety, their effects are not always known before a product comes to market. In the case of Roundup, recent research has discovered that the herbicide is highly toxic to amphibians, not because of the plant-killing chemical but because of the inert ingredients. It appears that the same properties that allow the penetration of leaves also allow the penetration of tadpole gill cells. The gills burst and the tadpoles suffocate. Unintended consequences such as these have roused interest in both Europe and North America to require that inert ingredients be tested for potentially harmful effects.

Pharmaceuticals and Hormones

While most people know that pesticides are commonly found in the environment, they often are surprised to learn that pharmaceutical drugs are also common. For example, the U.S. Geological Survey tested 139 streams across the United States for a variety of chemical contaminants. FIGURE 42.5 shows data for the frequency of detection. Among the different types of chemicals that were present at detectable levels, approximately 50 percent of all streams contained antibiotics and reproductive hormones, 80 percent contained nonprescription drugs, and 90 percent contained steroids. In most cases the concentrations of these chemicals are quite low and currently are not thought to pose a risk to environmental or human health. Some chemicals such as hormones, however, operate at very low concentrations inside the tissues of organisms and we have a poor understanding of their effects. As noted in our discussion of the Chesapeake

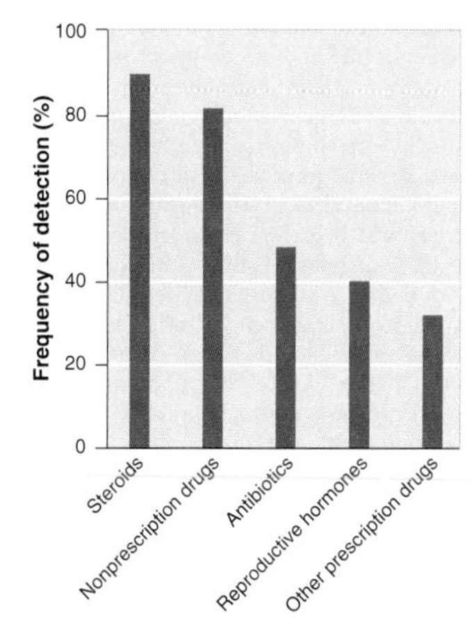

FIGURE 42.5 Contaminants in streams. Streams contain a wide variety of chemicals including pharmaceutical drugs and hormones. These come from a combination of wastewater inputs, agriculture, forestry, and industry. *(After D. W. Kolpin et al. 2002. Pharmaceuticals, hormones, and other organic wastewater contaminants in U.S. streams, 1999–2000: A national reconnaissance. Environmental Science & Technology 36: 1202–1211.)*

more **math** practice:
Calculating Percent Change

Many students struggle with how to calculate percent change. To calculate a percent change, first find the difference between the beginning and ending quantity. Then find the proportion of this change by dividing the difference between the beginning and ending quantity by the original beginning quantity. Finally, multiply this proportion by 100 to find the percent change.

Use the following problem to demonstrate the process of calculating a percent change:

In 2010, the South Anna River had a largemouth bass population of 25,000 individuals over a 10-mile stretch. In early 2011, a chemical spill occurred and the population of largemouth bass decreased to only 5,000 individuals over the same area. Calculate the percent change in largemouth bass in the South Anna River.

- **Step 1:** Find the difference between the number of individuals in the population before and after the chemical spill occurred.

  25,000 individuals − 5,000 individuals = 20,000 individuals lost

- **Step 2:** To find the proportion of individuals lost, divide the number lost by the original number

  $20{,}000 \div 25{,}000 = 0.8$

- **Step 3:** To find the percentage, multiply the proportion by 100

  $0.8 \times 100 = 80\%$

There was an 80 percent decrease in largemouth bass found in the South Anna River after the chemical spill.

▲ TEACHING with FIGURES

Show students Figure 42.5 and ask the following questions:

- How many streams contained detectable levels of antibiotics? *Approximately 50 percent of all streams contained antibiotics.*
- How many streams contained nonprescription drugs? *Approximately 80 percent of streams contained nonprescription drugs.*
- How many streams contained steroids? *Approximately 90 percent of streams contained steroids.*
- Does the figure tell us anything about quantities of contaminants found in streams? *This figure does not indicate whether the concentration of the sample was high or low. It reports only on the presence of the chemicals.*

Bay, it is now clear that low concentrations of pharmaceutical drugs that mimic estrogen are connected to male fish growing eggs in their testes. The extent of hormone effects on humans and wildlife around the world is currently unknown but environmental scientists are showing increased attention to these effects. We will talk more about the ways that chemicals can alter the endocrine systems of animals in Chapter 17.

Military Compounds

Perchlorates, a group of harmful chemicals used for rocket fuel, sometimes contaminate the soil in regions of the world where military rockets are manufactured, tested, or dismantled. Perchlorates come in many forms. The U.S. space shuttle, for example, used a booster rocket that contained 70 percent ammonium perchlorate. Perchlorates easily leach from contaminated soil into the groundwater, where they can persist for many years. Human exposure to perchlorates comes primarily through consumption of contaminated food and water. In the human body, perchlorates can affect the thyroid gland and reduce the production of hormones necessary for proper functioning of the human body.

Industrial Compounds

Industrial compounds are chemicals used in manufacturing. Unfortunately, it used to be common for manufacturers in the United States to dispose of industrial compounds directly into bodies of water. One of the most widely publicized consequences of this practice occurred in the Cuyahoga River of Ohio. For more than 100 years, industries along the river dumped industrial wastes that formed a slick of pollution along the surface, killing virtually all animal life. The problem became so bad that the river actually caught fire and burned several times over the decades (FIGURE 42.6). The fire on the river in 1969 garnered national attention and led to a movement to clean up America's waterways. Today, the Cuyahoga River and most other rivers in the United States are much cleaner because of legislation that substantially reduced the amount of industrial and other waste that can be legally dumped into waterways.

Polychlorinated biphenyls (PCBs) are a group of industrial compounds that were once used to manufacture plastics and insulate electrical transformers. PCBs have caused many environmental problems. Ingested PCBs are lethal and carcinogenic, or cancer-causing. Although the manufacture of PCBs ended in 1979 and they are no longer used in the United States, because of their long-term persistence they are still present in the environment. One particularly high-profile case involves two General Electric manufacturing plants in New York State that dumped 590,000 kg (1.3 million pounds) of PCBs into the Hudson River from 1947 to 1977. In 2002, the EPA ruled that General Electric must pay for the dredging and removal of approximately 2.03 million cubic meters (2.65 million cubic yards) of PCB-contaminated sediment from a 64-km (40-mile) stretch of the upper Hudson River in New York State. General Electric argued that the dredging should not occur because it would stir up the sediments and re-suspend the PCBs in the river water, causing further problems. The courts and EPA scientists disagreed and in 2009 the dredging of the PCBs finally began. As of 2013, the dredging was on schedule to be completed by 2015. As this case demonstrates, though it can take decades of scientific research and debate, in the end we can partially reverse the contamination of our water bodies.

While PCBs have long been a concern, there is a growing uneasiness over compounds known as PBDEs (polybrominated diphenyl ethers). PBDEs are most commonly known as flame retardants added to a wide variety of items that include construction materials, furniture, electrical components, and clothing. They make buildings and their contents considerably less flammable than they would be otherwise. Since the 1990s, however, scientists have been detecting PBDEs in some unexpected places, including fish, aquatic birds, and human breast milk. Exposure to some types of PBDEs can lead to brain damage, especially in children. As a result, the European Union and several states, including Washington and California, have banned the manufacture of several types of PBDEs.

FIGURE 42.6 **A river on fire.** In 1952, the polluted Cuyahoga River caught fire after a spark ignited the film of industrial pollution that was floating on the surface of the water. *(AP Photo)*

Perchlorates A group of harmful chemicals used for rocket fuel.

Polychlorinated biphenyls (PCBs) A group of industrial compounds used to manufacture plastics and insulate electrical transformers, and responsible for many environmental problems.

Teaching Tip: Discussion Starter

Ask students to name examples of synthetic compounds that have been found in the water supply and explain why they are a concern. *Examples of synthetic compounds found in the water supply include pesticides, pharmaceuticals, military compounds, and industrial compounds. These compounds are a concern because they can be toxic, can cause birth defects, and, in the case of compounds that resemble animal hormones, can interfere with growth and sexual development.*

module

42

REVIEW

In this module, we learned that a wide variety of chemicals in the water can have harmful effects on humans and other organisms. These chemicals include heavy metals, acids, and synthetic organic compounds. Through legislation and cleanup efforts, some of these compounds are now found less frequently in the environment. However, some chemicals continue to make their way into water bodies and others persist over time. In the next module, we will discuss the effects of oil pollution in bodies of water.

Module 42 AP® Review Questions

1. Arsenic is
   (a) found naturally in groundwater.
   (b) often discovered due to its rapid health effects.
   (c) inexpensive to remove from drinking water.
   (d) found in rainwater.
   (e) primarily found in water in North America.

2. Which is NOT true of acid deposition?
   (a) It is primarily due to the burning of coal.
   (b) It is treated using limestone.
   (c) It can occur as a result of mining.
   (d) It is lethal to many aquatic organisms.
   (e) It causes increased solubility of many ions.

3. What caused the eggshells of some birds to become thin and break?
   (a) Heavy metal concentrations in prey species
   (b) An inert ingredient in herbicides
   (c) A pharmaceutical in the water supply
   (d) An insecticide ingested by prey species
   (e) A pesticide used to control pest birds such as crows

4. Mercury
   (a) is harmless once converted into methylmercury.
   (b) exposure often occurs through shellfish.
   (c) is most concentrated in herbivores.
   (d) can be safely trapped during the production of concrete.
   (e) damages the immune system.

5. Which hazardous material is known to cause cancer?
   (a) Lead
   (b) DDT
   (c) Mercury
   (d) Perchlorates
   (e) PCBs

Answers to Module 42 AP® Review Questions

1. a
2. c
3. d
4. b
5. e

Teaching Tip: Beyond the Classroom

Ask your students to research the *Exxon Valdez* oil spill. In their report they should address the following questions:

- When and where did the *Exxon Valdez* spill occur? *The spill occurred on March 24, 1989, in Prince William Sound on the coast of Alaska.*
- How large was the tanker? How much oil could it carry? *The tanker could hold 1.48 million barrels of oil. Its dimensions were 301 m long, 51 m wide, and 26 m in depth.*
- What species have been affected? *Some of the wildlife that was affected were seabirds, sea otters, harbor seals, killer whales, and plant life including kelp forests. Many of these organisms died due to contamination from the crude oil.*
- Describe the water quality now, approximately 25 years after the spill. *Scientists have done yearly monitoring at the spill sites. They concluded that the harmed populations of many species have rebounded, including bald eagles and salmon. However, several species have not yet rebounded, including killer whales and sea otters. Nor has the oil been completely removed from the environment. Pits dug into the shoreline suggest that approximately 55,000 L of oil remain.*
- What new rules or regulations have been put into place now to help prevent such a spill from happening again? *All oil tankers today are required to be of a double-hull design where two steel walls contain any leaking oil. In Alaska, two tugboats must now escort large oil tankers through the channels.*

module 43

Oil Pollution

In Chapter 12 we noted that the pollution of Earth's oceans and shorelines from crude oil and other petroleum products is an ongoing problem. Petroleum products are highly toxic to many marine organisms, including birds, mammals, and fish, as well as to algae and microorganisms that form the base of the aquatic food chain. Oil is a persistent substance that can spread below and across the surface of the water for hundreds of kilometers and leave shorelines with a thick, viscous covering that is extremely difficult to remove. In this section, we will examine the many sources of oil pollution and then talk about some of the ways currently used to clean up oil and to reduce its harmful effects.

Learning Objectives

After reading this module you should be able to

- identify the major sources of oil pollution.
- explain some of the current methods to remediate oil pollution.

There are several sources of oil pollution

There are many different sources of oil pollution in water. Oil and other petroleum products can enter the oceans as spills from oil tankers. One of the best-known spills involved the tanker *Exxon Valdez* that ran aground off the coast of Alaska in 1989 (FIGURE 43.1). As we discussed in Chapter 12, the ship spilled 42 million liters (11 million gallons) of crude oil that spread across the surface of several kilometers of ocean and coastline. The spill killed 250,000 seabirds, 2,800 sea otters, 300 harbor seals, and 22 killer whales. Cleanup efforts have been going on for 2 decades.

In 2009, 20 years after the *Exxon Valdez* spill, scientists evaluated the state of the contaminated Alaskan ecosystem. They concluded that the harmed populations of many species have rebounded, including

FIGURE 43.1 **The *Exxon Valdez* oil spill.** In 1989, the oil tanker ran aground and spilled millions of liters of crude oil onto the shores of Alaska, where the oil killed thousands of animals and harmed many others such as this red-necked grebe (*Podiceps grisegena*) on Knight's Island, located about 35 miles from the spill. *(AP Photo)*

bald eagles and salmon. However, several species have not yet rebounded, including killer whales (*Orcinus orca*) and sea otters (*Enhydra lutris*). Nor has the oil been completely removed from the environment. Pits dug into the shoreline suggest that approximately 55,000 L (14,500 gallons) of oil remain. It is estimated that this oil will take more than 100 years to break down and the long-lasting effects will only become apparent over the coming decades.

For its part, Exxon has paid $1 billion for the cleanup and $500 million in damages. The company also changed the ship's name, although the ship has been banned from carrying oil in North America. The *Valdez* accident sparked new rules for oil tankers in North America. The *Exxon Valdez* had a single-hull design, but tankers must now have a double-hull design with two steel walls to contain leaking oil.

Offshore drilling is another source of oil pollution. There are approximately 5,000 offshore oil platforms in North America and another 3,000 worldwide. Drilling platforms often experience leaks. The best estimate for the amount of petroleum leaking into North American waters is 146,000 kg (322,000 pounds) per year. In other parts of the world, antipollution regulations are often less stringent. Estimates of the amount of petroleum leaking into the ocean annually from foreign oil platforms range from 0.3 million to 1.4 million kilograms (0.6 million to 3.1 million pounds).

One of the most famous oil leaks from an offshore platform occurred in 2010 on a BP operation in the Gulf of Mexico. In this case, an explosion on the *Deepwater Horizon* platform caused a pipe to break on the ocean floor nearly 1.6 km (1 mile) below the surface of the ocean. From the time of the explosion in April until the well was sealed in August 2010, the broken pipe released an estimated 780 million liters (206 million gallons) of crude oil into the Gulf of Mexico (see Figure 35.4 on page 412). This spill contaminated beaches, wildlife, and the estuaries that serve as habitats for the reproduction of commercially important fish and shellfish. The magnitude of the oil spill was nearly 20 times larger than that of the *Exxon Valdez*. However, because much of the oil spilled into the ocean, scientists may not be able to assess the full impact of the oil spill for several decades. The accident has the potential to become one of the largest environmental disasters in history.

In addition to oil spills, oil pollution in the ocean occurs naturally. In fact, the U.S. National Academy of Sciences recently estimated that natural releases of oil from seeps in the bottom of the ocean account for 60 percent of all oil in the waters surrounding North America and 45 percent of all oil in water worldwide. FIGURE 43.2 shows the proportion of different sources of oil in water for both North American and worldwide marine waters. In the waters controlled by the United States, the ocean seeps more than 270,000 L (70,000 gallons) of oil every day. This means that when we assess the environmental impact of oil in our oceans, we must consider the combination of both natural and anthropogenic releases of oil.

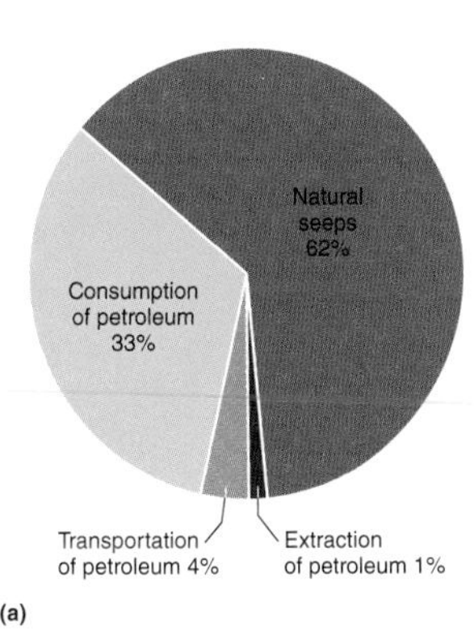

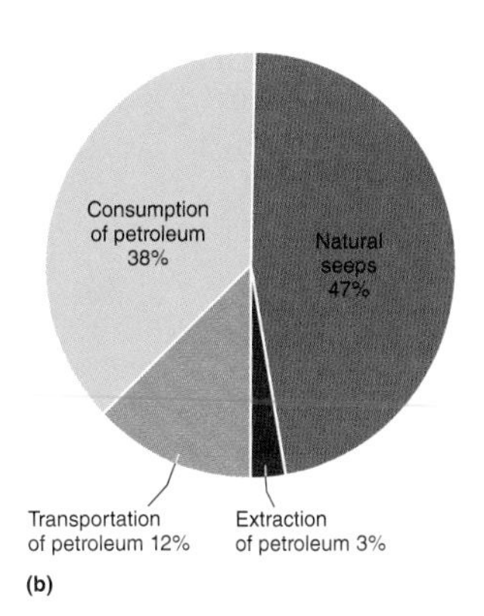

FIGURE 43.2 Sources of oil in the ocean. Oil contamination in the ocean, both (a) in North America and (b) worldwide, comes from a variety of sources including natural seeps, extraction of oil from underneath the ocean, transport of oil by tanker or pipeline, and consumption of petroleum-based products. *(After http://oceanworld.tamu.edu/resources/oceanography-book/contents.htm)*

Module 43 Oil Pollution 499

Teaching Tip: Video

Oil-Spill Cleanup Methods
This short, silent video, produced by the American Petroleum Institute, demonstrates sorbants, booms, skimmers, in-situ burning, and dispersants. After watching the video once, you can ask students to pick one of the oil-spill cleanup methods and write a narrative to accompany the film. Choose several students to share their scripts while you watch the film again.

The link to this video may be found by clicking the link buttons in the TE-book, opening the TRFD, or logging onto the book's companion website bcs.whfreeman.com/friedlandapes2e (teacher login required).

 **Chapter 14 Web Resources**

Teaching Tip: Review

Ask students the following questions:

- Name several ways in which oil gets into the ocean. *Oil gets into the ocean through natural seeps, extraction of oil from underneath the ocean, transport of oil by tanker or pipeline, and consumption of petroleum-based products.*
- Describe the effects of an oil spill. *Because petroleum products are highly toxic they can cause harm to many marine organisms, including birds, mammals, and fish, as well as the algae and microorganisms that form the base of the aquatic food chain.*
- What are three ways to remediate an oil spill? *The first approach is containment through the laying out of oil booms that consist of plastic barriers floating on the surface of the water and extending down into the water for several meters. This keeps the oil from spreading. A second approach is applying chemical dispersants that help break up oil on the surface before it hits the shoreline and is able to cause damage to coastal ecosystems. A third approach is using genetically engineered bacteria to consume the oil.*

There are ways to remediate oil pollution

Since the 1989 *Exxon Valdez* oil spill, researchers have been investigating how best to remediate oil spills. Contaminated mammals and waterfowl must be cleaned by hand. Bird feathers that are covered with oil, for example, become heavy and lose their ability to insulate. The best approach to cleaning up the spilled oil, however, is not always clear.

Oil spilled in the ocean can either float on the surface or remain far below in the form of underwater plumes. For oil floating on the surface of the open ocean, a common approach is to contain the oil within an area and then suck it off the surface of the water. Containment occurs by laying out oil containment booms that consist of plastic barriers floating on the surface of the water and extending down into the water for several meters. These plastic walls keep the floating oil from spreading further. Once the oil is contained, boats equipped with giant vacuums suck up as much oil as possible (FIGURE 43.3). In shallow areas and along the coastline, absorbent materials are used to suck up the spilled oil.

A second approach to treating oil floating on the surface is to apply chemicals that help break up and disperse the oil before it hits the shoreline and causes damage to the coastal ecosystems. Although the dispersants can be effective, they can also be toxic to marine life. Current research is examining ways to make chemical dispersants more environmentally friendly.

FIGURE 43.3 **Oil-spill containment.** Floating plastic walls can contain oil spills while the oil is sucked off the surface of the water. *(Matthew Hinton/The Times-Picayune/Landov)*

A third approach to cleaning up oil uses genetically engineered bacteria. Several years ago scientists discovered a naturally occurring bacterium that obtained its energy by consuming oil emerging from natural seeps. These bacteria were typically rare in the ocean, but were very abundant in areas where oil spills or seeps occurred. Scientists are currently trying to determine the genes that confer the bacteria's ability to consume oil and hope to insert copies of these genes into genetically modified bacteria to consume oil spills even faster.

In contrast to oil floating on the surface of the water, oil in underwater plumes persists as a mixture of water and oil, similar to the mixture of vinegar and oil in a salad dressing. In the case of the BP platform explosion in the Gulf of Mexico, scientists reported observing an oil plume moving approximately 1,000 m (3,000 feet) below the surface of the ocean. The plume was approximately 24 to 32 km (15–20 miles) long, 8 km (5 miles) wide, and hundreds of meters thick. There is currently no agreed-upon method of removing underwater plumes from the water.

When spilled oil comes to shore, the best solution in not always clear. For example, there is some debate over the best way to treat rocky coastlines after an oil spill. Scientists have been monitoring parts of Prince William Sound that were treated in different ways after the *Exxon Valdez* spill. Workers cleaned some areas with high-pressure hot water to remove the oil. Unfortunately, the hot water sprayers not only removed the oil, but also removed most of the marine life and, in some cases, the fine-grained sediments containing nutrients. Without the fine-grained sediment, many organisms were unable to recolonize the coast.

Other parts of the coastline received no human intervention. Over the years since the spill, the repeated action of waves and tides slowly removed much of the oil. However, the remaining oil existing in crevices of the rocky shoreline continues to have a negative effect on organisms that live among the rocks. Thus, leaving the oil on beaches also poses problems. At present, there is no clear consensus on the best way to respond to oil spills on coastlines.

module

43

REVIEW

In this module, we learned that oil pollution in water comes from both natural sources and human activity. Oil pollution can be devastating to plants and animals. We have developed several ways to help clean up oil spills. In the next module, we will examine non-chemical forms of pollution.

Module 43 AP® Review Questions

1. The *Exxon Valdez* oil spill
   (a) was cleaned up within 2 years of the spill.
   (b) lead to new regulations for oil tankers.
   (c) was larger than the *Deepwater Horizon* spill.
   (d) primarily affected aquatic organisms.
   (e) has not significantly changed the ecosystem 20 years later.

2. Approximately what percent of the oil in marine waters worldwide is due to natural causes?
   (a) 5 percent
   (b) 20 percent
   (c) 35 percent
   (d) 45 percent
   (e) 60 percent

3. Which is NOT used in the cleanup of oil spills?
   (a) Genetically engineered bacteria.
   (b) Clumping agents applied to underwater plumes.
   (c) Chemical dispersants.
   (d) Large vacuums.
   (e) Absorbent materials along the coast.

Answers to Module 43 AP® Review Questions

1. b
2. d
3. b

module

44

Nonchemical Water Pollution

When we think about water pollution, we most commonly envision scenes of dirty water contaminated with toxic chemicals. Though such scenarios certainly receive a great deal of public attention, there are other, less familiar types of water pollution. In this module, we will examine four types of nonchemical water pollution: solid waste, sediment, heat, and noise.

Learning Objectives

After reading this module you should be able to

- identify the major sources of solid waste pollution.
- explain the harmful effects of sediment pollution.
- discuss the sources and consequences of thermal pollution.
- understand the causes of noise pollution.

Solid waste pollution includes garbage and sludge

Solid waste consists of discarded materials that do not pose a toxic hazard to humans and other organisms. Much solid waste is what we call garbage and the sludge produced by sewage treatment plants. In the United States, solid waste is generally disposed of in landfills, but in some cases it is dumped into bodies of water and can later wash up on coastal beaches. In 1997, scientists discovered a large area of solid waste, composed mostly of discarded plastics, floating in the North Pacific. This area, named the Great Pacific Garbage Patch, appears to collect much of the solid waste that is dumped into waters and concentrates it in the middle of the rotating currents in an area the size of Texas. Other ocean currents appear to also have the ability to concentrate garbage. Given the vastness of these areas, no one is certain how much solid waste is floating but current estimates are in the range of hundreds of millions of kilograms.

Garbage on beaches and in the ocean is not only unsightly; it is dangerous to both marine organisms and people. Plastic rings from beverage six-packs, for instance, can strangle many animals. Medical waste poses a threat to people on the beach, particularly children. During the 1970s and 1980s, the public turned its attention to garbage dumping off the coasts of the United States by both municipalities and cruise ships. The outcry grew when garbage was found to include medical waste such as used hypodermic needles. As a result, the practice of dumping garbage in the ocean was curtailed in the early 1980s. The problem remains in many developing countries where there is often no political mechanism to prevent dumping or it is economically difficult to manage proper garbage disposal (FIGURE 44.1).

Another major source of solid waste pollution is the coal ash and coal slag that remain behind when coal is burned. As we saw in Chapter 12, such waste contains a number of harmful chemicals including mercury, arsenic, and lead. The solid waste products from burning coal are typically dumped into landfills, ponds, or

FIGURE 44.1 **A river of garbage.** Environmental regulations have greatly reduced the amount of solid waste that is dumped in U.S. waters, but other parts of the world such as the Citarum River in Indonesia still face a major environmental challenge. *(DADANG TRI/ Reuters/Landov)*

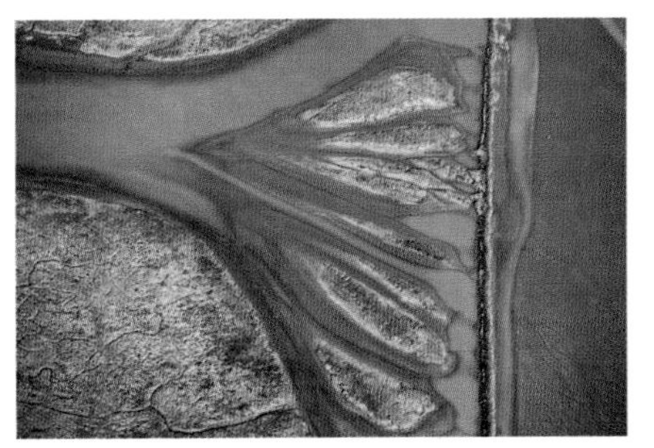

FIGURE 44.2 **Sediments carried by rivers.** Some rivers, such as the Fraser River Delta as it enters the Pacific Ocean, carry a large amount of sediment that gets emptied into lakes and oceans to form deltas. *(National Geographic Image Collection/Alamy)*

abandoned mines where it can contaminate groundwater. In the United States, the Environmental Protection Agency considers the waste from burning coal and other fossil fuels to be "special waste" that is exempt from federal regulations for the disposal of hazardous waste. In addition, most states have either no regulations or weak regulations governing the disposal of coal ash and coal slag.

Sediment pollution consists of soil particles that are carried downstream

As noted in the our discussion of the Chesapeake Bay, sediments are particles of sand, silt, and clay carried by moving water in streams and rivers that eventually settle out in another location where water movement is slowed, such as where streams empty into lakes and rivers empty into oceans forming deltas. The transport of sediments by streams and rivers is a natural phenomenon, but sediment pollution is the result of human activities that can substantially increase the amount of sediment entering natural waterways (FIGURE 44.2).

Numerous human activities lead to increased sedimentation. Construction of buildings, for example, requires digging up the soil, and the bare ground can lose some of its soil to erosion. As we saw in Chapters 8 and 10, plowed agricultural fields are susceptible to erosion from rain and wind. Sediments can also enter streams and rivers when natural vegetation is removed from the edge of a water body and replaced with either crops or domesticated animals that continually disturb the soil when they come to drink. According to recent estimates, 30 percent of all sediments in our waterways comes from natural sources while 70 percent comes from human activities.

What is the effect of increased sediment caused by human activities? Perhaps the most noticeable is that waterways become brown and cloudy due to the suspension of soil particles in the water. Increased sediment in the water column reduces the infiltration of sunlight, which can reduce the productivity of aquatic plants and algae. Because sediments can also clog gills, they sometimes hinder the ability of fish and other aquatic organisms to obtain oxygen. In locations where water moves slowly, the sediments settle out and accumulate on the bottom of the water body. This accumulation can clog the gills of bottom dwellers such as oysters or clams. Since the sediments can contain nutrients, this pollution may also contribute to increased nutrients coming into the ecosystem. In total, the U.S. Environmental Protection Agency estimates that sediment pollution costs $16 billion annually in environmental damage.

Thermal pollution causes substantial changes in water temperatures

A third type of nonchemical water pollution, **thermal pollution,** occurs when human activities cause a substantial change in the temperature of water. Thermal pollution is most common when an industry removes cold water from a natural supply, uses it to absorb heat

Thermal pollution Nonchemical water pollution that occurs when human activities cause a substantial change in the temperature of water.

AP® Exam Tip

The impact of sedimentation on waterways and coastlines is covered on the 2013 AP® Exam, Question 1. To answer this question, students must

- name two human activities that change the flow of sediments emptying into Gulf Coast ecosystems
- describe two ways that the loss of natural sediment harms Gulf Coast wetlands
- describe two environmental impacts on marine ecosystems when fertilizer runs into the Gulf of Mexico
- explain the economic impact of fertilizer runoff into the Gulf of Mexico
- describe a strategy for reducing the use of fertilizer that contributes to the runoff problem

Teaching Tip: Video

Toxic Sediments

This video describes problems with toxic sediments found in Sydney Harbour, Australia. Toxic sediment has caused plants and aquatic organisms to absorb high levels of contaminants. The two main sources of these contaminants are industrial processes that occurred in the past as well as current stormwater runoff.

The link to this video may be found by clicking the link buttons in the TE-book, opening the TRFD, or logging onto the book's companion website bcs .whfreeman.com/friedlandapes2e (teacher login required).

 **Chapter 14 Web Resources**

Teaching Tip: Debate the Issue

How should we regulate thermal pollution?

Have students research the subject of thermal pollution in preparation for a debate on how thermal pollution should be regulated. Students should be prepared to answer the following questions:

- What is thermal pollution?
- Should the EPA mandate that water sent back to the rivers and lakes be the same temperature as when it was originally taken?
- What are the possible economic consequences of regulating thermal pollution? How can they be addressed?

In class, select groups to debate the issue. Allow each side 2 minutes to prepare, 3 minutes to present, and 1 minute for rebuttal. Have the remainder of the class take notes, ask questions, and decide which side has the most persuasive argument.

FIGURE 44.3 **Thermal pollution.** Nuclear reactors, such as the Vermont Yankee nuclear plant, use water to generate steam. To cool this steam, they either use cooling towers or empty the water into holding ponds. In both cases, the water must be cooled before it is returned to the natural source of water such as a lake or river. In 2013 the owners of Vermont Yankee announced they would not seek to renew their license. *(CJ Gunther/epa/Corbis)*

that is generated in a manufacturing process, and returns the heated water back to the natural supply. For example, we saw in Chapter 12 that electric power plants use nearly half of all water extracted; they remove cold water from rivers, lakes, or oceans, cool the steam converted from water back into water, and then return the water to nature at temperatures that are 10°C to 15°C (18°F–27°F) warmer. A variety of other industries make use of water for cooling, including steel mills and paper mills that need to cool their machines.

Species in a given community are generally adapted to a particular natural range of temperatures. Therefore a dramatic change in temperature can kill many species, a phenomenon called **thermal shock.** High temperatures also cause organisms to increase their respiration rate. But because warmer water does not contain as much dissolved oxygen as cold water, an increase in respiration and a decrease in oxygen will cause many animals to suffocate. In recent years, steps have been taken to help reduce thermal pollution, including pumping the heated water into outdoor holding ponds where it can cool further before being pumped back into natural water bodies (FIGURE 44.3).

In the United States, the EPA regulates how much heated water can be returned to natural water bodies. Compliance can become a real challenge in the summer when high demand for electricity causes an increased demand for cooling water even though the water in rivers and lakes will then probably be at its lowest volume and at its highest temperature. One common solution to this problem has been the construction of cooling towers that release the excess heat into the atmosphere instead of into the water. A cooling tower relies on the cooling power of evaporation to reduce the temperature of the water, much as we depend on our own sweat and a breeze to cool ourselves on a hot day. Some industries have built closed systems in which they cool the hot water in a cooling tower and then recycle the water to be heated again. In this way, the industries do not extract water from natural water bodies, nor do they release any heated water back into nature.

Thermal shock A dramatic change in water temperature that can kill organisms.

Noise pollution may interfere with animal communication

It may strike you as odd to think of noise as a type of water pollution. Indeed, noise pollution has received the least amount of attention from environmental scientists and, as a result, we know the least about it. When we think of noise pollution, what often comes to mind is the sound of city traffic. However, noise pollution also occurs in the water. For example, sounds emitted by ships and submarines that interfere with animal communication are the major concern. Especially loud sonar could negatively affect species such as whales that rely on low-frequency, long-distance communication. Several instances of beached whales in the Bahamas, the Canary Islands, and the Gulf of California have been connected to the use of military sonar and the use of loud, underwater air guns by energy companies searching for oil deposits under the oceans.

In 2003 a federal judge rejected the U.S. Navy's request to install a network of long-range sonar systems across the ocean floor to detect incoming submarines because of suspected negative impacts on endangered whales and other species of marine animals. In 2008, however, the U.S. Supreme Court ruled that the president of the United States could exempt the navy from environmental laws that were related to potential sonar effects on ocean life. To better understand noise pollution in the ocean, the U.S. National Oceanic and Atmospheric Administration announced in 2012 that had it completed the first step in mapping the noises across different regions of the ocean. At the same time, an increased awareness of noise pollution in the ocean has inspired some ship builders to design ships equipped with quieter propellers.

Teaching Tip: Activity

Have students work in groups of two or three to fill out the table below on nonchemical water pollution. (Answers are provided in italics.) Once students have completed the table, suggest that they find a specific example of one of the four pollutants as a homework assignment. Have them write a short report that describes the example and explains the organisms affected as well as any steps being taken to address the problem.

Pollutant	Source	Effects	Solution
Solid waste	• *Garbage* • *Sludge* • *Plastics* • *Coal ash*	• *Can entangle ocean organisms* • *Medical waste poses threat to humans on beach*	• *Create laws and penalties for dumping* • *Fine those who violate the law* • *Public relations campaign*
Sediment	• *Construction sites* • *Poor farming practices*	• *Decreases photosynthetic rates* • *Clogs gills of fish and bottom dwellers*	• *Use silt fences at construction sites* • *Mandate riparian buffer zones near farms*
Thermal	• *Electric power plants* • *Industry*	• *Thermal shock, which causes organisms to increase respiration rate*	• *Monitor and regulate water temperature at power plants and manufacturing facilities* • *Require cooling towers or discharge lagoons*
Noise	• *Sound emitted by ships and submarines that interfere with animal communication*	• *Interferes with animal communication* • *Causes beaching of dolphins and whales*	• *Require large ships to use quieter propellers*

module

44

REVIEW

In this module, we learned that there are several nonchemical forms of water pollution. Solid waste pollution comes from garbage, sludge, and the ash and slag produced by burning coal. Sediment pollution consists of soil particles that erode from the land and are carried downstream in streams and rivers where they can reduce light penetration, clog the gills of aquatic organisms, and add nutrients to aquatic ecosystems. Thermal pollution typically happens when industrial processes take in cool water from water bodies, use the water to cool their equipment, and then return much warmer water to the water body. Finally, noise pollution from ships, sonar, and air guns exploring for energy deposits has the potential to interfere with the communication of aquatic animals in the ocean, including whales. In the next module, we will examine how water pollution laws have been designed to control some of the major sources of pollution.

Module 44 AP® Review Questions

1. Solid waste
   (a) sinks to the ocean floor.
   (b) dumping in the ocean is prohibited worldwide.
   (c) can include coal-burning byproducts.
   (d) is primarily problematic once washed up on beaches.
   (e) is rarely toxic to humans.

2. Sedimentation of water
   (a) decreases the solubility of oxygen.
   (b) increases the amount of photosynthesis.
   (c) clogs the gills of some aquatic animals.
   (d) results in decreased nutrient availability.
   (e) is primarily due to industrialization.

3. Thermal pollution
   (a) is primarily a problem in the winter.
   (b) is rarely lethal.
   (c) is primarily due to residential water use.
   (d) is not regulated in the United States.
   (e) has been reduced by the use of cooling towers.

4. The use of sonar
   (a) can reduce the productivity of some algae.
   (b) has little effect on aquatic ecosystems.
   (c) has decreased significantly.
   (d) has a positive effect on some fish species.
   (e) disrupts communication among whales.

Answers to Module 44 AP® Review Questions

1. e
2. c
3. e
4. e

Teaching Tip: Beyond the Classroom

Have your students research different types of water pollution, either individually or in groups. Instruct students to create a 5-minute class presentation. Each report should identify the problem, give examples, and discuss ways to address the problem. Possible research subjects include:

- offshore oil drilling and well abandonment
- sewage dumping from cruise ships
- sewage dumping in coastal areas of developing nations
- pollution in the Great Lakes
- pollution in Chesapeake Bay
- noise pollution
- thermal pollution
- algal bloom (red tide)
- oil spill

module

45 Water Pollution Laws

A country's water quality improves when its citizens demand it and the country is affluent enough to afford measures to clean up pollution and to take steps to prevent it in the future. In this module we will look at U.S. laws that protect water from pollution and ensure safe drinking water. We will also examine how laws in developing nations are changing to address water pollution.

Learning Objectives

After reading this module you should be able to

- explain how the Clean Water Act protects against water pollution.
- discuss the goals of the Safe Drinking Water Act.
- understand how water pollution legislation is changing in developing countries.

Teaching Tip: Journal Prompt

Ask students the following questions:

- What is the significance of a maximum contaminant level (MCL)? *The maximum contaminant levels for 77 different elements were set under the Safe Drinking Water Act (1974, 1986, 1996). These maximum concentrations consider both the concentration of each compound that can cause harm as well as the feasibility and cost of reducing the compound to such a concentration.*
- What are the two most important water pollution laws in the United States and what do they legislate? *In the United States, the most important water pollution laws are the Clean Water Act and Safe Drinking Water Act. The Clean Water Act supports the "protection and propagation of fish, shellfish, and wildlife and recreation in and on the water" by maintaining and, when necessary, restoring the chemical, physical, and biological properties of surface waters. The Safe Drinking Water Act is legislation that sets the national standards for safe drinking water. It established maximum contaminant levels for 77 different elements or substances in both surface water and groundwater.*

The Clean Water Act protects water bodies

As recently as the 1960s, water quality was very poor in much of the United States, but a growing awareness of the problem encouraged a series of laws to fight water pollution. The Federal Water Pollution Control Act of 1948 was the first major piece of legislation affecting water quality. In 1972, the act was expanded into the **Clean Water Act,** which supports the "protection and propagation of fish, shellfish, and wildlife and recreation in and on the water" by maintaining and, when necessary, restoring the chemical, physical, and biological properties of surface waters. Note that this objective does not include the protection of groundwater.

Clean Water Act Legislation that supports the "protection and propagation of fish, shellfish, and wildlife and recreation in and on the water" by maintaining and, when necessary, restoring the chemical, physical, and biological properties of surface waters.

Safe Drinking Water Act Legislation that sets the national standards for safe drinking water.

The Clean Water Act originally focused mostly on the chemical properties of surface waters. More recently, there has been an increased focus on ensuring that the biological properties of the waters also receive attention, including the abundance and diversity of various species. Most importantly, the Clean Water Act issued water quality standards that defined acceptable limits of various pollutants in U.S. waterways. To help enforce these limits, the act allowed the EPA and state governments to issue permits to control how much pollution industries can discharge into the water. Over time, more and more categories of pollutants have been brought under the jurisdiction of the Clean Water Act, including animal feedlots and storm runoff from municipal sewer systems.

The Safe Drinking Water Act protects sources of drinking water

In addition to the Clean Water Act, other legislation has been passed to regulate water pollution, including the **Safe Drinking Water Act** (1974, 1986, 1996), which sets the national standards for safe drinking water. Under the Safe Drinking Water Act, the EPA is

TABLE 45.1 The maximum contaminant levels (MCL) for a variety of contaminants in drinking water as determined by the U.S. Environmental Protection Agency, in parts per billion (ppb)

Contaminant category	Contaminant	Maximum contaminant level (ppb)
Microorganism	Giardia	0
Microorganism	Fecal coliform	0
Inorganic chemical	Arsenic	10
Inorganic chemical	Mercury	2
Organic chemical	Benzene	5
Organic chemical	Atrazine	3

Source: U.S. Environmental Protection Agency, http://www.epa.gov/safewater/contaminants/index.html.

TABLE 45.2 The current leading causes and sources of impaired waterways in the United States

	Causes of impairment	Sources of impairment
Streams and rivers	Bacterial pathogens, habitat alteration, oxygen depletion	Agriculture, water diversions, dam construction
Lakes, ponds, and reservoirs	Mercury, PCBs, nutrients	Atmospheric deposition, agriculture
Bays and estuaries	Bacterial pathogens, oxygen depletion, mercury	Atmospheric deposition, municipal discharges including sewage

Source: Data from U.S. Environmental Protection Agency. 2004. *National Water Quality Inventory: Report to Congress.*

responsible for establishing **maximum contaminant levels (MCL)** for 77 different elements or substances in both surface water and groundwater. This list includes some well-known microorganisms, disinfectants, organic chemicals, and inorganic chemicals (**TABLE 45.1**). These maximum concentrations consider both the concentration of each compound that can cause harm as well as the feasibility and cost of reducing the compound to such a concentration.

MCLs are somewhat subjective and are subject to political pressures. For example, despite the evidence that 50 ppb of arsenic caused harm in humans, the MCL for arsenic was kept at 50 ppb for many years because of concerns that many communities could not afford to reduce levels to 10 ppb. As noted earlier in this chapter, the MCL for arsenic was finally reduced to 10 ppb in 2001.

What has been the impact of these water pollution laws? In general, they have been very successful. The EPA defines bodies of water in terms of their designated uses, including aesthetics, recreation, protection of fish, and as a source of safe drinking water. The EPA then determines if a particular waterway fully supports all of the designated uses. According to the most recent EPA data, 56 percent of all streams, 35 percent of lakes and ponds, and 70 percent of bays and estuaries in the United States now fully support their designated uses. This is a large improvement from decades past but, as **TABLE 45.2** shows, we still have a lot of work to do to improve the remaining waterways. Today, the water in municipal water systems in the United States is generally safe. Water regulations have greatly reduced contamination of waters and nearly eliminated major point sources of water pollution. But nonpoint sources such as oil from parking lots and nutrients and pesticides from suburban lawns are not covered under existing regulations. In addition, the U.S.

Maximum contaminant level (MCL) The standard for safe drinking water established by the EPA under the Safe Drinking Water Act.

Module 45 Water Pollution Laws 507

FIGURE 45.1 The Tietê River in Brazil. The Tietê River, which passes through the large city of São Paolo, was badly polluted in the 1950s but is much cleaner today. *(Marcos Hirakawa/age footstock)*

government has exempted fracking for natural gas (see Chapter 1) from the Safe Drinking Water Act, despite the fact that fracking injects a suite of harmful chemicals deep into the ground.

Water pollution legislation is becoming more common in the developing world

If we look at water pollution legislation around the world, there is a clear difference between developed and developing countries. Developed countries, including those in North America and Europe, experienced tremendous industrialization many decades ago and widely polluted their air and water at that time. More recently they have addressed the problems of pollution by cleaning up polluted areas and by passing legislation to prevent pollution in the future. Developing countries are still in the process of industrializing. They are less able to afford water-quality improvements such as wastewater treatment plants or the costs associated with restrictive legislation. Moreover, political instability and corruption often make enforcement of legislation difficult. In some cases, contaminating industries move from developed countries to developing countries. Although the developing countries suffer from the additional pollution, they benefit economically from the additional jobs and spending that the new industries bring with them.

Water pollution problems are prevalent in many of the developing nations of Africa, Asia, Latin America, and eastern Europe. China and India, for example, have undergone rapid industrialization and have many areas of dense human population—a recipe for major water pollution problems. However, as a nation becomes more affluent, people often show more interest in the environment and resources available to address environmental issues. In Brazil, for example, industrialization began to take off in the 1950s. By the 1990s, the Tietê River, which passes through the large city of São Paulo, was badly polluted. More than a million Brazilians signed a petition in 1992 requesting that the government regulate the industrial and municipal pollution being dumped into the river. Today the Tietê River is much cleaner (FIGURE 45.1).

module

45

REVIEW

In this module, we looked at the two major water pollution laws in the United States. The Clean Water Act is designed to protect surface water from pollution, but it does not protect groundwater. In contrast, the Safe Water Drinking Act sets maximum levels for microorganisms, disinfectants, organic chemicals, and inorganic chemicals in drinking water. Whereas developed countries have gone through periods of large-scale water pollution as they industrialized, they have since passed many laws to improve water quality. Developing countries are currently experiencing industrialization and widespread water pollution. As these countries become more affluent, it is expected that they will also turn their attention to improving water quality.

Module 45 AP® Review Questions

1. The first legislation on water quality was
   (a) the Federal Water Pollution Control Act.
   (b) the Clean Water Act.
   (c) the Safe Drinking Water Act.
   (d) the Resource Conservation and Recovery Act.
   (e) the Water Quality Act.

2. Maximum contaminant levels for groundwater were set in
   (a) the Federal Water Pollution Control Act.
   (b) the Clean Water Act.
   (c) the Safe Drinking Water Act.
   (d) the Resource Conservation and Recovery Act.
   (e) the Water Quality Act.

3. Which water quality issue is not covered in existing legislation?
   (a) Water diversions
   (b) Organic chemicals
   (c) Offshore drilling
   (d) Nonpoint sources
   (e) Groundwater

4. Which does NOT contribute to poor water quality in developing countries?
   (a) High population density
   (b) Political instability
   (c) Rapid industrialization
   (d) High unemployment
   (e) Relocation of industry

Answers to Module 45 AP® Review Questions

1. a
2. c
3. d
4. d

working toward sustainability

Purifying Water for Pennies

As we discussed in this chapter, having access to clean drinking water is a worldwide problem. Nearly 1 billion people drink unsafe water that contains soil sediments, pesticides, heavy metals, and disease-causing organisms. The problem of unsafe drinking water is largely a problem of developing countries that lack the financial resources to build proper sewage and water treatment facilities. Any solution to this challenge would need to be both low cost and effective.

Nearly a decade ago, the Procter & Gamble Company teamed up with the Centers for Disease Control and Prevention to come up with a solution. Together they developed a powder with two components that purify water. The first component is known as a flocculant; it attaches to soil sediments, heavy metals, and pesticides that are in the water and forces them to settle out of the water. The second component is a chlorine compound that kills 99.99 percent of all harmful bacteria and viruses. The first step is to mix a small amount of the powder into 10 L of water and allow the flocculant to work for 5 minutes. Once the flocculant settles to the bottom, it is filtered out through a piece of cotton fabric. After an additional 20 minutes, the chlorine has killed bacteria and viruses. In short, for a small amount of effort, a person can create 10 L of safe drinking water in just 30 minutes.

Although Proctor & Gamble typically develops products designed to make a profit, this project was different. Purifying powder is needed in developing countries with populations that cannot afford to pay much even for something as essential as clean drinking water. Procter & Gamble decided to create a nonprofit organization, called the Children's Safe Drinking Water Program, that would supply the powder to people in developing countries at no profit. They created small powder packets that could clean 10 L of contaminated water but were not much bigger than a fast-food ketchup packet. The powder is inexpensive for Procter & Gamble to manufacture and the company is able to sell the packets at 3.5 cents each. Partnering with more than 100 governments and humanitarian groups including UNICEF, the company distributes the packets throughout the world. In September 2013, Procter & Gamble announced that they had distributed enough packets to filter 6 billion liters of water and estimated that this saved 32,000 lives.

As the water purification program became known around the world, Procter & Gamble realized that the product they created with no expectation for profit could be sold at a profit to wealthier people who need to purify water that they collect from streams and lakes when they go hiking and camping. In 2007, the company announced that it would begin selling the packets to consumers under the brand name PUR, for $2.50

Making safe drinking water. Nearly 1 billion people do not have access to safe drinking water. Procter & Gamble partnered with many other organizations to distribute its water purifying powder technology to people in developing countries. The powder technology kills bacteria and viruses and removes parasites and solid materials. In this photo, a family poses with a cup of typical dirty water in one hand and the same water after adding the power in the other hand. *(Courtesy P&G)*

each. This price yields a considerable profit, part of which supports the goals of the nonprofit program.

The Children's Safe Drinking Water Program is an excellent example of new technologies that can be developed to be effective, inexpensive solutions to major environmental problems. It also demonstrates that such solutions can ultimately be profitable while providing a positive corporate image.

Critical Thinking Questions

1. Why is it important for a water-purifying system to do more than just add chlorine to the water to kill harmful pathogens?
2. How might a company benefit when it produces a product that purifies water at no profit?

References

Deutsch, C. H. 2007. Procter & Gamble to benefit in all but name from water purifier. *New York Times,* July 23.

P&G Children's Safe Drinking Water. http://www.csdw.org/csdw/index.shtml.

Water-purification plant the size of a fast-food ketchup packet saves lives. 2013. *ScienceDaily*. September 9. http://www.sciencedaily.com/releases/2013/09/130909092343.htm.

Water purifying packet. The contents of each small packet can remove soil and kill harmful organisms in 10 L of water. *(AP Photo/Tom Uhlman)*

Suggested Answers to Critical Thinking Questions

1. Whereas chlorine is important for killing pathogens, water-purifying systems also need to include flocculants to remove soil, heavy metals, and pesticides.

2. The company will have a good image, making more people likely to purchase products that are profitable for it.

Teaching Tip: Measuring Your Impact

Gaining Access to Safe Water and Proper Sanitation
Measuring Your Impact 14 asks students to analyze data about access to safe drinking water and improved sanitation in different countries, and how this affects the rates of death among children due to diarrheal diseases. The link to this resource may be found by clicking on the link buttons in the TE-book, opening the TRFD, or logging onto the book's companion website bcs.whfreeman.com /friedlandapes2e (teacher login required).

Measuring Your Impact 14: Gaining Access to Safe Water and Proper Sanitation

Exploring the Literature

Evers, D. C., et al. 2007. Biological mercury hotspots in the northeastern United States and southeastern Canada. *BioScience* 57: 29–43.

Galbraith, K. 2012. If mercury pollution knows no borders, neither can its solution. *New York Times*, December 12. (http://www.nytimes.com/2012/12/13/business/energy-environment/if-mercury-pollution-knows-no-borders-neither-can-its-solution.html)

Henry, T., and K. Galbraith. 2013. As fracking proliferates, so do wastewater wells. *New York Times*, March 28. (http://www.nytimes.com/2013/03/29/us/wastewater-disposal-wells-proliferate-along-with-fracking.html)

Payne, J. R., et al. 2008. Long-term monitoring for oil in the *Exxon Valdez* spill region. *Marine Pollution Bulletin* 56: 2067–2081.

Revkin, A. C. 2009. Dredging of pollutants begins in Hudson. *New York Times,* May 16.

Download a printable version of this list by clicking the link buttons in the TE-book, opening the TRFD, or logging onto the book's companion website bcs.whfreeman.com/friedlandapes2e (teacher login required).

Exploring the Literature 14

chapter 14 REVIEW

In this chapter, we learned about the many sources of water pollution and their effects on humans and the environment. One prominent source of water pollution is the wastewater produced by humans and livestock. Wastewater can bring excessive nutrients to water bodies and be a source of disease-carrying organisms. Fortunately, a number of technologies exist to treat wastewater including septic tanks, sewage treatment facilities, and manure lagoons. Water pollution can also occur when heavy metals, pesticides, pharmaceuticals, and acids find their way into groundwater or surface waters. As we continue to extract energy from the ground, we face the risk of pollution from the pipelines and ships that carry petroleum products. Fortunately, new technologies present options for cleaning up some of the petroleum products that have leaked or spilled. We also face pollution risks from nonchemical sources including solid waste, sediments, heat, and noise.

Key Terms

Water pollution
Wastewater
Point source
Nonpoint source
Biochemical oxygen demand (BOD)
Dead zone
Eutrophication
Cultural eutrophication
Indicator species
Fecal coliform bacteria
Septic system
Septic tank
Sludge
Septage
Leach field
Manure lagoon
Acid deposition
Perchlorates
Polychlorinated biphenyls (PCBs)
Thermal pollution
Thermal shock
Clean Water Act
Safe Drinking Water Act
Maximum contaminant level (MCL)

Learning Objectives Revisited

Module 41 Wastewater from Humans and Livestock

- **Discuss the three major problems caused by wastewater pollution.**
  Because wastewater contains organic matter, decomposition by microbes can cause declines in the amount of oxygen available in water bodies and produce dead zones. Dead zones also occur where wastewater adds enough nutrients to a water body to cause eutrophication. Finally, wastewater can be a source of many harmful pathogens including viruses, bacteria, and protists.
- **Explain the modern technologies used to treat wastewater.**
  At low residential densities, septic systems can be used to treat sewage. At higher residential densities such as in cities, sewage treatment plants treat the collected sewage from thousands of people. For livestock operations that raise large numbers of animals, manure lagoons hold manure until it is broken down by bacteria and can be applied to agricultural fields.

Module 42 Heavy Metals and Other Chemicals

- **Explain the sources of heavy metals and their effect on organisms.**
  Among the heavy metals of most concern are lead, mercury, and arsenic. Lead can harm the nervous system and kidneys of fetuses; the major source of lead is old water pipes. Arsenic can cause cancers of the skin, lungs, kidneys, and bladder; it exists naturally in rocks that release arsenic into groundwater that people drink. Mercury is not particularly toxic, but bacteria in water convert it into methylmercury, which can damage the nervous systems of animals, including humans. Major sources of mercury include coal burning, garbage incineration, medical supplies, and cement manufacturing.

- **Discuss the sources and effects of acid deposition and acid mine drainage.**

  The major sources of acid deposition are industrial plants that burn coal and release sulfur dioxide and nitrogen dioxide into the atmosphere where they are converted into sulfuric acid and nitric acid. The major source of acid mine drainage is abandoned mines that contain pyrite rocks. When the mines become flooded with groundwater, rocks release hydrogen ions and any water that leaves the mine and goes into a stream can have a pH close to zero. Both sources of acid pollution cause a decline in aquatic pH to a point that many organisms cannot survive.

- **Explain how synthetic organic compounds can affect aquatic organisms.**

  Pesticides are designed to kill particular organisms, but their presence in water bodies can have lethal effects on a wide range of species. Pharmaceuticals including hormones can alter the physiology of organisms at very low concentrations and have a range of effects, including causing males to grow eggs inside their testes. Military compounds such as perchlorates used for rocket fuel and industrial compounds such as PCBs and PBDEs can also make their way into groundwater and water bodies.

Module 43 Oil Pollution

- **Identify the major sources of oil pollution.**

  The major sources of oil pollution are natural leaks from the ocean floor, oil spills from tankers, and drilling for undersea oil using offshore platforms.

- **Explain some of the current methods to remediate oil pollution.**

  When an oil spill occurs, several steps can be taken to remediate the problem. Contaminated wildlife are often washed to remove oil. Floating oil slicks can be surrounded by containment booms and then vacuumed off the surface of the water. Oil slicks can also be treated with chemicals that break up the oil before it can come to shore. Genetically modified bacteria are also being developed to consume oil in the water. There is currently no agreement on how to remove underwater plumes of oil.

Module 44 Nonchemical Water Pollution

- **Identify the major sources of solid waste pollution.**

  The major sources of solid waste are discarded materials from households and industries, sludge from wastewater treatment plants, and coal ash and coal slag that are produced when coal is burned.

- **Explain the harmful effects of sediment pollution.**

  Sediments that make their way into water bodies from natural sources, agricultural fields, and construction sites can reduce the ability of light to transmit into the water, which makes it difficult for plants and algae to grow. Sediments also can clog the gills of aquatic organisms.

- **Discuss the sources and consequences of thermal pollution.**

  Thermal pollution occurs when industries pump in cool water from a local water body, use the water for cooling purposes, and then return the much warmer water back to the water body. Many organisms cannot survive the shock of having the temperature of the environment substantially increase.

- **Understand the causes of noise pollution.**

  Noise pollution is primarily an issue of concern in oceans where noises come from the propellers of large ships, long-range sonar, and loud air guns that are used to explore for oil located below the ocean floor.

Module 45 Water Pollution Laws

- **Explain how the Clean Water Act protects against water pollution.**

  The Clean Water Act has a goal of maintaining and, when necessary, restoring the chemical, physical, and biological properties of surface waters. It does this by setting water quality standards for various pollutants.

- **Discuss the goals of the Safe Drinking Water Act.**

  The Safe Drinking Water Act sets maximum contaminant levels that can be found in sources of drinking water including groundwater. The contaminants include microorganisms, disinfectants, organic chemicals, and inorganic chemicals.

- **Understand how water pollution legislation is changing in developing countries.**

  Developed countries have gone through a period of industrialization and water pollution followed by steps to improve their water quality. Most developing countries are in the stage of industrializing and, as a result, they commonly have problems of polluted surface and groundwater. As developing countries rise in affluence, they begin to have the desire and the financial ability to improve the quality of their water.

Answers to Chapter 14 AP® Exam Multiple-Choice Questions

1. d
2. e
3. b
4. a
5. c
6. c
7. e
8. a
9. b

Answers to Chapter 14 AP® Exam

Download the full answers to the Chapter 14 AP® Exam by clicking on the link buttons in the TE-book, opening the TRFD, or logging onto the book's companion website bcs .whfreeman.com/friedlandapes2e (teacher login required).

Answers to Chapter 14 AP® Exam

Chapter 14 AP® Environmental Science Practice Exam

Section 1: Multiple-Choice Questions

Choose the best answer for questions 1–9.

1. Which of the following statements about nonpoint source (NPS) pollution is NOT correct?
   (a) NPS results from rain or snowmelt moving over or permeating through the ground.
   (b) NPS is more difficult to control, measure, and regulate than point source pollution.
   (c) NPS includes sediment from improperly managed construction sites as a pollutant.
   (d) NPS is water pollution that originates from a distinct source such as a pipe or tank.
   (e) NPS disperses pollutants over a large area, such as oil and grease in a parking lot.

2. Human wastewater results in which of the following water-pollution problems?
   I. Decomposition of organic matter reduces dissolved oxygen levels.
   II. Decomposition of organic matter releases great quantities of nutrients.
   III. Pathogenic organisms are carried to surface waters.

   (a) I only
   (b) II only
   (c) III only
   (d) I and III
   (e) I, II, and III

3. Which of the following indicates that a body of water is contaminated by human wastewater?
   (a) Low BOD and a fecal coliform bacteria count of zero
   (b) High levels of nutrients, such as nitrogen and phosphorus, and high BOD
   (c) Low BOD and low levels of nutrients, such as nitrogen and phosphorus
   (d) Low levels of nutrients, such as nitrogen and phosphorus, and a fecal coliform bacteria count of zero
   (e) A lack of dead zones

4. Both septic systems and sewage treatment plants utilize bacteria to break down organic matter. Where in each system does this process occur?
   (a) Septic tank and leach field; primary treatment and secondary treatment
   (b) Septic tank only; primary treatment and chlorination
   (c) Leach field only; secondary treatment only
   (d) Septic tank and leach field; secondary treatment only
   (e) Leach field only; secondary treatment and chlorination

5. Under which of the following circumstances is a sewage treatment plant legally permitted to bypass normal treatment protocol and discharge large amounts of sewage directly into a lake or river?
   (a) When the receiving surface water is designated for swimming only
   (b) When the population of the surrounding community surpasses the plant's capacity
   (c) When combined volumes of storm water and wastewater exceed the capacity of an older plant
   (d) When a permit to modernize the plant is denied by the Environmental Protection Agency
   (e) When an extended period of drought restricts water flow in a lake or river

6. Tertiary treatment of wastewater
   (a) removes pathogens.
   (b) reduces sediment.
   (c) reduces eutrophication.
   (d) removes heavy metals.
   (e) reduces the amount of sludge.

7. Which of the following inorganic substances is naturally occurring in rocks, soluble in groundwater, and toxic at low concentrations?
   (a) Mercury
   (b) Lead
   (c) PCBs
   (d) Copper
   (e) Arsenic

8. Acid mine drainage occurs when acidic water formed belowground makes its way to the surface; the acidic water is formed in flooded abandoned mines where the underground water
   (a) reacts with a type of rock, pyrite, which releases iron and hydrogen ions.
   (b) reacts with sulfur dioxide and nitrogen dioxide to form sulfuric and nitric acids.
   (c) flushes out the chemicals used in the mining process.
   (d) permeates a limestone layer that lowers the pH.
   (e) reacts with copper and aluminum to form pyrite rock and hydrogen ions.

9. All of the following are problems that result from the use of pesticides except
   (a) most pesticides are not target-specific and kill other related and nonrelated species.
   (b) pesticide runoff enters surface waters and increases the solubility of heavy metals.
   (c) pesticides affect nontarget organisms by changing community relationships.
   (d) pesticides target specific physiological functions, but also disrupt other functions.
   (e) most inert ingredients are not tested for safety and may pose unacceptable risks.

Additional Multiple-Choice AP® Practice Exam

To download an additional Multiple-Choice AP® Environmental Science Practice Exam for Chapter 14, click on the link buttons in the TE-book, open the TRFD, or log onto the book's companion website bcs.whfreeman.com/friedlandapes2e

Multiple-Choice AP® Practice Exam 14

Section 2: Free-Response Questions

Write your answer to each part clearly. Support your answers with relevant information and examples. Where calculations are required, show your work.

1. Answer questions a–d using the graph below, which contains data collected by the Maryland Department of Natural Resources and the Chesapeake Bay Program.

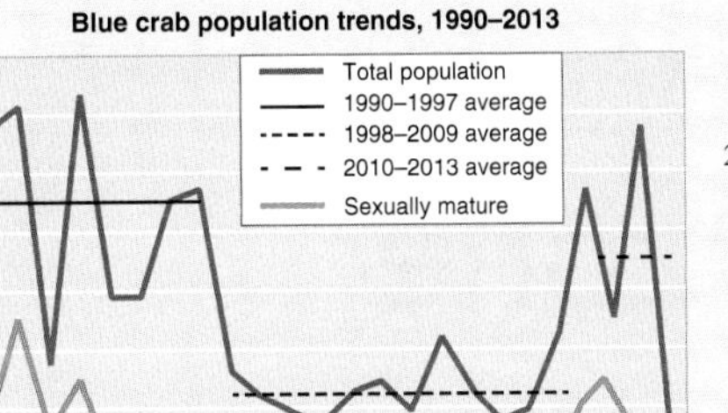

(a) Calculate the differences in crab population from 1990–1997, 1998–2009, and 2010–2013. Predict the average blue crab population for 2010–2020 and explain your answer. (4 points)

(b) Identify and explain three possible factors related to water pollution that could have contributed to the decline in the total blue crab population in the Chesapeake Bay. (3 points)

(c) Select one factor stated in (b) and describe how that source of water pollution could be managed and controlled. (2 points)

(d) What federal legislation would apply to the Chesapeake Bay and the blue crabs? (1 point)

2. The Food and Drug Administration (FDA) has developed guidelines for the consumption of canned tuna fish. These guidelines were developed particularly for children, pregnant women, or women who were planning to become pregnant, because mercury poses the most serious threat to these segments of society. However, the guidelines can be useful for everyone.

(a) Identify two major sources of mercury pollution and one means of controlling mercury pollution. (6 points)

(b) Explain how mercury is altered and how it finds its way into albacore tuna fish. (2 points)

(c) Identify two health effects of methylmercury on humans. (2 points)

Answers to the Chapter 14 Free-Response Questions can be found in the Answer Appendix on page ANS-11.

chapter

15 Air Pollution and Stratospheric Ozone Depletion

PD Overview

This chapter discusses air pollution, one of the most significant topics in the AP® Environmental Science course and a subject that students often find relevant to their lives. Air pollution is a problem that occurs in both developing and developed countries. Because air is a common resource across Earth, air pollution crosses many system boundaries. Human activity contributes to both outdoor and indoor air pollution. This chapter identifies the major air pollutants found around the globe and examines the specific air pollution situations that occur with photochemical smog and acid deposition. The chapter looks at a variety of air pollution control measures, examines stratospheric ozone depletion, and concludes with a discussion of indoor air pollution.

Many free-response questions from past exams have tested concepts covered in this chapter, so this is an especially good time to use those free-response questions for exam study and practice. Students who have not previously taken a class in chemistry may be directed to review concepts from Chapter 2, especially before covering Module 47 on photochemical smog, which relies on an understanding of chemical reactions.

Module 46: Major Air Pollutants and Their Sources

This module identifies and describes the major air pollutants. Air pollution is defined as the introduction of chemicals, particulate matter, or microorganisms into the atmosphere at concentrations high enough to harm plants, animals, and materials such as buildings, or to alter ecosystems. Generally, the term *air pollution* refers to pollution in the troposphere, the first 16 km (10 miles) of the atmosphere above the surface of Earth. This module also describes the sources of air pollution. Air pollution can occur naturally, from sources such as volcanoes and fires, or it can be anthropogenic, from sources such as automobiles and factories.

Module 47: Photochemical Smog and Acid Rain

The air quality in the United States has improved greatly in recent decades. However, some cities and even entire regions of the United States continue to experience intermittent air pollution, often related to smog formation. This module examines photochemical smog, which is a problem in the United States and elsewhere in the world, and acid rain, which is no longer a problem in the United States but has become a problem in Asia.

Module 48: Pollution Control Measures

During the discussion of energy and energy choices, students saw that sustainability is best achieved by considering conservation and efficiency first. Similarly, if we can address the problems of pollution by seeking ways to avoid creating it in the first place, we will require less energy and fewer resources to clean it up. Preventing pollution is usually much less expensive and energy intensive then controlling it. Unfortunately, pollution prevention is not always possible. This module examines strategies for controlling pollutants that enter the atmosphere. It also describes innovative pollution control measures.

Module 49: Stratospheric Ozone Depletion

Tropospheric or ground-level pollution has been shown to contribute to a number of problems in the natural world, to exacerbate asthma and breathing difficulties in humans, and to cause cancer. This module turns to the effects of certain pollutants in the stratosphere that have a substantial impact on the health of humans and ecosystems. In the troposphere, ozone is an oxidant that can harm respiratory systems in animals and damage a number of structures in plants. However, in the stratosphere ozone forms a necessary protective shield against radiation from the Sun; it absorbs ultraviolet light and prevents harmful ultraviolet radiation from reaching Earth. This module explains the benefits of stratospheric ozone and how it forms. It describes the depletion of stratospheric ozone and global efforts to reduce that depletion.

Module 50: Indoor Air Pollution

When we think of air pollution, we usually do not associate it with air inside our buildings, but indoor air pollution actually causes more deaths each year than does outdoor air pollution. Most of these deaths occur in the developing world. The amount of time one spends indoors depends on culture, climate, and economic situation. The quality of indoor air is highly variable and when polluting activities take place indoors, exposure to pollutants in a confined space can be a significant health risk. This module examines indoor air pollution in both developed and developing countries. It also looks at the subject of indoor air pollution, which particularly affects people in developing nations.

Alignment to AP® Environmental Science Course Description

AP® Outline	Module	
VI. Pollution (25–30%)		
A. Pollution Types	Module 46	Major Air Pollutants and Their Sources
	Module 47	Photochemical Smog and Acid Rain
	Module 50	Indoor Air Pollution
B. Impacts on the Environment and Human Health	Module 46	Major Air Pollutants and Their Sources
	Module 47	Photochemical Smog and Acid Rain
	Module 48	Pollution Control Measures
	Module 49	Stratospheric Ozone Depletion
	Module 50	Indoor Air Pollution
C. Economic Impacts	Module 46	Major Air Pollutants and Their Sources
	Module 48	Pollution Control Measures
	Module 49	Stratospheric Ozone Depletion
VII. Global Change (10–15%)		
A. Stratospheric Ozone	Module 49	Stratospheric Ozone Depletion

Chapter Learning Objectives

After completing this chapter students will be able to

- identify and describe the major air pollutants.
- describe the sources of air pollution.
- explain how photochemical smog forms and why it is still a problem in the United States.
- describe how acid deposition forms and why it has improved in the United States and become worse elsewhere.
- explain strategies and techniques for controlling sulfur dioxide, nitrogen oxides, and particulate matter.
- describe innovative pollution control measures.
- explain the benefits of stratospheric ozone and how it forms.
- describe the depletion of stratospheric ozone.
- explain efforts to reduce ozone depletion.
- explain how indoor air pollution differs in developing and developed countries.
- describe the major indoor air pollutants and the risks associated with them.

Chapter 15 Pacing Guide

This pacing guide is based on a schedule with 120 sessions of 50 minutes each before the AP® Exam. If you have a different number of sessions before the exam, you can modify the pacing to suit your needs. If you have additional time, consider incorporating quizzes, released AP® Environmental Science free-response and multiple-choice questions, or additional activities.

Module	Standard Schedule Days	Block Schedule Days
Module 46	2	1
Module 47	2	1
Module 48	1	½
Module 49	1	½
Module 50	1	½
Assessment	1	½

Chapter 15 Resources

The link to these resources can be found by clicking on the link buttons in the Teacher's e-Book (TE-book), opening the Teacher's Resource Flash Drive (TRFD), or logging onto the book's companion website bcs.whfreeman.com/friedlandapes2e (teacher login required).

- **PowerPoint Presentation 15**
- **Optimized Art PowerPoint and JPEG Files 15**
- **Do the Math Videos**
- **Measuring Your Impact 15:** Mercury Release from Coal
- **Lab 15.1:** Ozone Sampling
- **Chapter 15 Web Resources**
- **Exploring the Literature 15**
- **Answers to Chapter 15 AP® Exam**
- **Multiple-Choice AP® Practice Exam 15**

Free-Response Questions from Previous AP® Environmental Science Exams

Free-response questions from prior AP® Environmental Science Exams are available on the AP® course website: https://apstudent.collegeboard.org/apcourse/ap-environmental-science/exam-practice. Students should be able to answer all of the questions listed in the table to the right with material learned in this and previous chapters. When a question requires students to understand material from multiple chapters, the question will be listed in the last chapter required to complete the entire question. Questions marked with an asterisk (*) are from exams with released multiple-choice questions. You may want to save these questions until the end of the year so you can give your students a complete released exam for practice. Questions marked with double asterisks (**) require math calculations. Look for references to these questions throughout the chapter.

Year	Question	Content
1999	3**	• Historical concentration trends for ozone and lead • Sources of ozone and lead in the atmosphere • Physiological effect of ozone and lead on humans • Pollution reduction methods for particulates of gasses
2000	1**	• BTUs of heat needed to generate electricity in a coal-fired electric power plant • Calculation of coal needed to run an electric plant • Calculation of sulfur releases by a coal-fired electric plant • Calculation of the pounds of coal consumed each day; calculation of the sulfur content released by the power plant each day • EPA standards for sulfur emissions • Negative environmental impact of sulfur emissions
2001	3	• Indoor air pollution • Sick building syndrome
2002	1**	• Environmental benefits of using an electric car
2007	3	• Thinning of the stratospheric ozone layer • Effects of ground-level ozone on humans and on an ecosystem
2007	4	• Difference in temperatures in urban and suburban areas • Air pollution in urban and suburban areas • Differences in the hydrologic cycle in urban and rural areas
2009	1	• The nitrogen cycle • Nitrogen and photochemical smog
2013*	3	• Importance of stratospheric ozone layer • Chemical processes that lead to the composition and decomposition of ozone molecules • CFCs and the ozone layer • Formation of tropospheric ozone

PD Chapter 15 Overview

Watch the video overview of Chapter 15 (for teachers) by clicking on the link buttons in the TE-book, opening the TRFD, or logging onto the book's companion website bcs.whfreeman.com/friedlandapes2e (teacher login required).

TRM PowerPoint Presentation 15

Download the PowerPoint presentation for Chapter 15 by clicking on the link buttons in the TE-book, opening the TRFD, or logging onto the book's companion website bcs.whfreeman.com/friedlandapes2e (teacher login required).

Recent aerial view of Chattanooga, Tennessee, with Lookout Mountain clearly visible in the background. Four or five decades ago, Lookout Mountain was obscured by haze much of the time. (© Rock Creek Aviation)

chapter

15 Air Pollution and Stratospheric Ozone Depletion

Cleaning Up in Chattanooga

The summer of 2012 was not kind to Chattanooga, Tennessee. Dust storms in the West led to a number of poor air quality days in a city that is generally known for very good air quality. Chattanooga sits along the Tennessee River in a natural basin formed by the Appalachian Mountains, one of which—Lookout Mountain—rises 600 m (1,970 feet) over the city. After the Civil War, foundries, textile mills, and other industrial plants were quickly built and Chattanooga soon became one of the leading manufacturing centers in the nation.

The economic boom in Chattanooga had an environmental cost. Like Los Angeles and many other highly polluted cities, Chattanooga is located in a bowl formed by surrounding mountains. This geography traps pollutants that hover above the city. By 1957, Chattanooga had the third-worst particulate pollution in the country and rates of respiratory diseases were well above the national average. Over the next decade conditions worsened and by the 1960s, people were often unable to see Lookout Mountain even from a distance of a quarter mile. In 1969, a U.S. survey of the nation's air quality confirmed what many Chattanooga residents suspected: Their city topped the list of the worst cities in the United States for particulate air pollution.

By 1957, Chattanooga had the third-worst particulate pollution in the country and rates of respiratory diseases were well above the national average.

Obviously the poor quality of the air needed to be addressed. In 1969, Chattanooga, in conjunction with Hamilton County, created its own air pollution legislation by enacting the Air Pollution Control Ordinance.

Teaching Tip: Chapter Opening Case

The opening case, "Cleaning Up in Chattanooga," discusses how the city of Chattanooga, Tennessee, has improved its air quality. One part of this case discusses how ground-level ozone concentrations came to be extremely high in the late 1990s. It goes on to discuss how the private and public sectors had to work together to help Chattanooga attain a lower ozone level. Use this opening case to discuss some of the challenges cities face in cleaning up pollution and the importance of cooperation between government and industry.

Teaching Tip: Activity

In this activity, students look at several EPA websites to learn about the Air Quality Index (AQI) and to evaluate and compare air quality in various locations. Ask students to look at the websites and answer the following questions:

- What is the Air Quality Index (AQI) and what is its purpose? *The AQI is an index that basically reports the air quality of a certain region in the United States. Its purpose is to educate people on the health effects they might encounter after a few hours or days of breathing polluted air.*
- How many air pollutants are involved in calculating an air quality index? What are the major air pollutants subject to regulation? *The EPA uses five major air pollutants to calculate the AQI. The five major air pollutants, which are regulated by the Clean Air Act, are ground-level ozone, particle pollution (also known as particulate matter), carbon monoxide, sulfur dioxide, and nitrogen dioxide.*
- How does the AQI present air quality? *The AQI is divided into six categories; each of the six categories corresponds to a different level of health concern. The six levels are good, moderate, unhealthy for sensitive groups, unhealthy, very unhealthy, and hazardous. The AQI is presented in colors: green is good, yellow is moderate, orange is unhealthy for sensitive groups, red is unhealthy, purple is very unhealthy, and maroon is hazardous.*
- Pick a state on the East Coast, select the state's capital, and record the city's AQI for today and tomorrow. Next go to and find the AQI for Los Angeles. Does Los Angeles have a higher AQI? Would you expect it to have a higher AQI most of the time? Explain your answer. *The answer will vary depending on what location each student selects. Los Angeles will most likely have a higher AQI because it is a very populous city and its topography increases its chances of having high levels of air pollution. Weather conditions in Los Angeles also frequently cause higher than normal AQI readings.*

It controlled the emissions of sulfur oxides, allowed open burning by permit only, placed regulations on odors and dust, outlawed visible automobile emissions, capped the sulfur content of fuel at 4 percent, and limited visible emissions from industry. At the same time, the city and county governments put in place new pollution monitoring techniques to make sure the ordinance was being followed.

The city and county governments, along with private industry, poured approximately $40 million into the cleanup effort. Actions to improve air quality did not hinder business, as some people feared, but rather created new industrial opportunities related to the cleanup effort, such as the establishment of a local manufacturer of smokestack scrubbers. As a result of all these measures, in 1972—just 3 years after passage of the city ordinance and 2 years after the passage of the Federal Clean Air Act—Chattanooga achieved compliance with Clean Air Act air-quality standards.

The people of Chattanooga and the local governments recognized that keeping their air clean and maintaining economic sustainability would be an ongoing effort. To maintain their newly improved air quality, the city government and local businesses began several programs. One such program was a comprehensive recycling program, chosen as an alternative to a waste incinerator that would have added particles to the air. Public and private sectors successfully partnered to achieve both environmental and economic sustainability in creating the largest municipal fleet of electric buses in the United States, manufactured by another local business.

Unfortunately, while Chattanooga's efforts dramatically reduced the levels of particulate pollutants, the concentration of ozone, mostly from automotive pollutant precursors within and beyond the city limits, continued to climb. Ozone concentrations exceeded the 1997 standard of 0.08 parts per million by volume set by the Environmental Protection Agency. Chattanooga has responded to the new air pollution problem in the same way it faced the particulate pollutant problem of the 1960s—through a combined effort of government, the public, and local industries. The city and county governments formed an Early Action Compact with the EPA, agreeing to improve ozone concentrations ahead of EPA requirements in return for not being labeled a "nonattainment area," a designation that can result in the loss of federal highway funds and create a negative image that makes industrial recruitment and economic development more difficult. Like the 1969 Air Pollution Control Ordinance, the new Early Action Compact calls for a concerted effort by private and public sectors, and includes educating people on how they can limit ozone production on high-ozone days.

Chattanooga attained the 0.08 parts per million standard in 2007, two years ahead of schedule. However, national legislation has since lowered the ozone standard to 0.075 parts per million. In 2011, before the dust storms from the West, Chattanooga met the new, lower ozone standard. However, residents know that to achieve their goals of an economically vibrant city with clean air, they must continue to encourage cooperation among government, people, and business. They also need fewer dust storms from the West.

Sources:
Chattanooga Area Chamber of Commerce, *Summary of the Chattanooga Area Chamber of Commerce's Position on Strengthening the National Ambient Air Quality Standard for Ozone*, 2010, www.chattanoogachamber.com; National Ambient Air Quality Standards: *www.epa.gov/ttn/naaqs*.

Throughout this book, we have identified a number of systems that cover relatively small aspects of the natural and human worlds. Because air is a common resource across Earth, air pollution crosses many system boundaries. Human activity contributes to both outdoor and indoor air pollution. To understand air pollution and its effects, we need to look at all air pollutants, where they come from, and what happens to them after they are released into the atmosphere. In this chapter, we will identify the major air pollutants found around the globe and we will examine the specific air pollution situations that occur with photochemical smog and acid deposition. We will review a variety of air pollution control measures and examine stratospheric ozone depletion. We conclude the chapter with a discussion of indoor air pollution.

Find the links to the EPA websites by clicking on the link buttons in the TE-book, opening the TRFD, or logging onto the book's companion website bcs.whfreeman.com/friedlandapes2e (teacher login required).

TRM Chapter 15 Web Resources

module

46

Major Air Pollutants and Their Sources

Air pollution is defined as the introduction of chemicals, particulate matter, or microorganisms into the atmosphere at concentrations high enough to harm plants, animals, and materials such as buildings, or to alter ecosystems. Generally, the term air pollution refers to pollution in the troposphere, the first 16 km (10 miles) of the atmosphere above the surface of Earth. Tropospheric pollution is also sometimes called ground-level pollution. Air pollution can occur naturally, from sources such as volcanoes and fires, or it can be anthropogenic, from sources such as automobiles and factories.

In this module, we will examine the major air pollutants that occur in the troposphere and where they come from.

Learning Objectives

After reading this module, you should be able to

- identify and describe the major air pollutants.
- describe the sources of air pollution.

AP® Exam Tip

This is one of the most important chapters in the textbook. Over the years, at least 11 free-response questions have been devoted to air pollution concepts. Make sure students understand how primary and secondary pollutants are formed. Photochemical smog and acid deposition also frequently appear on the exam.

Air pollution is a global system

Since one of the major repositories for air pollutants is the atmosphere, which envelops the entire globe, we must think of the air pollution system as a global system. In fact, evidence appears to link air pollution across long distances. For example, in recent years, air pollution in Asia has been responsible for acidic rainfall on the West Coast of the United States (FIGURE 46.1).

The air pollution system has many inputs, which are the sources of pollution. It also has many outputs, which are components of the atmosphere and biosphere that remove air pollutants. It is difficult to conceptualize this system because the inputs do not originate from just one location. Air pollution inputs can come from automobiles on the ground, airplanes in the sky, or vegetation (tree leaves) 100 feet in the air. Similarly, air pollution can be removed or altered by vegetation, soil, and components of the atmosphere such as clouds, particles, or gases.

Air pollution The introduction of chemicals, particulate matter, or microorganisms into the atmosphere at concentrations high enough to harm plants, animals, and materials such as buildings, or to alter ecosystems.

Teaching Tip: Activity

Divide the class into groups of two. Give each group 20 index cards and the vocabulary terms in the list below. Have each team label one side of each card with the numbers 1 through 20. On the blank sides of cards 1 through 10, students should write one of the key terms listed below, using a new card for each term. Instruct students to shuffle the order of the cards; they should not duplicate the order of the list. On the blank sides of cards numbered 11 through 20, students should write the definition or some other information that connects the term to the definition of the terms on cards 1 through 10, making sure to keep an answer key of which numbered term card matches which numbered definition card. The groups should lay out their cards number side up in two rows—one with terms and one with definitions. Have the groups swap cards and answer keys.

The first player in each pair then chooses one card from each row. Read the cards to see if they are a match. If they are, that person keeps the cards and takes another turn. If they are not a match, the second member of the team has a turn. Each group continues playing until all the cards are gone. The person with the most cards is the winner.

- sulfur dioxide
- nitrogen oxides
- carbon oxide
- particulate matter
- lead
- ground-level ozone
- volatile organic compounds (VOC)
- mercury
- primary pollutant
- secondary pollutant

FIGURE 46.1 **Particulate pollution and visibility.** Particulates and sulfate aerosols are most responsible for causing pollution and reducing visibility in cities, such as this location in China. Pollution can also be transported long distances and cause problems far from the source. *(Natalie Behring/Panos Pictures)*

As a starting point for understanding the global air pollution system, we will identify the major pollutants and determine where they come from.

Classifying Pollutants

Even though air pollution has been with us for millennia, both the specific definition of pollution and the classification of a substance as a pollutant have evolved. The atmosphere is a public resource—in effect, a global commons—and consequently the science of air pollution is closely intertwined with political and social perspectives. In formulating the U.S. Clean Air Act of 1970 and subsequent amendments, legislators used information from environmental scientists and others on the most important air pollutants to monitor and control. The original act identified six pollutants that significantly threaten human well-being, ecosystems, and structures: sulfur dioxide, nitrogen oxides, carbon monoxide, particulate matter, tropospheric ozone, and lead. These were called criteria air pollutants because under the Clean Air Act, the EPA must specify allowable concentrations of each pollutant.

Although carbon dioxide was not included among the major air pollutants identified in the 1970s, today it is widely accepted that carbon dioxide is altering ecosystems in a substantial way. In 2007, the U.S. Supreme Court ruled that carbon dioxide should be considered an air pollutant under the Clean Air Act, and in 2012, a federal appeals court agreed that the EPA is required to impose limits on harmful greenhouse gas emissions, including carbon dioxide. In addition, volatile organic compounds and mercury, though not officially listed in the Clean Air Act, are commonly measured air pollutants that have the potential to be harmful.

The sources and effects of the major air pollutants, including the six criteria air pollutants, are summarized in **TABLE 46.1**. Let's take a closer look at each of these major air pollutants.

Sulfur Dioxide

Sulfur dioxide (SO_2) is a corrosive gas that comes primarily from combustion of fuels such as coal and oil. It is a respiratory irritant and can adversely affect plant tissue as well. Because all plants and animals contain sulfur in varying amounts, the fossil fuels derived from their remains contain sulfur. When these fuels are combusted, the sulfur combines with oxygen to form sulfur dioxide. Sulfur dioxide is also released in large quantities during volcanic eruptions and can be released, though in much smaller quantities, during forest fires.

Nitrogen Oxides

Nitrogen oxides are generically designated NO_X, with the X indicating that there may be either one or two oxygen atoms per nitrogen atom: nitrogen oxide (NO), a colorless, odorless gas; and nitrogen dioxide (NO_2), a pungent, reddish-brown gas, respectively. When we use the term *nitrogen oxides* in our discussion, we will be referring to either nitrogen oxide or nitrogen dioxide since they easily transform from one to the other in the atmosphere. The atmosphere is 78 percent nitrogen gas (N_2), and all combustion in the atmosphere leads to the formation of some nitrogen oxides. Motor vehicles and stationary fossil fuel combustion are the primary anthropogenic sources of nitrogen oxides. Natural sources include forest fires, lightning, and microbial action in soils. Atmospheric nitrogen oxides play a role in forming ozone and other components of smog. We will take a closer look at this process later in the chapter.

Carbon Oxides

Carbon monoxide (CO) is a colorless, odorless gas that is formed during incomplete combustion of most matter, and therefore is a common emission in vehicle exhaust and most other combustion processes. Carbon monoxide can be a significant component of air pollution in urban areas. It also can be a dangerous indoor air pollutant when exhaust systems on natural gas heaters malfunction. Carbon monoxide is a particular problem in developing countries, where people may cook with manure, charcoal, or kerosene within poorly ventilated structures.

Carbon dioxide (CO_2) is a colorless, odorless gas that is formed during the complete combustion of most matter, including fossil fuels and biomass. It is absorbed by plants during photosynthesis and is released during respiration. In general, the complete combustion of matter that produces carbon dioxide is more desirable than the incomplete combustion that

Teaching Tip: Remember

When teaching air pollution, help students remember the three main gases emitted during combustion of fossil fuels by using the saying "Socks, Nocks, and CO_2" written out on the board as "SO_x, NO_x, and CO_2." SO_x stands for *sulfur oxides*, such as *sulfur dioxide* (SO_2). NO_x stands for *nitrogen oxides* (NO and NO_2). CO_2 is *carbon dioxide*. SO_2 and NO_x are extremely important to remember because they are both primary pollutants that transform into secondary pollutants such as ground-level ozone and acid deposition. Students will need to know how ground-level ozone and acid deposition form.

TABLE 46.1 Major air pollutants

Compound	Symbol	Human-derived sources	Effects/Impacts
Criteria air pollutants			
Sulfur dioxide	SO_2	• Combustion of fuels that contain sulfur, including coal, oil, gasoline.	• Respiratory irritant, can exacerbate asthma and other respiratory ailments. • SO_2 gas can harm stomates and other plant tissue. • Converts to sulfuric acid in atmosphere, which is harmful to aquatic life and some vegetation.
Nitrogen oxides	NO_X	• All combustion in the atmosphere including fossil fuel combustion, wood, and other biomass burning.	• Respiratory irritant, increases susceptibility to respiratory infection. • An ozone precursor, leads to formation of photochemical smog. • Converts to nitric acid in atmosphere, which is harmful to aquatic life and some vegetation. • Contributes to overfertilizing terrestrial and aquatic systems.
Carbon monoxide	CO	• Incomplete combustion of any kind. • Malfunctioning exhaust systems and poorly ventilated cooking fires.	• Bonds to hemoglobin, thereby interfering with oxygen transport in the bloodstream. • Causes headaches at low concentrations. • Can cause death with prolonged exposure at high concentrations.
Particulate matter	PM_{10} (smaller than 10 micrometers) $PM_{2.5}$ (2.5 micrometers and less)	• Combustion of coal, oil, and diesel, and of biofuels such as manure and wood. • Agriculture, road construction, and other activities that mobilize soil, soot, and dust.	• Can exacerbate respiratory and cardiovascular disease and reduce lung function. • May lead to premature death. • Reduces visibility, and contributes to haze and smog.
Lead	Pb	• Gasoline additive, oil and gasoline, coal, old paint.	• Impairs central nervous system. • At low concentrations, can have measurable effects on learning and ability to concentrate.
Ozone	O_3	• A secondary pollutant formed by the combination of sunlight, water, oxygen, VOCs, and NO_X.	• Reduces lung function and exacerbates respiratory symptoms. • A degrading agent to plant surfaces. • Damages materials such as rubber and plastic.
Other air pollutants			
Volatile organic compounds	VOC	• Evaporation of fuels, solvents, paints. • Improper combustion of fuels such as gasoline.	• A precursor to ozone formation.
Mercury	Hg	• Coal, oil, gold mining.	• Impairs central nervous system. • Bioaccumulates in the food chain.
Carbon dioxide	CO_2	• Combustion of fossil fuels and clearing of land.	• Affects climate and alters ecosystems by increasing greenhouse gas concentrations.

produces carbon monoxide and other pollutants. However, burning fossil fuels has contributed additional carbon dioxide to the atmosphere and led to its becoming a major pollutant. It recently exceeded a concentration of 400 parts per million in the atmosphere and appears to be steadily increasing each year. This topic will be covered in more detail in Chapter 19 where we discuss the issue of climate change.

Particulate Matter

Particulate matter (PM), also called **particulates** or **particles,** is solid or liquid particles suspended in air. FIGURE 46.2 shows the sources of particulate matter and its effects. Particulate matter comes from the combustion of wood, animal manure and other biofuels, coal, oil, and gasoline. It is most commonly known as a class of pollutants released from the combustion of fuels such as coal and oil. Diesel-powered vehicles give off more particulate matter, in the form of black smoke, than do gasoline-powered vehicles. Particulate matter can also come from road dust and rock-crushing operations. Volcanoes, forest fires, and dust storms are important natural sources of particulate matter.

Particulate matter ranges in size from 0.01 micrometer (μm) to 100 μm (1 micrometer = 0.000001 m). For comparison, a human hair has a diameter of roughly 50 to 100 μm. Particulate matter larger than 10 μm is usually filtered out by the nose and throat; particulate matter of this size is not regulated by the EPA. Particles smaller than 10 μm are called Particulate Matter-10, written as PM_{10}, and are of concern to air pollution

Particulate matter (PM) Solid or liquid particles suspended in air. *Also known as* **Particulates; Particles.**

Teaching Tip: Discussion Starter

Ask students the following questions:

- What is particulate matter and where does it come from? *Particulate matter is solid or liquid particles suspended in air. Particulate matter comes from the combustion of wood, animal manure and other biofuels, coal, oil, and gasoline. Particulate matter can also come from road dust and rock-crushing operations. Volcanoes, forest fires, and dust storms are natural sources of particulate matter.*
- How does the size of particulate matter affect our concern about it? *Particulate matter ranges in size. Particulate matter larger than 10 micrometers is usually filtered out by the nose and throat; particulate matter of this size is not regulated by the EPA. Particles smaller than 10 micrometers are called Particulate Matter-10, written as PM_{10}, and are of concern to environmental scientists because they are not filtered out by the nose and throat and can be deposited deep within the respiratory tract. Particles of 2.5 μm and smaller, called $PM_{2.5}$, are an even greater health concern because they deposit deeply within the respiratory tract and they tend to be composed of more toxic substances than larger particles.*

Teaching Tip: Video

The Air We Breathe
This video describes how natural compounds in the air combine with pollutants from human activities to create air quality problems.

The link to this resource may be found by clicking on the link buttons in the TE-book, opening the TRFD, or logging onto the book's companion website bcs.whfreeman.com/friedlandapes2e (teacher login required).

Chapter 15 Web Resources

Teaching Tip: Video

Air Quality: The FRAPPÉ Field Campaign
This video describes the work and goals of the Front Range Air Pollution and Photochemistry Éxperiment (FRAPPÉ), which aims to characterize and understand summertime air quality in the Northern Front Range metropolitan area in Colorado. FRAPPÉ is a collaborative effort between the Colorado Department of Public Health and the Environment, the University of Colorado, Colorado State University, University of California, Berkeley, and other university collaborators.

The link to this resource may be found by clicking on the link buttons in the TE-book, opening the TRFD, or logging onto the book's companion website bcs.whfreeman.com/friedlandapes2e (teacher login required).

Chapter 15 Web Resources

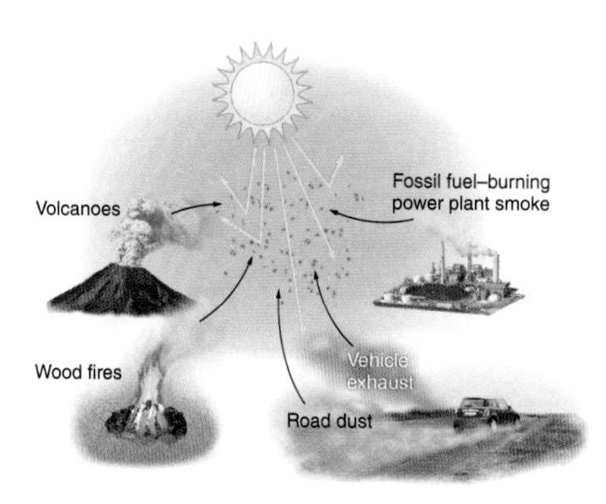

FIGURE 46.2 **The sources of particulate matter and its effect.** Particulate matter can be natural or anthropogenic. Particulate matter in the atmosphere ranges considerably in size and can absorb or scatter light, which creates a haze and reduces the light that reaches the surface of Earth.

scientists because they are not filtered out by the nose and throat and can be deposited deep within the respiratory tract. Particles of 2.5 μm and smaller, called $PM_{2.5}$, are an even greater health concern because they deposit deeply within the respiratory tract and they tend to be composed of more toxic substances than particles in larger size ranges.

Particulate matter also scatters and absorbs sunlight. If the atmospheric concentration of particulate matter is high enough, as it would be immediately following a large forest fire or a volcanic eruption, incoming solar radiation in the region will be reduced enough to affect photosynthesis. This happened in 1816, a year after a large volcanic eruption in Java released more than 150 million metric tons of particles that slowly spread around the globe. That year is commonly referred to as the year without a summer.

Haze Reduced visibility.

Photochemical oxidant A class of air pollutants formed as a result of sunlight acting on compounds such as nitrogen oxides.

Ozone (O_3) A secondary pollutant made up of three oxygen atoms bound together.

Smog A type of air pollution that is a mixture of oxidants and particulate matter.

Photochemical smog Smog that is dominated by oxidants such as ozone. *Also known as* **Los Angeles-type Smog; Brown smog.**

Sulfurous smog Smog dominated by sulfur dioxide and sulfate compounds. *Also known as* **London-type smog; Gray smog; Industrial smog.**

Reduced visibility, also known as **haze,** occurs primarily when particulate matter from air pollution scatters light. But as we will see in the next section, ozone and photochemical oxidants can also play an important indirect role in the formation of haze.

Photochemical Oxidants

Oxides are reactive compounds that remove electrons from other substances. **Photochemical oxidants** are a class of air pollutants formed as a result of sunlight (*photo*) acting on chemical compounds such as nitrogen oxides and sulfur dioxide. There are many photochemical oxidants, and they are generally harmful to plant tissue, human respiratory tissue, and construction materials. However, environmental scientists frequently focus on *ozone*. **Ozone (O_3)** is a secondary pollutant made up of three oxygen atoms bound together. It is harmful to both plants and animals and impairs respiratory function.

In the presence of sulfur and nitrogen oxides, photochemical oxidants can enhance the formation of certain particulate matter, which contributes to scattering light. The resulting mixture is called **smog,** a mixture of oxidants and particulate matter. The word is derived by combining *smoke* and *fog*. Smog is partly responsible for the hazy view and reduced sunlight observed in many cities. Smog can be divided into two categories. **Photochemical smog** is dominated by oxidants such as ozone and is sometimes called **Los Angeles–type smog** or **brown smog. Sulfurous smog** is dominated by sulfur dioxide and sulfate compounds and is sometimes called **London-type smog, gray smog,** or **industrial smog.**

Atmospheric brown cloud is a relatively new descriptive term that has been given to the combination of particulate matter and ozone. Derived primarily from combustion of fossil fuel and burning biomass, atmospheric brown clouds have been observed in cities and throughout entire regions, especially in Asia. The brownish tint that characterizes these clouds of pollution is typically caused by the presence of black or brown light absorbing carbon particles and/or nitrogen dioxide.

In addition to human health problems, particulate matter and photochemical oxidants also cause economic harm, since poor visibility in popular vacation destinations can reduce tourism revenues for recreation areas, such as lower incomes for hotels and restaurants in these areas. We will examine photochemical smog in greater detail in the next module.

Lead and Other Metals

Lead (Pb) is a trace metal that occurs naturally in rocks and soils. It is present in small concentrations in fuels including oil and coal. Lead compounds were added to gasoline for many years to improve vehicle performance. During that time, lead compounds released

AP® Exam Tip

Air pollution from ozone and lead is covered in the 1999 AP® Exam, Question 3. To answer this question, students must

- describe and compare the concentration trends for lead and ozone and calculate the percentage change in each, based on provided data
- identify the major source of either lead or ozone as a pollutant and describe its negative health effects in humans
- identify the source of either particulates or carbon monoxide and describe the most effective methods for reducing concentration in the atmosphere

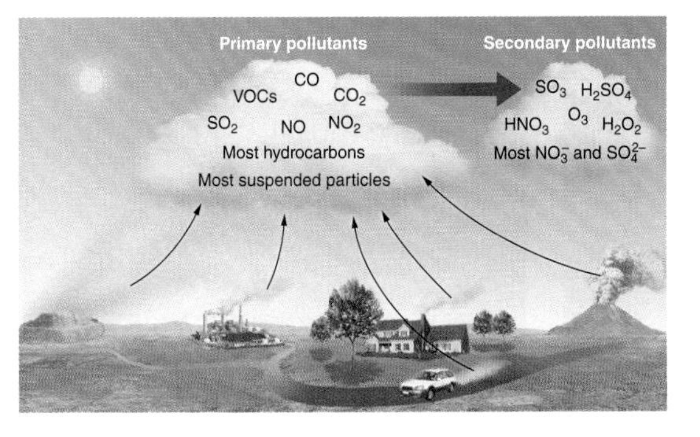

FIGURE 46.3 **Primary and secondary air pollutants.** The transformation from primary to secondary pollutant requires a number of factors including sunlight, water (clouds), and the appropriate temperature.

into the air traveled with the prevailing winds and were deposited on the ground by rain or snow. They became pervasive around the globe, including in polar regions far from combustion sources.

Lead was phased out as a gasoline additive in the United States between 1975 and 1996, and since then its concentration in the air has dropped considerably. A campaign to phase out lead use in gasoline globally is still underway. Another persistent source of lead is lead-based paint in older buildings. When the paint peels off, the resulting dust or chips can be toxic to the central nervous system and can affect learning and intelligence, particularly for young children who, attracted by the sweet taste, may ingest the dust or chips.

Mercury (Hg), another trace metal, is also found in coal and oil and, like lead, is toxic to the central nervous system of humans and other organisms. The EPA regulates mercury through its hazardous air pollutants program. As a result of the release of mercury into the air, primarily from the combustion of fossil fuels, especially coal, the concentrations of mercury in both air and water have increased dramatically in recent years. As we will see in Chapter 17, mercury concentrations in some fish have also increased. People who eat these fish increase their own mercury concentrations—an example of the interconnectedness of air pollution, air, water, aquatic health, and human health. Over the past 20 years, mercury emissions in the United States from waste incinerators have been reduced substantially. Because coal-fired electricity generation plants remain the largest uncontrolled source of mercury, emissions standards for coal plants will likely be the focus of future regulations.

Volatile Organic Compounds

Organic compounds that evaporate at typical atmospheric temperatures are called **volatile organic compounds (VOCs).** Many VOCs are hydrocarbons—compounds that contain carbon-hydrogen bonds, such as gasoline, lighter fluid, dry-cleaning fluid, oil-based paints, and perfumes. Compounds that give off a strong aroma are often VOCs since the chemicals are easily released into the air. VOCs play an important role in the formation of photochemical oxidants such as ozone. VOCs are not necessarily hazardous; many, such as VOCs given off by conifer trees, cause no direct harm. VOCs are not currently considered a criteria air pollutant, but because they can lead to the formation of photochemical oxidants, they have the potential to be harmful and are therefore of concern to air pollution scientists.

Primary and Secondary Pollutants

When trying to understand pollution, and when attempting to reduce pollution emissions, it is important to understand if a particular pollutant is coming directly from an emission source such as a smokestack or tailpipe, or if it has undergone transformations after emission. Thus, pollutants in the air can be categorized as *primary* or *secondary*.

Primary Pollutants

Primary pollutants are polluting compounds that come directly out of a smokestack, exhaust pipe, or natural emission source. As you can see in FIGURE 46.3, they include CO, CO_2, SO_2, NO_X, and most suspended particulate matter. Many VOCs are also primary pollutants. For example, as gasoline is burned in a car, it volatilizes from a liquid to a vapor, some of which is emitted

Volatile organic compound (VOC) An organic compound that evaporates at typical atmospheric temperatures.

Primary pollutant A polluting compound that comes directly out of a smokestack, exhaust pipe, or natural emission source.

Teaching Tip: Discussion Starter

Ask students the following questions:

- What are VOCs and how do they get into the atmosphere? *VOC stands for* volatile organic compound. *VOCs are organic compounds that evaporate at typical atmospheric temperatures.*
- Give several examples of VOCs. *Many VOCs are hydrocarbons—compounds that contain carbon-hydrogen bonds, such as gasoline, lighter fluid, dry-cleaning fluid, oil-based paints, and perfumes. Compounds that give off a strong aroma are often VOCs because the chemicals are easily released into the air.*
- Why are VOCs considered air pollutants? *VOCs are considered air pollutants because of their role in the formation of photochemical oxidants such as ozone.*

Teaching Tip: Activity

Have students work with a partner to fill in the table to the right. (Suggested answers are provided in italics.)

Pollutant	Definition	Examples	Formation
Primary pollutant	*A polluting compound that comes directly out of the smokestack, exhaust pipe, or natural emission source*	*CO, CO_2, SO_2, NO_x, VOCs, and particulate matter*	*Burning fossil fuels (such as from cars and factories), volcanoes, and forest fires*
Secondary pollutant	*A primary pollutant that has undergone transformation in the presence of sunlight, water, oxygen, or other compounds*	*SO_3, H_2SO_4, HNO_3, O_3, H_2O_2, most NO_3^- and SO_4^{2-}*	*Chemical transformation through interaction with sunlight, water (clouds), and the appropriate temperature*

Teaching Tip: Review

Ask students the following questions:

- Why is air pollution considered a global system? *Air pollution crosses many system boundaries. Human activities in one country can create pollution that affects the air quality in other countries.*
- What are the major air pollutants? *The major air pollutants are sulfur dioxide, nitrogen oxides, carbon oxides, particulate matter, lead, volatile organic compounds, mercury, and ground-level ozone.*
- What is the difference between a primary and a secondary pollutant? *A primary pollutant is a pollutant compound that comes directly out of a smokestack, exhaust pipe, or natural emission source. A secondary pollutant is a primary pollutant that has undergone transformation in the presence of sunlight, water, oxygen, or other compounds to form a new chemical or pollutant.*

(a)

(b)

FIGURE 46.4 **Natural sources of air pollution.** There are many natural sources of air pollution, including volcanoes, lightning strikes, forest fires, and plants. (a) A forest fire in Ojai, California, produces air pollution. (b) The Great Smoky Mountains were named for the natural air pollutants that reduce visibility and give the landscape a smoky appearance. *(a: R. Krubner/ClassicStock/The Image Works; b: Marvin Dembinsky Photo Associates/Alamy)*

from the exhaust pipe in an uncombusted form. The effect is more pronounced if the car is not operating efficiently. The resulting VOC becomes a primary air pollutant.

Secondary Pollutants

Secondary pollutants are primary pollutants that have undergone transformation in the presence of sunlight, water, oxygen, or other compounds. Because solar radiation provides energy for many of these transformations, and because water is usually involved, the conversion to secondary pollutants occurs more rapidly during the day and in wet environments.

Ozone is an example of a secondary pollutant. Ozone is formed in the atmosphere as a result of the emission of the primary air pollutants NO_X and VOCs in the presence of sunlight. The main components of acid deposition—sulfate (SO_4^{2-}) and nitrate (NO_3^-)—are also secondary pollutants. Both of these secondary pollutants will be discussed more fully below.

When trying to control secondary pollutants, it is necessary to consider the primary pollutants that create them, as well as factors that may lead to the breakdown or reduction in the secondary pollutants themselves. For example, when municipalities such as Chattanooga try to reduce ozone concentrations in the air, as described at the beginning of this chapter, they focus on reducing the compounds that lead to ozone formation—NO_X and VOCs—rather than on the ozone itself.

Secondary pollutant A primary pollutant that has undergone transformation in the presence of sunlight, water, oxygen, or other compounds.

Air pollution comes from both natural and human sources

Environmental scientists and concerned citizens direct much of their attention toward air pollution that comes from human activity. But human activity is not the only source of air pollution because processes in nature also cause air pollution. In this section we will examine both natural and human sources of air pollution.

Natural Emissions

Volcanoes, lightning, forest fires, and plants both living and dead all release compounds that can be classified as pollutants. Volcanoes release sulfur dioxide, particulate matter, carbon monoxide, and nitrogen oxides. Lightning strikes create nitrogen oxides from atmospheric nitrogen. Forest fires release particulate matter, nitrogen oxides, and carbon monoxide (FIGURE 46.4a). Living plants release a variety of VOCs, including ethylene and terpenes. The fragrant smell from conifer trees such as pine and fir and the smell from citrus fruits are mostly from terpenes; though we enjoy their fragrance, they can be precursors to photochemical smog. Long before anthropogenic pollution was common, the natural VOCs from plants gave rise to smog and photochemical oxidant pollution—hence the names of the forested mountain ranges in the southeastern United States, the Blue Ridge and the Smoky Mountains (Figure 46.4b). Large nonindustrial areas such as agricultural fields can give rise to particulate matter when they are plowed, as happened in the dustbowl of the 1930s. *The Encyclopedia of Earth,* using data in part from the Intergovernmental Panel on Climate Change, estimates that across the globe, sulfur dioxide emissions are 30 percent natural, nitrogen oxide emissions are 44 percent natural, and volatile organic compound emissions are 89 percent natural. However, in certain locations, such as North America, the anthropogenic contribution is much greater, perhaps as much as 95 percent for nitrogen oxides and sulfur dioxide.

The effects of these various compounds, especially when major natural events occur, depend in part on natural conditions such as wind direction. In May 1980, the prevailing westerly winds carried volcanic emissions from Mount St. Helens in Washington State and distributed particulate matter and sulfur oxides from the volcano across the United States. The rainfall pH was noticeably lower in the eastern United States that summer.

Anthropogenic Emissions

In contrast to natural emissions, emissions from human activity are monitored, regulated, and in many cases controlled. The EPA reports periodically on the emission sources of the criteria air pollutants for the entire United States, listing pollution sources in a variety of categories, such as on-road vehicles, power plants, industrial processes, and waste disposal. FIGURE 46.5 shows some of the most recent data. On-road vehicles, also referred to as the general category of *transportation,* are the largest sources of carbon monoxide and nitrogen oxides. Electricity generation, 40 percent of which, as we know from Chapter 12, is fueled by coal, is the major source of anthropogenic sulfur dioxide. Particulate matter comes from a variety of sources including natural and human-made fires, road dust, and the generation of electricity.

The Clean Air Act and its various amendments require that the EPA establish standards to control

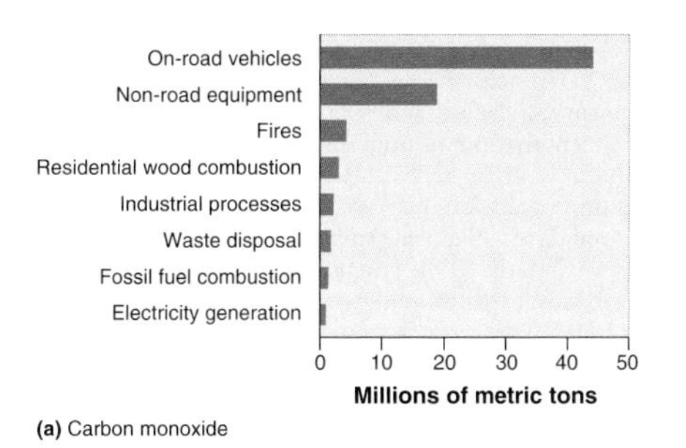

On-road vehicles
Non-road equipment
Electricity generation
Fossil fuel combustion
Industrial processes
Waste disposal
Fires
0 5 10
Millions of metric tons
(b) Nitrogen oxides

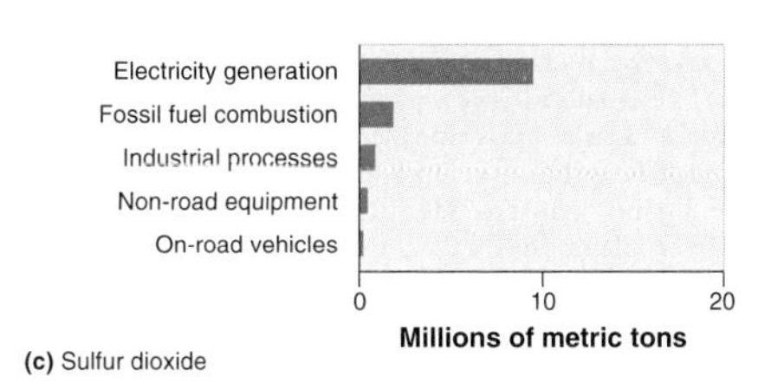

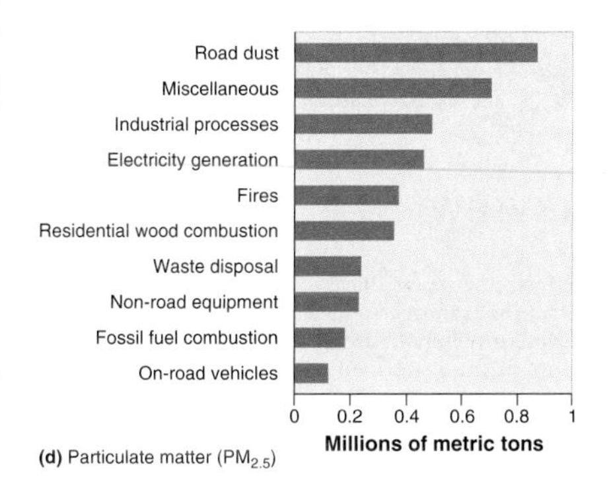

FIGURE 46.5 Emission sources of criteria air pollutants for the United States. Recent EPA data show that on-road vehicles, categorized as "transportation," are the largest source of (a) carbon monoxide and (b) nitrogen oxides. The major source of (c) anthropogenic sulfur dioxide is the generation of electricity, primarily from coal. Among the sources of (d) particulate matter are road dust, industrial processes, electricity generation, and natural and human-made fires. *(Data from http://www.epa.gov/air/emissions/index.htm)*

Teaching Tip: Warm-up

Ask students the following questions:

- What are the major natural sources of air pollution? *Volcanoes, lightning, forest fires, and both living and dead plants all release compounds that can be classified as pollutants.*
- What are the major sources of anthropogenic air pollution? *Sources of anthropogenic air pollution include on-road vehicles, power plants, industrial processes, and waste disposal.*
- How do the National Ambient Air Quality Standards (NAAQS) regulate anthropogenic emissions? *For each anthropogenic pollutant, the NAAQS set a concentration that should not be exceeded over a specified period of time. If a locality violates the air quality standard and does not make an attempt to improve air quality, it is subject to penalties.*

▲ TEACHING with FIGURES

Ask students to use Figure 46.5 to fill in the table to the right. (Answers are provided in italics.)

Major anthropogenic emissions	Top two sources
Carbon monoxide	• *On-road vehicles* • *Non-road equipment*
Nitrogen oxides	• *On-road vehicles* • *Non-road equipment*
Sulfur dioxide	• *Electricity generation* • *Fossil fuel combustion*
Particulate matter	• *Road dust* • *Miscellaneous*

TEACHING with FIGURES ▶

Show students Figure 46.6. Ask them to use the graph to answer the following questions:

- For which six criteria air pollutants has the EPA set concentration limits? *The EPA has set concentration limits for lead, ozone, particulate matter, carbon monoxide, nitrogen dioxide, and sulfur dioxide.*
- Which air pollutant had the largest decrease in the United States between 1990 and 2010? Why? *Lead showed the largest decrease in the United States between 1990 and 2010 because it is no longer added to gasoline. Regulations that prohibit leaded gasoline have caused lead concentrations in the atmosphere to drop significantly.*
- Which of the six criteria air pollutants did not change much between 1990 and 2010? Why? *Ground-level ozone has not changed much because it is a secondary pollutant that forms as a result of photochemical oxidants, sunlight, heat, and other primary pollutants.*

pollutants that are harmful to "human health and welfare." The term *human health* means the health of the human population and includes the elderly, children, and sensitive populations such as those with asthma. The term *welfare* refers to visibility, the status of crops, natural vegetation, animals, ecosystems, and buildings. Through the National Ambient Air Quality Standards (NAAQS), the EPA periodically specifies concentration limits for each air pollutant. For each pollutant the NAAQS note a concentration that should not be exceeded over a specified time period. Our discussion of air pollution in Chattanooga at the beginning of this chapter partially described the standard for ozone: For each locality in the United States, the average ozone concentration for any 8-hour period should not exceed 0.075 parts of ozone per million parts of air by volume more than 4 days per year, averaged over a 3-year period. If a locality violates the ozone air-quality standard and does not make an attempt to improve air quality, it is subject to penalties.

Each year, the EPA issues a report that shows the national level of the six criteria air pollutants relative to the published standards. FIGURE 46.6 shows that all criteria air pollutants have decreased in the United States over the last 2 decades. Only ozone and particulate matter concentrations exceed NAAQS on a regular basis. The cases where particulate matter exceeds national standards are not evident from Figure 46.6. Lead has decreased most significantly because it is no longer added to gasoline.

The situation is less positive in other parts of the world. Large areas in Germany, Poland, and the Czech Republic contain a great deal of "brown" coal or lignite that provides fuel for nearby coal-fired power plants and other industries. Emissions from combustion of this high-sulfur-content coal have caused this so-called Black Triangle to become one of the most polluted areas in the world. In addition to human health problems such as respiratory illnesses, forest ecosystems in this region have also been damaged in the last 25 years. In many parts of Asia, air quality has been so severely impaired by particulate matter and sulfates that visibility has been reduced by as much as 20 percent. A variety of nongovernmental and environmental organizations have prepared lists indicating the top 10 or top 20 most-polluted cities in the world. Cities in China and India usually dominate the list.

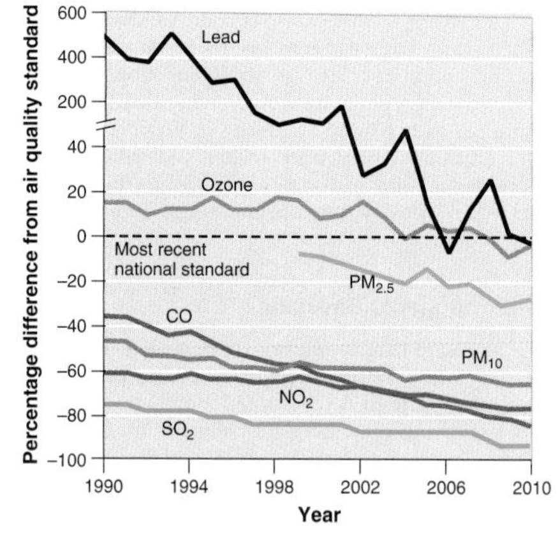

FIGURE 46.6 Criteria and other air pollutant trends. Trends in the criteria air pollutants in the United States between 1990 and 2010. All criteria air pollutants have decreased during this time period. The decrease for lead is the greatest. *(After http://www.epa.gov/airtrends/reports.html)*

module 46

REVIEW

In this module, we have seen that that air pollution occurs in a global system across the troposphere. The major air pollutants are sulfur dioxide, which comes from sulfur contained in coal and oil, nitrogen oxides, which come from combustion, and carbon monoxide, which comes from incomplete combustion. Combustion of coal and oil is the most common source of particulate matter, but it can also be released during the combustion of biofuels such as wood and animal manure. Photochemical oxidants are a class of pollutants that form in the presence of sunlight, nitrogen, and sulfur oxides. There are both natural and human-generated sources of air pollution. Presently, the most polluted cities in the world are found in China and India. These cities suffer from smog, which is the focus of the next module.

Teaching Tip: Activity

Write the following terms on the board:

- air pollution
- sulfur dioxide
- nitrogen oxides
- carbon oxides
- particulate matter
- haze
- photochemical oxidant
- ozone
- smog
- photochemical smog
- sulfurous smog
- lead
- mercury
- volatile organic compounds
- primary pollutants
- secondary pollutants
- natural emissions
- anthropogenic emissions
- volcanoes
- forest fires
- industry
- vehicle exhaust
- National Ambient Air Quality Standards (NAAQS)

In pairs or groups, have students use the words to create two true statements and two false statements about air pollution. Encourage them to use as many words from the board as possible. Have groups share their statements with the class. Award bonus points to groups that use the most terms from the list.

Sample answers:

- *Primary pollutants are pollutant compounds that come directly out of a smokestack, such as nitrogen dioxide, sulfur dioxide, and carbon dioxide. (True)*
- *Air pollutants that come from volcanoes and forest fires are considered a type of natural emission, while vehicle exhaust and pollution from industry are considered to be anthropogenic. (True)*
- *Air pollutants such as nitrogen dioxide and sulfur dioxide are considered secondary pollutants. (False)*
- *Mercury is considered a volatile organic compound because it is emitted when coal is burned. (False)*

Module 46 AP® Review Questions

1. Which is NOT a criteria air pollutant?
   (a) Sulfur dioxide
   (b) Lead
   (c) Carbon dioxide
   (d) Particulate matter
   (e) Tropospheric ozone

2. A secondary pollutant
   (a) forms in the stratosphere.
   (b) is transformed by sunlight or water.
   (c) cannot be directly tracked.
   (d) does not directly harm humans.
   (e) forms primarily from incomplete combustion.

3. Which is a source of sulfur dioxide found in nature?
   (a) Forest fires
   (b) Lightning strikes
   (c) Plant emissions
   (d) Volcanoes
   (e) Soil

4. Which size of particulate matter causes the greatest health concern?
   (a) $PM_{2.5}$
   (b) PM_{10}
   (c) PM_{100}
   (d) PM_{30}
   (e) PM_{15}

5. Carbon monoxide
   (a) increases lung cancer rates.
   (b) leads to the formation of photochemical smog.
   (c) is most problematic in rural areas.
   (d) does not directly cause deaths.
   (e) is produced by incomplete combustion.

Answers to Module 46 AP® Review Questions

1. c
2. b
3. d
4. a
5. e

module 47

Photochemical Smog and Acid Rain

The air quality in the United States has greatly improved in recent decades. However, some cities and even regions of the United States continue to experience intermittent air pollution, often related to smog formation. In this module, we will examine photochemical smog, which is a problem in the United States and elsewhere in the world, and acid rain, which is no longer a problem in the United States but has become a problem in Asia.

Learning Objectives

After reading this module, you should be able to

- explain how photochemical smog forms and why it is still a problem in the United States.
- describe how acid deposition forms and why it has improved in the United States and become worse elsewhere.

AP® Exam Tip

Past exams have asked students to describe the natural formation and destruction of ozone as discussed in this module. Students are encouraged to use chemical equations in their answers. Make sure students understand this process and can write the basic chemical equations for natural ozone accumulation and formation, as outlined on page 528 in Figure 47.1.

Teaching Tip: Review

Ask students: What is the difference between stratospheric ozone depletion and ground-level ozone? *Ozone occurs in two layers of the atmosphere. The layer closest to the Earth's surface is the troposphere. Here, ground-level or "bad" ozone is an air pollutant that is harmful to breathe and is damaging to crops, trees, and other vegetation. It is a main ingredient of urban smog. The troposphere generally extends to a level about 6 miles up, where it meets the second layer, the stratosphere. The stratosphere extends upward from about 6 to 30 miles. The stratospheric or "good" ozone protects life on Earth from the Sun's harmful ultraviolet (UV) rays.*

AP® Exam Tip

Ozone formation and destruction is covered on the 2013 AP® Exam, Question 1. To answer this question, students must

- name the type of solar radiation that is absorbed by stratospheric ozone and discuss a human health benefit of this absorption
- describe the chemical processes that lead to decomposition and formation of stratospheric ozone
- explain how CFCs contribute to the destruction of stratospheric ozone and why the destruction of stratospheric ozone has not ended quickly since CFCs were banned
- name a human activity that leads to the formation of tropospheric ozone as a secondary pollutant and explain why tropospheric ozone levels peak in the daytime
- name a negative ecological impact and a negative human health impact of tropospheric ozone

Photochemical smog remains an environmental problem in the United States

A recent headline read, "EPA Says Half of the United States Is Breathing Excessive Levels of Smog." You might think this was from a newspaper in the 1970s, before the Clean Air Act was fully in effect. But in December 2013, the EPA reported that 46 regions within the United States did not comply with the maximum allowable ozone concentration in the air of 0.075 parts of ozone per million parts of air over an 8-hour period. Although sulfur, nitrogen, and carbon monoxide pollution have been reduced well below the specified standards since the Clean Air Act was implemented, photochemical smog and ozone present especially difficult challenges. The reason lies in the chemistry of smog formation and the behavior of the atmosphere during changing weather conditions. These factors make smog formation very complex and difficult to predict.

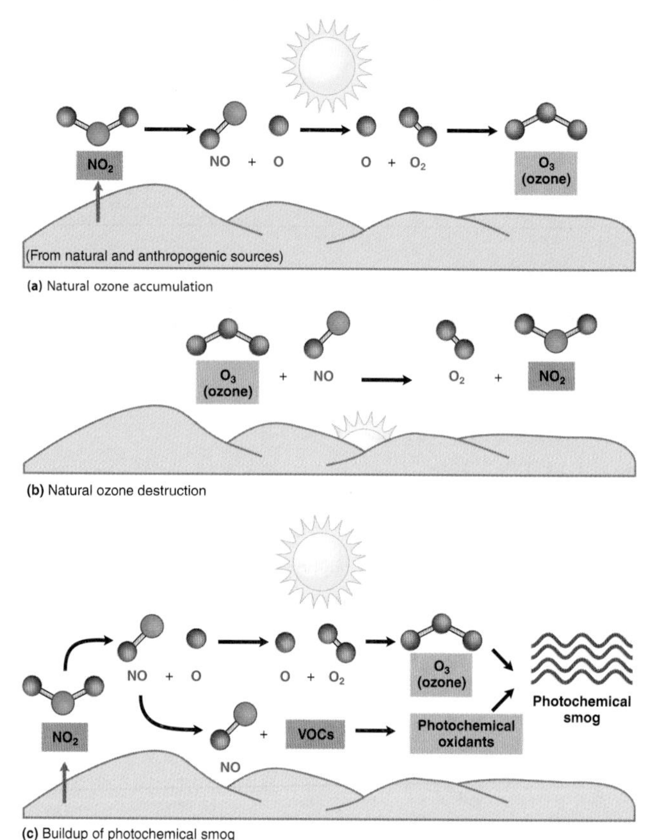

FIGURE 47.1 **Tropospheric ozone and photochemical smog formation.** (a) In the absence of VOCs, ozone will form during the daylight hours. (b) After sunset, the ozone will break down. (c) In the presence of VOCs, ozone will form during the daylight hours. The VOCs combine with nitrogen oxides to form photochemical oxidants, which reduce the amount of ozone that will break down later and contribute to prolonged periods of photochemical smog.

Teaching Tip: Video

Greenhouse Gases
This *National Geographic* video shows an experiment by scientists who seek to predict the effect of increased greenhouse gases on the forests of the future. It provides a good introduction to the subject of photochemical smog.

The link to this resource may be found by clicking on the link buttons in the TE-book, opening the TRFD, or logging onto the book's companion website bcs.whfreeman.com/friedlandapes2e (teacher login required).

lab
Ozone Sampling

This lab asks students to determine the amount of ozone present in different areas.

Download the lab by clicking on the link buttons in the TE-book, opening the TRFD, or logging onto the book's companion website bcs.whfreeman.com/friedlandapes2e (teacher login required).

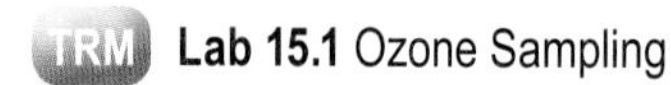

The Chemistry of Ozone and Photochemical Smog Formation

As we mentioned earlier, the term *smog* was originally used to describe the combination of smoke, fog, and sometimes sulfur dioxide that used to occur in cities that burned a lot of coal. Today, Los Angeles–type brown photochemical smog is still a problem in many U.S. cities. The formation of this photochemical smog is complex and still not well understood. A number of pollutants are involved and they undergo a series of complex transformations in the atmosphere that involve sunlight, water, and the presence of VOCs.

FIGURE 47.1 shows a portion of the chemical process that creates photochemical smog. The first part of the process, shown in Figure 47.1a, takes place during the day, in the presence of sunlight. When an abundance of nitrogen oxides are present in the atmosphere, with very few VOCs present, nitrogen dioxide (NO_2) splits to form nitrogen oxide (NO) and a free oxygen atom (O). In the presence of energy inputs from sunlight, this free oxygen atom combines with diatomic oxygen (O_2) to form ozone (O_3). With abundant nitrogen dioxide and abundant sunlight, ozone can accumulate in the atmosphere.

Figure 47.1b shows that a few hours later, when sunlight intensity decreases and with nitrogen oxide still present in the atmosphere, the ozone combines with nitrogen oxide (NO), and re-forms into O_2 + NO_2. This is referred to as ozone destruction and it is a natural process that happens in the latter part of the day and evening.

Volatile organic compounds come from human activity such as spilling of gasoline on pavement and from natural sources such as forests. When volatile organic compounds are absent or in small supply, the cycle of ozone formation and destruction generally takes place on a daily basis and relatively small amounts of photochemical smog form.

As shown in Figure 47.1c, a different scenario occurs when VOCs are present. The first part is the same: Sunlight causes nitrogen dioxide to break apart into nitrogen oxide and a free oxygen atom. The free oxygen atom combines with diatomic oxygen to form ozone. However, because VOCs have combined with nitrogen oxide in a strong bond, nitrogen oxide is no longer available to combine with ozone. Since the nitrogen oxide is not available to break down ozone by recombining with it, a larger amount of ozone accumulates. This explains, in part, the daytime accumulation of ozone in urban areas with an abundance of both VOCs and nitrogen dioxide.

Although smog is associated with urban areas, it is not limited to such areas. Trees and shrubs in rural areas produce VOCs that can contribute to the formation of photochemical smog, as do forest fires that begin naturally.

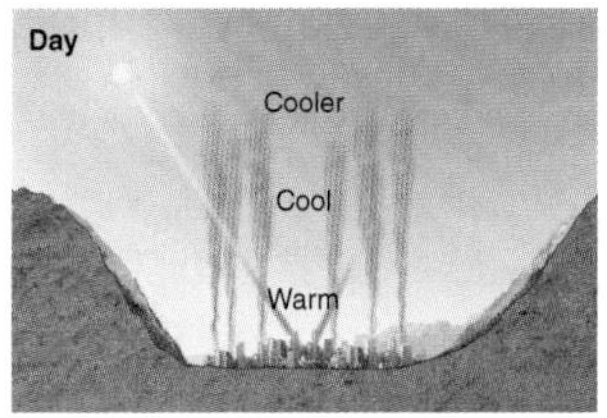

(a) Normal conditions

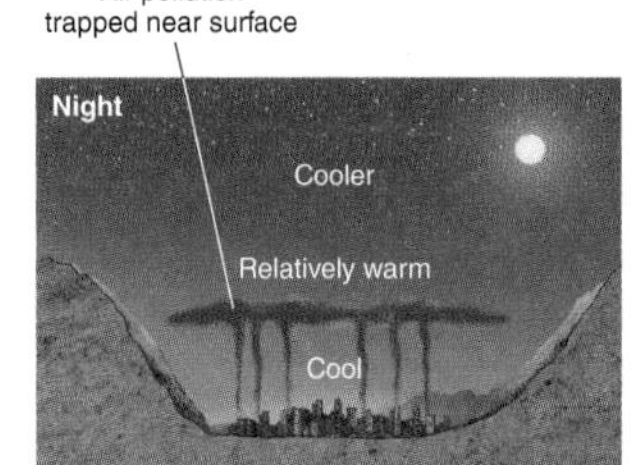

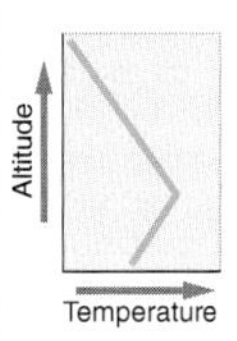

(b) Thermal inversion

FIGURE 47.2 **A thermal inversion.** (a) Under normal conditions, where temperatures decrease with increasing altitude, emissions rise into the atmosphere. (b) When a mid-altitude, relatively warm inversion layer blankets a cooler layer, emissions are trapped and accumulate.

Atmospheric temperature influences the formation of smog in several important ways. Emissions of VOCs from vegetation such as trees, as well as from evaporation of volatile liquids like gasoline, increase as the temperature increases. NO_X emissions from electric utilities are also greater with air-conditioning demands for electricity increasing on the hottest days. Moreover, many of the chemical reactions that form ozone and other photochemical oxidants proceed more rapidly at higher temperatures. These and other factors increase smog concentrations when temperatures are higher.

Thermal Inversions

Temperature also influences air pollution conditions in more complex ways. Normally, temperature decreases as altitude increases. As shown in FIGURE 47.2a, the warmest air is closest to Earth. This warm air, which is less dense than the colder air above it, can easily rise, dispersing pollutants into the upper atmosphere. This allows pollutants from the surface to be reduced or

◀ TEACHING with FIGURES

Ask students to examine Figure 47.2 and answer the following questions:

- What is the difference between Figure 47.2a and Figure 47.2b? *Figure 47.2a shows the normal atmosphere during the day; the temperature at the surface is warm and as altitude increases the temperature decreases. In Figure 47.2b, a relatively warm layer of air at mid-altitude covers a layer of cool, dense air below it. The warm layer of air trapped between the two cooler layers is known as an inversion layer.*
- Why does the inversion layer increase air pollutants at the surface? *Because the air closest to the surface of Earth is denser than the air above it, the cool air and the pollutants within it do not rise. Thus, the inversion layer traps emissions that then accumulate beneath it, and these emissions can cause a severe pollution event. These events typically occur in large cities where high concentrations of vehicle exhaust and industrial emissions are easily trapped.*

Teaching Tip: Video

Temperature Inversion

This video provides an overview of how a thermal inversion occurs. It also explains why a thermal inversion is greater when surrounding hills cause cold, dense air to drain down the slopes of the hills.

The link to this resource may be found by clicking on the link buttons in the TE-book, opening the TRFD, or logging onto the book's companion website bcs.whfreeman.com /friedlandapes2e (teacher login required).

Chapter 15 Web Resources

AP® Exam Tip

The formation of photochemical smog and the nitrogen cycle are covered in the 2009 AP® Exam, Question 1. To answer this question, students must

- describe an environmental problem caused by nitrogen-based fertilizers
- name a nitrogen-containing primary pollutant that contributes to the formation of photochemical smog
- name a secondary pollutant that is a component of photochemical smog, describe how that pollutant forms, and name a health effect and an environmental effect of the pollutant
- explain one chemical transformation that occurs in the nitrogen cycle and explain its importance to an ecosystem

Teaching Tip: Warm-up

Ask students: How does an inversion layer influence air pollution events? *A thermal inversion layer is created when a warm layer of air at mid-altitude covers a layer of cold, dense air below it. The warm layer of air trapped between the two cooler layers is known as an inversion layer. Because the air closest to the surface of Earth is denser than the air above it, the cool air and the pollutants within it do not rise. Thus, the inversion layer traps emissions that then accumulate beneath it, and these trapped emissions can cause a server pollution event.*

diluted by all of the atmosphere above. However, during a **thermal inversion**—shown in Figure 47.2b—a relatively warm layer of air at mid-altitude covers a layer of cold, dense air below it. The warm layer of air trapped between the two cooler layers is known as an **inversion layer.** Because the air closest to the surface of Earth is denser than the air above it, the cool air and the pollutants within it do not rise. Thus, the inversion layer traps emissions that then accumulate beneath it, and these trapped emissions can cause a severe pollution event. Thermal inversions that create pollution events are particularly common in some cities, where high concentrations of vehicle exhaust and industrial emissions are easily trapped by the inversion layer.

Thermal inversions can also lead to other forms of pollution. A striking example occurred in spring 1998 in the northern Chinese city of Tianjin. A cold spell that occurred after the city had shut off its central heating system for the season led many households to use individual coal-burning stoves for heat. A temperature inversion trapped the carbon monoxide and particulate matter from the coal used in these stoves, and caused over 1,000 people to suffer carbon monoxide poisoning, or respiratory ailments from the polluted air. Eleven people died.

Acid deposition has improved in the United States

All rain is naturally somewhat acidic; the reaction between water and atmospheric carbon dioxide lowers the pH of precipitation from neutral 7.0 to 5.6 (see Figure 4.7 on page 39). In Chapter 14 we described acid deposition, which refers to deposition with a pH lower than 5.6. Acid deposition is largely the result of human activity, although natural processes, such as volcanoes, may also contribute to its formation. In this section we will look at how acid deposition is formed, how it travels, and its effects.

How Acid Deposition Forms and Travels

FIGURE 47.3 shows how acid deposition forms. Nitrogen oxides (NO and NO_2) and sulfur dioxide (SO_2) are released into the atmosphere by natural and anthropogenic combustion processes. Through a series of reactions with atmospheric oxygen and water, these primary pollutants are transformed into the secondary pollutants nitric acid (HNO_3) and sulfuric acid (H_2SO_4). These latter compounds break down further, producing nitrate, sulfate—inorganic pollutants that we have discussed earlier—and hydrogen ions (H^+) that generate the acidity in acid deposition. These transformations occur over a number of days, and during this time, the pollutants may travel a thousand kilometers (600 miles) or more. Eventually, these secondary acidifying pollutants are washed out of the air and deposited either as precipitation or in dry form on vegetation, soil, or water.

Acid deposition has been reduced in the United States as a result of lower sulfur dioxide and nitrogen oxide emissions, as shown in Figure 46.6. Much of this improvement is a result of the Clean Air Act Amendments that were passed in 1990 and implemented in 1990 and 1995.

Studies have documented regional acid deposition in West Africa, South America, Japan, China, and many areas in eastern and central Europe. Acid deposition crosses the border between the United States and Canada and is carried from England, Germany, and the Netherlands to Scandinavia. Because of this mobility, the precursors to acid deposition emitted in one region may have a significant impact on another region or another country. For example, over the years, there have been legislative and legal attempts to restrict emissions from coal-burning power plants in the midwestern United States that fall as acid deposition in Canada. Most recently, acidic deposition documented along the West Coast of the United States is believed to be the result of coal combustion in China; sulfur and nitrogen oxides are released in China and elsewhere in Asia and are carried by the prevailing westerlies across the Pacific Ocean.

Effects of Acid Deposition

As we saw in Chapter 14, acid deposition in the United States increased substantially from the 1940s through the 1990s due to human activity. It had a variety of effects on materials, on agricultural lands, and on both aquatic and terrestrial natural habitats. Newspaper headlines in the United States and Europe in the 1980s contained frequent reports about adverse effects of acid deposition on forests, lakes, and streams.

Effects of acid deposition may be direct, such as a decrease in the pH of lake water, or indirect. It is often difficult to determine whether an effect is direct or indirect, making remediation challenging. The greatest effects of acid deposition have been on aquatic ecosystems. Lower pH of lakes and streams in areas of northeastern North America, Scandinavia, and the United Kingdom has caused decreased species diversity of aquatic organisms. As we saw in Chapter 6, many species are able to survive and reproduce only within

Thermal inversion A situation in which a relatively warm layer of air at mid-altitude covers a layer of cold, dense air below.

Inversion layer The layer of warm air that traps emissions in a thermal inversion.

Teaching Tip: Video

Appalachian Trail: Acid Rain: Invisible Menace This video from *National Geographic* describes problems with acid rain in the Appalachian Mountains of the eastern United States and the negative effects it has on the local aquatic life.

The link to this resource may be found by clicking on the link buttons in the TE-book, opening the TRFD, or logging onto the book's companion website bcs.whfreeman.com/friedlandapes2e (teacher login required).

TRM **Chapter 15 Web Resources**

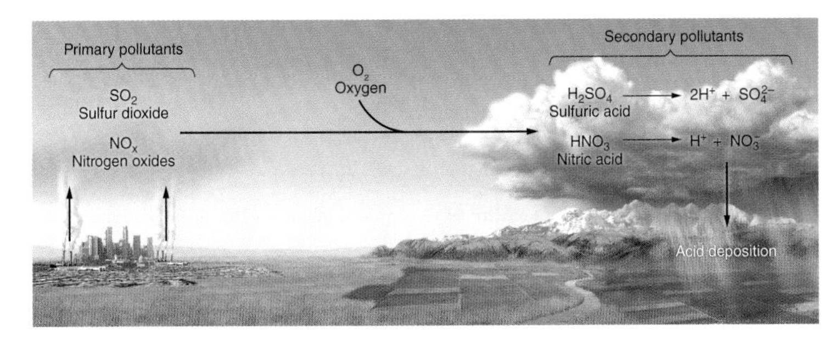

FIGURE 47.3 **Formation of acid deposition.** The primary pollutants sulfur dioxide and nitrogen oxides are precursors to acid deposition. After transformation to the secondary pollutants—sulfuric and nitric acid—dissociation occurs in the presence of water. The resulting ions—hydrogen, sulfate, and nitrate—cause the adverse ecosystem effects of acid deposition.

a narrow range of environmental conditions. Many amphibians, for instance, will survive when the pH of a lake is 6.5, but when the lake acidifies to pH 6.0 or 5.5, the same organism will begin to have developmental or reproductive problems. In water below pH 5.0, most salamander species cannot survive.

Lower pH can also lead to mobilization of metals, an indirect effect. When this happens, metals bound in organic or inorganic compounds in soils and sediments are released into surface water. Because metals such as aluminum and mercury can impair the physiological functioning of aquatic organisms, exposure can lead to species loss. Decreased pH can also affect the food sources of aquatic organisms, creating indirect effects at several trophic levels. On land, at least one species of tree, the red spruce (*Picea rubens*), at high elevations of the northeastern United States was shown to have been harmed by acid deposition. It is likely that these trees have been harmed by both the acidity of the deposition as well as by the nitrate and sulfate ions.

People are not harmed by direct contact with precipitation at the acidities commonly experienced in the United States or elsewhere in the world because human skin is a sufficiently robust barrier. Human health is more affected by the precursors to acid deposition such as sulfur dioxide and nitrogen oxides.

Acid deposition can, however, harm human-built structures such as statues, monuments, and buildings. For example, buildings from ancient Greece such as those on and near the Acropolis, many of which have stood for approximately 2,000 years, have been seriously eroded over the last half century by acid deposition (FIGURE 47.4). The damage happens because acid deposition reacts with building materials. When the hydrogen ion in acid deposition interacts with limestone or marble, the calcium carbonate reacts with H^+ and gives off Ca_2^+. In the process, the calcium carbonate material is partially dissolved. The more acidic the precipitation, the more hydrogen ions there are to interact with the calcium carbonate. In the case of the Acropolis and some other stone structures, other components of acidic deposition, including gaseous sulfur dioxide (SO_2) or sulfuric acid vapor, have contributed to the deterioration. Acid deposition also erodes many exposed painted surfaces, including automobile finishes.

FIGURE 47.4 **Material damage from acid deposition.** Hadrian's Arch, near the Acropolis in Athens, Greece, has been damaged by acid deposition. It is made of marble, which contains calcium carbonate, and is susceptible to deterioration from acid deposition and acids in the air. *(Sites & Photos/HIP/The Image Works)*

Teaching Tip: Review

Ask students the following questions:

- What is acid deposition? *Acid deposition is any type of precipitation that has a pH lower than 5.6. Acid deposition is largely the result of human activity, although natural phenomena, such as volcanoes, may also contribute to its formation.*
- What are the two primary pollutants that lead to the formation of acid deposition? *Two primary pollutants that lead to the formation of acid deposition are sulfur dioxide and nitrogen dioxide.*
- How do primary pollutants transform into secondary pollutants or acid deposition? *Through a series of reactions with atmospheric oxygen and water, primary pollutants are transformed into the secondary pollutants nitric acid (HNO_3) and sulfuric acid (H_2SO_4). These latter compounds break down further, producing nitrate, sulfate, and hydrogen ions (H^+) that generate the acidity in acid deposition.*

Teaching Tip: Activity

Have students fill out a table like the one to the right describing the effects of acid deposition in different biotic and abiotic ecosystems. (Suggested answers are provided in italics.)

Area affected by acid deposition	Effects
Agricultural land	*Acid deposition can lower the pH of soil on farmland. If soil is too acidic, most crops will not grow well. Farmers must add large amounts of lime to neutralize the acid.*
Aquatic ecosystem (streams and lakes)	*Aquatic organisms cannot tolerate waters with low pH.*
Terrestrial ecosystem (forests and grasslands)	*Trees, shrubs, and grasses can be harmed by rains that are very acidic. If plant life is adversely affected, then animals and the entire habitat can change.*
Human construction (historic statues and buildings)	*Historical structures like Hadrian's Arch (Figure 47.4), are damaged by acid deposition. Statues or buildings made of marble, which contains calcium carbonate, are susceptible to damage from acid deposition.*

Answers to Module 47 AP® Review Questions

1. d
2. c
3. b
4. d
5. b

module

47

REVIEW

In this module, we have seen that that photochemical smog accumulates in the troposphere in the presence of sunlight when nitrogen oxides and volatile organic compounds are present. Despite many improvements in air quality in the United States, photochemical smog is still a problem in a number of urban and some relatively rural areas. Air pollution is intensified by thermal inversions, when pollutants in a layer of cold air are trapped beneath a layer of warm air. Acid deposition forms when the oxides of sulfur and nitrogen undergo transformations in the atmosphere in the presence of water. It travels hundreds of kilometers and is deposited in precipitation. Acid deposition can adversely affect material structures, vegetation, soils, and aquatic systems. In the next module, we shall look at how acid deposition has been reduced significantly in the United States in recent decades.

Module 47 AP® Review Questions

1. High levels of photochemical smog are due to
   I. nitrogen dioxide.
   II. sulfur dioxide.
   III. VOCs.

   (a) I only
   (b) I and II only
   (c) II and III only
   (d) I and III only
   (e) I, II, and III

2. Recent increases in acid deposition in the western United States are due to
   (a) increased emissions in the United States.
   (b) decreased precipitation due to climate change.
   (c) increased emissions in China.
   (d) increased precipitation due to climate change.
   (e) increased emissions in Europe.

3. Thermal inversions
   (a) increase the rate of smog formation.
   (b) trap high concentrations of pollution at ground level.
   (c) result in increased levels of acid deposition.
   (d) are caused by high levels of precipitation.
   (e) reduce the negative effects of VOCs.

4. Acid deposition forms as a result of
   (a) nitrogen oxides and photochemical smog.
   (b) carbon monoxide and VOCs.
   (c) sulfur dioxide and VOCs.
   (d) nitrogen oxides and sulfur dioxide.
   (e) carbon monoxide and photochemical smog.

5. The effects of acid deposition include
   (a) increased cancer rates in aquatic organisms.
   (b) decreased aquatic biodiversity.
   (c) decreased mobilization of metals.
   (d) increased pH of lake water.
   (e) decreased human health in impacted areas.

module

48

Pollution Control Measures

In our discussion of energy and energy choices, we saw that sustainability is best achieved by considering conservation and efficiency first. Similarly, if we can address the problems of pollution by seeking ways to avoid creating it in the first place, we will require less energy and fewer resources to clean it up. Preventing pollution is usually much less expensive and energy intensive then controlling it. Unfortunately, pollution prevention is not always possible.

Learning Objectives

After reading this module, you should be able to

- explain strategies and techniques for controlling sulfur dioxide, nitrogen oxides, and particulate matter.
- describe innovative pollution control measures.

Pollution control includes prevention, technology, and innovation

As with other types of pollution, the best way to decrease air pollution emissions is to avoid them in the first place. This can be achieved through the use of fuels that contain fewer impurities. Coal and oil, for example, both occur naturally with different sulfur concentrations and are available for purchase at a variety of sulfur concentrations. In addition, during refining and processing, the concentration of sulfur can be reduced in both fuels.

Use of a low-sulfur coal or oil is certainly one of the best means of controlling air pollution, although typically low-sulfur coal or oil is more expensive to purchase than coal or oil containing higher sulfur concentrations. As we discussed in Chapters 12 and 13, other ways to reduce air pollution include increased efficiency and conservation: Use less fuel and you will produce less air pollution. While these measures reduce emissions by a certain amount, wherever fuel is combusted, pollution will be emitted. So ultimately, most attempts to reduce air pollution will depend on the control of pollutants after combustion. There are a number of approaches to controlling emissions, as we will discuss in the next section.

Control of Sulfur and Nitrogen Oxide Emissions

Sulfur and nitrogen oxides are common air pollutants in the United States and they cause a variety of environmental problems, including acid deposition that we described in the previous module. A substantial number of air pollution control measures have been directed toward sulfur and nitrogen oxides. Sulfur dioxide emissions from coal exhaust can be reduced by a process known as fluidized bed combustion. In this process, granulated coal is burned in close proximity to calcium carbonate. The heated calcium carbonate absorbs sulfur dioxide and produces calcium sulfate, which can be used in the production of gypsum wallboard, also known as sheetrock, for houses.

AP® Exam Tip

Government action to reduce outdoor air pollution is one of the topics covered on the 2007 AP® Exam, Question 4. To answer this question, students must

- describe how urban temperatures differ from the temperatures of surrounding rural areas
- identify and describe two differences between urban and rural areas that contribute to temperature differences
- identify a feature of urban areas that leads to higher levels of air pollution and explain why that feature leads to greater levels of air pollution
- suggest two actions that a local government can take to reduce air pollution
- identify two differences in the hydrologic cycles of urban and rural areas

Note that this free-response question uses material from Chapters 3, 10, and 15. This makes it a particularly good question to use as a cumulative assessment.

Teaching Tip: Concept Map

Write the words "Pollution Control Device" in the center of a square on the board. Ask students to create a definition. When you are satisfied with the definition, ask students to identify four different types of pollution control devices and briefly describe how each removes the pollutant. Have students create a concept map like the one shown to the right. Remind your class that avoiding pollution is almost always the most desirable option.

Fluidized bed combustion
Burns granulated coal in close proximity to calcium carbonate; the heated calcium carbonate absorbs SO_2 and produces calcium sulfate

Wet scrubbers
Use a combination of water and air to actually separate and remove particles

Pollution control device
A device that is used after combustion to remove harmful air pollutants such as NO_x, SO_x, and particulate matter.

Baghouse filters
Allow gas to pass through them but remove particluate matter

Electrostatic precipitators
Use an electrical charge to make particles coalesce so they can be removed

AP® Exam Tip

The effects of coal-fired electric power are covered in the 2000 AP® Exam, Question 1. To answer this question, students must

- use provided data to calculate the BTUs of energy needed to generate the electricity produced by the power plant each day
- use provided data to calculate the pounds of coal consumed by the plant
- use the provided data to calculate the amount of sulfur released by the plant
- determine whether the plant meets the EPA standards for sulfur release
- describe two ways the plant can reduce its sulfur emissions
- discuss why sulfur emissions harm the environment and describe one negative effect of sulfur emissions on an ecosystem

Note that students earned points for setting up the calculation problems correctly. Throughout the year encourage students to take the time to set up problems clearly so they will gain points on the exam even if they make calculation errors.

Some of the sulfur oxide that does escape the combustion process can be captured by other methods after combustion.

The atmosphere of Earth is 78 percent nitrogen gas and, as a result, nitrogen oxides are produced in virtually all combustion processes. Hotter burning conditions and the presence of oxygen allow proportionally more nitrogen oxide to be generated per unit of fuel burned. In order to reduce nitrogen oxide emissions, burn temperatures must be reduced and the amount of oxygen must be controlled—a procedure that is sometimes utilized in factories and power plants by means of certain air pollution control technologies. However, lowering temperatures and oxygen supply can result in less-complete combustion, which reduces the efficiency of the process and increases the amount of particulates and carbon monoxide. Finding the exact mix of air, temperature, oxygen, and other factors is a significant challenge.

Nitrogen oxide emissions from automobiles have also been reduced significantly in the United States over the last 35 years. Beginning in 1975, all new automobiles sold in the United States were required to include a catalytic converter, which reduces the nitrogen oxide and carbon monoxide emissions. In order to operate properly, the precious metals in the catalytic converter (mostly platinum and palladium) cannot be exposed to lead. Therefore gasoline could no longer contain lead. As Figure 46.6 illustrates, the change in gasoline formulation caused a significant reduction in emissions of lead from automobiles and in lead concentrations in the atmosphere. At the same time, improvements in the combustion technology of power plants and factories also reduced emissions of nitrogen oxides.

Control of Particulate Matter

The removal of particulate matter is the most common means of pollution control. Sometimes the process of removing particulate matter also removes sulfur. There are a variety of methods used to remove particulate matter. The simplest is gravitational settling, which relies on gravity as the exhaust travels through the smokestack. The particles simply settle out to the bottom. The ash residue that accumulates must be disposed of in a landfill. Depending on the fuel that was burned, the ash may contain sufficiently high concentrations of metals that require special disposal. This subject will be covered in more detail in Chapter 16 on solid waste.

Pollution control devices remove particulate matter and other compounds after combustion. Each has its advantages and disadvantages and all of them use energy—most commonly electricity—which generates additional pollution. Fabric filters are a type of filtration device that allow gases to pass through them but remove particulate matter. Often called baghouse filters, certain fabric filters can remove almost 100 percent of the particulate matter emissions. Electrostatic precipitators use an electrical charge to make particles coalesce so they can be removed. Polluted air enters the precipitator and the electrically charged particles within are attracted to negative or positive charges on the sides of the precipitator. The particles collect and relatively clean gas exits the precipitator. A scrubber, shown in FIGURE 48.1, uses a combination of water and air that actually separates and removes particles. Particles are removed in the scrubber in a liquid or sludge form and clean gas exits. Borrowing from the concept utilized in the electrostatic precipitator, particles are sometimes ionized before entering the scrubber to increase its efficiency. Scrubbers are also used to reduce the emissions of sulfur dioxide. All three types of pollution control devices, because they use additional energy and increase resistance to air flow in the factory or power plant, require the use of more fuel and result in increased carbon dioxide emissions.

Devices such as the electrostatic precipitator and the scrubber have helped reduce pollution significantly before it is released into the atmosphere. It is much harder—if not impossible—to remove pollutants from the environment after they have been dispersed over a wide area.

Smog Reduction

We have seen that many cities in the United States and around the world continue to have smog problems. Because the main component of photochemical smog—ozone—is a secondary pollutant, control efforts must be directed toward reducing the precursors, or primary pollutants. Historically, most local smog reduction measures have been directed primarily at reducing emissions of VOCs in urban areas. As noted earlier, with fewer VOCs in the air there are fewer compounds to interact with nitrogen oxides, and thus more nitrogen oxide will be available to recombine with ozone. More recently, regional efforts to control ozone have focused on reducing nitrogen oxide emissions, which appears to be a more effective method of controlling smog in areas away from urban centers.

Around the world, people are implementing innovative pollution control measures

A number of cities around the world, including those in China, Mexico, and England, have taken innovative and often controversial measures to reduce smog levels. Municipalities have passed measures, for example,

Teaching Tip: Beyond the Classroom

In the United States coal-fired electricity plants can be found in every state except Vermont and Rhode Island. Have students write a report or create a presentation about the use of coal-fired electricity in one state. Assign each student a state to research. Students should start with the following questions:

- How many coal-fired electricity plants are there in the state?
- What is the overall MW capacity of the state's coal-fired electricity plants?
- What percentage of electricity generated in the state is from coal-fired electricity?

In the second part of the report ask students to select a plant from the state they researched and find the following details:

- Owner
- Parent company
- Length of time in operation
- Most recent available emissions data:

CO_2 emissions
SO_2 emissions
NO_x emissions
Hg emissions

A link to a website with data for this project may be found by clicking on the link buttons in the TE-book, opening the TRFD, or logging onto the book's companion website bcs.whfreeman.com/friedlandapes2e (teacher login required).

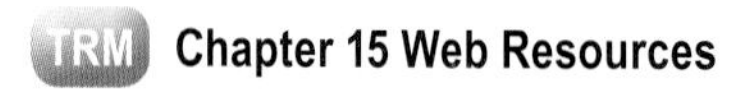
Chapter 15 Web Resources

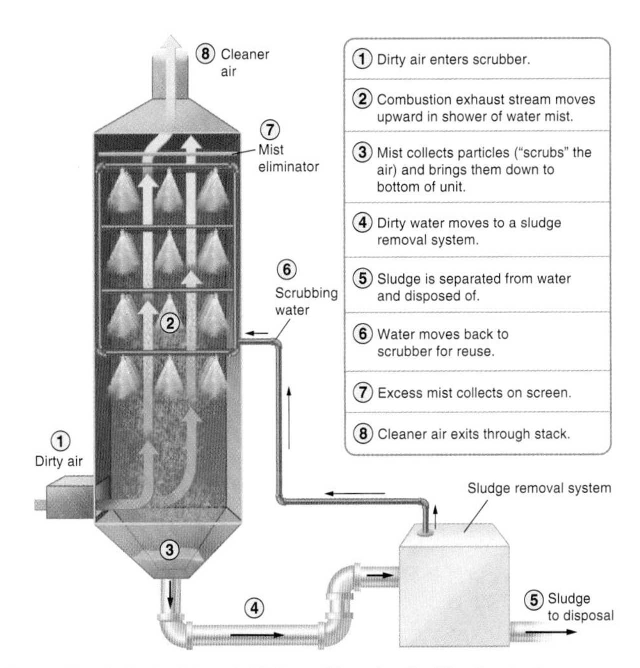

FIGURE 48.1 **The scrubber.** In this air pollution control device, particles are "scrubbed" from the exhaust stream by water droplets. A water-particle "sludge" is collected and processed for disposal.

to reduce the amount of gasoline spilled at gasoline stations, restrict the evaporation of dry-cleaning fluids, or restrict the use of lighter fluid (a VOC) for starting charcoal barbecues. Both urban and suburban areas have taken additional actions such as calling for a reduction in the use of wood-burning stoves or fireplaces that would reduce emissions of not only nitrogen oxide but also particulate matter, VOCs, and carbon monoxide. A number of California municipalities even discussed reducing the number of bakeries within certain areas, as the emissions from rising bread contain VOCs. This proposal was not very popular, as you can imagine, but emissions from bakeries along with many other businesses are sometimes regulated by local air-quality ordinances.

Since cars are responsible for large emissions of nitrogen oxides and VOCs in urban areas, and these two compounds are the major contributors to smog formation, some municipalities have tried to achieve lower smog concentrations by restricting automobile use. A number of cities, including Mexico City, have instituted plans permitting automobiles to be driven only every other day—for example, those with license plates ending in odd numbers may be used on one day and those with even-numbered license plates on alternate days. In China, during the 2008 Beijing Olympics, the government successfully expanded public transportation networks, imposed motor vehicle restrictions, and temporarily shut down a number of industries as a way to reduce photochemical smog and improve visibility (FIGURE 48.2).

Limiting automobile use has also helped to reduce other air pollutants. Carpool lanes, available in many areas, reduce the number of cars on the road by encouraging two or more people to share one vehicle. Improving the quality and accessibility of public transportation encourages people to leave their cars at home. A number of cities in England, including London, have been experimenting with charging individual user fees (tolls) for the use of roads at certain times of the day or within certain parts of a city as a way to reduce automobile traffic. Road user fees have been proposed for cities in the United States, including New York City, but none has yet been implemented.

In 1990 and again in 1995, scientists, policy makers, and academics collaborated on amendments to the Clean Air Act that would allow the free market to determine the least expensive ways to reduce emissions

Teaching Tip: Engage

Ask students: On days when high ozone levels are expected, what are four steps a person can take to help reduce ozone pollution levels? *Answers will vary. Some possible answers include:*

- *Share a ride to work or use public transportation*
- *Combine errands and reduce trips*
- *Walk to errands when possible*
- *Avoid excessive idling of your vehicle*
- *Refuel your car in the evening when it is cooler*
- *Conserve electricity and set air conditioners no lower than 78 degrees*
- *Defer lawn and gardening chores that use gasoline-powered equipment, or wait until evening*

Answer to Your Turn

Total percentage reduction:

22.6 million metric tons −
17.2 million metric tons =
5.4 million metric tons total reduction

5.4 million metric tons ÷
22.6 million metric tons × 100 = 23.9%

Annual percentage reduction:

$$\frac{23.9\%}{15\text{ years}} = 1.6\%/\text{year}$$

Teaching Tip: Journal Prompt

Ask students to describe ways that urban and suburban areas might reduce smog. *Because the main component of photochemical smog—ozone—is a secondary pollutant, control efforts must be directed toward reducing the precursors, or primary pollutants. Both urban and suburban areas have taken steps to reduce the use of wood-burning stoves or fireplaces in an attempt to reduce emissions of nitrogen oxides and VOCs. Restricting the use of lighter fluid, which is a VOC, can also make a difference. A number of large cities, including Mexico City, have instituted plans that restrict the use of automobiles to every other day.*

do the math

Calculating Annual Sulfur Reductions

Given the data presented below for total SO_2 emission reductions in the United States, calculate the total percentage reduction and the annual percentage reduction of SO_2 emissions.

23.5 million metric tons − 10.3 million metric tons
= 13.2 million metric tons total reduction

Divide the reduction by the original amount and multiply by 100 to obtain a percent reduction:

13.2 million metric tons ÷ 23.5 million metric tons × 100 = 56%

The total reduction was 56 percent.

To calculate the reduction per year, divide 56 percent by the number of years from beginning to end:

2008 − 1982 = 26 years

56% ÷ 26 years = 2.2%/year

Your Turn In the United States, nitrogen oxide emissions decreased from 22.6 million metric tons (25 million U.S. tons) in 1990 to 17.2 million metric tons (19 million U.S. tons) in 2005. Calculate the total percentage reduction and the annual percentage reduction for nitrogen oxides.

(a)

(b)

FIGURE 48.2 **Reducing photochemical smog.** The view in Beijing, China, (a) during the Beijing Olympics in 2008 and (b) after the pollution restrictions were removed. *(Peter Parks/AFP/Getty Images)*

of sulfur dioxide. The free-market program was implemented in two phases between 1995 and 2000, and approximately 3,000 power plants are now covered under the Acid Rain Program of the act. So far, each phase has led to significant reductions in sulfur emissions.

One of the most innovative aspects of the Clean Air Act amendments was the provision for the buying and selling of allowances that authorized the owner to release a certain quantity of sulfur. Each allowance authorizes a power plant or industrial source to emit one ton of SO_2 during a given year. Sulfur allowances are awarded annually to existing sulfur emitters proportional to the amounts of sulfur they were emitting before 1990, and the emitters are not allowed to emit more sulfur than the amount for which they have permits. At the end of a given year, the emitter must possess a number of allowances at least equal to its annual emissions. In other words, a facility that emits 1,000 tons of SO_2 must possess at least 1,000 allowances that are usable in that year. Facilities that emit quantities of SO_2 above their allowances must pay a financial penalty.

Sulfur allowances can be bought and sold on the open market by anyone. If emitters wanted to exceed their allowance level—say, because they intended to increase their industrial output—they would be required to purchase more allowances from another source. If, on the other hand, a company decreased its sulfur emissions more than it needed to in order to

AP® Exam Tip

Air pollution and electric vehicles are covered in the 2002 AP® Exam, Question 1. To answer this question, students must

- identify and describe two environmental benefits of using electric vehicles instead of gasoline-powered vehicles
- use provided data to calculate an estimate of the reduction in petroleum consumption that could be achieved if 10 percent of all cars in the United States were replaced with electric vehicles
- discuss the concern that electric vehicles shift emission of air pollutants from dispersed sources to point sources
- propose two government policies that might encourage more use of electric vehicles

comply with its allowance amount, it could sell any unused sulfur emission allowances. Over time, the number of allowances distributed each year has been gradually reduced: The total SO_2 emissions from all sources in the United States have declined from 23.5 million metric tons (26 million U.S. tons) in 1982 to 10.3 million metric tons (11.4 million U.S. tons) in 2008. "Do the Math: Calculating Annual Sulfur Reductions" shows you how to calculate these decreases as percentages. The overall economic cost for achieving these reductions has been about one-quarter of the original cost estimate. Global change researchers have used the sulfur allowance example as a model for the more recent experiments with buying and selling carbon dioxide allowances. We discuss this further in "Science Applied 8: Can We Solve the Carbon Crisis Using Cap-and-Trade?" following Chapter 20 on page 728.

Teaching Tip: Warm-up

Ask students to explain the purpose of sulfur allowances and how they work. *A sulfur allowance grants the holder the right to release a certain amount of sulfur during a given year. Allowances can be bought and sold. If emitters want to exceed their allowance level, they must purchase more allowances from another source. If a company decreases its sulfur emissions, it can sell unused emissions allowances. The ability to sell an unused allowance gives companies an incentive to emit less sulfur. Over time, the number of allowances distributed each year has been reduced, which makes the sulfur allowances even more valuable to those who can sell them.*

module 48 REVIEW

In this module, we have seen that the best approach for decreasing air pollution emissions is by avoiding them in the first place. This can be done by choosing fuels that contain fewer impurities or by increasing the efficiency of operations. After these options have been considered and implemented, a number of pollution control technologies are available to prevent pollutants from entering the air. The baghouse filter physically removes particles and the electrostatic precipitator uses an electrical charge to attract and remove pollutant particles. The scrubber literally scrubs the exhaust stream with water droplets, creating a sludge that can be collected and removed. Innovative techniques for reducing pollution include putting a price on the right to emit certain pollutants and allowing pollution permits bought and sold on the free market to bring down emissions in the most cost-efficient way possible. In the next module we will look at how pollutants caused a depletion of stratospheric ozone and the international agreement that led to a successful resolution of the problem.

Module 48 AP® Review Questions

1. Which method is used to reduce nitrogen oxide emissions?
   (a) Installation of catalytic converters
   (b) Increased temperature of combustion
   (c) The addition of oxygen to combustion processes
   (d) The use of fluidized bed combustion
   (e) The use of calcium carbonate to absorb the emissions

2. Which is NOT used to prevent the emission of particulate matter?
   (a) Gravitational settling
   (b) Fabric filters
   (c) Electrostatic precipitators
   (d) Catalytic converters
   (e) Scrubbers

3. Which pollution control method was proposed by an amendment to the Clean Air Act?
   (a) Allowing tolls to limit the use of automobiles
   (b) A market for sulfur emissions
   (c) The regulation of radon emissions
   (d) The use of catalytic converters
   (e) Chemicals developed to prevent the formation of ozone

4. If carbon monoxide emissions decreased from 145 million tons annually to 80 million tons annually, by what percentage have emissions been reduced?
   (a) 41 percent
   (b) 45 percent
   (c) 47 percent
   (d) 52 percent
   (e) 55 percent

Answers to Module 48 AP® Review Questions

1. a
2. d
3. b
4. b

module

49

Stratospheric Ozone Depletion

We have seen that tropospheric or ground-level pollution has been shown to contribute to a number of problems in the natural world, to exacerbate asthma and breathing difficulties in humans, and to cause cancer. Now we turn to the effects of certain pollutants in the stratosphere that have a substantial impact on the health of humans and ecosystems. In the troposphere, ozone is an oxidant that can harm respiratory systems in animals and damage a number of structures in plants. However, in the stratosphere ozone forms a necessary, protective shield against radiation from the Sun; it absorbs ultraviolet light and prevents harmful ultraviolet radiation from reaching Earth.

Learning Objectives

After reading this module, you should be able to

- explain the benefits of stratospheric ozone and how it forms.
- describe the depletion of stratospheric ozone.
- explain efforts to reduce ozone depletion.

AP® Exam Tip

The AP® Exam regularly asks questions about the difference between stratospheric ozone and ground level ozone. Make sure students are easily able to differentiate the two.

Stratospheric ozone is beneficial to life on Earth

The Sun radiates energy at many different wavelengths, including the ultraviolet range (see Figure 5.1 on page 45). The ultraviolet wavelengths are further classified into three groups: UV-A, or low-energy ultraviolet radiation, and the shorter, higher-energy UV-B and UV-C wavelengths. UV radiation of all types can damage the tissues and DNA of living organisms. Exposure to UV-B radiation increases the risk of skin cancer and cataracts, and suppresses the immune system in humans. Exposure to UV-B is also harmful to the cells of plants, and it reduces their ability to convert sunlight into usable energy. UV-B exposure can therefore harm entire biological communities. For example, losses of phytoplankton—the microscopic algae that form the base of many marine food chains—will cause the depletion of fisheries.

As we saw in Chapter 4, a layer of ozone in the stratosphere (see Figure 9.1 on page 106) absorbs ultraviolet radiation, filtering out harmful UV rays from the Sun. It is easy to confuse stratospheric ozone with tropospheric, or ground-level, ozone that we discussed earlier in this chapter because it *is* the same gas, O_3. However, stratospheric ozone occurs higher in the atmosphere where its ability to absorb ultraviolet radiation and thereby shield the surface below makes stratospheric ozone critically important to life on Earth. In this section we will examine the formation of stratospheric ozone and look at how it breaks down.

Formation of Stratospheric Ozone

When solar radiation strikes O_2 in the stratosphere, 16 to 50 km (10–31 miles) above Earth's surface, a series of chemical reactions begins that produces a new molecule: ozone (O_3).

In the first step, UV-C radiation breaks the molecular bond holding an oxygen molecule together:

$$O_2 + \text{UV-C} \rightarrow O + O$$

This happens to only a few oxygen molecules at any given time. The vast majority of the oxygen in the atmosphere remains in the form O_2.

In the second step, a free oxygen atom (O) produced in the first reaction encounters an oxygen molecule, and they form ozone.

$$O + O_2 \rightarrow O_3$$

Both UV-B and UV-C radiation can break a bond in this new ozone molecule, forming molecular oxygen and a free oxygen atom once again:

$$O_3 + \text{UV-B or UV-C} \rightarrow O_2 + O$$

Thus formation of ozone in the presence of sunlight and its subsequent breakdown is a cycle that can occur indefinitely as long as there is UV energy entering the atmosphere. Under normal conditions, the amount of ozone in the stratosphere remains at steady state.

Breakdown of Stratospheric Ozone

We rely on refrigeration to keep our foods safe and edible, and on air conditioning to keep us comfortable in hot weather. For many years, the same chemicals that made refrigeration and air conditioning possible were also used in a host of other consumer items, including aerosol spray cans and products such as Styrofoam. These chemicals, called chlorofluorocarbons, or CFCs, were considered essential to modern life, and producing them was a multibillion-dollar industry. CFCs were considered "safe" because they are both nontoxic and nonflammable. But it turned out that these chemicals had adverse effects in one part of the upper atmosphere, the stratosphere, by promoting the breakdown of ozone.

CFCs introduce chlorine (Cl) into the stratosphere. When chlorine is present, it can attach to an oxygen atom in an ozone molecule, thereby breaking the bond between that atom and the molecule and forming chlorine monoxide (ClO) and O_2:

$$O_3 + Cl \rightarrow ClO + O_2$$

Subsequently, the chlorine monoxide molecule reacts with a free oxygen atom, which pulls the oxygen from the ClO to produce free chlorine again:

$$ClO + O \rightarrow Cl + O_2$$

When we consider these reactions together, we see that chlorine starts out and ends up as a free Cl atom. In contrast, an ozone molecule and a free oxygen atom are converted into two oxygen molecules. A substance that aids a reaction but does not get used up itself is called a catalyst. A single chlorine atom can catalyze the breakdown of as many as 100,000 ozone molecules until finally one chlorine atom finds another and the process is stopped. In the process, the ozone molecules are no longer available to absorb incoming UV-B radiation. As a result, the UV-B radiation can reach Earth's surface and cause harm to biological organisms.

Humans have contributed to significant destruction of the ozone layer

Ozone formation and ozone destruction have occurred for many years. But the use of CFCs as refrigerants starting in the 1920s and their increased use since that time led to the more rapid destruction of stratospheric ozone. After decades of CFC use, its effects became apparent.

Depletion of the Ozone Layer

In the mid-1980s, atmospheric researchers noticed that stratospheric ozone in Antarctica had been decreasing each year, beginning in about 1979. Since the late 1970s, global ozone concentrations had decreased by more than 10 percent. Depletion was greatest at the poles, but occurred worldwide. One study from Switzerland showed an erratic but clearly decreasing trend of ozone concentrations since 1970. The graph in FIGURE 49.1 shows the results of this study.

Researchers also determined that, in the Antarctic, ozone depletion was seasonal: Each year the depletion occurred from roughly August through November (late winter through early spring in the Southern Hemisphere).The depletion caused an area of severely reduced ozone concentrations over most of Antarctica, creating what has come to be called the "ozone hole." A depletion of ozone also occurs over the Arctic in January through April, but it is not as severe, varies more from year to year, and does not cause a "hole" as in the Antarctic.

The cause of the formation of the ozone hole, which has received a great deal of media attention and has been studied intensively, is complex. It appears that extremely cold weather conditions during the polar winter cause a buildup of ice crystals mixed with nitrogen oxide. This in turn provides the perfect surface for the formation of the stable molecule Cl_2,

Module 49 Stratospheric Ozone Depletion 539

AP® Exam Tip

The thinning of the ozone layer and the harmful effects of ground-level ozone are covered in the 2007 AP® Exam, Question 3. To answer this question, students must

- identify the class of chemical compounds that is responsible for ozone layer thinning
- describe how the compounds responsible for ozone layer thinning destroy ozone molecules
- identify the major environmental consequence of stratospheric ozone depletion and describe two harmful effects on ecosystems or human health
- explain how ground-level ozone affects ecosystems and human health

which accumulates as atmospheric chlorine interacts with the ice crystals. When the Sun reappears in the spring, UV radiation breaks down this molecule into Cl again, which in turn catalyzes the destruction of ozone as described above. Because almost no ozone forms in the dark of the polar winter, a large "hole" occurs. Only after the temperatures warm up and the chlorine gets diluted by air coming from outside the polar region does the hole diminish. In contrast, the overall global trend of decreasing stratospheric ozone concentration is not related to temperature, but is caused by the breakdown reactions described earlier that result from increased concentrations of chlorine in the atmosphere.

Decreased stratospheric ozone has increased the amount of UV-B radiation that reaches the surface of Earth. A United Nations study showed that in mid-latitudes in North America, UV radiation at the surface of Earth increased about 4 percent between 1979 and 1992, although the increase is greater closer to the poles. For plants, both on land and in water, increasing exposure to UV-B radiation can be harmful to cells and can reduce photosynthetic activity, which could have an adverse impact on ecosystem productivity, among other things. In humans, particularly those with light skin, increasing exposure to UV-B radiation is correlated with increased risks of skin cancer, cataracts, and other eye problems, and with a suppressed immune system. Significant increases in skin cancers have already been recorded, especially in countries near the Antarctic ozone hole such as Chile and Australia.

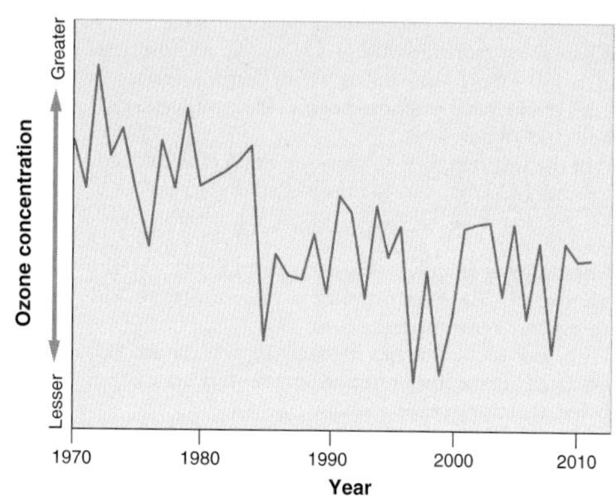

FIGURE 49.1 **Stratospheric ozone concentration.** This data for one area of Switzerland shows a generally decreasing trend from 1970 to 2011. *(Data from https://www.mfe.govt.nz/environmental-reporting/atmosphere/levels-stratospheric-ozone-indicator/report-card-2012.html)*

Efforts to reduce ozone depletion have been effective

In response to the decrease in stratospheric ozone, 24 nations in 1987 signed the Montreal Protocol on Substances That Deplete the Ozone Layer. This was a commitment to reduce CFC production by 50 percent by the year 2000. It was the most far-reaching environmental treaty to date, in which global CFC exporters like the United States appeared in some ways to prioritize the protection of the global biosphere over their short-term economic self-interest. More than 180 countries eventually signed a series of increasingly stringent amendments that required the elimination of CFC production and use in the developed world by 1996. In total, the protocol addressed 96 ozone-depleting compounds.

Because of these efforts, the concentration of chlorine in the stratosphere has stabilized at about 5 ppb (parts per billion) and should fall to about 1 ppb by 2100. The chlorine concentration reduction process is slow because CFCs are not easily removed from the stratosphere and in some recent years ozone depletion has continued to reach record levels. However, with the leveling off of chlorine concentrations, stratospheric ozone depletion should decrease in subsequent decades. The number of additional skin cancers should eventually decrease as well, although this effect will take some time due to the long time it takes for these cancers to develop.

Teaching Tip: Activity

Have students take a few minutes to fill in a table like the one below to compare stratospheric ozone and ground-level ozone. (Answers are provided in italics.)

	Location	Environmental effects	Significant chemical involvement	Current status
Stratospheric ozone	*Upper to middle stratosphere*	*Breakdown causes increased exposure to harmful increased UV-B radiation that damages cells and reduces photosynthetic activity in plants and causes health problems in humans, including skin cancer, cataracts, and immune system suppression*	*CFCs introduce Cl, which causes the reaction* $O_3 + Cl \rightarrow ClO + O_2$	*Efforts to reduce ozone depletion through restricting the use of CFCs have been effective but Cl reduction in the stratosphere is slow*
Ground-level ozone	*Lower troposphere*	*Reduces lung function and exacerbates respiratory symptoms, harms plant surfaces, and damages materials such as rubber and plastic*	*A secondary pollutant formed by the combination of sunlight, water, oxygen, VOCs, and* NO_x.	*Increased concentrations of ground-level ozone have been detected in many U.S. cities; in December 2013, the EPA reported that 46 regions in the country did not comply with the maximum allowable ozone concentration in the air of 0.075 parts of ozone per million parts of air over an 8-hour period*

module

49

REVIEW

In this module, we have seen that there is a natural process of ozone formation and ozone destruction in the stratosphere. As a result, with relatively constant ozone concentrations, harmful ultraviolet radiation is absorbed in the upper atmosphere and does not reach ground level where it could be harmful to plants and animals, including humans. However, the introduction of human-synthesized chlorofluorocarbons led to increased amounts of chlorine in the stratosphere, leading to destruction of stratospheric ozone. This caused an increase in UV-B radiation in certain locations on Earth. Since nations signed the Montreal Protocol on Substances That Deplete the Ozone Layer, stratospheric ozone depletion has begun to slow. In the next module we will turn from outdoor air pollution to indoor air pollution that occurs in houses and other living environments.

Module 49 AP® Review Questions

1. The formation of ozone begins when an O_2 molecule is split by
   (a) UV-A radiation.
   (b) UV-B radiation.
   (c) UV-C radiation.
   (d) UV-A or UV-B radiation.
   (e) UV-B or UV-C radiation.

2. How many ozone atoms can a single chlorine atom break down?
   (a) 5,000
   (b) 10,000
   (c) 50,000
   (d) 100,000
   (e) 500,000

3. Which is NOT a result of the reduction in stratospheric ozone?
   (a) Reduced photosynthetic activity
   (b) Increased skin cancer
   (c) Increased eye problems
   (d) Increased birth defects
   (e) Suppressed immune systems

4. Through international cooperation, the concentration of chlorine in the atmosphere
   (a) has already decreased below 3 ppb.
   (b) will stabilize around 3 ppb by 2030.
   (c) has stabilized at 5 ppb.
   (d) is expected to stabilize at 5 ppb by 2100.
   (e) is expected to fall below 3 ppb by 2020.

5. When is the Antarctic ozone hole largest?
   (a) Early spring
   (b) Late summer
   (c) Mid-summer
   (d) Late fall
   (e) Early winter

Answers to Module 49 AP® Review Questions

1. c
2. d
3. d
4. c
5. a

Teaching Tip: Activity

In the last 10 minutes of class, ask students to write a test question based on material covered in this chapter. Have students write their question on the top half of a sheet of paper and their answer on the bottom half of the same page. Collect their work and save the sheets for use the next day. Make extra copies of questions or add some of your own so you have a question for each student. At the beginning of the next class, pass out the questions and answers. Have students quiz each other and discuss their answers with a partner.

AP® Exam Tip

Indoor air pollution and sick building syndrome are covered on the 2001 AP® Exam, Question 3. To answer these questions, students must

- identify and discuss two specific indoor air pollutants, naming the likely location of the pollutant, the source, its effects on human health, and prevention or remediation strategies
- define the term *sick building syndrome* and describe the criteria used to determine whether a building has sick building syndrome

module 50

Indoor Air Pollution

When we think of air pollution, we usually don't associate it with air inside our buildings, but indoor air pollution actually causes more deaths each year than does outdoor air pollution. Most of these deaths occur in the developing world. The amount of time one spends indoors depends on culture, climate, and economic situation. The quality of indoor air is highly variable and when polluting activities take place indoors, exposure to pollutants in a confined space can be a significant health risk.

Learning Objectives

After reading this module you should be able to

- explain how indoor air pollution differs in developing and developed countries.
- describe the major indoor air pollutants and the risks associated with them.

Indoor air pollution is a significant hazard in developing and developed countries

Although it generally receives less attention than outdoor air pollution, indoor air pollution is a hazard all over the world. The reasons for indoor air pollution and its characteristics differ between the developing world and the developed world.

Developing Countries

In Chapter 13 we saw that around the world, more than 3 billion people use wood, animal manure, or coal indoors for heat and cooking. Biomass and coal are usually burned in open-pit fires that lack the proper mix of fuel and air to allow complete combustion. Usually, there is no exhaust system and little or no ventilation available in the home, which makes indoor air pollution from carbon monoxide and particulates a particular hazard in developing countries (FIGURE 50.1). Exposure to indoor air pollution from cooking and heating increases the risk of acute respiratory infections, pneumonia, bronchitis, and even cancer. The World Health Organization estimates that indoor air pollution is responsible for approximately 4 million deaths annually worldwide, and that

FIGURE 50.1 **Indoor air pollution in the developing world.** This photo shows a woman and her child in their home in Zimbabwe. *(Crispin Hughes/Panos Pictures)*

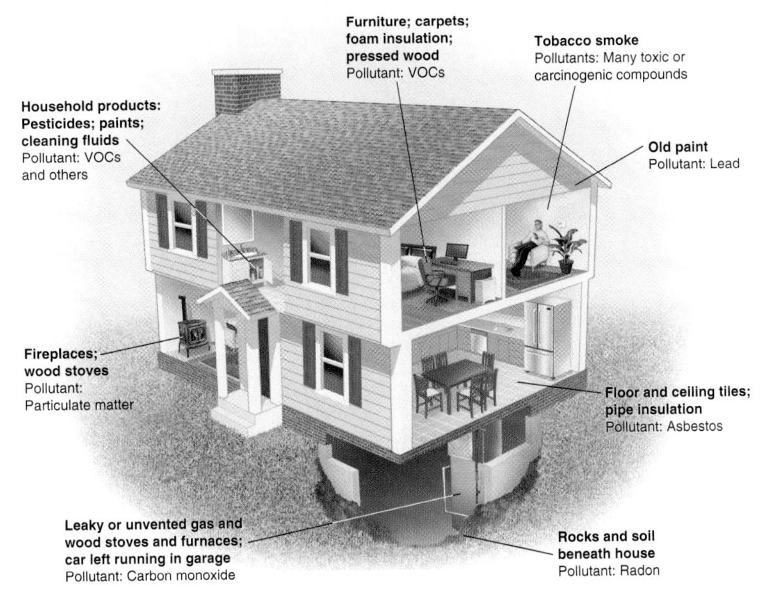

FIGURE 50.2 **Some sources of indoor air pollution in the developed world.** A typical home in the United States may contain a variety of chemical compounds that could, under certain circumstances, be considered indoor air pollutants. *(After U.S. EPA http://www.epa.gov/iaq/)*

more than 50 percent of those deaths occur among children less than 5 years of age. Ninety percent of deaths attributable to indoor air pollution are in developing countries.

Developed Countries

There are a number of factors that have caused the quality of air in homes in developed countries to take on greater importance in recent decades. First of all, people in much of the developed world have begun to spend more and more time indoors. Although improved insulation and tightly sealed building envelopes reduce energy consumption, these tightly sealed buildings also keep existing air in contact with the inhabitants of homes, schools, and offices for greater amounts of time. Finally, an increasing number of materials in the home and office are made from plastics and other petroleum-based materials that can give off chemical vapors. As FIGURE 50.2 shows, all of these factors combine to allow many possible sources of indoor air pollution to impact occupants.

Most indoor air pollutants differ from outdoor air pollutants

Because a house is a closed system with an abundance of manufactured materials, there is ample opportunity for indoor air pollutants to accumulate and for the occupants of that house to come into contact with harmful substances. Indoor air pollutants are for the most part different from outdoor pollutants, although, as we will see, carbon monoxide is one pollutant that causes problems both indoors and outdoors.

Carbon Monoxide

We have already described carbon monoxide as an outdoor air pollutant, but it can be even more dangerous as an indoor air pollutant. It occurs as a result of malfunctioning exhaust systems on household heaters, most typically natural gas heaters. When the exhaust system malfunctions, exhaust air escapes into the living space of the house. Because natural gas burns relatively

Teaching Tip: Activity

Ask students to draw a schematic of their house or apartment similar to the one in Figure 50.2. For homework, ask students to evaluate any potential indoor air pollution hazard and indicate it on the schematic.

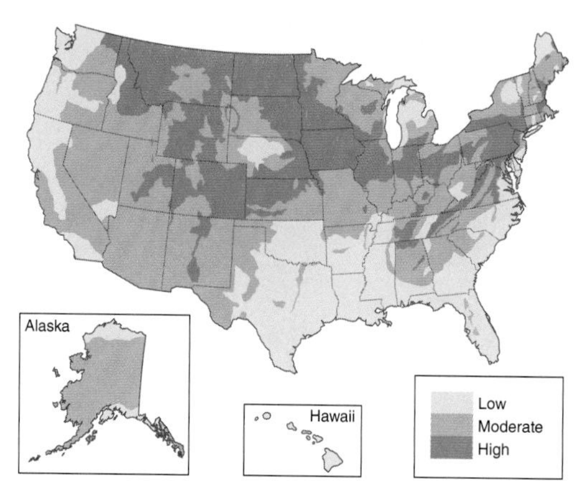

FIGURE 50.3 Potential radon exposure in the United States. Depending on the underlying bedrock and soils, the potential for exposure to radon exists in houses in certain parts of the United States. *(After U.S. Geological Survey and http://www.epa.gov/radon/zonemap.html)*

cleanly with little odor, a malfunctioning natural gas burner allows the colorless, odorless carbon monoxide to build up in a house without the occupants noticing, particularly if they are asleep. In the body, carbon monoxide binds with hemoglobin more efficiently than oxygen, thereby interfering with oxygen transport in the blood. Extended exposure to high concentrations of carbon monoxide in air can lead to oxygen deprivation in the brain and, ultimately, death.

Asbestos

Asbestos is a long, thin, fibrous silicate mineral with insulating properties. For many years it was used as an insulator on steam and hot water pipes and in shingles for the siding of buildings. The greatest health risks from asbestos have been respiratory diseases such as asbestosis and lung cancer found at very high rates among those who have mined asbestos. In manufactured form, asbestos is relatively stable and not dangerous until it is disturbed. When insulating materials become old or are damaged or disrupted, however, the fine fibers can become airborne and can enter the respiratory tract. In the United States, asbestos is no longer used as an insulating material, but it can still be found in older buildings, including schools. Removal of asbestos insulation must be done under tightly controlled conditions so that the fibers, typically less than 10 microns in diameter, cannot enter the air inside the building. Some studies have shown that when asbestos removal is complete, the concentration of asbestos in the air of the remediated building can be greater in the year after removal than during the year before removal. For this reason, it is absolutely necessary that asbestos removal be carefully done by qualified asbestos abatement personnel.

Asbestos A long, thin, fibrous silicate mineral with insulating properties, which can cause cancer when inhaled.

Radon

Radon-222, a radioactive gas that occurs naturally from the decay of uranium, exists in granitic and some other rocks and soils in many parts of the world. The map in FIGURE 50.3 shows areas for potential radon exposure in the United States. Humans can receive significant exposure to radon if it seeps into a home through cracks in the foundation, or from underlying rock, soil, or groundwater. Radon-222 decays within 4 days to a radioactive daughter product, Polonium-210. Either the radon or the polonium can attach to dust and other particles in the air and then be inhaled by the inhabitants of the home.

The EPA, the federal agency most responsible for identifying, measuring, and addressing environmental risks, estimates that about 21,000 people die each year from radon-induced lung cancer. This is 15 percent of yearly lung cancer deaths, and makes radon the second leading cause of lung cancer, after smoking. The EPA suggests that people test their homes for airborne radon. If radon levels are determined to be high, it is important to increase ventilation in the home. Other relatively inexpensive actions, such as sealing cracks in the basement, can be beneficial if radon is coming from underlying soil and bedrock.

VOCs in Home Products

Many volatile organic compounds are used in building materials, furniture, and other home products such

Teaching Tip: Activity

Ask students to compare asbestos and radon by filling in the table below. (Answers are provided in italics.)

Type of indoor pollutant	Appearance	Source	Prevention or remediation	Human health effects
Asbestos	*Long, thin, fibrous silicate mineral, micron-sized*	*Ceiling and floor tiles prior to the 1980s, insulation in older buildings*	*Professional encapsulation or removal*	*Respiratory diseases such as asbestosis and lung cancer*
Radon	*Invisible gas*	*Decaying uranium in granite, other rocks, and soils*	*Seal cracks and drains in the basement, install ventilation system*	*Lung cancer*

as glues and paints. One of the most toxic of these compounds is formaldehyde, which is used widely to manufacture a variety of building products such as particle board and carpeting glue. Formaldehyde is common in new homes and new products made from pressed wood, such as cabinets. The pungent smell that you may have noticed in a new home or one with new carpeting comes from formaldehyde, which is volatile and emits gases over time. A high enough concentration in a confined space can cause a burning sensation in the eyes and throat, and breathing difficulties and asthma in some people. There is evidence that people develop a sensitivity to formaldehyde over time; though they may not be very sensitive at first, with continued exposure they can experience irritation from ever-smaller exposures. Formaldehyde has been shown to cause cancer in laboratory animals and has recently been suspected of being a human carcinogen.

Many other consumer products such as detergents, dry-cleaning fluids, deodorizers, and solvents may contain VOCs and can be harmful if inhaled. Plastics, fabrics, construction materials, and carpets may also release VOCs over time.

Sick Building Syndrome

In newer buildings in developed countries in the temperate zone, more and more attention is given to insulation and prevention of air leaks in order to reduce the amount of heating or cooling necessary for a comfortable existence. This reduces energy use but may have the unintended side effect of allowing the buildup of toxic compounds and pollutants in an airtight space. In fact, such a phenomenon has been observed often enough in new or renovated buildings to be given a name: **sick building syndrome,** which describes a buildup of toxic pollutants in airtight spaces such as in newer buildings. Because new buildings contain many products made with synthetic materials and glues that may not have fully dried out, a significant amount of off-gassing occurs, which usually means that the indoor levels of VOCs, hydrocarbons, and other potentially toxic materials are quite high. Sick building syndrome has been observed particularly in office buildings, where large numbers of workers have reported a variety of maladies such as headaches, nausea, throat or eye irritations, and fatigue.

The EPA has identified four specific reasons for sick building syndrome: inadequate or faulty ventilation; chemical contamination from indoor sources such as glues, carpeting, furniture, cleaning agents, and copy machines; chemical contamination in the building from outdoor sources such as vehicle exhaust transferred through building air intakes; and biological contamination from inside or outside, such as molds and pollen.

Sick building syndrome A buildup of toxic pollutants in an airtight space, seen in newer buildings.

Teaching Tip: Review

Ask students the following questions:

- What are the main sources of indoor pollution in the developing world? *The main sources of indoor pollution in the developing world are the burning of wood, animal manure, or coal indoors for heat and cooking. Many homes in the developing world have no exhaust system so the carbon monoxide and particulate matter build up in the home.*
- List common sources of indoor air pollution in the developed world. *People in developed countries spend a lot of time indoors. New homes have improved insulation and their tightly sealed building envelopes reduce energy consumption. These tightly sealed buildings also keep existing air in contact with the inhabitants of homes, schools, and offices. Also, many materials in the home and office are made from plastics and other petroleum-based materials that can give off chemical vapors.*

module 50 REVIEW

In this module, we have seen that indoor air pollution is a problem around the world, but that its manifestations in the developing and the developed worlds differ. Carbon monoxide from combustion for cooking is the biggest problem in the developing world. In the developed world, houses sealed up for greater energy efficiency are often the causes of exposure to radon, carbon monoxide, and volatile organic compounds for people who spend substantial amounts of time indoors.

Module 50 AP® Review Questions

1. What percentage of worldwide deaths due to indoor air pollution occurs in developing nations?
   (a) 30 percent
   (b) 50 percent
   (c) 60 percent
   (d) 70 percent
   (e) 90 percent
2. Which outdoor air pollutant is also a significant indoor air pollutant?
   (a) Sulfur dioxide
   (b) Nitrogen oxides
   (c) Carbon monoxide
   (d) Lead
   (e) Ozone

Answers to Module 50 AP® Review Questions

1. e
2. c
3. d
4. c
5. a

Teaching Tip: Concept Map

Write the words "Sick Building Syndrome" in the center of a square on the board. Ask students to create a definition. When you are satisfied with the definition, ask your students to identify three specific reasons for sick building syndrome and create a concept map like the one shown on the right.

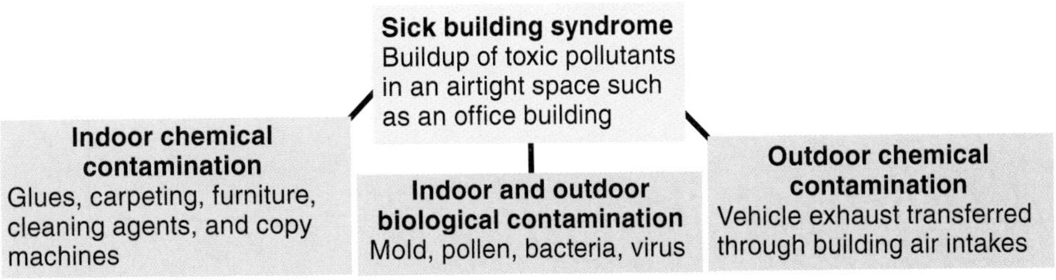

Sick building syndrome
Buildup of toxic pollutants in an airtight space such as an office building

Indoor chemical contamination
Glues, carpeting, furniture, cleaning agents, and copy machines

Indoor and outdoor biological contamination
Mold, pollen, bacteria, virus

Outdoor chemical contamination
Vehicle exhaust transferred through building air intakes

3. The primary source of radon is
   (a) electronics.
   (b) indoor fires.
   (c) household chemical fumes.
   (d) rocks and soils.
   (e) construction materials.

4. Sick building syndrome
   (a) occurs most often in old buildings.
   (b) is a primary cause of lung cancer.
   (c) is a result of off-gassing.
   (d) can be prevented by renovations.
   (e) occurs most often in wet tropical areas.

5. Asbestos
   (a) is used for insulation.
   (b) can be easily removed and treated.
   (c) can be a problem in new construction.
   (d) causes skin irritation, nausea, and fatigue.
   (e) is commonly used in furniture.

Teaching Tip: Beyond the Classroom

To learn more about the BioLite CampStove and how it works, have your students go to the BioLite website.

The link to this resource may be found by clicking on the link buttons in the TE-book, opening the TRFD, or logging onto the book's companion website bcs.whfreeman.com/friedlandapes2e (teacher login required).

Chapter 15 Web Resources

working toward  sustainability

A New Cook Stove Design

In China, India, and sub-Saharan Africa, people in 80 to 90 percent of households cook food using wood, animal manure, and crop residues as their fuel. Since women do most of the cooking, and young children are with the women of the household for much of the time, it is the women and young children who receive the greatest exposure to carbon monoxide and particulate matter. When biomass is used for cooking, concentrations of particulate matter in the home can be 200 times higher than the exposure limits recommended by the EPA. A wide range of diseases has been associated with exposure to smoke from cooking. Earlier in this chapter, we described that indoor air pollution is responsible for 4 million deaths annually around the world, and indoor cooking is a major source of indoor air pollution.

There are hundreds of projects underway around the world to enable women to use more efficient cooking stoves, ventilate cooking areas, cook outside whenever possible, and change customs and practices that will reduce their exposure to indoor air pollution. The use of an efficient cook stove will have the added benefit of consuming less fuel. This improves air quality and reduces the amount of fuel needed, which has environmental benefits and also reduces the amount of time that a woman must spend searching for fuel.

Increasing the efficiency of the combustion process requires the proper mix of fuel and oxygen. One effective method of ensuring a cleaner burn is the use of a small fan to facilitate greater oxygen delivery. However, because most homes in developing countries with significant indoor air pollution problems do not have access to electricity, some sort of internal source of energy for the fan is needed.

Two innovators from the United States developed a cook stove for backpackers and other outdoor enthusiasts who needed to cook a hot meal with little impact on the environment. They described their stove as needing no gasoline and no batteries, both desirable features for people carrying all their belongings on their backs. They soon realized that their stove, which could burn wood, animal manure, or crop residue, could make an important contribution in the developing world. This stove, called BioLite, physically separates the solid fuel from the gases that form when the fuel is burned and allows the stove to burn the gases. In addition, a small electric fan, located inside the stove, harnesses energy from the heat of the

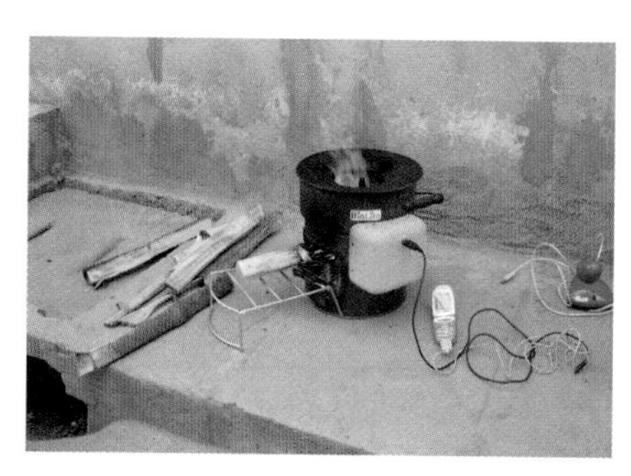

BioLite cookstove. This small stove, and others like it, has the potential to reduce the amount of firewood needed to cook a meal, and lower the amount of indoor air pollution emitted as well. *(Jonathan den Hartog; courtesy of Jonathan Cedar, www.BioLiteStove.com)*

fire and moves air through the stove at a rate that ensures complete combustion. The result is a more efficient burn, less fuel use, and less release of carbon monoxide and particulate matter. The stove weighs 0.7 kg (1.6 pounds).

How did the innovators manage to generate the electricity? They added a small semiconductor that generates electricity from the heat of the stove. All components of the stove except the semiconductor could be manufactured or repaired in a developing country. The BioLite stove won an international competition in early 2009 for the lowest emission stove. It was also the only stove in the competition that required no additional electricity inputs to operate. The BioLite stove is commercially available. One review of it stated that "it charges your phone while cooking your dinner."

There are many possible hurdles for those who are trying to introduce cleaner, more efficient cooking apparatus to the developing world. Manufacturing costs might make the stove difficult to afford for many. There has been some concern about possible reluctance to accept a different kind of cooking appliance. However, a number of studies in the developing world suggest that most households are quite receptive to using efficient stoves because of the benefits of improved air quality and reduced time spent obtaining fuel. Other promising ways to reduce fuel use and improve indoor air quality include the solar cooker shown in Figure 39.2 on page 451.

Critical Thinking Questions

1. Why are women and children often the ones most exposed to indoor air pollution in developing countries?
2. How can technology offer solutions to cooking over open fires?

References

Bilger, B. 2009. Annals of Invention, Hearth Surgery, *The New Yorker*, December 21, p. 84; http://www.newyorker.com/reporting/2009/12/21/091221fa_fact_bilger#ixzz0sMCn DR00.

www.biolitestove.com, homepage of BioLite stove.

Suggested Answers to Critical Thinking Questions

1. Women and children are more involved in cooking, and cook fires or stoves are the primary source of indoor air pollution in developing countries.

2. Several new products such as solar cookers and the BioLite stove offer alternatives to cooking over open fires.

chapter 15 REVIEW

In this chapter, we examined the major air pollutants and their natural and anthropogenic sources. We found that photochemical smog and acidic deposition are two air pollution problems that have had different outcomes, at least for now. Smog is still a problem in many locations around the world while acidic deposition has become less of a problem in North America and Europe. There are a variety of measures for controlling air pollution including pollution prevention and devices that remove pollutants from smokestacks before it is released into the atmosphere. Stratospheric ozone depletion has occurred because of the release of chlorofluorocarbons (CFCs) from refrigeration and air-conditioning units. Due to an international agreement, the Montreal Protocol on Substances That Deplete the Ozone Layer, there was a significant reduction in the use of CFCs and stratospheric ozone depletion has been reduced. Indoor air pollution is a problem that occurs around the world, although with causes and pollutants that differ between developing and developed countries.

Key Terms

Air pollution
Particulate matter (PM)
Particulates
Particles
Haze
Photochemical oxidant
Ozone (O_3)
Smog
Photochemical smog
Los Angeles–type smog
Brown smog
Sulfurous smog
London-type smog
Gray smog
Industrial smog
Volatile organic compound (VOC)
Primary pollutant
Secondary pollutant
Thermal inversion
Inversion layer
Asbestos
Sick building syndrome

Teaching Tip: Measuring Your Impact

Mercury Release from Coal
Measuring Your Impact 15 asks students to calculate mercury emissions from coal-burning power plants and examine ways that emissions can be reduced. The link to this resource may be found by clicking on the link buttons in the TE-book, opening the TRFD, or logging onto the book's companion website bcs.whfreeman.com/friedlandapes2e (teacher login required).

Measuring Your Impact 15:
Mercury Release from Coal

Exploring the Literature

Baird, C., and M. Cann. 2012. *Environmental Chemistry.* 5th ed. W. H. Freeman.

Manahan, S. E. 2009. *Environmental Chemistry.* 9th ed. CRC Press.

Ramanathan, V. M., et al. 2008. *Atmospheric Brown Clouds: Regional Assessment Report with Focus on Asia.* United Nations Environment Programme. U.S. Environmental Protection Agency. Section on Air: http://www.epa.gov/ebtpages/air.html.

Spiro, T. G., K. L. Purvis-Roberts, and W. M. Stifliani. 2012. *Chemistry of the Environment.* 3rd ed. University Science Books.

Download a printable version of this list by clicking on the link buttons in the TE-book, opening the TRFD, or logging onto the book's companion website bcs.whfreeman.com/ friedlandapes2e (teacher login required).

 **Exploring the Literature 15**

Learning Objectives Revisited

Module 46 Major Air Pollutants and Their Sources

- **Identify and describe the major air pollutants.**
  Sulfur oxides, nitrogen oxides, carbon monoxide, carbon dioxide, particulates, and ozone are some of the major ground-level air pollutants. Carbon dioxide was not originally considered one of the major pollutants but, because it alters ecosystems, it is now considered an air pollutant.

- **Describe the sources of air pollution.**
  Air pollution comes from both natural and human sources. Human activities that release these pollutants or their precursors include transportation, generation of electricity, space heating, and industrial processes.

Module 47 Photochemical Smog and Acid Rain

- **Explain how photochemical smog forms and why it is still a problem in the United States.**
  Smog forms when sunlight, nitrogen oxides, and volatile organic compounds are present. The secondary pollutant ozone is a major component of photochemical smog. Sulfur is the dominant ingredient in sulfurous smog. Smog impairs respiratory function in human beings. Because of an abundance of both nitrogen oxides and volatile organic compounds, smog occurs during daylight hours in many parts of the world, including the United States.

- **Describe how acid deposition forms and why it has improved in the United States and become worse elsewhere.**
  Acidic deposition, which is composed of hydrogen, sulfate, and nitrate ions, forms from both sulfur dioxide and nitrogen oxides. Acid deposition is harmful to aquatic organisms and can reduce forest productivity in sensitive ecosystems. Due to reductions in sulfur emissions in the United States, acidic deposition is much less of a problem than it used to be. Asia is a location today where acidic deposition is an environmental problem.

Module 48 Pollution Control Measures

- **Explain strategies and techniques for controlling sulfur dioxide, nitrogen oxides, and particulate matter.**
  Air pollution is best controlled by increasing the efficiency of processes that cause pollution, thereby reducing emissions, or by removing pollutants from fuel before combustion. After combustion occurs, filters and scrubbers remove pollutants from the exhaust stream before they can be released into the environment. The use of filters and scrubbers is preferable to trying to remove pollutants after they have been distributed throughout the environment.

- **Describe innovative pollution control measures.**
  Some innovative approaches to pollution control include targeting specific sources of certain kinds of pollution. Allowing for the buying and selling of permits to emit sulfur has been an effective means of reducing sulfur dioxide emissions and subsequent acidic deposition.

Module 49 Stratospheric Ozone Depletion

- **Explain the benefits of stratospheric ozone and how it forms.**
  Although ozone is the same gas that is a component of photochemical smog in the troposphere, in the stratosphere it is an important gas that absorbs harmful ultraviolet radiation. Its presence allows for plant and animal life to exist on the surface of Earth.

- **Describe the depletion of stratospheric ozone.**
  Chlorine-containing compounds such as CFCs that were part of refrigeration and air-conditioning systems lead to a reduction in stratospheric ozone. This reduction drew attention from scientists and the media around the world.

- **Explain efforts to reduce ozone depletion.**
  International efforts to reduce CFC emissions have helped stratospheric ozone levels to recover. The Montreal Protocol on Substances That Deplete the Ozone Layer is regarded as one of the most successful international agreements of modern times.

Module 50 Indoor Air Pollution

- **Explain how indoor air pollution differs in developing and developed countries.**
  Indoor air pollution is a different kind of problem in developing than in developed countries. Cooking over open fires with inadequate ventilation exposes occupants to carbon monoxide in the developing world. Relatively well-insulated living spaces in the developed world expose occupants to higher concentrations of carbon monoxide, radon, and certain volatile organic compounds.

- **Describe the major indoor air pollutants and the risks associated with them.**
  Carbon monoxide and particulates are the most harmful aspects of indoor air pollution and account for 4 million deaths annually worldwide. Fifty-six percent of those deaths occur among children less than 5 years of age. Cooking over open fires in the developing world exposes women and children to particulate matter and carbon monoxide pollution.

Chapter 15 AP® Environmental Science Practice Exam

Section 1: Multiple-Choice Questions

Choose the best answer for questions 1–19.

Questions 1–4 refer to the selections (a)–(e) below. Match the lettered item with the numbered descriptor.

(a) CO
(b) CH_4
(c) NO_2
(d) SO_2
(e) PM_{10}

1. A pungent reddish-brown gas often associated with photochemical smog
2. A corrosive gas from burning coal often associated with industrial smog
3. A dangerous indoor air pollutant
4. Emitted from both diesel and burning wood
5. All of the following are examples of primary air pollutants except
   (a) sulfur dioxide.
   (b) carbon dioxide.
   (c) tropospheric ozone.
   (d) nitrogen oxide.
   (e) particulates.
6. The greatest emission of sulfur dioxide comes from
   (a) on-road vehicles.
   (b) biofuels.
   (c) industrial processes.
   (d) electricity generation.
   (e) fires.
7. The least amount of nitrogen oxide emissions comes from
   (a) on-road vehicles.
   (b) fossil fuel combustion.
   (c) industrial processes.
   (d) electricity generation.
   (e) fires.
8. The accumulation of tropospheric ozone at night depends mainly upon the atmospheric concentration of
   (a) nitrogen dioxide.
   (b) volatile organics.
   (c) chlorofluorocarbons.
   (d) sulfates and nitrates.
   (e) nitric acid.
9. Under natural conditions the pH of rainfall is closest to
   (a) 8.5.
   (b) 7.1.
   (c) 5.6.
   (d) 4.5.
   (e) 3.1.
10. The effects of acid deposition include all of the following except
    (a) mobilization of metal ions from the soil into surface water.
    (b) increased numbers of salamanders in ponds and streams.
    (c) reduced food sources for aquatic organisms.
    (d) erosion of marble buildings and statues.
    (e) erosion of painted automobile finishes and metals.
11. The World Health Organization estimates that over half of the deaths worldwide due to indoor air pollution occur among
    (a) children less than 5 years old.
    (b) elderly people over 65 years of age.
    (c) people who work in office buildings.
    (d) workers in the smelting industry.
    (e) workers who manufacture asbestos.
12. Two major factors involved in the conversion of primary pollutants into secondary pollutants are
    (a) sunlight and water.
    (b) sulfates and sunlight.
    (c) water and volatile organics.
    (d) nitrogen oxides and sulfates.
    (e) sulfur dioxide and sulfuric acid.
13. The pollutant least likely to be emitted from a smokestack would be
    (a) carbon monoxide.
    (b) carbon dioxide.
    (c) ozone.
    (d) sulfur dioxide.
    (e) particulates.

Answers to Chapter 15 AP® Exam Multiple-Choice Questions

1. c
2. d
3. a
4. e
5. c
6. d
7. e
8. b
9. c
10. b
11. a
12. a
13. c
14. c
15. e
16. b
17. b
18. c
19. d

Answers to Chapter 15 AP® Exam

Download the full answers to the Chapter 15 AP® Exam by clicking on the link buttons in the TE-book, opening the TRFD, or logging onto the book's companion website bcs.whfreeman.com/friedlandapes2e (teacher login required).

Answers to Chapter 15 AP® Exam

Additional Multiple-Choice AP® Practice Exam

Download an additional Multiple-Choice AP® Environmental Science Practice Exam for Chapter 15 by clicking on the link buttons in the TE-book, opening the TRFD, or logging onto the book's companion website bcs.whfreeman.com/friedlandapes2e (teacher login required).

Multiple-Choice AP® Practice Exam 15

14. How might the increased use of insulation in a home affect air pollution?
    (a) Some common insulation off-gasses sulfur dioxide.
    (b) The production of insulation releases large amounts of particulate matter.
    (c) Increased insulation can create indoor air pollution problems.
    (d) Insulation prevents problems with radon in homes.
    (e) Good insulation can help prevent sick building syndrome.

15. The EPA identifies all of the following as reasons for sick building syndrome except
    (a) faulty ventilation systems.
    (b) emissions from carpets and furniture.
    (c) contamination from outdoor air.
    (d) contamination from molds and pollen.
    (e) high levels of radon in the basement.

16. Which statement regarding the decreased levels of stratospheric ozone is correct?
    (a) Increased photosynthetic activity has been measured in the phytoplankton around Antarctica.
    (b) The largest decrease in the level of stratospheric ozone over the Arctic region occurs between January and April.
    (c) Although the Montreal Protocol led to a reduction in the use of CFCs, it will have little effect on stratospheric ozone levels in the long term.
    (d) There is no correlation between the incidence of cataracts and skin cancers and the lower levels of stratospheric ozone.
    (e) The global crop yields of wheat, rice, and corn have increased since the reduction in CFC use.

17. Which is most strongly absorbed by ozone in the stratosphere?
    (a) UV-A
    (b) UV-B
    (c) UV-C
    (d) UV-B and UV-C
    (e) UV-A and UV-B

18. Natural sources of air pollution include all of the following except
    (a) forest fires.
    (b) volcanic eruptions.
    (c) waterfalls.
    (d) dust storms.
    (e) conifer trees.

19. A thermal inversion
    (a) rarely occurs in cities but is common in rural areas.
    (b) helps remove pollutants from the atmosphere.
    (c) leads to decreased amounts of ground-level smog.
    (d) occurs when a warm air layer overlies a cooler layer.
    (e) occurs when a cool air layer overlies a warmer layer.

550 Chapter 15 Air Pollution and Stratospheric Ozone Depletion

Section 2: Free-Response Questions

Write your answer to each part clearly. Support your answers with relevant information and examples. Where calculations are required, show your work.

1. The table below shows the ambient air data collected for Pittsburgh, Pennsylvania. Examine the data and answer the following questions.
   (a) Based on the National Ambient Air Quality Standards (NAAQS), the 2008 standard for ozone states that the average ozone levels are not to exceed 0.075 ppm (75 ppb) in any 8-hour period. Is Pittsburgh in compliance with this standard? Discuss how this NAAQS may not truly reflect the overall air quality. (2 points)
   (b) Ozone is classified as a secondary pollutant. Identify the primary pollutants necessary for its formation and describe how tropospheric ozone is formed. (2 points)
   (c) Identify two relationships between the data presented. Apply these relationships to your answer in (b) to explain the pattern of ozone levels in Pittsburgh. (4 points)
   (d) Explain how the same ozone that is harmful in the troposphere is beneficial in the stratosphere. (2 points)

2. The following appeared in *Medical News Today* on November 28, 2003.

Ultraviolet light in ventilation system cures sick building syndrome

So-called sick building syndrome (SBS) is caused by the increasing use of air conditioning in modern offices. Bacteria can build up in the ventilation systems and leave office workers suffering from a range of problems, such as breathing difficulties, headaches, sore throats, stuffy noses and itchy eyes....

Researchers from the Montreal Chest Institute in Canada tested three offices in the city where people suffered from SBS. They installed UVGI (ultraviolet germicidal irradiation) in the ventilation systems of the buildings and compared reports of illness when the lights were turned on with sickness rates when they were off.

The lights resulted in a 20 percent reduction in all symptoms. There were 40 percent fewer complaints about respiratory problems and a 30 percent reduction in people complaining of stuffy noses. Muscular complaints also halved, according to the study.

Scientists said the UVGI killed the bacteria and molds in ventilation systems. Installing the systems would be relatively cheap and could save companies millions by reducing rates of sickness...

Source: *Excerpted from http://www.medicalnewstoday.com/releases/4757.php*

   (a) According to the article, what is the cause of sick building syndrome and what are the advantages of this cure? (3 points)
   (b) What other sources of indoor pollution contribute to sick building syndrome? Will ultraviolet light be effective against these sources? Suggest one additional control measure that would be effective against SBS. (4 points)
   (c) Will the cure mentioned in the article be effective against indoor pollution in developing countries? Discuss why or why not. (3 points)

2008 Monthly Ambient Air Monitoring Report, Pittsburgh, PA

Month	Monthly maximum ozone levels (ppb)	Monthly average ozone levels (ppb)	Monthly average solar radiation (watts/m^3)
January	37	14	65
February	49	15	63
March	56	23	86
April	76	31	81
May	75	27	152
June	77	32	208
July	95	31	215
August	92	27	204
September	89	20	153
October	48	14	109
November	57	12	64
December	30	14	45

Source: *http://www.ahs.dep.pa.gov/aq_apps/aadata/*

Answers to the Chapter 15 Free-Response Questions can be found in the Answer Appendix on page ANS-12.

chapter

16 Waste Generation and Waste Disposal

PD Overview

Chapter 16 engages students in a topic that is probably somewhat familiar to them: solid waste. Humans are the only organisms to generate waste that other organisms cannot use. The chapter begins by looking at how humans generate waste. Total municipal solid waste (MSW) and per capita waste generation in the United States have recently started to decrease. The chapter moves from how humans generate waste to ways in which we can divert material from the solid waste stream: reduction, reuse, recycling, and composting. The chapter also looks closely at the environmental costs and benefits of landfills and incineration. Hazardous waste is a special category of waste that must be handled and disposed of with particular care. National legislation that addresses hazardous waste includes the U.S. Resource Conservation and Recovery Act (RCRA) and the Comprehensive Environmental Response, Compensation, and Liability Act (CERCLA). The chapter continues with a detailed discussion of life-cycle analysis, which uses a holistic approach to study the entire waste stream from the creation of materials through their use and ultimate disposal. Finally, the chapter explores integrated waste management, which offers multiple approaches for reducing the waste stream.

Module 51: Only Humans Generate Waste

This module begins the discussion of solid waste by considering the inputs and outputs of materials that end up becoming MSW. It defines solid waste and examines the contents of the waste stream and waste generation trends. Only a fraction of MSW is reused, recycled, or composted. Of the 227 million metric tons of MSW that could be disposed of, a little more than one-third ends up being diverted from landfills for reuse, recycling, or composting. The module concludes with an examination of electronic waste and the specific challenges that such waste raises. E-waste is a relatively small component of MSW, but because of the toxicity of the metals it often contains, its disposal is a significant problem.

Module 52: The Three Rs and Composting

Almost every schoolchild in the United States has heard the phrase "reduce, reuse, and recycle." In recent years, composting has been added to the list of actions one should take before adding material to the waste stream. This module examines the three Rs and composting. It is important to observe the "three Rs" in order—reduce, reuse, and recycle—because each action uses less energy than the next action. Composting is another pathway for diverting material from the waste stream. The diversion of compostable materials from the landfill reduces the production of methane gas and produces a desirable end product, decomposed organic matter.

Module 53: Landfills and Incineration

Beginning in the 1930s in the United States, as public opposition to open dumps was growing, the most convenient locations for disposal of MSW became holes in the ground created by the removal of soil, sand, or other earth material that was then used for construction purposes. When people filled those holes with waste, the sites became known as landfills. Though open dumps are now rare in developed countries, they still exist in the developing world, where they pose a considerable health hazard. This module examines the basic design and construction of landfills and incinerators in the United States. Landfills are designed to keep MSW dry and isolated so that the material in the landfill does not contaminate the surrounding environment. Leachate can contaminate nearby land and waterways if it escapes. Anaerobic decomposition in landfills can lead to the formation of methane, a potent greenhouse gas. This module also examines incineration as an alternative to landfills. Incinerators can generate a variety of air pollutants. Any metals contained within incinerated MSW may be released into the air. Bottom ash is particulate matter that accumulates underneath the combustion furnace. Sometimes ash from incinerators can be considered toxic and must be removed using special precautions.

Module 54: Hazardous Waste

When solid waste material is deemed toxic or otherwise harmful to people or natural ecosystems, it should not be placed in a MSW landfill or incinerated. In these cases, special means of handling and disposal are required. This module discusses the proper treatment of hazardous waste, much of which is composed of byproducts from industrial processes. In these cases, special means of handling and disposal are required. This module also looks at some of the legislation concerning hazardous waste including RCRA and CERCLA. In addition, the module considers international consequences of shipping hazardous waste to other countries.

Module 55: New Ways to Think About Solid Waste

Each method of managing solid waste has both benefits and drawbacks. Because every method of waste disposal will have adverse environmental effects, the challenge is to find the least detrimental option. Because there is no obvious best method and because waste is a pervasive fact of contemporary life, the problem of waste disposal seems overwhelming. This module demonstrates how lifecycle analysis and a holistic approach can help us determine what we should do with our solid waste. It also explores how integrated waste management can reduce the volume of waste entering the waste stream.

Alignment to AP® Environmental Science Course Description

AP® Outline	Module
VI. Pollution (25–30%)	
A. Pollution Types	Module 51 Only Humans Generate Waste
	Module 52 The Three Rs and Composting
	Module 53 Landfills and Incineration
	Module 54 Hazardous Waste
	Module 55 New Ways to Think About Solid Waste
B. Impacts on Environment and Human Health	Module 53 Landfills and Incineration
	Module 54 Hazardous Waste
C. Economic Impacts	Module 55 New Ways to Think About Solid Waste

Chapter Learning Objectives

After completing this chapter students will be able to

- explain why we generate waste and describe recent waste disposal trends.
- describe the content of the solid waste stream in the United States.
- describe the three Rs.
- understand the process and benefits of composting.
- describe the goals and function of a solid waste landfill.
- explain the design and purpose of a solid waste incinerator.
- define hazardous waste and discuss the issues involved in handling it.
- describe regulations and legislation regarding hazardous waste.
- explain the purpose of life-cycle analysis.
- describe alternative ways to handle waste and waste generation.

Chapter 16 Pacing Guide

This pacing guide is based on a schedule with 120 sessions of 50 minutes each before the AP® Exam. If you have a different number of sessions before the exam, you can modify the pacing to suit your needs. If you have additional time, consider incorporating quizzes, released AP® Environmental Science free-response and multiple-choice questions, or additional activities.

Module	Standard Schedule Days	Block Schedule Days
Module 51	1	½
Module 52	1	½
Module 53	1	½
Module 54	1	½
Module 55	1	½
Assessment	1	½

Chapter 16 Resources

The link to these resources can be found by clicking on the link buttons in the Teacher's e-Book (TE-book), opening the Teacher's Resource Flash Drive (TRFD), or logging onto the book's companion website bcs.whfreeman.com/friedlandapes2e (teacher login required).

- **PowerPoint Presentation 16**
- **Optimized Art PowerPoint and JPEG Files 16**
- **Do the Math Videos**
- **More Math Practice 16.1:** Graphing MSW Generation, Recycling, and Disposal
- **Measuring Your Impact 16:** Understanding Household Solid Waste
- **Lab 16.1:** What's in Our Trash?
- **Chapter 16 Web Resources**
- **Exploring the Literature 16**
- **Answers to Chapter 16 AP® Exam**
- **Multiple-Choice AP® Practice Exam 16**

Free-Response Questions from Previous AP® Environmental Science Exams

Free-response questions from prior AP® Environmental Science Exams are available on the AP® course website: https://apstudent.collegeboard.org/apcourse/ap-environmental-science/exam-practice. Students should be able to answer all of the questions listed below with material learned in this and previous chapters. When a question requires students to understand material from multiple chapters, the question will be listed in the last chapter required to complete the entire question. Questions marked with an asterisk (*) are from exams with released multiple-choice questions. You may want to save these questions until the end of the year so you can give your students a complete released exam for practice. Questions marked with double asterisks (**) require math calculations. Look for references to these questions throughout the chapter.

Year	Question	Content
2000	2	• Evaluating the benefits and costs of recycling newspaper • Comparing recycling of aluminum and recycling of newspaper • Challenges of implementing a community recycling program
2006	3	• Decontamination of a hazardous waste site • Social and environmental benefits of brownfield reclamation • Reduction and disposal of hazardous waste
2008*	2**	• Calculation of yearly water volume flowing into a landfill and leachate treated • Calculation of cadmium released yearly from a landfill • Cost of treating leachate • Strategies for reducing cadmium flowing into a landfill

PD **Chapter 16 Overview**

Watch the video overview of Chapter 16 (for teachers) by clicking on the link buttons in the TE-book, opening the TRFD, or logging onto the book's companion website bcs.whfreeman.com /friedlandapes2e (teacher login required).

TRM **PowerPoint Presentation 16**

Download the PowerPoint presentation for Chapter 16 by clicking on the link buttons in the TE-book, opening the TRFD, or logging onto the book's companion website bcs.whfreeman.com /friedlandapes2e (teacher login required).

An earthmover prepares solid waste for compaction and capping in a landfill in Boise, Idaho. *(David R. Frazier/DanitaDelimont.com)*

chapter 16

Waste Generation and Waste Disposal

Paper or Plastic?

Polystyrene is a plastic polymer that has high insulation value. More commonly known by its trade name, Styrofoam, it is particularly useful for food packaging because it minimizes temperature changes in both food and beverages. Polystyrene is lighter, insulates better, and is less expensive than the alternatives. However, a number of years ago, polystyrene was deemed harmful to the environment because, like all plastics, it is made from petroleum and because it does not decompose in landfills. In response to public sentiment, most food businesses greatly reduced or eliminated their use of polystyrene. All over the country, schools, businesses, and public institutions have purged their cafeterias of polystyrene cups and most have replaced them with disposable paper cups.

> Today, there is still no definitive answer as to whether the paper cup or the polystyrene cup causes less harm to the environment.

But was the elimination of polystyrene actually an environmental victory? It is hard to quantify the exact environmental benefits and costs of using a paper cup versus using a polystyrene cup. For example, because a paper cup does not insulate as well as a Styrofoam cup, paper cups filled with hot drinks are usually too hot to hold and vendors often wrap them in a cardboard band that becomes additional waste. To fully quantify the environmental costs and

Teaching Tip: Chapter Opening Case

The chapter opening case, "Paper or Plastic?" reinforces the idea that there are no easy answers when it comes to solid waste and recycling. Although students may be familiar with some of the concepts in this chapter and have ideas about the benefits of recycling, for many students this will be the first time they approach the issues of solid waste in a scientific way. Use the opening case to emphasize that waste generation and waste disposal are subjects that require all the tools of environmental science.

Teaching Tip: Activity

Before reading the opening case, ask students which they believe is better for the environment: a paper cup or polystyrene cup. Have students create a chart listing the negatives of polystyrene and paper. (Suggested answers are provided in italics.) After the exercise, ask students to make a choice between polystyrene and paper. Then read the chapter opening case and ask students if they have changed their position.

Polystyrene cup	Paper cup
• *Does not decompose in landfills* • *Production uses 3 grams of petroleum (but no wood)* • *May be used many times but is usually thrown away after one use* • *May leach chemicals into coffee and harm human health* • *Toxic emissions are created in the production process*	• *Additional paper must be used for insulation* • *Production requires 2 grams of petroleum and 33 grams of wood* • *Uses twice as much energy and much more water for production* • *Heavier, so requires more energy to transport* • *Because more energy is used, more air pollution is created* • *Used only once or twice* • *May be composted but is usually thrown away* • *Bleach used in manufacturing harms aquatic life* • *Will degrade in a landfill and produce methane*

benefits of each type of cup, one must create a list of inputs and outputs related to their manufacture, use, and disposal. This input-output analysis of all energy and materials is also called a *cradle-to-grave,* or *life-cycle, analysis.* When we make a list of all the materials and all the energy required to produce and then dispose of each type of cup, we find that it is not easy to determine which choice is better for the environment.

One study found that making a paper cup requires approximately 2 grams of petroleum along with 33 grams of wood and bark, which are renewable materials. A polystyrene cup requires 3 grams of petroleum, a nonrenewable material, but no wood or bark. Manufacturing the paper cup requires about twice as much energy, and much more water. A paper cup of the exact same size as a polystyrene cup is substantially heavier, which means it requires more energy to transport a paper cup to the location where it will be used. Air emissions are different in the manufacturing of each type of cup and it is difficult to say which emissions are more harmful to the environment. Since more energy is needed to make and transport a paper cup, it is reasonable to assume that using it generates more air pollution. A paper cup is normally used once or at most a few times while the polystyrene cup can, at least in theory, be reused many times. However, both types of cups are usually thrown away after one use. There has been concern among some scientists—but no consensus—that a polystyrene cup might leach chemicals from the plastic into the coffee; if this is true, using a paper cup could pose less risk to human health. However, without proper disposal, the bleach used to make paper cups in a paper mill, along with small amounts of the associated by-product, dioxin, can cause harm to aquatic life when the water is discharged into rivers and streams. Incineration of both types of cup could yield a small amount of energy. In a landfill, the paper cup will degrade and eventually produce methane gas, while the polystyrene cup, because it is made of an inert material, will remain there for a very long time.

Weighing these and other factors, one study concluded that a polystyrene cup is more desirable than a paper cup for one-time use. However, critics of that study felt that the author did not consider the toxicity of emissions from making polystyrene, the exposure of workers to those emissions, the impact of both cups on global carbon dioxide emissions, or the possibility of making the cup from materials other than petroleum or paper. Today, there is still no definitive answer as to whether the paper cup or the polystyrene cup causes less harm to the environment.

These types of studies illustrate that analyzing the environmental effects of the products we use is complex since it involves the synthesis of many aspects of environmental studies. Not only does it include science, ethics, and social judgments, it also necessitates a systems-based understanding of waste generation, waste reduction, and waste disposal. There is widespread agreement that paper and Styrofoam are not the only options. Reusable mugs are a possibility but they would require consideration of a host of different life-cycle issues such as greater inputs for manufacturing, and energy consumption, as well as the water and other resources needed to clean them after each use.

Sources:
M. B. Hocking, Paper versus polystyrene: A complex choice, *Science* 251 (1991): 504–505, DOI: 10.1126/science.251.4993.504; M. Brower and L. Warren. *The Consumer's Guide to Effective Environmental Choices* (Three Rivers Press, 1999).

As life in many countries has become increasingly dependent on disposable items, the generation of solid waste has become more of a problem for both the natural and human environments. In this chapter, we examine solid waste, something that only humans generate. We examine methods of reducing waste and ways of recycling. Then we describe the principal methods of getting rid of waste: landfills and incineration. We also consider the most toxic forms of waste: hazardous waste. We conclude the chapter with a discussion of innovative ways to think about solid waste from a systems perspective.

module

51

Only Humans Generate Waste

Throughout this book, we have examined systems in terms of inputs, outputs, and internal changes. We will begin our discussion of solid waste in the same way—by considering the inputs and outputs of materials that end up becoming solid waste. We will define solid waste and examine the contents of the waste stream and waste generation trends. We will conclude the module with an examination of electronic waste and the specific challenges that it raises.

Learning Objectives

After reading this module, you should be able to

- explain why we generate waste and describe recent waste disposal trends.
- describe the content of the solid waste stream in the United States.

Humans generate waste that other organisms cannot use

Humans are the only organisms that produce waste others cannot use. To explore this further, we need to learn why materials generated by humans become waste and what that waste contains. Although this seems like a trivial and simple question, it touches upon recent U.S. history, human behavior, and many other topics, including some that go beyond environmental science. We need to establish that waste can be viewed as a system, just like other materials. And it is necessary to describe how we got here—what features of U.S. society allowed us to generate the quantities of waste that we do.

Waste as a System

In an ecological system, plant materials, nutrients, water, and energy are the inputs. In a human system, inputs are very similar but contain materials manufactured by humans as well as natural materials. Within this system, humans use these inputs and materials to produce goods. And, as in any system, outputs are generated. We call these outputs **waste,** which is defined as material outputs that are not useful or are not consumed. Energy waste is also an output. FIGURE 51.1 shows a diagram of the relationship between inputs and outputs in a human system.

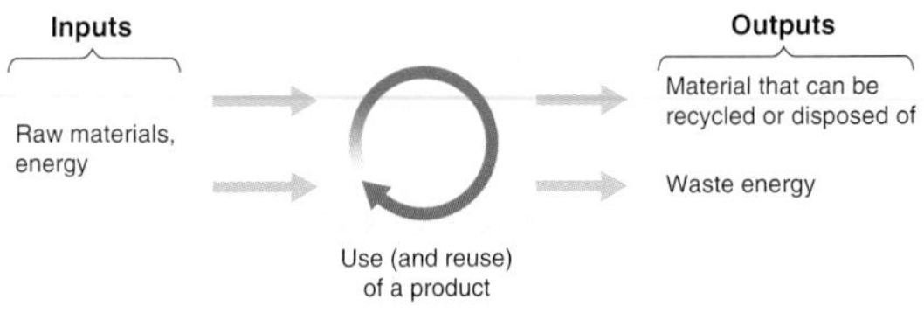

Inputs

Raw materials, energy

Outputs

Material that can be recycled or disposed of

Waste energy

Use (and reuse) of a product

FIGURE 51.1 **The solid waste system.** Waste is a component of a human-dominated system in which products are manufactured, used, and eventually disposed of (arrows are not proportional). At least some of the waste of this system may become the input of another system.

Waste Material outputs from a system that are not useful or consumed.

Teaching Tip: Video

The Story of Stuff
This 20-minute video describes the inputs and outputs of production, use, and disposal systems in developed nations.

The link to this video may be found by clicking the link buttons in the TE-book, opening the TRFD, or logging onto the book's companion website bcs.whfreeman.com/friedlandapes2e (teacher login required).

 **Chapter 16 Web Resources**

▲ TEACHING with FIGURES

Show students Figure 51.1 and note the direction of the green arrows. Ask students to predict a shortcoming of this system with regard to sustainability. Ask them to describe the direction of the blue arrow, and how following that pathway may reduce the use of limited resources. *The green arrows are linear. If the raw material inputs are always converted to waste, then our limited resources may be depleted. The blue arrow is circular and suggests that products will be reused. If more resources are reused rather than discarded, fewer raw materials and less energy will be needed and less waste will be created.*

If waste is the nonuseful output of a system, how do we determine what is useful? The detritivores we described in Chapter 3 recycle waste from animals and plants; they use the energy and nourishment they obtain and turn the remainder into compost or humus that nourishes other organisms. Dung beetles, for example, live on the energy and nutrients contained within the dung of elephants and other animals; in the natural world, this is not waste, it is food (FIGURE 51.2). Even humans make use of animal waste—for fertilizer, heat, and cooking fuel. In most situations, the waste of one organism becomes a source of energy for another.

FIGURE 51.2 **A dung beetle.** This dung beetle is using elephant waste as a resource. The waste of most organisms in the natural world ends up being a resource for other organisms. *(john michael evan potter/Shutterstock)*

The Throw-Away Society

Until a society becomes relatively wealthy, it generates little waste. Every object that no longer has value for its original purpose becomes useful for another purpose. In 1900 in the United States, virtually all metal, wood, and glass materials were recycled, although no one called it recycling back then. Those who collected recyclables were called junk dealers, or scrap metal dealers. For example, if a wooden bookcase broke and it could not be repaired, the pieces could be used to make a step stool. When the step stool broke, the wood was burned in a wood stove to heat the house. After World War II and with the rapid population growth that occurred in the United States, consumption patterns changed. The increasing industrialization and wealth of the United States, as well as cultural changes, made it possible for people to purchase household conveniences that could be used and then thrown away. Families were large, and people were urged to buy "labor-saving" household appliances and to dispose of them as soon as a new model was available. Planned obsolescence—designing a product so that it will need to be replaced within a few years—became a typical characteristic of everything from toasters to cars. TV dinners, throw-away napkins, and disposable plates and forks became common. In the 1960s disposable diapers became widely available and eventually replaced reusable cloth diapers. The components of household materials also changed. Objects were more likely to contain mixtures of different materials, which makes them harder to use for another purpose and difficult to recycle. The United States became the leader of what came to be known as the "throw-away society."

Refuse collected by municipalities from households, small businesses, and institutions such as schools, prisons, municipal buildings, and hospitals is known as **municipal solid waste (MSW).** The Environmental Protection Agency (EPA) estimates that approximately 60 percent of MSW comes from residences and 40 percent from commercial and institutional facilities. Other kinds of waste generated in the United States in addition to MSW include agricultural waste, mining waste, and industrial waste. Waste other than MSW is typically deposited and processed on-site rather than transferred to a different location for disposal. Although some of these other categories generate a much greater percentage of yearly total solid waste, this chapter focuses on MSW.

Municipal solid waste (MSW) Refuse collected by municipalities from households, small businesses, and institutions.

FIGURE 51.3 shows the trend toward greater generation of MSW both overall and on a per capita basis from 1960 to 2011. In the first 47 years of this period, the total amount of MSW generated in the United States increased from 80 million metric tons (88 million U.S. tons) to 227 million metric tons (250 million U.S. tons) per year. In the last several years for which there are data, the total amount of MSW actually decreased by a small amount. The increase for all but the last several years can be explained in part because of population growth and in part because individuals have been generating increasing amounts of MSW. In 2011, average waste generation was 2.0 kg (4.4 pounds) of MSW per person per day. Waste generation varies by season of the year, socioeconomic status of the individual generating the waste, and even geographic location within the country.

Waste generation in much of the rest of the world stands in contrast to the United States. In Japan, for example, each person generates an average of 1.1 kg (2.4 pounds) of MSW each day. The 2010 UN-HABITAT estimate for the developing world is 0.55 kg (1.2 pounds) per person per day. The estimate for the developed world ranges from 0.8 to 2.2 kg (1.8–4.8 pounds) per person per day. Some indigenous people create virtually no waste per day, with as much as 98 percent of MSW being used for something by

Teaching Tip: Journal Prompt

Ask students the following questions:

- What does it mean to be a throw-away society? *A throw-away society uses products that are made to be used for a short time and then discarded.*
- How does planned obsolescence contribute to waste generation? *Planned obsolescence perpetuates increased waste generation because products are designed so that they will need to be replaced within a few years. Greater amounts of materials are added to the waste stream with planned obsolescence.*

Teaching Tip: Discussion Starter

Ask students: What are the main sources of waste in the United States? *The main sources of waste in the United States are MSW, agriculture, mining, and industry.*

Teaching Tip: Warm-up

Ask students to identify the sources of MSW. *The sources of MSW include refuse collected by municipalities from households, schools, small businesses, municipal buildings, prisons, and hospitals.*

Teaching Tip: Engage

Ask students to suggest some reasons why MSW generation has started to decrease in recent years. *Answers will vary but may include increased public opposition to unnecessary packaging and more reuse of items.*

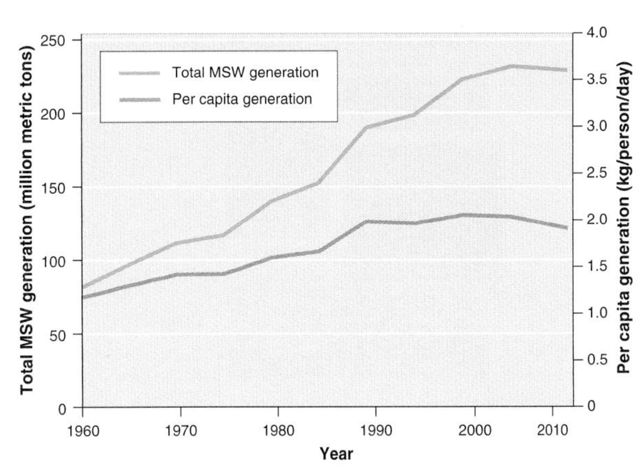

FIGURE 51.3 **Municipal solid waste generation in the United States, 1960–2011.** Total MSW generation and per capita MSW generation had been increasing from 1960 through 2008. They have recently started to decrease. *(Data from U.S. Environmental Protection Agency,* MSW Generation, Recycling and Disposal in the United States: Facts and Figures for 2011)

FIGURE 51.4 **A large dump in Manila, Philippines.** Throughout the world, impoverished people scavenge dumps. *(Digital Vision)*

someone. The remaining 2 percent ends up in a landfill or waste pile. Even there, impoverished people scavenge and reuse some of the discarded material (FIGURE 51.4).

Developing countries have become responsible for a greater portion of global MSW, in part because of their growing populations. In addition, as developing countries produce more of the goods used in the developed world, they generate more waste in the production process for these goods. For example, computers sold to consumers in the United States are assembled in such places as Taiwan, Singapore, and China, and the waste products generated are disposed of at the manufacturing location.

The solid waste stream contains materials from many sources

We have seen that MSW is made up of the things we use and then throw away. The goods that we use are generally a combination of organic items, fibers, metals, and plastics made from petroleum. A certain amount of waste is generated during any manufacturing process. Waste is also generated from the packaging and transporting of goods.

Depending on the particular materials, products, and goods that consumers use, such items can remain in the consumer-use system for a long time. For example, a ceramic plate or drinking mug might last

◀ TEACHING with FIGURES

Show students Figure 51.3 and ask them to describe the overall trend in MSW generation in the United States. To answer the question they should compare the total MSW generation with per capita generation. *Both total MSW generation and per capita MSW generation increased until 2008. More recently, both have started to decrease. Per capita generation began to level off and decline before total MSW generation began to decline.*

lab
What's in Our Trash?

This lab offers students the opportunity to see the types and amounts of waste they generate in a day.

Download this resource by clicking on the link buttons in the TE-book, opening the TRFD, or logging onto the book's companion website bcs.whfreeman.com /friedlandapes2e (teacher login required).

 **Lab 16.1** What's in Our Trash?

TEACHING with FIGURES ▶

Show students Figure 51.5 and ask them the following questions:

- What materials comprise MSW? Identify them from the largest percentage to the smallest. *The following materials comprise MSW, from the largest to the smallest percentage: paper; food scraps; yard waste; plastics; metals; rubber, leather, and textiles; wood; glass; and other materials.*
- Which group of materials could be removed from the waste stream most easily and would make the greatest impact on MSW reduction? *Compostable materials, which include yard trimmings, wood, and food scraps, could be removed and composted rather than being disposed of in a landfill. Removal of these components could reduce MSW by 34 percent.*

Total: 227 million metric tons (250 million US tons)

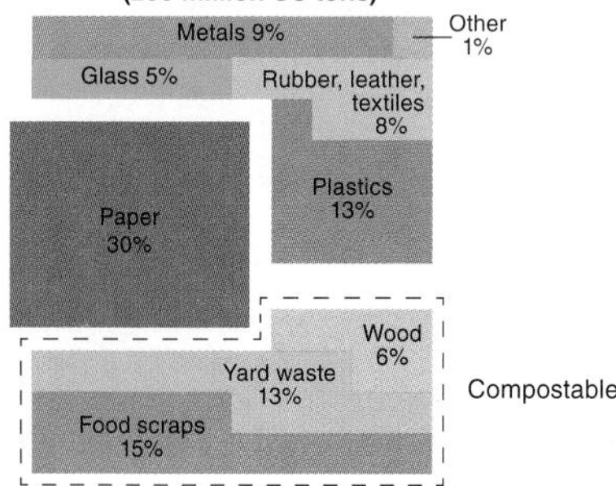

(a) Original waste stream

Total recovered: 79 million metric tons (87 million US tons)

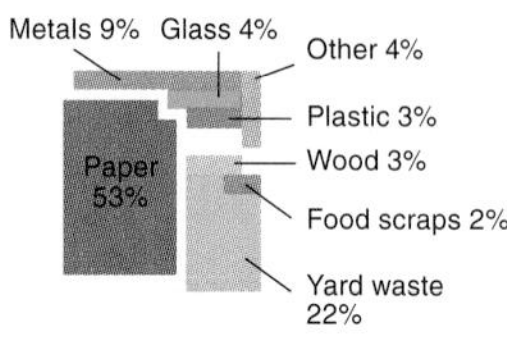

Total discarded: 148 million metric tons (163 million US tons)

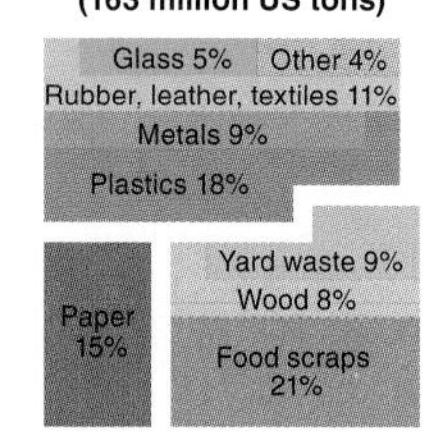

(b) After recovery and disposal

FIGURE 51.5 **Composition and sources of municipal solid waste (MSW) in the United States.** (a) The composition, by weight, of MSW in the United States in 2011 before recycling. Paper, food, and yard waste make up more than half of the MSW by weight. (b) The breakdown of the material that is recovered and the material that is discarded. Paper makes up more than half of the material that is recovered. Food and yard waste make up almost one-third of material that is discarded. *(After U.S. Environmental Protection Agency,* MSW Generation, Recycling and Disposal in the United States: Facts and Figures for 2011)

for 5 to 10 years. In most cases, a disposable paper cup leaves the system within minutes or hours after it is used. Ultimately, all products wear out, lose their value, or are discarded. At this point they enter the **waste stream**—the flow of solid waste that is recycled, incinerated, placed in a solid waste landfill, or disposed of in another way. In this section we will look at the composition of MSW and then explore e-waste in more detail.

Waste stream The flow of solid waste that is recycled, incinerated, placed in a solid waste landfill, or disposed of in another way.

Composition of Municipal Solid Waste

FIGURE 51.5a shows the data for MSW composition in the United States in 2011 by category before any recycling has occurred. The category "paper," which includes newsprint, office paper, cardboard, and boxboard such as cereal and food boxes, made up 30 percent of the 227 million metric tons (250 million U.S. tons) of waste generated before recycling. The fraction of paper in the solid waste stream has been decreasing; less than a decade ago it was 40 percent of MSW. Organic materials other than paper products make up another large category, with yard waste and food scraps together making up 28 percent of MSW. Wood, which includes construction debris, accounts for another 6 percent. So, not including paper products, which are more easily recycled, roughly 34 percent of current MSW could be composted, although some wood construction debris is difficult to compost because of its size and thickness. The combination of all plastics makes up approximately 13 percent of MSW.

As Figure 51.5b shows, 35 percent of the material that could potentially end up in a landfill or an incinerator is recovered. More than half of the recovered material is paper. Yard trimmings account for another large portion of recovered material, 22 percent.

After roughly one-third of our MSW is recycled, the resulting 148 million tons (164 million U.S. tons) that do end up in the landfill or incinerator have a different composition. As Figure 51.5b shows, paper represents a much smaller part of the diagram than in Figure 51.5a, largely because paper is so easily and frequently recycled. Food waste becomes a large part of the diagram because there are fewer composting programs available in the United States. Plastic is another component that increases, from 13 percent before recycling to 18 percent after materials have been recycled, in part because of the difficulty of recycling certain plastics.

Long-term viability is another way to consider MSW: Durable goods will last for years, nondurable goods are disposable, and compostable goods are those largely made up of organic material that can decompose under proper conditions. Containers and packaging make up 30 percent of MSW and are typically intended for one use. Food and yard waste account for 28 percent, and nondurable goods such as newspaper, white paper, printed products like telephone books, clothing, and plastic items like utensils and cups are 21 percent of the solid waste stream. Durable goods such as appliances, tires, and other manufactured products make up 20 percent of the waste stream. In addition to considering waste by weight, there is sometimes merit in considering waste by volume, especially when considering how much can be transported per truckload and how much will fit in a particular landfill.

E-Waste

Electronic waste, or e-waste, is one component of MSW that is small by weight but very important and rapidly increasing. Consumer electronics, including televisions, computers, portable music players, and cell phones, account for roughly 2 percent of the waste stream. This may not sound like a large amount but the environmental effect of these discarded objects is far greater than their weight. The older-style cathode-ray-tube (CRT) television or computer monitor contains 1 to 2 kg (2.2–4.4 pounds) of the heavy metal lead as well as other toxic metals such as mercury and cadmium. These metals may eventually leach out of the bottom of the landfill into groundwater or surface water. The toxic metals and other components can be extracted, but at present there is little formalized infrastructure or incentive to recycle them. However, many communities have begun voluntary programs to divert e-waste from landfills.

In the United States, most electronic devices are not designed to be easily dismantled after they are discarded. It generally costs more to recycle a computer than to put it in a landfill. The EPA estimates that approximately 27 percent of televisions and computer products discarded in 2010 were sent to recycling facilities. Unfortunately, much e-waste from the United States is exported to China, where adults as well as some children separate valuable metals from other materials using fire and acids in open spaces with no protective clothing and no respiratory gear (FIGURE 51.6). So even when consumers do send electronic products to be recycled, there is a good chance that the recycling will not be done properly.

FIGURE 51.6 **Electronic waste recycling in China.** Much of the recycling is done without protective gear and respirators that would typically be used in the United States. In addition, children are sometimes part of the recycling workforce in China. *(Peter Essick/Aurora Photos/Alamy)*

Teaching Tip: Beyond the Classroom

Have students read the article "High Tech Trash" from *National Geographic*. View the images in the photo gallery together in class and complete the e-waste quiz to increase awareness surrounding the disposal of electronic items. Have students design a campaign to educate others at your school about discarded electronics. Students may research how their community deals with electronic waste and whether it is shipped overseas.

The link to this resource may be found by clicking on the link buttons in the TE-book, opening the TRFD, or logging onto the book's companion website bcs.whfreeman.com/friedlandapes2e (teacher login required).

 **Chapter 16 Web Resources**

Teaching Tip: Video

The Story of Electronics
This video describes the production and disposal of electronic devices. There are 25 million tons of e-waste as well as issues of human health that are introduced as external costs often overlooked when considering the life of electronic devices. The video challenges designers to make long-lasting, non-toxic products that are fully and easily recyclable.

The link to this video may be found by clicking the link buttons in the TE-book, opening the TRFD, or logging onto the book's companion website bcs.whfreeman.com/friedlandapes2e (teacher login required).

TRM **Chapter 16 Web Resources**

Teaching Tip: Warm-up

Ask students: Why does electronic waste have a greater impact on the environment than its percentage of MSW by weight might imply? *E-waste contains heavy metals that are toxic and may leach into groundwater or surface water systems. Even though electronic waste may not comprise much of a landfill by weight, it has the potential to be especially harmful.*

module

51

REVIEW

In this module, we have seen that creating waste that is unusable by other organisms is a uniquely human characteristic. The refuse we dispose of is called municipal solid waste. The generation of MSW, both total and on a per capita basis, has increased steadily in the last 45 years although in recent years it has leveled off and decreased slightly. Paper, food waste, and yard waste make up 58 percent of MSW. A fraction of each is reused, recycled, or composted. Of the 227 million metric tons of MSW that could be disposed of, a little more than one-third ends up being diverted from the landfill for reuse, recycling, or composting. E-waste is a relatively small component of MSW but because of the toxicity of the metals it often contains, its disposal is a significant problem. In the next module, we will examine diversion from the landfill more closely.

Module 51 AP® Review Questions

1. Which played an important role in the development of the "throw-away" society?
   (a) The increased use of glass and metals
   (b) Objects made of many materials
   (c) Attitude changes after World War I
   (d) The shift in manufacturing to developing nations
   (e) The rejection of planned obsolescence

2. On average, how much municipal solid waste is generated per person each day in the United States?
   (a) 0.1 kg
   (b) 0.5 kg
   (c) 1 kg
   (d) 2 kg
   (e) 4 kg

3. The material that makes up the highest proportion of MSW is
   (a) plastic.
   (b) rubber, leather, and textiles.
   (c) paper and paperboard.
   (d) food.
   (e) metals.

4. Electronic waste
   (a) accounts for over 10 percent of the waste stream.
   (b) is almost always recycled.
   (c) is only a problem in developed nations.
   (d) contains few toxic components.
   (e) is more expensive to recycle than to put in a landfill.

5. Approximately how much MSW is recovered before it enters a landfill or incinerator?
   (a) 15 percent
   (b) 20 percent
   (c) 35 percent
   (d) 45 percent
   (e) 60 percent

Answers to Module 51 AP® Review Questions

1. b
2. d
3. c
4. e
5. c

module

52

The Three Rs and Composting

Almost every schoolchild in the United States has heard the phrase "reduce, reuse, and recycle." In recent years, composting has been added to the list of actions one should take before adding material to the waste stream. In this module we will examine the *three Rs* and composting.

Learning Objectives

After reading this module, you should be able to

- describe the three Rs.
- understand the process and benefits of composting.

Teaching Tip: Warm-up

Ask students: What is source reduction? Give some examples of source reduction. *Source reduction is an approach to waste management that seeks to cut waste by reducing the use of potential waste materials in the early stages of design and manufacturing. Some examples of source reduction are: using both sides of the paper when making copies; using packaging that provides the same protection with fewer materials; and giving workers reusable mugs instead of using paper cups.*

The three Rs divert materials from the waste stream

Starting in the 1990s, people in the United States began to promote the idea of diverting materials from the waste stream with a popular phrase **"Reduce, Reuse, Recycle,"** also known as **the three Rs.** The phrase incorporates a practical approach to the subject of solid waste management, with the techniques presented from the most environmentally beneficial to the least (FIGURE 52.1).

FIGURE 52.1 **Reduce, Reuse, Recycle.** This popular slogan emphasizes the actions to take in the proper order, from the most environmentally effective to the least. *(Christopher Steer/iStockphoto.com)*

Reduce

"Reduce" is the first choice among the three Rs because reducing inputs is the optimal way to achieve a reduction in solid waste generation. This strategy is also known as waste minimization and waste prevention. If the input of materials to a system is reduced, the outputs will also be reduced; in terms of waste, this means that when less material is used, there will be less material to discard. One approach, known as **source reduction,** seeks to cut waste by reducing the use of potential waste materials in the early stages of design and manufacture. In many cases, source reduction also increases energy efficiency because it means that manufacturing produces less waste to begin with and can minimize disposal processes. Since fewer resources are being expended, source reduction also provides economic benefits.

Source reduction can be implemented both on individual and on corporate or institutional levels. For

Reduce, reuse, recycle A popular phrase promoting the idea of diverting materials from the waste stream. *Also known as* **the three Rs.**

Source reduction An approach to waste management that seeks to cut waste by reducing the use of potential waste materials in the early stages of design and manufacture.

COMMON MISCONCEPTIONS
The three Rs

Students sometimes believe that "reduce," "reuse," and "recycle" all have the same impact. This is not true. Reducing means to generate less. In terms of reducing energy use, using less is the most desirable option. Reusing something is the next most desirable option. Recycling uses the most energy of the three but is often a more efficient option than throwing an item away.

Teaching Tip: Journal Prompt

Challenge students to keep a record of items that they reuse over the week. *Examples could include reusing a polystyrene cup, donating items to the thrift store, or using the back of a piece of paper instead of throwing it away.*

example, if an instructor has two pages of handout material for a class, she could reduce her paper use by 50 percent if she provides her students with double-sided photocopies. A copy machine that can automatically make copies on both sides of the page might use more energy and require more materials in manufacturing than a copy machine that only prints on one side, but with up to half the number of copies needed, the overall energy used to produce them over time will probably be less. Further source reduction could be achieved if the instructor did not hand out any sheets of paper at all but sent copies to the class electronically, and the students refrained from printing the documents.

Source reduction in manufacturing can happen in several ways. If the company creates new packaging that provides the same amount of protection to the product with less material, successful source reduction has occurred. Consider the incremental source reduction that occurred with purchasing music. Music compact discs were packaged in large plastic sleeves that were three times the size of the CD. Today, most CDs are wrapped with a small amount of plastic material that just covers the CD case. Many people no longer purchase CDs at all and instead download their music from the Web. Less wrapping on CDs is an example of source reduction on the corporate level. Not purchasing CDs at all is an example of source reduction on the individual level.

Source reduction can also be achieved by material substitution. In an office where workers drink water and coffee from paper cups, providing every worker with a reusable mug will reduce MSW. In some categorization schemes, this could be considered reuse rather than source reduction. Nevertheless, cleaning the mugs will require water, energy to heat the water, soap, and processing of wastewater. The break-even point, beyond which there are gains achieved by using a ceramic mug (though this depends on a variety of factors), might be approximately 50 uses. The break-even point is shorter for a reusable plastic mug, in part because less energy is required to manufacture the plastic mug and, because it is lighter, less energy is used to transport it. Source reduction may also involve substituting less toxic materials or products in situations where manufacturing utilizes or generates toxic substances. For example, switching from an oil-based paint that contains toxic petroleum derivatives to a relatively nontoxic latex paint is a form of source reduction.

All of these examples achieve a reduction in material use and, ultimately, material waste, without an additional expenditure of materials or energy. For that reason, reduction is the first "R"—and it is the most environmentally beneficial.

Reuse

Reuse of a product or material that would otherwise be discarded, rather than disposal, allows a material to cycle within a system longer before it becomes an output. In other words, its mean residence time in the system is greater. Optimally, no additional energy or resources are needed for the object to be reused. For example, a mailing envelope can be reused by covering the first address with a label and writing the new address over it. Here we are increasing the residence time of the envelope in the system and reducing the waste disposal rate. Or we could reuse a disposable polystyrene cup more than once, though reuse might involve cleaning the cup, which would add some energy cost and generate some wastewater. Sometimes reuse may involve repairing an existing object, which costs time, labor, energy, and materials.

Energy may also be required to prepare or transport an object for reuse by someone other than the original user. For example, certain companies reuse beverage containers by shipping them to the bottling factory where they are washed, sterilized, and refilled. Although energy is involved in the transport and preparation of the containers, it is still less than the energy that would be required for recycling or disposal.

We have noted that reuse is still common in many countries and that it was common practice in the United States before we became a "throw-away society." However, it is still practiced in many ways that we might not think of as reuse. For example, people often reuse newspapers for animal bedding or art projects. Many businesses and universities have surplus equipment agents who help find a home for items no longer needed. Flea markets, swap meets, and even popular websites such as eBay, craigslist, and Freecycle are all agents of reuse. Reuse sometimes involves the expenditure of energy and generates waste. For example, in order to transport, wash, and sterilize the beverage bottles in the example previously cited, energy is expended and wastewater is generated. So reuse does have environmental costs that typically exceed reducing the use of something, but it is preferable to using new material.

Reuse Using a product or material that was intended to be discarded.

Recycling The process by which materials destined to become municipal solid waste (MSW) are collected and converted into raw material that is then used to produce new objects.

Recycle

The third "R" is **recycling,** the process by which materials destined to become MSW are collected and converted into raw materials that are then used to produce new objects. We divide recycling into two categories: *closed-loop* and *open-loop*. FIGURE 52.2 shows

562 CHAPTER 16 ■ Waste Generation and Waste Disposal

Teaching Tip: Activity

Ask students to complete a table like the one below arranging the three Rs in order from most beneficial to least beneficial. Explain how each approach affects solid waste management. Give potential economic benefits to each practice and the potential disadvantages. (Sample answers are provided in italics.)

	Reduce	*Reuse*	*Recycle*
Effect on solid waste management	*If inputs are decreased, then output will also decrease*	*Materials cycle within a system for a longer time with no additional energy or resources needed*	*Fewer raw materials must be extracted, which reduces pollution, environmental degradation, and energy use*
Potential economic benefits	*Companies can save money when they use less packaging; people can manage on a smaller budget if they use less*	*Money can be saved by reusing an item rather than purchasing a new one; used items can be acquired at reduced prices*	*Recycling cuts the cost of extracting or harvesting raw materials, which can save money in the production process*
Potential disadvantages	*Sometimes using less can be considered a lower standard of living*	*Energy may be required to clean and transport an item for reuse; water may be required to clean an item for reuse and wastewater may be generated*	*Recycling materials requires time for processing, cleaning, transporting, and possible modification before material is usable*

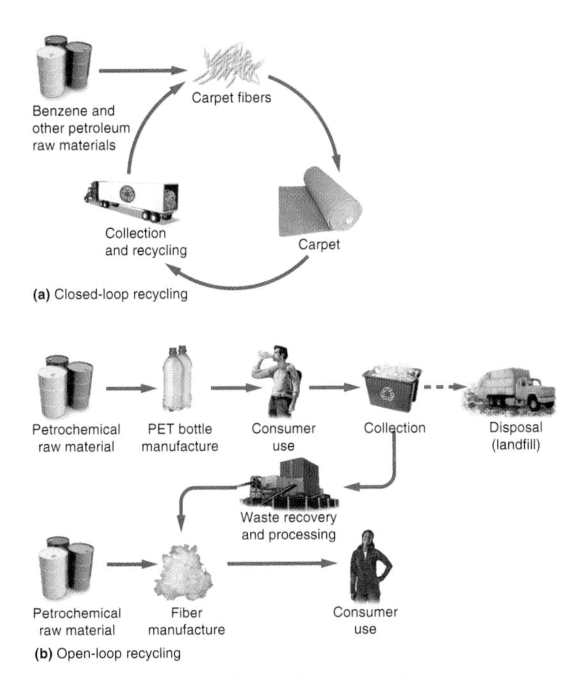

FIGURE 52.2 **Closed- and open-loop recycling.** (a) In closed-loop recycling, a discarded carpet can be recycled into a new carpet, although some additional energy and raw material are needed. (b) In open-loop recycling, a material such as a beverage container is used once and then recycled into something else, such as a fleece jacket.

the process for each. **Closed-loop recycling,** shown in Figure 52.2a, is the recycling of a product into the same product. Aluminum cans are a familiar example; they are collected, brought to an aluminum plant, melted down, and made into new aluminum cans. This process is called a closed loop because in theory it is possible to keep making aluminum cans from only old aluminum cans almost indefinitely; the process is thus similar to a closed system. In **open-loop recycling,** shown in Figure 52.2b, one product, such as plastic soda bottles, is recycled into another product, such as polar fleece jackets. Although recycling plastic bottles into other materials avoids sending the plastic bottles to a landfill, it does not reduce demand for the raw material—in this case petroleum—to make plastic for new bottles.

Recycling is not new in the United States, but over the past 25 years it has been embraced enthusiastically by both individuals and municipalities in the belief that it measurably improves environmental quality. The graph in FIGURE 52.3 shows both the increase in the weight of MSW in the United States from 1960 to 2011 and the increase in the percent of waste that was recycled over the same period of time. Recycling rates have increased in the United States since 1975, and today we recycle roughly one-third of MSW. In Japan, recycling rates are more than 50 percent, and some colleges and universities in the United States report recycling rates of 60 percent for their campuses.

Extracting resources from Earth requires energy, time, and usually a considerable financial investment. As we have seen, these processes generate pollution. Therefore, on many levels, it makes sense for manufacturers to utilize resources that have already been extracted. Today, many communities are adopting

Closed-loop recycling Recycling a product into the same product.

Open-loop recycling Recycling one product into a different product.

Teaching Tip: Warm-up

Ask students to differentiate between closed-loop and open-loop recycling. *Closed-loop recycling is the recycling of a product into the same product, while open-loop recycling recycles a product into a different product.*

Teaching Tip: Activity

Have students read "Science Applied 7: Is Recycling Always Good for the Environment?" on page 627. This article investigates the costs and benefits of recycling various materials and shows that not all recycled materials result in equal environmental benefits.

AP® Exam Tip

The economics of recycling is covered in the 2000 AP® Exam, Question 2. To answer this question, students must

- consider arguments for and against recycling in general and create arguments for and against recycling newspaper
- describe two pieces of scientific information that would be needed to evaluate the environmental benefits and environmental costs of recycling
- recommend either recycling aluminum or newspaper to a community that wishes to start a recycling program and explain the recommendation
- name and describe two difficulties a community might face in beginning a recycling program

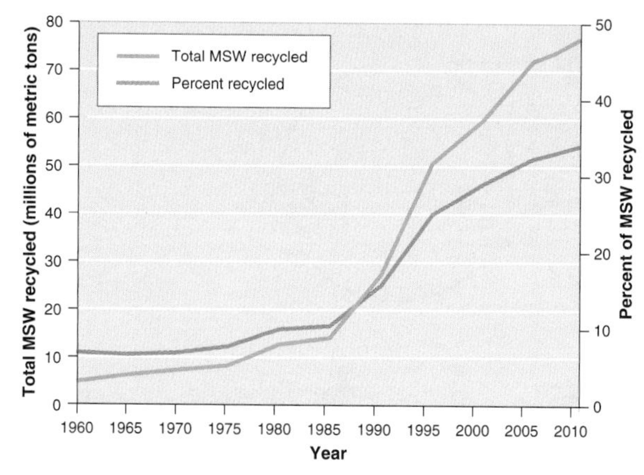

FIGURE 52.3 **Total weight of municipal solid waste recycled and percent of MSW recycled in the United States over time.** Both the total weight of MSW that is recycled and the percentage of MSW that is recycled have increased over time. *(After U.S. Environmental Protection Agency,* MSW Generation, Recycling and Disposal in the United States: Facts and Figures for 2011)

zero-sort recycling programs. These programs allow residents to mix all types of recyclables in one container that they deposit on the curb outside the home or bring to a transfer station. This saves time for residents who were once required to sort materials. At the sorting facility, workers sort the materials destined for recycling into whatever categories are in greatest demand at a given time and offer the greatest economic return (FIGURE 52.4). The markets for glass and paper are highly volatile. While there is always a demand for metals such as aluminum and copper, demand for recycled newspapers fluctuates. When newspaper is in low demand, the single-stream sorting facility might pull out newsprint to sell or give to local stables for use as horse bedding. At other times, when demand for paper is higher, the newsprint might be kept with other paper to be recycled into new paper products.

FIGURE 52.4 **A mixed single-stream solid waste recycling facility in San Francisco, California.** With single-stream recycling, also called no-sort or zero-sort recycling, consumers no longer have to worry about separating materials. *(Justin Sullivan/Getty Images)*

Nevertheless, because recycling requires time, processing, cleaning, transporting, and possible modification before the waste is usable as raw material, it does require more energy than reducing or reusing materials. Such costs caused New York City to make a controversial decision in 2002 to suspend glass and plastic recycling. This was a major shift in policy for the city, which had been encouraging as much recycling as possible, including collection of mixed recyclables—glass, plastic, newspapers, magazines, and boxboard—at the same time as waste materials. After collection, the recyclables were sent to a facility where they were sorted. According to city officials, the entire process of sorting glass and plastic from other recyclables and selling the material was not cost effective. In 2004, the recycling of all materials was reinstated in New York City. Today, the goal in most recycling programs is to maximize diversion from the landfill, even if that means collecting materials that have little economic value. However, many communities periodically have difficulty finding buyers for glass and plastic since the price paid for these materials fluctuates widely.

The New York City case is just one example of why recycling is the last choice among the three Rs.

Teaching Tip: Video

Seas of Plastic

This video from TED-Ed introduces Captain Charles Moore, who first discovered the Great Pacific Garbage Patch—an endless floating waste of plastic trash in the Pacific Ocean. In this video, he draws attention to the growing problem of plastic debris in the oceans.

The link to this video may be found by clicking the link buttons in the TE-book, opening the TRFD, or logging onto the book's companion website bcs.whfreeman.com/friedlandapes2e (teacher login required).

Chapter 16 Web Resources

This doesn't mean that we should abandon recycling programs. Not only does it work well for materials such as paper and aluminum, but it also encourages people to be more aware of the consequences of their consumption patterns. Nevertheless, in terms of the environment, source reduction and reuse are preferable. The environmental implications of recycling are considered further in "Science Applied 7: Is Recycling Always Good for the Environment?" that follows Chapter 17 on page 625.

Composting is becoming more popular

While diversion from the landfill is usually referred to as the three Rs, there is one more diversion pathway that is equally important, if not more so. Organic materials such as food and yard waste that end up in landfills cause two problems. Like any material, they take up space, but unlike glass and plastic that are chemically inert, organic materials are also unstable. As we will see later in this chapter, the absence of oxygen in landfills causes organic material to decompose anaerobically, which produces methane gas, a much more potent greenhouse gas than carbon dioxide.

An alternate way to treat organic waste is through *composting*. **Composting** creates organic matter (humus) that has decomposed under controlled conditions to produce an organic-rich material that enhances soil structure, cation exchange capacity, and fertility. Vegetables and vegetable by-products, such as cornstalks, grass, animal manure, yard wastes (like leaves and branches), and paper fiber not destined for recycling are suitable for composting. Normally, meat and dairy products are not composted because they do not decompose as easily, produce foul odors and are more likely to attract unwanted visitors such as rats, skunks, and raccoons.

Outdoor compost systems can be as simple as a pile of food and yard waste in the corner of a yard, or as sophisticated as compost boxes and drums that can be rotated to ensure mixing and aeration. From the decomposition process described in Chapter 3, we are already familiar with the process that takes place during composting. In order to encourage rapid decomposition, it is important to have the ratio of carbon to nitrogen (C:N) that will best support microbial activity—about 30:1. While it is possible to calculate the carbon and nitrogen content of each material you put into a compost pile, most compost experts recommend layering dry material such as leaves or dried cut grass—normally brown material—with wet material such as kitchen vegetables—normally green material. This will provide the correct carbon to nitrogen ratio for optimal composting. Frequent turning or agitation is usually necessary to ensure that decomposition processes are aerobic and to maintain appropriate moisture levels—otherwise the compost pile will produce methane and associated gases and emit a foul odor. If the pile becomes particularly dry, water needs to be added. Although many people assume that a compost pile must smell bad, with proper aeration and not too much moisture, the only odor will be that of fresh compost in 2 to 3 months' time (FIGURE 52.5).

FIGURE 52.5 **Composting.** Good compost has a pleasant smell and will enhance soil quality by adding nutrients to the soil and by improving moisture and nutrient retention. In a compost pile that is turned frequently, compost can be ready to use in a few months. *(Amanda Koster/Corbis)*

Large-scale composting facilities currently operate in many municipalities in the United States. Some facilities are indoors, but most employ the same basic process we have described, though on a much larger scale. FIGURE 52.6 shows the process. Organic material is piled up in long, narrow rows of compost. The material is turned frequently, exposing it to a combination of air and water that will speed natural aerobic decomposition. As with household composting, the organic material must include the correct combination of green (fresh) and brown (dried) organic material so that the ratios of carbon and nitrogen are optimal for bacteria. Various techniques are used to turn the organic material over periodically, including the use of rotating blades that move through the piles of organic material or a front loader that turns over the piles. The respiration activity of the microbes generates enough heat to kill any pathogenic bacteria that may be contained in food scraps, which is typically a concern only in large municipal composting systems. If the pile becomes too hot, it should be turned more frequently.

Composting Creation of organic matter (humus) by decomposition under controlled conditions to produce an organic-rich material that enhances soil structure, cation exchange capacity, and fertility.

Teaching Tip: Warm-up

Ask students to explain the benefits of including composting in a waste management strategy. *Organic materials in a landfill can cause two problems: They take up space and are unstable. In the anaerobic conditions that exist in a landfill, organic materials decompose and produce methane gas as a byproduct. Methane is a potent greenhouse gas. Composting removes organic waste from the landfill and converts it into highly productive humus that can be used to enhance soil structure and increase plant productivity.*

Teaching Tip: Journal Prompt

Ask students to give some general guidelines for maintaining a composting system. *Vegetables and vegetable by-products, leaves, grass, and paper fiber can be added to the compost. Meat and dairy products should not be added. Rapid decomposition is supported by a 30:1 carbon to nitrogen ratio. Layer dry materials such as leaves or cut grass with wet materials such as kitchen vegetables for optimal composting. Frequent turning or agitation allows for aeration. Water should be added as needed in dry conditions.*

Teaching Tip: Activity

If there is an appropriate place on campus, have students build and maintain a compost pile. If an outdoor location is unavailable, create a small-scale composting system in an aquarium. If possible create a composting program in partnership with the school cafeteria.

more **math** practice

The Dowling family is looking at ways to reduce its solid waste footprint. Each of the four members of the family produces 3 pounds of solid waste per day. If the family decides to compost all organic materials (food scraps, yard waste, etc.), they would reduce their solid waste footprint by 20 percent.

What is the total solid waste that the Dowling family produces in 1 year?

In 1 year the family produces:

(3.0 lb per person) × (4 people) × (365 days) = 4,380 lb/year

What is the family's total solid waste footprint after they have implemented composting?

***Step 1:** Composting would reduce the solid waste footprint by 20 percent. Find composted solid waste in pounds.*

(4,380 lb/year) × (0.20 composted) = 876 lb solid waste composted

***Step 2:** Subtract the composted amount from the total quantity:*

4,380 lb/year − 876 lb composted = 3,504 lb

The Dowling family's solid waste footprint after they have implemented composting would be 3,504 pounds per year.

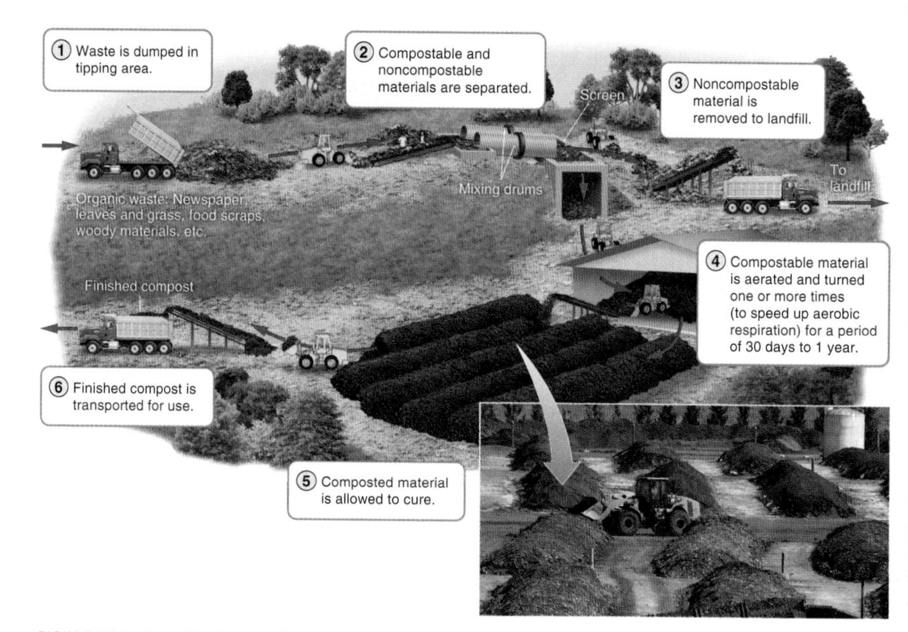

FIGURE 52.6 **A municipal composting facility.** A typical facility collects almost 100,000 metric tons of food scraps and paper per year and turns it into usable compost. Most facilities have some kind of mechanized system to allow mixing and aeration of the organic material, which speeds conversion to compost. *(Justin Sullivan/Getty Images)*

If the pile doesn't become hot enough, operators should check to make sure their C:N ratio is optimal, or they should slow the turnover rate. Within a matter of weeks, the organic waste becomes compost. Large-scale municipal composting systems with relatively little mechanization and labor may take up to a year to create a finished compost.

It is not necessary to have an outdoor space to compost household waste; composting is possible even in a city apartment or a dorm room. It is even possible to set up a composting system in a kitchen or basement. The very popular book *Worms Eat My Garbage: How to Set Up and Maintain a Worm Composting System* by Mary Appelhof has encouraged thousands of individuals across the country to compost kitchen waste using red wiggler worms. A small household recycling bin is large enough to serve as a worm box. As with an outdoor compost pile, a properly maintained worm box does not give off bad odors.

The composting process does take time and space. Source separation can be an inconvenience or, in some situations, not possible. Also, in certain environments, storing materials before they are added to the compost pile can attract flies or vermin. Finally, the compost pile itself can attract unwanted animals such as rats, skunks, raccoons, and even bears. But because compost is high in organic matter, which has a high cation exchange capacity and contains nutrients, it enhances soil quality when added to agricultural fields, gardens, and lawns.

Teaching Tip: Video

Vermicomposting: How Worms Can Reduce Our Waste
This video clip from TED-Ed introduces vermicomposting and details the steps we can all take to vermicompost at home.

The link to this video may be found by clicking the link buttons in the TE-book, opening the TRFD, or logging onto the book's companion website bcs.whfreeman.com/friedlandapes2e (teacher login required).

TRM **Chapter 16 Web Resources**

more **math** practice
Graphing MSW Generation, Recycling, and Disposal

Using the information provided by the EPA, students will graph and evaluate MSW generation, recycling, and disposal in the United States.

The link to this activity may be found by clicking on the link buttons in the TE-book, opening the TRFD, or logging onto the book's companion website bcs.whfreeman.com/friedlandapes2e (teacher login required).

 **More Math Practice 16.1:** Graphing MSW Generation, Recycling, and Disposal

module

52

REVIEW

In this module, we have seen that reduce, reuse, recycle, and compost are appropriate actions to take before sending a solid waste material into the waste stream. It is important to observe the "three Rs" in the order stated here, because each action uses less energy than the next action. Composting is another pathway for diverting material from the waste stream. The diversion of compostable materials from the landfill reduces methane production and produces a desirable end product, decomposed organic matter. In the next module, we will see what happens to material that is not diverted from the solid waste stream.

Module 52 AP® Review Questions

1. The correct order of the three Rs is
   (a) Reduce, Recycle, Reuse.
   (b) Reduce, Reuse, Recycle.
   (c) Recycle, Reuse, Reduce.
   (d) Recycle, Reduce, Reuse.
   (e) Reuse, Reduce, Recycle.

2. Which is NOT a form of source reduction?
   (a) Printing double-sided instead of single-sided
   (b) Purchasing digital versions of music instead of CDs
   (c) Replacing plastic mugs with disposable paper cups
   (d) Substituting a nontoxic material for something toxic
   (e) Designing a smaller box for packaging

3. Which material usually uses closed-loop recycling?
   (a) Aluminum
   (b) Glass
   (c) Paper
   (d) Cardboard
   (e) Plastic

4. Organic matter in landfills is a problem primarily because
   (a) there is a high risk of toxic leaching.
   (b) it contains bacteria that spreads disease.
   (c) as it breaks down it can dissolve the containment.
   (d) it creates excessive heat.
   (e) it produces methane gas.

5. For composting to work effectively, the compost
   (a) must be kept very wet.
   (b) must be mixed.
   (c) must be anaerobic.
   (d) must be protected from high temperatures.
   (e) must contain more nitrogen than carbon.

Answers to Module 52 AP® Review Questions

1. b
2. c
3. a
4. e
5. b

module

Landfills and Incineration

53

Beginning in the 1930s in the United States, as public opposition to open dumps was growing, the most convenient locations for disposal of MSW became holes in the ground created by the removal of soil, sand, or other earth material that was then used for construction purposes. When people filled those holes with waste, the sites became known as landfills. Though open dumps are now rare in developed countries, they still exist in the developing world, where they pose a considerable health hazard.

In this module, we will examine the basic design and implementation of landfills and incinerators in the United States.

Learning Objectives

After reading this module, you should be able to

- describe the goals and function of a solid waste landfill.
- explain the design and purpose of a solid waste incinerator.

AP® Exam Tip

Landfills and calculations associated with leachate are covered in the 2008 AP® Exam, Question 2. To answer this question, students must

- calculate the yearly volume of water flowing into a landfill and the leachate treated per year based on data provided
- calculate the weight of cadmium released into surrounding soil per year based on data provided
- calculate the annual cost of treating leachate from the landfill drainage system based on provided data
- describe two viable strategies for reducing the amount of cadmium entering the municipal waste input
- explain a shortcoming with at least one of the proposed methods for reducing the amount of cadmium entering the municipal waste input

This question is mathematically challenging. The question provides an opportunity to practice conversions between units. Students need to remember that m^3 is a unit of volume. Teach students to look for the unit(s) that a problem solution requires and then work backward looking for units in the given information that will help them reach the end point.

Landfills are the primary destination for MSW

As FIGURE 53.1 shows, in the United States a third of our waste is recovered through reuse and recycling, while more than half is discarded. The remainder is converted into energy through incineration. In the United States, MSW that is not diverted from the waste stream ends up becoming either buried in landfills or incinerated.

Initially, there were few concerns about what material went into a landfill. Those responsible for collecting and disposing of solid waste in landfills did not recognize the many problems associated with landfills, such as components of the MSW generating harmful runoff and *leachate*. **Leachate** is the water that leaches through the solid waste and removes various chemical compounds with which it comes into contact. Nor did they recognize the harm a landfill could cause when it

Leachate Liquid that contains elevated levels of pollutants as a result of having passed through municipal solid waste (MSW) or contaminated soil.

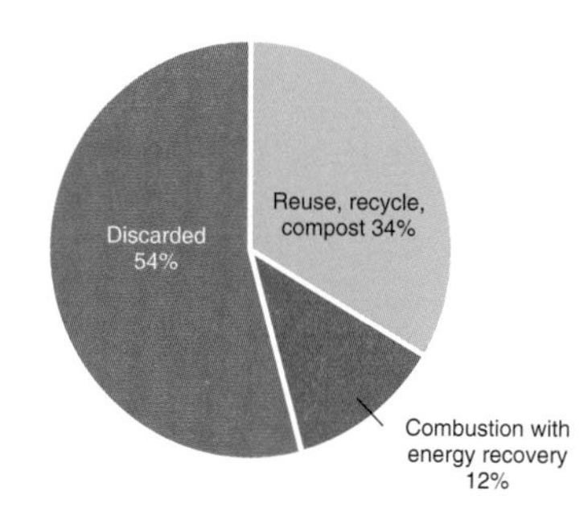

FIGURE 53.1 **The fate of municipal solid waste in the United States.** The majority of MSW is disposed of in landfills. (*After U.S. Environmental Protection Agency,* MSW Generation, Recycling and Disposal in the United States: Facts and Figures for 2011)

▲ **TEACHING with FIGURES**

Show students Figure 53.1 and ask: What is the destination of most MSW? *Most MSW ends up in a landfill.*

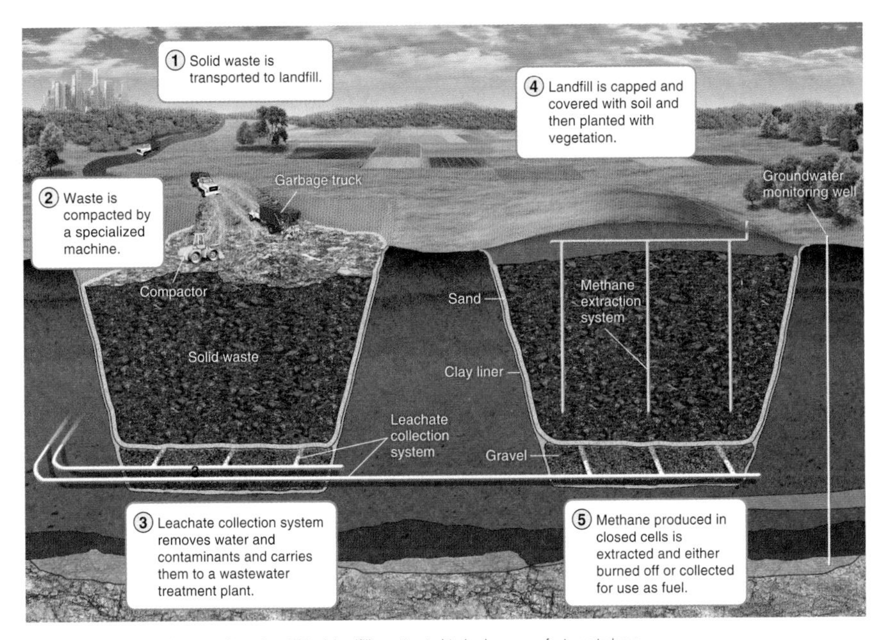

FIGURE 53.2 **A modern sanitary landfill.** A landfill constructed today has many features to keep components of the solid waste from entering the soil, water table, or nearby streams. Some of the most important environmental features are the clay liner, the leachate collection system, the cap—which prevents additional water from entering the landfill—and, if present, the methane extraction system.

was located near sensitive features of the landscape such as aquifers, rivers, streams, drinking-water supplies, and human habitation.

Today, some environmental scientists believe we should not use landfills at all; in later sections of this chapter we will discuss alternative means of waste disposal. Although landfills are still a component of solid waste management in the United States, we can make them much less harmful than they have been in the past. In this section we will examine landfill basics, how a landfill is sited, and the problems of using landfills.

Landfill Basics

Many people believe that a great deal of biological and chemical activity occurs in landfills. They imagine that organic matter breaks down and that landfills shrink over time as the material in them is converted into carbon dioxide or methane and released to the atmosphere. Professor William Rathje, from the University of Arizona, has challenged these misconceptions by using archaeological tools to examine modern-day garbage. Rathje and his colleagues have found newspapers with headlines still legible 40 years after being deposited in landfills, proving that little decomposition had taken place. Today it is widely accepted that decomposition takes place only in those areas of a landfill where the correct mixtures of air, moisture, and organic material are present. Because most areas do not contain this necessary mixture, a landfill will probably remain the same size it was when capped.

Given the lack of decomposition that occurs in a landfill, design and construction must proceed with care. In the United States today, repositories for MSW, known as **sanitary landfills,** are engineered ground facilities designed to hold MSW with as little contamination of the surrounding environment as possible. These facilities, like the one illustrated in FIGURE 53.2, generally utilize a variety of technologies that safeguard against the problems of traditional dumps.

Sanitary landfill An engineered ground facility designed to hold municipal solid waste (MSW) with as little contamination of the surrounding environment as possible.

AP® Exam Tip

Many past AP® Exams have asked students to describe and explain landfill design and function. Students should know how a landfill works. They should understand how leachate and methane are collected. They should be able to explain that landfills are designed in layers to prevent the spread of contamination. To reinforce knowledge of all components of a landfill, ask students to draw Figure 53.2 in their journals.

Teaching Tip: Journal Prompt

Ask students to describe the features of a modern sanitary landfill. How does a modern landfill compare to the open dumps of the past? *Construction of a modern sanitary landfill begins with a clay or plastic liner being placed along the bottom of the landfill that impedes the flow of water and can retain positively charged metals. Systems of pipes are constructed below the landfill to collect leachate. Finally a cover of soil and clay caps the landfill when it reaches capacity. Open dumps allow pollution to seep into ground and water systems. Sanitary landfills are designed to limit environmental impacts on the surrounding area. Open dumps are simply holes in the ground where contaminated leachate, methane gas, and debris can pollute the surrounding areas.*

Teaching Tip: Warm-up

Ask students to list the items that should not go into a landfill. *Aluminum and other metals, toxic materials such as oil-based paints and household cleaners, consumer electronics, appliances, batteries, and organic materials should never be put in a landfill.*

FIGURE 53.3 **Reclamation of a landfill.** The Fresh Kills landfill in Staten Island, New York, was open from 1947 to 2001. It is currently in the process of being converted into a 890-ha (2,200-acre) park. The mounds of landfill have been capped and planted with vegetation. *(Richard Levine/Alamy)*

Sanitary landfills are constructed with a clay or plastic lining at the bottom. Clay is often used because it can impede water flow and retain positively charged ions, such as metals. A system of pipes is constructed below the landfill to collect leachate, which is sometimes recycled back into the landfill. Finally, a cover of soil and clay, called a cap, is installed when the landfill reaches capacity.

Rainfall and other water inputs are minimized because excess water in the landfill causes a greater rate of anaerobic decomposition and consequent methane release. Also, with a large amount of water entering the landfill from both MSW and rainfall, there is a greater likelihood that some of that water will leave the landfill as leachate. Leachate that is not captured by the collection system may leach into nearby soils and groundwater. Leachate is tested regularly for its toxicity and if it exceeds certain toxicity standards, the landfill operators could be required to collect it and treat it as a toxic waste.

Once the landfill is constructed, it is ready to accept MSW. Perhaps the most important component of operating a safe modern-day landfill is controlling inputs. The materials destined for a landfill are those least likely to cause environmental damage through leachate or by generating methane. Composite materials made of plastic and paper, such as juice boxes for children, are good candidates for a landfill because they are difficult to recycle, while aluminum and other metals such as copper may contribute to leaching. In addition, metals are valuable as recyclables. Therefore, aluminum and copper should never go into a landfill.

Tipping fee A fee charged for disposing of material in a landfill or incinerator.

Glass and plastics are both chemically inert, making them suitable for a landfill when reuse or recycling is not possible. Toxic materials (such as household cleaners, oil-based paints, and automotive additives like motor oil and antifreeze), consumer electronics, appliances, batteries, and anything that contains substantial quantities of metals should not be deposited in landfills. All organic materials, such as food and garden scraps and yard waste, are potential sources of methane and should not be placed in landfills.

The MSW added to a landfill is periodically compacted into "cells," which reduce the volume of solid waste, thereby increasing the capacity of the landfill. The cells are covered with soil, which minimizes the amount of water that enters them and so reduces odor. When a landfill is full, it must be closed off from the surrounding environment so the input and output of water are reduced or eliminated. Once a landfill is closed and capped, the MSW within it is entombed. Some air and water may enter from the outside environment, but this should be minimal if the landfill was well-designed and properly sealed. The design and topography of the landfill cap, which is a combination of soil, clay, and sometimes plastic, encourage water to flow off to the sides, rather than into the landfill. Closed landfills can be reclaimed, meaning that some sort of herbaceous, shallow-rooted vegetation can be planted on the topsoil layer, both for aesthetic reasons and as a way to reduce soil erosion. Construction on the landfill is normally restricted for many years, although parks, playgrounds, and even golf courses have been built on reclaimed landfills (FIGURE 53.3).

A municipality or private enterprise constructs a landfill at a tremendous cost. These costs are recovered by charging a fee for waste delivered to the landfill, called a **tipping fee** because each truckload is put

Teaching Tip: Journal Prompt

Ask students to describe how a landfill is closed once it reaches capacity. *When a landfill reaches capacity it must be closed off to the surrounding environment so that the input and output of water is reduced or eliminated. The landfill is capped with soil and clay and sometimes plastic. Topography is engineered to encourage water to flow off to the sides rather than into the landfill. Shallow rooted vegetation can be planted for aesthetic reasons and to reduce erosion. Construction on a landfill is limited for many years but the land may be converted to parks, playgrounds, and golf courses.*

on a scale and, after the MSW is weighed, it is tipped into the landfill. Tipping fees at solid waste landfills average $50 per ton in the United States, although in certain regions, such as the Northeast, fees can be twice that much. These fees create an economic incentive to reduce the amount of waste that goes to the landfill. Many localities accept recyclables at no cost but charge for disposal of material destined for a landfill. This practice encourages individuals to separate recyclables. Some localities mandate that recyclable material be removed from the waste stream and disposed of separately. However, if tipping fees become too high, and regulations too stringent, a locality may inadvertently encourage illegal dumping of waste materials in locations other than the landfill and recycling center.

Choosing a Site for a Sanitary Landfill

Many considerations go into **siting** a landfill, which means designating its location. A landfill should be located in a soil rich in clay to reduce migration of contaminants. It should be located away from rivers, streams, and other bodies of water and drinking-water supplies. A landfill should also be sufficiently far from population centers so that trucks transporting the waste and animal scavengers such as seagulls and rats present minimal risks to people. However, the energy needed to transport MSW must also be considered in siting; as distance from a population center increases, so does the amount of energy required to move MSW to the landfill. Regional landfills, though, are becoming more common because sending all waste to a single location often offers the greatest economic advantage.

A landfill siting is always highly controversial and sometimes politically charged. Because landfills are unsightly and smell bad they are not considered desirable neighbors. Landfill siting has been the source of considerable environmental injustice. People with financial resources or political influence often adopt what has been popularly called a "not-in-my-backyard," or NIMBY, attitude about landfill sites. Because of this, a site may be chosen not because it meets the safety criteria better than other options but because its neighbors lack the resources to mount an effective opposition.

In Fort Wayne, Indiana, for example, the Adams Center Landfill was located in a densely populated, low-income, and predominantly minority neighborhood. A University of Michigan environmental justice study quotes Darrell Leap, a hydrogeologist and professor at Purdue University, as saying that on a scale of 1 to 10, with 10 being a geologically ideal site, the Adams Center Landfill would rate a "3, possibly 4" because the site held a substantial risk of water contamination. When the communities surrounding the Adams Center Landfill learned of the report and this danger, they protested the renewal of the federal permit and fought expansion of the landfill at both the state and local levels. Ultimately, a 1997 decision by the Indiana Department of Environmental Management closed the landfill.

Environmental Consequences of Landfills

Though sanitary landfills, as we have seen, are an improvement over open dumps, they present many problems. Locating landfills near populations that do not have the resources to object is a global problem. No matter how careful the design and engineering, there is always the possibility that leachate from a landfill will contaminate underlying and adjacent waterways. The EPA estimates that virtually all landfills in the United States have had some leaching. Even after a landfill is closed, the potential to harm adjacent waterways remains. The amount of leaching, the substances that have leached out, and how far they will travel are impossible to know in advance. To get an idea of how much leachate is generated from a landfill and how much might be collected, see "Do the Math: How Much Leachate Might Be Collected?" on page 572.

The risk to humans and ecosystems from leachate is uncertain. Public perception is that landfill contaminants pose a great threat to human health, though the EPA has ranked this risk as fairly low compared with other risks such as global climate change and air pollution. But methane and other organic gases generated from decomposing organic material in landfills do release greenhouse gases.

When solid waste is first placed in a landfill, some aerobic decomposition may take place, but shortly after the waste is compacted into cells and covered with soil, most of the oxygen is used up. At this stage, anaerobic decomposition begins, a process that generates methane and carbon dioxide—both greenhouse gases—as well as other gaseous compounds. The methane also creates an explosion hazard. For this reason, landfills are vented so that methane does not accumulate in highly explosive quantities. In recent years, more and more landfill operators are collecting the methane the landfill produces and using it to generate heat or electricity. An even more desirable environmental choice would be keeping organic material out of landfills entirely by using it to make compost.

Siting The designation of a landfill location, typically through a regulatory process involving studies, written reports, and public hearings.

Teaching Tip: Discussion Starter

Ask students: Why are landfill sites often the source of environmental injustice? *People who lack financial resources and political influence are often unable to effectively fight the construction of a landfill in their neighborhood. People in the area of the landfill may be exposed to water pollution, unsightly landscapes, and foul odors. The site may be chosen because people in the area lack the resources to oppose development rather than for the safety criteria of the chosen site.*

Teaching Tip: Journal Prompt

Ask students to list some parameters used when choosing a landfill site. *A landfill should be located in a soil rich in clay and located away from rivers, streams, and other bodies of water used for drinking-water supplies. A landfill should be sufficiently far away from populated areas to protect human health but not so far that transport costs are economically unfeasible.*

Answer to Your Turn

Calculate the volume of water in cubic meters (m^3) that infiltrates the landfill per year.

100 mm/year = 0.1 m/year
0.1 m/year $\times$ 5,000 m^2 $\times$ 30% = 150 m^3

So the volume of leachate that is treated in a year, in m^3, is:

150 m^3 $\times$ 90% = 135 m^3

Teaching Tip: Journal Prompt

Ask students to describe the process of incineration. When or why might incineration be more desirable than a landfill? *Incineration burns waste materials to reduce the volume and creates energy in the process. At the incinerator site, recyclables are sorted out of the waste stream. The remaining material is dumped onto a platform where metals are removed. The waste is dumped into a furnace. After incineration, ash is tested for toxicity, and if the toxicity is low enough, it may be sent to a landfill or used in road construction or in cement applications. If deemed toxic, the ash is sent to a landfill designed for toxic substances. From the combustion waste stream, filters and scrubbers remove sulfur dioxides, nitrogen oxides, and other air pollutants. When land is in short supply, incineration may be preferable to a landfill. Incineration reduces the volume of waste up to 90 percent. Energy harnessed from incineration can be used to generate electricity.*

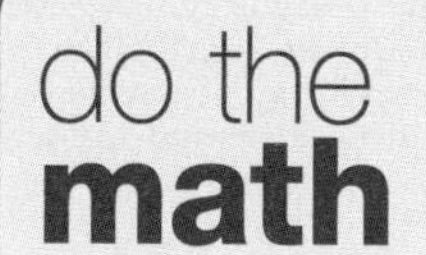

How Much Leachate Might Be Collected?
Annual precipitation at a landfill in the town of Fremont is 100 mm per year, and 50 percent of this water runs off the landfill without infiltrating the surface. The landfill has a surface area of 5,000 m^2. Underneath the landfill, the town installed a leachate collection system that is 80 percent effective. Any leachate not collected by the system enters the surrounding soil and groundwater. This leachate contains cadmium and other toxic metals.

Calculate the volume of water in cubic meters (m^3) that infiltrates the landfill per year.

100 mm/year = 0.1 m/year

0.1 m/year $\times$ 5,000 m^2 $\times$ 50% = 250 m^3

So the volume of leachate in m^3 that is treated per year is: 250m^3 $\times$ 80% = 200m^3

Your Turn In a neighboring landfill, 70 percent of annual precipitation runs off without infiltrating the surface and the leachate collection system is 90 percent effective. Calculate in cubic meters (m^3) the volume of leachate that is treated each year.

Incineration is another way to treat waste materials

Given all the problems of landfills, people have turned to a number of other means of solid waste disposal, including *incineration*. **Incineration** is the process of burning waste materials to reduce their volume and mass and, sometimes, to generate electricity or heat. More than three-quarters of the material that constitutes municipal solid waste is easily combustible. Because paper, plastic, and food and yard waste are composed largely of carbon, hydrogen, and oxygen, they are excellent candidates for incineration. An efficient incinerator operating under ideal conditions may reduce the volume of solid waste by up to 90 percent and the weight of the waste by approximately 75 percent, although the reductions vary greatly depending on the incinerator and the composition of the MSW.

Incineration The process of burning waste materials to reduce volume and mass, sometimes to generate electricity or heat.

Ash The residual nonorganic material that does not combust during incineration.

Bottom ash Residue collected at the bottom of the combustion chamber in a furnace.

Fly ash The residue collected from the chimney or exhaust pipe of a furnace.

Incineration Basics

FIGURE 53.4 shows a mass-burn municipal solid waste incinerator. Typically at such an incinerator, MSW is sorted and certain recyclables are diverted to recycling centers. The remaining material is dumped from a refuse truck onto a platform where certain materials such as metals are identified and removed. A moving grate or other delivery system transfers the waste to a furnace. Combustion rapidly converts much of the waste into carbon dioxide and water, which are released into the atmosphere along with heat.

Particulates, more commonly known as *ash* in the solid waste industry, are an end product of combustion. **Ash** is the residual nonorganic material that does not combust during incineration. Residue collected underneath the furnace is known as **bottom ash** and residue collected beyond the furnace is called **fly ash.**

Because incineration often does not operate under ideal conditions, ash typically fills roughly one-quarter the volume of the precombustion material. Disposal of this ash is determined by its concentration of toxic metals. The ash is tested for toxicity by leaching it with a weak acid. If the leachate is relatively low in concentration of contaminants such as lead and cadmium, the ash can be disposed of in a landfill. Ash deemed safe can also be used for other purposes such as fill in road construction or as an ingredient in cement blocks and cement flooring. If deemed toxic, the ash goes to a special ash landfill designed specifically for toxic substances.

Metals and other toxins in the MSW may be released to the atmosphere or may remain in the ash, depending on the pollutant, the specific incineration

Teaching Tip: Video

Turning Trash into Energy
This video from the Science Channel explains how New York City uses its trash to generate electricity.

The link to this video may be found by clicking the link buttons in the TE-book, opening the TRFD, or logging onto the book's companion website bcs.whfreeman.com/friedlandapes2e (teacher login required).

TRM **Chapter 16 Web Resources**

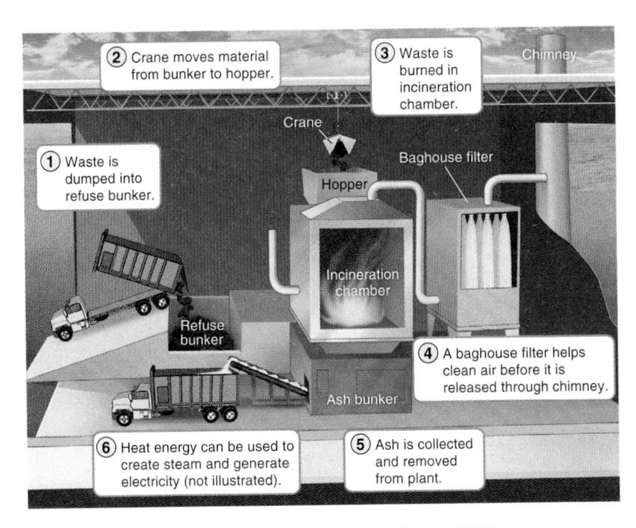

FIGURE 53.4 **A municipal mass-burn waste-to-energy incinerator.** In this plant, MSW is combusted and the exhaust is filtered. Remaining ash is disposed of in a landfill. The resulting heat energy is used to make steam, which turns a generator that generates electricity in the same manner as was illustrated in Figure 34.7 on page 406.

process, and the type of technology used. Exhaust gases from the combustion process, such as sulfur dioxide and nitrogen oxides, move through collectors and other devices that reduce their emission to the atmosphere. These collectors are similar in design to those described in Chapter 15 on air pollution. Acidic gases such as hydrogen chloride (HCl), which results from the incineration of certain materials including plastic, are recovered in a scrubber, neutralized, and sometimes treated further before disposal in a regular landfill or ash landfill.

Incineration also releases a great deal of heat energy, which is often used in a boiler immediately adjacent to the furnace either to heat the incinerator building or to generate electricity, using a process similar to that of a coal, natural gas, or nuclear power plant. When heat generated by incineration is used rather than released to the atmosphere, it is known as a **waste-to-energy** system. Although energy generation is a positive benefit of incineration, as we shall see, there are a number of environmental problems with incineration as a method of waste disposal.

Environmental Consequences of Incineration

Though incinerators address some of the problems of landfills, they also have shortcomings. To cover the costs of construction and operation, incineration facilities also charge tipping fees. Tipping fees are generally higher at incinerators than at landfills; national averages are around $70 per U.S. ton. We have seen that an incinerator may release air pollutants such as organic compounds from the incomplete combustion of plastics and metals contained in the solid waste that was burned. Some environmental scientists believe that incinerators are a poor solution to solid waste disposal because they produce ash that is more concentrated and thus more toxic than the original MSW. Therefore, the siting of an incinerator raises NIMBY and environmental justice issues similar to those of landfill siting.

In addition, because incinerators are typically large and expensive to build and operate, they require large quantities of MSW on a daily basis in order to burn efficiently and to be profitable. As a means of supporting these costs, communities that use incinerators may be less likely to encourage recycling. One solution is to use rate structures and other programs to encourage MSW reduction and diversion, with the goal of using incineration only as a last resort. However, in order to be successful, this usually must be a community effort.

Incinerators may not completely burn all the waste deposited in them. Plant operators can monitor and

Waste-to-energy A system in which heat generated by incineration is used as an energy source rather than released into the atmosphere.

Teaching Tip: Activity

Complete the following chart to compare the advantages and disadvantages of landfills and incineration. (Answers are provided in italics.)

Landfill		Incineration	
Advantages	**Disadvantages**	**Advantages**	**Disadvantages**
• *Engineered to hold large amounts of MSW with little contamination of the surrounding environment* • *Less expensive to build and maintain than incineration*	• *Sites may lead to environmental injustice issues* • *Leachate can contaminate underlying adjacent waterways* • *Anaerobic decomposition leads to the production of the greenhouse gases methane and carbon dioxide* • *Methane creates an explosion hazard if not properly vented*	• *Waste reduction up to 90%* • *Energy may be generated for electricity and heat*	• *Toxic ash must be disposed of in a designated landfill* • *Metals and other toxins may be released into the atmosphere* • *Tipping fees are higher at incineration facilities than at landfills* • *Incinerators are expensive to build and require large quantities of MSW in order to burn efficiently and to be profitable*

modify the oxygen content and temperature of the burn, but because the contents of MSW are extremely variable and lumped all together, it is difficult to have a uniform burn. Consider a truckload of MSW from your neighborhood. The same load may contain food waste with high moisture content and, right next to that, packaging and other dry, easily burnable material. It is difficult for any incinerator—even a state-of-the-art modern facility—to burn all of these materials uniformly.

Inevitably, MSW contains some toxic material. The concentration of toxics in MSW is generally quite low relative to all the paper, plastic, glass, and organics in the waste. However, rather than being dissipated to the atmosphere, most metals remain in the bottom ash or are captured in the fly ash. As we have already mentioned, incinerator ash that is deemed toxic must be disposed of in a special landfill for toxic materials.

As we have seen, there is no ideal choice for waste disposal. Sometimes the decision between whether to construct a landfill or build an incinerator is in part a decision about the kind of pollution a community prefers. The best choice is the production of less material for either the landfill or the incinerator.

module

53

REVIEW

In this module, we have seen that landfills and incinerators are the two ways to dispose of material not diverted from the waste stream. Landfills are designed to keep MSW dry and isolated so that the material in the landfill does not contaminate the surrounding environment. Leachate can contaminate nearby land and waterways if it escapes. Anaerobic decomposition in landfills can lead to the formation of methane, a potent greenhouse gas. Incinerators can generate a variety of air pollutants. Any metals contained within MSW that is incinerated may be released into the air. Bottom ash is particulate matter that accumulates underneath the combustion furnace. Sometimes, ash from incinerators can be considered toxic and must be removed using special precautions. Hazardous waste is the subject of our next module.

Module 53 AP® Review Questions

1. Which material, when placed in a landfill, is most likely to cause problems as a result of leaching?
   (a) Plastic
   (b) Paper
   (c) Food matter
   (d) Glass
   (e) Aluminum

2. A 500 m^2 landfill experiences 150 mm of rain each year and 60 percent of the rain is runoff. If the landfill has a 90 percent effective leachate collection system, how much leachate escapes each year?
   (a) 3 m^3
   (b) 5 m^3
   (c) 16 m^3
   (d) 27 m^3
   (e) 41 m^3

3. Incineration of waste is primarily used
   (a) to generate heat or electricity.
   (b) to reduce waste volume and mass.
   (c) to eliminate heavy metals.
   (d) when there is no other option.
   (e) if the waste is mostly organic material.

4. NIMBY describes
   (a) the fact that materials cannot be easily incinerated.
   (b) the costs associated with the use of incinerators.
   (c) factors contributing to leaching from landfills.
   (d) attempts to develop better landfills.
   (e) an attitude about the placement of landfills.

5. Which is NOT a detriment of waste incineration compared with landfills?
   (a) Increased toxicity of waste
   (b) Increased cost to dispose of waste
   (c) Increased space taken up by solid waste
   (d) Increased air pollution
   (e) Decreased incentive to recycle

Answers to Module 53 AP® Review Questions

1. e
2. a
3. b
4. e
5. c

module

54

Hazardous Waste

When solid waste material is deemed toxic or otherwise harmful to people or natural ecosystems, it should not be placed in a MSW landfill or incinerated. In these cases, special means of handling and disposal are required. In this module we will discuss the proper treatment of hazardous waste and some of the legislation concerning it.

Learning Objectives

After reading this module, you should be able to

- define hazardous waste and discuss the issues involved in handling it.
- describe regulations and legislation regarding hazardous waste.

Teaching Tip: Beyond the Classroom

Many household cleaners are toxic to the environment and may negatively affect human health. Ask students to research nontoxic homemade cleaning products and have them create a recipe card for 5 to 10 products they could use to replace commercial cleaning products at home. An example is provided below:

Heavy Duty Floor Cleaner

- *1 cup white vinegar*
- *1 tablespoon liquid dishwashing soap*
- *1 cup baking soda*
- *2 gallons warm tap water*

Mix ingredients well in a large 3- to 4-gallon bucket and use immediately to clean the floor.

Hazardous waste requires proper handling and disposal

Hazardous waste is liquid, solid, gaseous, or sludge waste material that is harmful to humans or ecosystems. According to the EPA in 2011, over 16,000 hazardous waste generators in the United States produce about 34 million metric tons (38 million U.S. tons) of hazardous waste each year. Only about 4 percent of that waste is recycled. The majority of hazardous waste is the by-product of industrial processes such as textile production, cleaning machinery, and manufacturing computer equipment, but it is also generated by small businesses such as dry cleaners, automobile service stations, and small farms. Even individual households generate hazardous waste—1.5 million metric tons (1.7 million U.S. tons) per year in the United States, including materials such as oven cleaners, batteries, and lawn fertilizers. All of these materials have a much greater likelihood of causing harm to humans and ecosystems than other materials and should not be disposed of in regular landfills.

Most municipalities do not have regular collection sites for hazardous waste or household hazardous waste. Rather, homeowners and small businesses are asked to keep their hazardous waste in a safe location until periodic collections are held (FIGURE 54.1).

FIGURE 54.1 **A typical household hazardous waste collection site in Seattle, Washington.** Residents are encouraged to keep hazardous household waste separate from their regular household waste. Collections are held periodically. *(David R. Frazier/Photolibrary, Inc./Alamy)*

Every aspect of the treatment and disposal of hazardous waste is more expensive and more difficult than the disposal of ordinary MSW. Households are filled

Hazardous waste Liquid, solid, gaseous, or sludge waste material that is harmful to humans or ecosystems.

Teaching Tip: Warm-up

Ask students: What is the definition of hazardous waste, and what are its main sources? *Hazardous waste is liquid, solid, gaseous, or sludge waste material that is harmful to humans or ecosystems. The majority of hazardous waste is the by-product of industrial processes. It can also be generated by small businesses such as dry cleaners, small farms, and automobile service stations. Households can also be a source of hazardous waste from items such as oil-based paints, motor oil, and chemical cleaners.*

Teaching Tip: Journal Prompt

Love Canal is one of the best-known Superfund sites. Ask students to describe what happened at Love Canal. *Love Canal, New York, was a hazardous waste landfill. Love Canal was covered with fill and topsoil and used as a site for a school and a housing development. In 1978 and 1980, known cancer-causing wastes such as benzene (a solvent), dioxin (a by-product of chemical manufacturing), and trichloroethylene (a degreasing agent) were found in the basements of homes in the area. When it became clear that a disproportionately large number of illnesses, possibly connected to the chemical waste, had been diagnosed in the local population, the situation attracted national attention. The contamination was so bad that in 1983 Love Canal was listed as a Superfund site and the inhabitants of the area were evacuated.*

AP® Exam Tip

Make sure the students know the Superfund Act (CERCLA) and the U.S. Resource Conservation and Recovery Act (RCRA). These laws are frequently tested in multiple-choice questions on the AP® Exam.

with numerous substances, such as oil-based paints, motor oil, or chemical cleaners, that are easy to purchase but become regulated hazardous waste as soon as the municipality collects them. Hazardous waste must be treated before disposal. Treatment, according to the EPA definition, means making it less environmentally harmful. To accomplish this, the waste must usually be altered through a series of chemical procedures.

Collection sites are designated as hazardous waste collection facilities that must be staffed with specially trained personnel. Sometimes the materials gathered are unlabeled and unknown and must be treated with extreme caution. Ultimately, the wastes may be sorted into a number of categories, such as fuels, solvents, and lubricants, for example. Some items, such as paint, may be reused, while others may be sent to a special facility for treatment.

As with other waste, there are no truly good options for disposing of hazardous waste. Source reduction is the most beneficial and one of the least expensive ways to approach hazardous waste: Don't create the waste in the first place. In the case of household hazardous waste, many community groups and municipalities encourage consumers to substitute products that are less toxic or to use as little of the toxic substances as possible.

Legislation oversees and regulates the treatment of hazardous waste

Because of the dangers that hazardous waste presents, disposal has received much public attention. At times in the past few decades, hazardous waste sites have become commonly known household names. Because lawmakers and the general public have mandated that hazardous waste be regulated, there are a number of laws and acts that specifically cover hazardous waste. We will discuss two of them in this section.

Regulation and Oversight of Handling Hazardous Waste

Regulation and handling of hazardous waste in the United States falls under two pieces of federal legislation, the U.S. Resource Conservation and Recovery Act (RCRA) and the Comprehensive Environmental Response, Compensation, and Liability Act (CERCLA).

In 1976, RCRA expanded previous solid waste laws. Its main goal was to protect human health and the natural environment by reducing or eliminating the generation of hazardous waste. Under RCRA's provision for "cradle-to-grave" tracking, the EPA maintains lists of hazardous wastes and works with businesses and state and local authorities both to minimize hazardous waste generation and to make sure that the waste is tracked until proper disposal. In 1984, RCRA was modified with the federal Hazardous and Solid Waste Amendments (HSWA), which encouraged waste minimization and phased out the disposal of hazardous wastes on land. The amendments also increased law enforcement authority in order to punish violators.

CERCLA, usually referred to as the **Superfund Act,** is a 1980 U.S. federal act that imposes a tax on the chemical and petroleum industries, funds the cleanup of abandoned and nonoperating hazardous waste sites, and authorizes the federal government to respond directly to the release or threatened release of substances that may pose a threat to human health or the environment. The Superfund Act is well known because of a number of sensational cases that have fallen under its jurisdiction. Originally passed in 1980 and amended in 1986, this legislation has several parts. First, it imposes a tax on the chemical and petroleum industries. The revenue from this tax is used to fund the cleanup of abandoned and nonoperating hazardous waste sites where a responsible party cannot be established. The name *Superfund* came from this provision. CERCLA also authorizes the federal government to respond directly to the release or threatened release of substances that may pose a threat to human health or the environment.

Under Superfund, the EPA maintains the National Priorities List (NPL) of contaminated sites that are eligible for cleanup funds. For a long time, very little Superfund money was disbursed and few sites underwent remediation. The map in FIGURE 54.2 shows the location of the NPL sites. As of mid-2014, there were 1,321 Superfund sites—at least 1 in every state except North Dakota. New Jersey, with 113 Superfund sites, has the most. California has the next highest number with 97 Superfund sites, followed by Pennsylvania with 94. New York has 86 sites.

Perhaps the best-known Superfund site is Love Canal, New York (FIGURE 54.3). Originally a hazardous waste landfill, Love Canal was covered with fill and topsoil and used as a site for a school and a housing development. In 1978 and 1980, known cancer-causing wastes such as benzene (a solvent), dioxin (a by-product of chemical manufacturing), and trichloroethylene (a degreasing agent) were found in the basements of

Superfund Act The common name for the Comprehensive Environmental Response, Compensation, and Liability Act (CERCLA); a 1980 U.S. federal act that imposes a tax on the chemical and petroleum industries, funds the cleanup of abandoned and nonoperating hazardous waste sites, and authorizes the federal government to respond directly to the release or threatened release of substances that may pose a threat to human health or the environment.

Teaching Tip: Activity

Ask students to complete the table below, which differentiates between the two hazardous waste laws. (Answers are provided in italics.)

	RCRA	CERCLA
Also known as	*"Cradle-to-Grave"*	*"Superfund Act"*
Goal	*To protect human health and the natural environment by reducing or eliminating the generation of hazardous waste*	*A tax on the chemical and petroleum industries to fund the cleanup of abandoned hazardous waste sites*
Authorizes	*EPA maintains lists of hazardous waste from generation through disposal*	*Allows the federal government to respond directly to the release of hazardous substances*

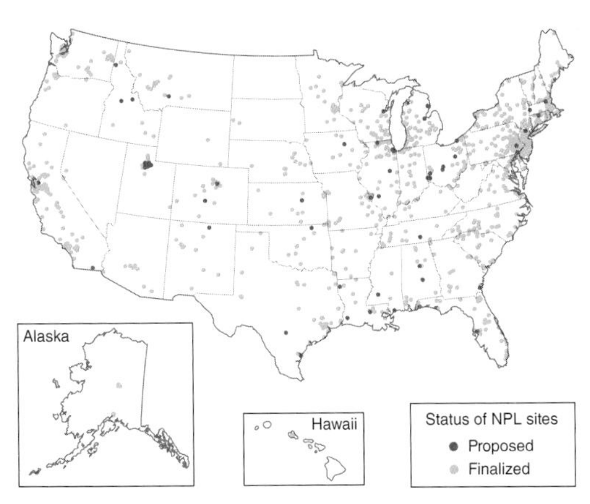

FIGURE 54.2 **Distribution of NPL (Superfund) sites in the United States.** Under Superfund, the EPA maintains the National Priorities List (NPL) of contaminated sites that are eligible for cleanup funds. *(After http://www.epa.gov/superfund/sites/products/nplmap.pdf)*

homes in the area. When it became clear that a disproportionately large number of illnesses, possibly connected to the chemical waste, had been diagnosed in the local population, the situation attracted national attention. The contamination was so bad that in 1983 Love Canal was listed as a Superfund site and the inhabitants of the area were evacuated. In 1994, the EPA removed Love Canal from the National Priorities List because the physical cleanup had been completed and the site was no longer deemed a threat to human health.

Since its inception, many have observed that CERCLA has not had enough funding to clean up the numerous hazardous waste sites around the country. The listing of NPL sites in a state does not necessarily include all contaminated sites within that state. For example, the Department of Environmental Protection in New Jersey, a state that has approximately 113 listed Superfund sites, believes that there are more than 9,000 sites where soil or groundwater is contaminated with hazardous chemicals.

Brownfields

The Superfund designation is reserved strictly for those locations with the highest risk to public health, and Superfund sites are managed solely by the federal government. In 1995, the EPA created the *Brownfields* Program. **Brownfields,** like Superfund sites, are contaminated industrial or commercial sites that may require environmental cleanup before they can be redeveloped or expanded. The Brownfields Program assists state and local governments in cleaning up contaminated industrial and commercial land that did not achieve conditions necessary to be in the Superfund category. Old factories, industrial areas and waterfronts, dry cleaners, gas stations, landfills, and rail yards are common examples of brownfield sites. Brownfields legislation has prompted the revitalization of several sites throughout the country. One notable instance is Seattle's Gasworks Park. In 1962, the city of Seattle purchased the land, which had housed a coal gasification plant, to rehabilitate the site into a park. After undergoing chemical abatement and environmental cleanup, the park has become a distinctive landmark for the city and is the site of many public events throughout the year.

The Brownfields Program has been criticized as an inadequate solution to the estimated 450,000 contaminated locations throughout the country. Since the cleanup is managed entirely by state and local governments, brownfields management can vary widely from region to region. Furthermore,

FIGURE 54.3 **Love Canal, New York.** Love Canal became a symbol of hazardous chemical pollution in the United States in the 1970s. Lois Gibbs, with no prior experience in either environmental science or activism, became a spokesperson for the neighborhood of Love Canal and is widely credited with bringing national attention to her community, including the elementary school that had been constructed on top of large quantities of hazardous chemical waste. Gibbs is currently the executive director of the Center for Health, Environment and Justice. *(Harry Scull, Jr./Getty Images)*

Brownfields Contaminated industrial or commercial sites that may require environmental cleanup before they can be redeveloped or expanded.

AP® Exam Tip

Hazardous waste disposal and brownfields are covered in the 2006 AP® Exam, Question 3. To answer this question, students must

- describe two problems caused by removing contaminated soil from a brownfield
- explain how vegetation can be used to decontaminate soil and the advantages and disadvantages of this method
- explain one environmental benefit and one social benefit of brownfield reclamation
- list and explain one method currently used to reduce hazardous waste production and one legal method of hazardous waste disposal

Teaching Tip: Warm-up

Ask students to differentiate between a Brownfield site and a Superfund site. *The Superfund designation is reserved strictly for those locations with the highest risk to public health and Superfund sites are managed by the EPA. Brownfield sites are contaminated industrial or commercial sites that may require a cleanup before they can be redeveloped and are managed by local and state governments. Brownfield legislation lacks legal liability to compel polluters to rehabilitate their properties.*

Teaching Tip: Discussion Starter

Ask students to discuss how the international community is involved in the disposal of hazardous waste. *Municipalities and industries in the United States have tried to send hazardous waste to countries with less-stringent regulations in exchange for cash payments. The United States has also accepted hazardous waste from other countries because it has the facilities to treat certain types of hazardous waste.*

the brownfields legislation lacks legal liability controls to compel polluters to rehabilitate their properties. Without legal recourse, many brownfield sites remain unused and contaminated, posing a continued risk to public health.

International Consequences

Because of the difficulties involved in disposing of hazardous waste, municipalities and industries sometimes try to send the waste to countries with less stringent regulations. There are numerous reports of garbage and ash barges that travel the oceans looking for a developing country willing to accept hazardous waste from the United States in exchange for a cash payment. Perhaps the most famous is the story of the cargo vessel *Khian Sea,* which left Philadelphia in 1986 with almost 13,000 metric tons of hazardous ash from an incinerator. It traveled to a number of countries in the Caribbean in search of a dumping place. Some of the ash was dumped in Haiti, and some was dumped in the ocean. In 1996, the United States ordered that the ash dumped in Haiti be retrieved and returned to the United States. After being held at a dock in Florida, the ash was deemed nonhazardous by the EPA and other agencies and in 2002, the ash was placed in a landfill in Franklin County, Pennsylvania, not very far from its source of origin.

In 2003, a notable instance of a waste transfer in the other direction occurred. A Pennsylvania company that specializes in recovering mercury from a wide variety of products agreed to accept 270 metric tons of mercury waste generated by a company in the state of Tamil Nadu, India, during the manufacture of thermometers. India had no facilities for recycling mercury waste, and most of the thermometers manufactured at the plant were shipped to the United States and Europe, so the transfer of material seemed appropriate. The mercury was shipped from India to the United States, where the mercury was concentrated, purified, and then sold to industrial users of the metal.

There are currently incinerator operators in Vancouver, British Columbia; Plymouth, England; and elsewhere around the world that are seeking appropriate disposal sites for their ash, but they seem to have stopped looking for other countries to accept this material. Most incinerator operators are now identifying closer locations for ash disposal.

module 54 REVIEW

Hazardous wastes are a special category of waste material that is harmful to humans or ecosystems. Much of this waste is composed of byproducts from industrial processes. Hazardous wastes must be handled and treated separately from the MSW stream. There are a number of legislative acts that specifically address hazardous waste. Sometimes hazardous waste is disposed of in other countries, resulting in international consequences to the safe handling and treatment of hazardous waste.

Answers to Module 54 AP® Review Questions

1. d
2. a
3. c
4. a

Module 54 AP® Review Questions

1. Hazardous waste
   (a) costs much more to recycle than regular waste.
   (b) is primarily generated by individuals and small businesses.
   (c) should be disposed of in landfills separately from MSW.
   (d) includes many household items such as certain paints and oil.
   (e) cannot be treated to make safe.

2. Which legislation calls for listing hazardous waste to use in cradle-to-grave tracking?
   (a) Resource Conservation and Recovery Act
   (b) Comprehensive Environmental Response, Compensation, and Liability Act
   (c) Hazardous and Solid Waste Amendments
   (d) Superfund Act
   (e) National Priorities Act

3. The Brownfields Program
   (a) includes hazardous locations with the highest risk to public health.
   (b) attempts to prevent pollution in sites near agricultural lands.
   (c) is managed primarily by state and local governments.
   (d) has made significant progress in cleaning the 450,000 identified sites.
   (e) includes fines and legal consequences for polluters who do not contribute to cleaning.

4. Why might hazardous waste disposal in the United States be an international issue?
   (a) The lower costs of disposal elsewhere means municipalities and industries sometimes export waste.
   (b) Air pollution from waste treatment often crosses international boundaries.
   (c) The United States continues to dump its hazardous waste into oceans.
   (d) The United States produces the majority of the world's hazardous waste.
   (e) The majority of electronic waste is processed in the United States.

module 55

New Ways to Think About Solid Waste

Throughout this chapter we have described a variety of ways of managing solid waste, from creating less of it to burying it in a landfill to burning it. Each method has both benefits and drawbacks. Because there is no obvious best method and because waste is a pervasive fact of contemporary life, the problem of waste disposal seems overwhelming. How can an individual, a small business, an institution, or a municipality best deal with solid waste? The answer is highly specific to each case and varies by region. Every method of waste disposal will have adverse environmental effects; the challenge is to find the least detrimental option. How do we decide which choices are best? Life-cycle analysis and a holistic approach are two useful approaches to gaining insights into the question of what we should do with our solid waste.

Learning Objectives

After reading this module, you should be able to

- explain the purpose of life-cycle analysis.
- describe alternative ways to handle waste and waste generation.

Module 55 New Ways to Think About Solid Waste 579

Teaching Tip: Beyond the Classroom

Ask students to perform a life-cycle analysis on two comparable products or processes and present their findings to the class. Examples could include plastic versus aluminum soda containers, paper versus plastic shopping bags, incineration versus using landfills, and the production of polyester versus cotton fabric.

Teaching Tip: Journal Prompt

Ask students: How is integrated waste management different from other approaches to waste management? *Integrated waste management employs multiple approaches to the reduction, management, and disposal of waste in order to reduce the environmental impact of MSW. Options are available in any combination that is most appropriate at a given time.*

Life-cycle analysis considers materials used and released throughout the lifetime of a product

Recall the beginning of the chapter where we attempted to make an objective assessment of solid waste disposal options by comparing paper and polystyrene cups. This process, known as **life-cycle analysis,** and also known as **cradle-to-grave analysis,** is an important systems tool that looks at the materials used and released throughout the lifetime of a product—from the procurement of raw materials through their manufacture, use, and disposal. As we saw when we compared a paper cup with a polystyrene cup, the full inventory of a life-cycle analysis sometimes yields surprising results.

In theory, conducting a life-cycle analysis should help a community determine whether incineration is more or less desirable than using a landfill. However, such an analysis has limitations. For example, it can be difficult to determine the overall environmental impact of a specific material. It is not possible to know whether the particulates and nitrogen oxides released from incinerating food waste are better or worse for the environment than the amount of methane that might be released if the same food waste were placed in a landfill. So in the case of waste that contains food matter, it is not possible to directly compare the full environmental impact of disposal in a landfill versus incineration. Similarly, we saw that it is a challenge to compare the sulfur dioxide released when making a paper cup with the volatile organic compounds released when making a plastic cup. Although life-cycle analysis may not be able to determine absolute environmental impact, it can be very helpful in assessing other considerations, especially those related to economics and energy use.

In terms of economics, the municipality might compare the costs of different disposal methods. For example, a glass manufacturing plant might pay $5 per ton for green glass that it will recycle into new glass. Economically, a municipality might do better if it received $5 for a ton of glass from a bottle manufacturing plant than if it paid a $50 per ton tipping fee so it could throw the material into a landfill. But the municipality must also consider the lower cost of transporting the glass to a relatively close landfill rather than to a distant glass plant.

In some parts of the country, the cost of waste disposal is covered by local taxes; in other locations, municipalities, businesses, or households may have to pay directly for disposal of their solid waste. Whether direct or indirect, there is always a cost to waste disposal. Normally, disposal of recyclables costs less than material destined for the landfill because the landfill always involves a tipping fee while recyclables either incur a lower tipping fee or generate revenue. However, as we have seen, costs change depending on many factors, including market conditions. For example, in a particular year, there may not be a market for recycled newspapers in the United States, but in the following year a huge increase in Japanese demand for newsprint may cause the price to go up in the United States. It is therefore essential for municipalities to have many choices and to be able to modify these choices as market environments change.

From the perspective of energy use, a life-cycle analysis should also consider the energy content of gasoline or diesel fuel used and the pollution generated in trucking material to each destination, as well as the monetary, energy, and pollution savings achieved if the new glass is made from old glass rather than from raw materials (sand, potash, lime). Reconciling all these competing factors is very challenging and the ultimate decisions based on such analyses are often debatable.

Life-cycle analysis A systems tool that looks at the materials used and released throughout the lifetime of a product–from the procurement of raw materials through their manufacture, use, and disposal. *Also known as* **Cradle-to-grave analysis.**

Integrated waste management An approach to waste disposal that employs several waste reduction, management, and disposal strategies in order to reduce the environmental impact of MSW.

Integrated waste management is a more holistic approach

A more holistic method seeks to develop as many options as possible, emphasizing reduced environmental harm and cost. **Integrated waste management** employs several waste reduction, management, and disposal strategies in order to reduce the environmental impact of MSW. Such options include a major emphasis on source reduction and include any combination of recycling, composting, use of landfills, incineration, and whatever additional methods are appropriate to the particular situation. FIGURE 55.1 shows how a nation or a community could consider a series of steps, starting with source reduction during manufacturing and procurement of items. After that, behavior related to use and disposal can be considered and possibly altered in order to obtain the desired outcome: less generation of MSW. According to this approach, no community should be forced into any one method of waste disposal. If a region makes a large investment in an incinerator, for

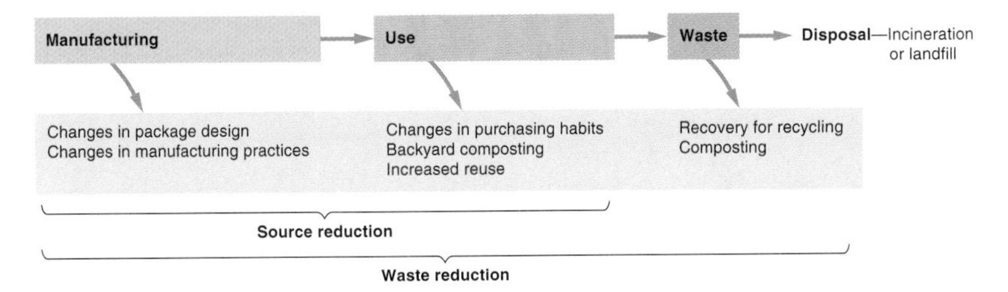

FIGURE 55.1 **A holistic approach to waste management.** Depending on the kind of waste and the geographic location, reducing waste can take much less time and money than disposing of it. Horizontal arrows indicate the waste stream from manufacture to disposal and curved arrows indicate ways in which waste can either be reduced or removed from the stream, thereby reducing the amount of waste incinerated or placed in landfills.

example, there is a risk that it would then need to attract large quantities of waste to pay for that incinerator, thereby reducing the incentive to recycle or use a landfill. Landfill space may be abundant or scarce, depending on the location of the community. If the municipality is free to consider all options, it can make the choice or choices that are efficient, cost effective, and least harmful to the environment.

The architect and former University of Virginia dean William McDonough looks at holistic waste management from a more far-reaching perspective. In the book *Cradle to Cradle*, McDonough and coauthor Michael Braungart propose a new approach to the manufacturing process. They argue that it is first necessary to assess existing practices in order to minimize waste generation before, during, and after manufacturing. Beyond that, manufacturers of durable goods such as automobiles, computers, appliances, and furniture should develop plans for disassembling goods when they are no longer useful so that parts or materials can be recycled with as little as possible becoming part of the waste stream.

Some industries are already developing new approaches to waste. Volkswagen, for example, manufactures some of its cars so that they can be easily taken apart and materials of different composition easily separated to allow recycling. Certain carpet manufacturers design their carpets so that when worn out they can be easily recycled into new carpeting (FIGURE 55.2). This is typically done by making a base that is extremely durable with a top portion of the carpet that can be changed when the color fades, is worn out, or is no longer desired. Finally, McDonough and Braungart point out that many organisms in the natural world, such as the turtle, produce very hard, impact-resistant materials, such as a shell, without producing any toxic waste. They suggest that humans should examine how a turtle creates such a hard shell without the production of toxic wastes. Humans can use this example as a goal for other kinds of production where no toxic wastes would be produced. More recently, McDonough and Braungart have introduced the term *upcycle* to describe the conversion of a waste material to something of higher quality and greater value than the original product.

FIGURE 55.2 **A recyclable carpet.** FLOR carpet tiles are designed to be easily replaced and easily recycled when the carpet wears out. *(Akron Beacon Journal/MCT/Landov)*

◀ TEACHING with FIGURES

Ask students to evaluate Figure 55.1 and list the methods that can be used along the manufacturing, use, and waste pathway to reduce or remove waste from the stream. *Methods that can reduce or remove waste from the stream include changes in packaging that require less material; changes in manufacturing processes that reduce resource use; changes in purchasing habits (for example, not buying certain products); backyard composting; and reuse of products and increased recycling.*

Teaching Tip: Engage

Ask students the following questions:

- What is upcycling? *Upcycling is taking a waste material or product and upgrading it to something of higher quality.*
- Can you think of a discarded item in your home that you might upcycle? *Answers will vary.*

Teaching Tip: Video

Creative Houses from Reclaimed Stuff
This video from TED introduces sustainable design. Dan Phillips describes how he uses recycled and reclaimed materials to build houses.

The link to this video may be found by clicking the link buttons in the TE-book, opening the TRFD, or logging onto the book's companion website bcs.whfreeman.com/friedlandapes2e (teacher login required).

Chapter 16 Web Resources

Answers to Module 55 AP® Review Questions

1. b
2. a
3. e

Although we can still greatly improve our handling of solid waste, there is some evidence that the nation as a whole has taken source reduction seriously; per capita waste generation has been level since 1990. It appears that recycling has been taken seriously, too, since recycling rates have increased since 1985. However, considering both the amount of waste generated in this country and the recycling rates found in other countries, we have a long way to go.

module
55
REVIEW

In this module, we have seen that there are new and innovative ways to think about and handle MSW. Life-cycle analysis allows the determination of all of the materials and energy that go into the manufacture, use, and disposal of a product, with the goal of minimizing environmental impact. Life-cycle analysis is sometimes called cradle-to-grave analysis because it tracks an object from the time it is manufactured or born (cradle) until it is disposed of (grave). Sometimes life-cycle analysis can inform decisions about existing manufacturing practices. In some cases, the manufacturing process can be modified so that less waste is generated and, in addition, after a useful lifetime, the manufactured product can be remanufactured into something else.

Module 55 AP® Review Questions

1. Life-cycle analysis
   (a) examines the total environmental impact an object will make when it is discarded.
   (b) examines the materials and energy associated with an object from extraction of materials to disposal.
   (c) is used to analyze the least environmentally harmful way to dispose of an object.
   (d) considers only the environmental costs associated with an object from extraction of materials to disposal.
   (e) is not used when considering the environmental impact of products that are already being recycled.

2. Integrated waste management
   (a) suggests that communities should have multiple options for waste disposal.
   (b) focuses on recycling waste.
   (c) is a method of educating individuals about the best method of waste disposal.
   (d) often ignores the benefits of composting.
   (e) focuses on the manufacturing process of new products.

3. Which is NOT part of the cradle to cradle concept proposed by William McDonough?
   (a) Planning for disassembling materials during design
   (b) Minimized waste generation during manufacturing
   (c) Increased use of composite materials
   (d) Mimicking the natural world in the production of hard materials
   (e) Upcycling waste materials

working toward sustainability

Recycling E-Waste in Chile

Electronic waste is a small part of the waste stream but it contains a large fraction of waste that ends up in landfills, dumps, and other locations. Toxic metals such as lead, mercury, and cadmium as well as carcinogenic organic compounds are common in electronic waste. Recycling rates for electronic waste vary around the world. In Switzerland, more than 80 percent of electronic waste is recycled. In the United States, the recycling rate for e-waste is about 20 percent. Some countries recycle almost none of their electronic waste. For e-waste that is collected separately, fully assembled electronic devices are usually exported for disassembly and recycling elsewhere. Not only does this practice export the pollution burden, it also uses unnecessary energy and space to export the entire device rather than just the components that need recycling.

Until the year 2000, the South American nation of Chile recycled less than 1 percent of its electronic waste. Businessman and entrepreneur Fernando Nilo decided he wanted to increase the e-waste recycling rate in Chile. He devised a business plan that won a local competition and subsequently obtained funding to launch his plan. In 2005, Recycla Chile of Santiago opened the first recycling facility in Latin America, which received electronic devices such as computers, cell phones, scanners, and televisions. While in China and other countries, workers, including children, pick apart e-waste without proper face and respiratory protection, at Recycla the workers are trained to dismantle and separate electronic components safely. Certain separated materials are sold to reputable dealers within Chile, while other materials are compacted to minimize shipping and energy costs and sent to environmentally certified metal smelters around the globe. These metal smelting companies safely recover valuable metals for recycling and use in new electronic devices. One of the innovations introduced by Recycla is the first "Green Seal" in Latin America. The manufacturer typically passes along the responsibility for recycling a product to the new owner. For example, when a customer purchases a printer, he or she is then responsible for proper disposal of that printer. Recycla introduced a program where the manufacturer contracted with Recycla to be responsible for the disposal of a product whenever the customer decides it is no longer needed. The firm that made the electronic device is financially or physically responsible for recycling that object. With this guarantee at the time of purchase, the object is much more likely to end up in a proper location at the end of its useful life, with no additional cost to the consumer.

Nilo and coworkers began a national advertising campaign both to generate revenue for their company and to educate leaders in their country and around the world about the importance of recycling electronic waste. Today Recycla promotes environmental sustainability in Chile and around the world, and provides jobs for marginalized people including prisoners and former prisoners. The company is profitable and has received many awards. When Recycla started, Chile was recycling less than 1 percent of its electronic waste, though it had a goal to recycle 10 percent within a few years. In 2009, Recycla reported that they had achieved a 5 percent recycling rate in Chile, well on the way to their goal and more than five times higher than the recycling rate when they started.

Critical Thinking Questions

1. What are some of the barriers to e-waste recycling?
2. Are you aware of any e-waste recycling activities in your community? Are they well publicized? Can you think of ways to enhance them?

References

Information Development Incubator Support Center. *Environmental Impact of IT Solutions: Recycling E-Waste in Chile.* http://www.infodev.org/highlights/network-member-client-companies-compete-ebn-smart-entrepreneurship-award.

Recycla Chile, SA. *Solving the E-Waste Problem* (company report), 2010.

A Recycla collection site in Chile. In this facility, electronic waste is deposited for recycling, thereby keeping metals and other harmful components of electronic products from the landfill or incinerator. *(imago stock&people/Newscom)*

Teaching Tip: Journal Prompt

Ask students: Which strategies differentiate Recycla Chile of Santiago from typical e-waste models? *E-waste is not exported but remains in Chile. Workers are trained to dismantle electronics safely. Some materials are sold to companies in Chile. To minimize energy and transportation costs, other materials are sold to environmentally certified smelters in other countries. The manufacturer of the product is responsible for returning the product to be recycled. The company ran a campaign to educate the public about their initiatives.*

Suggested Answers to Critical Thinking Questions

1. Due to the toxic nature of many parts of e-waste, the recycling can be hazardous if not done properly. It is also difficult to separate the many parts of e-waste from each other as well as from solid waste.

2. Responses will vary. Enhancements could include increased publicity, or methods for making recycling easier such as picking up e-waste from homes or businesses.

Teaching Tip: Measuring Your Impact

Understanding Household Solid Waste Measuring Your Impact 16 asks students to consider the MSW generated by a typical household in the United States and to calculate how much that household could save through composting and recycling. The link to this resource may be found by clicking on the link buttons in the TE-book, opening the TRFD, or logging onto the book's companion website bcs.whfreeman.com/friedlandapes2e (teacher login required).

 **Measuring Your Impact 16:** Understanding Household Solid Waste

Exploring the Literature

Appelhof, M. 2003. *Worms Eat My Garbage: How to Set Up and Maintain a Worm Composting System.* 2nd ed. Flower Press.

McDonough, W., and M. Braungart. 2002. *Cradle to Cradle: Remaking the Way We Make Things.* North Point Press.

McDonough, W., and M. Braungart. 2013. *The Upcycle: Beyond Sustainability—Designing for Abundance.* North Point Press.

UN-HABITAT. 2010. *Solid Waste Management in the World's Cities: Water and Sanitation in the World's Cities.* Earthscan.

U.S. Environmental Protection Agency. 2011. *Municipal Solid Waste Generation, Recycling and Disposal in the United States: Facts and Figures for 2011.* http://www.epa.gov/waste/nonhaz/municipal/msw99.htm

Download a printable version of this list by clicking the link buttons in the TE-book, opening the TRFD, or logging onto the book's companion website bcs.whfreeman.com/friedlandapes2e (teacher login required).

 **Exploring the Literature 16**

chapter 16 REVIEW

In this chapter, we examined municipal solid waste. We observed that waste is unique to human beings. Total MSW and per capita waste generation in the United States has leveled off in recent years. The diversion of material from the solid waste stream can occur from reduction, reusing, recycling, and composting. The total amount and per capita recycling had been increasing for a number of years but has leveled off since roughly 2008. Waste disposal relies on landfills or incineration, both of which have benefits and adverse consequences. Hazardous waste is a special category of waste that must be handled and disposed of with particular care. There is national legislation that addresses hazardous waste. Life-cycle analysis allows a holistic approach to studying the entire waste stream from the creation of materials through their use and ultimate disposal.

Key Terms

Waste
Municipal solid waste (MSW)
Waste stream
Reduce, reuse, recycle
The three Rs
Source reduction
Reuse
Recycling
Closed-loop recycling
Open-loop recycling
Composting
Leachate
Sanitary landfill
Tipping fee
Siting
Incineration
Ash
Bottom ash
Fly ash
Waste-to-energy
Hazardous waste
Superfund Act
Brownfields
Life-cycle analysis
Cradle-to-grave analysis
Integrated waste management

Learning Objectives Revisited

Module 51 Only Humans Generate Waste

- **Explain why we generate waste and describe recent waste disposal trends.**

  From an ecological and systems perspective, waste is composed of the nonuseful products of a system. Much of the solid waste problem in the United States stems from the attitudes of the "throw-away society" adopted after World War II. While the United States may be a major generator of solid waste, the lifestyle and goods disseminated around the world have made solid waste a global problem.

- **Describe the content of the solid waste stream in the United States.**

  Solid waste in the United States is composed of primarily paper, food, and yard waste. The total amount and per capita waste generation was steadily increasing in the United States until the last few years, during which there has been a slight decline.

Module 52 The Three Rs and Composting

- **Describe the three Rs.**

  The three Rs—Reduce, Reuse, and Recycle—divert materials from the waste stream. Reduce refers to activities that encourage a reduction in the use and disposal of materials. Reuse refers to using items multiple times whenever possible. Recycling refers to returning an object to a manufacturing plant where it is turned into the same or another object made from the same material. They are listed in order of least environmental impact to greatest environmental impact.

- **Understand the process and benefits of composting.**

  Composting is the diversion of organic material such as food and yard waste from the waste stream and allowing it to decompose into organic soil (humus). Composting, source reduction, and reuse generally have lower energy and financial costs than recycling, but all are important ways to minimize solid waste production.

Module 53 Landfills and Incineration

- **Describe the goals and function of a solid waste landfill.**

  Currently, most solid waste in the United States is buried in landfills. Contemporary landfills entomb the garbage and keep water and air from entering and leachate from escaping. The potential for toxic leachate to contaminate surrounding waterways is one major concern. The other concern is the generation of methane gas. Siting of landfills often raises issues of environmental justice.

- **Explain the design and purpose of a solid waste incinerator.**

  Incineration is an alternative to landfills. Its main benefit is that it reduces the waste material to roughly one-quarter of its original volume. In addition, waste-to-energy incineration often uses the excess heat produced to generate electricity. However, incineration generates air pollution and ash, which can sometimes contain a high concentration of toxic substances and require disposal in a special ash landfill.

Module 54 Hazardous Waste

- **Define hazardous waste and discuss the issues involved in handling it.**

  Hazardous waste is a special category of material that is especially toxic to humans and the environment. It includes industrial by-products and some household items such as batteries and oil-based paints, all requiring special means of disposal.

- **Describe regulations and legislation regarding hazardous waste.**

  Though a variety of regulations and legislation have been implemented to address issues of hazardous waste, many problems remain. CERCLA, also called the Superfund Act, is probably the most well-known regulation concerning hazardous waste because it provides for the cleanup of hazardous waste sites.

Module 55 New Ways to Think About Solid Waste

- **Explain the purpose of life-cycle analysis.**

  Life-cycle analysis tracks material "from cradle to grave." Using life-cycle analysis and integrated waste management—which draws on all the available treatment methods—we can make optimal decisions regarding our solid waste.

- **Describe alternative ways to handle waste and waste generation.**

  The best solution is to design products with a strategy for their ultimate reuse or their dismantling and recycling. This approach has become more common in recent years.

Chapter 16 Review 585

Answers to Chapter 16 AP® Exam Multiple-Choice Questions

1. b
2. d
3. e
4. d
5. e
6. b
7. a
8. b
9. e
10. a
11. d
12. a
13. c

Answers to Chapter 16 AP® Exam

Download the full answers to the Chapter 16 AP® Exam by clicking on the link buttons in the TE-book, opening the TRFD, or logging onto the book's companion website bcs .whfreeman.com/friedlandapes2e (teacher login required).

Answers to Chapter 16 AP® Exam

Chapter 16 AP® Environmental Science Practice Exam

Section 1: Multiple-Choice Questions

Choose the best answer for questions 1–13.

1. Which is NOT a reason to keep household batteries out of landfills?
   (a) They can leach toxic metals.
   (b) Their decomposition can contribute to greenhouse gas emissions.
   (c) They can be recycled, which would reduce the need for new raw materials.
   (d) They can be recycled, which would reduce the need for additional energy.
   (e) They take up space in landfills and we have a finite supply of landfill space.

2. All of the following are desired outcomes of MSW incineration except
   (a) extracting energy.
   (b) reducing volume.
   (c) prolonging the life of landfills.
   (d) increasing air pollution.
   (e) generating electricity.

3. In the last 15 years, MSW per capita in the United States has
   (a) decreased drastically.
   (b) decreased, then increased drastically.
   (c) increased drastically.
   (d) increased moderately, and then decreased drastically.
   (e) stayed the same.

4. The EPA estimates that approximately ____ percent of municipal solid waste comes from residences and ____ percent comes from commercial and institutional facilities.
   (a) 30/70
   (b) 40/60
   (c) 50/50
   (d) 60/40
   (e) 70/30

5. Which material constituted the largest component of municipal solid waste?
   (a) Metals
   (b) Yard waste
   (c) Food scraps
   (d) Discarded electronic devices
   (e) Paper

6. Increasing tipping fees can cause
   (a) decreased rates of recycling.
   (b) increases in illegal dumping.
   (c) incentives for proper waste disposal.
   (d) reduced use of hazardous material.
   (e) a lack of funding for landfills.

7. From an environmental waste perspective, which of the following is the most desirable?
   (a) Reduce
   (b) Reuse
   (c) Recycle
   (d) Compost
   (e) Incinerate

8. In the United States, how much of generated waste ends up being recycled?
   (a) $<$ 23 percent
   (b) $<$ 33 percent
   (c) $<$ 43 percent
   (d) $>$ 53 percent
   (e) $>$ 60 percent

9. Which legislation imposes a tax on the chemical and petroleum industries to generate funds to pay for the cleanup of hazardous substances?
   (a) RCRA
   (b) Cradle-to-Grave Act
   (c) The National Priorities List
   (d) HSWA
   (e) CERCLA

10. Of the following, which contributes most to the weight of MSW?
   (a) Packaging
   (b) E-waste
   (c) Plastics
   (d) Yard waste
   (e) Metals

For questions 11–13, refer to the following lettered choices and choose the compound that is most associated with each numbered statement.

(a) Benzene
(b) Dioxin
(c) Methane
(d) Hydrochloric acid

11. It may be present in the emissions from waste incinerators.

12. It contaminated the land and water near the housing development Love Canal in New York.

13. It is produced by anaerobic decomposition in landfills.

Additional Multiple-Choice AP® Practice Exam

Download an additional Multiple-Choice AP® Environmental Science Practice Exam for Chapter 16 by clicking on the link buttons in the TE-book, opening the TRFD, or logging onto the book's companion website bcs.whfreeman.com /friedlandapes2e (teacher login required).

Multiple-Choice AP® Practice Exam 16

Section 2: Free-Response Questions

Write your answer to each part clearly. Support your answers with relevant information and examples. Where calculations are required, show your work.

1. The total amount of municipal solid waste (MSW) generated in the United States increased from 80 million metric tons (88 million U.S. tons) in 1960 to 232 million metric tons (255 million U.S. tons) in 2007.
   (a) Describe reasons for this increase and explain how the United States became the leader of the "throw-away society." (2 points)
   (b) Explain why reducing is more favorable than reusing, which in turn is more favorable than recycling. (4 points)
   (c) Describe the process of composting and compare a home composting system with that of a large-scale municipal facility. (4 points)

2. In many developing countries garbage is still deposited in open dumps. Developed countries have phased out open dumps in favor of landfills or incinerators for handling garbage in the waste stream.
   (a) Explain why open dumps have been phased out in developed countries. (2 points)
   (b) Describe how a modern-day sanitary landfill is constructed and explain how it is an improvement on the earlier landfills. (2 points)
   (c) Discuss one benefit and one problem of landfilling waste. (2 points)
   (d) Describe how a waste incinerator works and how it could be used to generate electricity. (2 points)
   (e) Discuss one benefit and one problem of incinerating waste. (2 points)

Answers to the Chapter 16 Free-Response Questions can be found in the Answer Appendix on page ANS-13.

chapter

17 Human Health and Environmental Risks

PD Overview

Human health is a subject of great interest to students. This chapter examines human health and environmental risk from the perspective of an environmental scientist. The chapter begins with a look at historical and emergent infectious diseases and then considers toxicology and chemical risks. It investigates chemicals of major concern, which are grouped into five categories: neurotoxins, carcinogens, teratogens, allergens, and endocrine disruptors. Other topics introduced include dose-response studies and biomagnification. The chapter concludes with an in-depth look at risk analysis and the three major steps in the risk-analysis process: risk assessment, risk acceptance, and risk management. When teaching this chapter make sure to emphasize bioaccumulation and biomagnification as well as LD50 calculations. These topics frequently appear on the exam.

Module 56: Human Diseases

Humans face a wide range of potential diseases. This module explores different types of diseases and discusses the risk factors that make some people more likely to contract diseases. It then discusses a number of diseases that have been important in human history as well as the diseases that have become prevalent in recent decades.

Module 57: Toxicology and Chemical Risks

The complexity of the biological risks humans face is matched by the complexity of chemical risks. Our modern society has developed an incredible array of chemicals to improve human health and food production, including pharmaceuticals, insecticides, herbicides, and fungicides. Chemical by-products from manufacturing and the generation of energy can be harmful to humans and the environment. Even beneficial chemicals, when released into the environment, can harm humans and other organisms. Many pharmaceuticals, for example, have unexpected consequences when released into the environment. This module looks at the types of chemicals that can have harmful effects. It examines how scientists study these chemicals and their effect on humans.

Module 58: Risk Analysis

Most people face some kind of environmental hazard every day. The hazards we face may be voluntary, as when we make a decision to smoke tobacco, or they may be involuntary, as when we are exposed to air pollution. When assessing the risk of different environmental hazards, regulatory agencies, environmental scientists, and policy makers usually follow three steps for risk analysis: risk assessment, risk acceptance, and risk management. This module examines each of the three steps in risk analysis.

Alignment to AP® Environmental Science Course Description

AP® Outline	Module	
VI. Pollution (25–30%)		
B. Impacts on the Environment and Human Health	Module 56	Human Diseases
	Module 57	Toxicology and Chemical Risks
	Module 58	Risk Analysis

Chapter Learning Objectives

After completing this chapter students will be able to

- **identify the different types of human diseases.**
- **understand the risk factors for human chronic diseases.**
- **discuss the historically important human diseases.**
- **identify the major emergent infectious diseases.**
- **discuss the future challenges for improving human health.**
- **identify the major types of harmful chemicals.**
- **explain how scientists determine the concentrations of chemicals that harm organisms.**

- explain the processes of qualitative versus quantitative risk assessment.
- understand how to determine the amount of risk that can be tolerated.
- discuss how risk management balances potential harm against other factors.
- contrast the innocent-until-proven-guilty principle and the precautionary principle.

Chapter 17 Pacing Guide

This pacing guide is based on a schedule with 120 sessions of 50 minutes each before the AP® Exam. If you have a different number of sessions before the exam, you can modify the pacing to suit your needs. If you have additional time, consider incorporating quizzes, released AP® Environmental Science free-response and multiple-choice questions, or additional activities.

Module	Standard Schedule Days	Block Schedule Days
Module 56	2	1
Module 57	2	1
Module 58	1	½
Assessment	1	½

Chapter 17 Resources

The link to these resources can be found by clicking on the link buttons in the Teacher's e-Book (TE-book), opening the Teacher's Resource Flash Drive (TRFD), or logging onto the book's companion website bcs.whfreeman.com/friedlandapes2e (teacher login required).

- **PowerPoint Presentation 17**
- **Optimized Art PowerPoint and JPEG Files 17**
- **Do the Math Videos**
- **Measuring Your Impact 17:** How Does Risk Affect Your Life Expectancy?
- **Chapter 17 Web Resources**
- **Exploring the Literature 17**
- **Answers to Chapter 17 AP® Exam**
- **Multiple-Choice AP® Practice Exam 17**
- **Answers to Unit 7 AP® Exam**
- **Unit 7 Additional Free-Response Question**
- **Answer to Science Applied 7 Free-Response Question**

Free-Response Questions from Previous AP® Environmental Science Exams

Free-response questions from prior AP® Environmental Science Exams are available on the AP® course website: https://apstudent.collegeboard.org/apcourse/ap-environmental-science/exam-practice?envsci. Students should be able to answer all of the questions listed below with material learned in this and previous chapters. When a question requires students to understand material from multiple chapters, the question will be listed in the last chapter required to complete the entire question. Questions marked with an asterisk (*) are from exams with released multiple-choice questions. You may want to save these questions until the end of the year so you can give your students a complete released exam for practice. Questions marked with double asterisks (**) require math to calculate a problem. Look for references to these questions throughout the chapter.

Year	Question	Content
2002	3	• Toxicity testing and LD50
2004	1	• How human activities lead to mercury accumulation in seafood • Health risks associated with eating large predatory fish • Other toxic metals with acute sublethal effects
2005	1	• Transmission of infectious diseases and prevention in human populations • Environmental factors that contribute to the emergence or reemergence of infectious diseases • Importance of addressing infectious disease on a global scale
2010	1**	• Effects of industrial pollutants on human health • Reduction of toxic pollutants that threaten human health • Susceptibility of children to industrial pollutants • CO_2 release and cost of preventing it • Technology available to remove pollutants from waste stream of coal-burning power plants • Reasons for businesses to locate industry in developing nations

PD **Chapter 17 Overview**

Watch the video overview of Chapter 17 (for teachers) by clicking on the link buttons in the TE-book, opening the TRFD, or logging onto the book's companion website bcs.whfreeman.com /friedlandapes2e (teacher login required).

The citizens of Norco, Louisiana, live in the shadows of chemical plants and oil refineries. *(Mark Ludak/The Image Works)*

TRM **PowerPoint Presentation 17**

Download the PowerPoint presentation for Chapter 17 by clicking on the link buttons in the TE-book, opening the TRFD, or logging onto the book's companion website bcs.whfreeman.com /friedlandapes2e (teacher login required).

chapter 17

Human Health and Environmental Risks

Module **56** Human Diseases

Module **57** Toxicology and Chemical Risks

Module **58** Risk Analysis

Citizen Scientists

The neighborhood of Old Diamond in Norco, Louisiana, was composed of four city blocks located between a chemical plant and an oil refinery, both owned by the Shell Oil Company. There were approximately 1,500 residents in the neighborhood, largely lower-income African Americans. In 1973, a pipeline explosion blew a house off its foundation and killed two residents. In 1988, an accident at the refinery killed seven workers and sent more than 70 million kg (159 million pounds) of potentially toxic chemicals into the air. Nearly one-third of the children in Old Diamond suffered from asthma and there were many cases of cancer and birth defects. The unusually high rates of disease raised suspicions that the two nearby industrial facilities were harming the residents.

By 1989, local resident and middle school teacher Margie Richard had seen enough. Richard organized the Concerned Citizens of Norco. The primary goal of the group was to get Shell to buy the residents' properties at a fair price so they could move away from the industries that were putting their health at risk. Richard contacted environmental scientists and quickly learned that to make a solid case to the company and to the U.S. Environmental Protection Agency (EPA), she needed to be more than an organizer; she also needed to be a scientist.

> The unusually high rates of disease raised suspicions that the two nearby industrial facilities were harming the residents.

The residents knew that the local air had a foul smell, but they had no way of knowing which chemicals were present and in what concentrations. To determine whether they were being exposed to chemicals at concentrations that posed a health risk, the air had to be tested. Richard learned about specially built buckets that could collect air samples. She organized a

589

Teaching Tip: Chapter Opening Case

The chapter opening case, "Citizen Scientists," provides an opportunity to show the students that individuals can make a difference. Ask students to describe an environmental issue or issues in their community that concerns them. Have them propose a way to address the issue and find a solution.

Teaching Tip: Beyond the Classroom

In 2004, Margie Richard won the Goldman Environmental Prize. Have your students learn more about Margie Richard by exploring the Goldman Environmental Prize website.

The link to this website may be found by clicking on the link buttons in the TE-book, opening the TRFD, or logging onto the book's companion website bcs.whfreeman.com/friedlandapes2e (teacher login required).

Chapter 17 Web Resources

"Bucket Brigade" of volunteers and slowly collected the data she and her collaborators needed. As a result of these efforts, scientists were able to document that the Shell refinery was releasing more than 0.9 million kg (2 million pounds) of toxic chemicals into the air each year.

The fight against Shell met strong resistance from company officials and went on for 13 years. But in the end, Margie Richard won her battle. In 2002, Shell agreed to purchase the homes of the Old Diamond neighborhood. The company also agreed to pay an additional $5 million for community development and it committed to reducing air emissions from the refinery by 30 percent to help improve the air quality for those residents who remained in the area. In 2007, Shell agreed that it had violated air pollution regulations in several of its Louisiana plants and paid the state of Louisiana $6.5 million in penalties.

For her tremendous efforts in winning the battle in Norco, Margie Richard was the 2004 North American recipient of the Goldman Environmental Prize, which honors grassroots environmentalists. Since then, Richard has brought her message to many other minority communities located near large polluting industries. She teaches that protecting human health and the local environment requires community action—and learning how to be a citizen scientist.

Sources:
The Goldman Environmental Prize: Margie Richard. http://www.goldmanprize.org/2004/northamerica; M. Scallan, Shell, DEQ settle emission charges: Fines and upgrades to cost $6.5 million, *Times-Picayune* (New Orleans), March 15, 2007, http://royaldutchshellplc.com/2007/03/15/the-times-picayune-shell-deq-settle-emission-charges-fines-and-upgrades-to-cost-65-million/.

Teaching Tip: Warm-up

Ask students: What are the three major categories of risk for human health? Give an example of each. *Three major categories of risk that can harm human health are physical, biological, and chemical. Examples of physical risk include natural disasters and excessive exposure to UV light. Examples of biological risk include all diseases. Examples of chemical risk include being exposed to toxic chemicals such as arsenic or pesticides.*

The number of health risks that we face in our lives can feel overwhelming. Sometimes it seems that we hear new warnings every day. How do we evaluate and manage these risks? The first step in understanding health risks is to consider the three major categories of risk that can be detrimental to human health: physical, biological, and chemical. Physical risks include environmental factors, such as natural disasters, that can cause injury and loss of life. Physical risks also include less dramatic factors such as excessive exposure to ultraviolet radiation from the Sun, which causes sunburn and skin cancer, and exposure to radioactive substances such as radon, which we discussed in Chapter 15. Biological risks, which cause the most human deaths, are those risks associated with disease. Chemical risks are associated with exposure to chemicals ranging from those that occur naturally, such as arsenic, to those that are manufactured, such as synthetic chemicals and pesticides. In this chapter, we will focus on biological and chemical risks, and we will determine which of those risks are common and the current state of our understanding about each of them. We will look at how to assess and manage these risks. As we will see, although many health risks exist in both the developed and developing worlds, we can do a great deal to manage these risks and improve our lives.

module

56

Human Disease

Humans face a wide range of potential diseases. In this module, we will explore different types of diseases and discuss the risk factors that make some people more likely to contract diseases. We will then discuss a number of diseases that have been important in human history as well as the diseases that have become prevalent in recent decades.

Learning Objectives

After reading this module you should be able to

- identify the different types of human diseases.
- understand the risk factors for human chronic diseases.
- discuss the historically important human diseases.
- identify the major emergent infectious diseases.
- discuss the future challenges for improving human health.

There are different types of human diseases

Of all causes of human deaths worldwide, which are shown in FIGURE 56.1, approximately three-quarters stem from various types of diseases. A **disease** is any impaired function of the body with a characteristic set of symptoms. Moreover, *infectious diseases* cause about one-quarter of worldwide deaths. **Infectious diseases** are diseases caused by infectious agents, known as pathogens. Examples include pneumonia and sexually transmitted diseases. Noninfectious diseases are not caused by pathogens; these include cardiovascular diseases, respiratory and digestive diseases, and most cancers.

The pathogens that cause most infectious diseases include viruses, bacteria, fungi, protists, and a group of parasitic worms called helminths. However, only six types of disease cause 94 percent of all deaths attributed to infectious diseases. The three most common types of infectious disease are AIDS, diseases that cause respiratory infections, and diarrheal diseases. The next three are tuberculosis, malaria, and childhood diseases such as measles and tetanus. We will discuss many of these important infectious diseases later in the chapter.

All diseases can also be categorized as either *acute* or *chronic*. **Acute diseases** rapidly impair the functioning of a person's body. In some cases, such as a disease called Ebola hemorrhagic fever that we will discuss later in this chapter, death can come in a matter of days or weeks. In contrast, **chronic diseases** slowly impair the functioning of a person's body. Heart disease and most cancers, for example, are chronic diseases that develop over several decades.

Disease Any impaired function of the body with a characteristic set of symptoms.

Infectious disease A disease caused by a pathogen.

Acute disease A disease that rapidly impairs the functioning of an organism.

Chronic disease A disease that slowly impairs the functioning of an organism.

Teaching Tip: Warm-up

Ask students to create a table like the one shown to the right. Have them fill in the definitions and provide an example for each. (Suggested answers are provided in italics.)

	Definition	Example
Acute disease	*A disease that rapidly impairs the functioning of an organism*	*Ebola hemorrhagic fever*
Chronic disease	*A disease that slowly impairs functioning of an organism*	*Heart disease*

TEACHING with FIGURES ▶

Show students Figure 56.1 and ask the following questions:

- What are the top three leading causes of death in the world? *The top three leading causes of death in the world are cardiovascular diseases at 29 percent, infectious diseases at 26 percent, and cancers at 13 percent.*
- What are the top six infectious diseases found throughout the world? *The top six infectious diseases are respiratory infections at 30 percent, HIV/AIDS at 21 percent, diarrheal diseases at 14 percent, tuberculosis at 12 percent, malaria at 9 percent, and childhood diseases at 8 percent.*

TEACHING with FIGURES ▶

Show students Figure 56.2 and ask the following questions:

- What are the two leading health risks for high-income nations and low-income nations? *The two leading health risks for high-income nations are high blood pressure and tobacco use. The two leading health risks for low-income nations are childhood underweight and high blood pressure.*
- What accounts for the differences in health risks between high-income and low-income nations? *Affluence changes the major health risk factors for chronic disease. Because people in high-income countries can afford better nutrition, proper sanitation, and good medical care, fewer die at a young age from infectious diseases and diarrhea. Poverty and lack of education and medical care in low-income nations lead to an increase in risk factors such as unsafe sex, childhood malnutrition, unsafe water, and poor sanitation. Risk factors in high-income countries include increased availability of tobacco, and a combination of less-active lifestyles, poor nutrition, and overeating that leads to obesity and high blood pressure.*

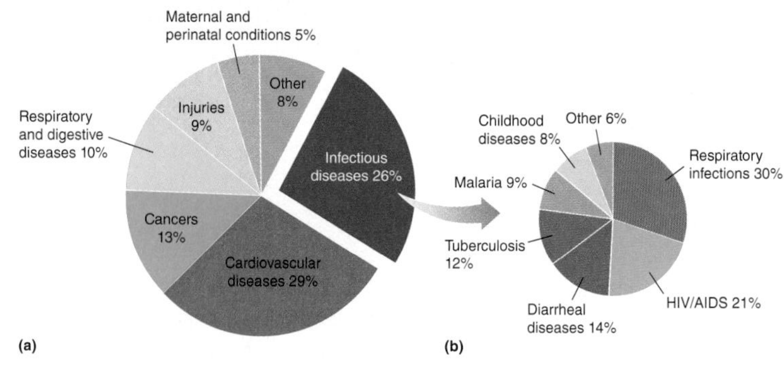

FIGURE 56.1 **Leading causes of death in the world.** (a) More than three-quarters of all world deaths are caused by diseases, including respiratory and digestive diseases, various cancers, cardiovascular diseases, and infectious diseases. (b) Among the world's deaths caused by infectious diseases, 94 percent are caused by only six types of diseases. *(Data from World Health Organization, 2004)*

Numerous risk factors exist for chronic disease in humans

Numerous factors cause people to be at a greater risk for chronic diseases such as cancer, cardiovascular diseases, diabetes, and chronic infectious diseases. The World Health Organization (WHO) has found that these risk factors differ substantially between low- and high-income countries. FIGURE 56.2 shows the WHO data for a variety of risks.

In low-income countries, the top risk factors leading to chronic disease are associated with poverty and include underweight children, unsafe drinking water,

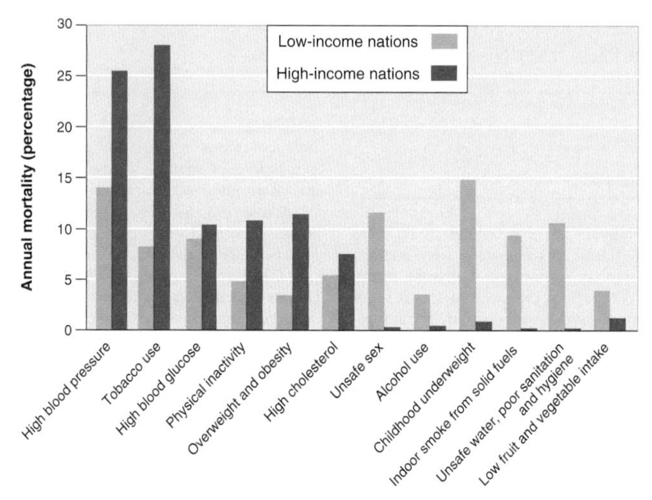

FIGURE 56.2 **Leading health risks in the world.** If we consider all deaths that occur and separate them into different causes, we can examine which categories cause the highest percentage of all deaths. The leading health risks for low-income countries include issues related to low nutrition and poor sanitation. The leading risks for high-income countries include issues related to tobacco use, inactivity, obesity, and urban air pollution. *(Data from World Health Organization, 2009)*

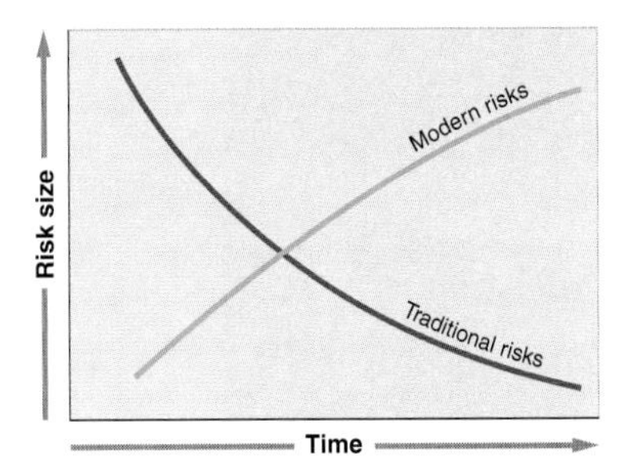

FIGURE 56.3 **The transition of risk.** As a nation becomes more developed over time and attains higher income levels, the risks of inadequate nutrition and sanitation decline while the risks of tobacco, obesity, and poor urban air quality rise. *(After World Health Organization, 2009)*

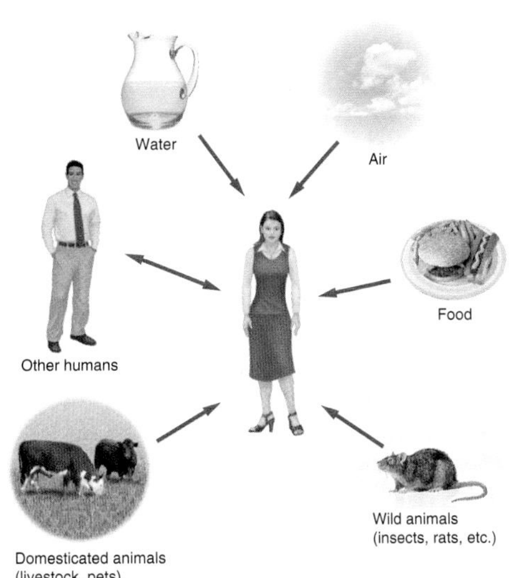

FIGURE 56.4 **Pathways of transmitting pathogens.** Pathogens have evolved a wide variety of ways to infect humans.

poor sanitation, and malnutrition. In Chapter 11 we looked at some of the factors that contribute to malnutrition, including political upheaval and high costs of fossil fuels that lead to increased food prices. As an example of poverty leading to chronic disease, nearly half of the children under the age of 5 who die from pneumonia succumb to the disease because they suffer from poor nutrition. Similarly, nearly three-quarters of children who die from diarrhea are also malnourished. If these children had improved nutrition, they would be better able to fight infectious diseases and many would survive.

Affluence changes the major health risk factors for chronic disease. Because people in high-income countries can afford better nutrition and proper sanitation, fewer die at a young age from diseases such as pneumonia and diarrhea. Risk factors for chronic disease in high-income countries include increased availability of tobacco, and a combination of less active lifestyles, poor nutrition, and overeating that leads to high blood pressure and obesity.

The change in risk factors between low- and high-income countries occurs over time as a given country becomes more affluent. The graph in FIGURE 56.3 illustrates how this transition in economic development affects health risk. A poor country initially faces the challenge of supplying food and proper sanitation to its citizens. As it begins to accumulate wealth, though, the health risks will change in a predictable fashion and the health care system of that country must change as well.

Some infectious diseases have been historically important

Although diseases can have genetic causes, environmental scientists are generally interested in diseases that have environmental causes, especially those caused by pathogens such as fungi, bacteria, and viruses. These pathogens have evolved a wide variety of pathways for infecting humans, including transmission through food, water, other humans, and other animals. FIGURE 56.4 illustrates some of these relationships.

Historically, disease-causing pathogens have taken a large toll on human health and mortality. When a pathogen causes a rapid increase in disease, we call it an **epidemic.** When an epidemic occurs over a large geographic region such as an entire continent, we call it a **pandemic.** Among the diversity of human diseases that have caused epidemics and pandemics, we will consider both those that have been historically important and those that have emerged recently. In Chapters 7 and 14 we looked at several diseases historically associated with poor sanitation and unsafe drinking water including cholera, hepatitis, and diarrheal diseases. Additional historic diseases that are passed between hosts include *plague, malaria,* and *tuberculosis.*

Epidemic A situation in which a pathogen causes a rapid increase in disease.

Pandemic An epidemic that occurs over a large geographic region.

Teaching Tip: Activity

Divide the class into groups of two. Give each group 20 index cards and the vocabulary terms in the list below. Have each team label one side of each card with the numbers 1 through 20. On the blank sides of cards 1 through 10, students should write one of the terms, using a new card for each term. Instruct students to shuffle the order of the cards; they should not duplicate the order of the list. On the blank sides of cards numbered 11 through 20, students should write the definition or some other information connected to each of the terms on cards 1 through 10, making sure to keep an answer key of which numbered term card matches which numbered definition card. The groups should lay out their cards number side up in two rows—one with terms and one with definitions. Have the groups swap cards and answer keys.

The first player in each pair then chooses one card from each row. Read the cards to see if they are a match. If the cards match, that player keeps the cards and takes another turn. If the cards are not a match, the second member of the team has a turn. Each group continues playing until all the cards are gone. The person with the most cards is the winner.

- acute disease
- chronic disease
- epidemic
- pandemic
- infectious disease
- plague
- malaria
- tuberculosis
- emergent infectious disease
- Acquired Immune Deficiency Syndrome (AIDS)

Teaching Tip: Warm-up

Ask students: What is the difference between an acute disease and a chronic disease? *An acute disease rapidly impairs the functioning of an organism. A chronic disease slowly impairs the functioning of an organism.*

AP® Exam Tip

The transmission of infectious diseases and environmental influences are covered on the 2005 AP® Exam, Question 1. To answer this question, students must

- choose an old disease and a new disease mentioned in an accompanying article and explain how these diseases are transmitted through a human population as well as an effective method to control the spread of the disease
- for one disease, identify an environmental factor that contributed to the emergence or reemergence of the disease and explain how that factor contributed to the spread of the disease
- explain why fighting an infectious disease must be a global effort

FIGURE 56.5 **The Black Death in Europe.** As depicted in Carlo Coppola's *The Marketplace in Naples During the Plague of 1656,* plague pandemics repeatedly swept through Europe from the 1300s through the 1800s and killed millions of people. Because the disease caused black sores on people's bodies, it also had the name *Black Death.* *(Roger-Viollet/The Image Works)*

Plague

Plague is caused by an infection from a bacterium (*Yersinia pestis*) that is carried by fleas. Fleas attach to rodents such as mice and rats, which gives the fleas tremendous mobility. One of the most well-known diseases of human history, plague also carries several historical names including *bubonic plague* and *Black Death.* When humans live in close contact with mice and rats, the bacterium can be transmitted either by flea bites or by handling the rodents. Individuals who become infected often experience swollen glands, black spots on their skin, and extreme pain. Plague is estimated to have killed hundreds of millions of people throughout history, including nearly one-fourth of the European population in the 1300s (FIGURE 56.5). The last major pandemic of plague occurred in Asia in the early 1900s. Today there are still occasional small outbreaks of plague around the world, including in the southwestern United States, but modern antibiotics are highly effective at killing the bacterium and preventing human death.

Plague An infectious disease caused by a bacterium (*Yersinia pestis*) that is carried by fleas.

Malaria An infectious disease caused by one of several species of protists in the genus *Plasmodium*.

Malaria

Malaria, caused by an infection from any one of several species of protists in the genus *Plasmodium,* is another widespread disease that has killed millions of people over the centuries. The malaria parasite spends one stage of its life inside a mosquito and another stage of its life inside a human. Infections cause recurrent flulike symptoms. Each year, 350 to 500 million people in the world contract the disease and 1 million people, mostly children under 5 years of age, die from it. The regions hardest hit include sub-Saharan Africa, Asia, the Middle East, and Central and South America. Since 1951, the malaria parasite has been eliminated from the United States by mosquito eradication programs. Although more than 1,000 cases of malaria are diagnosed in the United States each year, these are found in people who have returned from regions of the world where the malaria parasite lives.

The traditional approach to combating malaria was widespread spraying of insecticides such as DDT to eradicate the mosquitoes. Eradication efforts have proven to be ineffective in many parts of the world. Moreover, as we will see later in this chapter, the widespread use of many insecticides can create new problems. At the end of this chapter, "Working Toward Sustainability: The Global Fight Against Malaria" on page 616 examines the latest approaches toward combating malaria.

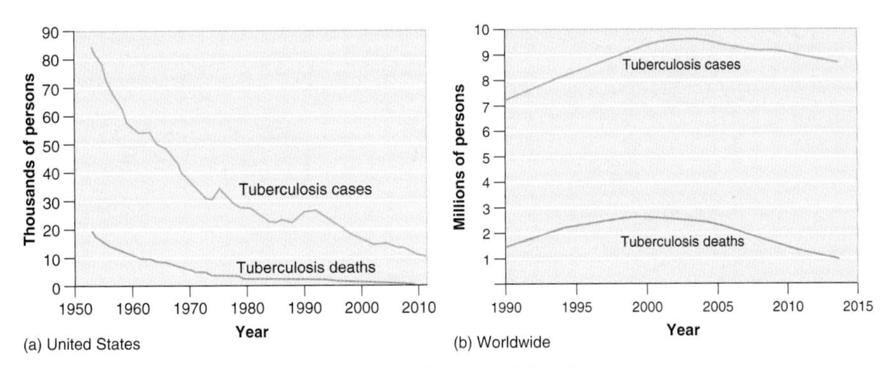

FIGURE 56.6 **Tuberculosis cases and deaths.** (a) Due to effective and available medicines, tuberculosis has gone from being one of the most deadly diseases in the United States to a disease that rarely kills. (b) Worldwide, however, tuberculosis has continued to infect and kill millions of people, especially in low- and middle-income countries.

Tuberculosis

Tuberculosis is a highly contagious disease caused by a bacterium (*Mycobacterium tuberculosis*) that primarily infects the lungs. Tuberculosis is spread when a person coughs and expels the bacteria into the air. The bacteria can persist in the air for several hours and infect a person who inhales them. Symptoms of an infection include feeling weak, sweating at night, and coughing up blood. As is the case with many pathogens, a person can be infected but not develop the tuberculosis disease. Indeed, it is estimated that one-third of the world's population is infected with tuberculosis. Each year 9 million people develop the disease and 2 million of them die.

Taking antibiotics for a year can easily treat most tuberculosis infections. In countries such as the United States, where such medicines are readily available, there has been a dramatic drop in both the number of new cases and the number of deaths from tuberculosis. FIGURE 56.6 shows the decline of tuberculosis in the United States since the mid-1950s. In fact, 95 percent of all tuberculosis deaths occur in low- and middle-income countries and it is the leading cause of death by disease in the developing world. In these countries, the medicines are not as available or affordable and those who receive the medicine sometimes do not take the prescribed dose for the full duration of time. When a patient stops taking antibiotics before the last bacteria have been killed, there are two consequences. First, the pathogen can quickly rebuild its population inside the person's body. Second, because the last few bacteria are generally the most drug-resistant, stopping the antibiotics before the bacteria are eradicated selects for drug-resistant strains. Drug-resistant strains of tuberculosis are becoming a major concern, particularly in parts of Africa and in Russia, where up to 20 percent of the people infected with tuberculosis carry a drug-resistant strain. Such strains are much harder to kill and therefore require newer antibiotics that can cost 100 times more than the traditional drugs.

Emergent infectious diseases pose new risks to humans

In recent decades, we have witnessed the appearance of many **emergent infectious diseases,** which are defined as infectious diseases that were previously not described or have not been common for at least the prior 20 years. FIGURE 56.7 locates some of the best known emergent infectious diseases. Since the 1970s, the world has observed an average of one emergent disease each year. Many of these new diseases have

Tuberculosis A highly contagious disease caused by the bacterium *Mycobacterium tuberculosis* that primarily infects the lungs.

Emergent infectious disease An infectious disease that has not been previously described or has not been common for at least 20 years.

◀ TEACHING with FIGURES

Have students study the trends from 1950 to 2010 in Figure 56.6a and the trends from 1990 to 2013 in Figure 56.6b. Then ask: Why has the number of tuberculosis cases in the United States dropped to record lows while the number of tuberculosis cases around the world has remained at very high levels? *In developed counties like the United States, where antibiotics are readily available, there has been a dramatic drop in both the number of new cases and the number of deaths from tuberculosis. In many developing countries medicines are not as available or affordable and those who receive the medicine sometimes do not take the prescribed dose for the full duration of time.*

Teaching Tip: Discussion Starter

Ask students the following questions:

- What is an emergent infectious disease? *An emergent infectious disease is an infectious disease that has not been previously described or has not been common for at least 20 years.*
- What are some recent emergent infectious diseases? *Recent emergent infectious diseases include HIV/AIDS, Ebola hemorrhagic fever, mad cow disease, swine flu, bird flu, SARS, and West Nile virus.*

Teaching Tip: Beyond the Classroom

Have students research a historic or an emergent infectious disease and create a presentation for the class. Students can work individually or you can break them up into small groups. The presentation should have six or seven slides or poster panels. Each slide or panel must have information related to the topic. Encourage students to find a graph or statistical information about the disease and include that information as part of their presentation. You may recommend the following outline:

Slide 1: Title with picture
Slide 2: What causes the disease?
Slide 3: How does the disease spread through the human population?
Slide 4: What are the symptoms or the effects on humans of the disease?
Slide 5: What is the best way to fight the disease?
Slide 7: Concluding statement

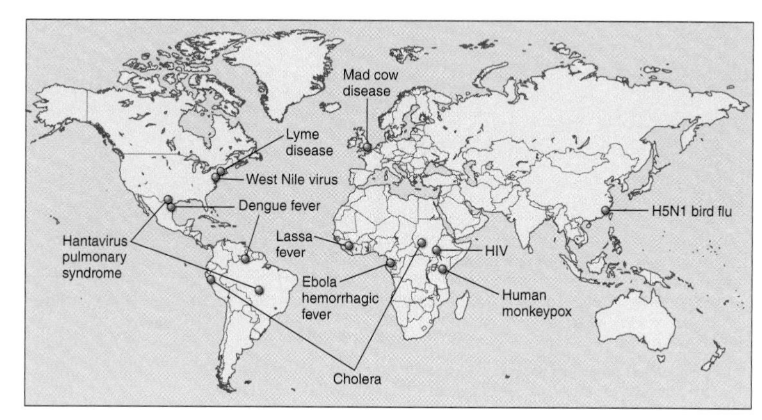

FIGURE 56.7 **The emergence of new diseases.** Since the 1970s, new diseases, or diseases that have been rare for more than 20 years, have been appearing throughout the world at a rate of approximately one per year. *(After http://ww3.niaid.nih.gov/NR/rdonlyres/DF9CBA79-F005-4550-857A-63D62FEFBCA8/0/emerging_diseases1.gif)*

come from pathogens that normally infect animal hosts but then unexpectedly jump to human hosts. This typically occurs because the diseases can mutate rapidly and eventually produce a genotype that can infect humans. Some of the most high-profile diseases that have jumped from animals to humans include HIV/AIDS, Ebola, mad cow disease, bird flu, SARS, and West Nile virus. These emerging diseases are of increasing concern because of the increased movement of people and cargo throughout the world during the past century. In fact, currently diseases can spread to nearly any place on Earth within 24 hours.

HIV/AIDS

In the late 1970s, rare types of pneumonia and cancer began appearing in individuals with weak immune systems. The condition responsible for the weakened immune systems was the disease **Acquired Immune Deficiency Syndrome (AIDS),** which is caused by a virus known as **Human Immunodeficiency Virus (HIV).** This virus spreads through sexual contact and blood transfusions, from mothers who pass it on to their fetuses, and among drug users who share unsanitized needles.

The origin of this new virus remained a mystery until 2006 when researchers found a genetically similar virus in a wild population of chimpanzees living in the African nation of Cameroon (FIGURE 56.8). The researchers hypothesized that local hunters were exposed to the virus when butchering or eating the chimps (a common practice in this part of the world). With this exposure, the virus was able to infect a new host, humans. Today, more than 33 million people in the world are infected with HIV and 25 million people have died from AIDS-related illnesses.

Fortunately, new antiviral drugs are able to maintain low HIV populations inside the human body and thereby substantially extend life. From the lessons learned in combating other diseases such as tuberculosis, combinations of antiviral drugs are being used to reduce the risk that the virus will evolve resistance to any single drug. Unfortunately, many of these drugs are expensive and most people living in low-income countries cannot afford them, although this is changing with wider availability and improved distribution of these drugs to the poor.

Acquired Immune Deficiency Syndrome (AIDS) An infectious disease caused by the human immunodeficiency virus (HIV).

Human Immunodeficiency Virus (HIV) A type of virus that causes Acquired Immune Deficiency Syndrome (AIDS).

Ebola hemorrhagic fever An infectious disease with high death rates, caused by the Ebola virus.

Ebola Hemorrhagic Diseases

In 1976, researchers first discovered **Ebola hemorrhagic fever,** an infectious disease with high death rates, caused by the Ebola virus. First discovered in the Democratic Republic of the Congo near the Ebola River, the virus has infected several hundred humans and a variety of other primates from several

Teaching Tip: Beyond the Classroom

Students can get current information about Ebola and current vaccination trials on the World Health Organization website. The website also contains details on how the disease is spread, the effects on humans, and survival rates as well as travel advisories, maps, and situation assessments.

The link to this website may be found by clicking the link buttons in the TE-book, opening the TRFD, or logging onto the book's companion website bcs.whfreeman.com/friedlandapes2e (teacher login required).

Chapter 17 Web Resources

FIGURE 56.8 **The source of HIV.** In 2006, researchers found that chimpanzees in Cameroon carried a virus that was genetically very similar to HIV. Thus, these chimps are the most likely source of this emerging human disease. *(Tim E White/Alamy)*

countries in central Africa. Infections have been sporadic since Ebola was first discovered, but there was a large outbreak in 2014 that infected thousands of people. The Ebola virus is of particular concern because it kills a large percentage of those infected. Infected individuals have suffered a 50 to 89 percent death rate from different outbreaks of the disease. Those infected quickly begin to experience fever, vomiting, and sometimes internal and external bleeding

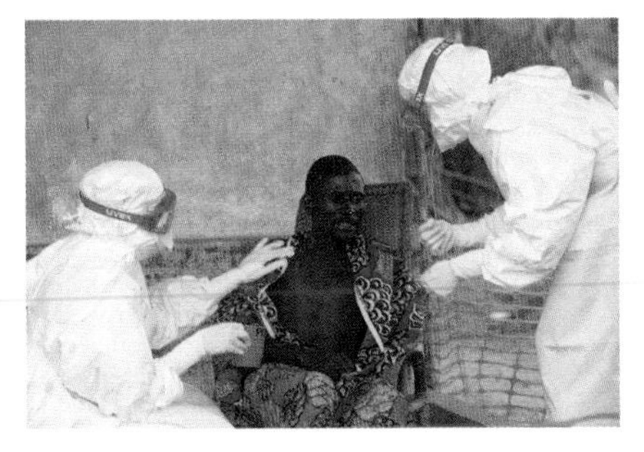

FIGURE 56.9 **Ebola hemorrhagic fever.** The Ebola virus is highly lethal to humans and there are only experimental drugs for treatment. When treating a person infected with the virus, researchers and medical workers have to exercise extreme caution. *(AP Photo/WHO/Christopher Black)*

FIGURE 56.10 **Mad cow disease.** Cows that have been fed the remains of dead cows and sheep can become infected with harmful prions. These prions damage the nervous system and cause the cows to develop glazed eyes, body tremors, and a loss of coordination, eventually leading to death. Humans who consume beef from infected cows can become infected and suffer a similar fate. *(C.E.V./Science Source)*

(FIGURE 56.9). Death occurs within 2 weeks, and currently only experimental drugs are available to fight the virus. Unlike the progress that has been made with identifying the origin of HIV, the natural source of the Ebola virus has been difficult to determine. Because the virus also kills other primates at high rates, leaving no primate hosts for the virus, primates are not a likely long-term source of the virus. In 2013, however, researchers discovered that fruit bats carried the Ebola virus and were likely the reservoir species that spread the virus to primates.

Mad Cow Disease

In the 1980s, scientists first described a neurological disease, later known as **mad cow disease,** in which prions mutate into deadly pathogens and slowly damage a cow's nervous system. The cow loses coordination of its body (a condition compared with a person going mad), and then dies (FIGURE 56.10). Scientists now know that small, beneficial proteins in brains of cattle, called **prions,** occasionally mutate into deadly proteins that act as pathogens and subsequently cause mad cow disease. Prions are not well understood and represent a new category of pathogen.

In 1996, scientists in Great Britain announced that mad cow disease, also known as bovine spongiform encephalopathy (BSE), could be transmitted to

Mad cow disease A disease in which prions mutate into deadly pathogens and slowly damage a cow's nervous system.

Prion A small, beneficial protein that occasionally mutates into a pathogen.

Teaching Tip: Video

Mad Cow Disease Found in California
This video from NBC discusses a case of mad cow disease found at an unidentified location in central California. The video describes mad cow disease and how it is spread.

The link to this video may be found by clicking the link buttons in the TE-book, opening the TRFD, or logging onto the book's companion website bcs.whfreeman.com/friedlandapes2e (teacher login required).

 **Chapter 17 Web Resources**

FIGURE 56.11 **Bird flu.** A farm worker feeds a large number of chickens on a farm in the Republic of Niger in western Africa. The virus that causes bird flu normally infects only birds. In 2006, however, the virus began jumping to human hosts where people and birds were in close contact. *(Pius Utomi Ekpei/AFP/Getty Images)*

Teaching Tip: Video

How Flu Viruses Attack
This *National Geographic* video shows students how a swine or bird flu virus can attack a host, mutate, and become contagious, perhaps resulting in an outbreak or even a pandemic.

The link to this video may be found by clicking the link buttons in the TE-book, opening the TRFD, or logging onto the book's companion website bcs.whfreeman.com/friedlandapes2e (teacher login required).

 **Chapter 17 Web Resources**

humans who ate meat from infected cattle. Unlike harmful bacteria that can be killed with proper cooking, prions are difficult to destroy by cooking. Infected humans developed variant Creutzfeldt-Jakob Disease (CJD) and suffered a fate similar to the infected cattle.

Mutant prions cannot be transmitted among cattle that only live together. Transmission requires an uninfected cow to consume the nervous system of an infected cow. As a result, when cattle feed on grass together in a pasture, a rare mutation in a prion would be restricted to a single cow and not spread to other cattle. In the 1980s, however, the diets of European cattle commonly included the ground-up remains of dead cattle as a source of additional protein. When these remains happened to contain a mutant prion, the prions spread rapidly through the entire cattle population and, in turn, infected humans who ate the beef. In Britain, as of 2009, a total of 180,000 cattle were infected and 166 people had died. It is estimated that several thousand people are currently infected, but the prions can exist in the human body for many years before they begin to cause symptoms of the disease. The European Union temporarily banned British beef imports in 1996 and the British government destroyed tens of thousands of cattle. Since that time, the disease has been found in several other countries including Canada and the United States, but only in a few cattle and no humans. Today, new rules exist that forbid the feeding of animal remains to cattle. As a result, the current risk of mad cow disease to humans has been greatly reduced and only a few cases were detected through 2014.

Swine Flu and Bird Flu

Humans commonly contract many types of flu viruses. As we saw in Chapter 5, the Spanish flu of 1918 killed up to 100 million people. Spanish flu was a type of influenza, known as **swine flu,** caused by the H1N1 virus. This virus is similar to a flu virus that humans normally contract, but H1N1 normally infects only pigs. Another pandemic of swine flu occurred around the world in 2009–2010 and it caused more than 18,000 deaths. The most recent outbreak occurred in Venezuela in 2013 and caused more than 1,000 people to become ill. Fortunately, only a small percentage of those people died thanks to the availability of drugs to combat the flu virus.

In 2006, reports emerged from Asia that a related virus, known as H5N1, or **bird flu,** had jumped from birds to people, primarily to people who were in close contact with birds (FIGURE 56.11). Infections are rarely deadly to wild birds but can frequently cause domesticated birds such as ducks, chickens, and turkeys to become very sick and die. Humans often contract a variety of flu viruses. Because humans have no evolutionary history with the H5N1 virus, they have few defenses against it. As of 2014, more than 600 people had become infected by H5N1 and over half of them died. Governments responded to this risk by destroying large numbers of infected birds. Currently the H5N1 virus is not easily passed among people, but if a future mutation makes transmission easier, scientists estimate that H5N1 has the potential to kill 150 million people.

Swine flu A type of flu caused by the H1N1 virus.
Bird flu A type of flu caused by the H5N1 virus.

SARS

In 2003, an unusual form of pneumonia was spreading through human populations in Southeast Asia that was eventually named **severe acute respiratory syndrome (SARS).** While some of the respiratory symptoms are similar to bird flu and swine flu, SARS is a disease caused by a different type of virus known as a coronavirus that can spread from one person to another. During this outbreak, there were more than 8,000 people infected and nearly 10 percent of them died.

West Nile Virus

The **West Nile virus** lives in hundreds of species of birds and is transmitted among birds by mosquitoes. Although the virus can be highly lethal to some species of birds, including blue jays (*Cyanocitta cristata*), American crows (*Corvus brachyrhynchos*), and American robins (*Turdus migratorius*), most species of birds survive the infection. During the latter half of the twentieth century there were increasing reports that the virus could sometimes infect horses and humans who had been bitten by mosquitoes. The first human case was identified in 1937 in the West Nile region of Uganda, thus giving the virus its name. In humans, the virus causes an inflammation of the brain leading to illness and sometimes death. In 1999, the virus appeared in New York and quickly spread throughout much of the United States. FIGURE 56.12 shows the history of the West Nile virus in the United States. The highest numbers of infections and deaths from the virus occurred in 2002 and 2003. Increased efforts to combat mosquito populations and protect against mosquito bites are causing a decline in the disease.

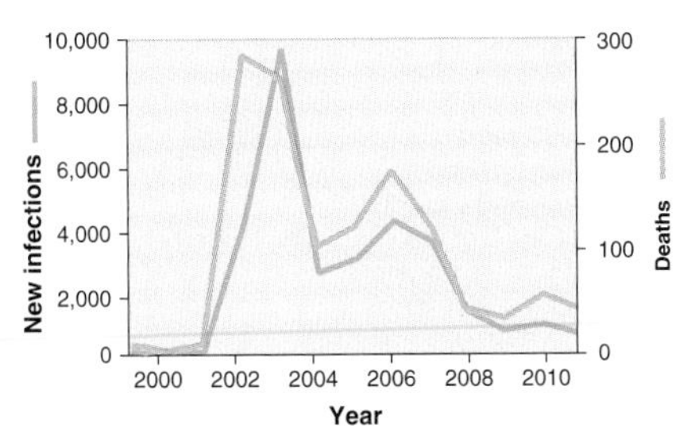

FIGURE 56.12 **West Nile virus in the United States.** Following the first appearance of West Nile virus in the United States in 1999, the number of human infections and deaths rapidly increased. Efforts to control populations of mosquitoes that carry the virus are helping to reduce the prevalence of the disease. *(Data from http://www.cdc.gov/ncidod/dvbid/westnile/surv&control.htm)*

Human health faces a number of future challenges

While humans face a large number of health risks, we have an excellent understanding of important risk factors and the ways to combat many historical and emerging infectious diseases. Combating diseases in low-income countries requires improvements in nutrition, wider availability of clean drinking water, and proper sanitation. In high-income countries, we need to promote healthier lifestyle choices such as increased physical activity, a balanced diet, and limiting excess food consumption and tobacco use. In all countries, continued education is needed to reduce the spread of diseases such as HIV and tuberculosis.

As we combat many diseases, an issue of growing concern is the ability of pathogens to evolve resistance. As we noted in the case of tuberculosis, patients often feel much better long before they complete the full year of prescribed medicine. Because they feel better or because they cannot afford a full year of drugs, they stop the drug treatment early. This allows a small fraction of highly resistant bacteria to survive in the body, reproduce, and then potentially spread to other people. When a pathogen evolves resistance to one drug, physicians often prescribe a different drug to combat the pathogen. Over time, however, some pathogens such as tuberculosis have evolved multiple drug resistance. Without new drugs, little can be done for patients with pathogen strains that possess multiple drug resistance.

A similar issue occurs with the increased use of antiseptic cleaners that are designed to kill microbes such as bacteria. As of 2014, there are more than 2,000 antiseptic cleaners being sold including a large number of antibacterial soaps; these products typically kill a large proportion of harmful pathogens, but not all of them. As a result of our efforts to wipe out pathogens, we are inadvertently selecting for pathogens that possess a stronger resistance to our efforts. Moreover, these products also move through wastewater and into streams, rivers, and lakes. Ironically, it is unclear that antiseptic products kill microbes any better than plain soap. Because of these concerns, in 2013 the U.S. Food and Drug Administration announced that they were considering new rules that would require manufacturers to conduct research to demonstrate that their antiseptic cleaners really are more effective than plain

Severe acute respiratory syndrome (SARS) A type of flu caused by a coronavirus.

West Nile virus A virus that lives in hundreds of species of birds and is transmitted among birds by mosquitoes.

Teaching Tip: Activity

From Birds to People: The West Nile Virus Story

The Howard Hughes Medical Institute has a "click and learn" activity on West Nile virus. The activity includes 25 slides with animations, videos, diagrams, and illustrations showing students how West Nile virus is spread from mosquito to bird or from mosquito to human. The website also offers a student worksheet that can be downloaded.

The link to this resource may be found by clicking the link buttons in the TE-book, opening the TRFD, or logging onto the book's companion website bcs.whfreeman.com/friedlandapes2e (teacher login required).

TRM **Chapter 17 Web Resources**

soap and water and that their products are safe for people to use on a daily basis.

Though many historical diseases are either currently under control or likely to be soon if the financial resources become available, emerging infectious diseases may present a greater challenge. New diseases often arise from new pathogens with which we have no experience. Since we cannot predict which diseases will emerge next, public health officials throughout the world must develop rapid response plans when a particular disease does appear. These include rapid worldwide notification of newly identified diseases and strategies to isolate infected persons, which will slow the spread of the disease and provide time for researchers to develop appropriate tactics to combat the threat.

module 56 REVIEW

In this module, we examined biological risks. We learned that human health risks include physical, chemical, and biological risks. Biological risks are those associated with diseases that are either noninfectious, such as heart disease, or infectious, such as diseases caused by pathogens. Diseases can also be categorized as either acute or chronic and the risk factors for chronic diseases differ between low- and high-income countries. Some of the infectious diseases that have a long history of harming humans include plague, malaria, and tuberculosis. Other diseases have emerged more recently and include HIV/AIDS, mad cow disease, swine flu, bird flu, SARS, and West Nile virus. Knowing the disease risk factors and the ways to combat these diseases will help to reduce their effect on humans, providing that we remember issues related to the evolution of drug resistance. In the next module, we will look to chemical risks.

Module 56 AP® Review Questions

1. An infectious disease is always
   (a) transmitted directly between humans.
   (b) caused by a virus.
   (c) transmitted between humans and animals.
   (d) treatable with antibiotics.
   (e) caused by a pathogen.

2. A disease that rapidly impairs a body's function is
   (a) infectious.
   (b) acute.
   (c) chronic.
   (d) pathogenic.
   (e) epidemic.

3. Tuberculosis
   (a) is transmitted by mosquitoes.
   (b) has been almost entirely eliminated in the world.
   (c) is caused by a virus.
   (d) has strains that have developed resistance to antibiotics.
   (e) is almost never fatal.

4. Prions are pathogens that are responsible for
   (a) AIDS.
   (b) Ebola hemorrhagic fever.
   (c) mad cow disease.
   (d) bird flu.
   (e) SARS.

5. Which is NOT a step to be taken in the future for combating diseases?
   (a) Improving nutrition
   (b) The use of many antiseptic cleaners
   (c) Increased education
   (d) Proper sanitation
   (e) Increased physical activity

Answers to Module 56 AP® Review Questions

1. e
2. b
3. d
4. c
5. b

module

57

Toxicology and Chemical Risks

The complexity of the biological risks humans face is matched by the complexity of chemical risks. Our modern society has developed an incredible array of chemicals to improve human health and food production, including pharmaceuticals, insecticides, herbicides, and fungicides. We have also seen that chemical by-products from manufacturing and the generation of energy can be harmful to humans and the environment. Even beneficial chemicals, when released into the environment, can harm humans and other organisms. Many pharmaceuticals, for example, have unexpected consequences when released into the environment. In this module we will look at the types of chemicals that can have harmful effects. We will see how scientists study these chemicals, and what effect the chemicals have on humans.

Learning Objectives

After reading this module you should be able to

- identify the major types of harmful chemicals.
- explain how scientists determine the concentrations of chemicals that harm organisms.

Many types of chemicals can harm organisms

Chemicals can have many different effects on organisms, and some of the most harmful are common in our environment; **TABLE 57.1** lists those of current concern. They can be grouped into five categories: *neurotoxins, carcinogens, teratogens, allergens,* and *endocrine disruptors.*

Neurotoxins

Neurotoxins are chemicals that disrupt the nervous systems of animals. Many insecticides, for example, are neurotoxins that interfere with an insect's ability to control its nerve transmissions. Insects and other invertebrates are highly sensitive to neurotoxin insecticides. These animals can become completely paralyzed, cannot obtain oxygen, and quickly die. Other important neurotoxins include lead and mercury. As we discussed in Chapters 14 and 15, lead and mercury are very harmful heavy metals that can damage the human kidneys, brain, and nervous system.

Neurotoxin A chemical that disrupts the nervous systems of animals.

Module 57 Toxicology and Chemical Risks 601

AP® Exam Tip

Heavy metals found in seafood is covered on the 2004 AP® Exam, Question 1. To answer this question, students must

- read an article on mercury in seafood and explain both how human activity releases mercury into the environment and how it is transported
- describe two ways that the amount of mercury released into the environment could be reduced
- explain why the health risks associated with eating large predatory fish are greater than the health risks associated with eating small nonpredatory fish
- name another toxic heavy metal that has a negative effect on human health, explain how it is introduced into the environment, and describe an acute sublethal effect on humans who are exposed

Table 57.1 Some chemicals of major concern

Chemical	Sources	Type	Effects
Lead	Paint, gasoline	Neurotoxin	Impaired learning, nervous system disorders, death
Mercury	Coal burning, fish consumption	Neurotoxin	Damaged brain, kidneys, liver, and immune system
Arsenic	Mining, groundwater	Carcinogen	Cancer
Asbestos	Building materials	Carcinogen	Impaired breathing, lung cancer
Polychlorinated biphenyls (PCBs)	Industry	Carcinogen	Cancer, impaired learning, liver damage
Radon	Soil, water	Carcinogen	Lung cancer
Vinyl chloride	Industry, water from vinyl chloride pipes	Carcinogen	Cancer
Alcohol	Alcoholic beverages	Teratogen	Reduced fetal growth, brain and nervous system damage
Atrazine	Herbicide	Endocrine disruptor	Feminization of males, low sperm counts
DDT	Insecticide	Endocrine disruptor	Feminization of males, thin eggshells of birds
Phthalates	Plastics, cosmetics	Endocrine disruptor	Feminization of males

As shown in FIGURE 57.1, since the federal government required a gradual elimination of lead in gasoline and paint in the 1970s, lead exposure in the United States has declined sharply. However, lead contamination in children remains a serious problem in low-income neighborhoods due to the presence of old lead paint in buildings. Mercury also remains a major problem.

Carcinogens

Carcinogens are chemicals that cause cancer. Carcinogens cause cell damage and lead to uncontrolled growth of these cells either by interfering with the normal metabolic processes of the cell or by damaging the genetic material of the cell. Carcinogens that cause damage to the genetic material of a cell are called **mutagens** (although not all mutagens are carcinogens). Some of the most well-known carcinogens include asbestos, radon, formaldehyde, and the chemicals found in tobacco.

Teratogens

Teratogens are chemicals that interfere with the normal development of embryos or fetuses. One of the most infamous teratogens was the drug thalidomide, prescribed to pregnant women during the late 1950s and early 1960s to combat morning sickness. Sadly, tens of thousands of these mothers around the world

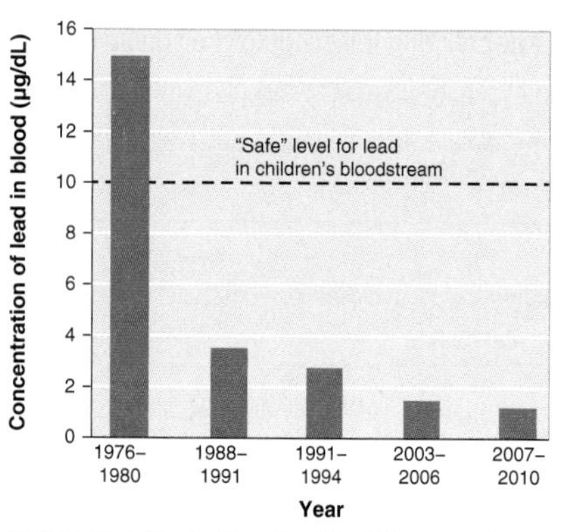

FIGURE 57.1 **The decline of lead in children over time.** Lead poses a particular risk to childhood development. Lead was gradually phased out of gasoline and paint in the 1970s. Since that time, the average concentration of lead in the bloodstream of children between 1 and 5 years of age has declined dramatically. *(Data from CDC NHANES survey of blood lead levels in children)*

Carcinogen A chemical that causes cancer.

Mutagen A type of carcinogen that causes damage to the genetic material of a cell.

Teratogen A chemical that interferes with the normal development of embryos or fetuses.

Teaching Tip: Activity

Numerous past free-response questions have asked students to identify the impact on humans of each of the five major types of harmful chemicals. To learn these effects, have students complete the table below. (Answers are provided in italics.)

Type of chemical	Selected sources	Examples	Effects
Neurotoxin	*Paint, gasoline, burning coal, fish consumption*	*Lead, mercury*	*Impaired learning, nervous system disorder, damage to brain, kidneys, liver*
Carcinogen	*Mines, groundwater, building materials, soil, industry*	*Arsenic, asbestos, PCBs, radon, vinyl chloride*	*Cancer, impaired breathing*
Teratogen	*Alcoholic beverages, pharmaceuticals, tobacco*	*Alcohol*	*Reduced fetal growth, brain and nervous system damage*
Allergen	*Food, latex, pharmaceuticals, pollutants*	*Peanuts, milk, shellfish, drugs*	*Breathing difficulties, hives*
Endocrine disruptor	*Herbicides, insecticides, plastics, cosmetics, pharmaceuticals*	*Atrazine, DDT, phthalates*	*Feminization of males, early onset of puberty, low sperm count, thin eggshells of birds*

FIGURE 57.2 **The effects of thalidomide.** Thalidomide was widely prescribed to pregnant women in the late 1950s to alleviate the symptoms of morning sickness, but it had the unanticipated effect of causing birth defects in tens of thousands of newborn children. *(Leonard McCombe/Life Magazine/Time & Life Pictures/Getty Images)*

gave birth to children with defects before the drug was taken off the market in 1961 (FIGURE 57.2). One of the most common modern teratogens is alcohol. Excessive alcohol consumption reduces the growth of the fetus and damages the brain and nervous system of the fetus, a condition known as fetal alcohol syndrome. This is why physicians recommend that women not consume alcoholic beverages while they are pregnant.

Allergens

Allergens are chemicals that cause allergic reactions. Although allergens are not pathogens, allergens are capable of causing an abnormally strong response from the immune system. In some cases, this response can cause breathing difficulties and even death. Typically, a given allergen only causes allergic reactions in a small fraction of people. Some common chemicals that cause allergic reactions include the chemicals naturally found in peanuts and milk and several drugs including penicillin and codeine.

Endocrine Disruptors

Endocrine disruptors are chemicals that interfere with the normal functioning of hormones in an animal's body. Hormones are normally manufactured in the endocrine system and released into the bloodstream in very low concentrations. As the hormones move through the body, they bind to specific cells. Binding stimulates the cell to respond in a way that regulates the functioning of the body including growth, metabolism, and the development of reproductive organs. As FIGURE 57.3 shows, an endocrine disruptor can bind to receptive cells and cause the cell to respond in ways that are not beneficial to the organism.

One high-profile example of endocrine disruptors in our environment is the group of reproductive hormones that can be found in wastewater. As we discussed in Chapter 14, wastewater may contain hormones from a variety of sources including animal-rearing facilities, human birth control pills, and pesticides that mimic animal hormones. In waterways exposed to hormones through wastewater, scientists are increasingly finding that male fish, reptiles, and amphibians are becoming feminized; males possess testes that have low sperm counts and, in some cases, testes that produce both eggs and sperm. Males normally convert the female hormone estrogen into the male chemical testosterone. Reproductive hormones in wastewater can interfere with the production of testosterone, which causes males to have higher concentrations of estrogen and lower concentrations of testosterone in their bodies. Such discoveries raise serious concerns about whether endocrine disruptors might affect the normal functioning of human hormones. These effects include low sperm counts in men and an increased risk of breast cancer in women.

Scientists can determine the concentrations of chemicals that harm organisms

To assess the risk a chemical poses, we need to know the concentrations that cause harm. Scientists have three techniques to determine harmful concentrations: dose-response studies, prospective studies, and retrospective studies.

Dose-Response Studies

Dose-response studies expose animals or plants to different amounts of a chemical and then look for a variety of possible responses including mortality or changes in

Allergen A chemical that causes allergic reactions.

Endocrine disruptor A chemical that interferes with the normal functioning of hormones in an animal's body.

Dose-response study A study that exposes organisms to different amounts of a chemical and then observes a variety of possible responses, including mortality or changes in behavior or reproduction.

Teaching Tip: Beyond the Classroom

Many chemicals used in homes are neurotoxins or carcinogens. Have students do a web search for neurotoxins or carcinogens found in household products. Common household products containing toxic chemicals include nail polish remover (acetone), metal polish (ethylene glycol), and insect repellent (DEET). Ask students to present a short report or a poster on one chemical and to suggest safer alternatives to the use of that chemical.

AP® Exam Tip

The effect of industrial pollutants on human health is covered on the 2010 AP® Exam, Question 1. To answer this question students must

- choose one pollutant discussed in a provided excerpt about industrial pollutants in China and describe a specific source, how the pollutant enters the human body, and its effects on human health; suggest two steps that can be taken to reduce the threat it poses
- explain why children are more vulnerable than adults to toxic pollutants
- use provided data to calculate the dollar cost of capturing the CO_2 generated in creating the energy needed to produce a single MP3 player
- offer two reasons a multinational company might decide to locate a manufacturing plant in China or India rather than in the United States or Europe

Teaching Tip: Warm-up

Ask students: What is the difference between an acute and a chronic study? *An acute study is an experiment that exposes organisms to an environmental hazard for a short duration, while a chronic study is an experiment that exposes organisms to an environmental hazard for a long duration.*

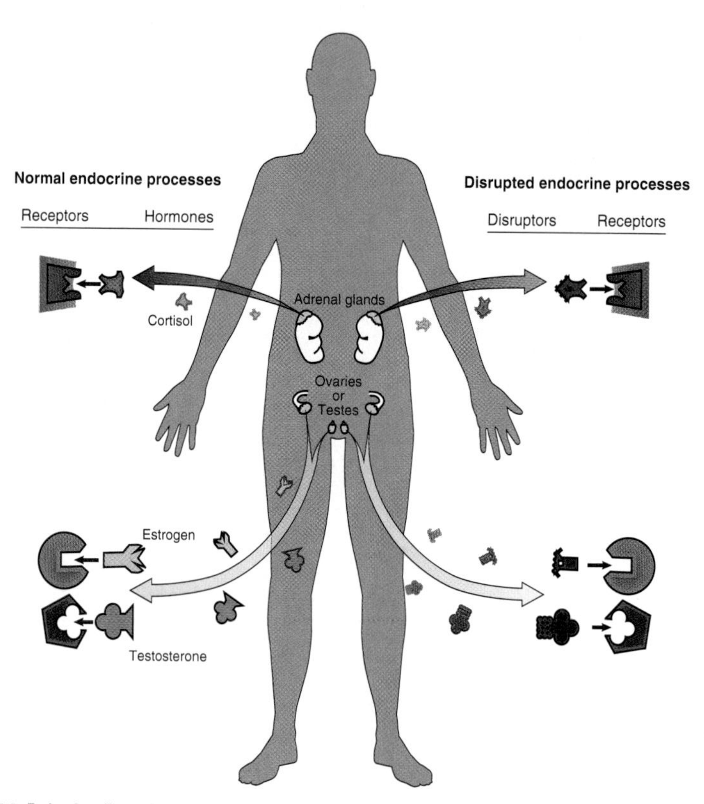

FIGURE 57.3 **Endocrine disruption.** In normal endocrine processes, hormones bind with receptors on cells to regulate the functioning of the body including growth, metabolism, and the development of reproductive organs. Hormone-disrupting chemicals mimic the hormones in the body and also bind to receptive cells and cause the cell to respond in ways that are not beneficial to the organism.

behavior or reproduction. For example, dose-response studies of aquatic animals such as tadpoles are used to determine the concentrations of various pesticides that cause 50 percent of the animals to die (FIGURE 57.4). The concentration of the chemicals being considered can be measured in air, water, or food. They can also be measured as the dose of a chemical, which is the amount that an organism absorbs or consumes. For reasons of efficiency, most dose-response studies only last for 1 to 4 days. Experiments that expose organisms to an environmental hazard for a short duration are called **acute studies.** Studies that are conducted for longer periods of time are called **chronic studies.**

Dose-response studies most commonly measure mortality as a response. At the end of a dose-response experiment, scientists count how many individuals die after exposure to each concentration. When the data are graphed, they generally follow an S-shaped curve, like the one in FIGURE 57.5. If you examine the purple curve, you will see that at the lowest dose no individuals die. At slightly higher doses, a few individuals die. The dose at which an effect can be detected is

Acute study An experiment that exposes organisms to an environmental hazard for a short duration.

Chronic study An experiment that exposes organisms to an environmental hazard for a long duration.

(a)

(b)

FIGURE 57.4 **Conducting dose-response experiments.** (a) Researchers determine how chemicals affect the mortality of animals using dose-response experiments in the laboratory. (b) In the experiment shown, researchers are examining the effects of different insecticide concentrations on the survival of tadpoles. *(a: Rick Relyea; b: Jason Hoverman)*

called the threshold. These individuals generally are in poorer health or genetically are not very tolerant to the chemical. As the dose is further increased, many more individuals begin to die. At the highest concentrations all individuals die.

To compare the harmful effects of different chemicals scientists measure the **LD50,** which is an abbreviation for the lethal dose that kills 50 percent of the individuals in a dose-response study. The LD50 value helps assess the relative toxicity of a chemical to a particular species. For example, scientists can compare the LD50 value of a new chemical with the LD50 value of thousands of previously tested chemicals to determine whether the new chemical is more or less lethal to a given organism than other chemicals.

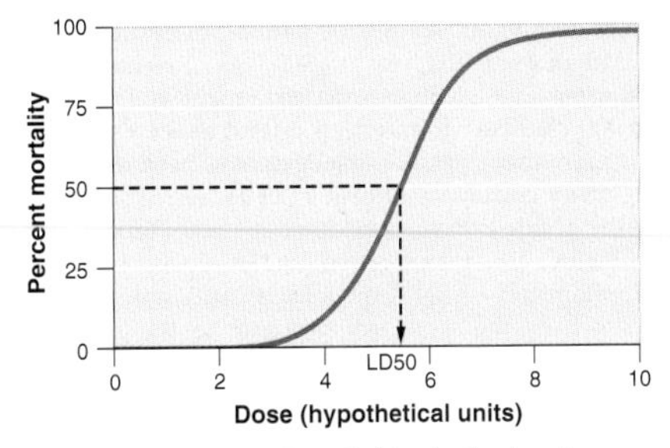

FIGURE 57.5 **LD50 studies.** To determine the dose of a chemical that causes a 50 percent death rate, scientists expose animals to different doses of a chemical and determine what proportion of the animals die at each dose. Such an experiment typically produces an S-shaped curve.

Although the vast majority of toxicology studies are only conducted for a few days, chronic studies will often last from the time an organism is very young to when it is old enough to reproduce. For some species such as fish, chronic experiments can take several months. The goal of chronic studies is to examine the long-term effects of chemicals, including how they affect survival and reproduction.

Not all dose-response experiments measure death as a response to chemicals. In many cases, scientists are interested in other harmful effects, including acting as a teratogen, carcinogen, or neurotoxin. When exposure to a chemical does not kill an organism but impairs its behavior, physiology, or reproduction, we say the chemical has **sublethal effects.** In these cases, the experiments are conducted to determine the **ED50,** which is the effective dose that causes 50 percent of the individuals in a dose-response study to display the harmful, but nonlethal, effect.

Testing Standards

In the United States, chemicals that affect humans and other species are regulated by the Environmental Protection Agency (EPA). The Toxic Substances

LD50 The lethal dose of a chemical that kills 50 percent of the individuals in a dose-response study.

Sublethal effect The effect of an environmental hazard that is not lethal, but which may impair an organism's behavior, physiology, or reproduction.

ED50 The effective dose of a chemical that causes 50 percent of the individuals in a dose-response study to display a harmful, but nonlethal, effect.

AP® Exam Tip

LD50 is covered on the 2002 AP® Exam, Question 3. To answer this question, students must

- read a short description of an experiment and use provided data to plot a log graph of a curve showing copper sulfate toxicity
- explain the meaning of LD50 and determine the LD50 concentration of copper sulfate in the experiment
- define "threshold of toxicity" and label it on the graph
- provide arguments for and against extending the experiment results to humans

Note that because this experiment involves killing brine shrimp, the question has raised ethical issues among teachers and students. Although the question provides students with an excellent opportunity to practice graphing skills and to make sure they understand the concept of LD50, you may wish to acknowledge the ethical challenges of animal testing and provide your students with an opportunity to express their opinions on it.

Teaching Tip: Activity

In pairs or groups, have students use the words from the list below to create two true statements and two false statements about human health and environmental risks. Encourage them to use as many words from the list as they can, as well as others as needed. Then have groups share their statements with the class. Award "bonus points" to the groups that use the most terms.

- neurotoxin
- carcinogen
- mutagen
- teratogen
- allergen
- endocrine disruptor
- dose response
- acute study
- chronic study
- sublethal effect
- LD50
- ED50
- retrospective study
- prospective study
- synergistic interaction
- route of exposure
- solubility
- bioaccumulation
- biomagnification
- persistence

Sample answers:

- *Recently an acute study was done on mice using neurotoxins such as mercury and lead. (True)*
- *The pregnant mother has been drinking alcohol, which is known as a teratogen; she also smokes cigarettes, which are carcinogens. Together they can have a synergistic interaction on the fetus. (True)*
- *The pesticide DDT is known as a carcinogen, which can cause an allergic reaction. (False)*

Control Act of 1976 gives the EPA the authority to regulate many chemicals, but does not include food, cosmetics, and pesticides. Pesticides are regulated under a separate law—the Federal Insecticide, Fungicide, and Rodenticide Act of 1996. Under this act, a manufacturer must demonstrate that a pesticide "will not generally cause unreasonable adverse effects on the environment."

Because no chemical can be tested on every one of the approximately 10 million species of organisms on Earth, scientists have devised a system of testing a few species—a bird, mammal, fish, and invertebrate—that are thought to be among the most sensitive in the world. The particular species tested from each of the four animal groups can vary, depending on which species is thought to be the most sensitive to a particular chemical. The reasoning for this is that regulations devised to protect the most sensitive species in a group will automatically protect all other species in that group. Since conducting LD50 studies on humans would be unethical, results from studies conducted on mice and rats are extrapolated to humans. For nonhuman animals, test results from mice and rats are used to represent all mammals, birds such as pigeons and quail are used to represent all birds, fish such as trout are used to represent all fish, and invertebrates such as water fleas are used to represent all invertebrates.

You might have noticed that the groups of tested animals do not include amphibians or reptiles. Unfortunately, the standards for testing chemicals were set up before there was much interest in protecting amphibians and reptiles. Currently, test results from fish are used to represent aquatic amphibians and reptiles, whereas test results from birds are used to represent terrestrial amphibians and reptiles. Because amphibians and reptiles are now experiencing population declines throughout the world, there is increased interest in requiring tests on species from these two groups as well.

Using the LD50 and ED50 values from dose-response experiments, regulatory agencies such as the EPA can determine the concentrations in the environment that should cause no harm. For most animals, a safe concentration is obtained by taking the LD50 value and dividing it by 10. The logic is that if the LD50 value causes 50 percent of the animals to die, then 10 percent of the LD50 value should cause few or no individuals to die.

Retrospective study A study that monitors people who have been exposed to an environmental hazard at some time in the past.

Prospective study A study that monitors people who might become exposed to harmful chemicals in the future.

The regulatory agencies, however, are much more conservative in setting concentrations for humans. Scientists determine the LD50 or ED50 values for rats or mice and then divide by 10 to determine a safe concentration for rats and mice. This value is divided by 10 again to reflect that rats and mice may be less sensitive to a chemical than humans. Finally, this value is often divided by 10 again to ensure an extra level of caution. In short, the LD50 and ED50 values obtained from rats and mice are divided by 1,000 to set the safe values for humans. "Do the Math: Estimating LD50 Values and Safe Exposures" on page 605 shows you how to make this calculation.

Retrospective versus Prospective Studies

Estimating the effects of chemicals on humans is a major challenge. We have seen that one approach is to conduct dose-response experiments on rats and mice and extrapolate the results to humans. An alternative approach is to examine large populations of humans or animals who are exposed to chemicals in their everyday lives and then determine whether these exposures are associated with any health problems. Such investigations fall within the study of epidemiology, a field of science that strives to understand the causes of illness and disease in human and wildlife populations. There are two ways of conducting this type of research: *retrospective studies* and *prospective studies*.

Retrospective studies monitor people who have been exposed to a chemical at some time in the past. In such studies, scientists identify a group of people who have been exposed to a potentially harmful chemical and a second group of people who have not been exposed to the chemical. Both groups are then monitored for many years to see if the exposed group experiences more health problems than the unexposed group. In 1984, for example, there was an accidental release of methyl isocyanate gas from a Union Carbide pesticide factory in Bhopal, India (FIGURE 57.6a). More than 36,000 kg (80,000 pounds) of hazardous gas spread through the city of 500,000 inhabitants. An estimated 2,000 people died that night and another 15,000 died later from effects related to the exposure. For more than 2 decades scientists have been monitoring many citizens of Bhopal to determine if survivors of the accident have developed any additional health problems. The retrospective studies have found that approximately 100,000 people are still suffering illnesses from the accidental exposure to the gas. The survivors have higher rates of genetic abnormalities, infant mortality, kidney failure, and learning disabilities. As shown in Figure 57.6b, they also have higher rates of respiratory problems and stillbirths.

In contrast to retrospective studies, **prospective studies** monitor people who might become exposed to harmful chemicals in the future. In this case, scientists might select a group of 1,000 participants and

Teaching Tip: Discussion Starter

Ask students: Why is it important to conduct retrospective and prospective studies? *Retrospective studies monitor people who have been exposed to an environmental hazard at some time in the past. Prospective studies monitor people who might become exposed to harmful chemicals in the future. Together, these studies help us understand how and why people become ill from environmental hazards.*

do the math

Estimating LD50 Values and Safe Exposures

Using our knowledge of how scientists conduct LD50 studies, we can consider an example. Let's imagine that you are a scientist charged with determining the safe levels for mammals of a pesticide in the environment. Using lab rats, you feed them a diet that contains different amounts of the pesticide, ranging from 0 to 4 mg of pesticide per kg of the rat's mass. After feeding them these diets for 4 days, you count how many rats are still alive. When you plot the data, you obtain the graph below.

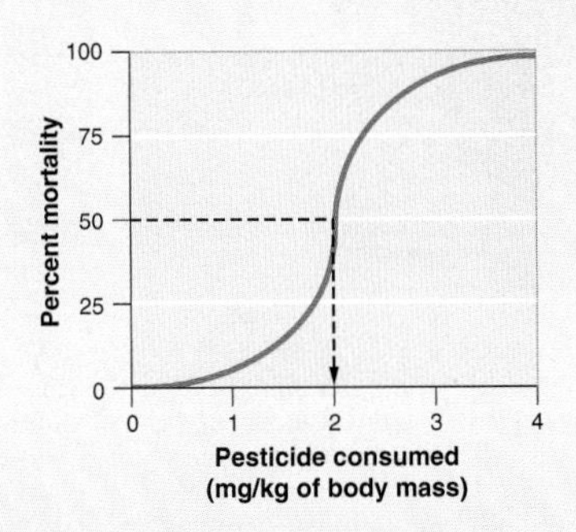

What is the LD50 value for lab rats? To determine this, we can draw a horizontal line at the point of 50 percent mortality on the *y* axis. Where this line intersects the purple line, we can draw another line straight down to the *x* axis. This second line crosses the *x* axis at 2 mg/kg of mass.

Based on this LD50 study, what amount of pesticide would be considered safe for mammals to ingest? Recall that we can calculate this number by taking the LD50 value and dividing it by 10. Thus the safe amount of pesticide for a rat is:

$$\frac{2 \text{ mg/kg of mass}}{10} = 0.2 \text{ mg/kg of mass}$$

Your Turn Using the same LD50 study, what amount of pesticide would be considered safe for a human to ingest?

ask them to keep track of the food they eat, the tobacco they use, and the alcohol they drink over a period of several decades. As time passes, the researchers can determine if the habits of the participants are associated with any future health problems. Prospective studies can be quite challenging because a participant's habits, such as tobacco use, can also be associated with many other risk factors, such as socioeconomic status. Of particular concern is when multiple risks cause **synergistic interactions,** in which two risks together cause more harm than expected based on the separate effects of each risk alone. For example, the health impact of a carcinogen such as asbestos can be much higher if an individual also smokes tobacco.

Studies of lead in children are often prospective. In one study, researchers at Harvard University looked at the effects of lead on children's intelligence by following 276 children in Rochester, New York, from 6 months to 5 years of age. IQ tests are reliable at the age of 5. In addition to lead exposure, the researchers also accounted for other factors that might affect childhood IQ including the mother's IQ, exposure to tobacco, and the intellectual environment of their homes. After controlling for these

Synergistic interaction A situation in which two risks together cause more harm than expected based on the separate effects of each risk alone.

Answer to Your Turn

To determine the "safe" level for humans, we divide the LD50 obtained from the experiment by 1,000.

LD50 $\div$ 1,000 = safe level for humans

2 mg/kg of body mass $\div$ 1,000 = 0.002 mg/kg of body mass

more math practice

LD50

Assume that for a certain pesticide, the LD50 dosage level for laboratory rats is determined to be 450 mg/kg of body mass.

1. Calculate the amount of the pesticide that would be considered safe for animals to ingest.
For most animals, a safe concentration is obtained by dividing the LD50 by 10. Therefore:

450 mg/kg $\div$ 10 = 45 mg/kg of body mass

2. What amount of pesticide would be considered safe for humans to ingest?
The LD50 values obtained from rats and mice are divided by 1,000 to set the safe values for humans. Therefore:

450 mg/kg $\div$ 1,000 = 0.45 mg/kg of body mass

Teaching Tip: Beyond the Classroom

Have your students research the 1984 chemical disaster in Bhopal, India. Ask them to write a short report or presentation based on the following questions:

- What did the disaster involve?
- What was the cause?
- What were the human health effects?
- What regulations changed as a result of the accident?
- What were some of the long-lasting effects from the accident?

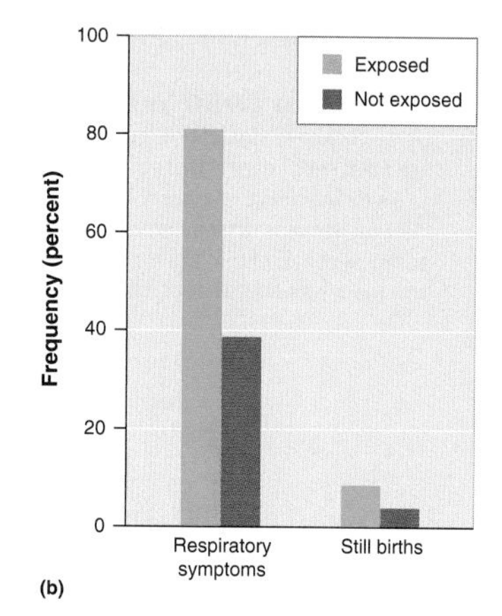

FIGURE 57.6 The chemical disaster in Bhopal, India. (a) In 1984, a massive release of methyl isocyanate gas killed and injured thousands of people. (b) Retrospective studies that followed the survivors of the accident have identified a large number of longer-term health effects from the accident. *(Data from P. Cullinan, S.D. Acquilla, and V.R. Dhara, Long-term morbidity in survivors of the 1984 Bhopal gas leak,* National Medical Journal of India, *Jan.–Feb. (1996) 9(1):5–10. Photo by AFP/ Getty Images.)*

other factors, the researchers found that among children who had been exposed to lead in the environment—primarily from breathing lead dust and consuming lead paint chips—those with higher lead exposures scored lower on subsequent IQ tests. Such prospective studies can help regulators determine acceptable levels of chemical exposure.

Factors That Determine the Concentrations of Chemicals That Organisms Experience

Knowing the concentrations of chemicals that can harm humans or other animals is important, but it is only useful when combined with information about the concentrations that an individual might actually experience in the environment. If a chemical is quite harmful at some moderate concentration but individuals only experience lower concentrations of that chemical, we might not be particularly concerned. Therefore, to identify and understand the effects of chemical concentrations that organisms experience, we need to know something about how the chemicals behave in the environment.

Routes of Exposure

The ways in which an individual might come into contact with a chemical are known as **routes of exposure.** As FIGURE 57.7 illustrates, the full range of possibilities is complex because it includes potential exposures from the air, from water used for drinking, bathing, or swimming, from food, and from the environments of places where people live, work, or visit. For any particular chemical, however, the major routes of exposure are usually limited to just a few of the many possible routes. For example, bisphenol A is a chemical used in manufacturing hard plastic items such as toys, food containers, and baby bottles. Recent research has raised concerns that bisphenol A may be responsible for early puberty and increased rates of cancer. While these

Route of exposure The way in which an individual might come into contact with an environmental hazard.

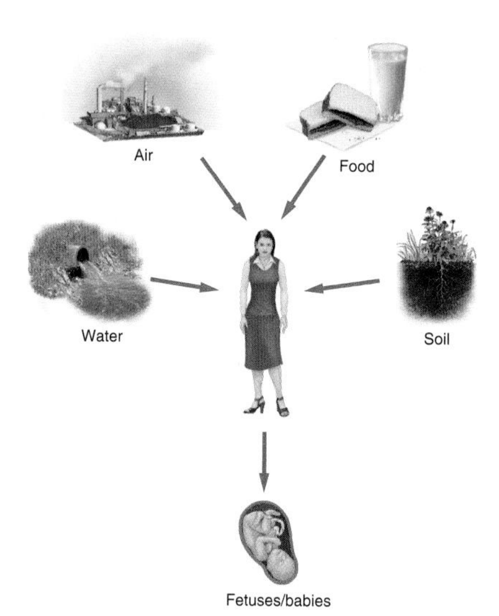

FIGURE 57.7 Routes of exposure. Despite a multitude of potential routes of exposure to chemicals, most chemicals have a limited number of major routes.

Teaching Tip: Engage

Ask students: What is solubility and how does it relate to bioaccumulation? *Solubility is a measure of how well a chemical can dissolve in a liquid. Some chemicals are readily soluble in water whereas others are much more soluble in fats and oils. Oil-soluble chemicals are also readily stored in the fat tissues of animals. Continued exposure to oil-soluble chemicals can cause bioaccumulation, which is the increased concentration of a chemical within an organism over time.*

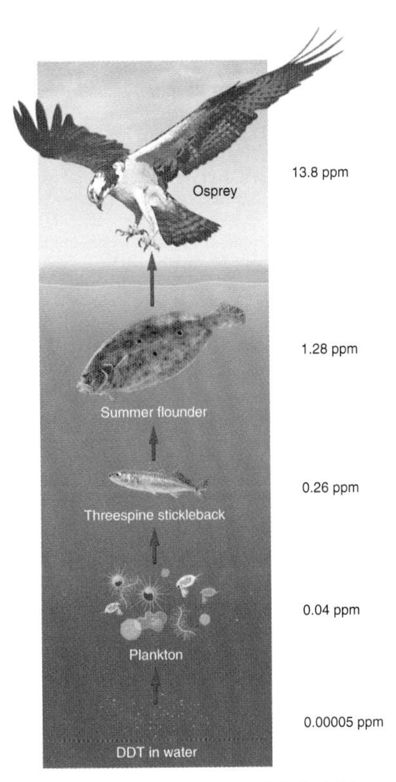

FIGURE 57.8 **The biomagnification of DDT.** The initial exposure is primarily in a low trophic group such as the plankton in a lake. Consumption causes the upward movement of the chemical where it is accumulated in the bodies at each trophic level. The combination of bioaccumulation at each trophic level and upward movement by consumption allows the concentration to magnify to the point where it can be substantially more concentrated in the top predator than it was in the water. *(Data from G. M. Woodwell, C. F. Wurster, Jr., and Peter A. Isaacson, DDT Residues in an East Coast Estuary: A Case of Biological Concentration of a Persistent Insecticide, Science, New Series, 156 (3776) (May 12, 1967): 821–824. http://www.jstor.org/stable/1722018)*

effects are being debated and investigated, it is clear that a child's routes of exposure to bisphenol A are limited to toys, food containers, and baby bottles.

Solubility of Chemicals, Bioaccumulation, and Biomagnifications

Once we know the potential routes of exposure, scientists can then determine the chemical's *solubility* and its potential for *bioaccumulation* and *biomagnification*. The movement of a chemical in the environment depends in part on its **solubility,** which is how well a chemical can dissolve in a liquid. For example, some chemicals such as herbicides are readily soluble in water whereas others such as insecticides are much more soluble in fats and oils. When a chemical is highly soluble in water, it can be washed off surfaces, percolate into groundwater, and run off into surface waters including rivers and lakes. In contrast, chemicals that are soluble in fats and oils are not very soluble in water so they tend not to be found percolating into the groundwater or running off into surface waters. Instead, they can be found in higher concentrations bound to soils, including the benthic soils that underlie bodies of water.

Chemicals that are soluble in fats and oils can also become stored in the fatty tissues of animals. For example, in Chapter 11 we mentioned that DDT accumulates in the fatty tissues of eagles and pelicans. This process, known as **bioaccumulation,** occurs when an organism increases the concentration of a chemical in its body over time. The process of bioaccumulation begins when an individual is exposed to small amounts of a chemical from the environment and incorporates the chemical into its tissues, typically its fat tissues. Fish, for example, are exposed to low concentrations of methyl mercury when they drink water, pass water over their gills to breathe, and consume food that contains mercury. A fish stores mercury in its fat tissues and, over time, the mercury accumulates. The rate of accumulation for any animal will depend on the concentration of the chemical in the environment, the rate at which the animal takes up each source of the chemical, the rate at which the chemical breaks down inside the animal, and the rate at which it is excreted by the animal.

Biomagnification is the increase in chemical concentration in animal tissues as the chemical moves up the food chain. In this way, the original concentration in the environment is magnified to occur at a much higher concentration in the top predator of the community. The classic example of biomagnification is the case of DDT, an insecticide that has been widely used to kill insect pests in agriculture and to kill the mosquitoes that carry malaria and other diseases. DDT is not soluble in water, so when sprayed over water it quickly binds to particulates in the water and the underlying soil or is quickly taken up by the tiny zooplankton that act as primary consumers on algae. As we see in FIGURE 57.8, the very low concentration of DDT in the water

Solubility How well a chemical dissolves in a liquid.

Bioaccumulation An increased concentration of a chemical within an organism over time.

Biomagnification The increase in chemical concentration in animal tissues as the chemical moves up the food chain.

COMMON MISCONCEPTION
Biomagnification and bioaccumulation

Students frequently confuse biomagnification and bioaccumulation. Bioaccumulation occurs within a single organism, and biomagnification refers to the accumulation of a toxin as it moves up the food chain.

Teaching Tip: Journal Prompt

Ask students: What do we mean when we talk about the persistence of a chemical? *The persistence of a chemical refers to how long the chemical remains in the environment. Harmful chemicals may cause even greater damage when they persist for many years. Today many chemicals are designed to break down much more rapidly than chemicals did in the past so that any unintended effects will be short-lived.*

bioaccumulates in the bodies of the zooplankton where it becomes approximately 1,000 times more concentrated. Small fish eat the zooplankton for many weeks or months and the DDT is further concentrated approximately sixfold. Large fish spend their lives eating the contaminated smaller fish and the DDT in the large fish is further concentrated approximately fivefold. Finally, fish-eating birds such as pelicans and eagles spend years eating the large fish and further magnify the DDT in their own bodies. Because of biomagnification along the food chain, the concentration of DDT in the birds is nearly 276,000 times higher than the concentration of DDT in the water. The concentrated DDT in the fish-eating birds causes them to produce thin-shelled eggs that often break when the parent birds try to incubate the eggs. This was a primary cause in the decline of these birds in the 1960s. Since DDT was banned in the United States in 1972, the populations of fish-eating birds have dramatically increased.

Persistence

The **persistence** of a chemical refers to how long the chemical remains in the environment. Persistence depends on a number of factors including temperature, pH, whether the chemical is in water or soil, and whether it can be degraded by sunlight or broken down by microbes. Scientists often measure persistence by observing the time needed for a chemical to degrade to half its original concentration, known as the half-life of the chemical. **TABLE 57.1** lists the persistence of various chemicals in the environment measured according to half-life. DDT, for example, has a half-life in soil of up to 30 years. Thus, even after DDT is no longer sprayed in an area, half of the chemical that was absorbed in the soil would still be present after 30 years, and one-fourth would be present after 60 years. Chemicals that cause harmful effects on humans and other organisms may become even larger risks when they persist for many years. For this reason, many modern chemicals are designed to break down much more rapidly so that any unintended effects will be short-lived.

Persistence The length of time a chemical remains in the environment.

Table 57.1 The persistence of various chemicals in the environment

Chemical	Source	Half-Life
Malathion	Insecticide	1 day
Radon	Rocks and soil	4 days in air
Vinyl chloride	Industry, water from vinyl chloride pipes	4.5 days in air
Phthalates	Plastics, cosmetics	2.5 days in water
Roundup	Herbicide	7 to 70 days in water
Atrazine	Herbicide	224 days in wetland soils
Polychlorinated biphenyls (PCBs)	Industry	8 to 15 years in water
DDT	Insecticide	30 years in soil

module

57

REVIEW

In this module, we learned there are many types of chemicals that can potentially cause harmful effects in humans. Neurotoxins disrupt the nervous systems of animals, carcinogens cause cancer, teratogens cause abnormal development in embryos and fetuses, allergens cause abnormally high immune responses, and endocrine disruptors interfere with the normal functioning of hormones. For each of these types of chemicals, scientists can determine the concentrations of chemicals that will harm organisms using short-term LD50 to assess lethal effects and ED50 studies to assess sublethal effects. Scientists can also use chronic studies that examine long-term effects of chemical exposure including both retrospective and prospective studies. Once we understand how different concentrations of a chemical can affect an organism, we can determine the concentrations that an organism could experience by examining solubility, bioaccumulation, biomagnification, and the persistence of chemicals in the environment. In the next module, we will examine how scientists analyze the risk that chemicals or any other environmental hazard poses to humans and other species.

Module 57 AP® Review Questions

1. Atrazine and DDT are examples of
   (a) neurotoxins.
   (b) carcinogens.
   (c) teratogens.
   (d) allergens.
   (e) endocrine disruptors.

2. Teratogens
   (a) interfere with embryo and fetus development.
   (b) disrupt the circulatory system.
   (c) alter the function of hormones.
   (d) suppress the immune system.
   (e) cause inflammation and cell damage.

3. If 1 mg/kg of mass of a pesticide is the LD50 for rats in an experiment, what would be considered the safe exposure for humans?
   (a) 10 mg/kg
   (b) 1 mg/kg
   (c) 0.1 mg/kg
   (d) 0.01 mg/kg
   (e) 0.001 mg/kg

4. Which is NOT a cause of high concentrations of DDT in fish-eating birds?
   (a) Bioaccumulation
   (b) Biomagnification
   (c) Persistence
   (d) Synergistic interactions
   (e) Solubility

5. A prospective study
   (a) determines synergistic interactions of toxins.
   (b) measures the effect of a particular event after it has occurred.
   (c) monitors individuals who might be exposed to harmful chemicals in the future.
   (d) determines the number of individuals who might be effected by a particular chemical.
   (e) determines the lethal dose of a chemical or toxin.

Answers to Module 57 AP® Review Questions

1. e
2. a
3. e
4. d
5. c

module

58

Risk Analysis

Most people face some kind of environmental hazard every day. The hazards we face may be voluntary, as when we make a decision to smoke tobacco, or they may be involuntary, as when we are exposed to air pollution. When assessing the risk of different environmental hazards, regulatory agencies, environmental scientists, and policy makers usually follow three steps for risk analysis: risk assessment, risk acceptance, and risk management. In this module, we will examine each of the three steps.

Learning Objectives

After reading this module you should be able to

- explain the processes of qualitative versus quantitative risk assessment.
- understand how to determine the amount of risk that can be tolerated.
- discuss how risk management balances potential harm against other factors.
- contrast the innocent-until-proven-guilty principle and the precautionary principle.

Risk assessment estimates potential harm

As illustrated in FIGURE 58.1, risk assessment is the first of the three steps involved in risk analysis. Risk analysis seeks to identify a potential hazard and determine the magnitude of the potential harm. There are two types of risk assessment—qualitative and quantitative. Each of us has some idea of the risk associated with different environmental hazards. For our purposes, an **environmental hazard** is anything in our environment that can potentially cause harm. Environmental hazards include substances such as pollutants or other chemical contaminants, human activities such as driving cars or flying in airplanes, or natural catastrophes such as volcanoes and earthquakes.

Environmental hazard Anything in the environment that can potentially cause harm.

We generally make qualitative judgments in which we might categorize our decisions as having low, medium, or high risks. When we choose to slow down on a wet highway or to buy a more expensive car because we feel it is safer, we are making qualitative judgments of the relative risks of various decisions. We are making judgments that are based on our perceptions but that are not based on actual data. It would be unusual for us to consider the actual probability—that is, the statistical likelihood—of an event occurring and the probability of that event causing us harm. Because our personal risk assessments are not quantitative, they often do not match the actual risk. For example, some people find air travel very stressful because they are afraid the plane might crash. These same people often prefer riding in a car, which they perceive to be much safer. Or, a person may be very cautious about safety while walking in an area with heavy traffic but never consider the health dangers of smoking or a lack of exercise. To manage our risk effectively, we need to ask how closely our perceptions of risk match the reality of actual risk.

Teaching Tip: Activity

Ask students to fill in a risk analysis flow chart, like the one below.

Risk assessment
1. Identify the hazard
2. Characterize the toxicity
3. Determine the extent of exposure

Risk acceptance
Determine acceptable level of risk balanced against social, economic, political considerations.

Risk management
Determine policy with input from private citizens, industry, interest groups.

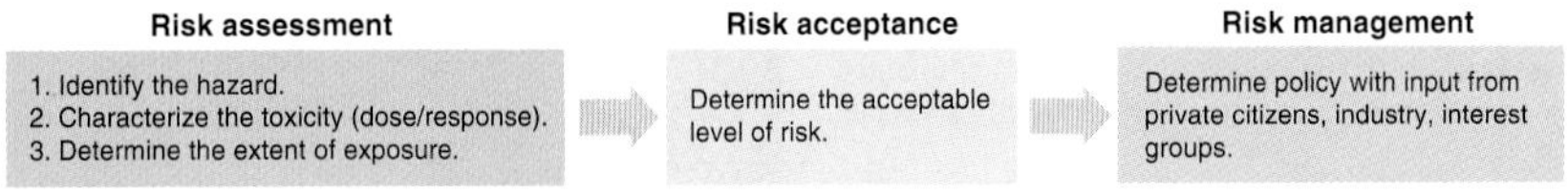

Risk assessment

1. Identify the hazard.
2. Characterize the toxicity (dose/response).
3. Determine the extent of exposure.

Risk acceptance

Determine the acceptable level of risk.

Risk management

Determine policy with input from private citizens, industry, interest groups.

FIGURE 58.1 The process of risk analysis.
Risk analysis involves risk assessment, risk acceptance, and risk management.

In the United States, the probability of death from various hazards can be calculated from data kept by the government. By looking at the total number of people who die in a year and their causes of death, researchers are able to determine the probability that an individual will die from a particular cause. FIGURE 58.2 provides current data on causes of death in the United States. Because these risk estimates are based on real data, they are quantitative rather than qualitative. If we examine this figure, we see that the probability of dying in an automobile is far greater than the probability of dying in an airplane. Similarly, the probability of dying from heart disease is monumentally greater than the risk of dying in a pedestrian accident. These numbers underscore the fact that our perceptions of risk can often be very different from the actual risk. Because a catastrophic event, such as a nuclear plant meltdown or a plane crash, can do a great deal of harm and receives great media attention, people believe that it is very risky to use nuclear reactors or to fly in airplanes. However, as these events rarely occur, the risk of harm is low. In contrast, we tend to downplay the risk of activities that provide us with cultural, political, or economic advantages such as drinking alcohol or working in a coal mine.

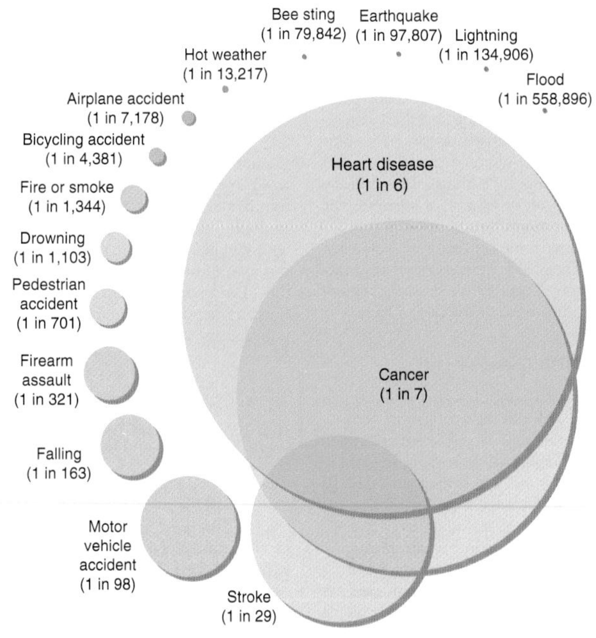

FIGURE 58.2 The probabilities of death in the United States. Some causes of death that people perceive as having a high probability of occurring, such as dying in an airplane crash, actually have a low probability of occurring. In contrast, some causes of death that people rate as having a low probability of occurring, such as dying from heart disease, actually have a very high probability of occurring. *(After National Geographic Society (2006); data from National Safety Council, 2012)*

Module 58 Risk Analysis 613

Quantitative Risk Assessment

The most common approach to conducting a quantitative risk assessment can be expressed with a simple equation:

Risk = probability of being exposed to a hazard × probability of being harmed if exposed

Using this equation, we could ask whether it is riskier in a year to fly on commercial airlines for 1,609 km (1,000 miles) per year or to eat 40 tablespoons of peanut butter, which contains tiny amounts of a carcinogenic chemical produced naturally by a fungus that sometimes occurs in peanuts that are used to make peanut butter. The risk of dying in a plane crash depends on the probability of experiencing a plane crash, which is very low, multiplied by the probability of dying if the plane does crash, which approaches 100 percent. The risk of dying of cancer from consuming peanut butter depends on your probability of eating peanut butter, which may be near 100 percent, multiplied by the probability that consuming peanut butter will cause you to develop lethal cancer, which is very low. It turns out that both behaviors produce a 1 in 1 million chance of dying. This example demonstrates a fundamental rule of risk assessment: The risk of a rare event that has a high likelihood of causing harm can be equal to the risk of a common event that has a low likelihood of causing harm.

Quantitative risk assessments bring together tremendous amounts of data. The estimates of harm can come from acute and chronic dose-response experiments, retrospective studies, and prospective studies. The estimates of which concentrations of a chemical an organism will experience in the environment incorporates the concentrations found in nature, routes of exposure, solubility, persistence, and the potential for the chemical to bioaccumulate or biomagnify. Together, these two groups of data can be used to estimate the probability of harm.

A Case Study in Risk Assessment

As we saw in our discussion of water pollution in Chapter 14, from the 1940s to the 1970s some companies manufacturing electrical components dumped PCBs (polychlorinated biphenyls) into rivers. Beginning in the 1960s, there was increasing evidence that PCBs might have harmful health effects on organisms that came into contact with them, including liver damage in animals and impaired learning in human infants.

Once the EPA identified PCBs as a potential hazard, it began a risk assessment. The agency brought together a range of data. First, scientists had to determine which concentrations of PCBs might cause cancer. To accomplish this objective, they examined dose-response studies on laboratory rats exposed to different concentrations of PCBs. They also examined retrospective studies of cancer cases in workers employed by industries that used PCBs. Next, they had to determine what concentrations

FIGURE 58.3 The outcome of a risk assessment of PCBs. Based on a risk assessment of humans consuming fish, the EPA determined that the fish living in the Hudson River in New York State and in Silver Lake in Massachusetts had unacceptably high concentrations of PCBs due to illegal dumping of PCBs by General Electric. As a result, anglers were not allowed to keep and consume the fish that they caught from those water bodies. *(AP Photo/Alan Solomon)*

people might experience. To accomplish this, scientists examined data on current concentrations in the air, soil, and water and considered the half-life of the chemical. Because PCBs were found throughout the environment and because they are very persistent, the probability of coming into contact with PCBs was considered relatively high. They also considered the potential routes of exposure: eating contaminated fish, drinking contaminated water, and breathing contaminated air.

The final result of the risk assessment on PCBs showed that the risk from eating contaminated fish is higher than the risk from drinking contaminated water and much higher than breathing contaminated air. As a result, signs were posted on the Hudson River and at Silver Lake instructing anglers not to consume any fish they caught (FIGURE 58.3). With limited fish consumption, the EPA concluded that the absolute risk of an individual developing cancer from PCB exposure was low. However, the risk was high enough to cause the EPA to recommend a dredging of the Hudson River to remove a large fraction of the PCBs that had settled at the bottom of the river (see Chapter 14).

Risk acceptance determines how much risk can be tolerated

Once the risk assessment is completed, the second step in risk analysis is to determine risk acceptance—the level of risk that can be tolerated. Risk acceptance may be the most difficult of the three steps in the risk-analysis process. No amount of information on the extent of the risk will overcome the conflict between those who are willing to live with some amount of risk and those who are not. Even among those people who are willing to accept some risk, the precise amount of acceptable risk is open to heated disagreement. For example, according to the EPA, a risk of 1 in 1 million is acceptable for most environmental hazards. Some people believe this is too high. Others feel that a risk

Teaching Tip: Warm-up

Ask students: Why is risk acceptance the most difficult of the three steps in risk analysis? *Some people are willing to live with more risk than others. Even among those people who are willing to accept some risk, the precise amount of acceptable risk is open to heated disagreement. For example, according to the EPA, a risk of 1 in 1 million is acceptable for most environmental hazards. Some people believe this is too high. Others feel that a risk such as a 1 in 1 million chance of death from a radiation leak is a small price to pay for electricity generated by nuclear energy.*

such as a 1 in 1 million chance of death from radiation leaks is a small price to pay for the electricity generated by nuclear energy. While personal preferences will always complicate the determination of risk acceptance, environmental scientists, economists, and others can help us weigh the options as objectively as possible by providing accurate estimates of the costs and benefits of activities that affect us and the environment.

Risk management balances potential harm against other factors

Risk management, the third step of the risk-analysis process, seeks to balance possible harm against other considerations. Risk management integrates the scientific data on risk assessment and the analysis of acceptable levels of risk with a number of additional factors including economic, social, ethical, and political issues. Whereas risk assessment is the job of environmental scientists, risk management is a regulatory activity that is typically carried out by local, national, or international government agencies.

The regulation of arsenic in drinking water provides an excellent example of the difference between risk assessment and risk management. As we saw in Chapter 14, despite the fact that scientists knew that 50 ppb of arsenic could cause cancer in people, from 1942 to 1999 the federal government set the acceptable concentration of arsenic at 50 ppb. In 1999, the EPA announced it was lowering the maximum concentration of arsenic in drinking water to 10 ppb, which matched the standards set by the European Union and the World Health Organization. This regulation threatened to place a large financial burden on mining companies that produced arsenic as a by-product of mining, and an economic burden on several municipalities in western states with naturally high concentrations of arsenic in their drinking water. Both groups lobbied hard against the lower arsenic limits. In 2001, weeks before the new lower limits were to go into effect, the EPA announced that it would return to the 50 ppb. The agency argued that further risk assessments needed to be conducted and any risk assessment had to be balanced by economic interests. Later in 2001, the National Academy of Sciences concluded that the acceptable amount of arsenic was a mere 5 ppb, which was lower than some previous estimates. This new risk assessment played a key role in striking a balance between the scientific data and economic interests and the EPA revised its ruling, ultimately setting the safe arsenic concentration at 10 ppb.

Worldwide standards of risk can be guided by two different philosophies

There are currently about 80,000 regulated chemicals in the world but they are not regulated the same way around the globe. A key factor determining the type of chemical regulation is whether the regulations are guided by the *innocent-until-proven-guilty principle* or the *precautionary principle,* both illustrated in FIGURE 58.4.

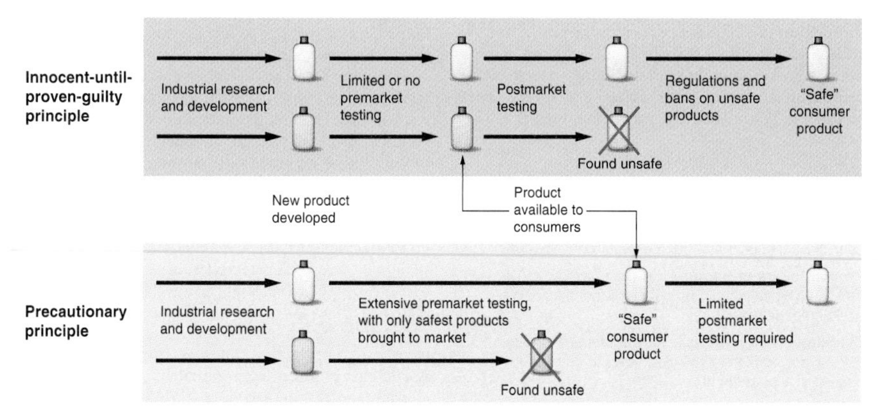

FIGURE 58.4 **The two different approaches to managing risk.** The innocent-until-proven-guilty principle requires that researchers prove harm before the chemical is restricted or banned. The precautionary principle requires that when there is scientific evidence that demonstrates a plausible risk, the chemical must then be further tested to demonstrate it is safe before it can continue to be used.

Teaching Tip: Discussion Starter

Ask students: What is the difference between the innocent-until-proven-guilty principle and the precautionary principle? *The innocent-until-proven-guilty principle is based on the belief that a potential hazard should not be considered an actual hazard until scientific data definitively demonstrate that it actually causes harm. The precautionary principle is based on the belief that action should be taken against a plausible environmental hazard.*

Teaching Tip: Discussion Starter

Ask students the following questions:

- What does the acronym REACH stand for? *REACH stands for registration, evaluation, authorization, and restriction of chemicals. It was an agreement among all 27 nations of the European Union in 2007.*
- How is REACH related to the precautionary principle? *REACH embraces the precautionary principle by putting more responsibility on chemical manufacturers to confirm that chemicals used in the environment pose no risk to people or the environment. This regulation was enacted because many chemicals used for decades in the European Union had not been subjected to rigorous risk analyses.*

The **innocent-until-proven-guilty principle** is based on the belief that a potential hazard should not be considered a real hazard until the scientific data definitively demonstrates that it actually causes harm. This strategy allows beneficial chemicals to be sold sooner. The downside is that harmful chemicals can affect humans or wildlife for decades before sufficient scientific evidence accumulates to confirm that they are harmful.

In contrast, the **precautionary principle** is based on the belief that when a hazard is plausible but not yet certain, we should take actions to reduce or remove the hazard. The plausibility of the risk cannot be speculation; it must have a scientific basis. In addition, the intervention should be in proportion to the potential harm that might be caused by the hazard. This approach allows fewer harmful chemicals to enter the environment. However, if the initial assessment indicates a plausible risk and the chemical ultimately proves harmless but beneficial, its introduction can be delayed for many years. Moreover, the slower pace of approval can reduce the financial motivation of manufacturers to invest in research for new chemicals. In short, there is a trade-off between greater safety with slower introduction of beneficial chemicals versus greater potential risk with a greater rate of discovery of helpful chemicals. Use of the precautionary principle has been growing throughout many parts of the world and was instituted by the European Union in 2000. The United States, however, continues to use the innocent-until-proven-guilty principle.

The benefit of the precautionary principle can be illustrated using the case of asbestos. Asbestos is a white, fibrous mineral that is very resistant to burning. This made asbestos a popular building material throughout much of the twentieth century. It is now widely accepted that dust from asbestos can cause a number of deadly diseases including asbestosis (a painful inflammation of the lungs) and several types of cancer. When asbestos was first mined in 1879, there was no evidence that it harmed humans. The first report of deaths in humans was in 1906 and the first experiment showing harmful effects in rats was conducted in 1911. In 1930, it was reported that 66 percent of workers in an asbestos factory suffered from asbestosis. In 1955, researchers found that asbestos workers had a higher risk of lung cancer than other groups. In 1965, a study linked a rare form of cancer with workers who were exposed to asbestos dust. Despite all of the growing scientific evidence that asbestos was harming human health, little was done to reduce the exposure of workers and the public. Indeed, it was not until 1998 that the European Union banned asbestos. Today, workers go to great lengths not to be exposed to asbestos dust such as by wearing protective suits and using respirators (FIGURE 58.5). A study in the Netherlands estimated that had asbestos been banned in 1965 when the harm to health became clear, the country would have had 34,000 fewer deaths from asbestos and would have saved approximately $25 billion in cleanup and compensation costs. Because the effects of asbestos can take several decades to harm a person's health, the European Union estimates that from 2005 to 2040 it will have 250,000 to 400,000 additional people die as a result of past exposures to

FIGURE 58.5 **The risks of asbestos dust.** Despite nearly a century of studies on the risks of asbestos dust to human health, only recently have workers been required to go to great lengths to prevent exposure. Today, they dress in chemical suits and respirators when removing asbestos from a building. Applying the precautionary principle would have required protection of workers many decades earlier and saved hundreds of thousands of lives. *(Ashley Cooper/Alamy)*

Innocent-until-proven-guilty principle A principle based on the belief that a potential hazard should not be considered an actual hazard until the scientific data definitively demonstrate that it actually causes harm.

Precautionary principle A principle based on the belief that action should be taken against a plausible environmental hazard.

Stockholm Convention A 2001 agreement among 127 nations concerning 12 chemicals to be banned, phased out, or reduced.

REACH A 2007 agreement among the nations of the European Union about regulation of chemicals; the acronym stands for registration, evaluation, authorization, and restriction of chemicals.

asbestos. Had the European Union been using the precautionary principle decades earlier, the number of deaths would have been considerably less.

International Agreements on Hazardous Chemicals

In 2001 a group of 127 nations gathered in Stockholm, Sweden, to reach an agreement on restricting the global use of some chemicals. The agreement, known as the **Stockholm Convention,** produced a list of 12 chemicals to be banned, phased out, or reduced. These 12 chemicals came to be known as the "dirty dozen" and included pesticides such as DDT, industrial chemicals such as PCBs, and certain chemicals that are by-products of manufacturing processes. All of the chemicals were known to be endocrine disruptors, and a number of them had already been banned or were experiencing declining use in many countries. However, bringing countries together in a forum to discuss controlling the most harmful chemicals was the great achievement of the Stockholm Convention. By 2013, an additional 11 chemicals were added to the original list of 12; several more have been suggested for future listing.

In 2007, the 27 nations of the European Union put into effect an agreement on how chemicals should be regulated within the European Union. Known as **REACH,** an acronym for registration, evaluation, authorization, and restriction of chemicals, the agreement embraces the precautionary principle by putting more responsibility on chemical manufacturers to confirm that chemicals used in the environment pose no risk to people or the environment. This regulation was enacted because many chemicals used for decades in the European Union had not been subjected to rigorous risk analyses. The new regulations are being phased in over 11 years to permit sufficient time for chemical manufacturers to complete the required testing.

module

58 REVIEW

In this module, we learned that we can examine the environmental hazards faced by humans by using the process of risk analysis. Risk analysis begins with a risk assessment, which can be either qualitative or quantitative. Quantitative risk assessments are typically preferred because a person's perception of risk can be quite different from the actual risk. Once we assess the level of risk posed by a hazard, we need to determine how much risk humans are willing to accept. With this information on risk assessment and risk tolerance from environmental scientists, government regulators balance the risk of a particular environmental hazard against numerous other factors including economic, social, ethical, and political issues. Balancing this risk can be done using the innocent-until proven guilty principle or the precautionary principle, with each principle having different costs and benefits. In recent years, there has been worldwide agreement on a number of hazardous chemicals that will be banned, phased out, or reduced.

Module 58 AP® Review Questions

1. Which is NOT an environmental hazard?
   (a) Air pollutants
   (b) Driving a car
   (c) An earthquake
   (d) Smoking tobacco
   (e) Cancer

2. What does the EPA consider the limit of acceptable risk for environmental hazards?
   (a) 1 in 10,000
   (b) 1 in 100,000
   (c) 1 in 1,000,000
   (d) 1 in 10,000,000
   (e) 1 in 100,000,000

3. The Stockholm Convention
   (a) was an international agreement to ban a number of endocrine disruptors.
   (b) led to the REACH agreement on chemical evaluation.
   (c) was a European Union agreement to use the precautionary principle.
   (d) was an international agreement on asbestos and other hazardous materials.
   (e) was an agreement between the United States and the European Union to ban many carcinogens.

Teaching Tip: Beyond the Classroom

In 2001, 127 nations came together in Stockholm, Sweden, to reach an agreement on restricting the worst hazardous chemicals. The Stockholm Convention produced a list of 12 chemicals to be banned, phased out, or reduced. Ask students to research one of the following 12 chemicals:

- aldrin
- chlordane
- DDT
- dieldrin
- endrin
- heptachlor
- hexachlorobenzene
- mirex
- toxaphene
- polychlorinated biphenyl (PCB)
- polychlorinated dibenzofuran (PCDF)
- polychlorinated dibenzodioxins (PCDD)

You may suggest students use the following questions to guide their research:

- What is the chemical?
- What is the chemical used for? Is it still in use?
- What happens to the chemical when it enters the environment?
- How might a person be exposed to the chemical?
- How can the chemical enter and leave a human body?
- How can the chemical affect human health?
- How can we reduce the risk of becoming exposed to the chemical?
- Is there a medical test to determine whether a person has been exposed to the chemical?

Answers to Module 58 AP® Review Questions

1. e
2. c
3. a
4. a
5. a

4. Which risk has the highest probability of death in the United States?
   (a) Motor vehicle accident
   (b) Drowning
   (c) Fire
   (d) Firearm assault
   (e) Earthquake

5. The precautionary principle
   (a) decreases financial incentives for chemical development.
   (b) was used in considering the use of asbestos.
   (c) causes faster adoption of new products.
   (d) is primarily used in the United States.
   (e) increases the risk of harmful chemicals being used.

Teaching Tip: Beyond the Classroom

Many organizations are involved with the global fight against malaria. Have your students research organizations that provide mosquito nets. As a community service project they may wish to raise funds for one or more mosquito nets. Alternatively, have them create posters about the global fight against malaria with information about the organizations they researched.

Teaching Tip: Video

Ghana: Mosquito Net Partnership
This video describes efforts to prevent malaria in Ghana through education and insecticide-treated mosquito nets.

The link to this resource may be found by clicking the link buttons in the TE-book, opening the TRFD, or logging onto the book's companion website bcs.whfreeman.com/friedlandapes2e (teacher login required).

Chapter 17 Web Resources

working toward  sustainability

The Global Fight Against Malaria

Bill Gates is best known as the founder of Microsoft, the computer software company, but he is also an active philanthropist. In 2007 he stood up in front of a large group of scientists in Seattle, Washington, and declared that the world needed to eradicate malaria. In challenging the scientists of the world, he asked, "Why would anyone want to follow a long line of failures by becoming the umpteenth person to declare the goal of eradicating malaria?"

Bill Gates knew the history of malaria. People have died from this disease for thousands of years. In modern times, 350 million to 500 million people are infected each year and 1 million of them die. Most malaria cases are in Africa and most of those who die are children. Several eradication efforts have been attempted over the past 6 decades, mostly focused on eliminating the mosquitoes that carry the malaria pathogen. In the United States, eradication was achieved in 1951 through widespread spraying of the insecticide DDT as well as numerous other public health measures. The spraying program became controversial in the 1960s and 1970s because DDT was found to be widely distributed around the globe and it was linked to the thinning of egg shells in large birds of prey due to bioaccumulation and biomagnification. DDT is still sprayed in many parts of the world to assist in the eradication of malaria, but malaria persists.

Combating malaria. By distributing medicine and nets impregnated with insecticide to households in Africa, childhood death from malaria has declined by as much as 60 percent. *(Paula Bronstein/Getty Images)*

Malaria is difficult to combat for a number of reasons. First, mosquito populations that are reduced by spraying insecticides can rebound quickly. In Sri Lanka, for example, consistent spraying to kill mosquitoes reduced the number of malaria cases from 1 million to a mere 18. Because of this great success, the spraying program was stopped, but within a few years, malaria cases rapidly increased to a half million. In short, the spraying program was ended before the job was done. Moreover, if one country is spraying to kill mosquitoes and neighboring countries are not, mosquitoes will continue to enter from the neighboring countries. Mosquitoes can rapidly evolve resistance to insecticides such as DDT. The malaria pathogen can rapidly evolve resistance to antimalarial drugs. Finally, eradicating malaria is expensive. Typically, countries use multiple strategies including insecticide spraying, antimalarial drugs, and the distribution of mosquito tents in which people can sleep and avoid being bitten during the night. Collectively, these strategies can carry a price tag that many low-income countries cannot afford. Additionally, other social and economic priorities of these countries, as well as social disruption, have precluded or curtailed malaria control programs.

So in 2007, after so many past failures, why did Bill Gates think there was now a possibility of eradicating malaria? Earlier that year, scientists had reported that a new drug, combined with a new style of mosquito net, produced large reductions in malaria cases—for example, as much as 97 percent in Uganda. The new nets, which were impregnated with more modern insecticides, could last 3 to 5 years. This was a big improvement from the earlier nets, which only lasted no more than 3 months. In addition to having a new drug and longer-lasting nets, the key to the success of the Ugandan program was to pay for and distribute the drug and nets to everyone who needed them. Employing this strategy around the world is an expensive endeavor, and that's where Bill Gates comes in.

The Bill and Melinda Gates Foundation funds projects that have been historically underfunded. Equally important, by throwing its prominent name and financial resources behind a cause like malaria eradication, the foundation can rally significant financial support from other foundations and from developed countries. Western governments joined the movement and increased malaria funding from $50 million to $1.1 billion. This gave new hope to the declared goal of eradicating malaria from the globe within 50 years.

Many challenges remain. One of the largest is simply organizing distribution systems to hand out the drugs and millions of mosquito tents. In some regions, there are no roads into the villages and the items must be brought there by foot or by boat. There is also the challenge to continue research into new strategies against the pathogen and the mosquito. Currently, a new antimalaria drug from China has proven very effective against the pathogen and is quite inexpensive to manufacture. The manufacturer of this drug agreed to sell it at less than the cost of manufacturing it, making the drug a very attractive option for low-income countries. The company estimates that this decision cost it $253 million in profits but saved 550,000 lives and brought the company very positive publicity.

Another possibility is the development of a vaccine that would provide immunity to malaria infections. For nearly 25 years, the pharmaceutical company GlaxoSmithKline spent more than $350 million to develop a malaria vaccine, but the company was reluctant to fully fund a study of the effectiveness of the vaccine in African children. The Gates Foundation provided $200 million to help fund the study and the results have been very encouraging. In 2013, it was announced that children vaccinated between 5 and 17 months of age experienced 46 percent fewer cases of malaria than similar children who were not vaccinated. The researchers conducting the study will continue to follow these children as part of a prospective study to determine if the vaccination continues to protect the children throughout their lives.

Today there is tremendous hope that Bill Gates's dream of eradicating malaria is gaining ground. Most experts agree that malaria cases could be reduced by at least 85 percent in most African countries. The reduction in illness and death would also be highly beneficial to the economies of these low-income countries by reducing health costs and creating a healthier, and therefore more productive, workforce. The success of the global fight against malaria critically depends on sustained financial support from foundations and governments, continued discovery of new drugs and vaccines, and the recognition that we cannot stop fighting malaria until the job is done.

Critical Thinking Questions

1. If mosquitoes vary in their resistance to insecticides, what might you predict about the long-term success of trying to eliminate all of the mosquitoes with insecticide as a way to eliminate malaria?
2. How might a reduction in illness and death from malaria affect the economies of low-income countries in terms of future health costs and the health and productivity of their workforce?

References

Editorial Board. 2013. Hope for a malaria vaccine. *New York Times*,October 13.http://www.nytimes.com/2013/10/14/opinion/hope-for-a-malaria-vaccine.html.

Kingsbury, K. 2009. A better deal on malaria. *Time*, February 26. http://content.time.com/time/magazine/article/0,9171,1881988,00.html.

McNeil, D., Jr. 2008. Nets and new drug make inroads against malaria. *New York Times*, February 1. http://www.nytimes.com/2008/02/01/health/01malaria.html.

McNeil, D., Jr. 2008. Eradicate malaria? Doubters fuel debate. *New York Times*, March 4. http://www.nytimes.com/2008/03/04/health/04mala.html.

Suggested Answers to Critical Thinking Questions

1. Eventually mosquitoes will become resistant to any insecticide. Therefore, we cannot eliminate all mosquitoes using only insecticides. While insecticides are effective in the short term, any given insecticide will gradually lose its effectiveness as resistance develops.

2. The reduction in illness and death will reduce health care costs and increase the productivity of the workforce by increasing the number of healthy adults.

Teaching Tip: Measuring Your Impact

How Does Risk Affect Your Life Expectancy?
Measuring Your Impact 17 asks students to examine risky behaviors by determining how they can affect life expectancy. The link to this resource may be found by clicking on the link buttons in the TE-book, opening the TRFD, or logging onto the book's companion website bcs.whfreeman.com/friedlandapes2e (teacher login required).

 **Measuring Your Impact 17:** How Does Risk Affect Your Life Expectancy?

chapter 17

REVIEW

In this chapter, we learned about human diseases and chemicals that can affect human health and how we analyze the risk of environmental hazards. Human diseases can be categorized as either acute or chronic and can be infectious or not. We reviewed many of the historically important infectious diseases and then discussed the modern problem of emerging infectious diseases. In addition to these biological risks, we also need to consider chemical risks to humans and other species. Chemical risks are assessed by experiments that determine the LD50 or ED50 for various species and by following a large sample of individuals using prospective and retrospective studies. Such risk assessments can be combined with data on risk tolerance to help in risk management, which weighs the assessed risk against social, economic, and political considerations. In conducting risk management, regulators in some regions of the world use the precautionary principle while regulators in other countries, including the United States, use the guilty-until-proven-innocent principle.

Key Terms

Disease
Infectious disease
Acute disease
Chronic disease
Epidemic
Pandemic
Plague
Malaria
Tuberculosis
Emergent infectious disease
Acquired Immune Deficiency Syndrome (AIDS)
Human Immunodeficiency Virus (HIV)
Ebola hemorrhagic fever
Mad cow disease
Prion
Swine flu
Bird flu
Severe acute respiratory syndrome (SARS)
West Nile virus
Neurotoxin
Carcinogen
Mutagen
Teratogen
Allergen
Endocrine disruptor
Dose-response study
Acute study
Chronic study
LD50
Sublethal effect
ED50
Retrospective study
Prospective study
Synergistic interaction
Route of exposure
Solubility
Bioaccumulation
Biomagnification
Persistence
Environmental hazard
Innocent-until-proven-guilty principle
Precautionary principle
Stockholm Convention
REACH

Learning Objectives Revisited

Module 56 Human Diseases

- **Identify the different types of human diseases.**
  Human diseases can be categorized as either infectious or noninfectious. Infectious diseases are caused by pathogens such as viruses, bacteria, fungi, protists, and helminths. Human diseases can also be categorized as either acute, which means they rapidly impair a body's functions, or chronic, which means they slowly impair a body's functions.

- **Understand the risk factors for human chronic diseases.**
  Risk factors for human health differ between low- and high-income countries. In low-income countries, the top risk factors include unsafe drinking water, poor sanitation, and malnutrition. In high-income countries, the top risk factors include tobacco use, less active lifestyles, poor nutrition, and overeating that leads to high blood pressure and obesity.

- **Discuss the historically important human diseases.**

  Among the historically important infectious diseases, plague is caused by a bacterium that is carried by fleas, malaria is caused by several different species of protists, and tuberculosis is caused by a bacterium that primarily infects the lungs.

- **Identify the major emergent infectious diseases.**

  Among the emerging infectious diseases, HIV/AIDS is caused by a virus that weakens the immune system, Ebola hemorrhagic fever is caused by a highly lethal virus, and mad cow disease is caused by a prion that damages the nervous system. In addition, the viruses that cause bird flu and swine flu are easily spread and sometimes lethal, SARS is a type of pneumonia caused by a virus, and West Nile virus normally infects birds but can be transmitted to humans by mosquitoes.

- **Discuss the future challenges for improving human health**

  The future challenges for improving human health including the improvement of nutrition and sanitation in low-income regions of the world and promoting healthier lifestyles in high-income regions of the world. We also need to educate people about the importance of taking the full duration of medicines that combat pathogens to prevent the evolution of drug-resistant strains of pathogens. Finally, we need to continue to develop rapid response to emerging infectious diseases to reduce the probability of them spreading worldwide.

Module 57 Toxicology and Chemical Risks

- **Identify the major types of harmful chemicals.**

  The major types of harmful chemicals are neurotoxins, carcinogens, teratogens, allergens, and endocrine disruptors. Neurotoxins disrupt the nervous systems of animals and they include insecticides, lead, and mercury. Carcinogens are cancer-causing chemicals that include asbestos, formaldehyde, radon, and chemicals from tobacco. Teratogens interfere with the normal development of embryos or fetuses and include thalidomide and alcohol. Allergens cause abnormally strong immune responses and a given allergen typically only affects a small fraction of people. Endocrine disruptors such as hormones from animal rearing facilities interfere with the normal functioning of hormones in organisms.

- **Explain how scientists determine the concentrations of chemicals that harm organisms.**

  Scientists can conduct LD50 experiments to determine lethal effects of chemicals and ED50 experiments to determine sublethal effects of chemicals. They can also follow large groups of individuals backward in time using retrospective studies or forward in time using prospective studies. Once we know the chemical concentrations that can cause harm, we also need to determine the routes of exposure by which an individual may come in contact with the chemical as well as the chemical's solubility and its potential to bioaccumulate and biomagnify.

Module 58 Risk Analysis

- **Explain the processes of qualitative versus quantitative risk assessment.**

  For a given environmental hazard, we can qualitatively categorize risks as relatively low, medium, or high. However, the actual risk of a given hazard may be quite different from our qualitative assessments. Quantitative assessments use actual data to determine the actual probability of various risks, either based on government death statistics or by calculating the probability of being exposed to a hazard multiplied by the probability of being harmed if exposed.

- **Understand how to determine the amount of risk that can be tolerated.**

  Individuals differ in how much risk they are willing to tolerate. For most environmental hazards, we often set risk tolerance at 1 in 1 million.

- **Discuss how risk management balances potential harm against other factors.**

  Understanding the level of risk is important, but we must also assess the effects of trying to reduce the risk. Such effects include economic, political, and social considerations that collectively can come to some compromise that balances all of these factors. Whereas risk assessment is conducted by environmental scientists, risk management is typically conducted by local, national, or international government agencies.

- **Contrast the innocent-until-proven-guilty principle and the precautionary principle.**

  According to the innocent-until-proven-guilty principle, a potential hazard should not be considered harmful until it can definitively be demonstrated to cause harm. According to the precautionary principle, when a hazard is plausible but not yet certain, we should reduce or remove the hazard.

Exploring the Literature

Newman, M. C., and M. A. Unger. 2009. *Fundamentals of Ecotoxicology.* 3rd ed. CRC Press.

Quammen, D. 2012. *Spillover: Animal Infections and the Next Human Pandemic.* W.W. Norton.

World Health Organization. 2009. *Global Health Risks: Mortality and Burden of Disease Attributable to Selected Major Risks.* WHO Press. http://www.who.int/entity/healthinfo/global_burden_disease/GlobalHealthRisks_report_full.pdf

Download a printable version of this list by clicking the link buttons in the TE-book, opening the TRFD, or logging onto the book's companion website bcs.whfreeman.com/friedlandapes2e (teacher login required).

Exploring the Literature 17

Answers to Chapter 17 AP® Exam Multiple-Choice Questions

1. a
2. d
3. e
4. c
5. e
6. a
7. b
8. e
9. b
10. b
11. b

Chapter 17 AP® Environmental Science Practice Exam

Section 1: Multiple-Choice Questions

Choose the best answer for questions 1–11.

1. Which statement is true regarding human health risks?
   (a) More people die from infectious diseases than from noninfectious diseases.
   (b) More people die from accidents than from any other cause.
   (c) More people die from chemical risks than from physical or biological risks.
   (d) More people die from cancer than from any other cause.
   (e) More people die from heart disease than from any other cause.

2. Which statement is true regarding the relationship between health risks and income?
   (a) A major risk in high-income countries is a lack of food.
   (b) A major risk in high-income countries is poor sanitation.
   (c) A major risk in low-income countries is obesity.
   (d) A major risk in low-income countries is a lack of food.
   (e) The major risks in high- and low-income countries are similar.

3. Which statement about historical infectious diseases is NOT true?
   (a) Plague is a disease that is carried by fleas attached to rodents.
   (b) Malaria is a disease that is carried by rodents.
   (c) Tuberculosis is a disease that is transmitted through the air.
   (d) The pathogen that causes tuberculosis can become drug-resistant.
   (e) Historically important infectious diseases still pose a health risk.

4. Which statement about emerging infectious diseases is NOT true?
   (a) HIV is a virus that most likely came from chimps.
   (b) Ebola hemorrhagic fever causes a high rate of death.
   (c) Mad cow disease is spread when cows are fed grain in large feedlots.
   (d) Bird flu is a virus that jumps from birds to people.
   (e) West Nile virus is a virus that comes from birds.

5. Which is NOT an example of an infectious disease?
   (a) AIDS
   (b) Pneumonia
   (c) Tetanus
   (d) Malaria
   (e) Leukemia

6. Which statement about toxins is correct?
   (a) Neurotoxins impair the nervous system.
   (b) Carcinogens cause birth defects.
   (c) Teratogens cause cancer.
   (d) Allergens mimic naturally occurring hormones.
   (e) Endocrine disruptors cause allergic reactions.

7. Which statement about dose-response studies is NOT true?
   (a) Dose-response studies test chemicals across a range of concentrations.
   (b) Dose-response studies only test for lethal effects.
   (c) Dose-response studies can last for days or months.
   (d) LD50 values are divided by 10 to determine safe concentrations for wildlife.
   (e) LD50 values are divided by 1,000 to determine safe concentrations for humans.

8. Which statement about retrospective and prospective toxicity studies is correct?
   (a) Retrospective studies are not conducted on humans.
   (b) Prospective studies are only conducted on wild animals.
   (c) Retrospective studies monitor health effects from future chemical exposures.
   (d) Prospective studies monitor health effects from future chemical exposures.
   (e) Prospective studies monitor health effects from past chemical exposures.

9. The concentration of chemical exposure does NOT depend on
   (a) the persistence of the chemical.
   (b) the solubility of the chemical.
   (c) the ability of the chemical to bioaccumulate.
   (d) the ability of the chemical to biomagnify.
   (e) the LD50 value of the chemical.

Answers to Chapter 17 AP® Exam

Download the full answers to the Chapter 17 AP® Exam by clicking on the link buttons in the TE-book, opening the TRFD, or logging onto the book's companion website bcs.whfreeman.com/friedlandapes2e (teacher login required).

Answers to Chapter 17 AP® Exam

10. Which statement is NOT correct?
    (a) Risk assessment quantifies the potential harm that a chemical poses.
    (b) Risk assessment does not include social, political, and economic considerations.
    (c) Risk acceptance determines the amount of tolerated risk.
    (d) Risk management includes social, political, and economic considerations.
    (e) Risk management does not consider the potential harm that a chemical poses.

11. What is NOT true about the two approaches to regulating chemicals?
    (a) The innocent-until-proven-guilty principle assumes chemicals are safe unless harm can be demonstrated.
    (b) The precautionary principle is used in the United States.
    (c) The precautionary principle assumes chemicals are harmful unless safety can be demonstrated.
    (d) The innocent-until-proven-guilty principle allows rapid approval of chemicals by regulatory agencies but increases the risk that harmful chemicals will be approved.
    (d) The innocent-until-proven-guilty principle allows rapid approval of chemicals by regulatory agencies but increases the risk that harmful chemicals will be approved.
    (e) The precautionary principle can cause delays in the use of beneficial chemicals but reduces the risk of harmful chemicals being approved.

Section 2: Free-Response Questions

Write your answer to each part clearly. Support your answers with relevant information and examples. Where calculations are required, show your work.

1. You are an employee of the Environmental Protection Agency. You are given the task of conducting risk management for spraying insecticides to kill the mosquitoes that carry West Nile virus.
    (a) How might you determine the proper concentration needed to kill mosquitoes? (2 points)
    (b) How might you determine whether the concentration used to kill mosquitoes might also kill other species of insects? (2 points)
    (c) If you knew the LD50 value of the insecticide for humans, what concentration would be considered the safe upper limit for humans? (2 points)
    (d) Given the information you have accumulated as part of your risk assessment, describe the factors that might be important in the risk management of spraying insecticides to kill the mosquitoes that carry West Nile virus. (4 points)
2. Given the differences in health risks that exist between low- and high-income countries, consider the following issues.
    (a) What strategies might you use to reduce the health risks of low-income countries? (3 points)
    (b) What strategies might you use to reduce the health risks of high-income countries? (3 points)
    (c) Suppose a low-income country discovers oil and is projected to become a high-income country within a decade. What changes in the country's health care system might you suggest? (4 points)

Additional Multiple-Choice AP® Practice Exam

Download an additional Multiple-Choice AP® Environmental Science Practice Exam for Chapter 17 by clicking on the link buttons in the TE-book, opening the TRFD, or logging onto the book's companion website bcs.whfreeman.com/friedlandapes2e (teacher login required).

Multiple-Choice AP® Practice Exam 17

Answers to Chapter 17 AP® Exam Free-Response Questions

1. (a) One would take a sample of the mosquitoes and expose them to varying concentrations of the chemical insecticide, introduced using an aerosol or spray. For each concentration, measure the response, the number of mosquitoes that succumbed to the insecticide. From the data find the LD50, which is the concentration required to produce 50 percent mortality in a population. To kill all mosquitoes would obviously require some multiple, maybe 2 or 2.5 times, of this LD50 value.
(b) A companion study might be done using the insecticide(s) of interest for some species of beneficial insect, such as a ladybug. Ideally we want an insecticide that kills mosquitoes but does not harm ladybugs. The ideal substance would have a low LD50 for mosquitoes and a high LD50 for ladybugs.
(c) Such a value is rarely known, but if it were, a safe dose or concentration should be between 0.1 and 1.0 percent of the LD50 value.
(d) In risk management, we first quantitatively measure risk and then consider economic, political, and social factors that might influence the decision of whether it should be applied to the environment.

2. (a)
- Provide means for good childhood nutrition and prenatal care
- Provide sources of clean water
- Provide clean-burning fuels for cooking and heating
- Provide high-quality medical care

(b)
- Provide information about good nutrition and wellness
- Reduce air pollution
- Encourage exercise and discourage use of tobacco products
- Emphasize the importance of a well-managed, safe workplace

(c) There should be more emphasis on wellness, good nutrition, and timely medical checkups.

Answers to Unit 7 AP® Exam Multiple-Choice Questions

1. c
2. e
3. b
4. a
5. c
6. e
7. c
8. e
9. a
10. b
11. b
12. e
13. e
14. c
15. a
16. d
17. c
18. b
19. b
20. c

Unit 7 AP® Environmental Science Practice Exam

Section 1: Multiple-Choice Questions

Choose the best answer for questions 1–20.

1. The owner of a rural home poured several liters of concentrated bleach into a sink drain connected to a buried septic system. Which of the following is the owner most likely to notice after a few weeks?
   (a) Settled sludge will be more sterile and dispersed into the leach field.
   (b) Sludge has started to accumulate in the leach field.
   (c) Sludge has started to accumulate faster in the septic tank.
   (d) There is a lower abundance of pathogens in groundwater.
   (e) The septage released from the septic tank is cleaner and safer for the environment.

2. Which of the following sources of mercury pollution may harm humans?
   I. Mercury-based compounds that are absorbed by marine plankton
   II. Inorganic or synthetic mercury-based compounds
   III. Wetlands with mercury concentrations below the legal limit set by the EPA

   (a) I only
   (b) II only
   (c) I and II
   (d) II and III
   (e) I, II, and III

3. To remediate an oil spill in the ocean, the use of dispersants
   (a) is preferred over the use of hot water sprayers.
   (b) can prevent oil from reaching the shorelines.
   (c) removes the oil from the ocean.
   (d) is less environmentally friendly than the use of oil-consuming, genetically modified bacteria.
   (e) is an effective way to slow the damage of oil plumes.

4. Although thermal and sediment pollution have different sources, they both can
   (a) lead to respiratory problems in aquatic animals.
   (b) contaminate waterways with heavy metals.
   (c) decrease the transparency of water.
   (d) lower the temperature of water.
   (e) harm fish in the open ocean.

5. Which of the following is NOT regulated by EPA Clean Water Act or Safe Drinking Water Act?
   (a) Levels of arsenic in drinking water
   (b) Septage disposal from sewage treatment plants
   (c) Nonpoint sources of oil pollution
   (d) Levels of giardia in headwater streams
   (e) Inorganic chemicals in groundwater

6. Which of the following groups consists entirely of secondary pollutants?
   (a) Carbon dioxide, methane, lead
   (b) Particulate matter, nitrogen oxide, nitrogen dioxide
   (c) Nitrogen dioxide, sulfur dioxide, ozone
   (d) VOC, mercury, carbon dioxide
   (e) Nitrate, ozone, sulfate

7. __________ is NOT likely to lead to the formation of chemical smog.
   (a) Primary production
   (b) Combustion of gasoline
   (c) Industrial release of CO_2
   (d) A thermal inversion
   (e) Intense sunlight

8. Sulfur dioxide emissions can be reduced by
   (a) selling sulfur emission credits.
   (b) increasing the temperature at which coal is combusted.
   (c) installing catalytic converters in cars.
   (d) installing fabric filters in smokestacks.
   (e) using electrostatic precipitators.

Question 9 refers to the following equations:

$$\text{i} + O_2 \rightarrow O_3 + \text{UV-C} \rightarrow O + \text{ii}$$
$$\text{iii} + 2O + O_2 \rightarrow \text{iv}$$

9. Roman numerals i, ii, iii, and iv refer to which of the following compounds?
   (a) O, O_2, chlorofluorocarbon, $2O_2$
   (b) O, 2O, UV-C, O + O_3
   (c) O, O_2, energy, O_3
   (d) Chlorofluorocarbon, O_2, chlorofluorocarbon, O_3
   (e) O, O_2 chlorofluorocarbon, O + O_3

10. Which of the following is NOT likely to reduce the risk of sick building syndrome?
    (a) Replacing asbestos insulation with more modern insulating material
    (b) Installing energy efficient windows
    (c) Using hardwood flooring instead of carpeting
    (d) Installing carbon monoxide detectors
    (e) Using paint with low amounts of volatile organic carbon

11. Which of the following factors is most likely to decrease the total amount of noncompostable municipal solid waste?
    (a) Packaging products in recyclable material
    (b) Increasing production of reusable materials
    (c) Lowering the cost of disposable products
    (d) Manufacturing products that are less durable
    (e) Decreasing the production of composite materials

Answers to Unit 7 AP® Exam

Download the full answers to the Unit 7 AP® Exam by clicking on the link buttons in the TE-book, opening the TRFD, or logging onto the book's companion website bcs .whfreeman.com/friedlandapes2e (teacher login required).

Answers to Unit 7 AP® Exam

12. Open-loop recycling refers to
   (a) the recycling of a product into the same product.
   (b) the recycling of a product into a product that will enter a different waste stream.
   (c) the recycling of a product into compost.
   (d) a process where only part of the waste is ultimately recycled.
   (e) a process where one product is recycled into a different product.

13. Which factor is least likely to influence the placement of a landfill?
   (a) NIMBY politics
   (b) Distance of landfill from homes
   (c) Local hydrology
   (d) Major type of consumer waste
   (e) NPP of habitat

Question 14 refers to the following table on the costs associated with landfills and incinerators

Landfill		**Incinerator**	
Transporting material to landfill	\$10/ton	Transporting material to incinerator	\$10/ton
Sorting waste	\$40/ton	Sorting waste	\$40/ton
Compacting waste	\$30/ton	Burning waste	\$160/ton
Dumping waste	\$10/ton	Removing ash	\$30/ton

14. Assume an incinerator can produce 500 kWh of energy by incinerating 1 ton of waste, and can sell a single kWh for \$0.15. To become economically beneficial, how much more waste would an incinerator have to collect and burn relative to a landfill?
   (a) 1 ton
   (b) 1.5 tons
   (c) 2 tons
   (d) 4 tons
   (e) 6 tons

15. The Comprehensive Environmental Response, Compensation, and Liability Act is referred to as "Superfund" because it
   (a) collects taxes from chemical and petroleum industries to fund the cleanup of abandoned waste sites where responsible parties cannot be established.
   (b) collects taxes from chemical and petroleum industries to regulate the proper disposal of hazardous waste by small, local businesses.
   (c) provides a fund to help both large and small industries to train workers properly on methods of disposing hazardous substances.
   (d) provides a fund to help industries design environmental responses to emergencies involving the accidental release of hazardous waste.
   (e) distributes money from varying government agencies to chemical and petroleum industries.

16. Which of the following is an example of an integrated waste management strategy?
   (a) Integrating the waste management goals of several small townships into a regional waste management plan.
   (b) Developing a nationwide program to handle hazardous waste materials.
   (c) Training waste management workers on how to sort recyclables from nonrecyclables.
   (d) Simultaneously developing several waste management strategies to reduce the amount of waste that must be incinerated or placed in landfills.
   (e) Collecting state taxes to fund the development of more landfills.

17. HIV, H1N1, and mad cow disease are all similar in that
   (a) they can be sexually transmitted.
   (b) they have been largely eradicated from developed nations.
   (c) they originated from an animal other than humans.
   (d) humans infected with these diseases cannot be cured.
   (e) they are all coronaviruses.

Questions 18 and 19 refer to the following graph:

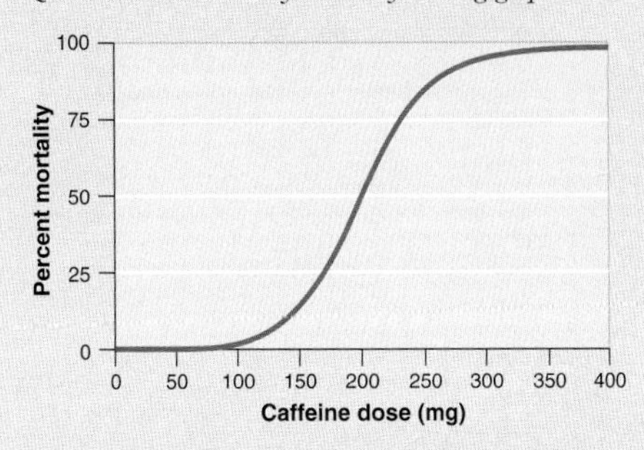

18. Given the mortality curve in the figure for a 0.5-kg rat, what would be the safe dosage of caffeine for a rat?
   (a) 10 mg
   (b) 20 mg
   (c) 40 mg
   (d) 100 mg
   (e) 200 mg

19. What would be the safe dosage of caffeine for a human?
   (a) 0.1
   (b) 0.2
   (c) 0.3
   (d) 0.5
   (e) 0.6

20. Which factor is the most difficult to predict when quantifying the risk of human exposure to a given contaminant?
   (a) Route of contaminant exposure
   (b) Solubility of a contaminant in water
   (c) Synergistic interactions with other contaminants
   (d) Biomagnification of a contaminant
   (e) Persistence of a contaminant in water

Unit 7 AP® Environmental Science Practice Exam 625

Answers to the Unit 7 Free-Response Questions can be found in the Answer Appendix on page ANS-15.

Teaching Tip: Unit 7 Additional Free-Response Question

An additional practice free-response question covering concepts from Unit 7 is available. The question presents a short discussion of the process of carbon fractionation. To answer this question students must:

- evaluate provided statistics on recycling systems to determine which promote the most recycling
- determine which system results in the greatest average percentage of recyclables
- determine which recycling system generates the greatest profit per ton
- compare the effect of an increase in recycled material on single-system and dual-system recycling
- compare the level of impurities in the two recycling systems
- consider how public education may alter the costs of single- and dual-stream recycling

Download this resource by clicking on the link buttons in the TE-book, opening the TRFD, or logging on to the book's companion website bcs.whfreeman.com/friedlandapes2e (teacher login required).

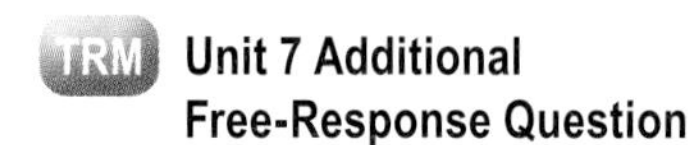

Unit 7 Additional Free-Response Question

Section 2: Free-Response Questions

Write your answer to each part clearly. Support your answers with relevant information and examples. Where calculations are required, show your work.

1. Concentrated Animal Feeding Operations (CAFOs) house thousands of cows that are concentrated in close living quarters. Manure generated at CAFOs is often washed into massive manure lagoons where it decomposes anaerobically until the manure is sprayed over fields as fertilizer. This practice has several problematic consequences. First, the anaerobic decomposition of manure releases sulfur gases and methane. Second, manure from lagoons can potentially leak into surrounding soils, streams, and lakes. Third, to make room for additional manure in the lagoons, more manure is often sprayed on agricultural fields than is needed for plant growth.
   (a) Explain two reasons why the release of sulfur gases poses an environmental and health risk. (2 points)
   (b) Name two reasons the leakage of manure from lagoons or the overspraying of manure on agricultural fields poses environmental risks for nearby soils, streams, and lakes. (2 points)
   (c) Suggest a retrospective study that could be conducted to determine the effect of manure lagoons on nearby residents. (2 points)
   (d) List two possible ways to determine if there is leakage in the soil surrounding a manure lagoon. (2 points)
   (e) Which two legislative acts require remediation of groundwater and stream water in cases where a manure lagoon leak occurs? (2 points)

2. During the first half of the twentieth century, residents in Los Angeles were allowed to have "backyard incinerators" in which they could burn their trash. Although these incinerators reduced the amount of trash going to a landfill, they also created a substantial amount of air pollution that led to severe smog. Initially, Los Angeles attempted to reduce smog by limiting the use of backyard incinerators to the hours between 4 AM and 7 AM. Backyard incinerators were ultimately banned in 1957.
   (a) Describe the chemical process that generates smog, and how backyard incinerators are likely to contribute to that process. (2 points)
   (b) How would limiting the incineration of trash to the hours between 4 AM and 7 AM potentially reduce the production of smog? (2 points)
   (c) Define a thermal inversion and describe how it affects smog. (2 points)
   (d) Much of the ash from backyard incinerators was either used in gardens or disposed of in landfills. What is the environmental risk of this practice? (2 points)
   (e) How do modern incinerators control the release of both bottom and fly ash? (2 points)

pryzmat/Shutterstock

7

scienceapplied

Is Recycling Always Good for the Environment?

One of the three ways to reduce solid waste is to recycle. As we discussed in Chapter 16, when we recycle items such as paper, plastic, bottles, and cans, less material ends up in landfills and fewer natural resources need to be extracted to produce these items in the future. In 2012, the EPA estimated that Americans recycle 60 million metric tons (65 million U.S. tons) of trash. This represents about 34 percent of all the trash that we generate. At first glance, recycling appears to make a lot of sense both economically and environmentally. Indeed, many state and local governments have encouraged or required recycling programs and the public generally associates recycling with being good for the environment (FIGURE SA7.1). But what do the data tell us? When we decide to recycle, what are the measurable benefits for the environment? How do these benefits compare with benefits from other decisions we make, such as the type of car we drive? The answers to these questions may surprise you.

FIGURE SA7.1 **Recycling.** There is increasing interest in recycling many materials. From a perspective of energy savings, some items are more important to recycle than others. *(David R. Frazer/ The Image Works)*

How do we begin to assess the benefits of recycling?

To determine the overall effect of recycling any type of waste, we need to consider the full range of costs and benefits of recycling and then compare these with the costs and benefits of manufacturing the same item from raw materials. For example, to assess the benefits of recycling paper, we need to compare the cost of recycling old paper into new paper products versus the cost of manufacturing new paper products from trees.

As we saw in Chapter 16, the best way to answer these questions is to complete a life-cycle analysis. Let's look at two examples: aluminum cans and plastic containers. To compare the environmental and economic costs of transporting and manufacturing these items from recycled materials versus raw materials, we begin at the manufacturing facility.

The recycling of aluminum, primarily from aluminum cans, is widespread in the United States. According to the Aluminum Association, more than 60 billion cans were recycled in the United States in 2012, which represents nearly 70 percent of all aluminum cans that were manufactured in that year (FIGURE SA7.2). To manufacture aluminum cans from raw materials, aluminum ore or bauxite must be mined and processed into pure aluminum. Not only does mining have environmental impacts as discussed in Chapter 8, but this processing of aluminum from ore also takes a substantial amount of energy. In contrast, manufacturing aluminum cans from recycled cans requires only 5 percent of this energy. In

FIGURE SA7.2 **Recycling aluminum cans.** Converting old aluminum cans into new aluminum cans requires only 6 percent of the energy used to convert aluminum ore from a mine into new aluminum cans. *(Erik Isakson/age fotostock)*

short, making new cans from recycled cans saves a large amount of energy and therefore saves manufacturers a lot of money. In fact, according to the EPA, recycling 0.9 metric ton (1 U.S. ton) of aluminum cans saves about 32 barrels of oil. When this energy comes from burning fossil fuels, it also means that manufacturing recycled cans reduces the amount of carbon dioxide and other pollutants in the atmosphere. Moreover, aluminum can be recycled over and over again without any loss of quality.

The recycling of plastic containers is also widely practiced. According to the American Chemistry Council and the Association of Postconsumer Plastic Recyclers, the recycling of plastic continues to grow each year with 1.3 billion kilograms (2.8 billion pounds) of bottles recycled in 2012, representing 30 percent of all plastic bottles that are manufactured. The cost of energy required to make new plastic bottles from raw material—in this case petroleum—is substantially less than the cost of recycling plastic bottles. As a result, manufacturing plastic bottles from recycled plastic bottles results in much smaller energy savings; washing and reprocessing plastic bottles requires nearly 50 percent of the energy required in manufacturing plastic bottles from oil. In addition, recycled plastic typically degrades in quality, so plastic from recycled bottles is typically used for products such as carpets and insulation in jackets and sleeping bags rather than new plastic bottles. This means that the economic and environmental benefits of using recycled plastic are much smaller than the economic benefits of using recycled aluminum.

What other costs of recycling do we need to consider?

Regardless of the type of material that is being recycled, we have to remember that there are several additional costs of recycling beyond the cost of energy used in manufacturing. To understand these costs, let's start at your house. If you rinse out your cans and bottles before recycling, energy is needed to get the water to your sink, particularly if you use hot water. If you use hot water to rinse out the peanut butter from a plastic peanut butter jar, for example, you are likely using more energy to clean the jar than is saved when you recycle the jar.

After they are cleaned, the materials to be recycled must be transported to a central recycling facility. Depending on location, the homeowner must either set out recycled items on the curb for pickup by a collection truck (FIGURE SA7.3) or bring them to a central facility. Both scenarios require burning fossil fuels for transportation. Additional fossil fuels must be consumed to transport the recycled items from the collection facility to the manufacturing facility. Although transportation costs will vary among different towns and cities, in terms of energy consumed and pollutants produced, they reduce the benefits of recycling. However, we can easily compare the cost of transporting recycled materials to manufacturers against the cost of transporting raw materials from their source, such as an aluminum mine. In addition, when homeowners pay for transporting the items to the collection

FIGURE SA7.3 **Transportation costs.** Although early recycling programs had garbage trucks make separate trips to pick up trash and recycling materials, modern trucks have separate compartments that allow both garbage and recycled items to be picked up at the curb in a single trip. A single trip saves time and money, and reduces consumption of fossil fuels as well as the production of air pollutants. *(Rob Crandall/The Image Works)*

center through taxes or trash collection fees, they will avoid the costs of putting the waste in a landfill.

What other benefits of recycling do we need to consider?

The primary argument for recycling is that it reduces the need for raw materials and keeps solid waste out of landfills. During the 1990s, there was a growing concern that the United States was running out of landfill space and that recycling was critical to extending the life of existing landfills. While it is true that many landfills are nearing capacity, particularly in the northeastern United States, there is still a large amount of land throughout the country that could serve as landfill space if people in those areas agreed to the construction of new landfills.

Reducing the amount of solid waste going into landfills allows existing landfills to operate longer. This, in turn, reduces the costs of closing and monitoring existing landfills. It also reduces the costs of building more landfills in the future as well as the costs of trucking the waste to new landfills likely to be farther away. Increased trucking raises both the economic cost and the environmental impact.

When we consider how we can improve the environment it is often helpful to gather the scientific data to make objective comparisons rather than simply make decisions based on perceptions. In the case of recycling, the analysis of the data makes it clear that recycling certain materials will have much greater environmental and economic benefit than recycling other materials. This helps us understand why manufacturers might be much more inclined to promote the recycling of certain items such as aluminum cans. Identifying the full range of costs and benefits also helps us identify the complexity of the question. The energy costs of recycling, for example, are wide-ranging and include homeowner costs, transportation costs, manufacturing costs, and landfill costs. By identifying all of the costs and benefits, we can strive to design more efficient recycling programs.

Questions

1. Why is it much more beneficial to recycle aluminum than plastic?
2. Why is it important to take a life-cycle approach when considering the benefits of recycling?

Free-Response Question

Write your answer to each part clearly. Support your answers with relevant information and examples. Where calculations are required, show your work.

Suppose you are planning a party and want to determine the most environmentally friendly way to serve drinks to your friends. You can choose either one-use, recyclable plastic cups or glass cups that can be kept and reused.

(a) List four factors that are likely to increase or decrease the environmental benefits of using recyclable plastic cups. (2 points)

(b) Suggest three reasons why using plastic cups at a party may be less environmentally costly than using glass cups. (3 points)

(c) To reduce the amount of waste that ends up in landfills, engineers have developed plastic cups made with oxo-biodegradable plastic that are designed to be composted. However, many of them end up in landfills because people do not have compost piles or do not understand that they can be composted. Despite their biodegradability, these cups persist for many years in landfills. Why might biodegradable plastic persist in a landfill for a long time? (2 points)

(d) Suggest one method to reduce, one method to reuse, and one method to recycle waste at your party. (3 points)

References

Aluminum Association. 2012. *Aluminum continues leadership in sustainable packaging as most recycled beverage container.* October 24. http://www.aluminum.org/AM/Template.cfm?Section=Home&TEMPLATE=/CM/ContentDisplay.cfm&CONTENTID=35146.

Association of Postconsumer Plastic Recyclers. 2012. *2012 United States National Postconsumer Plastics Bottle Recycling.* http://plastics.americanchemistry.com/Education-Resources/Publications/2012-National-Post-Consumer-Plastics-Bottle-Recycling-Report.pdf.

Friedland, A., et al. 2003. Personal decisions and their impacts on energy use and the environment. *Environmental Science and Policy* 6:175–179.

Municipal Solid Waste Generation, Recycling, and Disposal in the United States: Facts and Figures for 2011. http://www.epa.gov/waste/nonhaz/municipal/pubs/MSWcharacterization_508_053113_fs.pdf.

The answer to the Science Applied 7 Free-Response Question can be found in the Answer Appendix on page ANS-16.

Answer to Science Applied 7 Free-Response Question

Download the full answer to the Science Applied 7 Free-Response Question by clicking on the link buttons in the TE-book, opening the TRFD, or logging onto the book's companion website bcs.whfreeman.com/friedlandapes2e (teacher login required).

Answer to Science Applied 7 Free-Response Question

Suggested Answers to Science Applied Questions

1. While recycling aluminum uses much less energy than mining new aluminum, recycling plastic bottles actually uses more energy than just making new bottles. Additionally, plastic degrades during recycling, while aluminum does not.

2. A life-cycle approach to the benefits of recycling provides complete information on the costs and benefits of recycling a material. This approach provides a more realistic assessment of the environmental and economic value in recycling and insures that the activity of recycling is in fact providing environmental benefits.

chapter

18 Conservation of Biodiversity

PD Overview

This chapter explores the reasons for the decline in biodiversity and various strategies for preserving biodiversity. The chapter begins by noting that extinctions are a natural process and have occurred regularly throughout Earth's history. However, in recent times the number of extinctions has increased as the result of human activity. The chapter goes on to look at the specific mechanisms for the loss of biodiversity, and finally examines efforts currently underway to conserve biodiversity.

Module 59: The Sixth Mass Extinction

The world has experienced five major extinctions during the past 500 million years. Many scientists have suggested that we may currently be in the midst of a sixth mass extinction event, the first mass extinction to occur since humans have been present on Earth. This module examines the declines in biodiversity on Earth at various levels of complexity, including genetic diversity, species diversity, and ecosystem function. In each case, it examines the roles that humans have played in the decline of biodiversity.

Module 60: Causes of Declining Biodiversity

Declines in biodiversity are happening around the globe. This module builds on the basics of population and community ecology from Chapter 5 to understand how a number of factors can affect biodiversity. It looks at the role of habitat loss, exotic species, and overharvesting. Finally, it considers pollution and climate change.

Module 61: The Conservation of Biodiversity

It is important that we consider how to protect and increase biodiversity because of the large number of factors that can reduce biodiversity. There are two general approaches to conserving biodiversity: the single-species approach and the ecosystem approach. This module explores each of these approaches.

Alignment to AP® Environmental Science Course Description

AP® Outline	Module	
VII. Global Change (10–15%)		
C. Loss of Biodiversity	Module 59	The Sixth Mass Extinction
	Module 60	Causes of Declining Biodiversity
	Module 61	The Conservation of Biodiversity

Chapter Learning Objectives

After completing this chapter students will be able to

- explain the global decline in the genetic diversity of wild species.
- discuss the global decline in the genetic diversity of domesticated species.
- identify the patterns of global decline in species diversity.
- explain the values of ecosystems and the global declines in ecosystem function.
- discuss how habitat loss can lead to declines in species diversity.
- explain how the movement of exotic species affects biodiversity.
- describe how overharvesting causes declines in populations and species.
- understand how pollution reduces populations and biodiversity.
- identify how climate change affects species diversity.
- identify legislation that focuses on protecting single species.
- discuss conservation efforts that focus on protecting entire ecosystems.

Chapter 18 Pacing Guide

This pacing guide is based on a schedule with 120 sessions of 50 minutes each before the AP® Exam. If you have a different number of sessions before the exam, you can modify the pacing to suit your needs. If you have additional time, consider incorporating quizzes, released AP® Environmental Science free-response and multiple-choice questions, or additional activities.

Module	Standard Schedule Days	Block Schedule Days
Module 59	1	½
Module 60	2	1
Module 61	1	½
Assessment	1	½

Chapter 18 Resources

The link to these resources can be found by clicking on the link buttons in the Teacher's e-Book (TE-book), opening the Teacher's Resource Flash Drive (TRFD), or logging onto the book's companion website bcs.whfreeman.com/friedlandapes2e (teacher login required).

- **PowerPoint Presentation 18**
- **Optimized Art PowerPoint and JPEG Files 18**
- **Do the Math Videos**
- **Measuring Your Impact 18:** How Large Is Your Home?
- **Chapter 18 Web Resources**
- **Exploring the Literature 18**
- **Answers to Chapter 18 AP® Exam**
- **Multiple-Choice AP® Practice Exam 18**

Free-Response Questions from Previous AP® Environmental Science Exams

Free-response questions from prior AP® Environmental Science Exams are available on the AP® course website: https://apstudent.collegeboard.org/apcourse/ap-environmental-science/exam-practice?envsci. Students should be able to answer all of the questions listed below with material learned in this and previous chapters. When a question requires students to understand material from multiple chapters, the question will be listed in the last chapter required to complete the entire question. Questions marked with an asterisk (*) are from exams with released multiple-choice questions. You may want to save these questions until the end of the year so you can give your students a complete released exam for practice. Questions marked with double asterisks (**) require math to calculate a problem. Look for references to these questions throughout the chapter.

Year	Question	Content
2000	3	• Endangered species and the vulnerability of species to extinction • Federal and international laws to protect endangered species
2003*	4	• The decline and resurgence of the whooping crane and California condor • Characteristics that contribute to the vulnerability of a species • Pros and cons of protecting endangered species
2010	3	• Invasive species

PD Chapter 18 Overview

Watch the video overview of Chapter 18 (for teachers) by clicking on the link buttons in the TE-book, opening the TRFD, or logging onto the book's companion website bcs.whfreeman.com/friedlandapes2e (teacher login required).

The Papahānaumokuākea Marine National Monument, designated in 2006, surrounds the northwestern Hawaiian Islands and protects m[ore] than 7,000 species of marine organisms, includ[ing] these Hawaiian squirrel fish (*Sargocentron xantherythrum*). (James D. Watt/Oceanstock/SeaPics.com)

TRM PowerPoint Presentation 18

Download the PowerPoint presentation for Chapter 18 by clicking on the link buttons in the TE-book, opening the TRFD, or logging onto the book's companion website bcs.whfreeman.com/friedlandapes2e (teacher login required).

chapter 18

Conservation of Biodiversity

Modern Conservation Legacies

The biodiversity of the world is currently declining at such a rapid rate that many scientists have declared that we are in the midst of a sixth mass extinction. There are many causes of this decline, but all are related to human activities ranging from habitat destruction to overharvesting plant and animal populations. In response to this crisis, there is growing interest in conserving biodiversity by setting aside areas that are protected from many human activities.

The conservation of biodiversity has a long history. The United States, for example, has been protecting habitats as national parks, national monuments, national forests, and wilderness areas for more than a century. Yellowstone National Park was the first national park in the United States, designated in 1872 by President Ulysses Grant. During the presidency of Theodore Roosevelt (1901–1909), nearly 93 million hectares (230 million acres) received federal protection. This included the creation of more than a hundred national forests, although much of this land was set aside to ensure a future supply of trees for lumber and therefore lacked complete protection.

> Efforts to protect marine habitats are relatively recent.

In contrast to the long history of protecting terrestrial habitats, efforts to protect marine habitats are relatively recent. One of the most expansive efforts in the United States was made during the administration of George W. Bush. From 2006 to 2009, President Bush designated a total of 95 million hectares (215 million acres) of marine habitats as protected around the northwestern Hawaiian Islands and other U.S. Pacific islands. In the northwestern Hawaiian Islands, 36 million hectares (90 million acres) of these marine habitats were set aside as the Papahānaumokuākea Marine National Monument. This protected region is immense, covering an area about the size of California.

The marine ecosystem that surrounds the Hawaiian Islands contains a great deal of biodiversity—more than 7,000 marine species, approximately

Teaching Tip: Chapter Opening Case

The chapter opening case, "Modern Conservation Legacies," introduces students to some of the important conservation work being done today. It describes the Papahānaumokuākea Marine National Monument and other areas set aside for the protection and enjoyment of the natural world. This case provides a good opportunity for you to discuss the different levels of conservation, from genetic diversity to habitat preservation. It is also important for students to understand the economic pressures involved in setting aside land and resources for protection as well as the economic benefits of preserving biodiversity. Past free-response questions have asked students to write about both the ecological and the economic pros and cons of species preservation.

Teaching Tip: Beyond the Classroom

Have students research the Papahānaumokuākea Marine National Monument discussed in the chapter opening case. Students might pick a marine organism found in this protected area and answer the following questions:

- What is the name of the organism and what does it look like?
- What are some of its unique features?
- What role does it play in the food web?
- What human activities could cause it harm?

Organisms found in the Papahānaumokuākea Marine National Monument include:

- green sea turtle
- Laysan finch
- Nihoa finch
- Nihoa millerbird
- Laysan duck
- Laysan albatross
- Pritchardia palms
- red pencil urchin
- Hawaiian squirrelfish

one-fourth of which are found nowhere else in the world. Unfortunately, in recent decades human activities have caused a decline in this diversity. The human causes of declining diversity are wide ranging. Although Hawaii has only 1.3 million residents, 7 million tourists visit each year. Individual anglers and commercial fishing operations have exploited marine life, including coral and fish. In addition to this exploitation, there are thousands of kilograms of old fishing equipment lying at the bottom of the ocean that sometimes wash up on shore, entangling wildlife in old fishing lines. Invasive species of algae also dominate some areas.

The Papahānaumokuākea monument presents an opportunity for improving the Hawaiian marine environment. As a national monument, the area is protected from fishing, harvesting of coral, and the extraction of fossil fuels. Large amounts of solid waste debris are being removed from the shorelines and coral reefs, and efforts are under way to clean out much of the invasive algae. It is expected that the biodiversity of the area will quickly respond to these efforts. As the populations of organisms increase in the protected areas, individuals will disperse and add to the populations in the larger surrounding area. In this way, the protected area can serve as a constant supply of individuals to help neighboring areas maintain their diversity of species.

In the United States and the rest of the world, conserving the biodiversity of marine areas by creating marine reserves is a relatively new activity for governments, but the idea is gaining ground. In the Galápagos Islands, where Charles Darwin studied the evolution of finches, the nation of Ecuador recently designated a marine reserve that extends 64 km (40 miles) into the ocean from the islands and allows only limited fishing. Marine reserves have also been designated by Russia, the United Kingdom, Australia, Canada, and Belize.

Efforts to protect critical wildlife habitats continue today. For example, in 2009 the Obama administration set aside more than 484,000 km^2 (187,000 square miles) of Alaska coastline and waters as critical habitat for polar bears. Although this does not prevent activities such as gas and oil drilling, it does mean that potential impacts on polar bears must now be considered when such activities are proposed in this area. As more countries develop marine reserves, we have to make sure these areas are large enough to allow long-term protection of local species and we must consider how each new reserve is positioned relative to other reserves so that individuals are able to move among them. Furthermore, countries must decide what human activities will be allowed in each reserve, perhaps protecting a core area and allowing tourism, fishing, or extraction of fossil fuels to occur in more distant areas of the reserve. These are exciting times that demonstrate that there is a great potential for conserving biodiversity in the twenty-first century.

Sources:
P. Thomas, President Bush to add marine reserves; not all are applauding, *Los Angeles Times*, January 6, 2009. http://latimesblogs.latimes.com/outposts/2009/01/news-flash-pres.html; U.S. government heads for row with big business after Obama sets aside land in Alaska for polar bear sanctuary, *Daily Mail*, November 25, 2010. http://www.dailymail.co.uk/news/article-1333008/Obama-sets-aside-land-Alaska-polar-bear-sanctuary.html.

Preserving habitats is one important way to protect against declines in the world's biodiversity. In this chapter, we will examine declines in biodiversity at multiple levels including declines in the genetic diversity of wild plants and animals, declines in the genetic diversity of domesticated plants and animals, and declines in the species of large taxonomic groups. We will also investigate the major causes of these declines, which include habitat loss, overharvesting, and the introduction of species from other regions of the world. To help curb the loss of biodiversity, we have a number of laws and international agreements that are in effect. Approaching these efforts with an understanding of the concepts of metapopulations, island biogeography, and biosphere reserves can help us succeed in protecting large ecosystems.

module

59

The Sixth Mass Extinction

In Chapter 5, we noted that the world has experienced five major extinctions during the past 500 million years. Many scientists have suggested that we may currently be in the midst of a sixth mass extinction event. In the most recent assessment made in 2014, scientists estimate that the world is currently experiencing approximately 1,000 species extinctions per year. This sixth mass extinction is unique because it is happening over a relatively short period of time and is the first mass extinction to occur since humans have been present on Earth.

In this module, we will examine the declines in biodiversity of Earth at various levels of complexity including genetic diversity, species diversity, and ecosystem function. In each case, we will examine the roles that humans have played in the decline of biodiversity.

Learning Objectives

After reading this module you should be able to

- explain the global decline in the genetic diversity of wild species.
- discuss the global decline in the genetic diversity of domesticated species.
- identify the patterns of global decline in species diversity.
- explain the values of ecosystems and the global declines in ecosystem function.

We are experiencing global declines in the genetic diversity of wild species

At the lowest level of complexity, environmental scientists are concerned about conserving genetic diversity. Populations with low genetic diversity are not well suited to surviving environmental change and they are prone to inbreeding depression, as we discussed in Chapter 6. Inbreeding depression by parents that each carry a harmful recessive mutation causes some of their offspring to receive two copies of the harmful mutation and, as a result, causes the offspring to have a poor chance of survival and later reproduction. High genetic diversity ensures that a wider range of genotypes is present, which reduces the probability that an offspring will receive the same harmful mutation from both parents. In addition, high genetic diversity improves the probability of surviving future change in the environment. This happens because high genetic diversity produces a wide range of phenotypes that survive and reproduce under different environmental conditions.

Some declines in genetic diversity have natural causes. Cheetahs, for example, possess very low genetic diversity. Researchers have determined that this condition is the result of a population bottleneck that occurred approximately 10,000 years ago (see Figure 15.10). Other

Teaching Tip: Warm-up

Ask students the following questions:

- Why is it important to have high genetic diversity? *High genetic diversity ensures that a wider range of genotypes is present, which reduces the probability that an offspring will receive the same harmful mutation from both parents. High genetic diversity also improves the probability of surviving future change in the environment.*
- Why should we be concerned about inbreeding? *Inbreeding depression by parents that each carry a harmful recessive mutation causes some of their offspring to receive two copies of the harmful mutation and, as a result, causes the offspring to have a poor chance of survival and later reproduction.*

declines in genetic diversity have human causes. For example, we discussed in Chapter 5 that the Florida panther once roamed throughout the southeastern United States (FIGURE 59.1). Because of hunting and habitat destruction, the population of the Florida panther shrank to only a small group in south Florida and this led to inbreeding. This inbreeding caused a number of harmful defects that caused the population to decline even further. After scientists released 8 panthers from Texas into Florida to add genetic diversity, the Florida panther population increased from 20 to nearly 100 individuals.

FIGURE 59.1 **Declines in genetic diversity.** The Florida panther was reduced to such a small population that it suffered severe effects of inbreeding. In recent years the introduction of new genotypes from a Texas population has allowed the Florida panther to rebound. *(Thomas & Pat Leeson)*

We are also experiencing global declines in the genetic diversity of domesticated species

Although declining genetic variation of plants and animals in the wild is of great concern to scientists, there are also major concerns about declining genetic variation in the domesticated species of crops and livestock on which humans depend. The United Nations notes that the majority of livestock species comes from seven species of mammals (donkeys, buffalo, cattle, goats, horses, pigs, and sheep) and four species of birds (chickens, ducks, geese, and turkeys). In different parts of the world, these species have been bred by humans for a variety of characteristics including adaptations that allow them to survive local climates. For example, humans have bred for a tremendous diversity of traits in cattle, as illustrated in FIGURE 59.2. This wide variety of adaptations, which is produced by a great deal of genetic variation, could be used for

FIGURE 59.2 **The genetic diversity of livestock.** Over thousands of years, humans have selected for numerous breeds of domesticated animals to thrive in local climatic conditions and to resist diseases common in their local environments. Modern breeding, which focuses on productivity, has caused the decline or extinction of many of these animal breeds.

FIGURE 59.3 **A global seed bank.** The Svalbard Global Seed Vault in northern Norway is an international storage area for many varieties of crop seeds from throughout the world. *(Jim Richardson/National Geographic Society/Corbis)*

adapting to changing environmental conditions in the future or resisting new diseases. Unfortunately, livestock producers have concentrated their efforts on the breeds that are most productive and much of this genetic variation is being lost. In Europe, for example, half of the breeds of livestock that existed in 1900 are now extinct. Of those that remain, 43 percent are currently at serious risk of extinction. Of the 200 breeds of domesticated animals that have been evaluated in North America, 80 percent of these breeds are either declining or are already facing extinction.

A similar story exists for crop plants. A century ago, most of the crops that humans consumed were composed of hundreds or thousands of unique genetic varieties. Each variety grew well under specific environmental conditions and was usually resistant to local pests. In addition, each variety often had its own unique flavor. As we saw in Chapter 11, the green revolution in agriculture focused on techniques that increased productivity. Farmers planted fewer varieties, concentrating on those with higher yields. Fertilizers and irrigation helped humans control many of the abiotic conditions, allowing fewer but higher-yielding varieties to be grown across large regions of the world. For example, at the turn of the twentieth century, farmers grew approximately 8,000 varieties of apples. Today, that number has been reduced to about 100, and considerably fewer are available in your local grocery store.

Planting only a few varieties leaves us open to crop loss if the abiotic or biotic environment changes. For example, in the 1970s, a fungus spread through cornfields of the southern United States and killed half the crop. Although the fungus was uncommon, the high-yielding variety of corn that most farmers planted turned out to be susceptible to it. Following this crisis, scientists modified this high-yielding corn by adding a gene from a variety that is resistant to the fungus. Had the resistant variety not been preserved, this gene would not have been available.

The nations of the world have recognized the problem of declining seed diversity and have responded by storing seed varieties in specially designed warehouses to preserve genetic diversity. In fact, there are currently more than 1,400 such storage facilities around the world. However, many of these facilities are at risk from war and natural disasters. In the past decade, nations and philanthropists have funded an international storage facility known as the Svalbard Global Seed Vault (FIGURE 59.3). This facility consists of a tunnel built into the side of a frozen mountain on an island in the Arctic region of northern Norway. It was designed to resist a wide range of possible calamities, including natural disasters and global warming. Should the environment change in future years, either in terms of abiotic conditions or because of emergent diseases, the seed bank will be available to help scientists address the challenge. The Svalbard facility opened in 2008 with a capacity of 14.5 million seed varieties. As of 2013, more than 700,000 seed samples had been sent to Svalbard for long-term storage.

Species diversity has declined around the world

Extinction occurs when the last member of a species dies. These major extinction events are characterized as a loss of at least 75 percent of all species within a period of 2 million years. Scientists estimate that as a result of

Teaching Tip: Activity

Divide the class into groups of two. Give each group 20 index cards and the vocabulary terms in the list below. Have each team label one side of each card with the numbers 1 through 20. On the blank sides of cards 1 through 10, students should write one of the key terms below, using a new card for each term. Instruct students to shuffle the order of the cards; they should not duplicate the order of the list. On the blank sides of cards numbered 11 through 20, students should write the definition or some other information that connects the term to the definition of the terms on cards 1 through 10, making sure to keep an answer key of which numbered term card matches which numbered definition card. The groups should lay out their cards number side up in two rows—one with terms and one with definitions. Have the groups swap cards and answer keys.

The first player in each pair then chooses one card from each row and reads the cards to see if they are a match. If they are, that person keeps the cards and takes another turn. If they are not a match, the second member of the team has a turn. Each group continues playing until all the cards are gone. The person with the most cards is the winner.

- extinction
- exotic species
- inbreeding depression
- endangered
- invasive species
- Red List
- native species
- alien species
- CITES
- Endangered Species Act

TEACHING with FIGURES ▶

Show students Figure 59.4 and ask: Which group of species has had the greatest percentage of global decline since the year 1500? *Amphibians have had the greatest global decline.*

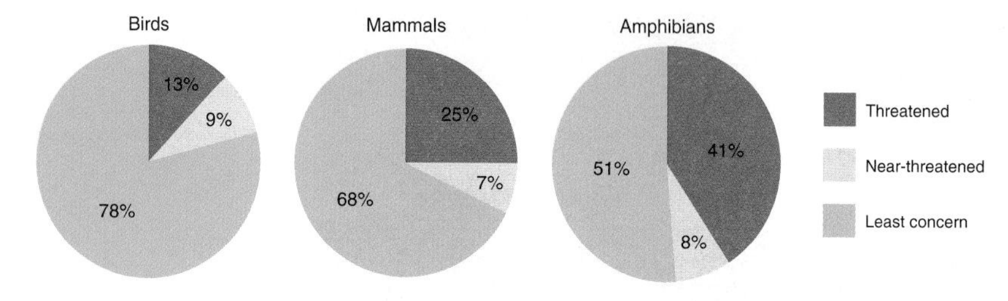

FIGURE 59.4 **The decline of birds, mammals, and amphibians.** Based on those species for which scientists have reliable data, 21 percent of birds, 32 percent of mammals, and 49 percent of amphibians are currently classified as threatened or near-threatened with extinction. *(After International Union for Conservation of Nature, 2009)*

these multiple mass extinctions and many minor extinctions, nearly 99 percent of the 4 billion species that have existed on Earth have gone extinct. However, because each of these mass extinction events has been followed by high rates of speciation that produced new species, we still have millions of species on Earth.

One way to assess the current extinction rate is by comparing the rate of extinction for groups of organisms, such as mammals, for which we have an excellent fossil record. Using this fossil record, we can compare the rate of species extinctions during the past 500 years to previous 500-year intervals. When we do this, we find that the rate of extinction during the past 500 years is higher than in previous 500-year periods. Indeed, the United Nations Convention on Biological Diversity estimates that the rate of extinction has been 1,000 times higher during the past 50 years than at any other time in human history and rivals the rates observed during the mass extinction event that eliminated the dinosaurs 65 million years ago.

To understand the current loss of species around the world, we can look at how particular groups of species are declining. When considering the status of a species, we use one of five categories defined by the International Union for Conservation of Nature (IUCN). Data-deficient species have no reliable data to assess their status; they may be increasing, decreasing, or stable. Species for which we have reliable data are placed in one of four categories. Extinct species are those that were known to exist as recently as the year 1500 but no longer exist today. The IUCN defines **threatened species** as those that have a high risk of extinction in the future and **near-threatened species** are very likely to become threatened in the future. **Least concern species** are widespread and abundant. These categories provide a mechanism for comparing the status of different groups of species.

Threatened species According to the International Union for Conservation of Nature (IUCN), species that have a high risk of extinction in the future.

Near-threatened species Species that are very likely to become threatened in the future.

Least concern species Species that are widespread and abundant.

Evaluating the status of different plant and animal groups presents several challenges. Many species fall under the category of data-deficient. At the same time, we are still discovering many new species, particularly in remote areas of the world. Since the number of species known to science constantly increases, it is not possible to evaluate every species and our estimates of what fraction of species are declining will constantly change. Finally, the work is expensive. Making an assessment for even one group of species, such as birds or mammals, requires thousands of scientists and millions of dollars.

Of the estimated 10 million species that currently live on Earth, ranging from bacteria to whales, only about 50,000 have been assessed to determine whether their populations are increasing, stable, or declining. Across all groups of organisms that have been assessed, nearly one-third are threatened with extinction. Given that some of the best data are for birds, mammals, and amphibians, we will examine these groups in more detail. FIGURE 59.4 shows the data for those species that are not yet extinct.

Since the year 1500, nearly 10,000 bird species have existed and 130 have become extinct. Today, 22 percent are threatened or near-threatened. Among the 800 species of birds living in the United States, nearly one-third are experiencing declining populations. These include 40 percent of bird species that live in grasslands and 30 percent of bird species that live in arid regions. Multiple threats, including reduced habitat and rising sea levels, have caused a growing concern for all species of birds that live on coastlines or on islands.

A similar pattern exists for mammals. Of the nearly 5,500 species of mammals known to have existed after 1500, 77 are extinct. Among the approximately 4,600 species for which there are reliable data, 25 percent are threatened and 32 percent are either threatened or near-threatened. This means that more than 1,400 species of mammals may be at risk of extinction.

Amphibians are experiencing the greatest global declines. Of the more than 6,300 species of amphibians, 34 species are extinct. However, a recent assessment of amphibian populations suggests that the number of extinctions may accelerate in the coming decades. Among the approximately 4,700 species for which reliable data exist, 49 percent are either threatened or near-threatened. This means that nearly 2,300 species of amphibians are declining around the world.

Many other groups of organisms are also experiencing large declines, but complete assessments have not yet been conducted because of the time and money required for each assessment. However, from the sample of species that have been assessed in each group, we see an emerging picture that is far from positive. For example, from this sample, approximately one-third of all reptiles, fish, and invertebrates are threatened with extinction. Similarly, one-fourth of plant species are threatened. These results suggest that when the assessments are complete, the news will most likely not be good.

Ecosystem values and the global declines in ecosystem function

Given that we rely on a relatively small number of the millions of species on Earth for our essential needs, why should we care about the millions of other species that live in various ecosystems? To understand the value of ecosystems, we can consider both *intrinsic values* and *instrumental values*.

Many people believe that ecosystems have **intrinsic value**—that is, that ecosystems are valuable independent of any benefit to humans. These beliefs may grow out of religious or philosophical convictions. People who believe that ecosystems are inherently valuable may argue that we have a moral obligation to preserve them. They may equate the obligation of protecting ecosystems with our responsibility toward people or animals that might need our help to survive. People who argue that ecosystems are valuable independent of any benefit to humans generally believe that environmental policy and the protection of ecosystems should be driven by this intrinsic value.

An ecosystem may also have **instrumental value,** meaning that it has worth as an instrument or tool that can be used to accomplish a goal. Instrumental values, which include the value of items such as crops, lumber, and pharmaceutical drugs, can be thought of in terms of how much economic benefit a species bestows. As noted in Chapter 1, we often refer to these instrumental values as ecosystem services. When calculating the instrumental value of various ecosystem services, we can consider five categories: *provisions, regulating services, support systems, resilience,* and *cultural services.*

Provisions

Goods produced by ecosystems that humans can use directly are called **provisions.** Examples include lumber, food crops, medicinal plants, natural rubber, and furs. Of the top 150 prescription drugs sold in the United States, about 70 percent come from natural sources. For example, Taxol, a potent anticancer drug, was originally discovered in the bark of the Pacific yew (*Taxus brevifolia*), a rare tree that grows in forests of the Pacific Northwest (FIGURE 59.5). Once approved by

FIGURE 59.5 **Provisions.** Scientists discovered that the bark of the Pacific yew contains a chemical that has anticancer properties. (Inga Spence/Science Source)

Intrinsic value Value independent of any benefit to humans.

Instrumental value Worth as an instrument or a tool that can be used to accomplish a goal.

Provision A good that humans can use directly.

Teaching Tip: Warm-up

Ask students the following questions:

- What is the difference between intrinsic value and instrumental value? *People who believe that an ecosystem has intrinsic value, believe that an ecosystem has value independent of any benefit to humans. People who believe that ecosystems are inherently valuable may argue that we have a moral obligation to protect them. The instrumental value of an ecosystem refers to its worth as an instrument or tool that can be used to accomplish a goal.*
- What is a provision? *A provision is a good that humans can use directly that is produced by an ecosystem. Some examples of provisions are lumber, food crops, medicinal plants, natural rubber, and furs.*

Teaching Tip: Discussion Starter

Ask students to give an example of a provision, a regulating service, and a support system. *An example of a provision would be a medicinal plant. An example of a regulating service would be natural ecosystems such as rainforests and oceans absorbing some of the excess carbon that humans add to the atmosphere. An example of a support system would be crop pollination provided by insects, birds, and bats.*

FIGURE 59.6 Regulating services. Tropical rainforests play a major role in regulating the amount of carbon in the atmosphere. *(John Pontier/Earth Scenes/Animals Animals)*

the FDA, the synthetic version of this single drug has had annual sales of over $1.5 billion. There is no way to estimate the potential value of natural pharmaceuticals that have yet to be discovered, but currently more than 800 natural chemicals have been identified as having potential uses to improve human health. Therefore, our best strategy may be to preserve as much biodiversity as we can to improve our chances of finding the next critical drug.

Regulating Services

Natural ecosystems help to regulate environmental conditions. For example, humans currently add about 8 gigatons of carbon to the atmosphere annually (1 gigaton = 1 trillion kilograms), but only about 4 gigatons of carbon remain there. The rest is removed by natural ecosystems, such as tropical rainforests and oceans, which provide us with more time to address climate change than we would otherwise have (FIGURE 59.6). As we have already seen, ecosystems also are important in regulating nutrient and hydrologic cycles.

Support Systems

Natural ecosystems provide numerous support services that would be extremely costly for humans to generate. One example is pollination of food crops (FIGURE 59.7). The American Institute of Biological Sciences estimates that crop pollination in the United States by native species of bees and other insects, hummingbirds, and bats is worth roughly $3.1 billion in added food production. In addition to providing habitat for animals that pollinate crops, ecosystems provide natural pest control services because they provide habitat for predators that prey on agricultural pests. Although organic farmers, who rarely use synthetic pesticides, gain the most from these pest controls, conventional agriculture benefits as well.

Healthy ecosystems also filter harmful pathogens and chemicals from water, leaving humans with water that requires relatively little treatment prior to drinking. Without these water-filtering services, humans would have to build many new water treatment facilities that use expensive filtration technologies. New York City, for example, draws its water from naturally clean reservoirs in the Catskill Mountains. But residential development and tourism in the area has threatened to increase contamination of the reservoirs with

FIGURE 59.7 Support systems. Pollinators such as this honeybee play an essential role in ensuring the pollination of food crops such as cherries. *(Steffan and Alexandria Sailer/Ardea/Earth Scenes/Animals Animals)*

silt and chemicals. Building a filtration plant adequate to address these problems would cost $6 billion to $8 billion. For this reason, New York City and the U.S. Environmental Protection Agency have been working to protect sensitive regions of the Catskills.

Resilience

We have already seen that resilience ensures an ecosystem will continue to exist in its current state, which means it can continue to provide benefits to humans. Resilience depends greatly on species diversity. For example, several different species may perform similar functions in an ecosystem but differ in their susceptibility to disturbance. If a pollutant kills one plant species that contains nitrogen-fixing bacteria, but does not kill all plant species that contain nitrogen-fixing bacteria, the ecosystem can still continue to fix nitrogen (FIGURE 59.8).

Cultural Services

Ecosystems provide cultural or aesthetic benefits to many people. The awe-inspiring beauty of nature has instrumental value because it provides an aesthetic benefit for which people are willing to pay (FIGURE 59.9). Similarly, scientific funding agencies may award grants to scientists for research that explores biodiversity with no promise of any economic gain. Nevertheless, the research itself has instrumental value because the scientists and others benefit from these activities by gaining knowledge. While intellectual gain and aesthetic satisfaction may be difficult to quantify, they can be considered cultural services that have instrumental value.

FIGURE 59.8 Species diversity as a component of resilience. This prairie ecosystem contains a high diversity of grasses and wildflowers, including many species of nitrogen-fixing wildflowers. If one nitrogen-fixing species is eliminated, the lost function can be compensated for by another nitrogen-fixing species. *(Mike MacDonald/ChicagoNature.com)*

FIGURE 59.9 Cultural services. Many natural areas, such as this scene from the Grand Tetons National Park, provide aesthetic beauty valued by humans. *(Buddy Mays/Corbis)*

Teaching Tip: Journal Prompt

Many natural areas or national parks have cultural services, meaning they have aesthetic benefits for many people. Ask students the following questions. (Answers will vary.)

- What are the cultural or aesthetic benefits of local, state, and national parks?
- Describe the last park or natural setting you visited and the aesthetic benefits you received from visiting it. It can be a private area like a farm, an orchard, or a lake.

Teaching Tip: Journal Prompt

Ask students: Is it possible to place a monetary value on ecosystem services? *Most economists believe that ecosystem services can be assigned a monetary value, and they are beginning to incorporate these values into their calculations of the economic costs and benefits of various human activities. However, assigning a dollar value is easier for some categories of ecosystem services than for others.*

The Monetary Value of Ecosystem Services

Most economists believe that the instrumental values of an ecosystem can be assigned monetary values, and they are beginning to incorporate these values into their calculations of the economic costs and benefits of various human activities. However, assigning a dollar value is easier for some categories of ecosystem services than for others. In 1997, a team of scientists and economists attempted to estimate the total value of ecosystem services to the human economy. They considered replacement value—the cost to replace the services provided by natural ecosystems. They also looked at other factors, such as how property values were affected by location relative to these services—for example, oceanfront housing. Finally, they considered how much time or money people were willing to spend to use these services—for example, whether they were willing to pay a fee to visit a national park. Using this method, researchers estimated that ecosystem services were worth over $30 trillion per year, or more than the entire global economy at that time.

The Decline of Ecosystem Services

Because species help determine the services that ecosystems can provide, we would expect declines in species diversity to be associated with declines in ecosystem function. In the Millennium Ecosystem Assessment conducted in 2005, the most recent assessment conducted, scientists from around the world examined the current state of 24 ecosystem functions, including food production, pollination, water purification, and the cycling of nutrients such as nitrogen and phosphorus. Of these 24 different ecosystem functions, 15 were found to be declining or used at a rate that cannot be sustained. If we want to improve ecosystem functions, we need to improve the fate of the species and ecosystems that provide these services.

module 59

REVIEW

In this module, we have seen that declines in biodiversity are happening at a rate that may indicate the start of a sixth mass extinction. At the genetic level, a loss of diversity is a concern because it places species under the threat of inbreeding depression that can cause population declines. We have observed declines in the genetic diversity of both wild species and domesticated species such as crops and livestock. At the species level, we have learned that major groups of organisms, including mammals, fish, and amphibians, have all experienced extinctions and contain large groups that are threatened or near-threatened. This global decline in species affects the functioning of ecosystems and reduces the intrinsic and instrumental values that these ecosystems provide. In the next module, we will examine the many causes of these declines in biodiversity.

Answers to Module 59 AP® Review Questions

1. d
2. a
3. a
4. d
5. e

Module 59 AP® Review Questions

1. In a major extinction event, what is the minimum percentage of species that goes extinct?
   (a) 25 percent
   (b) 40 percent
   (c) 50 percent
   (d) 75 percent
   (e) 90 percent

2. What factor has played the largest role in decreased diversity of domesticated species?
   (a) A focus on increased yields
   (b) The use of genetic engineering
   (c) The adaptation to specific growing environments
   (d) The decreased number of pests
   (e) Increased seed storage efforts

3. Which group of organisms has had the greatest number of extinctions since 1500?
   (a) Birds
   (b) Amphibians
   (c) Mammals
   (d) Reptiles
   (e) Fish

4. The intrinsic value of an ecosystem
   (a) is the total monetary worth of its features.
   (b) is based on the goods the ecosystem produces.
   (c) considers the potential benefits that could be discovered in the ecosystem.
   (d) is the value it has independent of humans.
   (e) is the function of an ecosystem that humans cannot replicate.

5. Which is NOT a category of instrumental value?
   (a) Regulating services
   (b) Resilience
   (c) Provisions
   (d) Cultural services
   (e) Diversity

module 60

Causes of Declining Biodiversity

We have seen that declines in biodiversity are happening around the globe. In "Science Applied 2: How Should We Prioritize the Protection of Species Diversity?" on page 184, we discussed biodiversity hotspots, which are areas around the world rich in biodiversity. Many organisms in these biodiversity hotspots face threats of extinctions due to human activities. However, threats to biodiversity exist throughout the world. In this module, we build on the basics of population and community ecology from Chapter 5 to understand how a number of factors can all affect biodiversity.

Learning Objectives

After reading this module you should be able to

- discuss how habitat loss can lead to declines in species diversity.
- explain how the movement of exotic species affects biodiversity.
- describe how overharvesting causes declines in populations and species.
- understand how pollution reduces populations and biodiversity.
- identify how climate change affects species diversity.

Teaching Tip: Warm-up

Ask students: How does habitat loss influence species extinction? *Many species thrive in a particular habitat within a narrow range of abiotic and biotic conditions. Species requiring such specialized habitats are particularly prone to population declines. As their favored habitat grows smaller, species are restricted to a smaller geographic area with resources that can support only a small population.*

Teaching Tip: Remember

Help your students remember the causes of biodiversity loss with the mnemonic HIPCO. HIPCO stands for:

- **H:** Habitat loss
- **I:** Invasive species
- **P:** Pollution
- **C:** Climate change
- **O:** Overharvesting

Habitat loss is the major cause of declining species diversity

For most species, the greatest cause of decline and extinction is habitat loss. In modern times, the primary cause of habitat loss is human development that removes natural habitats and replaces them with homes, industries, agricultural fields, shopping malls, and roads. Many species can only thrive in a particular habitat within a narrow range of abiotic and biotic conditions. Species requiring such specialized habitats are particularly prone to population declines, especially when their favored habitat is limited, which restricts their distribution to a specific geographic area suitable only for a small population.

Altering distinctive characteristics of a habitat, such as removing trees or damming streams, has an effect on the organisms that live in that habitat. For example, for thousands of years the northern spotted owl (*Strix occidentalis caurina*) lived in old-growth forests—those dominated by trees that are hundreds of years old—in the northwestern United States and southwestern Canada (see Figure 29.1c on page 332). This habitat provided the ideal sites for nesting, roosting, and catching small mammals to eat. The removal of the old trees, for lumber and housing developments, has transformed much of the former old-growth forest into a different habitat (FIGURE 60.1). This habitat alteration reduces the number of northern spotted owls because they have fewer trees in which to nest and less forest in which to find food.

The map in FIGURE 60.2 shows the changing face of forest habitats over the past few decades. As we saw in Chapter 6, much of the forest in the United States during the 1700s and 1800s was logged for lumber and cleared for agriculture. In recent decades, forested land has been increasing, although humans have often planted the new forests, which have a lower diversity of species than the original forests. At the same time, developing countries in South America, sub-Saharan Africa, and Southeast Asia are clearing their forests much as the United States and Europe did in years past. As a result, large declines in forest cover are occurring in developing countries that were once forested. It is currently unclear whether these countries will follow the pattern of Europe and North America and eventually allow their forested areas to increase.

Although deforestation receives a lot of attention, many other habitats are also being lost. According to the Millennium Ecosystem Assessment, approximately 70 percent of the woodland/shrubland ecosystem that borders the Mediterranean Sea has been lost. Similarly, across the globe we have lost nearly 50 percent of grassland habitats and 30 percent of desert habitats. Wetlands exhibit a mixed picture. Although the amount of wetland habitat is less than half of what

(a)

(b)

FIGURE 60.1 **Habitat loss.** Habitat loss caused by humans is the largest threat to biodiversity. (a) Old-growth forests, such as this one in the Mount Rainier National Park in Washington State, serve as habitat for spotted owls. (b) After clear-cutting an old-growth forest for timber, such as this site in the Olympic National Forest in Washington State, the forest provides a very different habitat. It may take hundreds of years before the habitat again becomes suitable for the spotted owl. *(a: Stephen Matera/DanitaDelimont.com; b: Kent Foster/Science Source)*

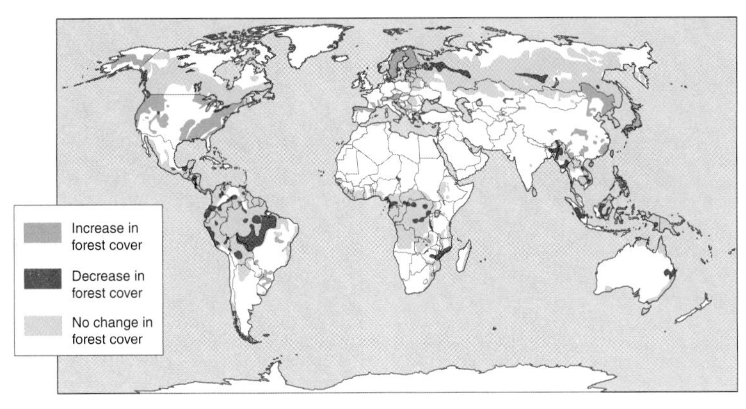

FIGURE 60.2 **Changing forests.** Some regions of the world experienced large declines in the amount of forested land from 1980 to 2000 while other regions have shown little change or have seen increases in forest cover. *(After Global Biodiversity Outlook 2, Convention on Biological Diversity, 2006)*

existed in the United States during the 1600s, from 1998 to 2004 the amount of freshwater wetland habitat actually increased. This overall growth occurred due to large increases in wetland habitat in the Great Lakes region. This growth offset a decline in coastal wetlands in the eastern United States and the Gulf of Mexico caused by growing human populations that built more roads, homes, and businesses.

In marine systems, there has been a sharp decline in the amount of living coral in the Caribbean Sea, as shown in FIGURE 60.3, from a high of 50 percent live coral in the 1970s to a mere 8 percent by 2012. Living coral provide habitat for thousands of other species, which makes them particularly vital to the persistence of marine habitats. The decline in coral is the result of human impacts including the warming of oceans (associated with global warming), increased pollution, and the removal of coral by collectors. This loss of coral habitat is occurring at a rapid rate throughout the world.

A species may decline in abundance or become extinct even without complete habitat destruction since a reduction in the size of critical habitat also can lead to extinctions through a variety of processes. As we saw in the case of the Florida panther, a smaller habitat supports a smaller population, reducing genetic diversity. Less habitat also reduces the variety of physical and climatic options available to individuals during periods of extreme conditions. The presence of cooler, high-altitude areas in a habitat, for example, allows animals a place to move during periods of hot weather. Also, loss of habitat can restrict the movement of migratory or highly mobile species. While many species can thrive in small habitats, other species, such as mountain lions, wolves, and tigers, require large tracts of relatively uninhabited, undisturbed land.

Smaller habitats can also cause increased interactions with other harmful species. For example, many songbirds in North America live in forests. When these birds make their nests near the edge of the forests—where the forest meets a field—they often have to contend with the brown-headed cowbird (*Molothrus ater*). The cowbird is a nest parasite—it does not build its own nest, but lays its eggs in the nests of several

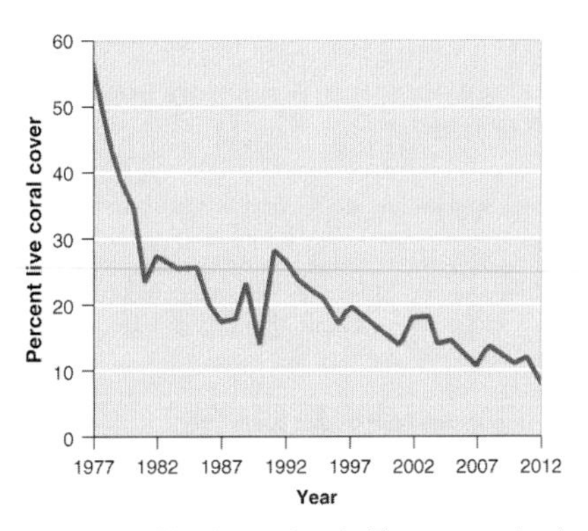

FIGURE 60.3 **Changing coral reefs.** The percentage of coral that remains alive in coral reefs has declined sharply in the Caribbean from 1977 to 2012. *(Tropical Americas Coral Reef Workshop, 2012)*

◀ TEACHING with FIGURES

Show students Figure 60.3 and ask: What are three causes for the decline in living coral reefs from the late 1970s to 2012?
Three main causes for the rapid decline in coral reefs found in the Caribbean are the warming of oceans, increased water pollution, and the removal of coral by collectors.

Teaching Tip: Beyond the Classroom

Have students create a "wanted" poster for an invasive species. Students should research an invasive species and create a poster that conveys the following information:

- What is the species and what does it look like?
- What problems does it cause?
- Where is it originally from and where is it considered invasive?
- How was it transported?
- What is being done to prevent its further spread?

Teaching Tip: Video

Invasion of the Earthworms!
This video from the National Science Foundation discusses the impact of invasive nonnative earthworms. Earthworms can cause a rapid incorporation of organic material into the soil, changing the chemistry and nutrient dynamic of the soil.

The link to this video may be found by clicking the link buttons in the TE-book, opening the TRFD, or logging onto the book's companion website bcs.whfreeman.com/friedlandapes2e (teacher login required).

Chapter 18 Web Resources

(a)

(b)

FIGURE 60.4 **Habitat fragmentation.** (a) Increased fragmentation of forests has caused forest songbirds to come into increasing contact with the brown-headed cowbird. (b) The cowbird does not make its own nest. Instead, it lays its brown, spotted eggs in the nests of other species, such as this nest containing four blue eggs of the chipping sparrow (*Spizella passerina*). *(a: Jim Zipp/Ardea/Animals Animals; b: Paul J. Fusco/Science Source)*

other species of birds (FIGURE 60.4). In this way the cowbird tricks forest birds into raising its offspring, which takes food away from the forest birds' own offspring. In some cases, the host bird will simply abandon the nest. As forests are broken up into smaller fragments, the proportion of forest near the edge increases and, therefore, the number of bird nests that are susceptible to brown-headed cowbirds increases. Over time, increased fragmentation has allowed brown-headed cowbirds to cause declines in many species of North American songbirds.

Exotic species are moving around the world

Native species are species that live in their historical range, typically where they have lived for thousands or millions of years. In contrast, **exotic species,** also known as **alien species,** are species that live outside their historical range. For example, honeybees (*Apis mellifera*) were introduced to North America in the 1600s to provide a source of honey for European colonists. Red foxes (*Vulpes vulpes*), now abundant in Australia, were introduced there in the 1800s for the purpose of fox hunts, which were popular in Europe at the time.

During the past several centuries, humans have frequently moved animals, plants, and pathogens around the world. Some species are also moved accidentally. For example, rats that have stowed away in shipping containers have ended up on distant oceanic islands. Because these islands never had rats or other ground predators, there had never been any natural selection against nesting on the ground, and numerous island bird species had evolved to nest on the ground. When the rats arrived, they found the eggs and hatchlings from ground nests an easy source of food, resulting in a high rate of extinction in ground-nesting birds in places such as Hawaii. Similar accidental movements have occurred for many pathogens, including exotic fungi that were introduced to North America nearly a century ago and have since killed nearly all American elm (*Ulmus americana*) and American chestnut (*Castanea dentata*) trees in eastern North America. Similarly, an exotic protist that causes avian malaria has driven many species of Hawaiian birds to extinction. Other movements of exotic species are intentional, such as exotic plants that are sold in greenhouses for houseplants and outdoor landscape plants, or exotic animals that are sold as pets or to game ranches that raise exotic species of large mammals for hunting.

Native species Species that live in their historical range, typically where they have lived for thousands or millions of years.

Exotic species A species living outside its historical range. *Also known as* **alien species.**

In most cases, exotic species fail to establish successful populations when they are introduced to a new region. For the small percentage of introductions that are successful, exotic species can live in their new surroundings and have no negative effect on the native species. In other cases, however, the exotic species rapidly increase in population size and cause harmful effects on native species. When exotic species spread rapidly across large areas, we call them **invasive species.** Rapid spread of invasive species is possible because invasive species, which have natural enemies in their native regions that act to control their population, often have no natural enemies in the regions where they are introduced. Two of the best-known examples of invasive exotic species in North America are the kudzu vine (*Pueraria lobata*) and the zebra mussel (*Dreissena polymorpha*).

The kudzu vine is native to Japan and southeast China but was introduced to the United States in 1876. Throughout the early 1900s, farmers in the southeastern states were encouraged to plant kudzu to help reduce erosion in their fields. By the 1950s, it became apparent that the southeastern climate was ideal for kudzu, with growth rates of the vine approaching 0.3 m (1 foot) *per day*. Because herbivores in the region do not eat kudzu, the species has no enemies and can spread rapidly. The vine grows up over most wildflowers and trees and shades them from the sunlight, causing the plants to die. Indeed, the vine grows over just about anything that does not move (FIGURE 60.5). Kudzu currently covers approximately 2.5 million hectares (7 million acres) in the United States.

The zebra mussel is native to the Black Sea and the Caspian Sea in eastern Europe and western Asia. Over the years, large cargo ships that traveled in these seas unloaded their cargo in the ports of the Black Sea and Caspian Sea and then pumped seawater into the holding tanks to ensure that the ship sat low enough in the water to remain stable. This water that is pumped into the ship is called ballast water. When the ships arrived in the St. Lawrence River and the Great Lakes, they loaded on new cargo and no longer needed the weight of the ballast water, which they pumped out of the ship into local waters. One consequence of transporting ballast water from Asia to North America is that many aquatic species from Asia, including zebra mussels, have been introduced into the aquatic ecosystems of North America. Because the St. Lawrence River and the Great Lakes provided an ideal ecosystem for the zebra mussel, and because a single zebra mussel can produce up to 30,000 eggs, the mussel spread rapidly through the Great Lakes ecosystem. On the positive side, because the mussels feed by filtering the water, they remove large amounts of algae and some contaminants, which, to some degree, counteracts the cultural eutrophication that has occurred in the Great Lakes ecosystem. On the negative side, the zebra mussels physically crowd out many native mussel species and the zebra mussels can consume so much algae that they negatively affect native species that also need to consume the algae. Moreover, the invasive mussels can achieve such high densities that they can clog intake pipes and impede the flow of water on which industries and communities rely.

FIGURE 60.5 **The spread of the exotic kudzu vine.** The fast-growing kudzu vine is native to Asia but was introduced to the United States to control erosion. It has since spread rapidly, growing over the top of nearly anything that does not move. *(Melissa Farlow/National Geographic Creative/Getty Images)*

A new threat to the Great Lakes is the silver carp (*Hypophthalmichthys molitrix*), a fish that is native to Asia but has been transported around the world in an effort to consume excess algae that accumulates in aquaculture operations and the holding ponds of sewage treatment plants. After being brought to the United States, some of the fish escaped and rapidly spread through many of the major river systems, including the Mississippi River. Over the years, the carp population has expanded northward, and by 2010 it approached a canal where the Mississippi River connects to Lake Michigan. Although researchers detected the DNA of the carp in water samples from the Great Lakes from 2009 to 2011, substantial netting efforts in the spring of 2013 failed to find any of the fish in the lakes. There are two major concerns about this invading fish. First, scientists worry that it will outcompete native species of fish that also consume algae. Second, the silver carp has an unusual behavior; it jumps out of

Invasive species A species that spreads rapidly across large areas.

AP® Exam Tip

Invasive species are covered in the 2010 AP® Exam, Question 3. To answer this question, students must

- explain why zebra mussels are located primarily in the eastern United States rather than in the western United States, based on a provided reading
- explain how zebra mussels are introduced into lakes and propose a viable method for preventing the spread of zebra mussels into isolated lakes
- identify and describe the impact zebra mussels have had on aquatic ecosystems
- describe another invasive species and one negative impact that species has had on an ecosystem
- give an example of using a nonnative species to control another nonnative species and discuss a negative impact of this strategy
- suggest two characteristics of invasive species that permit them to thrive in a nonnative environment

Module 60 Causes of Declining Biodiversity 645

Teaching Tip: Journal Prompt

Ask students the following questions:

- Why are scientists concerned about the introduction of invasive species? *Some possible answers are that introduced species compete for resources better than the native species, they can reproduce rapidly, they have no natural predators, and they can prey on native species.*
- List three things you can do to prevent the introduction of invasive species. *Ways to prevent the introduction of invasive species could include not taking seeds or plants from one area to another, inspecting goods as you travel from country to country, not transporting water, animals, or plants from one body of water to another, and only purchasing pets that you know are native.*

FIGURE 60.6 Silver carp invading the Mississippi River. The silver carp has been introduced into the Mississippi River from Asia and is quickly heading toward an invasion of the Great Lakes. *(Chris Olds/U.S. Fish & Wildlife Service)*

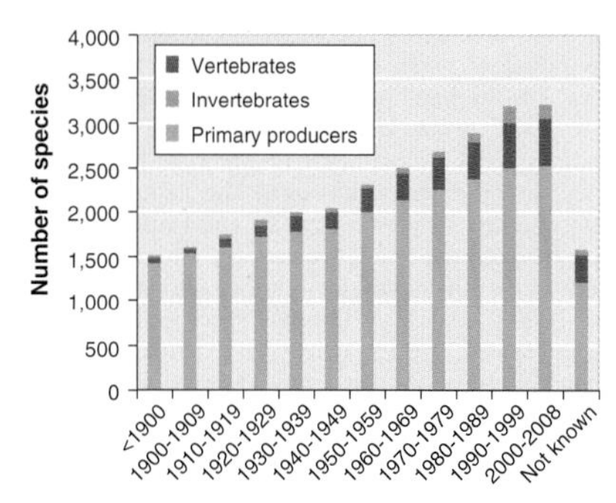

FIGURE 60.7 Exotic species. Over the decades, there has been a steady increase in the number of exotic species in Europe. This example shows the number of exotic species recorded in terrestrial ecosystems. *(Data from European Environment Agency,* EU 2010 Biodiversity Baseline, *2010)*

the water when startled by passing boats (FIGURE 60.6). Given that the carp can grow to 18 kg (40 pounds) and jump up to 3 m (10 feet) into the air, this poses a major safety issue to boaters.

Around the world, invasive exotic species pose a serious threat to biodiversity by acting as predators, pathogens, or superior competitors to native species. Some of the most complete data exist in Europe. As FIGURE 60.7 shows, during the past 100 years, Europe has experienced a steady increase of nearly 2,000 exotic species in terrestrial ecosystems. Additional species have been introduced into freshwater and marine ecosystems. A number of efforts are currently being used to reduce the introduction of invasive exotic species, including the inspection of goods coming into a country and the prohibition of wooden packing crates made of untreated wood that could contain insect pests.

Overharvesting causes declines in populations and species

Hunting, fishing, and other forms of harvesting are the most direct human influences on wild populations of plants and animals. Most species can be harvested to some degree, but a species is overharvested when individuals are removed at a rate faster than the population can replace them. In the extreme, overharvesting of a species can cause extinction. In the seventeenth century, for example, ships sailing from Europe stopped for food and water at Mauritius, an uninhabited island in the Indian Ocean. On Mauritius, the sailors would hunt the dodo (*Raphus cucullatus*), a large flightless bird that had no innate fear of humans because it had never seen humans during its evolutionary history (FIGURE 60.8). The dodo, unable to protect itself from human hunters and the rats (introduced by humans) that consumed dodo eggs and hatchlings, became extinct in just 80 years. This

FIGURE 60.8 Overharvesting. The dodo was a large flightless bird that served as an easy source of meat for sailors and settlers on the island of Mauritius. Because it evolved on an island with no large predators or humans, the dodo had no instinct to fear humans. *(Stock Montage/Getty Images)*

Teaching Tip: Video

Asian Carp in Bath, Illinois

This video shows an example of an invasive species gone wrong. The tiny town of Bath, Illinois, puts on an Asian carp fishing tournament each year where no rods or reels are needed. The contestants only need large fishing nets, because the fish jump high out of the water. The video discusses the harm the Asian carp has caused to local ecosystems.

The link to this video may be found by clicking the link buttons in the TE-book, opening the TRFD, or logging onto the book's companion website bcs.whfreeman.com/friedlandapes2e (teacher login required).

Chapter 18 Web Resources

same scenario appears to have taken place with many other large animal species as well. These animals include the giant ground sloths, mammoths, American camels of North and South America, and the 3.7 m (12 feet) tall moa birds of New Zealand. Each species became extinct soon after humans arrived, suggesting that the animals' demise may have been due to overharvesting.

Overharvesting has also occurred in the more recent past. In the 1800s and early 1900s, for example, market hunters slaughtered wild animals to sell their parts on such a scale that many species, including the American bison, declined dramatically. Bison were once abundant on the western plains, with estimates ranging from 60 to 75 million individuals. By the late 1800s fewer than 1,000 were left. This means that 99.999 percent of all bison were killed. Following enactment of legal protections, the bison population today has increased to more than 500,000, including both wild bison and bison raised commercially for meat.

Not all species harvested by market hunters fared as well as the American bison. The passenger pigeon was once one of the most abundant species of birds in North America. Population estimates range from 3 to 5 billion birds in the nineteenth century. In fact, during annual migrations, people observed continuous flocks of pigeons flying overhead for 3 days straight in densities that blocked out most of the sun. Breeding flocks could cover 40,000 ha (100,000 acres) with 100 nests built into each tree. With such high densities, market hunters could shoot or net the birds in very large numbers and fill train cars with harvested pigeons to be sold in eastern cities. This overharvesting, combined with the effects of forest clearing for agriculture, caused the passenger pigeon to decline quickly. The last passenger pigeon died in 1914 at the Cincinnati Zoo.

During the past century, regulations have been passed to prevent the overharvesting of plants and animals. In the United States, for example, state and federal regulations restrict hunting and fishing of game animals to particular times of the year and limit the number of animals that can be harvested. Similar agreements have been reached among countries through international treaties. In general, these regulations have proven very successful in preventing species declines caused by overharvesting. In some regions of the world, however, harvest regulations are not enforced and illegal poaching, especially of large, rare animals that include tigers, rhinoceroses, and apes, continues to threaten species with extinction. Harvesting of rare plants, birds, and coral reef dwellers for private collections has also jeopardized these species.

Plant and Animal Trade

For some species, the legal and illegal trade in plants and animals represents a serious threat to their ability to persist in nature. One of the earliest laws in the United States to control the trade of wildlife was the **Lacey Act.** First passed in 1900, the act originally prohibited the transport of illegally harvested game animals, primarily birds and mammals, across state lines. Over the years, a number of amendments have been added so that the Lacey Act today forbids the interstate shipping of all illegally harvested plants and animals.

At the international level, the United Nations **Convention on International Trade in Endangered Species of Wild Fauna and Flora,** also known as **CITES,** was developed in 1973 to control the international trade of threatened plants and animals. Today, CITES is an international agreement among 175 countries throughout the world. The IUCN maintains a list of threatened species known as the **Red List.** Each member country assigns a specific agency to monitor and regulate the import and export of animals on the list. For example, in the United States, the U.S. Fish and Wildlife Service conducts this oversight.

Despite such international agreements, much illegal plant and animal trade still occurs throughout the world. In 2008, a report by the Congressional Research Service estimated that illegal trade in wildlife was worth $5 billion to $20 billion annually. In some cases, animals are sold for fur or for body parts that are thought to have medicinal value. In other cases, rare animals are in demand as pets. For example, in 2001 a population of the Philippine forest turtle (*Siebenrockiella leytensis*), once thought to be extinct, was discovered on a single island in the Philippines (FIGURE 60.9). This animal, one of the most endangered species in the world, cannot be traded legally, but demand for it as a pet has caused it to be sold illegally and the last remaining population has declined sharply in only a few years. A single turtle sells for $50 to $75 in the Philippines and up to $2,500 in the United States and Europe. Similar cases of illegal trade occur in rare species of trees for lumber such as big-leaf mahogany (*Swietenia macrophylla*), rare species of plants for medicine such as goldenseal (*Hydrastis canadensis*), and many rare species of orchids for their beautiful flowers.

Lacey Act A U.S. act that prohibits interstate shipping of all illegally harvested plants and animals.

Convention on International Trade in Endangered Species of Wild Fauna and Flora (CITES) A 1973 treaty formed to control the international trade of threatened plants and animals.

Red List A list of worldwide threatened species.

Teaching Tip: Activity

Have your students take a few minutes to create a table to compare and contrast the Lacey Act and the Convention on International Trade in Endangered Species of Wild Fauna and Flora (CITES), both of which help prevent illegal plant and animal trade. (Answers are provided in italics.)

	Lacey Act	CITES
Description	*One of the earliest laws in the United States to control wildlife*	*A United Nations agreement among 175 countries throughout the world*
Year passed	*1900*	*1973*
Purpose	*The act originally prohibited the transport of illegally harvested game animals, primarily birds and mammals, across state lines. Subsequent amendments prohibited interstate shipping of all illegally harvested plants and animals.*	*The convention helps control international trade of threatened plants and animals. It also maintains the Red List, a list of threatened species worldwide.*

FIGURE 60.9 **Species declines due to the pet trade.** The only remaining population of this turtle lives on a single island in the Philippines. Although protected by law, illegal trade has caused a rapid decline of this species in the wild. *(NHPA/Photoshot)*

Sometimes even when trade in a particular species is legal, it can pose a potential long-term threat to species persistence. In the southwestern United States, for example, there is a growing movement to reduce water use by replacing grass lawns with desert landscapes. One of the unintended consequences is the increased demand for cacti and other desert plants that are collected from the wild. Sales are currently estimated to be $1 million annually. Given the slow growth of desert plants, this increased demand is causing heightened concern for these plant populations in the wild.

Pollution can have harmful effects on species

In Chapters 14 and 15, we saw how water and air pollution harm ecosystems. Threats to biodiversity come from toxic contaminants such as pesticides, heavy metals, acids, and oil spills. Other contaminants, such as endocrine disrupters, can have nonlethal effects that prevent or inhibit reproduction. Pollution sources that cause declines in biodiversity also include the release of nutrients that cause algal blooms and dead zones as well as thermal pollution that can make water bodies too warm for species to survive.

In 2010, for example, an oil platform in the Gulf of Mexico owned by BP and named the *Deepwater Horizon* exploded, causing a massive release of oil that lasted for several months. As we discussed in Chapter 14, the release of oil caused a tremendous amount of death across a wide range of animal species including sea turtles, pelicans, fish, and shellfish. In response to the massive oil spill, BP released hundreds of thousands of liters of oil dispersant, a chemical designed to break up large areas of oil into tiny droplets that can be consumed by specialized species of bacteria. However, the dispersant is also toxic to many species of animals. The total impact of the spilled oil and applied dispersants on the wildlife of the Gulf of Mexico may not be known for many years.

Climate change has the potential to affect species diversity

We have mentioned climate change in previous chapters and will discuss it in detail in Chapter 19. As a threat to biodiversity, the primary concern about climate change is its effect on patterns of temperature and precipitation in different regions of the world. In some regions, a species may be able to respond to warming temperatures and changes in precipitation by migrating to a place where the climate is well suited to the species niche. In other cases, this is not possible. For example, in southwestern Australia, a small woodland/shrubland peninsula exists on the edge of the continent with a much larger area of subtropical desert farther inland (see Figure 12.3 on page 123). Scientists expect conditions on the peninsula to become drier during the next 70 years. If this occurs, many species of plants in this small ecosystem will not have a nearby hospitable environment to which they can migrate, since the surrounding desert ecosystem is already too dry for them. An examination of 100 species of plants in the area (all from the genus *Banksia*) has led scientists to project that 66 percent of the species will decline in abundance and up to 25 percent will become extinct. As we will see in Chapter 19, many species in the world are expected to be affected by climate change.

module

60

REVIEW

In this module, we learned that the biodiversity of our planet is declining for a number of reasons. The primary causes of this decline are the loss of habitats and the fragmentation of habitats that species need to survive and reproduce. Exotic species are being moved around the world with the increased global movement of people and materials. Many populations of these exotic species remain small and cause no discernible harm, but some become invasive species that spread quickly and have harmful effects on native species. Overharvesting species can cause larger declines in population sizes and, in some cases, extinction. Current regulations within and among countries are designed to limit harvesting to sustainable levels, although these regulations are not always successfully enforced. Toxic compounds including pesticides, heavy metals, and spilled oil can also be detrimental to species either by direct lethal effects or by altering communities and ecosystems. Finally, climate change has the potential to alter populations and the long-term persistence of species, but more time is needed to determine if these predictions will come true. In the next module, we will examine past and current efforts to conserve biodiversity.

Module 60 AP® Review Questions

1. The most significant cause of species decline and extinction throughout the world is
   (a) habitat loss.
   (b) overharvesting.
   (c) pollution.
   (d) climate change.
   (e) invasive species.

2. Invasive species are
   (a) usually not a threat to biodiversity.
   (b) rare in island habitats.
   (c) successful due to a lack of natural enemies.
   (d) mostly specialist species.
   (e) often unable to compete effectively in the new environment.

3. Passenger pigeons were driven extinct primarily by
   (a) habitat loss.
   (b) overharvesting.
   (c) pollution.
   (d) climate change.
   (e) invasive species.

4. The Lacey Act
   (a) provides protected habitats for a number of threatened species.
   (b) forbids the interstate shipping of illegally harvested plants and animals.
   (c) provides harvesting quotas and prevents overharvesting.
   (d) prevents the spread of invasive species to the United States.
   (e) gives penalties for polluting ecosystems, especially water.

5. The primary impact of climate change on species diversity is expected to be
   (a) an increased number of extreme weather events.
   (b) an increased variability in weather.
   (c) decreased precipitation worldwide.
   (d) changes in available habitat because of changing temperatures.
   (e) the increased ability of species to disperse.

Answers to Module 60 AP® Review Questions

1. a
2. c
3. b
4. b
5. d

module

61

The Conservation of Biodiversity

It is important that we consider how to protect and increase biodiversity because of the large number of factors that can reduce biodiversity. There are two general approaches to conserving biodiversity: the single-species approach and the ecosystem approach. In this module, we explore each of these approaches.

Learning Objectives

After reading this module you should be able to

- identify legislation that focuses on protecting single species.
- discuss conservation efforts that focus on protecting entire ecosystems.

Conservation legislation often focuses on single species

The single-species approach to conserving biodiversity focuses our efforts on one species at a time. When a species declines significantly, the natural response is to encourage a population rebound by improving the conditions in which that species exists. This might be accomplished by providing additional habitat, reducing the harvest, or reducing the presence of a contaminant that is impairing survival or reproduction. When the population of a species has declined to extremely low numbers, sometimes the remaining few individuals will be captured and brought into captivity. Captive animals are bred with the intention of returning the species to the wild. A well-known example of captive breeding occurred with the California condor. As we discussed in Chapter 6, the condor had declined to a mere 22 birds in 1987. Thanks to captive breeding and several improvements in the condor's habitat, the population in 2013 was more than 400 birds. Programs such as these are a major function of zoos and aquariums around the world.

Marine Mammal Protection Act A 1972 U.S. act to protect declining populations of marine mammals.

The Marine Mammal Protection Act

In the United States, the single-species approach to conservation formed the foundation of the *Marine Mammal Protection Act* and the *Endangered Species Act*. The **Marine Mammal Protection Act** prohibits the killing of all marine mammals in the United States and prohibits the import or export of any marine mammal body parts. Only the U.S. Fish and Wildlife Service and the National Marine Fisheries Service are allowed to approve any exceptions to the act. The act was passed in 1972 in response to declining populations of many marine mammals, including polar bears, sea otters, manatees (*Trichechus manatus*), and California sea lions (*Zalophus californianus*) (FIGURE 61.1).

(a)

(b)

FIGURE 61.1 **Protected marine mammals.** The Marine Mammal Protection Act protects marine mammals in the United States from being killed. (a) Sea otter, (b) California sea lion. *(a: Hal Beral/V&W/The Image Works; b: Craig K. Lorenz/Science Source)*

AP® Exam Tip

Endangered species and the Endangered Species Act are covered on the 2000 AP® Exam, Question 3. To answer this question, students must

- identify one threatened or endangered species and describe the reasons it has declined
- describe three characteristics of organisms that could make them vulnerable to extinctions
- outline three arguments in favor of preserving biodiversity
- explain one U.S. federal law or one international treaty intended to protect species from extinction

The Endangered Species Act

In Chapter 10, we noted that the Endangered Species Act is a 1973 law designed to protect species from extinction. This act authorizes the U.S. Fish and Wildlife Service to determine which species can be listed as threatened species or *endangered species* and prohibits the harming of such species, including prohibitions on the trade of listed species, their fur, or their body parts. You might recall that earlier in this chapter we discussed the international definitions of threatened and near-threatened as used by the IUCN (see page 636): Threatened species have a high risk of extinction in the future and near threatened species are very likely to become threatened in the future. In the United States, an **endangered species** is defined as a species that is in danger of extinction within the foreseeable future throughout all or a significant portion of its range whereas a **threatened species** is defined as any species that is likely to become an endangered species within the foreseeable future throughout all or a significant portion of its range. As you can see, the U.S. definition of endangered is similar to the international definition for threatened and the U.S. definition of threatened is similar to the international definition of near-threatened.

The Endangered Species Act was first passed in 1973 and has been amended several times since then. From an international perspective, the act also implements the international CITES agreement that we discussed in the previous module. To assist in the conservation of threatened and endangered species, the act authorizes the government to purchase habitat that is critical to the conservation of these species and to develop recovery plans to increase the population of threatened and endangered species. This is often one of the most important steps in allowing endangered species to persist.

In 2013, the species that have been listed as threatened or endangered in the United States include 227 invertebrate animals, 394 vertebrate animals, and 815 plants. An additional 245 species are currently being considered for listing, a process that can take several years. Once listed, however, many threatened and endangered species have experienced stable or increasing populations. Indeed, some species have experienced sufficient increases in numbers to be removed from the endangered species list; these include the bald eagle (FIGURE 61.2), peregrine falcon, American alligator, and the eastern Pacific population of the gray whale (*Eschrichtius robustus*). Other species are currently increasing in number and may be taken off the list in the future. The gray wolf, for example, was reintroduced into Yellowstone National Park to help improve the species' abundance in the United States and it is now no longer endangered.

The Endangered Species Act has sparked a great deal of controversy in recent years because it permits restriction of certain human activities in areas where listed species live, including how landowners use their land. For example, some construction projects have been prevented or altered to accommodate threatened or endangered species. Organizations whose activities are restricted by the Endangered Species Act often try to pit the protection of listed species against the jobs of people in the region. In the

Endangered species A species that is in danger of extinction within the foreseeable future throughout all or a significant portion of its range.

Threatened species According to U.S. legislation, any species that is likely to become an endangered species within the foreseeable future throughout all or a significant portion of its range.

COMMON MISCONCEPTION
Once endangered, always endangered

Many students incorrectly believe that there are no success stories for endangered species. However, that is far from the truth. Endangered species that have been brought back from the edge of extinction include the bald eagle, peregrine falcon, grizzly bear, and red wolf. Many endangered species have seen population recoveries in recent years.

Teaching Tip: Activity

Have your students work in groups to create a 1-minute public service announcement that promotes an endangered species. The announcement should include suggestions on how to increase the population of the endangered species. Students should be sure to consider the following questions:

- Where does the endangered species live? What biome does it live in? Describe its habitat.
- Describe the species. What does it look like? What does it eat? What is its role in its native ecosystem?
- Does the endangered species have an economic benefit that the public should know about?
- Why is the species endangered?
- Is there a plan to help the species recover? What can be done?
- What are the economic and ecological pros and cons of recovery plans?

FIGURE 61.2 **Bringing back endangered species.** Habitat protection and reduced contaminants in the environment have allowed bald eagle populations to increase to the point where they could be taken off the Endangered Species List. *(Ron Niebrugge/Aurora Photos)*

1990s, for example, logging companies wanted to continue logging the old-growth forest of the Pacific Northwest. As we discussed earlier in this chapter, these forests are home to the threatened northern spotted owl and many other species that depend on old-growth forest. Automation had caused a large decline in the number of logging jobs over the preceding several decades, and many loggers perceived the Endangered Species Act as a further threat to their livelihood. They denounced the act because they said it placed more value on the spotted owl than it did on the humans who depended on logging. In the end, a compromise allowed continued logging on some of the old-growth forest while the rest became protected habitat.

During the past decade, several politicians have attempted to weaken the Endangered Species Act. However, strong support from the public and scientists has allowed it to retain much of its original power to protect threatened and endangered species. The biggest current challenge is a lack of sufficient funds and personnel required to implement the law.

The Convention on Biological Diversity

Protection of biodiversity is an international concern. In 1992, world nations came together and created the **Convention on Biological Diversity,** which is an international treaty to help protect biodiversity. The treaty had three objectives: conserve biodiversity, sustainably use biodiversity, and equitably share the benefits that emerge from the commercial use of genetic resources such as pharmaceutical drugs.

In 2002, the convention developed a strategic plan to achieve a substantial reduction in the worldwide rate of biodiversity loss by 2010. The nations that signed this agreement recognized both the instrumental and intrinsic values of biodiversity. In 2010, the convention evaluated the current trends in biodiversity around the world and concluded that the goal had not been met. They identified the following trends from 2002 to 2010:

- On average, species at risk of extinction have moved closer to extinction.
- One-quarter of all plant species are still threatened with extinction.
- Natural habitats are becoming smaller and more fragmented.
- The genetic diversity of crops and livestock is still declining.
- There is a widespread loss of ecosystem function.
- The causes of biodiversity loss have either stayed the same or increased in intensity.
- The ecological footprint of humans has increased.

Collectively, the message emerging from the convention is not very positive. From the perspectives of genetic diversity, species diversity, and ecosystem services, all of the trends during the 8-year period continue to move in the wrong direction.

Convention on Biological Diversity An international treaty to help protect biodiversity.

Some conservation efforts focus on protecting entire ecosystems

Awareness of a potential sixth mass extinction in which humans have played a major role has brought a growing interest in the ecosystem approach to conserving biodiversity. This approach recognizes the benefit of preserving particular regions of the world, such as biodiversity hotspots. Protecting entire ecosystems has been one of the major motivating factors in setting aside national parks and marine reserves. In some cases, these areas were originally protected for their aesthetic beauty, but today they are also valued for their communities of organisms. The amount of protected land has increased dramatically throughout the world since 1960. As an example of this increase, FIGURE 61.3 shows changes in the amount of protected land worldwide since 1900.

AP® Exam Tip

The pros and cons of protecting endangered species are covered on the 2003 AP® Exam, Question 4. To answer this question, students must

- name and describe two causes for the decline of the California condor and the whooping crane
- explain two measures that have been taken to protect these species
- describe two characteristics that would affect the ability of an endangered species to recover
- offer an ecological or economic argument in favor of protecting a particular endangered species and one ecological or economic argument against protecting it

Teaching Tip: Activity

In 1992, many nations met to create the Convention on Biological Diversity, which is an international treaty to help protect biodiversity. Later, in 2002, the convention developed a strategic plan to achieve a substantial reduction in the worldwide rate of biodiversity loss by 2010. In 2010, the convention evaluated the current trends in biodiversity around the world and concluded that the goal had not been met. Ask your students the following True/False questions:

- On average, species at risk of extinction have moved closer to extinction from 2002 to 2010. *True*
- One-half of all plant species are still threatened with extinction. *False. Only one quarter are threatened.*
- Natural habitats such as wetlands and rainforests are becoming slightly larger and less fragmented. *False. Natural habitats are becoming smaller and more fragmented.*
- The genetic diversity of crops and livestock is still declining. *True*
- The causes of biodiversity loss have either stayed the same or increased in intensity. *True*
- The ecological footprint of humans has decreased in the last 5 years. *False. The ecological footprint of humans is increasing.*

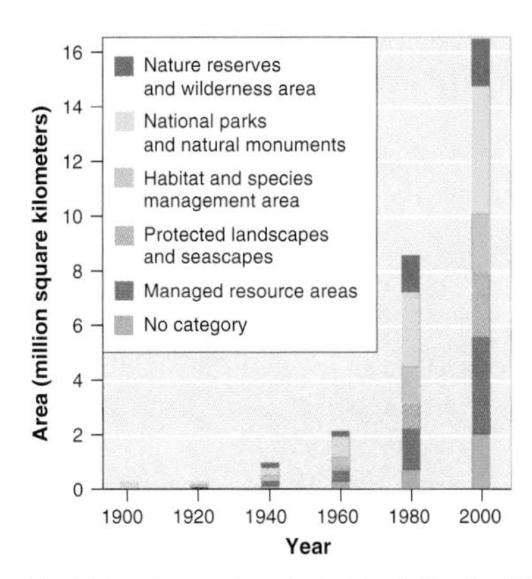

FIGURE 61.3 **Changes in protected land.** Since the 1960s, there has been a large increase in the amount of land that is under various types of protection throughout the world. *(Data from Global Biodiversity Outlook 2, Convention on Biological Diversity, 2006)*

When protecting ecosystems to conserve biodiversity, a number of factors must be taken into consideration including the size and shape of the protected area. We must also consider the amount of connectedness to other protected areas and how best to incorporate conservation while recognizing the need for sustainable habitat use for human needs.

The Size, Shape, and Connectedness of Protected Areas

A number of questions arise when we consider protecting areas of land or water. For example, how large should the designated area be? Should we protect a single large area or several smaller areas? Does it matter whether protected areas are isolated or if they are near other protected areas? To help us answer these questions, we can return to our discussion of the theory of island biogeography from Chapter 6.

As you may recall, the theory of island biogeography looks at how the size of islands and the distance between islands and the mainland affect the number of species that are present on different islands. Larger islands generally contain more species because they support larger populations of each species, which makes them less susceptible to extinction. Larger islands also contain more species because they typically contain more habitats and, therefore, provide a wider range of niches for different species to occupy. The distance between an island and the mainland, or between one island and another, is another crucial factor, since more species are capable of dispersing to close islands than to islands farther away.

Although the theory of island biogeography was originally applied to oceanic islands, it has since been applied to islands of protected areas in the midst of less hospitable environments. For example, we can think of all the state and national parks, natural areas, and wilderness areas as islands surrounded by environments subject to high levels of human activity, including agricultural fields, logged forests, housing developments, and cities (FIGURE 61.4). These areas provide habitats for species and places to stop and rest for migrating species. Applying the theory of island biogeography from this perspective gives us some idea of the best ways to design and manage protected areas. For example, when protected areas are far apart, it is less likely that species can travel among them. This means that when a species has been lost from one ecosystem, it will be harder for individuals of that species from other ecosystems to recolonize it. So when we create smaller areas, they should be close enough for species to move among them easily.

Decisions regarding the design of protected areas can also be informed by the concept of metapopulations. As we learned in Chapter 6, a metapopulation is a collection of smaller populations connected by occasional dispersal of individuals along habitat corridors. Each population fluctuates somewhat independently of the other populations and a population that declines or goes extinct, due to a disease for example, can be rescued by dispersers from a neighboring population. So if we set aside multiple protected areas, and recognize the need for connecting habitat corridors, a species is more likely to be protected from extinction by a decimating event such as a disease or natural disaster that could eliminate all individuals in a single protected area.

The concepts of island biogeography and metapopulations raise an interesting dilemma for conservation efforts. If we have limited resources to protect the biodiversity of a region, should we protect a single large area or several small areas? A single large area would support larger populations, but a species is more likely to survive a disease or natural disaster if it occupies several different areas. The debate over the best approach is known as SLOSS, which is an acronym for "single large or several small." While both approaches have their merits, in reality, human development and other factors often mean that only one of the two strategies is available. For example, due to human development of a region, there may simply not be a single large area available to protect, so the only available strategy is to protect several small areas. A final consideration regarding the size and shape of protected areas is the amount of *edge habitat* that an area contains.

◀ TEACHING with FIGURES

Show students Figure 61.3 and ask the following questions:

- How much did protected land increase from 1980 to 2000? *The amount of protected land increased from 8.5 million km² to 16 million km², an increase of 7.5 million km².*
- What were the two largest sectors of protected lands in the year 2000? *The largest two sectors were managed resource areas and national parks and natural monuments.*

Teaching Tip: Discussion Starter

Ask students: How does the theory of island biogeography help us decide how to protect species? *We can think of all the state and national parks, natural areas, and wilderness areas as islands surrounded by environments subject to high levels of human activity, including agricultural fields, logged forests, housing developments, and industrial parks. Applying the theory of island biogeography from this perspective gives us some idea of the best way to design and manage protected areas. For example, when protected areas are far apart, it is less likely that species can travel among them. This means that when a species has been lost from one ecosystem, it will be harder for individuals of that species from other ecosystems to recolonize it.*

Teaching Tip: Activity

In pairs or groups, have students use the words below to create two true statements and two false statements about conservation of biodiversity. Encourage them to use as many words on the page as they can, as well as others as needed. Have groups share their statements with the class. Award "bonus points" to the groups that use the most terms.

- extinction
- inbreeding
- endangered
- native species
- alien species
- exotic species
- Lacey Act
- CITES
- Red List
- Marine Mammal Protection Act
- Endangered Species Act
- Convention on Biological Diversity
- edge habitat
- biosphere reserves

Sample answers:

- *Sometimes invasive species can overpower native species and increase their chances of extinction. (True)*
- *The Lacey Act was first passed in 1900 and it later created the Red List, which is a list of the most threatened species found in the United States. (False)*

FIGURE 61.4 **Habitat islands provide habitat.** Central Park is an extreme example of an island of hospitable habitat surrounded by an urban environment that is not hospitable to most species. *(Michael S. Yamashita/National Geographic Stock)*

Edge habitat occurs where two different communities come together, typically forming an abrupt transition, such as where a grassy field meets a forest. While some species will live in either field or forest, others, like the brown-headed cowbird, specialize in living at the forest edge. So another challenge of protecting many small areas is the comparatively larger amount of edge habitat. When we protect several small forests, for example, the proliferation of species such as the cowbird in this larger amount of edge habitat can have a detrimental effect on songbirds that typically live farther inside a forest.

Biosphere Reserves

In Chapter 10 we saw that managing national parks and other protected areas so they serve multiple users can be a challenge. While we want to make places of great natural beauty available to everyone, when large numbers of people use an area for recreation, at least some degradation is very likely. To address this problem, the United Nations Educational, Scientific and Cultural Organization (UNESCO) developed the innovative concept of *biosphere reserves*. **Biosphere reserves** are protected areas consisting of zones that vary in the amount of permissible human impact. These reserves protect biodiversity without excluding all human activity. FIGURE 61.5 shows the different zones in a hypothetical biosphere reserve. The central core is an area that receives minimal human impact and is therefore the best location for preserving biodiversity. A buffer zone encircles the core area. Here, modest amounts of human activity are permitted, including tourism, environmental education, and scientific research facilities. Farther out is a transition area containing sustainable logging, sustainable agriculture, and residences for the local human population.

Designing reserves with these three zones represents an ideal scenario. In reality, biosphere reserves can take many forms depending on their location, though all attempt to maintain low-impact core areas. As of 2013 there are 610 biosphere reserves worldwide—47 in the United States—with a total of 117 nations participating.

Edge habitat Habitat that occurs where two different communities come together, typically forming an abrupt transition, such as where a grassy field meets a forest.

Biosphere reserve Protected area consisting of zones that vary in the amount of permissible human impact.

Teaching Tip: Discussion Starter

Ask students: What is a biosphere reserve and how does it help preserve biodiversity? *Biosphere reserves are protected areas consisting of zones that vary in the amount of permissible human impact. These reserves protect biodiversity without excluding all human activity. They help preserve biodiversity because they have core areas where no human contact can occur except to monitor or study the ecosystem.*

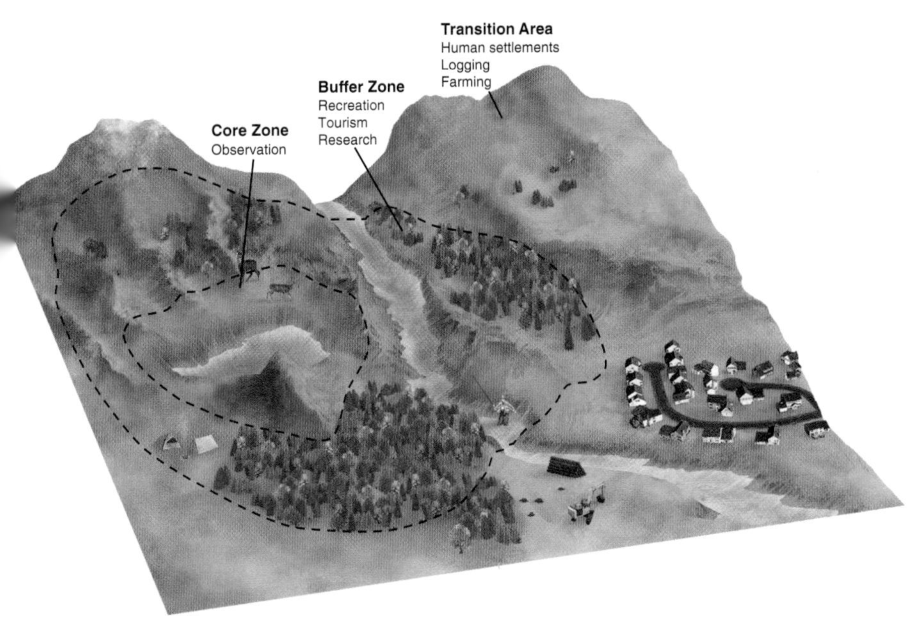

FIGURE 61.5 **Biosphere reserve design.** Biosphere reserves ideally consist of core areas that have minimal human impact and outer zones that have increasing levels of human impacts.

One well-known biosphere reserve is Big Bend National Park in Texas. The park itself serves as the core area and receives relatively little human impact, although hikers are permitted to walk through the beautiful desert landscapes and tree-covered mountain peaks. The park contains several dozen threatened and endangered plant and animal species. It also contains more than 1,000 species of plants and 400 species of birds, many of which pass through the Big Bend region during their annual spring and fall migrations. Outside the boundaries of the park is a region of increased human impact including tourist facilities, human settlements, and agriculture (FIGURE 61.6).

FIGURE 61.6 **Biosphere reserve.** Big Bend National Park, located in southwest Texas, serves as a low-impact core area of the Big Bend biosphere reserve. *(Tom Till/Alamy)*

Teaching Tip: Beyond the Classroom

In the United States there are 47 UNESCO biosphere reserves. These reserves ideally consist of a core zone designated as an area of minimal human impact and outer zones that have increasing levels of human impact. Have each student pick one of the 47 biosphere reserves and write a research paper. Ask students to answer the following questions:

- Where is the reserve located? What biome is it in?
- How large is the reserve?
- What is the history of the reserve?
- What are five species of plants or animals that are important to maintaining the biodiversity of the reserve?
- What are some ways the reserve is being monitored? Who is responsible for monitoring?
- How large is the core area designated for minimal human impact?
- How many people visit the reserve in a given year?

Find a link to a UNESCO map of the United States with information about biosphere reserves by clicking on the link buttons in the TE-book, opening the TRFD, or logging onto the book's companion website bcs.whfreeman.com/friedlandapes2e (teacher login required).

 **Chapter 18 Web Resources**

module

61

REVIEW

In this module, we learned that efforts to conserve biodiversity focus on either single species or entire ecosystems. The single-species approach is often the approach taken by conservation legislation. In the United States, single-species legislation includes the Marine Mammal Protection Act and the Endangered Species Act. Internationally, there are treaties that protect species, including CITES and the Convention on Biological Diversity.

A number of conservation efforts have focused on protecting entire ecosystems by considering the concepts of island biogeography and metapopulations. To reach a compromise between complete protection, on the one hand, and human use of habitats, on the other, scientists have developed the concept of biosphere reserves in which core areas receive greater protection while outer areas are allowed to have sustainable impacts.

Module 61 AP® Review Questions

1. The Marine Mammal Protection Act
   (a) allows states to make exceptions regarding the killing of marine mammals.
   (b) was passed primarily to protect whales.
   (c) prohibits the killing of all marine mammals.
   (d) allows the import of marine mammal body parts.
   (e) protects sharks as well as marine mammals.

2. Which is NOT true of the Endangered Species Act?
   (a) It is an example of the single-species approach to conservation.
   (b) It prohibits the hunting or harvesting of some listed species.
   (c) It includes the development of recovery plans for listed species.
   (d) It uses a different definition for threatened than the IUCN.
   (e) It has resulted in the delisting of several species after successful population growth.

3. Problems with protecting many small habitats include
   I. increased proportions of edge habitats.
   II. increased dispersal between populations.
   III. the need for corridors between some protected species.

   (a) I only
   (b) I and III only
   (c) II only
   (d) II and III only
   (e) III only

4. Which is a trend identified by the Convention on Biological Diversity between 2002 and 2010?
   (a) Over half of threatened species have moved away from extinction.
   (b) Very few plant species are at risk of extinction.
   (c) The genetic diversity of crops is increasing.
   (d) Marine species are affected most by recent biodiversity losses.
   (e) The human ecological footprint has increased.

5. According to the theory of island biogeography
   (a) when conservation areas are close to each other, more species will persist.
   (b) species on islands far from the mainland are at the least risk of extinction.
   (c) multiple small conservation areas will protect species better than one large area of the same size.
   (d) conservation areas should be connected with corridors to increase migration.
   (e) edge habitat is important to protect for increased diversity.

Answers to Module 61 AP® Review Questions

1. c
2. b
3. b
4. e
5. a

working toward sustainability

Swapping Debt for Nature

Preserving biodiversity is expensive. A case in point is the money required to set aside terrestrial or aquatic areas for protection. As an example, if the land is privately owned, it must be purchased. Indirect costs can also be high. Not using the land, water, or other natural resources—such as wood materials, metals, and fossil fuels—results in lost income. Finally, the costs of maintaining the protected area can be prohibitive, ranging from monitoring the biodiversity to hiring guards to prevent illegal activities such as poaching. Given the fact that preserving biodiversity is expensive, how can the developing nations of the world, which contain so much biodiversity but have such little wealth, afford it?

In 1984, Thomas Lovejoy from the World Wildlife Fund came up with an idea that would help protect large areas of land but at the same time improve the economic conditions of developing countries. Lovejoy observed that developing nations possessed much biodiversity but were often deep in debt to wealthier, developed countries. Developing countries borrowed large amounts for the purpose of improving economic conditions and political stability. While the developing countries were slowly repaying their loans with interest, some had fallen so far behind on these payments that it seemed unlikely the loans would ever be repaid in full. These debtor countries had little money left over for investment in an improved environment after they had paid their loans to developed countries. Lovejoy considered the possibility that the wealthy countries might be willing to let debtor nations swap their debt in exchange for investing in the conservation of the biodiversity of the debtor nations.

The "debt-for-nature" swap has been used several times in Central and South America. In these swaps, the United States government and prominent environmental organizations provide cash to pay down a portion of a country's debt to the United States. The debt is then transferred to environmental organizations within that country with the debtor government making payments to the environmental organizations rather than to the United States. This does not mean that the country is out of debt, just that it now sends its loan payments to the environmental organizations for the purpose of protecting the country's biodiversity. In short, the indebted country switches from sending its money out of the country to investing in its own environmental conservation.

One of the largest debt-for-nature swaps recently happened in the Central American country of Guatemala. The United States government paired with two conservation organizations to provide $17 million to Guatemala. Over a period of 15 years, this amount, with interest, would have grown to more than $24 million, or about 20 percent of Guatemala's debt to the United States. In exchange, Guatemala agreed to pay $24 million over 15 years to improve conservation efforts in four areas of the country, including the purchase of land, the prevention of illegal logging, and future grants to conservation organizations helping to document and preserve the local biodiversity. The four areas include two ecosystems—mangrove forests and tropical forests. Each forms a core area within a biosphere reserve that contains a large number of rare and endangered species including the jaguar (*Panthera onca*). More than twice the size of Yellowstone National Park in the United States, this reserve offers important protection to biodiversity while also preserving historic Mayan temples that are part of Guatemala's cultural heritage and allowing sustainable use of some of the forest by local people.

Since the program began in 1998, the United States has used the debt-for-nature swap to protect tropical forests in 15 countries from Central America to the Philippines. To take part in the swap program, the countries are required to have a democratically elected government, a plan for improving their economies, and an agreement to cooperate with the United States on issues related to combating drug trafficking and terrorism. The results of these agreements have been encouraging. In Belize, for example, a debt-for-nature swap allowed 9,300 ha (23,000 acres) to be protected and an

Swapping debt for nature in Guatemala. The Maya Biosphere Reserve is one of four areas of Guatemala that will be better protected under an agreement between the governments of the United States and Guatemala as well as several conservation organizations. *(Rob Crandall/The Image Works)*

Suggested Answers to Critical Thinking Questions

1. One goal of a debt-for-nature program is to help struggling economies grow stronger through debt relief. If the government does not have a viable plan for improvement, the nation may quickly sink back into debt.

2. The debt-for-nature program protects areas that have high biodiversity.

additional 109,000 ha (270,000 acres) to be managed for conservation. In Peru, a $10.6 million debt-for-nature swap led to the protection of more than 11 million ha (27 million acres) of tropical forest. Although these arrangements are only currently being applied to tropical forests, there is no inherent reason that this unique, modern-day conservation strategy would not also work in many other developing countries around the world.

Critical Thinking Questions

1. In debt-for-nature swaps, why might the United States require that developing countries receiving such assistance have a plan for improving their economies?
2. How might the debt-for-nature program promote the goals of the Convention on Biological Diversity?

References

How debt-for-nature swaps protect tropical forests. The Nature Conservancy. http://www.nature.org/ourinitiatives/regions/centralamerica/guatemala/guatemala-debt-for-nature-swap-is-a-win-for-tropical-forest-conservation.xml.

Lacey, M. 2006. U.S. to cut Guatemala's debt for not cutting trees. *New York Times*, October 2. http://www.nytimes.com/2006/10/02/world/americas/02conserve.html.

U.S.-Brazil debt for nature swap to protect forests. August 12, 2010. http://www.bbc.co.uk/news/world-latin-america-10958695.

chapter 18 REVIEW

In this chapter, we examined the state of the world's biodiversity. We learned that the genetic diversity of many wild and domesticated populations has declined substantially over the past century. In addition, the species diversity of most major taxonomic groups has also declined, with large proportions of birds, mammals, and amphibians being threatened or near-threatened. These declines in species diversity also lead to declines in the intrinsic and instrumental values of ecosystems. The causes underlying these declines in diversity are wide-ranging and may include any combination of habitat loss, intrusion of exotic species, overharvesting, pollution, and climate change. While legislation to reverse these declines has focused on single species, conservation efforts have applied the concepts of metapopulations, island biogeography, and biosphere reserves to protect large areas of habitat and thereby protect large ecosystems.

Key Terms

Threatened species (IUCN)
Near-threatened species
Least concern species
Intrinsic value
Instrumental value
Provision
Native species
Exotic species
Alien species
Invasive species
Lacey Act
Convention on International Trade in Endangered Species of Wild Fauna and Flora (CITES)
Red List
Marine Mammal Protection Act
Endangered species
Threatened species (U.S.)
Convention on Biological Diversity
Edge habitat
Biosphere reserve

Learning Objectives Revisited

Module 59 The Sixth Mass Extinction

- **Explain the global decline in the genetic diversity of wild species.**

  Declines in the abundance of individuals in a population can lead to reductions in genetic diversity that cause inbreeding depression. Inbreeding depression can cause offspring to inherit two copies of a harmful mutation and experience reduced survival and reproduction.

- **Discuss the global decline in the genetic diversity of domesticated species.**

  Humans have bred a wide variety of domesticated plants and animals, but in recent decades farmers have focused on the most productive varieties and many of the other varieties have disappeared over time. Such reductions in genetic diversity limit the options available to respond to new diseases or changing environmental conditions.

- **Identify the patterns of global decline in species diversity.**

  Of the estimated 10 million species on Earth, only about 50,000 have been assessed to determine whether their populations are increasing, stable, or declining. In examining those groups with the most complete data, scientists have found that every group has a substantial percentage of species that are threatened or near-threatened.

- **Explain the values of ecosystems and the global declines in ecosystem function.**

  Ecosystems can have intrinsic values, which are independent of any benefit to humans, or they can have instrumental values, which provide a benefit to humans and can be assigned a monetary value. Instrumental values include provisions, regulating services, support systems, resilience, and cultural services. Recent assessments of ecosystem function have found that more than half of those assessed are either declining or used at a rate that cannot be sustained.

Module 60 Causes of declining biodiversity

- **Discuss how habitat loss can lead to declines in species diversity.**

  The loss of habitat means that fewer individuals can be sustained in the habitat that remains. Smaller populations can then suffer from inbreeding depression. A reduction in habitat can also prevent the normal migration of species to important seasonal habitats and cause increased interactions with other species that have negative effects.

- **Explain how the movement of exotic species affects biodiversity.**

  Exotic species are those that are moved to new parts of the world where they are not native. Some of these species spread rapidly in their new locations and cause the demise of native species either as competitors, predators, herbivores, or pathogens.

- **Describe how overharvesting causes declines in populations and species.**

  Overharvesting plants and animals at rates that exceed the production of new individuals can cause population declines and even extinctions. Many of these extinctions have occurred due to unregulated harvesting in the past. However, in most parts of the world, governments have imposed harvest regulations to ensure that harvests occur in a sustainable manner.

- **Understand how pollution reduces populations and biodiversity.**

  Some pollutants can have direct lethal effects on species. Many other pollutants, however, can have sublethal effects that prevent or inhibit reproduction or alter ecosystems in ways that indirectly harm species.

- **Identify how climate change affects species diversity.**

  Climate change has the potential to alter the distribution of environmental conditions around the world. When conditions change and species are unable to move to more hospitable conditions, scientists predict that these species will either decline in abundance or go extinct.

Teaching Tip: Measuring Your Impact

How Large Is Your Home?

Measuring Your Impact 18 asks students to consider how logging affects habitat loss by having students calculate how much lumber is used for an average house in the United States. The link to this resource may be found by clicking on the link buttons in the TE-book, opening the TRFD, or logging onto the book's companion website bcs.whfreeman.com/friedlandapes2e (teacher login required).

Measuring Your Impact 18: How Large Is Your Home?

Exploring the Literature

Convention on Biological Diversity. 2010. *Global Biodiversity Outlook 3.* http://www.cbd.int/gbo/gbo3/doc/GBO3-Summary-final-en.pdf

Rosenthal, E. 2005. Food for thought: Crop diversity is dying. *New York Times*, August 18. http://www.nytimes.com/2005/08/17/world/europe/17iht-food.html

Download a printable version of this list by clicking the link buttons in the TE-book, opening the TRFD, or logging onto the book's companion website bcs.whfreeman.com/friedlandapes2e (teacher login required).

TRM **Exploring the Literature 18**

Answers to Chapter 18 AP® Exam Multiple-Choice Questions

1. e
2. b
3. e
4. d
5. a
6. c
7. b
8. e
9. d
10. a
11. b

- **Identify legislation that focuses on protecting single species.**

  The primary pieces of legislation in the United States to protect species are the Marine Mammal Protection Act and the Endangered Species Act. Internationally, nations created the Convention on Biological Diversity in order to conserve biodiversity, to use biodiversity sustainably, and to share equitably the benefits that emerge from the commercial use of biodiversity.

- **Discuss conservation efforts that focus on protecting entire ecosystems.**

  There has been a continual increase in the amount of aquatic and terrestrial habitats that have been protected around the world. When preserving such habitats, scientists consider the size, shape, and connectedness of these habitats as well as the presence of edge habitats. They have also incorporated the need to balance human use and habitat protection by designing biosphere reserves.

Chapter 18 AP® Environmental Science Practice Exam

Section 1: Multiple-Choice Questions

Choose the best answer for questions 1–11.

1. Which is a cause of declining global biodiversity?
   I. Pollution
   II. Habitat loss
   III. Overharvesting

   (a) I
   (b) I and II
   (c) I and III
   (d) II and III
   (e) I, II, and III

2. Which statement about global biodiversity is correct?
   (a) Species diversity is decreasing but genetic diversity is increasing.
   (b) Species diversity is decreasing and genetic diversity is decreasing.
   (c) Species diversity is increasing but genetic diversity is decreasing.
   (d) Declines in genetic diversity are occurring in wild plants but not in crop plants.
   (e) Declines in genetic diversity are occurring in crop plants but not in wild plants.

3. Which group of animals is declining in species diversity around the world?
   I. Fish and amphibians
   II. Birds and reptiles
   III. Mammals

   (a) I
   (b) I and II
   (c) I and III
   (d) II and III
   (e) I, II, and III

4. Which of the following species was historically overharvested?
   (a) Brown-headed cowbird
   (b) Honeybee
   (c) Kudzu vine
   (d) Dodo bird
   (e) Zebra mussel

5. Which statement is NOT correct regarding the genetic diversity of livestock?
   (a) The use of only the most productive breeds improves genetic diversity.
   (b) Livestock come from very few species.
   (c) The genetic diversity of livestock has declined during the past century.
   (d) Different breeds are adapted to different climatic conditions.
   (e) Different breeds are adapted to resist different diseases.

6. Which statement is NOT correct about invasive exotic species?
   (a) Their populations grow rapidly.
   (b) They often have no major predators or herbivores.
   (c) Most introduced species become established in new regions.
   (d) A well-known invasive exotic plant is the kudzu vine.
   (e) A well-known invasive exotic animal is the zebra mussel.

Additional Multiple-Choice AP® Practice Exam

Download an additional Multiple-Choice AP® Environmental Science Practice Exam for Chapter 18 by clicking on the link buttons in the TE-book, opening the TRFD, or logging onto the book's companion website bcs.whfreeman.com /friedlandapes2e (teacher login required).

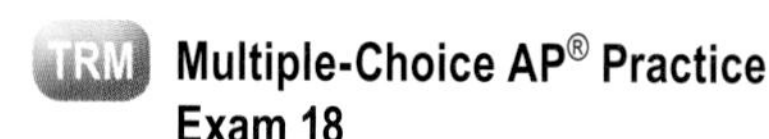

Answers to Chapter 18 AP® Exam

Download the full answers to the Chapter 18 AP® Exam by clicking on the link buttons in the TE-book, opening the TRFD, or logging onto the book's companion website bcs.whfreeman.com /friedlandapes2e (teacher login required).

Answers to Chapter 18 AP® Exam

7. Which is an example of the single-species approach to conservation?
   I. The Endangered Species Act
   II. The Marine Mammal Protection Act
   III. The Biosphere Reserve

   (a) I
   (b) I and II
   (c) I and III
   (d) II and III
   (e) I, II, and III

8. Which is NOT an example of how the Endangered Species Act can affect human activities?
   (a) It has been used to prevent new construction.
   (b) It has prevented logging in particular areas.
   (c) It prevents the killing of listed species.
   (d) It prevents the trade of any parts of listed species.
   (e) It prevents human use of biosphere reserves.

9. Which principle of island biogeography is NOT correct?
   (a) A larger protected area should contain more species.
   (b) Protected areas that are closer together should contain more species.
   (c) National parks can be thought of as islands of biodiversity.
   (d) A larger protected area will have fewer habitats.
   (e) Marine reserves can be thought of as islands of biodiversity.

10. In a biosphere reserve
   (a) sustainable agriculture and tourism are permitted in different zones.
   (b) human activities are allowed throughout the reserve.
   (c) human activities are restricted to the central core of the reserve.
   (d) no human activities are permitted in a biosphere reserve.
   (e) sustainable agriculture is permitted, but tourism is not.

11. Which statement is correct regarding swapping debt for nature?
   (a) Protecting land and water is typically not expensive.
   (b) Developing countries can pay part of their debt by investing in their own environment.
   (c) Developing countries pay their debt to the United States by investing in U.S. national parks.
   (d) Having a plan to improve the economy of a developing country is not important.
   (e) The only expense of protecting biodiversity is the purchase of an area.

Section 2: Free-Response Questions

Write your answer to each part clearly. Support your answers with relevant information and examples. Where calculations are required, show your work.

1. The conservation of biodiversity is an international problem.
   (a) Name and describe one U.S. law that is intended to prevent the extinction of species. (4 points)
   (b) Name and describe one international treaty that is intended to prevent the extinction of species. (4 points)
   (c) Explain the benefits of taking an ecosystem approach, as opposed to a single-species approach, to conserving biodiversity. (2 points)

2. Tropical rainforests are home to a tremendous diversity of species. You have been asked to develop a plan to protect this diversity.
   (a) Describe the advantages and disadvantages of protecting a single large area versus several small areas. (2 points)
   (b) How might increasing the amount of edge habitat affect species that typically live deep in the forest? (3 points)
   (c) Discuss the merits of preserving individual species that are threatened and endangered versus preserving the function of the ecosystem. (3 points)
   (d) Describe three characteristics of organisms that would make them particularly vulnerable to extinction. (2 points)

Answers to Chapter 18 AP® Exam Free-Response Questions

1. (a) The Endangered Species Act was enacted in 1973 and forbids government and private citizens from taking actions that would destroy endangered species or their habitats. It also prohibits trade in products made from these species. Another example is the Marine Mammal Protection Act of 1972, which forbids killing of marine mammals or importing their body parts.
(b) The United Nations Convention on International Trade in Endangered Species of Wild Fauna and Flora (CITES) controls the international trade of threatened plants and animals.
(c) The ecosystem approach sets aside protected areas as a means of conserving biodiversity. Factors to be considered are size, shape, and connectedness to other protected areas. Single-species approaches are focused on increasing the abundance of a particular species but these efforts often do not protect the entire biodiversity found in an ecosystem.

2. (a) The advantage of protecting a single large area is that the area will have more species and support larger populations of each species, both of which protect species from extinction. Larger areas also contain more species because they typically offer a greater variety of habitats and, therefore, provide a wider range of niches for different species to occupy. A disadvantage of protecting a single large area is that an event such as a disease or natural disaster could eliminate all individuals in the area.
(b) Increasing the amount of edge habitat will affect species that live deep in the forest, because some animals that thrive in the edge habitat can have detrimental effects on species living deep in the forest.
(c) Many people believe it is beneficial to focus conservation on specific endangered species to preserve biodiversity. However, awareness of a potential sixth mass extinction in which humans have played a major role has brought a growing interest in the ecosystem approach to conserving biodiversity. This approach recognizes the benefit of preserving particular regions of the world, such as biodiversity hotspots. Protecting entire ecosystems has been one of the major motivating factors in setting aside national parks and marine reserves.
(d) Possible answers may include:

- low reproduction rate (panda, polar bear)
- specialized feeding habits (black-footed ferret eats only prairie dogs)
- feeds at high trophic levels (tiger)
- large size (rhino, various species of whales)
- limited or specialized nesting or breeding areas (red-cockaded woodpecker)
- found in only one place (elephant seal)

chapter

19 Global Change

PD Overview

This chapter examines how humans have altered the world's climate. It explores the underlying causes of global change and their consequences. In this chapter, students have the opportunity to apply many of the themes developed throughout the course, including the interconnectedness of the systems on Earth, environmental indicators, and the interaction of environmental science and policy. This chapter provides several opportunities for reviewing the biogeochemical cycles covered in Chapter 3.

Module 62: Global Climate Change and the Greenhouse Effect

This module considers the distinctions among global change, global climate change, and global warming. It then explores the processes that underlie changes in global climates and, more specifically, global temperatures.

Module 63: The Evidence for Global Warming

This module explores the evidence that an increased concentration of greenhouse gases is causing Earth to become warmer. If Earth is becoming warmer, we can begin to evaluate whether or not this warming is caused by human activities that have released greenhouse gases. One way to make these assessments is by determining gas concentrations and temperatures from the past and comparing them to gas concentrations and temperatures in the present day. We can also use information about changes in gas concentrations and temperatures to predict future climate conditions. Finally, the module examines how greenhouse gas concentrations have changed over time and how these changes are linked to global warming.

Module 64: The Consequences of Global Climate Change

Numerous environmental indicators demonstrate that global warming is affecting global processes and contributing to overall global change. In many cases, we have clear evidence of how global warming is having an effect. In other cases, we can use climate models to make predictions about future changes. As with all predictions of the future, there is considerable uncertainty regarding the future effects of global warming. This module discusses how global warming is expected to affect the environment and organisms living on Earth.

Alignment to AP® Environmental Science Course Description

AP® Outline	Module	
VII. Global Change (10–15%)		
B. Global Warming	Module 62	Global Climate Change and the Greenhouse Effect
	Module 63	The Evidence for Global Warming
	Module 64	The Consequences of Global Climate Change

Chapter Learning Objectives

After completing this chapter students will be able to

- **distinguish among global change, global climate change, and global warming.**
- **explain the process underlying the greenhouse effect.**
- **identify the natural and anthropogenic sources of greenhouse gases.**
- **explain how CO_2 concentrations have changed over the past 6 decades and how emissions compare among the nations of the world.**
- **explain how temperatures have increased since records began in 1880.**
- **discuss how we estimate temperatures and levels of greenhouse gases over the past 500,000 years and into the future.**

- explain the role of feedbacks on the impacts of climate change.
- discuss how global climate change has affected the environment.
- explain how global climate change has affected organisms.
- identify the future changes predicted to occur with global climate change.
- explain the global climate change goals of the Kyoto Protocol.

Chapter 19 Pacing Guide

This pacing guide is based on a schedule with 120 sessions of 50 minutes each before the AP® Exam. If you have a different number of sessions before the exam, you can modify the pacing to suit your needs. If you have additional time, consider incorporating quizzes, released AP® Environmental Science free-response and multiple-choice questions, or additional activities.

Module	Standard Schedule Days	Block Schedule Days
Module 62	2	1
Module 63	2	1
Module 64	1	½
Assessment	1	½

Chapter 19 Resources

The link to these resources can be found by clicking on the link buttons in the Teacher's e-Book (TE-book), opening the Teacher's Resource Flash Drive (TRFD), or logging onto the book's companion website bcs.whfreeman.com/friedlandapes2e (teacher login required).

- PowerPoint Presentation 19
- Optimized Art PowerPoint and JPEG Files 19
- Do the Math Videos
- Measuring Your Impact 19: Carbon Produced by Different Modes of Travel
- Lab 19.1: Climate Change
- Chapter 19 Web Resources
- Exploring the Literature 19
- Answers to Chapter 19 AP® Exam
- Multiple-Choice AP® Practice Exam 19

Free-Response Questions from Previous AP® Environmental Science Exams

Free-response questions from prior AP® Environmental Science Exams are available on the AP® course website: https://apstudent.collegeboard.org/apcourse/ap-environmental-science/exam-practice?envsci. Students should be able to answer all of the questions listed below with material learned in this and previous chapters. When a question requires students to understand material from multiple chapters, the question will be listed in the last chapter required to complete the entire question. Questions marked with an asterisk (*) are from exams with released multiple-choice questions. You may want to save these questions until the end of the year so you can give your students a complete released exam for practice. Questions marked with double asterisks (**) require math to calculate a problem. Look for references to these questions throughout the chapter.

Year	Question	Content
2006	2 **	• Evaluation of historical data on temperature and CO_2 concentrations • Anthropogenic causes of increased greenhouse gasses
2009	3	• Costs and benefits of damming a river • Strategies for reducing agricultural water consumption • Environmental consequences of climate change on the hydrology or water cycle of the Colorado River
2010	4**	• Relationship of sea level and global mean atmospheric temperature • Environmental impact of a rise in sea level with respect to estuarine ecosystems • Economic impact of rising sea level on human populations living along coastlines

PD Chapter 19 Overview

Watch the video overview of Chapter 19 (for teachers) by clicking on the link buttons in the TE-book, opening the TRFD, or logging onto the book's companion website bcs.whfreeman.com/friedlandapes2e (teacher login required).

TRM PowerPoint Presentation 19

Download the PowerPoint presentation for Chapter 19 by clicking on the link buttons in the TE-book, opening the TRFD, or logging onto the book's companion website bcs.whfreeman.com/friedlandapes2e (teacher login required).

The polar ice cap currently melts 3 weeks earlier than it did 30 years ago, leaving polar bears less time to hunt for seals.
(Steven J. Kaziowski/GHG/Aurora Photos)

chapter 19

Global Change

Walking on Thin Ice

The polar bear is one of the best-known species of the Arctic. With a geographic distribution surrounding the North Pole, the polar bear plays an important role in this frozen ecosystem. Polar bears are voracious predators that specialize in eating several species of seals. The seals live under large expanses of ocean ice and come up for air where there are holes in this ice. The bears roam the ice in search of holes and pounce on seals when they come to the surface. In many cases, the bears will consume only seal blubber, which is a concentrated source of the energy critical for an organism living in such a cold environment. The portion of the seal carcass that remains is a significant food source for other animals, including the Arctic fox (*Vulpes lagopus*). The polar bear is also important to the indigenous people of the area who have long depended on its meat for food and its fur for clothing.

Over the last few decades, temperatures in the Arctic have risen much faster than in other parts of the world. Warmer air and ocean water have caused the polar ice cap to melt, raising concern among both scientists and the general public because of both the popular appeal of the polar bear and its importance to the ecosystem and native peoples.

> While the plight of the polar bear has drawn attention to the effects of global warming, it is only one of many indicators that our world is rapidly changing because of human activity.

Scientists have been taking satellite photos of the polar ice cap for over 30 years. In 1978, photos revealed that the ice cap extended from Russia and Norway to Greenland, Canada, and Alaska. Over the years, however, the ice cap has shrunk and retreated away from the land. During each year, a natural cycle of shrinking and expansion of the polar ice cap occurs; it becomes smaller during the summer and larger during the winter. Because of the rise in Arctic temperatures, the amount of summer shrinking of the ice has increased. To measure the extent of this shrinkage, scientists compare yearly measurements taken at the end of each summer. Compared with the average amount of ice present from 1979 to 2000, scientists found that there was 45 percent less ice each year from 2006 to 2012—a total area twice the size of Texas. In addition, the remaining ice is considerably thinner, making it more vulnerable to future melting. In 2013,

Teaching Tip: Chapter Opening Case

The chapter opening case, "Walking on Thin Ice," introduces students to the themes of this chapter using the example of polar bears. This is a good time to remind students that Earth's climate is a global system and that the actions of one nation can affect climate conditions many thousands of miles away.

Teaching Tip: Beyond the Classroom

Have your students research changing conditions at the North and South Poles. The National Snow and Ice Data Center provides current news and an overview of conditions at the poles. Have students look at the most current sea ice extent data and current trends. Ask them to write a paragraph describing how the sea ice extent has changed in the Arctic and Antarctic over the past 20 years.

Find a link to the NSIDC website by clicking on the link buttons in the TE-book, opening the TRFD, or logging onto the book's companion website bcs.whfreeman.com/friedlandapes2e (teacher login required).

Chapter 19 Web Resources

Teaching Tip: Discussion Starter

Ask students: How does global climate change threaten polar bears? *Global climate change has caused the Arctic sea ice to melt earlier in the year. This affects food availability, habitat, and breeding.*

the area of the polar ice cap at the end of summer was a bit larger, but still smaller than observed in past decades. In fact, the 10 years with the smallest areas of summer ice have all occured during the last 10 years.

As the polar ice cap melts, the polar bears have been losing habitat. During the summer, when the ice cap becomes unusually small and ice retreats far away from the land, polar bears can no longer reach the ice to hunt for seals. Today, the sea ice melts 3 weeks earlier than it did 30 years ago. Because of the shorter time they are able to hunt for seals, male polar bears near Hudson Bay in Canada currently weigh 67 kg (150 pounds) less than they weighed 30 years ago. Moreover, the population of polar bears living on the western side of Hudson Bay has declined by 10 percent during the past decade.

The status of polar bear populations is not easy to assess given their remote locations around the world. Of the 19 distinct populations that live in the Arctic, researchers at the International Union for the Conservation of Nature found that one is increasing, three are stable, and eight are decreasing; the remaining seven populations lack sufficient long-term data to determine their status. When considering all populations combined, scientists are predicting a 30 percent decline in the number of polar bears in the next 25 to 40 years.

A sharp decline or extinction of polar bears would have a wide range of effects on the ecosystem and on the indigenous people in the area. Seal populations could increase with the demise of their major predator, while other species, such as the Arctic fox, could decline because there would be fewer seal carcasses for them to feed on. In 2014, researchers working along the Hudson Bay found that the bears had begun to compensate for the reduced feeding on seals by changing their diets in the late summer; they started feeding more on snow geese, snow geese eggs, caribou, and even some plants. These changes in feeding patterns, in turn, should cause reverberations throughout the Arctic food web. Indigenous people would also be affected by polar bear declines, not only in terms of the food and clothing that the polar bear provides, but also from the perspective of their cultural and social identity. As the ecosystem changes with the decline of the polar bear, people of the Arctic may find their lives and livelihoods altered.

In 2008, the United States classified polar bears as a threatened species because of the decline in their ice habitat. If current conditions persist, the species will continue to decline. In 2009, the five nations with polar bear populations (Canada, Greenland, Norway, Russia, and the United States) agreed that the polar bear should be classified as threatened throughout its entire global range. While acknowledging that pollution and hunting contributed to the bears' bleak future, the nations agreed that the effect of global warming on the ice cap poses the greatest threat to polar bears. This means that the solution to the problem is not simply in the hands of those few nations that contain polar bear habitats, but is in the hands of everyone on Earth who contributes to global warming. In 2013, the United States and Russia went a step further by announcing an agreement to push for a CITES rule to ban the international trade in all products from polar bears including furs and teeth. While the plight of the polar bear has drawn attention to the effects of global warming, it is only one of many indicators that our world is rapidly changing because of human activity.

Sources:

A. Revkin, Nations near Arctic declare polar bears threatened by climate change, *New York Times*, March 20, 2009, http://www.nytimes.com/2009/03/20/science/earth/20bears.html; K. Snyder, Polar bear diet changes as sea ice melts. *American Museum of Natural History*, January 21, 2014, http://www.amnh.org/content/download/69060/1188269/file/Polar bear foraging.pdf.

The melting of the polar ice cap is just one example of many changes taking place on Earth. In this chapter, we will examine how humans have altered the world's climate and explore the underlying causes of these climate changes. We will also investigate the observed consequences of these changes for humans and other species, and consider predictions of future consequences. This chapter ties together many of the themes we have developed throughout the book: the interconnectedness of the systems on Earth, the environmental indicators that enable us to measure and evaluate the environmental status of Earth, and the interaction of environmental science and policy.

module

62

Global Climate Change and the Greenhouse Effect

In this module, we will consider the distinctions among *global change*, *global climate change*, and *global warming*. We will then explore the processes that underlie changes in global climates and, more specifically, global temperatures.

Learning Objectives

After reading this module you should be able to

- distinguish among global change, global climate change, and global warming.
- explain the process underlying the greenhouse effect.
- identify the natural and anthropogenic sources of greenhouse gases.

Global change includes global climate change and global warming

Throughout this book, we have highlighted a wide variety of ways in which the world has changed as a result of a rapidly growing human population. Human activity has placed increasing demands on natural resources such as water, trees, minerals, and fossil fuels. We have also emitted growing amounts of carbon dioxide, nitrogen compounds, and sulfur compounds into the atmosphere. Our agricultural methods depend on chemicals, including fertilizers and pesticides. Finally, a growing population faces challenges of waste disposal, sanitation, and the spread of human diseases.

Change that occurs in the chemical, biological, and physical properties of the planet is referred to as **global change.** As you can see in FIGURE 62.1, some types of global change are natural and have been occurring for millions of years. Global temperatures, for example, have fluctuated over millions of years. During periods of cold temperatures, Earth has experienced ice ages. In modern times, however, the rates of change have often been much higher than those that occurred historically. Many of these changes are the result of human activities, and they can have significant, sometimes cascading, effects. For example, as we saw in Chapter 17, emissions from coal-fired power plants and waste incinerators have increased the amount of mercury in the air and water, with concentrations roughly triple those of preindustrial levels. This mercury bioaccumulates in fish caught thousands of kilometers away from the sources of pollution. Because mercury has harmful effects on the nervous system of children, women who might become pregnant and children are advised to avoid eating top predator fish such as swordfish and tuna. Far-reaching effects on this scale were unimaginable just 50 years ago.

One type of global change of particular concern to scientists is **global climate change,** which refers to changes in the average weather that occurs in an area over a period of years or decades. Changes in climate

Global change Change that occurs in the chemical, biological, and physical properties of the planet.

Global climate change Changes in the average weather that occurs in an area over a period of years or decades.

Teaching Tip: Warm-up

Pretest students' knowledge of climate change with the questions below:

- Define the term "global change." *Global change is the change that occurs in the chemical, biological, and physical properties of the planet.*
- What role does the greenhouse effect play in global climate change? *The greenhouse effect occurs when atmospheric gases absorb infrared radiation and reradiate the energy back toward Earth. The more greenhouse gases found in the troposphere, the more the temperature will increase.*
- What determines the warming potential of a greenhouse gas? *Two factors determine the warming potential of a greenhouse gas: the amount of infrared energy that the gas can absorb and the length of time a molecule of the gas can persist in the atmosphere.*

Give students 5 to 10 minutes to answer the questions. When time is up, have students exchange papers and review their answers. Did anyone write correct answers to all three questions? When you are finished, ask students to review pages 665–669.

Teaching Tip: Activity

Divide the class into groups of two. Give each group 20 index cards and the vocabulary terms in the list below. Have each team label one side of each card with the numbers 1 through 20. On the blank sides of cards 1 through 10, students should write one of the key terms below, using a new card for each term. Instruct students to shuffle the order of the terms; they should not duplicate the order of the list. On the blank sides of cards numbered 11 through 20, student should write the definition or some other information that connects the term to the definition of the terms on cards 1 through 10, making sure to keep an answer key of which numbered term card matches which numbered definition card. The groups should lay out their cards number side up in two rows—one with terms and one with definitions. Have students exchange places and answer keys.

The first player in each pair then chooses one card from each row. Read the cards to see if they are a match. If they are, that person keeps the cards and takes another turn. If they are not a match, the second member of the team has a turn. Each group continues playing until all the cards are gone. The person with the most cards is the winner.

- global change
- global climate change
- global warming
- greenhouse effect
- greenhouse warming potential
- water vapor
- carbon dioxide
- methane
- nitrous oxide
- chlorofluorocarbon

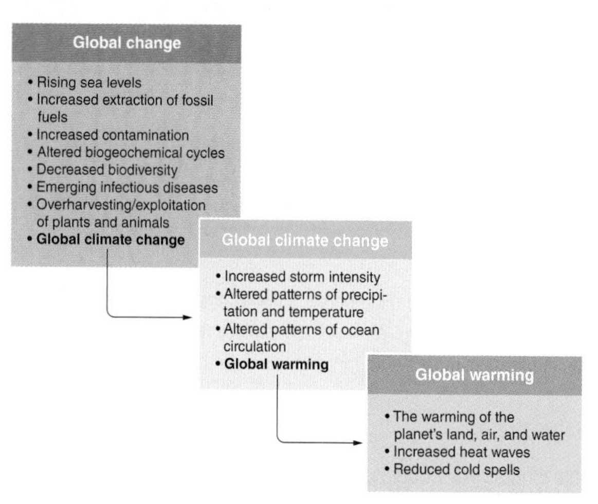

FIGURE 62.1 **Global change.** Global change includes a wide variety of factors that are changing over time. Global climate change refers to those factors that affect the average weather in an area of Earth. Global warming refers to changes in temperature in an area.

can be categorized as either natural or anthropogenic. For example, you might recall from Chapter 4 that El Niño events, which occur every 3 to 7 years, alter global patterns of temperature and precipitation (see Figure 11.3 on page 120). Anthropogenic activities such as fossil fuel combustion and deforestation also have major effects on global climates. **Global warming** refers to a specific aspect of climate change: the warming of the oceans, land masses, and atmosphere of Earth.

Solar radiation and greenhouse gases make our planet warm

The physical and biogeochemical systems that regulate temperature at the surface of Earth—the concentrations of gases, distribution of clouds, atmospheric currents, and ocean currents—are essential to life on our planet. It is therefore critical that we understand how the planet is warmed by the Sun and how the greenhouse effect contributes to the warming of Earth.

Global warming The warming of the oceans, land masses, and atmosphere of Earth.

The Sun-Earth Heating System

The ultimate source of almost all energy on Earth is the Sun. In the most basic sense, the Sun emits solar radiation that strikes Earth. As the planet warms, it emits radiation back toward the atmosphere. However, the types of energy radiated from the Sun and Earth are different (see Figure 5.1 on page 45). Because the Sun is very hot, most of its radiated energy is in the form of high-energy visible radiation and ultraviolet radiation—also known as visible light and ultraviolet light. When this radiation strikes Earth, the planet warms and radiates energy. Earth is not nearly as hot as the Sun, so it emits most of its energy as infrared radiation—also known as infrared light. We cannot see infrared radiation, but we can feel it being emitted from warm surfaces like the heat that radiates from an asphalt road on a hot day.

Differences in the types of radiation emitted by the Sun and Earth, in combination with processes that occur in the atmosphere, cause the planet to warm. Using FIGURE 62.2, we can walk through each step of this process. As radiation from the Sun travels toward Earth, about one-third of the radiation is reflected back into space. Although some ultraviolet radiation is absorbed by the ozone layer in the stratosphere, the remaining ultraviolet radiation, as well as visible light, easily passes through the atmosphere. Once it has passed through the atmosphere, this solar radiation strikes clouds and the surface of Earth. Some of this radiation is reflected from

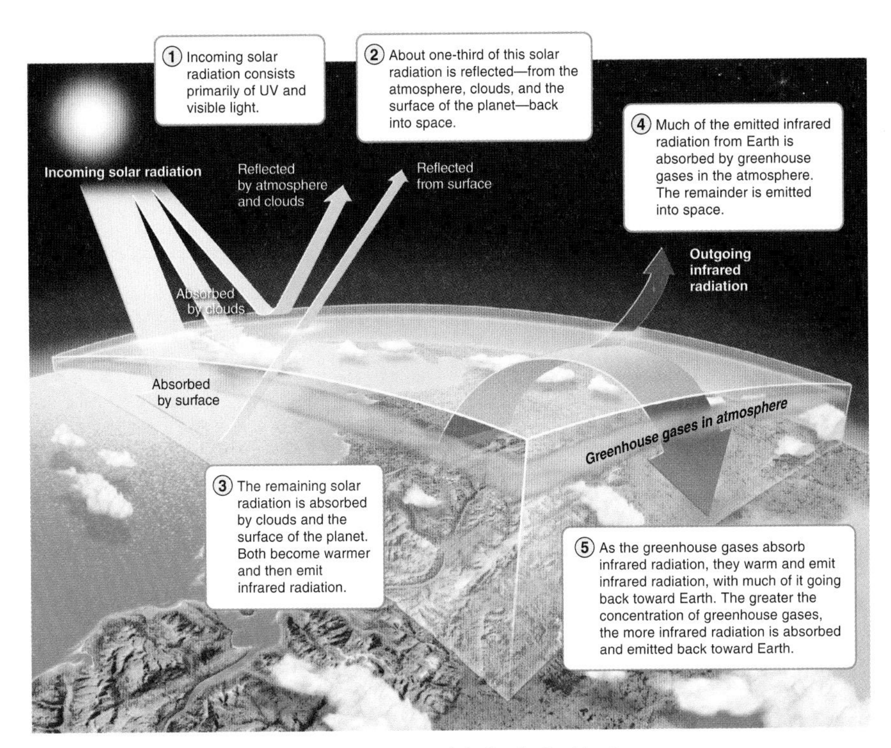

FIGURE 62.2 **The greenhouse effect.** When the high-energy radiation from the Sun strikes the atmosphere, about one-third is reflected from the atmosphere, clouds, and the surface of the planet. Much of the high-energy ultraviolet radiation is absorbed by the ozone layer where it is converted to low-energy infrared radiation. Some of the ultraviolet radiation and much of the visible light strikes the land and water of Earth where it is also converted into low-energy infrared radiation. The infrared radiation radiates back toward the atmosphere where it is absorbed by greenhouse gases that radiate much of it back toward the surface of Earth. Collectively, these processes cause warming of the planet.

the surface of the planet back into space. The remaining radiation is absorbed by clouds and the surface of Earth, which become warmer and begin to emit lower-energy infrared radiation back toward the atmosphere. Unlike ultraviolet and visible radiation, infrared radiation does not easily pass through the atmosphere. It is absorbed by gases, which causes theses gases to become warm. The warmed gases emit infrared radiation out into space and back toward the surface of Earth. The infrared radiation that is emitted toward Earth causes Earth's surface to become even warmer. This absorption of infrared radiation by atmospheric gases and reradiation of the energy back toward Earth is the **greenhouse effect.**

The greenhouse effect gets its name from the idea that solar radiation causes a gardener's greenhouse to become very warm. However, the process by which actual greenhouses are warmed by the Sun involves glass windows holding in heat whereas the process by which Earth is warmed involves greenhouse gases radiating infrared energy back toward the surface of the planet.

In the Sun-Earth heating system, the net flux of energy is zero; the inputs of energy to Earth equal the outputs from Earth. Over the long term—thousands or millions of years—the system has been in a steady state. However, in the shorter term—over years or decades—inputs can be slightly higher or lower than outputs. Factors that influence short-term fluctuations

Greenhouse effect Absorption of infrared radiation by atmospheric gases and reradiation of the energy back toward Earth.

◀ TEACHING with FIGURES

Show students Figure 62.2 and discuss the colors of the arrows. Yellow signifies UV light from the Sun and red signifies infrared light that is reradiated from Earth's surface. This infrared radiation is trapped by the greenhouse gases and warms Earth's atmosphere. Once you have discussed the figure, have your students draw their own version, making sure to include each of the five steps of the greenhouse effect.

Teaching Tip: Video

The Greenhouse Effect
This interactive video from *National Geographic* explores the greenhouse effect in five steps. When the class has finished watching the video, ask students to explain the process in a written paragraph.

The link to this resource may be found by clicking on the link buttons in the TE-book, opening the TRFD, or logging onto the book's companion website bcs.whfreeman.com/friedlandapes2e (teacher login required).

 **Chapter 19 Web Resources**

Teaching Tip: Discussion Starter

Ask students the following questions:

- How does the energy of the Sun cause Earth to warm? *As the planet warms, it emits radiation back toward the atmosphere. However, the types of energy radiated from the Sun and Earth are different. Because the Sun is very hot, most of its radiated energy is in the form of high-energy visible radiation and ultraviolet radiation, also known as visible light and ultraviolet light. When this radiation strikes Earth, the planet warms and radiates energy. Earth is not nearly as hot as the Sun, so it emits most of its energy as infrared radiation, also known as infrared light. We cannot see infrared radiation, but we can feel it being emitted as heat from warm surfaces such as an asphalt road on a hot day.*
- What is a greenhouse gas? *A greenhouse gas is a gas that can absorb infrared radiation emitted by the surface of the planet and radiate much of it back toward the surface.*
- Which greenhouse gases are the most common on Earth? *The most common greenhouse gas is water vapor. Water vapor absorbs more infrared radiation from Earth than any other compound, although a molecule of water vapor does not persist nearly as long as other greenhouse gases.*

include changes in incoming solar radiation from increased solar activity and changes in outgoing radiation from an increase in atmospheric gases that absorb infrared radiation. If incoming solar energy is greater than the sum of reflected solar energy and radiated infrared energy from Earth, then the energy accumulates faster than it is dispersed and the planet becomes warmer. If incoming solar energy is less than the sum of the two outputs, the planet becomes cooler. Such natural changes in inputs and outputs cause natural changes in the temperature of Earth over time.

The Gases That Cause the Greenhouse Effect

Throughout the book we have seen that certain gases in the atmosphere can absorb infrared radiation emitted by the surface of the planet and radiate much of it back toward the surface. As we have seen, gases in the atmosphere that absorb infrared radiation are known as greenhouse gases.

The two most common gases in the atmosphere, N_2 and O_2, compose 99 percent of the atmosphere. Because these two gases do not absorb infrared radiation, they are not greenhouse gases and do not contribute to the warming of Earth. This means that greenhouse gases make up a very small fraction of the atmosphere. The most common greenhouse gas is water vapor (H_2O). Water vapor absorbs more infrared radiation from Earth than any other compound, although a molecule of water vapor does not persist nearly as long as other greenhouse gases. Other important greenhouse gases include carbon dioxide (CO_2), methane (CH_4), nitrous oxide (N_2O), and ozone (O_3). All of these gases have been a part of the atmosphere for millions of years, and have kept Earth warm enough to be habitable. In the case of ozone, we have seen that its effects on Earth are diverse. Ozone in the stratosphere is beneficial because it filters out harmful ultraviolet radiation. In contrast, ozone in the lower troposphere acts as a greenhouse gas and can cause increased warming of Earth. It also is an air pollutant in the lower troposphere because it can cause damage to plants and human respiratory systems. There is one other type of greenhouse gas, chlorofluorocarbons (CFCs), which does not exist naturally. It occurs in the atmosphere exclusively due to production of CFCs by humans and, as we discussed in Chapter 15, these CFCs have contributed to a hole in the ozone layer over Antarctica.

Although we commonly think of the greenhouse effect as detrimental to our environment, without any greenhouse gases the average temperature on Earth would be approximately −18°C (0°F) instead of its current average temperature of 14°C (57°F). Concern about the danger of greenhouse gases is based on our understanding that an increase in the concentration of these gases—as has occurred due to human activities—can cause the planet to warm even more than usual.

The contribution of each gas to global warming depends in part on its *greenhouse warming potential*. The **greenhouse warming potential** of a gas estimates how much a molecule of any compound can contribute to global warming over a period of 100 years relative to a molecule of CO_2. In calculating this potential, scientists consider the amount of infrared energy that a given gas can absorb and how long a molecule of the gas can persist in the atmosphere. Because greenhouse gases can differ a great deal in these two factors, greenhouse warming potentials span a wide range of values. For example, water vapor has a lower potential compared with carbon dioxide. The remaining greenhouse gases have much higher values, either because they absorb more infrared radiation than a molecule of CO_2 or because they persist longer in the atmosphere than CO_2. **TABLE 62.1** shows the global warming potential for five common greenhouse gases. Compared with CO_2, the greenhouse warming potential is 25 times higher for methane (CH_4), nearly 300 times higher for nitrous oxide (N_2O), and up to 13,000 times higher for CFCs.

The effect of each greenhouse gas depends on both its warming potential and its concentration in the atmosphere. Although carbon dioxide has a relatively low warming potential, it is much more abundant than most other greenhouse gases, except for water vapor, which can have a concentration similar to carbon dioxide. While human activity appears to have little effect on the amount of water vapor in the atmosphere, it has caused substantial increases in the amount of the other greenhouse gases. Among these, carbon dioxide remains the greatest contributor to the greenhouse effect because its concentration is so much higher than any of the others. As a result, scientists and policy makers focus their efforts on ways to reduce carbon dioxide in the atmosphere.

Given what we now know about how greenhouse gases work, the concentrations of each gas, and how much infrared energy each gas absorbs, we can understand how changes in the concentrations of greenhouse gases can contribute to global warming. Increasing the concentration of any historically present greenhouse gas should cause more infrared radiation to be absorbed in the atmosphere, which will then radiate more energy back toward the surface of the planet and cause the planet to warm. Likewise, producing new greenhouse gases that can make their way into the atmosphere, such as CFCs, should also cause increased absorption of infrared radiation in the atmosphere and further cause the planet to warm.

Greenhouse warming potential An estimate of how much a molecule of any compound can contribute to global warming over a period of 100 years relative to a molecule of CO_2.

Teaching Tip: Activity

Have students name the five greenhouse gases and create a table like the one to the right. (Suggested answers are provided in italics.)

Greenhouse gas	Source (example)
Water vapor (H_2O)	*Evaporation*
Carbon dioxide (CO_2)	*Burning fossil fuels*
Methane (CH_4)	*Livestock manure*
Nitrous oxide (N_2O)	*Produced during the nitrogen cycle in the denitrification phase*
Chlorofluorocarbons	*Industrial chemicals used as refrigerants in air conditioners, refrigerators, and freezers*

TABLE 62.1 The major greenhouse gases

The major greenhouse gases differ in their ability to absorb infrared radiation and in the duration of time that they stay in the atmosphere. The units "ppm" are parts per million.

Greenhouse gas	Concentration in 2010	Global warming potential (over 100 years)	Duration in the atmosphere
Water Vapor	Variable with temperature	<1	9 days
Carbon Dioxide	390 ppm	1	Highly variable (ranging from years to hundreds of years)
Methane	1.8 ppm	25	12 years
Nitrous Oxide	.3 ppm	300	114 years
Chlorofluorocarbons	.9 ppm	1,600 to 13,000	55 to >500 years

Source: Data on concentration are from the National Oceanic and Atmospheric Administration. www.esrl.noaa.gov/gmd/aggi. Data on global warming potential are from the United Nations Framework Convention on Climate Change.

Sources of greenhouse gases are both natural and anthropogenic

As we have seen, greenhouse gases include a variety of compounds such as water vapor, carbon dioxide, methane, nitrous oxide, and CFCs. These gases have natural and anthropogenic sources. After reviewing the different sources of greenhouse gases, we will discuss the relative ranks of the different anthropogenic sources.

Natural Sources of Greenhouse Gases

Natural sources of greenhouse gases include volcanic eruptions, decomposition, digestion, denitrification, evaporation, and evapotranspiration.

Volcanic Eruptions

Over the scale of geologic time, volcanic eruptions can add a significant amount of carbon dioxide to the atmosphere. Other gases and the large quantities of ash released during volcanic eruptions can also have important, short-term climatic effects. A volcanic eruption emits a large quantity of ash into the atmosphere. The ash reflects incoming solar radiation back out into space, which has a cooling effect on Earth. In 1991, for example, Mount Pinatubo in the Philippines erupted and spewed millions of tons of ash into the atmosphere, as far as 20 km (12 miles) high (FIGURE 62.3). The large amount of ash in the atmosphere reduced the amount of radiation striking Earth, which caused a 0.5°C (0.9°F) decline in the temperature on the planet's surface. Because the ash and small particles eventually settle out of the atmosphere, such effects usually last only a few years.

Decomposition and Digestion

When decomposition occurs under high-oxygen conditions, the dead organic matter is ultimately converted into carbon dioxide. As we saw in our discussion of landfills in Chapter 16, methane is created when there is not enough oxygen available to produce carbon dioxide. This is a common occurrence at the bottom of wetlands where plants and animals decompose and oxygen is in low supply. Wetlands are the largest natural source of methane.

FIGURE 62.3 **Ash from volcanic eruptions.** Volcanic eruptions, such as this eruption of Mount Pinatubo in the Philippines in 1991, send millions of tons of ash into the atmosphere where it can absorb incoming solar radiation, reradiate it back to space, and cause Earth to cool. *(Exactostock/SuperStock)*

Teaching Tip: Review

This is a good time to review the carbon cycle (p. 83) and the nitrogen cycle (p. 84).

A similar situation occurs when certain animals digest plant matter. Animals that consume significant quantities of wood or grass, including termites and grazing antelopes, require gut bacteria to digest the plant material. Because the digestion occurs in the animal's gut, the bacteria do not have access to oxygen and methane is produced as a by-product. A single termite colony can contain more than a million termites (FIGURE 62.4). Termites are abundant throughout the world—especially in the tropics—and represent the second largest natural source of methane.

Denitrification

As we learned in Chapter 3, nitrous oxide (N_2O) is a natural component of the nitrogen cycle that is produced through the process of denitrification. Denitrification occurs in the low-oxygen environments of wet soils and at the bottoms of wetlands, lakes, and oceans. (Figure 7.3 on page 84 shows the nitrogen cycle.) In these environments, nitrate is converted to nitrous oxide gas, which then enters the atmosphere as a powerful greenhouse gas.

Evaporation and Evapotranspiration

As we stated earlier, water vapor is the most abundant greenhouse gas in the atmosphere and the greatest natural contributor to global warming. In Chapter 3 we examined the role of water vapor in the hydrologic cycle. (Figure 7.1 on page 80 shows the hydrologic cycle.) Water vapor is produced when liquid water from land and water bodies evaporates and by the evapotranspiration process of plants. Because the amount of evaporation into water vapor varies with climate, the amount of water vapor in the atmosphere can vary regionally.

FIGURE 62.4 **Termites and methane.** The bacteria that live in the anaerobic gut environment of herbivores such as termites produce methane as a by-product of their digestive activities. Because termite colonies, such as this one in Australia, can achieve population sizes of more than one million, collectively they can produce large amounts of methane. *(Anton Harder/Shutterstock)*

Anthropogenic Sources of Greenhouse Gases

As shown in FIGURE 62.5, there are many anthropogenic sources of greenhouse gases. The most significant of these are the burning of fossil fuels, agricultural practices, deforestation, landfills, and industrial production of new greenhouse chemicals.

Burning Fossil Fuels

Tens to hundreds of millions of years ago, organisms were sometimes buried without first decomposing into carbon dioxide. In Figure 7.2 on page 83 we outlined the process by which the carbon contained in these organisms, called fossil carbon, is slowly converted to fossil fuels deep underground. When humans burn these fossil fuels, we produce CO_2 that goes into the atmosphere. Because of the long time required to convert carbon into fossil fuels, the rate of putting carbon into the atmosphere by burning fossil fuels is much greater than the rate at which producers take CO_2 out of the air and both the producers and consumers contribute to the pool of buried fossil carbon.

Because fossil fuels differ in how they store energy, each type of fossil fuel produces different amounts of carbon dioxide. For a given amount of energy, burning coal produces the most CO_2. In comparison, burning oil produces 85 percent as much CO_2 as coal, and natural gas produces 56 percent as much. As we saw in Chapter 12, from the perspective of CO_2 emissions, natural gas is considered better for the environment than coal. The production of fossil fuels, such as the mining of coal, and the combustion of fossil fuels can also release methane and, in some cases, nitrous oxide.

Particulate matter may also play an important role in global warming. Although particulate matter, also known as black soot, may reflect solar radiation under some conditions, recent findings suggest that it may

Teaching Tip: Journal Prompt

Ask students: What impact does the burning of fossil fuels have on the carbon cycle? *Burning fossil fuels produces CO_2, which goes into the atmosphere. Because carbon is converted into fossil fuels so slowly, the rate of releasing carbon into the atmosphere by burning fossil fuels is much faster than the rate at which fossil fuels are produced. Therefore, burning fossil fuels leads to greater amounts of CO_2 in the atmosphere.*

FIGURE 62.5 **Anthropogenic sources of greenhouse gases.** Human activities are a major contributor of greenhouse gases including CO_2, methane, and nitrous oxide. These activities include the use of fossil fuels, agricultural practices, the creation of landfills, and the industrial production of new greenhouse gases.

be responsible for up to one-quarter of observed global warming during the past century. Particulates that fall on ice and snow in the higher latitudes absorb more energy of the Sun by lowering the albedo. As the snow and ice begin to melt, the particulates become more concentrated on the surface. The increased concentration raises the amount of solar radiation absorbed, which increases melting. This positive feedback system might help explain warming that occurred early in the last century, when atmospheric concentrations of greenhouse gases had not yet increased much but soot from the burning of coal was widespread.

Agricultural Practices

Agricultural practices can produce a variety of greenhouse gases. Agricultural fields that are overirrigated, or those that are deliberately flooded for cultivating crops such as rice, create low oxygen environments similar to wetlands and therefore can produce methane and nitrous oxide. Synthetic fertilizers, manures, and crops that naturally fix atmospheric nitrogen—for example, alfalfa—can create an excess of nitrates in the soil that are converted to nitrous oxide by the process of denitrification.

Raising livestock can also produce large quantities of methane. Many livestock such as cattle and sheep consume large quantities of plant matter and rely on gut bacteria to digest this cellulose. As we saw in the case of termites, gut bacteria live in a low-oxygen environment and digestion in this environment produces methane as a by-product. Manure from livestock operations will decompose to CO_2 under high-oxygen conditions, but in low-oxygen conditions, for example in manure lagoons that are not aerated, it will decompose to methane.

Deforestation

Each day, living trees remove CO_2 from the atmosphere during photosynthesis, and decomposing trees add CO_2 to the atmosphere. This part of the carbon cycle does not change the net atmospheric carbon because the inputs and outputs are approximately equal. However, when forests are destroyed by burning or decomposition and not replaced, as can happen during deforestation, the destruction of vegetation will contribute to a net increase in atmospheric CO_2. This is because the mass of carbon that made up the trees is added to the atmosphere by combustion or decomposition. The shifting agriculture described in Chapter 11, which involves clearing forests and burning the vegetation to make room for crops, is a major source of both particulates and a number of greenhouse gases, including carbon dioxide, methane, and nitrous oxide.

Landfills

As we saw in Chapter 16, landfills receive a great deal of household waste that slowly decomposes under layers of soil. When the landfills are not aerated properly, they create a low-oxygen environment, like wetlands, in which decomposition causes the production of methane as a by-product.

Industrial Production of New Greenhouse Chemicals

The creation of new industrial chemicals often has unintended effects on the atmosphere. In Chapter 15 we looked at CFCs, the family of chemicals that serves as refrigerants used in air conditioners, freezers, and refrigerators. CFCs were used in the past until scientists discovered that they were damaging the protective

◀ TEACHING with FIGURES

Show students Figure 62.5 and ask them to create a table listing the greenhouse gas produced by each human activity. (Answers are provided in italics.)

Human activity	Greenhouse gas produced
Use of fossil fuels	• *Burning fossil fuels releases atmospheric carbon dioxide*
Deforestation	• *Destruction of vegetation leads to an increase in atmospheric carbon dioxide*
Agriculture	• *Livestock produces large quantities of methane* • *Flooded fields can produce methane and nitrous oxide*
Landfills	• *Can create a low-oxygen environment in which decomposition causes the production of methane*
Industry	• *Many chemicals used in industrial processes have high greenhouse warming potentials*

Teaching Tip: Beyond the Classroom

Have your students research landfills in your state to determine which landfills collect methane to generate electricity, which burn off excess methane, and which release methane into the atmosphere. What are the regulations in your state? Begin by looking at the EPA website listing of municipal solid waste landfills for your state. If your state has a large number of landfills, assign two or three to each student. Have students create a table with the researched information.

The link to the EPA website may be found by clicking on the link buttons in the TE-book, opening the TRFD, or logging onto the book's companion website bcs.whfreeman.com/friedlandapes2e (teacher login required).

 **Chapter 19 Web Resources**

Teaching Tip: Journal Prompt

Ask students: Which of the three anthropogenic sources of greenhouse gases can humans decrease most easily and how? *Answers will vary.*

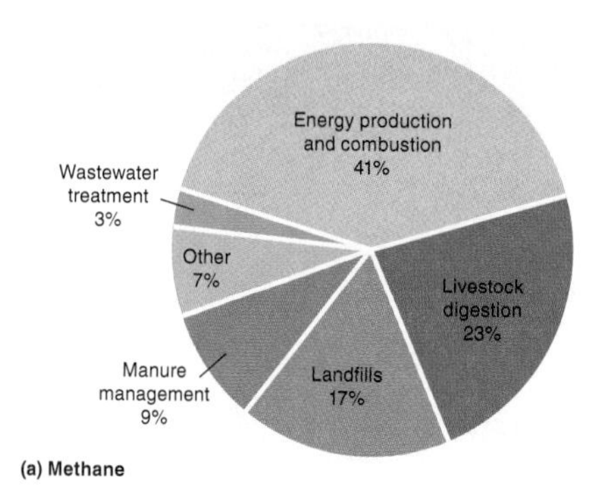

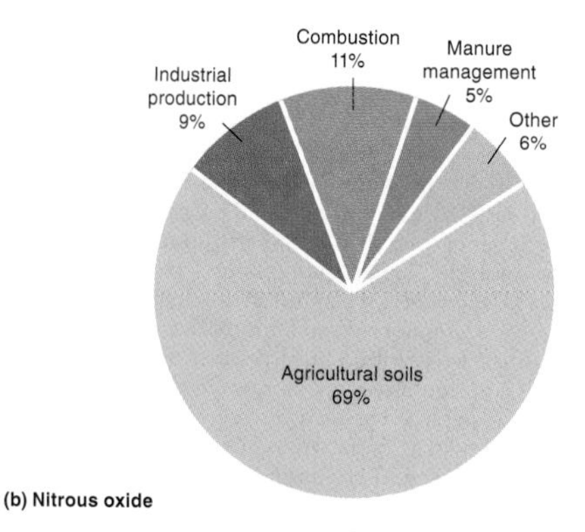

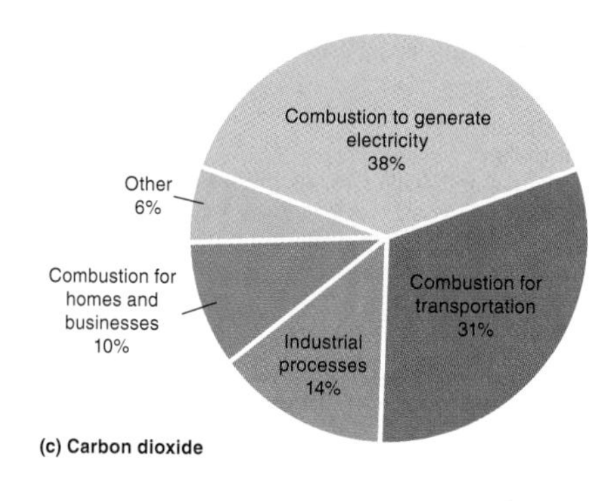

ozone layer. As we discussed in Chapter 15, the nations of the world joined together to sign the Montreal Protocol on Substances That Deplete the Ozone Layer, which phased out the production and use of CFCs by 1996. Unfortunately, many of the alternative refrigerants that are less harmful to the ozone layer, including a group of gases known as hydrochlorofluorocarbons (HCFCs), still have very high greenhouse warming potentials. As a result, developed countries will phase out the use of HCFCs by 2030.

Ranking the Anthropogenic Sources of Greenhouse Gases

We have seen that there are multiple anthropogenic sources of greenhouse gases. What is the relative contribution of each source? FIGURE 62.6 shows the major anthropogenic sources of greenhouse gases in the United States. Figure 62.6a shows that the three major contributors of methane in the atmosphere are the digestive processes of livestock, landfills, and the production of natural gas and petroleum products. The major contributor of nitrous oxide, shown in Figure 62.6b, is agricultural soil because they receive nitrogen from synthetic fertilizers, combustion, and industrial production of fertilizers and other products. Finally, looking at the numbers for carbon dioxide in Figure 62.6c, we see that approximately 94 percent of all CO_2 emissions come from industrial processes and the burning of fossil fuels.

FIGURE 62.6 **Anthropogenic sources of greenhouse gases in the United States.** (a) The largest contributions of methane in the atmosphere arise from gut bacteria that help many livestock species digest plant matter, landfills that experience decomposition in low-oxygen environments, and the production, storage, and transport of natural gas and petroleum products from which methane escapes. (b) The largest contributions of nitrous oxide in the atmosphere arise from the agricultural soils that obtain nitrogen from applied fertilizers, combustion, and industrial production of fertilizers and other products. (c) Nearly all anthropogenic CO_2 emissions come from the burning of fossil fuels. *(Data from http://www.epa.gov/climatechange/ghgemissions/gases/ch4.html, http://www.epa.gov/climatechange/ghgemissions/gases/n2o.html, http://www.epa.gov/climatechange/ghgemissions/gases/co2.html)*

TEACHING WITH FIGURES ▶

Show students Figure 62.6 and ask the following questions:

- What are the three largest anthropogenic sources of methane gas? *The three largest anthropogenic sources of methane gas are energy production and combustion at 41 percent, livestock digestion at 23 percent, and landfills at 17 percent.*
- Will the percentage of methane associated with livestock in the United States and worldwide increase in the future? *Although demand for meat in the United States has decreased slightly, worldwide demand for meat has increased. Therefore, the worldwide percentage of methane associated with livestock digestion will most likely increase.*

module

62

REVIEW

In this module, we considered global change, global climate change, and global warming. In a natural process known as the greenhouse effect, visible light and ultraviolet light from the Sun strike our planet and this energy is converted to infrared radiation that is emitted back to the atmosphere. A tiny percentage of gases in the atmosphere, known as greenhouse gases, absorb this infrared radiation and emit a portion of it back to Earth and this causes the planet to warm even more. Although most greenhouse gases have natural sources, human activities have increased the concentration of these gases in the atmosphere and produced new chemicals that are potent greenhouse gases. In the next module, we will examine the evidence that this increase in greenhouse gases has caused our planet to become warmer during the past 2 centuries.

Module 62 AP® Review Questions

1. The greenhouse gas with the highest greenhouse warming potential is
   (a) carbon dioxide.
   (b) methane.
   (c) water vapor.
   (d) chlorofluorocarbon.
   (e) nitrous oxide.

2. Methane is naturally produced by
   (a) decomposition.
   (b) volcanic eruptions.
   (c) denitrification.
   (d) evapotranspiration.
   (e) forest fires.

3. The greenhouse effect is due to
   (a) the absorption and reradiation of infrared radiation by the atmosphere.
   (b) the reflection of ultraviolet radiation by the atmosphere.
   (c) the absorption of ultraviolet radiation by the atmosphere.
   (d) the reflection of infrared radiation from Earth's surface.
   (e) the absorption of ultraviolet radiation by Earth.

4. Most nitrous oxide emissions are from
   (a) fossil fuel combustion.
   (b) agricultural practices.
   (c) refrigerants.
   (d) deforestation.
   (e) industrial processes.

5. Particulate matter can increase global warming by
   (a) reacting with chlorofluorocarbons.
   (b) reducing surface absorption of ultraviolet radiation.
   (c) producing additional nitrous oxides.
   (d) reflecting radiation.
   (e) lowering surface albedo.

Answers to Module 62 AP® Review Questions

1. d
2. a
3. a
4. b
5. e

AP® Exam Tip

Changes in atmospheric temperature and atmospheric concentration of CO_2 are covered in the 2006 AP® Exam, Question 2. To answer this question, students must

- answer several questions by evaluating graphs that show changes in temperature and CO_2 levels over time, including calculation of these changes, calculation of the ratio of these changes, and predicting future changes based on these ratios
- identify and describe two major causes for predicted future changes in the increase of atmospheric CO_2 levels
- identify two other gasses that contribute to the anthropogenic increase in mean global temperature and describe how human activity causes these gasses to be released

module 63

The Evidence for Global Warming

Now that we have reviewed greenhouse gases and their role in global warming, we can explore the evidence that an increased concentration of greenhouse gases is causing Earth to become warmer. If Earth is becoming warmer, we can begin to evaluate whether or not this warming is caused by human activities that have released greenhouse gases. One way to make these assessments is to determine gas concentrations and temperatures from the past and compare them to gas concentrations and temperatures in the present day. We can also use information about changes in gas concentrations and temperatures to predict future climate conditions. In this module, we will examine how greenhouse gas concentrations have changed over time and how these changes are linked to global warming.

Learning Objectives

After reading this module you should be able to

- explain how CO_2 concentrations have changed over the past 6 decades and how emissions compare among the nations of the world.
- explain how temperatures have increased since records began in 1880.
- discuss how we estimate temperatures and levels of greenhouse gases over the past 500,000 years and into the future.
- explain the role of feedbacks on the impacts of climate change.

CO_2 concentrations have been increasing for the past 6 decades

In 1988 the United Nations and the World Meteorological Organization created the Intergovernmental Panel on Climate Change (IPCC), a group of more than 3,000 scientists from around the world working together to assess climate change. Their mission is to understand the details of the global warming system, the effects of climate change on biodiversity and energy fluxes in ecosystems, and the economic and social effects of climate change. The IPCC enables scientists to assess and communicate the state of our knowledge and to suggest research directions that would improve our understanding in the future. This effort has produced an excellent understanding of how greenhouse gases and temperatures are linked.

674 CHAPTER 19 ■ Global Change

Measuring CO_2 Concentrations in the Atmosphere

Through the work of the IPCC, we now understand that CO_2 is an important greenhouse gas that can contribute to global warming, but we didn't always realize this. In the first half of the twentieth century, most scientists believed that if any excess CO_2 were being produced, it would be absorbed by the oceans and vegetation. In addition, because the concentration of atmospheric CO_2 was low compared to gases such as oxygen and nitrogen, it was difficult to measure accurately.

Charles David Keeling was the first to overcome the technical difficulties in measuring CO_2. When Keeling set out to measure the precise level of CO_2 in the atmosphere, most atmospheric scientists believed that two measurements several years apart would be sufficient to answer the question of whether human activities were causing increased concentrations of CO_2 in the atmosphere. Keeling did not agree and in 1958 he began collecting data throughout the year at the Mauna Loa Observatory in Hawaii. After just 1 year of work, Keeling found that CO_2 levels varied seasonally and that the concentration of CO_2 increased from year to year. His results prompted him to take measurements for several more years, and he and his students have continued this work into the twenty-first century. The results, shown in FIGURE 63.1, confirm Keeling's early findings; although CO_2 concentrations vary between seasons, there is a clear trend of rising CO_2 concentrations across the years. This increase over time is correlated to increased human emissions of carbon from the combustion of fossil fuels and net destruction of vegetation. "Do the Math: Projecting Future Increases in CO_2" gives you an opportunity to estimate the increase in CO_2 concentrations by the end of the century.

What causes the seasonal variation? Each spring, as deciduous trees, grasslands, and farmlands in the Northern Hemisphere turn green, they increase their absorption rates of CO_2 to carry out photosynthesis.

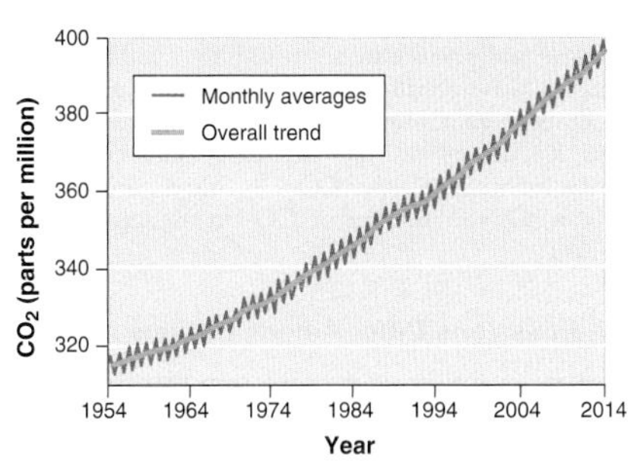

FIGURE 63.1 Changes in atmospheric CO_2 over time. Carbon dioxide levels have risen steadily since measurement began in 1958. *(Data from http://www.esrl.noaa.gov/gmd/ccgg/trends/)*

do the math

Projecting Future Increases in CO_2

Because Charles David Keeling and his colleagues began measuring CO_2 in 1958, we now have an excellent record of how CO_2 concentrations have changed in the atmosphere over time. From 1960 to 2010, the concentration of CO_2 in the atmosphere increased from 320 to 390 ppm (parts per million).

Based on these two points in time, what has been the average annual increase of CO_2 in the atmosphere?

$$\text{Time} = 2010 - 1960 = 50 \text{ years}$$

$$\text{Increase in } CO_2 = 390 \text{ ppm} - 320 \text{ ppm} = 70 \text{ ppm}$$

$$\text{Average annual increase in } CO_2 = 70 \text{ ppm} \div 50 \text{ years} = 1.4 \text{ ppm/year}$$

Your Turn

1. If the annual rate of CO_2 increase is 1.4 ppm, what concentration of CO_2 do you predict for the year 2100?
2. From 2000 to 2010, the rate of increase grew to 1.9 ppm per year. Based on this faster rate, what concentration of CO_2 do you predict for the year 2100?

◀ TEACHING with FIGURES

Show students Figure 63.1. Note that this graph has shown up on past exams. Ask: What is the reason for the yearly CO_2 fluctuation? *The yearly fluctuation in CO_2 can be linked to the seasonal photosynthetic activity of green plants. In the fall and winter, many trees drop their leaves and the photosynthetic rate decreases and CO_2 levels increase. In the spring and summer the trees have a much higher photosynthetic rate and they take in large amounts of CO_2.*

AP® Exam Tip

Do the Math, "Projecting Future Increases in CO_2," contains two problems that have appeared on past exams. Make sure your students understand how to calculate the average annual increase in CO_2 and how to predict the concentration of CO_2.

Answers to Your Turn

1. From the year 2010 (the last year of the data) until the year 2100 we have a 90-year span. Therefore we can multiply 90 years by 1.4 ppm per year to calculate the added CO_2:

$$90 \text{ years} \left(\frac{1.4 \text{ ppm } CO_2}{\text{year}}\right) = 126 \text{ ppm extra } CO_2$$

Predicted concentration 2100
= Concentration 2010 + calculated extra
= 390 ppm + 126 ppm
= 516 ppm CO_2

2. Again we have a 90-year span and calculate as before:

$$90 \text{ years} \left(\frac{1.9 \text{ ppm } CO_2}{\text{year}}\right) = 171 \text{ ppm extra } CO_2$$

Predicted concentration 2100
= Concentration 2010 + calculated extra
= 390 ppm + 171 ppm
= 561 ppm CO_2

At the same time, bodies of water begin to warm and the algae and plants also begin to photosynthesize. In doing so, these producers take up some of the CO_2 in the atmosphere. Conversely, in the fall, as leaves drop, crops are harvested, and bodies of water cool, the uptake of atmospheric CO_2 by algae and plants declines and the amount of CO_2 in the atmosphere increases.

CO_2 Emissions Differ Among Nations

Throughout this book we have seen that per capita consumption of fossil fuel and materials is greatest in developed countries. It is not surprising, then, that the production of carbon dioxide has also been greatest in the developed world. For many decades, the 20 percent of the population living in the developed world—roughly 1 billion people—produced about 75 percent of the carbon dioxide. However, these percentages are changing as some developing nations industrialize and acquire more vehicles that burn fossil fuels. In 2009, developing countries surpassed developed countries in the production of CO_2.

Development has been especially rapid in China and India, which together contain one-third of the world's population. From 2000 to 2009, China more than doubled its emissions of carbon dioxide as the country built many new coal-powered electrical plants that increased its ability to burn coal. FIGURE 63.2a shows that today China is the leading emitter of CO_2. China emits more than 7,200 million metric tons of CO_2, representing 24 percent of all global CO_2 emissions. The United States is in second place, emitting more than 5,300 million metric tons. This represents 18 percent of all global CO_2 emissions, yet the United States contains only 5 percent of the world's population. If we consider the amount of per capita CO_2 emissions, shown in Figure 63.2b, we obtain a very different picture of which countries produce the most CO_2. The United States and Australia are the leading per capita emitters of CO_2, followed by Saudi Arabia and Canada. Despite the fact that China and India rank among the top producers of CO_2, their per capita production ranks them sixteenth and twentieth, respectively, which reflects the fact that these two countries both have very large populations.

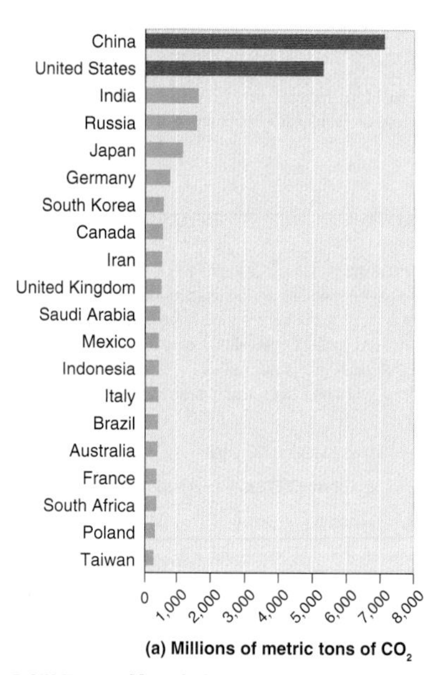

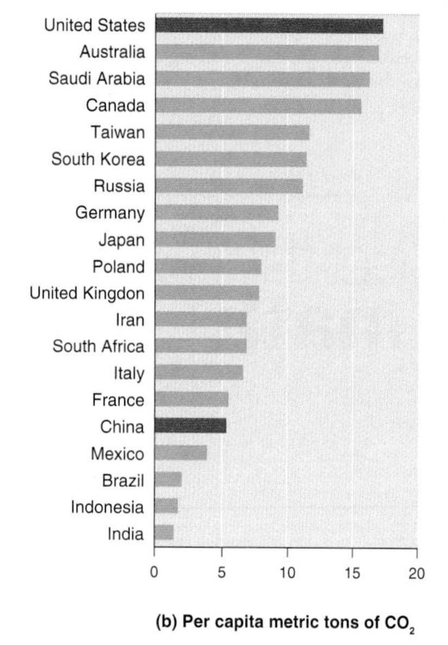

FIGURE 63.2 **CO_2 emissions by country in 2010.** (a) When we consider the total amount of CO_2 produced by a country, we see that the largest contributors are the developed and rapidly developing countries of the world. (b) On a per capita basis, some major CO_2 emitters have relatively low per capita CO_2 emissions. *(Data from International Energy Commission 2012)*

TEACHING with FIGURES ▶

Show students Figure 63.2 and ask the following questions:

- According to Figure 63.2a, how many metric tons of CO_2 does China produce? *As of 2010, China produced approximately 7,000 million metric tons of CO_2.*
- What are two reasons China is the leading CO_2 emitter in the world? *China is the leader in CO_2 emissions for many reasons. First, China has the largest population in the world. Second, it is a large user of coal to produce electricity. Third, the country is going through rapid industrialization and citizens now have access to more modern conveniences such as household appliances and cars.*
- According to Figure 63.2b, which three countries produce the highest per capita amounts of CO_2? *The three countries that produce the highest per capita amounts of CO_2 are the United States, Australia, and Canada.*
- Why are these countries the highest per capita producers of CO_2? *These countries are the highest per capita producers of CO_2 because they have among the highest GDP per capita and most people in these countries can afford to use fossil fuels for energy and transportation.*

Global temperatures have steadily increased since records began in 1880

Before we can determine if global temperature increases are a recent phenomenon and if these increases are unusual, we must establish how the temperatures of Earth have changed in the past. Since about 1880, there have been enough direct measurements of land and ocean temperatures that the NASA Goddard Institute for Space Studies has been able to generate a graph of global temperature change over time. This graph, updated monthly, is shown in FIGURE 63.3. Comprising thousands of measurements from around the world, the graph shows global temperatures have increased 0.8°C (1.4°F) from 1880 through 2013. In fact, the 10 warmest years on record since 1880 have all occurred between 1998 and 2013.

While an increase in average global temperature of 0.8°C (1.4°F) may not sound very substantial, it not evenly distributed around the globe. As the map in FIGURE 63.4 shows, some regions, including parts of Antarctica, have experienced cooler temperatures. Some regions, including areas of the oceans, have experienced no change in temperature. Finally, some regions, such as those in the extreme northern latitudes, have experienced increases of 1°C to 4°C (1.8°F –7.2°F). The substantial increases in temperatures in the northern latitudes have caused, among other problems, nearly 45 percent of the northern ice cap to melt, which has threatened polar bears and their ecosystem, as discussed at the beginning of this chapter.

The data collected by the NASA clearly demonstrate that the globe has been slowly warming during the past 120 years. However, it is possible that such changes in temperature are simply a natural phenomenon. If we want to know whether these changes are typical, we must examine a much longer span of time.

Scientists can estimate global temperatures and greenhouse gas concentrations for over 500,000 years

Since no one was measuring temperatures thousands of years ago, we must use indirect measurements. Common indirect measurements include changes in the species composition of organisms that have been preserved over millions of years and chemical analyses of air bubbles formed in ice long ago.

Changing Species Compositions

One commonly used biological measurement is the change in species composition of a group of small protists, called foraminifera. Foraminifera are tiny, marine organisms with hard shells that resist decay after death (FIGURE 63.5). In some regions of the ocean floor, the tiny shells have been building up in sediments for millions of years. The youngest sediment layers are near the top of the ocean floor whereas the oldest sediment layers are much deeper. Fortunately, different species of foraminifera prefer different water temperatures. As a result, when scientists identify the predominant species of foraminifera in a layer of sediment, they can infer the likely temperature of the ocean at the time the layer of sediment was deposited. By examining thousands of sediment layer samples, we can gain insights into temperature changes over millions of years.

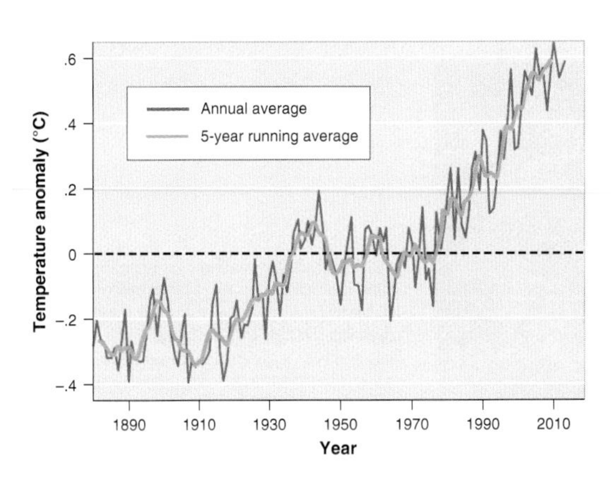

FIGURE 63.3 Changes in mean global temperatures over time. Although annual mean temperatures can vary from year to year, temperatures have exhibited a slow increase from 1880 to 2012. This pattern becomes much clearer when scientists compute the average temperature each year for the past 5 years. *(Data from http://data.giss.nasa.gov/gistemp/graphs_v3/Fig.A2.gif)*

Module 63 The Evidence for Global Warming 677

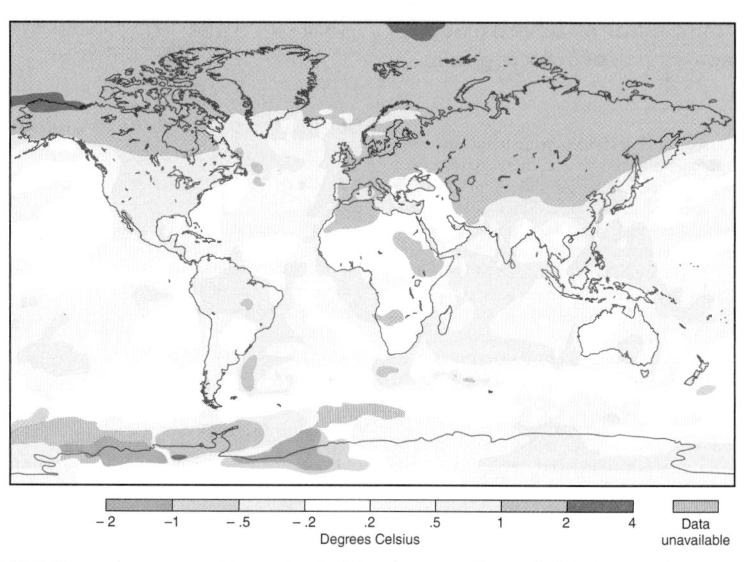

FIGURE 63.4 **Changes in mean annual temperature in different regions of the world.** In 2010, some regions became cooler, some regions had no temperature change, and the northern latitudes became substantially warmer than the long-term average temperature. The surface temperatures plotted on the map represent differences relative to the average temperature from 1951 to 1980. *(After http://www.nasa.gov/topics/earth/features/rapid-change-feature.html)*

Air Bubbles in Ancient Ice

Scientists can determine changes in greenhouse gas concentrations and temperatures over long periods of time by examining ancient ice. In cold areas such as Antarctica and at the top of the Himalayas, the snowfall each year eventually compresses to become ice. Similar to marine sediments, the youngest ice is near the surface and the oldest ice is much deeper. During the process of compression, the ice captures small air bubbles. These bubbles contain tiny samples of the atmosphere that existed at the time the ice was formed. Scientists have traveled to these

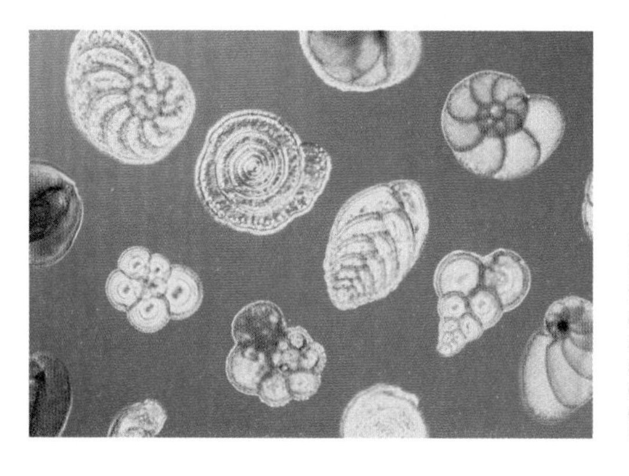

FIGURE 63.5 **Estimating past temperatures using the ancient shells of foraminifera.** The tiny shells of the protists become buried in layers of ocean sediments. By knowing the age of different ocean sediments and the preferred temperature of different species of foraminifera, scientists can indirectly estimate ocean temperature changes over time based on which species are found in each sediment layer. *(Astrid and Hanns-Frieder Michler/Science Source)*

Teaching Tip: Activity

Have your students research the ancient protists known as foraminifera. Ask them to describe how foraminifera help climate scientists estimate historical temperatures of oceans. *Foraminifera are tiny marine organisms with hard shells that resist decay after death. In some regions of the ocean floor, the tiny shells have built up in the sediment over millions of years. The youngest sediment layers are near the top of the ocean floor whereas the oldest sediment layers are much deeper below that. Different species of foraminifera prefer different water temperatures. As a result, when scientists identify the predominant species of foraminifera in a layer of sediment, they can infer the temperature of the ocean at the time the layer of sediment was deposited. By examining thousands of sediment layer samples, scientists can gain insights into temperature change over a very long time horizon.*

(a)

(b)

FIGURE 63.6 **Estimating past greenhouse gas concentrations and past temperatures using ice cores.** (a) Ice cores are extracted from very cold regions of the world such as this glacier on Mount Sajama in Bolivia. (b) Ice cores have tiny trapped air bubbles of ancient air that can provide indirect estimates of greenhouse gas concentrations and global temperatures. *(a: George Steinmetz/Corbis; b: Marc Steinmetz/Visum/The Image Works)*

frozen regions of the world to drill deep into the ice and extract long tubes of ice called ice cores (FIGURE 63.6). Samples of ice cores can span up to 500,000 years of ice formation. Scientists determine the age of different layers in the ice core and then melt the ice from a piece associated with a particular time period. When the piece of ice melts, air bubbles are released and scientists measure the concentration of greenhouse gases in the air when the bubbles were trapped in the ancient ice.

Oxygen atoms in melted ice cores can also be used to determine temperatures from the distant past. Oxygen atoms occur in two forms, or isotopes: light oxygen, also known as oxygen-16 (^{16}O), contains 8 protons plus 8 neutrons. In contrast, heavy oxygen, also known as oxygen-18 (^{18}O), contains 8 protons plus 10 neutrons. Ice formed during a period of warmer temperatures contains a higher percentage of heavy oxygen than ice formed during colder temperatures. By examining changes in the percentage of heavy oxygen atoms from different layers of the ice core, we can indirectly estimate temperatures from hundreds of thousands of years ago.

Combining data from different biological and physical measurements, researchers have created a picture of how the atmosphere and temperature of Earth have changed over hundreds of thousands of years. FIGURE 63.7 shows the pattern of atmospheric CO_2. Notice that for over 400,000 years, the atmosphere

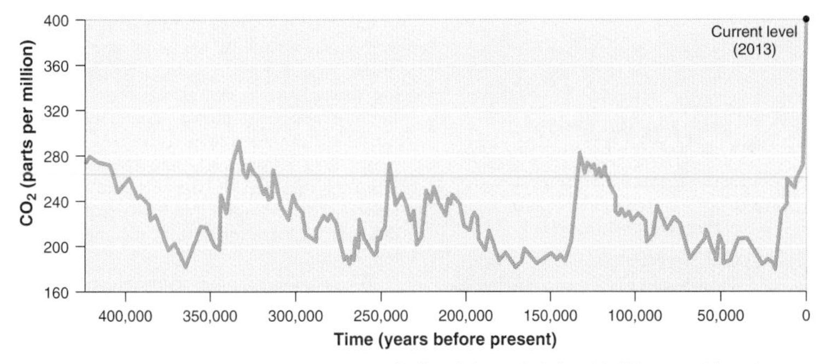

FIGURE 63.7 **Historic CO_2 concentrations.** Using a variety of indirect indicators including air bubbles trapped in ancient ice cores, scientists have found that for more than 400,000 years CO_2 concentrations never exceeded 300 ppm. After 1950, CO_2 concentrations have sharply increased to their current level of nearly 400 ppm. *(After http://climate.nasa.gov/evidence/)*

Teaching Tip: Warm-up

Ask students the following questions:

- How do ice cores from very cold regions of the world help climate scientists calculate historic CO_2 concentrations? *Ice core samples can span up to 500,000 years of ice formation. Scientists determine the age of different layers in the ice core and then melt the ice from a piece associated with a particular time period. When the piece of ice melts, air bubbles are released and scientists measure the concentration of greenhouse gases in the air at the time the bubbles were trapped in the ancient ice.*
- How do scientists use ice cores to examine oxygen isotopes, and what do they learn from this study? *Oxygen atoms occur in two forms, or isotopes. Light oxygen, known as oxygen-16, contains 8 protons plus 8 neutrons. In contrast, heavy oxygen, also known as oxygen-18, contains 8 protons plus 10 neutrons. Ice formed during a period of warmer temperatures contains a higher percentage of heavy oxygen than ice formed during colder temperatures. Scientists can indirectly estimate temperatures from hundreds of thousands of years ago.*

never contained more than 300 ppm of CO_2. In contrast, from 1958 to 2013 the concentration of CO_2 in the atmosphere has rapidly climbed from 310 to 400 ppm. This means that the rise of CO_2 in the atmosphere during the past 50 years is unprecedented.

During the past 10,000 years, CO_2 is not the only greenhouse gas whose concentration has increased. As you can see in FIGURE 63.8, methane and nitrous oxide

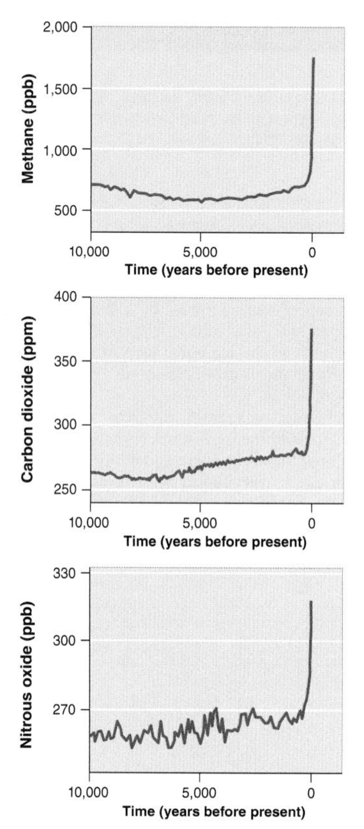

FIGURE 63.8 **Historic concentrations of CO_2, methane, and nitrous oxide.** Using samples from ice cores and modern measurements of the atmosphere, scientists have demonstrated that the concentrations of all three greenhouse gases have increased dramatically during the past 200 years. *(Data from IPCC, 2007: Summary for Policymakers. In* Climate Change 2007: The Physical Science Basis. *Contribution of Working Group I to the Fourth Assessment Report of the Intergovernmental Panel on Climate Change, ed. S.D. Solomon, et al. Cambridge University Press.)*

show a pattern of increase that is similar to the pattern we saw for CO_2. For all three gases, there was little change in concentration for most of the previous 10,000 years. After 1800, however, concentrations of the three gases all rose dramatically. Given what we now know about the anthropogenic sources of greenhouse gases, this increase in greenhouse gases occurred because this time period marks the start of the Industrial Revolution when humans began burning large amounts of fossil fuel and producing a variety of greenhouse gases.

FIGURE 63.9 charts historic temperatures and CO_2 concentrations. Looking at the blue line, we see that temperatures have changed dramatically over the past 400,000 years. Most of these rapid shifts occurred during the onset of an ice age or during the transition from an ice age to a period of warm temperatures after an ice age. Because these changes occurred before humans could have had an appreciable effect on global systems, scientists suspect the changes were caused by small, regular shifts in the orbit of Earth. The path of the orbit, the amount of tilt on Earth's axis, and the position relative to the Sun all change regularly over hundreds of thousands of years. These changes alter the amount of sunlight that hits high northern latitudes in the winter, the amount of snow that can accumulate, and the way the albedo effect keeps energy from being absorbed and converted to heat. These changes could give rise to fairly regular shifts in temperature over a long period of time.

The more important insight from Figure 63.9 is the close correspondence between historic temperatures and CO_2 concentrations. But the graph does not tell us the nature of this relationship. Did periods of increased CO_2 cause increased temperature; did periods of increased temperature cause increased production of CO_2; or is another factor at work? Scientists believe that the relationship between fluctuating levels of CO_2 and the temperature is complex and that both factors play a role. As we know, the increase of CO_2 in the atmosphere causes a greater capacity for warming through the greenhouse effect. However, when Earth experiences higher temperatures, the oceans warm and cannot contain as much CO_2 gas and, as a result, they release CO_2 into the atmosphere. What ultimately matters is the net movement of CO_2 between the atmosphere and the oceans and how these different feedback loops work together to affect global temperatures.

We can also examine temperatures over somewhat shorter time periods. For example, scientists have used indirect and direct measures to determine global temperatures during the past 2,000 years. They then selected the average temperature from 1901 to 2000 as a baseline against which all other years can be compared. FIGURE 63.10 shows that average temperature

680 Chapter 19 Global Change

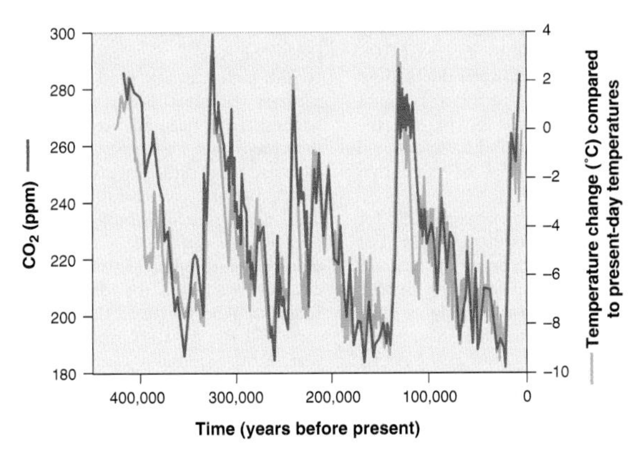

FIGURE 63.9 **Historic temperature and CO_2 concentrations.** Ice cores used to estimate historic temperatures and CO_2 concentrations indicate that the two factors vary together. *(Data from http://www.ncdc.noaa.gov/paleo/globalwarming/temperature-change.html)*

from 0 CE to 800 CE was cooler than the average measured during the twentieth century. Temperatures then tended to warm during 800 CE to 1200 CE and cooler from 1500 to 1800. It is particularly striking that the change in temperature during the past century has increased rapidly and that the average temperatures that we are experiencing today are higher than those experienced in the past 2,000 years.

Greenhouse Gases versus Increased Solar Radiation

We have seen that the surface temperature of Earth has increased roughly 0.8°C (1.4°F) over the past 120 years. But larger changes in temperature have occurred over the past 400,000 years without human influence. How can we tell if the recent changes are anthropogenic? One explanation for warming temperatures during the

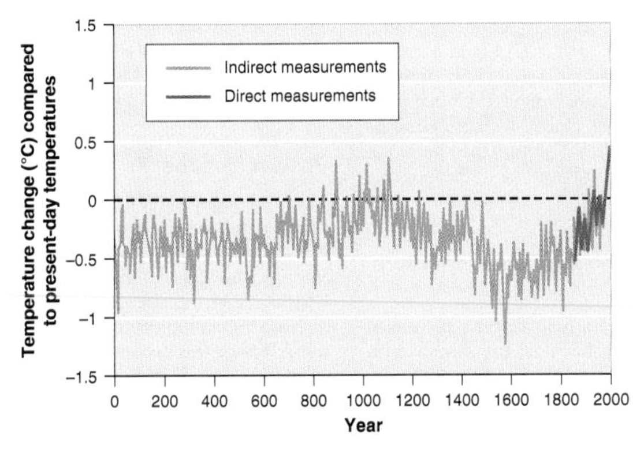

FIGURE 63.10 **Temperature changes in the Northern Hemisphere for the past 2,000 years.** By combining indirect measures of temperatures with direct measures after 1880, we can gain a longer-range perspective on how current global warming compares with past global warming. The average temperature from 1901 to 2000 serves as a baseline against which all other years are compared. As a baseline, the average temperature from 1901 to 2000 is set at "zero temperature change." *(Data from Moberg, A., et al. 2005. Highly variable Northern Hemisphere temperatures reconstructed from low- and high-resolution proxy data. Nature 433:613–617.)*

lab
Climate Change

Lab 19.1 has students observe the greenhouse effect and analyze its impact on Earth, using models with bottles, light, and paper.

Download the lab by clicking on the link buttons in the TE-book, opening the TRFD, or logging onto the book's companion website bcs.whfreeman.com/friedlandapes2e (teacher login required).

TRM **Lab 19.1** Climate Change

past century is an increase in solar radiation. Another possibility is that warming is caused by increased CO_2 in addition to warming caused by natural fluctuations in solar radiation. In other words, both factors might be important. Unfortunately, simply looking at temperature and CO_2 data averages around the globe will not allow us to determine if either of these two possibilities is correct.

One way to approach the problem is to look for more detailed patterns in temperature changes. For example, if increased CO_2 concentrations caused global warming by preventing heat loss, then periods of elevated CO_2 would be associated with higher temperatures more commonly in winter than in summer, at night rather than during the day, and in the Arctic rather than in warmer latitudes. These three scenarios are all associated with colder temperatures, so reducing heat loss would have a bigger impact on temperature than in scenarios in which the temperatures were already quite warm. In fact, we already observed this when we examined the changes in temperature around the world in Figure 63.4—the Arctic regions are experiencing the greatest amount of warming.

On the other hand, if increased solar radiation were the cause of global warming, periods of elevated solar radiation would be associated with higher temperatures more commonly when the Sun is shining more—namely in the summer, during the day, and at low latitudes. These times and locations on Earth receive the greatest amount of sunlight, so an increased intensity of solar radiation would cause these times and places on Earth to warm more than other times and places. When scientists make these types of detailed comparisons, they find that the patterns in temperature change are strongly consistent with increased greenhouse gases such as CO_2 and not consistent with increased solar radiation. This body of evidence led the IPCC to conclude in 2007 that most of the observed increase in global average temperatures since the mid-twentieth century has been the result of increased concentrations of anthropogenic greenhouse gas.

Climate Models and the Prediction of Future Global Temperatures

Just as indirect indicators can help us get a picture of what the temperature has been in the distant past, computer models can help us predict future climate conditions. Researchers can determine how well a model approximates real-world processes by applying it to a time in the past for which we have accurate data on conditions such as air and ocean temperatures, CO_2 concentration, extent of vegetation, and sea ice coverage at the poles. Modern models reproduce recent temperature fluctuations well over large spatial scales. From this work modelers are fairly confident that climate models capture the most significant features of today's climate.

Although climate models cannot forecast future climates with total accuracy, as the models improve scientists have been able to place more confidence in their predictions of temperature change, although they have had more difficulty predicting changes in precipitation. Because assumptions vary among different climate models, when multiple models predict similar changes, we can have increased confidence that the predictions are robust. FIGURE 63.11 shows recent predictions based on the data from climate models. Scientists generally agree that average global temperatures will rise by 1.8°C to 4°C (3.2°F–7.2°F) by the year 2100, depending on whether CO_2 emissions experience slow, moderate, or high growth over time.

Feedbacks can increase or decrease the impact of climate change

The global greenhouse system is made up of several interconnected subsystems with many potential positive and negative feedbacks. As we saw in Chapter 2 with population systems, positive feedbacks amplify changes. Because of this, positive feedback often leads to an unstable situation in which small fluctuations in inputs lead to large observed effects. On the other hand, negative feedbacks dampen changes. When we think about how anthropogenic greenhouse gases will affect Earth, we must ask whether positive or negative feedbacks will predominate. We do not currently have enough evidence to settle this question conclusively, but we can examine some of the feedback cycles and the way they influence temperatures on Earth.

Positive Feedbacks

There are many ways that a rise in temperatures could create a positive feedback. For example, global soils contain more than twice as much carbon as the amount currently in the atmosphere. As shown in FIGURE 63.12a, higher temperatures are expected to increase the biological activity of decomposers in these soils. Because this decomposition leads to the release of additional CO_2 from the soil into the atmosphere, the temperature change will be amplified even more.

A similar, but more troubling, scenario is expected in tundra biomes containing permafrost. As atmospheric concentrations of CO_2 from anthropogenic sources increase, the Arctic regions become substantially warmer and the frozen tundra begins to thaw. As it thaws, the tundra develops areas of standing water with little oxygen available under the water as the

Teaching Tip: Warm-up

Ask students the following questions:

- What are positive and negative feedbacks? *Positive feedbacks amplify changes or lead to an unstable situation in which small fluctuations in inputs cause large observed effects. Negative feedbacks dampen changes.*
- What is an example of a positive feedback in climate change? *If higher temperatures continue, an increase in decomposers in the soils will occur. This increased decomposition leads to the release of additional CO_2 from the soil into the atmosphere. With more CO_2 in the atmosphere, the temperature will continue to rise.*
- What is an example of a negative feedback in climate change? *Because CO_2 is required for photosynthesis, an increase in CO_2 can stimulate plant growth. More plants will cause more CO_2 to be removed from the atmosphere. This negative feedback reduces atmospheric CO_2 and causes temperature increases to be smaller than they otherwise would have been.*

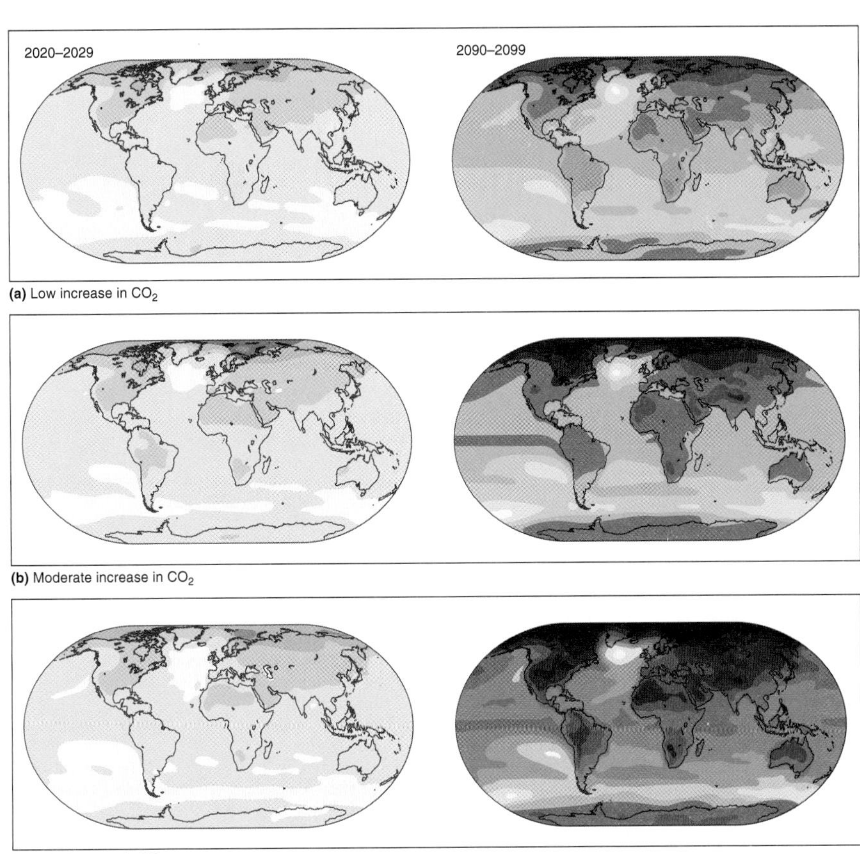

FIGURE 63.11 **Predicted increase in global temperatures by 2100.** The predictions depend on whether we expect (a) low, (b) moderate, or (c) high increases in how much CO_2 the world emits during the current century. These changes in temperature are relative to the mean temperatures from 1961 to 1990. *(Data from http://www.epa.gov/climatechange/science/futuretc.html)*

thick organic layers of the tundra begin to decompose. As a result, the organic material experiences anaerobic decomposition that produces methane, a stronger greenhouse gas than CO_2, which should lead to even more global warming.

Negative Feedbacks

One of the most important negative feedbacks occurs as plants respond to increases in atmospheric carbon. Figure 63.12b shows this cycle. Because carbon dioxide is required for photosynthesis, an increase in CO_2 can stimulate plant growth. The growth of more plants will cause more CO_2 to be removed from the atmosphere. This negative feedback, which causes carbon dioxide and temperature increases to be smaller than they otherwise would have been, appears to be one of the reasons why only about half of the CO_2 emitted into the atmosphere by human activities has remained in the atmosphere.

Module 63 The Evidence for Global Warming 683

Teaching Tip: Review

Before you discuss the text example of negative feedbacks in the oceans, review the hydrologic cycle on p. 80.

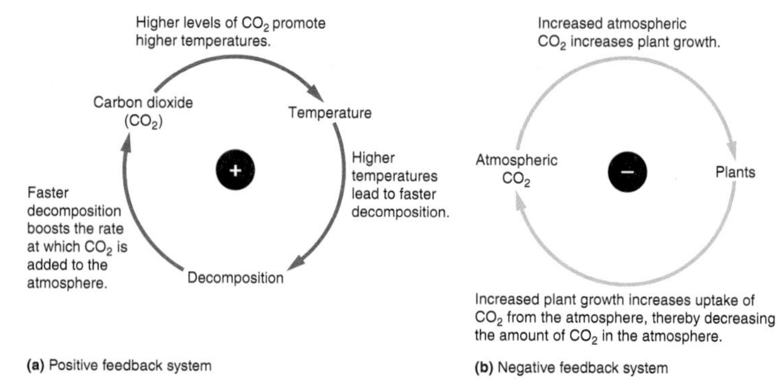

FIGURE 63.12 **Global change feedback systems.** (a) Temperature and CO_2 represent a positive feedback system. When the concentration of CO_2 increases in the atmosphere, it can cause global temperatures to increase. This in turn can cause more rapid decomposition, thereby releasing even more CO_2 into the atmosphere. (b) Carbon dioxide and producers represent a negative feedback system. Increased CO_2 in the atmosphere from anthropogenic sources can be partially removed by increased photosynthesis by producers.

A second negative feedback exists in the oceans. As CO_2 concentrations increase in the atmosphere, more CO_2 is absorbed by the oceans. Although this is beneficial because it reduces CO_2 in the atmosphere, it causes harmful effects to the oceans. When CO_2 dissolves in water, much of it combines with water molecules to form carbonic acid (H_2CO_3). Since this is an equilibrium reaction, an increase in ocean CO_2 causes more CO_2 to be converted to carbonic acid, which lowers the pH of the water in a process known as **ocean acidification.** Ocean acidification is of particular concern for the wide variety of species that build shells and skeletons made of calcium carbonate, including corals, mollusks, and crustaceans. As pH decreases, the calcium carbonate in these organisms can begin to dissolve and the ocean's saturation point for calcium carbonate declines, which makes it harder for organisms to acquire the material they need to build their shells and skeletons.

The Limitations of Feedbacks

Most of the feedbacks we have discussed are limited by features of the systems in which they take place. For example, the soil-carbon feedback is limited by the amount of carbon in soils. While warming soils could add large amounts of CO_2 to the atmosphere for a time, eventually soil stocks will become so low that biological activity falls back to earlier rates. The enhanced CO_2 uptake by plants is also limited: Studies indicate that only some plants benefit from CO_2 fertilization, and often the growth is enhanced only until another factor becomes limiting, such as water or nutrients.

Ocean acidification The process by which an increase in ocean CO_2 causes more CO_2 to be converted to carbonic acid, which lowers the pH of the water.

The magnitude and direction of many feedbacks are complex. For example, water vapor has both positive and negative feedbacks and there are limits to each. As temperatures increase, water can evaporate into the atmosphere more easily. Because water vapor is an important greenhouse gas, this increased evaporation will lead to further warming. There is a limit, however, to the amount of water vapor that can exist in the air. As we discussed in Chapter 4, air can become saturated with water vapor and the amount of saturation changes with temperature. Increased water vapor in the atmosphere can lead to the formation of clouds that can shield the surface of Earth from solar radiation, leading to a negative feedback. In short, the net effect of water vapor on global temperatures depends on several simultaneously occurring processes that make predictions difficult.

module

63

REVIEW

In this module, we learned that the concentrations of greenhouse gases have been steadily increasing in the past century and that the production of these gases differs among nations. At the same time, global temperatures have also increased since 1880, when direct measurements were first made. We also saw how scientists can estimate temperatures and greenhouse gas concentrations for over 500,000 years using changes in species composition of foraminifera and examining the concentrations of gases in bubbles of air that are frozen in ancient ice. Together, these data suggest a close relationship between changes in CO_2 and changes in global temperatures for more than 400,000 years. Using climate models, we can predict future changes in global temperatures under different scenarios of small, medium, or high increases in CO_2 concentrations. In the next module, we will examine the consequences of current and future changes in global temperatures.

Module 63 AP® Review Questions

1. What is the evidence that solar radiation is less important to global warming than the increase in greenhouse gases?
   (a) Temperatures have increased more in the summer than the winter.
   (b) Temperatures have increased more in the winter than the summer.
   (c) Temperature increases have been uneven across the globe.
   (d) There is an historic correlation of CO_2 and temperature.
   (e) There is a lack of temperature change in some areas.

2. If the annual rate of CO_2 increase is 1.5 ppm and the concentration in 2010 is 390 ppm, what concentration would you expect in 2100?
   (a) 420 ppm
   (b) 480 ppm
   (c) 505 ppm
   (d) 525 ppm
   (e) 540 ppm

3. How much has the average global temperature increased in the last 120 years?
   (a) 0.6°F
   (b) 1.0°F
   (c) 1.4°F
   (d) 2.7°F
   (e) 3.2°F

4. Which data are used to estimate historic temperatures and carbon dioxide concentrations?
   I. Marine organism fossils
   II. Air trapped in ice
   III. Glacier depth

   (a) I only
   (b) I and II only
   (c) I and III only
   (d) II and III only
   (e) I, II, and III

5. Which is an example of a negative feedback?
   (a) Higher air temperatures cause increased ocean evaporation.
   (b) Higher air temperatures cause increased decomposition in soils.
   (c) Lower concentrations of CO_2 cause increased absorption of oceanic CO_2.
   (d) Low albedo causes decreased reflection of sunlight.
   (e) Higher CO_2 concentrations cause increased photosynthesis.

Answers to Module 63 AP® Review Questions

1. a
2. d
3. c
4. b
5. e

AP® Exam Tip

Climate change and environmental issues associated with water diversion are considered in the 2009 AP® Exam, Question 3. To answer this question, students must

- name two benefits (other than agriculture and recreation) of large-scale dams
- list two environmental consequences of damming a river
- provide two conservation strategies for reducing the use of agricultural water usage
- list two possible consequences of climate change in the hydrology of the Colorado River
- describe two consequences of climate change on coastal ecosystems

module 64

Consequences of Global Climate Change

A wide range of environmental indicators demonstrate that global warming is affecting global processes and contributing to overall global change. In many cases, we have clear evidence of how global warming is having an effect. In other cases, we can use climate models to make predictions about future changes. As with all predictions of the future, there is a fair degree of uncertainty regarding the future effects of global warming. In this module, we will discuss how global warming is expected to affect the environment and organisms living on Earth.

Learning Objectives

After reading this module you should be able to

- discuss how global climate change has affected the environment.
- explain how global climate change has affected organisms.
- identify the future changes predicted to occur with global climate change.
- explain the global climate change goals of the Kyoto Protocol.

Global climate change is already affecting the environment

Warming temperatures are expected to have a wide range of impacts on the environment. Many of these effects are already happening, including melting of polar ice caps, glaciers, and permafrost and rising sea levels. Other effects are predicted to occur in the future, including an increased frequency of heat waves, fewer and less-intense cold spells, altered precipitation patterns and storm intensity, and shifting ocean currents.

Polar Ice

As we have seen, the Arctic has already warmed by 1°C to 4°C (1.8°F–7.2°F). FIGURE 64.1 illustrates the extent of the reduction in the size of the ice cap that surrounds the North Pole. These data are collected in September of each year, which is about the time that the sea ice has reached its minimum extent. As you can see, the extent of sea ice fluctuates, but there is an overall trend of a 14 percent decline per decade from 1979 to 2013.

Over the next 70 years, the Arctic is predicted to warm by an additional 4°C to 7°C (7°F–13°F) compared to the mean temperatures experienced from 1980 to 1990. If this prediction is accurate, large openings in sea ice will continue to expand and the ecosystem of the Arctic region will be negatively affected. At the same time, though, there may also be benefits to humans. For example, the opening in the polar ice cap could create new shipping lanes that would reduce by thousands of kilometers the distance some ships have to travel. Also, it is estimated that

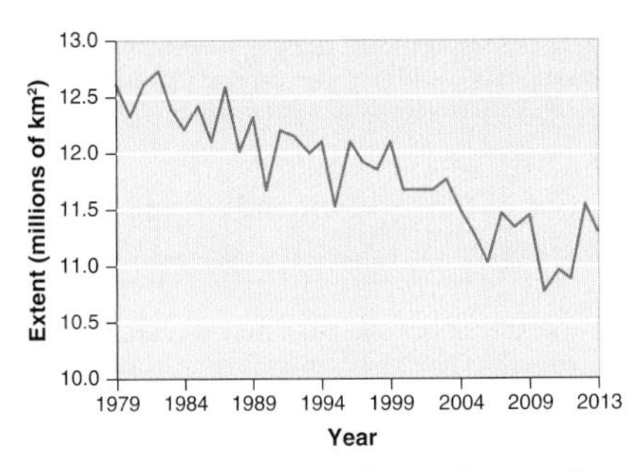

FIGURE 64.1 **The melting polar ice cap.** Because northern latitudes have experienced the greatest amount of global warming, the extent of the ice cap near the North Pole has been declining over the past 3 decades. The polar ice cap reaches its minimum late in the summer of each year, so we can look for a trend by examining the extent of ice each September. From 1979 to 2013, the polar ice has declined an average of 14 percent per decade. *(Data from http://nsidc.org/arcticseaicenews/2013/10/)*

nearly one-fourth of all undiscovered oil and natural gas lies under the polar ice cap and a melted polar ice cap might make these fossil fuels more easily obtainable. However, the combustion of these fossil fuels would further facilitate global warming, representing another example of a positive feedback.

In addition to the polar ice cap in the Arctic, Greenland and Antarctica have also experienced melting. As you can see in FIGURE 64.2, sea ice mass has been measured in Antarctica and Greenland from 2000 to 2013. During this time, Antarctica has lost more than 1,300 gigatons (about 3,000 trillion pounds) of ice while Greenland lost more than 3,000 gigatons (6,600 trillion pounds) of ice. The polar ice cap in Antarctica is particularly interesting. It has shown a small increase in total area over the past three decades. However, the mass of the ice cap includes both area and thickness. Because the ice cap is losing thickness due to melting, its overall mass has been reduced. Current evidence shows that the melting rate of these ice-covered regions is continuing to increase. As we will see, such large amounts of melted ice has caused sea levels to rise.

Glaciers

As we discussed in Chapter 9, global warming has caused the melting of many glaciers around the world. Glacier National Park in northwest Montana, for example, had 150 individual glaciers in 1850 but has only 25 glaciers today. It is estimated that by 2030 Glacier National Park will no longer have any glaciers. The loss of glaciers is not simply a loss of an aesthetic natural wonder. In many parts of the world the melting of glaciers starting each spring provides a critical source of water for many communities. Historically, these glaciers partially melted during the spring and summer and then grew back to their full size during the winter. However, as summers become warmer, glaciers are melting faster than they can grow back in the winter, which leaves some people without a reliable water supply.

Permafrost

As warmer temperatures cause ice caps and glaciers to melt, it is perhaps not surprising that areas of permafrost are also melting. You may recall from the discussion on biomes in Chapter 4 that permafrost is permanently frozen ground that exists in the cold regions of high altitudes and high latitudes, which include the tundra and boreal forest biomes. About 20 percent of land on Earth contains permafrost; in some places it can be as much as 1,600 m (1 mile) thick. Melting of the permafrost causes overlying lakes to become smaller as the lake water

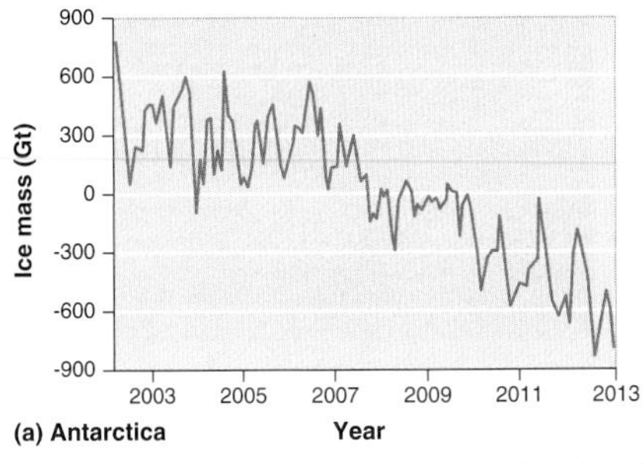

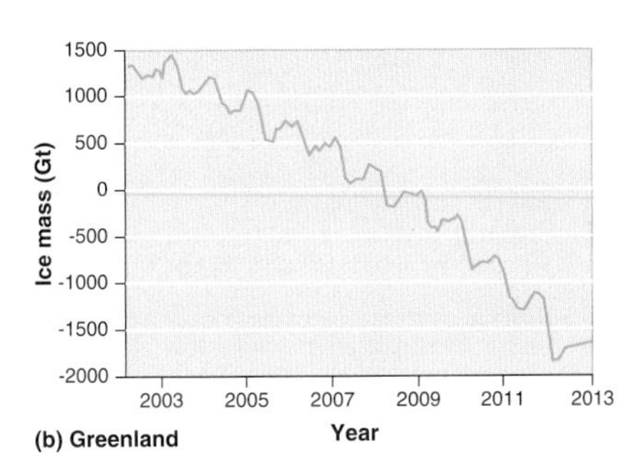

FIGURE 64.2 **Declining ice in Antarctica and Greenland.** Measurements of ice mass from 2002 to 2013 have detected decline in both (a) Antarctica and (b) Greenland. *(Data from http://climate.nasa.gov/key_indicators)*

Teaching Tip: Video

Melting Mountain Glaciers

In this video from the National Science Foundation and NBC Learn, scientists discuss melting glaciers and the effects of climate change, with a focus on Mount Kilimanjaro in Tanzania.

The link to this video may be found by clicking the link buttons in the TE-book, opening the TRFD, or logging onto the book's companion website bcs.whfreeman.com/friedlandapes2e (teacher login required).

Chapter 19 Web Resources

Teaching Tip: Video

Thawing Permafrost

This video from the *Changing Planet* series by the National Science Foundation and NBC Learn discusses how permafrost has been thawing over the past 3 decades and the risk this may pose to the environment.

The link to this video may be found by clicking the link buttons in the TE-book, opening the TRFD, or logging onto the book's companion website bcs.whfreeman.com/friedlandapes2e (teacher login required).

Chapter 19 Web Resources

AP® Exam Tip

The relationship between global mean atmospheric temperature and the sea level are considered in the 2010 AP® Exam, Question 4. To answer this question, students must

- use provided data to calculate the expected increase in sea level during the next 50 years
- identify the two phenomena that result from an increase in global temperature and explain how they affect sea level
- describe two environmental impacts of a rising sea level on estuarine ecosystems
- identify one negative economic impact of a rise in sea level on human populations living along a coast and describe two viable government strategies to discourage people from settling in coastal areas

drains deeper down into the ground. Melting can also cause substantial problems with human-built structures that are anchored into the permafrost, including houses and oil pipelines. As the frozen ground melts, it can subside and slide away.

Melting permafrost also means that the massive amounts of organic matter contained in the tundra will begin to decompose. Because this decomposition would be occurring in wet soils under low-oxygen conditions, it would release substantial amounts of methane, increasing the concentration of this potent greenhouse gas. This chain of events could produce a positive feedback in which the warming of Earth melts the permafrost, releasing more methane that causes further global warming.

Sea Levels

The rise in global temperatures affects sea levels in two ways. First, the water from melting glaciers and ice sheets on land adds to the total volume of ocean water. Second, as the water of the oceans becomes warmer, it expands. FIGURE 64.3 shows that, as a result of both these effects, sea levels have risen 220 mm (9 inches) since 1870. Scientists predict that by the end of the twenty-first century, sea levels could rise an additional 180 to 590 mm above 1999 levels (7–23 inches). This could endanger coastal cities and low-lying island nations by making them more vulnerable to flooding, especially during storms, with more saltwater intrusion into aquifers and increased soil erosion. Currently, 100 million people live within 1 m (3 feet) of sea level. The actual impact on these areas of the world will depend on the steps taken to mitigate these effects. For example, as we saw in Chapter 9, some countries may be able to build up their shorelines with dikes to prevent inundation from rising sea levels. Countries possessing less wealth are not expected to be able to respond as effectively to coastal flooding.

Global climate is already affecting organisms

The warming of the planet not only is affecting polar ice caps and sea levels but also is affecting living organisms. These effects range from temperature-induced changes in the timing of plant flowering and animal behavior to the ability of plants and animals to disperse to more hospitable habitats.

During the last decade, the IPCC reviewed approximately 2,500 scientific papers that reported the effects of warmer temperatures on plants and animals. The panel concluded that over the preceding 40 years, the growing season for plants had lengthened by 4 to 16 days in the Northern Hemisphere, with the greatest increases occurring in higher latitudes. Indeed, scientists are finding that in the Northern Hemisphere many species of plants now flower earlier, birds arrive at their breeding grounds earlier, and insects emerge earlier. At the same time, the ranges occupied by different species of plants, birds, insects, and fish have shifted toward both poles.

Rapid temperature changes have the potential to cause harm if organisms do not have the option of moving to more hospitable climates and do not have sufficient time to evolve adaptations. Historically, organisms have migrated in response to climatic changes. This ability to migrate is one reason that temperature shifts have not been catastrophic over the past few million years. Today, however, fragmentation of

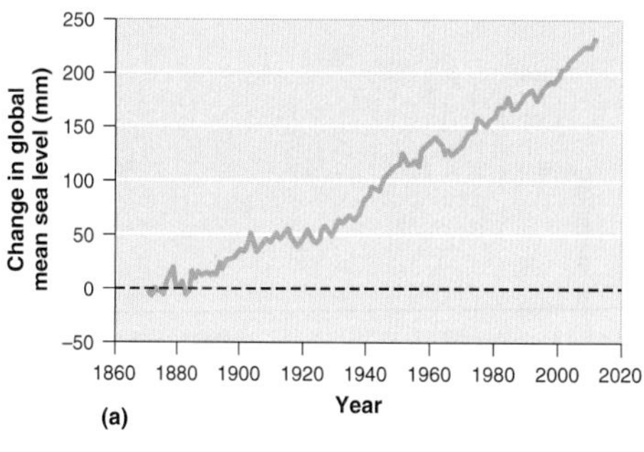

FIGURE 64.3 **Rising sea levels.** (a) Since 1870, sea levels have risen by 220 mm (9 inches). Future sea level increases are predicted to be 180 to 590 mm (7 to 23 inches) above 1999 levels by the end of this century. (b) Nearly 100 million people live within 1 m (3 feet) of sea level, such as on this island in the Maldives in the Indian Ocean. *(a: Data from http://www.cmar.csiro.au/sealevel/index.html; b: WO/Shutterstock)*

Teaching Tip: Video

Rising Sea Levels

This video from the *Changing Planet* series by the National Science Foundation and NBC Learn discusses how sea levels are rising due to global climate change.

The link to this video may be found by clicking the link buttons in the TE-book, opening the TRFD, or logging onto the book's companion website bcs.whfreeman.com/friedlandapes2e (teacher login required).

TRM **Chapter 19 Web Resources**

FIGURE 64.4 **Effects of global warming on the pied flycatcher.** Due to global warming in the Netherlands, the bird's main food source, a caterpillar, now becomes abundant 2 weeks earlier. However, the time when eggs hatch for the flycatcher has not changed. As a result, the birds hatch after the caterpillar population has begun to decline. *(Kats Edwin/age footstock)*

certain habitats by roads, farms, and cities has made movement much more difficult. In fact, this fragmentation may be the primary factor that allows a warming climate to cause the extinction of species.

The pied flycatcher (*Ficedula hypoleuca*) is a bird that provides an interesting example. In the Netherlands, the pied flycatcher has evolved to synchronize the time that its chicks hatch with the time of peak abundance of caterpillars, a major source of food for the newly hatched birds (FIGURE 64.4). In 1980, the date of hatching preceded the peak in caterpillar abundance by a few days, so there was plenty of food for the new chicks. Twenty years later, warmer spring temperatures have caused trees to produce leaves earlier in the spring. Because the caterpillars feed on tree leaves, the peak abundance of caterpillars now occurs about 2 weeks earlier than it did in 1980. However, the hatching date of the pied flycatcher has not been affected by temperature change. Thus, by the time the chicks hatch, the caterpillars are no longer abundant and the hatchlings lack a major source of food, causing flycatcher populations in these areas to decline by 90 percent. There are other similar examples of the effects of rapid temperature change throughout the natural world.

Corals are one group of organisms that are particularly sensitive to global warming because their range of temperature tolerance is quite small. Many corals worldwide are currently undergoing "bleaching." Coral bleaching occurs when stressed corals eject their mutualistic algae, which provide corals with energy. The loss of algae causes the coral to turn white. The underlying causes of coral bleaching appear to be a combination of warming oceans, pollution, and sedimentation. Bleaching is sometimes temporary, but if it lasts for more than a short time, the corals die. While new corals should colonize regions at higher latitudes, it will take centuries before major new reef systems can be built. More coral bleaching is expected from global warming even if climate changes are kept relatively small.

Global climate change is predicted to have additional effects in the future

Much of the controversy about global warming is not related to how the planet is already being affected, but rather what is predicted to happen in the future. Predicted future changes have some amount of uncertainty because they are based on computer models of complicated systems, such as the world's climates. It will take several decades to determine whether these predictions come true. In this section, we examine a number of these predicted effects including the frequency of heat waves, cold spells, precipitation patterns, storm intensity, and changes in ocean currents.

Heat Waves

As temperatures increase, long periods of hot weather—known as heat waves—are likely to become more frequent. Heat waves cause an increased energy demand for cooling the homes and offices where people live and work. For people who lack air conditioning in their homes, especially the elderly, heat waves increase the risk of death. Heat waves also cause heat and drought damage to crops, prompting the need for greater amounts of irrigation. The increased energy required for irrigation would raise the cost of food production.

Cold Spells

With global temperatures rising, minimum temperatures are expected to increase over most land areas, with fewer extremely cold days and fewer days below

Teaching Tip: Video

Withering Crops
This video from the *Changing Planet* series by the National Science Foundation and NBC Learn discusses the impact climate change may have on the world's crops.

The link to this video may be found by clicking the link buttons in the TE-book, opening the TRFD, or logging onto the book's companion website bcs.whfreeman.com/friedlandapes2e (teacher login required).

Chapter 19 Web Resources

freezing. Such conditions would have two major positive effects for humans: fewer deaths due to cold temperatures and a decrease in the risk of crop damage from freezing temperatures. It may also make new areas available for agriculture that are currently too cold to grow crops. In addition, warmer temperatures would decrease the energy needed to heat buildings in the winter. However, a decrease in freezing temperatures that normally would cause the death of some pest species might allow these pest species to expand their range. The hemlock wooly adelgid (*Adelges tsugae*), for example, is an invasive insect from Asia that is causing the death of hemlock trees in North America by feeding on sap. Researchers have found that the range of the species is limited by cold temperatures. Warmer conditions in future decades are expected to allow this pest to expand its range and kill hemlock trees over a much larger area.

Precipitation Patterns

Because warmer temperatures should drive increased evaporation from the surface of Earth as part of the hydrologic cycle (see Figure 7.1 on page 80), global warming is projected to alter precipitation patterns. As mentioned earlier, climate models do not make consistent predictions about precipitation. Current models predict that some areas will experience increased rainfall, but the models differ in predicting which regions of the world would be affected. Regions receiving increased precipitation would benefit from an increased recharge to aquifers and higher crop yields, but they could also experience more flooding, landslides, and soil erosion. In contrast, other regions of the world are predicted to receive less precipitation, making it more difficult to grow crops and requiring greater efforts to supply water.

Storm Intensity

Although it is impossible to link any single weather event to climate change because of the multiple factors that are always involved, ocean warming may be increasing the intensity of Atlantic storms. For example, in 2005, hurricanes Katrina and Rita devastated coastal areas in Texas, Louisiana, and Mississippi. These hurricanes appear to have become as powerful as they did because waters in the Gulf of Mexico were unusually warm. Scientists at the National Center for Atmospheric Research concluded that climate change was responsible for at least half of this warming. As temperatures increase, such conditions should become more frequent, and hurricanes are likely to become more common farther north. The devastation in New Orleans did not come as a surprise; for many years scientists had warned that a strong hurricane could flood the city because of its position below sea level. Unfortunately, scientists caution that New Orleans is not the only American city that could be devastated by a powerful storm. Other cities at risk include New York, Miami, and Tampa.

Ocean Currents

Global ocean currents may shift as a result of more fresh water being released from melting ice. In Chapter 4 we saw that ocean currents have major effects on the climate of nearby continents. If the currents change, the distribution of heat on the planet could be disrupted. Scientists are particularly concerned about the thermohaline circulation, which, as we saw in Chapter 4, is a deep ocean circulation driven by water that comes out of the Gulf of Mexico and moves up to Greenland where it becomes colder and saltier and sinks to the ocean floor. This sinking water mixes with the deep waters of the ocean basin, resurfaces near the equator, and eventually makes its way back to the Gulf of Mexico (see Figure 11.2 on page 119). This circulating water moves the warm water from the Gulf of Mexico up toward Europe and moves cold water from the North Atlantic down to the equator. However, increased melting from Greenland and the northern polar ice cap could dilute the salty ocean water sufficiently to stop the water from sinking near Greenland and thereby shut off the thermohaline circulation. If this occurrs, much of Europe would experience significantly colder temperatures.

Effects on Humans

Global warming and climate change could also affect many aspects of our lives. For example, some people may have to relocate from such vulnerable areas as coastal communities and some ocean islands. Poorer communities close to or along coastlines might not have the resources to rebuild on higher ground. If these communities do not obtain financial assistance, they will face severe consequences from flooding and saltwater intrusion. On the other hand, certain areas that have not been suitable for human habitation might become more hospitable if they become warmer, although other factors, such as water availability, might still limit their habitability.

Climate change has the potential to affect human health. Continued warming of the planet could affect the geographic range of temperature-limited disease vectors. For example, the mosquitoes that carry West Nile virus and malaria could spread beyond their current geographic range and bring health threats to regions that were once relatively untouched. As the climate changes, heat waves could cause more deaths to the very young, the very old, and those without access to air conditioning. Infectious diseases and bacterial and fungal illnesses might extend over a wider range than they do at present.

FIGURE 64.5 **Adjusting tourism to climate change.** In the French Alps, global warming has meant that many ski resorts are receiving less snow in the winter. In response, some ski resorts are altering their mountaintop facilities to cater to other types of tourists, including those who come in the summer because they want to leave hot cities and recreate in the cooler mountains. *(Patrice Schreyer/ Rapsodia/Aurora Photos)*

Climate changes will also have economic consequences. In northern locations, for example, warmer temperatures and shorter winters would drastically alter the character of northern communities that depend on snow for tourism. In the Alps, for example, many ski resorts are already adjusting to reduced snow on the mountains by catering to new groups of tourists who are more interested mountain biking than skiing (FIGURE 64.5). In warmer regions, the damage to corals reefs would negatively affect tourism as well. The economic impact on these types of tourist attractions depends on the rate of climate change and the ability of the tourism industry to adjust.

Assessing Uncertainty

How much controversy really exists regarding climate change? This question has been portrayed in many ways by various interest groups. Advocates in the environmental community talk of a "scientific consensus" on the topic of global warming while opponents of government regulation often speak of the global warming "controversy." The fundamental basis of climate change—that greenhouse gas concentrations are increasing and that this will lead to global warming—is not in dispute among the vast majority of scientists. Increases in greenhouse gases have been documented with real data and the ability of greenhouse gases to absorb and emit infrared radiation is a simple application of physics. Furthermore, the fact that the globe is warming is not in dispute; as we have seen, the data have clearly demonstrated increased global temperatures since direct measurements began in 1880, declines in the polar ice cap since measurements began in 1979, declines in Greenland and Antarctica ice sheets since measurements began in 2002, and rises in sea level since measurements began in 1880. As we discussed, what remains unclear is the likelihood that other changes in our climate have already started to happen.

In their 2007 report, the IPCC attempted to address some of the uncertainty by listing the likelihood that various types of climate change are already occurring, the likelihood that humans contributed to the changes, and the likelihood that these trends will continue through the twenty-first century. Their results are shown in **TABLE 64.1**. For example, the panel concluded that a decline in cold days and an increase in warm days very likely occurred in most land areas, that these changes likely had a human contribution, and that it was virtually certain that these trends would continue through the twenty-first century. They also concluded that heat waves, droughts, heavy precipitation, and hurricanes have likely increased since 1960, that these changes probably were influenced by human activities, and that these trends were likely to continue through the twenty-first century.

The Kyoto Protocol addresses climate change at the international level

Awareness of global change is a relatively recent phenomenon. In the past, most environmental issues could be dealt with at the national, state, or even local level. Global change is different because the scale of impact is so much larger and because the people and ecosystems affected can be extremely distant from the cause. In the case of climate change, many of the adverse effects are expected to be in the developing world, which has received disproportionately fewer benefits from the use of fossil fuels that led to the change in the first place. It would be impossible for

Teaching Tip: Review

Ask students the following questions:

- What are the causes of global climate change? *Global climate change has occurred because of increases in the amounts of the five greenhouse gases: water vapor, carbon dioxide, methane, nitrous oxide, and chlorofluorocarbons.*
- How does global climate change affect the environment? *The environmental effects of global climate change include melting of polar ice and glaciers, melting of permafrost, increasing sea levels around the world, and changing habitats.*
- What changes might we expect to see in the future as a result of climate change? *Future change might include increased heat spells, change in precipitation patterns, greater storm intensity, and changes in ocean currents.*

Teaching Tip: Discussion Starter

Ask students the following questions:

- What is the Kyoto Protocol? *The Kyoto Protocol is an international agreement that set a goal for global emissions of greenhouse gases from all industrialized countries to be reduced by 5.2 percent below their 1990 levels by 2012.*
- How is the Kyoto Protocol an example of the precautionary principle? *The main argument for the Kyoto Protocol is grounded in the precautionary principle, which states that in the face of scientific uncertainty we should behave cautiously. In the case of climate change, this means that since there is sufficient evidence to suggest human activities are altering the global climate, we should take measures to stabilize greenhouse gas concentrations either by reducing emissions or by removing the gases from the atmosphere.*

TABLE 64.1 The 2007 assessment of global change by the Intergovernmental Panel on Climate Change (IPCC)

The scientists considered the likelihood that specific changes have occurred, the likelihood that humans contributed to the change, and the likelihood that current trends will continue.

Definitions: More likely than not = more than 50% certain; Likely = more than 60% certain; Very likely = more than 90% certain; Virtually certain = more than 99% certain.

Phenomenon and direction of trend	Likelihood that trend occurred in the late 20th century (typically post-1960)	Likelihood of a human contribution to observed trend	Likelihood of future trends based on projections for 21st century from *Special Report on Emissions Scenarios*
Warmer and fewer cold days and nights over most land areas	Very Likely	Likely	Virtually Certain
Warmer and more frequent hot days and nights over most land areas	Very Likely	Likely (nights)	Virtually Certain
Warm spells/heat waves. Frequency increases over most land areas	Likely	More likely than not	Very Likely
Heavy precipitation events. Frequency (or proportion of total rainfall from heavy falls) increases over most areas	Likely	More likely than not	Very Likely
Area affected by drought increases	Likely in many regions since 1970	More likely than not	Likely
Intense tropical cyclone activity increases	Likely in many regions since 1970	More likely than not	Likely
Increased incidence of extreme high sea level (excludes tsunamis)	Likely	More likely than not	Likely

Source: IPCC, 2007: Summary for Policymakers. In *Climate Change 2007: The Physical Science Basis. Contribution of Working Group I to the Fourth Assessment Report of the Intergovernmental Panel on Climate Change* [Solomon, S., D. Qin, M. Manning, Z. Chen, M. Marquis, K.B. Averyt, M.Tignor and H.L. Miller (eds.)]. Cambridge University Press, Cambridge, United Kingdom and New York, NY, USA.

one nation to pass legislation allowing it to avoid the impacts of climate change. To address the problem of global warming, the nations of the world must work together.

In 1997, representatives of the world's nations convened in Kyoto, Japan, to discuss how best to control the emissions contributing to global warming. At this meeting, they drew up the **Kyoto Protocol,** an international agreement which set a goal for global emissions of greenhouse gases from all industrialized countries to be reduced by 5.2 percent below their 1990 levels by 2012. Due to special circumstances and political pressures, countries agreed to different levels of emission restrictions, including a 7 percent reduction for the United States, an 8 percent reduction for the countries of the European Union, and a 0 percent reduction for Russia. Developing nations, including China and India, did not have emission limits imposed by the protocol. These developing nations argued that different restrictions on developed and developing countries are justified because developing countries are unfairly exposed to the consequences of global warming that in large part come from the developed nations. Indeed, the poorest countries in the world have only contributed to 1 percent of historic carbon emissions but are still affected by global warming. Thus, the approach was to have the countries that have historically emitted the most CO_2 pay most of the costs of reducing CO_2.

The main argument for the Kyoto Protocol is grounded in the precautionary principle, which, as we saw in Chapter 17, states that in the face of scientific evidence that contains some uncertainty we should behave cautiously. In the case of climate change, this means that since there is sufficient evidence to suggest human activities are altering the global climate, we should take measures to stabilize greenhouse gas concentrations either by reducing emissions or by removing the gases from the atmosphere. The first option includes trying to increase fuel efficiency or switching from coal and oil to energy sources that emit less or no

Kyoto Protocol An international agreement that sets a goal for global emissions of greenhouse gases from all industrialized countries to be reduced by 5.2 percent below their 1990 levels by 2012.

CO_2 such as natural gas, solar energy, wind-powered energy, or nuclear energy. The second option includes **carbon sequestration**—an approach that involves taking CO_2 out of the atmosphere. Methods of carbon sequestration might include storing carbon in agricultural soils or retiring agricultural land and allowing it to become pasture or forest, either of which would return atmospheric carbon to longer-term storage in the form of plant biomass and soil carbon. Researchers are also working on cost-effective ways of capturing CO_2 from the air, from coal-burning power plants, and from other emission sources. This captured CO_2 can then be compressed and pumped into abandoned oil wells or the deep ocean. Such technologies are still being developed, so their economic feasibility and potential environmental impacts are not yet known.

In developed countries, reductions in CO_2 emissions would require major changes to manufacturing, agriculture, or infrastructure at significant expense and economic impact. In 1997, before the Kyoto Protocol was finalized, the U.S. Senate voted unanimously (95–0) that the United States should not sign any international agreement that lacked restrictions on developing countries or any agreement that would harm the economy of the United States. Despite this vote, the vice president at the time, Al Gore, was instructed by President Bill Clinton (1993–2001) to sign the protocol on behalf of the United States. President Clinton, however, never sent the agreement to the Senate for ratification since it was clear the Senate would not vote in favor. Since it was never ratified, the Kyoto Protocol is not legally binding on the United States.

In 2001, the Kyoto Protocol was modified to convince more developed nations to ratify it. At that time, the United States, under the administration of President George W. Bush (2001–2009), argued that there was too much uncertainty in global warming predictions to justify ratification of the protocol. The administration also argued that the costs of controlling carbon dioxide emissions would unfairly disadvantage businesses in the United States while businesses in China and India—two developing countries that were not significant emitters of greenhouse gases prior to 1990 but are significant emitters today—have no reduction requirements. Ultimately, the United States argued that efforts to limit greenhouse gases should wait until there is more scientific evidence for global warming and the effects it produces before accepting the costs that the protocol would entail. Furthermore, the United States argued that all countries should be subject to emissions limits. Otherwise polluting factories in developed countries could simply relocate to the developing countries, and little good would have been accomplished. Based on similar arguments, Canada, Russia, and Japan have since withdrawn targets for reduced CO_2 emissions. Proponents of reducing greenhouse gas emissions argue that profits gained from manufacturing new pollution-control technologies and savings in fuel costs through greater efficiency will offset any costs or decrease in short-term profits.

More recently, the U.S. government has taken stronger steps to regulate CO_2 emissions. In 2007, the U.S. Supreme Court ruled that the U.S. Environmental Protection Agency not only had the authority to regulate greenhouse gases as part of the Clean Air Act, but that it was required to do so. As a result, in 2009 the EPA announced it would begin regulating greenhouse gases for the first time. In 2010, the U.S. EPA began to look more closely at possible ways to regulate emissions of carbon dioxide. One proposal embraced by auto manufacturers has been to increase the fuel efficiency of vehicles. In 2012, the Obama Administration announced that the average fuel efficiency of cars and light trucks would increase from the current 12 km per liter (29 miles per gallon) to 16 km per liter (37 miles per gallon) by 2017 and 23 km per liter (55 miles per gallon) by 2025. This increase in efficiency is projected to cause a 50 percent reduction in CO_2 and other greenhouse gases from vehicles by 2025. The more fuel-efficient cars are expected to cost an extra $1,000, but the reduced consumption of gasoline is expected to save the average driver $3,000 over the lifetime of the vehicle. This would allow the United States to reduce greenhouse gases, invest in new automotive technology, reduce its consumption of fossil fuels, and save money.

As of 2014, 192 countries have ratified the Kyoto Protocol, including most developed and developing countries, although more than 100 developing countries are exempt from any limits on CO_2 emissions including China and India. The United States is the only developed country that has not yet ratified the agreement. Changes in CO_2 emissions in various countries have been mixed. For example, the total emissions of greenhouse gases from 1990 to 2012 have decreased by 25 percent for Germany and 34 percent for Russia. In contrast, greenhouse gas emissions have increased by 30 percent for Australia, 30 percent for Canada, and 8 percent for the United States. However, emissions in the United States have been declining since 2007 for a variety of reasons including a slower economy, increased fuel efficiency, and an increase in the use of natural gas rather than coal for power generation. Globally, total greenhouse gases increased by 34 percent between 1990 and 2013.

Carbon sequestration An approach to stabilizing greenhouse gases by removing CO_2 from the atmosphere.

Teaching Tip: Beyond the Classroom

Ask students to research different methods of carbon sequestration and write a one-page paper describing them.

Here are some questions for students to consider when they research the subject:

- What is the history of the idea to sequester carbon?
- Is there a small model being tested currently? Where is this carbon sequestration model being used?
- What are the economic feasibility and the potential environmental impacts of this idea?
- What is your opinion on the feasibility of carbon sequestration?

module

64

REVIEW

In this module, we learned that the warming of Earth due to human activities has had a number of effects. In regard to the environment, global warming has caused declines in the polar ice cap, declines in the ice masses of Greenland and Antarctica, melting of glaciers, thawing of permafrost, and increases in sea level. In regard to organisms, global warming has caused changes in the dates of flowering, bird migrations, insect emergence, and the length of the growing season. Scientists predict a number of additional effects with various levels of certainty, including more extreme temperatures, changes in global patterns of precipitation, more intense storms, and altered ocean currents. These predicted changes could affect many aspects of human life including where humans can live as well as their health.

Answers to Module 64 AP® Review Questions

1. a
2. d
3. d
4. e
5. b

Module 64 AP® Review Questions

1. Global warming might limit the availability of fresh water in many areas because of
   (a) melting glaciers.
   (b) increased sea level.
   (c) thawing permafrost.
   (d) more frequent heat waves.
   (e) changes in storm intensity.

2. Effects of climate change on organisms include all of the following except
   (a) increased growing seasons.
   (b) disruption of animal life cycles.
   (c) coral bleaching.
   (d) decreased species ranges in temperate areas.
   (e) increased dispersal of plants and animals.

3. What is NOT a potential negative effect of climate change on agriculture?
   (a) Increased range of pests
   (b) Increased risk of droughts
   (c) Increased damage from severe weather events
   (d) Increased rates of photosynthesis
   (e) Decreased growing season in western Europe

4. How might climate change increase the range of pests?
   (a) Higher intensity weather events
   (b) Changing precipitation patterns
   (c) Increasing number of heat waves
   (d) Disruption of oceanic currents
   (e) Decreased duration of cold spells

5. In 1997 the United States did not ratify the Kyoto Protocol because
   (a) it required the same emission reductions from all signatories.
   (b) it did not limit the emissions of developing nations.
   (c) the predictions regarding the severity of climate change were too uncertain.
   (d) current technology was unable to meet the required goals.
   (e) the target emission reductions were seen as impossible to achieve.

working toward sustainability

Cities, States, and Businesses Lead the Way to Reduce Greenhouse Gases

Although the United States signed the original Kyoto Protocol, the U.S. Congress never ratified the agreement so the protocol has never been legally binding on the United States. The administration of President George W. Bush argued that there was no scientific consensus on global warming and that the costs of reducing greenhouse gases were simply too high. However, many state and local governments felt they had waited long enough for change at the federal level. In 2005, mayors from 141 cities and both major political parties gathered in San Francisco to organize their own efforts to reduce the causes and consequences of global warming. Their goal was to reduce greenhouse emissions in their own cities by the same 7 percent that the United States had agreed to in the Kyoto Protocol.

As of 2014, a total of 1,060 out of 1,139 mayors of U.S. cities had signed the U.S. Conference of Mayors Climate Protection Agreement. Among the reasons the mayors cited for supporting this agreement were concerns in their communities over increasing droughts, reduced supplies of fresh water due to melting glaciers, and rising sea levels in coastal cities. "The United States inevitably will have to join this effort," Seattle mayor Greg Nickels said. "Ultimately we will make it impossible for the federal government to say no. They will see that it can be done without huge economic disruption and that there's support throughout the country to do this."

Similar actions are being taken at the state level. In 2005, then-governor of California Arnold Schwarzenegger stated at a press conference, "The debate is over . . . and we know the time for action is now." In 2006, Governor Schwarzenegger signed the California Global Warming Solutions Act. The goal of the act was to bring California into compliance with the Kyoto Protocol by 2020, an effort that would require a 25 percent reduction in greenhouse gases for a state that, if a country, would be the tenth largest producer of greenhouse gases in the world. At the signing ceremony, the governor stated, "I say unquestionably it is good for businesses." Indeed, a cost analysis by the California Air Resources Board in 2008 indicated that the law would add $27 billion to the economy of the state and add 100,000 jobs.

The California effort is gaining popularity around the country. In the northeastern United States, for example, nine states have joined together collectively to form the Regional Greenhouse Gas Initiative to control regional production of greenhouse gases. A similar group emerged in western North America when seven western states and four Canadian provinces joined together in 2007 to form the Western Climate Initiative. For both groups, the goal was to to regulate greenhouse emissions. By 2014, northeastern group continued to work together while the western group had a reduced membership that included only California and the four Canadian provinces.

A number of large businesses are also joining in efforts to reduce greenhouse gases. General Electric, for example, announced in 2014 that it had reduced its greenhouse emissions by 34 percent since 2004. In addition, the company has invested $12 billion for research and development of technologies that can reduce greenhouse gases and is planning to invest a total of $25 billion by 2020. In 2011, General Electric announced that its technology generated more than $100 billion in revenues, which confirmed that creating technology that would reduce greenhouse emissions was a profitable thing to do.

In 2013, the *New York Times* reported that a growing number of companies including Microsoft, ExxonMobil, and Google have developed long-term financial plans that include the cost of producing greenhouse gases. These companies recognize that the scientific evidence of human-caused global climate change continues to grow and that they will increasingly need to factor the costs of emissions into their budgets. Those companies

Innovative technologies to reduce greenhouse emissions. Companies such as General Electric have invested billions of dollars into technologies that are more energy efficient and produce less greenhouse gas. The Evolution® Series Tier 3 locomotive uses less fuel and has 40 percent lower emissions than previous locomotives. *(Courtesy of GE Transportation)*

Teaching Tip: Beyond the Classroom

Have your students research the U.S. Conference of Mayors Climate Protection Agreement. At the website, students can look at a map of the United States to see if their city is one that is presently involved with the Mayors Climate Protection Agreement. Agreement cities are committed to participate in three actions. Ask students to list the three actions.

The three actions are:

- *Strive to meet or beat the Kyoto Protocol targets in their own communities through actions ranging from antisprawl land-use policies to urban forest restoration projects to public information campaigns.*
- *Urge their state governments and the federal government to enact policies and programs to meet or beat the greenhouse gas emission reduction target suggested for the United States in the Kyoto Protocol.*
- *Urge the U.S. Congress to pass the bipartisan greenhouse gas reduction legislation, which would establish a national emission trading system.*

Find a link to the U.S. Conference of Mayors Climate Protection Agreement website by clicking the link buttons in the TE-book, opening the TRFD, or logging onto the book's companion website bcs.whfreeman.com/friedlandapes2e (teacher login required).

Chapter 19 Web Resources

Suggested Answers to Critical Thinking Questions

1. Data on the increase in CO_2 concentrations and correlation to human activities (primarily the burning of fossil fuel) and historic correlations between CO_2 and temperature can help support the assertion that humans are causing global warming.

2. Because global warming is an Earth systems problem, it will be more effective to address the problem on the largest scale possible. Many economic actions like a carbon tax or subsidies for renewable or efficient technologies will have a much larger impact if they occur over large areas.

that include plans to accommodate and reduce these costs are likely to profit from such planning.

From these stories, it is clear that progress on reducing greenhouse gases that cause global warming does not have to wait for national and international agreements to take effect. The public overwhelmingly understands that Earth is warming, states and cities are pushing forward with solutions that save money, and large corporations understand that reducing emissions can reduce costs and improve profits over the long term. In short, curbing greenhouse gases and global warming is not only good for humans and the environment, it can be good for business as well.

Critical Thinking Questions

1. What data might city mayors use to support their assertion that humans are causing global warming?
2. Why is it more effective for states and provinces to create regional partnerships to combat global warming rather than doing so alone?

References

Barringer, F. 2013. States group calls for 45% cut in amount of carbon emissions allowed. *New York Times*, February 7. http://www.nytimes.com/2013/02/08/business/energy-environment/states-group-calls-for-45-cut-in-amount-of-carbon-emissions-allowed.html.

Davenport, C. 2013. Large companies prepared to pay price on carbon. *New York Times*, December 5, 2013. http://www.nytimes.com/2013/12/05/business/energy-environment/large-companies-prepared-to-pay-price-on-carbon.html.

Peer, M. 2009. GE's green goals. *Forbes*, June 11. http://www.forbes.com/2009/06/11/general-electric-energy-markets-equities-alternative.html.

chapter 19 REVIEW

In this chapter, we learned the distinctions among global change, global climate change, and global warming. Global warming is an inherently natural process whereby a tiny percentage of gases in the atmosphere, known as greenhouse gases, absorb infrared radition from Earth and emit some of this energy back to Earth. This warms the planet. Environmental scientists are concerned that human activities have caused a higher concentration of greenhouse gases in the atmosphere that is responsible for a gradual warming of the planet above that observed a century ago. The average amount of warming around the world has been relatively small at 0.8°C but some regions such as high northern latitudes have experienced increases of up to 4°C. This warming has caused a decline in polar ice, a decline in glaciers, an increased thawing of permafrost, and an increase in sea level. It has also affected the timing of plant flowering, bird migration, insect emergence, and the length of growing seasons. While these effects have already been observed, other climate changes are also predicted to occur including more extreme temperatures, more instense storms, changing patterns of precipitation, and altered ocean current. The Kyoto Protocol was designed to reduce the global emissions of greenhosue gases, but the goal set by nations around the world has not yet been achieved.

Key Terms

Global change
Global climate change
Global warming
Greenhouse effect
Greenhouse warming potential
Ocean acidification
Kyoto Protocol
Carbon sequestration

Teaching Tip: Measuring Your Impact

Carbon Produced by Different Modes of Travel

Measuring Your Impact 19 asks students to calculate their personal yearly carbon emission by looking at how much fuel is used by different forms of travel. Then, students will be asked to consider whether their calculations provide a good estimate of global carbon emissions.

The link to this resource may be found by clicking on the link buttons in the TE-book, opening the TRFD, or logging onto the book's companion website bcs.whfreeman.com/friedlandapes2e (teacher login required).

TRM **Measuring Your Impact 19:** Carbon Produced by Different Modes of Travel

Teaching Tip: Beyond the Classroom

Ask students to write a letter or send an email to a local official asking his or her personal position on global climate change. Sample questions include:

- Do you support efforts to address climate change?
- Are there any plans in your area to help reduce greenhouse gases?
- Do you think mayors in our state should participate in the U.S. Conference of Mayors Climate Protection Agreement?

Learning Objectives Revisited

Module 62 Global Climate Change and the Greenhouse Effect

- **Distinguish among global change, global climate change, and global warming.**

  Global change refers to changes in the chemical, biological, and physical properties of the planet. One aspect of this is global climate change, which refers more specifically to the average weather that occurs in an area over a period of years or decades. One aspect of global climate change is global warming, which refers to the warming of the oceans, land masses, and atmosphere.

- **Explain the process underlying the greenhouse effect.**

  The greenhouse effect occurs when high-energy visible and ultraviolet light strike Earth and are emitted back from Earth as infrared radiation. Some of this infrared radiation is absorbed by greenhouse gases in the atmosphere, which make up a very small percentage of all atmospheric gases. The greenhouse gases subsequently emit infrared radiation, some of which head back to Earth and this causes the planet to become warmer.

- **Identify the natural and anthropogenic sources of greenhouse gases.**

  The natural and anthropogenic sources of greenhouse gases include water vapor, carbon dioxide, methane, and nitrous oxide. Chlorofluorocarbons are a type of greenhouse gas that only has an anthropogenic source.

Module 63 The Evidence for Global Warming

- **Explain how CO_2 concentrations have changed over the past 6 decades and how emissions compare among the nations of the world.**

  CO_2 concentrations have continually increased since atmospheric measurements first began in 1958. These concentrations have increased from 310 ppm in 1958 to nearly 400 ppm in 2013. The major producers of CO_2 include large developed countries, such as the United States and Russia, and rapidly growing developing countries, such as China and India. When we consider the per capita production of CO_2, we see that some of the largest producers of CO_2 have relatively low per capita production of CO_2.

- **Explain how temperatures have increased since records began in 1880.**

  Direct measurements of land and sea temperatures have taken place since 1880. Averaged across the globe, mean annual temperatures have increased by 0.8°C. However, some regions have become a bit cooler while other regions have become much warmer with increases up to 4°C.

- **Discuss how we estimate temperatures and greenhouse gases over the past 500,000 years and into the future.**

  To estimate changes in global temperatures over 500,000 years, we can use changes in the species composition of foraminifera that are found in ocean sediments. We can also examine the ratio of ^{16}O and ^{18}O atoms in the air bubbles that are preserved in ancient ice. These air bubbles can also be used to determine the concentrations of various greenhouse gases from different time periods in the past. We can predict future climate changes using climate models designed to understand processes that affect climate, such as air and ocean temperatures, CO_2 concentration, extent of vegetation, and sea ice coverage at the poles. The models use estimates about how changes in these factors will change in the future to predict how climates will change.

- **Explain the role of feedbacks on the impacts of climate change.**

  Feedbacks in the environment can be either positive or negative. Positive feedbacks, such as global warming that causes higher rates of soil decomposition, can amplify the effects of global warming. Negative feedbacks, such as plants responding to increased CO_2 concentrations, can reduce the effects of global warming.

Module 64 The Consequences of Global Climate Change

- **Discuss how global climate change has affected the environment.**

  Global warming has already caused a decline in the ice mass of the Arctic polar ice cap and in Antarctica and Greenland. It has also caused a decline in glaciers around the world, an increase in the thawing of permafrost, and an increase in sea level.

Exploring the Literature

Houghton, J. 2004. *Global Warming: The Complete Briefing*. 3rd ed. Cambridge University Press.

Intergovernmental Panel on Climate Change. *Climate Change 2007: Synthesis Report. Fourth Assessment Report of the Intergovernmental Panel on Climate Change*. http://www.ipcc.ch/pdf/assessment-report/ar4/syr/ar4_syr.pdf

National Research Council. 2002. *Abrupt Climate Change: Inevitable Surprises*. National Academies Press.

U.S. Global Change Research Program. 2009. *Global Climate Change Impacts in the United States*. http://downloads.globalchange.gov/usimpacts/pdfs/climate-impacts-report.pdf

Download a printable version of this list by clicking the link buttons in the TE-book, opening the TRFD, or logging onto the book's companion website bcs.whfreeman.com/friedlandapes2e (teacher login required).

Exploring the Literature 19

- **Explain how global climate change has affected organisms.**

  Warmer global temperatures have caused longer growing seasons. Warmer temperatures have also caused many species of plants to flower earlier, and they have changed the times when animals breed and insects emerge.

- **Identify the future changes predicted to occur with global climate change.**

  Future global changes include longer periods of cold and warm temperatures, more intense storms, changes in global precipitation patterns, and the alteration of ocean currents. These changes may not only affect wild plants and animals but also may influence where humans can live and how humans are affected by diseases.

- **Explain the global climate change goals of the Kyoto Protocol.**

  The Kyoto Protocol is an international agreement to reduce the concentrations of greenhouse gases by 5.2 percent below 1990 levels by 2012. Different developed countries agreed to different emission limits and the protocol did not impose emission limits for developing countries. Today, some developed countries have accomplished substantial reductions in the emissions whereas others have not. Collectively, the global goal has not been met.

Answers to Chapter 19 AP® Exam Multiple-Choice Questions

1. a
2. d
3. e
4. e
5. a
6. a
7. c
8. e
9. c
10. d
11. c

Chapter 19 AP® Environmental Science Practice Exam

Section 1: Multiple-Choice Questions

Choose the best answer for questions 1–11.

1. Which of the following activities causes a cooling of Earth?
   (a) Volcanic eruptions
   (b) Emissions of anthropogenic greenhouse gases
   (c) Evaporation of water vapor
   (d) Combustion of fossil fuels
   (e) Deforestation

2. In regard to the greenhouse effect, which statement is NOT true?
   (a) Ultraviolet and visible radiation are converted to infrared radiation at the surface of Earth.
   (b) Approximately one-third of the radiation of the Sun does not enter the atmosphere of Earth.
   (c) Infrared radiation is absorbed by greenhouse gases.
   (d) Greenhouse gases were not historically present in the atmosphere.
   (e) Ultraviolet radiation is absorbed by ozone.

3. Which of the following is NOT a greenhouse gas?
   (a) Carbon dioxide
   (b) Water vapor
   (c) Methane
   (d) Nitrous oxide
   (e) Nitrogen

4. Of the following factors, which are important when considering the effect of a greenhouse gas on global warming?
   I. The amount of infrared radiation the gas can absorb
   II. How long the gas remains in the atmosphere
   III. The concentration of the gas in the atmosphere

   (a) I
   (b) I and II
   (c) I and III
   (d) II and III
   (e) I, II, and III

5. Which greenhouse gas is NOT correctly paired with one of its sources?
   (a) Nitrous oxide : landfills
   (b) Methane : termites
   (c) Water vapor : evaporation
   (d) Nitrous oxide : automobiles
   (e) CO_2 : deforestation

6. Carbon sequestration
   (a) is a process to remove CO_2 from the atmosphere.
   (b) is a method for preventing carbon emissions from landfills.
   (c) causes increased coral bleaching.
   (d) is a method for emissions reduction that focuses on improved efficiency.
   (e) is the release of carbon from soils due to warming.

Answers to Chapter 19 AP® Exam

Download the full answers to the Chapter 19 AP® Exam by clicking on the link buttons in the TE-book, opening the TRFD, or logging onto the book's companion website bcs.whfreeman.com/friedlandapes2e (teacher login required).

Answers to Chapter 19 AP® Exam

Additional Multiple-Choice AP® Practice Exam

Download an additional Multiple-Choice AP® Environmental Science Practice Exam for Chapter 19 by clicking on the link buttons in the TE-book, opening the TRFD, or logging onto the book's companion website bcs.whfreeman.com/friedlandapes2e (teacher login required).

Multiple-Choice AP® Practice Exam 19

7. Which statement about global warming is true?
   (a) The planet is not warming.
   (b) The planet is warming, but humans have not played a role.
   (c) The planet has had many periods of warming and cooling in the past.
   (d) Greenhouse gases compose only a small fraction of the atmosphere, so they cannot be important in causing global warming.
   (e) Small increases in average global temperatures are not likely to affect polar bears.

8. Which sources of data have been used to assess changes in global CO_2 and temperature?
   I. Air bubbles in ice cores from glaciers
   II. Thermometers placed around the globe
   III. CO_2 sensors placed around the globe

   (a) I
   (b) I and II
   (c) I and III
   (d) II and III
   (e) I, II, and III

9. Which is a true statement about feedback loops that occur with climate change?
   (a) All feedback loops are positive.
   (b) All feedback loops are negative.
   (c) Increased soil decomposition under warmer temperatures represents a positive feedback loop.
   (d) Increased evaporation under warmer temperatures represents a negative feedback loop.
   (e) Increased plant growth under higher CO_2 concentrations represents a positive feedback loop.

10. Which predicted consequence of global warming has not yet occurred?
   (a) Melting ice caps
   (b) Rising sea levels
   (c) Melting permafrost
   (d) Shutting down the thermohaline circulation of the ocean
   (e) Altered breeding times and flowering times of animals and plants

11. Which statement regarding the Kyoto Protocol is true?
   (a) All nations agreed to reduce their emission of greenhouse gases.
   (b) All nations agreed to stop emitting greenhouse gases.
   (c) Developed nations agreed to different levels of emission reductions.
   (d) Developing nations agreed to reduce their emission of greenhouse gases.
   (e) Developing nations agreed to stop their emission of greenhouse gases.

Section 2: Free-Response Questions

Write your answer to each part clearly. Support your answers with relevant information and examples. Where calculations are required, show your work.

1. During a debate on climate change legislation in 2009, a U.S. congressman declared that human-induced global warming was a "hoax" and that "there is no scientific consensus."
   (a) If you were a member of Congress, what points might you raise in the debate to demonstrate that global warming is real? (4 points)
   (b) What points might you raise to demonstrate that global warming has been influenced by humans? (4 points)
   (c) What are some human health and economic effects that could occur because of global warming? (2 points)

2. Given what you have learned about global warming and global climate change,
   (a) what actions might you propose in the United States to reduce CO_2 emissions? (3 points)
   (b) what actions might you propose in the United States to reduce methane emissions? (3 points)
   (c) what actions might you propose in the United States to reduce nitrous oxide emissions? (3 points)
   (d) what evidence have scientists used to support the assertion that global warming is happening? (1 point)

Answers to the Chapter 19 Free-Response Questions can be found in the Answer Appendix on page ANS-16.

chapter

20 Sustainability, Economics, and Equity

PD Overview

This chapter explores sustainability in the context of economics. It describes how sustainability can be achieved through the use of sound economic and business practices as well as through effective environmental regulations and laws. The chapter looks at the work of agencies such as the United Nations Environment Programme as well as a number of U.S. agencies, including the Environmental Protection Agency that oversees all governmental efforts related to the environment. The chapter describes the role of different worldviews in approaches to sustainability. Finally, the chapter considers the importance of reducing poverty as part of achieving sustainability.

Module 65: Sustainability and Economics

This module reviews the definition of sustainability and explains why economic analysis is essential to achieving a sustainable existence. The module describes environmental and ecological economics, which are economic subfields that attempt to place value on benefits offered by the natural environment. Finally, the module considers cradle-to-grave and cradle-to-cradle systems analyses, which are tools used to merge sustainability with life-cycle analysis.

Module 66: Regulations and Equity

This module looks at how different worldviews influence approaches to environmental protection and regulation. A variety of world agencies such as the United Nations, the World Bank, and the World Health Organization have environmental programs. In the United States, the Environmental Protection Agency and the Department of Energy are two of the most significant environmental agencies. A variety of measures are used to assess sustainability and environmental well-being, including the human development index and the human poverty index. The triple bottom line maintains that sustainability can be achieved at the intersection of the three factors—economic, environmental, and social—that influence most development endeavors. Finally, the module explores challenges in achieving sustainability: poverty, inequality, and environmental justice.

Alignment to AP® Environmental Science Course Description

AP® Outline	Module	
IV. Land and Water Use (10–15%)		
G. Global Economics	Module 65	Sustainability and Economics
	Module 66	Regulations and Equity

Chapter Learning Objectives

After completing this chapter students will be able to

- **explain why efforts to achieve sustainability must consider both sound environmental science and economic analysis.**
- **describe how economic health depends on the availability of natural capital and basic human welfare.**
- **explain the role of agencies and regulations in efforts to protect our natural and human capital.**
- **describe the approaches to measuring and achieving sustainability.**
- **discuss the relationship among sustainability, poverty, personal action, and stewardship.**

Chapter 20 Pacing Guide

This pacing guide is based on a schedule with 120 sessions of 50 minutes each before the AP® Exam. If you have a different number of sessions before the exam, you can modify the pacing to suit your needs. If you have additional time, consider incorporating quizzes, released AP® Environmental Science free-response and multiple-choice questions, or additional activities.

Module	Standard Schedule Days	Block Schedule Days
Module 65	1	½
Module 66	1	½
Assessment	1	½

Chapter 20 Resources

The link to these resources can be found by clicking on the link buttons in the Teacher's e-Book (TE-book), opening the Teacher's Resource Flash Drive (TRFD), or logging onto the book's companion website bcs.whfreeman.com/friedlandapes2e (teacher login required).

- **PowerPoint Presentation 20**
- **Optimized Art PowerPoint and JPEG Files 20**
- **Do the Math Videos**
- **Measuring Your Impact 20:** GDP and Footprints
- **Chapter 20 Web Resources**
- **Exploring the Literature 20**
- **Answers to Chapter 20 AP® Exam**
- **Multiple-Choice AP® Practice Exam 20**
- **Answers to Unit 8 AP® Exam**
- **Unit 8 Additional Free-Response Question**
- **Answer to Science Applied 8 Free-Response Question**

PD Chapter 20 Overview

Watch the video overview of Chapter 20 (for teachers) by clicking on the link buttons in the TE-book, opening the TRFD, or logging onto the book's companion website bcs.whfreeman.com/friedlandapes2e (teacher login required).

TRM PowerPoint Presentation 20

Download the PowerPoint presentation for Chapter 20 by clicking on the link buttons in the TE-book, opening the TRFD, or logging onto the book's companion website bcs.whfreeman.com/friedlandapes2e (teacher login required).

Poverty and environmental degradation are evident along the border between Mexico and the United States, as seen here in Ciudad Juárez, Mexico. *(Stephen Sharnoff)*

chapter

20 Sustainability, Economics, and Equity

Module **65** Sustainability and Economics

Module **66** Regulations and Equity

Assembly Plants, Free Trade, and Sustainable Systems

Although citizens of Ciudad Juárez, Mexico, call the border with the United States la línea, or "the line," the border, which stretches nearly 3,200 km (2,000 miles), is more a network of passageways than a division. Trade among people and cultures across this international boundary clearly affects both countries. The link between the Mexican and American economies, strengthened by globalization and increased trade, is exemplified by the maquiladora, or assembly plant, industry.

Established in the 1960s, this industry allows international companies to import materials and equipment free of tariffs to Mexican maquiladoras, and then to export the finished product to markets in other countries. In 1994, the United States, Canada, and Mexico passed the North American Free Trade Agreement (NAFTA), which was intended to increase trade among the three countries by reducing tariffs and other taxes as well as regulations. After NAFTA, the use of maquiladoras increased significantly, with the export of assembled products tripling between 1995 and 2000. Maquiladoras, which export 90 percent of their products to the United States, currently constitute 80 percent of the economy in the northern border region and a quarter of Mexico's total GDP. And while the jobs have been welcomed in these economically depressed areas, there have been many negative consequences as well, including industrial pollution, poor working conditions, and discrimination. In addition, the maquiladoras raise questions of social justice because much of the profit is sent to other countries.

> If we could give equal attention to economic profit, environmental integrity, and human welfare, could we ultimately create more sustainable development?

In terms of the environment, maquiladora operations often contaminate the border region with toxic industrial waste. Environmental regulations are lenient or nonexistent, and the majority of companies do not comply with mandates that maquiladora waste be shipped to the company's home country. Disposal of toxic chemicals and heavy metals into the local environment causes

Teaching Tip: Chapter Opening Case

The chapter opening case, "Assembly Plants, Free Trade, and Sustainable Systems," introduces students to the economic, social, and environmental implications of NAFTA, the international trade agreement with the United States, Mexico, and Canada. Students will consider the benefits to all three countries, the environmental and social consequences for Mexican community members, and the balance that must exist among economic profit, environmental integrity, and human welfare in order to establish sustainable development.

Teaching Tip: Video

Demand a Fair Trade Cell Phone
In this video from TED, Bandi Mbubi illuminates the practice of mining for coltan in the Congo. Coltan is used in cell phones, laptops, and game systems. The valuable mineral has led to war, slavery, and environmental degradation. Mbubi asks his audience to demand better mining practices from their cell phone manufacturers so that technology is a positive force in the world and not reliant on negative externalities.

The link to this video may be found by clicking on the link buttons in the TE-book, opening the TRFD, or logging onto the book's companion website bcs.whfreeman.com/friedlandapes2e (teacher login required).

Chapter 20 Web Resources

Teaching Tip: Activity

After reading the opening case, have your students make a chart of the benefits and costs of the maquiladora industry. (Suggested answers are provided in italics.)

Benefits	Costs
Increased trade among Canada, Mexico, and the United States benefits the economies of all three countries	*Industrial pollution due to limited environmental regulations*
Provides one-quarter of Mexico's GDP	*Poor working conditions*
Provides 80 percent of the border's economy	*Increased population numbers have led to problems with sanitation and MSW disposal*
Availability of jobs in an economically depressed area	*Social injustices include firing pregnant women, using underage workers, and a lack of "right to know" laws*
Facilitates international trade in developing nations	*Profits from factories often go to home countries rather than to Mexico*

When students have completed the table, ask them to describe ways in which they might be able to increase the benefits and decrease the costs that they listed. *Possible answers include amending NAFTA to include more environmental and social laws and improving environmental conditions to improve social and economic conditions for those living in Mexico.*

groundwater and surface water pollution and significant harm to human health. Many maquiladora employees are women of reproductive age, a population that is particularly vulnerable to toxic chemicals.

In addition to pollution from the manufacturing processes, an increase in the human population in maquiladora areas has added greatly to other environmental problems. Many municipalities in which maquiladoras are situated do not have sewage treatment facilities or trash collection capabilities. The solid waste pollutes water sources, and seasonal floods spread garbage throughout the areas.

Social abuses also occur in this system. Employers often test women for pregnancy before they are hired, and those who become pregnant may be illegally fired. Managers employ underage workers. Factory conditions are hazardous, and employees are often unaware of risks because of the lack of "right to know" laws and an absence of warning signs in Spanish. Companies exploit the poverty of the region by offering wages that barely support employee needs. An average maquiladora worker earns the equivalent of a few dollars per day, and these wages have remained stagnant for years even as living costs have risen.

Sometimes the profits from these factories do not enter the Mexican economy, but rather go to the home countries of the companies that run the plants. Many observers believe that northern Mexico pays the social and environmental prices for the maquiladora industry, while foreign corporations reap the benefits.

Free trade and globalization agreements like NAFTA are designed to enhance developing economies by facilitating international business. However, in northern Mexico, increased free trade has stimulated an industry that in some cases may sacrifice social well-being and environmental health. Nevertheless, because many people are employed in the maquiladora industry, money does enter the local economy and helps individuals. Environmental scientists interested in human social welfare and the well-being of the environment look at situations such as these and ask: If we could give equal attention to economic profit, environmental integrity, and human welfare, could we ultimately create more sustainable development?

Sources:

J. Carrillo and R. Zarate, The evolution of maquiladora best practices: 1965–2008, *Journal of Business Ethics* (2009) 88: 335–348; J. G. Samstad and S. Pipkin, Bringing the firm back in: Local decision making and human capital development in Mexico's maquiladora sector, *World Development* (2005) 33: 805–822.

Throughout this book we have seen that economic development, social justice, and sustainable environmental practices are often in conflict. In recent years, environmental scientists have begun to address these relationships by using the tools from economics and other fields to help find ways in which we can achieve a sustainable, equitable, and prosperous existence for all inhabitants on Earth. However, it is difficult to expect people to be concerned about the welfare of the planet on which they live if they have not met their own basic needs of water, food, health, and housing. Thus in recent years, environmental well-being and human well-being have become linked. In this chapter we will begin to explore some of these connections.

module

65

Sustainability and Economics

Sustainability is a relatively new and evolving concept in contemporary environmental science. We have seen that something is sustainable when it meets the needs of the present generation without compromising the ability of future generations to meet their own needs. Although human needs can be defined in various ways, for our purposes we identify the basic necessities as access to food, water, shelter, education, and a healthy, disease-free existence. In order for these five necessities to be available, there must be functioning environmental systems that provide us with breathable air, drinkable water, and productive land for growing food, fiber, and other raw materials—the ecosystem services that we have described in this book.

The quest to obtain resources and increase **well-being**—the status of being healthy, happy, and prosperous—has caused individuals and nations to exploit and degrade natural resources such as air, land, water, wildlife, minerals, and even entire ecosystems. To address questions of sustainability, we need to be able to understand where human well-being and the condition of environmental systems are in conflict. To do this we will consider economic analysis, ecological economics and ecosystem services, and the role of regulatory agencies in bringing about environmental regulation and protection.

Learning Objectives

After reading this module you should be able to

- explain why efforts to achieve sustainability must consider both sound environmental science and economic analysis.
- describe how economic health depends on the availability of natural capital and basic human welfare.

Well-being The status of being healthy, happy, and prosperous.

Teaching Tip: Discussion Starter

Ask students: What is sustainability? *Sustainability is meeting the needs of the current generation without compromising the needs of future generations.*

Teaching Tip: Journal Prompt

Ask students: How do the laws of supply and demand influence the price of a good or service? *When a good is in great demand and wanted by many people, producers are typically unable to provide enough to meet demand. Price is the method that producers and consumers use to communicate the value of an item and to allocate it.*

COMMON MISCONCEPTIONS

The principles of economics apply only to business

Students often believe that the principles of economics are only applicable to business decision making. Economics is the study of scarcity and choice and provides a useful framework for making environmental decisions. Some universities now offer degree programs in environmental economics and ecological economics.

Achieving sustainability requires both sound environmental science and economic analysis

In an attempt to reduce environmental harm, researchers and policy makers have experimented with a variety of techniques to encourage consumers to change their behavior in ways that would benefit the environment. We explored some of these techniques in Chapter 10 where we discussed externalities and in Chapter 15 where we discussed the buying and selling of air pollution allowances as well as charging a fee or tax for the use of certain resources or for the emission of certain pollutants. **Economics** is the study of how humans allocate scarce resources in the production, distribution, and consumption of goods and services.

Throughout this text we have already applied many concepts from the field of economics. When we looked at the problem of externalities and pollution, we were using economic theory. Life-cycle analysis is very similar to the cost-benefit analysis that economic policy makers use. In this section we will look at some basic economic concepts and learn how they can be applied to environmental issues.

Supply, Demand, and the Market

In today's world, most economies are market economies. In the simplest sense, a market occurs wherever people engage in trade. In a market economy, the cost of a good is determined by supply and demand. When a good is in great demand and wanted by many people, producers are typically unable to provide an unlimited supply of that good. Price is the method that producers and consumers use to communicate the value of an item and to allocate the scarce item.

The graph shown in FIGURE 65.1 illustrates the relationship between supply, demand, and price. The supply curve (S) shows how many units that suppliers of a given product or service—for example, T-shirts—are willing to provide at a particular price. Factors that influence supply of a good include input prices (the cost of the resources used to produce the item), technology, expectations about future prices, and the number of people selling the product. For example, if you are the only person selling T-shirts and many people want them, you will be willing to make the investment required to produce many T-shirts. However, if a new T-shirt seller comes along, because you will be concerned that you will not sell as many,

Economics The study of how humans allocate scarce resources in the production, distribution, and consumption of goods and services.

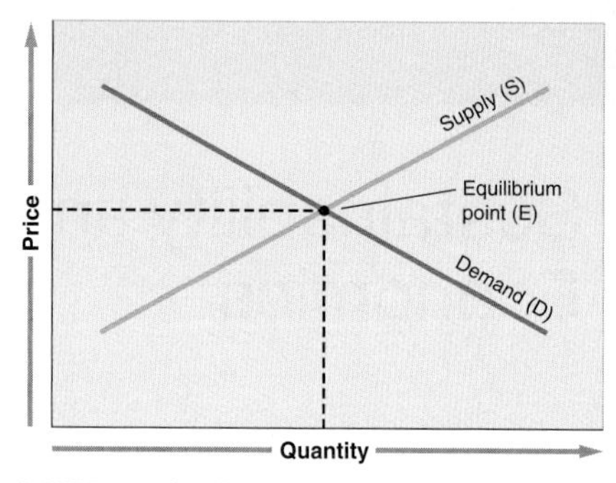

FIGURE 65.1 **Supply and demand.** A manufacturer will supply a certain number of units of an item based on the revenue that will be received. A consumer will demand a certain number of units of that item based on the price paid. The intersection of the supply and demand curves determines the market equilibrium point for that item.

you will decrease your production because you now must share the market with another supplier.

The demand curve (D) shows how much of a good consumers want to buy. Factors that influence demand include income, prices of related goods, tastes, expectations, and the number of people who want the good. For example, if your boss gives you a raise, you may feel like you can afford that T-shirt you have been wanting to buy.

Notice that the demand curve slopes downward. In other words, as the price of T-shirts rises, the demand for them declines. This illustrates the law of demand, which states that when the price of a good rises, the quantity demanded falls and when the price falls, the quantity demanded rises. Conversely, the supply curve slopes upward. This reflects the law of supply, which states that when the price of a good rises, the quantity supplied of that good will rise and when the price of a good falls, the quantity supplied will fall.

The laws of supply and demand make intuitive sense. After all, if you are selling T-shirts and you find that your profits have shrunk, you are more likely to use your resources to produce and sell something more popular, and more profitable. If you are a consumer of T-shirts, the less expensive they are, the more you are inclined to buy.

With these different interests, how do demand and supply ever meet? In a market system, without any restrictions such as taxes or other regulations, the price of a good will come to an equilibrium point (E) where the two curves on the graph intersect. Here the quantity demanded and the quantity supplied are exactly equal. At this price, suppliers find it worthwhile to supply exactly as many T-shirts as consumers are willing to buy.

Unfortunately, markets—composed of many buyers and sellers—do not always take all costs of production into account. We have already seen that this is the case in situations of land degradation where people, organizations, or even governments deplete or damage a natural resource because they do not bear any direct costs for doing so. As we saw in Chapter 10, the cost of using a resource that is not included in the purchase price is called an externality. When we pollute air or water without directly paying for it, that is also an externality. When we account for the costs of externalities created by manufacturing a good or offering a service, the price changes. This, in turn, affects demand and supply.

Let's look at the example of coal. The dollar cost of coal-generated electricity includes the cost of the coal, the cost of paying people to operate the power plant, and the cost of electricity distribution to customers. However, the cost to the environment of emitting sulfur dioxide, carbon dioxide, and other waste products, all of which are negative externalities, is largely missing from the price customers pay. However, these negative externalities certainly add costs, both financially and in terms of the well-being of people living downwind from the power plant. For example, someone with a respiratory ailment could incur greater medical expenses because of increased sulfur dioxide and particulates in the air. There may be provisions requiring polluters to pay some of the costs related to these emissions, but often these payments are not sufficient to cover the total cost of the pollution. In addition, they often do not reach the affected individuals or groups.

If the dollar cost of a good included externalities such as the expenses incurred by emitting pollutants into the air, or the expenses related to removing the pollutants before they were emitted, then the cost for most items produced would be greater. This could only occur if a tax were imposed by a regulatory agency. When the cost of production rises due to this tax, the supply curve shifts to the left, from S to S_1 as shown in FIGURE 65.2. The new market equilibrium (E_1) is at a higher cost and, as a result, fewer items are manufactured and purchased. In other words, including the externalities raises the price and lowers the demand. Therefore the price that includes externalities is more reflective of the true cost of the item.

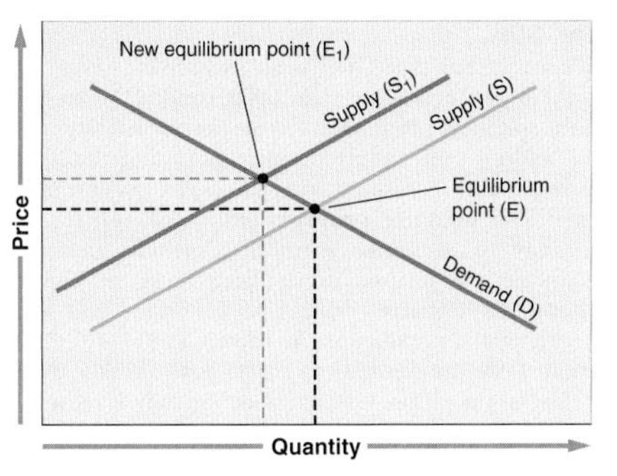

FIGURE 65.2 **Supply and demand with externalities.** When the cost of emitting pollutants is included in the price of a good, for any given quantity of items, the price increases. This causes the supply curve to shift to the left, from S to S_1. Since the law of demand states that when the price of a good goes up, demand falls, the amount demanded falls, and the market reaches a new equilibrium, E_1.

Measuring Wealth and Productivity

There are a variety of ways of measuring wealth and productivity. While there is no consensus on which method is most accurate and each has shortcomings, researchers do agree that measuring wealth and productivity can be a useful way of examining the health of an economy. In this section we will look at the most common and widely accepted measure of wealth and productivity and then look at alternatives.

GDP

Economists use different national economic measurements to gauge the economic wealth of a country in terms of its productivity and consumption. Most of them do not take externalities into account. The most common of these measurements is the gross domestic product (GDP), which refers to the value of all products and services produced in a year in a given country. GDP includes four types of spending: consumer spending, investments, government spending, and exports minus imports. As a measure of well-being, GDP has been criticized for a number of reasons. Because costs for health care contribute to a higher GDP, a society that has a great deal of illness would have a higher GDP than an equivalent society without a great deal of illness. Such an inclusion does not appear to be an accurate reflection of the "wealth" or "well-being" of a society. And because externalities such as pollution and land degradation are not included in GDP, measurement of GDP does not reflect the true cost of production.

Some social scientists maintain that the best way to improve the global environment is to increase the GDP in the less developed world. In Chapter 7 we examined the relationship between rising income and falling birth rates; as GDP increases, population growth slows. This, in turn, should lead to a reduction in anthropogenic environmental degradation. Wealthier, developed countries are able to purchase goods and services that will lead to environmental improvements—for example, pollution control devices like catalytic converters—and to use their resources more efficiently. On the other hand, as we have seen, developed countries use many more resources than developing countries, which leads to more environmental degradation.

◀ **TEACHING with FIGURES**

Show students Figure 65.2 and ask what happens to the equilibrium point as externalities are added to the price of a product through taxes. *When taxes are added to the price of a product, the price increase causes the quantity demanded to fall.*

AP® Exam Tip

Make sure the students are familiar with the concepts of GDP and externalities. These topics have come up frequently on past exams.

Teaching Tip: Warm-up

Ask students: What is an externality? How might the inclusion of negative externalities affect the cost of a good? *The cost of using a resource that is not included in the purchase price is called an externality. Including negative externalities will increase the price of a product and may lead to a reduction in environmental degradation.*

TEACHING with FIGURES ▶

Show students Figure 65.3 and ask: What does the difference between GDP and GPI for the United States indicate? *Over the years GDP has increased while GPI has remained level since 1980. This indicates that although economic progress has occurred, environmental protection, health of the population, and income distribution have not increased at the same rate.*

Teaching Tip: Journal Prompt

Ask students: How do GPI and the Kuznets curve address the shortcomings of GDP as a measurement of a country's well-being? *Each of these models considers social and environmental factors in addition to economic factors. GPI is a measure of economic status that includes personal consumption, income distribution, levels of higher education, resource depletion, pollution, and health. The Kuznets curve suggests that as per capita income in a country increases, environmental degradation first increases and then decreases.*

The GPI

We have seen that GDP is an incomplete measurement of the economic status of a country because it only considers production. Some researchers attempt to address this shortcoming by using another measurement that is known as the *genuine progress indicator*. The **genuine progress indicator (GPI)** is a measure of economic status that includes personal consumption, income distribution, levels of higher education, resource depletion, pollution, and the health of the population. As shown in FIGURE 65.3, while GDP in the United States rose steadily from 1950 through 2004, GPI has been virtually level since about 1980. A number of countries, including England, Germany, and Sweden, have recalculated their GDP using the GPI. They have found that their overall wealth, when human and environmental welfare are included, has steadily declined over the last 3 decades.

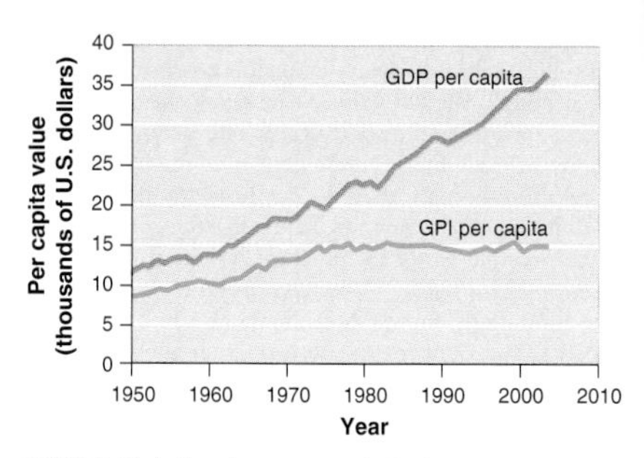

FIGURE 65.3 **Genuine progress indicator versus gross domestic product, per capita, for the United States from 1950 to 2004.** While gross domestic product measures the value of all products and services a country produces, the genuine progress indicator attempts to include the level of education, personal consumption, income distribution, resource depletion, pollution, and the health of the population. *(Data from http://genuineprogress.net)*

The Kuznets Curve

To address some of the shortcomings of GDP as a measurement of wealth, some environmental economists and scientists advocate using a model known as the Kuznets curve. The Kuznets curve, shown in FIGURE 65.4, suggests that as per capita income in a country increases, environmental degradation first increases and then decreases. The model is controversial because it is not easily applicable to all situations. For example, despite the increasing affluence of developed countries, carbon dioxide emissions and municipal solid waste (MSW) generation have both continued to increase. It is possible that these developed countries are not yet wealthy enough to deal with these problems effectively, but it is also possible that there are certain problems that cannot be solved simply with greater wealth. For example, as countries become wealthier, residents tend to use more fossil fuel for travel, to consume more resources, and to generate more waste.

Sometimes less developed countries experience technological leaps without going through each phase of technological development. These kinds of changes may influence the shape of the Kuznets curve or influence how well it characterizes a given situation.

Technology transfer happens when less developed countries adopt technological innovations that were developed in wealthy countries. For example, in many less developed countries, a significant proportion of the population uses cell phones without ever having had access to a network of landline telephones. A situation in which less developed countries use new technology without first using the precursor technology is known as **leapfrogging.** Leapfrogging occurs whenever new technology develops in a way that makes the older technology unnecessary or obsolete. This allows the developing nations to take advantage of the expensive research, development, and experience of the more developed nations.

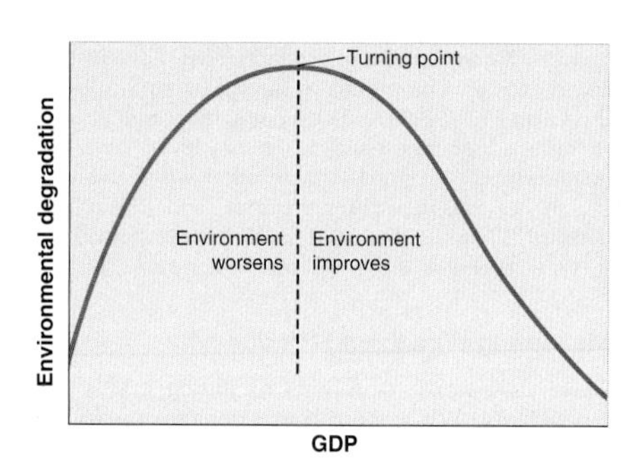

FIGURE 65.4 **The Kuznets curve.** This model suggests that as per capita income in a country increases, environmental degradation first increases and then decreases. In many respects, China is on the first part of this curve while the United States is on the second part of the curve.

Genuine progress indicator (GPI) A measure of economic status that includes personal consumption, income distribution, levels of higher education, resource depletion, pollution, and the health of the population.

Technology transfer The phenomenon of less developed countries adopting technological innovations developed in wealthy countries.

Leapfrogging The phenomenon of less developed countries using new technology without first using the precursor technology.

▲ TEACHING with FIGURES

Show students Figure 65.4 and ask them to describe why environmental degradation caused by a society typically increases as a country develops and then decreases as it becomes wealthy. *As per capita income increases, environmental degradation increases due to increased consumption with limited pollution control. As a country becomes wealthier, it acquires the financial resources to implement and enforce pollution laws and to protect the environment.*

FIGURE 65.5 **Solar panels in Africa.** In areas where the electrical grid is not established and electricity supply lines are not present, the installation and use of photovoltaic solar cells may be less expensive and less environmentally disruptive than a traditional electrical infrastructure. *(Pallava Bagla/Corbis)*

Solar energy is a particularly good example of leapfrogging. In industrialized nations, solar electricity has not been cost-competitive with gas- or coal-generated electricity. However, it has been very successful in nations in Africa, Asia, and South America that lack the resources to build a reliable electrical distribution grid. Solar energy is a small-scale energy source not dependent on outside connections to an electrical grid (see Chapter 13). In fact, it is possible that many less developed countries will continue to increase their use of solar energy and skip the step of building a nationwide electrical grid, much like what has happened with cell phones versus landlines for telephone service. Solar energy allows developing countries to produce and distribute their own electricity without investment in the massive infrastructure of an electrical distribution grid that would be needed in a developed country (FIGURE 65.5).

Economic health depends on the availability of natural capital and basic human welfare

Capital, or the totality of our economic assets, is typically divided into three categories: natural, human, and manufactured. **Natural capital** refers to the resources of the planet, such as air, water, and minerals. **Human capital** refers to human knowledge and abilities. **Manufactured capital** refers to all goods and services that humans produce. While economists usually base their assessment of national wealth on productivity and consumption, environmental scientists point out that all economic systems require a foundation of natural capital. Without natural capital, humans would not be able to produce very much and would probably not survive.

Environmental and Ecological Economics

Some advocates of a purely free-market system believe that as long as market forces are left alone, human work and creativity will find solutions to problems of natural resource degradation and depletion. But as we have seen, externalities are not assessed appropriately if the cost of environmental degradation is not charged to the individuals responsible for that degradation. A **market failure** occurs when the economic system does not account for all costs. Among those economic thinkers who have sought ways to respond to market failures, many have become part of the discussion in the fields of *environmental economics* and *ecological economics*. **Environmental economics** is a subfield of economics that examines the costs and benefits of various policies and regulations that seek to regulate or limit air and water pollution and other causes of environmental degradation. **Ecological economics** is the study of economics as a component of ecological systems rather than as a distinctly separate field of study. Ecological economics is a method of understanding and managing the economy as a subsystem of both natural and human systems. It has as a goal the preservation of natural capital, the goods and services related to the natural world.

Environmental and ecological economists attempt to assign monetary value to intangible benefits and natural capital, a practice known as **valuation.** For example, they have developed methods for assessing the monetary value of a pristine nature preserve, a spotted owl, or a scenic view. One method is to calculate the revenue generated by people who pay for the benefit—for example, the amount tourists pay to visit a nature preserve

Natural capital The resources of the planet, such as air, water, and minerals.

Human capital Human knowledge and abilities.

Manufactured capital All goods and services that humans produce.

Market failure When the economic system does not account for all costs.

Environmental economics A subfield of economics that examines the costs and benefits of various policies and regulations that seek to regulate or limit air and water pollution and other causes of environmental degradation.

Ecological economics The study of economics as a component of ecological systems.

Valuation The practice of assigning monetary value to intangible benefits and natural capital.

Teaching Tip: Discussion Starter

Ask students the following questions:

- What is leapfrogging? *Leapfrogging occurs whenever new technology develops in a way that makes the older technology unnecessary or obsolete. This allows developing nations to take advantage of the expensive research, development, and experience of more-developed nations. Less-developed countries may go from minimal electrification directly to solar PV electricity sources and skip the step of building a nationwide electrical grid.*
- How are PV panels an example of leapfrogging and how have less-developed nations benefited from this phenomenon? *PV panels allow developing countries to produce and distribute their own electricity without investment in the massive infrastructure of an electrical distribution grid that is commonly found in a developed country.*

Teaching Tip: Concept Map

Ask students to complete the concept map below to differentiate between the categories of capital. (Answers are provided in italics.)

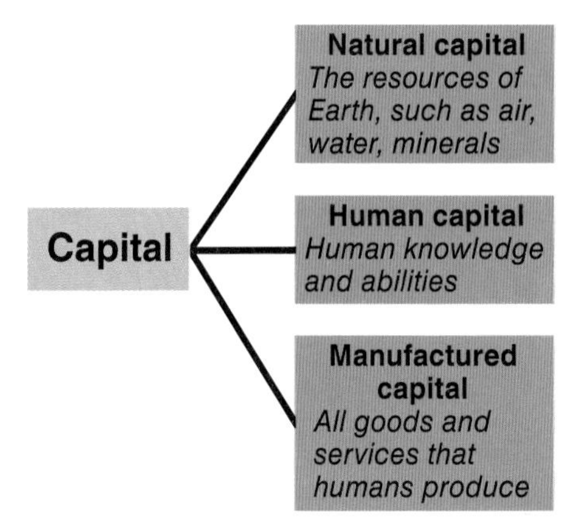

would represent the dollar value of the preserve. Another method is to use surveys. They might ask a number of people how much they are willing to pay just to know that spotted owls exist, even if they are unlikely ever to see one. The most extensive assessments have attempted to determine the value of ecosystem services such as oxygen that plants produce or pollination that insects do. Although estimates vary, global ecosystem services might have a dollar value of approximately $30 trillion per year. The 2004 Millennium Ecosystem Assessment categorized the variety of services that ecosystems provide for the benefit of humans. In many cases, it is possible to estimate the cost of a particular service if it were provided using technology rather than naturally. For example, as we discussed in Chapter 3, New York City could have constructed a massive water purification system at a known cost. Instead, it chose to protect watersheds in the Catskill Mountains, a region north of New York City that supplies the city with water, so the water would not need expensive purification. Accordingly, the ecosystem service of water purification has a known value, which can be used to help calculate the total dollar value of ecosystem services.

Given all of the natural capital and ecological services distributed around the world, it is quite likely that human activities will generate multiple negative externalities. Economic tools can be used to incorporate the dollar cost of the externalities in the price of goods and services. We have seen examples of this with our discussions of charging for allowances to allow sulfur dioxide and carbon dioxide emissions. These economic tools can be used in many other ways as well. Typically, a tax or regulation calls for reducing externalities through a market-driven system. This system calls for the incorporation of negative environmental impacts of a commodity or service in its cost of production. For example, a car manufacturer would include in the cost of production for each car not only the cost of labor and natural resources, such as steel and water, but also the cost of the air pollution caused by the manufacturing process. Viewed this way, the cost of production of a car will immediately increase. Typically, the manufacturer would want to distribute at least part of this additional cost to the consumer by raising the price of the car. Calculating the full costs of a commodity or service by internalizing externalities will likely cause consumers to buy fewer items with high negative impacts because those impacts will be reflected in higher prices. The most obvious way for the costs of externalities to be included is by requiring the producer to pay them. This could be achieved through regulation, imposition of a tax, or some sort of public action mandating reparation for externalities or making it difficult for the company to produce its product in a way that pollutes. Much of the debate in environmental and ecological economics revolves around how best to impose the dollar cost on the producer.

Sustainable Economic Systems

Critics of our current economic system maintain that it is based on maximizing the utilization of resources, energy, and human labor. This encourages the extraction of large amounts of natural resources and does not provide any incentives that would reduce the amount of waste generated. A system analysis of the current economic situation, shown in FIGURE 65.6a, suggests that continuing with such a system is not sustainable.

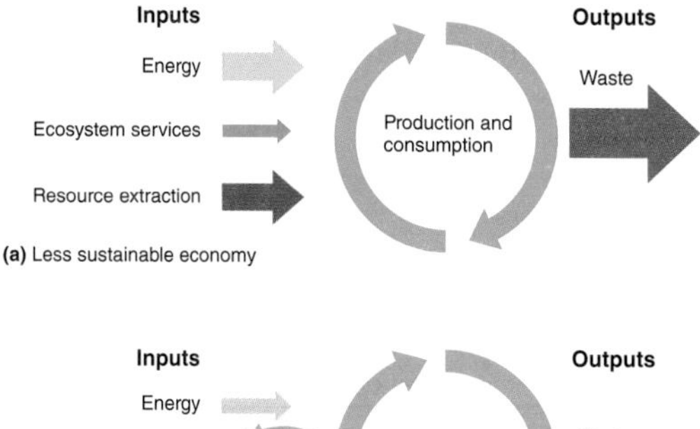

FIGURE 65.6 Systems diagrams of two economic systems. (a) A less sustainable system, like our current economy, is based on maximizing the utilization of resources and results in a fairly large waste stream. (b) A more sustainable system is based on greater use of ecosystem services, less resource extraction, and minimizing the waste stream.

Teaching Tip: Journal Prompt

Ask students: What is the role of valuation in protecting natural capital? *Valuation is the practice of assigning monetary value to intangible benefits and natural capital. For example, placing value on an ecosystem service such as flood control provided by a wetland allows us to make more informed decisions about management and protection in light of the costs that would be incurred if the service were removed.*

Teaching Tip: Warm-up

Ask students: What is an example of how externalities have been included in the cost of a good or service? *Charging for sulfur dioxide and carbon dioxide emission allowances is one example of including externalities in the cost of a good.*

Teaching Tip: Journal Prompt

Ask students: What are the features of a sustainable economic system? *A sustainable economic system relies more on ecosystem services and reuse of existing manufactured materials and less on resource extraction and energy use. It also takes some of the waste stream and reuses it in the production and consumption cycle. Therefore, the cycle would use more renewable energy, reduce negative externalities, and reuse more of the products that were destined for the waste stream.*

In the current system, large amounts of extracted resources and energy and relatively small amounts of ecosystem services are the inputs and large amounts of waste are the outputs.

A sustainable economic system, depicted in Figure 65.6b, will rely more on ecosystem services and reuse of existing manufactured materials and less on resource extraction. In this system, there is greater reliance on ecosystem services and less reliance on resource extraction that requires energy. It also takes some of the waste stream and reuses it in the production and consumption cycle, as indicated by the arrow labeled "Waste stream recycling." Therefore, the cycle in Figure 65.6b would use more renewable energy, lessen negative externalities, and reuse more of the products that were destined for the waste stream. This model has led architects, environmental scientists, and engineers to a collaborative discussion of the optimal way to design, manufacture, use, and dispose of objects such as automobiles, houses, and consumer goods. Currently, a consumer purchases an object, such as an automobile or computer, and when that object has reached the end of its useful life, the consumer is responsible for its disposal. As we pointed out in Chapter 16, because the responsibility for the object rests with the consumer, there is no incentive for the manufacturer to make it easy to reuse or recycle the object. Some observers of the current situation have noted the irony that a can of a chemical oven cleaner purchased for $5 may cost $20 for disposal. These kinds of discussions have led to the cradle-to-grave and cradle-to-cradle analysis described in Chapter 16.

Because the cradle-to-cradle system includes human capital, resource, and energy inputs as well as a redirection of the waste stream, it gains even greater importance when we consider the entire economic system. FIGURE 65.7 shows an alternative approach to the previous diagram. In this case, while there is some waste disposal, a good fraction of materials that are used up contribute to the raw materials for new items. The ultimate goal is to produce a good that at the end of its

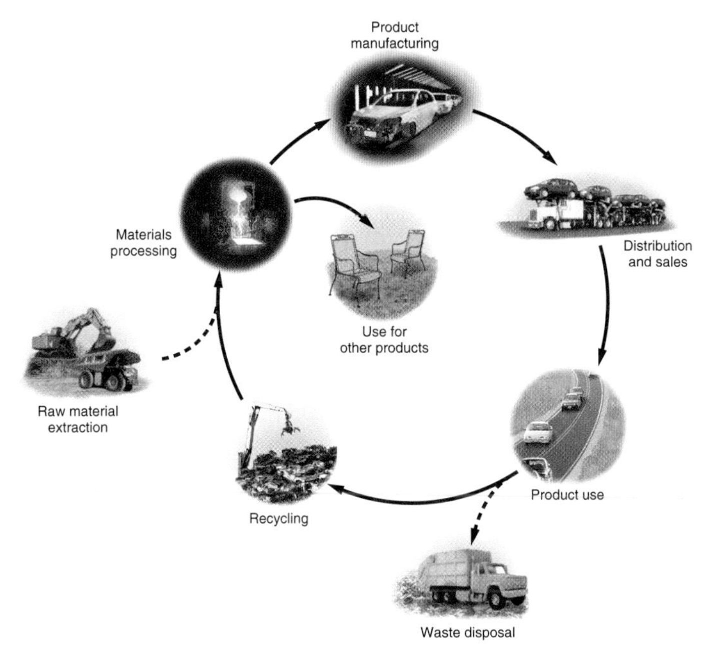

FIGURE 65.7 **A cradle-to-cradle system for material use and waste recycling.** The manufacture of automobiles serves as one example. Products made at a factory use recycled materials whenever possible. Products are designed and manufactured with the goal of recycling as much of the automobile as possible when its useful life is over. Energy costs in manufacturing, distribution, and use are all taken into consideration when designing the automobile and the distribution network.

Module 65 Sustainability and Economics 709

useful life—as it approaches its "grave"—can be easily reused to make a new product. That is, most or all of the parts of an old product will become the "cradle"—the beginning of life—for the same or other products. An automobile would be ideal for a cradle-to-cradle system because each individual car contains a ton or more of steel and other metals as well as rubber, plastic, and a host of other materials that can be reused.

module 65 REVIEW

In this module, we have seen that economic analysis is essential to achieving a sustainable existence. Market economics must be balanced with different measures of wealth, productivity, and other measures of the economic status of a country. Innovative techniques are sometimes needed to promote successful economic strategies along with successful protection of the environment. Environmental and ecological economics are subfields in economics that attempt to place value on benefits offered by the natural environment, some of which are hard to quantify. Cradle-to-grave and cradle-to-cradle systems analyses are tools used to merge sustainability with life-cycle analysis. In conjunction with applying economics to sustainability efforts, regulations are sometimes used as well. This is the focus of the first part of our next module.

Module 65 AP® Review Questions

1. How might the inclusion of an externality affect the supply and demand of a product?
   (a) It will increase price and decrease quantity demanded.
   (b) It will increase price and increase quantity demanded.
   (c) It will decrease price and increase quantity demanded.
   (d) It will decrease price and decrease quantity demanded.
   (e) Price will remain the same and quantity demanded will decrease.

2. What is NOT included in the calculation of gross domestic product?
   (a) Costs of health care
   (b) Government spending
   (c) Spending on durable goods
   (d) Earnings from investments
   (e) The costs of externalities

3. The use of cell phones in the developing world is an example of
   (a) the Kuznets curve.
   (b) a positive externality.
   (c) natural capital.
   (d) leapfrogging.
   (e) a negative externality.

4. Human capital includes
   (a) the goods that humans produce.
   (b) services that humans provide.
   (c) human knowledge and skills.
   (d) assets directly related to human survival.
   (e) services and processes that use manual labor.

5. Which is NOT a goal of a sustainable economic system?
   (a) Give priority to ecosystem health.
   (b) Use nonrenewable resources.
   (c) Place value on ecosystems.
   (d) Rely on ecosystem services rather than resource extraction.
   (e) Make manufacturers responsible for the disposal of products.

Answers to Module 65 AP® Review Questions

1. a
2. e
3. d
4. c
5. b

module

Regulations and Equity

66

We have looked at the economic system and the roles of natural capital and human capital. This understanding provides us with the tools we need to evaluate different options for monitoring and managing human systems in a way that will result in the least amount of harm to the natural environment. Regulatory tools are also used to bring about the least environmental harm. Ultimately, the goal is freedom from exposure to environmental harm for all the people in the world. This is the final topic of the book: environmental equity.

Learning Objectives

After reading this module you should be able to

- explain the role of agencies and regulations in efforts to protect our natural and human capital.
- describe the approaches to measuring and achieving sustainability.
- discuss the relationship among sustainability, poverty, personal action, and stewardship.

Agencies, laws, and regulations are designed to protect our natural and human capital

Many different techniques and approaches are used to influence how we treat the environment. Sometimes laws and regulations help achieve a certain outcome. Before examining some of the major laws and regulations in the United States and the world, we must familiarize ourselves with an important factor that shapes the way nations approach policy making—how people look at the world.

Environmental Worldviews and Regulatory Approaches

We have seen that the approach a nation takes to the regulation of economic activity and the environment depends in part on that nation's stage of development. In addition, worldview and attitude toward risk also shape a nation's approach to economics and the environment.

Worldviews

An **environmental worldview** is a worldview that encompasses how one thinks the world works; how one views his or her role in the world; and what one believes to be proper environmental behavior. Three

Environmental worldview A worldview that encompasses how one thinks the world works; how one views one's role in the world; and what one believes to be proper environmental behavior.

Teaching Tip: Engage

Write the three environmental worldviews (anthropocentric, biocentric, ecocentric) on the board. Have students place a sticky-note under the title they feel is most representative of their viewpoint. This activity creates a quick visual of the differences in how individuals see their role in the world.

Teaching Tip: Activity

Read each statement listed below. For each statement, direct students to stand if they agree and to stay seated if they disagree. Have one student from each position defend his or her position. This activity is meant to demonstrate that there are many different opinions and personal views that determine the response to a dilemma. Because there are no clear answers to these questions, students should understand that there are many factors and points of consideration when making environmental decisions.

- Extraction of natural gas should be banned if it threatens the quality of our water.
- Environmental regulations should be limited in less-developed nations because such regulations prevent the economy from developing.
- Electric companies should be required to pay a fee for the pollutants they release into the air even if it raises the cost of electricity.
- Governments should regulate the number of children a family can have in order to limit environmental degradation and to protect the population from starvation.

types of environmental worldviews dominate: human-centered, life-centered, and Earth-centered.

The **anthropocentric worldview** is a worldview that focuses on human welfare and well-being. In other words, nature has an instrumental value to provide for our needs. There are variations on this human-centered worldview. For example, those who favor a free-market approach to economics are optimistic about the results of unlimited competition and minimal government intervention. The planetary management school, while optimistic that we can solve resource depletion issues with technological innovations, believes that nature requires protection and that government intervention is at times necessary to provide this protection. **Stewardship,** a subset of the anthropocentric worldview, supports the careful and responsible management and care for Earth and its resources. The stewardship school of thought considers that while the natural world requires protection, it is also our ethical responsibility to be good managers of Earth.

The **biocentric worldview** is life-centered and holds that humans are just one of many species on Earth, all of which have equal intrinsic value. At the same time, this worldview considers that the ecosystems in which humans live have an instrumental value. There are various positions within the life-centered approach. While some consider that it is our obligation to protect a species, others consider that it is our obligation to protect every living creature.

The **ecocentric worldview** is Earth-centered. It places equal value on all living organisms and the ecosystems in which they live, and it demands that we consider nature free of any associations with our own existence. This worldview takes various forms. The environmental wisdom school, for example, believes that since resources on Earth are limited, we should adapt our needs to nature rather than adapt nature to our needs. The deep ecology school, meanwhile, insists that humans have no right to interfere with nature and its diversity. Our worldviews determine the decisions we make about our lives, our work, and the way we treat the planet.

Anthropocentric worldview A worldview that focuses on human welfare and well-being.

Stewardship The careful and responsible management and care for Earth and its resources.

Biocentric worldview A worldview that holds that humans are just one of many species on Earth, all of which have equal intrinsic value.

Ecocentric worldview A worldview that places equal value on all living organisms and the ecosystems in which they live.

Environmental worldviews can play a significant role in the policies a nation considers and how it implements them. For example, a nation that operates on an anthropocentric worldview might not concern itself with how economic activity affects the natural environment. A country with an ecocentric worldview might carefully regulate economic activity in order to protect ecosystems and the species within them. In practice, the policies and regulations of most nations represent a variety of worldviews depending on the particular nation and the specific resource or region of the biosphere that is being affected.

The Precautionary Principle

A nation's approach to environmental policy and regulation may also be influenced by whether or not it tends to follow the precautionary principle. In Chapter 17 we discussed the precautionary principle, which states that when the results of an action are uncertain—such as the effects caused by the introduction of a compound or chemical—it is better to choose an alternative known to be harmless. In many situations, scientific uncertainty complicates the estimation of the comparative risks of different actions. This is an important part of environmental decision making. In the United States, environmental law and policy has at times treated scientific uncertainty as a reason to discount or downplay scientific evidence of problems in the environment. Industrial and business groups have also used scientific uncertainty as a reason to avoid implementing expensive measures that would mitigate future environmental harms. Those who favor using the precautionary principle argue that if we wait for widespread scientific consensus about the adverse effects of a particular compound or action, we run the risk of creating an environmentally unsustainable and inequitable future.

Critics of the precautionary principle maintain that economic progress and human well-being will be hindered if we wait to use something until we verify that it is completely safe for the environment. There may also be an additional economic cost to waiting, or for choosing alternative means of achieving a goal.

In 1994, the International Union for Conservation of Nature, an organization composed of over 800 government and nongovernmental wildlife organizations, strengthened the 1992 Convention on Biological Diversity by publishing guidelines that included applying the precautionary principle as a tool in reaching decisions about the sustainable use of plant and animal species. The guidelines emphasize using the "best science available" in deciding whether to list a species as endangered and whether to ban any activity that could jeopardize that species.

The 1987 Montreal Protocol on Substances That Deplete the Ozone Layer is an example of the precautionary principle applied to global change. When the

Teaching Tip: Journal Prompt

Ask students the following questions:

- What is the precautionary principle? *The precautionary principle states that when the results of an action are uncertain, it is better to choose an alternative known to be harmless.*
- What do proponents of the precautionary principle argue? *Those who favor using the precautionary principle argue that if we wait for widespread scientific consensus about the adverse effects of a particular compound or action, we run the risk of creating an environmentally unsustainable and inequitable future.*
- What do critics of the precautionary principle argue? *Critics of the precautionary principle maintain that economic progress and human well-being will be hindered if we wait to use something until we verify that it is completely safe for the environment.*

protocol was adopted, there was still some scientific uncertainty about the evidence for the effect of CFCs on ozone depletion. You may recall from Chapter 14 that CFCs are chemicals that were used for refrigeration and other commercial applications. Despite the uncertainty about the effect of CFCs on the atmosphere, the protocol recommended eliminating their use. In this case, economic and political factors were balanced with the scientific findings to reach an agreement that CFCs should be phased out over a period of decades rather than immediately. Part of the success of the Montreal Protocol has been credited to the availability of an affordable and fairly effective replacement for CFCs. It has proven much more difficult to find affordable and effective replacements for the fossil fuels we currently use. Therefore, it is less likely that a similar scenario will unfold with respect to a reduction of greenhouse gases.

The precautionary principle is a relatively new and important part of environmental policy. It does not recommend or require any specific actions, but it does provide a reminder to environmental policy makers and managers that, in many cases, absolute scientific certainty may come too late when dealing with potentially serious environmental harms.

World Agencies

By considering the variety of worldviews presented earlier, and to the extent a particular country or agency subscribes to the precautionary principle, we can begin to understand more about the decision-making process that influences the various world agencies that have jurisdiction over global environmental issues. Global, national, or personal situations may prompt key beneficial decisions out of a sense of necessity and urgency. After World War II (1939 –1945), leaders of the allied nations agreed to found the **United Nations (UN),** a global institution dedicated to promoting dialogue among countries with the goal of maintaining world peace. When its charter was ratified in 1945, the UN had 51 member countries; by 2011, it had grown to 193, which is the number of member countries today. Since its establishment, the UN has created many internal agencies and institutions. Four of the many important UN organizations relating to the environment are the United Nations Environment Programme, the World Bank, the World Health Organization, and the United Nations Development Programme.

The United Nations Environment Programme

The **United Nations Environment Programme (UNEP)** is a program of the United Nations responsible for gathering environmental information, conducting research, and assessing environmental problems. Headquartered in Nairobi, Kenya, UNEP is also the international agency responsible for negotiating certain environmental treaties. In particular, the Convention on Biological Diversity, the Convention on International Trade in Endangered Species (CITES), and the Montreal Protocol on Substances That Deplete the Ozone Layer are three important international treaties UNEP negotiated. The Global Environment Outlook (GEO) reports are prepared under the auspices of UNEP.

The World Bank

The **World Bank** is a global institution that provides technical and financial assistance to developing countries with the objectives of reducing poverty and promoting growth, especially in the poorest countries. The World Bank cites four goals for economic development: (1) educating government officials and strengthening governments; (2) creating infrastructure; (3) developing financial systems, from microcredit to much larger systems; and (4) combating corruption. Critics of the World Bank maintain that there is too little consideration of environmental and ecological impacts when projects are evaluated and approved.

The World Health Organization

Headquartered in Geneva, Switzerland, the **World Health Organization (WHO)** is a global institution dedicated to the improvement of human health by monitoring and assessing health trends and providing medical advice to countries (FIGURE 66.1). It is the group within the UN responsible for human health, including combating the spread of infectious diseases, such as those that are exacerbated by global climate changes. This organization is also responsible for health issues in crises and emergencies created by storms and other natural disasters. The five key objectives of the WHO are: (1) promoting development, which should lead to improved health of individuals;

United Nations (UN) A global institution dedicated to promoting dialogue among countries with the goal of maintaining world peace.

United Nations Environment Programme (UNEP) A program of the United Nations responsible for gathering environmental information, conducting research, and assessing environmental problems.

World Bank A global institution that provides technical and financial assistance to developing countries with the objectives of reducing poverty and promoting growth, especially in the poorest countries.

World Health Organization (WHO) A global institution dedicated to the improvement of human health by monitoring and assessing health trends and providing medical advice to countries.

Teaching Tip: Video

Poverty, Money, and Love
This video from TED introduces the co-founder of Kiva.org who talks about how her work with microloans has brought new power to people who live on a few dollars a day.

The link to this resource may be found by clicking on the link buttons in the TE-book, opening the TRFD, or logging onto the book's companion website bcs.whfreeman.com/friedlandapes2e (teacher login required).

Chapter 20 Web Resources

Teaching Tip: Journal Prompt

Ask students the following questions:

- What is the human development index? *The human development index is a measurement index that combines three basic measures of human status:*
  - *life expectancy*
  - *knowledge and education, as shown in adult literacy rates and educational attainment*
  - *standard of living, as shown in per capita GDP and individual purchasing power*
- What is the human poverty index? *The human poverty index is a measurement index developed by the United Nations to investigate the proportion of a population suffering from deprivation in a country with a high HDI.*
- What might a high HDI value along with a high HPI value indicate about a country? *A high HDI value and a high HPI value might indicate a developed nation in which a substantial percentage of the population lives in poverty and lacks access to clean drinking water and health services.*

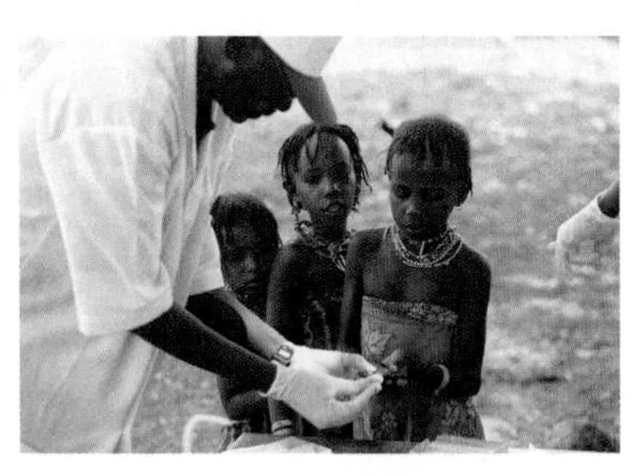

FIGURE 66.1 **World Health Organization workers.** A WHO worker in Chad draws blood from children for disease testing. *(Patrick Robert/Corbis)*

(2) fostering health security to defend against outbreaks of emerging diseases; (3) strengthening health care systems; (4) coordinating and synthesizing health research, information, and evidence; and (5) enhancing partnerships with other organizations.

The United Nations Development Programme

The **United Nations Development Programme (UNDP)** is an international program that operates in 166 countries around the world to advocate change that will help people obtain a better life through development. Headquartered in New York City, UNDP has a primary mission of addressing and facilitating issues of democratic governance, poverty reduction, crisis prevention and recovery, environment and energy issues, and prevention of the spread of HIV/AIDS. UNDP prepares an annual Human Development Report (HDR) that is an extremely useful measurement tool for the status of the human population.

Other Agencies

There are also a great number of nongovernmental organizations (NGOs) that work on worldwide environmental issues. These include Greenpeace, the International Union for Conservation of Nature, World Wide Fund for Nature (formerly World Wildlife Fund), and Friends of the Earth International.

U.S. Agencies

In January 1969, offshore oil platforms 10 kilometers (6 miles) from Santa Barbara, California, began to leak oil. Roughly 11.4 million liters (3 million gallons) of oil spilled out over the next 11 days and the leak continued throughout the year. This was not the first oil spill during the 1960s, nor the largest, but its proximity to the southern California coast resulted in something new—vast media attention. Daily television news reports of dead seabirds, fish, and marine mammals, as well as large stretches of oil-soaked beaches, shocked the American public and government officials. The Santa Barbara oil spill caused a major shift in federal policy toward incorporating an awareness of how human society affects the environment.

The first Earth Day, April 22, 1970, was partially the result of public reaction to the Santa Barbara oil spill and to other environmental problems that surfaced during the 1960s, such as those documented by Rachel Carson in her book *Silent Spring* (FIGURE 66.2). Earth Day 1970 is the symbolic birthday of the modern expression of the view that the natural environment and human society are inextricably connected. It also signals the beginning of contemporary environmental policy. Before 1970,

United Nations Development Programme (UNDP) An international program that works in 166 countries around the world to advocate change that will help people obtain a better life through development.

FIGURE 66.2 **The first Earth Day, New York City, 1970.** Large numbers of people gathered at many locations around the United States on April 22, 1970, to bring attention to the condition of Earth. *(Julian Wasser/Time & Life Pictures/Getty Images)*

Teaching Tip: Journal Prompt

Have students complete the table below to familiarize themselves with the agencies and organizations listed in the left column. (Answers are provided in italics.)

Organization	Description
United Nations Environment Programme (UNEP)	*A program of the United Nations responsible for gathering environmental information, conducting research, and assessing environmental problems*
World Bank	*A global institution that provides technical and financial assistance to developing countries with the objectives of reducing poverty and promoting growth, especially in the poorest countries*
World Health Organization (WHO)	*A global institution dedicated to the improvement of human health by monitoring and assessing health trends and providing medical advice to countries*
United Nations Development Programme (UNDP)	*An international program that works in 166 countries around the world to advocate change that will help people obtain a better life through development*
Environmental Protection Agency (EPA)	*The U.S. organization that oversees all governmental efforts related to the environment, including science, research, assessment, and education*
Occupational Safety and Health Administration (OSHA)	*An agency of the U.S. Department of Labor, responsible for the enforcement of health and safety regulations*
Department of Energy (DOE)	*The U.S. organization that advances the energy and economic security of the United States*

environmental policy focused primarily on biological and physical systems as economic resources for an industrial society. After 1970, sound environmental policy expanded to include the idea that economic benefits must be balanced by environmental science, environmental equity, and intergenerational equity—the interests of future generations in a healthy environment.

Since the early 1970s, several important U.S. agencies have been created to monitor human impact on the environment as well as to promote environmental and human health.

The Environmental Protection Agency

In 1970, President Richard Nixon signed the bill authorizing the creation of the **Environmental Protection Agency (EPA),** which oversees all governmental efforts related to the environment including science, research, assessment, and education. Headquartered in Washington, D.C., the EPA also writes and develops regulations and works with the Department of Justice and Department of State and U.S. Native American governments to enforce those regulations.

The Occupational Safety and Health Administration

Also in 1970, President Nixon signed the act creating the **Occupational Safety and Health Administration (OSHA),** an agency of the U.S. Department of Labor that is responsible for the enforcement of health and safety regulations. Its main mission is to prevent injuries, illnesses, and deaths in the workplace. OSHA conducts inspections, workshops, and education efforts to achieve its goals. Limiting exposure to chemicals and pollutants in the workplace is one way that OSHA is involved in environmental protection.

The Department of Energy

In 1977, President Jimmy Carter signed an act creating the **Department of Energy (DOE),** which advances the energy and economic security of the United States. Among its top goals are scientific discovery, innovation, and environmental responsibility. Within the DOE, the Energy Information Agency gathers data on the use of energy in the United States and elsewhere.

There are several approaches to measuring and achieving sustainability

Just as there are agencies, laws, and regulations designed to initiate and enhance sustainability, there are also a number of lenses through which to view the world, and a number of measurements or indexes used to evaluate sustainability. This section introduces some of these measurements and views. Eventually, some or all of these indices may become more directly involved in the measurement and assessment of sustainability.

Measuring Human Status

Despite the variety of economic indicators that are used around the world, there is still a call for a measurement that reports on the status of human beings with the specific goal of covering some of the noneconomic parameters such as levels of health and education. A variety of these are used and we describe two of them here.

The Human Development Index

The **human development index (HDI)** is a measurement index that combines three basic measures of human status: life expectancy; knowledge and education, as shown in adult literacy rate and educational attainment; and standard of living, as shown in per capita GDP and individual purchasing power. HDI was developed in 1990 by economists from Pakistan, England, and the United States, and it has been used since then by the UNDP in its annual HDR. As an index, HDI serves to rank countries in order of development and determine whether they are developed, developing, or underdeveloped. FIGURE 66.3 shows the range of HDI values and the distribution among countries. As you might expect, most of the developed countries have the highest HDI values.

The Human Poverty Index

The **human poverty index (HPI)** is a measurement index developed by the United Nations to investigate the proportion of a population suffering from deprivation

Environmental Protection Agency (EPA) The U.S. organization that oversees all governmental efforts related to the environment, including science, research, assessment, and education.

Occupational Safety and Health Administration (OSHA) An agency of the U.S. Department of Labor, responsible for the enforcement of health and safety regulations.

Department of Energy (DOE) The U.S. organization that advances the energy and economic security of the United States.

Human development index (HDI) A measurement index that combines three basic measures of human status: life expectancy; knowledge and education.

Human poverty index (HPI) A measurement index developed by the United Nations to investigate the proportion of a population suffering from deprivation in a country with a high HDI.

Module 66 Regulations and Equity 715

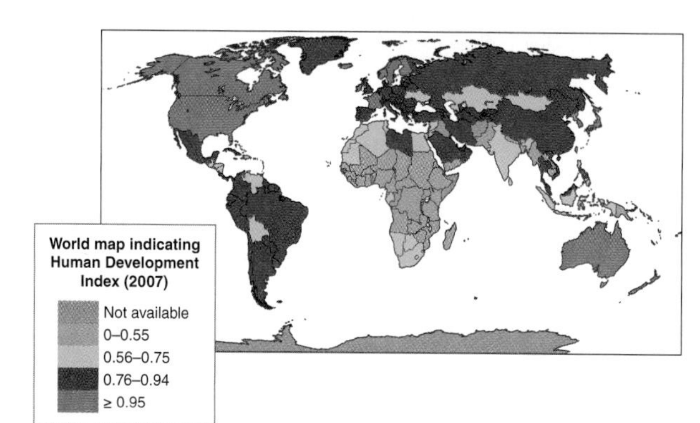

FIGURE 66.3 **The human development index.** The HDI is an index of well-being proposed by some as an alternative to GDP. Higher values indicate greater development.

Teaching Tip: Activity

Have students read "Science Applied 8: Can We Solve the Carbon Crisis Using Cap-and-Trade?" on page 730. Have students complete the free-response question on page 733 in preparation for a class discussion considering the benefits and costs of this approach.

in a country with a high HDI. This index measures three things: longevity, as indicated by the percentage of the population not expected to live past 40; knowledge, as measured by the adult illiteracy rate; and standard of living, as indicated by the proportion of the population without access to clean water and health services, as well as the percentage of children under 5 years of age who are underweight.

The Policy Process in the United States

To be fair and effective, environmental policies should be based on scientific indicators that suggest a certain behavior or action will be best for the environment. When policy makers believe there is adequate understanding of the science, and there is a course of preferred action for states or individuals, they begin a process to develop a policy.

The five basic steps in a policy cycle are problem identification, policy formulation, policy adoption, policy implementation, and policy evaluation. FIGURE 66.4 depicts this process as a circular or reiterative process. As a policy is evaluated, the need for amendment might arise. When an amendment is initiated, it follows roughly the same steps. Many good environmental policies have had numerous amendments. For example, the Clean Air Act has been amended twice, and even the original Clean Air Act of 1970 was actually a modification of earlier clean air legislation.

Command-and-control approach A strategy for pollution control that involves regulations and enforcement mechanisms.

Incentive-based approach A strategy for pollution control that constructs financial and other incentives for lowering emissions based on profits and benefits.

Legislative Approaches to Encourage Sustainability

United States governmental agencies have tried many ways to protect the environment, promote human safety and welfare and, in some cases, internalize externalities. The **command-and-control approach** is a strategy for pollution control that involves regulations and enforcement mechanisms. The **incentive-based approach** constructs financial and other incentives for lowering emissions based on profits and benefits. A combination of both approaches is likely to generate the maximum amount of desired changes.

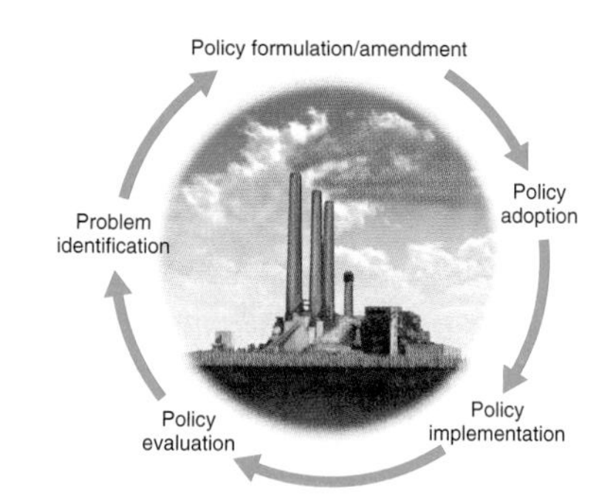

FIGURE 66.4 **The environmental policy cycle.** After an environmental problem is identified, environmental policy is formulated or modified. After a policy is adopted and implemented, it is evaluated and, if necessary, adjustments to the policy are made.

Taxation is a major deterrent used to discourage companies from producing pollution and generating other negative impacts. A **green tax** is a tax placed on environmentally harmful activities or emissions in an attempt to internalize some of the externalities that may be involved in the life cycle of those activities or products. However, a tax alone may not be sufficient to achieve the desired results. Sometimes rebates or tax credits are given to individuals and businesses purchasing certain items such as energy-efficient appliances or building materials such as windows and doors. Another technique, known as cap-and-trade, is discussed in "Science Applied: Can We Solve the Carbon Crisis Using Cap-and-Trade" on page 738.

In 1996, President Clinton's Council on Sustainable Development declared that "the essence of sustainable development is the recognition that the pursuit of one set of goals affects others and that we must pursue policies that integrate economic, environmental, and social goals." The **triple bottom line** is an approach to sustainability that considers three factors—economic, environmental, and social—when making decisions about business, the economy, and development. FIGURE 66.5 shows that the intersection of these three factors is sustainability. There are many organizations and businesses that place one of these three factors at the top of a priority list. Some businesses strive for economic well-being—a sound financial bottom line—to the exclusion of human welfare or the environment. They may be regarded as successful within certain communities, but the triple bottom line concept emphasizes that to be a true success, there must be adequate treatment of both humans and environment. Paul Hawken, the author of *Natural Capitalism*, states the objective as, "Leave the world better than you found it, take no more than you need, try not to harm life or the environment, and make amends if you do."

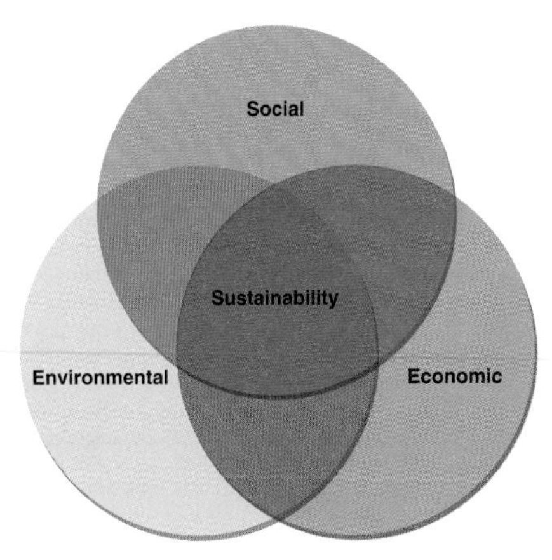

FIGURE 66.5 **The triple bottom line.** Sustainability is believed to be achievable at the intersection of the social, economic, and environmental factors that influence most development endeavors.

U.S. Policies for Promoting Sustainability

Of the many regulations that have been established in the last 50 years or so in the United States, there are at least seven important pieces of legislation that may help move the United States toward sustainability. All of these regulations have been discussed in other chapters and are summarized in **TABLE 66.1**.

Two major challenges of our time are reducing poverty and stewarding the environment

The classic environmental dichotomy is "jobs versus the environment." Those primarily concerned with human well-being ask how we can make demands for environmental improvements when there is so much poverty and injustice in the world. Those primarily concerned with the environment ask how we can focus exclusively on human suffering when an impoverished environment cannot support human health and well-being.

Poverty and Inequity

Approximately one-sixth of the human population—more than one billion people—lives in unsanitary conditions in informal settlements, slums, and shantytowns. Roughly one-sixth of the human population earns less than $1 a day, and one-third earns less than $2 a day. In the last 100 years, as developed countries have increased their GDPs and as many countries have modernized and developed their economies, the disparities between the rich and the poor have become greater. Poverty is simultaneously an issue of human

Green tax A tax placed on environmentally harmful activities or emissions in an attempt to internalize some of the externalities that may be involved in the life cycle of those activities or products.

Triple bottom line An approach to sustainability that considers three factors—economic, environmental, and social—when making decisions about business, the economy, and development.

Teaching Tip: Warm-up

Ask students to explain how sustainability is addressed by incorporating the idea of the triple bottom line. *The triple bottom line is an approach to sustainability that considers three factors—economic, environmental, and social—when making decisions about business, the economy, and development. A sound financial bottom line is balanced by adequate treatment of both humans and the environment.*

Teaching Tip: Engage

Ask students: What are some challenges that keep the international community from reaching the triple bottom line? *Answers may vary but can include: Placing limited monetary value on environmental resources and human rights; inequality among nations and within nations that limit fair use or distribution of resources; powerful corporations or governments that determine policy based on economic growth.*

Teaching Tip: Journal Prompt

Have students re-create Table 66.1 in their journals. Students should be familiar with each of these laws in order to answer multiple-choice questions and, occasionally, free-response questions on the AP® Exam.

TABLE 66.1 Major U.S. legislation for promoting sustainability

Act	Abbreviation	Year enacted	Purpose	Prime example of a success
National Environmental Policy Act	NEPA	1970	Enhance environment; monitor with a tool: the Environmental Impact Assessment	Protection of coral formation and sea turtles has occurred.
Occupational Safety and Health Act	OSHA	1970	Prevent occupational injury, illness, death from work-related exposure to physical and chemical harm	Worker training and knowledge of toxins has increased.
Endangered Species Act	ESA	1973	Protect animal and plant species from extinction	Bald eagle, peregrine falcon, and gray wolf populations have recovered
Clean Air Act	CAA	1970	Promote clean air	Sulfur dioxide reductions from cap-and-trade have occurred.
Clean Water Act	CWA	1972	Promote clean water	Swimmable and fishable rivers across the United States have increased.
Resource Conservation and Recovery Act	RCRA	1976	Govern tracking and disposal of solid and hazardous waste	Numerous brownfields and contaminated lands have been cleaned up.
Comprehensive Environmental Response, Compensation, and Liability Act	CERCLA, also called Superfund	1980	Force and/or implement the cleanup of hazardous waste sites	Dozens of Superfund sites have been cleaned up around the United States.

rights, economics, and the environment. Every human has a basic right to survival, well-being, and happiness—all directly threatened by poverty. Indebted individuals and nations are often unable to pay what they owe. In 2005, the 8 major industrial countries of the world, known as the G8, canceled the debt of the 18 poorest countries. From an environmental standpoint, poverty increases overuse of the land, degradation of the water, and incidence of disease.

In 2000, the United Nations offered an eight-point resolution listing what its member countries agreed were pressing issues that the world could no longer ignore. The member countries committed to reaching these Millennium Development Goals (MDGs), as outlined in the United Nations' Millennium Declaration, by 2015:

- Eradicate extreme poverty and hunger
- Achieve universal primary education
- Promote gender equality and empower women
- Reduce child mortality
- Improve maternal health
- Combat HIV/AIDS, malaria, and other diseases
- Ensure environmental sustainability
- Develop a global partnership for development

As of this writing, some countries are well on their way to meeting these goals while others lag far behind. As with environmental laws and their implementation, the distance from resolutions to results is immense, and not all developed countries have committed the resources that they had promised.

One proponent of the UN MDGs was Nobel Peace Prize Laureate Dr. Wangari Maathai (1940–2011) from Kenya (FIGURE 66.6). Dr. Maathai was the founder of the Green Belt Movement, a Kenyan and international environmental organization that empowers women by paying them to plant trees, some of which can be harvested for firewood after a few years. The Green Belt Movement is credited with replanting large expanses of land in East Africa, thereby reducing erosion and improving soil conditions and moisture retention. In addition, the trees that have been replanted, provided that they are not overharvested, offer a renewable source of fuel for cooking. The Green Belt Movement is considered a global sustainability success story promoting both individual human and environmental well-being. Dr. Maathai was also involved in environmental activism to achieve her goals, which sometimes caused her difficulties with certain governmental organizations.

Environmental Justice

The typical North American uses many more resources than the average person in many other parts of the world. This situation is not equitable. The subject of fair distribution of the resources of Earth, known as environmental equity, has received increasing international

Teaching Tip: Journal Prompt

Ask students to list the Millennium Development Goals developed by the United Nations.

- *Eradicate extreme poverty and hunger*
- *Achieve universal primary education*
- *Promote gender equality and empower women*
- *Reduce child mortality*
- *Improve maternal health*
- *Combat HIV/AIDS, malaria, and other diseases*
- *Ensure environmental sustainability*
- *Develop a global partnership for development*

Teaching Tip: Warm-up

Ask students: What is the connection between poverty and inequity? *People living in poverty are more likely to face exposure to increased pollution and to receive fewer environmental resources than people with higher socioeconomic status.*

FIGURE 66.6 **Wangari Maathai.** Dr. Maathai was the founder of the Green Belt Movement in Kenya. (Micheline Pelletier/Corbis)

attention in recent years. Beyond moral objections to inequity, there are concerns about sustainability. We have seen how increased resource use usually increases harm to the environment. As more and more people develop a legitimate desire for better living conditions, the resources of Earth may not be able to support continued consumption at such high levels. Closely related to the equity of resource allocation are questions of the inequitable distribution of pollution and of environmental degradation with their adverse effects on humans and ecosystems. All of these topics fall under the subject of environmental equity.

As we discussed in Chapter 16 and elsewhere, African Americans and other minorities in the United States are more likely than Caucasians to live in an area with solid waste incinerators, chemical production plants, and other so-called "dirty" industries. In a number of studies in the 1980s and later, investigators used the distribution of minority residents by postal zip code to relate race and class to the location of hazardous sites. In Atlanta, 83 percent of the African American population lived within the same zip code area as the 94 uncontrolled toxic waste sites, while 60 percent of the whites lived in those areas. In Los Angeles, roughly 60 percent of Hispanics lived in the same areas as the toxic waste sites, while only 35 percent of the white population lived in those areas. One study in five southern states compared the size of specific landfill facilities with the percentage of minorities in the zip code area in which the landfill was located. The study concluded that the largest landfills are located in areas that have the greatest percentage of minorities. An important issue that has not been entirely resolved, and that can vary from case to case, is whether the affected population or the hazardous facility came first to a given area. By knowing which came first, people and organizations attempting to remedy the situation will have a better idea of how to modify existing legislation and regulations to reduce the number of people who live in degraded environments.

More recently, it has become clear that the subjects of disproportionate exposure to environmental hazards were not only African Americans, but all races in lower income brackets. Moreover, the problem was not limited to the United States. The concept was broadened and named environmental justice, which is both a social movement and an academic field of study. Those involved in environmental justice examine whether there is equal enforcement of environmental laws and elimination of disparities—intended or unintended—in the exposure to pollutants and other environmental harms affecting different ethnic and socioeconomic groups within a society. Delegates to the First National People of Color Environmental Leadership Summit in 1991 established 17 principles of environmental justice. Professor Robert Bullard of Texas Southern University has published books and papers in the academic area of environmental justice and has been involved in the social movement as well. He is probably best known for his 1990 book *Dumping in Dixie*, which demonstrated that minority and lower socioeconomic groups were often the recipients of pollution from dumping of MSW and hazardous wastes. More recently, Professors Paul Mohai of the University of Michigan and Robin Saha of the University of Montana reassessed the unequal distribution of hazardous waste dumping in the United States and found that the situation is actually worse than previously reported. In particular, they believe they have resolved the issue of whether the hazardous waste facility or the lower-socioeconomic and minority population came first to an area. The authors maintain that the minority community in many cases was present first and that the hazardous waste facility, which came later, was specifically targeted to that community.

Individual and Community Action

There are a fair number of people who believe that whether or not governments and private agencies are able to achieve their goals, individuals can and must act to further their own goals of sustaining human existence on the planet. These individuals have begun to make attempts to live a sustainable existence without government incentives, taxes, or other measures. They have begun activities such as calculating their own ecological footprint, carbon footprint, energy footprint, and other metrics to determine how much of an impact they are making on Earth. From this starting point, they have begun to make changes in their consumption, behavior, and lifestyle to reduce that impact. Some people act on their own while others act through groups and organizations. They have adopted a philosophy represented by the saying, "If the people lead, the leaders will follow." Some individuals have

Teaching Tip: Journal Prompt

Ask students: What is environmental justice? Give an example of an environmental injustice. *Environmental justice is both a social movement and an academic field of study. Those involved in environmental justice examine whether there is equal enforcement of environmental laws and elimination of disparities—intended or unintended—in the exposure to pollutants and other environmental harms affecting different ethnic and socioeconomic groups within a society. One example of environmental injustice is that minorities in the United States are more likely to live in an area with solid waste incinerators, chemical production plants, and other so-called "dirty" industries.*

Teaching Tip: Engage

Ask students: How did Dr. Maathai's Green Belt Movement meet the goals of the triple bottom line?

- ***Economic:*** *Women are paid to plant trees, increasing the economic status of women.*
- ***Social:*** *Women are empowered to provide for their families because the trees that have been planted—provided that they are not overharvested—offer a renewable source of fuel for cooking.*
- ***Environmental:*** *Replanting trees reduces erosion and improves soil conditions and moisture retention.*

Teaching Tip: Video

A Teacher Growing Green in the South Bronx

This video from TED showcases the efforts of Stephen Ritz who has inspired his students to create a better community. Through his educational efforts he has given his students "green job" skills that cannot be outsourced.

The link to this resource may be found by clicking on the link buttons in the TE-Book, opening the TRCD, or logging onto the book's companion website bcs.whfreeman.com/friedlandapes2e (teacher login required).

Chapter 20 Web Resource

joined together to organize communities centered around philosophies of sustainability.

Van Jones, a graduate of Yale Law School, was a community organizer in San Francisco working on civil rights and human justice issues when he decided to combine concerns about the environment and global climate change with the need for creating jobs in cities (FIGURE 66.7). He founded an organization called Green for All and in 2008 published a book titled *The Green Collar Economy: How One Solution Can Fix Our Two Biggest Problems*. The two problems, as he sees it, are global warming and urban poverty. He believes that creating green jobs, such as insulating buildings, constructing wind turbines and solar collectors, and building and operating mass transit systems, will improve the living conditions of some of the poorest people in the nation and reduce our impact on the environment. For Van Jones, this is a win-win solution—people will be employed and our emissions of global greenhouse gases will decrease.

Majora Carter exemplifies another model of how to participate in community activities. She was born and raised in the South Bronx section of New York City and is a Wesleyan University and New York University graduate. She founded a not-for-profit environmental justice organization before forming her own private sector firm. She advocates improving health and quality of life in communities by promoting economic development in a sustainable and environmentally sound way. She has attracted a great deal of attention by creating gardens and greenways in the South Bronx, while at the same time creating all types of employment opportunities, including green jobs.

FIGURE 66.7 A gathering of Green for All supporters in Oakland, California. Over the decades, individual and community action have influenced achievements in environmental quality and sustainability. *(©greenforall.org)*

module 66

REVIEW

In this module, we have seen that a series of worldviews can be used to approach environmental protection and regulation. A variety of world agencies such as the United Nations, the World Bank, and the World Health Organization have numerous environmental programs. In the United States, the Environmental Protection Agency and the Department of Energy are two of a number of environmental agencies. A variety of measures are used to assess sustainability and environmental well-being, including the human development index and the human poverty index. The triple bottom line maintains that sustainability can be achieved at the intersection of the three factors—economic, environmental, and social—that influence most development endeavors. Reducing poverty and taking sound care of the environment are two challenges that are essential for sustainability. Achieving these goals involves addressing poverty and inequality, environmental justice, and individual and community action.

Answers to Module 66 AP® Review Questions

1. d
2. c
3. b
4. a
5. d

Module 66 AP® Review Questions

1. Which worldview considers ecosystems to have intrinsic value?
   I. Anthropocentric
   II. Biocentric
   III. Ecocentric

   (a) I only
   (b) I and III only
   (c) II only
   (d) II and III only
   (e) III only

2. Enacting legislation that restricts a chemical suspected of being harmful while there is still scientific uncertainty about that chemical is an example of
   (a) an anthropocentric worldview.
   (b) a biocentric worldview.
   (c) the precautionary principle.
   (d) technology leapfrogging.
   (e) an incentive-based approach.

3. In what way does the Occupational Safety and Health Administration (OSHA) contribute to environmental protection?
   (a) It develops regulations to limit emissions.
   (b) It limits human exposure to chemicals and pollutants.
   (c) It improves the quality of water.
   (d) It provides funds to clean contaminated sites.
   (e) It protects rare habitats from industrial development.

4. The triple bottom line
   (a) is an approach to sustainability that considers economic, environmental, and social factors.
   (b) is a method for encouraging sustainability that includes incentives, regulations, and penalties.
   (c) consists of three measures of human status used in the Human Development Index.
   (d) is used by United Nations Development Programme to determine if a program is successful.
   (e) is an application of the precautionary principle to identify a chemical for further study.

5. Which is an example of the command-and-control approach to encourage sustainability?
   (a) A cap-and-trade system for carbon emissions
   (b) A rebate for energy-efficient products
   (c) Funding for solar energy projects
   (d) Regulations that limit sulfur emissions and include a provision for fines
   (e) Voluntary standards such as fair trade

working toward  sustainability

Reuse-A-Sneaker

In the 1990s, athletic shoe manufacturer Nike, based in Beaverton, Oregon, drew a great deal of negative publicity for the conditions of its factories and treatment of its workers overseas. The prevalence of child labor, unsafe working conditions, and inadequate wages highlighted the social costs that often accompany commercial success. Environmental degradation was the most pressing unseen effect of a very profitable shoe manufacturing company. However, in subsequent years, Nike imposed a system of factory standards and inspections that, along with recently passed labor laws in countries like China, have considerably improved conditions for workers.

Unlike heavy industry and energy generation, the shoe industry is not often part of the environmental discussion in the United States. However, in Asian and South American countries where the factories are located (one-third of Nike shoes are produced in China), the environmental impacts of shoe manufacturing are far more visible. Many aspects of production reduce the quality of air, water, and soil both near and far from production plants, including emissions from energy use and transportation of materials and products, solid waste from all levels of production, and resource extraction prior to manufacturing. Besides these threats, workers are often exposed to the toxic solvents, adhesives, and rubber used in shoe fabrication. The evaporation of these substances contributes to the concentration of volatile organic compounds (VOCs) in the atmosphere. VOCs also cause respiratory impairment in factory employees and contribute to hazardous waste. As shoe companies like Nike continue to increase production, the environmental effects are compounded. However, Nike is making significant efforts to lessen the environmental cost of a successful business.

In order to improve the sustainability of its industry, Nike has developed a comprehensive cradle-to-grave program that addresses environmental impacts in every stage of raw material extraction, product fabrication, sale, and disposal. The "Nike Considered" program offers more sustainable products and provides an index that evaluates each product and assigns it a score based on its environmental impact. Factors considered include efficiency of design, solvent use, and waste creation. Essentially, Nike has created a life-cycle analysis of materials including extraction processes, energy and water use, manufacturing practices, and recyclability. Products labeled "Considered" must meet standards of sustainability significantly higher than the average Nike

Teaching Tip: Activity

Ask students working in pairs to create a table that compares the negative implications of Nike's manufacturing processes and the sustainable practices they have employed to improve these practices. (Answers are provided in italics.)

Negative practices	Sustainable practices
• *Child labor* • *Poor working conditions* • *Inadequate wages*	*Nike imposed a system of factory standards and inspections that, along with recently passed labor laws in China, have considerably improved conditions for workers.*
Many aspects of production reduce the quality of air, water, and soil, including emissions from energy use and transportation of materials, solid waste from all levels of production, and resource extraction prior to manufacturing.	*Nike has developed a comprehensive cradle-to-grave program that addresses environmental impacts in every stage of raw material extraction, product fabrication, sale, and disposal. Nike's Reuse-A-Shoe program and recycling of scrap materials left over from manufacturing are incorporated into sports surfaces and new Nike products. The program reduces landfill waste and reduces the need for extraction of raw materials for athletic facility surfaces.*
The evaporation of toxic solvents contributes to the concentration of volatile organic compounds (VOCs) in the atmosphere. VOCs also cause respiratory impairment in factory employees and contribute to hazardous waste.	*Some products are made using water-based adhesives instead of solvent-based adhesives, thereby reducing VOC evaporation. Soles of the shoes consist of recycled and less toxic rubber. The design of the shoe as well as the manufacturing process must demonstrate efficiency.*

Teaching Tip: Beyond the Classroom

Ask students to research local companies that incorporate sustainable practices. Have them create a presentation to share with the class. Alternatively, invite a member of the company to give a presentation about these practices.

Suggested Answers to Critical Thinking Questions

1. One significant obstacle can be economic. While some forms of sustainability, like improvements in efficiency and materials, can often save money in the long run, other sustainability measures can increase the costs by internalizing externalities like pollution and disposal. These increased costs make it harder for the business to compete unless there is sufficient consumer interest in supporting sustainable business and paying the additional cost.

2. Answers will vary. Sustainable practices could include recycling programs, energy efficiency efforts, renewable energy use, or pollution reduction.

Nike Grind. Making this athletic playing surface from recycled sneakers and sneaker manufacturing waste saved resources and energy. *(Courtesy of NIKE, Inc.)*

product. These products are made using water-based adhesives instead of solvent-based adhesives, thereby reducing VOC evaporation. Soles of the shoes consist of recycled and less toxic rubber. The design of the shoe as well as the manufacturing process must demonstrate efficiency. Many Nike products already follow the "Considered" standards. Nike intends for all products made by it and its affiliates to be manufactured following the "Considered" standards, at which point waste will decrease by 17 percent and use of environmentally preferable materials will increase by 20 percent. As part of this program, Nike has created a Sustainable Business Performance Summary.

In addition to designing and fabricating more sustainable shoes and supporting organic agriculture, Nike reduces waste and positively affects communities through its Reuse-A-Shoe program. The company encourages the public to recycle used athletic shoes so that they can be broken down into a substance called Nike Grind. Nike Grind is the raw material that results from the recycling of athletic shoes collected through Nike's Reuse-A-Shoe program and from the recycling of scrap materials left over from the manufacture of Nike footwear. By recycling old shoes and manufacturing scrap, and incorporating this recycled material into sports surfaces and new Nike products, the program reduces landfill waste and reduces the need for extraction of raw materials for athletic facility surfaces. Additionally, Nike athletic shoes are now designed for easier breakdown, which decreases the energy needed to create Nike Grind. Nike has partnered with market-leading sports surfacing companies that use Nike Grind in the manufacture of tennis and basketball courts, running tracks, synthetic turf, children's playgrounds, and fitness room flooring, doubly benefiting the environment and the local population. Nike estimates that more than 25 million pairs of shoes have been recycled in the Reuse-A-Shoe program.

In April of 2009, volunteers used Nike Grind to build the first community playground in the Poncey-Highland neighborhood of Atlanta, Georgia. The rubber surface consisted of 92 percent recycled materials, and equipment was constructed from recycled milk jugs and other reused products. By using cradle-to-cradle practices to reduce environmental damage and support social programs, Nike is working to achieve a more ethical and sustainable business model. Presumably, over time, more companies and industries will develop innovative sustainable business practices. Sustainable practices are an essential part of being a consumer, but they may also become part of business and industry.

Critical Thinking Questions

1. What are some of the obstacles that prevent businesses from adopting sustainability goals and actually practicing sustainability in the workplace?
2. Can you identify and describe a sustainable business in your community?

References

Locke, R. M., et al. 2009. Nike Considered: Getting Traction on Sustainability. *MIT Sloan Teaching Innovation Resources (MSTIR)*. https://mitsloan.mit.edu/MSTIR/sustainability/NikeConsidered/Pages/default.aspx.

Sharma, A. 2013. Swoosh and sustainability: Nike's emergence as a global sustainable brand. Sustainable Brands www.sustainablebrands.com/news_and_views (May 17, 2013).

chapter

20 REVIEW

In this chapter, we identified sustainability as one of the ultimate goals of environmental science. It is often achieved through the use of sound economic and business practices as well as effective environmental regulations and laws. There are international agencies such as the United Nations Environment Programme and a number of U.S. agencies to enforce regulations and laws. The Environmental Protection Agency oversees all governmental efforts related to the environment. Different worldviews affect how people approach the goal of sustainability. Proper stewardship of the natural world and a reduction of poverty are challenges that must be met if people are to achieve sustainability.

Key Terms

Well-being
Economics
Genuine progress indicator (GPI)
Technology transfer
Leapfrogging
Natural capital
Human capital
Manufactured capital
Market failure
Environmental economics
Ecological economics
Valuation
Environmental worldview
Anthropocentric worldview
Stewardship
Biocentric worldview
Ecocentric worldview
United Nations (UN)
United Nations Environment Programme (UNEP)
World Bank
World Health Organization (WHO)
United Nations Development Programme (UNDP)
Environmental Protection Agency (EPA)
Occupational Safety and Health Administration (OSHA)
Department of Energy (DOE)
Human development index (HDI)
Human poverty index (HPI)
Command-and-control approach
Incentive-based approach
Green tax
Triple bottom line

Learning Objectives Revisited

Module 65 Sustainability and Economics

- **Explain why efforts to achieve sustainability must consider both sound environmental science and economic analysis.**

Sustainable environmental systems must allow for maintaining air, water, land, and biosphere systems and must also maintain human well-being—the status of being healthy, happy, and prosperous. Sustainability will not be achieved if certain groups are exposed to a disproportionate share of dirty jobs or waste material in the home or workplace.

- **Describe how economic health depends on the availability of natural capital and basic human welfare.**

Sustainable systems must include a consideration of externalities. Gross domestic product (GDP) is the value of all products and services produced in a year in a given country. Genuine progress indicator (GPI) includes measures of personal consumption, income distribution, levels of higher education, resource depletion, pollution, and health of the population. Economic assets, or capital, can come from the natural systems on Earth, from humans, or from the manufactured products made by humans. Valuing all three kinds of capital is essential to systems that are sustainable.

Module 66 Regulations and Equity

- **Explain the role of agencies and regulations in efforts to protect our natural and human capital.**

Once a society believes it has enough scientific information to act with the intent of protecting or reducing harm to the environment, it must determine the rules and regulations it wishes to enact. A group of government agencies in the

Teaching Tip: Measuring Your Impact

GDP and Footprints

Measuring Your Impact 20 offers students the opportunity to calculate and determine the relationship between ecological footprint and GDP for China, India, Japan, and the United States. Students will discuss environmental equity between a developed nation and a less-developed nation. Finally, students will identify two goals of the United Nations Millennium Declaration that would assist a less-developed nation in environmental equity.

Download this resource by clicking on the link buttons in in the TE-book, opening the TRFD, or logging onto the book's companion website bcs.whfreeman.com/friedlandapes2e (teacher login required).

Measuring Your Impact 20: GDP and Footprints

Exploring the Literature

Bullard, R. 2009. *Race, Place, and Environmental Justice after Hurricane Katrina.* Westview Press.

Ehrenfeld, J.R., and A.J. Hoffman. 2013. *Flourishing: A Frank Conversation About Sustainability.* Stanford Business Books.

Goodstein, E. 2010. *Economics and the Environment.* 6th ed. John Wiley & Sons.

Hawken, Paul. 1994. *The Ecology of Commerce.* Collins.

Jones, Van. 2012. *Rebuild the Dream.* Nation Books.

Download a printable version of this list by clicking on the link buttons in the TE-book, opening the TRFD, or logging onto the book's companion website bcs.whfreeman.com/friedlandapes2e (teacher login required).

Exploring the Literature 20

Additional Multiple-Choice AP® Practice Exam

Download an additional Multiple-Choice AP® Environmental Science Practice Exam for Chapter 20 by clicking on the link buttons in the TE-book, opening the TRFD, or logging onto the book's companion website bcs.whfreeman.com /friedlandapes2e (teacher login required).

Multiple-Choice AP® Practice Exam 20

Answers to Chapter 20 AP® Exam Multiple-Choice Questions

1. e
2. b
3. c
4. d
5. d
6. a
7. b
8. a
9. d
10. b
11. e
12. c
13. e
14. a
15. d

United States handles the areas that offer protection to the environment and humans. Policies are enacted through passage and modification of laws.

- **Describe the approaches to measuring and achieving sustainability.**

The Human Development Index combines life expectancy, knowledge and education, and standard of living as a measure of human status. The Human Poverty Index measures the percentage of population in a country that is suffering from deprivation. A green tax can be used to internalize externalities or reduce environmental harm. The triple bottom line accounts for three factors—economic, environmental, and social—when making decisions about the environment and development. These ideas have led to a variety of policies in the United States for promoting sustainability.

- **Discuss the relationship among sustainability, poverty, personal action, and stewardship.**

One-sixth of the world population has inadequate housing and inadequate income. People will need access to food, housing, clean water, and adequate medical care before they can be concerned about environmental sustainability. The UN Millennium Development Goals have established objectives for improving the status of people and the sustainability of the environment. The human-centered worldview maintains that humans have intrinsic value and nature provides for our needs. The life-centered worldview holds that humans are one of many species on Earth, all of which have value. The Earth-centered worldview places equal value on both all living organisms and ecosystems. Individual and community action can lead to sustainable actions occurring at a greater level worldwide.

Chapter 20 AP® Environmental Science Practice Exam

Section 1: Multiple-Choice Questions

Choose the best answer for questions 1–15.

1. Based on the supply and demand curve below, which of the following can be inferred?

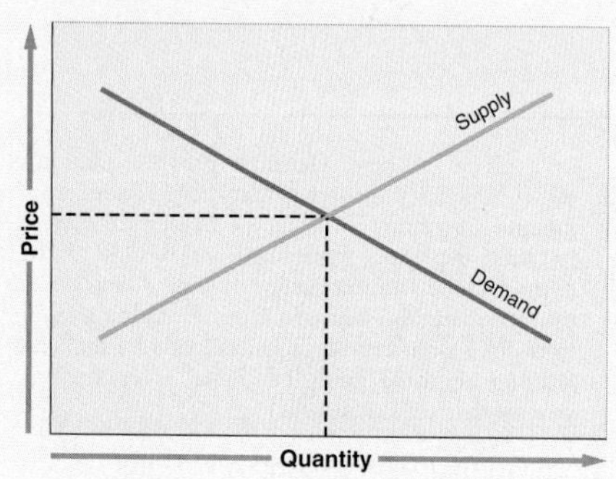

I. A lower price results in a greater demand.
II. A higher price results in a greater supply.
III. Price changes as supply and demand fluctuate.

(a) I only
(b) II only
(c) III only
(d) I and II
(e) I, II, and III

2. All of the following are examples of negative externalities except
(a) global climate change as a result of greenhouse gas emissions from burning coal, oil, and gasoline.
(b) increased pollination rates of crop plants as a result of local beekeeping.
(c) a pulp mill that produces paper and pollutes the surrounding water and air.
(d) runoff of pesticides and fertilizers from a farm into a nearby river.
(e) acid deposition in the Adirondacks as a result of coal-burning power plants in the Midwest.

3. The genuine progress indicator (GPI) is more representative measure of the wealth and well-being of a country than gross domestic product (GDP) because GPI
(a) measures productivity and consumption without taking externalities into account.
(b) has risen in the United States while the GDP has remained fairly constant since 1970.
(c) includes resource depletion, pollution, and health of the population in its calculation.
(d) can be increased by higher health care costs and a greater incidence of illnesses.
(e) does not reflect personal consumption, income distribution, or levels of higher education.

Answers to Chapter 20 AP® Exam

Download the full answers to the Chapter 20 AP® Exam by clicking on the link buttons in the TE-book, opening the TRFD, or logging onto the book's companion website bcs .whfreeman.com/friedlandapes2e (teacher login required).

Answers to Chapter 20 AP® Exam

4. Economic assets are the sum total of
   I. Natural capital
   II. Human capital
   III. Manufactured capital

   (a) II only
   (b) III only
   (c) I and III
   (d) I, II, and III

5. Valuation, according to environmental and ecological economics, would include all of the following except
   (a) the revenue generated from tourists visiting a national park.
   (b) the cost of wastewater treatment provided by a natural wetland.
   (c) the benefits derived from medicinal plants found in tropical rainforests.
   (d) the profits realized from hiring more employees to increase production.
   (e) the cost of converting animal wastes into reusable organic matter by detritivores.

6. Full cost pricing by the internalization of externalities could result in which of the following?
   (a) Higher prices and a reduction in the consumption of items with high negative impacts
   (b) Lower prices and an increase in the consumption of items with high negative impacts
   (c) Greater consumer demand for products with high negative impacts
   (d) Lower production costs due to diminishing natural resources
   (e) Reduced production of environmentally friendly goods and services

7. Cradle-to-cradle and cradle-to-grave analyses of manufactured goods can best be described as the study of the
   (a) changes in the use of a product from one generation to the next.
   (b) life cycle of a product from its production to use to ultimate disposal.
   (c) use of resource extraction over the use of ecosystem services.
   (d) options for the disposal of solid waste generated by the product.
   (e) natural and human resources required for production.

8. Recently the Los Angeles Unified School District adopted a new policy on the use of pesticides in schools. This policy assumes that the use of pesticides constitutes a risk to the health of children and the environment. Pesticides will be employed only after nonchemical methods have been explored. The pest control measure that is the least harmful will be implemented. This is an example of
   (a) the precautionary principle.
   (b) ecosystem services.
   (c) a market-driven approach.
   (d) sustainable use.
   (e) full cost pricing.

9. Which is a United Nations organization concerned with the environment?
   (a) World Resources Institute (WRI)
   (b) Occupational Safety and Health Administration (OSHA)
   (c) Department of Energy (DOE)
   (d) World Health Organization (WHO)
   (e) Environmental Protection Agency (EPA)

10. Which U.S. law contributes to sustainability by governing the tracking and disposal of solid and hazardous waste?
    (a) National Environmental Policy Act (NEPA)
    (b) Resource Conservation and Recovery Act (RCRA)
    (c) Clean Water Act (CWA)
    (d) Comprehensive Environmental Response, Compensation, and Liability Act (CERCLA)
    (e) Occupational Safety and Health Act (OSHA)

11. Strategies to implement environmental laws and regulations include all of the following except
    (a) standards for emission levels with fines when these levels are exceeded.
    (b) green taxes on environmentally harmful activities or emissions.
    (c) buying and selling of pollution permits.
    (d) an incentive-based approach based on profits.
    (e) banning the cap-and-trade practice.

12. Which is a harmful effect of poverty?
    (a) Decrease in unsanitary conditions
    (b) Greater access to clean drinking water
    (c) Increased overuse of the land
    (d) Decreased malnutrition
    (e) Lower infant mortality rates

13. The United Nations Millennium Declaration proposes to meet which of the following goals?
    I. Reduce environmental sustainability through economic development
    II. Eliminate extreme poverty and hunger and reduce child mortality
    III. Empower women and improve maternal health

    (a) I only
    (b) II only
    (c) IIII only
    (d) I and III
    (e) II and III

Answers to the Chapter 20 Free-Response Questions can be found in the Answer Appendix on page ANS-17.

14. The following is a summary report for the Distribution of Environmental Burdens for Allegheny County in Pennsylvania.

Population categories	Number of facilities emitting criteria air pollutants per square mile
Minorities	11
Whites	4.5
Low-income families	8
High-income families	3.9
Families below poverty threshold	8.9
Families above poverty threshold	4.1
Non-high school graduates	6.9
High school graduates	4.7

The information in this table reflects
(a) an environmental equity issue.
(b) an anthropocentric worldview.
(c) a biocentric worldview.
(d) an ecocentric worldview.
(e) a stewardship school issue.

15. The idea that all people regardless of ethnic or socioeconomic status deserve equal environmental conditions is a central principle of
(a) the triple bottom line.
(b) the National Environmental Policy Act.
(c) the United Nations Environment Programme.
(d) environmental justice.
(e) the Resource Conservation and Recovery Act.

Section 2: Free-Response Questions

Write your answer to each part clearly. Support your answers with relevant information and examples. Where calculations are required, show your work.

1. Use the following information about gasoline consumption in the United States to answer the questions below.
   - In 2008 the United States consumed approximately 138 billion gallons of gasoline.
   - The current federal tax on gasoline is 18.4 cents per gallon.
   - The national average cost of a gallon of regular unleaded gasoline in June 2008 was $4.00 per gallon.
   - 80 percent of the federal gasoline tax is used to subsidize road construction.

   (a) Calculate the total amount of money spent in the United States on the purchase of gasoline in 2008 (when gasoline cost $4.00 per gallon). (2 points)
   (b)
   (i) What percent of the cost per gallon is the gasoline tax?
   (ii) How much revenue was generated by the gasoline tax in 2008?
   (iii) How much was used in 2008 to subsidize road construction? (3 points)
   (c) Does the federal tax on gasoline qualify as a green tax? Explain your answer. (1 point)
   (d) Advocates of raising the gasoline tax suggest that the tax be increased to 80 cents per gallon. Identify two economic effects and two environmental effects of raising this tax. (4 points)

2. In 1997, the ecological economist Robert Costanza and his associates published a report titled *The Value of the World's Ecosystem Services and Natural Capital.* They estimated that if all the ecosystem services provided worldwide had to be paid for, the cost would average $33 trillion per year with a range from $16 trillion to $54 trillion. In that same year the global gross national product (GNP) was $18 trillion.
   (a) What is meant by ecosystem or ecological services? Give three specific examples and identify which United Nations organization might oversee these services. (4 points)
   (b) Define the term valuation. What would the worldwide consequences be if the world actually had to pay for ecosystem services and natural capital? (2 points)
   (c) Explain how this report could be used to develop a sustainable economic system. (2 points)
   (d) Which environmental worldview is most consistent with the concerns of environmental economics? Explain. (2 points)

Unit 8 AP® Environmental Science Practice Exam

Section 1: Multiple-Choice Questions

Choose the best answer for questions 1–20.

1. In a small town, local residents recently discussed a proposal to fill in a wetland near the planned site of a new apartment complex. A farmer objected to this proposal because the wetland provides a habitat for crop pollinators. A business owner objected because she was worried it would harm the town's image and reduce business. Which categories of instrumental values do these arguments consider?
   (a) The farmer is considering the intrinsic value of the wetland; the business owner is considering the instrumental value of the wetland.
   (b) Both are considering the intrinsic value of the wetland.
   (c) The farmer is considering the regulating services of the wetland; the business owner is considering the support services of the wetland.
   (d) The farmer is considering the support services of the wetland; the business owner is considering the cultural services of the wetland.
   (e) Both are considering the regulating services of the wetland.

2. Which of these organizations is NOT an international organization?
   (a) IUCN
   (b) USEPA
   (c) WWF
   (d) UNEP
   (e) UNDP

3. Which of the following phenomenon is NOT likely to be a potential consequence of habitat fragmentation?
   (a) A decrease in the proportion of edge habitat
   (b) An increase in the rate of inbreeding depression
   (c) Loss of genetic diversity within populations
   (d) A decrease in the range of animal movement
   (e) A decline in the abundance of endemic species

4. In North America, honeybees (*Apis mellifera*) should be considered
   (a) a native species.
   (b) an exotic species.
   (c) an invasive species.
   (d) a threatened species.
   (e) an endemic species.

5. In the United States, the definitions of *endangered* and *threatened* are
   (a) similar to IUCN definitions of threatened and near-threatened.
   (b) identical to the IUCN definitions of threatened and near-threatened.
   (c) similar to definitions set by UNEP.
   (b) designed to classify more species as endangered.
   (e) defined by the Lacey Act.

6. Which of the following factors is least likely to be a consideration when applying the concept of SLOSS (single large or several small) to the conservation of a population of a particular species?
   (a) The monetary cost of protecting suitable habitat
   (b) The genetic diversity of the population
   (c) The ability of individuals in the population to move between suitable habitats
   (d) The amount of edge habitat surrounding a suitable habitat
   (e) The IUCN risk categorization of the species

7. Overharvesting a species for sport, medicinal, or industrial purposes may alter the ________ associated with that species.
   I. intrinsic value
   II. instrumental value
   III. ecological interactions

   (a) I only
   (b) II only
   (c) III only
   (d) II and III
   (e) I, II, and III

Question 8 refers to the following figure:

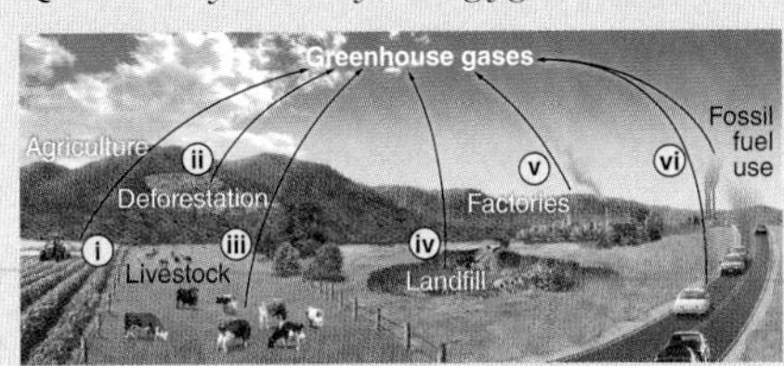

8. In the diagram of greenhouse gas sources, which refers to significant sources of methane (CH_4)?
   (a) i, ii, iv
   (b) i, ii, vi
   (c) ii and iii
   (d) iii and iv
   (e) i, ii, iii, iv, v, and vi

Answers to Unit 8 AP® Exam Multiple-Choice Questions

1. d
2. b
3. a
4. b
5. a
6. e
7. c
8. d
9. b
10. d
11. b
12. b
13. c
14. d
15. c
16. d
17. e
18. b
19. c
20. c

Answers to Unit 8 AP® Exam

Download the full answers to the Unit 8 AP® Exam by clicking on the link buttons in the TE-book, opening the TRFD, or logging onto the book's companion website bcs .whfreeman.com/friedlandapes2e (teacher login required).

Answers to Unit 8 AP® Exam

9. Black soot causes global warming by
   (a) reflecting solar radiation.
   (b) lowering the albedo of snow.
   (c) releasing volatile organic compounds.
   (d) interacting with existing greenhouse gases.
   (e) increasing transmission of infrared radiation to Earth's surface.

10. From 1980 to 1990, the concentration of atmospheric CO_2 increased from 335 to 350 ppm. Based on these two time points, predict the concentration of atmospheric CO_2 in 2030 assuming that the rate of CO_2 increase remains the same.
   (a) 360 ppm
   (b) 395 ppm
   (c) 405 ppm
   (d) 410 ppm
   (e) 430 ppm

11. Which of the following factors is responsible for annual fluctuations in atmospheric CO_2 concentrations?
   (a) Melting of glacial ice in the Arctic
   (b) Primary production during the spring and summer
   (c) The El Niño–Southern Oscillation
   (d) Deforestation
   (e) Combustion of fossil fuels during the summer

12. Scientists can analyze the __________ of foraminifera in ocean sediments to detect past changes in __________.
   (a) species composition/ CO_2 concentrations
   (b) species composition/water temperatures
   (c) quantity/global temperatures
   (d) quantity/water temperatures
   (e) quantity/CO_2 concentrations

13. Which of the following is most likely to cause a negative feedback loop with regard to global climate change?
   (a) Ocean acidification
   (b) Melting permafrost
   (c) Primary production
   (d) Evapotranspiration
   (e) Anaerobic decomposition

14. Which of the following is a potential consequence of glacial melting due to global warming?
   I. Europe and North America experiencing warmer temperatures
   II. Loss of the thermohaline circulation
   III. Global rises in sea level

   (a) I only
   (b) II only
   (c) I and II
   (d) II and III
   (e) I, II, and III

15. Ocean acidification is primarily caused by
   (a) acid rain.
   (b) the precipitation of base compounds.
   (c) increased atmospheric CO_2.
   (d) the melting of glaciers.
   (e) increases in average global temperatures.

16. Which of the following best describes the U.S. justification for not signing the Kyoto Protocol after it was modified in 2001?
   (a) The U.S. government argued that the protocol imposed unreasonably stringent regulations on developing nations.
   (b) It is impossible to reduce carbon emissions without causing substantial harm to the U.S. economy.
   (c) The U.S. Senate voted unanimously that the United States should not sign any international agreement that lacked restrictions on developing countries.
   (d) The agreement did not require sufficient reductions in the carbon emissions of China and India despite their large population sizes.
   (e) Signing the agreement was unnecessary for the United States, because the Supreme Court had already ruled that the U.S. Environmental Protection Agency has the authority to regulate greenhouse gases as part of the Clean Air Act.

Question 17 refers to the following figure:

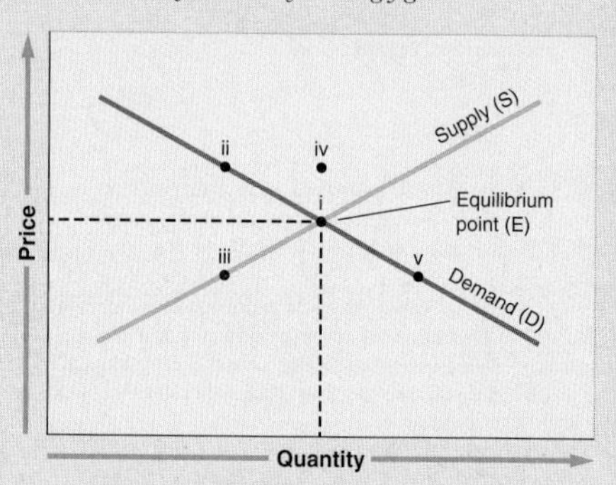

17. The graph depicts the supply and demand curves for production of energy-efficient light bulbs in the United States. Currently, the equilibrium point of the two curves is at *i*. Suppose that government subsidies are given to the light bulb manufacturers in order to decrease the overall cost of production and increase the quantity of bulbs produced. Where is the new equilibrium point most likely to lie?
   (a) i
   (b) ii
   (c) iii
   (d) iv
   (e) v

728 Unit 8 AP® Environmental Science Practice Exam

18. Which of the following phenomenon challenges the conceptual basis of the Kuznets curve?
    (a) Technology transfer between developed nations
    (b) Leapfrogging by undeveloped nations
    (c) Increasing environmental degradation with increasing gross domestic product
    (d) Decreasing environmental degradation with increasing per capita income
    (e) Improved education standards with increasing environmental degradation

19. The biocentric worldview believes that
    (a) we should adapt to nature rather than adapt nature to our needs.
    (b) we can solve resource depletion with technological innovation but nature does require some protection.
    (c) humans are one of many species on Earth, and each has equal intrinsic value.
    (d) it is the ethical responsibility of humans to care for all species on the planet.
    (e) government intervention is always necessary to protect the environment.

20. The United Nations Environment Programme
    (a) seeks to improve human health by assessing health trends among countries.
    (b) provides technical and financial assistance to developing countries.
    (c) is responsible for gathering environmental information and conducting research.
    (d) oversees the quality of working conditions for workers in developing nations.
    (e) encourages the elimination of poverty through ecotourism.

Section 2: Free-Response Questions

Write your answer to each part clearly. Support your answers with relevant information and examples. Where calculations are required, show your work.

1. In 1997, tropical nations joined together in a massive plan to create a continuous terrestrial corridor that would conserve endemic species and allow dispersal of species between North and South America. Currently, the project is still underway and each nation is continually adding fragments of land to the corridor. Developers of this initiative have faced several problems. For example, many indigenous tribes exist throughout the corridor. Displacement from their land would mean the loss of the culture and heritage associated with those tribes. There are also challenges in purchasing and connecting fragmented land, as well as monitoring the success of the corridor.
    (a) Identify one environmental worldview that advocates for allowing these tribes to persist on their land. Justify your answer by defining the worldview. (2 points)
    (b) Describe how the concept of SLOSS can be used to overcome the challenges of connecting fragmented land. (4 points)
    (c) Describe two ecosystem services that could be monitored to evaluate the health of protected land in the corridor. (2 points)
    (d) Define the IUCN Red List and suggest how we can use the Red List to monitor the success of the corridor. (2 points)

2. Evidence indicates that atmospheric carbon dioxide has greatly increased over the past century and that this increase is associated with changes in average temperatures. Although these changes are likely to alter the current range of species, ecologists are concerned that these changes will also alter the timing and placement of ecological interactions.
    (a) As temperatures warm during the spring, many species of butterfly migrate to northern latitudes to find food and breeding spots. As temperatures cool during the fall, they migrate south to warmer conditions.
        (i) Describe two ways in which migrating butterflies might suffer as a result of changes in global temperature. (2 points)
        (ii) Suggest one way that migrating butterflies might rapidly evolve over a few generations to cope with changes in global temperatures. (2 points)
    (b) Describe how we can use ice cores from the Antarctic to determine if the abundance of carbon dioxide in the atmosphere has recently increased. (3 points)
    (c) Describe three ways in which global warming might alter weather patterns in North America. (3 points)

Teaching Tip: Unit 8 Additional Free-Response Question

An additional practice free-response question covering concepts from Unit 8 is available. To answer this question, students must

- discuss how flowering time is likely to shift with climate change
- explain how we might combine our current data with Thoreau's observations over many years to predict the effects of climate change on flowering time
- describe and justify the pattern that a naturalist in the 1880s would have seen in CO_2 levels over the four seasons of each year
- explain how an earlier flowering time and longer growing season result in a negative feedback loop with respect to global warming

Download this resource by clicking on the link buttons in the TE-book, opening the TRFD, or logging onto the book's companion website bcs.whfreeman.com/friedlandapes2e (teacher login required).

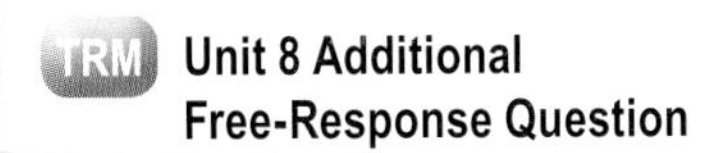

Unit 8 Additional Free-Response Question

Answers to the Unit 8 Free-Response Questions can be found in the Answer Appendix on page ANS-18.

8

Gemenacom/Shutterstock

science applied

Can We Solve the Carbon Crisis Using Cap-and-Trade?

In Chapter 19, we saw that the nations of the world have become concerned about the increasing amount of CO_2 in the atmosphere of Earth. While the Kyoto Protocol called for developed nations to reduce total CO_2 production, there has been a great deal of debate regarding the most effective way to achieve this reduction.

What are our options for controlling CO_2 emissions?

Although many different options for controlling CO_2 emissions have been considered, most approaches can be categorized as either command-and-control or *cap-and-trade*. As discussed in Chapter 20, with a command-and-control approach, a government regulates the amount of pollution that can be emitted by different industries. For example, in 2010 the U.S. federal government announced substantially higher fuel-efficiency standards for cars and light-duty trucks and in 2014 it announced higher standards for medium- and heavy-duty trucks such as delivery trucks and tractor trailers (FIGURE SA8.1). These mandates would reduce the consumption of gasoline and thereby lower the output of CO_2 emitted into the atmosphere. The new standards, which ultimately received support from the major automobile manufacturers, require new technology that would make the vehicles more expensive for consumers, but this higher initial expense is more than offset by reduced fuel costs over the life of each vehicle.

FIGURE SA8.1 **The command-and-control approach.** When the federal government sets minimum standards for fuel efficiency, it is using a command-and-control approach to pollution. *(AP Images/David Zalubowski)*

Cap-and-trade An approach to controlling CO_2 emissions, where a cap places an upper limit on the amount of pollutant that can be emitted and trade allows companies to buy and sell allowances for a given amount of pollution.

While a command-and-control approach means the government sets a single mandatory standard to which all companies must adhere, a **cap-and-trade** approach uses a different philosophy. As the name suggests, there are two elements to this approach. First, there is a *cap*, or upper limit, that is placed on the amount of a pollutant that can be emitted. Instead of placing a limit on individual companies or industries, the limit is placed on the total amount of the pollutant produced in a region, such as a state, country, or group of cooperating countries. In the case of CO_2, for example, a cap could be placed on the total amount of CO_2 that a state or country could emit. To meet the goals of the Kyoto Protocol, for example, the United States agreed

FIGURE SA8.2 **Carbon cap-and-trade approach.** Many different industries emit large amounts of CO_2, including coal-fired electricity-generating power plants, automobiles, airplanes, and steel manufacturers. Under a cap-and-trade approach, all of these industries could be involved in buying and selling permits for CO_2 emission. *(top left: David Parsons/iStockphoto.com; bottom left: Elena Elisseeva/age fotostock; right: Digital Vision/Getty Images)*

to a 7 percent reduction in CO_2 from 1990 levels. This value could serve as a cap for the United States.

The *trade* portion of the cap-and-trade approach should be familiar from Chapter 15. It allows companies to buy and sell permits for a given amount of pollution. A government can either require all companies that emit pollutants to buy permits or it can distribute a large number of free permits and then slowly reduce the number of permits available to be traded. Allowing companies to buy and sell pollution permits provides an incentive for companies to invest in pollution reduction, since they would gain income by selling permits they no longer need to companies that continue to pollute. Companies still generating large amounts of pollution will also be motivated to reduce their emissions since they would have to spend more to buy the additional permits, assuming that the cost of the permit is higher than the cost of not polluting (FIGURE SA8.2).

Using this strategy, over time, the government would gradually reduce the number of pollution permits available. By the laws of supply and demand, reducing the supply of permits (lowering the cap) causes an increased demand for the remaining permits and results in higher prices for the permits. This price increase further motivates polluting companies to reduce their emissions, resulting in a decline in the total amount of pollution. The cap-and-trade approach avoids the complex regulations that occur with the command-and-control approach and gives companies the freedom to choose from a wide range of possible solutions the approach that best suits them. At the same time, it offers an economic incentive to reduce pollution. The rate at which the pollution declines would depend on how quickly the government reduced the number of pollution permits.

What are the concerns about controlling carbon by using cap-and-trade?

There is currently a great deal of debate concerning the use of cap-and-trade as a means of controlling CO_2 emissions. One of the major issues is that most cap-and-trade proposals for CO_2 include an additional aspect known as *carbon offsets*. **Carbon offsets** are methods of promoting global CO_2 reduction that do not involve a direct reduction in the amount of CO_2 actually emitted by a company. For example, a company could pay for the reforestation of an area that could absorb CO_2 or pay a landowner not to log a forest, thereby keeping the carbon locked up in the biomass of the trees. By paying these costs, a company could avoid paying a potentially higher cost of reducing its own CO_2 emissions. Opponents contend that while these efforts would allow a particular piece of land to sequester carbon, it does not decrease the total amount of land needed to grow lumber or crops. As a result, the deforestation simply occurs in another location and the carbon offset does nothing to reduce the global production of CO_2. Furthermore, even if a section of land is allowed to sequester carbon, changes in government regulations or political control, or even a natural event such as a forest fire, could rapidly convert sequestered carbon into atmospheric carbon dioxide.

Another concern about the cap-and-trade approach is the exemption of certain companies or facilities. Most experts agree that such exemptions

Carbon offsets Methods of promoting global CO_2 reduction that do not involve a direct reduction in the amount of CO_2 actually emitted by a company.

Science Applied 8 Can We Solve the Carbon Crisis Using Cap-and-Trade? 731

are likely to occur. As we saw in Chapter 15, the cap-and-trade approach was used for the control of sulfur dioxide emissions from coal-powered plants in 1990 to reduce acid precipitation. Many of the older power plants were exempt from the program because at the time it was believed that these plants would soon be retired. However, nearly two-thirds of the coal-powered plants in operation today were constructed before 1975. For cap-and-trade to be a fair system, all polluting companies will need to pay for polluting permits.

Has cap-and-trade ever worked successfully?

The concept of cap-and-trade, developed by economists in the 1970s, has experienced limited use in the decades since. As noted above, it was first implemented to reduce sulfur dioxide emissions that caused acid precipitation as part of the U.S. Clean Air Act of 1990. Following a suggestion from the Environmental Defense Fund, the federal government developed a cap-and-trade program for coal-powered plants. Motivated by the new cost of buying pollution permits, the companies owning these plants had to retrofit existing facilities, build new and less polluting facilities, or buy pollution permits from other companies that had already reduced their emissions. Even though some of the older plants were exempt from certain provisions of the act, from 1990 to 2012 there has been a 72 percent reduction in sulfur emissions. The cap-and-trade system motivated investment in new technologies, which is one of the underlying reasons for the success of this program. For example, General Electric developed a system that converted sulfur into gypsum. Because gypsum is a valuable product used in construction, the economic incentive to reduce sulfur emissions motivated new research and development that produced a profitable solution.

In 2005, a carbon cap-and-trade system was instituted in the European Union. Critics of the European system point out that the national governments in the European Union have overestimated the baseline amount of CO_2 that their industries were producing and have given away most of the permits to the polluters for free. As a result, these companies have surplus CO_2 permits that they can sell at a profit. In rebuttal, proponents of the system point out that this strategy was necessary at the beginning to gain the support of companies. Starting the system with a surplus of permits and then gradually reducing the number of permits available allowed for a gradual adjustment by the companies that emitted CO_2. In 2010, the European Union started to lower the cap and more companies, including electricity-generating plants and airlines, began paying for the permits. Some industries, such as steel and cement manufacturers, however, have argued that they needed to continue having free permits if they are to remain competitive with companies that operate outside the European Union. In 2014, the European Union projected that industries covered by the cap-and-trade system would reduce CO_2 emissions by 21 percent in 2020 compared with emissions in 2005.

At this stage, it is too early to tell if the cap-and-trade system will be a successful strategy to reduce CO_2 emissions. Cap-and-trade was successful in reducing sulfur emissions, but the sulfur program had no offsets that could be used as credit and allow continued pollution by a company. If carbon cap-and-trade is used in the United States, a major focus will be placed on energy companies, including those that produce petroleum products and those that generate electricity. A study by Point Carbon, a market research company, found that the major oil companies, including ExxonMobil and Chevron, would likely pay hundreds of millions of dollars in CO_2 permits while electricity-generating companies that have diversified to include hydroelectric and nuclear energy will pay substantially less. Ultimately, if cap-and-trade is successful in the European Union, it could serve as an effective model for other nations and help reduce the growing problem of CO_2 and the global changes that come with it.

References

Hansen, J. 2009. Cap and fade. *New York Times*, December 7. http://www.nytimes.com/2009/12/07/opinion/07hansen.html.

Krugman, P. 2009.Unhelpful Hansen. *New York Times*, December 7. http://krugman.blogs.nytimes.com/2009/12/07/unhelpful-hansen/.

Reed, S. 2013. European lawmakers support carbon trading system. *New York Times*, December 10. http://www.nytimes.com/2013/12/11/business/energy-environment/european-lawmakers-vote-to-support-carbon-trading.html.

The European Union Emissions Trading System http://ec.europa.eu/clima/policies/ets/index_en.htm.

732 Science Applied 8 Can We Solve the Carbon Crisis Using Cap-and-Trade?

Key Terms

Cap-and-trade
Carbon offsets

Questions

1. What is the difference between a command-and-control and a cap-and-trade approach to regulating CO_2?
2. If a cap-and-trade approach were to be used, what is the expected effect on CO_2-emitting factories as a nation's government begins reducing the number of CO_2 emission permits?

Free-Response Question

Write your answer to each part clearly. Support your answers with relevant information and examples. Where calculations are required, show your work.

Carbon taxes provide an alternative method to cap-and-trade for the federal regulation of greenhouse gas emissions. Unlike cap-and-trade, there is no cap on the amount of emissions that can be produced. However, there is a tax on every unit of emissions that is produced. Economists have debated which system is better.

(a) Considering the laws of supply and demand, how is the equilibrium value of pollution permits likely to change? Justify your answer by describing supply and demand curves. Assume a fixed number of permits are given. (2 points)

(b) If a fixed number of pollution permits are distributed in a single year and no more are offered in subsequent years, describe why new companies have a disadvantage under this system. (2 points)

(c) Which of the two methods of regulating greenhouse gas emissions is better if

  (i) the environment is highly sensitive to small changes in greenhouse gas emissions? (1 point)

  (ii) the economy is rapidly growing, and the environment is not highly sensitive to changes in gas emissions? (1 points)

(d) What are carbon offsets and why are they a potential concern in implementing a cap-and-trade system? (2 points)

(e) If a company produces more pollution than allowed by their number of acquired pollution permits, they must pay a fine. Why is the size of the fine important to consider when discussing the potential benefits of cap-and-trade? (2 points)

Suggested Answers to Science Applied Questions

1. A command-and-control approach sets a regulation that all companies in an industry must follow, such as phasing out and banning incandescent bulbs in many countries. A cap-and-trade system has a country- or statewide cap, but allows companies to trade their allowances with each other, so if one company is more efficient than required, it can sell the permits to a company that is less efficient.

2. The reduced number of permits will increase the value of the remaining permits, making it more affordable for companies to switch to practices that produce less CO_2 rather than continuing to buy permits.

Answer to Science Applied 8 Free-Response Question

(a) If we assume that a fixed number of permits are given, then the supply of permits cannot change. Hence, the supply curve is simply a vertical line. If demand increases, then the demand curve is pushed to the right and the equilibrium value of pollution permits will increase. If demand decreases, the demand curve is pushed to the left and the equilibrium value of pollution permits will decrease.

(b) If a fixed number of pollution permits are distributed in a single year, they will only be distributed to existing companies. As new companies enter the market, they must buy permits from the preexisting companies. This makes it difficult for newer companies to start.

(c) (i) If the environment is highly sensitive to small changes in greenhouse gas emissions, it is better to use a cap-and-trade system, since there is an absolute cap on the amount of gases that are produced.

(ii) If the economy is rapidly growing and the environment is not highly sensitive to changes in gas emissions, then it is better to use a carbon tax system. This system allows starting companies to predict their future expenses and compete in the market.

(d) Carbon offsets are methods of promoting CO_2 reduction that do not involve a direct reduction in the amount of CO_2 produced. Instead, companies pay for projects that can remove carbon from the atmosphere, such as reforestation projects. Although these efforts allow a particular piece of land to sequester carbon, it does not decrease the total amount of land needed to grow lumber or crops. As a result, the deforestation simply occurs in another location and the carbon offset does nothing to reduce the global production of CO_2. Furthermore, even if a section of land is allowed to sequester carbon, changes in government regulations or political control, or even a natural event such as a forest fire, could rapidly convert sequestered carbon to atmospheric CO_2.

(e) If the size of the fine is too small, companies are likely to ignore regulations that limit carbon pollution. This will be particularly true if the size of the fine is similar to the cost of a pollution permit, which is determined by supply and demand. In such cases, companies have little incentive to limit their pollution, since the cost of the fine will be identical to purchasing a pollution permit. Hence, the size of the fine must be substantially larger than the cost of a pollution permit.

Answer to Science Applied 8 Free-Response Question

Download the full answer to the Science Applied 8 Free-Response Question by clicking on the link buttons in the TE-book, opening the TRFD, or logging onto the book's companion website bcs .whfreeman.com/friedlandapes2e (teacher login required).

Answer to Science Applied 8 Free-Response Question

cumulative AP® environmental science practice exam

Section 1: Multiple-Choice Questions

Choose the best answer for questions 1–100.

1. Primary production is an example of
   I. an ecosystem service.
   II. an environmental indicator.
   III. heterotrophic activity.

   (a) I only
   (b) II only
   (c) I and II
   (d) II and III
   (e) I, II, and III

2. Which of the following is likely to increase biodiversity within a biome?
   (a) Landscape fragmentation
   (b) Introduction of an invasive species
   (c) Immigration of humans
   (d) Speciation
   (e) A disease epidemic

3. The United States produces 8 million tons of oranges in a single year. However, many orange crops are succumbing to a deadly invasive bacteria. If 10,000 hectares of orange cropland are lost in a year to this bacteria, and a single acre can produce 20 tons of oranges, what percentage of the total orange crop is lost to the disease in a year? (Note that 1 hectare = 2.5 acres.)
   (a) 2 percent
   (b) 6 percent
   (c) 10 percent
   (d) 20 percent
   (e) 24 percent

Questions 4 and 5 refer to the following experiment:

A group of scientists wanted to test the effects of increased greenhouse gas concentrations on plant growth. They hypothesized that elevated levels of CO_2 would increase plant biomass after 2 weeks, whereas elevated levels of N_2O and CH_4 would have no effect. To test this hypothesis, they placed red maple (*Acer rubrum*) tree saplings in incubators, and then subjected each sapling to one of three treatments. The treatments included 10 ppm of CO_2, N_2O, or CH_4 gas above ambient concentrations. Each treatment had four replicates. After 2 weeks, they measured plant biomass.

4. Which is a flaw of this experiment?
   (a) The experiment lacks a control treatment.
   (b) 10 ppm is a negligible increase of CO_2 relative to ambient concentrations.
   (c) The hypothesis is actually a null hypothesis.
   (d) The measured response variable does not relate to the hypothesis.
   (e) N_2O gas is not a greenhouse gas.

5. As hypothesized, the researchers found that plants exposed to elevated CO_2 had increased biomass after 2 weeks, whereas plants exposed to elevated N_2O and CH_4 did not exhibit any change in biomass. Which would be a deductive statement based solely on these results?
   (a) Elevated levels of CO_2 are due to global climate change.
   (b) Reduced levels of CO_2 due to global climate change will decrease red maple production.
   (c) An observed increase in red maple production is probably due to elevated levels of CO_2.
   (d) Increases in red maple production in nature are probably not due to elevated levels of N_2O or CH_4.
   (e) CH_4 and N_2O are not likely to be biologically important greenhouse gases for tree growth.

6. For radioactive elements, the transformation between a parent and daughter atom involves
   (a) the creation of ionic bonds.
   (b) a release of neutrons and energy.
   (c) an increase in total energy.
   (d) the transformation of chemical energy to potential energy.
   (e) the transformation of heat energy to kinetic energy.

7. Which group of compounds is listed in order of increasing pH?
   (a) OH^-, H_2O, $CaCO_3$
   (b) $CaCl$, $LiCl$, HCl
   (c) $NaOH$, BaO, OH^-
   (d) $NaOH$, H_2O, H_2SO_4
   (e) HF, $NaCl$, $NaOH$

8. You have installed a solar-charged battery that can provide 4 MJ of electrical energy each day. Approximately how many 50 W bulbs can you run on the battery if each bulb is on for an average of 1 hour per day?
   (a) 3
   (b) 10
   (c) 22
   (d) 32
   (e) 45

Answers to Cumulative AP® Environmental Science Practice Exam Multiple-Choice Questions

1. c
2. d
3. b
4. a
5. c
6. b
7. e
8. c

Answers to Cumulative AP® Exam

Download the full answers to the Cumulative AP® Exam by clicking on the link buttons in the TE-book, opening the TRFD, or logging onto the book's companion website bcs.whfreeman.com /friedlandapes2e (teacher login required).

Answers to Cumulative AP® Exam

Answers

9. e
10. c
11. d
12. a
13. e
14. d
15. b
16. b

9. Which of the following is involved in generating electricity through nuclear fission?
   I. A positive feedback loop
   II. Creation of chemical energy
   III. Release of kinetic energy

   (a) I only
   (b) II only
   (c) III only
   (d) I and II
   (e) I and III

10. Calculate the energy efficiency of converting coal into heated water if the efficiency of turning coal into electricity is 35 percent, the efficiency of transporting the electricity is 90 percent, and the efficiency of hot water heaters is 90 percent.
    (a) 10 percent
    (b) 17 percent
    (c) 28 percent
    (d) 32 percent
    (e) 60 percent

Question 11 refers to the following graphs:

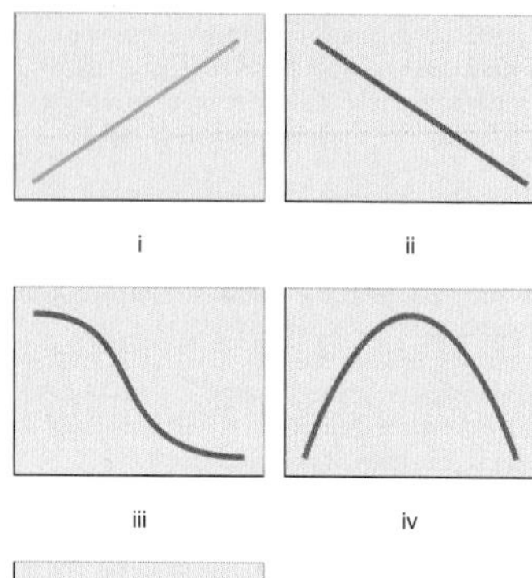

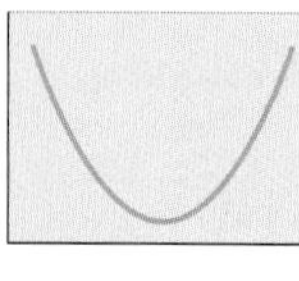

11. According to the intermediate disturbance hypothesis, which graph best represents the effects of predator density (x axis) on prey species richness (y axis)?
    (a) i
    (b) ii
    (c) iii
    (d) iv
    (e) v

12. Which describes a key difference between photosynthesis and aerobic respiration?
    (a) Photosynthesis generates organic compounds whereas aerobic respiration generates inorganic compounds.
    (b) Photosynthesis releases energy whereas aerobic respiration consumes energy.
    (c) Photosynthesis uses energy derived from external sources whereas aerobic respiration uses energy derived from internal sources.
    (d) Aerobic respiration has a higher energy efficiency than photosynthesis.
    (e) Photosynthesis consumes chemical energy whereas aerobic respiration generates chemical energy.

13. In an acre of grassland, net primary productivity is best measured as
    (a) carbon taken up by plants in sunlight plus carbon released by plants in the dark.
    (b) carbon taken up by plants in sunlight.
    (c) aboveground biomass of plants.
    (d) aboveground and belowground biomass of plants.
    (e) carbon taken up by plants in sunlight minus carbon released by plants in the dark.

14. Ocean acidification represents a key component of the
    (a) nitrogen cycle.
    (b) hydrologic cycle.
    (c) phosphorus cycle.
    (d) carbon cycle.
    (e) sulfur cycle.

15. Which would best estimate the resilience of a forest ecosystem?
    (a) Measurement of tree biomass once before and immediately after a forest fire.
    (b) Measurement of tree biomass once before a drought and annually for the decade following the drought.
    (c) Measuring LD_{50} values in various herbivores exposed to a common herbicide.
    (d) Daily measurement of stream discharge and sediment loads.
    (e) Measurement of net primary productivity before and after the introduction of an invasive herbivore.

16. When does the South Pole experience 24 hours of sunlight per day?
    (a) From the March equinox to the September equinox
    (b) From the September equinox to the March equinox
    (c) Only on the March and September equinoxes
    (d) Only on the December and June solstices
    (e) From the December solstice to the June solstice

17. Heavy precipitation over the intertropical convergence zone (ITCZ) and the windward side of mountain ranges are both caused by
   (a) the Coriolis effect.
   (b) prevailing winds.
   (c) adiabatic heating.
   (d) reduced air pressure.
   (e) the El Niño–Southern Oscillation.

18. Which of the following is NOT true regarding an El Niño–Southern Oscillation (ENSO) event?
   (a) An ENSO event relies on thermohaline circulation.
   (b) An ENSO event brings warmer air temperatures to South America.
   (c) Glacial melting could eliminate thermohaline circulation.
   (d) An ENSO event typically starts around Christmas.
   (e) ENSO events are driven by a reversal of trade winds.

Question 19 refers to the following graph:

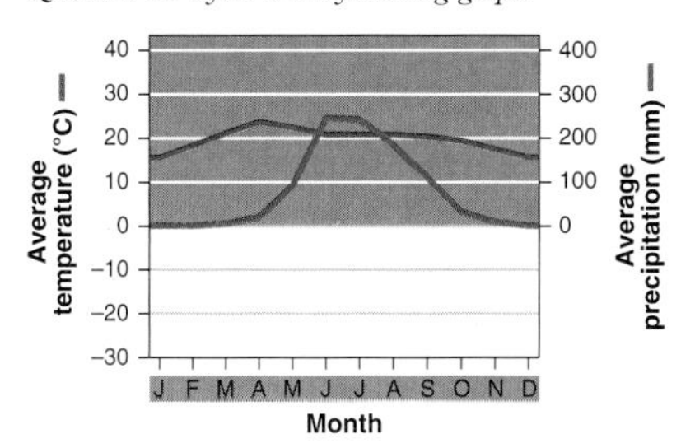

19. Which biome does this graph represent?
   (a) Boreal forest
   (b) Temperate rainforest
   (c) Temperate grassland
   (d) Temperate seasonal rainforest
   (e) Tropical seasonal forest

20. The limnetic zone is defined as the area of a lake where
   (a) algae and emergent plants grow.
   (b) sunlight does not reach.
   (c) nutrient levels are primarily oligotrophic.
   (d) the thermocline exists.
   (e) phytoplankton are the only photosynthetic organisms.

Question 21 refers to the following table, which contains the number of different spider species found in a survey of 1-m^2 plots from five gardens. Empty spaces indicate that no spiders of that species were found in the garden.

	Gardens				
	A	B	C	D	E
Grass spider	2	7	3	2	2
Orb weaver	1		3	2	4
Jumping spider		8	3		4
Wolf spider	2	2	2	2	6
Sac spider	7		3		2

21. Which garden has the highest Shannon's Index?
   (a) Garden A
   (b) Garden B
   (c) Garden C
   (d) Garden D
   (e) Garden E

22. Which of the following is NOT a key idea of Darwin's theory of natural selection?
   (a) Individuals differ in their phenotypes.
   (b) Differences in traits can be passed on from parents to offspring.
   (c) The genotype of an organism determines its phenotype.
   (d) Not all offspring can survive.
   (e) Differences in traits are associated with differences in the ability to survive and reproduce.

23. _____ is a change in the genetic composition of a population as a result of descending from a small number of colonizing individuals.
   (a) The founder effect
   (b) Inbreeding depression
   (c) Gene flow
   (d) The bottleneck effect
   (e) Genetic drift

24. Which human activity is most likely to decelerate the pace of evolution?
   (a) Monocropping corn of the same genotype
   (b) Fragmenting habitat
   (c) Overharvesting fisheries
   (d) Enforcing size limits on hunting
   (e) Connecting lakes through an artificial canal

Answers

17. d
18. a
19. e
20. e
21. c
22. c
23. a
24. a

Answers

25. e
26. e
27. b
28. a
29. c
30. e
31. c
32. c
33. a
34. e

25. Over long time periods, global climate change may change the __________ of a species.
I. realized niche
II. fundamental niche
III. extinction rate

(a) I only
(b) II only
(c) I and II
(d) I and III
(e) I, II, and III

26. Which is least likely to affect the carrying capacity of a habitat for a particular species?
(a) Abundance of a limiting nutrient
(b) Limited food
(c) Density-dependent factors
(d) Abundance of competing species
(e) Population distribution

27. Suppose that a population of squirrels can increase by 70 percent each year under ideal conditions. For a population of 30 squirrels living in a temperate forest, what is the most likely number of squirrels after 3 years of population growth?
(a) $30 \times e^{0.5 \times 3}$
(b) $30 \times e^{0.7 \times 3}$
(c) $70 \times e^{30 \times 3}$
(d) $0.7 \times e^{30 \times 3}$
(e) $0.5 \times e^{30 \times 3}$

28. Which is a requirement for the existence of a metapopulation?
I. A system of habitat corridors that link individual populations
II. The presence of both *r*- and *K*-selected species
III. Variation in the size of individual populations

(a) I only
(b) II only
(c) I and II
(d) II and III
(e) I, II, and III

29. Which is NOT true regarding keystone species?
(a) Keystone species can be mutualists with other species.
(b) Keystone species are often rare individuals in a community.
(c) Keystone species are most likely to be commensalists.
(d) Keystone species can be highly lethal to several species in a community.
(e) Keystone species may interact with only a single other species.

30. During the last ice age, glaciers covered much of the higher latitudes in North America. Considering the factors that influence the diversity of a community, which of the following is probably NOT true?
(a) At northern latitudes, genetic diversity of a given fish species is greater than at southern latitudes.
(b) At northern latitudes, large lakes contain more fish species than small lakes.
(c) At northern latitudes, large lakes have larger populations of each species than small lakes.
(d) At northern latitudes, large lakes have experienced fewer species extinctions than small lakes.
(e) At northern latitudes, small lakes were colonized by new species faster than large lakes.

31. In 2012, there were 2 births per 100 people and 5 deaths per 1,000 people in Peru. What is the approximate doubling time of Peru's population? Assume equal rates of immigration and emigration.
(a) 5 years
(b) 15 years
(c) 47 years
(d) 72 years
(e) 86 years

32. Which is most likely to be the replacement-level fertility for a developing nation in the third phase of demographic transition?
(a) 1.8
(b) 2.0
(c) 2.2
(d) 3.0
(e) 3.5

33. Which is most likely to result in an increase in pollution?
(a) An increase in GDP
(b) Switching from inland to offshore oil drilling
(c) A shift from subsistence energy sources to natural gas
(d) An increase in affordable health care
(e) An increase in the use of family planning

34. A population with an age distribution shaped like a pyramid
I. exhibits a stable population size.
II. is likely to exhibit a high population momentum.
III. has a high crude birth rate.

(a) I only
(b) III only
(c) I and II
(d) I and III
(e) II and III

35. The four factors represented in the IPAT equation include
 (a) impact on the environment, population size, affluence, and total fertility rate.
 (b) infant mortality rate, population size, affluence, and total fertility rate.
 (c) impact on the environment, population size, affluence, and technology.
 (d) industry, population growth, architecture, and technology.
 (e) industry, population growth, affluence, and technology.

36. Divergent plate boundaries are responsible for creation of new rock
 (a) along transform fault boundaries.
 (b) at the bottom of the ocean.
 (c) through volcanic activity.
 (d) above geological hotspots.
 (e) adjacent to subduction zones.

37. Which of the following statements about earthquakes is NOT true?
 (a) An earthquake of magnitude 8.0 on the Richter scale occurs about once every year.
 (b) Earthquakes occur when rocks of the asthenosphere rupture along a fault.
 (c) Earthquakes are most common along fault zones.
 (d) The Ring of Fire describes a geological circle on Earth where earthquakes are likely to occur.
 (e) An earthquake of magnitude 4 on the Richter scale is 100 times greater than an earthquake of magnitude 2.

38. Which contributes to the physical weathering of rocks?
 (a) Acid rain
 (b) Burial in sediment
 (c) Volcanic activity
 (d) Growth of plant roots
 (e) Deposition

39. A field containing soil that is 60 percent clay can hold a tremendous amount of nutrients but often has poor crop growth. What is the most likely explanation for this phenomenon?
 (a) Excessive concentrations of nutrients in soil can become toxic to many types of plants.
 (b) Clay can hold heavy metals that are toxic to plants.
 (c) Soil that is 60 percent clay can deprive plant roots of oxygen.
 (d) The base saturation of clay soil is too high for most plants.
 (e) Earthworms are unable to penetrate clay soils.

40. Which type of mining is potentially the most harmful to human health?
 (a) Open-pit mining
 (b) Mountaintop removal
 (c) Strip mining
 (d) Subsurface mining
 (e) Placer mining

41. Saltwater intrusion is likely to occur when
 (a) a cone of depression is created near a coastal ecosystem.
 (b) humans drill into a confined aquifer near the coast.
 (c) farmers irrigate their crops with an excess of saline water.
 (d) an artesian well is created near the coast.
 (e) the soil near a coastline allows for rapid water drainage.

42. The Dead Sea, which lies on the border between Jordan and Israel, is one of the saltiest lakes on Earth. Suppose you are a geologist exploring the area. What geological factors would you expect to see that might explain this salinity?
 I. High concentrations of uranium in or around the Dead Sea.
 II. Lack of outlets for water to escape
 III. Relatively impermeable soil at the bottom of the lake

 (a) I only
 (b) II only
 (c) III only
 (d) I and II
 (e) II and III

43. Which is a positive externality that results from the construction of a hydroelectric dam?
 (a) Construction of a fish ladder to help migrating fish pass the dam
 (b) Millions of fish killed when they are sucked into the dam turbines
 (c) Creation of valuable lake shoreline that promotes housing and tourism
 (d) Loss of seasonal flooding that promotes downstream fish diversity
 (e) Changes in the temperature profile of the river before and after the dam

44. Which of the following describes a limitation of drip irrigation systems?
 (a) Drip irrigation requires far more work for annual crops than perennial crops.
 (b) Drip irrigation is efficient, but encourages weed growth.
 (c) When considering the total cost of materials and water, drip irrigation is one of the most expensive irrigation methods.
 (d) Drip irrigation cannot supply sufficient water for most types of crops.
 (e) Drip irrigation uses a tremendous amount of energy.

Answers

35. c
36. b
37. b
38. d
39. c
40. d
41. a
42. b
43. c
44. a

Answers

45. c
46. d
47. b
48. a
49. c
50. b
51. c
52. c

45. Which represents water consumption at a power plant?
    (a) The plant withdraws cold water from a stream and later returns warm water to the stream.
    (b) The plant contaminates water with by-products of coal combustion.
    (c) The plant withdraws cold water from a stream and later releases water vapor from a cooling tower.
    (d) In the plant, water passes through a hydroelectric turbine.
    (e) Water is condensed after it has been converted to steam.

46. Which is NOT likely to lead to a tragedy of the commons?
    (a) Power plants emitting air pollution
    (b) Farmers siphoning water from a county aqueduct for irrigation
    (c) Office staff using a common coffee machine
    (d) A ranch charging farmers to board their horses
    (e) Eliminating laws that regulate permits for sport hunting

Question 47 refers to the following graph:

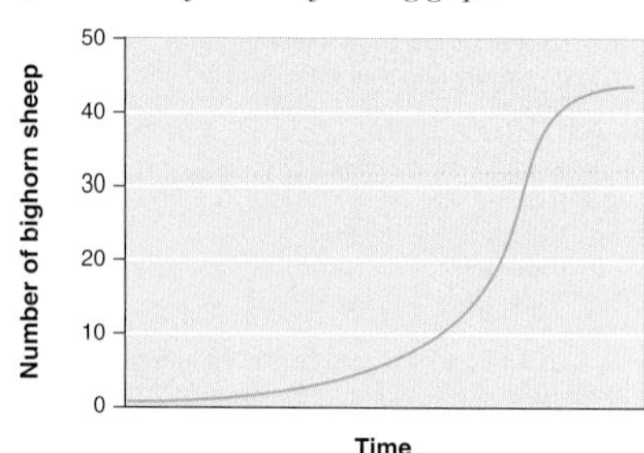

47. The graph shows the population growth of bighorn sheep through time for a 10-km^2 plot without predators. In a more realistic scenario with both sheep and wolf predators, how many bighorn sheep can hunters harvest without compromising the future of the sheep population?
    (a) 0
    (b) 10
    (c) 30
    (d) 35
    (e) 45

48. Which of the following is NOT an outcome of a prescribed forest burn?
    (a) Dead biomass is retained in the forest.
    (b) Nutrients in dead material are liberated for primary producers.
    (c) The risk of forest fires is minimized.
    (d) Habitat is provided for early-succession species.
    (e) The burning of the forest can stimulate the growth of other plant species.

49. Exurban communities are similar to
    (a) urban communities with low population densities.
    (b) suburban communities with low population densities.
    (c) suburban communities not connected to any central city.
    (d) urban communities whose services are no longer funded by a major city's government.
    (e) urban communities that are now converted into suburbs.

50. One U.S. city recently held a town hall meeting to decide on the best way to address growing problems of urban sprawl. Several suggestions were considered. Which of the following suggestions does NOT follow the EPA's principles of smart growth?
    (a) Fund individual communities within the city to develop a theme that makes each community distinct from adjacent neighborhoods.
    (b) Provide public transportation for routes that circle the city in order to encourage driving of personal cars within the city.
    (c) Involve neighborhood residents in making decisions regarding regulation of local services and development of new infrastructure.
    (d) Develop apartments and condominiums in the same areas.
    (e) Avoid the development of businesses in green spaces.

51. Which contributes significantly to obesity in the United States?
    (a) Insufficient access to vegetables due to diversion of crops for animal feed
    (b) Lack of nutritional information
    (c) High availability of inexpensive food with low nutrient content
    (d) High rates of iron deficiency among people
    (e) High caloric expenditure among people

52. To produce 10,000 kg of cattle, a farmer uses 50 acres of corn as cattle feed. If a single acre produces 4,000 kg of corn, what is the energy subsidy of beef? Assume mass can be used as a substitute for energy.
    (a) 0.4
    (b) 2.5
    (c) 20
    (d) 80
    (e) 200

53. Which of the following describes how intercropping alters the ecology of a farm?
   I. Intercropping increases the carrying capacity of insect herbivores more than monocropping.
   II. Intercropping promotes plant mutualisms.
   III. Relative to monocropping, intercropping increases the prevalence of disease outbreaks among crops.

   (a) I only
   (b) II only
   (c) III only
   (d) I and II
   (e) I, II, and III

54. Which of the following statements about no-till agriculture is NOT true?
   (a) No-till agriculture reduces wind and water erosion.
   (b) No-till soil undergoes less oxidation.
   (c) Organic farms are less likely to use no-till agriculture.
   (d) No-till agriculture increases crop production relative to conventional tilling methods.
   (e) No-till agriculture releases less CO_2 than conventional tilling methods.

55. Which provides protection against the risks that genetically modified crops pose to global crop biodiversity?
   (a) No-till agriculture
   (b) Integrated pest management
   (c) Broad-spectrum herbicides
   (d) Use of a pesticide treadmill
   (e) Buffer zones around crops

Question 56 refers to the following table:

Fuel source	EROEI
Biodiesel	1
Liquid petroleum	20
Ethanol from corn	2
Ethanol from sugarcane	5
Natural gas	10

56. All of the fuel sources listed in the table can be used in personal vehicles. Assuming that each vehicle achieves approximately the same MJ per passenger-mile, which type of fuel is likely to contribute most to overall operation efficiency?
   (a) Biodiesel
   (b) Liquid petroleum
   (c) Ethanol from corn
   (d) Ethanol from sugarcane
   (e) Natural gas

57. The typical home in the United States uses 1.25 kW of electricity. One wind farm has 50 wind turbines, each of which produces 2 MW of electricity when the wind is blowing. If the capacity factor of the wind farm is 0.2, how many homes can this wind farm supply with energy?
   (a) 10,000
   (b) 16,000
   (c) 21,000
   (d) 22,400
   (e) 25,100

58. Which two nonrenewable energy sources are typically extracted together?
   (a) Liquid coal and oil sands
   (b) Petroleum and oil sands
   (c) Petroleum and coal
   (d) Petroleum and natural gas
   (e) Uranium and coal

59. Which statement best explains why overall energy use has remained relatively constant even though energy intensity has decreased?
   (a) We are nearing the peak of the Hubbert curve.
   (b) The average number of electrical devices per human has increased steadily.
   (c) Power plants are becoming more efficient and are now producing more electricity.
   (d) We have switched to less energy intense forms of fossil fuels that have a smaller impact on the environment.
   (e) We continue to find more oil reserves, so the supply of energy has increased while demand has remained relatively steady.

60. In the following equation, which describes the fission of uranium, what do i, ii, and iii represent?

$$\text{i} + {}^{235}\text{Uranium} \rightarrow \text{ii} + {}^{91}\text{Krypton} + \text{iii}$$

   (a) 1 neutron, ^{142}Boron, heat energy
   (b) 1 neutron, ^{142}Barium, ^{135}Cesium
   (c) 1 neutron, ^{142}Barium, kinetic energy
   (d) Kinetic energy, ^{142}Boron, potential energy
   (e) Potential energy, ^{142}Barium, kinetic energy

61. Which of the following will NOT result in energy savings?
   (a) Increasing insulation of walls and ceilings in a residence
   (b) Replacing LED bulbs with CFL bulbs
   (c) Using a passive solar design in the construction of a new building
   (d) Installing sensors to turn off lights when a room is vacant
   (e) Implementing a financial incentive to encourage less driving

Answers

53. d
54. d
55. e
56. b
57. b
58. d
59. b
60. c
61. b

Cumulative AP® Environmental Science Practice Exam EXAM-7

Answers

62. e
63. a
64. b
65. d
66. c
67. e
68. b
69. c
70. b

62. Biomass is created through the conversion of __________ energy into __________ energy, which can then be used to generate electricity. In contrast, tidal energy involves the conversion of __________ energy into electricity.
    (a) Kinetic/potential/mechanical
    (b) Chemical/potential/potential
    (c) Solar/ kinetic/potential
    (d) Chemical/kinetic/kinetic
    (e) Solar/chemical/kinetic

63. Concentrated solar thermal systems implement
    (a) active solar technology.
    (b) photovoltaic cell technology.
    (c) passive solar technology.
    (d) smart grid technology.
    (e) Energy Star appliances.

64. Which of the following factors should NOT be considered when determining the EROEI of hydrogen fuel cell vehicles?
    (a) Efficiency of isolating hydrogen from water or natural gas
    (b) By-products of generating electricity from hydrogen
    (c) Energy produced by the reaction between hydrogen and oxygen
    (d) Efficiency of electric engines
    (e) Energy required to compress or liquefy hydrogen for transport

65. The primary purpose of a smart grid is to
    (a) improve the efficiency of energy production.
    (b) improve the efficiency of energy transportation through power lines.
    (c) improve the capacity factor of power plants.
    (d) streamline energy distribution and reduce excess energy capacity on electrical grids.
    (e) generate more affordable energy for consumers.

66. In the northeastern United States, during what time(s) of the year are lakes located near agricultural fields in temperate ecosystems most likely to experience the highest biochemical oxygen demand?
    (a) Throughout the entire growing season
    (b) Primarily during the spring
    (c) Primarily during the summer
    (d) Throughout the year
    (e) During the fall

67. You want to conduct a natural experiment to test whether leakage from a manure lagoon has an adverse effect on a nearby stream. What is an appropriate null hypothesis for your experiment?
    (a) There will be a higher abundance of an indicator species in the stream before the lagoon began leaking than after it began leaking.
    (b) There will be a higher evenness of an indicator species in the stream before the lagoon began leaking than after it began leaking.
    (c) There will be a lower abundance of an indicator species in the stream before the lagoon began leaking than after it began leaking.
    (d) There will be no difference in the abundance of phytoplankton in the stream before the lagoon began leaking and after it began leaking.
    (e) There will be no difference in the abundance of an indicator species in the stream before the lagoon began leaking and after it began leaking.

68. Which of the following minerals is ultimately responsible for the yellow color of streams that drain from abandoned mines?
    (a) Salt
    (b) Pyrite rock
    (c) Aluminum
    (d) Coal
    (e) Sulfur

69. Which of the following is NOT an example of a consequence of biomagnification?
    (a) Incidences of human cancer associated with spraying crops with pesticides
    (b) Formation of ovaries in male fish
    (c) Suffocation of fish as a result of inert pesticide ingredients
    (d) Thinning of bird egg shells
    (e) Increasing concentrations of iron at higher levels of a food chain

70. A group of researchers monitored the health of all streams and rivers in a single watershed. Several streams in a part of the watershed were found to have a high content of silt, large boulders, and a low diversity of insects and fish. What would be the most direct inference from this description?
    (a) Most of the sediment load in the stream is probably from a natural source.
    (b) The surrounding area probably has a high amount of impervious surface.
    (c) The primary cause of low insect and fish diversity is industrial pollution.
    (d) There is probably a high biochemical oxygen demand in the benthos of the stream.
    (e) The streams have very low resilience.

71. Atmospheric brown cloud results from a combination of
    I. primary and secondary pollutants.
    II. $PM_{2.5}$ and O_3.
    III. volatile organic compounds and O_3.

    (a) I only
    (b) II only
    (c) III only
    (d) I and II
    (e) II and III

72. Which location has been experiencing greater amounts of acid rain in recent years?
    (a) The central United States and the Mississippi River
    (b) The West Coast of the United States
    (c) The northeastern United States
    (d) Areas east of the Rocky Mountains
    (e) The southeastern United States

Question 73 refers to the following figure:

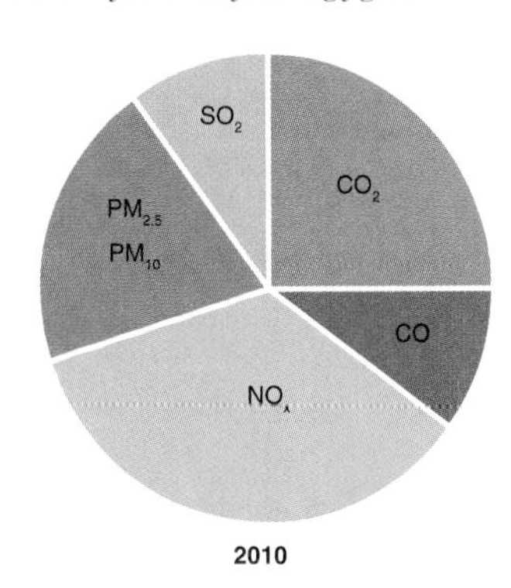

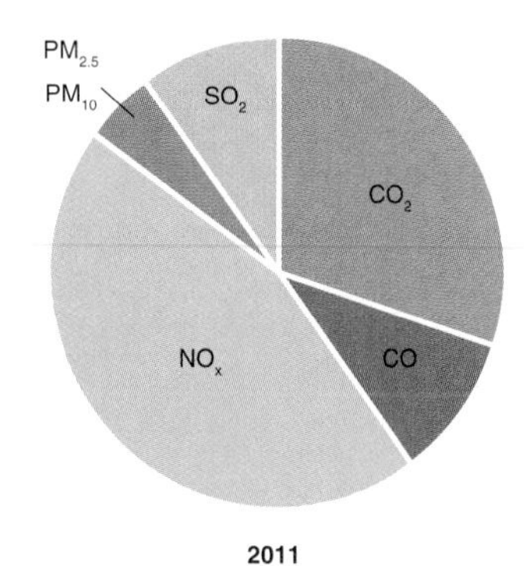

73. One city recently conducted a multiyear survey of the percentage of various atmospheric pollutants from commercial and industrial practices. Based on the pie charts, what practice was most likely implemented between 2010 and 2011?
    (a) The city instituted a cap-and-trade policy on the emissions of sulfur dioxide.
    (b) Several power plants reduced the temperature at which fuel was combusted.
    (c) Automobile use was limited by enforcing a toll on major roadways.
    (d) The city began enforcing policies that mandated every car have a working catalytic converter.
    (e) Power plants began installing electrostatic precipitators in their smokestacks.

74. Identify i, ii, iii, and iv in the following equations, which show the interaction between CFCs and ozone:

$$\text{i} + \text{ii} \rightarrow 2\text{O}$$
$$\text{iii} + \text{CFC} \rightarrow \text{iv} + O_2$$

    (a) UV-C radiation, O_2, O_3, ClO
    (b) UV-C radiation, O_2, O_3, O
    (c) UV-B radiation, O_2, 3O, ClO
    (d) O_2, CFC, O_3, ClO
    (e) UV-B radiation, O_2, O_3, ClO

75. The release of radon-222 can result from
    (a) the use of asbestos in building insulation.
    (b) radioactive decay of uranium and polonium-210.
    (c) the use of granitic rock in construction material.
    (d) the use of electronic equipment, such as copy machines.
    (e) poor ventilation in buildings.

76. More than half of the material in the solid waste stream that is recovered before ending up in a landfill or incinerator is
    (a) yard waste.
    (b) paper products.
    (c) plastic products.
    (d) glass products.
    (e) food scraps.

77. Which statement regarding composting is NOT true?
    (a) Composting can be a way to divert material from the landfill.
    (b) Efficient composting requires the material to be mixed frequently.
    (c) Composting produces a substantial amount of carbon dioxide.
    (d) Efficient composting requires an abundance of anaerobic bacteria.
    (e) Meat and dairy products are typically not composted.

Answers

71. d
72. b
73. e
74. a
75. c
76. b
77. d

Answers

78. c
79. b
80. b
81. e
82. c
83. b
84. d

Questions 78–80 refer to the following scenario:

In one particular landfill that has been capped, 80 percent of the annual precipitation runs off without infiltrating the surface. The remaining 20 percent infiltrates and leaches through the soil. There is no leachate collection system, so any water that infiltrates the surface eventually enters the groundwater. Suppose that this 50-acre landfill receives a total of 200 million L of rainfall each year. Also suppose that the total mass of aluminum in the landfill is 100,000 kg and 1 percent of that is leached each year.

78. What is the concentration of aluminum in the water that leaches from the landfill?
(a) 20 mg/L
(b) 25 g/L
(c) 25 mg/L
(d) 250 mg/L
(e) 2,500 g/L

79. To determine the effect of aluminum leachate on human health, researchers conducted a series of toxicity tests on rats. They found 20 percent mortality among rats given water containing 2 g/L of aluminum, 50 percent mortality among rats given water containing 5 g/L of aluminum, and 90 percent mortality of rats given water containing 10 g/L of aluminum. What is the safe concentration of aluminum in drinking water for humans?
(a) 2 mg/L
(b) 5 mg/L
(c) 20 mg/L
(d) 50 mg/L
(e) 500 mg/L

80. As one moves away from the landfill, the concentration of aluminum in the groundwater decreases by half every 1 km of distance due to adsorption of aluminum by soil and groundwater dilution. At what minimum distance from the landfill would it be safe to drink the groundwater?
(a) 1 km
(b) 3 km
(c) 10 km
(d) 15 km
(e) 20 km

81. Which statement best describes the difference between the Superfund and Brownfield programs?
(a) Superfund sites are currently hazardous to human health whereas Brownfields are potentially hazardous to humans in the future.
(b) Superfund sites are limited to cities whereas Brownfields can occur in both urban and rural locations.
(c) The Superfund program only funds the cleanup of wastes deemed hazardous under the Resource Conservation and Recovery Act, whereas the Brownfield program funds the cleanup of all hazardous wastes.
(d) Superfund sites are generally reclaimed for human uses (e.g., recreation) whereas Brownfields are cleaned but never reclaimed.
(e) Superfund sites are federally funded and reserved for locations with high public health risk whereas Brownfields are state funded and designated in less hazardous locations.

82. Which of the following is NOT likely to provide an incentive for a township to initiate a recycling program?
(a) Tipping fees that exceed the value of recycled goods
(b) A stable demand, but low supply of recycled cardboard
(c) Construction of a landfill with a low tipping fee
(d) Initiation of an integrated waste management program
(e) Initiation of an upcycling program

83. The majority of emergent infectious diseases arise
(a) when people generate resistant bacteria by not completing antibiotic dosages.
(b) when a disease unexpectedly jumps from an animal to human host.
(c) from prion mutations.
(d) from consumption of uncooked meat.
(e) from sexual transmission.

84. Which could be a set of treatment and control groups for a retrospective study on the effects of endocrine disruptors on humans?
(a) A group of farmers that are planning to spray crops with atrazine and a group of farmers that are planning to use organic pesticides
(b) A group of construction workers that have been exposed to asbestos and a group of construction workers that have been exposed to PCBs
(c) A group of children with mothers exposed to PCBs during pregnancy and a group of children who were exposed to DDT after they were born
(d) Members of a town that was downwind of a commercial farm that sprayed pesticides and members of a town that was downwind of an organic farm
(e) Humans raised when DDT was used and humans raised after DDT was no longer used

85. Which of the following is an application of the innocent-until-proven-guilty principle?
   (a) Policies enacted as a result of the Stockholm Convention
   (b) Allowing the use of a pesticide that has a low risk to human health
   (c) Banning the use of certain nanoparticles until further research has assessed their risk to human health
   (d) Dividing the LD_{50} of a chemical by 1,000 to determine safe levels of human consumption
   (e) Laws that permit the sale of drugs for human use only if the drug has been tested for toxicity

86. After spraying pesticide on his crops, a farmer notices that the population of amphibians in a nearby wetland has begun to decrease. The concentration of pesticide in the water is well below the LD_{50} value for amphibians. Based on this information, which is the most likely explanation for the decrease in amphibian population?
   (a) There was a lack of nearby source populations to recolonize the wetland.
   (b) There were synergistic interactions between the pesticide and environmental factors.
   (c) An emerging infectious disease was killing susceptible individuals.
   (d) The pesticide does not biomagnify in the amphibians.
   (e) Wetland conditions allowed for a longer persistence time of the pesticide than laboratory studies revealed.

87. Suppose it was discovered that the genetic diversity of a bacteria species responsible for disease in mice was declining. What is a likely outcome(s) of this decline?
   I. Loss of ecosystem resilience
   II. Increase in disease prevalence
   III. Loss of regulating services

   (a) II only
   (b) III only
   (c) I and II
   (d) I and III
   (e) I, II, and III

88. Which is NOT associated with smaller habitat sizes?
   (a) A decrease in genetic diversity
   (b) A decrease in edge habitat
   (c) Lower resiliency
   (d) Loss of habitat diversity
   (e) Faster extinction rates

89. Which agreement regulates the shipping of endangered or threatened species within the United States?
   (a) The Convention on International Trade in Endangered Species
   (b) The Lacey Act
   (c) The Endangered Species Act
   (d) Resource Conservation and Recovery Act
   (e) National Environmental Policy Act

90. The Galápagos Islands, located off the coast of Ecuador, contain a rich diversity of endemic plants and animals. Several of the islands are entirely off-limits to tourists, whereas other islands allow tourists on the outer edges and on designated trails. As a whole, which of the following best describes the Galápagos Islands?
   (a) A wildlife refuge
   (b) A national park
   (c) A biosphere reserve
   (d) Multiple-use lands
   (e) A managed resource protection area

91. Which is true of invasive species?
   (a) Invasive species are defined as species that come from other countries.
   (b) By definition, all exotic species are invasive.
   (c) All invasive species spread rapidly.
   (d) Invasive species typically have few ecological interactions.
   (e) Invasive species must come from another continent.

92. Which is NOT true regarding greenhouse gases?
   (a) Greenhouse gases with high global warming potential persist in the atmosphere for tens to hundreds of years.
   (b) Greenhouse gases absorb infrared radiation reflected by Earth's surface.
   (c) Greenhouse gases with high warming potential include compounds derived from both natural and anthropogenic sources.
   (d) Some greenhouse gases filter out harmful ultra-violet radiation.
   (e) Greenhouse gases that contribute to warming of Earth are located in the stratosphere.

93. Where is denitrification NOT likely to occur?
   (a) Topsoil of freshly tilled croplands
   (b) Sediments of a deep lake
   (c) Well-fertilized croplands
   (d) Compacted, wet soils under cattle farms
   (e) Recently flooded cropland

Answers

85. a
86. b
87. d
88. b
89. b
90. c
91. c
92. e
93. a

Answers

94. d
95. a
96. a
97. c
98. b
99. d
100. c

Question 94 refers to the following graph:

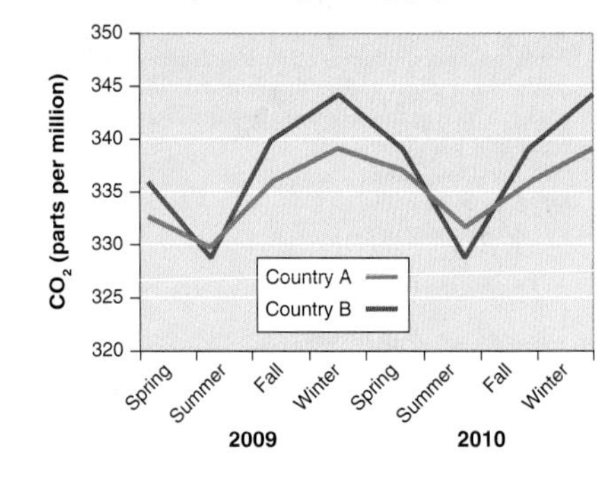

94. The graph shows average CO_2 concentrations during each season for two countries from 2009 to 2010. Based only on the information in the graph, which statement best explains what the graph shows?
   (a) Country B has higher per-capita CO_2 emissions.
   (b) Country B experiences warmer fall and winter temperatures than country A.
   (c) Average CO_2 concentrations for country A are increasing relative to average concentrations for country B.
   (d) Country B experiences higher levels of spring and summer primary production.
   (e) Country B is more industrial than country A.

95. Which phenomenon is part of a negative feedback with regard to global warming?
   (a) Increased absorption of atmospheric CO_2 by the ocean
   (b) Methane production from peatland decomposition
   (c) Increased evapotranspiration from plant growth
   (d) CO_2 release from forest fires during dry spells
   (e) Reduced albedo resulting from melting ice

96. Which statement is NOT true regarding U.S. carbon emission output and politics?
   (a) The United States was the only developed nation that did not sign the Kyoto protocol.
   (b) Total carbon emissions in the United States have increased over the past two decades.
   (c) A switch from coal to natural gas has led to a major decrease in carbon emissions.
   (d) The U.S. EPA is legally required to regulate greenhouse gases.
   (e) Total carbon emissions in the United States have decreased since 2007.

Question 97 refers to the following figure:

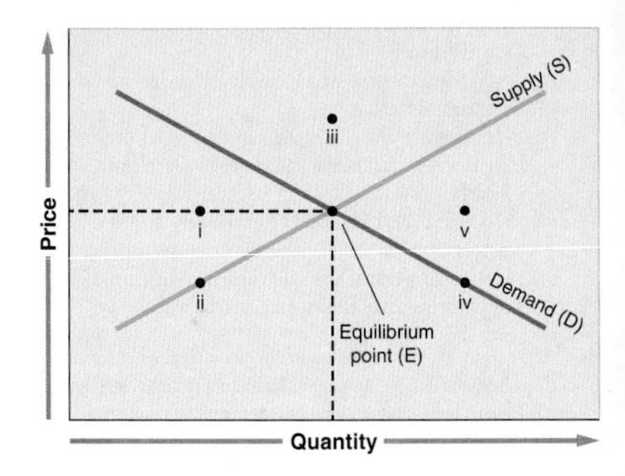

97. Where will the equilibrium point be if demand for a good increases, and the negative environmental externalities are internalized into the price of the good?
   (a) i
   (b) ii
   (c) iii
   (d) iv
   (e) v

98. In the United States, which organization is most likely to address issues related to sick building syndrome in the work environment?
   (a) EPA
   (b) OSHA
   (c) DOE
   (d) WHO
   (e) UNDEP

99. Which of the following are incentive-based approaches to control environmental degradation?
   I. Cap-and-trade on carbon emissions
   II. Green taxes
   III. Required reclamation of open-pit mines

   (a) I only
   (b) II only
   (c) III only
   (d) I and II
   (e) II and III

100. Which is most likely to discourage environmental inequity?
   (a) NIMBY politics
   (b) Higher carbon emissions in developed nations
   (c) Technology transfer
   (d) Racial segregation
   (e) A higher abundance of landfills in economically poor areas

Section 2: Free-Response Questions

Write your answer to each part clearly. Support your answers with relevant information and examples. Where calculations are required, show your work.

1. The City of Philadelphia recently replaced one out of every 10 trash bins with solar-powered trash compactors. The compactor is an enclosed unit with a door that opens for trash disposal. The compactor automatically detects when the bin is full and uses a solar-powered mechanical crusher to compact the contents. When the compactor needs to be emptied, it sends an electronic signal. Use of solar-powered compactors has increased the capacity of public trash bins and has reduced the number of trash collection visits to each bin from 17 times per week to 5 times per week.
   (a) Describe four positive externalities of installing solar-powered trash compactors. (2 points)
   (b) Describe six cradle-to-grave components of solar-powered trash compactors. (2 points)
   (c) Suggest one way that the installation of solar-powered trash compactors can reverse the effects of urban blight. (2 points)
   (d) The price of a regular trash bin is \$300, and it has a lifespan of 20 years. The price of a solar-powered trash compactor is \$4,000, and it has a lifespan of 10 years; it also requires approximately \$150 in maintenance costs each year. On average, a trash collection visit costs \$5 in fuel and \$20 in employee salary. Based on this information, are solar-powered trash compactors economically beneficial? (2 points)
   (e) Describe two ways that you might determine if solar-powered trash compactors are environmentally beneficial. (2 points)

2. The country of Costa Rica has an abundance of climactic, geographic, and biological diversity. However, in the last century intensive farming and population growth have led to a 75 percent reduction in its forests. In the 1980s, the government of Costa Rica began to address concerns about the loss of forest with a series of political and environmental programs. These programs, designed to generate more sustainable economic development, include land protection and conservation of biodiversity.
   (a) Costa Rica lies just north of the equator and contains a series of mountain ranges that run the entire length of the country.
      (i) Given its geographic location, what is likely to be the prevailing wind pattern across the country? (1 point)
      (ii) Describe how mountain ranges contribute to the climactic, geographic, and biological diversity observed in Costa Rica. (1 point)
   (b) Given that Costa Rica is bordered by the Atlantic and Pacific Oceans, how are weather patterns in the country likely to be affected by the El Niño–Southern Oscillation (ENSO)? (2 points)
   (c) Describe four ecosystem services that are provided through the protection of land and how the Costa Rican government may profit from each of them. (4 points)
   (d) To promote economic sustainability, a large proportion of land was protected through debt-for-nature programs. Describe debt-for-nature programs and why they are effective. (2 points)

Answers to the Cumulative Practice Exam Free-Response Questions can be found in the Answer Appendix on page ANS-19.

Cumulative AP® Environmental Science Practice Exam EXAM-13

3. To fully understand how inorganic and organic material flows through ecosystems, environmental scientists often create energy and nutrient budgets. These budgets are similar to a financial budget that you might create for yourself; just as you account for money that goes into and out of a bank account, environmental scientists account for energy and nutrients that flow into and out of an ecosystem. Understanding energy and nutrient budgets for farmlands is particularly useful, because it allows farmers to improve crop yield and farm sustainability. However, accounting for all inputs and outputs is often very difficult.
   (a) List three possible inputs and three outputs of organic material to a cropland. (2 points)
   (b) List three possible inputs and three outputs of inorganic material to a cropland. (2 points)
   (c) Define gross primary productivity (GPP) and describe how a farmer could measure GPP on a cropland. (2 points)
   (d) For most crop species, farmers only harvest aboveground primary production. All below-ground plant material is left to decompose.
      (i) Describe the processes of ammonification and dentrification. (2 points)
      (ii) Discuss how the spraying of fungicides might alter the nitrogen budget of a farmland. (2 points)

4. On April 26, 1986, a series of power surges occurred during a routine systems test at the Chernobyl nuclear power plant in Russia. These surges resulted in the most catastrophic nuclear accident in history, and released 80 petabecquerel of radioactive material on 200,000 km^2 of surrounding land (1 petabecquerel = 10^{15} becquerel). This created an immediate, severe, and lasting health hazard for humans and wildlife. Although most wildlife died without producing viable offspring, a recent press release suggested that animals may be adapting to low levels of chronic radiation:

   Exposure to radiation generates health problems through the formation of free radicals that damage DNA and other organic molecules. Twenty-five years after the disaster at Chernobyl, findings indicate that several bird species in the area have started producing a greater abundance of antioxidants that can remove free radicals. However, birds that produce these antioxidants demonstrate lower reproduction.

   (a) Describe the processes that could have led to changes in antioxidant production in birds around Chernobyl. (2 points)
   (b) Define the bottleneck effect, describe how it may apply to wildlife populations around Chernobyl, and discuss potential problems associated with the bottleneck effect. (2 points)
   (c) Discuss the importance of having intact bird habitat in areas surrounding Chernobyl that were not exposed to radiation. (2 points)
   (d) In general, birds that produce high levels of antioxidants exhibit lower levels of reproduction than birds that produce lower levels of antioxidants. However, researchers hypothesize that birds that produce high levels of antioxidants in the area around Chernobyl are likely to have a higher overall fitness than birds in the area without this trait. Design an experiment to test this hypothesis. In your experiment, define
      (i) a null hypothesis. (1 point)
      (ii) treatments and controls. (1 point)
      (iii) a description of the experimental procedure. (2 points)

EXAM-14 Cumulative AP® Environmental Science Practice Exam

appendix

Reading Graphs

Environmental scientists often use graphs to display the data they collect. Unlike a table that contains rows and columns of data, graphs help us visualize patterns and trends in the data and they communicate our results more clearly. Because graphs are such a fundamental tool of environmental science as well as most other sciences, we have designed this appendix to help you become familiar with the major types of graphs that environmental scientists use; we also discuss how to create each type of graph and how to interpret the data that are presented in the graphs.

Scientists use graphs to present data and ideas

A graph is a tool that allows scientists to visualize data or ideas. Organizing information in the form of a graph can help us understand relationships more clearly. Throughout your study of environmental science you will encounter many different types of graphs. In this section we will look at the most common types of graphs that environmental scientists use.

Scatter Plot Graphs

Although many of the graphs in this book may look different from each other, they all follow the same basic principles. Let's begin with an example in which researchers who were investigating a possible relationship between per capita income and total fertility rate plotted data points for 11 countries. We can examine this relationship by creating a *scatter plot graph,* as shown in FIGURE A.1. In the simplest form of a scatter plot graph, researchers look at two variables; they put the values of one variable on the x axis and the values of the other variable on the y axis. By convention, the place where the two axes converge in the bottom left corner, called the origin, represents a value of 0 for each variable. The units of measurement tend to get larger as we move from left to right on the x axis, and from bottom to top on the y axis. In our example, per capita income is on the x axis and total fertility rate is on the y axis. As you can see in the graph, as income increases the fertility rate decreases.

When two variables are graphed using a scatter plot, we can draw a line through the middle of the data points that describes the general trend of these data points—as you can see in Figure A.1. Because such a line is drawn in a way that fits the general trend of the data, we call it *the line of best fit*. The line of best fit allows us to visualize a general trend. In our graph, the addition of a line of best fit makes it easier to identify a trend in the data; as we move from low to high income we observe lower fertility. This is known as a negative relationship between the two variables because as one variable gets larger the other variable gets smaller. When graphing data using a scatter plot graph, the line of best fit may be either straight or curved.

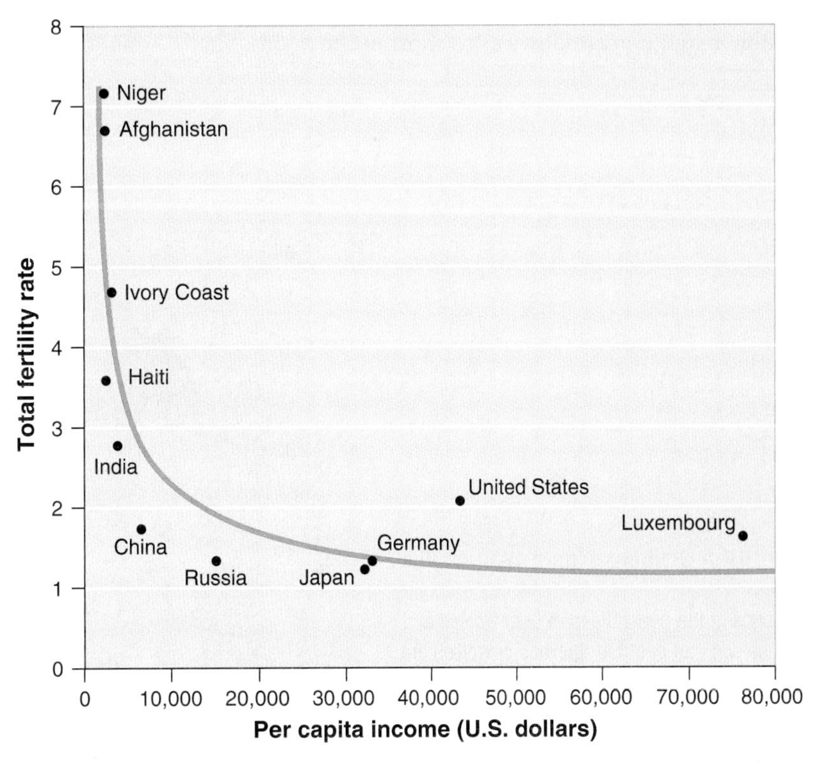

FIGURE A.1 **(book Figure 23.2) Scatter plot graph.** In this graph, we place data points that coincide with the income of different countries on the x axis and the corresponding fertility rate of each country on the y axis. *(Data from http://www.gapminder.org)*

Line Graphs

A *line graph* displays data that occur as a sequence of measurements over time or space. For example, scientists have estimated the number of humans living on Earth from 8,000 years ago to the present time. Using all of the available data points, a line graph can be used to connect each data point over time, as shown in FIGURE A.2. In contrast to a line of best fit that fits a straight or curved line through the middle of all data points, a line graph connects one data point to another, so it can be straight or curved, or it can move up and down as it follows the movement of the data points.

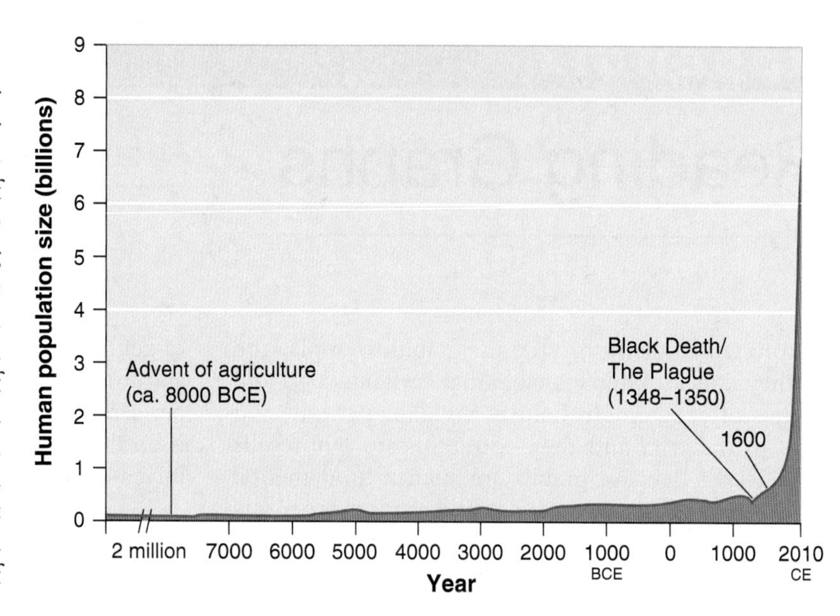

FIGURE A.2 **(book Figure 22.1) Line graph.** In this graph, a line is used to track the change in human population size over time. Note that this graph extends far back in time and there is only a small increase in population size from 2 million BCE to 7000 BCE, so a break in the *x* axis is used to allow a shorter axis. This allows us to focus on the period of rapid population growth that happened during the past 7,000 years.

When a graph includes data points with a very large range of values, the size of the graph can become cumbersome. To keep the graph from becoming too large, we can use a break in the axis. For example in our graph of human population growth, we see that the size of the population varied little from 7000 BCE to 2 million BCE. Showing the data for those years would not provide much additional useful information but it would make the graph a lot wider. The break in the *x* axis between 7000 BCE to 2 million BCE, indicated by the double hatch marks, allows us to shorten the *x* axis. The double hatch marks indicate that we are condensing the middle part of the *x* axis.

Line graphs can also illustrate how several different variables change over time. When two variables contain different units or a different range of values, we can use two *y* axes. For example, FIGURE A.3 presents data on changes in the population sizes of two

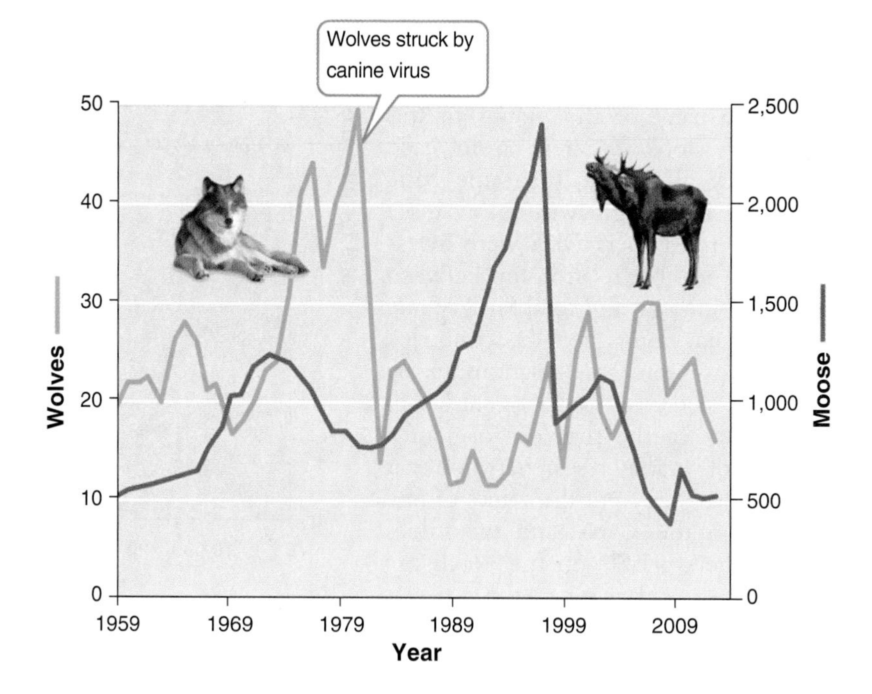

FIGURE A.3 **(book Figure 19.6) A line graph with two sets of data.** By graphing changes in the populations of both wolves and moose, we can see that declines in wolves are associated with increases in moose. *(Data from J. A. Vucetich and R. O. Peterson,* Ecological Studies of Wolves on Isle Royale: Annual Report 2007–2008, *School of Forest Resources and Environmental Science, Michigan Technological University.)*

different animals on Isle Royale in the years 1955 to 2011. The left y axis represents the population changes in the wolf population whereas the right y axis represents the population changes in the moose population during the same time span.

Bar Graphs

A *bar graph* plots numerical values that come from different categories. For example, in FIGURE A.4, the x axis contains categories that represent regions of the world. The y axis represents a numerical value—the number of people infected with HIV. The visual impact of the different bar heights provides a dramatic comparison of the incidence of HIV in different regions.

A bar graph is a very flexible tool and can be altered in several ways to accommodate data sets of different sizes or even several data sets that a researcher wishes to compare. When scientists measured the net primary productivity of different ecosystems, as shown in FIGURE A.5, they put the categories—various ecosystems—on the y axis and the plotted values—net primary productivity—on the x axis. This orientation makes it easier to accommodate the relatively large amount of text needed to name each ecosystem.

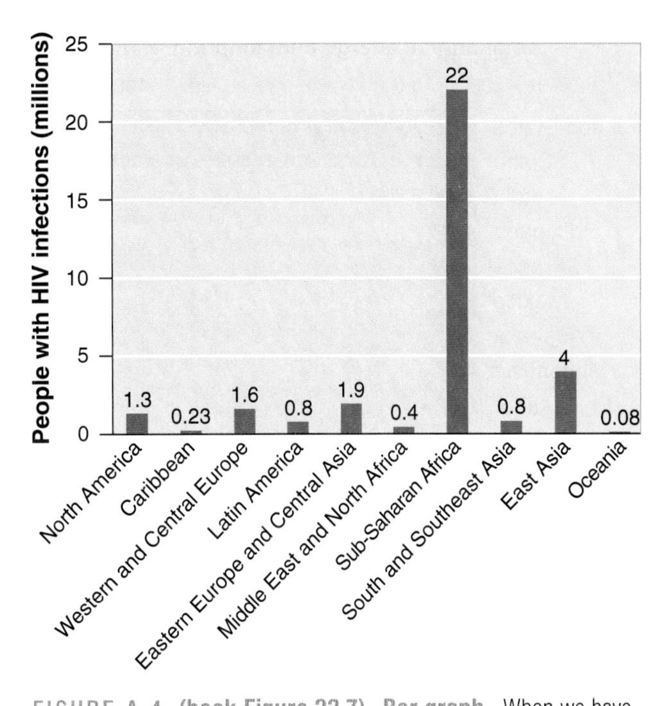

FIGURE A.4 **(book Figure 22.7) Bar graph.** When we have numerical values that come from different categories we can use a bar graph to plot data. In this example, we can plot the number of people infected with HIV in several countries around the world. *(Data from World Health Organization, UNAIDS)*

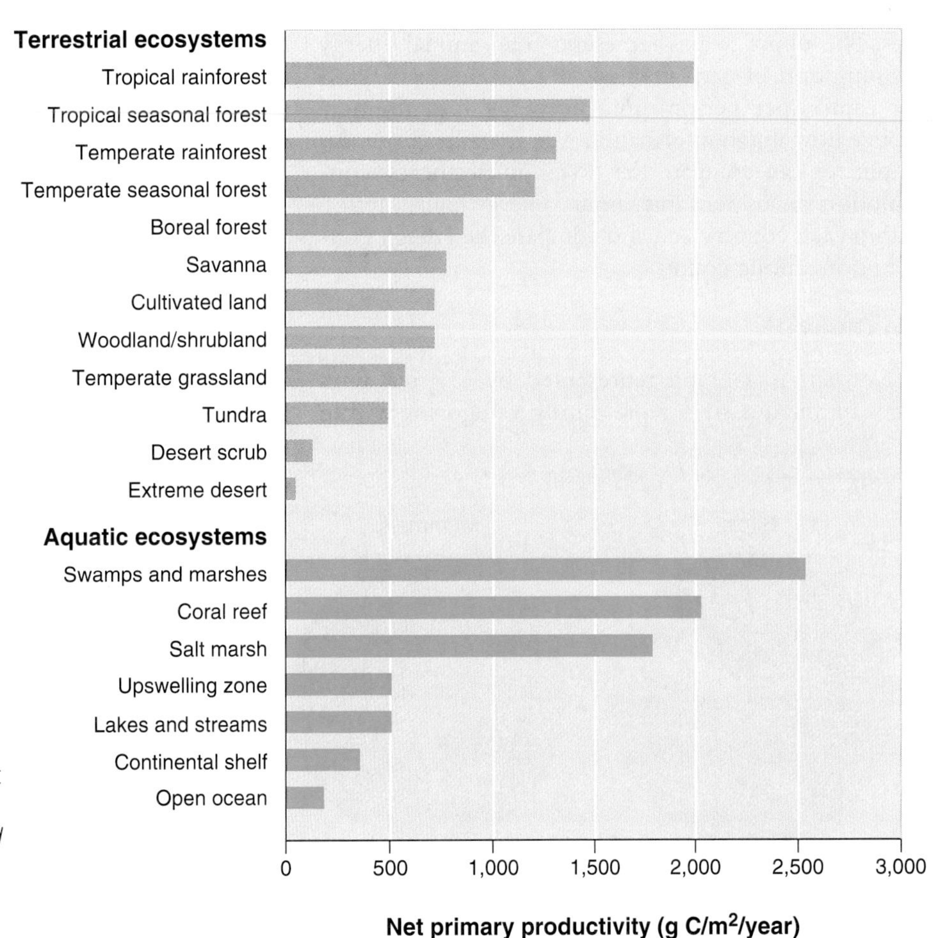

FIGURE A.5 **(book Figure 6.8) A rotated bar graph.** Bar graphs can place the categories on either the x axis, as in Figure A.4, or on the y axis, as in this figure that plots the net primary productivity of different ecosystems. *(After R. H. Whittaker and G. E. Likens, Primary production: The biosphere and man,* Human Ecology *1 (1973): 357–369.)*

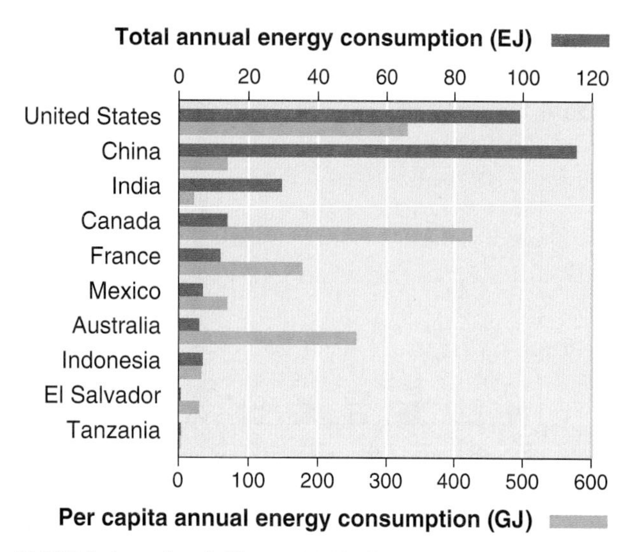

FIGURE A.6 **(book Figure 34.2) A rotated bar graph with two sets of data.** In this bar graph, we have nations as our categories and two sets of numerical data: the total energy consumed by each country and the per capita annual energy consumed. *(Data from the U.S. Department of Energy, Energy Information Administration, 2012)*

FIGURE A.6 shows an example of a bar graph that presents two sets of data for each category. In this example, the bar graph is rotated such that the categories are on the *y* axis and the numeric data are on two *x* axes. The upper *x* axis represents total annual energy consumption of each country. The lower *x* axis plots per capita (per person) annual energy consumption. Notice how much information we can gather from this graph; we can compare the total annual energy consumption versus per capita annual energy consumption within each country and also compare the energy consumption among countries.

Pie Charts

A *pie chart* is a graph represented by a circle with slices of various sizes representing categories within the whole pie. The entire pie represents 100 percent of the data and each slice is sized according to the percentage of the pie that it represents. For example, FIGURE A.7 shows the percentage of birds, mammals, and amphibians from around the world that have been categorized as threatened, near-threatened, or of least concern from a conservation point of view. For each group of animals, each slice of the pie represents the percentage of species that fall within each conservation category.

Two special types of graphs are used by environmental scientists

While scatter plots, line graphs, bar graphs, and pie charts are used by many different types of scientists, environmental scientists also use two types of graphs that are not common in most other fields of science: *climate diagrams* and *age structure diagrams*. Although these two types of graphs are discussed within the text, we provide them here for review.

Climate Graphs

Climate diagrams are used to illustrate the annual patterns of temperature and precipitation that help to determine the productivity of biomes on Earth. FIGURE A.8 shows two hypothetical biomes. By graphing the average monthly temperature and precipitation of a biome, we can see how conditions in a biome vary during a typical year. We can also observe the specific time period when the temperature is warm enough for plants to grow. In the biome illustrated in Figure A.8a, the growing season—indicated by the shaded region on the *x* axis—is mid-March through mid-October. In Figure A.8b, the growing season is mid-April through mid-September.

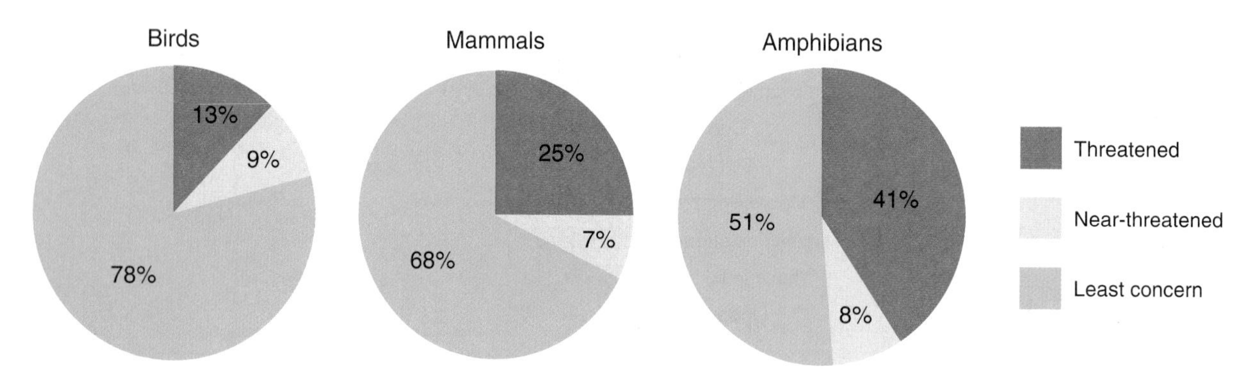

FIGURE A.7 **(book Figure 59.4) Pie chart.** Pie charts plot data that are percentages and collectively add up to 100 percent. These pie charts illustrate the percentages of birds, mammals, and amphibians of the world that are categorized as either threatened, near-threatened, or least concern. *(After International Union for Conservation of Nature, 2009)*

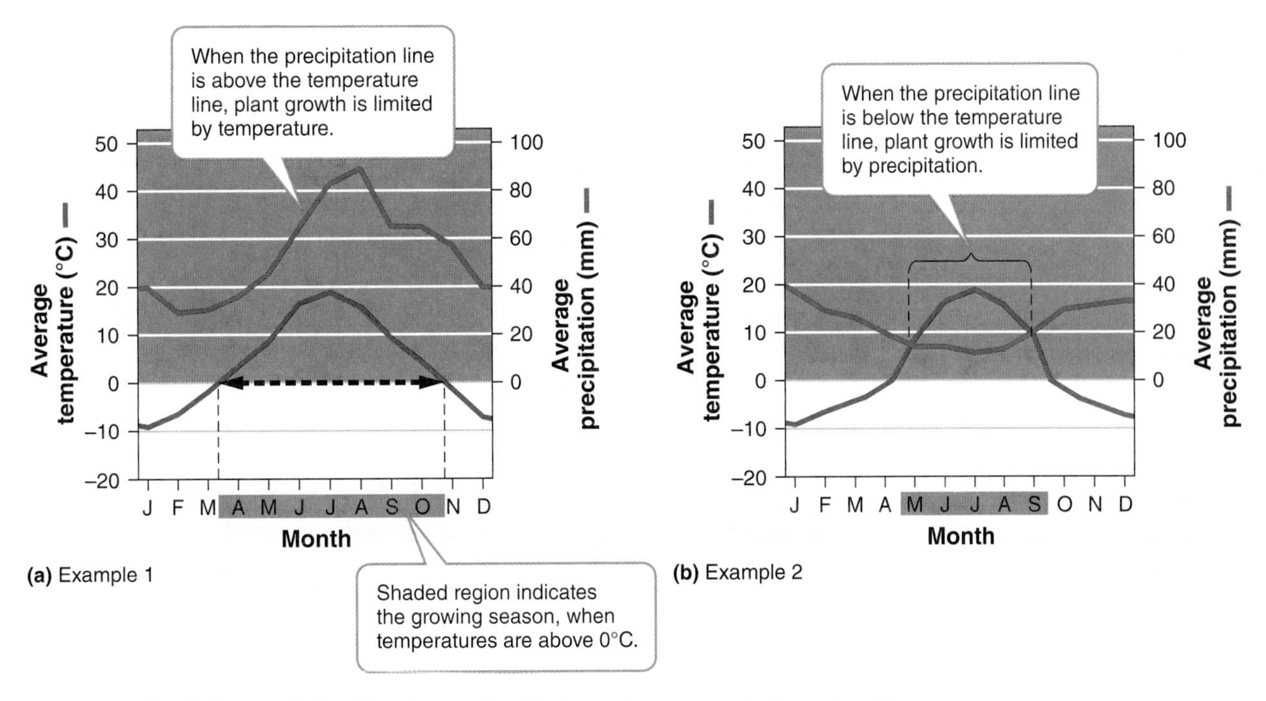

FIGURE A.8 **(book Figure 12.4) Climate graph.** Climate graphs are a special type of graph that plots monthly temperatures and monthly precipitation in a way that tells us whether plant growth is more limited by temperature or water. These diagrams help us understand the productivity of different biomes.

In addition to identifying the growing season, climate diagrams show the relationships among precipitation, temperature, and plant growth. In Figure A.8a, the precipitation line is above the temperature line in every month. This means that water supply exceeds demand, so plant growth is more constrained by temperature than by precipitation throughout the entire year. In Figure A.8b, the precipitation line intersects the temperature line. At this point, the amount of precipitation available to plants equals the amount of water lost by plants through evapotranspiration. When the precipitation line falls below the temperature line from May through September, water demand exceeds supply and plant growth will be constrained more by precipitation than by temperature.

Age Structure Diagrams

Age structure diagrams are visual representations of age distribution for both males and females in a country. FIGURE A.9 presents four examples. Each horizontal bar of the diagram represents a 5-year age group and the length of a given bar represents the number of males or females in that age group.

While every nation has a unique age structure, we can group countries very broadly into three categories. Figure A.9a shows a country with many more young people than older people. The age structure diagram of a country with this population will be in the shape of a pyramid, with its widest part at the bottom, moving toward the smallest at the top. Age structure diagrams with this shape are typical of countries in the developing world, such as Venezuela and India.

A country with a smaller difference between the number of individuals in the younger and older age groups has an age structure diagram that looks more like a column. With fewer individuals in the younger age groups, we can deduce that the country has little or no population growth. Figure A.9b shows the age structure of people in the United States, which is similar to the age structure of people in Canada, Australia, Sweden, and many other developed countries. Panels (c) and (d) show countries with a proportionally larger number of older people. This age structure diagram resembles an inverted pyramid. Such a country has a decreasing number of males and females within each younger age range and that number will continue to shrink. Italy, Germany, Russia, and a few other developed countries display this pattern. In recent years China has also begun to show this pattern.

As you can see, we can use many different types of graphs to display data collected in environmental studies. This makes it easier to view patterns in our data and to reach the correct interpretation. With this knowledge of graph making, you are now well prepared to interpret the graphs throughout the book.

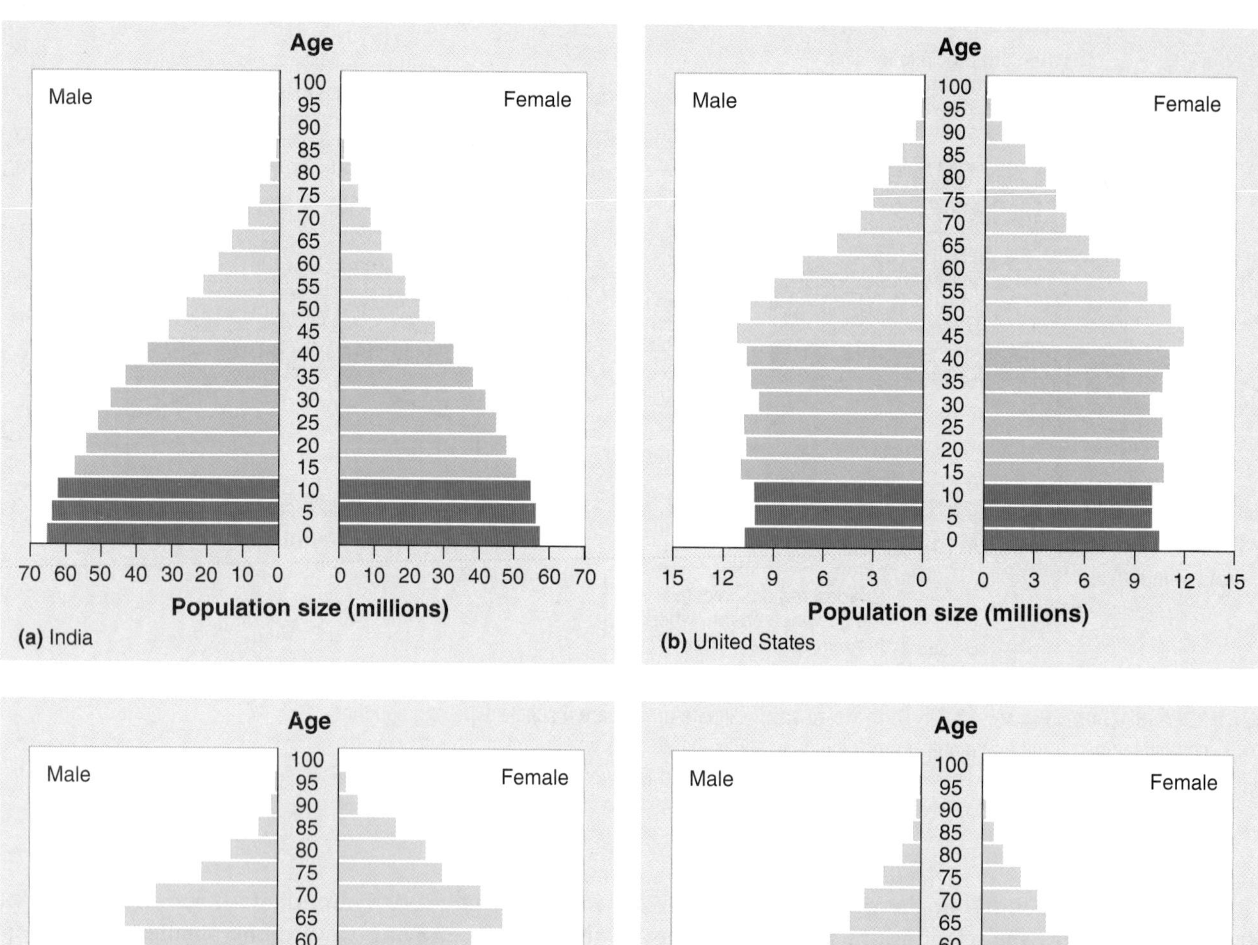

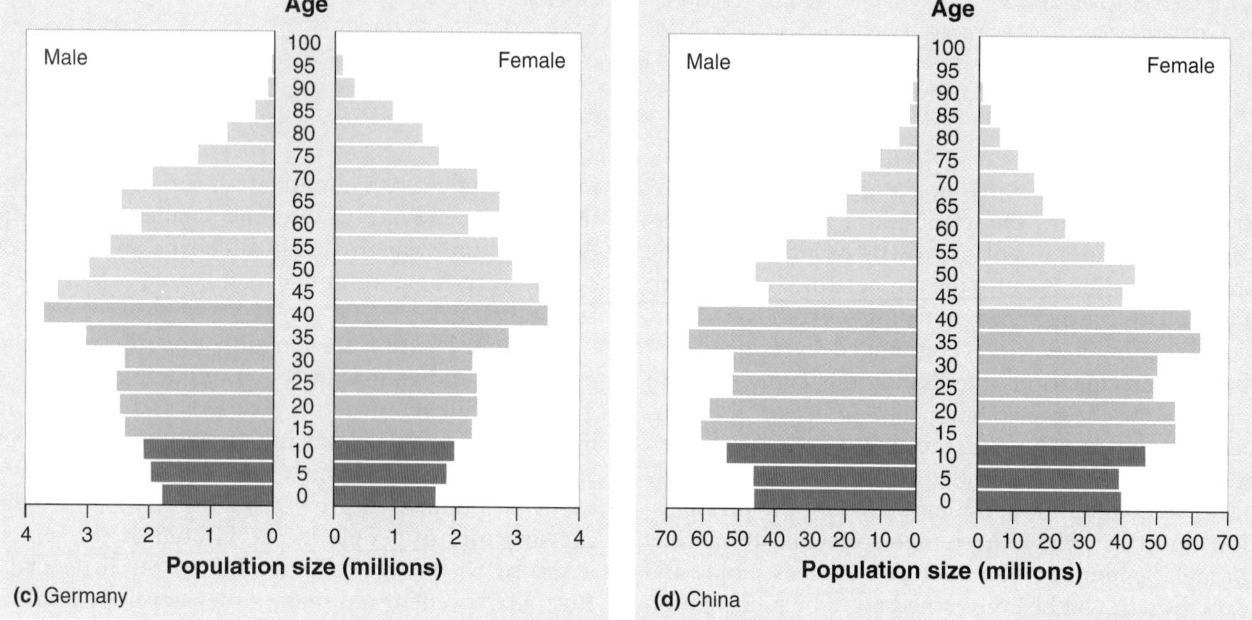

FIGURE A.9 **(book Figure 22.8) Age-structure diagrams.** These graphs allow us to understand the relative number of males and females in different age classes. In doing so, these graphs illustrate whether a population is likely to grow, stay stable, or shrink in future years. *(Data from http://www.census.gov/ipc/www/idb/pyramids.html)*

Answer Appendix

Chapter 1 Free-Response Questions (p. 29)

2. (a) Possible answers could include cutting down trees, which leads to massive erosion. Other answers could be loss of soil, which causes food production to decrease, and overuse of needed water resources.

(b) Possible answers could include practices such as conserving and finding alternatives to nonrenewable resources as well as protecting the capacity of the environment so that it continues to supply renewable resources. Recycling and reuse are other possible answers with a suitable discussion of the strategy.

(c) Answers can include a discussion of how environmental scientists believe that there are limits to the supply of clean air and water, nutritious foods, and other life-sustaining resources our environment provides. They also feel there is a point at which Earth will no longer be able to maintain a stable climate.

Chapter 2 Free-Response Questions (p. 60)

2. (a) Possible inputs:
- water
- sunlight
- seeds
- fertilizers
- pesticides
- gasoline (to run machinery)
- land (space)
- soil

Possible outputs:
- food
- seeds
- air pollution
- water pollution

(b) There are many possible answers; two examples are provided in the figures below.

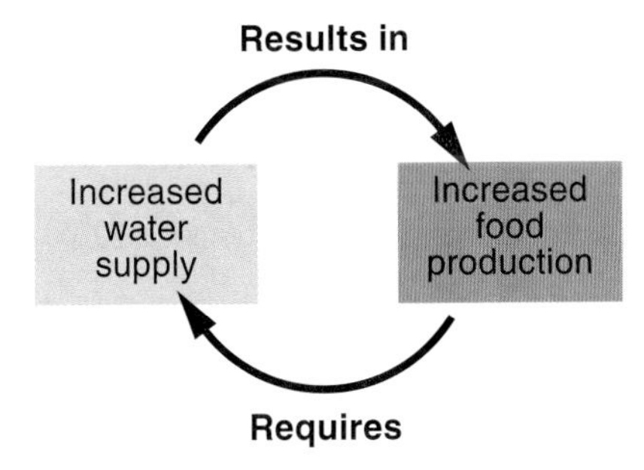

Increased water supply results in increased food production, which requires more water.

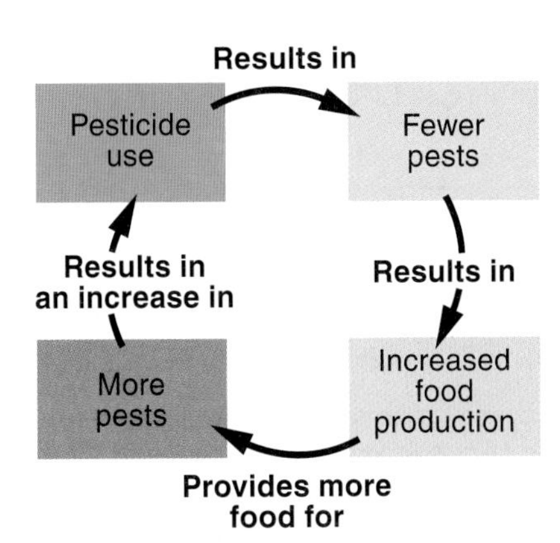

Pesticide use reduces the number of pests, which results in an increase in crop production, which in turn provides more food for more pests and necessitates the use of more pesticides.

(c) Possible adaptive management strategies:

- plant fewer crops
- replant specific areas with prairie grasses for ground cover to prevent erosion
- divert water from other areas not affected by the drought
- import water (e.g., truck it in from distant areas)
- use water that has been stored from previous nondrought years
- switch to a crop that requires less water
- leave land fallow

(d) kcal consumed by male from wheat

$= 2{,}500\ \text{kcal/day} \times (365\ \text{days} \times 0.30)$

$= 2.74 \times 10^5\ \text{kcal}$

kg wheat required per year

$= (2.74 \times 10^5\ \text{kcal} \div 2.5\ \text{kcal/g}) \times$

$(1\ \text{kg}/1{,}000\text{g}) = 110\ \text{kg per year}$

hectares needed

$= (1\ \text{ha}/3{,}000\ \text{kg}) \times (110\ \text{kg}) = 0.037\ \text{ha}$

Unit 1 Free-Response Questions (p. 63)

1. (a) The area of natural habitat that will be lost is the sum of forest and stream area minus the area of wetland area. To conduct this calculation in terms of hectares, it is necessary to know that a hectare is an area of land that is 100 m by 100 m, or 10,000 m^2, and that 1 km is equal to 1,000 m. An area of 1 km^2 is 1,000 m by 1,000 m, and contains 1,000,000 m^2.

To convert 4 km^2 of forest habitat to hectares:

$$(4\ \text{km}^2) \times \left(\frac{1{,}000{,}000\ \text{m}^2}{1\ \text{km}^2}\right) \times \left(\frac{1\ \text{ha}}{10{,}000\ \text{m}^2}\right) = 400\ \text{ha}$$

To convert 5,000 m^2 of stream habitat to hectares:

$$(5{,}000\ \text{m}^2) \times \left(\frac{1\ \text{ha}}{10{,}000\ \text{m}^2}\right) = 0.5\ \text{ha}$$

To convert 0.75 km^2 of wetland habitat to hectares:

$$(0.75\ \text{km}^2) \times \left(\frac{1{,}000{,}000\ \text{m}^2}{1\ \text{km}^2}\right) \times \left(\frac{1\ \text{ha}}{10{,}000\ \text{m}^2}\right) = 75\ \text{ha}$$

Hence, the total area of natural habitat lost as a result of development will be

$$400 + 0.5 - 75 = 325.5\ \text{ha}$$

(b) Ecosystem services are processes by which life-supporting resources are produced for humans. Important services that will be lost as a result of burying the stream may include:

- cycling of nutrients through downstream transport of organic matter, decomposition, and consumption of energy and nutrients by stream organisms
- fertilization of streamside (i.e., riparian) land
- flood prevention by carrying surface water runoff out of the local area (Note that natural streams also maintain slower flow rates relative to culvert streams, which further prevents flooding.)
- slowing sediment transport (Since natural streams maintain slower flow rates relative to culvert streams, sediment from surface-water

runoff is often trapped in small streams for extended periods of time. This results in less buildup of sediment downstream, which can fill up reservoirs, damage fisheries, increase water filtration costs, and harm downstream aquatic habitats.)
- filtering harmful contaminants from water (Biological processes in streams often convert excess nutrients from fertilizer and other environmental contaminants into less harmful substances. By burying the stream, many of these biological processes will be lost and lead to harmful effects downstream.)
- aesthetic and recreational services (Streams provide habitat for recreational activities such as sport fishing, and they also provide calm areas for people to relax.)
- production of biodiversity (Stream ecosystems serve as habitat and breeding grounds for myriad organisms that humans depend on, such as amphibians for mosquito control and flying insects for pollination.)

Important services that will be gained as a result of building a wetland include many of the same services that will be lost as a result of burying a stream. For example, services provided by a wetland include

- cycling of nutrients via chemical and biological processes
- fertilization of nearby vegetation (Although note that as water drains, into a wetland, water and nutrients are often absorbed into nearby soil and provide resources for plants.)
- flood control and prevention by retaining water from surface runoff
- trapping sediment carried by streams
- filtering contaminants zfrom water (Note that wetland plants are particularly good at absorbing excess nutrients from water.)
- aesthetic and recreational services
- production of biodiversity

(c) (i) A suitable hypothesis for this experiment is one that requires observations of the change in water quality before and after development. For example, the hypothesis may concern the change in water quality of the stream before and after burial, the quality of water in the stream before burial versus the quality of water in the wetland, or the quality of groundwater before and after construction. A poor hypothesis would simply state that change will happen. A better hypothesis would indicate the direction of this change (e.g., water quality in the stream will decrease after burial), and the best hypothesis would provide an explanation (e.g., water quality in the stream will decrease after burial because biological processes that filter the water will decrease). The student may also choose to define water quality prior to the hypothesis, which is also acceptable (e.g., water quality in the stream, defined as the concentration of pollutants in the water, will decrease after burial).

(ii) The data to be collected should specifically address the hypothesis as well as the definition of water quality. If the hypothesis regards the quality of water in the stream before and after burial, the most appropriate measurements will be in-stream measurements—not groundwater or wetland measurements. Although water quality lacks a precise definition, good water quality should be something that increases ecosystem services. For example, the textbook commonly refers to water quality as the concentration of contaminants, but the term can also refer to the suitability of water for organisms such as pH levels and nutrient concentrations.

The best answer will refer to data that directly measure water quality with a specific variable or set of variables. The measure of water quality should *explicitly* address the definition of water quality. For example, the student may choose to measure water quality by measuring a related ecosystem process, such as the number of insect species found in the stream. Although this does indirectly address water quality, it is not the best answer since diversity does not explicitly describe water quality. The student should also indicate the frequency of measurements (e.g., once before and after development, weekly during development, etc.).

An example of a suitable answer is, "I will conduct a natural experiment in which I will measure the concentration of pesticides in stream water before development begins and after development is finished."

(iii) Although students should not be expected to know details of experimental methods (e.g., how to analyze water for pesticides), students should be able to identify treatments, a control, and replication. Treatments are the object of measurement before and after development, such as water at the outflow of the stream before and after development. The control should demonstrate thought about the influence of external factors on the treatments that might influence interpretation. For example, contaminants in the stream may increase after development as a result of contamination from a source upstream of the development site. Hence, it would be appropriate to measure water quality both upstream and downstream of the development site, where the upstream measurement is the control. Replication should occur during the sampling procedure (e.g., taking more than one sample of water) so that the quality of a single sample does not bias the results. If the answer includes laboratory analysis of water quality, the answer should also include replication of analytical procedures to ensure that human error does not bias the results.

(iv) Results that validate a hypothesis should explicitly address the hypothesis. For example, if the hypothesis is that sediment in the water will increase after development, results that validate this hypothesis would be an equal amount of sediment in the control water before and after development, and a greater amount of sediment will be in the downstream water after development.

2. (a) To calculate the energy efficiency of growing corn and then converting corn into milk, we must first know how much corn the average cow consumes to produce a single liter of milk. If 8 million cows produce 72 billion L of milk, then

$$\left(\frac{72{,}000{,}000{,}000 \text{ L milk}}{8{,}000{,}000 \text{ cows}}\right) \times \left(\frac{1 \text{ cow}}{13{,}500 \text{ kg corn}}\right) = \left(\frac{0.6667 \text{ L milk}}{1 \text{ kg of corn}}\right)$$

This can be restated as taking 1.5 kg of corn to produce 1 L of milk. Hence, the amount of corn energy it takes to produce 1 L of milk is 20 MJ $\times$ 1.5 = 30 MJ.

$$\left(\frac{20 \text{ MJ}}{1 \text{ kg corn}}\right) \times \left(\frac{1.5 \text{ kg corn}}{1 \text{ L milk}}\right) = \left(\frac{30 \text{ MJ corn energy}}{1 \text{ L milk}}\right)$$

The energy efficiency of producing corn energy is

$$\left(\frac{20 \text{ MJ energy in corn}}{40 \text{ MJ to produce corn}}\right) \times 100 = 50\%$$

The energy efficiency of converting corn energy into milk energy is

$$\left(\frac{15 \text{ MJ energy in milk}}{1 \text{ L milk}}\right) \times \left(\frac{1 \text{ L milk}}{30 \text{ MJ corn energy to produce 1 L of milk}}\right) \times 100 = 50\%$$

Hence, the total energy efficiency of converting corn into milk is

$$50\% \times 50\% = 25\%$$

(b) Since no process of energy conversion is 100 percent efficient, any answer that relates to the process of milk production will be correct. However, the best answers will be ones that directly describe how energy efficiency is reduced. For example, appropriate answers could be:

- To produce cattle feed, farmers must use large machines that plow and plant fields. Motors in these machines are often highly inefficient at converting liquid fuel to kinetic energy.
- To grow feed for cattle, farmers must apply large amounts of fertilizer to fields. Only a fraction of this fertilizer is used by the plants; the remaining fertilizer is leached from the ground by groundwater. Consequently, any kinetic energy that was used to apply this wasted fertilizer creates inefficiency in converting the chemical energy of fertilizer into the chemical energy of plants.
- The process of harvesting plants is highly inefficient; farmers must use a lot of fuel to cut and grind the feed. Motors in these machines are inefficient at converting liquid fuel to kinetic energy.
- When cows consume cattle feed, they must use a large amount of ingested energy for their own metabolism. This chemical energy is converted to body heat and kinetic energy. Some of the material consumed is never used by the body and is egested.

(c) (i) Methane (CH_4) is the main chemical gas produced by the decomposition of cow manure that can be used as a potent source of fuel.

(ii) Energy is released from methane when the covalent bonds between carbon and hydrogen are broken. These bonds are typically broken in the presence of oxygen gas (O_2): $CH_4 + 2O_2 \rightarrow CO_2 + 2H_2O + \text{energy}$

(iii) As determined from answer (a), 50% of all energy ingested by cows is not used to produce milk. If 10% of this energy can be recovered from gas released by manure, then the total efficiency of converted corn feed into milk will be $50\% + 5\% = 55\%$. Hence, the total efficiency of growing corn and converting corn into milk will be: $50\% \times 55\% = 27.5\%$

Unit 2 Free-Response Questions (p. 183)

1. (a) The nitrogen cycle is initially altered when the bacteria and fungi die, and their internal contents are leached into the soil. Following this event, nitrogen cycling in temperate forests may be affected in several ways:

- Rainwater can transport leached nutrients away from the point of leaching. This reduces the total amount of nitrogen that will be present in the forest during the growing season. Plant growth will be reduced, and the number of consumers will also be reduced.
- If nutrients are not carried away, then they will be available for plant nutrition after they are mineralized. The death of bacteria and fungi releases a tremendous amount of nutrients. However, those nutrients will be in organic form and they must be converted into inorganic forms such as ammonia through the process of mineralization. The mineralization of nitrogen is also called ammonification.
- Bacteria and fungi are the organisms primarily responsible for mineralization of nutrients. Hence, the loss of bacteria and fungi may reduce the amount of nitrogen that is mineralized in the soil. Less mineralization will reduce the amount of nitrogen available for primary production. As a result, primary production will decrease.

(b) The carbon cycle is intimately linked with the growth, respiration, and metabolism of bacteria and fungi in a forest. Following the death of bacteria and fungi, the carbon cycle in temperate forests may be affected in several ways:

- Bacteria and fungi are the organisms primarily responsible for the decomposition and mineralization of organic materials, such as leaf litter. Loss of bacteria and fungi will reduce decomposition and mineralization, and increase burial of organic material.
- A reduction in decomposition and mineralization will also reduce the amount of carbon and nutrients available for primary production. Hence, the loss of bacteria and fungi will likely reduce photosynthesis.
- Like any aerobic organism, many bacteria and fungi respire CO_2 through cellular respiration. Loss of bacteria and fungi will decrease the amount of CO_2 respired in the forest.

(c) In temperate forests, stream and pond food webs receive a substantial amount of energy and nutrients from the surrounding forests in the form of runoff and dead organic material such as leaf litter. The loss of bacteria and fungi could affect stream and pond food webs in several ways:

- Leached nutrients from bacteria and fungi will likely enter stream and pond food webs via runoff. This influx of nutrients will be mineralized in the aquatic ecosystem and will provide more nutrients for aquatic primary production. Nutrients that are not mineralized may be consumed by decomposers.
- Reduced decomposition of leaf litter on the forest floor provides more material that may enter streams and ponds, for example through runoff or wind. This influx of energy and nutrients can increase the biomass of organisms in aquatic systems.
- Similarly, reduced mineralization of organic compounds in the forest soil provides more material that may enter streams and ponds, for example through runoff or wind.
- Reduced primary production or consumer biomass in ponds and streams will decrease the abundance of secondary and tertiary consumers in these ecosystems.
- The loss of bacteria and fungi can have delayed consequences for stream and pond food webs. The loss of bacteria and fungi will result in the reduction of nutrient mineralization in the soil and lead to a reduction in primary production. Less primary production means fewer leaves and therefore less leaf litter. A reduction in leaf litter will decrease the amount of organic material that enters the aquatic system during the following autumn.

(d) Exposure of the soil to extreme temperatures allows only individuals who are buried deeper in the soil to survive. This will reduce the local genetic diversity of wood frogs, but also select for individuals that are better adapted to more extreme temperatures because they bury themselves deeper. Subsequent generations of wood frogs will therefore be better adapted for extreme temperatures. An extreme reduction in genetic diversity would be an example of a bottleneck effect. As a result of reduced genetic diversity, populations that have experienced a bottleneck are often less able to adapt to future changes in the environment. If the genetic diversity of a local wood frog population is drastically reduced, future generations may not be able to cope with other challenges in the environment such as extremely high temperatures or disease.

2. (a)

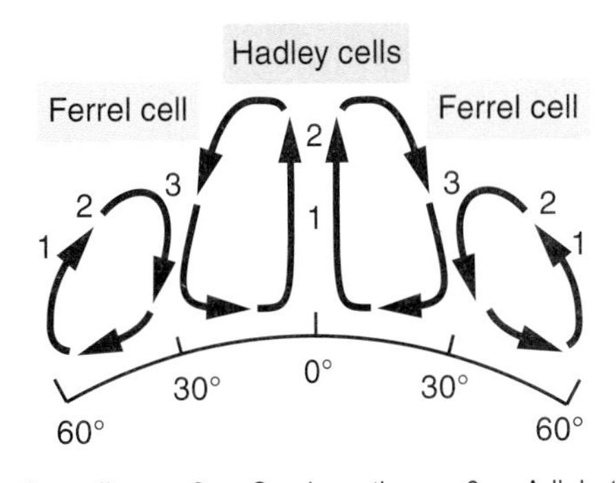

(b) The western slopes of the Colorado Rockies would experience greater precipitation than the eastern slopes because the eastern slopes are in a rain shadow. Air moving inland from the ocean contains water vapor that is carried up the western slopes of the Rockies. As the air rises, it experiences adiabatic cooling, which causes the water vapor to condense and precipitate. Condensation causes latent heat release, which causes the air to continue moving up the western slopes and over the top of the mountains. The air that moves over the mountains is much drier and capable of delivering far less precipitation.

(c) **(i)** Gross primary productivity (GPP) is the total amount of solar energy that the producers in an ecosystem capture via photosynthesis over a given amount of time. Hence, the equation for GPP is:

$$\text{GPP} = \text{net primary productivity} + \text{respiration by producers}$$

Note that GPP differs from net primary productivity (NPP), which is the energy captured by photosynthesis minus the energy respired by producers (NPP = GPP – respiration by producers).

(ii) To determine GPP, we must determine NPP and the rate of respiration. A common approach is to measure the compounds that participate in these processes. The most common approach is to measure the movement of CO_2, which is a compound required by photosynthesis and produced by respiration. First, we measure the production of CO_2 in the dark. Since photosynthesis does not occur in the dark, this provides a measurement of CO_2 released by respiration. Then, we measure the uptake of CO_2 in sunlight, which provides the net movement of CO_2 when respiration and photosynthesis are both occurring. By adding the amount of CO_2 lost in sunlight and the amount of CO_2 gained in the dark, we can determine the gross amount of CO_2 that is taken up during photosynthesis. Note that both NPP and GPP are rates, so the measurements of CO_2 lost and gained must be expressed as quantities over time.

(iii) A null hypothesis for this study would be that there is no difference in GPP on the western slopes of the Colorado Rockies between ENSO years and normal (non-ENSO) years.

Science Applied 2 Free-Response Question (p. 187)

(a) For an area to be considered a hotspot, Conservation International requires (1) that the area contain at least 1,500 endemic plant species, and (2) the area must have lost more than 70 percent of the vegetation that contains those endemic plant species. Coral reefs are an interesting problem in this regard, as the diversity of reefs is spread among multiple trophic layers. Consequently, coral reefs may not have enough plant species to be considered a hotspot, although they have many species of primary, secondary, and tertiary consumers. Although Conservation International does work to protect coral reefs, the reefs would not technically be classified as hotspots under the definition of endemic plant species. However, even if the requirement of endemic species were extended to organisms across all trophic levels, it is possible that Florida Reef does not have at least 1,500 truly endemic species to be considered a hotspot. It is also possible that an insufficient proportion of the reef is threatened with bleaching to warrant classification as a hotspot.

(b) Ecosystem services provided by a coral reef may include:

- Aesthetic value: Humans frequently visit coral reefs to admire the scenery and diversity.
- Recreational value: Coral reefs provide opportunities for snorkeling and diving. In addition, reefs provide a source population for species that may be valued in sport fishing and in the aquarium trade.
- Food and natural products: Coral reefs provide a source population and habitat for species that are valued in the seafood industry and for many species that are used by local communities.
- Materials for medicine: Scientists continue to find medicinal uses for compounds derived from newly discovered species that reside on or around coral reefs.
- Production of biodiversity: Coral reefs are home to a vast array of endemic species that are critical for ecosystem processes, such as water filtration and nutrient cycling.
- CO_2 assimilation: Coral reef food webs store a tremendous amount of carbon in the form of plant and animal life. This has the potential to offset carbon emissions generated by humans.
- Shoreline protection: By existing between the open ocean and the mainland, coral reefs often provide a buffer to slow wave action before it strikes coastlines. This reduces shoreline erosion.
- Monitoring of pollution and climate: Reefs are highly sensitive to disturbances and provide scientists with a means of identifying past and current sources of pollution and climate change.

(c) Coral bleaching is a phenomenon in which symbiotic algae inside the corals die. Corals are tiny animals that live inside a hollow tube made of limestone. Because corals often live in water that is relatively poor in nutrients and food, they rely on symbiotic relationships with algae that live inside coral tissue. When a coral digests food, it releases CO_2 and nutrients that the algae absorb. In exchange, the algae release sugars that the corals use for food. When the algae dies—often as a consequence of disease and environmental changes such as ocean acidification and temperature rise—the coral dies soon after. When the coral dies, it leaves behind the white limestone tube.

(d) The most common hypothesis to explain this pattern is the intermediate disturbance hypothesis. The intermediate disturbance hypothesis states that ecosystems experiencing intermediate levels of disturbance will favor a higher diversity of species than those with high or low disturbance levels. Ecosystems in which disturbances are rare experience intense competition among species, and populations of only a few highly competitive species typically dominate the system. When disturbances are frequent, the only species to persist are those with fast growth or those that can tolerate the disturbance. In ecosystems that experience intermediate disturbances—for example those with occasional hurricanes—many types of species persist.

Chapter 7 Free-Response Questions (p. 251)

1. **(a)** See Figure 23.1.

(b) During phase 1, a country is preindustrial. There are high birth rates but also high death rates, so the population remains stable. High birth rates reflect a lack of both contraception and education about how to prevent pregnancy, as well as a need for children to contribute to family income. The high death rates reflect a lack of proper medical care, harsh living conditions, poor sanitation, and inadequate nutrition. In these countries, life expectancy for adults is relatively short due to difficult and often dangerous working conditions. Infant mortality is high because of disease, lack of health care, and poor sanitation. As a result, many families have more children because they expect many will not live to adulthood. In a subsistence economy, where most people are farmers, having numerous children is an asset. Children can do jobs such as collecting firewood, tending crops, watching livestock, and caring for younger siblings. Parents also count on children to care for them in their old age.

During phase 2, the country is in the transitional phase. Births are still high, but the death rate begins to fall. The populations of countries in phase 2 grow at a very fast rate. The drop in the death rate is because medical care becomes more available and living conditions and sanitation improve, although the country is still developing and poverty remains high. As the country modernizes, better sanitation, clean drinking water, increased access to food and goods, and access to health care (including childhood vaccinations) all reduce infant mortality and CDR. However, the birth rate does not markedly decline. Couples continue to have large families because it takes at least one generation, if not more, for people to notice the decline in infant mortality and adjust to it. It also takes time to implement educational systems and birth control measures.

During phase 3, the country is now industrialized. Birth rates fall and death rates remain low. The population growth slows and may reach zero. The birth rates drop because contraception is available, and women and girls are more educated and participate in the workforce. The death rate remains low because medical care is widespread and sanitation has dramatically improved. A country enters phase 3 as its economy and educational system improve. In general, as family income increases, people have fewer children. Phase 3 is typical of many developed countries, including the United States and Canada. As societies transition from subsistence farming to more complex economic specializations, having large numbers of children may become a financial burden rather than an economic benefit. Relative affluence, more time spent pursuing education, and the availability of birth control increase the likelihood that people will choose to have smaller families. However, it is important to note that cultural, societal, and religious norms may also play a role in birth rates. As birth rates and death rates decrease in phase 3, the system returns to a steady state. Population growth levels off during this phase, and population size does not change very quickly, because low birth rates and low death rates cancel each other out.

During phase 4, the country is in the postindustrial phase. The birth rate is low and the death rate is low. The death rate may even exceed the birth rate. This country's population is either stable or decreasing. The birth rate remains low because there is available contraception, women and girls are educated, and jobs are available. The death rate remains low because medical care is widespread and sanitation has dramatically improved. Phase 4 is characterized by a relatively high level of affluence and economic development. Japan, the United Kingdom, Germany, and Italy are phase 4 countries, with crude birth rates well below crude death rates. The declining population in phase 4 means fewer young people and a higher proportion of elderly people. This demographic shift can have important social and economic effects. With fewer people in the labor force and more people retired or working part-time, the ratio of dependent elderly to wage earners increases, and pension programs and social security services put a greater tax burden on each wage earner. There may be a shortage of health care workers to care for an aging population. Governments may encourage immigration as a source of additional workers. In some countries, such as Japan, the government provides economic incentives to encourage families to have more children in order to offset the demographic shift.

(c) The government could write laws that strictly enforce birth control, or the government could begin to educate the people on how to use and obtain proper birth control. Some countries provide birth control, while other countries tax additional children or enforce penalties for families that do not stick to the official population program. The country could also educate its people on the benefits of a smaller family. Depending on the country's political situation, there could be different ways this plan would work. The country may force its program on the people or it might encourage the program but allow the people to choose whether they will participate. The drawback to this plan is that people who need or want more children may decide to ignore the country's suggestions and continue to have large families. Another drawback could be that the plan will work so well that the country's population begins to decrease, causing adverse economic or societal issues. These problems include a larger elderly population with no younger, working-age people to help support these individuals.

2. (a) (i) This country's age structure is roughly equal at all ages, with a slighter lower number of the elderly. The number of adults and children are about the same.

(ii) This country has a low total fertility rate because the parents have had just enough children to replace themselves.

(iii) Life expectancy in this country would be generally high, with many people surviving into old age.

(iv) The growth rate of this country will be low and the doubling time will be very high.

(b) The socioeconomic status of this country is high. There are plenty of jobs, the GDP is high, and there is advanced technology and industry. In fact, there may be more available jobs than the population can fill, so immigration may be required to drive country A's economy.

(c) (i) Country B has a fast-growing population with many more children and infants than adults. The country's population is growing exponentially. The country has a high growth rate now and it will continue to grow rapidly because of the large number of people currently beginning their reproductive years.

(ii) The infant mortality rate in this country is high because health care is not easily available or is of low quality.

(iii) The rate of population growth is very high.

Unit 3 Free-Response Questions (p. 254)

2. (a) A possible answer could be: A growing elderly population increases the ratio of dependent elderly to wage earners. Consequently, money and time may be diverted from economic growth to care for the elderly. For individual families, this means that wage earners will save less money and invest less into the economy, which will slow economic growth. Diverting money toward care of the elderly may also reduce the health of wage earners, who will have less money to spend on their own personal health care. This may serve to slow overall population growth. Diverting money and time toward care of the elderly may also delay or eliminate decisions to have offspring, which can severely slow population growth.

(Other answers may be correct; correct answers should clearly demonstrate how the diversion of time, money, or manpower toward elderly care will alter overall population growth.)

(b) If elderly in need of health care move to urban areas, they will have better access to modern health care, which may reduce the burden on individual families. In addition, condensing populations into urban areas generally has the effect of reducing per capita ecological footprints. However, condensing the elderly into urban areas also exposes them to a higher amount of pollutants and disease, which could shorten average lifespan.

(c) To calculate the crude death rate of a nation, we can use the following equation:

$$\text{National population growth rate} = \frac{(\text{crude birth rate} + \text{immigration}) - (\text{crude death rate} + \text{emigration})}{10}$$

From the problem, we know that the national population growth rate is 0.50 percent and the crude birth rate is 12 births per 1,000 individuals. We also know that the number of emigrants from China is 700,000 and the number of immigrants to China is 200,000. These numbers must be expressed per 1,000 people. To do this, we first divide the numbers by the total population size and multiple by 1,000:

$$\text{emigration} = \frac{700{,}000}{1{,}400{,}000{,}000} \times 1{,}000 = 0.5$$

$$\text{immigration} = \frac{200{,}000}{1{,}400{,}000{,}000} \times 1{,}000 = 0.14$$

Next, we insert the numbers into the equation for national population growth rate and solve for crude death rate:

$$0.50 = \frac{(12 - 0.14) - (\text{crude death rate} + 0.5)}{10}$$

$$5.0 = 11.86 - \text{crude death rate} - 0.5$$

$$\text{crude death rate} = 6.36 \text{ death per 1,000 individuals}$$

(d) If all families in China obeyed the one-child policy rule, the total fertility rate would be 1.

Science Applied 3 Free-Response Questions (p. 257)

(a) Density-dependent factors that could influence the population of white-tailed deer could be:

- Competition: As population density increases, the amount of competition for limiting resources also increases. This results in fewer per capita resources, and lower chances of survival or reproduction.
- Predation: As population density increases, there will likely be a parallel rise in the number of deer predators (e.g., coyotes), which may reduce deer population size. If the number of predators does not increase, then per capita predation rate will decrease and population size will continue to increase. As population density decreases, it becomes harder for predators to find prey, which will eventually allow the deer population to increase in size.
- Parasitism: As population density increases, there are more hosts available for diseases and infection rates will increase.
- Reproduction: If population density decreases to a point where it is difficult for individuals to find mates, lower rates of reproduction and a further decrease in population size may result.
- Sport hunting: The number of deer hunted is higher when deer abundance is higher because hunting is regulated. Also the ability for hunters to actually find deer affects the number of deer that are taken.

Density-independent factors that could influence that population of white-tailed deer could be:

- Weather disturbances: Natural events such as droughts, floods, tornadoes, hurricanes, and fires can reduce population sizes. Alternatively, weather patterns that increase resource availability can increase population sizes.
- Human activity: Human practices such as logging, fire suppression, and damming rivers can change resource and habitat availability, leading to changes in population growth.
- Seasonal cycles: Seasonal changes in weather patterns lead to fluctuations in population growth that are often independent of population density.

(b) At low abundance, deer can accelerate the process of ecological succession. For example, by grazing some species of plants, deer may reduce competition for other plant species. Deer may also promote the growth of grazing-resistant plant species. Deer may also disperse seeds of plants; many herbivores consume seeds but do not digest them. Instead, the seeds pass through the gut and are egested onto the ground far from their parental plant.

(c) (i) Population growth rate can be calculated using the following equation:

$$\text{Population growth rate} = \frac{\text{crude birth rate} - \text{crude death rate}}{10}$$

However, we must take into account the presence of infant mortality, which should be subtracted from the crude birth rate. For the deer population in this study, the crude birth rate is 40, the crude death rate is 20, and infant mortality rate is 10. Hence, the population growth rate would be $(40 - 10 - 20) \div 10 = 1$ percent.

(ii) Given a population growth rate of 1 percent, we can calculate the number of individuals that are added to the population in the following year:

$$100{,}000 + (100{,}000 \times 0.01) = 101{,}000$$

In order to keep the population size stable, hunters must kill 1,000 deer per year.

Chapter 8 Free-Response Questions (p. 291)

2. (a) (i) Earthworms are part of the macrofauna of the soil community, along with other species such as insects. They are essential for mixing soil and for shredding detritus. Detritus broken down into smaller pieces can be chemically degraded and release nutrients and minerals into the soil. The burrowing earthworms also help with the aeration and water infiltration of the soil. Having an abundant number of earthworms in the soil community helps the sustainability of the soil community. Without earthworms, the soil community would become less stable and degrade more quickly.

(ii) Organic material enhances the physical and chemical properties of soil. Depending on the soil texture, organic matter or humus can improve water and aeration properties and workability. As organic matter is broken down, it releases into the soil mineral nutrients and other material that benefit plants. Having a good supply of organic matter in the soil enhances the stability of the soil and has a positive effect on sustainability. Soils that lose organic material are prone to degradation; mineralization of the soil can take place and lead to desertification.

(iii) When animal feed grain is continually grown on agricultural land, inorganic fertilizers have to be applied to the fields in order to attain good crop yields. Inorganic fertilizers contain nitrates, phosphates, and potassium compounds that are obtained by mining rocks rich in mineral resources. This can have a negative impact on the rock cycle and the environment, and is clearly a nonsustainable practice. The nitrates from the application of inorganic fertilizer can dissolve in rainwater, infiltrate the soil, and build up in the groundwater below. Often more inorganic fertilizer is applied than needed. When animals are allowed to graze in pastureland, it is unnecessary to apply large amounts of inorganic fertilizers, and therefore fewer nitrates will be found in the groundwater. Fertilizer in the pastureland may be in the form of animal dung, which is organic.

(iv) The stream quality improved as a result of less fertilizer runoff, which reduced eutrophication. When tested, the stream water will probably have lower amounts of nitrates and phosphates than would be found on land continually used for grain production. As less soil erosion takes place from pastureland, the sediment load in the stream will be lower and not cover the streambed, allowing larvae to build up on the rocks and provide habitat for small aquatic organisms. With lower amounts of sediments in the stream, light penetration will enable benthic plants to survive. Grain production requires the application of pesticides and herbicides, and a switch to pasture grazing will result in lower amounts of these chemical pollutants in the runoff into the stream. All of these changes will lead to improved stream quality.

(b) A farmer who feeds beef cattle a diet of grain will generally be able to raise a high number of animals in a relatively small area. This could be in a building designed to house cattle in stalls. Such diets often allow a farmer to bring the animals to market sooner. A farmer may not want to switch to permanent pasture grazing because not as many cattle can be raised with that method, and the farmer's profits will decrease.

(c) There are several possibilities here. Animal manure contains nutrients that would diminish water quality of the stream. That might mean eutrophication or an increase in pathogens such as *E. coli*, which would make the downstream water unfit for uses such as drinking. Another possibility is that the animals might eat vegetation around the stream, eroding the soil of the bank and making flooding more likely.

Chapter 9 Free-Response Questions (p. 323)

2. (a) An earthquake is caused by movement of Earth's crust along a fault, which is a fracture in rock. Fault zones form in the upper lithosphere where two plates meet or slide past each other. The rock along a fault is jagged and resists movement of the plates. Eventually, the mounting pressure of the plates overcomes the resistance, and the plates slip past each other. The plates can move up to several meters in just a few seconds, and the result is an earthquake. The epicenter of an earthquake is the exact point on the surface of Earth above the location where the rock ruptures.

(b) The potential for an earthquake is generated by the buildup of potential energy in rocks surrounding a fault. Once enough potential energy is accumulated, that energy is converted to kinetic energy of rock movement. The potential energy in rocks will be expressed as increased pressure on the compounds found within the rocks. The pressure will be greatest near the epicenter. To predict earthquakes, we could set up electric charge sensors along a fault zone and monitor the amount of charge generated from the piezoelectric forces. As the charge increases, there is an elevated probability for the location of an epicenter. However, earthquakes can occur at all magnitudes, and it would be difficult to predict how much electric charge would be required before an earthquake actually occurred.

(c) The Richter scale measures the largest ground movement that occurs during an earthquake on a logarithmic scale. If an earthquake of magnitude 4.0 on the Richter scale causes ground movement of 10 mm, then an earthquake of magnitude 6.0 on the Richter scale will cause ground movement that is 100 times that movement, or 1,000 mm.

(d) An earthquake of magnitude 6.0 on the Richter scale will cause ground movement of 1,000 mm, or 1 m. If an earthquake of magnitude 6.0 occurs every 40 years, we can use the equation rate $\times$ time $=$ distance to determine the time (in years) it will take for an area of land to move 10 m:

$$\frac{1\text{ m}}{40\text{ years}} \times \text{time} = 10\text{ m}$$

$$\frac{0.025\text{ m}}{\text{year}} \times \text{time} = 10\text{ m}$$

$$\text{time} = 10\text{ m} \div \frac{0.025\text{ m}}{\text{year}} = 400\text{ years}$$

Chapter 10 Free-Response Questions (p. 355)

1. (a) Depending on how the land is to be used, all four major public land management agencies might be involved. Additionally, the land may be shared by several agencies with overlapping responsibilities, or divided among agencies. Students should recognize that, as presented in the photograph, all four agencies could be involved.

- **Bureau of Land Management:** The major responsibility of this agency is to oversee grazing practices; since this property appears to include open grasslands, BLM may be involved. Additionally, off-road driving and forests are also pictured, which fall under BLM jurisdiction.
- **U.S. Forest Service:** USFS oversees the timber program and builds and maintains logging roads. Recreational use and the maintenance of the forest biodiversity also are the responsibilities of this agency.
- **National Park Service:** The large building in the center of the land might suggest lodgings for tourists or an information center. The NPS is responsible for recreation and wildlife habitat/species protection.
- **U.S. Fish and Wildlife Service:** The forested area may represent an intact forest ecosystem. There is no evidence of roads. (Road building is banned in national wilderness preservation areas.) The existing building may predate the wilderness designation. The land may be used for wildlife conservation, hunting, and/or recreation.

(b) Some of the principles of smart growth are economic and social in nature, and therefore are not appropriate for this question, which specifically asks for minimization of environmental impacts. Possible answers might include:

- **Promote mixed land uses:** As opposed to traditional zoning, mixed land uses containing business, residential, retail, educational, and recreational facilities promote walking and bicycling as opposed to automobile use. This reduces noise, air pollution, and energy consumption.
- **Create walkable neighborhoods:** Positioning of buildings and parking lots works in conjunction with mixed land use to reduce noise, air pollution, and energy consumption.
- **Take advantage of compact building design:** Through the use of multistory buildings for residences, businesses, and parking garages, the amount of impervious surfaces can be reduced. This reduces runoff and the risk of flooding, and increases groundwater recharge.
- **Preserve farmland, open space, and critical environmental areas:** Farmland can provide a local source of fresh produce and other agricultural products, thereby reducing the need to import food. Preserving open space and critical environmental areas can enhance these areas in the federally owned land, thereby reducing habitat fragmentation.
- **Provide a variety of transportation choices:** With viable options for walking, bicycling, and mass transit, energy consumption can be reduced and air quality improved.

(c) An environmental impact statement (EIS) typically outlines the scope and purpose of a project, describes the environmental context, suggests alternative approaches to the project, and analyzes the environmental impact of each alternative. NEPA requires an environmental assessment of all projects involving federal money or federal permits. An EIS might be required under any of the following conditions:

- construction of an interstate highway
- construction of an airport
- construction of a wastewater treatment plant
- construction of an electrical power plant
- implementation of oil/gas leasing, exploration, and development
- construction of a bridge over an existing waterway
- construction of a dam

2. (a) The tragedy of the commons involves the degradation of a common resource that is accessible by everyone. In this case, the rangelands (grasslands) comprise the common resource used by ranchers, conservationists, and developers. The prairie dog populations and the populations of wildlife that depend on them for food have declined due to habitat loss, shooting, and poisoning. Ranchers in particular consider the prairie dogs detrimental to the grassland because they destroy grass and promote erosion. The BLM has exempted energy development companies from complying with rules that protect prairie dog colonies and habitat. Interspersed land ownership (public and private) prevents consistency in land management policies.

(b) Possible arguments for preserving the grasslands as a habitat for prairie dogs include: It is necessary to preserve the grassland in order to preserve the prairie dog. Prairie dogs are a keystone species, important to maintaining the integrity of the grasslands biome. In addition to eating grasses, prairie dogs also are known to eat insects. They may be instrumental in controlling insect pests. Their burrows aid in soil aeration and improve water penetration. Their nitrogen-rich dung improves soil fertility.

Possible arguments for maintaining the grasslands for livestock grazing include: Grazing on rangelands is less expensive than raising cattle in feedlots, which reduces the cost of beef for the consumer. Livestock provide a valuable food source to the public. The livestock industry provides employment for local residents.

(c) The Bureau of Land Management could raise (or lower) grazing fees to discourage (or encourage) grazing by cattle ranchers. It could mitigate the damages caused by cattle grazing so that prairie dogs could coexist. The U.S. Fish and Wildlife Service could designate the various species of prairie dogs as endangered species, thereby limiting the activities of cattle ranchers. It could establish periodic hunting seasons for prairie dogs, to control their numbers. It could designate specific areas of the grasslands as national wildlife refuges for the protection of the prairie dogs, where grazing would be banned.

Unit 5 Free-Response Questions (p. 391)

1. (a) The 10 principles related to the strategy of smart growth are:

- Smart growth mixes residential, retail, educational, recreational, and business land uses. For Detroit, this could mean mixing parks, shops, schools, and homes in the same area. The inclusion of a park and safe areas for children to play produces an area that is highly desirable for residential purposes.
- Smart growth provides housing for people of all income levels, which counters rising poverty levels, allows more people to find jobs near where they live, improves schools, and generates support for local businesses. For Detroit, this could mean the development of apartments, condominiums, and houses in similar areas.
- Smart growth creates walkable neighborhoods that decrease traffic by encouraging people to walk to shops and schools. This requires increasing the safety of the neighborhood, which would necessitate a larger police force.
- Smart growth encourages community and stakeholder collaboration in development decisions. This encourages community members to take part in developing their own living areas and encourages the care and upkeep of communities.
- Smart growth takes advantage of compact building designs. This may be particularly important for Detroit due to its decreased population size. Compact building designs may encourage people to live closer together, which reduces the need to police and care for large areas of the city that are sparsely inhabited.
- Smart growth fosters distinctive, attractive communities with a strong sense of place. This may be quite easy for Detroit, which is a center of American history that brought people from many races and ethnicities to live together.
- Smart growth preserves open farmland, natural beauty, and critical environmental areas. Even in a city, there are opportunities to highlight attractive natural features. For example, many cities are built around lakes or streams, and preserving these areas can make them a focal point.
- Smart growth provides a variety of transportation choices through taxis, buses, trains, and sidewalks. By investing money in transportation, people can more easily move between their homes and workplaces within the city.
- Smart growth directs development toward existing communities, which helps to invigorate urban neighborhoods that are caught in a vicious cycle of depopulation and blight.
- Smart growth makes development decisions predictable, fair, and cost-effective. This encourages rapid growth of individualized housing and business, which also promotes a sense of place.

(b) The positive feedback loop of urban blight involves several steps. As people leave the city, city revenue from property, sales, and service taxes begins to shrink. Cities are then forced to reduce services, raise taxes, or both. Crime rates begin to increase because police services decrease, and because conditions for low-income residents decline. Infrastructure begins to deteriorate, which makes the city less attractive, and affluent people move away. Jobs and services follow this migration, and commuting patterns develop around cities rather than into and out of them. Urban retail stores begin to lose customers and are forced to close or relocate. As stores close, people have fewer reasons to live in the city, leading to further population reduction.

(c) Negative externalities of park development include any costs that are not included in the actual expense of developing the park. Note that construction and landscaping costs to build the area would not be externalities.

Possible negative externalities would be:

- potential loss of business due to the use of land for recreation instead of industrial or retail development
- increased demand for public services to maintain the area, such as police, landscaping, and sanitation services
- increased wildlife that may result in automobile collisions or health problems

Possible positive externalities of park development would be:

- aesthetic appeal that eases stress and anxiety among citizens

- increased traffic that draws people to local businesses
- increased area appeal that leads to increased growth of residential areas
- festivals held on park grounds that bring nonresidents to the area and expand local business growth
- increased outdoor activity among local residents that decreases health problems and reduces the costs of health care
- increased sense of place and community involvement that increases the development of local businesses

(d) To determine the maximum sustainable yield of a fishery, we must know many different parameters related to the growth and death rates of fish. These parameters include:

- birth rate of fish
- growth rate of fish
- death rate of fish
- diversity of aquatic life
- abundance of resources
- prevalence of disease that might kill living fish
- fertility of fish species
- location of fish
- likelihood of poaching (i.e., how many fish will be caught without permission)
- ecological interactions among fish species that might change with variation in population sizes

2. (a) The decomposition of plant litter is important on farmlands because it recycles the energy and nutrients from dead organic material. After harvest, a substantial amount of plant material still remains above and below ground (e.g., roots). This material contains valuable energy and nutrients for crops that are planted during the next season. If this material did not decompose, farmers would have to apply more fertilizer to their crops.

(b) The use of Bt-positive crops may influence future crop production in several ways. Leaching of toxic enzymes into the soil may inhibit the growth of soil microbes and other decomposers, which will result in reduced decomposition of dead plant material. This will mean slower turnover of energy and nutrients into the soil, and slower growth of future crops. The release of toxic enzymes may also directly inhibit the germination of crops, particularly if the enzymes are exuded through the roots that remain in the ground. Additionally, the use of Bt-positive crops may deter beneficial insects from coming to the farm, which may decrease pollination and other important ecosystem services.

(c) The use of Bt-positive crops may alter the evolution of pest species by causing selection for Bt-resistant strains. Similar problems are generated when farmers spray their crops with pesticides. Most pests die, but a few survive. Those that survive are likely to have a genetic variation that makes them less sensitive to the pesticide or the Bt-toxic enzyme. Survivors will mate and pass that gene variant along to future offspring. Ultimately, this suggests that current methods of pest deterrence will become useless in future years unless scientists develop novel strains of Bt-crops and pesticides.

(d) First, calculate the energy needed to grow 1 kg of nontransgenic potato crop:

1,100 kcal fuel to plant crops

+ 700 kcal of fuel to spray pesticides

+ 200 kcal to harvest = 2,000 kcal

The energy subsidy for nontransgenic potato crop is:

$$\frac{2{,}000 \text{ kcal}}{850 \text{ kcal}} = 2.4$$

Second, calculate the energy needed to grow 1 kg of nontransgenic potato crop:

1,100 kcal fuel to plant crops + 200 kcal to harvest = 1,300 kcal

The energy subsidy for transgenic potato crop is:

$$\frac{1{,}300 \text{ kcal}}{850 \text{ kcal}} = 1.5$$

The difference in energy subsidy between transgenic and nontransgenic potato crop is 2.4 – 1.5 = 0.9.

Chapter 12 Free-Response Questions (p. 429)

2. (a) Enriched uranium-235 is processed into pellets and put into fuel rods. Up to 100 fuel rod assemblies are placed into the reactor core. The uranium undergoes nuclear fission, splitting into other atoms and releasing a great deal of heat. The heat from the water in the reactor core is transferred via a heat exchanger into another water system, where the heat then transforms water into steam. The steam passes into a turbo generator that produces electricity as it spins. The electricity is then passed into the grid system. The rate at which electricity is generated is altered by raising or lifting the control rods, which adjust the amount of radioactive decay taking place in the reactor core.

(b) On March 28, 1979, a partial meltdown occurred in one of the two reactors at Three Mile Island, Pennsylvania. Human error involving a cooling valve led to a loss of coolant in the core, causing it to overheat and release radiation from the plant. On April 26, 1986, an explosion and fire damaged the nuclear power plant at Chernobyl in the Ukraine. The accident was the result of operators conducting an unauthorized test by disconnecting emergency cooling systems and removing the control rods from the core. The core overheated as the radioactive decay process became unstable and out of control, causing an explosion and the graphite control rods to catch fire. Over 30 people lost their lives, and many more may succumb to cancer in the coming years from the radioactive fallout throughout central and western Europe.

(c) Nuclear power plants do not produce air pollution (including carbon dioxide) when they are operating, and so nuclear power can be considered "clean" energy. On the other hand, coal-burning power plants emit sulfur dioxide, nitrogen oxides, and carbon dioxide. The first two are problematic in acid deposition and the latter is a potent greenhouse gas. Additionally, toxic heavy metals such as mercury and arsenic are also released into the environment from burning coal. Although there is some land disruption and pollution from the mining of uranium ore, there is much greater disruption of the land from the mining of coal. Coal mining is also a dangerous occupation; the toll on human life is higher than in other energy industries.

(d) The three types of radioactive waste produced by nuclear power plants are:

- high-level waste from fuel rods
- low-level waste from contaminated protective clothing, tools, rags, and other items used in operating and maintaining the nuclear power plant
- uranium mine tailings, which are the residue left over after the ore is mined and enriched

When humans are exposed to ionizing radiation, there is a chance they will develop cellular mutations that can lead to cancer. The higher the level of exposure, the greater the risk. Exposure to a very high level of radiation can result in acute radiation poisoning and death. Exposure to lower levels of radiation over a longer time period can also result in cancers and other detrimental health effects.

(e) It often takes a long time for radioactive isotopes to decay to a level where they are considered safe to handle and where they pose a minimal risk to humans. High-level radioactive wastes can take many thousands of years to decay to a safe level, due to their long half-lives. Some high-level radioactive wastes may have to be stored for over 100,000 years before they will no longer be considered dangerous.

Because radioactive waste cannot be incinerated or made inert through chemical reactions, other solutions have to be investigated for its disposal. Many suggestions about how to dispose of radioactive waste have been put forward. Suggestions have included sending it into space, dumping it into the ocean, or burying it in a deep ocean trench. All of these pose threats to the environment. Until recently, the United States had proposed the long-term underground storage of radioactive waste at a site at Yucca Mountain, Nevada. This site would have been the repository for the country's spent nuclear fuel and other high-level waste. The waste would have been stored deep underground in specialized vaults. At present, the Yucca Mountain plan has been canceled.

Chapter 13 Free-Response Questions (p. 471)

2. (a) The total amount of electricity generated by photovoltaic solar cells has fluctuated around a mean between 2002 and 2007. Some years there was a decrease (2002–2003; 2004–2005; 2005–2006), while others show an increase (2003–2004; 2006–2007).
The approximate change from 2002–2007 is:

$$\frac{(600{,}000 \times 10^3 \text{ kWh}) - (550{,}000 \times 10^3 \text{ kWh})}{550{,}000 \times 10^3 \text{ kWh}}$$

$$= \frac{50{,}000}{550{,}000} = \frac{1}{11} \times 100 = 9\%$$

The use of photovoltaic solar cells to generate electricity has increased by around 9 percent from 2002 to 2007.

(b) The total amount of electricity generated by wind has gradually increased between 2002 and 2007. The largest single yearly increase in terms of thousand MWh took place between 2005 and 2006 ($8{,}779 \times 10^3$ MWh). The approximate change from 2002 to 2007 is:

$$\frac{(32{,}000{,}000 \times 10^3 \text{ kWh}) - (10{,}000{,}000 \times 10^3 \text{ kWh})}{10{,}000{,}000 \times 10^3 \text{ kWh}}$$

$$= \frac{22{,}000{,}000}{10{,}000{,}000} = 2.2 \times 100 = 220\%$$

The use of wind to generate electricity has increased by around 220 percent from 2002 to 2007. In other words, the amount of electricity generated by wind has more than tripled over the 5-year period.

(c) Years ago, many areas used windmills as a source of energy to grind corn or pump water. Obtaining energy from wind is making a comeback and is widely used to generate electricity in the United States, where the generating capacity of wind power is the largest in the world. Wind is the fastest growing major source of electricity in the world. Because prices for installation and maintenance make the cost of generating electricity from wind power cost-effective, it is not surprising that the amount of electricity generated by wind has more than doubled from 2002 to 2007. However, solar power has seen only modest growth. Although photovoltaic cells have declined in cost in recent years, the initial installation cost remains a drawback, and it may take many years to attain financial payback from an installed PV system. This also makes the cost per kWh of electricity generated by photovoltaic solar cells higher than some other renewable sources. Solar power also shows marked variations in cost around the country. Perhaps when the price of PV cells drops further, pushing down the price of installation, then we may see a greater percentage increase in PV cell usage.

(d) Photovoltaic solar cells
Argument for: PV cell panels can be installed on the roof or can be freestanding on the side of a house that faces south. Solar power is sustainable, renewable, and free. There is no air or water pollution from the use of solar panels, and when they are installed and are operating they do not emit carbon dioxide. They work best on sunny days when demand for electricity is at its highest. There is no cost involved after initial installation, and tax credits may offset the overall cost. Excess electricity generation could be sold to the utility company.

Argument against: Nevada may be too hot for the optimal operation of a PV device. The cost to install a PV system is high and it may take many years to repay the initial investment. Overall, the cost per kWh of generated electricity is higher than that produced by wind. The production of PV cells has an environmental impact such as the high use of metals and water in the manufacturing process. Also the PV cells cannot be recycled, and can only be used during the day; batteries are needed to store electricity for nighttime use.

Wind
Argument for: Wind energy is sustainable, renewable, and free. There are low upfront costs for a home wind turbine, which can be mounted on the roof or be freestanding. There is no cost after the initial investment, and tax credits may be available. Initial investment can be recouped after a few years. No air or water pollution is emitted from a wind turbine. The overall cost per kWh of generated electricity is lower than that produced by PV cells. Electricity can be generated both during the day and at night.

Argument against: Wind generators make noise, which may be annoying to the homeowner and/or the neighbors. The blades may cause the death of birds or bats. The smaller wind turbines designed for home use may be unattractive.

(e) Energy conservation simply means using less energy (e.g., by turning off lights or not using the car as much). Energy efficiency means getting the same amount or more of work from a lesser input of energy. By installing either a photovoltaic or wind system, the homeowner would be making improvements in the energy efficiency of the home. There would be less electricity being used from the grid where energy losses occur in many steps of both the generation and transmission processes. Generating electricity at the site of use and from renewable sources such as the Sun or wind is more energy efficient than using electricity generated from nonrenewable fossil fuels in a polluting power plant many miles away.

Unit 6 Free-Response Questions (p. 475)

1. (a) There are several reason why hybrid and electric vehicles might be less energy efficient than gas-powered cars:

- The process of mining and conversion of fossil fuel into electricity is only about 35 percent efficient, which severely reduces the overall efficiency of an electric-powered car.

- Hybrid and electric vehicles require more expensive parts and may involve a more labor-intensive construction process. The energy used to build the car must be factored into the overall efficiency of the car.
- Hybrid and electric vehicles may require more frequent replacement of engine components, which require energy to produce and manufacture.
- Hybrid and electric vehicles may encourage more driving due to their low fuel costs, which would quickly eliminate any savings in energy efficiency.
- Batteries and other parts of hybrid and electric vehicles may require disposal or may not be recyclable. If the batteries or other parts last a shorter time than those of a conventional car, it will take more energy to run the electric or hybrid vehicle.
- Hybrid and electric vehicles may get fewer miles over the lifetime of the vehicle than gas-powered cars; because the cars may need to be replaced more often, the energy needed to produce new cars may increase.

(b) Cars require energy to build, operate, and recycle. If the total lifetime of a hybrid or electric car is shorter than that of the average conventional car, then the owner of the vehicle will generate higher energy demand to recycle and build new cars. In addition, parts such as batteries may not be fully recyclable, and could harm the environment if not disposed of properly.

(c) (i) Since the new vehicle costs \$40,000 and you received \$2,000 for your old vehicle, you must save \$38,000 in fuel costs to make your purchase cost-effective. With the price of gas at \$4.00 per 20 MJ, you can determine the total number of MJ that \$38,000 can buy:

$$\$38{,}000 \times \frac{20\text{ MJ}}{\$4} = 190{,}000\text{ MJ}$$

You save $4 - 1.5 = 2.5$ MJ per passenger-mile driven. So, the purchase becomes cost-effective after $190{,}000\text{ MJ} \div 2.5\text{ MJ} = 76{,}000$ miles.

(ii) A smart grid coordinates energy use with energy availability. Smart grids are efficient, self-regulating electricity distribution networks that accept any source of electricity and distribute it automatically to end users. End users have computers that coordinate with the distribution network to reduce energy use during peak demand, and increase energy use during times of low demand when there is excess energy capacity. By using smart grid technology with electric vehicles, owners could increase the efficiency of energy production and use. Such increased efficiency could reduce the ecological footprint of the vehicles by decreasing the amount of fossil fuel used to produce energy and reducing the number of new energy production plants that must be built.

2. (a) Any activity that does not change atmospheric CO_2 concentrations is referred to as carbon neutral. Although the combustion of fuels created by algae does release CO_2 gas into the atmosphere, growth of algae also sequesters CO_2 gas. In theory, if production and use of fuels derived from algae remain constant over time, then it is carbon neutral.

(b) (i) The second law of thermodynamics tells us that when energy is transformed, the quantity of energy never increases, and that its ability to do work diminishes.

(ii) The second law of thermodynamics suggests that the during the conversion of CO_2 into energy via plant photosynthesis some of that CO_2 is lost to plant respiration and metabolism. Although the production of fuel by algae is one of the most sustainable options for energy production, over many decades there will still be a net loss of energy. To compensate, we will have to grow slightly more algae each year, or reduce our energy demand.

(c) Similarities between the production of corn and algae biofuel include the following:

- Both are potentially renewable fuels.
- Both fuel sources have the potential to reduce our dependence on fossil fuels.
- Both fuel sources are potentially more environmentally friendly than fossil fuels.
- Both fuel sources require vast amounts of area for production.
- Both fuel sources emit CO_2 when combusted.
- Both require substantial amounts of water and sunlight.

Differences between the production of corn and algae biofuel include the following:

- Production of algae biofuels is not likely to increase the cost of food, unlike production of corn ethanol.
- Production of algae biofuels does not require the use of cropland, and they may be grown on marginal lands, in brackish water, or even on rooftops.
- Algae may produce 15 to 300 times more fuel per area than conventional crops.
- Algae grows very quickly and may be grown year-round, unlike corn, which has a relatively slower growth rate and is an annual crop.

(d) To grow and reproduce, algae require water, sunlight, and CO_2. The best location for an algae biofuel generating plant would be a location where all of these components are found in high abundance. One good starting point would be to place biofuel plants next to energy plants that combust fossil fuels, which emits CO_2 that the algae can use. However, if the location of these plants is not near a source of fresh water, then water will have to be pumped to the plant. Pumping water will reduce the efficiency of algae biofuel production, as well as deplete valuable water resources used for crop irrigation and human consumption. Hence, the best place for an algae biofuel plant would be next to an energy plant that emits fossil fuels and is located in a sunny region with a plentiful water supply.

(e) Although genetically engineered algae pose several risks to the environment, there are ways to mitigate potential harm:

- Biologists can engineer algal strains to be less competitive than natural strains, so that engineered algae would die in natural environments.
- Saltwater algae could be grown separately from purely freshwater environments, so that the release of an engineered strain would not survive in surrounding environments. Alternatively, freshwater algae could be grown around purely saltwater environments (e.g., offshore).
- Algae could be grown in indoor bioreactors to reduce the risk that strains would be released. However, this would require vast amounts of energy to generate light and building materials.

Chapter 14 Free-Response Questions (p. 515)

1. (a) Crab population in 1990–1997 was approximately 650 million.

Crab population in 1998–2009 was approximately 330 million.

Crab population in 2010–2013 was approximately 550 million.

Between the 1990–1997 period and the 1998–2009 period, the crab population fell by 650 million − 330 million = 320 million.

Between the 1998–2009 period and the 2010–2013 period, the crab population rose by 550 Million − 330 million = 220 million.

The average population of blue crab for 2010–2020 will be greater than the 1998–2009 average for the following reasons:

- The total population is increasing and the number of sexually mature crabs is rising.
- As water pollution sources are controlled, the water quality of the Chesapeake Bay will improve; as water quality improves, the blue crab survivorship will most likely improve as well.

(b) Possible answers may include the following:

- The introduction of the nutrients nitrogen and phosphorus causes an algal bloom; huge numbers of algae die and are decomposed by aerobic bacteria, which reduces the amount of dissolved oxygen in the water. Blue crabs are aquatic organisms that depend on dissolved oxygen to survive. Eutrophication causes a decline in the number of blue crabs.
- The introduction of sediments may have caused a decline in the number of blue crabs. The smaller sediment particles remain suspended in the water, making it cloudy and reducing the amount of sunlight reaching the grasses in the bay. This causes a decline in the population of the grasses, which are an important habitat for the blue crabs, and therefore causes a decline in the population of the crabs as well. Also, sediments can clog the gills of the crabs and lead to their eventual suffocation.
- Pesticides introduced into the Chesapeake Bay could be directly toxic to blue crabs, or they could enter the aquatic food chain and later end up in the blue crabs.
- Pharmaceutical drugs introduced into the bay could be toxic to crabs or could affect the sexual development of crabs, as is the case with largemouth and smallmouth bass.
- The introduction of heavy metals may have caused a decline in the number of blue crabs. Heavy metals can bioaccumulate in crabs, especially in their muscle tissue, and can thereby reduce their survivability.
- Thermal pollution could be harmful to the crab population in several ways. Thermal shock could result if the increase in temperature is above the tolerance level of the blue crab. Higher temperatures result in lower levels of dissolved oxygen, which could stress, weaken, or even kill the blue crabs. Higher temperatures also facilitate the growth of a dinoflagellate, Pfiesteria, which occurs naturally in the coastal waters of the east coast (outbreaks occur more frequently at 75° latitude or higher) and releases a toxin that is harmful or fatal to blue crabs.

(c) The method of management or control must be appropriate for the factor selected. Passing laws or levying fines are not acceptable answers, since they are regulatory and do not specifically explain how the source of pollution would be managed or controlled. Answers may include:

- The introduction of nitrogen and phosphorus could be managed by constructing manure lagoons in animal feedlots, creating a wetland where denitrifying bacteria can remove nitrogen from the water, or by adding tertiary treatment to remove nitrogen and phosphorus from the wastewater effluent of sewage treatment plants.
- The introduction of sediments could be controlled by planting vegetation to curtail erosion, or by installing barriers such as hay bales or stones to curtail erosion.
- The harmful effects of pesticides could be reduced by using biological or mechanical methods of controlling pests instead of chemicals, or by using chemicals that quickly degrade to harmless substances.
- The introduction of pharmaceutical drugs could be managed by adding tertiary treatment to remove pharmaceuticals from the wastewater effluent of sewage treatment plants.
- The introduction of heavy metals could be controlled by reducing emissions at the source, or by removing the metals by fine membrane filtration, distillation, or reverse osmosis.
- Thermal pollution could be reduced by making sure that water used for cooling during industrial processes is cooled again in holding ponds or cooling towers before it is returned to the source of the water.

(d) The Clean Water Act appropriately applies to the Chesapeake Bay and the blue crabs.

2. (a) Major sources of mercury pollution:

- burning fossil fuels (particularly coal)
- incinerating garbage, hazardous waste, and medical and dental supplies
- heating raw materials that are used to manufacture cement
- escaping wastewater from wells that occurs during petroleum exploration

Means of controlling mercury pollution:

- reducing emissions at the source
- filtration
- reverse osmosis
- granular activated carbon

(b) Mercury emissions are transported through water or air; in lakes and wetlands, bacteria convert inorganic mercury to methylmercury. Aquatic organisms absorb the methylmercury as water passes over their gills and as they feed. It then bioaccumulates in their tissues, reaching concentrations greater than the surrounding water. The concentration of methylmercury increases at each trophic level, resulting in the highest concentrations in the top predators—biomagnification.

(c) Health effects of methylmercury on humans:

- damages the central nervous system, especially in developing fetuses and young children
- impairs coordination
- impairs senses of touch, taste, and sight
- impairs speech
- impairs cognitive abilities

Chapter 15 Free-Response Questions (p. 551)

1. (a) Pittsburgh seems to be in compliance since the NAAQS standard is an average, and all of the monthly average ozone concentrations are below the 75 ppb standard. Since the NAAQS standard is an average, it is possible for the ozone concentrations to exceed the standard and still be in compliance, as evidenced by the monthly maximum ozone concentrations where the standard 75 ppb is exceeded five times.

(b) Primary pollutants necessary for the formation of ozone:

- NO_x
- VOCs (Volatile Organic Compounds)

Formation of ground-level ozone:

- NO_2 + sunlight = NO + O
- NO + O + VOC = photochemical oxidants + O_3

(c) Possible relationships between data:

- The higher the monthly average solar radiation, the higher the monthly ozone concentration (either maximum or average). Or the converse: The lower the monthly average solar radiation, the lower the monthly ozone concentration (either maximum or average).
- The higher the monthly maximum ozone concentration, the higher the monthly average ozone concentration. Or the converse: The lower the monthly maximum ozone concentration, the lower the monthly average ozone concentration.
- The monthly average solar radiation is lower in the fall and winter months and higher in the spring and summer months.
- The ozone concentrations (either maximum or average) are lower in the fall and winter months and higher in the spring and summer months.

Since the source of energy for this photochemical reaction is sunlight, the greater the amount of solar radiation (or the greater the duration of the solar radiation as indicated by seasonally longer or shorter days), the greater the amount of ozone produced. The data seem to indicate that ozone concentrations fluctuate seasonally in Pittsburgh.

(d) Possible explanations may include:

- Tropospheric and stratospheric ozone are chemically identical: At ground level it is a respiratory irritant; in the stratosphere, no humans are exposed.
- In the stratosphere, ozone is the only substance that absorbs the energy from UV-B radiation and prevents most of it from reaching the surface of Earth.
- Harmful effects of UV-B radiation include reduction in photosynthetic activity, reduction in ecosystem productivity, increase in skin cancer, and increase in cataracts and other eye problems.

2. (a) The cause of sick building syndrome, according to the article, is "the increasing use of air conditioning in modern offices. Bacteria can build up in the ventilation system and leave office workers suffering from a range of problems, such as breathing difficulties, headaches, sore throats, stuffy noses and itchy eyes."
Advantages of the cure:

- 20 percent reduction in all symptoms
- 40 percent reduction in complaints of respiratory problems
- 30 percent reduction in complaints of stuffy noses
- 50 percent reduction in muscular complaints
- killed bacteria and mold in ventilation systems
- system is relatively cheap
- reduced worker absences due to building related illnesses
- cost effective

(b) Other sources of indoor pollution that contribute to sick building syndrome:

- Chemical contamination from indoor sources such as glues, carpeting, furniture, cleaning agents, and copy machines
- Chemical contamination in the building from outdoor sources such as vehicle exhaust emitted near the air intakes of the building
- Long-term effects could come from asbestos in older buildings

Ultraviolet light will not be effective against these additional sources.
Additional control measures:

- increased ventilation
- filters in ventilating system (must be properly maintained)
- houseplants can reduce contaminant levels
- a ban on smoking
- sealing cracks in basement (for radon)
- removal of asbestos by qualified asbestos abatement personnel

(c) Ultraviolet light would not be effective in developing countries.
The major source of indoor pollution in developing countries is the burning of biomass and coal for heating and cooking. These fuels are usually burned in open-pit fires that are inefficient and lack the proper mixture of fuel and air for complete combustion. This results in the production of carbon monoxide and particulates, which would not be controlled or eliminated by ultraviolet light.

Chapter 16 Free-Response Questions (p. 587)

1. (a) The waste increased because the population of the United States grew and, as a consequence, so did the amount of trash produced. Another reason is that years ago objects were made to last longer, and when they were no longer needed, they were taken away by scrap metal merchants or junk dealers. Then after the Second World War, industrialization and increased birth rates (often referred to as the baby boom) resulted in a different type of product being developed. These new products were labor-saving devices, and many of them were designed to be discarded or thrown away after use. As time went by, many more of these types of products entered into the lifestyles of the American public. Many of the new disposable products could not be recycled and ended up entering an ever-growing waste stream. America gained the reputation of leading the world as a "throw-away society" due to this increase in affluence.

(b) Reducing the input of materials into the system will most likely reduce the output amount or waste that must be disposed of. This is often referred to as source reduction, where the use, avoidance of use, design, or manufacture of materials will result in smaller amounts of MSW being generated. As less waste is being generated to begin with, there are often associated increases in energy efficiency. Also, fewer resources will be used, which can have economic benefits. Source reduction can be done on an individual basis or within a household, and can be applied equally well at the point of product manufacture. For these reasons, reducing is the most favorable of the 3 Rs.

Once an item is purchased, reusing it rather than discarding it enables the item to remain in the system longer before it becomes an output or part of the waste stream. The longer an item is in the system, the less energy or resources are required for new items, and less waste disposal will occur. Some items (interoffice envelopes, for example) can be reused over and over. Sometimes reusing may involve the expenditure of energy, for example in the reuse of glass milk bottles. Energy is needed to transport, wash, and refill them. Reusing is a common practice throughout the world and still occurs in America at flea markets, swap meets, and online auction sites. Reusing an item for as long as possible is more desirable from an environmental perspective than using it once and throwing it away or recycling it.

Recycling is a way of removing items from the generated MSW. After an item is collected, it can be converted back into raw materials in order to produce new items. Closed-loop recycling is where an item is converted back into the original item that was recycled (for example, aluminum cans into aluminum cans). Open-loop recycling takes the recycled item and converts it into a totally different product, as exemplified by plastic bottles being turned into polar fleece products. This outcome is better than sending the plastic bottles to the landfill, but it does not negate the fact that petroleum has to be used to make new plastic bottles. The United States has embraced recycling over the past few decades. Recycling means that less energy, money, and fewer resources are expended. However, recycling does use more energy than either reducing or reusing.

(c) Most organic material can be removed from the waste stream and decomposed under controlled conditions to produce an organic-rich material (compost) that can be applied to soil to enhance its fertility. Meat and dairy products do not normally get composted, but vegetables and vegetable by-products such as potato peelings, along with grass cuttings, yard waste, animal manure, and paper products, can be treated in this way. The composting material has to be mixed on a regular basis to allow for the aerobic decomposition process to occur optimally and to reduce smells and methane production. The heat that is generated in the decomposition process is sufficient to kill any pathogenic bacteria in the material being composted. With the correct aeration and moisture content, the composting process can occur in a matter of weeks to months. The carbon to nitrogen ratio must be checked to ensure it is optimal for the decomposition process—usually around 30:1.

A home composting system can be very simple. A corner of the yard or garden can be the location of a pile of food scraps or yard waste. Sometimes a compost box or drum can be used. The compost has to be turned and mixed frequently. It takes a couple of months to produce fresh compost in this way. If the home does not have a garden or yard, it is possible to compost materials inside even a small apartment or dorm room. Red wriggler worms can be used to compost kitchen waste in a small worm box. A number of small countertop composters are also available commercially.

Although composting can be effectively accomplished on a small scale, many areas around the United States have established large-scale facilities to serve the whole community. Some of the facilities are indoors, while others are outside. The principle is still the same as the home composter but on a much bigger scale. The organic material is piled up in windrows—long narrow rows—where it is often turned either by rotating blades or a specially equipped tractor. After a few weeks fresh compost is obtained, which is then sold or given away to the people in the community.

2. (a) Open dumps are health hazards wherever they are located. They attract many pests, such as rats and scavenging birds, which can transmit diseases to humans. They can pollute the air and produce a significant odor. They can pollute both groundwater supplies and aboveground streams and rivers. Often open dumps were set alight to burn the trash and reduce its volume, which added to the problems of air pollution. All kinds of trash were left at the dump, including food and yard waste, car tires and batteries, and old appliances such as stoves and refrigerators. Open dumps have been closed down or converted into landfills, which are considered to pose less of a hazard to humans.

(b) Modern landfills are an improvement on the landfills that initially replaced open dumps. The early landfills were not much more than holes in the ground that were filled with trash. Not much thought was given to what entered the landfill, or what might be in the runoff and leachate. The location of the early landfills was not a consideration at the time and many were placed near streams, rivers, drinking water supplies, or where people lived.

Today landfills are high-tech facilities that are referred to as sanitary landfills and are designed to safely dispose of waste and minimize environmental hazards. The location of a sanitary landfill remains a controversial issue. Because people typically prefer not to have landfills close to where they live, they are often located in areas where residents have less political power or financial resources, including densely populated, low-income, minority neighborhoods. However, modern landfills are situated away from rivers and other bodies of water, including sources of drinking water. The soil is ideally rich in clay to minimize the danger from leachate and the site should not be close to centers of population in order to minimize the risks from transporting the waste and attracting pests such as rats and seagulls. After the necessary permits have been issued, which take into account what type of MSW can be deposited, the site is lined with clay and plastic. Monitoring wells are established close by that regularly check on the health of the aquifer. Leachate removal systems are built into the landfill to minimize contamination. Depending on the needs of the community, certain materials may be restricted for disposal in the landfill. Generally, toxic materials, car batteries, car tires, and other harmful substances are no longer allowed in landfills. Organic materials such as yard wastes may be restricted since decomposition can release methane. A methane collection system can pipe the gas either for useful combustion at a nearby power plant or vent it into the atmosphere to prevent explosions. At the end of each day the landfill cells are covered with soil to reduce water percolation.

(c) Benefits include:

- less odor; no open dumps or trash burning
- minimum groundwater pollution when correctly sited and built
- lower operating costs and, hence, lower tipping fees than at an incinerator
- building it can be quicker than constructing an incinerator
- large amounts of MSW can be processed
- methane can be collected and used to generate heat/electricity
- once filled and capped, landfill can be used for other purposes

Problems include:

- leachate leaks could pollute groundwater or other bodies of water
- waste-reduction strategies are not encouraged
- methane, carbon dioxide, hydrogen sulfide are generated
- noise, dust, air pollution from trucks transporting waste
- slow rate of decomposition

(d) MSW contains easily combustible materials. The MSW is trucked into the facility and dumped into a pit. A crane lifts the waste into the furnace. Some incinerators presort the waste, while others burn everything in the waste; the latter are known as mass-burn incinerators. Most incineration facilities remove metals from the waste stream before transfer to the furnace occurs. Bottom ash from the furnace is removed and tested for toxicity. Based on the results, it will be transferred to one of a variety of landfills. The hot gas from the combustion process passes out of the furnace. If the incinerator is equipped with a waste-to-energy system, the gas will pass around a boiler that produces steam, driving a turbo generator and producing electricity. The exhaust gas then continues on through a scrubber and electrostatic precipitator system; the residual ash (fly ash) is removed and taken to a landfill and the exhaust gas emitted to the atmosphere through the smokestack.

(e) Benefits include:

- less residual volume (ash) goes into the landfill, extending the life of landfills or minimizing construction of new landfills
- less water pollution than that generated by landfills
- recovery of resources such as metals before incineration in some facilities
- hazardous/toxic materials can be concentrated into the ash

- air pollution devices installed onto the exit gas smokestack can prevent or reduce emissions
- some incinerators can produce electricity from burning trash

Problems include:

- discourages local waste reduction strategies
- reduces amount of recycling
- expensive to build; siting a problems due to NIMBY
- higher tipping fees than landfills; may have to haul the trash further distances
- some air pollution, posing a particular problem at older facilities without modern scrubbers

Unit 7 Free-Response Questions (p. 626)

1. (a) Sulfur gases such as sulfur dioxide are reactive with organic tissues and atmospheric compounds, which creates several environmental and health risks. Sulfur gases emitted into the atmosphere can:

- create photochemical smog following the breakdown of sulfur dioxide in the atmosphere
- generate acid rain following interaction of sulfur dioxide with atmospheric oxygen and water; this interaction produces sulfate, which can produce photochemical smog
- lead to human respiratory ailments such as asthma
- reduce plant health by damaging stomates and other plant tissues

(b) Leakage or overspraying of manure from lagoons poses several environmental risks to surrounding soils, streams, and lakes:

- When manure leaks from a lagoon, harmful microbial diseases may also leak into the environment. These diseases can infiltrate groundwater used by humans and harm the biota of soils, streams, and lakes.
- Excessive spraying of manure can lead to runoff of nutrients into nearby bodies of water, which may generate algal blooms that harm aquatic biota.
- Leakage of manure can increase the nutrient content of soil. In turn, this can increase the nutrient concentrations of nearby streams or lakes and lead to algal blooms that can harm aquatic biota.
- Antibiotics that are given to livestock may be present in manure lagoons; leakage releases these antibiotics into the environment where they can harm microflora and larger organisms.
- Overspraying may release airborne pathogens that can pose health risks.

(c) Retrospective studies monitor organisms that have been exposed to a treatment (e.g., chemical, pollutant, disease) at some time in the past, and compare them to organisms that have not been exposed to the treatment. Hence, a retrospective study to determine the effect of manure lagoons on nearby residents would compare the health of residents near a manure lagoon to the health of residents far from a manure lagoon. Ideally, the two groups of people will reside in areas with a similar environment, except for the presence of a manure lagoon. Because this is nearly impossible, one should compare the average health of residents from several communities with and without manure lagoons.

(d) To determine if there is leakage in the soil surrounding a manure lagoon, researchers can:

- monitor soil samples for the presence of indicator species such as *E. coli*; high levels of *E. coli* might indicate leakage
- monitor soil samples for increased nitrogen or phosphorus levels; high levels of these nutrients might indicate leakage
- monitor the abundance of bacteria in soil surrounding the lagoon (If the abundance of bacteria changes over time, this might be an indication that there is leakage. However, it would be ideal to compare soil samples to long-term records, since bacterial abundance naturally changes through time.)
- monitor changes in plant growth over time (An increase in plant growth may indicate higher nutrient levels in the soil due to a leakage. In contrast, a decrease in plant growth may indicate the presence of contaminants from a lagoon leak.)

(e) The Clean Water Act of 1972 supports the "protection and propagation of fish, shellfish, and wildlife and recreation in and on the water" by maintaining and, when necessary, restoring the chemical, physical, and biological properties of surface waters. Although this act does not protect groundwater, the Safe Drinking Water Acts of 1974, 1986, and 1996 establish maximum levels for many contaminants in both surface water and groundwater.

2. (a) The generation of smog is a multistep process that requires the presence of sunlight and several chemicals. When there is an abundance of nitrogen oxides, nitrogen dioxide splits to form nitrogen oxide (NO) and a free oxygen atom (O). In the presence of sunlight, O combines with diatomic oxygen (O_2) to form ozone (O_3). When volatile organic compounds (VOCs) are low in abundance and as sunlight intensity decreases, O_3 combines with NO and re-forms into $NO_2 + O_2$. However, when VOCs are high in abundance, the NO binds to VOCs instead of O_3. As a result, O_3 accumulates in the atmosphere as photochemical smog. Backyard incinerators can potentially release substantial amounts of VOCs and other pollutants, which will decrease the abundance of atmospheric NO that can bind with O_3. In turn, the use of backyard incinerators can lead to an increase in photochemical smog.

(b) Sunlight provides the energy to combine O and O_2 into O_3. Hence, the highest accumulation rate of O_3 occurs during the daytime. By limiting backyard incineration to between the hours of 4 AM and 7 AM, VOCs may dissipate before production of O_3 begins. As a result, O_3 can combine with NO to form $NO_2 + O_2$ during the daytime.

(c) During a thermal inversion, a relatively warm layer of air at mid-altitude covers a layer of cold, dense air below it. The warm layer of air trapped between the two cooler layers is known as an inversion layer. Because the air closest to the surface of Earth is denser than the air above it, the cool air and the pollutants within it do not rise. Thus, the inversion layer traps VOCs that then accumulate beneath it. The trapped VOCs can severely increase the amount of smog by leading to an accumulation of O_3.

(d) Ash from incinerators can contain toxic compounds and high concentrations of metals, such as cadmium and lead, that are toxic to the wildlife. If ash is not properly disposed of, these toxic compounds can leach into the environment following rainfall. Hence, placing incinerator ash with toxic metals into gardens or landfills can lead to contamination of groundwater and may pose a hazard to human health and wildlife.

(e) Bottom ash is the ash that accumulates underneath an incinerator. Fly ash is the ash that is collected beyond the furnace in filters or other collection devices. Modern incinerators monitor the concentration of contaminants in both types of ash. If the ash has a low concentration of contaminants, then it can be disposed of in a landfill. Such ash can also be used for other purposes such as fill in road construction or as an ingredient in cement blocks and cement flooring. If the ash has a high concentration of contaminants, it is transported to a landfill designed for toxic substances. Due to the high cost of disposal, some hazardous ash is transported and sold to other countries with less stringent disposal regulations. These countries can potentially harvest the metals or use the ash in construction materials, but with high environmental and human health risks.

Science Applied 7 Free-Response Question (p. 629)

(a) There are many factors that can increase or decrease the environmental benefits of recycling plastic cups:

- The environmental benefits of using recyclable plastic cups increase if the cups are reused multiple times during the party. However, if cups are used once and discarded—followed by the use of more cups—the environmental benefits decrease.
- Using water to wash the cups before their disposal can decrease their environmental benefits. However, if the cups are washed and reused, this may increase their environmental benefits.
- Overall, recyclable cups have a shorter lifetime than glasses. Hence, more of them must be made, which requires energy and may decrease their environmental benefit.
- The energy and chemicals used in the recycling process may harm the environment and decrease their environmental benefit.
- Pollutants may also be generated during the recycling process that can harm the environment.
- Plastic cups are lighter than glass cups, so they may require less energy for transport than glass cups. However, more plastic cups must be transported, which will increase their overall weight and the energy needed to transport them.
- It is very likely that many people will not know that the cups can be recycled, so they will end up in the trash. This severely decreases the environmental benefits of recyclable plastic cups.
- There may be a substantial cost in transporting the cups to a recycling center if the center is far from the party. This will consume energy that may decrease the environmental benefits of using recyclable plastic cups.
- Sorting the recyclable cups from other types of recyclable material requires time and energy, which decreases their environmental benefit.
- The use of recyclable cups may decrease the amount of trash in landfills, which increases their environmental benefit.

(b) Overall, using recyclable cups may be less environmentally costly than using glass cups for several reasons:

- Glass cups are likely to break at a party and end up in a landfill. In contrast, plastic cups are less likely to break and—if disposed of properly—will always end up as recycled material.
- Glass cups may require more energy to manufacture than plastic cups.
- The production of glass cups may produce more pollutants than the production of plastic cups.
- Glass cups are heavier and require more fuel to transport than plastic cups.
- Glass cups will need to be washed after use, which will require the use of water and sanitization of that water.

(c) Composting any material requires oxygen for microbes. When oxo-biodegradable material is composted, the compost pile must be mixed and aerated multiple times to ensure complete decomposition. When oxo-biodegradable material ends up in a landfill, it often is buried under tons of trash. As a result, bacteria are deprived of oxygen and the material will not decompose. In fact, oxo-biodegradable material may persist as long as conventional plastics in a landfill.

(d) Methods to reduce waste at a party can include any strategy that encourages partygoers to use less disposable material. For example, persons can be encouraged to mark their disposable cups so that they do not lose them. You might also encourage the use of glassware and other reusable material. You could also request that people attending the party bring their own cups, so that they are more likely to save and reuse them through the night.

Methods to reuse waste at a party can include any strategy that encourages attendees to continue using material throughout the night without throwing anything away. Reusing waste can also be accomplished by upcycling material for other purposes, such as using plastic cups as planters after the party. Composting is a form of reusing that converts paper products and other compostable material into inorganic material for plant growth.

Methods for recycling waste at a party can include any strategy that encourages attendees to properly dispose of recyclable materials. This could be accomplished by creating a separate waste bin for recyclable materials. Since there is often confusion regarding which materials can be recycled, one could also place pictures or descriptions of recyclable materials next to the waste bin.

Chapter 19 Free-Response Questions (p. 699)

1. (a) Students could point out that CO_2 concentrations are today at unprecedented levels for the past 400,000 years, that 9 of the 10 hottest years on record have been during the past decade, that ice caps are melting, that the permafrost is thawing, and that the seasonal timing of events for many plants and animals has become earlier.

(b) Students might explain that when scientists look at all the data, they find that the patterns in temperature change are strongly consistent with increased greenhouse gases such as CO_2 and are not consistent with increased solar radiation. This body of evidence led the IPCC to conclude in 2007 that most of the observed increase in global average temperatures since the mid-twentieth century is the result of an increase in anthropogenic greenhouse gas concentrations.

(c) Climate change has the potential to affect human health. Continued warming of the planet could affect the geographic distribution of temperature-limited disease vectors. The mosquitoes that carry West Nile virus and malaria, for example, could expand the geographic range over which they live, introducing novel health threats to regions that were once relatively untouched. Heat waves could cause more deaths to the very young, the very old, and those without access to air conditioning. Infectious diseases and bacterial and fungal illnesses might extend over a wider range than at present. Climate change will also have economic consequences. In northern locations, for example, warmer temperatures and shorter winters may sound appealing at first, with reduced heating bills and less damage from winter storms. However, they would alter the character of northern communities that depend on snow for tourism, such as ski resorts. In the Swiss Alps, for example, many ski resorts are already adjusting to reduced snow on the mountains by catering to new groups of tourists who are more interested in warmer weather activities. In warmer regions, the damage to coral reefs would negatively affect tourism as well. The economic impact on these types of tourist attractions depends on the rate of climate change and the ability of entrepreneurs to adjust so they can continue attracting tourists.

2. (a) Students might talk about reducing carbon emissions by using more energy-efficient technologies and by having more fuel-efficient cars.

(b) The major producers of methane in the United States are livestock, landfills, and the production of natural gas and petroleum products. Possible examples of how to lower these methane levels would be to reduce meat consumption, follow the three Rs (reduce, reuse, and recycle), lower the amount of waste being added to landfills, and switch to alternative fuels.

(c) The major contributors of nitrous oxide are nitrogen from synthetic fertilizers, applications of manure as an organic fertilizer, and growing of nitrogen-fixing crops such as alfalfa. Agricultural fields that are overirrigated or deliberately flooded for cultivating crops such as rice create low-oxygen environments that can also produce nitrous oxide by the process of denitrification. With this information, students might discuss changing farming practices and switching to crops that produce less nitrous oxide and methane.

(d) One commonly used biological measurement has been the change in species composition of a small protist, called foraminifera, over millions of years. The foraminifera are tiny marine protists that have hard shells resistant to decay after the organism dies. In some regions of the ocean floor, the tiny shells have been building up in sediments for millions of years. Because different species of foraminifera prefer different water temperatures, identifying the predominant species of foraminifera in a layer of sediment can indicate the ocean temperature when each layer of sediment was deposited.

A commonly used physical measurement is the examination of ancient ice. In cold areas such as Antarctica and the top of the Himalayas, snow falls each year and eventually packs down to become ice. When it does so, small bubbles of air are captured in the ice. These bubbles contain tiny samples of the atmosphere that existed at the time the ice was formed. Scientists can drill deep into the ice and extract ice cores that represent up to 500,000 years of ice formation. By knowing the age of different ice depths, one can remove pieces of ice from different depths, melt the layer, and measure the concentration of CO_2 that was in the trapped air bubbles. The ice cores can also be used to estimate past temperatures by examining the oxygen atoms contained in the water of the melted ice. Oxygen atoms occur in two forms or isotopes: an atom of light oxygen contains 8 neutrons whereas an atom of heavy oxygen contains 10 neutrons. Scientists have known for some time that ice formed during warmer temperatures contains a higher percentage of heavy oxygen whereas ice formed during colder temperatures contains a lower percentage of heavy oxygen. By examining changes in the percentage of heavy oxygen atoms from different layers of the ice core, we can indirectly estimate the temperatures that existed for hundreds of thousands of years in the past.

Chapter 20 Free-Response Questions (p. 726)

1. (a) 138,000,000,000 gallons of gasoline × \$4.00 per gallon = \$552,000,000,000

(b) (i) \$0.184 tax per gallon ÷ \$4.00 cost per gallon × 100 = 4.60%

(ii) 138,000,000,000 gallons of gasoline × \$0.184 tax per gallon = \$25,392,000,000 in 2008 gasoline tax revenue

OR

0.046 × \$552,000,000,000 = \$25,392,000,000 in 2008 gasoline tax revenue

(iii) 0.80 × \$25,392,000,000 = \$20,313,600,000 spent to subsidize road construction

(c) No, the federal gasoline tax does not qualify as a green tax because it is primarily used to subsidize road construction. This actually encourages more driving, which results in an increase in the generation of carbon dioxide and nitrous oxide. Additionally, the natural habitats of the land are disrupted when roadways are constructed. A green tax by definition is an attempt to internalize these negative externalities; the gasoline tax is not channeled to pay for mitigation of land use or for the air pollution generated by driving.

(d) Possible economic effects:

- With an increase in gasoline prices due to an increased tax, consumers might use less gasoline, thereby reducing the total annual revenues to the oil industry.
- Reduced use by consumers also reduces the tax revenues and subsidies for road construction, thereby reducing the number of people employed in road construction and the materials purchased.
- With an increase in gasoline prices due to an increased tax, consumers might be inclined to purchase an automobile with higher fuel efficiency; this could economically increase the income of manufacturers of automobiles with higher gas mileage and reduce the income of manufacturers of automobiles with lower gas mileage.
- An increase in gasoline prices due to an increased tax might encourage consumers to use mass transit if such an alternative is available, therefore reducing the number of automobile sales and services.
- Increased costs may encourage the development of alternative fuels, thereby reducing the profits of the oil industry and increasing the profits for alternative energy companies.
- If gasoline consumption is not reduced, then tax revenues would be increased, as would the subsidies for road construction. This in turn would mean more employment and more materials used for the ensuing road construction.
- Such a tax tends to fall most heavily on individuals who have lower incomes and those who must drive long distances to get to work or to other activities.

Possible environmental effects:

- An increase in gasoline prices due to an increased tax might encourage consumers to drive less, resulting in a decrease in carbon dioxide and nitrogen oxide emissions.
- An increase in gasoline prices due to an increased tax might result in greater use of mass transit, thereby reducing automobile sales and the mining of raw materials and the disruption of the land needed to produce the automobiles.
- With fewer automobiles there would be fewer used cars entering the solid waste stream.
- An increase in gasoline prices due to an increased tax might encourage the development of alternative fuels that would decrease greenhouse gas emissions and reduce global warming. It might also increase the development of renewable energy sources and new means of transportation.
- An increase in gasoline prices due to an increased tax might encourage consumers to drive less, decreasing the amount of gasoline used. This in turn might reduce the amount of oil drilling (offshore and otherwise) and the resulting disruption of both land and sea.
- The demand for gasoline at present is relatively inelastic, so demand for more fuel-efficient cars, hybrid-electric cars, and, possibly, electric cars may increase. These have environmental impacts as well.

2. (a) Ecological services are those services provided by ecosystems throughout the world that benefit human beings either directly or indirectly. There are many possible examples, including the following:

- pollination by bees
- soil adsorption and filtration of chemicals
- detritivores' turning animal waste into organic material used by primary producers

The United Nations Environment Programme (UNEP) is a program of the United Nations responsible for gathering environmental information, conducting research, and assessing environmental problems.

(b) Valuation is a method by which a monetary value is placed on intangible benefits and on natural capital and physical entities provided by nature. Based on the above figures, the world could not afford these services; at the low end of the cost range ($16 trillion) the world could afford them and nothing else. Using the average figure of $33 trillion, the world would amass a debt of $15 trillion ($33 trillion to $18 trillion) per year for these services alone.

(c) A sustainable economic system would rely more on ecosystem services and less on resource extraction; it would also use more renewable energy sources, lessen negative externalities, and reuse more of the products destined to enter the solid waste stream. The details of this report would assist planners in identifying ecosystem services and natural capital to achieve a sustainable economic system.

(d) Environmental economics addresses the failure of free-market systems to allocate resources efficiently to improve the welfare of the greatest number of people. The anthropocentric worldview is most consistent with environmental economics because it maintains that human beings have intrinsic value and that nature has an instrumental value to provide for our needs. It also recognizes that resource depletion issues can be managed through technology and that at times nature requires protection that can be provided through the government.

Unit 8 Free-Response Questions (p. 729)

1. (a) Either the anthropocentric or biocentric worldview advocates for allowing tribes to persist on their land. The anthropocentric worldview focuses on human welfare and well-being, and it suggests that nature has an instrumental value to provide for our needs. Different schools of thought within this worldview suggest ways that humans can manage Earth for our needs. The biocentric worldview holds that humans are just one of many species on Earth, all of which have equal intrinsic value. By this reasoning, humans have as much intrinsic value as all other species, and all should be protected.

(b) According to the concept of SLOSS (single large or several small), patches of terrestrial land can act as islands for the movement and maintenance of species. As islands of land, they follow the principles of island biogeography. Large islands typically harbor more species than small islands. Also, individuals can more easily migrate to patches of land that are larger and closer to other islands. Hence, it may not be necessary to create a continuous patch of land as a corridor between North and South America; instead we can create a series of terrestrial islands that allow easy migration of species. Ideally, this patchwork of islands will include several large islands of land surrounded by many small islands. Unfortunately, this will not allow all species to migrate; species with limited movement may find it difficult to migrate between patches. However, species that can fly, jump, or be carried by the wind will benefit from a patchwork of terrestrial islands.

(c) Several types of ecosystem services can be used to monitor the success of protected land. Note that provisional services are not likely to be monitored, since humans would not be allowed to extract material from protected land.

- Regulating services are those that regulate environmental conditions. For example, functional ecosystems absorb atmospheric carbon and nutrients. To monitor the success of the corridor, we could monitor the amount of carbon and nutrient uptake in various parts of the corridor.
- Support services are those that support human activities, such as pollination, pest reduction, and pathogen removal. It is possible to monitor these services in lands surrounding protected parts of the corridor. For example, if pollination rates increase in croplands surrounding the corridor, this may be a sign that pollinators are increasing in abundance and diversity within the corridor.
- Resilience is a service that benefits both humans and natural ecosystems, and it is a measure of how susceptible the system is to disturbance. Resilience depends greatly on species diversity; systems with more species are typically more resilient. It is possible to monitor resilience of a system by monitoring production within the system before and after a major disturbance (e.g., a hurricane).
- Cultural services provide cultural or aesthetic benefits to people. Lands that harbor a greater number of species and a more pristine landscape are likely to provide more of these benefits. It may be possible to monitor cultural services by documenting changes in tourism, park revenues, and human health in areas surrounding protected land.

(d) The IUCN Red List is a list of threatened species around the world. The list is maintained by the International Union for Conservation of Nature. If threatened species increase in abundance within protected land along the corridor, it is likely that the corridor is functioning properly.

2. (a) (i) Following changes in global temperature, migrating butterflies are likely to be displaced from their normal range. Temperatures in their normal, southern resting spots may become too warm and they will be forced to find higher-latitude resting spots. Similarly, temperatures in northern feeding and breeding spots may become sufficiently warm so that butterflies can remain there year-round. In addition, the flowers on which a species relies for food and breeding are also likely to shift in range with global warming. Consequently, as temperatures warm, migrating butterflies may not find their normal source of food or habitat. This could decrease survival and the number of offspring. Alternatively, the migration patterns of the butterflies could shift northward with the range shift of the flowers, resulting in no overall harm to the butterflies.

(ii) To cope with changes in global climate, migrating butterflies may begin remaining at northern latitudes year-round. Alternatively, migrating butterflies may simply shift their normal migration locations northward with rising temperatures. Additionally, migrating butterflies may seek out new food sources as novel plants begin moving into migration territories.

(b) In cold areas (e.g., Antarctica), the yearly snowfall compresses into ice layers. During the process of compression, the ice captures small air bubbles that contain tiny samples of the atmosphere that exist at the time the ice is formed. Samples of ice cores can span up to 500,000 years of ice formation. Scientists determine the age of different layers in the ice core and then melt the ice from a piece associated with a particular time period. When the piece of ice melts, air bubbles are released and scientists measure the concentration of greenhouse gases in the air when the bubbles were formed in the ancient ice. Oxygen atoms in melted ice cores can also be used to determine past temperatures by measuring the percentage of heavy oxygen (^{18}O) in the air bubbles. Ice formed during a period of warmer temperatures contains a higher percentage of heavy oxygen than ice formed during colder temperatures.

(c) Global warming is likely to have several major impacts on weather patterns in North America:

- As temperatures increase, heat waves are predicted to become more frequent.
- As temperatures increase, cold spells are predicted to become less frequent.
- There is likely to be an increased intensity of Atlantic storms as a result of ocean warming.

Cumulative Practice Exam Free-Response Questions (p. EXAM-13)

1. (a) An externality is a cost or benefit of a good or service that is not included in the purchase price of that good or service. *(Note that incorrect answers would be anything that is already included in the cost of purchasing or installing the trash compactors, as well as paying or not paying people to collect the trash.)* Positive externalities of solar-powered trash compactors may include:

- less trash scattered on the street next to trash bins
- increased value of city property owing to cleaner streets and technological advancements of trash compactors
- fewer flies collecting around trash
- cleaner air resulting from fewer trash collection trips
- increased likelihood that people will properly dispose of garbage

(b) Cradle-to-grave components of solar-powered trash compactor units include:

- energy invested in mining of raw materials to produce the housing, mechanical parts, and electronics of the compactors
- energy invested in converting raw materials into actual components
- energy invested in constructing and installing the components
- fuel costs associated with delivering the trash compactors to the city
- labor and energy associated with installing the trash compactors
- labor and energy associated with cleaning and maintaining the solar panels and other electrical parts of the compactors
- labor and energy associated with cleaning and maintaining all mechanical components of the compactor
- fuel and labor costs associated with collecting the trash
- collecting each trash compactor at the end of its useful life
- properly disposing of each trash compactor at the end of its useful life, which involves the labor and energy associated with recycling and disposal of its various parts

(c) Urban blight is the degradation of the built and social environments of a city that often accompanies and accelerates migration of residents to the suburbs. Installation and use of solar trash compactors may reverse urban blight in numerous ways. Trash compactors reduce demand for trash collection and subsequently liberate time, energy, and financial resources for other city services or improvements. Since they can hold more trash, trash compactors are also likely to reduce the amount of waste on city streets. As a result, the streets will be more pleasant for residents and visitors. Additionally, solar-powered trash compactors provide a novelty that is likely to make living in a city or visiting it more appealing.

(d) To determine which type of trash receptacle is more economically beneficial, we need to determine the total price of trash bins and the cost of trash collection visits. It is easiest to do this for a 20-year period. For regular trash bins:

Price of bin over 20 years = \$300
Cost of a collection visit = \$5 fuel + \$20 employee salary = \$25
Collecting trash 17 times per week over 20 years =
(17 collections/week × \$25/collection) × (52 weeks/year)
× 20 years = \$442,000
Cost of maintenance over 20 years = \$0
Total cost of bin and trash collection over 20 years = \$442,000 + \$300
= \$442,300

In contrast, the price of solar-powered trash bins over 20 years can be calculated as:

Price of a bin over 10 years = \$4,000
Price of a bin over 20 years = \$4,000 × 2 = \$8,000
Cost of a collection visit = \$5 fuel + \$20 employee salary = \$25
Collecting trash 5 times per week over 20 years =
(5 collections/week × \$25/collection) × (52 weeks/year)
× 20 years = \$130,000/year
Cost of maintenance over 20 years = \$150 × 20 years = \$3,000
Total cost of bins and trash collection over 20 years = \$8,000
+ \$130,000 + \$3,000 = \$141,000

Hence, the total cost of a solar-powered trash compactor is one-third that of a regular trash bin and, thus, is more economically beneficial.

(e) There are several ways to determine if solar-powered trash bins are more environmentally friendly:

- The solar-powered trash bins may be more friendly to the environment if the total amount of fuel required to produce, ship, repair, maintain, and dispose of the bins is lower than the amount required for regular trash bins over the same period.
- Solar-powered trash bins may require the use of more precious metals, but this may not have much of an environmental impact (beyond mining for those metals) if those metals are properly disposed of or recycled.
- Solar-powered trash compactors may be more beneficial if they reduce the amount of trash that winds up in street gutters and in natural habitats.
- Solar-powered trash compactors may be more beneficial if they encourage people to handle and dispose of their waste more wisely.

2. (a) (i) Areas just north of the equator are dominated by northeast trade winds that blow from the northeast in a southwesterly direction toward the equator. These winds are often characterized as hot and dry winds that pick up moisture as they flow along the oceans. These trade winds meet southeast trade winds at the intertropical convergence zone and rise into the atmosphere.

(ii) As the trade winds flow over land and up mountain ranges, adiabatic cooling leads to precipitation on the side of mountains facing the Atlantic. In contrast, the leeward sides of the mountains are relatively dry. The height of the mountains, combined with their location through the center of the country, create several ecoregions that vary in temperature and precipitation along the altitudinal gradient of the mountains.

(b) During ENSO years, water along the Pacific coastline becomes much warmer than in other years. For the United States, this typically means wetter conditions in the southeast. This occurs because winds flow inward toward the Rocky Mountains. In contrast, winds across Central America flow toward the Pacific Ocean. As a result, conditions on the Pacific side become drier during ENSO years.

(c) In general, there are five categories of ecosystem services and each may be provided through the protection of land. Costa Rica can greatly profit from each of these services:

- Provisioning services refer to goods that humans use directly, such as timber, food crops, medicinal plants, natural rubber, and furs. If the government protects land from further degradation and promotes the growth of natural ecosystems, the land in Costa Rica can continue to provide provisioning services for residents.
- Regulating services are those that regulate environmental conditions such as weather, air quality, water quality, and temperature. Costa Rican forests pump oxygen into the atmosphere, purify and provide drinking water, and alter temperature through the generation of shade and humidity. These services protect wildlife and lead to stability in the provision of natural goods.
- Support services often improve the quality of human life and make it easier for humans to exist. For example, nature provides pollinators and pest predators. These services improve living conditions for residents, increase land values, and increase production of natural products.
- Resiliency services ensure that the environment continues to function in a stable state, which means that it can continue to provide benefits to humans. If Costa Ricans removed these services, it is very unlikely that the land would continue to produce the abundance and quality of goods and wildlife that it currently produces.
- Cultural services provide aesthetic and cultural benefits to humans. Above all other services, these services most directly benefit Costa Rica. Millions of tourists visit the country each year and spend money to see natural parks, stay at hotels, and eat at restaurants. Without the beauty of nature that currently exists in Costa Rica, few of these tourists would visit.

(d) The debt-for-nature program helps developing nations reduce national debt in exchange for protected land. Typically, environmental protection organizations based in developed nations (e.g., The Nature Conservancy) agree to pay off some of a developing nation's debt. In exchange, the developing nations agree to protect some amount of land. In a sense, the environmental organizations are purchasing land to protect it, but paying the money to the debtors instead of the landholders.

3. (a) Organic inputs and outputs are those that have carbon-carbon and carbon-hydrogen bonds. Possible organic inputs to a cropland include:

- organic fertilizer (e.g., manure) from animals or other plants; may include feces of birds and other animals that invade the cropland
- seeds from other farms or from previous crops
- insects from outlying natural habitats (e.g., worms, aphids)
- organic pesticides extracted from living sources
- beneficial soil bacteria

Possible organic outputs from a cropland include:

- insects, such as pests and pollinators
- plant material
- decaying or living organic material in the soil carried by runoff
- seeds that pest species may carry
- wind- or insect-carried pollen

(b) Inorganic inputs and outputs are those that either do not contain the element carbon or contain carbon that is bound to elements other than hydrogen. Possible inorganic inputs to a cropland include:

- water
- inorganic fertilizer (e.g., ammonia, nitrate, phosphate) from rainfall or anthropogenic sources
- carbon dioxide (CO_2)
- oxygen (O_2)
- nitrogen gas (N_2)
- inorganic components of animal egesta and excretia

Possible inorganic outputs from a cropland include:

- carbon dioxide
- oxygen
- methane (CH_4)
- water (from evapotranspiration)
- inorganic nutrients from runoff

(c) Gross primary production (GPP) is a measure of the total amount of solar energy that producers in a system capture via photosynthesis over a given amount of time. This includes the energy that is put into plant biomass as well as the energy that is respired by producers (i.e., net primary productivity + respiration). Determining GPP is often a challenge because plants rarely photosynthesize without respiring. A common method of determining GPP is to determine the rate of photosynthesis by measuring the compounds that participate in the reaction. For example, we can measure the rate at which CO_2 is taken up during photosynthesis and the rate at which CO_2 is produced during respiration. A common approach to measuring GPP is first to measure the production of CO_2 in the dark. Because no photosynthesis occurs in the dark, this measure eliminates CO_2 uptake by photosynthesis. Next, we measure the uptake of CO_2 in sunlight. This measure gives us the net movement of CO_2 when respiration and photosynthesis are both occurring. By adding the amount of CO_2 produced in the dark to the amount of CO_2 taken up in the sunlight, we can determine the gross amount of CO_2 that is taken up during photosynthesis:

CO_2 taken up during photosynthesis = CO_2 taken up in sunlight + CO_2 produced in the dark

A farmer could take these measurements for a few plants in various parts of his or her cropland. The average value of all plants could be used to determine total GPP for the cropland.

(d) (i) Ammonification is the process of converting organic nitrogen compounds into inorganic ammonium (NH_4^+). This process is often conducted by fungal and bacterial decomposers. It is also known as mineralization. Ammonium produced by this process can either be taken up by producers in the ecosystem or be converted into nitrite (NO_2^-) or nitrate (NO_3^-) through the process of nitrification. Denitrification is the conversion of nitrate (NO_3^-) in a series of steps into nitrous oxide gas (N_2O) and into nitrogen gas (N_2). Denitrification is conducted by specialized bacteria that live under anaerobic conditions.

(ii) When belowground crop biomass is left in the soil, the decomposition of this material can greatly improve the productivity of future crops by liberating soil nitrogen for plant uptake. Since most ammonification is conducted by bacterial and fungal decomposers, spraying fungicide on a farmland could severely reduce the quantity of belowground organic nitrogen that decomposes into inorganic nitrogen. In turn, this could cause a cropland to retain nitrogen in the soil instead of cycling it back into plants.

4. (a) Several processes could have led to changes in antioxidant production in birds around Chernobyl. A simple explanation may be that birds in the area are responding to their environment and have started to produce more antioxidants as a means of improving survival. However, this process does not imply evolution and does not fully explain why the trait of increased antioxidant production has increased in abundance over the years. Another possible explanation for this phenomenon is natural selection; birds that produce antioxidants exhibit greater survival than birds that do not produce antioxidants. If these birds go on to produce more offspring than the birds that do not survive as long, then selection will favor antioxidant production. Variation in antioxidant production could be produced by a variety of mechanisms, including mutation (occasional mistakes in the process of copying DNA), recombination (recombining genes through sexual reproduction), gene flow from surrounding populations, and random genetic drift.

(b) The bottleneck effect is a drastic reduction in the size of a population that reduces genetic variation. Having fewer individuals in the population means that there are fewer unique genotypes to provide the variation needed for adaptive natural selection. The bottleneck effect likely applies to populations around Chernobyl because population numbers were severely reduced following the explosion. Only a few individuals of each species were likely to have survived, which severely lowered the amount of genetic variation in each population. As a result, slight mutations could have widespread and deleterious effects on the remaining population. Additionally, low genetic variation could lead to inbreeding, which would create inbreeding depression and lower fitness.

(c) When the size of a population is severely reduced and genetic variation is diminished, gene flow from neighboring populations can increase the genetic variation of the diminished population. In turn, this could save the population from local extinction and even accelerate adaptation. Hence, having intact bird populations in areas near Chernobyl likely brought novel sources of genetic variation into diminished populations, allowing surviving individuals to breed and produce future generations that natural selection could act upon.

(d) There could be several experimental designs, but all of them should be fairly similar with similar treatments, controls, and expected results.

(e) **(i)** A null hypothesis would be that birds producing higher levels of antioxidants are not more likely to exhibit higher reproduction than birds in the area without the adaptation.

(ii) Appropriate treatments and controls for this experiment should allow one to specifically test for the adaptive effects of antioxidant production when challenged with high levels of radiation. Hence, treatments for the experiment could include one group of birds from the area around the explosion that exhibits high levels of antioxidant production and another group of birds from the area that exhibits low levels of antioxidant production. The control might be birds of the same species that are not from the area and do not exhibit high antioxidant production, but are handled in a similar manner to the two treatment groups.

However, it may not be possible to find any individuals in the surrounding area that do not produce high levels of antioxidants. Hence, another appropriate experiment might be a study where the treatment group consists of birds from the area surrounding the explosion, and another set of birds of the same species that were not exposed to radiation but are experimentally exposed to radiation. The control group would be a set of birds of the same species that are not exposed to any radiation and are handled in the same manner as the two control groups.

(iii) Experimental procedures are likely to vary among students, but all designs should aim specifically to address the hypothesis. Hence, the procedures should measure reproductive output of the birds from all treatments and controls. Reproductive output may be measured as the number of eggs produced per bird over a lifetime or in a single breeding season, or as the number of eggs that successfully hatch and are able to produce viable offspring. If birds are to be exposed to radiation, the answer should address the timing of radiation exposure (the length of time the bird should be exposed to radiation). Also, birds should be exposed to radiation prior to measurement of reproduction. Replication of treatments should also be mentioned, although no specific number of replicates is necessary. Students should also mention that control groups will be handled exactly the same way as treatment groups, although no treatment will be applied. Excellent answers will also conduct measurements of antioxidant production in each bird to ensure that individuals in both treatment and control groups are accurate representations of their groups.

Glossary

Glossário

A

A horizon Frequently the top layer of soil, a zone of organic material and minerals that have been mixed together. *Also known as* **topsoil.**

abiotic Nonliving.

accuracy How close a measured value is to the actual or true value.

acid A substance that contributes hydrogen ions to a solution.

acid deposition Acids deposited on Earth as rain and snow or as gases and particles that attach to the surfaces of plants, soil, and water.

acid precipitation Precipitation high in sulfuric acid and nitric acid from reactions between water vapor and sulfur and nitrogen oxides in the atmosphere. *Also known as* **acid rain.**

acid rain *See* **acid precipitation.**

Acquired Immune Deficiency Syndrome (AIDS) An infectious disease caused by the human immunodeficiency virus (HIV).

active solar energy Energy captured from sunlight with advanced technologies.

acute disease A disease that rapidly impairs the functioning of an organism.

acute study An experiment that exposes organisms to an environmental hazard for a short duration.

adaptation A trait that improves an individual's fitness.

adiabatic cooling The cooling effect of reduced pressure on air as it rises higher in the atmosphere and expands.

adiabatic heating The heating effect of increased pressure on air as it sinks toward the surface of Earth and decreases in volume.

aerobic respiration The process by which cells convert glucose and oxygen into energy, carbon dioxide, and water.

affluence The state of having plentiful wealth including the possession of money, goods, or property.

age structure A description of how many individuals fit into particular age categories in a population.

age structure diagram A visual representation of the number of individuals within specific age groups for a country, typically expressed for males and females.

horizonte A A menudo es la capa superficial del suelo, una zona de material orgánico y minerales que se han entremezclado. *También se conoce como* **manto.**

abiótico Sin vida.

exactitud Fidelidad que hay entre un valor medido y su valor real.

ácido Sustancia que aporta iones de hidrógeno a una solución.

depósitos ácidos Ácidos que se depositan sobre la Tierra en la forma de lluvia y nieve, o pueden ser gases y partículas que se adhieren a la superficie de plantas, suelos y agua.

precipitación ácida Precipitación con una concentración alta de ácido sulfúrico y ácido nítrico producto de reacciones entre el vapor de agua y los óxidos de nitrógeno y azufre en la atmósfera. *También se conoce como* **lluvia ácida.**

lluvia ácida *Ver* **acid precipitation.**

síndrome de inmunodeficiencia adquirida (SIDA) Enfermedad infecciosa causada por el virus de la inmunodeficiencia humana (VIH).

energía solar activa Energía captada de la luz solar con tecnologías de punta.

enfermedad aguda Enfermedad que rápidamente perjudica el funcionamiento de un organismo.

estudio agudo Experimento que expone a los organismos a un peligro ambiental durante un período de tiempo breve.

adaptación Característica que mejora la adecuación de un individuo.

enfriamiento adiabático Efecto enfriador al reducir la presión sobre el aire a medida que este asciende en la atmósfera y se expande.

calentamiento adiabático Efecto calentador al aumentar la presión sobre el aire a medida que este desciende hacia la superficie de la Tierra y su volumen disminuye.

respiración aeróbica Proceso mediante el cual las células convierten la glucosa y el oxígeno en energía, dióxido de carbono y agua.

abundancia Estado de contar con una plenitud de riqueza, inclusive poseer dinero, bienes o propiedades.

estructura de edades Descripción de cuántos individuos se ajustan a las distintas categorías de edad en una población.

diagrama de la estructura de edades Representación visual de la cantidad de individuos dentro de grupos de edades específicos por país; característicamente se expresa por separado para varones y para mujeres.

agribusiness *See* **industrial agriculture.**

agroforestry An agricultural technique in which trees and vegetables are intercropped.

air pollution The introduction of chemicals, particulate matter, or microorganisms into the atmosphere at concentrations high enough to harm plants, animals, and materials such as buildings, or to alter ecosystems.

albedo The percentage of incoming sunlight reflected from a surface.

algal bloom A rapid increase in the algal population of a waterway.

alien species *See* **exotic species.**

allergen A chemical that causes allergic reactions.

allopatric speciation The process of speciation that occurs with geographic isolation.

ammonification The process by which fungal and bacterial decomposers break down the organic nitrogen found in dead bodies and waste products and convert it into inorganic ammonium (NH_4^+).

anaerobic respiration The process by which cells convert glucose into energy in the absence of oxygen.

anemia A deficiency of iron.

annual plant A plant that lives only one season.

anthropocentric worldview A worldview that focuses on human welfare and well-being.

anthropogenic Derived from human activities.

aphotic zone The deeper layer of ocean water that lacks sufficient sunlight for photosynthesis.

aquaculture Farming aquatic organisms such as fish, shellfish, and seaweeds.

aquatic biome An aquatic region characterized by a particular combination of salinity, depth, and water flow.

aqueduct A canal or ditch used to carry water from one location to another.

aquifer A permeable layer of rock and sediment that contains groundwater.

artesian well A well created by drilling a hole into a confined aquifer.

asbestos A long, thin, fibrous silicate mineral with insulating properties, which can cause cancer when inhaled.

ash The residual nonorganic material that does not combust during incineration.

assimilation The process by which producers incorporate elements into their tissues.

asthenosphere The layer of Earth located in the outer part of the mantle, composed of semi-molten rock.

agroindustria *Ver* **industrial agriculture.**

agrosilvicultura Técnica agrícola en la que se combinan los cultivos de árboles y los de comestibles.

contaminación aérea La introducción de sustancias químicas, materias particuladas o microorganismos en la atmósfera a concentraciones que son lo suficientemente altas como para causar daño a las plantas, los animales o a materiales tales como edificios, o que alteran ecosistemas completos.

albedo Porcentaje de luz solar que llega y que se refleja a partir de una superficie.

eflorescencia de algas Crecimiento acelerado de la población de algas en una vía navegable.

especie introducida *Ver* **exotic species.**

alérgeno Sustancia química que produce reacciones alérgicas.

especiación alopátrida Proceso de especiación por aislamiento geográfico.

amonificación Proceso mediante el cual saprófitos fúngicos y bacterianos descomponen el nitrógeno orgánico que se encuentra en organismos muertos y en desechos, y lo convierte en amonio inorgánico (NH_4^+).

respiración anaeróbica Proceso mediante el cual las células convierten glucosa en energía en la ausencia de oxígeno.

anemia Deficiencia de hierro.

planta anual Planta que vive solamente una estación.

visión antropocéntrica del mundo Visión del mundo que se concentra en la felicidad y el bienestar humanos.

antropogénico Derivado de actividades humanas.

zona afótica Profundidad submarina en la que se carece de suficiente luz solar para que haya fotosíntesis.

acuacultura Cultivo de organismos acuáticos tales como peces, mariscos y algas marinas.

biomedio acuático Región acuática caracterizada por una combinación específica de salinidad, profundidad y flujo acuático.

acueducto Canal o trocha que se utiliza para transportar agua de una ubicación a otra.

acuífero Capa permeable de roca y sedimento que contiene aguas freáticas.

pozo artesiano Pozo que se crea taladrando un agujero en un acuífero cautivo.

asbesto Mineral silicato fibroso, delgado y largo con propiedades aislantes, que puede causar cáncer si se lo inhala.

ceniza Material residual no orgánico que no se quema durante la incineración.

asimilación Proceso mediante el cual los productores incorporan elementos en sus tejidos.

astenosfera Capa de la Tierra ubicada en la parte exterior del manto. Está compuesta por piedra semifundida.

atmospheric convection current Global patterns of air movement that are initiated by the unequal heating of Earth.

atom The smallest particle that can contain the chemical properties of an element.

atomic number The number of protons in the nucleus of a particular element.

autotroph *See* **producer.**

corriente de convección atmosférica Patrones globales de movimientos de aire iniciados por diferencias en el calentamiento de la Tierra.

átomo Partícula más pequeña que puede contener las propiedades químicas de un elemento.

número atómico Número de protones en el núcleo de un elemento dado.

autótrofo *Ver* **producer.**

B

B horizon A soil horizon composed primarily of mineral material with very little organic matter.

horizonte B Horizonte del suelo compuesto principalmente de material mineral con muy poco material orgánico.

background extinction rate The average rate at which species become extinct over the long term.

tasa de extensión en trasfondo Tasa promedio en la que las especies se extinguen a largo plazo.

base A substance that contributes hydroxide ions to a solution.

base Sustancia que aporta iones de hidróxido a una solución.

base saturation The proportion of soil bases to soil acids, expressed as a percentage.

saturación base Proporción de bases en el suelo, comparadas con los ácidos en el suelo. Se expresa en forma de porcentaje.

becquerel (Bq) Unit that measures the rate at which a sample of radioactive material decays; 1 Bq = decay of 1 atom or nucleus per second.

becquerelio (Bq) Unidad que mide la tasa de desintegración de una muestra de un material radiactivo; 1 Bq = desintegración de 1 átomo o núcleo por segundo.

benthic zone The muddy bottom of a lake, pond, or ocean.

zona béntica Fondo de lodo de un lago, una laguna o un océano.

bioaccumulation An increased concentration of a chemical within an organism over time.

bioacumulación Mayor concentración de una sustancia química en un organismo con el pasar del tiempo.

biocentric worldview A worldview that holds that humans are just one of many species on Earth, all of which have equal intrinsic value.

visión biocéntrica del mundo Visión del mundo en la que los seres humanos son solo una de muchas especies en la Tierra, y todas y cada una de dichas especies tienen el mismo valor intrínseco.

biochemical oxygen demand (BOD) The amount of oxygen a quantity of water uses over a period of time at specific temperatures.

demanda bioquímica de oxígeno Volumen de oxígeno que una cantidad de agua utiliza durante un período de tiempo a temperaturas específicas.

biodiesel A diesel substitute produced by extracting and chemically altering oil from plants.

biodiésel Sustituto del diésel producido de aceite extraído de plantas, que luego se altera químicamente.

biodiversity hotspot An area that contains a high proportion of all the species found on Earth.

punto caliente de biodiversidad Zona que contiene una proporción alta de todas las especies que se encuentran en la Tierra.

biodiversity The diversity of life forms in an environment.

biodiversidad Diversidad de las formas con vida en un entorno.

biofuel Liquid fuel created from processed or refined biomass.

biocombustible Combustible líquido creado a partir de biomasa procesada o refinada.

biogeochemical cycle The movements of matter within and between ecosystems.

ciclo biogeoquímico Movimientos de materia dentro de ecosistemas y entre ellos.

biomagnification The increase in chemical concentration in animal tissues as the chemical moves up the food chain.

biomagnificación Aumento en la concentración química en tejidos animales a medida que la sustancia química asciende por la cadena trófica.

biomass The total mass of all living matter in a specific area.

biomasa Masa total de materia viviente en una zona específica.

biophilia Love of life.

biofilia Amor por la vida.

biosphere reserve Protected area consisting of zones that vary in the amount of permissible human impact.

reserva de la biosfera Zona protegida que consta de extensiones que varían en la cantidad de impacto humano que permiten.

biosphere The region of our planet where life resides, the combination of all ecosystems on Earth.

biosfera Región de nuestro planeta en la que reside la vida, la combinación de todos los ecosistemas en la Tierra.

biotic Living.

biótico Con vida.

bird flu A type of flu caused by the H5N1 virus.

gripe aviaria Tipo de influenza causada por el virus H5N1.

bitumen A degraded petroleum that forms when petroleum migrates to the surface of Earth and is modified by bacteria.

betún Petróleo degradado que se forma cuando el petróleo migra a la superficie de la Tierra y se ve modificado por bacterias.

boreal forest A forest biome made up primarily of coniferous evergreen trees that can tolerate cold winters and short growing seasons.

selva boreal Biomedio boscoso compuesto primordialmente por coníferos de hoja perenne capaces de tolerar inviernos fríos y temporadas de crecimiento breves.

bottleneck effect A reduction in the genetic diversity of a population caused by a reduction in its size.

efecto cuello de botella Reducción en la diversidad genética de una población atribuible a una reducción en su volumen.

bottom ash Residue collected at the bottom of the combustion chamber in a furnace.

ceniza de fondo Residuo que se recoge en el fondo de la cámara de combustión de una caldera.

broad-spectrum pesticide A pesticide that kills many different types of pest.

plaguicida de espectro amplio Plaguicida que mata muchos tipos diferentes de plagas.

brown smog *See* **photochemical smog.**

esmog marrón *Ver* **photochemical smog.**

brownfields Contaminated industrial or commercial sites that may require environmental cleanup before they can be redeveloped or expanded.

zona contaminada Sitios comerciales o industriales que podrían precisar una limpieza ambiental antes de que se puedan volver a urbanizar o se puedan ampliar.

bycatch The unintentional catch of nontarget species while fishing.

pesca incidental Captura no intencional de especies que no son las anheladas cuando se pesca.

C

C horizon The least-weathered soil horizon, which always occurs beneath the B horizon and is similar to the parent material.

horizonte C Horizonte del suelo menos erosionado, que siempre ocurre debajo del horizonte B y es parecido al material parental.

capacity factor The fraction of time a power plant operates in a year.

factor de capacidad Fracción de tiempo que una central eléctrica funciona durante un año.

capacity In reference to an electricity-generating plant, the maximum electrical output.

capacidad Al referirse a centrales generadores de electricidad, la producción eléctrica máxima.

cap-and-trade An approach to controlling CO_2 emissions, where a cap places an upper limit on the amount of pollutant that can be emitted and trade allows companies to buy and sell allowances for a given amount of pollution.

sistema de control e intercambio Enfoque que propone el control de las emisiones de CO_2. El control consiste en limitar la cantidad máxima que se puede emitir de un agente contaminador y el intercambio les permite a las empresas comprar y vender cupos de una cantidad de contaminación.

capillary action A property of water that occurs when adhesion of water molecules to a surface is stronger than cohesion between the molecules.

capilaridad Propiedad del agua que ocurre cuando la adhesión de las moléculas de agua a la superficie es mayor que la cohesión entre las moléculas mismas.

carbohydrate A compound composed of carbon, hydrogen, and oxygen atoms.

carbohidrato Compuesto que consta de átomos de carbono, hidrógeno y oxígeno.

carbon cycle The movement of carbon around the biosphere.

ciclo de carbono Movimiento del carbono alrededor de la biosfera.

carbon neutral An activity that does not change atmospheric CO_2 concentrations.

emisión neutra de carbono Actividad que no cambia las concentraciones atmosféricas de CO_2.

carbon offsets Methods of promoting global CO_2 reduction that do not involve a direct reduction in the amount of CO_2 actually emitted by a company.

mitigación de emisiones de carbono Métodos para fomentar la reducción total de CO_2 que no implican una reducción directa de la cantidad de CO_2 que una empresa dada de hecho emana.

carbon sequestration An approach to stabilizing greenhouse gases by removing CO_2 from the atmosphere.

captura de carbono Enfoque para estabilizar la emisión de gases de efecto invernadero mediante la eliminación de CO_2 en la atmósfera.

carcinogen A chemical that causes cancer.

carcinógeno Sustancia química que causa cáncer.

carnivore A consumer that eats other consumers.

carnívoro Consumidor que se come otros consumidores.

carrying capacity (*K*) The limit of how many individuals in a population the environment can sustain.

capacidad poblacional (*K*) Límite del número de individuos de una población que el entorno puede sustentar.

cation exchange capacity (CEC) The ability of a particular soil to absorb and release cations.

capacidad de intercambio de cationes Capacidad de un suelo dado de absorber y de liberar cationes.

cell A highly organized living entity that consists of the four types of macromolecules and other substances in a watery solution, surrounded by a membrane.

célula Entidad viviente muy bien organizada que consta de cuatro tipos de macromoléculas y de otras sustancias en una solución acuosa, rodeadas por una membrana.

cellular respiration The process by which cells unlock the energy of chemical compounds.

respiración celular Proceso mediante el cual las células liberan la energía de compuestos químicos.

cellulosic ethanol An ethanol derived from cellulose, the cell wall material in plants.

etanol celulósico Etanol derivado de celulosa, el material de la pared celular en las plantas.

chemical energy Potential energy stored in chemical bonds.

energía química Energía potencial almacenada en los enlaces químicos.

chemical reaction A reaction that occurs when atoms separate from molecules or recombine with other molecules.

reacción química Reacción que ocurre cuando algunos átomos se separan de las moléculas o se recombinan con otras moléculas.

chemical weathering The breakdown of rocks and minerals by chemical reactions, the dissolving of chemical elements from rocks, or both.

erosión química Descomposición de rocas y minerales por reacciones químicas; disolución de elementos químicos de rocas, o ambas cosas.

chemosynthesis A process used by some bacteria in the ocean to generate energy with methane and hydrogen sulfide.

quimiosíntesis Proceso que utilizan algunas bacterias en el océano para generar energía valiéndose del metano y el ácido sulfhídrico.

child mortality The number of deaths of children under age 5 per 1,000 live births.

mortalidad infantil Cantidad de fallecimientos de niños menores de 5 años por millar de partos vivos.

chronic disease A disease that slowly impairs the functioning of an organism.

enfermedad crónica Enfermedad que lentamente va perjudicando las funciones de un organismo.

chronic study An experiment that exposes organisms to an environmental hazard for a long duration.

estudio crónico Experimento que expone los organismos a un peligro ambiental durante un período de tiempo prolongado.

Clean Water Act Legislation that supports the "protection and propagation of fish, shellfish, and wildlife and recreation in and on the water" by maintaining and, when necessary, restoring the chemical, physical, and biological properties of surface waters.

Ley de Agua Limpia Legislación estadounidense que fomenta la "protección y propagación de peces, moluscos y crustáceos, y vida silvestre, así como recreación en o sobre el agua" mediante el mantenimiento y, cuando se precise, la restauración de las propiedades químicas, físicas y biológicas de las aguas superficiales.

clear-cutting A method of harvesting trees that involves removing all or almost all of the trees within an area.

tala indiscriminada Método de cosechar árboles que implica la tala de todos o casi todos los árboles dentro de una zona.

climate The average weather that occurs in a given region over a long period of time.

clima Promedio del estado del tiempo que ocurre en una región dada durante un período de tiempo prolongado.

closed system A system in which matter and energy exchanges do not occur across boundaries.

sistema cerrado Sistema en el que no se dan intercambios de materia y energía a través de sus respectivas fronteras.

closed-loop recycling Recycling a product into the same product.

coal A solid fuel formed primarily from the remains of trees, ferns, and other plant materials preserved 280 million to 360 million years ago.

cogeneration The use of a fuel to generate electricity and produce heat. *Also known as* **combined heat and power.**

combined cycle A power plant that uses both exhaust gases and steam turbines to generate electricity.

combined heat and power *See* **cogeneration.**

command-and-control approach A strategy for pollution control that involves regulations and enforcement mechanisms.

commensalism A relationship between species in which one species benefits and the other species is neither harmed nor helped.

commercial energy source An energy source that is bought and sold.

community All of the populations of organisms within a given area.

community ecology The study of interactions between species.

competition The struggle of individuals to obtain a shared limiting resource.

competitive exclusion principle The principle stating that two species competing for the same limiting resource cannot coexist.

composting Creation of organic matter (humus) by decomposition under controlled conditions to produce an organic-rich material that enhances soil structure, cation exchange capacity, and fertility.

compound A molecule containing more than one element.

concentrated animal feeding operation (CAFO) A large indoor or outdoor structure designed for maximum output.

cone of depression An area lacking groundwater due to rapid withdrawal by a well.

confined aquifer An aquifer surrounded by a layer of impermeable rock or clay that impedes water flow.

consumer An organism that is incapable of photosynthesis and must obtain its energy by consuming other organisms. *Also known as* **heterotroph.**

contaminated water Wastewater from toilets, kitchen sinks, and dishwashers.

contour plowing An agricultural technique in which plowing and harvesting are done parallel to the topographic contours of the land.

reciclaje en circuito cerrado Reciclaje de un producto para formar el mismo producto.

carbón Combustible sólido formado principalmente de los residuos de árboles, helechos y otros materiales vegetales preservados hace entre 280 y 360 millones de años.

cogeneración Uso de un combustible para generar electricidad y producir calor. *También se conoce como* **energía y calor combinados.**

ciclo combinado Central eléctrica que utiliza tanto los gases de salida como turbinas de vapor para generar electricidad.

energía y calor combinados *Ver* **cogeneration.**

enfoque de mando y control Estrategia para el control de la contaminación que incluye reglamentaciones y mecanismos para verificar el acatamiento.

comensalismo Relación entre especies en la que una especie se beneficia y a la otra especie no se la ha perjudicado ni ayudado.

fuente comercial de energía Fuente de energía que se compra y se vende.

comunidad Poblaciones de todos los organismos que viven dentro de un área dada.

ecología de la comunidad Estudio de las interacciones entre las especies.

competencia Batalla entre individuos por obtener un recurso limitante compartido.

principio de la exclusión competitiva Principio que explica que dos especies que compiten por el mismo recurso limitante no pueden coexistir.

compostación Creación de materia orgánica (humus) mediante la descomposición en condiciones controladas con el fin de producir un material de riqueza orgánica que mejora la estructura del suelo, aumenta la capacidad de intercambio de cationes y fomenta la fertilidad.

compuesto Molécula que contiene más de un elemento.

operación concentrada para la alimentación de animales Estructura grande interior o al aire libre que se ha diseñado para generar una producción máxima.

cono de depresión Área que carece de aguas subterráneas debido a que estas han sido retiradas rápidamente por medio de un pozo.

acuífero cautivo Acuífero rodeado por una capa de roca o arcilla impermeable que impide el flujo de agua.

consumidor Organismo que es incapaz de hacer la fotosíntesis y debe obtener su energía consumiendo otros organismos. *También se conoce como* **heterótrofo.**

agua contaminada Aguas negras provenientes de inodoros, lavamanos y lavaplatos.

arar en curvas de nivel Técnica agrícola en la que se ara y se cosecha en paralelo a los contornos topográficos del terreno.

control group In a scientific investigation, a group that experiences exactly the same conditions as the experimental group, except for the single variable under study.

grupo de control En una investigación científica, grupo que experimenta exactamente las mismas condiciones que el grupo experimental, con la excepción de una variable que se está estudiando.

control rod A cylindrical device inserted between the fuel rods in a nuclear reactor to absorb excess neutrons and slow or stop the fission reaction.

vara de control Dispositivo cilíndrico que se inserta entre las barras de combustible en un reactor nuclear para absorber los neutrones que sobran y para enlentecer o detener las reacción de fisión.

Convention on Biological Diversity An international treaty to help protect biodiversity.

Convenio sobre la Diversidad Biológica Tratado internacional que busca proteger la biodiversidad.

Convention on International Trade in Endangered Species of Wild Fauna and Flora (CITES) A 1973 treaty formed to control the international trade of threatened plants and animals.

Convenio sobre el Comercio Internacional de Especies Amenazadas de Fauna y Flora Silvestres (CITES) Convenio promulgado en 1973 con el fin de controlar el comercio internacional de plantas y animales amenazados.

convergent plate boundary An area where plates move toward one another and collide.

borde de placas convergentes Zona en la que las placas tectónicas se acercan una a la otra y chocan una con otra.

coral bleaching A phenomenon in which algae inside corals die, causing the corals to turn white.

blanqueamiento de corales Fenómeno en el que las algas dentro de los corales fallecen, motivo por el cual los corales se ponen blancos.

coral reef The most diverse marine biome on Earth, found in warm, shallow waters beyond the shoreline.

arrecife de coral Biomedio más diverso de la Tierra, se encuentra en aguas tibias y de poca profundidad a cierta distancia del litoral.

core The innermost zone of Earth's interior, composed mostly of iron and nickel. It includes a liquid outer layer and a solid inner layer.

núcleo interno Zona interior de la Tierra, compuesta en su mayor parte de hierro y níquel. Incluye una capa externa líquida y una capa interna maciza.

Coriolis effect The deflection of an object's path due to the rotation of Earth.

efecto Coriolis Desviación de la trayectoria de un objeto debido a la rotación de la Tierra.

corridor Strips of natural habitat that connect populations.

corredor Franjas de hábitat natural que interconectan poblaciones.

covalent bond The bond formed when elements share electrons.

enlace covalente Enlace que se forma cuando los elementos comparten electrones.

cradle-to-grave analysis *See* **life-cycle analysis.**

análisis de la cuna a la tumba *Ver* **life-cycle analysis.**

crop rotation An agricultural technique in which crop species in a field are rotated from season to season.

rotación de cultivos Técnica agrícola en la que las especies de cultivos en un campo se rotan de estación a estación.

crude birth rate (CBR) The number of births per 1,000 individuals per year.

tasa bruta de nacimientos Cantidad de nacimientos por millar de individuos por año.

crude death rate (CDR) The number of deaths per 1,000 individuals per year.

tasa bruta de fallecimientos Cantidad de fallecimientos por millar de individuos por año.

crude oil Liquid petroleum removed from the ground.

petróleo crudo Petróleo líquido que ha sido extraído del suelo.

crust In geology, the chemically distinct outermost layer of the lithosphere.

corteza En geología, la capa más externa y distinguible en términos químicos de la litosfera.

crustal abundance The average concentration of an element in Earth's crust.

abundancia en la corteza Concentración promedio de un elemento en la corteza terrestre.

CTL (coal to liquid) The process of converting solid coal into liquid fuel.

carbón a líquido Proceso para convertir el carbón sólido en combustible líquido.

cultural eutrophication An increase in fertility in a body of water, the result of anthropogenic inputs of nutrients.

eutrofización cultural Aumento en la fertilidad de una extensión de agua como resultado del ingreso antropogenético de nutrientes.

curie A unit of measure for radiation; 1 curie = 37 billion decays per second.

curio Unidad con la que se mide la radiactividad; 1 curio = 37 mil millones de desintegraciones por segundo.

D

dam A barrier that runs across a river or stream to control the flow of water.

presa Barrera que atraviesa un río o arroyo. Se usa para controlar el flujo de agua.

dead zone In a body of water, an area with extremely low oxygen concentration and very little life.

zona muerta En una extensión de agua, zona con una concentración de oxígeno extremadamente baja y muy poca vida.

decomposers Fungi or bacteria that convert organic matter into small elements and molecules that can be recycled back into the ecosystem.

saprofitos Hongos o bacterias que convierten la materia orgánica en elementos y moléculas diminutos que se pueden reciclar y devolver al ecosistema.

demographer A scientist in the field of demography.

demógrafo Científico que trabaja en el campo de la demografía.

demography The study of human populations and population trends.

demografía Estudio de las poblaciones humanas y sus tendencias poblacionales.

denitrification The conversion of nitrate (NO_3^-) in a series of steps into the gases nitrous oxide (N_2O) and, eventually, nitrogen gas (N_2), which is emitted into the atmosphere.

desnitrificación Conversión, mediante una serie de pasos, de nitrato (NO_3^-) en los gases óxido nitroso (N_2O) y, con el tiempo, nitrógeno gaseoso (N_2), que se emite a la atmósfera.

density-dependent factor A factor that influences an individual's probability of survival and reproduction in a manner that depends on the size of the population.

factor densodependiente Factor que influye en la probabilidad de supervivencia y reproducción de una manera que depende del volumen de la población.

density-independent factor A factor that has the same effect on an individual's probability of survival and the amount of reproduction at any population size.

factor densoindependiente Factor que tiene el mismo efecto en la probabilidad de supervivencia y cantidad de reproducción de un individuo en una población de cualquier volumen.

Department of Energy (DOE) The U.S. organization that advances the energy and economic security of the United States.

Departamento de Energía (DOE) Dependencia del gobierno federal estadounidense que fomenta la seguridad energética y económica de Estados Unidos.

desalinization The process of removing the salt from salt water.

desalinización Proceso mediante el cual se le quita la sal al agua salobre.

desertification The transformation of arable, productive land to desert or unproductive land due to climate change or destructive land use.

desertificación Transformación de terrenos arables y productivos a suelos no productivos debido a cambios climáticos o al uso destructivo del suelo.

detritivore An organism that specializes in breaking down dead tissues and waste products into smaller particles.

detritívoro Organismo que se especializa en descomponer tejidos muertos y desechos en partículas más pequeñas.

developed country A country with relatively high levels of industrialization and income.

país desarrollado País con niveles relativamente altos de industrialización e ingresos.

developing country A country with relatively low levels of industrialization and income.

país en vías de desarrollo País con niveles relativamente bajos de industrialización e ingresos.

development Improvement in human well-being through economic advancement.

desarrollo Mejoras en el bienestar humano mediante adelantos económicos.

die-off A rapid decline in a population due to death.

mortandad Disminución rápida en una población debido a muertes.

dike A structure built to prevent ocean waters from flooding adjacent land.

dique Estructura construida para prevenir que las aguas del océano inunden los terrenos contiguos.

disease Any impaired function of the body with a characteristic set of symptoms.

enfermedad Toda función corporal disminuida que presenta un conjunto característico de síntomas.

distillation A process of desalinization in which water is boiled and the resulting steam is captured and condensed to yield pure water.

destilación Proceso de desalinización en el que el agua se hierve y el vapor resultante se capta y se condensa para producir agua pura.

distribution Areas of the world in which a species lives.

distribución Zonas del mundo en las cuales vive una especie.

disturbance An event, caused by physical, chemical, or biological agents, resulting in changes in population size or community composition.

perturbación Acontecimiento causado por agentes físicos, químicos o biológicos que produce cambios en el volumen de la población o en la composición de la comunidad.

divergent plate boundary An area beneath the ocean where tectonic plates move away from each other.

borde de placas divergentes Zona en la profundidad del océano en la que las placas tectónicas se separan y alejan una de la otra.

DNA (deoxyribonucleic acid) A nucleic acid, the genetic material that contains the code for reproducing the components of the next generation, and which organisms pass on to their offspring.

ADN (ácido desoxirribonucleico) Ácido nucleico, material genético que contiene el código para la reproducción de los componentes de la siguiente generación, y que los organismos pasan a sus descendientes.

dose-response study A study that exposes organisms to different amounts of a chemical and then observes a variety of possible responses, including mortality or changes in behavior or reproduction.

estudio de dosis-respuesta Estudio que expone organismos a diferentes cantidades de una sustancia química y luego observa una diversidad de respuestas posibles, las que incluyen la mortalidad o cambios en la conducta o reproducción.

doubling time The number of years it takes a population to double.

tiempo de duplicación Número de años que precisa una población para duplicarse.

E

E horizon A zone of leaching, or eluviation, found in some acidic soils under the O horizon or, less often, the A horizon.

horizonte E Zona de lixiviación o eluviación que se encuentra en algunos suelos acidizados debajo del horizonte O o, con menor frecuencia, el horizonte A.

earthquake The sudden movement of Earth's crust caused by a release of potential energy along a geologic fault and usually causing a vibration or trembling at Earth's surface.

terremoto El movimiento repentino de la corteza de la Tierra producido por la liberación de energía potencial a lo largo de una falla geológica y que por lo general produce una vibración o temblor en la superficie terrestre.

Ebola hemorrhagic fever An infectious disease with high death rates, caused by the Ebola virus.

fiebre hemorrágica del ébola Enfermedad infecciosa con tasas de mortalidad altas, causada por el virus del Ébola.

ecocentric worldview A worldview that places equal value on all living organisms and the ecosystems in which they live.

visión ecocéntrica del mundo Visión que le da el mismo valor a todos los organismos vivos y los ecosistemas en los que residen.

ecological economics The study of economics as a component of ecological systems.

economía ecológica Estudio de economía como componente de sistemas ecológicos.

ecological efficiency The proportion of consumed energy that can be passed from onetrophic level to another.

eficiencia ecológica Proporción de energía consumida que se puede pasar de un nivel trófico a otro.

ecological footprint A measure of how much an individual consumes, expressed in area of land.

presencia ecológica Medida de cuánto consume un individuo, expresada en extensión de tierra.

ecological succession The predictable replacement of one group of species by another group of species over time.

sucesión ecológica Reemplazo predecible de un grupo de especies por otro grupo de especies a través del tiempo.

ecologically sustainable forestry An approach to removing trees from forests in ways that do not unduly affect the viability of other trees.

silvicultura sostenible ecológicamente Enfoque de talar árboles en los bosques de maneras que no afecten indebidamente la viabilidad de los demás árboles.

economics The study of how humans allocate scarce resources in the production, distribution, and consumption of goods and services.

economía Estudio de cómo los seres humanos reparten recursos escasos en la producción, la distribución y el consumo de bienes y servicios.

economies of scale The observation that average costs of production fall as output increases.

economías de escala Observación de que los costos promedio de la producción caen en la medida en que aumenta la producción.

ecosystem A particular location on Earth with interacting biotic and abiotic components.

ecosistema Ubicación particular en la Tierra que tiene componentes bióticos y abióticos que interactúan entre sí.

ecosystem engineer A keystone species that creates or maintains habitat for other species.

ecosystem services The processes by which life-supporting resources such as clean water, timber, fisheries, and agricultural crops are produced.

ED50 The effective dose of a chemical that causes 50 percent of the individuals in a dose-response study to display a harmful, but nonlethal, effect.

edge habitat Habitat that occurs where two different communities come together, typically forming an abrupt transition, such as where a grassy field meets a forest.

El Niño–Southern Oscillation (ENSO) A reversal of wind and water currents in the South Pacific.

electrical grid A network of interconnected transmission lines that joins power plants together and links them with end users of electricity.

electrolysis The application of an electric current to water molecules to split them into hydrogen and oxygen.

electromagnetic radiation A form of energy emitted by the Sun that includes, but is not limited to, visible light, ultraviolet light, and infrared energy.

element A substance composed of atoms that cannot be broken down into smaller, simpler components.

emergent infectious disease An infectious disease that has not been previously described or has not been common for at least 20 years.

emigration The movement of people out of a country or region.

eminent domain A principle that grants government the power to acquire a property at fair market value even if the owner does not wish to sell it.

endangered species A species that is in danger of extinction within the foreseeable future throughout all or a significant portion of its range.

Endangered Species Act A 1973 U.S. act designed to protect species from extinction.

endemic species A species that lives in a very small area of the world and nowhere else.

endocrine disruptor A chemical that interferes with the normal functioning of hormones in an animal's body.

energy The ability to do work or transfer heat.

energy carrier Something that can move and deliver energy in a convenient, usable form to end users.

energy conservation Finding and implementing ways to use less energy.

energy efficiency The ratio of the amount of energy expended in the form you want to the total amount of energy that is introduced into the system.

ingeniero del ecosistema Especie clave que crea o mantiene el hábitat para otras especies.

servicios del ecosistema Procesos mediante los cuales se producen los recursos que sustentan la vida, tales como el agua limpia, la madera, las pesquerías y la agricultura.

ED50 Dosis efectiva de una sustancia química que causa que el 50 por ciento de los individuos en un estudio de dosis-respuesta presenten un efecto nocivo pero no letal.

hábitat de borde Hábitat que ocurre donde dos comunidades diferentes se unen. Característicamente forman una transición abrupta, como cuando un pastizal se topa con un bosque.

El Niño–oscilación austral Inversión de las corrientes de viento y de agua en el Pacífico Sur.

red eléctrica Red de líneas de transmisión interconectadas que unen varias centrales generadoras y las conectan con los usuarios finales de la electricidad.

electrólisis Aplicación de una corriente eléctrica a moléculas de agua para separarlas en hidrógeno y oxígeno.

radiación electromagnética Forma de energía emitida por el Sol que incluye, sin limitarse, luz visible, luz ultravioleta y energía infrarroja.

elemento Sustancia compuesta por átomos que no se pueden desintegrar en componentes más pequeños y sencillos.

enfermedad infecciosa emergente Enfermedad infecciosa que no se ha descrito antes o que no ha sido común durante al menos 20 años.

emigración Desplazamiento de personas que salen de un país o región.

derecho de expropiación Principio según el cual se le otorga al gobierno el poder de adquirir un inmueble al valor del mercado incluso si el propietario no desea venderlo.

especie en peligro de extinción Especie que está en peligro de desaparecer dentro de un futuro predecible en la totalidad o en una parte significativa de su hábitat.

Ley de Especies en Peligro de Extinción Ley promulgada en 1973 en Estados Unidos mediante la cual se protegen de extinción ciertas especies.

especie endémica Especie que vive en una zona muy reducida del mundo y en ningún otro sitio.

interruptor endocrino Sustancia química que interfiere con el funcionamiento normal de las hormonas en el organismo de un animal.

energía Capacidad de hacer trabajo o transferir calor.

transportador de electricidad Algo que puede transportar y entregar electricidad de una manera que sea conveniente y fácil para los usuarios finales.

conservación de energía Hallar y poner en práctica maneras de consumir menos energía.

eficiencia energética Cantidad de energía que se expende en la forma deseada, expresada como proporción de la cantidad de energía que se introduce en el sistema.

energy intensity The energy use per unit of gross domestic product.

energy quality The ease with which an energy source can be used for work.

energy subsidy The fossil fuel energy and human energy input per calorie of food produced.

entropy Randomness in a system.

environment The sum of all the conditions surrounding us that influence life.

environmental economics A subfield of economics that examines the costs and benefits of various policies and regulations that seek to regulate or limit air and water pollution and other causes of environmental degradation.

environmental hazard Anything in the environment that can potentially cause harm.

environmental impact statement (EIS) A document outlining the scope and purpose of a development project, describing the environmental context, suggesting alternative approaches to the project, and analyzing the environmental impact of each alternative.

environmental indicator An indicator that describes the current state of an environmental system.

environmental mitigation plan A plan that outlines how a developer will address concerns raised by a project's impact on the environment.

Environmental Protection Agency (EPA) The U.S. organization that oversees all governmental efforts related to the environment, including science, research, assessment, and education.

environmental science The field of study that looks at interactions among human systems and those found in nature.

environmental studies The field of study that includes environmental science and additional subjects such as environmental policy, economics, literature, and ethics.

environmental worldview A worldview that encompasses how one thinks the world works; how one views one's role in the world; and what one believes to be proper environmental behavior.

environmentalist A person who participates in environmentalism, a social movement that seeks to protect the environment through lobbying, activism, and education.

epicenter The exact point on the surface of Earth directly above the location where rock ruptures during an earthquake.

epidemic A situation in which a pathogen causes a rapid increase in disease.

erosion The physical removal of rock fragments from a landscape or ecosystem.

intensidad energética La energía por unidad del producto bruto interno.

calidad de la energía Facilidad con la cual una fuente de energía se puede aprovechar para hacer trabajo.

subsidio energético La energía de combustibles fósiles y la energía humana consumidas por caloría de alimento producida.

entropía Aleatoriedad en un sistema.

entorno Suma de todas las condiciones circundantes que influyen en la vida.

economía ambiental Subcampo de la economía en la que se estudian los costos y los beneficios de las normas y las reglamentaciones que regulan o limitan la contaminación del aire y del agua, así como otras causas de degradación ambiental.

peligro ambiental Toda cosa en el medioambiente que potencialmente pueda ser perjudicial.

declaración de impacto ambiental Documento que detalla el alcance y el propósito de un proyecto de desarrollo. Describe el contexto ambiental, sugiere enfoques alternativos que podría tomar el proyecto, y analiza el impacto ambiental de cada alternativa.

indicador ambiental Indicador que describe el estado actual de un sistema ambiental.

plan de mitigación ambiental Plan que detalla cómo el encargado tratará las inquietudes que surgen en torno al impacto ambiental de un proyecto.

Agencia de Protección Ambiental (EPA) Dependencia del gobierno federal estadounidense que mantiene un control de todos los esfuerzos del gobierno relacionados con el medio ambiente, incluidas las ciencias, las investigaciones, las valoraciones y la educación.

ciencia ambiental Campo en el que se estudian las interacciones entre sistemas humanos y sistemas que se hallan en la naturaleza.

estudios ambientales Campo de estudio que incluye la ciencia ambiental y otros temas tales como política ambiental, economía, literatura y ética.

visión ambiental del mundo Visión del mundo que abarca la manera como pensamos que funciona el mundo; cómo uno ve el papel que uno desempeña en el mundo; y lo que uno considera que es el comportamiento ecológico apropiado.

ambientalista Persona que participa en el ambientalismo, un movimiento social que aspira a proteger el medio ambiente por medio del cabildeo, el activismo y la educación.

epicentro Sitio exacto en la superficie terrestre directamente encima de la ubicación en la cual hay una ruptura de la roca durante un terremoto.

epidemia Situación en la que un patógeno causa la propagación rápida de una enfermedad.

erosión Retiro físico de fragmentos de roca de un paisaje o ecosistema.

ethanol Alcohol made by converting starches and sugars from plant material into alcohol and CO_2.

eutrophic Describes a lake with a high level of productivity.

eutrophication A phenomenon in which a body of water becomes rich in nutrients.

evapotranspiration The combined amount of evaporation and transpiration.

evolution A change in the genetic composition of a population over time.

evolution by artificial selection The process in which humans determine which individuals breed, typically with a preconceived set of traits in mind.

evolution by natural selection The process in which the environment determines which individuals survive and reproduce.

exotic species A species living outside its historical range. *Also known as* **alien species.**

exponential growth model ($N_t = N_0e^{rt}$) A growth model that estimates a population's future size (N_t) after a period of time (t), based on the intrinsic growth rate (r) and the number of reproducing individuals currently in the population (N_0).

externality The cost or benefit of a good or service that is not included in the purchase price of that good or service.

extinction The death of the last member of a species.

extrusive igneous rock Rock that forms when magma cools above the surface of Earth.

exurb An area similar to a suburb, but unconnected to any central city or densely populated area.

etanol Alcohol que se elabora convirtiendo almidones y azúcares de material vegetal en alcohol y CO_2.

eutrófico Se dice de un lago con un gran nivel de productividad.

eutrofización Fenómeno en el que una extensión de agua se enriquece de nutrientes.

evapotranspiración Cantidad combinada de evaporación y transpiración.

evolución Cambio en la composición genética de una población a través del tiempo.

evolución por selección artificial Proceso en el que los seres humanos determinan qué individuos se reproducen, característicamente teniendo en cuenta un conjunto de rasgos.

evolución por selección natural Proceso en el que el entorno determina qué individuos sobreviven y se reproducen.

especie exótica Especie que vive por fuera de su hábitat histórico. *También se conoce como* **especie introducida**.

modelo de crecimiento exponencial ($N_t = N_0e^{rt}$) Modelo de crecimiento que estima el volumen futuro (N_t) de una población después de un período de tiempo (t), con base en la tasa de crecimiento intrínseco (r) y la cantidad de individuos que se pueden reproducir en la actualidad en la población (N_0).

externalidad Costo o beneficio de un bien o servicio que no se incluye en el precio de compra de dicho bien o servicio.

extinción Muerte del último integrante de una especie.

roca ígnea extrusiva Roca que se forma cuando el magma se enfría en la superficie de la corteza terrestre.

exurbio Zona parecida a un suburbio, pero desconectada de la ciudad central o de la zona de densidad poblacional alta.

F

family planning The practice of regulating the number or spacing of offspring through the use of birth control.

famine The condition in which food insecurity is so extreme that large numbers of deaths occur in a given area over a relatively short period.

fault A fracture in rock caused by a movement of Earth's crust.

fault zone A large expanse of rock where a fault has occurred.

fecal coliform bacteria A group of generally harmless microorganisms in human intestines that can serve as an indicator species for potentially harmful microorganisms associated with contaminated sewage.

Ferrell cell A convection current in the atmosphere that lies between Hadley cells and polar cells.

planificación familiar Práctica de regular el espaciamiento de la descendencia gracias al uso de un método de control de la natalidad.

hambruna Condición en la que la inseguridad alimentaria es tan extrema que ocurren grandes cantidades de muertes en una zona dada o dentro de un período de tiempo relativamente breve.

falla Fractura en la roca causada por un movimiento de la corteza terrestre.

zona de falla Extensión amplia de roca en la cual ha ocurrido una falla.

bacteria fecal coliforme Grupo de microorganismos que en general no son nocivos en el intestino humano que pueden ser una especie indicadora de los microorganismos potencialmente dañinos que se asocian con aguas negras contaminadas.

célula de Ferrell Corriente de convección en la atmósfera que yace entre las células de Hadley y las células polares.

first law of thermodynamics A physical law which states that energy can neither be created nor destroyed but can change from one form to another.

fish ladder A stair-like structure that allows migrating fish to get around a dam.

fishery A commercially harvestable population of fish within a particular ecological region.

fishery collapse The decline of a fish population by 90 percent or more.

fission A nuclear reaction in which a neutronstrikes a relatively large atomic nucleus, which then splits into two or more parts, releasing additional neutrons and energy in the form of heat.

fitness An individual's ability to survive and reproduce.

flex-fuel vehicle A vehicle that runs on either gasoline or a gasoline/ethanol mixture.

floodplain The land adjacent to a river.

fly ash The residue collected from the chimney or exhaust pipe of a furnace.

food chain The sequence of consumption from producers through tertiary consumers.

food insecurity A condition in which people do not have adequate access to food.

food security A condition in which people have access to sufficient, safe, and nutritious food that meets their dietary needs for an active and healthy life.

food web A complex model of how energy and matter move between trophic levels.

forest Land dominated by trees and other woody vegetation and sometimes used for commercial logging.

fossil carbon Carbon in fossil fuels.

fossil fuel A fuel derived from biological material that became fossilized millions of years ago.

founder effect A change in the genetic composition of a population as a result of descending from a small number of colonizing individuals.

fracking Hydraulic fracturing, a method of oil and gas extraction that uses high-pressure fluids to force open cracks in rocks deep underground.

fracture In geology, a crack that occurs in rock as it cools.

freshwater wetlands An aquatic biome that is submerged or saturated by water for at least part of each year, but shallow enough to support emergent vegetation.

fuel cell An electrical-chemical device that converts fuel, such as hydrogen, into an electrical current.

primera ley de termodinámica Ley física que declara que la energía no se puede ni crear ni destruir, pero sí se puede transformar de una forma a otra.

escalera para peces Estructura en forma de escalera que facilita la migración de los peces que ha sido por un embalse.

pesquería Población de peces dentro de una región ecológica dada. Se puede cosechar con fines comerciales.

colapso de pesquería La caída en un 90 por ciento o más de una población de peces.

fisión Reacción nuclear en la que un neutrón bombardea un núcleo atómico relativamente grande, lo cual produce su rotura en dos o más partes, liberando otros neutrones y energía en forma de calor.

aptitud Capacidad que tiene un individuo para sobrevivir y reproducirse.

vehículo de combustible flexible Vehículo que funciona ya sea con gasolina o con una mezcla de gasolina y etanol.

terreno inundable Terreno contiguo a un río.

ceniza voladora Residuo que se recoge de la chimenea o del tubo de salida de una caldera.

cadena trófica Secuencia de consumo desde productores hasta consumidores terciarios.

inseguridad alimentaria Condición en la que las personas no cuentan con acceso adecuado a alimentos.

seguridad alimentaria Condición en la que las personas tienen acceso asegurado a alimentos nutritivos en cantidad suficiente para satisfacer las necesidades dietéticas que se precisan para llevar una vida activa y saludable.

red trófica Modelo complejo que explica cómo la energía y la materia se desplazan entre niveles tróficos.

bosque Terreno dominado por árboles y otra vegetación leñosa, que a veces se usa para la explotación forestal comercial.

carbono fósil Carbono en combustibles fósiles.

combustible fósil Combustible derivado de materiales biológicos que se fosilizaron hace millones de años.

efecto fundador Cambio en la composición genética de una población por descender de un número reducido de individuos colonizadores.

fraqueo Fracturación hidráulica, método para la extracción de petróleo y gas que se vale de líquidos a gran presión que abren a la fuerza fisuras en rocas que se encuentran a grandes profundidades subterráneas.

fractura En geología, fisura que ocurre en la roca cuando esta se enfría.

humedales de agua dulce Biomedio acuático sumergido o saturado de agua al menos una parte de cada año, pero suficientemente pando para dar sustento a vegetación emergente.

celda de combustible Dispositivo electroquímico que convierte un combustible, por ejemplo, el hidrógeno, en una corriente eléctrica.

fuel rod A cylindrical tube that encloses nuclear fuel within a nuclear reactor.

barra de combustible Tubo cilíndrico que contiene el combustible nuclear dentro de un reactor nuclear.

fundamental niche The suite of abiotic conditions under which a species can survive, grow, and reproduce.

nicho fundamental Conjunto de condiciones abióticas bajo las cuales una especie puede sobrevivir, crecer y reproducirse.

G

gene A physical location on the chromosomes within each cell of an organism.

gen Ubicación física en los cromosomas dentro de cada célula de un organismo.

gene flow The process by which individuals move from one population to another and thereby alter the genetic composition of both populations.

migración genética Proceso mediante el cual los individuos se desplazan de una población a otra y así alteran la composición genética de ambas poblaciones.

genetic diversity A measure of the genetic variation among individuals in a population.

diversidad genética Medida de la variación genética entre individuos en una población.

genetic drift A change in the genetic composition of a population over time as a result of random mating.

deriva genética Cambio en la composición genética de una población a través del tiempo producto del apareamiento aleatorio.

genetically modified organism (GMO) An organism produced by copying genes from a species with a desirable trait and inserting them into another species.

organismo transgénico Organismo producido al copiar genes de una especie con un rasgo deseado e insertarlos en otra especie.

genotype The complete set of genes in an individual.

genotipo Conjunto completo de genes de un individuo.

genuine progress indicator (GPI) A measure of economic status that includes personal consumption, income distribution, levels of higher education, resource depletion, pollution, and the health of the population.

indicador de avance genuino Medición del estado económico que incluye el consumo personal, la distribución de ingresos, los niveles de educación superior, el agotamiento de recursos, la contaminación y el estado de salud de la población.

geographic isolation Physical separation of a group of individuals from others of the same species.

aislamiento geográfico Separación física de un grupo de individuos de otros de la misma especie.

geothermal energy Heat energy that comes from the natural radioactive decay of elements deep within Earth.

energía geotérmica Energía térmica que proviene de la descomposición natural radiactiva de elementos en lo profundo de la Tierra.

global change Change that occurs in the chemical, biological, and physical properties of the planet.

cambio global Cambio que se da en las propiedades químicas, biológicas y físicas del planeta.

global climate change Changes in the average weather that occurs in an area over a period of years or decades.

cambio climático global Cambios en el promedio del estado del tiempo que se presentan en una región durante un período de años o décadas.

global warming The warming of the oceans, land masses, and atmosphere of Earth.

calentamiento global Calentamiento de los océanos, las masas terrestres y la atmósfera de la Tierra.

gray smog *See* **sulfurous smog.**

esmog gris *Ver* **sulfurous smog.**

gray water Wastewater from baths, showers, bathrooms, and washing machines.

aguas grises Aguas servidas provenientes de tinas, duchas, salas de baño y máquinas lavadoras de ropa.

Green Revolution A shift in agricultural practices in the twentieth century that included new management techniques, mechanization, fertilization, irrigation, and improved crop varieties, and that resulted in increased food output.

Revolución verde Modificación en el cultivo agrícola que se dio en el siglo XX, la cual incluyó nuevas técnicas de gestión, mecanización, fertilización, riego, así como cultivos mejorados, con lo cual se obtuvo un aumento en la producción alimentaria.

green tax A tax placed on environmentally harmful activities or emissions in an attempt to internalize some of the externalities that may be involved in the life cycle of those activities or products.

impuesto verde Gravamen que se impone a las actividades o emisiones que pueden ser perjudiciales para el medio ambiente con el fin de tratar de internalizar algunas de las externalidades que pueden estar implícitas en el ciclo de vida de dichas actividades o productos.

greenhouse effect Absorption of infrared radiation by atmospheric gases and reradiation of the energy back toward Earth.

efecto invernadero Absorción de radiación infrarroja por los gases atmosféricos y reirradiación de la energía de vuelta a la Tierra.

greenhouse gases Gases in Earth's atmosphere that trap heat near the surface.

gases de efecto invernadero Gases de la atmósfera terrestre que atrapan el calor y lo mantienen cerca de la superficie.

greenhouse warming potential An estimate of how much a molecule of any compound can contribute to global warming over a period of 100 years relative to a molecule of CO_2.

potencial de calentamiento por gases de efecto invernadero Estimación de cuánto una molécula de cualquier compuesto puede contribuir al calentamiento global durante un lapso de 100 años con relación a una molécula de CO_2.

gross domestic product (GDP) A measure of the value of all products and services produced in one year in one country.

producto bruto interno (PBI) Medición del valor de todos los productos y servicios producidos en un país en un año.

gross primary productivity (GPP) The total amount of solar energy that producers in an ecosystem capture via photosynthesis over a given amount of time.

productividad bruta primaria (PBP) Monto total de energía solar que los productores de un ecosistema captan por fotosíntesis durante un lapso de tiempo dado.

ground source heat pump A technology that transfers heat from the ground to a building.

bomba térmica de origen terrestre Tecnología mediante la cual se transfiere el calor de la tierra al interior de un edificio.

groundwater recharge A process by which water percolates through the soil and works its way into an aquifer.

recarga de aguas freáticas Proceso mediante el cual el agua se filtra a través del suelo y avanza hasta llegar al acuífero.

gyre A large-scale pattern of water circulation that moves clockwise in the Northern Hemisphere and counterclockwise in the Southern Hemisphere.

giro oceánico Sistema a gran escala de la circulación del agua que se desplaza en el sentido de las manecillas del reloj en el hemisferio septentrional (Norte) y en el sentido contrario al de las manecillas del reloj en el hemisferio austral (Sur).

H

Hadley cell A convection current in the atmosphere that cycles between the equator and 30° N and 30° S.

célula de Hadley Corriente de convección en la atmósfera entre el ecuador y 30° N y 30° S.

half-life The time it takes for one-half of an original radioactive parent atom to decay.

hemivida Tiempo necesario para que se desintegre la mitad de un átomo radiactivo original.

hazardous waste Liquid, solid, gaseous, or sludge waste material that is harmful to humans or ecosystems.

residuos nocivos Material líquido, sólido, gaseoso o fangoso que es perjudicial para los seres humanos o para algún ecosistema.

haze Reduced visibility.

bruma Visibilidad reducida.

herbicide A pesticide that targets plant species that compete with crops.

herbicida Plaguicida que se enfoca en especies de plantas que compiten con los cultivos.

herbivore A consumer that eats producers. *Also known as* **primary consumer.**

herbívoro Consumidor que come productores. *También se conoce como* **consumidor primario.**

herbivory An interaction in which an animal consumes a producer.

herbivorio Interacción en la que un animal consume un productor.

heterotroph *See* **consumer.**

heterótrofo *Ver* **consumer.**

Highway Trust Fund A U.S. federal fund that pays for the construction and maintenance of roads and highways.

Fondo Fideicomisario para el Transporte Fondo federal estadounidense que paga por la construcción y el mantenimiento de carreteras y autopistas.

horizon A horizontal layer in a soil defined by distinctive physical features such as texture and color.

horizonte Capa horizontal en un suelo que se define por sus rasgos físicos característicos tales como textura y color.

hot spot In geology, a place where molten material from Earth's mantle reaches the lithosphere.

punto caliente En geología, sitio en el que el material en estado líquido del manto terrestre alcanza la litosfera.

Hubbert curve A bell-shaped curve representing oil use and projecting both when world oil production will reach a maximum and when the world will run out of oil.

curva de Hubbert Curva en forma de campana que representa tanto el aprovechamiento de petróleo como proyecciones de cuándo la producción mundial de petróleo alcanzará un máximo y cuándo se agotará el petróleo en todo el mundo.

human capital Human knowledge and abilities.

capital humano Conocimientos y habilidades de los seres humanos.

human development index (HDI) A measurement index that combines three basic measures of human status: life expectancy; knowledge and education.

índice de desarrollo humano Índice de medición que combina tres mediciones básicas del estatus del ser humano: expectativa de vida, conocimientos y educación.

Human Immunodeficiency Virus (HIV) A type of virus that causes Acquired Immune Deficiency Syndrome (AIDS).

virus de la inmunodeficiencia humana (VIH) Tipo de virus que causa el síndrome de inmunodeficiencia adquirida (sida).

human poverty index (HPI) A measurement index developed by the United Nations to investigate the proportion of a population suffering from deprivation in a country with a high HDI.

índice de pobreza humana Índice de medición creado por Naciones Unidades para investigar la proporción de una población que sufre de privación en un país con un índice de desarrollo humano alto.

hydroelectricity Electricity generated by the kinetic energy of moving water.

hidroelectricidad Electricidad generada por la energía cinética del agua en movimiento.

hydrogen bond A weak chemical bond that forms when hydrogen atoms that are covalently bonded to one atom are attracted to another atom on another molecule.

enlace de hidrógeno Enlace químico débil que se forma cuando los átomos de hidrógeno que tienen un enlace covalente con un átomo son atraídos por otro átomo en otra molécula.

hydrologic cycle The movement of water through the biosphere.

ciclo hidrológico Movimiento del agua por la biosfera.

hydroponic agriculture The cultivation of plants in greenhouse conditions by immersing roots in a nutrient-rich solution.

agricultura hidropónica Cultivo de plantas en condiciones de invernadero mediante la inmersión de las raíces en una solución rica en nutrientes.

hypothesis A testable conjecture about how something works.

hipótesis Conjetura comprobable acerca de cómo funciona algo.

hypoxic Low in oxygen.

hipóxico De poco oxígeno.

I

igneous rock Rock formed directly from magma.

roca ígnea Roca que se formó directamente del magma.

immigration The movement of people into a country or region, from another country or region.

inmigración Desplazamiento de personas a un país o región, provenientes de otro país o región.

impermeable surface Pavement or buildings that do not allow water penetration.

superficie impermeable Pavimento o edificios que no permiten que el agua penetre.

inbreeding depression When individuals with similar genotypes—typically relatives—breed with each other and produce offspring that have an impaired ability to survive and reproduce.

depresión por endogamia Cuando individuos con genotipos parecidos — característicamente parientes — se aparean entre sí y producen descendientes a quienes les es imposible sobrevivir y reproducirse.

incentive-based approach A strategy for pollution control that constructs financial and other incentives for lowering emissions based on profits and benefits.

enfoque basado en incentivos Estrategia para controlar la contaminación que propone incentivos económicos y de otra índole por disminuir las emisiones. Se basa en utilidades y beneficios.

incineration The process of burning waste materials to reduce volume and mass, sometimes to generate electricity or heat.

incineración Proceso de quemar desechos para reducir su volumen y masa. A veces se usa para generar electricidad o calor.

indicator species A species that indicates whether or not disease-causing pathogens are likely to be present.

especie indicadora Especie que señala si es o no probable la presencia de patógenos que causan enfermedades.

individual transferable quota (ITQ) A fishery management program in which individual fishers are given a total allowable catch of fish in a season that they can either catch or sell.

cuota transferible individual Programa de gestión de pesquerías en el que a cada pescador individual se le otorga un volumen de pesca para la temporada. Dicha cuota se puede aprovechar pescándola o vendiéndola.

induced demand The phenomenon in which an increase in the supply of a good causes demand to grow.

demanda inducida Fenómeno en el que un aumento en la oferta de un bien da por resultado un aumento de la demanda.

industrial agriculture Agriculture that applies the techniques of mechanization and standardization. *Also known as* **agribusiness.**

agricultura industrial Agricultura que aplica las técnicas de mecanización y normalización. *También se conoce como* **agroindustria.**

industrial smog *See* **sulfurous smog.**

esmog industrial *Ver* **sulfurous smog.**

infant mortality The number of deaths of children under 1 year of age per 1,000 live births.

mortalidad infantil Número de muertes de niños menores de un año por millar de partos vivos.

infectious disease A disease caused by a pathogen.

enfermedad infecciosa Enfermedad causada por un patógeno.

infill Development that fills in vacant lots within existing communities.

aprovechamiento Desarrollo de lotes baldíos dentro de comunidades establecidas.

innocent-until-proven-guilty principle A principle based on the belief that a potential hazard should not be considered an actual hazard until the scientific data definitively demonstrate that it actually causes harm.

principio de ser inocente hasta que se compruebe la culpabilidad Principio que se basa en la convicción de que un posible peligro no debe ser considerado un peligro real sino hasta que se cuente con datos científicos que demuestren de manera definitiva que es perjudicial.

inorganic compound A compound that does not contain the element carbon or contains carbon bound to elements other than hydrogen.

compuesto inorgánico Compuesto que no contiene el elemento carbono ni contiene carbono enlazado a elementos que no sean el hidrógeno.

inorganic fertilizer *See* **synthetic fertilizer.**

fertilizante inorgánico *Ver* **synthetic fertilizer.**

input An addition to a system.

entrada Algo que se le agrega al sistema.

insecticide A pesticide that targets species of insects and other invertebrates that consume crops.

insecticida Plaguicida que se enfoca en especies de insectos y otros invertebrados que consumen cultivos.

instrumental value Worth as an instrument or a tool that can be used to accomplish a goal.

valor instrumental Valor como instrumento o recurso que se puede emplear para lograr un cometido.

integrated pest management (IPM) An agricultural practice that uses a variety of techniques designed to minimize pesticide inputs.

gestión integrada de plagas Técnica agrícola que emplea una diversidad de técnicas diseñadas para minimizar el uso de plaguicidas.

integrated waste management An approach to waste disposal that employs several waste reduction, management, and disposal strategies in order to reduce the environmental impact of MSW.

gestión integrada de desechos Manera de enfocar la disposición de desechos que emplea varias estrategias para la reducción, la gestión y la eliminación de desechos con el fin de mitigar el impacto de los desechos sólidos municipales.

intercropping An agricultural method in which two or more crop species are planted in the same field at the same time to promote a synergistic interaction.

cultivos asociados Técnica agrícola en la que dos o más especies de cultivos se siembran en el mismo terreno con el fin de fomentar una interacción de sinergia.

intermediate disturbance hypothesis The hypothesis that ecosystems experiencing intermediate levels of disturbance are more diverse than those with high or low disturbance levels.

hipótesis de la perturbación intermedia Hipótesis que explica que los ecosistemas experimentan niveles intermedios de perturbaciones que son más diversos que aquellos con niveles de disturbios altos o bajos.

intertidal zone The narrow band of coastline between the levels of high tide and low tide.

zona intermareal Banda delgada en el litoral entre los niveles conocidos de marea alta y marea baja.

intertropical convergence zone (ITCZ) The latitude that receives the most intense sunlight, which causes the ascending branches of the two Hadley cells to converge.

zona de convergencia intertropical Latitud que recibe la luz solar más intensa, lo que lleva a que las ramas ascendentes de la dos células de Hadley se junten.

intrinsic growth rate (*r*) The maximum potential for growth of a population under ideal conditions with unlimited resources.

tasa de crecimiento intrínseco (*r*) Máximo potencial de crecimiento de una población en condiciones ideales con recursos ilimitados.

intrinsic value Value independent of any benefit to humans.

valor intrínseco Valor independiente de todo beneficio para los seres humanos.

intrusive igneous rock Igneous rock that forms when magma rises up and cools in a place underground.

roca ígnea intrusiva Roca ígnea que se forma cuando el magma asciende y se enfría en un sitio subterráneo.

invasive species A species that spreads rapidly across large areas.

especie invasora Especie que se disemina rápidamente a través de zonas extensas.

inversion layer The layer of warm air that traps emissions in a thermal inversion.

capa de inversión Capa de aire cálido que atrapa las emisiones en una inversión térmica.

ionic bond A chemical bond between two ions of opposite charges.

enlace iónico Enlace químico entre dos iones con cargas opuestas.

IPAT equation An equation used to estimate the impact of the human lifestyle on the environment: Impact = population × affluence × technology.

ecuación IPAT Ecuación que se usa para estimar el impacto del estilo de vida humano en el medio ambiente: Impacto = población × afluencia × tecnología.

isotopes Atoms of the same element with different numbers of neutrons.

isótopos Átomos del mismo elemento con diferentes cantidades de neutrones.

J

joule (J) The amount of energy used when a 1-watt electrical device is turned on for 1 second.

julio (J) Cantidad de energía que se usa cuando se enciende un dispositivo eléctrico de 1 vatio por un segundo.

J-shaped curve The curve of the exponential growth model when graphed.

curva con forma de J Curva del modelo de crecimiento exponencial cuando se representa de manera gráfica.

K

keystone species A species that plays a far more important in its community than its relative abundance might suggest.

especie clave Especie que desempeña un papel mucho más importante en su comunidad que lo que indicaría su abundancia relativa.

kinetic energy The energy of motion.

energía cinética Energía del movimiento.

***K*-selected species** A species with a low intrinsic growth rate that causes the population to increase slowly until it reaches carrying capacity.

especie de selección *K* Especie con una tasa de crecimiento intrínseco baja que es motivo de que la población crezca lentamente hasta que llegue a su capacidad poblacional.

Kyoto Protocol An international agreement that sets a goal for global emissions of greenhouse gases from all industrialized countries to be reduced by 5.2 percent below their 1990 levels by 2012.

protocolo de Kyoto Convenio internacional que fija una meta para el total de emisiones mundiales de gases de efecto invernadero provenientes de todos los países industrializados. Para el año 2012, dichos gases se habrían de reducir por 5.2 por ciento a los niveles que tenían en 1990.

L

Lacey Act A U.S. act that prohibits interstate shipping of all illegally harvested plants and animals.

Ley Lacey Ley federal estadounidense que prohíbe el despacho interestatal de plantas y animales obtenidos de manera ilícita.

latent heat release The release of energy when water vapor in the atmosphere condenses into liquid water.

liberación térmica latente Liberación de energía cuando el vapor de agua en la atmósfera se condensa y forma agua líquida.

law of conservation of matter A law of nature stating that matter cannot be created or destroyed; it can only change form.

ley de conservación de la materia Ley natural que declara que la materia no se puede crear ni destruir; sólo puede cambiar de forma.

LD50 The lethal dose of a chemical that kills 50 percent of the individuals in a dose-response study.

LD50 Dosis letal de una sustancia química que mata el 50 por ciento de los individuos en un estudio de dosis-respuesta.

leach field A component of a septic system, made up of underground pipes laid out below the surface of the ground.

campo de lixiviación Componente de un sistema séptico compuesto por tuberías subterráneas dispuestas en un diseño planificado.

leachate Liquid that contains elevated levels of pollutants as a result of having passed through municipal solid waste (MSW) or contaminated soil.

lixiviado Líquido que contiene niveles elevados de contaminantes como resultado de haber pasado por desechos sólidos municipales o por suelos contaminados.

leaching The transportation of dissolved molecules through the soil via groundwater.

lixiviación Transporte de moléculas disueltas a través de la tierra por medio de aguas freáticas.

leapfrogging The phenomenon of less developed countries using new technology without first using the precursor technology.

brincos superadores Fenómeno en países menos desarrollados que se valen de tecnologías nuevas sin tener que usar primero la tecnología precursora.

least concern species Species that are widespread and abundant.

especies menos preocupantes Especies generalizadas y abundantes.

levee An enlarged bank built up on each side of a river.

dique de contención Orilla agrandada que se amplía a cada costado de un río.

life expectancy The average number of years that an infant born in a particular year in a particular country can be expected to live, given the current average life span and death rate in that country.

expectativa de vida Cantidad promedio de años que se puede prever que viva un bebé nacido en un año dado y en un país específico, dada la media de vida o la tasa de fallecimientos en dicho país.

life-cycle analysis A systems tool that looks at the materials used and released throughout the lifetime of a product—from the procurement of raw materials through their manufacture, use, and disposal. *Also known as* **cradle-to-grave analysis.**

análisis de ciclo de vida Recurso en sistemas que estudia los materiales utilizados y emanados durante la vida completa de un producto (es decir, desde que se compra la materia prima hasta que se fabrica, utiliza y tira). *También se conoce como* **análisis de la cuna a la tumba.**

limiting nutrient A nutrient required for the growth of an organism but available in a lower quantity than other nutrients.

nutriente limitante Nutriente que se precisa para el crecimiento de un organismo, pero que está disponible en cantidad menor que otros nutrientes.

limiting resource A resource that a population cannot live without and that occurs in quantities lower than the population would require to increase in size.

recurso limitante Recurso sin el que una población no puede vivir y que existe en cantidades menores de las que la población precisaría para aumentar en volumen.

limnetic zone A zone of open water in lakes and ponds.

zona limnética Espacio de aguas abiertas en lagos y lagunas.

lipid A smaller organic biological molecule that does not mix with water.

lípido Molécula biológica orgánica pequeña que no se mezcla con agua.

lithosphere The outermost layer of Earth, including the mantle and crust.

litosfera Capa exterior de la Tierra que incluye el manto y la corteza.

littoral zone The shallow zone of soil and water in lakes and ponds where most algae and emergent plants grow.

zona litoral El espacio pando de tierra y agua en lagos y lagunas, en los que crece la mayoría de las algas y las plantas emergentes.

logistic growth model A growth model that describes a population whose growth is initially exponential, but slows as the population approaches the carrying capacity of the environment.

modelo de crecimiento logístico Modelo de crecimiento que describe una población cuyo crecimiento inicialmente es exponencial, pero que pierde velocidad a medida que se aproxima a la capacidad poblacional del entorno.

London-type smog *See* **sulfurous smog.**

esmog tipo londinense *Ver* **sulfurous smog.**

Los Angeles–type smog *See* **photochemical smog.**

esmog tipo angelino o de Los Ángeles *Ver* **photochemical smog.**

M

macroevolution Evolution that gives rise to new species, genera, families, classes, or phyla.

macroevolución Evolución de la que surgen nuevas especies, géneros, familias, clases o filos.

macronutrient One of six key elements that organisms need in relatively large amounts: nitrogen, phosphorus, potassium, calcium, magnesium, and sulfur.

macronutriente Uno de seis elementos clave que los organismos necesitan en cantidades relativamente grandes: nitrógeno, fósforo, potasio, calcio, magnesio y azufre.

mad cow disease A disease in which prions mutate into deadly pathogens and slowly damage a cow's nervous system.

enfermedad de las vacas locas Enfermedad en la que priones mutan en patógenos mortales que lentamente perjudican el sistema nervioso de las vacas.

magma Molten rock.

magma Roca fundida.

malnourished Having a diet that lacks the correct balance of proteins, carbohydrates, vitamins, and minerals.

desnutrido Que cuenta con un régimen alimenticio que carece del equilibrio adecuado de proteínas, carbohidratos, vitaminas y minerales.

mangrove swamp A swamp that occurs along tropical and subtropical coasts, and contains salt-tolerant trees with roots submerged in water.

mantle The layer of Earth above the core, containing magma.

manufactured capital All goods and services that humans produce.

manure lagoon Human-made pond lined with rubber built to handle large quantities of manure produced by livestock.

Marine Mammal Protection Act A 1972 U.S. act to protect declining populations of marine mammals.

market failure When the economic system does not account for all costs.

mass A measurement of the amount of matter an object contains.

mass extinction A large extinction of species in a relatively short period of time.

mass number A measurement of the total number of protons and neutrons in an element.

matter Anything that occupies space and has mass.

maximum contaminant level (MCL) The standard for safe drinking water established by the EPA under the Safe Drinking Water Act.

maximum sustainable yield (MSY) The maximum amount of a renewable resource that can be harvested without compromising the future availability of that resource.

meat Livestock or poultry consumed as food.

mesotrophic Describes a lake with a moderate level of productivity.

metal An element with properties that allow it to conduct electricity and heat energy, and to perform other important functions.

metamorphic rock Rock that forms when sedimentary rock, igneous rock, or other metamorphic rock is subjected to high temperature and pressure.

metapopulation A group of spatially distinct populations that are connected by occasional movements of individuals between them.

microevolution Evolution below the species level.

mineralization The process by which fungal and bacterial decomposers break down the organic matter found in dead bodies and waste products and convert it into inorganic compounds.

mining spoils Unwanted waste material created during mining. *Also known as* **tailings.**

modern carbon Carbon in biomass that was recently in the atmosphere.

molecule A particle that contains more than one atom.

manglar Pantano que se da en las costas tropicales y subtropicales, que contiene árboles que toleran la sal y cuyas raíces permanecen sumergidas en el agua.

manto Capa de la Tierra que está sobre el núcleo interno y que contiene el magma.

capital manufacturado Todos los bienes y servicios que producen los seres humanos.

lago anaeróbico o de estiércol Estanque artificial con forro de caucho o hule construido para manejar grandes cantidades de estiércol producido por ganado.

Ley de Protección de los Mamíferos Marinos Ley federal estadounidense promulgada en 1972 con el fin de proteger las poblaciones de mamíferos marinos en disminución.

falla del mercado Ocurre cuando el sistema económico no tiene en cuenta todos los costos.

masa Medición de la cantidad de materia que contiene un objeto.

extinción masiva Extinción terminal de una especie en un período de tiempo relativamente corto.

número másico Medición de la cantidad total de protones y neutrones que tiene un elemento.

materia Todo aquello que ocupa espacio y tiene masa.

nivel máximo de contaminantes La norma que define el agua potable, establecida por la EPA conforme a la Ley de Agua Potable Segura.

producción sostenida máxima Cantidad máxima de un recurso renovable que se puede cosechar sin comprometer la disponibilidad futura de dicho recurso.

carne Ganado o aves que se consumen como alimento.

mesotrópico Se dice de un lago con un nivel moderado de productividad.

metal Elemento con propiedades que le permiten conducir energía eléctrica y térmica, y realizar otras funciones importantes.

roca metamórfica Roca que se forma cuando una roca sedimentaria, una roca ígnea o cualquier otra roca metamórfica se somete a temperatura y presión alta.

metapoblación Grupo de poblaciones separadas espacialmente que están conectadas entre sí por desplazamientos ocasionales de individuos entre ellas.

microevolución Evolución por debajo del nivel de especie.

mineralización Proceso mediante el cual los saprofitos fúngicos y bacterianos desintegran la materia orgánica que se encuentra en los cuerpos muertos y los desechos, y la convierten en compuestos inorgánicos.

relaves de minería Material de desecho creado durante el proceso de minería. *También se conoce como* **colas.**

carbono moderno Carbono en forma de biomasa que hace poco estuvo en la atmósfera.

molécula Partícula que contiene más de un átomo.

monocropping An agricultural method that utilizes large plantings of a single species or variety.

monocultivo Método agrícola en el que se siembran grandes plantaciones de una sola especie o variedad.

mountaintop removal A mining technique in which the entire top of a mountain is removed with explosives.

remoción de cumbres Técnica de minería en la que la cumbre completa de una montaña se elimina con explosivos.

multiple-use lands A U.S. classification used to designate lands that may be used for recreation, grazing, timber harvesting, and mineral extraction.

terrenos de usos múltiples Clasificación que se usa en Estados Unidos para designar terrenos que se pueden usar para fines recreativos, de pastoreo, de silvicultura y para la extracción de minerales.

multi-use zoning A zoning classification that allows retail and high-density residential development to coexist in the same area.

zonificación de múltiples usos Clasificación territorial que autoriza que coexistan en la misma zona tiendas detallistas y urbanizaciones residenciales de alta densidad.

municipal solid waste (MSW) Refuse collected by municipalities from households, small businesses, and institutions.

desechos sólidos municipales Basura recogida por el municipio de los hogares, la pequeña empresa y las instituciones.

mutagen A type of carcinogen that causes damage to the genetic material of a cell.

mutágeno Tipo de agente carcinógeno que causa daños a los materiales genéticos de una célula.

mutation A random change in the genetic code produced by a mistake in the copying process.

mutación Cambio aleatorio en el código genético producido por un error en el proceso de copia.

mutualism An interaction between two species that increases the chances of survival or reproduction for both species.

mutualismo Interacción entre dos especies que aumenta las posibilidades de supervivencia o reproducción para ambas especies.

N

National Environmental Policy Act (NEPA) A 1969 U.S. federal act that mandates an environmental assessment of all projects involving federal money or federal permits.

Ley Normativa Nacional de Integración Ambiental Ley de EE. UU., promulgada en 1969, que obliga a realizar una evaluación ambiental en todo proyecto en el que se usen dineros federales o que precisen una autorización federal.

national wilderness area An area set aside with the intent of preserving a large tract of intact ecosystem or a landscape.

área silvestre nacional Área que se ha separado con la intención de conservar un terreno de mucha extensión preservando intacto un ecosistema o paisaje.

national wildlife refuge A federal public land managed for the primary purpose of protecting wildlife.

refugio silvestre nacional Terreno de propiedad del gobierno federal administrado para el efecto principal de proteger la vida silvestre.

native species Species that live in their historical range, typically where they have lived for thousands or millions of years.

especies nativas Especies que viven dentro de sus hábitats históricos, generalmente donde se han arraigado durante miles o millones de años.

natural capital The resources of the planet, such as air, water, and minerals.

capital natural Recursos del planeta, tales como aire, agua y minerales.

natural experiment A natural event that acts as an experimental treatment in an ecosystem.

experimento natural Acontecimiento natural que hace las veces de tratamiento experimental en un ecosistema.

near-threatened species Species that are very likely to become threatened in the future.

especies casi amenazadas Especies que con mucha probabilidad van a verse amenazadas en un futuro.

negative feedback loop A feedback loop in which a system responds to a change by returning to its original state, or by decreasing the rate at which the change is occurring.

circuito de realimentación negativa Realimentación en la que un sistema responde a un cambio mediante el regreso a su estado original o mediante la disminución de la tasa a la cual sucede el cambio.

net migration rate The difference between immigration and emigration in a given year per 1,000 people in a country.

tasa neta de migración Diferencia entre inmigración y emigración en un año dado por cada millar de personas que viven en un país.

net primary productivity (NPP) The energy captured by producers in an ecosystem minus the energy producers respire.

productividad primaria neta Energía captada por productores en un ecosistema menos la energía que respiran los productores.

net removal The process of removing more than is replaced by growth, typically used when referring to carbon.

eliminación neta Proceso de eliminar más de lo que se reemplaza con crecimiento. Típicamente se usa el término al referirse al carbono.

neurotoxin A chemical that disrupts the nervous systems of animals.

neurotoxina Sustancia química que altera el sistema nervioso de los animales.

niche generalist A species that can live under a wide range of abiotic or biotic conditions.

generalista del nicho Especie que puede vivir bajo una amplia gama de condiciones abióticas o bióticas.

niche specialist A species that is specialized to live in a specific habitat or to feed on a small group of species.

especialista del nicho Especie especializada para vivir en un hábitat específico o para alimentarse de un grupo pequeño de especies.

nitrification The conversion of ammonia (NH_4^+) into nitrite (NO_2^-) and then into nitrate (NO_3^-).

nitrificación Conversión de amonio (NH_4^+) a nitrito (NO_2^-) y luego a nitrato (NO_3^-).

nitrogen cycle The movement of nitrogen around the biosphere.

ciclo de nitrógeno Movimiento del nitrógeno alrededor de la biosfera.

nitrogen fixation A process by which some organisms can convert nitrogen gas molecules directly into ammonia.

fijación de nitrógeno Proceso mediante el cual algunos organismos pueden convertir moléculas de nitrógeno gaseoso directamente en amonio.

nomadic grazing The feeding of herds of animals by moving them to seasonally productive feeding grounds, often over long distances.

pastoreo nomádico Alimentación de manadas de animales mudándolas cada temporada de un terreno de alimentación productiva a otro, a menudo cubriendo grandes distancias.

nondepletable An energy source that cannot be used up.

inagotable Fuente de energía que no se puede agotar ni consumir en su totalidad.

nonpersistent pesticide A pesticide that breaks down rapidly, usually in weeks or months.

plaguicida no persistente Plaguicida que se descompone rápidamente, generalmente en semanas o meses.

nonpoint source A diffuse area that produces pollution.

fuente sin punto Zona indefinida que produce contaminación.

nonrenewable energy resource An energy source with a finite supply, primarily the fossil fuels and nuclear fuels.

recurso energético no renovable Fuente de energía con un abastecimiento finito, primordialmente combustibles fósiles y combustibles nucleares.

no-till agriculture An agricultural method in which farmers do not turn the soil between seasons as a means of reducing topsoil erosion.

agricultura sin arar Método agrícola en el que los agricultores no voltean los suelos entre estaciones a fin de minimizar la erosión de la capa superior o manto.

nuclear fuel Fuel derived from radioactive materials that give off energy.

combustible nuclear Combustible derivado de materiales radiactivos que producen energía.

nuclear fusion A reaction that occurs when lighter nuclei are forced together to produce heavier nuclei.

fusión nuclear Reacción que ocurre cuando se amontonan núcleos ligeros para producir núcleos más pesados.

nucleic acid Organic compounds found in all living cells.

ácido nucleico Compuestos orgánicos que se encuentran en todas las células vivientes.

null hypothesis A prediction that there is no difference between groups or conditions, or a statement or an idea that can be falsified, or proved wrong.

hipótesis nula Predicción de que no hay diferencia entre los grupos o las condiciones, o declaración o concepto que puede ser falsificado o que se puede demostrar que es erróneo.

O

O horizon The organic horizon at the surface of many soils, composed of organic detritus in various stages of decomposition.

horizonte O Horizonte orgánico en la superficie de muchos suelos, compuesto por desperdicios en diversas etapas de descomposición.

Occupational Safety and Health Administration (OSHA) An agency of the U.S. Department of Labor, responsible for the enforcement of health and safety regulations.

Administración de Seguridad y Salud Ocupacional (OSHA) Dependencia del gobierno federal estadounidense. Forma parte del Departamento del Trabajo y se responsabiliza de hacer cumplir las reglamentaciones de salud y seguridad.

ocean acidification The process by which an increase in ocean CO_2 causes more CO_2 to be converted to carbonic acid, which lowers the pH of the water.

oil sands Slow-flowing, viscous deposits of bitumen mixed with sand, water, and clay.

oligotrophic Describes a lake with a low level of productivity.

open ocean Deep ocean water, located away from the shoreline where sunlight can no longer reach the ocean bottom.

open system A system in which exchanges of matter or energy occur across system boundaries.

open-loop recycling Recycling one product into a different product.

open-pit mining A mining technique that uses a large visible pit or hole in the ground.

ore A concentrated accumulation of minerals from which economically valuable materials can be extracted.

organic agriculture Production of crops without the use of synthetic pesticides or fertilizers.

organic compound A compound that contains carbon-carbon and carbon-hydrogen bonds.

organic fertilizer Fertilizer composed of organic matter from plants and animals.

output A loss from a system.

overnutrition Ingestion of too many calories and a lack of balance of foods and nutrients.

overshoot When a population becomes larger than the environment's carrying capacity.

oxygenated fuel A fuel with oxygen as part of the molecule.

ozone (O_3) A secondary pollutant made up of three oxygen atoms bound together.

P

pandemic An epidemic that occurs over a large geographic region.

parasitism An interaction in which one organism lives on or in another organism.

parasitoid A specialized type of predator that lays eggs inside other organisms—referred to as its host.

parent material The rock material from which the inorganic components of a soil are derived.

particles *See* **particulate matter.**

particulate matter (PM) Solid or liquid particles suspended in air. *Also known as* **particulates; particles.**

particulates *See* **particulate matter.**

passive solar design Construction designed to take advantage of solar radiation without active technology.

pathogens Parasites that cause disease in their host.

acidificación de los océanos Proceso mediante el cual un aumento en el CO_2 en el océano puede convertirse en ácido carbónico, el cual baja el pH del agua.

arenas petrolíferas Depósitos bituminosos viscosos de flujo lento mezclados con arena, agua y arcilla.

oligotrópico Se dice de un lago con un nivel bajo de productividad.

mar abierto Aguas profundas del océano, ubicadas a una distancia del litoral donde la luz solar ya no puede llegar hasta el fondo.

sistema abierto Sistema en el que los intercambios de materia o energía se dan a través de las fronteras del sistema.

reciclaje en circuito abierto Reciclaje de un producto para terminar con otro producto diferente.

minería a cielo abierto Técnica de minería en la que se usa un tajo o agujero grande y visible en el suelo.

mena Acumulación concentrada de minerales de los que se pueden extraer materiales de valor económico.

agricultura orgánica Producción de cultivos sin usar plaguicidas ni fertilizantes sintéticos.

compuesto orgánico Compuesto que contiene enlaces de carbono con carbono y carbono con hidrógeno.

fertilizante orgánico Fertilizante hecho de materia orgánica tomada de plantas y animales.

salida Pérdida en un sistema.

sobrenutrición Ingestión de demasiadas calorías y la falta de equilibrio entre alimentos y nutrientes.

extralimitación Cuando una población crece más que la capacidad poblacional del entorno.

combustible oxigenado Combustible que incluye oxígeno como parte de la molécula.

ozono (O_3) Contaminante secundario compuesto por tres átomos de oxígeno ligados entre sí.

pandemia Epidemia que cuando ocurre abarca una región geográfica amplia.

parasitismo Interacción en la que un organismo vive sobre o en otro organismo.

parasitoide Tipo especializado de depredador que pone huevos dentro de otros organismos, a los que se los designa huéspedes.

material parental Material rocoso a partir del cual se derivan los componentes inorgánicos del suelo.

partículas *Ver* **particulate matter.**

materia particulada Partículas sólidas o líquidas suspendidas en el aire. *También se conoce como* **particulados; partículas.**

particulados *Ver* **particulate matter.**

diseño solar pasivo Construcción diseñada para aprovechar la radiación solar sin tecnología activa.

patógenos Parásitos que pueden enfermar a su huésped (u hospedador).

peak demand The greatest quantity of energy used at any one time.

demanda pico Cantidad máxima de energía que se utiliza simultáneamente.

peak oil The point at which half the total known oil supply is used up.

pico petrolero Punto en el cual se ha consumido la mitad del total conocido de las existencias de petróleo.

perchlorates A group of harmful chemicals used for rocket fuel.

percloratos Grupo de sustancias químicas nocivas que se utilizan de combustible en cohetes.

perennial plant A plant that lives for multiple years.

planta perenne Planta que vive múltiples años.

periodic table A chart of all chemical elements currently known, organized by their properties.

tabla periódica Cuadro de todos los elementos químicos conocidos a la fecha, organizados según sus propiedades.

permafrost An impermeable, permanently frozen layer of soil.

permafrost Capa del suelo impermeable y permanentemente congelada.

persistence The length of time a chemical remains in the environment.

persistencia Tiempo que una sustancia química permanece en el medio ambiente.

persistent pesticide A pesticide that remains in the environment for a long time.

plaguicida persistente Plaguicida que permanece en el medio ambiente mucho tiempo.

pesticide A substance, either natural or synthetic, that kills or controls organisms that people consider pests.

plaguicida Sustancia, sea natural o sintética, que mata o controla organismos que las personas consideran plagas.

pesticide resistance A trait possessed by certain individuals that are exposed to a pesticide and survive.

resistencia a plaguicidas Rasgo de ciertos individuos que se exponen a un plaguicida y sobreviven.

pesticide treadmill A cycle of pesticide development, followed by pest resistance, followed by new pesticide development.

círculo vicioso de plaguicidas Ciclo de la creación de un plaguicida, seguida de resistencia al mismo de las plagas, seguida de la creación de un nuevo plaguicida.

petroleum A fossil fuel that occurs in underground deposits, composed of a liquid mixture of hydrocarbons, water, and sulfur.

petróleo Combustible fósil que se da en yacimientos subterráneos, compuesto por una mezcla líquida de hidrocarburos, agua y azufre.

pH The number that indicates the relative strength of acids and bases in a substance.

pH Cifra que indica la potencia relativa de ácidos y bases en una sustancia.

phenotype A set of traits expressed by an individual.

fenotipo Conjunto de rasgos que se manifiestan en un individuo.

phosphorus cycle The movement of phosphorus around the biosphere.

ciclo de fósforo Movimiento de fósforo alrededor de la biosfera.

photic zone The upper layer of ocean water in the ocean that receives enough sunlight for photosynthesis.

zona fótica Capa superior del agua del océano en la que el océano recibe suficiente luz solar para la fotosíntesis.

photochemical oxidant A class of air pollutants formed as a result of sunlight acting on compounds such as nitrogen oxides.

oxidante fotoquímico Clase de contaminantes aéreos que se forman como resultado de la acción de la luz solar sobre compuestos tales como óxidos de nitrógeno.

photochemical smog Smog that is dominated by oxidants such as ozone. *Also known as* **Los Angeles–type smog; brown smog.**

esmog fotoquímico Esmog dominado por oxidantes tales como ozono. *También se conoce como* **esmog angelino o de Los Ángeles** o como **esmog marrón.**

photon A massless packet of energy that carries electromagnetic radiation at the speed of light.

fotón Paquete de energía sin masa que transporta radiación electromagnética a la velocidad de la luz.

photosynthesis The process by which producers use solar energy to convert carbon dioxide and water into glucose.

fotosíntesis Proceso mediante el cual los productores aprovechan la energía solar para convertir el dióxido de carbono y el agua en glucosa.

photovoltaic solar cell A system of capturing energy from sunlight and converting it directly into electricity.

celda solar fotovoltaica Sistema para captar la energía del sol y convertirla directamente en electricidad.

phylogeny The branching pattern of evolutionary relationships.

filogenia Diseño de ramas de las interrelaciones evolucionarias.

physical weathering The mechanical breakdown of rocks and minerals.

desgaste físico Descomposición mecánica de rocas y minerales.

phytoplankton Floating algae.

fitoplancton Algas flotantes.

pioneer species A species that can colonize new areas rapidly and grow well in full sunshine.

especie pionera Especie que puede colonizar áreas nuevas rápidamente y crecer bien cuando hay plenitud de luz solar.

placer mining The process of looking for minerals, metals, and precious stones in river sediments.

minería de arenal Proceso mediante el cual se buscan minerales, metales y piedras preciosas en los sedimentos de ríos.

plague An infectious disease caused by a bacterium (*Yersinia pestis*) that is carried by fleas.

peste Enfermedad infecciosa causada por una bacteria (*Yersinia pestis*) transportada por pulgas.

plate tectonics The theory that the lithosphere of Earth is divided into plates, most of which are in constant motion.

tectónica de placas Teoría que explica que la litosfera de la Tierra está dividida en placas, la mayoría de las cuales está en constante movimiento.

point source A distinct location from which pollution is directly produced.

fuente de punto Ubicación definida en la que se produce la contaminación directamente.

polar cell A convection current in the atmosphere, formed by air that rises at 60° N and 60° S and sinks at the poles, 90° N and 90° S.

célula polar Corriente de convección en la atmósfera, formada por aire que asciende a 60° N y 60° S y se hunde en los polos, 90° N y 90° S.

polar molecule A molecule in which one side is more positive and the other side is more negative.

molécula polar Molécula en la que un lado es más positivo y el otro es más negativo.

polychlorinated biphenyls (PCBs) A group of industrial compounds used to manufacture plastics and insulate electrical transformers, and responsible for many environmental problems.

bifenilos policlorados Grupo de compuestos industriales que se usan para fabricar plásticos y aislar transformadores eléctricos. Son responsables de muchos problemas medioambientales.

population density The number of individuals per unit area at a given time.

densidad poblacional Cantidad de individuos en un área dada en un momento dado.

population distribution A description of how individuals are distributed with respect to one another.

distribución de la población Descripción de cómo los individuos se distribuyen con respecto a los demás.

population ecology The study of factors that cause populations to increase or decrease.

ecología poblacional Estudio de los factores que llevan a que las poblaciones aumenten o disminuyan.

population growth models Mathematical equations that can be used to predict population size at any moment in time.

modelos de crecimiento poblacional Ecuaciones matemáticas que se pueden utilizar para predecir el volumen poblacional en un momento dado.

population growth rate The number of offspring an individual can produce in a given time period, minus the deaths of the individual or its offspring during the same period.

tasa de crecimiento poblacional La cantidad de descendientes que un individuo puede producir en un período de tiempo dado, menos los fallecimientos del individuo o su descendencia durante ese mismo período de tiempo.

population momentum Continued population growth after growth reduction measures have been implemented.

momento poblacional Continuación del crecimiento de la población después de implementarse medidas para reducir la población.

population pyramid An age structure diagram that is widest at the bottom and smallest at the top, typical of developing countries.

pirámide poblacional Diagrama de estructuras de edad que es más ancho en la parte inferior y más pequeño en la cima, aspecto característico de los países en vías de desarrollo.

population size (*N*) The total number of individuals within a defined area at a given time.

volumen poblacional (*N*) Número total de individuos dentro de un área definida en un momento dado.

population The individuals that belong to the same species and live in a given area at a particular time.

población Individuos que pertenecen a la misma especie y conviven en un área dada en un momento dado.

positive feedback loop A feedback loop in which change in a system is amplified.

circuito de realimentación positiva Realimentación en la que se amplifica cualquier cambio en un sistema.

potential energy Stored energy that has not been released.

energía potencial Energía almacenada que no se ha liberado.

potentially renewable An energy source that can be regenerated indefinitely as long as it is not overharvested.

power The rate at which work is done.

precautionary principle A principle based on the belief that action should be taken against a plausible environmental hazard.

precision How close the repeated measurements of a sample are to one another.

predation An interaction in which one animal typically kills and consumes another animal.

prescribed burn A fire deliberately set under controlled conditions in order to reduce the accumulation of dead biomass on a forest floor.

primary consumer *See* **herbivore.**

primary pollutant A polluting compound that comes directly out of a smokestack, exhaust pipe, or natural emission source.

primary succession Ecological succession occurring on surfaces that are initially devoid of soil.

prion A small, beneficial protein that occasionally mutates into a pathogen.

producer An organism that uses the energy of the Sun to produce usable forms of energy. *Also known as* **autotroph.**

profundal zone A region of water where sunlight does not reach, below the limnetic zone in very deep lakes.

prospective study A study that monitors people who might become exposed to harmful chemicals in the future.

protein A critical component of living organisms made up of a long chain of nitrogen-containing organic molecules known as amino acids.

provision A good that humans can use directly.

potencialmente renovable Fuente de energía que se puede regenerar indefinidamente siempre y cuando no se sobreaproveche.

potencia Ritmo al cual se trabaja.

principio precaucionario Principio basado en la convicción de que se debe poner en práctica alguna acción contra un peligro ambiental plausible.

precisión Cercanía que existe entre las reiteradas mediciones de una muestra.

depredación Interacción en la que un animal característicamente mata y consume otro animal.

quema prescrita Incendio que se comienza a propósito y en condiciones controladas a fin de reducir la acumulación de biomasa muerta en un suelo forestal.

consumidor primario *Ver* **herbivore.**

contaminante primario Compuesto contaminante que proviene directamente de una chimenea, un tubo de escape o una fuente de emisión natural.

sucesión primaria Sucesión ecológica que se da en las superficies que inicialmente carecen de tierra.

prión Proteína pequeña y benigna que a veces muta en un patógeno.

productor Organismo que aprovecha la energía del Sol para producir formas utilizables de energía. *También se conoce como* **autótrofo.**

zona de profundidad Región del agua a la que no llega la luz solar, debajo de la zona limnética en lagos muy profundos.

estudio prospectivo Estudio en el que se vigilan personas que en un futuro podrían quedar expuestas a sustancias nocivas.

proteína Componente crítico de los organismos vivientes compuesto por una larga cadena de moléculas orgánicas que contienen nitrógeno conocidas como aminoácidos.

provisión Bien que los seres humanos pueden aprovechar directamente.

R

radioactive decay The spontaneous release of material from the nucleus of radioactive isotopes.

radioactive waste Nuclear fuel that can no longer produce enough heat to be useful in a power plant but continues to emit radioactivity.

rain shadow A region with dry conditions found on the leeward side of a mountain range as a result of humid winds from the ocean causing precipitation on the windward side.

range of tolerance The limits to the abiotic conditions that a species can tolerate.

rangeland A dry open grassland.

descomposición radiactiva Liberación espontánea de material desde el núcleo de los isótopos radiactivos.

residuos radiactivos Combustible nuclear que ya no puede producir suficiente calor para ser útil en una planta generadora pero que sigue emitiendo radiactividad.

sombra pluviométrica Región de condiciones secas que se encuentra en el lado de sotavento de una cordillera como resultado de los vientos húmedos que provienen del océano que producen precipitación en el lado de barlovento.

gama de tolerancia Límites de las condiciones abióticas que puede tolerar una especie.

pastizal Terreno seco y despejado con abundancia de pasto.

REACH A 2007 agreement among the nations of the European Union about regulation of chemicals; the acronym stands for registration, evaluation, authorization, and restriction of chemicals.

REACH Convenio logrado en el año 2007 entre los países de la Unión Europea en el que se controlan las sustancias químicas. La sigla representa en inglés registro, evaluación, autorización y restricción de las sustancias químicas.

realized niche The range of abiotic and biotic conditions under which a species actually lives.

nicho realizado Gama de condiciones abióticas y bióticas bajo las cuales vive una especie.

recombination The genetic process by which one chromosome breaks off and attaches to another chromosome during reproductive cell division.

recombinación Proceso genético en el cual un cromosoma se separa y se adhiere a otro cromosoma durante la división reproductiva de células.

recycling The process by which materials destined to become municipal solid waste (MSW) are collected and converted into raw material that is then used to produce new objects.

reciclaje Proceso en el que se recogen los materiales destinados a convertirse en desechos sólidos municipales y se convierten en materia prima que luego se aprovecha para producir objetos nuevos.

Red List A list of worldwide threatened species.

Lista Roja Lista de las especies amenazadas en todo el mundo.

reduce, reuse, recycle A popular phrase promoting the idea of diverting materials from the waste stream. *Also known as* **the three Rs.**

reducir, reutilizar, reciclar Frase de gran acogida que fomenta el concepto de extraer materiales del flujo de desechos. *También se conoce como* **las tres erres.**

renewable In energy management, an energy source that is either potentially renewable or nondepletable.

renovable En la gestión de energía, fuente energética que tiene el potencial de ser renovable o inagotable.

replacement-level fertility The total fertility rate required to offset the average number of deaths in a population in order to maintain the current population size.

fertilidad a nivel de reemplazo Tasa de fertilidad total requerida para compensar la cantidad promedio de fallecimientos en una población a fin de mantener el volumen poblacional actual.

replication The data collection procedure of taking repeated measurements.

replicación Procedimiento de recolección de datos en el que se repiten las mediciones.

reproductive isolation The result of two populations within a species evolving separately to the point that they can no longer interbreed and produce viable offspring.

aislamiento reproductor Lo que resulta cuando dos poblaciones dentro de una especie evolucionan separadamente hasta el punto de que ya no pueden aparearse ni producir descendientes viables.

reserve In resource management, the known quantity of a resource that can be economically recovered.

reserva En la gestión de recursos, la cantidad conocida de un recurso que se puede recuperar en términos económicos.

reservoir The water body created by a damming a river or stream.

estanque Extensión de agua creada cuando se le coloca una presa a un río o arroyo.

resilience The rate at which an ecosystem returns to its original state after a disturbance.

resiliencia Ritmo al cual un ecosistema regresa a su estado original después de una perturbación.

resistance A measure of how much a disturbance can affect flows of energy and matter in an ecosystem.

resistencia Medición del efecto que una perturbación puede tener respecto de los flujos de energía y materia en un ecosistema.

resource conservation ethic The belief that people should maximize use of resources, based on the greatest good for everyone.

ética de conservación de recursos Noción de que las personas deben aprovechar al máximo los recursos, basándose en el mayor bienestar para todos.

resource partitioning When two species divide a resource based on differences in their behavior or morphology.

partición de recursos Cuando dos especies se dividen un recurso según diferencias en su comportamiento o morfología.

restoration ecology The study and implementation of restoring damaged ecosystems.

ecología de restauración Estudio de ecosistemas perjudicados y de cómo restaurarlos.

retrospective study A study that monitors people who have been exposed to an environmental hazard at some time in the past.

estudio retrospectivo Estudio en el que se vigilan personas que han sido expuestas a un peligro medioambiental en algún momento en el pasado.

reuse Using a product or material that was intended to be discarded.

reutilizar Hacer uso de algún producto o material que se iba a tirar.

reverse osmosis A process of desalination in which water is forced through a thin semipermeable membrane at high pressure.

ósmosis inversa Proceso de desalinización en el que el agua se pasa por una membrana delgada y semipermeable bajo gran presión.

Richter scale A scale that measures the largest ground movement that occurs during an earthquake.

escala de Richter Escala que mide los movimientos telúricos de mayor escala que ocurren durante un terremoto.

RNA (ribonucleic acid) A nucleic acid that translates the code stored in DNA, which makes possible the synthesis of proteins.

ARN (ácido ribonucleico) Ácido nucleico que traduce el código almacenado en el ADN, que posibilita la síntesis de proteínas.

rock cycle The geologic cycle governing the constant formation, alteration, and destruction of rock material that results from tectonics, weathering, and erosion, among other processes.

ciclo rocoso Ciclo geológico que dirige la constante formación, alteración y destrucción de material rocoso que es resultado de la tectónica, la desintegración y la erosión, entre otros procesos.

route of exposure The way in which an individual might come into contact with an environmental hazard.

ruta de exposición Manera en que un individuo puede entrar en contacto con un peligro medioambiental.

***r*-selected species** A species that has a high intrinsic growth rate, which often leads to population overshoots and die-offs.

especies de selección *r* Especie que tiene una alta tasa de crecimiento intrínseco, lo cual a menudo conduce a extralimitarse en la población y a mortandades.

run off Water that moves across the land surface and into streams and rivers.

escorrentía Agua que se desplaza a través de la superficie terrestre y desemboca en ríos y arroyos.

run-of-the-river Hydroelectricity generation in which water is retained behind a low dam or no dam.

flujo del agua Generación hidroeléctrica en la que el agua se retiene detrás de una presa pequeña o se aprovecha sin presa.

S

Safe Drinking Water Act Legislation that sets the national standards for safe drinking water.

Ley de Agua Potable Segura Legislación federal estadounidense que fija las normas para el agua potable segura.

salinization A form of soil degradation that occurs when the small amount of salts in irrigation water becomes highly concentrated on the soil surface through evaporation.

salinización Forma de degradación del suelo que ocurre cuando una cantidad reducida de sales en el agua de riego termina con una concentración alta en la superficie del suelo por medio de la evaporación.

salt marsh A marsh containing nonwoody emergent vegetation, found along the coast in temperate climates.

marisma salina Marisma que contiene vegetación emergente no leñosa que se encuentra en el litoral en climas templados.

saltwater intrusion An infiltration of salt water in an area where groundwater pressure has been reduced from extensive drilling of wells.

intrusión de agua salada Infiltración de agua salada en una zona en la que la presión de las aguas freáticas se ha visto reducida por la extensa perforación de pozos.

sample size (*n*) The number of times a measurement is replicated in data collection.

tamaño de la muestra (*n*) Cantidad de veces que se replica una medición en una recolección de datos.

sanitary landfill An engineered ground facility designed to hold municipal solid waste (MSW) with as little contamination of the surrounding environment as possible.

basurero sanitario Instalación terrestre diseñada para retener desechos sólidos municipales con el mínimo posible de contaminación del entorno ambiental.

saturation point The maximum amount of water vapor in the air at a given temperature.

punto de saturación Cantidad máxima de vapor de agua en el aire a una temperatura dada.

scavenger An organism that consumes dead animals.

carroñero Organismo que consume animales muertos.

scientific method An objective method to explore the natural world, draw inferences from it, and predict the outcome of certain events, processes, or changes.

método científico Método objetivo para explorar el mundo natural, trazar inferencias del mismo y predecir el resultado de ciertos eventos, procesos o cambios.

seafloor spreading The formation of new ocean crust as a result of magma pushing upward and outward from Earth's mantle to the surface.

ensanchamiento del fondo marino Formación de una nueva corteza marítima como resultado de la presión hacia arriba y hacia fuera del magma sobre el manto terrestre hacia la superficie.

second law of thermodynamics The physical law stating that when energy is transformed, the quantity of energy remains the same, but its ability to do work diminishes.

secondary consumer A carnivore that eats primary consumers.

secondary pollutant A primary pollutant that has undergone transformation in the presence of sunlight, water, oxygen, or other compounds.

secondary succession The succession of plant life that occurs in areas that have been disturbed but have not lost their soil.

sedimentary rock Rock that forms when sediments such as muds, sands, or gravels are compressed by overlying sediments.

seismic activity The frequency and intensity of earthquakes experienced over time.

selective cutting The method of harvesting trees that involves the removal of single trees or a relatively small number of trees from among many in a forest.

selective pesticide A pesticide that targets a narrow range of organisms.

sense of place The feeling that an area has a distinct and meaningful character.

septage A layer of fairly clear water found in the middle of a septic tank.

septic system A relatively small and simple sewage treatment system, made up of a septic tank and a leach field, often used for homes in rural areas.

septic tank A large container that receives wastewater from a house as part of a septic system.

severe acute respiratory syndrome (SARS) A type of flu caused by a coronavirus.

sex ratio The ratio of males to females in a population.

shifting agriculture An agricultural method in which land is cleared and used for a few years until the soil is depleted of nutrients.

sick building syndrome A buildup of toxic pollutants in an airtight space, seen in newer buildings.

siltation The accumulation of sediments, primarily silt, on the bottom of a reservoir.

siting The designation of a landfill location, typically through a regulatory process involving studies, written reports, and public hearings.

sludge Solid waste material from wastewater.

segunda ley de termodinámica Ley física que manifiesta que cuando la energía se transforma, la cantidad de energía sigue igual pero su capacidad de hacer trabajos disminuye.

consumidor secundario Carnívoro que se come consumidores primarios.

contaminante secundario Contaminante primario que se ha transformado ante la presencia de luz solar, agua, oxígeno u otros compuestos.

sucesión secundaria Sucesión de flora que ocurre en áreas que han sido perturbadas pero que no han perdido su suelo.

roca sedimentaria Roca que se forma cuando sedimentos tales como lodos, arenas o gravas se comprimen sobre sedimentos sobreyacentes.

actividad sísmica Frecuencia e intensidad de terremotos que ocurren en un lapso de tiempo.

tala selectiva Método de tala de árboles que implica retirar un solo árbol o una cantidad relativamente reducida de árboles entre los muchos que hay en un bosque o selva.

plaguicida selectivo Plaguicida que se enfoca en una gama estrecha de organismos.

sentido de sitio Sentido de que una zona tiene un carácter distintivo y significativo.

residuos sépticos Capa de agua relativamente clara que se encuentra en el centro del tanque séptico.

sistema séptico Sistema relativamente pequeño y sencillo para el tratamiento de aguas residuales. Consta de un tanque séptico y un campo de lixiviación. A menudo se utilizan en hogares en zonas rurales.

tanque séptico Recipiente grande que forma parte de un sistema séptico y que recibe aguas negras del hogar.

síndrome respiratorio agudo grave (SRAG) Tipo de gripe causada por uno de los virus de la corona.

proporción sexual Proporción de machos a hembras en una población.

agricultura de desplazamiento Método agrícola en el que el terreno se despeja y se usa unos pocos años hasta que se agoten los nutrientes del suelo.

síndrome de edificio enfermo Acumulación de contaminantes tóxicos en un espacio hermético. Se registra en los edificios más recientes.

sedimentación Acumulación de sedimentos, primordialmente cieno, en el fondo de un embalse.

ubicación Designación del sitio donde se ubicará un basurero, por lo general mediante un proceso reglamentado que incluye estudios, informes escritos y audiencias públicas.

fangos residuales Material de desecho sólido que proviene de aguas negras.

smart grid An efficient, self-regulating electricity distribution network that accepts any source of electricity and distributes it automatically to end users.

red inteligente Red de distribución eléctrica eficiente, que se autorregula, y que recibe cualquier fuente de electricidad y la distribuye automáticamente a los usuarios finales.

smart growth A set of principles for community planning that focuses on strategies to encourage the development of sustainable, healthy communities.

crecimiento inteligente Conjunto de principios de planificación comunitaria que se concentra en estrategias que fomentan la creación de comunidades sostenibles y saludables.

smog A type of air pollution that is a mixture of oxidants and particulate matter.

esmog Tipo de contaminación del aire que es una mezcla de oxidantes y materias particuladas.

soil degradation The loss of some or all of a soil's ability to support plant growth.

degradación de los suelos Pérdida parcial o total de la capacidad que tiene el suelo para sustentar el crecimiento vegetal.

solubility How well a chemical dissolves in a liquid.

solubilidad La facilidad con que una sustancia química se disuelve en un líquido.

source reduction An approach to waste management that seeks to cut waste by reducing the use of potential waste materials in the early stages of design and manufacture.

reducción de fuentes Manera de enfocar la gestión de desechos mediante la cual se procura reducir el uso de materiales que posiblemente se tengan que desechar en las etapas iniciales de diseño y fabricación.

speciation The evolution of new species.

especiación Evolución de especies nuevas.

species A group of organisms that is distinct from other groups in its morphology (body form and structure), behavior, or biochemical properties.

especie Grupo de organismos que se distinguen de otros grupos en su morfología (forma y estructura del cuerpo), conducta o propiedades bioquímicas.

species diversity The number of species in a region or in a particular type of habitat.

diversidad de especies Número de especies en una región o en un tipo de hábitat en particular.

species evenness The relative proportion of individuals within the different species in a given area.

uniformidad de especies Proporción relativa de individuos dentro de especies diferentes en un área dada.

species richness The number of species in a given area.

riqueza de especies Cantidad de especies en un área dada.

spring A natural source of water formed when water from an aquifer percolates up to the ground surface.

manantial Fuente natural de agua que se forma cuando el agua de un acuífero se filtra hasta la superficie.

S-shaped curve The shape of the logistic growth model when graphed.

curva en forma de S Representación gráfica de la forma del modelo de crecimiento logístico.

stakeholder A person or organization with an interest in a particular place or issue.

parte interesada Persona u organización con un interés en un sitio o asunto específico.

standing crop The amount of biomass present in an ecosystem at a particular time.

cultivo activo Cantidad de biomasa presente en un ecosistema a una hora dada.

steady state A state in which inputs equal outputs, so that the system is not changing over time.

régimen estacionario Régimen en el que las entradas son equivalentes a las salidas, de manera que con el pasar del tiempo el sistema no cambia.

stewardship The careful and responsible management and care for Earth and its resources.

conciencia fiduciaria La gestión y cuidado esmerados y responsables de la Tierra y sus recursos.

Stockholm Convention A 2001 agreement among 127 nations concerning 12 chemicals to be banned, phased out, or reduced.

Convenio de Estocolmo Convenio suscrito en 2001 por 127 países referente al tratamiento de 12 sustancias químicas que se han de prohibir, eliminar de manera paulatina o reducir.

stratosphere The layer of the atmosphere above the troposphere, extending roughly 16 to 50 km (10–31 miles) above the surface of Earth.

estratosfera Capa de la atmósfera encima de la troposfera, que se extiende aproximadamente 16 a 50 km (de 10 a 31 millas) sobre la superficie terrestre.

strip mining The removal of strips of soil and rock to expose ore.

minería a cielo abierto Remoción de tajos de suelo y roca para dejar expuesto un mineral.

subduction The process of one crustal plate passing under another.

subducción Proceso en el que una placa de la corteza pasa debajo de otra.

sublethal effect The effect of an environmental hazard that is not lethal, but which may impair an organism's behavior, physiology, or reproduction.

efecto subletal Efecto de un peligro medioambiental que no es letal pero que puede perjudicar la conducta, fisiología o reproducción de un individuo.

subsistence energy source An energy source gathered by individuals for their own immediate needs.

fuente de energía de subsistencia Fuente energética recogida por individuos para satisfacer sus necesidades inmediatas.

subsurface mining Mining techniques used when the desired resource is more than 100 m (328 feet) below the surface of Earth.

minería subsuperficial Técnicas de minería que se emplean cuando el recurso deseado se encuentra a más de 100 m (328 pies) por debajo de la superficie terrestre.

subtropical desert A biome prevailing at approximately 30° N and 30° S, with hot temperatures, extremely dry conditions, and sparse vegetation.

desierto subtropical Biomedio predominante a aproximadamente 30° N y 30° S, con temperaturas calientes, condiciones extremadamente secas y vegetación escasa.

suburb An area surrounding a metropolitan center, with a comparatively low population density.

suburbio Zona circundante de un centro metropolitano, con una densidad poblacional comparativamente baja.

sulfur cycle The movement of sulfur around the biosphere.

ciclo de azufre Desplazamiento del azufre alrededor de la biosfera.

sulfurous smog Smog dominated by sulfur dioxide and sulfate compounds. *Also known as* **London-type smog; gray smog; industrial smog.**

esmog sulfuroso Esmog dominado por el dióxido de azufre y compuestos sulfatados. *También se conoce como* **esmog londinense; esmog gris; esmog industrial.**

Superfund Act The common name for the Comprehensive Environmental Response, Compensation, and Liability Act (CERCLA); a 1980 U.S. federal act that imposes a tax on the chemical and petroleum industries, funds the cleanup of abandoned and non-operating hazardous waste sites, and authorizes the federal government to respond directly to the release or threatened release of substances that may pose a threat to human health or the environment.

Ley *Superfund* Nombre popular de la Ley Integral de Respuesta, Indemnización y Responsabilidad Medioambiental (CERCLA); ley federal estadounidense promulgada en 1980 que impone un gravamen a las industrias químicas y petroleras, fondos para la limpieza de sitios con desechos nocivos que se hayan abandonado o que hayan dejado de funcionar, y que autoriza al gobierno federal a responder directamente a la liberación o posible liberación de sustancias que podrían atentar contra la salud humana o contra el medioambiente.

surface tension A property of water that results from the cohesion of water molecules at the surface of a body of water and that creates a sort of skin on the water's surface.

tensión superficial Propiedad del agua que resulta de la cohesión de moléculas de agua en la superficie de una extensión de agua y que crea una especie de piel sobre la superficie del agua.

survivorship curve A graph that represents the distinct patterns of species survival as a function of age.

curva de supervivencia Representación gráfica de los distintos patrones de supervivencia de las especies como función de la edad.

sustainability Living on Earth in a way that allows humans to use its resources without depriving future generations of those resources.

sostenibilidad Vivir en la Tierra de una manera que les permite a los humanos aprovechar sus recursos sin privar a generaciones futuras de tales recursos.

sustainable agriculture Agriculture that fulfills the need for food and fiber while enhancing the quality of the soil, minimizing the use of nonrenewable resources, and allowing economic viability for the farmer.

agricultura sostenible Agricultura que satisface la necesidad de alimento y fibra al mismo tiempo que mejora la calidad del suelo, minimizando el uso de recursos no renovables y facilitando la viabilidad económica del agricultor.

sustainable development Development that balances current human well-being and economic advancement with resource management for the benefit of future generations.

desarrollo sostenible Desarrollo en el que se equilibran el bienestar y el progreso económico de los seres humanos de hoy con la gestión de los recursos por el bien de las generaciones futuras.

swine flu A type of flu caused by the H1N1 virus.

gripe porcina Tipo de gripe causada por el virus H1N1.

symbiotic relationship The relationship between two species that live in close association with each other.

relación simbiótica Relación entre dos especies que viven en asociación de proximidad una con la otra.

sympatric speciation The evolution of one species into two, without geographic isolation.

especiación simpátrida Evolución de una especie en dos, sin que medie aislamiento geográfico.

synergistic interaction A situation in which two risks together cause more harm than expected based on the separate effects of each risk alone.

interacción sinérgica Situación en la que dos riesgos se unen para producir más perjuicios que lo previsto con base en los efectos separados de cada riesgo por sí solo.

synthetic fertilizer Fertilizer produced commercially, normally with the use of fossil fuels. *Also known as* **inorganic fertilizer.**

fertilizante sintético Fertilizante producido de manera comercial, generalmente a partir de combustibles fósiles. *También se conoce como* **fertilizante inorgánico**.

systems analysis An analysis to determine inputs, outputs, and changes in a system under various conditions.

análisis de sistemas Análisis para determinar las entradas, las salidas y los cambios en un sistema en diversas condiciones.

T

tailings *See* **mining spoils.**

colas *Ver* **mining spoils.**

technology transfer The phenomenon of less developed countries adopting technological innovations developed in wealthy countries.

transferencia de tecnología Fenómeno de los países menos desarrollados que adoptan innovaciones tecnológicas creadas en los países adinerados.

tectonic cycle The sum of the processes that build up and break down the lithosphere.

ciclo tectónico Suma de los procesos que componen y descomponen la litosfera.

temperate grassland/cold desert A biome characterized by cold, harsh winters, and hot, dry summers.

pradera templada o desierto frío Biomedio caracterizado por inviernos fríos y agrestes, y veranos calientes y secos.

temperate rainforest A coastal biome typified by moderate temperatures and high precipitation.

bosque lluvioso templado Biomedio litoral caracterizado por temperaturas moderadas y precipitación abundante.

temperate seasonal forest A biome with warm summers and cold winters with over 1 m (39 inches) of precipitation annually.

bosque templado estacional Biomedio con veranos cálidos e inviernos fríos con más de 1 metro (39 pulgadas) de precipitación al año.

temperature The measure of the average kinetic energy of a substance.

temperatura Medición de la energía cinética promedio de una sustancia.

teratogen A chemical that interferes with the normal development of embryos or fetuses.

teratógeno Sustancia química que interfiere con el desarrollo normal del embrión o feto.

terrestrial biome A geographic region categorized by a particular combination of average annual temperature, annual precipitation, and distinctive plant growth forms on land.

biomedio terrestre Región geográfica categorizada por una combinación dada de los promedios de temperatura anual y de precipitación anual, así como por las plantas terrestres características que se distinguen en la zona.

tertiary consumer A carnivore that eats secondary consumers.

consumidor terciario Carnívoro que come consumidores secundarios.

the three Rs *See* **reduce, reuse, recycle.**

las tres erres *Ver* **reduce, reuse, recycle.**

theory A hypothesis that has been repeatedly tested and confirmed by multiple groups of researchers and has reached wide acceptance.

teoría Hipótesis que múltiples grupos de investigadores han comprobado y confirmado en repetidas ocasiones y que ha logrado amplia aceptación.

theory of demographic transition The theory that as a country moves from a subsistence economy to industrialization and increased affluence it undergoes a predictable shift in population growth.

teoría de transición demográfica Teoría que explica que en la medida en que un país pasa de una economía de subsistencia a la industrialización y a una mayor abundancia, experimenta un desplazamiento predecible en su crecimiento poblacional.

theory of island biogeography A theory that demonstrates the dual importance of habitat size and distance in determining species richness.

teoría de biogeografía de islas Teoría que demuestra la importancia doble de la extensión del hábitat y la distancia para determinar su riqueza en especies.

thermal inversion A situation in which a relatively warm layer of air at mid-altitude covers a layer of cold, dense air below.

inversión térmica Situación en la que una capa de aire relativamente cálida a media altura cubre una capa de aire frío y denso más abajo.

thermal mass A property of a building material that allows it to maintain heat or cold.

masa térmica Propiedad de un material de construcción que le permite mantenerse caliente o frío.

thermal pollution Nonchemical water pollution that occurs when human activities cause a substantial change in the temperature of water.

contaminación térmica Contaminación del agua sin sustancias químicas que ocurre cuando ciertas actividades humanas causan un cambio sustancial en la temperatura del agua.

thermal shock A dramatic change in water temperature that can kill organisms.

choque térmico Cambio drástico en la temperatura del agua que puede matar organismos.

thermohaline circulation An oceanic circulation pattern that drives the mixing of surface water and deep water.

circulación termohalina Patrón de circulación oceánica en el que se mezclan las aguas superficiales con las aguas profundas.

threatened species According to the International Union for Conservation of Nature (ICUN), species that have a high risk of extinction in the future; according to U.S. legislation, any species that is likely to become an endangered species within the foreseeable future throughout all or a significant portion of its range.

especies amenazadas Según la Unión Internacional para al Conservación de la Naturaleza (ICUN), las especies que presentan un riesgo grave de extinción en el futuro; según la legislación estadounidense, toda especie que presenta una probabilidad de ser una especie en peligro de extinción dentro de un futuro previsible en todo o en una parte significativa de su hábitat.

tidal energy Energy that comes from the movement of water driven by the gravitational pull of the Moon.

energía mareomotriz Energía que se genera del movimiento del agua impulsado por la atracción gravitatoria de la Luna.

tiered rate system A billing system used by some electric companies in which customers pay higher rates as their use goes up.

sistema tarifario escalonado Sistema de facturación utilizado por algunas compañías eléctricas en la que los clientes pagan tarifas más altas a medida que aumenta su consumo de energía.

tipping fee A fee charged for disposing of material in a landfill or incinerator.

tarifa de carga Cuota que se cobra por enviar material a un basurero o incinerador.

topsoil *See* **A horizon.**

manto *Ver* **A horizon.**

total fertility rate (TFR) An estimate of the average number of children that each woman in a population will bear throughout her childbearing years.

tasa de fertilidad total Estimación del número promedio de niños que cada mujer en una población dará a luz durante sus años en edad reproductiva.

tragedy of the commons The tendency of a shared, limited resource to become depleted because people act from self-interest for short-term gain.

tragedia de los comunes Tendencia de un recurso compartido y limitado a agotarse debido a personas que actúan por interés propio con el fin de obtener una ganancia inmediata.

transform fault boundary An area where tectonic plates move sideways past each other.

borde de una falla de transformación Zona en la que las placas tectónicas se desplazan lateralmente, una pasando la otra.

transit-oriented development (TOD) Development that attempts to focus dense residential and retail development around stops for public transportation, a component of smart growth.

urbanización orientada por el tránsito Desarrollo urbano que trata de concentrar la creación de zonas residenciales y de venta detallista alrededor de paradas del transporte público. Es uno de los componentes del crecimiento inteligente.

transpiration The release of water from leaves during photosynthesis.

transpiración Liberación de agua desde las hojas durante la fotosíntesis.

tree plantation A large area typically planted with a single rapidly growing tree species.

plantación de árboles Zona extensa en la que característicamente se han sembrado especies de árboles de crecimiento rápido.

triple bottom line An approach to sustainability that considers three factors—economic, environmental, and social—when making decisions about business, the economy, and development.

triple fin de cuentas Manera de enfocar la sostenibilidad en la que se tienen en cuenta tres factores: el económico, el medioambiental y el social, cuando se toman decisiones sobre los negocios, la economía y el desarrollo.

trophic levels The successive levels of organisms consuming one another.

niveles tróficos Niveles sucesivos de organismos que se consumen unos a otros.

trophic pyramid A representation of the distribution of biomass, numbers, or energy among trophic levels.

pirámide trófica Representación gráfica de la distribución de biomasa, cifras o energía entre los niveles tróficos.

tropical rainforest A warm and wet biome found between 20° N and 20° S of the equator, with little seasonal temperature variation and high precipitation.

bosque lluvioso tropical Biomedio cálido y húmedo que se encuentra entre 20° N y 20° S del ecuador, con poca variación estacional de temperatura y niveles altos de precipitación.

tropical seasonal forest/savanna A biome marked by warm temperatures and distinct wet and dry seasons.

bosque/sabana tropical estacional Biomedio que se distingue por sus temperaturas cálidas y sus estaciones de lluvia y sequía claramente distinguibles.

troposphere A layer of the atmosphere closest to the surface of Earth, extending up to approximately 16 km (10 miles).

troposfera Capa de la atmósfera contigua a la superficie de la Tierra que se extiende en forma ascendente unos 16 km (10 millas).

tuberculosis A highly contagious disease caused by the bacterium *Mycobacterium tuberculosis* that primarily infects the lungs.

tuberculosis Enfermedad muy contagiosa causada por la bacteria *Mycobacterium tuberculosis.* Primordialmente afecta los pulmones.

tundra A cold and treeless biome with low-growing vegetation.

tundra Biomedio frío y carente de vegetación arbórea con vegetación de poco crecimiento.

turbine A device with blades that can be turned by water, wind, steam, or exhaust gas from combustion that turns a generator in an electricity-producing plant.

turbina Dispositivo con álabes que pueden ser girados por agua, viento, vapor o gases de escape de la combustión, que hacen girar un generador en una planta de producción eléctrica.

Type I survivorship curve A pattern of survival over time in which there is high survival throughout most of the life span, but then individuals start to die in large numbers as they approach old age.

curva de supervivencia tipo I Patrón de supervivencia temporal en el que existe un nivel de supervivencia alto durante la mayor parte de la duración de la vida, pero luego los individuos comienzan a fallecer en gran volumen a medida que se aproximan a la vejez.

Type II survivorship curve A pattern of survival over time in which there is a relatively constant decline in survivorship throughout most of the life span.

curva de supervivencia tipo II Patrón de supervivencia temporal en el que existe una disminución relativamente constante en la supervivencia durante la mayoría de la duración de la vida.

Type III survivorship curve A pattern of survival over time in which there is low survivorship early in life with few individuals reaching adulthood.

curva de supervivencia tipo III Patrón de supervivencia temporal en el que existe un bajo nivel de supervivencia temprano en la vida y son pocos los individuos que llegan a la adultez.

U

uncertainty An estimate of how much a measured or calculated value differs from a true value.

incertidumbre Estimación de la diferencia que hay entre un valor medido o calculado y el valor real.

unconfined aquifer An aquifer made of porous rock covered by soil out of which water can easily flow.

acuífero no cautivo Acuífero hecho de roca porosa, cubierto por suelo del cual fácilmente puede fluir agua.

undernutrition The condition in which not enough calories are ingested to maintain health.

subnutrición Condición producto de no haber ingerido suficientes calorías para mantener el estado de salud.

United Nations (UN) A global institution dedicated to promoting dialogue among countries with the goal of maintaining world peace.

Organización de Naciones Unidas (ONU) Institución global dedicada a fomentar el diálogo entre países con el objetivo de mantener la paz en el mundo.

United Nations Development Programme (UNDP) An international program that works in 166 countries around the world to advocate change that will help people obtain a better life through development.

Programa de Naciones Unidas para el Desarrollo (UNDP) Programa internacional que funciona en 166 países del mundo con el fin de promover cambios que les ayudarán a las personas a lograr una mejor vida gracias al desarrollo.

United Nations Environment Programme (UNEP) A program of the United Nations responsible for gathering environmental information, conducting research, and assessing environmental problems.

Programa de Naciones Unidas para el Medio Ambiente (UNEP) Programa de Naciones Unidades responsable por recoger información, realizar investigaciones y evaluar problemas de índole ambiental y ecológica.

upwelling The upward movement of ocean water toward the surface as a result of diverging currents.

surgencia Movimiento hacia arriba del agua oceánica hacia la superficie como resultado de corrientes divergentes.

urban area An area that contains more than 385 people per square kilometer (1,000 people per square mile).

zona urbana Extensión que contiene más de 385 personas por kilómetro cuadrado (1,000 personas por milla cuadrada).

urban blight The degradation of the built and social environments of the city that often accompanies and accelerates migration to the suburbs.

deterioro urbano Degradación del entorno construido y del ambiente social de la ciudad que a menudo acompaña y acelera la migración hacia los suburbios.

urban growth boundary A restriction on development outside a designated area.

límite del crecimiento urbano Limitación de la urbanización por fuera de una zona designada.

urban sprawl Urbanized areas that spread into rural areas, removing clear boundaries between the two.

expansión urbana Zonas urbanizadas que se desparraman hacia las zonas rurales, eliminando los límites claros entre las dos.

valuation The practice of assigning monetary value to intangible benefits and natural capital.

valoración Asignación de un valor monetario a los beneficios intangibles y al capital natural.

volatile organic compound (VOC) An organic compound that evaporates at typical atmospheric temperatures.

compuesto volátil orgánico Compuesto orgánico que se evapora a las temperaturas atmosféricas características.

V

volcano A vent in the surface of Earth that emits ash, gases, or molten lava.

volcán Escape en la superficie terrestre por el que se emiten cenizas, gases o lava fundida.

W

waste Material outputs from a system that are not useful or consumed.

desecho Efluentes materiales de un sistema. Se trata de efluentes que ni son útiles ni se pueden consumir.

waste stream The flow of solid waste that is recycled, incinerated, placed in a solid waste landfill, or disposed of in another way.

flujo de desechos El flujo de desechos sólidos que se reciclan, incineran, tiran en un basurero o eliminan de alguna otra manera.

waste-to-energy A system in which heat generated by incineration is used as an energy source rather than released into the atmosphere.

desecho a energía Sistema en el que se genera calor mediante incineración. Dicho calor se utiliza como fuente de energía en lugar de liberarlo a la atmósfera.

wastewater Water produced by livestock operations and human activities, including human sewage from toilets and gray water from bathing and washing of clothes and dishes.

aguas residuales Aguas producidas por operaciones de ganado y actividades humanas, incluidas aguas negras humanas de inodoros y aguas grises de baños y del lavado de ropas y trastes.

water footprint The total daily per capita use of fresh water.

consumo de agua El uso total diario per cápita de agua dulce.

water impoundment The storage of water in a reservoir behind a dam.

embalse de agua Almacenamiento de agua en un reservorio, detrás de una presa.

water pollution The contamination of streams, rivers, lakes, oceans, or groundwater with substances produced through human activities.

contaminación del agua Contaminación de arroyos, ríos, lagos, océanos o aguas freáticas con sustancias producidas mediante actividades humanas.

water table The uppermost level at which the water in a given area fully saturates rock or soil.

nivel freático El nivel más alto en el que el agua satura totalmente la roca o el suelo.

waterlogging A form of soil degradation that occurs when soil remains under water for prolonged periods.

saturación Forma de degradación del suelo que ocurre cuando este permanece bajo agua por lapsos de tiempo prolongados.

watershed All land in a given landscape that drains into a particular stream, river, lake, or wetland.

cuenca hidrográfica Todo el terreno en un paisaje dado que desemboca en un arroyo, río, lago o humedal dado.

weather The short-term conditions of the atmosphere in a local area, which include temperature, humidity, clouds, precipitation, and wind speed.

estado del tiempo Condiciones de la atmósfera a corto plazo en una zona local; incluye la temperatura, la humedad, la nubosidad, la precipitación y la velocidad del viento.

well-being The status of being healthy, happy, and prosperous.

bienestar Estado de ser feliz, y de gozar de buena salud y prosperidad.

West Nile virus A virus that lives in hundreds of species of birds and is transmitted among birds by mosquitoes.

virus del Nilo Occidental Virus que vive en centenares de especies de aves y que es transmitido entre aves por mosquitos.

wind energy Energy generated from the kinetic energy of moving air.

energía eólica Energía generada a partir de la energía cinética del aire en movimiento.

wind turbine A turbine that converts wind energy into electricity.

turbina aerogeneradora Turbina que convierte la energía del viento en electricidad. *También se conoce como* **torre eólica.**

woodland/shrubland A biome characterized by hot, dry summers and mild, rainy winters.

terreno de árboles y arbustos Biomedio caracterizado por veranos calurosos y secos, e inviernos suaves y lluviosos.

World Bank A global institution that provides technical and financial assistance to developing countries with the objectives of reducing poverty and promoting growth, especially in the poorest countries.

Bando Mundial Institución de alcance mundial que ofrece asesoría técnica y asistencia económica a los países en vías de desarrollo, con los objetivos de reducir la pobreza y fomentar el crecimiento, sobre todo en los países menos adinerados.

World Health Organization (WHO) A global institution dedicated to the improvement of human health by monitoring and assessing health trends and providing medical advice to countries.

Organización Mundial de la Salud (OMS) Institución de alcance mundial que se dedica a mejorar la salud humana mediante el monitoreo y la evaluación de tendencias sanitarias. Además, ofrece consejos médicos a los países que los soliciten.

Z

zoning A planning tool used to separate industry and business from residential neighborhoods.

zonificación Recurso de planificación que se utiliza para separar la industria y el comercio de los barrios residenciales.**horizonte A** A menudo es la capa superficial del suelo, una zona de material orgánico y minerales que se han entremezclado. *También se conoce como* **manto.**

Index

Note: Page numbers in **boldface** indicate definitions; those followed by f indicate figures; those followed by t indicate tables.

International System of Units (Metric System) and Common U.S. Unit Conversions

Measurement	Unit Abbreviation	Metric Equivalent	Metric to U.S.	U.S. to Metric
Area	1 square meter (m^2)	= 10,000 square centimeters	1 m^2 = 1.1960 square yards	1 square yard = 0.8361 m^2
			1 m^2 = 10.764 square feet	1 square foot = 0.0929 m^2
	1 square centimeter (cm^2)	= 100 square millimeters	1 cm^2 = 0.155 square inch	1 square inch = 6.4516 cm^2
Length	1 kilometer (km)	= 1,000 (10^3) meters	1 km = 0.62 mile	1 mile = 1.61 km
	1 meter (m)	= 100 (10^2) centimeters	1 m = 1.09 yards	1 yard = 0.914 m
		= 1,000 millimeters	1 m = 3.28 feet	1 foot = 0.305 m
			1 m = 39.37 inches	
	1 centimeter (cm)	= 0.01 (10^{-2}) meter	1 cm = 0.394 inch	1 inch = 2.54 cm
	1 millimeter (mm)	= 0.001 (10^{-3}) meter	1 mm = 0.039 inch	
Mass	1 metric tonne (t)	= 1,000 kilograms	1 t = 1.103 ton (referred to in text as "U.S. ton")	1 ton = 0.907 t
	1 kilogram (kg)	= 1,000 grams	1 kg = 2.205 pounds	1 pound = 0.4536 kg
	1 gram (g)	= 1,000 milligrams	1 g = 0.0353 ounce	1 ounce = 28.35 g
	1 milligram (mg)	= 0.001 gram		
Volume–solids	1 cubic meter (m^3)	= 1,000,000 cubic centimeters	1 m^3 = 1.3080 cubic yards	1 cubic yard = 0.7646 m^3
			1 m^3 = 35.315 cubic feet	1 cubic foot = 0.0283 m^3
	1 cubic centimeter (cm^3 or cc)	= 0.000001 cubic meter	1 cm^3 = 0.0610 cubic inch	1 cubic inch = 16.387 cm^3
	1 cubic millimeter (mm^3)	= 1 milliliter		
		= 0.000000001 cubic meter		
Volume–liquids and gases	1 liter (L)	= 1,000 milliliters	1 L = 0.264 gallons	1 quart = 0.946 L
	1 kiloliter (kL)	= 1,000 liters	1 kL = 264.17 gallons	1 gallon = 3.785 L
			1 L = 1.057 quarts	
	1 milliliter (ml)	= 0.001 liter	1 ml = 0.034 fluid ounce	1 quart = 946 ml
		= 1 cubic centimeter	1 ml = approximately 1/5 teaspoon	1 pint = 473 ml
				1 fluid ounce = 29.57 ml
				1 teaspoon = approx. 5 ml
Time	1 millisecond (ms)	= 0.001 second		
Temperature	Degrees Celsius (°C)		°C = 5/9 (°F − 32)	°F = 9/5 °C + 32
Energy and Power	1 kilowatt-hour	= 3,413 BTUs		
		= 860,421 calories		
	1 watt	= 3.413 BTU/hr		
		= 14.34 calorie/min		
	1 calorie	= amount of energy needed to raise the temperature of 1 gram (1 cm^3) of water by 1 degree Celsius		
	1 horsepower	= 7.457×10^2 watts		
	1 joule	= 9.481×10^{-4} BTU		
		= 0.239 cal		
		= 2.778×10^{-7} kilowatt-hour		